Handbook of Nonconvex Analysis and Applications

Handbook of Nonconvex Analysis and Applications

edited by
David Yang Gao and Dumitru Motreanu

International Press
www.intlpress.com

Handbook of Nonconvex Analysis and Applications

Editors:
David Yang Gao (University of Ballarat, Australia)
Dumitru Motreanu (Université de Perpignan, France)

2010 Mathematics Subject Classification. 35Jxx, 35Kxx, 47Jxx, 47Nxx, 49Jxx, 49Mxx, 58Exx, 58Jxx, 90Cxx.

ISBN 978-1-57146-298-5

Paperback re-issue 2014.
Previously published in 2010 under ISBN 978-1-57146-200-8 (hardcover).

Printed in the United States of America.

David Gao's work was supported in part by the U.S. Air Force Office of Scientific Research, under grant number FA9550-09-1-0285.

Preface

Nonconvex analysis is a rapidly developing, multi-disciplinary field of research, comprehending theoretical analysis in mathematical modelling of natural systems, bifurcation and chaos in dynamical systems, finite deformation theory, nonlinear partial differential equations, global optimization, calculus of variation, numerical methods, and scientific computations. The field of nonconvex analysis has undergone considerable development in a remarkably short time – with extensive applications to theoretical physics, material science, modern mechanics, complex systems, and scientific computations.

The present volume, *Handbook of Nonconvex Analysis and Applications*, was proposed by Shing-Tung Yau, a world-renowned mathematician at Harvard University, and we are honored to have been invited to edit it. The *Handbook* will serve as a much-needed reference work for the dynamic and ever-growing field of nonconvex analysis and its applications.

The *Handbook* consists of thirteen chapters written by notable experts in the field, addressing essential recent developments in nonconvex analysis and its applications, and keeping a balance between major areas of theory, methods, and applications. Each chapter provides an illuminating exposition of state-of-the-art approaches to a specific topic, with discussions of the central contributions, and pointers to some basic references. A variety of topics regarding nonconvex analysis and its applications are discussed: nonconvex variational principles; comparison principles; nonlinear eigenvalue problems; critical point theory; boundary value problems; topological methods, including Morse theory; nonlinear elliptic equations; evolution problems; difference equations; inequality problems; geometric properties of functions and spaces; and applications in mechanics.

We thank all the authors for their valuable work, and Mr. Brian J. Bianchini, production editor at International Press, for his efficient and generous support on behalf of the publisher.

David Gao's work was supported in part by the U.S. Air Force Office of Scientific Research, under grant number FA9550-09-1-0285.

David Gao
Dumitru Motreanu

Contents

Nonlinear difference equations through variational methods

Gabriele Bonanno and Pasquale Candito

ABSTRACT. A careful arrangement of some recent results on the existence and multiplicity of solutions for nonlinear difference equations involving the discrete p-Laplacian is given. Main tools are critical points theorems presented below.

CONTENTS

1. Introduction

Difference equations are an important field of research both from a mathematical point of view and for the great numbers of applications where such mathematical model appears naturally as, for instance, computer science, mechanical engineering, control systems, artificial or biological neural networks, economics and many others. The theory of linear difference equations has been extensively studied. Moreover, in the last years, many authors have widely developed various methods and techniques, such as fixed points theorems or upper and lower solutions methods, to study nonlinear discrete problems see, for instance, [**CJ**], [**JCOA**] and [**HT**]. On these subjects and methods used to investigate linear and nonlinear difference equations, a complete overview is given, for instance, in the monographs [**A**], [**ARW**] and [**KP**] and in the references given therein.

2000 *Mathematics Subject Classification.* Primary 39A05, 47J30; Secondary 39A10, 39-01.

Key words and phrases. Nonlinear difference equations; Discrete nonlinear boundary value problems; Multiple solutions; Infinitely many solutions; Variational methods.

Here, we will be concerned with nonlinear discrete boundary value problems by using variational methods. By starting from the seminal papers [**APR1**] and [**APR2**], this approach has been recently adopted, for instance, in [**BX**], [**CG**], [**FI**], [**JZ1**], [**JZ2**], [**LW**], [**ZL**] and [**ZZL**]. In particular, we will focus on the results established in [**CG**] [**BC4**] and [**BC5**]. To be precise, at the best of our knowledge, here we will give a survey on recent results obtained by variational methods for the following problem

$$(P_\lambda^{f,q})\begin{cases} -\Delta(\phi_p(\Delta u_{k-1}))+q_k\phi_p(u_k)=\lambda f(k,u_k), & k\in[1,N],\\ u_0=u_{N+1}=0, \end{cases}$$

where, N is a positive integer, $[1, N]$ denotes the discrete interval $\{1,\ldots,N\}$, $f: [1,N]\times\mathbb{R}\to\mathbb{R}$ is a continuous function, $\Delta u_{k-1} := u_k - u_{k-1}$ is the forward difference operator, $q_k \in \mathbb{R}_0^+$ for all $k\in[1,N]$, $\phi_p(s) := |s|^{p-2}s$, $1 < p < +\infty$, and $\lambda\in\mathbb{R}^+$.

Roughly speaking, the chapter is divided into three parts: the first part is de-dicated to critical point theorems (Sections 2–5), the second one to variational formulation and some basic results which are useful to the study of problem $(P_\lambda^{f,q})$ (Section 6) and the third part to the main results on existence and multiplicity of solutions for problem $(P_\lambda^{f,q})$ (Section 7).

In relation to the critical point theory, we have arranged in a careful way some recent results included in different papers as [**AB1**], [**AB2**], [**AB3**], [**B1**], [**B2**], [**B3**], [**BC1**], [**BC2**], [**BC3**], [**BD**], [**BM**], [**BL**], [**BO**], [**BMo**] and [**S**]. Although most of them work also for locally Lipschitz functionals perturbed with a convex, proper and lower semicontinuous term, to achieve our goal it has been enough to present only suitable smooth versions of such results, reducing the non smooth theory to the aspects needed for introducing the Palais-Smale condition cut off at M (briefly $(P.S.)_c^M$, Section 2). More precisely, let X be a reflexive real Banach space, let $\Phi,\Psi : X\to\mathbb{R}$ be two continuously Gâteaux differentiable functions, with Φ coercive and sequentially weakly lower semicontinuous and Ψ sequentially weakly upper semicontinuous, we point out several kinds of critical points theorems for functionals of type $\Phi-\lambda\Psi$. In particular, about the existence and localization of a critical point we present three results: two of these are in the spirit of the so called directed methods of the Calculus of Variations and they give, in addition to the classical version (Theorem 1.2 of [**St**]), a critical point in the set $[\Phi < r]$ (Theorem 5) and one in $[\Phi > r]$ (Theorem 7). The third one (Theorem 9) is a special version of the Mountain Pass Lemma of Ambrosetti-Rabinowitz [**AR**].

In order to have two critical points for each λ belonging to a well determined interval of parameters, the building blocks are: condition (i_1) of Theorem 12 below, for functionals possibly unbounded; condition (i_5) of Theorem 24 for coercive functionals. Then, the third critical point can be obtained, for instance, through the use of $(P.S.)_c^M$, without requiring any information on the behavior of the functional at infinity and in this case we also have a localization of the third critical point (Theorems 15, 18 see also Theorem 22); assuming some usual Palais-Smale type conditions (Remark 14) but in this case, with respect to the previous one, we lose the localization of the critical point (see also Theorem 24).

To the purpose to obtain infinitely many critical points, the main tool is Theorem 28 ([**BMo**, Theorem 2.1]).

Moreover, since the functional associated to the above problem is defined on a finite dimensional Banach space, we also give suitable versions of such results in this context (Section 5).
For general references on variational methods for regular functionals we refer the reader to the monographs [**R**], [**St**] and [**MP**], [**MR**] for the non-smooth case.

The variational formulation of problem $(P_\lambda^{f,q})$ is obtained in a standard way and some interesting inequalities and a strong maximum principle (Theorem 43) and its variant (Theorem 44) are presented (Section 6).

Obviously, the main results achieved for problem $(P_\lambda^{f,q})$ don't fill up all those we could obtain by using the critical point theorems presented here, nevertheless in our opinion, they are enough to test the efficiency of variational methods also in this framework.

In the last Section we present our main results, in particular, arguing in a standard way, we obtain at least one solution, requiring that the potential F, i.e. $F(k,t) = \int_0^t f(k,s)\,ds$ for each $(k,t) \in [1,N] \times \mathbb{R}$, is p-sublinear at infinity (Theorems 46 see also Remark 51). Moreover, if F is even, p-sublinear at infinity and superlinear at zero, problem $(P_1^{f,q})$ admits at least N distinct pairs of critical points (Theorem 48). Furthermore, if F is p-sublinear at infinity and in addition it has a rapid growth on a bounded interval, we achieve at least three solutions (Theorems 49, 56). On the other hand, if we have $sf(k,s) \geq 0$ for every $(k,s) \in [1,N] \times \mathbb{R}$, we get a precise estimate of an interval Λ such that, for each $\lambda \in \Lambda$, problem $(P_\lambda^{f,q})$ has at least three solutions which are uniformly bounded in norm with respect to λ (Theorem 60, see also Corollary 63 and Remark 64). While, if F is p-superliner at infinity we get at least three solutions for $(P_\lambda^{f,q})$ for suitable λ is large enough (Theorem 65).

With regard to the existence of infinitely many solutions, two types of results are given: either the existence of an unbounded sequence of solutions (Theorems 70 and its consequences, that is, Theorems 72, Corollaries 76 and 78, see also Theorem 79); or a sequence of pairwise distinct non zero solutions which converges to 0 (Theorem 71 and Remark 82), according as the nonlinear term has a suitable oscillating behavior, respectively, at infinity or at zero. In particular, at infinity (as well as at zero) we distinguish two situations. In the first case, Theorem 72 and Corollary 76, under an appropriate oscillating behavior at infinity of the function $\sum_{k=1}^{N} \frac{F(k,t)}{t^p}$ (see (h_5') in Remark 73), establish infinitely many solutions for each λ belonging to a precise real interval. In the second case, under different assumptions, Theorem 75 and Corollary 78 ensure the same conclusion for each λ lying in a well defined half-line.

Finally, several examples of functions that satisfy preceding theorems are also presented.

2. Preliminaries and auxiliary results

Let $(X, \|\cdot\|)$ be a real Banach space, let $I : X \to \mathbb{R}$ be a function, and let $r < s$ be a pair of real numbers, we set

$$[I < r] := \{u \in X : I(u) < r\},$$

$$[I > s] := \{u \in X : I(u) > s\},$$
$$[r < I < s] := \{u \in X : r < I(u) < s\}.$$

The sets $[I \le r]$, $[I \ge s]$, and $[r \le I \le s]$ are defined similarly.

A function I is upper semicontinuous if $[I < r]$ is open for all $r \in \mathbb{R}$. A function I is lower semicontinuous if $[I > r]$ is open for all $r \in \mathbb{R}$. A function I is continuous if and only if it is both upper and lower semicontinuous.

Taking into account that X is a Banach space, it is well known, see [**Z**], that a function I is upper semicontinuous iff I is sequentially upper semicontinuous, *i.e. for every $u \in X$ and for each sequence (u_n) in X such that $u_n \to u$ one has*

$$\limsup_{n \to +\infty} I(u_n) \le I(u).$$

If X is endowed with the weak topology and the previous inequality holds whenever (u_n) converges to u with respect to the weak topology, then I is said sequentially weakly upper semicontinuous. Bearing in mind that I is upper semicontinuous if and only if $-I$ is lower semicontinuous it is clear what's the meaning when we write that I is sequentially (weakly) lower semicontinuous.

Moreover, I is called coercive whenever

$$\lim_{\|u\| \to +\infty} I(u) = +\infty.$$

While, I is said locally Lipschitz when for every $u \in X$ there corresponds a neighborhood U of u and a constant $L \ge 0$ such that

$$|I(v) - I(w)| \le L\|v - w\|,$$

for all $v, w \in U$. As usual, X^* denotes the dual space of X and $< \cdot, \cdot >$ stands for the duality pairing between X^* and X. The generalized directional derivative of a locally Lipschitz function I at the point u along the direction v is defined as follows:

$$I^0(u; v) = \limsup_{w \to u \; t \to 0^+} \frac{I(w + tv) - I(w)}{t}.$$

Moreover, the generalized gradient of I at u is the following set:

$$\partial I(u) = \{u^* \in X^* :< u^*, v > \le I^0(u; v) \quad \forall v \in X\}.$$

We recall that if I is continuously Gâteaux differentiable at u, then I is locally Lipschitz at u and $\partial I(u) = \{I'(u)\}$. Further, a point $u \in X$ is said a (generalized) critical point of the locally Lipschitz function I if $0_{X^*} \in \partial I(u)$, namely

$$I^0(u; v) \ge 0,$$

for all $v \in X$. Clearly, if I is a continuously Gâteaux differentiable at u, then u becomes a (classical) critical point of I, that is

$$I'(u) = 0_{X^*}.$$

In this framework, by a simple calculation, it is easy to show that the classical Fermat's Theorem assume the following form

THEOREM 1. *If I is a locally Lipschitz function on X which attains a local minimum or maximum at $u \in X$, then u is a critical point for I.*

Finally, we say that a locally Lipschitz function I satisfies the Palais-Smale condition $(P.S.)_c$, $c \in \mathbb{R}$, if each sequence $\{u_n\}$ such that

(a_1) $I(u_n) \to c$,
(a_2) $\min_{v \in \partial I(u_n)} \|v\|_{X^*} \to 0$,

possesses a convergent subsequence.

It is well known that, condition (a_2) and

(a_2') $I^0(u_n, v - u_n) \geq -\epsilon_n \|v - u_n\| \ \forall v \in X$, where $\epsilon_n \to 0^+$,

are equivalent. Moreover, when I is a continuously Gâteaux differentiable function, the $(P.S.)_c$ condition is reduced to classical one, namely each sequence $\{u_n\}$ such that

(a_1) $I(u_n) \to c$,
(a_2) $\|I'(u_n)\|_{X^*} \to 0$,

possesses a convergent subsequence.

Let $\Phi, \Psi : X \to \mathbb{R}$ be two locally Lipschitz functions. Fix $M \in \mathbb{R}$ and put

$$\Psi_M(u) = \begin{cases} \Psi(u) & \text{if} \quad \Psi(u) \leq M \\ M & \text{if} \quad \Psi(u) > M. \end{cases}$$

It is a simple computation to show that also Ψ_M is a locally Lipschitz function. Now, let us to introduce a new type of Palais-Smale condition which it will be very usefull to find critical point for $\Phi - \Psi$ with Φ and Ψ belonging to $C^1(X)$ as critical point of the non smooth functional $\Phi - \Psi_M$.

DEFINITION 2. We say that the function $\Phi - \Psi$ satisfies the Palais-Smale condition cut off at M (briefly $(P.S.)_c^M$) if the function $\Phi - \Psi_M$ satisfies the Palais-Smale condition.

It is easy to verify that the $(P.S.)_c^M$ condition and the $(P.S.)_c$ condition on $\Phi - \Psi$ are mutually independent.

Now, we emphasize the following theorem that guarantees the $(P.S.)_c^M$ condition for all $M \in \mathbb{R}$ for functions of class C^1.

THEOREM 3. *Let $\Phi, \Psi : X \to \mathbb{R}$ be two continuously Gâteaux differentiable functions, with Φ coercive. Assume that $\Phi' : X \to X^*$ admits a continuous inverse operator on X^*, and $\Psi' : X \to X^*$ is compact. Then, the function $\Phi - \Psi$ satisfies the $(P.S.)_c^M$ condition for all $M \in \mathbb{R}$.*

PROOF. Fix $M \in \mathbb{R}$ and let $\{u_n\}$ be a sequence such that

(1) $\Phi(u_n) - \Psi_M(u_n) \to c$, $c \in \mathbb{R}$,
(2) $\min_{\xi \in \partial(\Phi - \Psi_M)(u_n)} \|\xi\|_{X^*} \to 0$.

Since Φ is coercive, also $\Phi - \Psi_M$ is coercive and, hence, from (1) we obtain that $\{u_n\}$ is bounded. Moreover, taking into account that

$$\partial(\Phi - \Psi_M)(u_n) = \partial\Phi(u_n) - \partial\Psi_M(u_n) = \Phi'(u_n) - \partial\Psi_M(u_n)$$

and

$$\partial\Psi_M(u_n) = \begin{cases} \Psi'(u_n) & \text{if } \Psi(u_n) < M \\ 0 & \text{if } \Psi(u_n) > M \\ \{r\Psi'(u_n) : r \in [0,1]\} & \text{if } \Psi(u_n) = M \end{cases}$$

(see [**Cl**, Corollary 2, p. 39; Proposition 2.2.4, p.33; Proposition 2.3.12, p.47]), one has

$$\begin{aligned} \min_{\xi \in \partial(\Phi - \Psi_M)(u_n)} \|\xi\|_{X^*} &= \min_{\xi \in \partial\Phi(u_n) - \partial\Psi_M(u_n)} \|\xi\|_{X^*} = \min_{\eta \in \partial\Psi_M(u_n)} \|\Phi'(u_n) - \eta\|_{X^*} \\ &= \min_{r \in [0,1]} \|\Phi'(u_n) - r\Psi'(u_n)\|_{X^*} = \|\Phi'(u_n) - r_n\Psi'(u_n)\|_{X^*}. \end{aligned}$$

From the compactness of Ψ' and by the fact that $\{r_n\} \subseteq [0,1]$, there are two sequences $\{r_{n_k}\}$ and $\{u_{n_k}\}$ such that $\{r_{n_k}\Psi'(u_{n_k})\}$ converges. From (2) and the previous equalities, one has $\|\Phi'(u_{n_k}) - r_{n_k}\Psi'(u_{n_k})\|_{X^*} \to 0$. Hence, $\{\Phi'(u_{n_k})\}$ also converges and, owing to our assumption on Φ', $\{u_{n_k}\}$ converges. □

REMARK 4. If in the previous theorem, instead of Φ coercive, we assume $\Phi - \Psi$ coercive, then the function $\Phi - \Psi$ satisfies the classical $(P.S.)_c$ condition (see, for instance, [**Z**, Example 38.25, page 162]).

3. Existence and localization of a critical point

Let X be a reflexive real Banach space, let $\Phi, \Psi : X \to \mathbb{R}$ be two continuously Gâteaux differentiable functions, with Φ sequentially weakly lower semicontinuous and Ψ sequentially weakly upper semicontinuous. Assume that Φ is also a coercive functional. Put

$$\varphi^{(1)}(r) = \inf_{u \in [\Phi < r]} \frac{\sup_{[\Phi<r]} \Psi - \Psi(u)}{r - \Phi(u)},$$

we have the following theorem.

THEOREM 5. *Let $r > \inf_X \Phi$. Then, for each $\lambda \in \left]0, \frac{1}{\varphi^{(1)}(r)}\right[$ the restriction of the functional $\Phi - \lambda\Psi$ to $[\Phi < r]$ has at least a critical point (a global minimum).*

PROOF. Fix $\lambda \in \left]0, \frac{1}{\varphi^{(1)}(r)}\right[$. Then, there is $\overline{u}$ such that $\Phi(\overline{u}) < r$ and $\Phi(\overline{u}) - \lambda\Psi(\overline{u}) < r - \lambda \sup_{[\Phi<r]} \Psi$. Moreover, put $M = \frac{r - \Phi(\overline{u})}{\lambda} + \Psi(\overline{u})$. Clearly, one has

$$\sup_{[\Phi<r]} \Psi < M. \tag{1}$$

Finally, put

$$\Psi_M(u) = \begin{cases} \Psi(u) & \text{if } \Psi(u) \leq M \\ M & \text{if } \Psi(u) > M. \end{cases}$$

It is a simple computation to show that Ψ_M is sequentially weakly upper semicontinuous, $\Phi - \lambda\Psi_M$ is a sequentially weakly lower semicontinuous and coercive functional. Therefore (see Theorem 1.2 of [**St**]), it admits a global minimum u_0. If $\Phi(u_0) - \lambda\Psi_M(u_0) = \Phi(\overline{u}) - \lambda\Psi_M(\overline{u})$, the point $\overline{u}$ is the global minimum requested. Otherwise, we claim that $\Psi(u_0) < M$. In fact, from $\Phi(u_0) - \lambda\Psi_M(u_0) < \Phi(\overline{u}) - \lambda\Psi_M(\overline{u})$, it follows, $\Phi(u_0) < \lambda\Psi_M(u_0) + \Phi(\overline{u}) - \lambda\Psi(\overline{u}) \leq \lambda M + \Phi(\overline{u}) - \lambda\Psi(\overline{u}) = r$ and, from (4) one has $\Psi(u_0) < M$. Therefore, $\Phi(u_0) - \lambda\Psi(u_0) = \Phi(u_0) - \lambda\Psi_M(u_0) \leq \Phi(u) - \lambda\Psi_M(u)$ for all $u \in X$ and, taking into account again (4), $\Phi(u_0) - \lambda\Psi(u_0) \leq \Phi(u) - \lambda\Psi(u)$ for all $u \in [\Phi < r]$. Hence, by Theorem 1, the conclusion is proved. □

REMARK 6. A careful reading of the proof of Theorem 5 shows that its conclusion holds also if Φ and Ψ are merely two locally Lipschitz maps. In [**MM2**, Theorem 1.1 (a)] the same result for each $\lambda \in \left]0, \frac{1}{\underline{\varphi}^{(1)}(r)}\right[$ was proved, where

$$\underline{\varphi}^{(1)}(r) = \inf_{u \in [\Phi < r]} \frac{\sup_{\overline{([\Phi<r])}^{w}} \Psi - \Psi(u)}{r - \Phi(u)},$$

and $\overline{([\Phi < r])}^{w}$ is the closure of $[\Phi < r]$ in the weak topology on X.

Since $\varphi^{(1)}(r) \leq \underline{\varphi}^{(1)}(r)$, our result, with a completely different proof, improves [**MM2**, Theorem 1.1 (a)] and [**Ri1**, Theorem 2.5 (a)].

Now, for all $r \in \left]\inf_X \Phi, \sup_X \Phi\right[$, put

$$\varphi^{(2)}(r) := \sup_{u \in [\Phi > r]} \frac{\Psi(u) - \sup_{[\Phi \leq r]} \Psi}{\Phi(u) - r}.$$

We have the following existence result which is a particular case of ([**BD**, Theorem 2.1]). It is worthwhile noticing that it works without requiring that the function Φ is coercive.

THEOREM 7. *Assume that $\Phi - \Psi$ is coercive. If there exists $r \in \left]\inf_X \Phi, \sup_X \Phi\right[$ such that $\varphi^{(2)}(r) > 1$. Then, the functional $\Phi - \Psi$ admits a critical point (the restriction of the function $\Phi - \Psi$ to $[\Phi > r]$ has a global minimum).*

PROOF. Define $\Phi^r : X \to \mathbb{R}$ by setting, for every $u \in X$,

$$\Phi^r(u) := \begin{cases} r & \text{if } \Phi(u) \leq r; \\ \Phi(u) & \text{otherwise.} \end{cases}$$

From our assumptions, it follows that Φ^r is sequentially weakly lower semicontinuous and that $\Phi^r - \Psi$ is coercive. Therefore, owing to [**St**, Theorem 1.2], the function $\Phi^r - \Psi$ admits a global minimum u_1 in X, that is

$$\Phi^r(u_1) - \Psi(u_1) \leq \Phi^r(u) - \Psi(u), \tag{2}$$

for all $u \in X$. Now, we claim that $u_1 \in [\Phi > r]$. Arguing by a contradiction, assume that $u_1 \in [\Phi \leq r]$). From (2) one has

$$r - \Psi(u_1) \leq \Phi(u) - \Psi(u),$$

for all $u \in [\Phi > r]$. Therefore, for all $u \in [\Phi > r]$, one has

$$r - \Phi(u) \leq \Psi(u_1) - \Psi(u) \leq \sup_{[\Phi \leq r]} \Psi - \Psi(u).$$

that is,

$$1 \geq \frac{\Psi(u) - \sup_{[\Phi \leq r]} \Psi(u)}{\Phi(u) - r}.$$

It follows that $1 \geq \varphi^{(2)}(r)$, this is an absurd and our claim is proved. From (2) one has

$$\Phi(u_1) - \Psi(u_1) \leq \Phi(u) - \Psi(u),$$

for all $u \in [\Phi > r]$ and the conclusion follows at once. □

REMARK 8. The existence of a critical point for $\Phi - \Psi$ follows directly from classical results (see, for instance, [**St**, Theorem 1.2]), while, Theorem 7, in addition, guarantees that a critical point u satisfies the condition $\Phi(u) > r$.

It is well known that a very important result on the existence of a critical point for smooth functionals is the famous Mountain Pass Theorem of Ambrosetti-Rabinowitz, see [**AR**]. Here, adopting the $(P.S.)_c^M$ condition instead of the usual one, we present a version of this classical result for functionals of type $\Phi - \Psi$ which in addition gives also a localization of a critical point.

In this direction, here and in the sequel of this section, we require the following assumptions on Φ and Ψ:

(A_1) Φ is convex;

(A_2) for every u_1, u_2 such that $\Psi(u_1) \geq 0$ and $\Psi(u_2) \geq 0$ one has

$$\inf_{t \in [0,1]} \Psi(tu_1 + (1-t)u_2) \geq 0.$$

(A_3) $\inf_X \Phi = \Phi(0) = \Psi(0) = 0$.

Let $u \in X$ and $s > 0$, put $B(u,s) = \{v \in X : \|v - u\| < s\}$ and $\partial B(u,s) = \{v \in X : \|v - u\| = s\}$, we have

THEOREM 9. *Let s and ρ be two positive real numbers. Let u_1, $u_2 \in X$, with $\max\{\Phi(u_1), \Phi(u_2)\} < \rho$, $\min\{\Psi(u_1), \Psi(u_2)\} \geq 0$, $\|u_2 - u_1\| > s$ and $\overline{B}(u_1, s) \subseteq [\Phi < \rho]$. Put, $J(u) = \Phi(u) - \Psi(u)$ for all $u \in X$, assume that*

(M_1) $a = \inf_{\partial B(u_1,s)} J \geq \max\{J(u_1), J(u_2)\}$.
There exists $M > 0$ such that

(M_2) $\sup_{[\Phi < \rho + M])} \Psi < M$.

(M_3) *J satisfies the $(P.S.)_c^M$ condition for all $c \in \mathbb{R}$.*

Then, the function J has a (classical) critical point u_3 distinct from u_1 and u_2 such that

$$J(u_3) \geq \max\{J(u_1); J(u_2)\}$$

and

$$\Phi(u_3) < \rho + M.$$

PROOF. We may suppose, without loss of generality, $u_1 = 0$ and $J(0) = 0$. Moveover, we put

$$\Psi_M(u) = \begin{cases} \Psi(u) & \text{if} \quad \Psi(u) \leq M \\ M & \text{if} \quad \Psi(u) > M, \end{cases}$$

and $J_M(u) = \Phi(u) - \Psi_M(u)$ for all $u \in X$. Clearly, J_M is a locally Lipschitz function, taking into account (M_1) and (M_2), it has the geometry of the mountain pass theorem and, by (M_3), satisfies the Palais-Smale condition $(P.S.)_c$, $c \in \mathbb{R}$. Hence, all the assumptions of the Mountain Pass Theorem for non-differentiable functions (see, for instance, Theorem 2.2 of [**MM1**]) are verified. Therefore, there exists a (generalized) critical point u_3 of J_M such that

$$J_M(u_3) = \inf_{\gamma \in \Gamma} \sup_{t \in [0,1]} J_M(\gamma(t)),$$

where $\Gamma = \{\gamma \in C^0([0,1], X) : \gamma(0) = 0, \gamma(1) = u_2\}$. Moreover, one has

$$J_M(u_3) \geq \max\{J_M(0); J_M(u_2)\}. \tag{3}$$

We claim that u_3 is a (classical) critical point of J. In fact, by choosing as γ the segment with endpoints 0 and u_2, and owing to (A_1) and (A_2), one has $J_M(u_3) \leq \sup_{t \in [0,1]} J_M((1-t)u_2) = \sup_{t \in [0,1]} \Phi((1-t)u_2) \leq \Phi(u_2) < \rho$, namely $J_M(u_3) < \rho$. Hence, owing to (M_2) one has

$$\Phi(u_3) < \rho + M.$$

So, it follows at once that J_M is a continuously Gâteaux differentiable at u_3 and our claim is proved. Finally, it is easy to verify that, from (M_1), we obtain $J(u_3) \geq \max\{J(u_1); J(u_2)\}$. Hence, the proof is completed. □

The next useful consequence of Theorem 9 is a version of [**PS**, Corollary 2.1] for functional $\Phi - \Psi$ obtained by using $(PS)_c^M$ condition instead of the usual $(PS)_c$ where in addition with respect to such classical result we obtain an information about the localization of the third critical point.

THEOREM 10. *Assume that the functional $\Phi - \Psi$ admits two distinct local minima u_1, $u_2 \in X$. Put $\rho \in \mathbb{R}$ such that $\max\{\Phi(u_1), \Phi(u_2)\} < \rho$ and assume that $\min\{\Psi(u_1), \Psi(u_2)\} \geq 0$. Further, suppose that there exists $M > 0$ such that conditions (M_2) and (M_3) of Theorem 9 hold. Then, the function $\Phi - \Psi$ has a third (classical) critical point u_3 distinct from u_1 and u_2 such that*

$$\Phi(u_3) < \rho + M.$$

PROOF. Without loss of generality, we can assume $(\Phi - \Psi)(u_2) \leq (\Phi - \Psi)(u_1)$. Since $u_1 \in [\Phi < \rho]$, which is an open set, there is $r > 0$ such that $B(u_1, r) \subseteq [\Phi < \rho]$. By choosing $s > 0$ such that $s < \min\{\|u_2 - u_1\|; r\}$, the condition (M_1) is easily verified. Hence, the conclusion follows at once from Theorem 9. □

Finally, at once from Theorems 3 and 10 we point out one of the main results of this Section.

THEOREM 11. *Assume that Φ is coercive, $\Phi' : X \to X^*$ admits a continuous inverse operator on X^*, $\Psi' : X \to X^*$ is compact, and $\Phi - \Psi$ admits two distinct local minima u_1, $u_2 \in X$. Put $\rho \in \mathbb{R}$ such that* $\max\{\Phi(u_1), \Phi(u_2)\} < \rho$ *and assume that* $\min\{\Psi(u_1), \Psi(u_2)\} \geq 0$ *and there is $M > 0$ such that condition (M_2) of Theorem 9 holds. Then, the function $\Phi - \Psi$ has a third critical point u_3 distinct from u_1 and u_2 such that*

$$\Phi(u_3) < \rho + M.$$

4. Multiple critical points

Now, provided $r, r_1, r_2 > \inf_X \Phi$, $r_2 > r_1$, we can define

$$\varphi_1(r_1, r_2) = \max\left\{\varphi^{(1)}(r_1); \varphi^{(1)}(r_2)\right\}$$

$$\varphi_2(r_1, r_2) = \inf_{u \in [\Phi < r_1]} \sup_{v \in [r_1 \leq \Phi < r_2]} \frac{\Psi(v) - \Psi(u)}{\Phi(v) - \Phi(u)}.$$

We have the following result [**BC3**, Theorem 3.1] see also [**B1**, Theorem 2.1].

THEOREM 12. *Assume that there are $r_1, r_2 \in \mathbb{R}$, with* $\inf_X \Phi < r_1 < r_2$, *such that*

$$(i_1) \qquad \varphi_1(r_1, r_2) < \varphi_2(r_1, r_2).$$

Then, for each $\lambda \in \Lambda_{r_1, r_2} := \left]\frac{1}{\varphi_2(r_1, r_2)}, \frac{1}{\varphi_1(r_1, r_2)}\right[$ the functional $\Phi - \lambda\Psi$ admits al least two critical points. To be precise, the restriction of the functional $\Phi - \lambda\Psi$ to $[\Phi < r_1]$ admits a global minimum u_1 and the restriction of the functional $\Phi - \lambda\Psi$ to $[\Phi < r_2]$ admits a global minimum $u_2 \notin [\Phi < r_1]$.

PROOF. Fix $\lambda \in \Lambda_{r_1, r_2}$, Theorem 5 ensures the existence both of a global minimum u_1 of the restriction of functional $\Phi - \lambda\Psi$ to $[\Phi < r_1]$ and of a global minimum u_2 of the restriction of the functional $\Phi - \lambda\Psi$ to $[\Phi < r_2]$. Therefore the conclusion follows at once if we show that $u_2 \notin [\Phi < r_1]$. Indeed, if $u_2 \in [\Phi < r_1]$, from $\frac{1}{\lambda} < \varphi_2(r_1, r_2)$ there is $\overline{v} \in [r_1 \leq \Phi < r_2]$ such that $\frac{1}{\lambda} < \frac{\Psi(\overline{v}) - \Psi(u_2)}{\Phi(\overline{v}) - \Phi(u_2)}$, that is $\Phi(\overline{v}) - \lambda\Psi(\overline{v}) < \Phi(u_2) - \lambda\Psi(u_2)$ and this is absurd since u_2 is a global minimum in $[\Phi < r_2]$. Hence, again by Theorem 1, the proof is completed. □

REMARK 13. We also observe that Theorem 12 works also for non-smooth functionals as one can see in [**BC3**, Theorem 3.1].

REMARK 14. If in Theorem 12, in addition, we assume the $(PS)_c$ condition, owing to [**PS**, Corollary 1], see also [**MM1**, Corollary 2.1] in the non smooth case, the functional $\Phi - \lambda\Psi$, for each $\lambda \in \Lambda_{r_1, r_2}$, admits at least three distinct critical points. On this subject, we refer also to [**BM**, Theorem 2.5], where the authors give a three critical point theorem under the assumptions of Theorem 12 whenever Φ and Ψ turn out to be two locally Lipschitz functions on X which satisfy a more general Palais-Smale condition, the so called Cerami's condition. However, nothing is known about the localization of the third point.

Given $r_3 > 0$, we define

$$\varphi^{(3)}(r_2, r_3) = \frac{\sup_{[\Phi<r_2+r_3]} \Psi}{r_3}$$

and

$$\varphi_3(r_1, r_2, r_3) = \max\{\varphi_1(r_1, r_2); \varphi^{(3)}(r_2, r_3)\}.$$

A useful three critical points result in order to obtain a localization of the third critical point is the following [**BC3**, Theorem 3.3].

THEOREM 15. *Suppose that conditions* (A_1), (A_2) *and* (A_3) *above hold. Assume that there are three positive constants* r_1, r_2, r_3, *with* $r_1 < r_2$, *such that*

(i_2) $\qquad \varphi_3(r_1, r_2, r_3) < \varphi_2(r_1, r_2)$.

Assume also that for each $\lambda \in \Lambda_{r_1,r_2,r_3} := \left]\frac{1}{\varphi_2(r_1,r_2)}, \frac{1}{\varphi_3(r_1,r_2,r_3)}\right[$ *one has*

(M_λ) *the functional* $\Phi - \lambda\Psi$ *fulfills* $(PS)_c^{\frac{r_3}{\lambda}}$, $c \in \mathbb{R}$.

Then, for each $\lambda \in \Lambda_{r_1,r_2,r_3}$ *the functional* $\Phi - \lambda\Psi$ *admits at least three critical points* u_1, u_2, u_3 *such that* $u_1 \in [\Phi < r_1]$, $u_2 \in [r_1 \leq \Phi < r_2]$ *and* $u_3 \in [\Phi < r_2 + r_3]$.

PROOF. Fix $\lambda \in \Lambda_{r_1,r_2,r_3}$. Owing to Theorem 12 the restriction of $\Phi - \lambda\Psi$ to $[\Phi < r_1]$ admits a global minimum that we call u_1, which is a local minimum of $\Phi - \lambda\Psi$, and the restriction of $\Phi - \lambda\Psi$ to $[\Phi < r_2]$ admits a global minimum that we call u_2, which is a local minimum of $\Phi - \lambda\Psi$ belonging to $[\Phi < r_2]$. We explicitly observe that

$$(\Phi - \lambda\Psi)(u_2) < (\Phi - \lambda\Psi)(u_1).$$

In fact, from $\frac{1}{\lambda} < \varphi_2(r_1, r_2)$ there is $\overline{v} \in [r_1 \leq \Phi < r_2]$ such that $\Phi(\overline{v}) - \lambda\Psi(\overline{v}) < \Phi(u_1) - \lambda\Psi(u_1)$, so that $(\Phi - \lambda\Psi)(u_2) \leq (\Phi - \lambda\Psi)(\overline{v}) < (\Phi - \lambda\Psi)(u_1)$. We may suppose, without loss of generality, $u_1 = 0$ and $(\Phi - \lambda\Psi)(u_1) = 0$. Moreover, since $\lambda < \dfrac{1}{\varphi^{(3)}(r_2, r_3)}$, one has

$$\sup_{[\Phi<r_2+r_3]} \lambda\Psi < r_3,$$

that is condition (M_2) of Theorem 9 holds. Hence, bearing in mind (M_λ), our conclusion follows at once by applying Theorem 10 to the functional $\Phi - \lambda\Psi$. □

REMARK 16. Taking into account the proof of Theorems 10 and 15, the assumption (A_2) can be replaced by the following more general condition:

(A_2') *for each* $\lambda \in \Lambda_{r_1,r_2,r_3}$ *and for every* u_1, u_2, *which are local minima for the functional* $\Phi - \lambda\Psi$, *and such that* $\Psi(u_1) \geq 0$ *and* $\Psi(u_2) \geq 0$ *one has*

$$\inf_{t\in[0,1]} \Psi(tu_1 + (1-t)u_2) \geq 0.$$

REMARK 17. With respect to classical results, Theorem 15 works with $(PS)_c^M$ condition instead of $(PS)_c$ and, in addition, since Φ is coercive, there exists $\sigma > 0$ such that

$$[\Phi < \rho_2) \subseteq B(0_X, \sigma).$$

Therefore, the conclusion of the previous theorem ensures that the three critical points are uniformly bounded in norm with respect to λ, that is

$$\|u_i\| < \sigma,$$

$i = 1,2,3$, for all $\lambda \in \Lambda_{r_1,r_2,r_3}$. With respect to [**MM1**, Theorem 3.1 and B] (see also [**BC2**, Theorem 2.1]), the coercivity of $\Phi - \lambda\Psi$ is not requested and a precisely determined interval of parameters λ for which the functional has three critical points, is obtained. We also observe that the key assumption of [**MM1**, Theorem 3.1] is a suitable minimax inequality which is equivalent to an inequality of similar type to (i_1) (see [**B1**], [**B3**]). We also observe that Theorem 15 gives back also [**AB2**, Theorem 5.1].

We point out one of the consequences of Theorem 15.

THEOREM 18. *Assume that $\Phi' : X \to X^*$ admits a continuous inverse operator on X^*, and $\Psi' : X \to X^*$ is compact. Assume that there are two positive constants ρ_1, ρ_2 and $\overline{v} \in X$, with $\rho_1 < \Phi(\overline{v}) < \rho_2/2$, such that*

$$(i_3) \qquad \frac{\sup\limits_{[\Phi<\rho_1]} \Psi}{\rho_1} < \frac{1}{2}\frac{\Psi(\overline{v})}{\Phi(\overline{v})},$$

$$(i_4) \qquad \frac{\sup\limits_{[\Phi<\rho_2]} \Psi}{\rho_2} < \frac{1}{4}\frac{\Psi(\overline{v})}{\Phi(\overline{v})}.$$

Then, for each

$$\lambda \in \Lambda'_{\rho_1,\rho_2,\overline{v}} := \left] \frac{2\Phi(\overline{v})}{\Phi(\overline{v})}, \min\left\{ \frac{\rho_1}{\sup\limits_{[\Phi<\rho_1]} \Psi}; \frac{\rho_2/2}{\sup\limits_{[\Phi<\rho_2]} \Psi} \right\} \right[,$$

the functional $\Phi - \lambda\Psi$ admits at least three critical points u_1, u_2, u_3 which lie in $[\Phi < \rho_2]$.

PROOF. Put $r_1 = \rho_1$, $r_2 = r_3 = \rho_2/2$. Clearly, one has

$$\varphi^{(1)}(r_1) \le \frac{\sup_{[\Phi<\rho_1]} \Psi}{\rho_1}, \quad \varphi^{(1)}(r_2) \le \frac{\sup_{[\Phi<\rho_2/2]} \Psi}{\rho_2/2} \le \frac{\sup_{[\Phi<\rho_2]} \Psi}{\rho_2/2},$$

$$\varphi^{(3)}(r_2, r_3) = \frac{\sup_{[\Phi<\rho_2]} \Psi}{\rho_2/2}.$$

Hence,

$$(4) \qquad \varphi_3(r_1, r_2, r_3) \le \max\left\{ \frac{\sup_{[\Phi<\rho_1]} \Psi}{\rho_1}; 2\frac{\sup_{[\Phi<\rho_2]} \Psi}{\rho_2} \right\}.$$

On the other hand, taking into account that from (i_3) one has $\sup\limits_{[\Phi<\rho_1]} \Psi < \Psi(\overline{v})$, and since $\Phi(u) \ge 0$ for all $u \in X$, we obtain

$$\varphi_2(r_1, r_2) \ge \frac{\Psi(\overline{v}) - \sup\limits_{[\Phi<\rho_1]} \Psi}{\Phi(\overline{v})} \ge \frac{\Psi(\overline{v})}{\Phi(\overline{v})} - \frac{\sup\limits_{[\Phi<\rho_1]} \Psi}{\rho_1}.$$

Hence, from (i_4), one has

$$\varphi_2(r_1,r_2) \geq \frac{1}{2}\frac{\Psi(\overline{v})}{\Phi(\overline{v})}. \tag{5}$$

Therefore, taking (i_3) and (i_4) into account, Theorem 15 ensures our conclusion. □

REMARK 19. The same reasoning made in the previous proof shows that when condition (A_3) above holds the following conditions

(i_2') For some r_1, r_2, $w \in X$ one has $r_1 < \Phi(w) < r_2$ such that

$$\frac{\sup_{[\Phi<r_1]}\Psi}{r_1} < \frac{\Psi(w)}{\Phi(w)}, \qquad \frac{\sup_{[\Phi<r_2]}\Psi}{r_2} < \frac{\Psi(w)}{\Phi(w)}.$$

forces (i_1) of Theorem 12.

REMARK 20. Put

$$\chi(r) = \sup_{[\Phi<r])} \Psi$$

for all $r > \inf_X \Phi$ and assume that condition (A_3) holds, that is

$$\inf_X \Phi = \Phi(0) = \Psi(0) = 0. \tag{6}$$

As seen in the proof of Theorem 18, the inequality

$$\frac{\chi(r_1)}{r_1} < \frac{1}{2}\frac{\Psi(\overline{v})}{\Phi(\overline{v})},$$

with $\Phi(\overline{v}) > r_1$ implies $\varphi^{(1)}(r_1) < \varphi_2(r_1,r_2)$; the inequality

$$\frac{\chi(r_2)}{r_2} < \frac{1}{2}\frac{\Psi(\overline{v})}{\Phi(\overline{v})},$$

with $r_1 < \Phi(\overline{v}) \leq r_2$, implies $\varphi^{(1)}(r_2) < \varphi_2(r_1,r_2)$; and, finally, the inequality

$$\frac{\chi(r_2+r_3)}{r_3} < \frac{1}{2}\frac{\Psi(\overline{v})}{\Phi(\overline{v})},$$

with $r_1 < \Phi(\overline{v}) \leq r_2$ and $r_3 > 0$, implies $\varphi^{(3)}(r_2,r_3) < \varphi_2(r_1,r_2)$. Therefore, to apply Theorem 15, to investigate nonlinear differential and difference problems, it is enough to estimate an upper bound of the function $\chi(r)$ for some $r > 0$.

REMARK 21. We explicitly observe that the constants in inequalities (i_3) and (i_4) of Theorem 18 can be improved when ρ_1 is significantly less than $\Phi(\overline{v})$. For instance, if $n\rho_1 < \Phi(\overline{v}) < \rho_2/2$ with $n \in \mathbb{N}$, arguing in the same way, we can write $\frac{n}{n+1}$ instead of $\frac{1}{2}$ in (i_3) and $\frac{1}{2}\frac{n}{n+1}$ instead of $\frac{1}{4}$ in (i_4), while the first endpoint of $\Lambda'_{\rho_1,\rho_2,\overline{v}}$ becomes $\frac{n+1}{n}\frac{\Phi(\overline{v})}{\Psi(\overline{v})}$. In particular, if

(i_3') $\dfrac{\sup_{[\Phi<\rho]}\Psi}{\rho} < \dfrac{\Psi(\overline{v})}{\Phi(\overline{v})}$ for all ρ small enough;

$$(i_4') \qquad \frac{\sup\limits_{[\Phi<\rho_2]} \Psi}{\rho_2} < \frac{1}{2}\frac{\Psi(\overline{v})}{\Phi(\overline{v})} \text{ with } 0 < \Psi(\overline{v}) < \rho_2/2,$$

then the first endpoint of the interval in the conclusion of Theorem 18 becomes $\frac{\Phi(\overline{v})}{\Psi(\overline{v})}$.
In this connection we mention also the following [**B1**, Theorem 2.2]

THEOREM 22. *Assume that there exist three constants r_1, r_2 and r_3 such that*

(k) $\inf\limits_X \Phi < r_1 < r_2 < r_3$*;*

(kk) $\max\{\varphi^{(1)}(r_1),\ \varphi^{(1)}(r_2),\ \varphi^{(1)}(r_3)\} < \min\{\varphi_2(r_1, r_2), \varphi_4(r_1, r_2, r_3\}$,

where

$$\varphi_4(r_1, r_2, r_3\} = \inf_{u\in[r_1\leq\Phi<r_2]} \sup_{v\in[r_2\leq\Phi<r_3]} \frac{\Psi(v)-\Psi(u)}{\Phi(v)-\Phi(u)}.$$

Then, there exists a positive σ such that for each

$$\lambda \in \left]\max\left\{\frac{1}{\varphi_2(r_1, r_2)}, \frac{1}{\varphi_4(r_1, r_2, r_3\}}\right\}; \min\left\{\frac{1}{\varphi^{(1)}(r_1)}, \frac{1}{\varphi^{(1)}(r_2)}, \frac{1}{\varphi^{(1)}(r_3)}\right\}\right[,$$

the functional $\Phi - \lambda\Psi$ admits at least three critical point (three local minimum) whose norms are less than σ.

PROOF. Fix

$$\lambda \in \left]\max\left\{\frac{1}{\varphi_2(r_1, r_2)}, \frac{1}{\varphi_4(r_1, r_2, r_3\}}\right\}; \min\left\{\frac{1}{\varphi^{(1)}(r_1)}, \frac{1}{\varphi^{(1)}(r_2)}, \frac{1}{\varphi^{(1)}(r_3)}\right\}\right[.$$

From Theorem 12, the functional $\Phi - \lambda\Psi$ has two local minima, namely u_1 and u_2 which lie in $[\Phi < r_1]$ and $[r_1 \leq \Phi < r_2]$, respectively. Moreover, since $\frac{1}{\lambda} > \varphi^{(1)}(r_3)$ Theorem 5 ensures that the restrictions of the functional $\Phi - \lambda\Psi$ to $[\Phi < r_3]$ has a global minimum, say u_3. We claim that $u_3 \notin [\Phi < r_2]$. First, taking into account that $\frac{1}{\lambda} < \varphi_2(r_1, r_2)$ and arguing as in the proof of Theorem 5 we obtain that $u_3 \notin [\Phi < r_1]$. Next, since one has $\frac{1}{\lambda} < \varphi_4(r_1, r_2, r_3)$ and arguing again as in the proof of Theorem 5 we obtain that $u_3 \notin [r_1 \leq \Phi < r_2]$. So, our claim is proved. The conclusion follows at once by choosing σ such that $\|u\| \leq \sigma$ for each $u \in [\Phi < r_3]$ and taking into account that u_1, u_2 and u_3 are three distinct critical point for the functional $\Phi - \lambda\Psi$. □

REMARK 23. Recently, three critical points theorems were established, [**B1**] and [**B2**]. In [**B2**, Theorem 2.1] (which is based on [**Ri2**, Theorem 3]) was established the existence of an interval $\Lambda \subseteq [0, +\infty[$ such that, for each $\lambda \in \Lambda$, the function $\Phi - \lambda\Psi$ has at least three critical points which are uniformly bounded in norm with respect to λ. However, there, only an upper bound of the interval Λ was guaranteed. On the other hand, in [Theorem B] (which is based on the variational principle of B. Ricceri in [**Ri1**]) a precise interval of parameters λ for which the function $\Phi - \lambda\Psi$ has three critical points was established, however losing the uniform bounded-ness in norm. For a more precise comparison between Theorem 2.1 of [**B2**] and Theorem B of we refer to [**B3**]. Here, we explicitly observe that the conclusions of previous theorems ensure both a precise interval of parameters λ for which the function

$\Phi - \lambda\Psi$ has at least three critical points and the uniform bounded-ness in norm of the three critical points with respect to λ. Further, we point out that one of the key assumptions both in Theorem 2.1 of [**B2**] and in Theorem B is

$$\lim_{\|x\|\to+\infty} (\Phi(x) - \lambda\Psi(x)) = +\infty$$

that is not requested in Theorems 15 and 18. On the other hand, in those theorems the assumption

$$\varphi^{(3)}(r_2, r_3) < \varphi_2(r_1, r_2)$$

is not requested. We also observe that the assumptions of Theorem 22 are mutually independent from those of Theorem 15 and, in that case, the third critical point is actually a local minimum point.

In [**BM**] a three critical points result for locally Lipschitz and coercive functionals has been given. Here we give a version of such result which it is more simple to apply in studying difference equations.

THEOREM 24. *Assume that $\Phi' : X \to X^*$ admits a continuous inverse operator on X^*, and $\Psi' : X \to X^*$ is compact. Suppose that*

(i_5) *There exist $r \in \mathbb{R}$, $u_0, u_1 \in X$ fulfilling $\Phi(u_0) < r < \Phi(u_1)$ and, moreover,*

$$\sup_{[\Phi\le r]} \Psi < \frac{(\Phi(x_1) - r)\Psi(u_0) + (r - \Phi(u_0))\Psi(x_1)}{\Phi(x_1) - \Phi(u_0)}.$$

Put,

$$\Lambda_r := \left] \frac{\Phi(u_1) - r}{\Phi(u_1) - \sup_{[\Psi\le r]} \Psi}, \frac{r - \Phi(u_0)}{\sup_{[\Psi\le r]} \Psi - \Psi(u_0)} \right[,$$

assume that

(i_6) *for each $\lambda \in \Lambda_r$, $\Phi - \lambda\Psi$ is a coercive functional.*

Then, for each $\lambda \in \Lambda_r$ the functional $\Phi - \lambda\Psi$ admits at least three critical points.

PROOF. Fix $\lambda \in \Lambda_r$, simple calculation guarantee that

$$\frac{1}{\varphi^{(2)}(r)} < \lambda < \frac{1}{\varphi^{(1)}(r)}.$$

Hence, Theorem 5 ensures the existence of a local minimum u_1 of $\Phi - \lambda\Psi$ which lies in $[\Phi < r]$, while, since $\Phi - \lambda\Psi$ is coercive, Theorem 7 furnishes a local minimum u_2 of $\Phi - \lambda\Psi$ belonging to $[\Phi > r]$. Taking into account the properties of Φ' and Ψ' and arguing as in the proof of Theorem 3, see also [**St**], the coercivity of the functional $\Phi - \lambda\Psi$ guarantees that the (PS) condition holds. Therefore, the conclusion follows by [**PS**, Corollary 2.1], see also Remark 14. □

REMARK 25. In particular, if $\Phi(0) = \Psi(0) = 0$ then (i_5) obviously reduces to
(i_6') *For some $r > 0$, $u_1 \in X$ one has $r < \Phi(u_1)$ as well as*

$$\sup_{[\Phi \le r]} \Psi < r\frac{\Psi(u_1)}{\Phi(u_1)}.$$

In this case, by Theorem 24, the function $\Phi - \lambda\Psi$ possesses at least three critical points provided

$$\frac{\Phi(u_1)}{\Psi(u_1)} < \lambda < \frac{r}{\sup_{[\Phi_r \le r]} \Psi}. \tag{7}$$

Evidently, this latter inequality is less restrictive than (i_3) and (i_4) of Theorem 18, i.e. Theorem 24 allows a wider range of parameter λ. However, in Theorem 15 the coercivity of $\Phi - \lambda\Psi$ is not requested and the three critical point are uniformly bounded in norm with respect to λ. Instead, Theorem 24 furnishes only a localization of at least a critical point in $[\Phi < r]$ and of other one in $[\Phi > r]$.

REMARK 26. Basically, we have three different kinds of Theorems for functionals like $\Phi - \lambda\Psi$ for which it is possible to guarantee the existence of at least three critical points. To be precise:

(I_1) For functionals that satisfy condition (i_1) of Theorem 12 and the Palais-Smale condition. In this case, our conclusion holds for each $\lambda \in \Lambda_{r_1,r_2}$ (see Remark 14). However, we loss the localization of the third critical point.

(I_2) For functionals possibly unbounded and satisfying Palais-Smale condition cut off at M, for which condition (i_2) of Theorem 15 holds or conditions (i_3) and (i_4) of Theorem 18 are verified (see also Theorem 22). In this case, the three critical points are uniformly bounded in norm for each $\lambda \in \Lambda_{r_1,r_2,r_3}$ with $\Lambda_{r_1,r_2,r_3} \subset \Lambda_{r_1,r_2}$.

(I_3) For coercive functionals fulfilling condition (i_5) of Theorem 24. In this case, with respect the previous cases the conclusion holds for each $\lambda \in \Lambda_r$ with $\Lambda_{r_1,r_2} \subset \Lambda_r$.

REMARK 27. Clearly, whenever a functional turns out to be bounded below, (I_1) is a special case of (I_3), see for instance [**MR**, Corollary 1.3].

Moreover, provided $r > \inf_X \Phi$, we can define

$$\gamma := \liminf_{r\to+\infty} \varphi^{(1)}(r), \qquad \delta := \liminf_{r\to(\inf_X \Phi)^+} \varphi^{(1)}(r).$$

The next result is devoted to the existence of infinitely many critical points for functional of the type $\Phi - \lambda\Psi$, which is based on Theorem 5.

THEOREM 28. ([**BMo**, Theorem 2.1]) *Under the above assumptions on X, Φ and Ψ, one has*

(a) *For every $r > \inf_X \Phi$ and every $\lambda \in \left]0, \frac{1}{\varphi^{(1)}(r)}\right[$, the restriction of the functional $\Phi - \lambda\Psi$ to $[\Phi < r]$ admits a global minimum, which is a critical point (local minimum) of $\Phi - \lambda\Psi$ in X.*

(b) *If* $\gamma < +\infty$ *then, for each* $\lambda \in]0, \frac{1}{\gamma}[$, *the following alternative holds: either*

(b_1) $\Phi - \lambda\Psi$ *possesses a global minimum,*

or

(b_2) *there is a sequence* $\{u_n\}$ *of critical points (local minima) of* $\Phi - \lambda\Psi$ *such that* $\lim_{n\to+\infty} \Phi(u_n) = +\infty$.

(c) *If* $\delta < +\infty$ *then, for each* $\lambda \in]0, \frac{1}{\delta}[$, *the following alternative holds: either*

(c_1) *there is a global minimum of* Φ *which is a local minimum of* $\Phi - \lambda\Psi$,

or

(c_2) *there is a sequence* $\{u_n\}$ *of pairwise distinct critical points (local minima) of* $\Phi - \lambda\Psi$, *with* $\lim_{n\to+\infty} \Phi(u_n) = \inf_X \Phi$, *which weakly converges to a global minimum of* Φ.

PROOF. Clearly (a) is exactly the conclusion of Theorem 5. Let us prove (b) and put $I_\lambda = \Phi - \lambda\Psi$. Pick $\lambda \in]0, \frac{1}{\gamma}[$ and assume that (b_1) is not true. We will show that when $\gamma < +\infty$ and I_λ does not posses a global minimum in X, then I_λ admits a sequence of critical points. Let $a \in \mathbb{R}$ such that $\lambda < a < \frac{1}{\gamma}$. From $\liminf_{r\to+\infty} \varphi(r) < \frac{1}{a}$ there exists a sequence $\{r_n\}$ such that $\lim_{n\to+\infty} r_n = +\infty$ and $\varphi(r_n) < \frac{1}{a}$ for all $n \in \mathbb{N}$. Put $r_{n_1} = r_1$. Owing to (a), one can find $u_1 \in [\Phi < r_{n_1}])$ such that u_1 is a local minimum for I_λ. From our assumption, u_1 is not a global minimum for I_λ. Therefore, there exists $\overline{u}_1 \in X$ such that $I_\lambda(\overline{u}_1) < I_\lambda(u_1)$. Hence, $\overline{u}_1 \notin [\Phi < r_{n_1}]$. Let $r_{n_2} \in \{r_n\}$ such that $r_{n_2} > \Phi(\overline{u}_1)$. Again from (a), there is $u_2 \in [\Phi < r_{n_2}]$ such that u_2 is a local minimum for the functional I_λ. Taking into account that u_2 is a global minimum in $[\Phi < r_{n_2}]$, we have that $I_\lambda(u_2) \leq I_\lambda(\overline{u}_1)$ and $I_\lambda(\overline{u}_1) < I_\lambda(u_1)$. Hence, $\Phi(u_2) > r_{n_1}$. Reasoning inductively we obtain a sequence $\{u_k\}$ of distinct critical points such that $\Phi(u_{k+1}) > r_{n_k}$ for all $k \in \mathbb{N}$. Hence, (b_2) holds.

Finally, we prove (c). Fix $\lambda \in]0, \frac{1}{\delta}[$ and let $\overline{u} \in X$ such that $\Phi(\overline{u}) = \inf_X \Phi$. Assume that ($c_1$) does not hold, that is $\overline{u}$ is not a local minimum for I_λ. Consider $a \in \mathbb{R}$ such that $\lambda < a < \frac{1}{\delta}$. By our assumption, $\liminf_{r\to(\inf_X \Phi)^+} \varphi(r) = \delta < \frac{1}{a}$, hence there exists a decreasing sequence $\{r_n\}$ such that $\lim_{n\to+\infty} r_n = \inf_X \Phi$ and $\varphi(r_n) < \frac{1}{a}$ for all $n \in \mathbb{N}$. Put $r_{n_1} = r_1$. Owing to (a), there exists $u_1 \in [\Phi < r_{n_1}])$, which is a local minimum for I_λ. Therefore, $u_1 \neq \overline{u}$. Then, $\Phi(\overline{u}) < \Phi(u_1)$. Let $r_{n_2} \in \{r_n\}$ such that $\Phi(\overline{u}) < r_{n_2} < \Phi(u_1)$. Again from (a), there is $u_2 \in [\Phi < r_{n_2}]$ which is a local minimum for the functional I_λ, with $\Phi(\overline{u}) < \Phi(u_2) < \Phi(u_1)$. Reasoning inductively we obtain a sequence $\{u_k\}$ of distinct local minima for I_λ such that $\Phi(\overline{u}) < \Phi(u_k) < r_{n_k}$ for all $k \in \mathbb{N}$. Hence, $\lim_{k\to+\infty} \Phi(u_k) = \inf_X \Phi$. Moreover, since $\{u_k\} \subseteq [\Phi < r_{n_1}]$ and Φ is coercive, then it is bounded. Since X is reflexive, taking a subsequence if necessary, $\{u_k\}$ weakly converges to $u^* \in X$. From the weak sequential lower semicontinuity, one has $\Phi(u^*) \leq \lim_{k\to+\infty} \Phi(u_k) = \inf_X \Phi$, that is $\Phi(u^*) = \inf_X \Phi$. Hence, the conclusion is obtained. □

REMARK 29. We explicitly observe that the proof here outlined is different from that proposed by Ricceri in [**Ri1**] and Marano and Motreanu in [**MM2**]. Further we do not use the weak closure of the sub-levels $[\Phi < r]$, for $r > \inf_X \Phi$.

5. Critical point theory in finite-dimensional spaces

Here and in the sequel X denotes a finite dimensional real Banach space and we assume the following structural hypothesis

(Λ) $\Phi, \Psi : X \to \mathbb{R}$ are two functionals of class C^1 on X with Φ coercive, and λ is a positive parameter.

In this framework, an immediately consequence of Theorem 24 is the following

THEOREM 30. *Assume that* $\Phi(0) = \Psi(0) = 0$ *and*

(j_1) *There exist* $r > 0$, $u \in X$ *with* $r < \Phi(u)$ *such that*

$$\sup_{[\Phi \le r]} \Psi < r \frac{\Psi(u)}{\Phi(u)}.$$

Put,

$$\Lambda'_r := \left] \frac{\Phi(u)}{\Psi(u)}, \frac{r}{\sup_{[\Psi \le r]} \Psi} \right[,$$

assume that

(j_2) *for each* $\lambda \in \Lambda'_r$, $\Phi - \lambda\Psi$ *is a coercive functional.*

Then, for each $\lambda \in \Lambda'_r$ *the functional* $\Phi - \lambda\Psi$ *admits at least three critical points with at least one belonging to* $[\Phi < r]$ *and an other one in* $[\Phi > r]$.

A finite dimensional variant of Theorem 18 is the following

THEOREM 31. *Assume that the conditions* (A_1), (A'_2) *and* (A_3) *introduced in the previous section hold. Further, assume that there are two positive constants* ρ_1, ρ_2 *and* $\overline{v} \in X$, *with* $\rho_1 < \Phi(\overline{v}) < \rho_2/2$, *such that*

(j_3) $$\frac{\sup_{[\Phi<\rho_1]} \Psi}{\rho_1} < \frac{1}{2}\frac{\Psi(\overline{v})}{\Phi(\overline{v})}$$

(j_4) $$\frac{\sup_{[\Phi<\rho_2]} \Psi}{\rho_2} < \frac{1}{4}\frac{\Psi(\overline{v})}{\Phi(\overline{v})}.$$

Then, for each $\lambda \in \left] \frac{2\Phi(\overline{v})}{\Psi(\overline{v})}, \min\left\{ \frac{\rho_1}{\sup_{[\Phi<\rho_1]} \Psi}; \frac{\rho_2/2}{\sup_{[\Phi<\rho_2]} \Psi} \right\} \right[$ *the functional* $\Phi - \lambda\Psi$ *admits at least three distinct critical points* u_1, u_2, u_3 *such that* $u_1 \in [\Phi < \rho_1]$, $u_2 \in [\rho_1 \le \Phi < \frac{\rho_2}{2}]$ *and* $u_3 \in [\Phi < \rho_2]$.

PROOF. Hence, taking into account that X is finite dimensional, it is enough to observe that the functional $\Phi - \lambda\Psi$ satisfies all the assumptions of Theorem 15 and the conclusion follows. □

For all $r > \inf_X \Phi$, put

$$\lambda^* := \frac{1}{\inf_{\{r > \inf_X \Phi\}} \varphi^{(1)}(r)},$$

where we read $\frac{1}{0} := +\infty$ if this case occurs.

We point out the following result [**BC4**, Theorem 2.2].

THEOREM 32. *Assume that, for each $\lambda \in]0, \lambda^*[$, one has:*

(j_5) *there exists $x_1 \in X$ such that $\lim_{t\to+\infty} \Phi(tx_1) - \lambda\Psi(tx_1) = -\infty$.*

(j_6) *$\Phi - \lambda\Psi$ satisfies the $(PS)_c$ condition.*

Then, for each $\lambda \in]0, \lambda^*[$, $\Phi - \lambda\Psi$ admits at least two distinct critical points.

PROOF. Fix $\lambda \in]0, \lambda^*[$, and define $I_\lambda = \Phi - \lambda\Psi$. We observe that there exists $r > \inf_X \Phi$ such that $\lambda < 1/\varphi^{(1)}(r)$. Hence, arguing as in the proof of Theorem 5, the functional I_λ admits a local minimum, say u_1. Therefore, taking (j_5) into account, there exist $\rho > 0$ and $t > 0$ large enough, such that

$$I_\lambda(tx_1) < I_\lambda(u_1) \leq \inf_{\|u-u_1\|=\rho} I_\lambda(u).$$

Hence, by (j_6) the Mountain Pass Theorem ensures the existence of a critical point u_2, which is distinct from u_1. □

REMARK 33. We explicitly observe that whenever X turns out to be an infinite dimensional Banach space, assuming in addition that X is reflexive and Φ and $-\Psi$ are sequentially weakly lower semicontinuous, Theorem 32 again holds. We also observe that Theorem 32 again holds even if Φ and Ψ are non-smooth functionals, that is, they are only locally Lipschitz functionals without being of class C^1 on X. The proof is the same as Theorem 32 by using Mountain Pass Theorem for non-smooth functionals (see, for instance, [**MM1**, Theorem 2.2]). Clearly, in this case, generalized critical points are obtained.

REMARK 34. When either X is a finite dimensional Banach space or Φ and Ψ are functionals regular enough ($\Phi' : X \to X^*$ admits a continuous inverse operator on X^*, and $\Psi' : X \to X^*$ is compact), the following condition

(j_7) $\lim_{\|u\|\to\infty} \Phi(u) - \lambda\Psi(u) = -\infty$

implies (j_5) and (j_6). When X is finite dimensional, the following [**BC4**, Theorem 2.3] shows that condition (j_7) actually guarantees three critical points.

THEOREM 35. *Assume (j_7) for each $\lambda \in]0, \lambda^*[$. Then, for each $\lambda \in]0, \lambda^*[$ the functional $\Phi - \lambda\Psi$ admits at least three distinct critical points.*

PROOF. Fix $\lambda \in]0, \lambda^*[$. Arguing as above, the functional $\Phi - \lambda\Psi$ has a local minimum. Since X is a finite dimensional space, from (j_7), $\Phi - \lambda\Psi$ admits a global maximum, which clearly is a local maximum. Moreover, again from (j_7) and the finite dimension of X, $\Phi - \lambda\Psi$ satisfies the Palais-Smale condition. Therefore, if $dim(X) \geq 2$, Corollary 2.11 of [**F**] ensures the third critical point. Otherwise, condition (j_7), by standard arguments, ensures the same conclusion. □

Moreover, in this framework we recall a finite version for even functional of Clarke's Theorem see also [**R**, Theorem 9.1]

THEOREM 36. *Assume that $\Phi - \Psi$ is a coercive functional such that $\Phi(0) = \Psi(0)$. In addition, suppose that there exist a set $K \subset S$ homeomorphic to $S^{N-1} = \{u \in S : \|u\| = 1\}$ such that $\sup_K \Phi < \inf_K \Psi$. Then, the functional $\Phi - \Psi$ admits at least N distinct pairs of critical points.*

Finally, for reader convenience we recall a finite version of Theorem 28

THEOREM 37. *Assume that condition* (Λ) *holds, one has*

(a) *For every* $r > \inf_X \Phi$ *and every* $\lambda \in \left]0, \frac{1}{\varphi(1)(r)}\right[$, *the restriction of the functional* $\Phi - \lambda\Psi$ *to* $[\Phi < r]$ *admits a global minimum, which is a critical point (local minimum) of* $\Phi - \lambda\Psi$ *in* X.

(b) *If* $\gamma < +\infty$ *then, for each* $\lambda \in \left]0, \frac{1}{\gamma}\right[$, *the following alternative holds: either*

(b_1) $\Phi - \lambda\Psi$ *possesses a global minimum,*

or

(b_2) *there is a sequence* $\{u_n\}$ *of critical points (local minima) of* $\Phi - \lambda\Psi$ *such that* $\lim_{n\to+\infty} \Phi(u_n) = +\infty$.

(c) *If* $\delta < +\infty$ *then, for each* $\lambda \in \left]0, \frac{1}{\delta}\right[$, *the following alternative holds: either*

(c_1) *there is a global minimum of* Φ *which is a local minimum of* $\Phi - \lambda\Psi$,

or

(c_2) *there is a sequence* $\{u_n\}$ *of pairwise distinct critical points (local minima) of* $\Phi - \lambda\Psi$, *with* $\lim_{n\to+\infty} \Phi(u_n) = \inf_X \Phi$, *which converges to a global minimum of* Φ.

6. Difference equations and variational framework

The end of this section is to point out the variational formulation of problem $(P_\lambda^{f,q})$ as well as some useful auxiliaries results on difference equations.

Let N be a positive integer, denote with $[1, N]$ the discrete interval $\{1, 2, ..., N\}$, we consider the following problem

$$(P_\lambda^{f,q}) \begin{cases} -\Delta(\phi_p(\Delta u_{k-1})) + q_k\phi_p(u_k) = \lambda f(k, u_k), & k \in [1, N], \\ u_0 = u_{N+1} = 0, \end{cases}$$

where, $f : [1, N] \times \mathbb{R} \to \mathbb{R}$ is a continuous function, $\Delta u_{k-1} = u_k - u_{k-1}$ is the forward difference operator, $q_k \in \mathbb{R}_0^+$ for all $k \in [1, N]$, $\phi_p(s) := |s|^{p-2}s$, $1 < p < +\infty$, and $\lambda \in \mathbb{R}^+$.

On the N-dimensional Banach space $S = \{u : [0, N+1] \to \mathbb{R} : u_0 = u_{N+1} = 0\}$ we introduce the norm, (see Lemma 40 below),

$$\|u\| := \left(\sum_{k=1}^{N+1} |\Delta u_{k-1}|^p + q_k|u_k|^p \right)^{1/p}.$$

Moreover, for every $u \in S$, we put

$$\Phi(u) := \frac{\|u\|^p}{p}, \quad \text{and} \quad \Psi(u) := \sum_{k=1}^{N} F(k, u_k), \tag{8}$$

where

$$F(k, t) := \int_0^t f(k, \xi)d\xi,$$

for every $(k,t) \in [1,N] \times \mathbb{R}$.

Now we give, see also [**APR2**] and [**JZ2**], the following.

LEMMA 38. *The above functionals Φ and Ψ are two continuously Gâteaux differentiable functionals with, for every $u,\ v \in S$,*

$$\Phi'(u)(v) = \sum_{k=1}^{N+1} (\phi_p(\Delta u_{k-1})\Delta v_{k-1} + q_k \phi_p(u_k) v_k),$$

and

$$\Psi'(u)(v) = \sum_{k=1}^{N} f(k, u_k) v_k.$$

PROOF. By the Lagrange's Mean Value Theorem it easy to prove that for every $x,\ y \in \mathbb{R}$ and $t \in \mathbb{R} \setminus \{0\}$, there exists $s \in]0,1[$ such that

$$\frac{|x+ty|^p}{p} - \frac{|x|^p}{p} = ty\phi_p(x+sty).$$

From this, a simple computation show that, for every $u,\ v \in S$ and $k \in [1,N]$ there exist $\theta_k,\ \eta_k,\ \rho_k \in]0,1[$ such that

$$\frac{\Phi(u+tv)-\Phi(u)}{t} = \sum_{k=1}^{N+1} \phi_p(\Delta u_{k-1} + t\theta_k \Delta v_{k-1})\Delta v_{k-1} + \sum_{k=1}^{N} q_k \phi_p(u_k + t\eta_k v_{k-1}) v_k,$$

and

$$\frac{\Psi(u+tv)-\Psi(u)}{t} = \sum_{k=1}^{N} f(k, u_k + t\rho_k v_k) v_k.$$

Hence, takin into account the continuity of the functions ϕ_p and f, it is clear as to achieve our conclusions. □

LEMMA 39. *Let u be in S. Then u is a solution of problem $(P_\lambda^{f,q})$ iff u is a critical point for the functional $\Phi - \lambda\Psi$.*

PROOF. Let $u \in S$ be a solution of problem $(P_\lambda^{f,q})$ and we show that u is a critical point of $\Phi - \lambda\Psi$. As usual let v be an arbitrary element of S, we multiply the difference equation in problem $(P_\lambda^{f,q})$ by v_k and summing, we have

$$-\sum_{k=1}^{N} \Delta(\phi_p(\Delta u_{k-1}))v_k + \sum_{k=1}^{N} q_k\phi_p(u_k)v_k = \lambda \sum_{k=1}^{N} f(k,u_k)v_k.$$

From this, bearing in mind the previous Lemma, it is easy to see that our claim follows if we verify that

$$-\sum_{k=1}^{N} \Delta(\phi_p(\Delta u_{k-1}))v_k = \sum_{k=1}^{N+1} \phi_p(\Delta u_{k-1})\Delta v_{k-1}.$$

Indeed, one has

$$\begin{aligned}\sum_{k=1}^{N}\Delta(\phi_p(\Delta u_{k-1}))v_k &= \sum_{k=1}^{N}(\phi_p(\Delta u_k)v_k - \phi_p(\Delta u_{k-1})v_k)\\ &= -\sum_{k=1}^{N}\phi_p(\Delta u_{k-1})\Delta v_{k-1} + \phi_p(\Delta u_N)v_N - \phi_p(\Delta u_0)v_0 = -\sum_{k=1}^{N+1}\phi_p(\Delta u_{k-1})\Delta v_{k-1}.\end{aligned}$$

Obviously, if u is a critical point for $\Phi - \lambda\Psi$, arguing backward, also the viceversa holds and the proof is completed. □

Denote with q the conjugate number of p, that is $1/p + 1/q = 1$, in the sequel, we will use the following discrete Hölder inequality,

LEMMA 40. *For every u, $v \in S$ and $p > 1$ one has*

$$\sum_{k=1}^{N}|u_k v_k| \leq \left(\sum_{k=1}^{N}|u_k|^p\right)^{1/p}\left(\sum_{k=1}^{N}|v_k|^q\right)^{1/q}. \tag{9}$$

PROOF. Without loss of generality we can assume $u \neq 0$ and $v \neq 0$. Taking into account Young Inequality (Cauchy-Schwartz inequality when $p = q = 2$), see for instance [**D**, Proposition 2.1], for every a, $b \in \mathbb{R}$ one has

$$|ab| \leq \frac{1}{p}|a|^p + |b|^q.$$

Hence, for every $k \in [1, N]$, we get

$$\frac{|u_k|}{\left(\sum_{k=1}^{N}|u_k|^p\right)^{1/p}}\frac{|v_k|}{\left(\sum_{k=1}^{N}|v_k|^q\right)^{1/q}} \leq \frac{1}{p}\frac{|u_k|^p}{\left(\sum_{k=1}^{N}|u_k|^p\right)} + \frac{1}{q}\frac{|v_k|^q}{\left(\sum_{k=1}^{N}|v_k|^q\right)}.$$

From this, adding for $k = 1, 2, ..., N$, one has

$$\sum_{k=1}^{N}\frac{|u_k|}{\left(\sum_{k=1}^{N}|u_k|^p\right)^{1/p}}\frac{|v_k|}{\left(\sum_{k=1}^{N}|v_k|^q\right)^{1/q}} \leq 1,$$

that is our conclusion. □

LEMMA 41. *For every $u \in S$ and $p > 1$, the following inequalities hold*

$$\| u \|_\infty := \max_{k\in[1,N]}|u_k| \leq \frac{(N+1)^{(p-1)/p}}{2}\|u\|; \tag{10}$$

$$\sum_{k=1}^{N}|u|^s \leq N\left(\frac{(N+1)^{(p-1)/p}}{2}\|u\|\right)^s \quad \textit{for every } s \geq 0; \tag{11}$$

$$\frac{\|u\|^p}{(2^p(N+1)+\parallel q \parallel_\infty N)\, N^{p-1}} \leq \sum_{k=1}^{N} |u_k|^p, \tag{12}$$

where $\parallel q \parallel_\infty := \max_{k\in[1,N]} |q_k|$,

$$\frac{\|u\|}{(2^p(N+1)+\parallel q \parallel_\infty N)^{1/p}} \leq \parallel u \parallel_\infty . \tag{13}$$

PROOF. Without loss of generality we can assume $q_k = 0$, for every $k \in [1, N]$. Let $\overline{k} \in [1, N]$ such that

$$u_{\overline{k}} = \max_{k\in[1,N]} |u_k|.$$

by (40), one has

$$u_{\overline{k}} = \left| \sum_{k=1}^{\overline{k}} \Delta u_{k-1} \right| \leq \sum_{k=1}^{\overline{k}} |\Delta u_{k-1}| \leq \overline{k}^{1/q} \left(\sum_{k=1}^{\overline{k}} |\Delta u_{k-1}|^p \right)^{1/p},$$

$$u_{\overline{k}} = \left| \sum_{k=\overline{k}+1}^{N+1} \Delta u_{k-1} \right| \leq \sum_{k=\overline{k}+1}^{N+1} |\Delta u_{k-1}| \leq (N-\overline{k}+1)^{1/q} \left(\sum_{k=\overline{k}+1}^{N+1} |\Delta u_{k-1}|^p \right)^{1/p}.$$

Hence, taking into account that the real function $x^{1/q}$, with $x \geq 0$, is increasing, since $\max\{\overline{k}, N-\overline{k}+1\} \leq (N+1)$, one has $2u_{\overline{k}} \leq (N+1)^{1/q}\|u\|$, that is (10). Obviously, since for every $0 \leq s$ and for every $k \in [1, N]$ one has $|u_k|^s \leq\parallel u \parallel_\infty^s$, (11) is a consequence of the previous one inequality. Now, we observe that

$$\begin{aligned} \|u\|^p &\leq (2^p(N+1)+\parallel q \parallel_\infty N)\, \|u\|_\infty^p \leq \\ &\leq (2^p(N+1)+\parallel q \parallel_\infty N) \left(\sum_{k=1}^{N} |u_k| \right)^p \leq \\ &\leq (2^p(N+1)+\parallel q \parallel_\infty N)\, N^{p-1} \sum_{k=1}^{N} |u_k|^p. \end{aligned}$$

Clearly, from this follows (12) and by product (13). □

REMARK 42. For $Q = 0$, about (10) see also [**FI**] and [**JZ2**].

Finally, we point out two results that we will use to obtain positive solutions for our problem. The first is the following strong maximum principle.

THEOREM 43. *Fix* $u \in S$ *such that*

$$-\Delta(\phi_p(\Delta u_{k-1})) + q_k|u_k|^{p-2}u_k \geq 0 \quad \text{for all} \quad k \in [1, N]. \tag{14}$$

Then, either $u > 0$ *in* $[1, N]$ *or* $u \equiv 0$.

PROOF. If $q_k = 0$ for all $k \in [1, N]$, from [**APR2**, Lemma 2.3] the conclusion follows. Assume that there is $k \in [1, N]$ such that $q_k \neq 0$. First, we prove that $u \geq 0$. Put $u_j = \min_{k\in[1,N]} u_k$. We can assume $q_j \neq 0$. In fact, if $q_j = 0$, from (14) one has

$$-|\Delta u_j|^{p-2}(u_{j+1} - u_j) + |\Delta u_{j-1}|^{p-2}(u_j - u_{j-1}) \geq 0,$$

so, taking into account that u_j is the minimum, $|\Delta u_j|^{p-2}\Delta u_j = |\Delta u_{j-1}|^{p-2}\Delta u_{j-1}$. Since ϕ_p is strictly monotone and u_j is the minimum, one has $u_{j+1} - u_j = u_j - u_{j-1} = 0$. Hence, if $q_{j+1} \neq 0$ or $q_{j-1} \neq 0$ we pick as minimum u_{j+1} or u_{j-1}, otherwise we repeat as already done. Now, again from (14) we obtain

$$q_j|u_j|^{p-2}u_j \geq |\Delta u_j|^{p-2}(u_{j+1} - u_j) - |\Delta u_{j-1}|^{p-2}(u_j - u_{j-1}) \geq 0,$$

so $u_j \geq 0$, hence $u \geq 0$. Finally, arguing as in the proof of [**APR2**, Lemma 2.2], we observe that if $u_k = 0$, $u_{k-1} \geq 0$ and $u_{k+1} \geq 0$ satisfy (14), then $u_{k-1} = u_{k+1} = 0$. To be precise, again from (14) and nonnegativity of u_{k-1}, u_{k+1} one has

$$0 \leq -|\Delta u_k|^{p-2}(u_{k+1}) + |\Delta u_{k-1}|^{p-2}(-u_{k-1}) \leq 0,$$

so $u_{k-1} = u_{k+1} = 0$. Hence, the conclusion follows. □

The second result is the following theorem that guarantees the same conclusion of the preceding strong maximum principle, but without any condition of sign on the operator.

THEOREM 44. *Fix $u \in S$ such that*

$$\text{if } u_k \leq 0 \text{ then } \quad -\Delta(\phi_p(\Delta u_{k-1})) + q_k|u_k|^{p-2}u_k = 0. \tag{15}$$

Then, either $u > 0$ in $[1, N]$ or $u \equiv 0$.

PROOF. Let u be a non trivial function satisfying our assumptions. From (15), if $u_j \leq 0$ then

$$|\Delta u_j|^{p-2}(u_{j+1} - u_j) = |\Delta u_{j-1}|^{p-2}(u_j - u_{j-1}) + q_j|u_j|^{p-2}u_j. \tag{16}$$

Now, we claim that $u_1 > 0$. Arguing by contradiction, we assume that $u_1 \leq 0$. Applying (16) with $j = 1$ we obtain $u_2 \leq u_1$, so we have $u_2 \leq 0$ and we can again apply (16) with $j = 2$ obtaining $u_3 \leq u_2$. So, we have $u_N \leq u_{N-1} \leq ... \leq u_3 \leq u_2 \leq u_1 \leq 0$. If $u_N = 0$ then u is trivial and this is absurd. If $u_N < 0$, from (16) with $j = N$ we obtain $0 < |\Delta u_N|^{p-2}(-u_N) = |\Delta u_{N-1}|^{p-2}(u_N - u_{N-1}) + q_N|u_N|^{p-2}u_N \leq 0$ and also this is absurd, so our claim is proved. The same conclusion for u_2 holds. Indeed, arguing by contradiction, if $u_2 < 0$, as before we obtain $u_N \leq u_{N-1} \leq ... \leq u_3 \leq u_2 < 0$, that is, $u_N < 0$ and this is absurd. Repeating the same computation we obtain $u_3 \geq 0$, $u_4 \geq 0$,... $u_{N-1} \geq 0$. Finally, we prove that also $u_N \geq 0$. Assuming $u_N < 0$, again from (16) for $j = N$, we obtain $0 < |-u_N|^{p-2}(-u_N) = |u_N - u_{N-1}|^{p-2}(u_N - u_{N-1}) + q_N|u_N|^{p-2}u_N < 0$ since $u_N < 0 \leq u_{N-1}$. To sum up then, we have $u_1 > 0$ and $u_j \geq 0$ for all $j \in [2, N]$. Now we prove that $u_2 > 0$. If $u_2 = 0$, from (16) with $j = 2$ we obtain $0 \leq |\Delta u_2|^{p-2}u_3 = |\Delta u_1|^{p-2}(-u_1) < 0$ that is absurd. Arguing in this way we obtain $u_3 > 0$, $u_4 > 0$,... $u_{N-1} > 0$. Finally, if $u_N = 0$, from (16) for $j = N$, we obtain $0 = |\Delta u_N|^{p-2}(u_{N+1} - u_N) = |\Delta u_{N-1}|^{p-2}(-u_{N-1}) < 0$. Hence, $u > 0$ and our proof is complete. □

REMARK 45. If $f : [1, N] \times \mathbb{R} \to \mathbb{R}$ is a nonnegative function, then, owing to Theorem 43, all solutions of Problem $(P_\lambda^{f,q})$ are either zero or positive.
Now, let $f : [1, N] \times \mathbb{R} \to \mathbb{R}$ be such that $f(k, 0) = 0$ for all $k \in [0, N]$. Put

$$f^*(k,t) = \begin{cases} f(k,t), & \text{if } t > 0 \\ 0 & \text{if } t \leq 0. \end{cases}$$

Clearly, also f^* is a continuous function. Owing to Theorem 44, all solutions of Problem $(P_\lambda^{f^*,q})$ are either zero or positive and, hence, are also solutions for $(P_\lambda^{f,q})$. Hence, we emphasize that when $(P_\lambda^{f^*,q})$ admits nontrivial solutions, then Problem $(P_\lambda^{f,q})$ admits positive solutions, independently of the sign of f.

Finally, denoted by λ_1 the first eigenvalue of the problem

$$\begin{cases} -\Delta(\phi_p(\Delta u_{k-1})) = \lambda \phi_p(u_k), & k \in [1, N], \\ u_0 = u_{N+1} = 0, \end{cases}$$

one has that

$$\lambda_1 = \min_{u \in S \setminus \{0\}} \frac{\sum_{k=1}^{N+1} |\Delta u_{k-1}|^p}{\sum_{k=1}^{N} |u_k|^p}.$$

7. Main results

First, we give a classical existence result. we have the following

THEOREM 46. *Assume that there exist two constants μ and c, with $\mu < \lambda_1$, such that for every $(k, t) \in [1, N] \times \mathbb{R}$ one has*

$$F(k,t) < \frac{\mu}{p}|t|^p + c.$$

Then, problem $(P_1^{f,q})$ admits at least one solution.

PROOF. Clearly, for every $u \in S$, bearing in mind that

$$\sum_{k-1}^{N} |u_k|^p \leq \frac{\sum_{k=1}^{N+1} |\Delta u_{k-1}|^p}{\lambda_1},$$

one has

$$\Psi(u) \leq \frac{\mu}{p} \sum_{k=1}^{N} |u_k|^p + cN \leq \frac{1}{p} \frac{\mu}{\lambda_1} \|u\|^p + cN.$$

Therefore, for every $u \in S$, we have

$$\Phi(u) - \Psi(u) \geq \frac{1}{p}\left(1 - \frac{\mu}{\lambda_1}\right) \|u\|^p - cN,$$

which implies that the functional $\Phi - \Psi$ turns out to be coercive. Therefore, by standard arguments presented in the previous section it is trivial to achieve our conclusion. □

REMARK 47. A sufficient condition that implies the assumptions of Theorem 46 is the following

$$\limsup_{|t|\to+\infty} \frac{f(k,t)}{|t|^{p-2}t} < \lambda_1.$$

In [**APR2**], requiring that f satisfies also the following condition:

There exist two non negative and non trivial function $\alpha, \beta : [1,N] \to \mathbb{R}$ and two positive constants γ and t_0, such that for every $(k,t) \in [1,N] \times (0,t_0)$, one has

$$\alpha_k \leq f(k,t)\beta_k t^{-\gamma},$$

among other results, the authors have been proved that problem $(P_1^{f,0})$ has at least two positive solutions.

Now we give an multiplicity result for $(P_1^{f,q})$ provided that f is an odd function. We remind that a simple calculation ensures that F is even. In this direction, let s be a real number, put

$$l(N) = \max_{k\in[1,N]} \limsup_{t\to+\infty} \frac{F(k,t)}{t^s},$$

we have the following

THEOREM 48. *Let $f : [1,N] \times \mathbb{R} \to \mathbb{R}$ be an odd continuous function such that $f(k,0) \neq 0$ for some $k \in [1,N]$. Assume that there exist four positive constant s, r, η and μ, with $1 < s \leq p$, $\eta > 1$ and $r^{\eta-p} > \frac{[2^\eta(N+1)+\|q\|_\infty N]N^{\eta-1}}{\mu p}$ such that*

(b_{10}) If $s < p$, one has $l(N) < +\infty$ while if $s = p$ one has $l(N) < \frac{2^p}{pN(N+1)^{p-1}}$.

(b_{11}) For every $k \in [1,N]$ and $t \in \left[0, \frac{(N+1)^{(p-1)/p}}{2}r\right]$, one has

$$F(k,t) \geq \mu t^\eta.$$

Then, problem $(P_1^{f,q})$ admits at least N distinct pairs of critical points.

PROOF. First, we explicitly observe that conditions (b_{10}) and (b_{11}) hold also for t going to $-\infty$ and $t \in \left[-\frac{(N+1)^{(p-1)/p}}{2}r, 0\right]$. By (b_{10}), it is easy to see that there exist a positive constant C such that, for every $(k,t) \in [1,N] \times \mathbb{R}$, it results

$$F(k,t) \leq l(N)|t|^s + C.$$

From this, for every $u \in S$, we have

$$\begin{aligned}\Phi(u) - \lambda\Psi(u) &\geq \frac{\|u\|^p}{p} - l(N)\sum_{k=1}^{N} |u|^s - NC \\ &\geq \left(\frac{1}{p} - l(N)\frac{N(N+1)^{s-1}}{2^s}\|u\|^{s-p}\right)\|u\|^p - NC.\end{aligned}$$

Hence, taking into account (b_{10}) the functional $\Phi - \lambda\Psi$ turns out to be coercive. On the other hand, take $u \in S$ with $\|u\| = r$. Bearing in mind (10), for every $k \in [1, N]$, we have

$$|u_k| \leq \frac{(N+1)^{(p-1)/p}}{2} r.$$

Moreover, takin into account (12), by (b_{11}), we get

$$\inf_{\|u\|=r} \sum_{k=1}^{N} F(k, u_k) \geq \mu \inf_{\|u\|=r} \sum_{k=1}^{N} |u_k|^{\eta} \geq \frac{\mu r^{\eta}}{[2^{\eta}(N+1) + \|q\|_{\infty} N] N^{\eta-1}}.$$

Therefore, take $K = \{u \in S : \|u\| = r\}$ we have

$$\sup_K \Phi - \inf_K \Psi \leq \frac{r^p}{p} - \frac{\mu r^{\eta}}{[2^{\eta}(N+1) + \|q\|_{\infty} N] N^{\eta-1}} < 0.$$

Bering in mind Theorem 36 our conclusion follows. □

Put $Q = \sum_{k=1}^{N} q_k$ and let c and d be two positive constants, we write

$$\Theta(c) := \frac{\sum\limits_{k=1}^{N} \sup\limits_{|\xi| \leq c} F(k, \xi)}{c^p} \quad \text{and} \quad \Gamma(\mathrm{d}) := \frac{\sum_{\mathrm{k=1}}^{\mathrm{N}} \mathrm{F}(\mathrm{k}, \mathrm{d})}{\mathrm{d}^{\mathrm{p}}}.$$

Now, we give

THEOREM 49. *Assume that there exist four positive constants a, c, d and s with $c < d$ and $s < p$ such that*

(b_1) $\Theta(c) < \frac{2^p}{(N+1)^{p-1}(2+Q)} \Gamma(d)$;

(b_2) $F(k, \xi) \leq a(1 + |\xi|^s)$ *for all* $(k, \xi) \in [1, N] \times \mathbb{R}$.

Then, for every $\lambda \in \left] \frac{2+Q}{p\Gamma(d)}, \frac{2^p}{p\Theta(c)(N+1)^{p-1}} \right[$, *problem* $(P_{\lambda}^{f,q})$ *admits at least three solutions, one with* $\|u\|_{\infty} \leq c$ *and one with*

$$\|u\|_{\infty} > \frac{2c}{[2^p(N+1)^p + \| q \|_{\infty} N(N+1)^{p-1}]^{1/p}}.$$

PROOF. In applying Theorem 30, choice $X = S$, and let Φ and Ψ be as in (8). Clearly, for each $\lambda > 0$, Φ and Ψ satisfy condition (Λ) and picked,

$$r = \frac{(2c)^p}{p(N+1)^{p-1}}, \tag{17}$$

taking in account (10) an easy computation ensures that

$$\frac{\sup_{[\Phi<r]} \Psi}{r} \leq \frac{\sup_{\|u\| \leq (pr)^{1/p}} \sum_{k=1}^{N} F(k, u_k)}{r} \leq \frac{p(N+1)^{p-1}}{2^p} \Theta(c). \tag{18}$$

Further, since one has $c < \left(\frac{N+1}{2}\right)^{(p-1)/p}(2+Q)^{1/p}d$, it results that $\|v\| > (pr)^{1/p}$ where,

$$v_k = \begin{cases} d & \text{if } k \in [1, N], \\ 0 & \text{otherwise.} \end{cases}$$

Hence, we get

$$\frac{\Psi(v)}{\Phi(v)} = \frac{\sum_{k=1}^{N} F(k,d)}{d^p(2+Q)} = \frac{p}{2+Q}\Gamma(d). \tag{19}$$

Combining (18) and (19), it is clearly that condition (j_1) is verified and, owing to (b_2) and by (12), for every $u \in S$ and $\lambda > 0$, we have

$$\begin{aligned} \Psi(u) - \lambda\Psi(u) &\geq \frac{\|u\|^p}{p} - \lambda \sum_{k=1}^{N} a(1 + |u_k|^s) \\ &\geq \frac{\|u\|^p}{p} - \frac{a\lambda N(N+1)^{(s-1)/p}}{2}\|u\|^s - a\lambda N, \end{aligned}$$

which clearly ensures that $\Phi - \lambda\Psi$ turns out to be coercive. So, the assumptions of Theorem 30 are satisfied and our conclusion follows taking into account (17), (10) and (13). □

Remark 50. [**CG**, Theorem 3.1] also ensures the existence of at least three solutions for problem $(P_\lambda^{f,0})$ under conditions (b_1) and (b_2). However, there $\Gamma(d)$ is given by

$$\Gamma(d) = \frac{\sum_{k=1}^{N} F(k,d) - \sup\limits_{|\xi|\leq c} F(k,\xi)}{d^p}.$$

Therefore, it clear that here we obtain a more precise estimate from below for the interval of parameter λ for which problem $(P_\lambda^{f,q})$ admits at least three solutions. Moreover, here we also give an estimate on the maximum for two of such solutions.

Remark 51. In many situations it is also important to obtain at least one solutions to $(P_\lambda^{f,q})$. In this order of ideas, a careful reading of the above proof reveals that condition (b_2) ensures the existence of at least one solution to $(P_\lambda^{f,q})$ for every $\lambda > 0$ and if $\lambda > \frac{2+Q}{p\Gamma(d)}$ one has $\|u\|_\infty > \frac{2c}{[2^p(N+1)^p + \| q \|_\infty N(N+1)^{p-1}]^{1/p}}$. Whereas, fix $c > 0$, arguing again as above, but taking into account the proof of Theorem 5, we can show that, for every $\lambda \in \left]0, \frac{2^p}{p\Theta(c)(N+1)^{p-1}}\right[$, problem $(P_\lambda^{f,q})$ has al least one solution u with $\|u\|_\infty \leq c$.

Let $g : \mathbb{R} \to \mathbb{R}$, $\alpha : [1, N] \to \mathbb{R}$ be two nonnegative functions with $A := \sum_{k=1}^{N} \alpha_k > 0$. Put, for every $t \in \mathbb{R}$, $G(t) := \int_0^t g(\xi)d\xi$, a simple consequence of the previous result is the following

Theorem 52. *Assume that there exist four positive constants ρ, c, d and s with $c < d$ and $s < p$ such that*

(b_3) $\frac{G(c)}{c^p} < \frac{2^p}{(2+Q)(N+1)^{p-1}} \frac{G(d)}{d^p}$;
(b_2') $G(t) \leq \rho(1+|t|^s)$, *for all* $t \in \mathbb{R}$.

Then, for each $\lambda \in \left]\frac{2+Q}{pA}\frac{d^p}{G(d)}, \frac{2^p}{pAQ(N+1)^{p-1}}\frac{c^p}{G(c)}\right[$, *the problem*

$$\begin{cases} -\Delta(\phi_p(\Delta u(k-1))) + q_k\phi_p(u_k) = \lambda\alpha_k g(u_k), & k \in [1,N], \\ u(0) = u(N+1) = 0, \end{cases}$$

admits at least three nonnegative solutions, one with $\|u\|_\infty \leq c$ *and one with*

$$\|u\|_\infty > \frac{2c}{[2^p(N+1)^p + \| q \|_\infty N(N+1)^{p-1}]^{1/p}}.$$

PROOF. We see at once that our conclusion follows form Theorem 49 by choosing $f(k,t) = \alpha_k g(t)$ for each $(k,t) \in [1,N] \times \mathbb{R}$ and $a = \rho Q$. □

REMARK 53. If $g(0) > 0$, then in Theorem 52 we can replace the word *nonnegative* with *positive*.

EXAMPLE 54. Write, for each $k \in [1,N]$, $\alpha_k = k$ and let $g : \mathbb{R} \to \mathbb{R}$ be defined by putting

$$g(t) = \begin{cases} e^t, & \text{if } t \leq 12; \\ e^{12}, & \text{if } t > 12. \end{cases}$$

By choosing for instance $\rho = e^{12}$, $c = 1$, $d = 12$, $s = 1$ and $p = 3$, the assumptions of Theorem 52 are satisfied. Therefore, for each $\lambda \in \left]\frac{157}{10^6}, \frac{344}{10^6}\right[$, the problem

$$\begin{cases} -\Delta(|\Delta u(k-1)|\Delta u(k-1))) = \lambda k g(u_k), & k \in [1,9], \\ u(0) = u(10) = 0, \end{cases}$$

has at least three positive solutions.

REMARK 55. Intentionally, to make easy the comparison with [**CG**, Theorem 3.2], in the above example, we take $q_k = 0$ for every $k \in [1,9]$. Indeed, with the same date this later result gives at least three positive solutions for $\lambda \in \left]\frac{164}{10^6}, \frac{344}{10^6}\right[$. Moreover, here, we have that at least one solution satisfying $\| u \|_\infty \leq 1$ and at least an other with $\| u \|_\infty > \frac{1}{\sqrt[3]{10}}$.

Now we discuss the case $s = p$ inside the growth condition (b_2).

THEOREM 56. *Assume that there exist three positive constants* a, c, d *with* $c < d$ *such that* (b_1) *holds and in addition suppose that*

(b_4) $F(k,\xi) \leq a(1+|\xi|^p)$ *for all* $(k,\xi) \in [1,N]\times\mathbb{R}$, *with* $a < \frac{2^p}{(2+Q)N(N+1)^{p-1}}\Gamma(d)$.

Then, for every $\lambda \in \left]\frac{2+Q}{p\Gamma(d)}, \frac{2^p}{p(N+1)^{p-1}}\min\left\{\frac{1}{\Theta(c)}, \frac{1}{aN}\right\}\right[$, *problem* $(P_\lambda^{f,q})$ *admits at least three solutions, one with* $\|u\|_\infty \leq c$ *and an other with*

$$\|u\|_\infty > \frac{2c}{[2^p(N+1)^p + \| q \|_\infty N]^{1/p}}.$$

PROOF. Arguing as in the proof of Theorem 49 it is clear that our conclusion follows again by Theorem 30 if we show that $\Phi - \lambda\Psi$ turns out to be coercive. Indeed, fix λ as above, since for every $u \in S$, by (10), it results

$$|u_k|^p \leq \frac{(N+1)^{(p-1)}}{2^p}\|u\|^p, \qquad \forall\, k \in [1, N],$$

(b_4) ensures that,

$$\Phi(u) - \lambda\Psi(u) \geq \left(\frac{1}{p} - \lambda\frac{aN(N+1)^{p-1}}{2^p}\right)\|u\|^p - \lambda a N.$$

Thus, begin $\lambda < \frac{2^p}{aN(N+1)^{p-1}p}$, our claim holds and the proof is completed. □

REMARK 57. Evidently, arguing as in the above proof, we get at least one solution to $(P_1^{f,q})$ if (b_4) holds with $a < \frac{2^p}{pN(N+1)^{(p-1)}}$.

Now, we give

THEOREM 58. *Assume that there exist four positive constants a, c, d and s with $c < d$ and $s < p$ such that (b_2) holds and moreover, suppose that*

(b_5) $\max_{|\xi|\leq c} F(k,\xi) \leq 0$ *for all* $k \in [1, N]$;

(b_6) *there exists* $\overline{k} \in [1, N]$ *such that* $\int_0^d f(\overline{k},\xi)d\xi > 0$.

Then, for every $\lambda \in \left]\frac{(2+q_{\overline{k}})d^p}{p\int_0^d f(\overline{k},\xi)d\xi}, +\infty\right[$, *problem (P_λ) admits at least two positive solutions.*

PROOF. In applying Theorem 30 take, as above, and $r = \frac{(2c)^p}{p(N+1)^{p-1}}$. For every $u \in S$ with $\|u\| \leq (pr)^{1/p}$, by (b_5) one has that $\Psi(u) \leq 0$. From this, it follows that

$$\sup_{[\Phi\leq r]} \Psi \leq \sup_{\|u\|\leq (pr)^{1/p}} \Psi \leq \Psi(0) = 0.$$

On the other hand it is easy to see that, put

$$w_k = \begin{cases} d & \text{if } k = \overline{k}; \\ 0 & \text{otherwise,} \end{cases}$$

one has $\|w\| > (pr)^{1/p}$. Hence, by (b_6), we get

$$\frac{\Phi(w)}{\Psi(w)} = \frac{\int_0^d f(\overline{k},\xi)d\xi}{(2+q_{\overline{k}})d^p} > 0.$$

So, since (b_5) implies $f(k, 0) = 0$, for every $k \in [1, N]$, Theorem 30 furnishes at least two nontrivial solutions, say v and w, to $(P_\lambda^{f,q})$. Moreover, a simple computation shows that v and w turn out to be also two solutions of the following problem

$$\begin{cases} -\Delta(\phi_p(\Delta u_{k-1})) + q_k\phi_p(u_k) = \lambda\widehat{f}(k, u_k), & k \in [1, N], \\ u_0 = u_{N+1} = 0, \end{cases}$$

where, $\widehat{f} : [1, N] \times \mathbb{R} \to \mathbb{R}$ is defined by putting

$$\widehat{f}(k,t) = \begin{cases} f(k,t) & \text{if } t \geq 0; \\ 0 & \text{otherwise.} \end{cases}$$

We claim that v and w turn out to be positive. By contradiction, suppose that at least one, said v, is non-positive. Thus, we have that v fulfills the following conditions

$$-\Delta(\phi_p(\Delta v_{k-1}) + q_k\phi_p(v_k) = 0, \quad k \in [1, N], \quad v_0 = v_{N+1} = 0.$$

On the other hand, bearing in mind that v is nontrivial as well as Lemma 44, one has that v is positive, which clearly is a contradiction. □

EXAMPLE 59. For every N and for every $p \geq 4$, the following class of problems

$$\begin{cases} -\Delta(\phi_p(\Delta u_{k-1})) + q_k\phi_p(u_k) = \lambda k u_k[(2+k)u_k - 4], & k \in [1, N], \\ u_0 = u_{N+1} = 0, \end{cases}$$

has at least two positive solutions provided that

$$\lambda \in \left] \frac{3^{p-2}}{p} \min_{k\in[1,N]} \left(\frac{2+q_k}{k^2}\right), +\infty\right[.$$

Indeed, since we have $F(k, t) = (kt)^2(t-2)$ for every $(k, t) \in [1, N] \times \mathbb{R}$, it is easy to see that by choosing $c = 1$, $d = 3$ and $s = 3$ all the assumptions of Theorem 58 are satisfied.

Now we give

THEOREM 60. *Let f be a positive continuous function in $[1, N] \times [0, +\infty]$. Assume that there exist three positive constants c_1, d, c_2, with*

$$c_1 < d < \left(\frac{1}{2+Q}\right)^{(1/p)} \left(\frac{2}{N+1}\right)^{(p-1)/p} c_2,$$

such that

(b_7) $\max\{\Gamma(c_1), 2\Gamma(c_2)\} < \frac{1}{2+Q}\left(\frac{2}{N+1}\right)^{p-1}\Gamma(d)$.

Then, for each $\lambda \in \left] \frac{2(2+Q)}{p}\frac{1}{\Gamma(d)}, \frac{2}{p}\left(\frac{2}{N+1}\right)^{p-1}\frac{1}{\max\{\Gamma(c_1),2\Gamma(c_2)\}}\right[$, problem $(P_\lambda^{f,q})$ admits at least three distinct positive solutions u_i, $i = 1, 2, 3$, such that

$$\|u_i\|_\infty < c_2,$$

for all $k \in [1, N]$, $i = 1, 2, 3$.

PROOF. Without loss of generality, we can assume $f(k, t) = f(k, 0)$ for all $(k, t) \in [1, N]\times]-\infty, 0[$. Our aim is to apply Theorem 31. Fix λ as in the conclusion and let Φ and Ψ be as in (8). Clearly, (A_1) and (A_3) are satisfied. Moreover, let u_1 and u_2 be two local minima for $\Phi - \lambda\Psi$. They are two solutions for $(P_\lambda^{f,q})$ and, owing to Lemma 43, one has $ru_1(t) + (1-r)u_2(t) \geq 0$ for all $t \in [1, N]$ and for all $r \in [0, 1]$. Hence, $\Psi(ru_1 + (1-r)u_2) \geq 0$ for all $r \in [0, 1]$ and, (A_2') is verified. Now, put

$$\rho_1 = \frac{(2c_1)^p}{p(N+1)^{(p-1)}} \quad \text{and} \quad \rho_2 = \frac{(2c_2)^p}{p(N+1)^{(p-1)}}.$$

From

$$\max_{k\in[1,N]} |u_k| \leq \frac{(N+1)^{(p-1)/p}}{2}\|u\|, \tag{20}$$

for every $u \in X$, we obtain

$$\max_{k\in[1,N]} |u_k| \leq \frac{(N+1)^{(p-1)/p}}{2}(p\rho_1)^{\frac{1}{p}} = c_1,$$

for all $u \in S$ such that $\|u\| \leq (p\rho_1)^{\frac{1}{p}}$, and

$$\max_{k\in[1,N]} |u_k| \leq \frac{(N+1)^{(p-1)/p}}{2}(p\rho_2)^{\frac{1}{p}} = c_2,$$

for all $u \in S$ such that $\|u\| \leq (p\rho_2)^{\frac{1}{p}}$.
Therefore, one has

$$\sup_{[\Phi<\rho_1]} \frac{\Psi}{\rho_1} = \frac{\sup_{\|u\|<(p\rho_1)^{1/p}} \sum_{k=1}^{N} F(k,u_k)}{\rho_1} \leq \frac{\sum_{k=1}^{N} F(k,c_1)}{\rho_1} = \frac{p(N+1)^{p-1}}{2^p}\Gamma(c_1) \tag{21}$$

as well as

$$\sup_{[\Phi<\rho_2]} \frac{\Psi}{\rho_2} \leq \frac{p(N+1)^{p-1}}{2^p}\Gamma(c_2). \tag{22}$$

On the other hand, pick $\overline{v} \in S$, defined as follows

$$\overline{v}(k) = \begin{cases} d & \text{if } k \in [1, N] \\ 0 & \text{if } k = 0, \quad k = N+1. \end{cases}$$

Clearly, one has $\Phi(\overline{v}) = \frac{2}{p}d^p$. Hence, from $c_1 < d$ we obtain $\rho_1 < \Phi(\overline{v})$ and from $d < \left(\frac{1}{(2+Q)}\right)^{(1/p)} \left(\frac{2}{N+1}\right)^{(p-1)/p} c_2$ we obtain $\Phi(\overline{v}) < \rho_2/2$. Moreover, one has

$$\frac{\Psi(\overline{v})}{\Phi(\overline{v})} = \frac{p}{2+Q}\sum_{k=1}^{N+1} \frac{F(k,d)}{d^p} = \frac{p}{2+Q}\Gamma(d). \tag{23}$$

Hence, taking (21), (22) and (23) into account, assumptions (j_3) and (j_4) of Theorem 31 follow from (b_7). Further, again from (21), (22) and (23), one has that

$$\lambda \in \left]\frac{2\Phi(\overline{v})}{\Psi(\overline{v})}, \min\left\{\frac{\rho_1}{\sup_{\Phi>\rho_1]} \Psi}; \frac{\rho_2/2}{\sup_{\Phi<\rho_2]} \Psi}\right\}\right[.$$

Therefore, the functional $\Phi - \lambda\Psi$ admits at least three critical points u_i, $i = 1, 2, 3$, which are solutions of $(P_\lambda^{f,q})$ and, owing to Lemma 43, are positive functions. Finally, from (20), for $i = 1, 2, 3$, one has

$$\|u_i\|_\infty \leq \frac{(N+1)^{(p-1)/p}}{2}\|u_i\| < \frac{(N+1)^{(p-1)/p}}{2}(p\rho_2)^{\frac{1}{p}} = c_2,$$

and the proof is complete. □

REMARK 61. Whenever $Q = 0$ the previous results give back [**BC4**, Theorem 3.1]

REMARK 62. Since the range of solutions obtained from Theorem 65 is in $[0, c_2]$, the conclusion still holds if we assume that $f(k,t) > 0$ only for each $(k,t) \in [1,N] \times [0, c_2]$. Indeed, it is enough to apply it to the function $\widetilde{f}$ defined as $\widetilde{f}(k,t) = f(k,t)$ if $t \leq c_2$ and $\widetilde{f}(k,t) = f(k, c_2)$ if $t \geq c_2$, for every $k \in [1,N]$. Hence, for $t > c_2$, f may be highly discontinuous and change sign and, in addition, f has no asymptotic condition.

Moreover, whenever f is nonnegative in $[1,N] \times [0, c_2]$, the proof of Theorem 60 ensures at least two positive solutions to $(P_\lambda^{f,q})$, or again at least three positive solutions if $f(k,0) \neq 0$ for some $k \in [1,N]$.

Among the consequences of Theorem 60 we point out here the following result.

COROLLARY 63. *Let $f : [1,N] \times \mathbb{R} \to \mathbb{R}$ be a continuous function such that $f(k,t)t > 0$ for all $(k,t) \in [1,N] \times (\mathbb{R} \setminus \{0\})$. Assume that*

(b_8) $\displaystyle\lim_{t\to 0} \frac{f(k,t)}{|t|^{p-1}} = \lim_{|t|\to\infty} \frac{f(k,t)}{|t|^{p-1}} = 0$ *for all $k \in [1,N]$.*

Then, for each $\lambda > \overline{\lambda}^*$, where

$$\overline{\lambda}^* = \max\left\{\inf_{d>0} \frac{d^p}{\sum_{k=1}^{N}\int_0^d f(k,t)dt}; \inf_{d<0} \frac{(-d)^p}{\sum_{k=1}^{N}\int_0^d f(k,t)dt}\right\},$$

problem $(P_\lambda^{f,q})$ admits at least four distinct nontrivial solutions.

PROOF. Put

$$f_1(k,t) = \begin{cases} f(k,t) & \text{if} \quad (k,t) \in [1,N] \times [0, +\infty[\\ 0 & \text{otherwise} \end{cases}$$

and

$$f_2(k,t) = \begin{cases} -f(k,-t) & \text{if} \quad (k,t) \in [1,N] \times [0, +\infty[\\ 0 & \text{otherwise.} \end{cases}$$

Moreover, define $F_1(k,t) := \int_0^t f_1(k,\xi)d\xi$ for every $(k,t) \in [1,N] \times \mathbb{R}$ and $\Gamma_1(c) :=$

$\frac{\sum_{k=1}^{N} F_1(k,c)}{c^p}$ for all $c > 0$. Fix $\lambda > \overline{\lambda}^*$. Then, there is $d > 0$ such that $\frac{1}{\lambda} < \Gamma_1(d)$. From $\lim_{t\to 0^+} \frac{f_1(k,t)}{t^{p-1}} = 0$ there is $c_1 < d$ such that $\Gamma_1(c_1) < \frac{1}{(2+Q)}\left(\frac{2}{N+1}\right)^{p-1}\frac{1}{\lambda}$ and from $\lim_{t\to +\infty} \frac{f_1(k,t)}{t^{p-1}} = 0$ there is $c_2 > (2+Q)^{\frac{1}{p}}\left(\frac{N+1}{2}\right)^{(p-1)/p} d$ such that $\Gamma_1(c_2) < \frac{1}{2}\frac{1}{2+Q}\left(\frac{2}{N+1}\right)^{p-1}\frac{1}{\lambda}$. Then, (b_7) of Theorem 60 is satisfied and

$$\lambda \in \left]\frac{2(2+Q)}{p}\frac{1}{\Gamma_1(d)}, \frac{2}{p}\left(\frac{2}{N+1}\right)^{p-1}\frac{1}{\max\{\Gamma_1(c_1), 2\Gamma_1(c_2)\}}\right[.$$

Hence, problem $(P_\lambda^{f_1,q})$ admits two positive solutions u_1, u_2, which are positive solutions of $(P_\lambda^{f,q})$. Next, arguing in the same way, from $\lim_{t\to 0^+} \frac{f_2(k,t)}{t^{p-1}} = \lim_{t\to +\infty} \frac{f_2(k,t)}{t^{p-1}} = 0$ we obtain that problem $(P_\lambda^{f_2,q})$ admits two positive solutions u_3, u_4. Clearly, $-u_3$, $-u_4$ are negative solutions of $(P_\lambda^{f,q})$ and the conclusion is achieved. □

REMARK 64. We explicitly observe that in Corollary 63 we have no symmetric assumption on f. However, whenever f is an odd continuous nonzero function such that $f(k,t) \geq 0$ for all $(k,t) \in [1,N] \times [0,+\infty[$ we can substitute (b_8) with

$$(b_8') \quad \lim_{t\to 0^+} \frac{f(k,t)}{t^{p-1}} = \lim_{t\to\infty} \frac{f(k,t)}{t^{p-1}} = 0 \quad \text{for all } k \in [1,N],$$

obtaining at least four distinct nontrivial solutions to problem $(P_\lambda^{f,q})$ for each

$$\lambda > \inf_{d>0} \frac{1}{\Gamma(d)}.$$

Now we present the following result.

THEOREM 65. *Let $f : [1,N] \times \mathbb{R} \to \mathbb{R}$ be a continuous function such that $f(k,0) \neq 0$ for some $k \in [1,N]$. Assume that there exist four constants c_1, c_2, η_1, and η_2, with $c_1 > 0$, $\eta_1 > p$ and $0 \leq \eta_2 < \eta_1$, such that*

(b_9) $F(k,t) \geq c_1|t|^{\eta_1} - c_2|t|^{\eta_2}$ *for every $(k,t) \in [1,N] \times \mathbb{R}$.*

Then, for each

$$\lambda \in \left]0, \frac{2^p}{p(N+1)^{p-1}} \sup_{d>0} \frac{1}{\Theta(d)}\right[$$

problem $(P_\lambda^{f,q})$ admits at least three nontrivial solutions.

PROOF. Our aim is to apply Theorem 35. Fix λ as in the conclusion and let Φ and Ψ be as in (8). Then, there is $\overline{d} > 0$ such that

$$\lambda < \frac{2^p}{p(N+1)^{p-1}} \frac{1}{\Theta(\overline{d})}.$$

Setting $\overline{r} = \frac{(2\overline{d})^p}{p(N+1)^{p-1}}$ and arguing as in the proof of Theorem 60 (see (15)), one has

$$\frac{1}{\lambda^*} \leq \varphi^{(1)}(\overline{r}) \leq \frac{\sup_{[\Phi<\overline{r}]} \Psi}{\overline{r}} \leq \frac{p(N+1)^{p-1}}{2^p}\Theta(\overline{d}) < \frac{1}{\lambda},$$

that is $\lambda < \lambda^*$. Moreover, bearing in mind (11) and (12), for each $u \in S$, one has

$$\sum_{k=1}^{N} |u_k|^{\eta_1} \geq \frac{1}{[2^{\eta_1}(N+1) + \|q\|_\infty N]\, N^{\eta_1 - 1}} \|u\|^{\eta_1} \tag{24}$$

and

$$\sum_{k=1}^{N} |u_k|^{\eta_2} \leq N \left[\frac{(N+1)^{(p-1)\eta_2/p}}{2^{\eta_2}} \right] \|u\|^{\eta_2} \tag{25}$$

Hence, from (b_9), for each $u \in S$, we get

$$\Phi(u) - \lambda\Psi(u) \leq \frac{\|u\|^p}{p} - \frac{\lambda c_1 \|u\|^{\eta_1}}{[2^{\eta_1}(N+1) + \|q\|_\infty N]\, N^{\eta_1 - 1}} + \frac{\lambda c_2 N (N+1)^{(p-1)\eta_2/p}}{2^{\eta_2}} \|u\|^{\eta_2}.$$

Therefore, since $\eta_1 > p$ and $\eta_1 > \eta_2$, condition (j_7) is verified. Hence, from Theorem 35, the functional $\Phi - \lambda\Psi$ admits at least three critical points, which are three solutions for $(P_\lambda^{f,q})$. Since $f(k,0) \neq 0$ for some $k \in [1,N]$, they are nontrivial solutions and our conclusion is proved. □

REMARK 66. If $f(k,0) = 0$ for all $k \in [1,N]$ the same proof of Theorem 65 ensures, for each $\lambda \in]0, \overline{\lambda}[$, the existence of at least two nontrivial solutions for problem $(P_\lambda^{f,q})$.

Now, we present some easy examples of applications of the previous results.

EXAMPLE 67. Let $\alpha : [1,7] \to \mathbb{R}$ be a positive function and let $g : [0, +\infty[\to \mathbb{R}$ be the function defined as follows

$$g(t) := \begin{cases} 1 & \text{if } 0 \leq t < 1; \\ t^{11} & \text{if } 1 \leq t < 2; \\ (4-t)^{11} & \text{if } 2 \leq t < 3; \\ 1, & \text{if } 3 \leq t < 11; \\ \overline{g}(t), & \text{if } 11 \leq t, \end{cases}$$

where $\overline{g} : [11, +\infty[\to \mathbb{R}$ is an arbitrary function.
The function $f(k,t) = \alpha_k g^*(t)$, $(k,t) \in [1,7] \times [0,+\infty[$, where g^* is a continuous function in $[0,+\infty[$ which coincides with g in $[0,11[$ (see also Remark 62), satisfies all the assumptions of Theorem 49 by choosing, for instance, $c_1 = 1$, $d = 2$, $c_2 = 11$ and by taking $p = 3$ and $N = 7$. Then, for each $\lambda \in \left] \frac{A}{32}, \frac{A}{24} \right[$, where $A := \frac{1}{\sum_{k=1}^{7} \alpha_k}$, the problem

$$\begin{cases} -\Delta(\phi_3(\Delta u_{k-1})) = \lambda \alpha_k g(u_k), & k \in [1,7], \\ u_0 = u_8 = 0, \end{cases}$$

admits at least three distinct positive solutions u_i, $i = 1,2,3$, such that, for each $k \in [1,7]$ one has $\|u_i\|_\infty < 11$, $i = 1,2,3$.

EXAMPLE 68. Owing to Corollary 63 (see also Remark 64), for each $\lambda > \frac{3}{4}$, the problem

$$\begin{cases} -\Delta(\phi_5(\Delta u_{k-1})) + q_k\phi_5(u_k) = \lambda k u^5 \exp(-ku^6), & k \in [1,10], \\ u(0) = u(11) = 0, \end{cases}$$

admits at least four nontrivial solutions.

EXAMPLE 69. Owing to Theorem 65, for each $\lambda \in \left]0, \frac{4}{3N(N+1)^2}\right[$, the problem

$$\begin{cases} -\Delta(\phi_3(\Delta u_{k-1})) + q_k\phi_3(u_k) = \lambda(4u^3 - 1), & k \in [1,N], \\ u(0) = u(N+1) = 0, \end{cases}$$

admits at least three nontrivial solutions.

Now we look for infinitely many solutions for problem $(P_\lambda^{f,q})$. Put

$$\Gamma^\infty := \limsup_{t\to+\infty} \Gamma(t).$$

Our main result is the following.

THEOREM 70. *Assume that there exist two real sequences* $\{a_n\}$ *and* $\{b_n\}$, *with* $\lim_{n\to+\infty} b_n = +\infty$, *such that*

$$(h_1) \qquad |a_n| < \frac{2}{(Q+2)^{1/p}(N+1)^{(p-1)/p}}\, b_n \qquad \textit{for each } n \in \mathbb{N},$$

$$(h_2)\ \mathcal{A}_\infty := \lim_{n\to+\infty} \frac{\sum_{k=1}^N \max_{|t|\le b_n} F(k,t) - \sum_{k=1}^N F(k,a_n)}{(2b_n)^p - (Q+2)\,(N+1)^{(p-1)}\,|a_n|^p} < \frac{\Gamma^\infty}{(Q+2)(N+1)^{p-1}}.$$

Then, for each $\lambda \in \left]\frac{2+Q}{p\Gamma^\infty}, \frac{1}{p(N+1)^{p-1}\mathcal{A}_\infty}\right[$, *problem* $(P_\lambda^{f,q})$ *admits an unbounded sequence of solutions.*

PROOF. Fix λ in $\left]\frac{2+Q}{p\Gamma^\infty}, \frac{1}{p(N+1)^{p-1}\mathcal{A}_\infty}\right[$, take the real Banach space S as defined in Section 3, and put Φ, Ψ, as in (8). Our aim is to apply Theorem 37 part (b) to function $\Phi - \lambda\Psi$, since critical points of $\Phi - \lambda\Psi$ are solutions to our problem. Clearly (Λ) holds. Therefore, our conclusion follows provided that $\gamma < +\infty$ as well as that $\Phi - \lambda\Psi$ turns out to be unbounded from below. To this end, write

$$r_n := \frac{(2b_n)^p}{p(N+1)^{p-1}}$$

for every $n \in \mathbb{N}$. Since, owing to (20), if $\|u\| \le (pr_n)^{1/p}$ then $|u_k| \le b_n$ for every $k \in [1,N]$, we obtain

$$\varphi^{(1)}(r_n) \le \inf_{\|u\|^p < pr_n} \frac{\sum_{k=1}^N \max_{|t|\le b_n} F(k,t) - \sum_{k=1}^N F(k,u_k)}{\frac{(2b_n)^p}{p(1+N)^{p-1}} - \frac{\|u\|^p}{p}}.$$

Now, for each $n \in \mathbb{N}$, let w_n be defined by $(w_n)_k := a_n$ for every $k \in [1, N]$, $(w_n)_0 = (w_n)_{N+1} = 0$. Clearly, $w_n \in S$ and $\|w_n\|^p = (2 + Q)|a_n|^p$. Moreover, from (h_1) one has $\|w_n\|^p < pr_n$. Therefore, one has

$$\varphi^{(1)}(r_n) \leq p(N+1)^{p-1} \frac{\sum_{k=1}^{N} \max_{|t| \leq b_n} F(k,t) - \sum_{k=1}^{N} F(k, a_n)}{(2b_n)^p - (2+Q)\,(N+1)^{(p-1)}\,|a_n|^p}.$$

Hence, bearing in mind (h_2), $\gamma \leq \lim_{n \to +\infty} \varphi(r_n) \leq p(N+1)^{p-1}\mathcal{A}_\infty < +\infty$ follows.

Now, we verify that $\Phi - \lambda\Psi$ is unbounded from below. First, assume that $\Gamma^\infty = +\infty$. Accordingly, fix M such that $M > \frac{Q+2}{p\lambda}$ and let $\{c_n\}$ be a sequence of positive numbers, with $\lim_{n \to +\infty} c_n = +\infty$, such that

$$\sum_{k=1}^{N} F(k, c_n) > M c_n^p, \qquad \forall\, n \in \mathbb{N}.$$

Thus, taking in S the sequence $\{s_n\}$ that, for each $n \in \mathbb{N}$, is given by $(s_n)_k := c_n$ for every $k \in [1, N]$, $(s_n)_0 = (s_n)_{N+1} = 0$, one has

$$\begin{aligned}(\Phi - \lambda\Psi)(s_n) &= \frac{\|s_n\|^p}{p} - \lambda \sum_{k=1}^{N} F(k, c_n) \\ &= \frac{1}{p}(2+Q)c_n^p - \lambda \sum_{k=1}^{N} F(k, c_n) < \left(\frac{2+Q}{p} - \lambda M\right) c_n^p,\end{aligned}$$

that is, $\lim_{n \to +\infty}(\Phi - \lambda\Psi)(s_n) = -\infty$.

Next, assume that $\Gamma^\infty < +\infty$. Since $\lambda > \frac{2+Q}{p\Gamma^\infty}$, we can fix $\varepsilon > 0$ such that $\varepsilon < \Gamma^\infty - \dfrac{Q+2}{p\lambda}$. Therefore, also calling $\{c_n\}$ a sequence of positive numbers such that $\lim_{n \to +\infty} c_n = +\infty$ and

$$(\Gamma^\infty - \varepsilon)c_n^p < \sum_{k=1}^{N} F(k, c_n) < (\Gamma^\infty + \varepsilon)c_n^p, \quad \forall\, n \in \mathbb{N},$$

arguing as before and by choosing $\{s_n\}$ in S as above, one has

$$(\Phi - \lambda\Psi)(s_n) < \left(\frac{2+Q}{p} - \lambda(\Gamma^\infty - \varepsilon)\right) c_n^p.$$

So, $\lim_{n \to +\infty} \Phi - \lambda\Psi(s_n) = -\infty$.

Hence, in both cases $\Phi - \lambda\Psi$ is unbounded from below and the proof is complete. □

Now, put

$$\Gamma^0 := \limsup_{t \to 0^+} \Gamma(t).$$

Arguing as in the proof of Theorem 70 and applying part (c) of Theorem 37, we get

THEOREM 71. *Assume that there exist two real sequences* $\{c_n\}$ *and* $\{d_n\}$, *with* $\lim_{n\to+\infty} d_n = 0$, *such that*

$$(h_3) \qquad |c_n| < \frac{2}{(Q+2)^{1/p}(N+1)^{(p-1)/p}}\, d_n \quad \text{for each } n \in \mathbb{N},$$

$$(h_4) \qquad \mathcal{A}_0 := \lim_{n\to+\infty} \frac{\sum_{k=1}^{N} \max_{|t|\le d_n} F(k,t) - \sum_{k=1}^{N} F(k,c_n)}{(2d_n)^p - (Q+2)\,(N+1)^{(p-1)}\,|c_n|^p} < \frac{\Gamma^0}{(Q+2)(N+1)^{p-1}}.$$

Then, for each $\lambda \in \left]\frac{Q+2}{p\Gamma^0}, \frac{1}{p(N+1)^{p-1}\mathcal{A}_0}\right[$, *problem* $(P_\lambda^{f,q})$ *admits a sequence of non zero solutions which converges to zero.*

Now, we point out two consequences of Theorem 70. The first is the following theorem. Setting

$$\Theta_\infty := \liminf_{t\to+\infty} \Theta(t),$$

we get

THEOREM 72. *Assume that*

$$(h_5) \qquad \Theta_\infty < \frac{2^p}{(Q+2)(N+1)^{p-1}}\Gamma^\infty.$$

Then, for each $\lambda \in \left]\frac{Q+2}{p\Gamma^\infty}, \frac{2^p}{p(N+1)^{p-1}\Theta_\infty}\right[$, *problem* $(P_\lambda^{f,q})$ *admits an unbounded sequence of solutions.*

PROOF. Let $\{b_n\}$ be a sequence of positive numbers which goes to infinity such that

$$\lim_{n\to+\infty} \Theta(b_n) = \Theta_\infty.$$

Taking $a_n = 0$ for every $n \in \mathbb{N}$, from Theorem 70 the conclusion follows. □

REMARK 73. When f is a nonnegative function, condition (h_5) becomes (h_5')

$$\Theta_\infty = \liminf_{t\to+\infty} \Gamma(t) < \frac{2^p}{(Q+2)(N+1)^{p-1}} \limsup_{t\to+\infty} \Gamma(t) = \Gamma^\infty.$$

In this case, owing to Theorem 72, (h_5') ensures an unbounded sequence of positive solutions to Problem $(P_\lambda^{f,q})$ for each $\lambda \in \left]\frac{Q+2}{p\Gamma^\infty}, \frac{2^p}{p(N+1)^{p-1}\Theta_\infty}\right[$.

REMARK 74. If $f(k,0) = 0$ for all $k \in [0,N]$, owing to Theorem 44 (see also Remark 45), the solutions in the conclusion of Theorem 65 are positive.

The next result is the second consequence of Theorem 70 and guarantees the existence of infinitely many solutions to $(P_\lambda^{f,q})$ for each λ which lies in a precise half-line.

THEOREM 75. *Assume that there exist two real sequences* $\{a_n\}$ *and* $\{b_n\}$, *with* $\lim_{n\to+\infty} b_n = +\infty$, *such that* (h_1) *holds and*

$$(h_6) \qquad \sum_{k=1}^{N} F(k,a_n) = \sum_{k=1}^{N} \max_{|t|\leq b_n} F(k,t) \qquad \textit{for each } n \in \mathbb{N}.$$

If $\Gamma^\infty > 0$, *then for each* $\lambda > \frac{Q+2}{p\Gamma^\infty}$, *problem* $(P_\lambda^{f,q})$ *admits an unbounded sequence of solutions.*

PROOF. From (h_6) we obtain $\mathcal{A}_\infty = 0$. Hence, since $\Gamma^\infty > 0$, (h_2) of Theorem 70 holds and the proof is complete. □

Now we point out two corollaries of Theorem 72 and Theorem 75 that, owing to Theorem 44, ensure infinitely many positive solutions without assuming the positivity of the nonlinearity.

The following is a consequence of Theorem 72.

COROLLARY 76. *Let* $\alpha : [1,N] \to \mathbb{R}$ *be a nonnegative and non zero function and let* $g : [0,+\infty) \to \mathbb{R}$ *be a continuous function such that* $g(0) = 0$. *Assume that*

$$(h_5'') \ \liminf_{t\to+\infty} \frac{\max_{0\leq\xi\leq t} \int_0^\xi g(s)\,ds}{t^p} < \frac{2^p}{(Q+2)(N+1)^{p-1}} \limsup_{t\to+\infty} \frac{\int_0^t g(s)\,ds}{t^p}.$$

Then, for each

$$\lambda \in \frac{1}{p\sum_{k=1}^N \alpha_k} \left] \frac{Q+2}{\limsup_{t\to+\infty} \frac{\int_0^t g(s)\,ds}{t^p}}, \frac{2^p}{(N+1)^{p-1}} \frac{1}{\liminf_{t\to+\infty} \frac{\max_{0\leq\xi\leq t}\int_0^\xi g(s)\,ds}{t^p}} \right[,$$

the problem

$$(P_\lambda^{\alpha g,q}) \begin{cases} -\Delta(\phi_p(\Delta u_{k-1})) + q_k|u_k|^{p-2}u_k = \lambda\alpha_k g(u_k), & k \in [1,N], \\ u_0 = u_{N+1} = 0, \end{cases}$$

admits an unbounded sequence of positive solutions.

PROOF. Put

$$f(k,t) = \begin{cases} \alpha_k g(t) & \text{if } t \geq 0 \\ 0 & \text{if } t < 0 \end{cases}$$

for every $k \in [1,N]$ and $t \in \mathbb{R}$. The conclusion follows from Theorem 72 owing to (h_5'') and taking into account Theorem 44. □

In particular one has

THEOREM 77. *Let $g : \mathbb{R} \to \mathbb{R}$ be a nonnegative and continuous function. Assume that*

$$\liminf_{t\to+\infty} \frac{\int_0^t g(\xi)\, d\xi}{t^p} < \frac{2^p}{(N+2)(N+1)^{p-1}} \limsup_{t\to+\infty} \frac{\int_0^t g(\xi)\, d\xi}{t^p}.$$

Then, for each

$$\lambda \in \frac{1}{pN} \left] \frac{N+2}{\limsup_{t\to+\infty} \frac{\int_0^t g(\xi)\, d\xi}{t^p}}, \left(\frac{2}{N+1}\right)^{p-1} \frac{1}{\liminf_{t\to+\infty} \frac{\int_0^t g(\xi)\, d\xi}{t^p}} \right[,$$

the problem

$$\begin{cases} -\Delta(\phi_p(\Delta u_{k-1})) + |u_k|^{p-2}u_k = \lambda g(u_k), & k \in [1, N], \\ u_0 = u_{N+1} = 0, \end{cases}$$

admits an unbounded sequence of positive solutions.

Instead, a consequence of Theorem 75 is

COROLLARY 78. *Let α and g be two functions as in Corollary 76. Let $\{a_n\}$ and $\{b_n\}$ be two real sequences, with $\lim_{n\to+\infty} b_n = +\infty$, such that $a_n < b_n$ for each $n \in \mathbb{N}$. Assume that*

(l_1) $\lim_{n\to+\infty} \frac{b_n}{a_n} > \frac{(Q+2)(N+1)^{(p-1)/p}}{2^p}$;

(l_2) $\max_{a_n \le t \le b_n} g(t) \le 0$;

(l_3) $\limsup_{t\to+\infty} \frac{\int_0^t g(x)\, dx}{t^p} > 0.$

Then, for each $\lambda > \frac{Q+2}{p\sum_{k=1}^N \alpha_k} \frac{1}{\limsup_{t\to+\infty} \frac{\int_0^t g(x)\, dx}{t^p}}$ *the conclusion of Corollary 76 holds.*

PROOF. Since (l_2) implies (h_6), our goal is achieved owing to Theorem 75 taking Theorem 44 into account. □

Next result establishes positive solutions even if the nonlinear term changes sign.

THEOREM 79. *Let $g : \mathbb{R} \to \mathbb{R}$ be a continuous function such that $g(0) = 0$ and $\liminf_{t\to+\infty} g(t) < 0$. Assume that $N + 1 < 2^p$ and* $\limsup_{t\to+\infty} \frac{\int_0^t g(x)\, dx}{t^p} = +\infty.$
Then, the problem

$$\begin{cases} -\Delta(\phi_p(\Delta u_{k-1})) = g(u_k), & k \in [1, N], \\ u_0 = u_{N+1} = 0, \end{cases}$$

admits an unbounded sequence of positive solutions.

PROOF. From $\liminf_{t\to+\infty} g(t) < 0$ there is a strictly increasing sequence of positive real numbers $\{t_n\}_{n\in\mathbb{N}}$ such that $g(t_n) < 0$ for all $n \in \mathbb{N}$. Now, fix $\delta_n > 0$ such that $g(t) < 0$ for all $t \in [t_n - \delta_n, t_n + \delta_n]$ and

$$t_n + \delta_n < t_{n+1} - \delta_{n+1}$$

for all $n \in \mathbb{N}$. Therefore, taking into account that

$$\frac{(Q+2)(N+1)^{(p-1)/p}}{2^p} = \frac{(N+1)^{(p-1)/p}}{2^{p-1}} < 1,$$

and by choosing, for instance, $b_n = t_n$ and $a_n = t_n - \frac{\delta_n}{2}$, one has

$$\lim_{n\to+\infty} \frac{b_n}{a_n} \geq 1 > \frac{(Q+2)(N+1)^{(p-1)/p}}{2^p},$$

and $\max_{a_n\leq t\leq b_n} g(t) < 0$ for all $n \in \mathbb{N}$. Hence, since $\limsup_{t\to+\infty} \frac{\int_0^t g(x)\,dx}{t^p} = +\infty$, Corollary 78 ensures our conclusion. □

Finally, we give two easy examples of applications of our results.

EXAMPLE 80. Let N, p and Q be such that

$$\frac{(Q+2)(N+1)^{p-1}}{2^p} < \frac{2+\sqrt{2}}{2-\sqrt{2}}. \tag{26}$$

Then, for each $\lambda \in \left]\frac{Q+2}{N}\frac{2p}{2+\sqrt{2}}, \frac{2^p}{N(N+1)^{p-1}}\frac{2p}{(2-\sqrt{2})}\right[$ the problem

$$\begin{cases} -\Delta(|\Delta u_{k-1}|^{p-2}\Delta u_{k-1})) + q_k|u_k|^{p-2}u_k = \lambda\frac{2}{p}u_k^{p-1}\cos^2(\frac{p}{2}\ln(u_k)), & k \in [1, N], \\ u_0 = u_{N+1} = 0, \end{cases}$$

admits an unbounded sequence of positive solutions.
It is enough to apply Corollary 76 by choosing $g(t) = \frac{2}{p}t^{p-1}\cos^2(\frac{p}{2}\ln(t))$, $t > 0$, $g(0) = 0$, $\alpha_k = 1$, $k \in [1, N]$. In fact,

$$\liminf_{t\to+\infty} \frac{\int_0^t \frac{2}{p}s^{p-1}\cos^2(\frac{p}{2}\ln(s))ds}{t^p} = \liminf_{t\to+\infty} \frac{\frac{t^p}{p^2}\left(1 + \frac{\sin(p\ln t) + \cos(p\ln t)}{2}\right)}{t^p} = \frac{2-\sqrt{2}}{2p^2}.$$

and

$$\limsup_{t\to+\infty} \frac{\int_0^t \frac{2}{p}s^{p-1}\cos^2(\frac{p}{2}\ln(s))ds}{t^p} = \frac{2+\sqrt{2}}{2p^2}.$$

Hence, from (26) one has

$$\liminf_{t\to+\infty} \frac{\int_0^t \frac{2}{p}s^{p-1}\cos^2(\frac{p}{2}\ln(s))ds}{t^p} < \frac{2^p}{(Q+2)(N+1)^{p-1}}\limsup_{t\to+\infty} \frac{\int_0^t \frac{2}{p}s^{p-1}\cos^2(\frac{p}{2}\ln(s))ds}{t^p},$$

that is, (h_5'') holds.

In particular, for instance, the problem

$$\begin{cases} -\Delta^2(u_{k-1}) = \frac{1}{2}u_k\cos^2(\ln(u_k)), & k \in [1,5], \\ u_0 = u_6 = 0, \end{cases}$$

admits an unbounded sequence of positive solutions.

EXAMPLE 81. Let N, p, Q be such that

$$\frac{(Q+2)(N+1)^{(p-1)/p}}{2^p} < e^\pi. \tag{27}$$

Then, for each $\lambda > 0$ the problem

$$\begin{cases} -\Delta(\phi_p(\Delta u_{k-1})) + q_k|u_k|^{p-2}u_k = \lambda u_k^p \sin(\ln(u_k)), & k \in [1,N], \\ u_0 = u_{N+1} = 0, \end{cases}$$

admits an unbounded sequence of positive solutions.
Indeed, put $g(t) = t^p \sin(\ln(t))$, $t > 0$, $g(0) = 0$, $\alpha_k = 1$, $k \in [1,N]$. By choosing $a_n = e^{\pi(2n+1)}$ and $b_n = e^{2(n+1)\pi}$ for every $n \in \mathbb{N}$, one has $\frac{b_n}{a_n} = e^\pi$ for every $n \in \mathbb{N}$. Therefore, from (27) (l_1) holds. Moreover, a simple matter shows that $\max_{a_n \le t \le b_n} g(t) \le 0$, that is, (l_2). Finally, one has

$$\limsup_{t\to+\infty} \frac{\int_0^t s^p \sin(\ln(s))ds}{t^p} = \limsup_{t\to+\infty} \frac{\frac{t^{p+1}}{p^2+2p+2}\Big((p+1)\sin(\ln t) - \cos(\ln t)\Big)}{t^p} = +\infty.$$

Hence, (l_3) holds and Corollary 78 ensures our claim.

In particular, for instance, fixed $N \le 2 \cdot 10^3$, the problem

$$\begin{cases} -\Delta^2(u_{k-1}) = u_k^2 \sin(\ln(u_k)), & k \in [1,N], \\ u_0 = u_{N+1} = 0, \end{cases}$$

admits an unbounded sequence of positive solutions.

REMARK 82. From Theorem 71 we obtain the same consequences of Theorem 70. Namely, replacing $\xi \to +\infty$ with $\xi \to 0^+$, statements such as Theorems 72 and 75, Corollaries 76 and 78 can be established.

Acknowledgments. The authors would like to thank Dr. Giuseppina D'Aguì for a careful reading of the Chapter.

Bibliography

[A] R. P. Agarwal, *Difference Equations and Inequalities: Theory, Methods and Applications*, Marcel Dekker, New York, Basel, 2000.

[APR1] R. P. Agarwal, K. Perera and D. O'Regan, *Multiple positive solutions of singular and nonsingular discrete problems via variational methods*, Nonlinear Anal., **58** (2004), 69–73.

[APR2] R. P. Agarwal, K. Perera and D. O'Regan, *Multiple positive solutions of singular discrete p-Laplacian problems via variational methods*, Adv. in Difference Equations, **2** (2005), 93–99.

[ARW] R. P. Agarwal, D. O'Regan and P. J. Y. Wong, *Positive Solutions of Differential, Difference and Integral Equations*, Kluwer Academic Publishers, Dordrecht, 1999.

[AR] A. Ambrosetti and P. H. Rabinowitz, *Dual variational methods in critical point theory and applications*, J. Funct. Anal. **14** (1973), 349–381.

[AB1] D. Averna and G. Bonanno, *A three critical points theorem and its applications to the ordinary Dirichlet problem*, Topol. Methods Nonlinear Anal., **22** (2003), 93–103.

[AB2] D. Averna and G. Bonanno, *A mountain pass theorem for a suitable class of functions*, Rocky Mountain J. Math., **39** (2009), 707–727.

[AB3] D. Averna and G. Bonanno, *Three solutions for a quasilinear two point boundary value problem involving the one-dimensional p-Laplacian*, Proc. Edinb. Math. Soc. **47** (2004), 257–270.

[B1] G. Bonanno, *Multiple critical points theorems without the Palais-Smale condition*, J. Math. Anal. Appl., **299** (2004), 600–614.

[B2] G. Bonanno, *Some remarks on a three critical points theorem*, Nonlinear Anal. **54** (2003), 651–665.

[B3] G. Bonanno, *A critical points theorem and nonlinear differential problems*, J. of Global Optim. **28** (2004), 249–258.

[BC1] G. Bonanno and P. Candito, *Three solutions to a Neumann problem for elliptic equations involving the p-Laplacian*, Arch. Math. (Basel), **80** (2003), 424–429.

[BC2] G. Bonanno and P. Candito, *On a class of nonlinear variational-hemivariational inequalities*, Appl. Anal. **83** (2004), 1229–1244.

[BC3] G. Bonanno and P. Candito, *Non-differentiable functions with applications to elliptic equations with discontinuous nonlinearities*, J. Differential Equations, **244** (2008) 3031–3059.

[BC4] G. Bonanno and P. Candito, *Nonlinear difference equations investigated via critical methods*, Nonlinear Anal. **70** (2009), 3180–3186.

[BC5] G. Bonanno and P. Candito, *Infinitely many solutions for a class of nonlinear discrete boundary value problems*, Appl. Anal. **88** (2009), 605–616.

[BD] G. Bonanno and G. D'Aguì, *A critical point theorem and existence results for a nonlinear boundary value problem*, Nonlinear Anal., **72** (2010), 1977–1982.

[BL] G. Bonanno and R. Livrea, *Multiplicity theorems for the Dirichlet problem involving the p-Laplacian*, Nonlinear Anal., **54** (2003), 1–7.

[BM] G. Bonanno and S. A. Marano, *On the structure of the critical set of non-differentialble functions with a weak compactness condiction*, App. Anal., **89** (2010), 1–10.

[BMo] G. Bonanno and G. Molica Bisci, *Infinitely many solutions for a boundary value problem with discontinuous nonlinearities*, Bound. Value Probl., (2009) 2009, 20 pages.

[BO] G. Bonanno and D. O'Regan, *A boundary value problem on the half-line via critical point methods*, Dynam. Systems Appl., **15** (2006), 395–407.

[BX] D. Bai and Y. Xu, *Nontrivial solutions of boundary value problems of second-order difference equations*, J. Math. Anal. Appl. **326** (2007), 297–302.

[CG] P. Candito and N. Giovannelli, *Multiple solutions for a discrete boundary value problem*, Comput. Math. Appl., **56** (2008), 959–964.

[Ch] K. C. Chang, *Variational methods for non-differentiable functionals and their applications to partial differential equations*, J. Math. Anal. Appl., **80** (1981), 102–129.

[CJ] J. Chu, D. Jiang, *Eigenvalues and discrete boundary value problems for the one-dimensional p-Laplacian*, J. Math. Anal. Appl. **305** (2005), 452–465.

[Cl] F. H. Clarke, *Optimization and Nonsmooth Analysis*, Classics Appl. Math. **5**, SIAM, Philadelphia, 1990.

[D] E. DiBenedetto, *Real Analysis*, Birkhäuser, Boston, 2002.

[F] G. Fang, *On the existence and the classification of critical points for non-smooth functionals*, Can. J. Math. **47**(4) (1981), 684–717.

[FI] F. Faraci and A. Iannizzotto, *Multiplicity theorems for discrete boundary problems*, Aequationes Math. **74** (2007), 111–118.

[HT] J. Henderson and H. B. Thompson, *Existence of multiple solutions for second order discrete boundary value problems*, Comput. Math. Appl. **43** (2002), 1239–1248.

[JCOA] D. Jiang, J. Chu, D. O'Regan and R. P. Agarwal, *Positive solutions for continuous and discrete boundary value problems to the one-dimensional p-Laplacian*, Math. Inequal. Appl., **7** (2004), 523–534.

[JZ1] L. Jiang and Z. Zhou, *Existence of nontrivial solutions for discrete nonlinear two point boundary value problems*, Appl. Math. and Comput. **180** (2006), 318–329.

[JZ2] L. Jiang and Z. Zhou, *Three solutions to Dirichlet boundary value problems for p-Laplacian difference equations*, Advance in Difference Equations **2008** (2008), 1–10.

[KP] W. G. Kelly and A. C. Peterson, *Difference Equations, An Introduction with Applications*, Academic Press, San Diego, New York, 1991.

[LW] H. Liang and P. Weng, *Existence and multiple solutions for a second-order difference boundary value problem via critical point theory*, J. Math. Anal. Appl. **326** (2007), 511–520.

[MM1] S. A. Marano and D. Motreanu, *On a three critical points theorem for non differentiable functions and applications to non linear boundary value problems*, Nonlinear Anal. **48** (2002), 37–52.

[MM2] S. A. Marano and D. Motreanu, *Infinitely many critical points of non-differentiable functions and applications to a Neumann type problem involving the p-Laplacian*, J. Differential Equations **182** (2002), 108–120.

[MP] D. Motreanu and P. D. Panagiotopoulos, *Minimax Theorems and Qualitative Properties of the Solutions of Hemivariational Inequalities*, Nonconvex Optim. Appl. **29** Kluwer, Dordrecht, 1998.

[MR] D. Motreanu and V. Rădulescu, *Variational and non-variational methods in nonlinear analysis and boundary value problems*, Kluwer, Dordrecht 2003.

[PS] P.Pucci and J.Serrin, *A mountain pass theorem*, J. Differential Equations **63** (1985), 142–149.

[R] P. H. Rabinowitz, *Minimax meshods in critical Point Theory with Applications to Differential Equations*, CBMS Reg. Conferences in Math., vol.**65** , Amer. Math. Soc. Providence, RI, 1686.

[Ri1] B. Ricceri, *A general variational principle and some of its applications*, J. Comput. Appl. Math. **113** (2000), 901–410.

[Ri2] B.Ricceri, *On a three critical points theorem*, Arch. Math. (Basel) **75** (2000), 220–226.

[S] R. Salvati, *Multiple solutions for a mixed boundary value problem*, Math. Sci. Res. J. **7** (2003), no.7, 275–983.

[St] W. Struwe, *Variational methods*, (1996) Springer, Berlin.

[Z] E. Zeidler, *Nonlinear functional analysis and its applications*, Vol. III. Berlin-Meidelberg-New York 1985.

[ZL] G. Zhang and S. Liu, *On a class of semipositone discrete boundary value problems*, J. Math. Anal. Appl. **325** (2007), 175–182.

[ZZL] G. Zhang, W. Zhang and S. Liu, *Existence of 2^n nontrivial solutions for a discrete two-point boundary value problems*, Nonlinear Anal. **59** (2004), 1181–1187

Gabriele Bonanno, Department of Science for Engineering and Architecture (Mathematics Section), Engineering Faculty, University of Messina, 98166 - Messina (Italy)

E-mail address: `bonanno@unime.it`

Pasquale Candito, Dipartimento di Informatica, Matematica, Elettronica e Trasporti Facoltà di Ingegneria, Università degli Studi Mediterranea di Reggio Calabria Via Graziella (Feo di Vito), 89100 Reggio Calabria (Italy)

E-mail address: `pasquale.candito@unirc.it`

Sub-supersolution method for multi-valued elliptic and evolution problems

Siegfried Carl and Dumitru Motreanu

ABSTRACT. We provide a general framework of the sub-supersolution method for multi-valued variational inequalities and variational-hemivariational inequalities, as well as for multi-valued periodic-parabolic problems and periodic-parabolic hemivariational inequalities, and we prove the equivalence between the latter. The obtained existence and comparison results are finally applied to study the existence of multiple solutions for a semilinear periodic-parabolic problem, which generalizes earlier results by others in this direction that now can be proved in a much simpler way.

CONTENTS

The method of sub-supersolution has been proved to be a powerful and fruitful tool in the qualitative and computational theory of elliptic and parabolic boundary value problems. In a recent monograph by Carl-Le-Motreanu (see [**7**]) this method has been extended in a systematic and unified way to both nonsmooth stationary and evolutionary variational problems including variational-hemivariational inequalities. Since the appearance of that monograph new results due to the authors have been obtained which allow to weaken the hypotheses which were needed in [**7**] to establish the sub-supersolution method. In particular, a certain one-sided growth

1991 *Mathematics Subject Classification.* Primary 35J85, 35R70, 47J20; Secondary 35B05, 35J60, 47H04.

Key words and phrases. sub-supersolution, variational inequality, hemivariational inequality.

condition for Clarke's generalized gradient of some locally Lipschitz function which was crucial in the development of the theory in [**7**] can now completely be dropped.

Based on the authors' recent joint work the main purpose of this article is to provide a unified and comprehensive exposition of the sub-supersolution method for both multi-valued variational inequalities involving Clarke's generalized gradient and related variational-hemivariational inequalities. By a number of examples we are going to demonstrate that much of the idea of the sub-supersolution method that has been known for various elliptic and parabolic boundary value problems can be developed in a rather general nonsmooth setting. An important result, which is based on the established method of sub-supersolution, is the equivalence of multi-valued variational inequalities involving Clarke's generalized gradient and variational-hemivariational inequalities.

1. Introduction

Let $V = W^{1,p}(\Omega)$ and $V_0 = W_0^{1,p}(\Omega)$, with $1 < p < \infty$, denote the usual Sobolev spaces, where $\Omega \subset \mathbb{R}^N$, $N \geq 1$, is a bounded domain with Lipschitz boundary $\partial\Omega$. We denote by V^* and V_0^* the dual spaces corresponding to V and V_0, respectively, and by $\langle \cdot, \cdot \rangle$ the duality pairing. Let K be a closed, convex subset of V, and let $j_1 : \Omega \times \mathbb{R} \to \mathbb{R}$ and $j_2 : \partial\Omega \times \mathbb{R} \to \mathbb{R}$ be functions that are only supposed to be measurable in their first and locally Lipschitz continuous with respect to their second argument. Our main goal is to extend the idea of the sub-supersolution method in a natural and systematic way to quasilinear multi-valued elliptic variational inequalities on the one hand and variational-hemivariational inequalities on the other hand, and, moreover, we are going to prove the equivalence of the two problems. Let q denote the Hölder conjugate to p, i.e., q satisfies $1/p + 1/q = 1$. We deal with the following problems.

Multi-valued elliptic variational inequality: Find $u \in K$, $\eta \in L^q(\Omega)$, and $\xi \in L^q(\partial\Omega)$ such that

$$(1) \quad \begin{cases} \eta(x) \in \partial j_1(x, u(x)), \text{ a.e. } x \in \Omega, \quad \xi(x) \in \partial j_2(x, \gamma u(x)), \text{ a.e. } x \in \partial\Omega, \\ \langle Au - h, v - u \rangle + \displaystyle\int_\Omega \eta\,(v - u)\,dx + \int_{\partial\Omega} \xi\,(\gamma v - \gamma u)\,d\sigma \geq 0, \ \forall\, v \in K. \end{cases}$$

where

- the multi-valued functions $s \mapsto \partial j_k(x, s)$ are given by Clarke's generalized gradient of the locally Lipschitz functions $s \mapsto j_k(x, s)$, $k = 1, 2$, defined by

$$\partial j_k(x, s) := \{\zeta \in \mathbb{R} : j_k^o(x, s; r) \geq \zeta r, \ \forall\, r \in \mathbb{R}\}$$

  for a.a. $x \in \Omega$ in case of $k = 1$, and for a.a. $x \in \partial\Omega$ in case of $k = 2$, with $j_k^o(x, s; r)$ denoting the generalized directional derivative of $s \mapsto j_k(x, s)$ at s in the direction r given by

$$j_k^o(x, s; r) = \limsup_{y \to s,\ t \downarrow 0} \frac{j_k(x, y + t\,r) - j_k(x, y)}{t},$$

  (cf., e.g., [**12**, Chap. 2]),

- A is a second-order quasilinear elliptic differential operator of the form

$$Au(x) = -\sum_{i=1}^{N} \frac{\partial}{\partial x_i} a_i(x, \nabla u(x)), \quad \text{with } \nabla u = \left(\frac{\partial u}{\partial x_1}, \dots, \frac{\partial u}{\partial x_N}\right),$$

- $\gamma : V \to L^p(\partial\Omega)$ denotes the trace operator which is known to be linear and compact from V into $L^p(\partial\Omega)$,
- $h \in V^*$.

Variational-hemivariational inequality: Find $u \in K$ such that

$$(2)\ \langle Au - h, v - u\rangle + \int_\Omega j_1^o(x, u; v - u)\, dx + \int_{\partial\Omega} j_2^o(x, \gamma u; \gamma v - \gamma u)\, d\sigma \geq 0,\ \forall\, v \in K.$$

REMARK 1. Applying the definition of Clarke's generalized gradient of j_k, $k = 1, 2$, one readily can see that any solution of the multi-valued variational inequality (1) must be a solution of the variational-hemivariational inequality (2) as well. A sufficient condition for the reverse to hold true is the regularity in the sense of Clarke (see [**12**, Chap. 2]) of the locally Lipschitz functions $s \mapsto j_k(\cdot, s)$. One of the important consequences of the sub-supersolution method to be developed here is that we are able to show the equivalence of (1) and (2) without assuming regularity in the sense of Clarke or any other additional hypotheses on j_k provided the closed convex set K satisfies the lattice structure (14) below.

Before we discuss various special cases of the general setting of the multi-valued variational inequality (1), let us formulate the assumptions on the operator A. We assume the following hypotheses of Leray–Lions type on the coefficient functions a_i, $i = 1, \dots, N$, of the operator A:

(A1) Each $a_i : \Omega \times \mathbb{R}^N \to \mathbb{R}$ satisfies the Carathéodory conditions, i.e., $a_i(x, \zeta)$ is measurable in $x \in \Omega$ for all $\zeta \in \mathbb{R}^N$, and continuous in ζ for a.a. $x \in \Omega$. There exist a constant $c_0 > 0$ and a function $k_0 \in L^q(\Omega)$ such that

$$|a_i(x, \zeta)| \leq k_0(x) + c_0\, |\zeta|^{p-1},$$

for a.a. $x \in \Omega$ and for all $\zeta \in \mathbb{R}^N$.

(A2) For a.a. $x \in \Omega$, and for all $\zeta, \zeta' \in \mathbb{R}^N$ with $\zeta \neq \zeta'$ the following monotonicity holds:

$$\sum_{i=1}^{N} (a_i(x, \zeta) - a_i(x, \zeta'))(\zeta_i - \zeta_i') > 0.$$

(A3) There is some constant $\nu > 0$ such that for a.a. $x \in \Omega$ and for all $\zeta \in \mathbb{R}^N$ the inequality

$$\sum_{i=1}^{N} a_i(x, \zeta)\zeta_i \geq \nu|\zeta|^p - k_1(x)$$

is satisfied for some function $k_1 \in L^1(\Omega)$.

A particular case of the operator A satisfying (A1)–(A3) is the negative p-Laplacian, i.e., $A = -\Delta_p$, which is obtained if

$$a_i(x,\zeta) = |\zeta|^{p-2}\zeta_i, \quad i = 1,\dots,N, \quad \zeta = (\zeta_1,\dots,\zeta_N).$$

In view of (A1), (A2) the operator A defined by

$$\langle Au, \varphi\rangle := \int_\Omega \sum_{i=1}^N a_i(x,\nabla u)\frac{\partial\varphi}{\partial x_i}\,dx, \qquad \forall\,\varphi\in V_0$$

is known to provide a continuous, bounded, and monotone (resp. strictly monotone) mapping from V (resp. V_0) into V_0^*.

To demonstrate the general frame-work that is provided by (1) let us consider a few important special cases.

Example 2. Let $f : \Omega\times\mathbb{R}\to\mathbb{R}$ be a Carathéodory function. Consider its primitive given by

$$j(x,s) := \int_0^s f(x,t)\,dt.$$

Then the function $s \mapsto j(x,s)$ is continuously differentiable, and thus Clarke's gradient reduces to a singleton, i.e.,

$$\partial j(x,s) = \{\partial j(x,s)/\partial s\} = \{f(x,s)\}.$$

If we set $\hat{j}_1(x,s) = j(x,s) + j_1(x,s)$ with j_1 given above, then $\hat{j}_1$ satisfies the same qualitative properties as j_1, i.e., $x\mapsto \hat{j}_1(x,s)$ is measurable, and $s\mapsto\hat{j}_1(x,s)$ is locally Lipschitz continuous. Therefore we can replace j_1 in (1) by $\hat{j}_1$, which, due to the calculus of Clarke's gradient (see [**12**, Chap.2])

$$\partial\hat{j}_1(x,s) = f(x,s) + \partial j_1(x,s)$$

yields

$$\langle Au + F(u) - h, v-u\rangle + \int_\Omega \eta\,(v-u)\,dx + \int_{\partial\Omega}\xi\,(\gamma v - \gamma u)\,d\sigma \ge 0, \quad \forall\, v\in K, \tag{3}$$

where F is the Nemytskij operator generated by f through $F(u)(x) = f(x,u(x))$. Thus problems with elliptic operators $\hat{A}u = Au + F(u)$ involving additional lower order terms of the form $a_0(x,u) = f(x,u)$ are already included in (1).

Example 3. Let $f_1 : \Omega\times\mathbb{R}\to\mathbb{R}$ and $f_2 : \partial\Omega\times\mathbb{R}\to\mathbb{R}$ be measurable functions, which are locally bounded with respect to their second argument. If j_k are the primitives of f_k, i.e.,

$$j_k(x,s) := \int_0^s f_k(x,t)\,dt. \tag{4}$$

then $s\mapsto j_k(x,s)$ are locally Lipschitz and their generalized Clarke's gradients are given by

$$\partial j_k(x,s) = [\underline{f_k}(x,s), \overline{f_k}(x,s)], \tag{5}$$

where

$$\underline{f_k}(x,t) := \lim_{\delta\to 0^+} \operatorname*{ess\,inf}_{|\tau-t|<\delta} f(x,\tau), \quad \overline{f_k}(x,t) := \lim_{\delta\to 0^+} \operatorname*{ess\,sup}_{|\tau-t|<\delta} f(x,\tau),$$

EXAMPLE 4. If $K = V$, $f_1 : \Omega\times\mathbb{R}\to\mathbb{R}$ and $f_2 : \partial\Omega\times\mathbb{R}\to\mathbb{R}$ are Carathéodory functions, and if j_k, $k = 1, 2$, are the primitives of f_k given by (4), then (1) reduces to the following quasilinear elliptic BVP

$$(6) \qquad \langle Au - h, v\rangle + \int_\Omega f_1(x,u)\, v\, dx + \int_{\partial\Omega} f_2(x,\gamma u)\,\gamma v\, d\sigma = 0, \quad \forall\, v\in V,$$

which is the formulation of the weak solution of the BVP

$$(7) \qquad Au + f_1(x,u) = h \;\text{ in }\; \Omega, \quad \frac{\partial u}{\partial\nu} + f_2(x,u) = 0 \quad\text{on}\quad \partial\Omega,$$

where $\partial/\partial\nu$ denotes the outward pointing conormal derivative associated with A.

EXAMPLE 5. If $K = V_0$, and j_k as in Example 4, then (1) is equivalent to

$$(8) \qquad u\in V_0 : \quad \langle Au - h, v\rangle + \int_\Omega f_1(x,u)\, v\, dx = 0, \quad \forall\, v\in V_0,$$

which is nothing but the weak formulation of the homogeneous Dirichlet problem

$$(9) \qquad Au + f_1(x,u) = h \;\text{ in }\; \Omega, \quad u = 0 \quad\text{on}\quad \partial\Omega.$$

EXAMPLE 6. If $K = V_0$ or $K = V$, then (1) reduces to elliptic inclusion problems, which for $K = V_0$ yields the following Dirichlet problem

$$(10) \qquad Au + \partial j_1(x,u) \ni h \;\text{ in }\; \Omega, \quad u = 0 \quad\text{on}\quad \partial\Omega,$$

and for $K = V$ the elliptic inclusion problem

$$(11) \qquad Au + \partial j_1(x,u) \ni h \;\text{ in }\; \Omega, \quad \frac{\partial u}{\partial\nu} + \partial j_2(x,u) \ni 0 \quad\text{on}\quad \partial\Omega.$$

We note that nonhomogeneous Dirichlet conditions that are given by the trace $\gamma\varphi$ with $\varphi\in V$ can always be transformed to a homogeneous Dirichlet problem. Moreover, by an appropriate choice of K also BVP with mixed boundary conditions can be seen to be included as special case of the general formulation (1). To this end let Γ_1 and Γ_2 be relatively open subsets of $\partial\Omega$ satisfying $\overline{\Gamma_1}\cup\overline{\Gamma_2} = \partial\Omega$ and $\Gamma_1\cap\Gamma_2 = \emptyset$. Then we obtain the following special case of (1).

EXAMPLE 7. If $K\subseteq V$ is the closed subspace given by

$$K = \{v\in V : \gamma v = 0 \;\text{ on }\; \Gamma_1\},$$

then (1) reduces to

$$(12) \qquad Au + \partial j_1(x,u) \ni h \;\text{ in }\; \Omega, \quad \frac{\partial u}{\partial\nu} + \partial j_2(x,u) \ni 0 \quad\text{on}\quad \Gamma_2, \quad u = 0 \;\text{ on }\; \Gamma_1.$$

EXAMPLE 8. If $K \subseteq V$, and $j_k = 0$, then (1) is equivalent to the usual variational inequality of the form

$$u \in K: \quad \langle Au - h, v - u\rangle \geq 0, \quad \forall\, v \in K.$$

In Sects. 2 and 3 we present the sub-supersolution method for the multi-valued variational inequality (1) and the variational-hemivariational inequality (2), respectively, and prove various topological and order related properties of the solution sets of these problems. Based on the results of Sects. 2 and 3, recently in [**4**] it has been proved that both problems (1) and (2) are in fact equivalent. For the latter no additional regularity assumptions on the functions j_k are needed. To the best of the authors' knowledge, this equivalence result, fills a gap in the literature, where both problems are treated as different. In Sect. 4 the sub-supersolution method is established for multi-valued periodic parabolic problems, and the equivalence to an associated periodic parabolic hemivariational is proved. Finally, in the last section the results of Sect. 4 are applied to study the existence of multiple solutions for a semilinear periodic parabolic problem, cf. [**5**].

2. Multi-valued variational inequalities

The main goal of this section is to establish the method of sub-supersolution for the multi-valued variational inequality (1). Our presentation is motivated by and relies on results recently obtained in [**2, 7, 3, 10, 9, 8, 11**]. It should be noted that the results we are going to present here include most of the results in the above works, which is mainly due to the weakened assumptions on the nonlinearities j_k, $k = 1, 2$, whose generalized gradients $s \mapsto \partial j_k(x, s)$ need only satisfy a local growth condition with respect to an ordered pair of sub-supersolutions whose definition will be given next. In particular, we should emphasize that unlike in earlier works by the authors, such as in [**7**], the following one-sided growth on Clarke's generalized gradients $s \mapsto \partial j_k(x, s)$ in the form: $\exists\ c_k \geq 0$ such that

$$\zeta_1 \leq \zeta_2 + c_k(s_2 - s_1)^{p-1}, \quad \forall\, \zeta_i \in \partial j_k(x, s_i), \quad \forall\, s_1, s_2 \text{ with } s_1 < s_2, \tag{13}$$

can completely be dropped in our treatment.

For functions w, z and sets W and Z of functions defined on Ω or $\partial\Omega$ we use the notations: $w \wedge z = \min\{w, z\}$, $w \vee z = \max\{w, z\}$, $W \wedge Z = \{w \wedge z : w \in W,\ z \in Z\}$, $W \vee Z = \{w \vee z : w \in W,\ z \in Z\}$, and $w \wedge Z = \{w\} \wedge Z$, $w \vee Z = \{w\} \vee Z$. The following lattice property of the closed convex subset $K \subseteq V$:

$$K \wedge K \subseteq K, \quad K \vee K \subseteq K \tag{14}$$

will play a crucial role in the proof of the existence of extremal solutions.

Our basic notion of sub- and supersolution of the multi-valued variational inequality (1) reads as follows.

DEFINITION 9. A function $\underline{u} \in V$ is called a **subsolution** of (1) if there is an $\underline{\eta} \in L^q(\Omega)$ and a $\underline{\xi} \in L^q(\partial\Omega)$ satisfying

(i) $\underline{u} \vee K \subseteq K$,

(ii) $\underline{\eta}(x) \in \partial j_1(x, \underline{u}(x))$, a.e. $x \in \Omega$, $\underline{\xi}(x) \in \partial j_2(x, \gamma\underline{u}(x))$, a.e. $x \in \partial\Omega$,

(iii) $\langle A\underline{u} - h, v - \underline{u}\rangle + \int_\Omega \underline{\eta}\,(v - \underline{u})\,dx + \int_{\partial\Omega} \underline{\xi}\,(\gamma v - \gamma\underline{u})\,d\sigma \geq 0,$
for all $v \in \underline{u} \wedge K$.

DEFINITION 10. A function $\overline{u} \in V$ is called a **supersolution** of (1) if there is an $\overline{\eta} \in L^q(\Omega)$ and a $\overline{\xi} \in L^q(\partial\Omega)$ satisfying

(i) $\overline{u} \wedge K \subseteq K$,

(ii) $\overline{\eta}(x) \in \partial j_1(x, \overline{u}(x))$, a.e. $x \in \Omega$, $\overline{\xi}(x) \in \partial j_2(x, \gamma\overline{u}(x))$, a.e. $x \in \partial\Omega$,

(iii) $\langle A\overline{u} - h, v - \overline{u}\rangle + \int_\Omega \overline{\eta}\,(v - \overline{u})\,dx + \int_{\partial\Omega} \overline{\xi}\,(\gamma v - \gamma\overline{u})\,d\sigma \geq 0,$
for all $v \in \overline{u} \vee K$.

REMARK 11. Note that the notions for sub- and supersolution defined in Definition 9 and Definition 10 have a symmetric structure, i.e., one obtains the definition for the supersolution $\overline{u}$ from the definition of the subsolution by replacing $\underline{u}$ in Definition 9 by $\overline{u}$, and interchanging $\vee$ by $\wedge$. Furthermore, the lattice condition (14) readily implies that any solution of the multi-valued variational inequality (1) is both a subsolution and a supersolution for (1).

To justify that Definitions 9 and 10 are in fact natural extensions of the usual notions of sub-supersolutions for elliptic boundary value problems let us discuss several special cases.

EXAMPLE 12. Consider Example 4, i.e., $K = V$, $f_1 : \Omega \times \mathbb{R} \to \mathbb{R}$ and $f_2 : \partial\Omega \times \mathbb{R} \to \mathbb{R}$ are Carathéodory functions, and j_k, $k = 1, 2$, are the primitives of f_k given by (4), then Clarke's generalized gradient ∂j_k reduces to a singleton, i.e.,

$$\partial j_k(x, s) = \{f_k(x, s)\},$$

and (1) reduces to the quasilinear elliptic BVP (7). If $\underline{u} \in V$ is a subsolution according to Definition 9, then the first condition (i) is trivially satisfied. The second condition (ii) of Definition 9 means that

$$\underline{\eta}(x) = f_1(x, \underline{u}(x)), \text{ a.e. } x \in \Omega, \ \underline{\xi}(x) = f_2(x, \gamma\underline{u}(x)), \text{ a.e. } x \in \partial\Omega.$$

Since $K = V$, any $v \in \underline{u} \wedge V$ has the form $v = \underline{u} \wedge \varphi = \underline{u} - (\underline{u} - \varphi)^+$ with $\varphi \in V$, where $w^+ = \max\{w, 0\}$, condition (iii) becomes

$$\begin{aligned} &\langle A\underline{u} - h, -(\underline{u} - \varphi)^+\rangle + \int_\Omega f_1(\cdot, \underline{u})\,(-(\underline{u} - \varphi)^+)\,dx \\ &\qquad + \int_{\partial\Omega} f_2(\cdot, \gamma\underline{u})\,(-(\gamma\underline{u} - \gamma\varphi)^+)\,d\sigma \geq 0, \end{aligned} \tag{15}$$

for all $\varphi \in V$. Since $\underline{u} \in V$, we have

$$M = \{(\underline{u} - \varphi)^+ : \varphi \in V\} = V \cap L^p_+(\Omega),$$

where $L^p_+(\Omega)$ is the positive cone of $L^p(\Omega)$, and thus we obtain from inequality (15)

$$(16)\quad \langle A\underline{u} - h, \chi\rangle + \int_\Omega f_1(x, \underline{u})\,\chi\, dx + \int_{\partial\Omega} f_2(x, \gamma\underline{u})\,\gamma\chi\, d\sigma \le 0, \quad \forall\, \chi \in V \cap L^p_+(\Omega),$$

which is nothing but the usual notion of a (weak) subsolution for the BVP (7). Similarly, one verifies that $\overline{u} \in V$ which is a supersolution according to Definition 10 is equivalent with the usual supersolution of the BVP (7).

Example 13. In case that $K = V_0$, and j_k as in Example 12, then (1) is equivalent to the BVP (8) (resp. (9)). Let us consider the notion of subsolution in this case given via Definition 9. For $\underline{u} \in V$ condition (i) means $\underline{u} \vee V_0 \subseteq V_0$. This last condition is satisfied if and only if

$$(17)\qquad \gamma\underline{u} \le 0 \quad \text{i.e.,} \quad \underline{u} \le 0 \ \text{ on } \ \partial\Omega,$$

and condition (ii) means, as above,

$$\underline{\eta}(x) = f_1(x, \underline{u}(x)), \ \text{ a.e. } x \in \Omega.$$

(Note, the boundary integral vanishes since $\gamma v = 0$ for $v \in V_0$.) Since any $v \in \underline{u} \wedge V_0$ can be represented in the form $v = \underline{u} - (\underline{u} - \varphi)^+$ with $\varphi \in V_0$, from (iii) of Definition 9 we obtain

$$(18)\qquad \langle A\underline{u} - h, -(\underline{u} - \varphi)^+\rangle + \int_\Omega f_1(\cdot, \underline{u})\,(-(\underline{u} - \varphi)^+)\, dx \ge 0, \quad \forall \varphi \in V_0.$$

Set $\chi = (\underline{u} - \varphi)^+$, then (18) results in

$$(19)\qquad \langle A\underline{u} - h, \chi\rangle + \int_\Omega f_1(\cdot, \underline{u})\,\chi\, dx \le 0, \quad \forall\, \chi \in M_0,$$

where $M_0 := \{\chi \in V : \chi = (\underline{u} - \varphi)^+,\ \varphi \in V_0\} \subseteq V_0 \cap L^p_+(\Omega)$. One can prove that the set M_0 is a dense subset of $V_0 \cap L^p_+(\Omega)$ (cf. [**7**]), which shows that (19) together with (17) is nothing but the weak formulation for the subsolution of the Dirichlet problem (8) (resp. (9)). Similarly, $\overline{u} \in V$ given by Definition 10 is shown to be a supersolution of the Dirichlet problem (8) (resp. (9)).

In the same way one can verify that the notion of sub- and supersolution defined via Definition 9 and Definition 10, respectively, turns out to be equivalent with the well established notion of sub-supersolution for all other special cases as well.

Assume the existence of an ordered pair $(\underline{u}, \overline{u})$ of sub-supersolutions of the multi-valued variational inequality (1) satisfying $\underline{u} \le \overline{u}$. With respect to this ordered pair we impose the following hypotheses on the nonlinearities j_k, $k = 1, 2$.

(j1) $j_1 : \Omega \times \mathbb{R} \to \mathbb{R}$ satisfies

- (i) $x \mapsto j_1(x, s)$ is measurable in Ω for all $s \in \mathbb{R}$, and $s \mapsto j_1(x, s)$ is locally Lipschitz continuous in $\mathbb{R}$ for a.e. $x \in \Omega$.

(ii) There exists a function $k_\Omega \in L^q_+(\Omega)$ such that for a.e. $x \in \Omega$ and for all $s \in [\underline{u}(x), \overline{u}(x)]$ the growth condition

$$|\eta| \le k_\Omega(x), \quad \forall\, \eta \in \partial j_1(x,s)$$

is fulfilled.

(j2) $j_2 : \partial\Omega \times \mathbb{R} \to \mathbb{R}$ satisfies

(i) $x \mapsto j_2(x,s)$ is measurable in $\partial\Omega$ for all $s \in \mathbb{R}$, and $s \mapsto j_2(x,s)$ is locally Lipschitz continuous in $\mathbb{R}$ for a.e. $x \in \partial\Omega$.

(ii) There exists a function $k_{\partial\Omega} \in L^q_+(\partial\Omega)$ such that for a.e. $x \in \partial\Omega$ and for all $s \in [\gamma\underline{u}(x), \gamma\overline{u}(x)]$ the growth condition

$$|\xi| \le k_{\partial\Omega}(x), \quad \forall\, \xi \in \partial j_2(x,s)$$

is fulfilled.

REMARK 14. We note that by the growth condition (ii) of (j1) and (j2) only a local L^q-boundedness condition on Clarke's generalized gradient ∂j_k is assumed, which is trivially satisfied, in particular, if we suppose the following natural growth condition on ∂j_k: There exist $c > 0$, $k_\Omega \in L^q(\Omega)$ and $k_{\partial\Omega} \in L^q_+(\partial\Omega)$ such that

$$|\eta| \le k_\Omega(x) + c|s|^{p-1}, \quad \forall\, \eta \in \partial j_1(x,s),$$

for a.a. $x \in \Omega$ and for all $s \in \mathbb{R}$, and

$$|\xi| \le k_{\partial\Omega}(x) + c|s|^{p-1}, \quad \forall\, \xi \in \partial j_2(x,s),$$

for a.e. $x \in \partial\Omega$ and for all $s \in \mathbb{R}$.

Our goal is to prove the existence of solutions of the multi-valued variational inequality (1) within the ordered interval $[\underline{u}, \overline{u}]$ of the given sub- and supersolution of (1). Therefore, truncation and comparison techniques will play an important role. We next provide some preliminaries in this direction.

Preliminaries

Let us first briefly recall the definition of pseudomonotonicity for a multi-valued operator $\mathcal{A} : X \to 2^{X^*}$ defined on a real reflexive Banach space X, see, e.g., [**18**], or [**7**, Chap. 2].

DEFINITION 15. Let X be a real reflexive Banach space with dual space X^*. The operator $\mathcal{A} : X \to 2^{X^*}$ is called **pseudomonotone** if the following conditions hold:

(i) The set $\mathcal{A}(u)$ is nonempty, bounded, closed and convex for all $u \in X$;

(ii) $\mathcal{A}$ is upper semicontinuous from each finite dimensional subspace of X to X^* equipped with the weak topology;

(iii) If $(u_n) \subset X$ with $u_n \rightharpoonup u$, and if $u_n^* \in \mathcal{A}(u_n)$ is such that $\limsup \langle u_n^*, u_n - u\rangle \le 0$, then to each element $v \in X$ there exists $u^*(v) \in \mathcal{A}(u)$ with

$$\liminf \langle u_n^*, u_n - v\rangle \ge \langle u^*(v), u - v\rangle.$$

The following proposition provides sufficient conditions for an operator $\mathcal{A} : X \to 2^{X^*}$ to be pseudomonotone.

PROPOSITION 16. *Let X be a real reflexive Banach space, and assume that $\mathcal{A} : X \to 2^{X^*}$ satisfies the following conditions:*

(i) *For each $u \in X$ we have that $\mathcal{A}(u)$ is a nonempty, closed and convex subset of X^*;*
(ii) *$\mathcal{A} : X \to 2^{X^*}$ is bounded;*
(iii) *If $u_n \rightharpoonup u$ in X and $u_n^* \rightharpoonup u^*$ in X^* provided one has that $u_n^* \in \mathcal{A}(u_n)$ and $\limsup\langle u_n^*, u_n - u\rangle \le 0$, then $u^* \in \mathcal{A}(u)$ and $\langle u_n^*, u_n\rangle \to \langle u^*, u\rangle$.*

Then the operator $\mathcal{A} : X \to 2^{X^*}$ is pseudomonotone.

As for the proof of Proposition 16 we refer, e.g., to [**18**, Chap. 2]. In the proof of our main result we make use of the following surjectivity result for multi-valued pseudomonotone mappings perturbed by maximal monotone operators in reflexive Banach spaces (cf., e.g., [**18**, Theorem 2.12]).

THEOREM 17. *Let X be a real reflexive Banach space with dual space X^*, $\Phi : X \to 2^{X^*}$ a maximal monotone operator, and $v_0 \in dom(\Phi)$. Let $\mathcal{A} : X \to 2^{X^*}$ be a pseudomonotone operator, and assume that either $\mathcal{A}_{v_0}$ is quasi-bounded or Φ_{v_0} is strongly quasi-bounded. Assume further that $\mathcal{A} : X \to 2^{X^*}$ is v_0-coercive, i.e., there exists a real-valued function $a : \mathbb{R}_+ \to \mathbb{R}$ with $a(r) \to +\infty$ as $r \to +\infty$ such that for all $(u, u^*) \in graph(\mathcal{A})$ one has $\langle u^*, u - v_0\rangle \ge a(\|u\|_X)\|u\|_X$. Then $\mathcal{A} + \Phi$ is surjective, i.e., range$(\mathcal{A} + \Phi) = X^*$.*

REMARK 18. The operators $\mathcal{A}_{v_0}$ and Φ_{v_0} that appear in the theorem above are defined by $\mathcal{A}_{v_0}(v) := \mathcal{A}(v_0 + v)$ and similarly for Φ_{v_0}. As for the notion of *quasi-bounded* and *strongly quasi-bounded* we refer to [**18**, p.51]. In particular, any bounded operator is quasi-bounded and strongly quasi-bounded as well.

The proof of our main existence and comparison result strongly relies on an appropriate modification of the functions j_k outside the interval $[\underline{u}, \overline{u}]$ formed by the given sub- and supersolutions. Let $(\underline{u}, \underline{\eta}, \underline{\xi}) \in V \times L^q(\Omega) \times L^q(\partial\Omega)$ and $(\overline{u}, \overline{\eta}, \overline{\xi}) \in V \times L^q(\Omega) \times L^q(\partial\Omega)$ satisfy the conditions of Definition 9 and Definition 10, respectively, with $\underline{u} \le \overline{u}$. Then we define the following modifications $\tilde{j}_k$ of the given j_k:

$$\tilde{j}_1(x, s) = \begin{cases} j_1(x, \underline{u}(x)) + \underline{\eta}(x)(s - \underline{u}(x)) & \text{if } s < \underline{u}(x), \\ j_1(x, s) & \text{if } \underline{u}(x) \le s \le \overline{u}(x), \\ j_1(x, \overline{u}(x)) + \overline{\eta}(x)(s - \overline{u}(x)) & \text{if } s > \overline{u}(x). \end{cases} \tag{20}$$

and similarly

$$\tilde{j}_2(x, s) = \begin{cases} j_2(x, \gamma\underline{u}(x)) + \underline{\xi}(x)(s - \gamma\underline{u}(x)) & \text{if } s < \gamma\underline{u}(x), \\ j_2(x, s) & \text{if } \gamma\underline{u}(x) \le s \le \gamma\overline{u}(x), \\ j_2(x, \gamma\overline{u}(x)) + \overline{\xi}(x)(s - \gamma\overline{u}(x)) & \text{if } s > \gamma\overline{u}(x). \end{cases} \tag{21}$$

The following two lemmas list essential properties of the modified functions $\tilde{j}_k$.

LEMMA 19. *Let hypotheses (j1) be satisfied. Then the function $\tilde{j}_1$ has the following properties:*

(P1) *$\tilde{j}_1 : \Omega \times \mathbb{R} \to \mathbb{R}$ satisfies*
  (i) *$x \mapsto \tilde{j}_1(x, s)$ is measurable in Ω for all $s \in \mathbb{R}$, and $s \mapsto \tilde{j}_1(x, s)$ is Lipschitz continuous in $\mathbb{R}$ for a.e. $x \in \Omega$.*

(ii) *Let* $\partial\tilde{j}_1$ *denote Clarke's generalized gradient of* $s \mapsto \tilde{j}_1(x,s)$, *then for a.e.* $x \in \Omega$ *and for all* $s \in \mathbb{R}$ *the growth*

$$|\eta| \le k_\Omega(x), \quad \forall\, \eta \in \partial\tilde{j}_1(x,s)$$

is fulfilled.

(iii) *Clarke's generalized gradient of* $s \mapsto \tilde{j}_1(x,s)$ *is given by*

$$\partial\tilde{j}_1(x,s) = \begin{cases} \underline{\eta}(x) & \text{if } s < \underline{u}(x), \\ \partial\tilde{j}_1(x,\underline{u}(x)) & \text{if } s = \underline{u}(x), \\ \partial j_1(x,s) & \text{if } \underline{u}(x) < s < \overline{u}(x), \\ \partial\tilde{j}_1(x,\overline{u}(x)) & \text{if } s = \overline{u}(x), \\ \overline{\eta}(x) & \text{if } s > \overline{u}(x), \end{cases} \tag{22}$$

and the inclusions $\partial\tilde{j}_1(x,\underline{u}(x)) \subseteq \partial j_1(x,\underline{u}(x))$ *and* $\partial\tilde{j}_1(x,\overline{u}(x)) \subseteq \partial j_1(x,\overline{u}(x))$ *hold true.*

Proof. The proof follows immediately from the definition (20) of $\tilde{j}_1$, and using the assumptions (j1) on j_1 as well as from the fact that Clarke's generalized gradient $\partial j_1(x,s)$ is a convex set. □

Lemma 20. *Let hypotheses (j2) be satisfied. Then the function* $\tilde{j}_2$ *has the following properties:*

(P2) $\tilde{j}_2 : \partial\Omega \times \mathbb{R} \to \mathbb{R}$ *satisfies*

(i) $x \mapsto \tilde{j}_2(x,s)$ *is measurable in* $\partial\Omega$ *for all* $s \in \mathbb{R}$, *and* $s \mapsto \tilde{j}_2(x,s)$ *is Lipschitz continuous in* $\mathbb{R}$ *for a.e.* $x \in \partial\Omega$.

(ii) *Let* $\partial\tilde{j}_2$ *denote Clarke's generalized gradient of* $s \mapsto \tilde{j}_2(x,s)$, *then for a.e.* $x \in \partial\Omega$ *and for all* $s \in \mathbb{R}$ *the growth*

$$|\xi| \le k_{\partial\Omega}(x), \quad \forall\, \xi \in \partial\tilde{j}_2(x,s)$$

is fulfilled.

(iii) *Clarke's generalized gradient of* $s \mapsto \tilde{j}_2(x,s)$ *is given by*

$$\partial\tilde{j}_2(x,s) = \begin{cases} \underline{\xi}(x) & \text{if } s < \gamma\underline{u}(x), \\ \partial\tilde{j}_2(x,\gamma\underline{u}(x)) & \text{if } s = \gamma\underline{u}(x), \\ \partial j_2(x,s) & \text{if } \gamma\underline{u}(x) < s < \gamma\overline{u}(x), \\ \partial\tilde{j}_2(x,\gamma\overline{u}(x)) & \text{if } s = \gamma\overline{u}(x), \\ \overline{\xi}(x) & \text{if } s > \gamma\overline{u}(x), \end{cases} \tag{23}$$

and the inclusions $\partial\tilde{j}_2(x,\gamma\underline{u}(x)) \subseteq \partial j_2(x,\gamma\underline{u}(x))$ *and* $\partial\tilde{j}_2(x,\gamma\overline{u}(x)) \subseteq \partial j_2(x,\gamma\overline{u}(x))$ *hold true.*

Proof. The proof is similar as for Lemma 19. □

By means of $\tilde{j}_1$ and $\tilde{j}_2$ we introduce integral functionals $\tilde{J}_1$ and $\tilde{J}_2$ defined on $L^p(\Omega)$ and $L^p(\partial\Omega)$, respectively, and given by

$$\tilde{J}_1(u) = \int_\Omega \tilde{j}_1(x,u(x))\,dx, \quad u \in L^p(\Omega),$$

$$\tilde{J}_2(v) = \int_{\partial\Omega} \tilde{j}_2(x,v(x))\,d\sigma, \quad v \in L^p(\partial\Omega).$$

Due to (P1)(ii) and (P2) (ii), and applying Lebourg's mean value theorem (see [**7**, Theorem 2.177]) the functionals $\tilde{J}_1 : L^p(\Omega) \to \mathbb{R}$ and $\tilde{J}_2 : L^p(\partial\Omega) \to \mathbb{R}$ are well-defined and Lipschitz continuous, so that Clarke's generalized gradients $\partial\tilde{J}_1 : L^p(\Omega) \to 2^{(L^p(\Omega))^*}$ and $\partial\tilde{J}_2 : L^p(\partial\Omega) \to 2^{(L^p(\partial\Omega))^*}$ are well-defined too. Moreover, Aubin–Clarke theorem (cf. [**12**, p. 83]) provides the following characterization of the generalized gradients. For $u \in L^p(\Omega)$ we have

$$\tilde{\eta} \in \partial\tilde{J}_1(u) \Longrightarrow \tilde{\eta} \in L^q(\Omega) \quad \text{with } \tilde{\eta}(x) \in \partial\tilde{j}_1(x, u(x)) \quad \text{for a.e. } x \in \Omega, \tag{24}$$

and similarly for $v \in L^p(\partial\Omega)$

$$\tilde{\xi} \in \partial\tilde{J}_2(v) \Longrightarrow \tilde{\xi} \in L^q(\partial\Omega) \quad \text{with } \tilde{\xi}(x) \in \partial\tilde{j}_2(x, v(x)) \quad \text{for a.e. } x \in \partial\Omega. \tag{25}$$

By means of Clarke's generalized gradient $\partial\tilde{J}_k$ we introduce the following multi-valued operators:

$$\Phi_1(u) := (i^* \circ \partial\tilde{J}_1 \circ i)(u), \quad \Phi_2(u) := (\gamma^* \circ \partial\tilde{J}_2 \circ \gamma)(u), \quad u \in V, \tag{26}$$

where $i^* : L^q(\Omega) \to V^*$ and $\gamma^* : L^q(\partial\Omega) \to V^*$ denote the adjoint operators of the embedding $i : V \hookrightarrow L^p(\Omega)$ and the trace operator $\gamma : V \to L^p(\partial\Omega)$, respectively, defined by: If $\eta \in L^q(\Omega)$ and $\xi \in L^q(\partial\Omega)$ then

$$\langle i^*\eta, \varphi\rangle = \int_\Omega \eta\,\varphi\,dx, \quad \langle \gamma^*\xi, \varphi\rangle = \int_{\partial\Omega} \xi\,\gamma\varphi\,d\sigma, \quad \forall\varphi \in V. \tag{27}$$

The operators Φ_k, $k = 1, 2$, possess the following properties.

LEMMA 21. *The operators* $\Phi_k : V \to 2^{V^*}$, $k = 1, 2$, *are bounded and pseudomonotone.*

PROOF. As for $\Phi_1 : V \to 2^{V^*}$ we refer to [**2**, Lemma 3.1], and for $\Phi_2 : V \to 2^{V^*}$ the proof is given in [**2**, Lemma 3.2] □

Let b be the cut-off function defined as follows

$$b(x,s) = \begin{cases} (s - \overline{u}(x))^{p-1} & \text{if } s > \overline{u}(x) \\ 0 & \text{if } \underline{u}(x) \le s \le \overline{u}(x) \\ -(\underline{u}(x) - s)^{p-1} & \text{if } s < \underline{u}(x). \end{cases}$$

Apparently, $b : \Omega \times \mathbb{R} \to \mathbb{R}$ is a Carathéodory function satisfying the growth condition

$$|b(x,s)| \le k_2(x) + c_1|s|^{p-1} \tag{28}$$

for a.e. $x \in \Omega$ and for all $s \in \mathbb{R}$, where $c_1 > 0$ is a constant and $k_2 \in L^q(\Omega)$. Moreover, one has the following estimate

$$\int_\Omega b(x, u(x))\,u(x)\,dx \ge c_2\|u\|^p_{L^p(\Omega)} - c_3, \quad \forall u \in L^p(\Omega), \tag{29}$$

for some constants $c_2 > 0$ and $c_3 > 0$. Let B denote the Nemytskij operator associated with b, i.e.,

$$B(u)(x) = b(x, u(x)).$$

In view of (28) the Nemytskij operator $B : L^p(\Omega) \to L^q(\Omega)$ is continuous and bounded, and thus due to the compact embedding $V \hookrightarrow L^p(\Omega)$, it follows that the composed operators $\tilde{B} := i^* \circ B \circ i : V \to V^*$ is completely continuous.

Consider the following multi-valued operator $\mathcal{A}$ defined by

$$\mathcal{A}(u) = Au + \tilde{B}(u) + \Phi_1(u) + \Phi_2(u). \tag{30}$$

LEMMA 22. *The operator $\mathcal{A} : V \to 2^{V^*}$ is bounded, pseudomonotone, and v_0-coercive for $v_0 \in K$.*

PROOF. Hypotheses (A1)–(A2) imply that $A : V \to V^*$ is continuous, bounded, and monotone, and thus, in particular, pseudomonotone, see, e.g., [**7**, Theorem 2.109]. The cut-off operator $\tilde{B} : V \to V^*$ is bounded and completely continuous. Therefore, the (single-valued) operator $A + \tilde{B} : V \to V^*$ is continuous, bounded and pseudomonotone. Due to Lemma 21, the multi-valued operators $\Phi_k : V \to 2^{V^*}$, $k = 1, 2$, are bounded and pseudomonotone. Since pseudomonotonicity is invariant under addition (see [**18**, Chap. 2]), it follows that $\mathcal{A} : V \to 2^{V^*}$ is bounded and pseudomonotone, and so it remains to show that $\mathcal{A}$ is v_0-coercive for $v_0 \in K$, i.e., we need to verify that there exists a real-valued function $a : \mathbb{R}_+ \to \mathbb{R}$ with $a(r) \to +\infty$ as $r \to +\infty$ such that for all $(u, u^*) \in$ graph $(\mathcal{A})$ one has

$$\langle u^*, u - v_0 \rangle \geq a(\|u\|_V)\|u\|_V. \tag{31}$$

To check (31) let $u^* \in \mathcal{A}(u)$, i.e., u^* has the form

$$u^* = Au + \tilde{B}(u) + i^*\tilde{\eta} + \gamma^*\tilde{\xi},$$

where $\tilde{\eta} \in L^q(\Omega)$ with $\tilde{\eta}(x) \in \partial\tilde{j_1}(x, u(x))$, for a.e. $x \in \Omega$, and $\tilde{\xi} \in L^q(\partial\Omega)$ with $\tilde{\xi}(x) \in \partial\tilde{j_2}(x, \gamma u(x))$, for a.e. $x \in \partial\Omega$. Hypotheses (A1), (A2), and the estimates (28) and (29), as well as the uniform boundedness of $\partial\tilde{j_1}$ and $\partial\tilde{j_2}$ in view of (P1) (ii) of Lemma 19 and (P2) (ii) of Lemma 20, respectively, allows us to estimate as follows:

$$|\langle Au + \tilde{B}(u) + i^*\tilde{\eta} + \gamma^*\tilde{\xi}, v_0 \rangle| \leq c(1 + \|u\|_V^{p-1}), \quad \forall\, u \in V, \tag{32}$$

for some constant $c > 0$, and

$$\begin{aligned} \langle u^*, u \rangle &= \langle Au + \tilde{B}(u) + i^*\tilde{\eta} + \gamma^*\tilde{\xi}, u \rangle \\ &\geq \nu\|\nabla u\|_{L^p(\Omega)}^p - c_4 + c_2\|u\|_{L^p(\Omega)}^p - c_3 - c_5\,\|u\|_{L^p(\Omega)} - c_6\,\|\gamma u\|_{L^p(\partial\Omega)} \\ &\geq c_7\,\|u\|_V^p - c_8\,\|u\|_V - c_9. \end{aligned} \tag{33}$$

From (32) and (33) we get the estimate

$$\langle Au + \tilde{B}(u) + i^*\tilde{\eta} + \gamma^*\tilde{\xi}, u - v_0 \rangle \geq c_7\,\|u\|_V^p - c\,\|u\|_V^{p-1} - c_8\,\|u\|_V - c_{10},$$

for some positive constants c_i, which proves the v_0-coercivity. □

We are now in a position to prove the main existence and comparison result for the multi-valued variational inequality (1).

THEOREM 23 (Sub-Supersolution Method). *Let (A1)–(A3) be satisfied. Assume the existence of sub- and supersolutions* $\underline{u}$ *and* $\overline{u}$*, respectively, of the multi-valued variational inequality (1) with* $\underline{u} \le \overline{u}$ *such that (j1)–(j2) is fulfilled. Then there exist solutions of (1) within the ordered interval* $[\underline{u}, \overline{u}]$.

PROOF. Let $I_K : V \to \mathbb{R} \cup \{+\infty\}$ be the indicator function related to the given closed convex set $K \neq \emptyset$, i.e.,

$$I_K(u) = \begin{cases} 0 & \text{if } u \in K, \\ +\infty & \text{if } u \notin K, \end{cases}$$

which is known to be proper, convex, and lower semicontinuous. Consider the following multi-valued operator

$$\mathcal{A} + \partial I_K : V \to 2^{V^*}, \tag{34}$$

where $\mathcal{A}$ is given by (30), and ∂I_K is the usual subdifferential of I_K which is known to be a maximal monotone operator, cf., e.g., [**20**]. Taking into account Lemma 22, we may apply the surjectivity result of Theorem 17, which implies the existence of $u \in K$ such that $h \in \mathcal{A}(u) + \partial I_K(u)$. By definition of $\mathcal{A}$ and ∂I_K the latter inclusion implies the existence of $\eta^* \in \Phi_1(u)$, $\xi^* \in \Phi_2(u)$, and $\theta^* \in \partial I_K(u)$ with $\eta^* = i^*\tilde{\eta}$ and $\xi^* = \gamma^*\tilde{\xi}$ such that equation

$$Au + \tilde{B}(u) + i^*\tilde{\eta} + \gamma^*\tilde{\xi} + \theta^* = h, \quad \text{in } V^* \tag{35}$$

holds, where in view of (24)–(25) we have $\tilde{\eta} \in L^q(\Omega)$ with $\tilde{\eta}(x) \in \partial\tilde{j}_1(x, u(x))$, for a.e. $x \in \Omega$, and $\tilde{\xi} \in L^q(\partial\Omega)$ with $\tilde{\xi}(x) \in \partial\tilde{j}_2(x, \gamma u(x))$, for a.e. $x \in \partial\Omega$. By using the definition of $\partial I_K(u)$, the solution u of equation (35) is seen to be a solution of the following multi-valued variational inequality

$$u \in K : \quad \langle Au - h + \tilde{B}(u) + i^*\tilde{\eta} + \gamma^*\tilde{\xi}, v - u\rangle \ge 0, \quad \forall\, v \in K, \tag{36}$$

which is equivalent to

$$\begin{aligned} u \in K : \quad \langle Au - h, v - u\rangle + \int_\Omega b(x,u)\,(v-u)\,dx + \int_\Omega \tilde{\eta}\,(v-u)\,dx \\ + \int_{\partial\Omega} \tilde{\xi}\,(\gamma v - \gamma u)\,d\sigma \ge 0, \quad \forall\, v \in K. \end{aligned} \tag{37}$$

We are going to show that any solution u of (37) is a solution of the multi-valued variational inequality (1) satisfying $u \in [\underline{u}, \overline{u}]$. To this end we first show that u indeed satisfies $\underline{u} \le u \le \overline{u}$. To prove the inequality $u \le \overline{u}$ we apply the special test function $v = \overline{u} \vee u = \overline{u} + (u - \overline{u})^+$ in Definition 10 (iii), and $v = \overline{u} \wedge u = u - (u - \overline{u})^+ \in K$ in (37), and get by adding the resulting inequalities the following:

$$\begin{aligned} \langle A\overline{u} - Au, (u-\overline{u})^+\rangle - \int_\Omega b(x,u)\,(u-\overline{u})^+\,dx + \int_\Omega (\overline{\eta} - \tilde{\eta})\,(u-\overline{u})^+\,dx \\ + \int_{\partial\Omega} (\overline{\xi} - \tilde{\xi})\,(\gamma u - \gamma\overline{u})^+\,d\sigma \ge 0. \end{aligned} \tag{38}$$

With the notation $\{u > \overline{u}\} = \{x \in \Omega : u(x) > \overline{u}(x)\}$ and $\{\gamma u > \gamma \overline{u}\} = \{x \in \partial\Omega : \gamma u(x) > \gamma \overline{u}(x)\}$, and by applying the results (P1)(iii) of Lemma 19 and (P2)(iii) of Lemma 20 we obtain

$$\int_\Omega (\overline{\eta} - \tilde{\eta})\,(u - \overline{u})^+\,dx = \int_{\{u>\overline{u}\}} (\overline{\eta} - \tilde{\eta})\,(u - \overline{u})\,dx = 0, \tag{39}$$

because $\tilde{\eta}(x) = \overline{\eta}(x)$ for $x \in \{u > \overline{u}\}$, and

$$\int_{\partial\Omega} (\overline{\xi} - \tilde{\xi})\,(\gamma u - \gamma\overline{u})^+\,d\sigma = \int_{\{\gamma u>\gamma\overline{u}\}} (\overline{\xi} - \tilde{\xi})\,(\gamma u - \gamma\overline{u})\,d\sigma = 0, \tag{40}$$

because $\tilde{\xi}(x) = \overline{\xi}(x)$ for $x \in \{\gamma u > \gamma\overline{u}\}$. Taking the definition of the cut-off function b into account we get

$$\int_\Omega b(x,u)\,(u - \overline{u})^+\,dx = \int_\Omega \left((u - \overline{u})^+\right)^p\,dx, \tag{41}$$

and by means of (A2) we obtain the estimate

$$\langle A\overline{u} - Au, (u - \overline{u})^+\rangle = -\langle Au - A\overline{u}, (u - \overline{u})^+\rangle \le 0. \tag{42}$$

Applying the results (39)–(42) to (38) finally yields

$$\int_\Omega \left((u - \overline{u})^+\right)^p\,dx = 0,$$

which implies $(u - \overline{u})^+ = 0$, and thus $u \le \overline{u}$. The proof for $\underline{u} \le u$ can be done in a similar way. So far we have shown that any solution u of the auxiliary multi-valued variational inequality (37) belongs to the interval $[\underline{u}, \overline{u}]$, and thus satisfies: $u \in K$, $b(x, u(x)) = 0$, $\tilde{\eta} \in L^q(\Omega)$, $\tilde{\xi} \in L^q(\partial\Omega)$ and

$$\tilde{\eta}(x) \in \partial\tilde{j_1}(x, u(x)),\ \text{a.e. } x \in \Omega,\quad \tilde{\xi}(x) \in \partial\tilde{j_2}(x, \gamma u(x)),\ \text{a.e. } x \in \partial\Omega, \tag{43}$$

$$\langle Au - h, v - u\rangle + \int_\Omega \tilde{\eta}\,(v - u)\,dx + \int_{\partial\Omega} \tilde{\xi}\,(\gamma v - \gamma u)\,d\sigma \ge 0, \quad \forall\, v \in K. \tag{44}$$

From (P1)(iii) of Lemma 19 we see that $\partial\tilde{j_1}(x, u(x)) \subseteq \partial j_1(x, u(x))$ for any $u \in [\underline{u}, \overline{u}]$, and from (P2)(iii) of Lemma 20 we see that $\partial\tilde{j_2}(x, \gamma u(x)) \subseteq \partial j_2(x, \gamma u(x))$ for $\gamma u \in [\gamma\underline{u}, \gamma\overline{u}]$, and therefore we also have

$$\tilde{\eta}(x) \in \partial j_1(x, u(x)),\ \text{a.e. } x \in \Omega,\quad \tilde{\xi}(x) \in \partial j_2(x, \gamma u(x)),\ \text{a.e. } x \in \partial\Omega,$$

which shows that the solution $u \in [\underline{u}, \overline{u}]$ of the auxiliary problem is in fact a solution of the original multi-valued variational inequality (1). This completes the proof. □

REMARK 24. Without difficulties, the existence and comparison result of Theorem 23 can be extended to more general elliptic operators A of Leray–Lions type such as

$$Au(x) = -\sum_{i=1}^N \frac{\partial}{\partial x_i} a_i(x, u, \nabla u(x)) + a_0(x, u, \nabla u(x)),$$

where the coefficients $a_i : \Omega \times \mathbb{R} \times \mathbb{R}^N \to \mathbb{R}$, $i = 0, 1, \dots, N$ satisfy the following conditions:

(A1') Each $a_i(x, s, \zeta)$ satisfies Carathéodory conditions, i.e., is measurable in $x \in \Omega$ for all $(s, \zeta) \in \mathbb{R} \times \mathbb{R}^N$ and continuous in (s, ζ) for a.e. $x \in \Omega$. There exist a constant $c_0 > 0$ and a function $k_0 \in L^q(\Omega)$ so that

$$|a_i(x, s, \zeta)| \le k_0(x) + c_0(|s|^{p-1} + |\zeta|^{p-1})$$

for a.e. $x \in \Omega$ and for all $(s, \zeta) \in \mathbb{R} \times \mathbb{R}^N$, with $|\zeta|$ denoting the Euclidean norm of the vector ζ.

(A2') The coefficients a_i satisfy a monotonicity condition with respect to ζ in the form

$$\sum_{i=1}^{N} (a_i(x, s, \zeta) - a_i(x, s, \zeta'))(\zeta_i - \zeta_i') > 0$$

for a.e. $x \in \Omega$, for all $s \in \mathbb{R}$, and for all $\zeta, \zeta' \in \mathbb{R}^N$ with $\zeta \neq \zeta'$.

(A3')

$$\sum_{i=1}^{N} a_i(x, s, \zeta)\zeta_i \ge \nu|\zeta|^p - k(x)$$

for a.e. $x \in \Omega$, for all $s \in \mathbb{R}$, and for all $\zeta \in \mathbb{R}^N$ with some constant $\nu > 0$ and some function $k \in L^1(\Omega)$.

The proof of Theorem 23 under these more general assumptions can be done in just the same way by appropriately modifying the multi-valued operator $\mathcal{A}$ defined by (30) in the following way:

$$\mathcal{A}(u) = A_T u + \lambda \tilde{B}(u) + \Phi_1(u) + \Phi_2(u), \tag{45}$$

where A_T is given by

$$A_T u(x) = -\sum_{i=1}^{N} \frac{\partial}{\partial x_i} a_i(x, Tu, \nabla u(x)) + a_0(x, Tu, \nabla Tu(x)),$$

and T is the following truncation mapping

$$(Tu)(x) = \begin{cases} \overline{u}(x) & \text{if } u(x) > \overline{u}(x) \\ u(x) & \text{if } \underline{u}(x) \le u(x) \le \overline{u}(x) \\ \underline{u}(x) & \text{if } u(x) < \underline{u}(x). \end{cases}$$

Then for $\lambda > 0$ sufficiently large one can prove that the modified operator $\mathcal{A} : V \to 2^{V^*}$ is bounded, pseudomonotone, and v_0-coercive for $v_0 \in K$.

2.1. Directedness of solution set. Let $\mathcal{S}$ denote the set of all solutions of the multi-valued variational inequality (1) that belong to the ordered interval $[\underline{u}, \overline{u}]$ formed by the given sub- and supersolution. In view of Theorem 23, we have that $\mathcal{S} \neq \emptyset$. The main goal of this subsection is to show that $\mathcal{S}$ is a *directed set*.

DEFINITION 25 (directed set). Let $(\mathcal{P}, \leq)$ be a partially ordered set. A subset $\mathcal{C}$ of $\mathcal{P}$ is said to be **upward directed** if for each pair $x, y \in \mathcal{C}$ there is a $z \in \mathcal{C}$ such that $x \leq z$ and $y \leq z$. Similarly, $\mathcal{C}$ is **downward directed** if for each pair $x, y \in \mathcal{C}$ there is a $w \in \mathcal{C}$ such that $w \leq x$ and $w \leq y$. If $\mathcal{C}$ is both upward and downward directed it is called **directed**.

Our main result is given by the following theorem.

THEOREM 26. *Let the hypotheses of Theorem 23 and the lattice condition (14) be satisfied. Then the solution set $\mathcal{S}$ of all solutions of (1) within the interval $[\underline{u}, \overline{u}]$ equipped with the natural partial ordering of functions introduced by the order cone $L^p_+(\Omega)$ is a directed set.*

PROOF. We are going to show that $\mathcal{S}$ is upward directed only, since the proof for $\mathcal{S}$ being downward directed can be done in a similar way. To this end let u_1, $u_2 \in \mathcal{S}$. Our goal is to prove the existence of an element $u \in \mathcal{S}$ such that $u \geq u_k$, $k = 1, 2$. The proof will be done in several steps, and crucially relies on an appropriately designed auxiliary problem.
Step 1: Auxiliary Problem
Let $u_k \in \mathcal{S}$, $k = 1, 2$, i.e., $u_k \in K$, and there exist $\eta_k \in L^q(\Omega)$, $\xi_k \in L^q(\partial\Omega)$ such that $\eta_k(x) \in \partial j_1(x, u_k(x))$, for a.e. $x \in \Omega$, $\xi_k(x) \in \partial j_2(x, \gamma u_k(x))$ for a.e. $x \in \partial\Omega$, and the following variational inequality is satisfied:

$$\langle Au_k - h, v - u_k\rangle + \int_\Omega \eta_k\,(v - u_k)\,dx + \int_{\partial\Omega} \xi_k\,(\gamma v - \gamma u_k)\,d\sigma \geq 0, \quad \forall\, v \in K. \tag{46}$$

Let us define $u_0 := \max\{u_1, u_2\}$, and η_0 as follows

$$\eta_0(x) = \begin{cases} \eta_1(x) & \text{if } x \in \{u_1 \geq u_2\}, \\ \eta_2(x) & \text{if } x \in \{u_2 > u_1\}, \end{cases}$$

as well as define ξ_0 by

$$\xi_0(x) = \begin{cases} \xi_1(x) & \text{if } x \in \{\gamma u_1 \geq \gamma u_2\}, \\ \xi_2(x) & \text{if } x \in \{\gamma u_2 > \gamma u_1\}. \end{cases}$$

By the definition of η_0, and ξ_0 we readily see that $\eta_0 \in L^q(\Omega)$, and $\xi_0 \in L^q(\partial\Omega)$, and

$$\eta_0(x) \in \partial j_1(x, u_0(x)) \text{ for a.e. } x \in \Omega, \xi_0(x) \in \partial j_2(x, \gamma u_0(x)) \text{ for a.e. } x \in \partial\Omega. \tag{47}$$

By means of η_0, ξ_0, u_0, and $\overline{\eta}$, $\overline{\xi}$, $\overline{u}$ of Definition 10 we introduce the following modifications $\tilde{j_1} : \Omega \times \mathbb{R} \to \mathbb{R}$ and $\tilde{j_2} : \partial\Omega \times \mathbb{R} \to \mathbb{R}$ of the given j_1 and j_2, respectively:

$$\tilde{j_1}(x, s) = \begin{cases} j_1(x, u_0(x)) + \eta_0(x)(s - u_0(x)) & \text{if } s < u_0(x), \\ j_1(x, s) & \text{if } u_0(x) \leq s \leq \overline{u}(x), \\ j_1(x, \overline{u}(x)) + \overline{\eta}(x)(s - \overline{u}(x)) & \text{if } s > \overline{u}(x). \end{cases} \tag{48}$$

and similarly

$$\tilde{j_2}(x, s) = \begin{cases} j_2(x, \gamma u_0(x)) + \xi_0(x)(s - \gamma u_0(x)) & \text{if } s < \gamma u_0(x), \\ j_2(x, s) & \text{if } \gamma u_0(x) \leq s \leq \gamma\overline{u}(x), \\ j_2(x, \gamma\overline{u}(x)) + \overline{\xi}(x)(s - \gamma\overline{u}(x)) & \text{if } s > \gamma\overline{u}(x). \end{cases} \tag{49}$$

In view of hypotheses (j1) and (j2) the modified function $\tilde{j}_1$ and $\tilde{j}_2$ given by (48) and (49) enjoy the properties of Lemma 19 and Lemma 20, respectively, with the only difference that Clarke's gradient of $\tilde{j}_1$ and $\tilde{j}_2$ is now given by:

$$\partial\tilde{j}_1(x,s)=\begin{cases}\eta_0(x) & \text{if } s<u_0(x),\\ \partial\tilde{j}_1(x,u_0(x)) & \text{if } s=u_0(x),\\ \partial j_1(x,s) & \text{if } u_0(x)<s<\overline{u}(x),\\ \partial\tilde{j}_1(x,\overline{u}(x)) & \text{if } s=\overline{u}(x),\\ \overline{\eta}(x) & \text{if } s>\overline{u}(x),\end{cases} \tag{50}$$

and the inclusions $\partial\tilde{j}_1(x,u_0(x))\subseteq\partial j_1(x,u_0(x))$ and $\partial\tilde{j}_1(x,\overline{u}(x))\subseteq\partial j_1(x,\overline{u}(x))$ hold true, and similarly,

$$\partial\tilde{j}_2(x,s)=\begin{cases}\xi_0(x) & \text{if } s<\gamma u_0(x),\\ \partial\tilde{j}_2(x,\gamma u_0(x)) & \text{if } s=\gamma u_0(x),\\ \partial j_2(x,s) & \text{if } \gamma u_0(x)<s<\gamma\overline{u}(x),\\ \partial\tilde{j}_2(x,\gamma\overline{u}(x)) & \text{if } s=\gamma\overline{u}(x),\\ \overline{\xi}(x) & \text{if } s>\gamma\overline{u}(x),\end{cases} \tag{51}$$

with the inclusions $\partial\tilde{j}_2(x,\gamma u_0(x))\subseteq\partial j_2(x,\gamma u_0(x))$ and $\partial\tilde{j}_2(x,\gamma\overline{u}(x))\subseteq\partial j_2$ $(x,\gamma\overline{u}(x))$. Furthermore, we introduce functions $g_{1,i}:\Omega\times\mathbb{R}\to\mathbb{R}$, $g_{2,i}:\partial\Omega\times\mathbb{R}\to\mathbb{R}$ related to η_i, ξ_i and u_i, $i=0,1,2$, and defined by:

$$g_{1,0}(x,s)=\begin{cases}\eta_0(x) & \text{if } s\le u_0(x),\\ \eta_0(x)+\dfrac{\overline{\eta}(x)-\eta_0(x)}{\overline{u}(x)-u_0(x)}(s-u_0(x)) & \text{if } u_0(x)<s<\overline{u}(x),\\ \overline{\eta}(x) & \text{if } s\ge\overline{u}(x),\end{cases} \tag{52}$$

and for $k=1,2$

$$g_{1,k}(x,s)=\begin{cases}\eta_k(x) & \text{if } s\le u_k(x),\\ \eta_k(x)+\dfrac{\eta_0(x)-\eta_k(x)}{u_0(x)-u_k(x)}(s-u_k(x)) & \text{if } u_k(x)<s<u_0(x),\\ g_{1,0}(x,s) & \text{if } s\ge u_0(x),\end{cases} \tag{53}$$

as well as

$$g_{2,0}(x,s)=\begin{cases}\xi_0(x) & \text{if } s\le\gamma u_0(x),\\ \xi_0(x)+\dfrac{\overline{\xi}(x)-\xi_0(x)}{\gamma\overline{u}(x)-\gamma u_0(x)}(s-\gamma u_0(x)) & \text{if } \gamma u_0(x)<s<\gamma\overline{u}(x),\\ \overline{\xi}(x) & \text{if } s\ge\gamma\overline{u}(x),\end{cases} \tag{54}$$

and for $k=1,2$
(55)

$$g_{2,k}(x,s)=\begin{cases}\xi_k(x) & \text{if } s\le\gamma u_k(x),\\ \xi_k(x)+\dfrac{\xi_0(x)-\xi_k(x)}{\gamma u_0(x)-\gamma u_k(x)}(s-\gamma u_k(x)) & \text{if } \gamma u_k(x)<s<\gamma u_0(x),\\ g_{2,0}(x,s) & \text{if } s\ge\gamma u_0(x).\end{cases}$$

Apparently, the functions $g_{1,i}:\Omega\times\mathbb{R}\to\mathbb{R}$, $g_{2,i}:\partial\Omega\times\mathbb{R}\to\mathbb{R}$, $i=0,1,2$, are Carathéodory functions which are piecewise linear with respect to s and uniformly $L^q(\Omega)$ and $L^q(\partial\Omega)$-bounded, respectively. Finally, define the cut-off function

$\hat{b}$ related to the pair u_0 and $\overline{u}$ by:

$$\hat{b}(x,s) = \begin{cases} (s-\overline{u}(x))^{p-1} & \text{if } s > \overline{u}(x) \\ 0 & \text{if } u_0(x) \le s \le \overline{u}(x) \\ -(u_0(x)-s)^{p-1} & \text{if } s < u_0(x), \end{cases} \tag{56}$$

which qualitatively satisfies the same estimates as b in (28) and (29). The function $\hat{b} : \Omega \times \mathbb{R} \to \mathbb{R}$ is a Carathéodory function, and its associated Nemytskij operator $\hat{B} : L^p(Q) \to L^q(Q)$ defined by $\hat{B}u(x,t) = \hat{b}(x,t,u(x,t))$ is continuous and bounded. With $i^* : L^q(\Omega) \hookrightarrow V^*$ being the adjoint operator of the embedding $i : V \hookrightarrow L^p(\Omega)$, the operator $\hat{\mathcal{B}} = i^* \circ \hat{B} \circ i : V \to V^*$ is completely continuous and bounded, due to the compact embedding $V \hookrightarrow L^p(\Omega)$. By means of the Carathéodory functions $g_{k,i}$ introduced in (52)–(55) we define the functions $g_1 : \Omega \times \mathbb{R} \to \mathbb{R}$ and $g_2 : \partial\Omega \times \mathbb{R} \to \mathbb{R}$ as follows:

$$g_j(x,s) = \sum_{l=1}^{2} |g_{j,l}(x,s) - g_{j,0}(x,s)|, \quad j = 1,2, \tag{57}$$

which are Carathéodory functions in their respective domains of definition, and which are uniformly $L^q(\Omega)$ and $L^q(\partial\Omega)$-bounded, respectively. Thus, the associated Nemytskij operators $G_1 : L^p(\Omega) \to L^q(\Omega)$ and $G_2 : L^p(\partial\Omega) \to L^q(\partial\Omega)$ are continuous and bounded which due to the compact embedding $V \hookrightarrow L^p(\Omega)$ and the compactness of the trace operator γ implies that

$$\mathcal{G}_1 = i^* \circ G_1 \circ i : V \to V^*, \quad \mathcal{G}_2 = \gamma^* \circ G_2 \circ \gamma : V \to V^* \tag{58}$$

is bounded and completely continuous. Further, in a similar way as in the Preliminaries, we let $\tilde{J}_1 : L^p(\Omega) \to \mathbb{R}$ and $\tilde{J}_2 : L^p(\partial\Omega) \to \mathbb{R}$ be the integral functionals, here associated to $\tilde{j}_1$ and $\tilde{j}_2$ defined by (48) and (49), respectively. With Clarke's generalized gradients $\partial\tilde{J}_k$ of $\tilde{J}_k$ we define the multi-valued functions Φ_k as in (26), i.e.,

$$\Phi_1(u) := (i^* \circ \partial\tilde{J}_1 \circ i)(u), \quad \Phi_2(u) := (\gamma^* \circ \partial\tilde{J}_2 \circ \gamma)(u), \quad u \in V.$$

Further, define the multi-valued operator $\mathcal{A}$ as

$$\mathcal{A}(u) = Au + \hat{\mathcal{B}}(u) + \Phi_1(u) + \Phi_2(u) - \mathcal{G}_1(u) - \mathcal{G}_2(u). \tag{59}$$

Since $\mathcal{A}$ given by (59) differs qualitatively from (30) basically only by uniformly bounded and completely continuous perturbations $\mathcal{G}_k$, we readily observe that $\mathcal{A}$ given by (59) fulfills Lemma 22, i.e., $\mathcal{A} : V \to 2^{V^*}$ is bounded, pseudomonotone, and v_0-coercive for $v_0 \in K$. Therefore, we may apply the surjectivity result of Theorem 17 to the multi-valued operator

$$\mathcal{A} + \partial I_K : V \to 2^{V^*},$$

where I_K is the indicator function related to K. As a consequence of Theorem 17 we obtain the existence of a $u \in K$ such that $h \in \mathcal{A}(u) + \partial I_K(u)$, which by definition (59) of $\mathcal{A}$ and ∂I_K results in the existence of $\eta^* \in \Phi_1(u)$, $\xi^* \in \Phi_2(u)$, and $\theta^* \in \partial I_K(u)$ with $\eta^* = i^*\tilde{\eta}$, $\xi^* = \gamma^*\tilde{\xi}$ such that equation

$$Au + \hat{\mathcal{B}}(u) + i^*\tilde{\eta} + \gamma^*\tilde{\xi} - \mathcal{G}_1(u) - \mathcal{G}_2(u) + \theta^* = h, \quad \text{in } V^* \tag{60}$$

holds, where $\tilde{\eta} \in L^q(\Omega)$ with $\tilde{\eta}(x) \in \partial\tilde{j}_1(x, u(x))$, for a.e. $x \in \Omega$, and $\tilde{\xi} \in L^q(\partial\Omega)$ with $\tilde{\xi}(x) \in \partial\tilde{j}_2(x, \gamma u(x))$, for a.e. $x \in \partial\Omega$. By using the definition of $\partial I_K(u)$, the solution u of equation (60) is seen to solve the following auxiliary multi-valued variational inequality

$$(61)\quad u \in K : \langle Au - h + \hat{\mathcal{B}}(u) + i^*\tilde{\eta} + \gamma^*\tilde{\xi} - \mathcal{G}_1(u) - \mathcal{G}_2(u), v - u\rangle \geq 0, \ \forall\, v \in K,$$

which is equivalent to

$$\begin{aligned} u \in K : \langle Au - h, v - u\rangle &+ \int_\Omega \hat{b}(x, u)\,(v - u)\,dx + \int_\Omega (\tilde{\eta} - g_1(x, u))\,(v - u)\,dx \\ &+ \int_{\partial\Omega} (\tilde{\xi} - g_2(x, \gamma u))\,(\gamma v - \gamma u)\,d\sigma \geq 0, \ \forall\, v \in K. \end{aligned} \tag{62}$$

Step 2: Comparison
Here we are going to show that any solution u of the auxiliary multi-valued variational inequality (62) satisfies $u_0 \leq u \leq \overline{u}$.

Let us first verify $u \leq \overline{u}$. To this end we recall Definition 10, and take as special test function in Definition 10 (iii) $v = \overline{u} \vee u = \overline{u} + (u - \overline{u})^+$, and in (62) the special test function $v = \overline{u} \wedge u = u - (u - \overline{u})^+ \in K$ is applied. By adding the resulting inequalities we obtain the following:

$$\begin{aligned} \langle A\overline{u} - Au, (u - \overline{u})^+\rangle &- \int_\Omega \hat{b}(x, u)\,(u - \overline{u})^+\,dx \\ &+ \int_\Omega (\overline{\eta} - \tilde{\eta} + g_1(x, u))\,(u - \overline{u})^+\,dx \\ &+ \int_{\partial\Omega} (\overline{\xi} - \tilde{\xi} + g_2(x, \gamma u))\,(\gamma u - \gamma\overline{u})^+\,d\sigma \geq 0. \end{aligned} \tag{63}$$

We recall that $\tilde{\eta}(x) \in \partial\tilde{j}_1(x, u(x))$ and $\tilde{\xi}(x) \in \partial\tilde{j}_2(x, \gamma u(x))$ with $\partial\tilde{j}_1$ and $\partial\tilde{j}_2$ given by (50) and (51), respectively. Therefore, if $x \in \{u > \overline{u}\}$ then $\overline{\eta}(x) = \tilde{\eta}(x)$, and due to (53) $g_{1,l}(x, u(x)) = g_{1,0}(x, u(x))$, for $l = 1, 2$, which results in $g_1(x, u(x)) = 0$, and thus

$$(64)\quad \int_\Omega (\overline{\eta} - \tilde{\eta} + g_1(x, u))\,(u - \overline{u})^+\,dx = \int_{\{u > \overline{u}\}} (\overline{\eta} - \tilde{\eta} + g_1(x, u))\,(u - \overline{u})\,dx = 0.$$

If $x \in \{\gamma u > \gamma\overline{u}\}$ then $\overline{\xi}(x) = \tilde{\xi}(x)$, and due to (55) we have $g_{2,l}(x, \gamma u(x)) = g_{2,0}(x, \gamma u(x))$, for $l = 1, 2$, which results in $g_2(x, \gamma u(x)) = 0$, and thus

$$\int_{\partial\Omega} (\overline{\xi} - \tilde{\xi} + g_2(x, \gamma u))\,(\gamma u - \gamma\overline{u})^+\,d\sigma = 0. \tag{65}$$

Taking the definition of the cut-off function $\hat{b}$ into account we get

$$\int_\Omega \hat{b}(x, u)\,(u - \overline{u})^+\,dx = \int_\Omega \left((u - \overline{u})^+\right)^p\,dx, \tag{66}$$

and by means of (A2) we obtain the estimate

$$\langle A\overline{u} - Au, (u - \overline{u})^+\rangle = -\langle Au - A\overline{u}, (u - \overline{u})^+\rangle \leq 0. \tag{67}$$

Applying the results (64)–(67) to (63) finally yields

$$\int_\Omega \Big((u-\overline{u})^+\Big)^p\,dx = 0,$$

which implies $(u-\overline{u})^+ = 0$, and thus $u \le \overline{u}$.

In order to show that $u \ge u_0 = \max\{u_1, u_2\}$, we show that $u \ge u_k$, $k = 1, 2$. We recall that $u_k \in \mathcal{S}$ means that u_k is a solution of (1) within the interval $[\underline{u}, \overline{u}]$, i.e., there exist $\eta_k \in L^q(\Omega)$, $\xi_k \in L^q(\partial\Omega)$ such that $\eta_k(x) \in \partial j_1(x, u_k(x))$, for a.e. $x \in \Omega$, $\xi_k(x) \in \partial j_2(x, \gamma u_k(x))$ for a.e. $x \in \partial\Omega$, and the variational inequality (46) holds. If we test (46) with $v = u_k \wedge u = u_k - (u_k - u)^+ \in K$, and the auxiliary multi-valued variational inequality (62) with $v = u \vee u_k = u + (u_k - u)^+ \in K$ we get by adding the resulting inequalities the following estimate:

$$\begin{aligned}
&\langle Au - Au_k, (u_k-u)^+\rangle + \int_\Omega \hat{b}(x,u)\,(u_k-u)^+\,dx \\
&\quad + \int_\Omega (\tilde{\eta} - g_1(x,u) - \eta_k)\,(u_k-u)^+\,dx \\
&\quad + \int_{\partial\Omega} (\tilde{\xi} - g_2(x,\gamma u) - \xi_k)\,(\gamma u_k - \gamma u)^+\,d\sigma \ge 0.
\end{aligned} \tag{68}$$

We estimate the terms on the left-hand side of the last inequality individually. By means of (A2) we get

$$\langle Au - Au_k, (u_k-u)^+\rangle = -\langle Au_k - Au, (u_k-u)^+\rangle \le 0. \tag{69}$$

For $x \in \{u_k > u\}$, from (52) it follows $g_{1,0}(x,u(x)) = \eta_0(x)$, and (53) yields $g_{1,k}(x,u(x)) = \eta_k(x)$, hence we obtain by using the definition of g_1 given in (57) the equation (with $l \ne k$, and $l, k \in \{1,2\}$)

$$g_1(x,u(x)) = |\eta_k(x) - \eta_0(x)| + |g_{1,l}(x,u(x)) - g_{1,0}(x,u(x))|, \quad x \in \{u_k > u\}. \tag{70}$$

Further, for $x \in \{u_k > u\} \subseteq \{u_0 > u\}$ from (50) it follows $\tilde{\eta}(x) \in \partial\tilde{j}_1(x,u(x)) = \eta_0(x)$, which in conjunction with (70) results in the following estimate of the third term on left-hand side of (68): $l \ne k$

$$\begin{aligned}
&\int_\Omega (\tilde{\eta} - g_1(x,u) - \eta_k)\,(u_k-u)^+\,dx = \int_{\{u_k>u\}} (\tilde{\eta} - g_1(x,u) - \eta_k)\,(u_k-u)\,dx \\
&\quad = \int_{\{u_k>u\}} \Big(\eta_0 - \eta_k - |\eta_k - \eta_0| - |g_{1,l}(x,u) - g_{1,0}(x,u)|\Big) \\
&\quad\quad \times (u_k - u)\,dx \le 0.
\end{aligned} \tag{71}$$

Similarly, for $x \in \{\gamma u_k > \gamma u\}$, from (54) it follows $g_{2,0}(x,\gamma u(x)) = \xi_0(x)$, and (55) yields $g_{2,k}(x,\gamma u(x)) = \xi_k(x)$, hence we obtain by using the definition of g_2 given in (57) the equation (with $l \ne k$, and $l, k \in \{1,2\}$)

$$g_2(x,\gamma u(x)) = |\xi_k(x) - \xi_0(x)| + |g_{2,l}(x,\gamma u(x)) - g_{2,0}(x,\gamma u(x))| \tag{72}$$

for all $x \in \{\gamma u_k > \gamma u\}$. Further, for $x \in \{\gamma u_k > \gamma u\} \subseteq \{\gamma u_0 > \gamma u\}$ from (51) it follows $\tilde{\xi}(x) \in \partial\tilde{j}_2(x,\gamma u(x)) = \xi_0(x)$, which in conjunction with (72) implies the following estimate of the 4th term on left-hand side of (68): $l \neq k$

$$\begin{aligned}
&\int_{\partial\Omega} (\tilde{\xi} - g_2(x,\gamma u) - \xi_k)\,(\gamma u_k - \gamma u)^+\, d\sigma \\
&= \int_{\{\gamma u_k > \gamma u\}} (\tilde{\xi} - g_2(x,\gamma u) - \xi_k)\,(\gamma u_k - \gamma u)\, d\sigma \\
&= \int_{\{\gamma u_k > \gamma u\}} \Big(\xi_0 - \xi_k - |\xi_k - \xi_0| - |g_{2,l}(x,\gamma u) - g_{2,0}(x,\gamma u)|\Big) \\
&\quad \times (\gamma u_k - \gamma u)\, d\sigma \le 0.
\end{aligned} \tag{73}$$

Using (69), (71), and (73) we get from (68) the estimate

$$\int_\Omega \hat{b}(x,u)\,(u_k - u)^+\, dx \ge 0,$$

which by applying the definition of $\hat{b}$ finally results in

$$\begin{aligned}
0 &\le \int_\Omega \hat{b}(x,u)\,(u_k - u)^+\, dx = \int_{\{u_k > u\}} \hat{b}(x,u)\,(u_k - u)\, dx \\
&= -\int_{\{u_k > u\}} (u_0 - u)^{p-1}\,(u_k - u)\, dx \le -\int_{\{u_k > u\}} (u_k - u)^p\, dx \\
&\le -\int_\Omega ((u_k - u)^+)^p\, dx \le 0.
\end{aligned}$$

The last inequality implies

$$\int_\Omega ((u_k - u)^+)^p\, dx = 0,$$

and hence it follows $(u_k - u)^+ = 0$, i.e., $u_k \le u$, $k = 1, 2,,$ which completes the comparison.

Step 3: $\mathcal{S}$ is Upward Directed

In this final step we complete the proof of Theorem 26 showing that $\mathcal{S}$ is indeed upward directed. To this end we only need to show that a solution u of the auxiliary multi-valued variational inequality (62), which due to Step 2 above satisfies $u_0 \le u \le \overline{u}$, belongs to $\mathcal{S}$. As $u \in [u_0, \overline{u}]$, it follows that $\hat{b}(x,u) = 0$, and $\gamma u \in [\gamma u_0, \gamma\overline{u}]$. Moreover, (52)–(55) imply $g_1(x,u) = 0$ and $g_2(x,\gamma u) = 0$, and hence the solution u of (62) satisfies: $u \in K$ and

$$\langle Au - h, v - u\rangle + \int_\Omega \tilde{\eta}\,(v - u)\, dx + \int_{\partial\Omega} \tilde{\xi}\,(\gamma v - \gamma u)\, d\sigma \ge 0, \quad \forall\, v \in K. \tag{74}$$

where $\tilde{\eta}(x) \in \partial\tilde{j}_1(x, u(x))$, and $\tilde{\xi} \in \partial\tilde{j}_2(x, \gamma u(x))$. Because $u \in [u_0, \overline{u}]$, from (50) and (51) it follows that $\partial\tilde{j}_1(x,u(x)) \subseteq \partial j_1(x,u(x))$ and $\partial\tilde{j}_2(x,\gamma u(x)) \subseteq \partial j_2(x,\gamma u(x))$ which proves that u is in fact a solution of our original multi-valued variational inequality (1). This completes the proof. □

REMARK 27. If $K = V$, or $K = V_0$, or K is the closed subspace of V given in Example 7, then the multi-valued variational inequality (1) reduces to a multi-valued variational equation. In this case Theorem 26 on the directedness of $\mathcal{S}$

remains true also if the elliptic operator A is replaced by the more general Leray–Lions operator as in Remark 24, i.e.,

$$Au(x) = -\sum_{i=1}^{N} \frac{\partial}{\partial x_i} a_i(x, u, \nabla u(x)) + a_0(x, u, \nabla u(x)),$$

where the coefficients $a_i : \Omega \times \mathbb{R} \times \mathbb{R}^N \to \mathbb{R}$, $i = 0, 1, \ldots, N$ satisfy assumptions (A1')–(A3') as in Remark 24, and the coefficients a_i with $i = 1, \ldots, N$ satisfy, in addition, hypothesis (A4'):

(A4') There is a function $k_2 \in L^q_+(\Omega)$ and a continuous function $\omega : \mathbb{R}_+ \to \mathbb{R}_+$ such that

$$|a_i(x, s, \zeta) - a_i(x, s', \zeta)| \leq [k_2(x) + |s|^{p-1} + |s'|^{p-1} + |\zeta|^{p-1}]\omega(|s - s'|),$$

holds for a.a. $x \in \Omega$, for all s, $s' \in \mathbb{R}$, and for all $\zeta \in \mathbb{R}^N$, where ω satisfies, for each $\varepsilon > 0$,

$$\int_0^{\varepsilon} \frac{dr}{\omega(r)} = +\infty. \tag{75}$$

Relation (75) means that the integral diverges near zero. Hypothesis (A4') includes, for example, $\omega(r) = cr$, for $c > 0$, and for all $r \geq 0$, which means that the coefficients $a_i(x, s, \zeta)$ are Lipschitz continuous with respect to s. The proof of the directedness makes use of a special test function technique analogous to that for multi-valued evolution problems as will be demonstrated in Sect. 4.4.

2.2. Extremal solutions. In this subsection, the solution set $\mathcal{S}$ of all solutions of the multi-valued variational inequality (1) within the ordered interval $[\underline{u}, \overline{u}]$ is shown to possess *extremal solutions*, i.e., $\mathcal{S}$ has a *smallest solution* u_* and a *greatest solution* u^* with respect to the natural underlying partial ordering induced by the order cone $L^p_+(\Omega)$. A first step toward this goal is the following topological characterization of $\mathcal{S}$.

LEMMA 28. *The solution set $\mathcal{S}$ is a compact set in $K \subseteq V$.*

PROOF. We first prove that $\mathcal{S}$ is bounded in V. Let $c > 0$ be a generic constant that may change size, but which is independent of u. As $\mathcal{S} \subseteq [\underline{u}, \overline{u}]$, for any $u \in \mathcal{S}$ one has $\gamma u \in [\gamma\underline{u}, \gamma\overline{u}]$, and thus

$$\|u\|_{L^p(\Omega)} \leq c, \quad \|\gamma u\|_{L^p(\partial\Omega)} \leq c, \quad \|u\|_V \leq c(1 + \|\nabla u\|_{L^p(\Omega)}), \quad \forall\, u \in \mathcal{S}. \tag{76}$$

Therefore, the boundedness of $\mathcal{S}$ in V is proved provided that $\|\nabla u\|_{L^p(\Omega)}$ is bounded for all $u \in \mathcal{S}$. Let $v_0 \in K$ be fixed, and $u \in \mathcal{S}$. Then u, in particular, satisfies: $u \in K$ and

$$\langle Au - h, v_0 - u\rangle + \int_\Omega \eta\,(v_0 - u)\, dx + \int_{\partial\Omega} \xi\,(\gamma v_0 - \gamma u)\, d\sigma \geq 0, \tag{77}$$

where $\eta(x) \in \partial j_1(x, u(x))$, for a.e. $x \in \Omega$, and $\xi(x) \in \partial j_2(x, \gamma u(x))$, for a.e. $x \in \partial\Omega$. By using (j1) and (j2) we immediately get

$$\left| \int_\Omega \eta\,(v_0 - u)\,dx \right| \le c, \quad \left| \int_{\partial\Omega} \xi\,(\gamma v_0 - \gamma u)\,d\sigma \right| \le c, \quad \forall\, u \in \mathcal{S}. \tag{78}$$

Assumption (A1) implies

$$|\langle Au - h, v_0\rangle| \le c(1 + \|\nabla u\|_{L^p(\Omega)}^{p-1}), \quad \forall\, u \in \mathcal{S}. \tag{79}$$

Further we have

$$|\langle h, u\rangle| \le \|h\|_{V^*}\|u\|_V \le c(1 + \|\nabla u\|_{L^p(\Omega)}) \quad \forall\, u \in \mathcal{S}. \tag{80}$$

From (77), we obtain by using (78)–(80) as well as (A3), the following estimate

$$\nu\,\|\nabla u\|_{L^p(\Omega)}^p \le c\,(1 + \|\nabla u\|_{L^p(\Omega)} + \|\nabla u\|_{L^p(\Omega)}^{p-1}), \quad \forall\, u \in \mathcal{S}, \tag{81}$$

which proves $\|\nabla u\|_{L^p(\Omega)} \le c$, and thus

$$\|u\|_V \le c, \quad \forall\, u \in \mathcal{S}. \tag{82}$$

Let $(u_n) \subseteq \mathcal{S}$. Since V is reflexive, there exists a weakly convergent subsequence (again denoted by (u_n)), which due to the compact embedding $V \hookrightarrow L^p(\Omega)$ and the compactness of the trace operator $\gamma : V \to L^p(\partial\Omega)$, has the following convergence properties

$$\begin{aligned} u_n &\rightharpoonup u \quad \text{in } V,\\ u_n &\to u \quad \text{in } L^p(\Omega) \text{ and a.e. in } \Omega,\\ \gamma u_n &\to \gamma u \quad \text{in } L^p(\partial\Omega) \text{ and a.e. in } \partial\Omega. \end{aligned} \tag{83}$$

Since $(u_n) \subset K$, and K is closed in V and convex, so K is weakly closed, and thus $u \in K$. We are going to show that $u_n \to u$ (strongly) in V and that $u \in \mathcal{S}$, which completes the proof of the lemma. Testing the variational inequality (1) with $v = u$, then each u_n satisfies

$$\langle Au_n - h, u - u_n\rangle + \int_\Omega \eta_n\,(u - u_n)\,dx + \int_{\partial\Omega} \xi_n\,(\gamma u - \gamma u_n)\,d\sigma \ge 0, \tag{84}$$

where $\eta_n(x) \in \partial j_1(x, u_n(x))$, for a.e. $x \in \Omega$, and $\xi_n(x) \in \partial j_2(x, \gamma u_n(x))$, for a.e. $x \in \partial\Omega$. Note $u_n \in \mathcal{S}$, which, in particular, implies $u_n \in [\underline{u}, \overline{u}]$, and therefore, by (j1) and (j2) the sequences (η_n) and (ξ_n) are uniformly bounded in $L^q(\Omega)$ and $L^q(\partial\Omega)$, respectively. The latter in conjunction with (84) and the convergence properties (83) result in

$$\limsup_{n\to\infty} \langle Au_n, u_n - u\rangle \le 0. \tag{85}$$

Taking into account that the operator A enjoys the (S_+)-property (see, e.g., [**7**, Chap. 2]), $u_n \rightharpoonup u$ and (85) imply the strong convergence $u_n \to u$ in V. Due to

$\|\eta_n\|_{L^q(\Omega)} \le c$ and $\|\xi_n\|_{L^q(\partial\Omega)} \le c$, we get by passing to a further subsequence if necessary (again denoted by (η_n) and (ξ_n))

$$\eta_n \rightharpoonup \eta \text{ in } L^q(\Omega), \quad \xi_n \rightharpoonup \xi \text{ in } L^q(\partial\Omega). \tag{86}$$

The strong convergence $u_n \to u$ in V and (86) allow to pass to the limit in

$$\langle Au_n - h, v - u_n\rangle + \int_\Omega \eta_n\,(v - u_n)\,dx + \int_{\partial\Omega} \xi_n\,(\gamma v - \gamma u_n)\,d\sigma \ge 0, \quad v \in K$$

which yields for the strong limit $u \in K$

$$\langle Au - h, v - u\rangle + \int_\Omega \eta\,(v - u)\,dx + \int_{\partial\Omega} \xi\,(\gamma v - \gamma u)\,d\sigma \ge 0, \quad v \in K. \tag{87}$$

Apparently $u \in [\underline{u}, \overline{u}]$. To complete the proof we need to verify that $\eta(x) \in \partial j_1(x, u(x))$, for a.e. $x \in \Omega$, and $\xi(x) \in \partial j_2(x, \gamma u(x))$, for a.e. $x \in \partial\Omega$, which together with (87) proves that $u \in \mathcal{S}$. Let us check $\eta(x) \in \partial j_1(x, u(x))$. For this purpose, one only needs to prove that

$$\eta(x) r \le j_1^o(x, u(x); r) \tag{88}$$

for all $r \in \mathbb{R}$ and a.e. $x \in \Omega$. Let r be a fixed real number and E be any measurable subset of Ω. From the upper semicontinuity of $(s, r) \mapsto j_1^o(x, s; r)$ (cf. e.g. [**12**]), and $u_n(x) \to u(x)$ for a.e. $x \in \Omega$ we get

$$j_1^o(x, u(x); r) \ge \limsup j_1^o(x, u_n(x); r),$$

for a.e. $x \in \Omega$, and thus

$$\int_E j_1^o(x, u(x); r)dx \ge \int_E \limsup j_1^o(x, u_n(x); r)dx. \tag{89}$$

On the other hand, since

$$j_1^o(x, s; r) = \max\{\zeta r : \zeta \in \partial j_1(x, s)\},$$

it follows from (j1) that

$$|j_1^o(x, s; r)| \le |r|\, k_\Omega(x), \ \forall s \in [\underline{u}(x), \overline{u}(x)], \tag{90}$$

Together with $u_n \in [\underline{u}, \overline{u}]$, this implies that

$$|j_1^o(x, u_n(x); r)| \le |r|\, k_\Omega(x) \tag{91}$$

for a.e. $x \in \Omega$. Since the right-hand side is, in particular, in $L^1(\Omega)$, one can apply Fatou's lemma to get

$$\int_E \limsup j_1^o(x, u_n(x); r)dx \ge \limsup \int_E j_1^o(x, u_n(x); r)dx.$$

This estimate, together with (89), yields

$$\int_E j_1^o(x, u(x); r)dx \ge \limsup \int_E j_1^o(x, u_n(x); r)dx. \tag{92}$$

From the weak convergence (86) of η_n to η it follows

$$\int_E \eta_n(x) r dx \to \int_E \eta(x) r dx \quad \text{as} \quad n \to \infty. \tag{93}$$

For each n, we have $\eta_n(x) \in \partial j_1(x, u_n(x))$ and consequently $j_1^o(x, u_n(x); r) \geq \eta_n(x) r$ for a.e. $x \in \Omega$. Hence,

$$\int_E j_1^o(x, u_n(x); r) dx \geq \int_E \eta_n(x) r dx, \ \forall \ n. \tag{94}$$

Letting $n \to \infty$ in (94) and making use of (93) and (92), we obtain

$$\int_E j_1^o(x, u(x); r) dx \geq \int_E \eta(x) r dx.$$

Since this inequality holds for all measurable subsets E of Ω, we must have (88) for a.e. $x \in \Omega$. The proof of $\xi(x) \in \partial j_2(x, \gamma u(x))$ can be done in an obvious similar way, which completes our compactness proof. □

Theorem 26 on the directedness of $\mathcal{S}$ and Lemma 28 on the compactness of $\mathcal{S}$ allow us to prove the following extremality property of $\mathcal{S}$.

THEOREM 29. *Let the hypotheses of Theorem 26 be satisfied. Then the solution set $\mathcal{S}$ has the greatest element u^* and the smallest element u_*, i.e., there exist greatest and smallest solutions u^* and u_*, respectively, of the multi-valued variational inequality (1) within the ordered interval $[\underline{u}, \overline{u}]$ in the sense that if u is any solution of (1) in $[\underline{u}, \overline{u}]$ then it satisfies $u_* \leq u \leq u^*$.*

PROOF. Let us focus on the existence of the greatest element of $\mathcal{S}$. Since V is separable, it follows that $\mathcal{S} \subset V$ is separable too, so there exists a countable, dense subset $Z = \{z_n : n \in \mathbb{N}\}$ of $\mathcal{S}$. Since $\mathcal{S}$ is, in particular, upward directed, so we can construct an increasing sequence $(u_n) \subseteq \mathcal{S}$ as follows. Let $u_1 = z_1$. Select $u_{n+1} \in \mathcal{S}$ such that

$$\max\{z_n, u_n\} \leq u_{n+1} \leq \overline{u}.$$

The existence of u_{n+1} is due to the directedness of $\mathcal{S}$ by Theorem 26. From the compactness of $\mathcal{S}$ according to Lemma 28, there exists a subsequence of (u_n), denoted again (u_n), and an element $u \in \mathcal{S}$ such that $u_n \to u$ in V, and $u_n(x) \to u(x)$ a.e. in Ω. This last property of (u_n) combined with its increasing monotonicity implies that the entire sequence is convergent in V and, moreover, $u = \sup_n u_n$. By construction, we see that

$$\max\{z_1, z_2, \dots, z_n\} \leq u_{n+1} \leq u, \quad \text{for all } n \in \mathbb{N},$$

thus $Z \subseteq [\underline{u}, u]$. Since the interval $[\underline{u}, u]$ is closed in V, we infer

$$\mathcal{S} \subseteq \overline{Z} \subseteq \overline{[\underline{u}, u]} = [\underline{u}, u],$$

which in conjunction with $u \in \mathcal{S}$ ensures that $u = u^*$ is the greatest solution. The existence of the smallest solution u_* can be proved in a similar way using the fact that $\mathcal{S}$ is downward directed. □

REMARK 30. If $K = V$, or $K = V_0$, or K is the the closed subspace of V given in Example 7, then the multi-valued variational inequality (1) reduces to a multi-valued variational equation. In this case Theorem 29 on the existence of extremal solutions remains true also if the elliptic operator A is replaced by the more general Leray–Lions operator as in Remark 24, i.e.,

$$Au(x) = -\sum_{i=1}^{N} \frac{\partial}{\partial x_i} a_i(x, u, \nabla u(x)) + a_0(x, u, \nabla u(x)),$$

where the coefficients $a_i : \Omega \times \mathbb{R} \times \mathbb{R}^N \to \mathbb{R}$, $i = 0, 1, \ldots, N$ satisfy assumptions (A1')–(A3') as in Remark 24, and the coefficients a_i with $i = 1, \ldots, N$ satisfy, in addition, hypothesis (A4') of Remark 27.

3. Variational-hemivariational inequalities

Keeping the notations and hypotheses of Sect. 2, in this section we develop the sub-supersolution method for the variational-hemivariational inequality (2), i.e., we consider the problem: Find $u \in K$ such that

$$(95)\quad \langle Au-h, v-u\rangle + \int_{\Omega} j_1^o(x, u; v-u)\, dx + \int_{\partial\Omega} j_2^o(x, \gamma u; \gamma v-\gamma u)\, d\sigma \geq 0, \ \forall\, v \in K.$$

Our presentation is based on results obtained recently in [**3**] in conjunction with the construction of appropriate multi-valued operators as in Sect. 2. By specifying the closed convex set $K \subseteq V$ one can see in a similar way as in Sect. 2 that the variational-hemivariational includes various elliptic boundary value problems. Consider only two of them.

EXAMPLE 31. If $K = V$, $f_1 : \Omega\times\mathbb{R} \to \mathbb{R}$ and $f_2 : \partial\Omega\times\mathbb{R} \to \mathbb{R}$ are Carathéodory functions, and if j_k, $k = 1, 2$, are the primitives of f_k given by (4), then $j_k^o(x, s; r) = \frac{\partial j_k}{\partial s}(x, s)\, r = f_k(x, s) r$, and thus (95) reduces to the following quasilinear elliptic BVP

$$(96)\qquad \langle Au - h, v\rangle + \int_{\Omega} f_1(x, u)\, v\, dx + \int_{\partial\Omega} f_2(x, \gamma u)\, \gamma v\, d\sigma = 0, \quad \forall\, v \in V,$$

which is the formulation of the weak solution of the BVP

$$(97)\qquad Au + f_1(x, u) = h \ \text{ in } \ \Omega, \quad \frac{\partial u}{\partial \nu} + f_2(x, u) = 0 \quad \text{on } \ \partial\Omega,$$

where $\partial/\partial\nu$ denotes the outward pointing conormal derivative associated with A.

EXAMPLE 32. If $K = V_0$, and j_k as in Example 31, then (95) is equivalent to

$$(98)\qquad u \in V_0 : \quad \langle Au - h, v\rangle + \int_{\Omega} f_1(x, u)\, v\, dx = 0, \quad \forall\, v \in V_0,$$

which is nothing but the weak formulation of the homogeneous Dirichlet problem

$$(99)\qquad Au + f_1(x, u) = h \ \text{ in } \ \Omega, \quad u = 0 \quad \text{on } \ \partial\Omega.$$

Next let us introduce the notion of sub- and supersolution.

DEFINITION 33. A function $\underline{u} \in V$ is called a **subsolution** of (95) if the following holds:

(i) $\underline{u} \vee K \subseteq K$,

(ii) $\langle A\underline{u} - h, v - \underline{u}\rangle + \int_\Omega j_1^o(x, \underline{u}; v - \underline{u})\, dx + \int_{\partial\Omega} j_2^o(x, \gamma\underline{u}; \gamma v - \gamma\underline{u})\, d\sigma \geq 0$

for all $v \in \underline{u} \wedge K$.

DEFINITION 34. A function $\overline{u} \in V$ is called a **supersolution** of (95) if the following holds:

(i) $\overline{u} \wedge K \subseteq K$,

(ii) $\langle A\overline{u} - h, v - \overline{u}\rangle + \int_\Omega j_1^o(x, \overline{u}; v - \overline{u})\, dx + \int_{\partial\Omega} j_2^o(x, \gamma\overline{u}; \gamma v - \gamma\overline{u})\, d\sigma \geq 0$

for all $v \in \overline{u} \vee K$.

REMARK 35. Note that the notion for sub- and supersolution defined in Definition 33 and Definition 34 have a symmetric structure, i.e., one obtains the definition for the supersolution $\overline{u}$ from the definition of the subsolution by replacing $\underline{u}$ in Definition 33 by $\overline{u}$, and interchanging $\vee$ by $\wedge$. If we apply Definitions 33 and 34 to the Examples 31 and 32 we see in just the same way as in Examples 12 and 13, that the above definitions are in fact natural extensions of the usual notions of sub-supersolutions for elliptic boundary value problems.

3.1. Existence and comparison principle. We are now establishing the method of sub-supersolution for the variational-hemivariational inequality (95).

THEOREM 36 (Sub-supersolution method). *Let (A1)–(A3) be satisfied. Assume the existence of sub- and supersolutions $\underline{u}$ and $\overline{u}$, respectively, of the variational-hemivariational inequality (95) with $\underline{u} \leq \overline{u}$ such that (j1)–(j2) is fulfilled. Then there exist solutions of (95)) within the ordered interval $[\underline{u}, \overline{u}]$.*

PROOF. The proof is just a modification of the proof of Theorem 23 in Sect. 2. This modification is related to the definition of the functions $\tilde{j}_k$ given in (20)–(21), i.e., of

$$\tilde{j}_1(x,s) = \begin{cases} j_1(x, \underline{u}(x)) + \underline{\eta}(x)(s - \underline{u}(x)) & \text{if } s < \underline{u}(x), \\ j_1(x,s) & \text{if } \underline{u}(x) \leq s \leq \overline{u}(x), \\ j_1(x, \overline{u}(x)) + \overline{\eta}(x)(s - \overline{u}(x)) & \text{if } s > \overline{u}(x), \end{cases} \tag{100}$$

and similarly

$$\tilde{j}_2(x,s) = \begin{cases} j_2(x, \gamma\underline{u}(x)) + \underline{\xi}(x)(s - \gamma\underline{u}(x)) & \text{if } s < \gamma\underline{u}(x), \\ j_2(x,s) & \text{if } \gamma\underline{u}(x) \leq s \leq \gamma\overline{u}(x), \\ j_2(x, \gamma\overline{u}(x)) + \overline{\xi}(x)(s - \gamma\overline{u}(x)) & \text{if } s > \gamma\overline{u}(x). \end{cases} \tag{101}$$

While the functions $\underline{\eta}, \overline{\eta} \in L^q(\Omega)$, and $\underline{\xi}, \overline{\xi} \in L^q(\partial\Omega)$ which appear in (20)–(21) are given via the definition of sub- and supersolution of the multi-valued variational inequality (1), no such functions appear in the Definitions 33 and 34 of the variational-hemivariational inequality (95), and therefore we need to specify these functions in (100)–(101). Here we let $x \mapsto \underline{\eta}(x)$ and $x \mapsto \overline{\eta}(x)$, and $x \mapsto \underline{\xi}(x)$ and $x \mapsto \overline{\xi}(x)$ be the following measurable selection of $x \mapsto \partial j_1(x, \underline{u}(x))$ and $x \mapsto \partial j_1(x, \overline{u}(x))$, and of $x \mapsto \partial j_2(x, \gamma \underline{u}(x))$ and $x \mapsto \partial j_2(x, \gamma \overline{u}(x))$, respectively:

$$
\begin{aligned}
\underline{\eta}(x) &= \min\{\zeta : \zeta \in \partial j_1(x, \underline{u}(x))\} = -j_1^o(x, \underline{u}(x); -1), \\
\overline{\eta}(x) &= \max\{\zeta : \zeta \in \partial j_1(x, \overline{u}(x))\} = j_1^o(x, \overline{u}(x); 1), \\
\underline{\xi}(x) &= \min\{\zeta : \zeta \in \partial j_2(x, \gamma \underline{u}(x))\} = -j_2^o(x, \gamma \underline{u}(x); -1), \\
\overline{\xi}(x) &= \max\{\zeta : \zeta \in \partial j_2(x, \gamma \overline{u}(x))\} = j_2^o(x, \gamma \overline{u}(x); 1).
\end{aligned} \tag{102}
$$

By (j1)–(j2) we infer that $\underline{\eta}$, $\overline{\eta} \in L^q(\Omega)$, and $\underline{\xi}$, $\overline{\xi} \in L^q(\partial\Omega)$, and with these functions we define $\tilde{j}_k$ in just the same way as in (20)–(21). With exactly the same notation and arguments as in the proof of Theorem 23 we obtain the existence of a solution of the following variational inequality (see (37)):

$$
\begin{aligned}
u \in K: \quad \langle Au - h, v - u \rangle &+ \int_\Omega b(x, u)\, (v - u)\, dx + \int_\Omega \tilde{\eta}\, (v - u)\, dx \\
&+ \int_{\partial\Omega} \tilde{\xi}\, (\gamma v - \gamma u)\, d\sigma \geq 0, \quad \forall\, v \in K,
\end{aligned} \tag{103}
$$

where $\tilde{\eta} \in L^q(\Omega)$ with $\tilde{\eta}(x) \in \partial \tilde{j}_1(x, u(x))$, for a.e. $x \in \Omega$, and $\tilde{\xi} \in L^q(\partial\Omega)$ with $\tilde{\xi}(x) \in \partial \tilde{j}_2(x, \gamma u(x))$, for a.e. $x \in \partial\Omega$. Taking into account the definition of Clarke's generalized gradient $\partial \tilde{j}_k$, from (103) we readily obtain that a solution u of (103) satisfied the variational-hemivariational inequality

$$
\begin{aligned}
u \in K: \quad \langle Au - h, v - u \rangle &+ \int_\Omega b(x, u)\, (v - u)\, dx + \int_\Omega \tilde{j}_1^o(x, u; (v - u))\, dx \\
&+ \int_{\partial\Omega} \tilde{j}_2^o(x, \gamma u; (\gamma v - \gamma u)\, d\sigma \geq 0, \quad \forall\, v \in K.
\end{aligned} \tag{104}
$$

By means of the definitions of the sub-supersolutions we next show that any u solution of (104) satisfies $\underline{u} \leq u \leq \overline{u}$, i.e., u belongs to the interval $[\underline{u}, \overline{u}]$. Let us verify that $u \leq \overline{u}$. Taking in Definition 34 $v = \overline{u} \vee u = \overline{u} + (u - \overline{u})^+$ we get

$$
\langle A\overline{u} - h, (u - \overline{u})^+ \rangle + \int_\Omega j_1^o(x, \overline{u}; (u - \overline{u})^+)\, dx + \int_{\partial\Omega} j_2^o(x, \gamma \overline{u}; \gamma (u - \overline{u})^+)\, d\sigma \geq 0, \tag{105}
$$

and taking in (104) for v the special element $v = \overline{u} \wedge u = u - (u - \overline{u})^+ \in K$ we obtain

$$
\begin{aligned}
\langle Au - h, -(u - \overline{u})^+ \rangle &+ \int_\Omega b(x, u)\, (-(u - \overline{u})^+)\, dx + \int_\Omega \tilde{j}_1^o(x, u; -(u - \overline{u})^+)\, dx \\
&+ \int_{\partial\Omega} \tilde{j}_2^o(x, \gamma u; -\gamma (u - \overline{u})^+)\, d\sigma \geq 0.
\end{aligned} \tag{106}
$$

Adding (105) and (106) we arrive at

$$\begin{aligned} &\langle A\overline{u} - Au, (u - \overline{u})^+\rangle \\ &\quad + \int_\Omega \Big(j_1^o(x, \overline{u}; 1) + \tilde{j}_1^o(x, u; -1)\Big)(u - \overline{u})^+ \, dx \\ &\quad + \int_{\partial\Omega} \Big(j_2^o(x, \gamma\overline{u}; 1) + \tilde{j}_2^o(x, \gamma u; -1)\Big)\gamma(u - \overline{u})^+ \, d\sigma \\ &\geq \int_\Omega b(x, u)\,(u - \overline{u})^+ \, dx. \end{aligned} \tag{107}$$

Let us estimate the terms on the left-hand side of inequality (107) separately. From (A2) we readily obtain

$$\langle A\overline{u} - Au, (u - \overline{u})^+\rangle \leq 0.$$

As for the second term on the left-hand side of inequality (107) we need to calculate the values $\tilde{j}_1^o(x, u(x); -1)$ for those $x \in \Omega$ belonging to the set $\{u > \overline{u}\} = \{x \in \Omega : u(x) > \overline{u}(x)\}$ because otherwise the integral is zero. To this end we make use of the definitions (100) and (102) and get for $x \in \{u > \overline{u}\}$ the following:

$$\begin{aligned} &\tilde{j}_1^o(x, u(x); -1) \\ &= \limsup_{s \to u(x),\, t \downarrow 0} \frac{\tilde{j}_1(x, s - t) - \tilde{j}_1(x, s)}{t} \\ &= \limsup_{s \to u(x),\, t \downarrow 0} \frac{\tilde{j}_1(x, \overline{u}(x)) + \overline{\eta}(x)(s - t - \overline{u}(x)) - \tilde{j}_1(x, \overline{u}(x)) - \overline{\eta}(x)(s - \overline{u}(x))}{t} \\ &= -\overline{\eta}(x). \end{aligned} \tag{108}$$

Taking into account (102) and (108) we readily see that the second term on the left-hand side of inequality (107) becomes zero. In a similar way one shows that also the third term on the left-hand side of inequality (107) becomes zero. Thus, from (107) we obtain the following inequality:

$$0 \geq \int_\Omega b(x, u)\,(u - \overline{u})^+ \, dx,$$

which by applying the definition of of the cut-off function b results in

$$0 = \int_\Omega [(u - \overline{u})^+]^p \, dx,$$

and thus we conclude $(u - \overline{u})^+(x) = 0$ a.e. in Ω, i.e., $u \leq \overline{u}$. In a similar way one can show that $\underline{u} \leq u$ holds true. So far we have proved that any solution of (103) (which is also a solution of (104)) belongs to the interval $[\underline{u}, \overline{u}]$, which implies that $b(\cdot, u) = 0$, and therefore any solution of (103) satisfies: $u \in K$ and

$$u \in K: \ \langle Au - h, v - u\rangle + \int_\Omega \tilde{\eta}\,(v - u)\, dx + \int_{\partial\Omega} \tilde{\xi}\,(\gamma v - \gamma u)\, d\sigma \geq 0, \ \forall\, v \in K, \tag{109}$$

where $\tilde{\eta} \in L^q(\Omega)$ with $\tilde{\eta}(x) \in \partial\tilde{j}_1(x, u(x))$, for a.e. $x \in \Omega$, and $\tilde{\xi} \in L^q(\partial\Omega)$ with $\tilde{\xi}(x) \in \partial\tilde{j}_2(x, \gamma u(x))$, for a.e. $x \in \partial\Omega$. As $\partial\tilde{j}_1(x, u(x)) \subseteq \partial j_1(x, u(x))$ and

$\partial\tilde{j_2}(x, \gamma u(x)) \subseteq \partial j_2(x, \gamma u(x))$, we see by applying the definition of Clarke's generalized gradient that any solution of (103) belongs to $[\underline{u}, \overline{u}]$, and satisfies the variational-hemivariational inequality (95), which completes the proof. □

REMARK 37. The existence and comparison result of Theorem 36 can be extended to more general elliptic operators A of Leray–Lions type such as

$$Au(x) = -\sum_{i=1}^{N} \frac{\partial}{\partial x_i} a_i(x, u, \nabla u(x)) + a_0(x, u, \nabla u(x)),$$

where the coefficients $a_i : \Omega \times \mathbb{R} \times \mathbb{R}^N \to \mathbb{R}$, $i = 0, 1, \ldots, N$ satisfy hypotheses (A1')–(A3') given in Remark 24, see [**11**].

By appropriately modifying the proofs of Theorem 26 and Theorem 29 we can show the following theorem.

THEOREM 38. *Let the hypotheses of Theorem 36 and the lattice condition (14) be satisfied. Then the solution set $\mathcal{S}$ of all solutions of (95) within the interval $[\underline{u}, \overline{u}]$ is a directed set. Moreover, there exist greatest and smallest solutions (extremal solutions) of the variational-hemivariational inequality (95) within the ordered interval $[\underline{u}, \overline{u}]$*

REMARK 39. Similarly as in Remark 27 we note that if $K = V$, or $K = V_0$, then the assertion of Theorem 38 remains true also in case that the elliptic operator A is replaced by the more general Leray–Lions operator as in Remark 24, i.e.,

$$Au(x) = -\sum_{i=1}^{N} \frac{\partial}{\partial x_i} a_i(x, u, \nabla u(x)) + a_0(x, u, \nabla u(x)),$$

where the coefficients $a_i : \Omega \times \mathbb{R} \times \mathbb{R}^N \to \mathbb{R}$, $i = 0, 1, \ldots, N$ satisfy assumptions (A1')–(A3') as in Remark 24, and the coefficients a_i with $i = 1, \ldots, N$ satisfy, in addition, the hypothesis (A4') given in Remark 27.

4. Multi-valued periodic evolution problems

Let $\Omega \subset \mathbb{R}^N$ be a bounded domain with Lipschitz boundary $\partial\Omega$, $Q = \Omega \times (0, \tau)$, and $\Gamma = \partial\Omega \times (0, \tau)$, with $\tau > 0$. In this section we consider the periodic–Dirichlet problem of the following quasilinear parabolic inclusion:

$$u_t + Au + \partial j(\cdot, \cdot, u) \ni f \quad \text{in } Q, \tag{110}$$

$$u(\cdot, 0) = u(\cdot, \tau) \quad \text{in } \Omega, \qquad u = 0 \quad \text{on } \Gamma, \tag{111}$$

where A is a second order quasilinear differential operator of Leray-Lions type of the form

$$Au(x,t) = -\sum_{i=1}^{N} \frac{\partial}{\partial x_i} a_i(x, t, u(x,t), \nabla u(x,t)), \quad \text{with } \nabla u = \left(\frac{\partial u}{\partial x_1}, \ldots, \frac{\partial u}{\partial x_N}\right).$$

The function $j : Q \times \mathbb{R} \to \mathbb{R}$ is assumed to be measurable with respect to $(x, t) \in Q$ for all $s \in \mathbb{R}$, and locally Lipschitz in $s \in \mathbb{R}$ for almost all (a.a.) $(x, t) \in Q$.

The multi-valued function $s \mapsto \partial j(x,t,s)$ in (110) is given by Clarke's generalized gradient of the locally Lipschitz function $s \mapsto j(x,t,s)$, and is defined by

$$\partial j(x,t,s) := \{\zeta \in \mathbb{R} : j^o(x,t,s;r) \geq \zeta r, \ \forall\, r \in \mathbb{R}\}$$

for a.a. $(x,t) \in Q$, with $(s,r) \mapsto j^o(x,t,s;r)$ denoting the generalized directional derivative of $s \mapsto j(x,t,s)$ at s in the direction r given by

$$j^o(x,t,s;r) = \limsup_{y \to s,\ \varepsilon \downarrow 0} \frac{j(x,t,y+\varepsilon r) - j(x,t,y)}{\varepsilon}.$$

If j has the form of a primitive of some Carathéodory function $g : Q \times \mathbb{R} \to \mathbb{R}$, i.e.,

$$j(x,t,s) := \int_0^s g(x,t,r)\, dr$$

then the function $s \mapsto j(x,t,s)$ is continuously differentiable, and thus $\partial j(x,t,s) = \{\partial j(x,t,s)/\partial s\} = \{g(x,t,s)\}$, and (110) reduces to the (single-valued) parabolic periodic–Dirichlet problem

$$u_t + Au + g(\cdot,\cdot,u) = f \quad \text{in } Q, \tag{112}$$

$$u(\cdot,0) = u(\cdot,\tau) \ \text{ in } \Omega, \quad u = 0 \quad \text{on } \Gamma. \tag{113}$$

Existence and comparison results via sub-supersolutions for (112), (113) have been obtained in [**1**].

Our aim is to extend these existence and comparison results to weak solutions of the inclusion (110), (111) in terms of appropriately defined sub- and supersolutions in such a way that the single-valued case is included as a special case. Our approach in dealing with (110)–(111) is based on recent papers by the authors (see [**5, 8, 9**]) where initial-boundary value problems for parabolic inclusions in the form (110) have been treated.

As for the definition of a weak solution of (110), (111), let $W^{1,p}(\Omega)$ and $W_0^{1,p}(\Omega)$ denote the usual Sobolev spaces with dual spaces $(W^{1,p}(\Omega))^*$ and $W^{-1,q}(\Omega)$, respectively, where q is the conjugate real satisfying $1/p + 1/q = 1$. Only for the sake of simplicity we assume throughout this section $2 \leq p < \infty$. Then $W^{1,p}(\Omega) \subset L^2(\Omega) \subset (W^{1,p}(\Omega))^*$ forms an evolution triple with all the embeddings being continuous, dense and compact.

We set $X = L^p(0,\tau; W^{1,p}(\Omega))$, and introduce the space W defined by

$$W = \{u \in X : u' \in X^*\},$$

where the derivative $u' := \partial u/\partial t$ is understood in the sense of vector-valued distributions, and $X^* = L^q(0,\tau;(W^{1,p}(\Omega))^*)$ is the dual space to X. The space W endowed with the graph norm of the operator $\partial/\partial t$

$$\|u\|_W = \|u\|_X + \|u'\|_{X^*}$$

is a Banach space which is separable and reflexive due to the separability and reflexivity of X and X^*. Furthermore, it is well known that the embedding $W \hookrightarrow C([0,\tau], L^2(\Omega))$ is continuous. Finally, because $W^{1,p}(\Omega)$ is compactly embedded in $L^p(\Omega)$, we have by Aubin's lemma a compact embedding of $W \hookrightarrow L^p(Q)$.

Similarly, we set $X_0 = L^p(0,\tau;W_0^{1,p}(\Omega))$, whose dual space is given by $X_0^* = L^q(0,\tau;W^{-1,q}(\Omega))$, and introduce the space W_0 defined by

$$W_0 = \{u \in X_0 : u' \in X_0^*\}.$$

We use the notation $\langle\cdot,\cdot\rangle$ for any of the dual pairings between X and X^*, X_0 and X_0^*, $W^{1,p}(\Omega)$ and $(W^{1,p}(\Omega))^*$, and $W_0^{1,p}(\Omega)$ and $W^{-1,q}(\Omega)$. We assume for f in (110) that $f \in X_0^*$. Then the dual pairing, e.g., of f and $u \in X_0$ is given by

$$\langle f,u\rangle = \int_0^\tau \langle f(t),u(t)\rangle\,dt.$$

Let $L := \partial/\partial t$ and its domain of definition $D(L)$ be given by

$$D(L) = \{u \in X_0 : u' \in X_0^* \text{ and } u(\cdot,0) = u(\cdot,\tau)\}.$$

The linear operator $L : D(L) \subset X_0 \to X_0^*$ is closed, densely defined and maximal monotone, e.g., cf. [**20**, Chap. 32].

DEFINITION 40. A function $u \in D(L) \subset W_0$ is called a **solution** of (110), (111) if there is an $\eta \in L^q(Q)$ with $\eta(x,t) \in \partial j(x,t,u(x,t))$ for a.e. $(x,t) \in Q$ such that

$$\langle Lu + Au,\varphi\rangle + \int_Q \eta\,\varphi\,dx\,dt = \langle f,\varphi\rangle, \quad \forall\,\varphi \in X_0. \tag{114}$$

By the definition of Clarke's generalized gradient ∂j one immediately sees that any solution of (110), (111) is a solution of the following *periodic-evolution hemivariational inequality*:

$$u \in D(L): \quad \langle Lu + Au - f,\varphi\rangle + \int_Q j^o(\cdot,\cdot,u;\varphi)\,dx\,dt \ge 0, \quad \forall\,\varphi \in X_0, \tag{115}$$

provided the integral exists, while the reverse is not clear at all. However, by means of the method of sub-supersolution to be developed here we will be able to show that (110), (111) and (115) are in fact equivalent.

Differential inclusions (110) and their related hemivariational inequalities (115) have attracted increasing interest over the last decades, because, among others, they arise, e.g., in mechanical problems governed by nonconvex, possibly nonsmooth energy functionals which appear if nonmonotone, multi-valued constitutive laws are taken into account, see [**18, 19**]. Here we focus on comparison principles for solutions of (110), (111). Results in this direction for initial-boundary value problems have been obtained by the authors in [**6**] and in a recent monograph, see [**7**] and references therein. However, besides a natural growth condition the following additional structure restriction to Clarke's generalized gradient ∂j was required in previous works to get comparison results, such as in [**7**]:

$$\eta_i \in \partial j(x,t,s_i): \quad \eta_1 \le \eta_2 + c\,(s_2 - s_1)^{p-1}, \tag{116}$$

for a.a. $(x,t) \in Q$, for all $\eta_i \in \partial j(x,t,s_i)$, $i = 1,2$, and for all s_1, s_2 with $s_1 < s_2$.

As in the elliptic case, the main goal in this section is to establish a comparison principle for (110), (111) in terms of sub-supersolution without assuming the one-sided condition (116). It should be also mentioned that the method of proof in this work can be extended to deal with comparison results for more general periodic parabolic boundary value problems in the form

$$u_t + Au + \partial j_1(x,t,u) \ni f \quad \text{in } Q, \quad u(\cdot,0) = u(\cdot,\tau) \text{ in } \Omega,$$
$$u = 0 \quad \text{on } \Gamma_2, \quad \frac{\partial u}{\partial \nu} + \partial j_2(x,t,u) \ni g \quad \text{on } \Gamma_1,$$

where A may include also lower order terms, and the functions $(x,t,s) \mapsto j_i(x,t,s)$, $i = 1,2$ are measurable in $(x,t) \in Q$ and $(x,t) \in \Gamma_1$, respectively, for all $s \in \mathbb{R}$, and locally Lipschitz in $s \in \mathbb{R}$ for a.a. (x,t). Here the components Γ_1 and Γ_2 are relatively open subsets of Γ satisfying $\overline{\Gamma_1} \cup \overline{\Gamma_2} = \Gamma$ and $\Gamma_1 \cap \Gamma_2 = \emptyset$. Only for the sake of simplifying our presentation and in order to emphasize the main idea we confine our considerations on problem (110), (111) and assume without loss of generality homogeneous boundary data.

4.1. Notations, hypotheses, and preliminary results. We assume the following Leray–Lions conditions on the coefficient functions a_i, $i = 1, \dots, N$, entering the definition of the operator A.

(A1) $a_i : Q \times \mathbb{R} \times \mathbb{R}^N \to \mathbb{R}$ are Carathéodory functions, i.e. $a_i(\cdot,\cdot,s,\xi) : Q \to \mathbb{R}$ is measurable for all $(s,\xi) \in \mathbb{R} \times \mathbb{R}^N$ and $a_i(x,t,\cdot,\cdot) : \mathbb{R} \times \mathbb{R}^N \to \mathbb{R}$ is continuous for a.e. $(x,t) \in Q$. In addition, the following growth condition holds:

$$|a_i(x,t,s,\xi)| \le k_0(x,t) + c_0 \left(|s|^{p-1} + |\xi|^{p-1}\right)$$

for a.a. $(x,t) \in Q$ and for all $(s,\xi) \in \mathbb{R} \times \mathbb{R}^N$, for some constant $c_0 > 0$ and some function $k_0 \in L^q(Q)$.

(A2) For a.a. $(x,t) \in Q$, for all $s \in \mathbb{R}$ and for all $\xi, \xi' \in \mathbb{R}^N$ with $\xi \neq \xi'$ the following monotonicity holds:

$$\sum_{i=1}^{N} (a_i(x,t,s,\xi) - a_i(x,t,s,\xi'))(\xi_i - \xi'_i) > 0.$$

(A3) There is some constant $\nu > 0$ such that for a.a. $(x,t) \in Q$ and for all $(s,\xi) \in \mathbb{R} \times \mathbb{R}^N$ the inequality

$$\sum_{i=1}^{N} a_i(x,t,s,\xi)\xi_i \ge \nu|\xi|^p - k_1(x,t)$$

is satisfied for some function $k_1 \in L^1(Q)$.

By (A1) the operator A defined by

$$\langle Au, \varphi \rangle := \int_Q \sum_{i=1}^{N} a_i(x,t,u,\nabla u) \frac{\partial \varphi}{\partial x_i}\, dx\, dt, \quad \forall \varphi \in X_0,$$

is continuous and bounded from X (resp. X_0) into X_0^*.

We denote by $L^p_+(Q)$ the positive cone of nonnegative elements of $L^p(Q)$. A partial ordering in $L^p(Q)$ is defined by $u \le v$ if and only if $v - u \in L^p_+(Q)$. If

$\underline{u}, \overline{u} \in W$ with $\underline{u} \le \overline{u}$, we denote by

$$[\underline{u}, \overline{u}] = \{u \in W \,:\, \underline{u} \le u \le \overline{u}\}$$

the ordered interval formed by $\underline{u}$ and $\overline{u}$. The following definitions provide a natural extension of the notion of sub-supersolution for (single-valued) periodic parabolic problems to the multi-valued problem considered here.

DEFINITION 41. A function $\underline{u} \in W$ is called a **subsolution** of (110), (111) if there is an $\underline{\eta} \in L^q(Q)$ satisfying $\underline{\eta}(x,t) \in \partial j(x,t,\underline{u}(x,t))$ for a.a. $(x,t) \in Q$ such that $\underline{u}(x,0) \le \underline{u}(x,\tau)$ for a.a. $x \in \Omega$, $\underline{u}|_\Gamma \le 0$, and

$$\langle \underline{u}_t + A\underline{u}, \varphi\rangle + \int_Q \underline{\eta}\,\varphi\, dx\, dt \le \langle f, \varphi\rangle, \quad \forall\, \varphi \in X_0 \cap L^p_+(Q). \tag{117}$$

DEFINITION 42. A function $\overline{u} \in W$ is called a **supersolution** of (110), (111) if there is an $\overline{\eta} \in L^q(Q)$ satisfying $\overline{\eta}(x,t) \in \partial j(x,t,\overline{u}(x,t))$ for a.a. $(x,t) \in Q$ such that $\overline{u}(x,0) \ge \overline{u}(x,\tau)$ for a.e. $x \in \Omega$, $\overline{u}|_\Gamma \ge 0$, and

$$\langle \overline{u}_t + A\overline{u}, \varphi\rangle + \int_Q \overline{\eta}\,\varphi\, dx\, dt \ge \langle f, \varphi\rangle, \quad \forall\, \varphi \in X_0 \cap L^p_+(Q). \tag{118}$$

As for the function $j : Q \times \mathbb{R} \to \mathbb{R}$ we suppose the following hypothesis:

(H) The function $(x,t) \mapsto j(x,t,s)$ is measurable in Q for all $s \in \mathbb{R}$, and $s \mapsto j(x,t,s)$ is locally Lipschitz continuous in $\mathbb{R}$ for a.a. $(x,t) \in Q$. There exists a pair $(\underline{u}, \overline{u})$ of sub-and supersolution of (110), (111) with $\underline{u} \le \overline{u}$, and a $k \in L^q_+(Q)$ such that

$$|\eta| \le k(x,t)$$

for a.a. $(x,t) \in Q$, for all $\eta \in \partial j(x,t,s)$, and for all $s \in [\underline{u}(x,t), \overline{u}(x,t)]$.

REMARK 43. There are different ways to define sub- and supersolutions for the multi-valued problem considered here. However, any notion of sub-supersolution should include the well known notion of sub-supersolution for the corresponding single-valued problems. In particular, the notions given here imply that any solution of (110), (111) is a subsolution and a supersolution as well, which will allow us to obtain existence and comparison results that preserve the characteristic features of the corresponding single-valued problems. Moreover, by means of Definition 41 and 42 in conjunction with the comparison principle to be developed here, the equivalence of (110), (111) and the hemivariational inequality (115) can be proved. Note that by (H) we only require a local L^q-boundedness condition on Clarke's gradient ∂j, which is trivially satisfied, in particular, if we assume the following natural growth condition:

(G) The function $(x,t) \mapsto j(x,t,s)$ is measurable in Q for all $s \in \mathbb{R}$, and $s \mapsto j(x,t,s)$ is locally Lipschitz continuous in $\mathbb{R}$ for a.a. $(x,t) \in Q$. There exist $c > 0$ and $k \in L^q(Q)$ such that

$$|\eta| \le k(x,t) + c|s|^{p-1}$$

for a.a. $(x,t) \in Q$, for all $\eta \in \partial j(x,t,s)$, and for all $s \in \mathbb{R}$.

Next we briefly recall a general surjectivity result for multi-valued operators in a real reflexive Banach space Y, which will be used to deal with some auxiliary problems specified later. To this end we introduce the notion of multi-valued pseudomonotone operators with respect to the graph norm topology of the domain $D(L)$ (w.r.t. $D(L)$ for short) of some linear, closed, densely defined and maximal monotone operator $L : D(L) \subseteq Y \to Y^*$.

DEFINITION 44. Let $L : D(L) \subseteq Y \to Y^*$ be a linear, closed, densely defined and maximal monotone operator. The operator $\mathcal{A} : Y \to 2^{Y^*}$ is called **pseudo-monotone w.r.t. D(L)** if the following conditions are satisfied:

(i) The set $\mathcal{A}(u)$ is nonempty, bounded, closed and convex for all $u \in Y$.
(ii) $\mathcal{A}$ is upper semicontinuous from each finite dimensional subspace of Y to Y^* equipped with the weak topology.
(iii) If $(u_n) \subset D(L)$ with $u_n \rightharpoonup u$ in Y, $Lu_n \rightharpoonup Lu$ in Y^*, $u_n^* \in \mathcal{A}(u_n)$ with $u_n^* \rightharpoonup u^*$ in Y^* and $\limsup \langle u_n^*, u_n - u\rangle \le 0$, then $u^* \in \mathcal{A}(u)$ and $\langle u_n^*, u_n\rangle \to \langle u^*, u\rangle$.

DEFINITION 45. The operator $\mathcal{A} : Y \to 2^{Y^*}$ is called **coercive** iff either the domain of $\mathcal{A}$ denoted by $D(\mathcal{A})$ is bounded or $D(\mathcal{A})$ is unbounded and

$$\frac{\inf\{\langle v^*, v\rangle : v^* \in \mathcal{A}(v)\}}{\|v\|_Y} \to +\infty \quad \text{as} \quad \|v\|_Y \to \infty, \ v \in D(\mathcal{A}).$$

The following surjectivity result which will be used later can be found, e.g., in [**15**, Theorem 1.3.73, p. 62].

THEOREM 46. *Let Y be a real reflexive, strictly convex Banach space with dual space Y^*, and let $L : D(L) \subseteq Y \to Y^*$ be a linear closed, densely defined and maximal monotone operator. If the multi-valued operator $\mathcal{A} : Y \to 2^{Y^*}$ is pseudomonotone w.r.t. $D(L)$, bounded and coercive, then $L + \mathcal{A}$ is surjective, i.e.,* range $(L + \mathcal{A}) = Y^*$.

Let $\underline{u}$ and $\overline{u}$ be the sub- and supersolution of (110), (111) given in (H). We introduce the truncation mapping T and the cut-off function b as follows:

$$(Tu)(x,t) = \begin{cases} \overline{u}(x,t) & \text{if } u(x,t) > \overline{u}(x,t) \\ u(x,t) & \text{if } \underline{u}(x,t) \le u(x,t) \le \overline{u}(x,t) \\ \underline{u}(x,t) & \text{if } u(x,t) < \underline{u}(x,t), \end{cases}$$

$$b(x,t,s) = \begin{cases} (s - \overline{u}(x,t))^{p-1} & \text{if } s > \overline{u}(x,t) \\ 0 & \text{if } \underline{u}(x,t) \le s \le \overline{u}(x,t) \\ -(\underline{u}(x,t) - s)^{p-1} & \text{if } s < \underline{u}(x,t). \end{cases}$$

We have that $b : Q \times \mathbb{R} \to \mathbb{R}$ is a Carathéodory function satisfying the growth condition

$$|b(x,t,s)| \le k_2(x,t) + c_1|s|^{p-1} \tag{119}$$

for a.e. $(x,t) \in Q$ and for all $s \in \mathbb{R}$, where $c_1 > 0$ is a constant and $k_2 \in L^q(Q)$. Moreover, one has the following estimate

$$\int_Q b(x,t,u(x,t))\, u(x,t)\, dx\, dt \ge c_2\|u\|^p_{L^p(Q)} - c_3, \quad \forall u \in L^p(Q), \tag{120}$$

for some constants $c_2 > 0$ and $c_3 > 0$.

By (119), the Nemytski operator $B : L^p(Q) \to L^q(Q)$ defined by $Bu(x,t) = b(x,t,u(x,t))$ is continuous and bounded. Therefore, the operator $\hat{B} = i^* \circ B \circ i : X_0 \to X_0^*$ is continuous, bounded, and pseudomonotone w.r.t. $D(L)$ due to the compact embedding $W_0 \hookrightarrow L^p(Q)$. Let A_T denote the following operator:

$$\langle A_T u, \varphi \rangle := \int_Q \sum_{i=1}^N a_i(x,t,Tu,\nabla u) \frac{\partial \varphi}{\partial x_i}\, dx\, dt, \quad \forall \varphi \in X_0.$$

LEMMA 47. *Let hypotheses (A1)–(A3) be satisfied. Then the operator $A_T : X_0 \to X_0^*$ is bounded, continuous, and pseudomonotone w.r.t. $D(L)$.*

PROOF. As $T : X \to X$ is continuous and bounded, the assertion of the lemma follows readily from [**7**, Theorem 2.153]. □

4.2. Comparison principle. The proof of our main existence and comparison result strongly relies on an appropriate modification of the function $j : Q \times \mathbb{R} \to \mathbb{R}$ which satisfies (H). Let $\underline{u} \in W$ and $\overline{u} \in W$ be the sub- and supersolution given in (H) according to Definition 41 and Definition 42, respectively, with $\underline{\eta}(x,t) \in \partial j(x,t,\underline{u}(x,t))$ and $\overline{\eta}(x,t) \in \partial j(x,t,\overline{u}(x,t))$. Then we define the following modification $\tilde{j} : Q \times \mathbb{R} \to \mathbb{R}$ of the given j:

$$\tilde{j}(x,t,s) = \begin{cases} j(x,t,\underline{u}(x,t)) + \underline{\eta}(x,t)(s - \underline{u}(x,t)) & \text{if } s < \underline{u}(x,t), \\ j(x,t,s) & \text{if } \underline{u}(x,t) \le s \le \bar{u}(x,t), \\ j(x,t,\overline{u}(x,t)) + \overline{\eta}(x,t)(s - \overline{u}(x,t)) & \text{if } s > \overline{u}(x,t). \end{cases} \tag{121}$$

LEMMA 48. *Let j satisfy (H). Then the modified function $\tilde{j} : Q \times \mathbb{R} \to \mathbb{R}$ has the following properties:*

(P1) *$(x,t) \mapsto \tilde{j}(x,t,s)$ is measurable in Q for all $s \in \mathbb{R}$, and $s \mapsto \tilde{j}(x,t,s)$ is Lipschitz continuous in $\mathbb{R}$ for a.a. $(x,t) \in Q$.*

(P2) *Let $\partial\tilde{j}(x,t,s)$ denote Clarke's generalized gradient of $s \mapsto \tilde{j}(x,t,s)$. Then*

$$\eta \in \partial\tilde{j}(x,t,s) : \quad |\eta| \le k(x,t)$$

for a.a. $(x,t) \in Q$ and for all $s \in \mathbb{R}$.

(P3) *Clarke's generalized gradient of $s \mapsto \tilde{j}(x,t,s)$ is given by*

$$\partial\tilde{j}(x,t,s) = \begin{cases} \underline{\eta}(x,t) & \text{if } s < \underline{u}(x,t), \\ \partial\tilde{j}(x,t,\underline{u}(x,t)) & \text{if } s = \underline{u}(x,t), \\ \partial j(x,t,s) & \text{if } \underline{u}(x,t) < s < \bar{u}(x,t), \\ \partial\tilde{j}(x,t,\overline{u}(x,t)) & \text{if } s = \overline{u}(x,t), \\ \overline{\eta}(x,t) & \text{if } s > \overline{u}(x,t), \end{cases} \tag{122}$$

and the inclusions $\partial\tilde{j}(x,t,\underline{u}(x,t)) \subseteq \partial j(x,t,\underline{u}(x,t))$ and $\partial\tilde{j}(x,t,\overline{u}(x,t)) \subseteq \partial j(x,t,\overline{u}(x,t))$ hold true.

PROOF. The proof follows immediately from the definition (121) of $\tilde{j}$, and using the assumptions (H) on j as well as from the fact that Clarke's generalized gradient $\partial j(x,t,s)$ is a convex set. □

By means of $\tilde{j}$ we introduce the integral functional J given by

$$J(v) = \int_Q \tilde{j}(x,t,v(x,t))\,dxdt, \quad v \in L^p(Q).$$

Due to hypotheses (H), and applying Lebourg's mean value theorem (see [**7**, Theorem 2.177]) the functional $J : L^p(Q) \to \mathbb{R}$ is well defined and Lipschitz continuous, so that Clarke's generalized gradient $\partial J : L^p(Q) \to 2^{L^q(Q)}$ is well defined too. Moreover, Aubin–Clarke theorem (cf. [**12**, p. 83]) provides the following characterization of the generalized gradient. For $u \in L^p(Q)$ we have
(123)
$$\tilde{\eta} \in \partial J(u) \Longrightarrow \tilde{\eta} \in L^q(Q) \quad \text{with } \tilde{\eta}(x,t) \in \partial\tilde{j}(x,t,u(x,t)) \quad \text{for a.a. } (x,t) \in Q.$$

By means of Clarke's generalized gradient ∂J we further introduce the following multi-valued operator:

$$\Psi(u) := (i^* \circ \partial J \circ i)(u), \quad u \in X_0,$$

where $i^* : L^q(Q) \hookrightarrow X_0^*$ is the adjoint operator of the embedding $i : X_0 \hookrightarrow L^p(Q)$. As the operator Ψ is easily seen to be bounded due to hypothesis (H), we obtain in conjunction with [**7**, Lemma 6.22] the following result.

LEMMA 49. *Let hypothesis (H) be fulfilled. Then the operator* $\Psi : X_0 \to 2^{X_0^*}$ *is bounded and pseudomonotone w.r.t.* $D(L)$.

The main result of this subsection is the following theorem.

THEOREM 50. *Let hypotheses (A1)–(A3) and (H) be satisfied, and let* $\underline{u}$ *and* $\overline{u}$ *be the sub- and supersolution, respectively, given in (H). Then there exist solutions of (110), (111) within the ordered interval* $[\underline{u}, \overline{u}]$.

PROOF. The following auxiliary, multi-valued, periodic-boundary value problem plays a crucial role in the proof of the theorem:

$$u_t + A_T u + \hat{B}(u) + \partial\tilde{j}(\cdot,\cdot,u) \ni f \quad \text{in } Q, \tag{124}$$
$$u(\cdot,0) = u(\cdot,\tau) \ \text{ in } \Omega, \quad u = 0 \quad \text{on } \Gamma, \tag{125}$$

where A_T and $\hat{B}$ are defined in Subsection 4.1, and $\tilde{j}$ is given by (121). The weak formulation of problem (124), (125) is as follows: Find $u \in D(L)$ such that

$$\langle Lu + A_T u + \hat{B}(u), \varphi\rangle + \int_Q \tilde{\eta}\,\varphi\,dx\,dt = \langle f, \varphi\rangle, \ \forall\, \varphi \in X_0, \tag{126}$$

where $\tilde{\eta} \in L^q(Q)$ and $\tilde{\eta}(x,t) \in \partial\tilde{j}(x,t,u(x,t))$ a.e. on Q. The proof of the theorem is accomplished in two steps. First, we'll show the existence of solutions of (126). Secondly, we are going to verify that any solution of (126) belongs to the interval $[\underline{u}, \overline{u}]$, which completes the proof, because then $A_T u = Au$, $\hat{B}(u) = 0$, and $\tilde{\eta}(x,t) \in \partial\tilde{j}(x,t,u(x,t)) \subseteq \partial j(x,t,u(x,t))$. The latter is due to Lemma 48 (P3).

As for the existence of solutions of (126) let us consider the following multi-valued mapping

$$\mathcal{A}(u) := A_T u + \hat{B}(u) + \Psi(u). \tag{127}$$

By Lemma 49 and 47 the multi-valued operator $\mathcal{A} : X_0 \to 2^{X_0^*}$ of (127) is bounded and pseudomonotone w.r.t. $D(L)$ in the sense of Definition 44, and thanks to (A3) and (120) it is also coercive. Theorem 46 can be applied which yields that $L + \mathcal{A} : X_0 \to 2^{X_0^*}$ is surjective. This implies the existence of $u \in D(L)$ such that

$$f \in Lu + \mathcal{A}(u) \quad \text{in } X_0^*.$$

The latter means that there is an $\tilde{\eta} \in \partial J(i(u)) = \partial J(u)$ such that

$$Lu + A_T u + \hat{B}(u) + i^* \tilde{\eta} = f \quad \text{in } X_0^*. \tag{128}$$

From (123) and (128) we see that the solution u of (128) is in fact a solution of the auxiliary problem (126). To complete the proof it remains to verify that any solution of (126) belongs to the interval $[\underline{u}, \overline{u}]$.

Let us prove that $u \le \overline{u}$ holds, where $\overline{u}$ is the given supersolution. Subtracting (118) from (128) we obtain the following inequality:

$$\langle u_t - \overline{u}_t + A_T u - A\overline{u} + \hat{B}(u) + i^* \tilde{\eta} - i^* \overline{\eta}, \varphi \rangle \le 0, \quad \forall\, \varphi \in X_0 \cap L_+^p(Q), \tag{129}$$

where $\tilde{\eta}(x,t) \in \partial \tilde{j}(x,t,u(x,t))$ and $\overline{\eta}(x,t) \in \partial j(x,t,\overline{u}(x,t))$ for a.a. $(x,t) \in Q$. With the special test function $\varphi = (u - \overline{u})^+ \in X_0 \cap L_+^p(Q)$ in (129) we get for the individual terms on the left-hand side of inequality (129) the following estimates:

$$\langle u_t - \overline{u}_t, (u-\overline{u})^+ \rangle = \frac{1}{2}\Big(\|(u-\overline{u})^+(\cdot,\tau)\|_{L^2(\Omega)}^2 - \|(u-\overline{u})^+(\cdot,0)\|_{L^2(\Omega)}^2 \Big) \ge 0. \tag{130}$$

The last inequality follows from [**7**, Lemma 2.146] and by using

$$(u-\overline{u})(\cdot,0) \le (u-\overline{u})(\cdot,\tau),$$

which is due to the property $\overline{u}(\cdot,0) \ge \overline{u}(\cdot,\tau)$ of the supersolution. Further we have

$$\begin{aligned} &\langle A_T u - A\overline{u}, (u-\overline{u})^+ \rangle \\ &= \int_{\{u > \overline{u}\}} \sum_{i=1}^N \Big(a_i(x,t,\overline{u},\nabla u) - a_i(x,t,\overline{u},\nabla \overline{u}) \Big) \frac{\partial (u-\overline{u})}{\partial x_i}\, dx\, dt \ge 0, \end{aligned} \tag{131}$$

where $\{u > \overline{u}\} = \{(x,t) \in Q : u(x,t) > \overline{u}(x,t)\}$, and

$$\langle i^* \tilde{\eta} - i^* \overline{\eta}, (u-\overline{u})^+ \rangle = \int_{\{u > \overline{u}\}} (\tilde{\eta} - \overline{\eta})(u - \overline{u})\, dx\, dt = 0, \tag{132}$$

because for $(x,t) \in \{u > \overline{u}\}$ we get by Lemma 48 (P3) that $\tilde{\eta}(x,t) = \overline{\eta}(x,t)$. In view of (130)–(132) we get from (129) the inequality

$$\langle \hat{B}(u), (u-\overline{u})^+ \rangle = \int_Q b(x,t,u)(u-\overline{u})^+\, dx\, dt \le 0,$$

which by using the definition of b yields

$$0 \le \int_Q [(u-\overline{u})^+]^p\, dx\, dt \le 0, \tag{133}$$

and thus it follows $(u-\overline{u})^+ = 0$, i.e., $u \le \overline{u}$. The inequality $\underline{u} \le u$ can be shown analogously, which completes the proof of the theorem. □

4.3. Equivalence of parabolic inclusion and hemivariational inequality. Theorem 50 allows us to justify the equivalence of the parabolic inclusion (110)–(111) and the hemivariational inequality (115) in case that j satisfies only hypothesis (G).

THEOREM 51. *Under hypotheses (A1)–(A3) and (G) problem (110), (111) and the hemivariational inequality (115) are equivalent.*

PROOF. We already mentioned above that only by applying the definition of Clarke's generalized gradient we readily infer that any solution of (110), (111) must be a solution of (115). So we need to show the reverse. In [**7**, Sect. 6.3] a notion of sub- and supersolution has been introduced for evolution hemivariational inequalities under homogeneous initial and boundary conditions. In the spirit of the approach given there we introduce the following notion for sub-supersolution of the periodic evolution hemivariational inequality (115).

DEFINITION 52. A function $\underline{u} \in W$ is called a **subsolution** of (115) if $\underline{u}(x,0) \le \underline{u}(x,\tau)$ for a.a. $x \in \Omega$, $\underline{u}|_\Gamma \le 0$, and

$$(134)\quad \langle \underline{u}_t + A\underline{u}, \varphi - \underline{u}\rangle + \int_Q j^o(x,t,\underline{u};\varphi - \underline{u})\,dx\,dt \ge \langle f, \varphi - \underline{u}\rangle, \quad \forall\, \varphi \in \underline{u} \wedge X_0.$$

DEFINITION 53. A function $\overline{u} \in W$ is called a **supersolution** of (115) if $\overline{u}(x,0) \ge \overline{u}(x,\tau)$ for a.a. $x \in \Omega$, $\overline{u}|_\Gamma \ge 0$, and

$$(135)\quad \langle \overline{u}_t + A\overline{u}, \varphi - \overline{u}\rangle + \int_Q j^o(x,t,\overline{u};\varphi - \overline{u})\,dx\,dt \ge \langle f, \varphi - \overline{u}\rangle, \quad \forall\, \varphi \in \overline{u} \vee X_0.$$

Note that any solution of (115) is both a subsolution and a supersolution of (115). Next we are going to prove that $\underline{u}(\overline{u}) \in W$ is a subsolution (resp. supersolution) of (115) if and only if it is a subsolution (resp. supersolution) of the inclusion (110), (111). So, let $\underline{u} \in W$ be a subsolution of (110), (111) according to Definition 41. For $\underline{u}$ being a subsolution of (115), we only need to verify (134). From (117) we immediately get by applying the definition of Clarke's gradient the following inequality:

$$(136)\qquad \langle \underline{u}_t + A\underline{u}, -v\rangle + \int_Q j^o(x,t,\underline{u};-v)\,dx\,dt \ge \langle f, -v\rangle, \quad \forall\, v \in X_0 \cap L^p_+(Q).$$

In particular, the last inequality holds for $v = (\underline{u} - w)^+$ with $w \in X_0$. Setting $\varphi = \underline{u} - (\underline{u} - w)^+ = \underline{u} \wedge w$, we may rewrite (136) in the form (134), which shows that $\underline{u}$ is a subsolution of (115).

To prove the reverse, let $\underline{u}$ be a subsolution of (115) according to Definition 52. As any $\varphi \in \underline{u} \wedge X_0$ can be written in the form $\varphi = \underline{u} - (\underline{u} - w)^+$ with $w \in X_0$, we get

(137)

$$\langle \underline{u}_t + A\underline{u}, -(\underline{u} - w)^+\rangle + \int_Q j^o(x,t,\underline{u};-(\underline{u} - w)^+)\,dx\,dt \ge \langle f, -(\underline{u} - w)^+\rangle, \quad \forall\, w \in X_0.$$

By the density result [**7**, Lemma 5.36] we get from (137)

$$(138)\qquad \langle \underline{u}_t + A\underline{u}, -\phi\rangle + \int_Q j^o(x,t,\underline{u};-\phi)\,dx\,dt \ge \langle f, -\phi\rangle, \quad \forall\, \phi \in X_0 \cap L^p_+(Q),$$

which yields

$$(139) \qquad \langle \underline{u}_t + A\underline{u}, \phi\rangle - \int_Q j^o(x,t,\underline{u};-1)\,\phi\,dx\,dt \le \langle f, \phi\rangle, \quad \forall\, \phi \in X_0 \cap L^p_+(Q).$$

From the calculus of Clarke's generalized gradient we infer the existence of some function $\underline{\eta} : Q \to \mathbb{R}$ such that

$$\underline{\eta}(x,t) \in \partial j(x,t,\underline{u}(x,t)) \quad \text{and} \quad j^o(x,t,\underline{u}(x,t);-1) = \underline{\eta}(x,t)(-1).$$

From the definition of the generalized directional derivative j^o it follows that the function $(x,t) \mapsto j^o(x,t,\underline{u}(x,t);-1)$ is measurable, which implies that

$$\underline{\eta}(x,t) = -j^o(x,t,\underline{u}(x,t);-1)$$

is measurable. The growth assumption (G) yields $\underline{\eta} \in L^q(Q)$, which proves that $\underline{u}$ is in fact a subsolution (110), (111) according to Definition 41. The proof for the supersolution is analogous.

To complete the proof of Theorem 51 we need to show that any solution of (115) must be a solution of (110), (111). Since any solution u of (115) is both a subsolution and a supersolution of (115), it must be a subsolution and a supersolution of the periodic evolution problem (110), (111). This means that there are $\underline{\eta}, \overline{\eta} \in L^q(Q)$ such that

$$\underline{\eta}(x,t) \in \partial j(x,t,,u(x,t)) : \quad u_t + Au + i^*\underline{\eta} \le f \quad \text{in} \quad X_0^*,$$

and

$$\overline{\eta}(x,t) \in \partial j(x,t,,u(x,t)) : \quad u_t + Au + i^*\overline{\eta} \ge f \quad \text{in} \quad X_0^*.$$

(Note: $\underline{\eta}$ and $\overline{\eta}$ need not be the same). Applying Theorem 50 we obtain the existence of a solution v of (110), (111) within the interval $[u,u] = \{u\}$ which implies $v = u$ and the existence of $\eta \in L^q(Q)$ with $\eta(x,t) \in \partial j(x,t,u(x,t))$ such that

$$u_t + Au + i^*\eta = f \quad \text{in} \quad X_0^*,$$

which shows that u is a solution of (110), (111). □

4.4. Directedness of solution set. Theorem 50 ensures the existence of solutions of the periodic evolution inclusion problem (110)–(111) within the ordered interval $[\underline{u}, \overline{u}]$ formed by an ordered pair of sub- and supersolution. Let $\mathcal{S}$ denote the set of all solutions of (110)–(111) that belong to the ordered interval $[\underline{u}, \overline{u}]$. In view of Theorem 50, we have that $\mathcal{S} \neq \emptyset$.

Our aim of this subsection is to show that under the hypotheses of Theorem 50 and hypothesis (A4) given below, the solution set $\mathcal{S}$ is a directed set. As for the notion of a directed set we refer to Definition 25. It should be noted that in proving the directedness of $\mathcal{S}$ one is faced with the difficulty that the underlying solution space for (110)–(111), which is the Lions space W_0, does not possess lattice structure. This is in contrast to the situation of the corresponding elliptic inclusions, where the solution space is the Sobolev space $W_0^{1,p}(\Omega)$ (resp. $W^{1,p}(\Omega)$) which does possess lattice structure. The presence of a lattice structure usually allows one to prove that $\max\{u_1, u_2\}$ is a subsolution and $\min\{u_1, u_2\}$ is a supersolution whenever u_1 and u_2 are solutions, from which the directedness is a straightforward consequence. Unfortunately, this lattice structure is not available for the underlying solution space W_0 of (110)–(111).

In addition to the hypotheses of Theorem 50 we assume the following hypothesis:

(A4) There is a function $k_2 \in L^q_+(Q)$ and a continuous function $\omega : \mathbb{R}_+ \to \mathbb{R}_+$ such that

$$|a_i(x,t,s,\xi) - a_i(x,t,s',\xi)| \le [k_2(x,t) + |s|^{p-1} + |s'|^{p-1} + |\xi|^{p-1}]\omega(|s-s'|),$$

holds for a.a. $(x,t) \in Q$, for all s, $s' \in \mathbb{R}$, and for all $\xi \in \mathbb{R}^N$, where ω satisfies, for each $\varepsilon > 0$,

$$\int_0^\varepsilon \frac{dr}{\omega(r)} = +\infty. \tag{140}$$

REMARK 54. Relation (140) means that the integral diverges near zero. Hypothesis (A4) includes, for example, $\omega(r) = c\,r$, for $c > 0$, and for all $r \ge 0$, which means that the coefficients $a_i(x,t,s,\xi)$ are Lipschitz continuous with respect to s.

Our main result of this subsection is given by the following theorem.

THEOREM 55. *Let hypotheses (A1)–(A4) and (H) be satisfied, and let* $\underline{u}$ *and* $\overline{u}$ *be the sub- and supersolution, respectively, given in (H). Then the solution set* $\mathcal{S}$ *equipped with the natural partial ordering of functions introduced by the order cone* $L^p_+(Q)$ *is a directed set.*

PROOF. We are going to show that $\mathcal{S}$ is upward directed only, since the proof for $\mathcal{S}$ being downward directed can be done in a similar way. To this end let u_1, $u_2 \in \mathcal{S}$. Our goal is to prove the existence of an element $u \in \mathcal{S}$ such that $u \ge u_k$, $k = 1, 2$. The proof will be done in several steps, and crucially relies on an appropriately designed auxiliary problem.

Step 1: Auxiliary Problem

Let $u_k \in \mathcal{S}$, $k = 1, 2$, i.e., $u_k \in D(L) \subset W_0$, and there are $\eta_k \in L^q(Q)$ with $\eta_k(x,t) \in \partial j(x,t,u_k(x,t))$ for a.e. $(x,t) \in Q$ such that

$$\langle Lu_k + Au_k, \varphi\rangle + \int_Q \eta_k\,\varphi\,dx\,dt = \langle f, \varphi\rangle, \quad \forall\, \varphi \in X_0. \tag{141}$$

Let us define $u_0 := \max\{u_1, u_2\}$ and η_0 as follows

$$\eta_0(x,t) = \begin{cases} \eta_1(x,t) & \text{if} \quad (x,t) \in \{u_1 \ge u_2\}, \\ \eta_2(x,t) & \text{if} \quad (x,t) \in \{u_2 > u_1\}, \end{cases}$$

where $\{u_l \ge (>)\, u_k\}$ stands for $\{(x,t) \in Q : u_l(x,t) \ge (>)\, u_k(x,t)\}$. By the definition of η_0 we readily see that $\eta_0 \in L^q(Q)$, and

$$\eta_0(x,t) \in \partial j(x,t,u_0(x,t)) \quad \text{for a.e. } (x,t) \in Q. \tag{142}$$

By means of η_0, u_0, and $\overline{\eta}$, $\overline{u}$ of Definition 42 we introduce the following modification $\tilde{j} : Q \times \mathbb{R} \to \mathbb{R}$ of the given j:
(143)

$$\tilde{j}(x,t,s) = \begin{cases} j(x,t,u_0(x,t)) + \eta_0(x,t)(s - u_0(x,t)) & \text{if } s < u_0(x,t), \\ j(x,t,s) & \text{if } u_0(x,t) \le s \le \overline{u}(x,t), \\ j(x,t,\overline{u}(x,t)) + \overline{\eta}(x,t)(s - \overline{u}(x,t)) & \text{if } s > \overline{u}(x,t). \end{cases}$$

In view of hypothesis (H) the modified function $\tilde{j} : Q \times \mathbb{R} \to \mathbb{R}$ has the following properties:

(a) $(x,t) \mapsto \tilde{j}(x,t,s)$ is measurable in Q for all $s \in \mathbb{R}$, and $s \mapsto \tilde{j}(x,t,s)$ is Lipschitz continuous in $\mathbb{R}$ for a.a. $(x,t) \in Q$.

(b) Let $\partial\tilde{j}(x,t,s)$ denote Clarke's generalized gradient of $s \mapsto \tilde{j}(x,t,s)$. Then for all $\eta \in \partial\tilde{j}(x,t,s)$, we have that

$$|\eta| \le k(x,t) \tag{144}$$

for a.a. $(x,t) \in Q$ and for all $s \in \mathbb{R}$.

(c) Clarke's generalized gradient of $s \mapsto \tilde{j}(x,t,s)$ is given by

$$\partial\tilde{j}(x,t,s) = \begin{cases} \eta_0(x,t) & \text{if } s < u_0(x,t), \\ \partial\tilde{j}(x,t,u_0(x,t)) & \text{if } s = u_0(x,t), \\ \partial j(x,t,s) & \text{if } u_0(x,t) < s < \overline{u}(x,t), \\ \partial\tilde{j}(x,t,\overline{u}(x,t)) & \text{if } s = \overline{u}(x,t), \\ \overline{\eta}(x,t) & \text{if } s > \overline{u}(x,t). \end{cases} \tag{145}$$

Moreover, the inclusions

$$\partial\tilde{j}(x,t,u_0(x,t)) \subseteq \partial j(x,t,u_0(x,t)), \quad \partial\tilde{j}(x,t,\overline{u}(x,t)) \subseteq \partial j(x,t,\overline{u}(x,t)) \tag{146}$$

hold true due to the fact that Clarke's generalized gradient $\partial j(x,t,s)$ is a convex set.

Furthermore, we introduce the functions $g_i : Q \times \mathbb{R} \to \mathbb{R}$ related with η_i and u_i, for $i = 0,1,2$, and defined by:
(147)

$$g_0(x,t,s) = \begin{cases} \eta_0(x,t) & \text{if } s \le u_0(x,t), \\ \eta_0(x,t) + \dfrac{\overline{\eta}(x,t) - \eta_0(x,t)}{\overline{u}(x,t) - u_0(x,t)}(s - u_0(x,t)) & \text{if } u_0(x,t) < s < \overline{u}(x,t), \\ \overline{\eta}(x,t) & \text{if } s \ge \overline{u}(x,t), \end{cases}$$

and for $k = 1,2$
(148)

$$g_k(x,t,s) = \begin{cases} \eta_k(x,t) & \text{if } s \le u_k(x,t), \\ \eta_k(x,t) + \dfrac{\eta_0(x,t) - \eta_k(x,t)}{u_0(x,t) - u_k(x,t)}(s - u_k(x,t)) & \text{if } u_k(x,t) < s < u_0(x,t), \\ g_0(x,t,s) & \text{if } s \ge u_0(x,t). \end{cases}$$

Apparently, the functions $(x,t,s) \mapsto g_i(x,t,s)$ are Carathéodory functions which are piecewise linear with respect to s. Finally, define the cut-off function $\hat{b}$ related to the pair u_0 and $\overline{u}$ by:

$$\hat{b}(x,t,s) = \begin{cases} (s - \overline{u}(x,t))^{p-1} & \text{if } s > \overline{u}(x,t) \\ 0 & \text{if } u_0(x,t) \le s \le \overline{u}(x,t) \\ -(u_0(x,t) - s)^{p-1} & \text{if } s < u_0(x,t). \end{cases} \tag{149}$$

We have that $\hat{b} : Q \times \mathbb{R} \to \mathbb{R}$ is a Carathéodory function satisfying the growth condition

$$|\hat{b}(x,t,s)| \le k_3(x,t) + c_1|s|^{p-1} \tag{150}$$

for a.a. $(x,t) \in Q$ and for all $s \in \mathbb{R}$, where $c_1 > 0$ is a constant and $k_3 \in L^q(Q)$. Moreover, one has the following estimate

$$\int_Q \hat{b}(x,t,u(x,t))\, u(x,t)\, dx\, dt \ge c_2\|u\|^p_{L^p(Q)} - c_3, \quad \forall u \in L^p(Q), \tag{151}$$

for some constants $c_2 > 0$ and $c_3 > 0$. By (150), the associated Nemytskij operator $\hat{B} : L^p(Q) \to L^q(Q)$ defined by $\hat{B}u(x,t) = \hat{b}(x,t,u(x,t))$ is continuous and bounded. If $i^* : L^q(Q) \hookrightarrow X_0^*$ is the adjoint operator of the embedding $i : X_0 \hookrightarrow L^p(Q)$, then the operator $\hat{\mathcal{B}} = i^* \circ \hat{B} \circ i : X_0 \to X_0^*$ is continuous, bounded, and pseudomonotone w.r.t. $D(L)$ due to the compact embedding $W_0 \hookrightarrow L^p(Q)$. Now we are ready to introduce the following auxiliary problem which is crucial for the rest of the proof:

$$u_t + A_T u + \hat{B}(u) + \partial\tilde{j}(\cdot,\cdot,u) - \sum_{l=1}^{2} |g_l(x,t,u) - g_0(x,t,u)| \ni f \quad \text{in } Q, \tag{152}$$

$$u(\cdot,0) = u(\cdot,\tau) \ \text{ in } \Omega, \quad u = 0 \quad \text{on } \Gamma, \tag{153}$$

with $\tilde{j}$, g_i, and $\hat{B}$ given by (143), (147)–(149), and with the operator A_T defined by

$$\langle A_T u, \varphi\rangle := \int_Q \sum_{i=1}^{N} a_i(x,t,Tu,\nabla u)\frac{\partial\varphi}{\partial x_i}\, dx\, dt, \quad \forall\varphi \in X_0,$$

where T denotes the truncation operator given by

$$(Tu)(x,t) = \begin{cases} \overline{u}(x,t) & \text{if } u(x,t) > \overline{u}(x,t) \\ u(x,t) & \text{if } u(x,t) \le \overline{u}(x,t), \end{cases}$$

which is known to be continuous from X into X. In the steps to follow we are going to show that the auxiliary Problem (152)–(153) enjoys the following properties:

- (i) Existence: There exist solutions of Problem (152)–(153).
- (ii) Comparison: Any solution u of (152)–(153) satisfies $u_k \le u \le \overline{u}$, $k = 1, 2$.
- (iii) Upward Directedness: Any solution u of (152)–(153) is a solution of the original problem (110)–(111), which due to (ii) exceeds u_k, and thus $\mathcal{S}$ is upward directed.

Step 2: Existence of Solutions of the Auxiliary Problem
To prove the existence of solutions of (152)–(153) let

$$g(x,t,u) := \sum_{l=1}^{2} |g_l(x,t,u) - g_0(x,t,u)|. \tag{154}$$

By definitions (147)–(148), the functions $g_i : Q\times\mathbb{R} \to \mathbb{R}$ are Carathéodory functions that are uniformly $L^q(Q)$-bounded. Therefore, the associated Nemytskij operator $G : L^p(Q) \to L^q(Q)$ is continuous and bounded, and $\mathcal{G} = i^* \circ G \circ i : X_0 \to X_0^*$ is continuous, bounded, and pseudomonotone w.r.t. $D(L)$ due to the compact embedding $W_0 \hookrightarrow L^p(Q)$. Thus the weak formulation of problem (152)–(153) is as follows: Find $u \in D(L)$ such that

$$\langle Lu + A_T u + \hat{\mathcal{B}}(u) - \mathcal{G}(u), \varphi\rangle + \int_Q \tilde{\eta}\,\varphi\,dx\,dt = \langle f, \varphi\rangle, \quad \forall\, \varphi \in X_0, \tag{155}$$

where $\tilde{\eta} \in L^q(Q)$ and $\tilde{\eta}(x,t) \in \partial\tilde{j}(x,t,u(x,t))$ a.e. on Q. By means of $\tilde{j}$ we introduce the integral functional $\tilde{J}$ given by

$$\tilde{J}(v) = \int_Q \tilde{j}(x,t,v(x,t))\,dxdt, \quad v \in L^p(Q).$$

Due to hypotheses (H), and applying Lebourg's mean value theorem the functional $\tilde{J} : L^p(Q) \to \mathbb{R}$ is well defined and even Lipschitz continuous, so that Clarke's generalized gradient $\partial\tilde{J} : L^p(Q) \to 2^{L^q(Q)}$ is well defined too. Moreover, Aubin–Clarke's theorem (cf. [**12**, p. 83]) provides the following characterization of the generalized gradient. For $u \in L^p(Q)$ we have

$$\tilde{\eta} \in \partial\tilde{J}(u) \Longrightarrow \tilde{\eta} \in L^q(Q) \text{ with } \tilde{\eta}(x,t) \in \partial\tilde{j}(x,t,u(x,t)) \text{ for a.a. } (x,t) \in Q. \tag{156}$$

By means of Clarke's generalized gradient $\partial\tilde{J}$ we further introduce the following multi-valued operator:

$$\tilde{\Psi}(u) := (i^* \circ \partial\tilde{J} \circ i)(u), \quad u \in X_0.$$

As the operator $\tilde{\Psi}$ is bounded due to hypothesis (H), we obtain, in conjunction with [**7**, Lemma 6.22], that the multi-valued operator $\tilde{\Psi} : X_0 \to 2^{X_0^*}$ is bounded and pseudomonotone w.r.t. $D(L)$. To show the existence of solutions of (155) let us consider the following multi-valued mapping

$$\mathcal{A}(u) := A_T u + \hat{\mathcal{B}}(u) - \mathcal{G}(u) + \tilde{\Psi}(u). \tag{157}$$

As pseudomonotonicity w.r.t. $D(L)$ is invariant under addition, the operator $\mathcal{A} : X_0 \to 2^{X_0^*}$ of (157) is bounded and pseudomonotone w.r.t. $D(L)$ in the sense of Definition 44. By the definitions of the g_i we deduce from (147)–(148) that $\mathcal{G} : X_0 \to X_0^*$ is globally bounded. From (144) it follows that also $\tilde{\Psi} : X_0 \to 2^{X_0^*}$ is globally bounded. Since $\hat{b}$ defining the operator $\hat{\mathcal{B}}$ satisfies an estimate of the form (151), we infer from hypothesis (A3) that $\mathcal{A} : X_0 \to 2^{X_0^*}$ is coercive. This allows us to apply Theorem 46. Hence it follows that the operator $L + \mathcal{A} : X_0 \to 2^{X_0^*}$ is surjective, which implies the existence of $u \in D(L)$ such that

$$f \in Lu + \mathcal{A}(u) \quad \text{in } X_0^*.$$

The latter means that there is an $\tilde{\eta} \in \partial\tilde{J}(i(u)) = \partial\tilde{J}(u)$ such that

$$Lu + A_T u + \hat{\mathcal{B}}(u) - \mathcal{G}(u) + i^*\tilde{\eta} = f \quad \text{in } X_0^*, \tag{158}$$

which in view of (156) completes the existence proof of the auxiliary problem (152)–(153) in its weak form (155).

Step 3: Comparison

Next we show that any solution u of the auxiliary problem (152)–(153) satisfies $u_k \le u \le \overline{u}$. Let us prove first $u \le \overline{u}$. Subtracting the inequality (118) satisfied by the supersolution $\overline{u}$ from the equation (158), and testing the resulting inequality by $(u-\overline{u})^+ \in X_0 \cap L^p_+(Q)$ we obtain:

$$\langle u_t - \overline{u}_t + A_T u - A\overline{u} + \hat{\mathcal{B}}(u) - \mathcal{G}(u) + i^*\tilde{\eta} - i^*\overline{\eta}, (u-\overline{u})^+\rangle \le 0. \tag{159}$$

We estimate the individual terms on the left-hand side of (159) as follows:

$$\langle u_t - \overline{u}_t, (u-\overline{u})^+\rangle \ge 0, \quad \text{see (130)}. \tag{160}$$

In view of (A2) and due to the definition of A_T we get

$$\begin{aligned} &\langle A_T u - A\overline{u}, (u-\overline{u})^+\rangle \\ &= \int_{\{u>\overline{u}\}} \sum_{i=1}^N \Big(a_i(x,t,\overline{u},\nabla u) - a_i(x,t,\overline{u},\nabla\overline{u})\Big)\frac{\partial(u-\overline{u})}{\partial x_i}\,dx\,dt \ge 0, \end{aligned} \tag{161}$$

and by (147), (148)

$$\langle -\mathcal{G}(u), (u-\overline{u})^+\rangle = \int_{\{u>\overline{u}\}} -G(u)\,(u-\overline{u})\,dx\,dt = 0. \tag{162}$$

From (145) we see that

$$\langle i^*\tilde{\eta} - i^*\overline{\eta}, (u-\overline{u})^+\rangle = \int_{\{u>\overline{u}\}} (\tilde{\eta}-\overline{\eta})\,(u-\overline{u})\,dx\,dt = 0. \tag{163}$$

Taking into account (160)–(163), and (149), from (159) we finally get

$$\langle \hat{\mathcal{B}}(u), (u-\overline{u})^+\rangle = \int_{\{u>\overline{u}\}} \hat{b}(x,t,u)\,(u-\overline{u})\,dx\,dt = \int_{\{u>\overline{u}\}} (u-\overline{u})^p\,dx\,dt \le 0,$$

and thus $(u-\overline{u})^+ = 0$, i.e., $u \le \overline{u}$.

The proof of $u_k \le u$ for $k = 1,2$ requires more involved tools. The solutions $u_k \in \mathcal{S}$ satisfy (141). Subtracting the equation (155) (or equivalently (158)) satisfied by a solution u of the auxiliary problem from (141), and noting that $Tu = u$ by the previous comparison, yields

$$\langle (u_k-u)_t + Au_k - Au - \hat{\mathcal{B}}(u) + \mathcal{G}(u) + i^*\eta_k - i^*\tilde{\eta}, \varphi\rangle = 0, \quad \forall\, \varphi \in X_0. \tag{164}$$

Unlike in the previous comparison procedure, a more subtle test function technique is required here. This is because u has to be compared with both u_1 and u_2, and the operator A is not specified for one particular u_k. It is here where the assumption (A4) comes into play. For this purpose we construct a special test function φ for (164). By (A4), for any fixed $\varepsilon > 0$ there exists $\delta(\varepsilon) \in (0,\varepsilon)$ such that

$$\int_{\delta(\varepsilon)}^{\varepsilon} \frac{1}{\omega(r)}\,dr = 1.$$

We define the function $\theta_\varepsilon : \mathbb{R} \to \mathbb{R}_+$ as follows

$$\theta_\varepsilon(s) = \begin{cases} 0 & \text{if } s < \delta(\varepsilon) \\ \displaystyle\int_{\delta(\varepsilon)}^{s} \frac{1}{\omega(r)}\,dr & \text{if } \delta(\varepsilon) \le s \le \varepsilon \\ 1 & \text{if } s > \varepsilon. \end{cases}$$

We readily verify that, for each $\varepsilon > 0$, the function θ_ε is continuous, piecewise differentiable and the derivative is nonnegative and bounded. Therefore the function θ_ε is Lipschitz continuous and nondecreasing. In addition, it satisfies

$$\theta_\varepsilon \to \chi_{\{s>0\}} \quad \text{as } \varepsilon \to 0, \tag{165}$$

where $\chi_{\{s>0\}}$ is the characteristic function of the real half line $\{s \in \mathbb{R} : s > 0\}$, and one has

$$\theta_\varepsilon'(s) = \begin{cases} 1/\omega(s) & \text{if } \delta(\varepsilon) < s < \varepsilon \\ 0 & \text{if } s \notin [\delta(\varepsilon), \varepsilon]. \end{cases}$$

As our special test function we choose the composition of θ_ε with $(u_k - u)$, i.e.,

$$\varphi = \theta_\varepsilon(u_k - u) \in X_0 \cap L_+^p(Q), \tag{166}$$

and note that

$$\frac{\partial}{\partial x_i}\theta_\varepsilon(u_k - u) = \theta_\varepsilon'(u_k - u)\frac{\partial(u_k - u)}{\partial x_i}.$$

We use (166) to test (164), and estimate the terms on the left-hand side individually. Let Θ_ε be the primitive of the function θ_ε defined by

$$\Theta_\varepsilon(s) = \int_0^s \theta_\varepsilon(r)\,dr.$$

We obtain for the first term on the left-hand side of (164) by using [**7**, Lemma 2.146] that

$$\left\langle \frac{\partial(u_k - u)}{\partial t}, \theta_\varepsilon(u_k - u) \right\rangle = \int_\Omega \Big(\Theta_\varepsilon(u_k - u)(x,\tau) - \Theta_\varepsilon(u_k - u)(x,0) \Big)\,dx = 0. \tag{167}$$

Making use of (A2) and (A4) the term $\langle Au_k - Au, \theta_\varepsilon(u_k - u)\rangle$ can be estimated as follows:

$$\begin{aligned} &\langle Au_k - Au, \theta_\varepsilon(u_k - u)\rangle \\ &\quad = \int_Q \sum_{i=1}^N \Big(a_i(x,t,u_k,\nabla u_k) - a_i(x,t,u,\nabla u) \Big) \frac{\partial}{\partial x_i}\theta_\varepsilon(u_k - u)\,dx\,dt \\ &\quad = \int_Q \sum_{i=1}^N \Big(a_i(x,t,u_k,\nabla u_k) - a_i(x,t,u_k,\nabla u) \Big)\theta_\varepsilon'(u_k - u)\frac{\partial(u_k - u)}{\partial x_i}\,dx\,dt \\ &\qquad + \int_Q \sum_{i=1}^N \Big(a_i(x,t,u_k,\nabla u) - a_i(x,t,u,\nabla u) \Big)\theta_\varepsilon'(u_k - u)\frac{\partial(u_k - u)}{\partial x_i}\,dx\,dt \end{aligned} \tag{168}$$

$$\geq \int_Q \sum_{i=1}^{N} \Big(a_i(x,t,u_k,\nabla u) - a_i(x,t,u,\nabla u)\Big)\theta'_\varepsilon(u_k-u)\frac{\partial(u_k-u)}{\partial x_i}\,dx\,dt$$
$$\geq -N\int_Q (k_2+|u|^{p-1}+|u_k|^{p-1}+|\nabla u|^{p-1})$$
$$\times\,\omega(|u_k-u|)\theta'_\varepsilon(u_k-u)|\,\nabla(u_k-u)|\,dx\,dt$$
$$\geq -N\int_{\{\delta(\varepsilon)<u_k-u<\varepsilon\}} \gamma\,|\nabla(u_k-u)|\,dx\,dt,$$

where $\gamma = k_2+|u|^{p-1}+|u_k|^{p-1}+|\nabla u|^{p-1}\in L^q(Q)$. The term on the right-hand side of (168) tends to zero as $\varepsilon\to 0$. Next we estimate the term

$$\begin{aligned}&\langle \mathcal{G}(u)+i^*\eta_k - i^*\tilde{\eta},\theta_\varepsilon(u_k-u)\rangle\\ &\quad= \int_Q\Big(\sum_{l=1}^{2}|g_l(x,t,u)-g_0(x,t,u)|+\eta_k-\tilde{\eta}\Big)\theta_\varepsilon(u_k-u)\,dx\,dt.\end{aligned}\tag{169}$$

By using (165) and applying Lebesgue's dominated convergence theorem, and taking into account the definitions of g_l $(l=0,1,2)$ according to (147) and (148), from (165) we obtain

$$\begin{aligned}&\lim_{\varepsilon\to 0}\langle \mathcal{G}(u)+i^*\eta_k - i^*\tilde{\eta},\theta_\varepsilon(u_k-u)\rangle\\ &\quad= \int_Q\Big(\sum_{l=1}^{2}|g_l(x,t,u)-g_0(x,t,u)|+\eta_k-\tilde{\eta}\Big)\chi_{\{u_k-u>0\}}\,dx\,dt.\end{aligned}\tag{170}$$

If $u_k-u>0$, then $u<u_0$, and thus for $(x,t)\in\{u_k-u>0\}$ we get $\tilde{\eta}(x,t)=\eta_0(x,t)$, $g_0(x,t,u)=\eta_0(x,t)$, and $g_k(x,t,u)=\eta_k(x,t)$, which implies the following estimate of the right-hand side of (170) (with $l\neq k$, and $l,k\in\{1,2\}$)

$$\begin{aligned}&\int_Q\Big(\sum_{l=1}^{2}|g_l(x,t,u)-g_0(x,t,u)|+\eta_k-\tilde{\eta}\Big)\chi_{\{u_k-u>0\}}\,dx\,dt\\ &\quad=\int_Q\Big(|g_k(x,t,u)-g_0(x,t,u)|+\eta_k-\tilde{\eta}\Big)\chi_{\{u_k-u>0\}}\,dx\,dt\\ &\qquad+\int_Q |g_l(x,t,u)-g_0(x,t,u)|\,\chi_{\{u_k-u>0\}}\,dx\,dt\\ &\quad\geq \int_{\{u_k-u>0\}}\Big(|\eta_k-\eta_0|+\eta_k-\eta_0\Big)dx\,dt\geq 0.\end{aligned}\tag{171}$$

Using (167)–(171) we finally obtain from (164) as $\varepsilon\to 0$ the following inequality:

$$\lim_{\varepsilon\to 0}\langle -\hat{\mathcal{B}}(u),\theta_\varepsilon(u_k-u)\rangle = -\int_Q \hat{b}(x,t,u)\chi_{\{u_k-u>0\}}\,dx\,dt\leq 0.\tag{172}$$

As $u_k \le u_0$, and for $(x,t) \in \{u_k - u > 0\}$ the cut-off function becomes $\hat{b}(x,t,u) = -(u_0(x,t) - u(x,t))^{p-1}$, inequality (172) yields the estimate

$$\begin{aligned} 0 &\le \int_{\{u_k-u>0\}} (u_k(x,t) - u(x,t))^{p-1}\,dx\,dt \\ &\le \int_{\{u_k-u>0\}} (u_0(x,t) - u(x,t))^{p-1}\,dx\,dt \le 0, \end{aligned}$$

which implies

$$\int_Q \Big((u_k(x,t) - u(x,t))^+\Big)^{p-1}\,dx\,dt = 0,$$

and thus $(u_k - u)^+ = 0$, i.e., $u_k \le u$ in Q, for $k = 1, 2$.
Step 4: $\mathcal{S}$ is Upward Directed
From Step 1–Step 3 we know that any solution u of the auxiliary problem (152)–(153) (or equivalently (155)) satisfies $u_k \le u \le \overline{u}$, where $u_k \in \mathcal{S}$, $k = 1, 2$, and therefore $u \ge u_0$. The proof of the upward directedness is complete, if the solution u of the auxiliary problem in fact turns out to be a solution of the original problem (110)–(111). Because $u_k \le u \le \overline{u}$ it is clear that $u \in [\underline{u}, \overline{u}]$. Due to $u_0 \le u \le \overline{u}$, it follows that $A_T u = Au$, $\hat{B}(u) = 0$, and $G(u) = 0$, because then $g_0(x,t,u) = g_l(x,t,u)$, $l = 1, 2$, see (148). Hence, the solution of the auxiliary problem in fact fulfills: $u \in D(L)$ such that

$$(173) \qquad \langle Lu + Au, \varphi\rangle + \int_Q \tilde{\eta}\,\varphi\,dx\,dt = \langle f, \varphi\rangle, \quad \forall\, \varphi \in X_0,$$

where $\tilde{\eta} \in L^q(Q)$ and $\tilde{\eta}(x,t) \in \partial\tilde{j}(x,t,u(x,t))$ a.e. on Q. In view of (145)–(146) we see that $\partial\tilde{j}(x,t,u(x,t)) \subseteq \partial j(x,t,u(x,t))$, which shows that u is in fact a solution of the original problem (110)–(111). □

REMARK 56. The directedness of the solution set $\mathcal{S}$ ensured by Theorem 55 is the crucial step towards the existence of the smallest and greatest element of $\mathcal{S}$, which are the smallest and greatest solution of (110)–(111) within the interval $[\underline{u}, \overline{u}]$ (called also the *extremal solutions*). The proof of the extremal solutions follows standard arguments and can be omitted.

5. Application: multiple periodic solutions

Let $\Omega \subset \mathbb{R}^N$ be a bounded domain with smooth boundary $\partial\Omega$, $Q = \Omega \times (0,\tau)$, and $\Gamma = \partial\Omega \times (0,\tau)$, with $\tau > 0$. In this section we apply the results obtained in Section 4 to prove the existence of multiple periodic solutions of the following (single-valued) periodic-parabolic problem:

$$(174) \qquad u_t - \Delta u = g(u) + h(x,t) \quad \text{in } Q,$$

$$(175) \qquad u(\cdot,0) = u(\cdot,\tau) \ \text{ in } \Omega, \quad u = 0 \quad \text{on } \Gamma,$$

where $g : \mathbb{R} \to \mathbb{R}$ and $h : \overline{\Omega} \times \mathbb{R} \to \mathbb{R}$ are assumed to satisfy the hypotheses

(g) $g(s) = |s|^{\alpha-1}s, \quad s \in \mathbb{R}, \quad 0 < \alpha < 1.$

(h) $h : \overline{\Omega} \times \mathbb{R} \to \mathbb{R}$ is continuous, $h > 0$ on $\Omega \times \mathbb{R}$, and $t \mapsto h(x,t)$ is τ-periodic for each $x \in \overline{\Omega}$.

Under the hypotheses (g) and (h) the periodic parabolic problem (174)–(175) has been studied by Hirano and Kim in [**17**] who were able to prove the following multiplicity result under an additional hypothesis on the smallness of h with respect to the $C^1(\overline{Q})$-norm.

THEOREM 57 (Hirano-Kim [**17**]). *Let hypotheses (g) and (h) be satisfied, and assume in addition $h \in C^1(\overline{Q})$. Then there exists a constant $m_0 > 0$ such that for each $h \in C^1(\overline{Q})$ with $\|h\|_{C^1(\overline{Q})} \le m_0$ there exist at least two solutions of (174)–(175).*

In their proof of Theorem 57 the authors use an approximating process by smooth nonlinear functions for g which allows for applying semigroup theory, and adopt rather involved topological arguments based on the variational structure of the functionals corresponding to the stationary part of the approximating equations.

Unlike in the paper [**17**], our proof of multiple periodic solutions is completely different and is based on the sub-supersolution method developed in Section 4 which considerably simplifies the proof of multiple periodic solutions. Even more, we not only provide a much simpler treatment, but at the same time we are able to weaken the hypothesis on h. In our proof only the smallness of h with respect to $C(\overline{Q})$-norm is needed. The approach given here is based on the recent paper [**5**], where even more general problems are considered.

By assumption (g), the nonlinearity $g : \mathbb{R} \to \mathbb{R}$ is continuous and, due to Young's inequality, satisfies for any $\varepsilon > 0$ the growth condition (note $0 < \alpha < 1$)

$$|g(s)| = |s|^\alpha \le C(\varepsilon) + \varepsilon |s|, \quad \forall\, s \in \mathbb{R}. \tag{176}$$

The negative Laplacian $A = -\Delta$ satisfies all assumptions (A1)–(A4) with $p = 2$, where $a_i(x,t,s,\xi) = \xi_i$ (note, hypothesis (A4) is trivially fulfilled). Defining j by

$$j(s) = -\int_0^s g(\zeta)\, d\zeta, \tag{177}$$

then problem (174)–(175) is easily seen as a special case of the general periodic inclusion problem (110)–(111) for which the sub-supersolution method was developed in Section 4. Here, the given continuous function $h \in C(\overline{Q})$ is considered as an element of X_0^*. Due to (177) we have $\partial j(s) = \{j'(s)\} = -g(s)$, and (176) shows that even the more restrictive hypothesis (G) (see Subsection 4.1) is fulfilled. Thus we may apply Theorem 50 as well as Theorem 55 of the previous section to the periodic problem (174)–(175). Therefore, the proof of the existence of two solutions of (174)–(175) is accomplished provided we are able to construct ordered intervals of sub- and supersolutions which do not overlap. We first construct an ordered pair of nonnegative sub- and supersolutions.

Let λ_1 be the first Dirichlet eigenvalue of $-\Delta$ in $W_0^{1,2}(\Omega)$, and denote by φ_1 the corresponding eigenfunction satisfying $\varphi_1 > 0$ in Ω and normalized with respect to the $L^2(\Omega)$-norm. Fix $\lambda \in (0, \lambda_1)$. Then there is a constant $c_\lambda > 0$ such that

$$s^\alpha \le \lambda\, s + c_\lambda, \quad \forall\, s \ge 0. \tag{178}$$

Let $v \in W_0^{1,2}(\Omega)$ be the unique solution of the following homogeneous Dirichlet problem:

$$-\Delta v = \lambda v + c_\lambda + \|h\|_{C(\overline{Q})} \text{ in } \Omega, \quad v = 0 \text{ on } \partial\Omega. \tag{179}$$

Applying regularity results and the strong maximum principle to (179) (see, e.g., [**16**]) we find $v \in \mathrm{int}(C_0^1(\overline{\Omega})_+)$, where $\mathrm{int}(C_0^1(\overline{\Omega})_+)$ means the interior of the positive cone of $C_0^1(\overline{\Omega})$. One readily verifies that $\overline{u}(x,t) = v(x)$ is a positive supersolution. As v is time-independent and satisfies homogeneous Dirichlet data, we only need to check inequality (118) of the definition of a supersolution (note, $\overline{\eta} = -g(\overline{u})$). The latter, however, is an immediate consequence of (178) and (179). Furthermore, apparently $\underline{u}(x,t) \equiv 0$ is a subsolution. A positive subsolution can also be obtained in the form $\underline{u}(x,t) = \varepsilon\varphi_1(x)$ for $\varepsilon > 0$ small enough. To see this we note that if $\varepsilon > 0$ is sufficiently small we have $(\varepsilon\varphi_1)^{\alpha-1} > \lambda_1$ on Ω, which implies that $\varepsilon\varphi_1$ is a subsolution of problem (174)–(175). Indeed, since $h \ge 0$, it is seen that

$$-\Delta(\varepsilon\varphi_1) = \varepsilon\lambda_1\varphi_1 < (\varepsilon\varphi_1)^\alpha + h.$$

Moreover, since both $\overline{u}$ and φ_1 belong to int $(C_0^1(\overline{\Omega})_+)$, we can choose $\varepsilon > 0$ sufficiently small to have $\varepsilon\varphi_1 \le \overline{u}$ on Ω. Applying Theorem 50 we have thus obtained the following result.

LEMMA 58. *Under hypotheses (g) and (h) the periodic-parabolic problem (174)–(175) has at least one positive solution within the ordered interval $[\varepsilon\varphi_1, v]$ for $\varepsilon > 0$ small enough.*

For the construction of an ordered pair of negative sub- and supersolutions we make use of the so called anti-maximum principle, see [**13, 14**]. Let us give a short account on the anti-maximum principle. Consider the elliptic boundary value problem

$$-\Delta u = \lambda u + f \text{ in } \Omega, \quad u = 0 \text{ on } \partial\Omega, \tag{180}$$

where Ω is a bounded smooth domain in $\mathbb{R}^N$ as assumed throughout this section. Then, due to the maximum principle, it is well known that for $\lambda < \lambda_1$ (note: λ_1 is the first eigenvalue of the Dirichlet negative Laplacian) the following sign-preserving property holds: If $f \ge 0$ then $u \ge 0$. In [**13**] a sign-reversing property, called the anti-maximum principle, has been proved which reads as follows:

THEOREM 59 (Anti-Maximum Principle). *Let $f \in L^p(\Omega)$ with $0 \ne f \ge 0$, and $p > N$. Then there exists a $\lambda_f > \lambda_1$ with λ_f depending on f such that for $\lambda \in (\lambda_1, \lambda_f)$ the solution of (180) satisfies $u < 0$ in Ω.*

For the construction of a negative supersolution the following auxiliary elliptic boundary value problem plays a crucial role:

$$-\Delta e = \lambda e + 1 \text{ in } \Omega, \quad e = 0 \text{ on } \partial\Omega. \tag{181}$$

By Theorem 59 there is a $\hat{\lambda} > \lambda_1$ such that for $\lambda \in (\lambda_1, \hat{\lambda})$, the solution e of (181) satisfies $e < 0$ in Ω. As λ is not an eigenvalue e is uniquely defined due to the Fredholm alternative, see [**16**, Chap. 8]). For some positive constant β we are looking for a supersolution of the problem (174)–(175) in the form: $\overline{u}(x,t) = \beta\, e(x)$. This is readily seen as follows:

$$-\Delta\overline{u} = \lambda\overline{u} + \beta = \lambda\overline{u} - |\overline{u}|^{\alpha-1}\overline{u} + \beta + |\overline{u}|^{\alpha-1}\overline{u}. \tag{182}$$

As $\overline{u} < 0$ we get $|\overline{u}|^{\alpha-1}\overline{u} = -(-\overline{u})^{\alpha}$ which yields for $z = -\overline{u} \geq 0$

$$\lambda\overline{u} - |\overline{u}|^{\alpha-1}\overline{u} = z^{\alpha} - \lambda z \geq 0 \tag{183}$$

provided z is small enough, which can always be achieved by choosing $\beta > 0$ sufficiently small. Therefore, from (182)–(183) we get

$$-\Delta\overline{u} \geq \beta + |\overline{u}|^{\alpha-1}\overline{u} \geq |\overline{u}|^{\alpha-1}\overline{u} + h \tag{184}$$

provided that $\|h\|_{C(\overline{Q})} \leq \beta$.

To construct a negative subsolution we make use of the following elliptic Dirichlet problem:

$$-\Delta w = -1 \ \text{ in } \Omega, \quad w = 0 \ \text{ on } \partial\Omega, \tag{185}$$

which has a unique negative smooth solution $w \in -\mathrm{int}(C_0^1(\overline{\Omega})_+)$. We are looking for a negative subsolution of the form: $\underline{u}(x,t) = cw(x)$ for some $c > 0$, which leads to

$$-\Delta\underline{u} \leq -c - |\underline{u}|^{\alpha-1}\underline{u} + |\underline{u}|^{\alpha-1}\underline{u} + h,$$

and thus $\underline{u} = cw$ is a subsolution provided that the constant $c > 0$ can be selected in such a way that

$$-c - |\underline{u}|^{\alpha-1}\underline{u} \leq 0,$$

which is equivalent to

$$-c + c^{\alpha}(-w)^{\alpha} \leq 0. \tag{186}$$

Inequality (186) is certainly fulfilled provided that

$$c \geq \|w\|_{C(\overline{\Omega})}^{\frac{\alpha}{1-\alpha}}. \tag{187}$$

Hence, for any constant $c > 0$ satisfying (187), the function $\underline{u} = cw$ is a negative subsolution. Furthermore, selecting c large enough, one can always achieve that

$$\underline{u} = cw \leq \beta e = \overline{u},$$

which shows that the sub- and supersolution constructed above form an ordered pair. Applying Theorem 50 we have thus obtained the following result.

LEMMA 60. *Under hypotheses (g) and (h) the periodic-parabolic problem (174)–(175) has at least one negative solution within the ordered interval* $[cw, \beta e]$ *provided that* $\|h\|_{C(\overline{Q})} \leq \beta$ *with* $\beta > 0$ *small enough, where e and w are the solutions of (181) and (185), respectively, and c satisfying (187) is sufficiently large.*

Combining Lemma 58 and Lemma 60 we get the following multiplicity result.

THEOREM 61. *Let hypotheses (g) and (h) be satisfied. Then the periodic-parabolic problem* (174))–(175) *has at least one positive and one negative periodic solution provided* $\|h\|_{C(\overline{Q})}$ *is small enough.*

REMARK 62. Our Theorem 61 strongly improves the result obtained by Hirano and Kim (see Theorem 57), in that only a smallness of h with respect to the $C(\overline{Q})$-norm is required. At the same time the proof of our result is much simpler, because it is only based on the sub-supersolution method for nonlinear periodic-parabolic problems provided in the previous section. It should be noted that in the above construction of ordered pairs of sub-supersolutions, the anti-maximum principle plays a crucial role.

Bibliography

[1] S. Carl, *Existence of extremal periodic solutions for quasilinear parabolic equations*, Abstr. Appl. Anal. **2** (1997), 257–270.

[2] S. Carl, *Existence and comparison results for noncoercive and nonmonotone multivalued elliptic problems*, Nonlinear Anal. **65** (8)(2006), 1532–1546.

[3] S. Carl, *The sup-supersolution method for variational-hemivariational inequalities*, Nonlinear Anal. **69** (2008), 816–822.

[4] S. Carl, *Equivalence of some multi-valued elliptic variational inequalities and hemivariational inequalities*, Adv. Nonlinear Stud. (in press).

[5] S. Carl, *Multiple solutions of quasilinear periodic-parabolic inclusions*, Nonlinear Anal. **72** (2010), 2909–2922.

[6] S. Carl and D. Motreanu, *Extremal solutions of quasilinear parabolic inclusions with generalized Clarke's gradient*, J. Differential Equations, **191** (2003), 206–233.

[7] S. Carl, V.K. Le, and D. Motreanu, *Nonsmooth Variational Problems and Their Inequalities*, Springer Monographs in Mathematics, Springer, New York, (2007).

[8] S. Carl and D. Motreanu, *Directedness of solution set for some quasilinear multi-valued parabolic problems*, Appl. Anal. **89** (2010), 161–174.

[9] S. Carl and D. Motreanu, *Comparison principle for Quasilinear parabolic inclusions with Clarke's gradient*, Adv. Nonlinear Stud. **9** (2009), 69–80.

[10] S. Carl and D. Motreanu, *General comparison principle for quasilinear elliptic inclusions*, Nonlinear Anal. **70** (2009), 1105–1112.

[11] S. Carl and P. Winkert, *General comparison principle for variational-hemivariational inequalities*, J. Inequal. Appl. **2009** (2009), Article ID 184348, 29 pages.

[12] F.H. Clarke, *Optimization and Nonsmooth Analysis*, Society for Industrial and Applied Mathematics (SIAM) Philadelphia, PA (1990).

[13] Ph. Clément and L. A. Peletier, *An anti-maximum principle for second-order elliptic opera tors*, J. Differential Equations **34** (1979), 218–229.

[14] Ph. Clément and G. Sweers, *Uniform anti-maximum principle*, J. Differential Equations **164** (2000), 118–154.

[15] Z. Denkowski, S. Migórski, and N. S. Papageorgiou, *An Introduction to Nonlinear Analysis: Theory*, Kluwer Academic Publishers Boston, Dordrecht, London 2003.

[16] D. Gilbarg and N. S. Trudinger, *Elliptic Partial Differential Equations of Second Order*, Springer-Verlag, Berlin, 1983.

[17] N. Hirano and W. S. Kim, *Multiple existence of periodic solutions for a nonlinear parabolic problem with singular nonlinearities*, Nonlinear Anal. **54** (2003), 445–456.

[18] Z. Naniewicz and P. D. Panagiotopoulos, *Mathematical Theory of Hemivariational Inequalities and Applications*, Marcel Dekker, New York, 1995.

[19] P. D. Panagiotopoulos, *Hemivariational Inequalities and Applications in Mechanics and Engineering*, Springer-Verlag, New York, 1993.

[20] E. Zeidler, *Nonlinear Functional Analysis and Its Applications*, Vols. II A/B, Springer-Verlag, Berlin, 1990.

Siegfried Carl, Institute of Mathematics, University of Halle, D-06099 Halle, Germany

E-mail address: `siegfried.carl@mathematik.uni-halle.de`

Dumitru Motreanu, Département de Mathématiques, Université de Perpignan, 52 Avenue de Villeneuve, 66860 Perpignan, France

E-mail address: `motreanu@univ-perp.fr`

Prox-regular sets and applications

Giovanni Colombo and Lionel Thibault

Abstract. Prox-regular sets (or sets with positive reach according to Federer [**63**]), are closed subsets of a Hilbert space with the property of existence of a supporting ball of locally constant radius in all possible normal directions. This assumption entails several properties which have natural analogues for convex sets. In particular, the smoothness of the distance function and the uniqueness and Lipschitz continuity of the metric projection hold *locally* for such sets, while hold *globally* for convex sets. This chapter is devoted first to a systematic study of those properties (Sections 3, 4, and 5), several of them being actually equivalent to prox-regularity, both in finite and in infinite dimensions. Among them we also mention the (hypo)monotonicity of the metric projection as well as, in $\mathbb{R}^n$, the a.e. second order Taylor expansion of lower semicontinuous functions with prox-regular epigraph, i.e., a counterpart of Alexandrov's theorem for convex functions. Second (Section 6), we study in detail the sweeping process, which is a time-dependent evolution inclusion modeling the displacement of a point constrained to belong to a moving set $C(t)$, with velocity in the opposite of the normal cone to $C(t)$. The results are proved here for prox-regular moving sets, and generalize many known properties enjoyed by the sweeping process with a convex moving set, which was first introduced by Moreau in the Seventies. Finally, we describe several further applications of prox-regularity to optimal control, to isoperimetric inequalities, to Hamilton-Jacobi equations, and to set-valued analysis. In particular, continuous functions with prox-regular epigraph or hypograph appear to be a good regularity class for minimum time functions with a target general enough.

Contents

2000 *Mathematics Subject Classification.* Primary 49-02, 49J52, 49J53, 58C05, 34G25.
Key words and phrases. Sets with positive reach, φ-convex sets, proximally smooth sets, proximate retracts, $O(2)$-convexity, weakly convex sets, hypomonotonicity, distance function, metric projection, differentiability of non-Lipschitz functions, sweeping process, minimum time function.

1. Introduction

In his famous 1959 paper ("Curvature measures", see [**63**]), Federer observed that there were two classical theories which allowed to compute the volume of an r-neighborhood of a closed subset of $\mathbb{R}^n$ as an n-degree polynomial in the positive r-variable, with coefficients having a clear geometrical meaning. One theory was devoted to *convex sets*, while the other to *possibly nonconvex, but sufficiently smooth sets* (actually with $\mathcal{C}^2$- boundary). Then he stated that "the search for a general theory," which must be based neither on convexity nor on differentiability assumptions, "is an obvious challenge." In the above cases of sets, the equality defining the polynomial is known as the Steiner polynomial formula.

This challenge led to Federer's fundamental paper [**63**] where he introduced the concept of r_0-positively reached (closed) set C in $\mathbb{R}^n$ as a set for which any point $x \in \mathbb{R}^n$ with $d(x,C) < r_0$ has a unique nearest point $P_C(x)$ in C to x, and obtained the desired result of extending the Steiner formula to such sets.

Other different motivations and problems of several independent researches induced some authors to introduce many classes of appropriate sets under distinct names: *prox-regular* ([**106**]) or *φ-convex* ([**29**]) or *proximally smooth* ([**43**]) sets, $O(2)$-sets ([**110**]) etc. All these concepts are actually known to coincide.

The above class of sets is the topic of this chapter and the aim here is giving an overview of some recent developments and applications of this concept, both in finite and in infinite dimensional Hilbert spaces, mainly from the viewpoint of nonsmooth analysis. We present also several applications to various fields, including the sweeping process, the regularity of the minimum time function, viscosity solutions of some Hamilton-Jacobi partial differential equations, isoperimetric inequalities and set-valued analysis. Overall, we believe that this concept is both a natural and a very effective generalization of convexity.

Actually, prox-regularity is defined (see Definition 1 below) through a suitable *external sphere condition with locally uniform radius*, hence it is a second order condition. More precisely, prox-regularity is defined by postulating the existence of a supporting ball of locally constant radius in all possible normal directions. The concept of supporting hyperplane for a *convex set* being interpreted as the same external sphere condition, but with arbitrarily large radius, one may conjecture that several features of convexity have a good counterpart for prox-regularity. In fact, a large part of the rich properties enjoyed by prox-regular sets and by lower semicontinuous functions with prox-regular epigraph comes out from this remark. For example, the smoothness of the distance function and the uniqueness and Lipschitz continuity of the metric projection hold *locally* for such sets, while hold *globally* for convex sets.

Our environment will be mostly a possibly infinite dimensional Hilbert space. However, one section will be devoted to an analysis in $\mathbb{R}^n$. In this case it is possible to use basic techniques of geometric measure theory, and several other properties can be proved.

The content of the chapter is as follows: after setting up basic definitions of nonsmooth analysis, we devote Section 3 to the introduction of $\rho(\cdot)$-prox-regular sets, which are the subject of our analysis. Next, results containing several conditions equivalent to $\rho(\cdot)$-prox-regularity are proved in full detail. Those conditions involve in turn existence, uniqueness and Lipschitz continuity of the metric projection, or smoothness of the distance function, or hypomonotonicity of the normal cone, or finally a uniform second order contact between the set and its tangent cone. Some proofs and some results in this section are new. Let us remark that the metric projection onto a closed set, in infinite dimensional spaces, is in general empty, and in any case may not be a singleton. For $\rho(\cdot)$-prox-regular sets, instead, it is always a singleton, when restricted to a suitably small neighborhood of the set, whose size is determined only by $\rho(\cdot)$. The nature of "quadratic perturbation of convexity" which characterizes $\rho(\cdot)$-prox-regularity becomes particularly transparent in the proof of existence and uniqueness of the metric projection (see the proof of Theorem 3). In fact it is shown that every minimizing sequence for the distance from a $\rho(\cdot)$-prox-regular set is Cauchy, with a proof which is very close to one of the classical proofs of the existence and uniqueness of the metric projection onto a closed convex subset of a Hilbert space. Section 4 is devoted to operations on prox-regular sets, including intersections, direct and inverse image, and Hausdorff convergence. Next (Section 5), we switch to finite dimensional spaces, dealing with finer regularity properties of $\rho(\cdot)$-prox-regular sets, which involve measure theoretic estimates on the nonsmooth subset of the boundary. In the same section we study regularity properties of lower semicontinuous functions with $\rho(\cdot)$-prox-regular epigraph. It turns out that those functions, although not necessarily Lipschitz continuous, enjoy several regularity properties which are typical of (finite dimensional) convexity, such as an a.e. second order Taylor expansion. Moreover a representation formula of the proximal subgradient involving limits of ordinary gradients can be given, in the spirit of the classical gradient formula for the Clarke generalized gradient of Lipschitz continuous functions. Section 6 is devoted to the sweeping, or Moreau, process, when the moving set is $\rho(\cdot)$-prox-regular. This is an evolution differential inclusion, which was introduced and thoroughly studied by Moreau for convex moving sets, which describes the displacement of a point which is dragged by the boundary of a moving set $C(t)$, with the opposite of its velocity constrained to be normal to $C(t)$. Existence and uniqueness of a solution are shown. Again, the results are very similar to the classical convex ones. The introduction of the section contains a more detailed informal discussion of the results. Section 7 is devoted to further applications of $\rho(\cdot)$-prox-regularity, the main one being the regularity of the minimum time function under controllability conditions which do not imply its Lipschitz continuity. We state a result giving sufficient conditions for the minimum time to have $\rho(\cdot)$-prox-regular epi/hypograph, according to the shape of the target and to some conditions on the dynamics, thus providing new information on the structure and the regularity of such a function. The proof involves control theoretic arguments and is not presented here. Next we present more briefly other applications, including a result on viscosity solutions for a class of Hamilton–Jacobi partial differential equations, together with results on continuous or Lipschitz continuous selections from a $\rho(\cdot)$-prox-regular valued multifunction, and isoperimetric estimates. Section 8 is devoted to miscellaneous results, including relations among prox-regular sets and semiconvex functions, together with examples and counterexamples. The last section contains some historical remarks and comparisons with related concepts.

2. Preliminaries

Throughout the paper H is a *real Hilbert space* endowed with the inner product $\langle \cdot, \cdot \rangle$ and $\| \cdot \|$ is the associated norm. For any $x \in H$ and any real number $\delta \geq 0$ the closed (resp. open) ball centered at x with radius δ will be denoted by $B_H[x, \delta]$ (resp. $B_H(x, \delta)$). For $x = 0$ and $\delta = 1$ we will put $\mathbb{B}_H$ or $\mathbb{B}$ in place of $B_H[0, 1]$. For $\delta = +\infty$, we adopt the convention $\delta \mathbb{B}_H = B(x, \delta) = B[x, \delta] = H$ for all $x \in H$.

We give first some background material on variational analysis and preliminary results which will be used later. We state only the definitions and results which will be needed in our development. For more details on the subject, we refer to [**22, 42, 95, 109**]. Let $f : H \to \mathbb{R} \cup \{+\infty\}$ be a lower semicontinuous (lsc) function and $x \in H$ be a point where it is finite. A vector $v \in H$ is a *Fréchet subgradient* of f at x (see, e.g., [**95, 109**]) if for each number $\varepsilon > 0$ there exists some neighborhood U of x such that

$$\langle v, x' - x \rangle \leq f(x') - f(x) + \varepsilon \|x' - x\| \quad \text{for all } x' \in U. \tag{1}$$

The set $\partial^F f(x)$ of all Fréchet subgradients of f at x is the *Fréchet subdifferential* of f at x. The Fréchet subdifferential enjoys only (see, e.g.,[**95**]) fuzzy calculus rules, i.e., for any function $g : H \to \mathbb{R} \cup \{+\infty\}$ which is Lipschitz continuous near x, for $v \in \partial^F (f + g)(x)$ and $\varepsilon > 0$, there exist $x', x'' \in H$ such that

$$v \in \partial^F f(x') + \partial^F g(x'') + \varepsilon \mathbb{B}_H \tag{2}$$

and $\|x' - x\| + |f(x') - f(x)| < \varepsilon$. So, a limiting process is needed to obtain pointwise calculus rules. A vector $v \in H$ is a *limiting (Fréchet) subgradient* or (basic subgradient) of f at x if there exists a sequence $((x_n, f(x_n)))_n$ converging to $(x, f(x))$ and a sequence $(v_n)_n$ converging weakly to v such that $v_n \in \partial^F f(x_n)$. The set $\partial^L f(x)$ of all limiting subgradients of f at x is the Mordukhovich *limiting (basic) subdifferential* [1] of f at x. As usual, if $f(x)$ is not finite, one puts $\partial^F f(x) = \emptyset$ and $\partial^L f(x) = \emptyset$. If a function g is Lipschitz continuous near x, then by Theorem 3.36 in [**95**] one has

$$\partial^L (f + g)(x) \subset \partial^L f(x) + \partial^L g(x) \tag{3}$$

and by Corollary 3.43 in [**95**]

$$\partial^L (g \circ G)(u) \subset DG(u)^* (\partial^L g(x)) \tag{4}$$

for any mapping G from a Hilbert space E into H which is continuously differentiable near u with $G(u) = x$. In the second inclusion $DG(u)^*$ denotes the adjoint of $DG(u)$ and

$$DG(u)^* (\partial^L g(x)) := \{DG(u)^*(v) : v \in \partial^L g(x)\}.$$

[1]Here the Riesz isomorphism is used to define the Fréchet and Mordukhovich subdifferentials as subsets of H instead of the topological dual H^*.

When f is the indicator function ψ_C of a closed subset C of H (defined by $\psi_C(x') = 0$ if $x' \in C$ and $\psi_C(x') = \infty$ otherwise), $\partial^F \psi_C(x)$ and $\partial^L \psi_C(x)$ are the *Fréchet normal cone* $N^F(C;x)$ and the Mordukhovich *limiting (basic) normal cone* $N^L(C;x)$ of C at x. This means that a vector $v \in H$ is a Fréchet normal vector to C at $x \in C$ if and only if for every number $\varepsilon > 0$ there is a neighborhood U of x such that

$$\langle v, x' - x\rangle \leq \varepsilon \|x' - x\| \quad \text{for all } x' \in C \cap U, \tag{5}$$

and

$$N^L(C;x) = {}^{\text{seq}}\operatorname*{Lim\,sup}_{C \ni y \to x} N^F(C;y). \tag{6}$$

In the latter equality, the second member denotes the $\| \; \| \times w(H,H)$ sequential Painlevé-Kuratowski outer or upper limit, that is,

$$^{\text{seq}}\operatorname*{Lim\,sup}_{C \ni y \to x} N^F(C;y) := \left\{ \begin{array}{c} v \in H : \exists \text{ sequences } C \ni y_n \to x, v_n \xrightarrow{w} v, \\ \text{with } v_n \in N(C;y_n) \end{array} \right\}.$$

Let $d(\cdot, C)$ or $d_C(\cdot)$ be the usual distance function associated with C, i.e., $d(x,C) := \inf_{y \in C} \|x - y\|$. For any $x \in C$ the Fréchet normal cone (see, e.g., [**95**]) is related to the distance function through the equalities

$$N^F(C;x) \cap \mathbb{B}_H = \partial^F d(\cdot, C)(x) \quad \text{and} \quad N^F(C;x) = \mathbb{R}_+ \partial^F d(\cdot;C)(x), \tag{7}$$

and the Mordukhovich limiting normal cone as well (see [**112**]) through the inclusion and equality

$$\partial^L d(\cdot, C)(x) \subset N^L(C;x) \cap \mathbb{B}_H \quad \text{and} \quad N^L(C;x) = \mathbb{R}_+ \partial^L d(\cdot, C)(x), \tag{8}$$

where $\mathbb{R}_+$ denotes the set of non negative real numbers.

The proximal subdifferential of f can be defined in a similar way. A vector $v \in H$ is a *proximal subgradient* of f at x (see, e.g., [**42, 109**]) if there exist some real number $\sigma \geq 0$ and some neighborhood U of x such that

$$\langle v, x' - x\rangle \leq f(x') - f(x) + \sigma \|x' - x\|^2 \quad \text{for all } x' \in U. \tag{9}$$

The set $\partial^P f(x)$ of all proximal subgradients of f at x is the *proximal subdifferential* of f at x and one puts $\partial_P f(x) = \emptyset$ when f is not finite at x. If $\partial^P f(x) \neq \emptyset$ (resp. $\partial^F f(x) \neq \emptyset$), one says that f is ∂^P*-subdifferentiable* (resp. ∂^F*-subdifferentiable*) at the point x. The proximal subdifferential enjoys on H, like the Fréchet subdifferential, the same fuzzy rule (2). It also characterizes (see [**42, 115**]) the Lipschitz behavior of the lsc function $f : H \to \mathbb{R} \cup \{+\infty\}$, in the sense that f is Lipschitz continuous on an open convex set U of H with γ as Lipschitz constant if and only if

$$\partial^P f(x) \subset \gamma \mathbb{B}_H \quad \text{for all } x \in U. \tag{10}$$

Like above, a vector $v \in H$ is a *proximal normal vector* to the set C at $x \in C$ when it is a proximal subgradient of the indicator function of C, that is, when there exist a constant $\sigma \geq 0$ and a neighborhood U of x such that

$$\langle v, x' - x\rangle \leq \sigma \|x' - x\|^2 \quad \text{for all } x' \in U \cap C. \tag{11}$$

The set of such vectors is the *proximal normal cone* $N^P(C;x)$ of C at x. Obviously

$$N^P(C;x) \subset N^F(C;x) \subset N^L(C;x).$$

Nevertheless, the limiting process leads, in the Hilbert setting, to equalities (see, e.g., [**74, 95**]), that is,

$$N^L(C;x) = {}^{\mathrm{seq}}\operatorname*{Lim\,sup}_{C\ni y\to x} N^F(C;y) = {}^{\mathrm{seq}}\operatorname*{Lim\,sup}_{C\ni y\to x} N^P(C;y). \tag{12}$$

The proximal normal cone enjoys a geometrical characterization (see, e.g., [**42**]) given by the equality

$$N^P(C;x) = \{v \in H : \exists t > 0 \text{ s.t. } x \in \operatorname{Proj}_C(x+tv)\}, \tag{13}$$

where

$$\operatorname{Proj}_C(u) := \{y \in C : d(u,C) := \|u-y\|\}$$

is the set of nearest points of C to u. For $x \in C$, the proximal normal cone is also related to the distance function to C via the equalities (see, e.g., [**25, 42**])

$$N^P(C;x) \cap \mathbb{B}_H = \partial^P d(\cdot,C)(x) \quad \text{and} \quad N^P(C;x) = \mathbb{R}_+\partial^P d(\cdot,C)(x). \tag{14}$$

We observe that (13) means that a proximal normal v with $\|v\| = 1$ is realized by a t-ball, in the sense that the intersection $C \cap B_H(x+rtv, r) = \emptyset$. Our analysis will be focused on sets such that all proximal normals are realized by a locally constant ball, which will be called prox-regular sets (see Definition 1).

The remarkable properties of prox-regular sets will be established in Section 3 and their relevance to Optimization, Control theory, Nonlinear Analysis, Differential inclusions etc. will be studied in the next sections.

In our development of properties of prox-regular sets we will also need the concept of Clarke normal cone (see [**41**]). One of the best ways to define it is as the polar of a tangent cone. Recall that a vector $h \in H$ is in the Clarke tangent cone $T^{Cl}(C;x)$ to the set C at $x \in C$ when for every sequence $(x_n)_n$ in C converging to x and every sequence of positive numbers $(t_n)_n$ converging to 0, there exists some sequence $(h_n)_n$ in H converging to h such that $x_n + t_n h_n \in C$ for all integers n. This cone is closed and convex and its negative polar $(T^{Cl}(C;x))^0$ is the Clarke normal cone $N^{Cl}(C;x)$ of C at $x \in C$, that is,

$$N^{Cl}(C;x) = \{v \in H : \langle v,h\rangle \le 0 \ \ \forall h \in T^{Cl}(C;x)\}.$$

As usual, one puts $N^{Cl}(C;x) = \emptyset$ if $x \notin C$. In the Hilbert space H, the equality (see, e.g., [**95**])

$$N^{Cl}(C;x) = \overline{\mathrm{co}}\,(N^L(C;x))$$

(where $\overline{\mathrm{co}}$ denotes the closed convex hull) provides the relationship between the Clarke normal cone and the Mordukhovich limiting normal cone. Through that normal cone, the Clarke subdifferential of the function $f : H \to \mathbb{R} \cup \{+\infty\}$ is defined by

$$\partial^{Cl} f(x) := \{v \in H : (v,-1) \in N^{Cl}(\operatorname{epi} f, (x, f(x)))\},$$

where $\operatorname{epi} f := \{(y,r) \in H \times \mathbb{R} : f(y) \leq r\}$ is the epigraph of f. When the function f is finite and locally Lipschitz continuous around x, the Clarke subdifferential is characterized (see [**41**]) in the following simple and amenable way

$$\partial^{Cl} f(x) = \{v \in H : \langle v, h\rangle \leq f^0(x;h) \ \forall h \in H\}, \tag{15}$$

where

$$f^0(x;h) := \limsup_{(t,x') \to (0^+,x)} t^{-1}[f(x'+th) - f(x')] \tag{16}$$

is the generalized directional derivative of the locally Lipschitz function f at x in the direction h. The function $f^0(x;\cdot)$ is in fact the support function of $\partial^{Cl} f(x)$. That characterization easily yields that the Clarke subdifferential of any locally Lipschitz function enjoys the important property of upper semicontinuity. The equality (see [**41**])

$$N^{Cl}(C;x) = \operatorname{cl}(\mathbb{R}_+ \partial^{Cl} d(\cdot,C)(x)) \quad \text{for } x \in C, \tag{17}$$

giving the expression of the Clarke normal cone in terms of the distance function, will allow us to take advantage, in the next section, of the upper semicontinuity of the Clarke subdifferential of the particular Lipschitz function $d(\cdot,C)$. As usual, it will be convenient to write below $\partial^{Cl} d(x,C)$ in place of $\partial^{Cl} d(\cdot,C)(x)$. When f is locally Lipschitz continuous one also has (see,e.g., [**95**])

$$\partial^{Cl} f(x) = \overline{\text{co}}\,(\partial^L f(x)) \quad \text{for all } x \in H \tag{18}$$

and hence in particular for the distance function

$$\partial^{Cl} d(x,C) = \overline{\text{co}}\,(\partial^L d(x,C)) \quad \text{for all } x \in H. \tag{19}$$

Whenever the extended real-valued function f is convex, the Clarke subdifferential of f at x as well as the proximal, Fréchet and Mordukhovich subdifferentials coincide with the Fenchel subdifferential of convex analysis, i.e., with the set

$$\{v \in H : \langle v, x' - x\rangle + f(x) \leq f(x') \, \forall x' \in H\}.$$

In addition to the Clarke normal, the Bouligand tangent cone will be also sometimes considered in the paper. A vector $h \in H$ is in the Bouligand tangent (or contingent) cone $T^B(C;x)$ to C at $x \in C$ if there are a sequence of positive numbers $(t_n)_n$ converging to 0 and a sequence $(h_n)_n$ converging to h such that $x + t_n h_n \in C$ for all $n \in \mathbb{N}$. Obviously, one always has $T^{Cl}(C;x) \subset T^B(C;x)$. Like for normal cones, one puts $T^B(C;x) = T^{Cl}(C;x) = \emptyset$ for $x \notin C$.

All the above concepts of subdifferential, normal cone, and tangent cone define set-valued mappings from H into H. Recall that for a set-valued mapping $M : X \rightrightarrows Y$ between two nonempty sets X and Y, its effective domain $\operatorname{Dom} M$ is defined as $\operatorname{Dom} M := \{x \in X : M(x) \neq \emptyset\}$. Also, the effective domain of an extended real-valued function $f : X \to \mathbb{R} \cup \{+\infty\}$ is $\operatorname{dom} f : \{x \in X : f(x) < +\infty\}$.

We also recall that he *excess* of a set A over a set B in H is defined as

$$e(A,B) = \sup_{a \in A} d_B(a)$$

while $H(A,B) := \max\{e(A,B), e(B,A)\}$ is the Hausdorff distance between the sets A and B. We say that a sequence of sets $C_k \subset H$, $k \in \mathbb{N}$, Hausdorff converges to $C \subset H$ if $H(C_k, C) \to 0$.
A cone N is said pointed if $N \cap (-N) = \{0\}$. If H is finite dimensional, we say (see [**108**, p.163]) that a ray $\mathbb{R}^+\bar{\zeta}$, $\bar{\zeta} \neq 0$, of a convex cone N is *exposed* if there exists a linear functional $h = \langle \theta, \cdot \rangle$ which is zero on it and

$$\langle \theta, p \rangle = 0 \text{ and } p \in N \text{ imply that } p \in \mathbb{R}^+\bar{\zeta}. \tag{20}$$

Finally, throughout the paper, the conventions $1/0 = +\infty$, $1/\infty = 0$ will be adopted.

3. Prox-regularity

In this section we study prox-regular sets with varying thickness $\rho(\cdot)$ and with constant thickness r. The local property around a fixed point of the set is also discussed.

3.1. $\rho(\cdot)$-prox-regularity.

Quantifying (13) in requiring to the positive number t (for the unit proximal normal vectors to C) to depend continuously on the point $x \in C$ leads to the concept of $\rho(\cdot)$-prox-regularity.

DEFINITION 1. Let $C \subset H$ be closed and let $\rho : C \to]0, +\infty]$ be a continuous function with positive extended-real values. The set C is $\rho(\cdot)$-prox-regular provided that, for any $x \in C$ and for any positive real number $t \leq \rho(x)$, every proximal normal vector v to C at x with $\|v\| \leq 1$ can be realized by a t-ball, which can be translated into the fact that for all $x \in C$ and all $v \in N^P(C;x)$ with $\|v\| \leq 1$ one has

$$x \in \mathrm{Proj}_C(x + tv),$$

for any positive real number $t \leq \rho(x)$. By saying simply that a set C is prox-regular, we mean that there exists a continuous $\rho(\cdot)$ such that C is $\rho(\cdot)$-prox-regular.

As above $\mathrm{Proj}_C(u)$ denotes the (possibly empty) set of nearest points of C to u. Sometimes it will be convenient to write $\mathrm{Proj}(C;u)$ instead of $\mathrm{Proj}_C(u)$. When the set $\mathrm{Proj}_C(u)$ is a singleton, its unique element will be denoted by $P_C(u)$ or $P(C;u)$. So, the notation $P_C(u)$ or $P(C;u)$ means that $\mathrm{Proj}_C(u)$ is a nonempty set with a unique element and $\mathrm{Proj}_C(u) = \{P_C(u)\}$.

Throughout the remaining of the paper, $\rho(\cdot)$ is a *continuous function* from the closed set C into $]0, +\infty]$. We recall the convention $1/r = 0$ for $r = +\infty$.

We start our study of prox-regular sets with Theorem 3. In its statement we use for any positive $\gamma < 1$ the notation

$$O^{\gamma}_{\rho(\cdot)}(C) := \Big\{ u \in H : \exists \gamma' \in]0, \gamma[, \exists \eta > 0 \text{ such that } \forall x \in C \\ \text{with } \|x - u\| < \eta + d_C(u) \text{ one has } \frac{\|x-u\|}{\rho(x)} < \gamma' \Big\}$$

along with

$$O_{\rho(\cdot)}(C) := \Big\{ u \in H : \exists \gamma \in]0,1[, \exists \eta > 0 \text{ such that } \forall x \in C \\ \text{with } \|x-u\| < \eta + d_C(u), \text{ one has } \frac{\|x-u\|}{\rho(x)} < \gamma \Big\},$$

that is, $O_{\rho(\cdot)}(C) = \underset{0<\gamma<1}{\cup} O^{\gamma}_{\rho(\cdot)}(C)$. It is not difficult to see that

$$O^{\gamma}_{\rho(\cdot)}(C) = \left\{ u \in H : \limsup_{\substack{\|x-u\| \to d_C(u) \\ x \in C}} \frac{\|x-u\|}{\rho(x)} < \gamma \right\}, \tag{21}$$

and

$$O_{\rho(\cdot)}(C) = \left\{ u \in H : \limsup_{\substack{\|x-u\| \to d_C(u) \\ x \in C}} \frac{\|x-u\|}{\rho(x)} < 1 \right\}. \tag{22}$$

We will see later that the sets $O^{\gamma}_{\rho(\cdot)}(C)$ and $O_{\rho(\cdot)}(C)$ are open whenever C is $\rho(\cdot)$-prox-regular.

This section will be mainly devoted to establishing several necessary and sufficient conditions for prox-regularity, which will be divided into several steps. Before stating the first theorem on characterizations of prox-regularity, let us establish the following lemma which provides a general result which has its own interest. It concerns the set of nearest points and in its statement I denotes the identity mapping on H.

LEMMA 2. For any point $u \in H$ one has

$$\operatorname{Proj}_C(u) \subset \left(I + d_C(u)\mathbb{B}_H \cap N^P(C;\cdot)\right)^{-1}(u).$$

PROOF. Suppose that $\operatorname{Proj}_C(u)$ is nonempty and fix any $y \in \operatorname{Proj}_C(u)$. Then by definition of proximal normal we have $u - y \in N^P(C;y)$. Since $\|u-y\| = d_C(u)$ the latter inclusion assures us that

$$\frac{1}{d_C(u)}(u-y) \in \mathbb{B}_H \cap N^P(C;y) \quad \text{if } d_C(u) > 0$$

and hence $u \in y + d_C(u)\mathbb{B}_H \cap N^P(C;y)$ even if $d_C(u) = 0$. This means that in any case we have

$$y \in \left(I + d_C(u)\mathbb{B}_H \cap N^P(C;\cdot)\right)^{-1}(u),$$

which proves the desired inclusion. □

We can now establish the following theorem providing some first results concerning $\rho(\cdot)$-prox-regular sets.

THEOREM 3. Let C be a closed subset of H and $\rho(\cdot) : C \to]0, +\infty]$ be as above a continuous function. Consider the following assertions.

(a) The set C is $\rho(\cdot)$-prox-regular.

(b) For any $x, x' \in C$ and $v \in N^P(C; x)$ one has

$$\langle v, x' - x \rangle \leq \frac{1}{2\rho(x)} \|v\| \, \|x' - x\|^2.$$

(c) For any $v \in N^P(C; x) \cap \mathbb{B}_H$ and any nonnegative real number $t < \rho(x)$ one has $x = P_C(x + tv)$ (we recall that $P_C(y)$ denotes the unique point of $\mathrm{Proj}_C(y)$ when this latter is a singleton).

(d) For any $x_i \in C$, $v_i \in N^P(C; x_i)$ with $i = 1, 2$ one has

$$\langle v_1 - v_2, x_1 - x_2 \rangle \geq -\frac{1}{2} \left(\frac{\|v_1\|}{\rho(x_1)} + \frac{\|v_2\|}{\rho(x_2)} \right) \|x_1 - x_2\|^2.$$

(e) For any $u_i \in H$ and for any $y_i \in (I + N^P(C; \cdot))^{-1}(u_i)$ with $i = 1, 2$, one has

$$\langle y_1 - y_2, u_1 - u_2 \rangle \geq \left(1 - \frac{\|u_1 - y_1\|}{2\rho(y_1)} - \frac{\|u_2 - y_2\|}{2\rho(y_2)} \right) \|y_1 - y_2\|^2.$$

(f) For any positive $\gamma < 1$ and any $u \in O^{\gamma}_{\rho(\cdot)}(C)$ the set $\mathrm{Proj}_C(u)$ is a singleton, i.e., $P_C(u)$ exists, the equality

$$P_C(u) = \left(I + \gamma\rho(\cdot)\mathbb{B}_H \cap N^P(C; \cdot) \right)^{-1}(u)$$

holds, and the mapping P_C is Lipschitz continuous on $O^{\gamma}_{\rho(\cdot)}(C)$ with $(1 - \gamma)^{-1}$ as a Lipschitz constant, i.e.,

$$\|P_C(u_1) - P_C(u_2)\| \leq (1 - \gamma)^{-1} \|u_1 - u_2\| \quad \text{for all } u_1, u_2 \in O^{\gamma}_{\rho(\cdot)}(C).$$

(f') For any positive $\gamma < 1$ and any $u_i \in O^{\gamma}_{\rho(\cdot)}(C)$, $i = 1, 2$, $P_C(u_i)$ exist and

$$\begin{aligned} &\|(I - P_C)(u_1) - (I - P_C)(u_2)\|^2 + (1 - 2\gamma)\|P_C(u_1) - P_C(u_2)\|^2 \\ &\quad \leq \|u_1 - u_2\|^2, \end{aligned}$$

where I denotes the identity mapping on H. In particular, for $\gamma \leq 1/2$ the mapping $I - P_C$ is Lipschitz continuous on $O^{\gamma}_{\rho(\cdot)}(C)$ with 1 as Lipschitz constant.

(g) The mapping P_C is well defined on $O_{\rho(\cdot)}(C)$ and for all $u_1, u_2 \in O_{\rho(\cdot)}(C)$ one has

$$\|P_C(u_1) - P_C(u_2)\| \leq \left(1 - \frac{d_C(u_1)}{2\rho(P_C(u_1))} - \frac{d_C(u_2)}{2\rho(P_C(u_2))} \right)^{-1} \|u_1 - u_2\|.$$

Then all the assertions (a) – (d) are pairwise equivalent, each one implies (e), and (e) implies (f), (f′) and (g).

PROOF. $(a) \Leftrightarrow (b)$. Fix any $x \in C$ and $v \in N^P(C;x) \cap \mathbb{B}_H$. For any $x' \in H$ we have for any positive real number $t \leq \rho(x)$

$$\|x + tv - x'\|^2 = t^2\|v\|^2 - 2t\langle v, x' - x\rangle + \|x' - x\|^2.$$

Consequently

$$\|x + tv - x\|^2 \leq \|x + tv - x'\|^2 \quad \text{for all } x' \in C$$

if and only if

$$\langle v, x' - x\rangle \leq \frac{1}{2t}\|x' - x\|^2 \quad \text{for all } x' \in C.$$

This yields that $x \in \mathrm{Proj}_C(x + tv)$ if and only if the latter inequality is fulfilled, and this means that the equivalence between (a) and (b) holds.
$(a) \Leftrightarrow (c)$. This equivalence is easily verified because it is not difficult to see that $x = P_C(x + tv)$ whenever $x \in \mathrm{Proj}_C(x + t'v)$ for some $t' > t$.
$(b) \Rightarrow (d)$. Let $v_i \in N^P(C; x_i)$ with $i = 1, 2$. By (b) we have

$$\langle v_1, x_1 - x_2\rangle \geq -\frac{\|v_1\|}{2\rho(x_1)}\|x_1 - x_2\|^2 \text{ and } \langle -v_2, x_1 - x_2\rangle \geq -\frac{\|v_2\|}{2\rho(x_2)}\|x_2 - x_1\|^2$$

and hence adding the two inequalities yields the desired inequality of (d).
$(d) \Rightarrow (b)$. For $x, x' \in C$ and $v \in N^P(C;x)$ it is enough to take $v_1 = v$, $x_1 = x$, $x_2 = x'$ and $v_2 = 0$ in the inequality of (d) to obtain that of (b).

So the equivalence between the assertions $(a) - (d)$ is established.
$(d) \Rightarrow (e)$. Suppose that (d) holds and fix any $u_i \in H$ and any $y_i \in (I + N^P(C;\cdot))^{-1}(u_i)$ for $i = 1, 2$. Then (d) allows us to write

$$\begin{aligned}\langle y_1 - y_2, u_1 - u_2\rangle &= \langle y_1 - y_2, (u_1 - y_1) - (u_2 - y_2)\rangle + \|y_1 - y_2\|^2 \\ &\geq -\frac{1}{2}\left(\frac{\|u_1 - y_1\|}{\rho(y_1)} + \frac{\|u_2 - y_2\|}{\rho(y_2)}\right)\|y_1 - y_2\|^2 + \|y_1 - y_2\|^2,\end{aligned}$$

that is,

$$\langle y_1 - y_2, u_1 - u_2\rangle \geq \left(1 - \frac{\|u_1 - y_1\|}{2\rho(y_1)} - \frac{\|u_2 - y_2\|}{2\rho(y_2)}\right)\|y_1 - y_2\|^2.$$

So, (e) is fulfilled.
$(e) \Rightarrow ((f) \text{ and } (g))$. Fix any positive $\gamma < 1$. For any $u_i \in H$ and any $y_i \in (I + \gamma\rho(\cdot)\mathbb{B}_H \cap N^P(C;\cdot))^{-1}(u_i)$ with $i = 1, 2$, the inclusion $u_i \in y_i + \gamma\rho(y_i)\mathbb{B}_H \cap N^P(C; y_i)$ assures us that $\|u_i - y_i\| \leq \gamma\rho(y_i)$ and hence it follows from the inequality of (c) that

$$\|y_1 - y_2\| \leq (1 - \gamma)^{-1}\|u_1 - u_2\|, \tag{23}$$

which entails that the set-valued operator $(I + \gamma\rho(\cdot)\mathbb{B}_H \cap N^P(C;\cdot))^{-1}(\cdot)$ is, on its domain, single valued and Lipschitz continuous with $(1 - \gamma)^{-1}$ as a Lipschitz constant. So, we are left to show that the domain of $(I + \gamma\rho(\cdot)\mathbb{B}_H \cap N^P(C;\cdot))^{-1}(\cdot)$ contains $O^{\gamma}_{\rho(\cdot)}(C)$, i.e., we want to prove that for any $u \in O^{\gamma}_{\rho(\cdot)}(C)$ we have

$$\mathrm{Proj}_C(u) \neq \emptyset \quad \text{and} \quad d_C(u) \leq \gamma\rho(y) \quad \text{forall } y \in \mathrm{Proj}_C(u). \tag{24}$$

The desired properties being obvious for $u \in C$, fix any $u \in O^{\gamma}_{\rho(\cdot)}(C) \setminus C$. Choose a sequence of positive numbers $(\varepsilon_n)_{n\in\mathbb{N}}$ with $\varepsilon_n \downarrow 0$ and a sequence $(z_n)_{n\in\mathbb{N}}$ in C such that $f(z_n) < \inf_C f + \varepsilon_n^2$, where $f(x) := \frac{1}{2}\|x-u\|^2$ for all $x \in H$, and observe that such a sequence $(z_n)_n$ is bounded (because of the definition of f). According to the Ekeland variational principle, for each integer n there exists some $y_n \in C$ such that

$$\|y_n - z_n\| < \varepsilon_n \quad \text{and} \quad f(y_n) = \inf_{x\in C}[f(x) + \varepsilon_n\|x - y_n\|]. \tag{25}$$

On one hand, the sequence $(y_n)_n$ is bounded and by the equality in (25) we have for n large enough

$$\frac{1}{2}\|y_n - u\|^2 \le \frac{1}{2}\|z_n - u\|^2 + \varepsilon_n\|z_n - y_n\|$$

and this ensures that

$$\|y_n - u\| \to d_C(u). \tag{26}$$

On the other hand, by the equality in (25) again and by the fuzzy calculus rule for the proximal subdifferential, there are

$$y'_n \in C \text{ with } \|y'_n - y_n\| < \varepsilon_n \quad \text{and} \quad y''_n \in H \text{ with } \|y''_n - y_n\| < \varepsilon_n \tag{27}$$

such that

$$0 \in \partial f(y''_n) + \varepsilon_n\mathbb{B}_H + N^P(C; y'_n),$$

that is, for some $b_n \in \mathbb{B}_H$ we have $u - y''_n + \varepsilon_n b_n \in N^P(C; y'_n)$. Set $u_n := u + y'_n - y''_n + \varepsilon_n b_n$ and observe that, by (26) and (27), $\lim_{n\to\infty} \frac{\|u_n - y'_n\|}{\|u - y'_n\|} = 1$. Since we see by definition of $O^{\gamma}_{\rho(\cdot)}(C)$ (see (21)) that $\limsup_{n\to\infty} \frac{\|u-y'_n\|}{\rho(y'_n)} < \gamma$, we obtain $\limsup_{n\to\infty} \frac{\|u_n - y'_n\|}{\rho(y'_n)} < \gamma$. Deleting a finite number of integers if necessary, we may then suppose that for all n we have $\|u_n - y'_n\| \le \gamma\rho(y'_n)$. As $u_n - y'_n \in N^P(C; y'_n)$ we obtain $y'_n \in (I + \gamma\rho(\cdot)\mathbb{B}_H \cap N^P(C; \cdot))^{-1}(u_n)$. It follows from (23) that

$$\|y'_n - y'_m\| \le (1-\gamma)^{-1}\|u_n - u_m\|,$$

and this implies that $(y'_n)_n$ is a Cauchy sequence because $u_n \to u$. Denoting by y the limit of the sequence $(y'_n)_n$, we see that $(y_n)_n$ also converges to y and it is not difficult to derive that $y \in \operatorname{Proj}_C(u)$. Further, the inequality $\|u_n - y'_n\| < \gamma\rho(y'_n)$ and the continuity of ρ yield $d_C(u) = \|u - y\| \le \gamma\rho(y)$. The properties in (24) are then proved.

Fixing any $u \in O^{\gamma}_{\rho(\cdot)}(C)$ and $y \in \operatorname{Proj}_C(u)$ we have $u - y \in N^P(C; y)$, and combining this (24) gives $d_C(u) \le \gamma\rho(y)$. So, we have

$$u \in y + d_C(u)\mathbb{B}_H \cap N^P(C; y) \subset y + \gamma\rho(y)\mathbb{B}_H \cap N^P(C; y)$$

and according to (24)

$$O^{\gamma}_{\rho(\cdot)}(C) \subset \operatorname{Dom}\operatorname{Proj}_C \text{ and } O^{\gamma}_{\rho(\cdot)}(C) \subset \operatorname{Dom}(I + \gamma\rho(\cdot)\mathbb{B} \cap N^P(C; \cdot))^{-1}.$$

Because of the single valuedness of $(I + \gamma\rho(\cdot)\mathbb{B}_H \cap N^P(C;\cdot))^{-1}(\cdot)$ on its domain and of Lemma 2, it follows that the mapping P_C is well defined on $O^\gamma_{\rho(\cdot)}(C)$,

$$\{P_C(u)\} = \mathrm{Proj}_C(u) = (I + \gamma\rho(\cdot)\mathbb{B}_H \cap N^P(C;\cdot))^{-1}(u) \quad \text{for all } u \in O^\gamma_{\rho(\cdot)}(C),$$

and by (23)

$$\|P_C(u) - P_C(u')\| \le (1-\gamma)^{-1}\|u - u'\| \quad \text{for all } u, u' \in O^\gamma_{\rho(\cdot)}(C),$$

that is, P_C is Lipschitz continuous on $O^\gamma_{\rho(\cdot)}(C)$ with $(1-\gamma)^{-1}$ as a Lipschitz constant. All the properties in (f) are then established.

To obtain (g), we see that for $u_i \in O_{\rho(\cdot)}(C)$ with $i = 1, 2$ we have by (f) that $P_C(u_i)$ exists. Then the inequality of (g) easily follows from (e) by taking $y_i = P_C(u_i)$ in (e).

$(e) \Rightarrow (f')$ Suppose (e) and take $u_i \in O^\gamma_{\rho(\cdot)}(C)$. Since $(e) \Rightarrow (f)$ holds, $P_C(u_i)$ exist. Therefore by (e) we have

$$\langle P_C(u_1) - P_C(u_2), u_1 - u_2\rangle \ge (1-\gamma)\|u_1 - u_2\|^2.$$

This yields

$$\begin{aligned}\|(I - P_C)(u_1) - (I - P_C(u_2))\|^2 &= \|u_1 - u_2\|^2 + \|P_C(u_1) - P_C(u_2)\|^2 \\ &\quad - 2\langle u_1 - u_2, P_C(u_1) - P_C(u_2)\rangle \\ &\le \|u_1 - u_2\|^2 + (2\gamma - 1)\|P_C(u_1) - P_C(u_2)\|^2,\end{aligned}$$

that is, (f'). So, the proof of the theorem is complete. □

Before going further let us establish the openness of the sets $O_{\rho(\cdot)}(C)$ and $O^\gamma_{\rho(\cdot)}(C)$ and compare them with the similar sets

$$U_{\rho(\cdot)}(C) = \left\{ u \in H : \liminf_{\substack{\mathrm{Dom\,Proj}_C \ni u' \to u \\ x \in \mathrm{Proj}_C(u')}} \frac{\|x - u\|}{\rho(x)} < 1 \right\} \tag{28}$$

and for positive $\gamma < 1$

$$U^\gamma_{\rho(\cdot)}(C) = \left\{ u \in H : \liminf_{\substack{\mathrm{Dom\,Proj}_C \ni u' \to u \\ x \in \mathrm{Proj}_C(u')}} \frac{\|x - u\|}{\rho(x)} < \gamma \right\}. \tag{29}$$

Observe that the conditions under which the liminf is taken in (28) and (29) make sense because Lau's theorem (see [**80**]) tells us that $\mathrm{Dom\,Proj}_C$ is dense in the space H.

Obviously, one always has

(30)
$$O^\gamma_{\rho(\cdot)}(C) \subset U^\gamma_{\rho(\cdot)}(C) \text{ and } \{u \in H : \exists y \in \mathrm{Proj}_C(u),\ d_C(u) < \gamma\rho(y)\} \subset U^\gamma_{\rho(\cdot)}(C),$$

and similar inclusions for $U_{\rho(\cdot)}(C)$. The next proposition shows that the three sets in (30) coincide whenever C is $\rho(\cdot)$-prox-regular.

PROPOSITION 4. Assume that the closed set C is $\rho(\cdot)$-prox regular. Then the following properties hold.

(1) One has

$$O_{\rho(\cdot)}(C) = \{u \in H : \exists y \in \operatorname{Proj}_C(u),\, d_C(u) < \rho(y)\} = U_{\rho(\cdot)}(C)$$

and for all positive $\gamma < 1$

$$O^{\gamma}_{\rho(\cdot)}(C) = \{u \in H : \exists y \in \operatorname{Proj}_C(u),\, d_C(u) < \gamma\rho(y)\} = U^{\gamma}_{\rho(\cdot)}(C).$$

(2) The sets $O_{\rho(\cdot)}(C)$ and $O^{\gamma}_{\rho(\cdot)}(C)$ for all positive $\gamma < 1$ are open.

(3) The mapping P_C is well-defined on the open set $O_{\rho(\cdot)}(C)$ and locally Lipschitz continuous on this open set.

PROOF. Observe first that for any $u' \in \operatorname{Dom}\operatorname{Proj}_C$, any $y' \in \operatorname{Proj}_C(u')$, and any $z \in C$, writing

$$\|u' - z\|^2 = \|u' - y'\|^2 + \|y' - z\|^2 - 2\langle u' - y', z - y'\rangle,$$

the inclusion $u' - y' \in N^P(C; y')$ and (b) of Theorem 3 yield

$$\|u' - z\|^2 \geq \|u' - y'\|^2 + \left(1 - \frac{\|u' - y'\|}{\rho(y')}\right) \|z - y'\|^2. \tag{31}$$

Fixing now any positive $\gamma < 1$ and any $u \in O^{\gamma}_{\rho(\cdot)}$ we know by Theorem 3 that $y := P_C(u)$ exists and the definition of $O^{\gamma}_{\rho(\cdot)}(C)$ ensures that $\frac{\|u-y\|}{\rho(y)} < \gamma$, that is, $d_C(u) < \gamma\rho_C(y)$, establishing the inclusion of $O^{\gamma}_{\rho(\cdot)}(C)$ in the second member of the equalities of the proposition concerning $U^{\gamma}_{\rho(\cdot)}(C)$. The second member of those equalities being always included in the third member (as observed above), let us show the inclusion of the third member into the first one. Take any u in that third member $U^{\gamma}_{\rho(\cdot)}(C)$ with $u \notin C$. Choose a positive $\gamma' < \gamma$, a sequence $(u_n)_n$ of $\operatorname{Dom}\operatorname{Proj}_C$ converging to u, and $x_n \in \operatorname{Proj}_C(u_n)$ such that

$$\lim_n \frac{\|u - x_n\|}{\rho(x_n)} = \liminf_{\substack{\operatorname{Dom}\operatorname{Proj}_C \ni u' \to u \\ x \in \operatorname{Proj}_C(u')}} \frac{\|x - u\|}{\rho(x)} < \gamma' < \gamma.$$

Writing $d_C(u) \leq \|u - x_n\| + d_C(u_n)$ (as in the proof of Theorem 3) we see that $\lim_n \|u - x_n\|/\|u_n - x_n\| = 1$ and hence we also have $\lim_n \frac{\|u_n - x_n\|}{\rho(x_n)} < \gamma'$. Deleting a finite number of n if necessary, we may suppose that $\|u_n - x_n\|/\rho(x_n) < \gamma'$ for all n. By (31) we then have for all integers m, n

$$\begin{aligned}(1 - \gamma')\|x_m - x_n\|^2 &\leq \|u_m - x_n\|^2 - \|u_m - x_m\|^2 \\ &\leq (\|u_m - u_n\| + \|u_n - x_n\|)^2 - \|u_m - x_m\|^2 \\ &= (\|u_m - u_n\| + d_C(u_n))^2 - d_C^2(u_m),\end{aligned}$$

and hence $\|x_m - x_n\|^2 \to 0$ as $m \to \infty$ and $n \to \infty$. Let $x \in C$ be the limit of the Cauchy sequence $(x_n)_n$. Since $\|u_n - x_n\| \to \|u - x\|$ and $\|u_n - x_n\| = d_C(u_n) \to d_C(u)$, we see that $x \in \mathrm{Proj}_C(u)$ and $\|u - x\|/\rho(x) \le \gamma'$ according to the continuity of $\rho(\cdot)$.

Consider now any sequence $(y_n)_n$ in C with $\|u - y_n\| \to d_C(u)$ as $n \to \infty$. By (31) again we have

$$\left(1 - \frac{\|u - x\|}{\rho(x)}\right) \|x - y_n\|^2 \le \|u - y_n\|^2 - d_C^2(u).$$

Therefore, $y_n \to x$ (because $\|u - x\|/\rho(x) < 1$) and thanks to the continuity of $\rho(\cdot)$ we obtain

$$\lim_n \frac{\|u - y_n\|}{\rho(y_n)} = \frac{\|u - x\|}{\rho(x)} \le \gamma'.$$

It then follows that $u \in O^{\gamma}_{\rho(\cdot)}(C)$ according to (21), proving the desired inclusion. The equalities concerning $O^{\gamma}_{\rho(\cdot)}(C)$ are then established and the ones concerning $O_{\rho(\cdot)}(C)$ are direct consequences of those equalities.

The openness of $O_{\rho(\cdot)}(C)$ being a direct consequence of that of $O^{\gamma}_{\rho(\cdot)}(C)$ for $0 < \gamma < 1$, it remains to show that $O^{\gamma}_{\rho}(C)$ is open for $0 < \gamma < 1$. Fix $\bar{u} \in O^{\gamma}_{\rho(\cdot)}(C)$. By (f) in Theorem 3 the point $\bar{y} := P_C(\bar{u})$ exists, and for some positive $\gamma' < \gamma$ we have $\|\bar{u} - \bar{y}\| < \gamma'\rho(\bar{y})$. By continuity of ρ we may choose some $\eta > 0$ such that

$$\frac{\|u - z\|}{\rho(y)} < \gamma' \quad \text{for all } u \in B(\bar{u}, \eta),\ \ z \in B(\bar{y}, \eta),\ \ y \in C \cap B(\bar{y}, \eta). \tag{32}$$

Choose a positive $\varepsilon < \eta$ such that for all $u \in B(\bar{u}, \varepsilon)$

$$\left(1 - \frac{\|\bar{u} - \bar{y}\|}{\rho(\bar{y})}\right)^{-1} \left((d_C(u) + \varepsilon)^2 - d_C(\bar{u})\right) < \eta.$$

Fix any $u \in B(\bar{u}, \varepsilon)$ and consider a sequence $(y_n)_n$ in C such that $\lim_n \|y_n - u\| = d_C(u)$ and

$$\limsup_{\substack{\|y - u\| \to d_C(u) \\ y \in C}} \frac{\|y - u\|}{\rho(y)} = \lim_n \frac{\|y_n - u\|}{\rho(y_n)}. \tag{33}$$

For each n the equality

$$\|y_n - \bar{y}\|^2 = \|y_n - \bar{u}\|^2 - \|\bar{u} - \bar{y}\|^2 + 2\langle \bar{u} - \bar{y}, y_n - \bar{y}\rangle$$

and the inclusion $\bar{u} - \bar{y} \in N^P(C; \bar{y})$ combining with (b) of Theorem 3 assure us that

$$\|y_n - \bar{y}\|^2 \le \|y_n - \bar{u}\|^2 - \|\bar{u} - \bar{y}\|^2 + \frac{\|\bar{u} - \bar{y}\|}{\rho(\bar{y})} \|y_n - \bar{y}\|^2,$$

which implies

$$\left(1 - \frac{\|\bar{u} - \bar{y}\|}{\rho(\bar{y})}\right) \|y_n - \bar{y}\|^2 \le (\|y_n - u\| + \|u - \bar{u}\|)^2 - \|\bar{u} - \bar{y}\|^2.$$

The second member of the latter inequality converging to $(d_C(u) + \|u - \bar{u}\|)^2 - d_C(\bar{u})$ when $n \to \infty$, we obtain for n large enough that $\|y_n - \bar{y}\| < \eta$. Therefore for n sufficiently large we have by (32) that $\|u - y_n\|/\rho(y_n) < \gamma'$ and hence $\lim_n \frac{\|y_n - u\|}{\rho(y_n)} \le \gamma' < \gamma$. According to (33) and to (21), we see that $u \in O^{\gamma}_{\rho(\cdot)}(C)$ and hence $B(\bar{u}, \varepsilon) \subset O^{\gamma}_{\rho(\cdot)}(C)$. This concludes the openness of $O^{\gamma}_{\rho(\cdot)}(C)$.

Finally, the local Lipschitz continuity of P_C is a direct consequence of (f) in Theorem 3 and of the openness of $O^{\gamma}_{\rho(\cdot)}(C)$ for any positive $\gamma < 1$. □

In the definition of prox-regularity as well as in its characterizations $(b) - (d)$ in Theorem 3 only points inside the closed set C are considered. The next proposition provides two additional characterizations taking into account outside points. We need first the following lemma.

LEMMA 5. Let u be any point of H.

(a) One has the inclusions

$$d_C(u)\partial^P d_C(u) = \partial^P \left(\frac{1}{2} d_C^2\right)(u) \quad \text{and} \quad d_C(u)\partial^F d_C(u) = \partial^F \left(\frac{1}{2} d_C^2\right)(u).$$

(b) If d_C is ∂^P-subdifferentiable at u, then $P_C(u)$ exists and

$$d_C(u)\partial^P d_C(u) = \partial^P \left(\frac{1}{2} d_C^2\right)(u) = \{u - P_C(u)\}.$$

(c) If d_C is ∂^F-subdifferentiable at u, then $P_C(u)$ exists and

$$d_C(u)\partial^F d_C(u) = \partial^F \left(\frac{1}{2} d_C^2\right)(u) = \{u - P_C(u)\}.$$

(d) The function d_C is Fréchet differentiable at a point $u \in H \setminus C$ if and only if it is ∂^F-subdifferentiable at that point. In that case $\nabla^F d_C(u) = \frac{1}{d_C(u)}(u - P_C(u))$.

(e) The square distance function d_C^2 is Fréchet differentiable at a point $u \in H$ if and only if it is ∂^F-subdifferentiable at that point. In that case $\nabla^F(\frac{1}{2}d_C^2)(u) = u - P_C(u)$.

PROOF. Suppose that the left member of the first relation of (a) is nonempty and fix any $v \in \partial^P d_C(u)$. By definition of proximal subgradient there exist some real number $\sigma > 0$ and some neighborhood U of u such that for all $y \in U$

$$\langle v, y - u\rangle \le d_C(y) - d_C(u) + \sigma\|y - u\|^2$$

and hence

$$\langle d_C(u)v, y - u\rangle \le d_C(u)d_C(y) - d_C^2(u) + \sigma d_C(u)\|y - u\|^2.$$

Observing that the second member of the latter inequality is equal to

$$\frac{1}{2}d_C^2(y) - \frac{1}{2}d_C^2(u) - \frac{1}{2}(d_C(y) - d_C(u))^2 + \sigma d_C(u)\|y-u\|^2,$$

we obtain for all $y \in U$

$$\langle d_C(u)v, y-u\rangle \le \frac{1}{2}d_C^2(y) - \frac{1}{2}d_C^2(u) + \sigma d_C(u)\|y-u\|^2,$$

which says that $d_C(u)v \in \partial^P(\frac{1}{2}d_C^2)(u)$. So the inclusion of the first member of (a) into the second one is established.

Take now any $v \in \partial^P(\frac{1}{2}d_C^2)(u)$ and suppose that $u \notin C$. There exists some real number $\sigma > 0$ and some neighborhood U of u such that for all $y \in U$

$$\langle v, y-u\rangle \le \frac{1}{2}d_C^2(y) - \frac{1}{2}d_C^2(u) + \sigma\|y-u\|^2,$$

which yields

$$\begin{aligned} \left\langle \frac{v}{d_C(u)}, y-u\right\rangle &\le \frac{d_C^2(y) - d_C^2(u)}{2d_C(u)} + \frac{\sigma}{d_C(u)}\|y-u\|^2 \\ &= d_C(y) - d_C(u) + \frac{(d_C(y)-d_C(u))^2}{2d_C(u)} + \frac{\sigma}{d_C(u)}\|y-u\|^2 \qquad (34)\\ &\le d_C(y) - d_C(u) + \frac{1+2\sigma}{2d_C(u)}\|y-u\|^2. \end{aligned}$$

So, $\partial^P(\frac{1}{2}d_C^2)(u) \subset d_C(u)\partial^P d_C(u)$ and this inclusion still holds for $u \in C$ because for $u \in C$ it is easily seen that both members are equal to $\{0\}$. The first equality of (a) is then established, and the second one is obtained in a similar way.

(b). Assume that $\partial^P(\frac{1}{2}d_C^2)(u)$ is nonempty and fix any v in it. Theorem 11 in [**18**] ensures that $\mathrm{Proj}_C(u) \neq \emptyset$. Choosing some $\sigma > 0$ and some neighborhood U of u such that for all $y \in U$

$$\langle v, y-u\rangle \le \frac{1}{2}d_C^2(y) - \frac{1}{2}d_C^2(u) + \sigma\|y-u\|^2,$$

we see that for each $w \in \mathrm{Proj}_C(u)$

$$\langle v, y-u\rangle \le \frac{1}{2}\|y-w\|^2 - \frac{1}{2}\|u-w\|^2 + \sigma\|y-u\|^2$$

for all $y \in U$ and hence $v \in \partial^P(\frac{1}{2}\|\cdot - w\|^2)(u) = \{u-w\}$. This yields that the set $u - \mathrm{Proj}_C(u)$ is a singleton that coincides with $\partial^P(\frac{1}{2}d_C^2)(u)$. Therefore, taking into account the first equation of (a), we obtain that $P_C(u)$ exists and

$$d_C(u)\partial^P d_C(u) = \partial^P\left(\frac{1}{2}d_C^2\right)(u) = \{u - P_C(u)\}$$

whenever $\partial^P d_C(u) \neq \emptyset$.

(c). The proof of (c) is similar because Theorem 11 in [**18**] also says that $\mathrm{Proj}_C(u)$ is nonempty whenever $\partial^F d_C(u) \neq \emptyset$.

(d) and (e). Fix any $u \in H \setminus C$. Obviously, d_C is ∂^F-subdifferentiable at u whenever it is Fréchet differentiable at u. Suppose now that $\partial^F d_C(u) \neq \emptyset$. To obtain the Fréchet differentiability of d_C at u, it suffices to prove that of $f := \frac{1}{2}d_C^2$ at u. By (c) we know that $P_C(u)$ exists and that for $v := u - P_C(u)$ we have $\partial^F f(u) = \{v\}$. Fix any $\varepsilon > 0$. By definition of Fréchet subgradient, there exists some $\delta > 0$ such that for any $t \in]0, \delta[$ and any $y \in \mathbb{B}_H$

$$\langle v, ty \rangle \leq f(u + ty) - f(u) + \varepsilon t,$$

that is,

$$t^{-1}[f(u + ty) - f(u)] - \langle v, y \rangle \geq -\varepsilon. \tag{35}$$

On the other hand, taking the above δ satisfying $\delta < 2\varepsilon$ we also have for all $y \in \mathbb{B}_H$ and $t \in]0, \delta[$

$$\begin{aligned} f(u + ty) - f(u) &\leq \frac{1}{2}\|u + ty - P_C(u)\|^2 - \frac{1}{2}\|u - P_C(u)\|^2 \\ &= \langle u - P_C(u), ty \rangle + \frac{1}{2}t^2\|y\|^2 \end{aligned}$$

and hence

$$t^{-1}[f(u + ty) - f(u)] - \langle v, y \rangle \leq \varepsilon.$$

Combining the latter with (35) yields the Fréchet differentiability of f at u with $\nabla^F f(u) = v$.

Finally, it is easily seen that at any point $u \in C$ the function f is Fréchet differentiable with $\nabla^F f(u) = 0$, i.e., the equality $\nabla^F f(u) = u - P_C(u)$ still holds. The proof of the lemma is then complete. □

Proposition 6. Anyone of the following properties can be added to the list of equivalences $(a) - (d)$ in Theorem 3.

(α) For any $u \in \operatorname{Dom} \partial^P d_C$ and any $v \in \partial^P d_C(u)$ one has

$$d_C(u) + \langle v, x - u \rangle \leq \frac{1}{2\rho(P_C(u))}\|x - P_C(u)\|^2 \quad \forall x \in C.$$

(β) For any $u \in \operatorname{Dom} \partial^P d_C$ and any $v \in \partial^P d_C(u)$ one has

$$\langle v, x - u \rangle \leq \frac{1}{2\rho(P_C(u))}\|x - P_C(u)\|^2 \quad \forall x \in C.$$

(γ) For any $x \in C$ and any $v \in N^P(C; x) \cap \mathbb{B}_H$ one has

$$\langle v, u - x \rangle \leq d_C(u) + \frac{1}{2\rho(x)}\|P_C(u) - x\|^2 \quad \forall u \in \operatorname{Dom} P_C.$$

(δ) For any $x \in C$ and any $v \in N^P(C; x) \cap \mathbb{B}_H$ one has

$$\langle v, u - x \rangle \leq d_C(u) + \frac{1}{2\rho(x)}\left(d_C(u) + \|u - x\|\right)^2 \quad \forall u \in H.$$

PROOF. Recalling that $\partial^P d_C(u) = N^P(C;u) \cap \mathbb{B}_H$ whenever $u \in C$, we see that (b) of Theorem 3 holds under anyone of the assumptions of (α), (β), (γ), (δ).

Now assume that (b) in Theorem 3 holds. On one hand, fix any $u \in \mathrm{Dom}\, \partial^P d_C$ and $v \in \partial^P d_C(u)$. By Proposition 6 we have that $P_C(u)$ exists. If we suppose $u \notin C$, by (b) in Lemma 5 we have that $v = \frac{1}{\|u - P_C(u)\|}(u - P_C(u))$ and hence

$$d_C(u) + \langle v, x - u\rangle = \langle v, u - P_C(u)\rangle + \langle v, x - u\rangle = \langle v, x - P_C(u)\rangle.$$

The same equality $v = \frac{1}{\|u - P_C(u)\|}(u - P_C(u))$ saying that $v \in N^P(C;x) \cap \mathbb{B}_H$, it follows according to (b) of Theorem 3 that

$$d_C(u) + \langle v, x - u\rangle \leq \frac{1}{2\rho(P_C(u))}\|x - P_C(u)\|^2,$$

that is, (b) in Theorem 3 implies (α) holds. Further, (α) obviously implies (β).

On the other hand, fix in a similar way any $x \in C$ and $v \in N^P(C;x) \cap \mathbb{B}_H$. Then for any $u \in \mathrm{Dom}\, P_C$ we have by (b) in Theorem 3 and the inequality $\|v\| \leq 1$ that

$$\langle v, u - x\rangle = \langle v, u - P_C(u)\rangle + \langle v, P_C(u) - x\rangle \leq d_C(u) + \frac{1}{2\rho(x)}\|P_C(u) - x\|^2,$$

which ensures (γ).

Finally, suppose that (γ) is fulfilled and fix any $x \in C$ and $v \in N^P(C;x) \cap \mathbb{B}_H$. For every $u \in \mathrm{Dom}\, P_C$, writing

$$\|P_C(u) - x\| \leq \|P_C(u) - u\| + \|u - x\| = d_C(u) + \|x - u\|$$

we obtain by (γ) that

$$\langle v, u - x\rangle \leq d_C(u) + \frac{1}{2\rho(x)}\left(d_C(u) + \|u - x\|\right)^2,$$

and this inequality still holds for any $u \in H$ by continuity because $\mathrm{Dom}\, P_C$ is dense in H according to Lau's theorem. So, the implication $(\gamma) \Rightarrow (\delta)$ holds.

So, each one of (α), (β), (γ), (δ) is equivalent to (b) in Theorem 3. □

Before stating other characterizations (in terms of the distance function) of $\rho(\cdot)$-prox regularity, let us observe the following tangential and normal regularities of such sets. Furthermore, we also show for such sets that the graph of the normal cone $N^P(C;\cdot)$ is sequentially $\|\cdot\| \times w(H,H)$-closed. This is a significant property since Example 1.7, p. 11, in [**95**] provides a closed set in $l^2(\mathbb{N})$ whose limiting normal cone at some point is not weakly sequentially closed and even not strongly closed.

PROPOSITION 7. For any closed subset C of H the following facts hold.

(a) If C is $\rho(\cdot)$-prox-regular, then the graph of the normal cone $N^p(C;\cdot)$ is strongly $\times$ weakly sequentially closed.

(b) If C is $\rho(\cdot)$-prox-regular, then the normal regularity

$$N^P(C;x) = N^F(C;x) = N^L(C;x) = N^{Cl}(C;x) \quad \text{for all } x \in C$$

is fulfilled along with the tangential regularity

$$T^B(C;x) = T^{Cl}(C;x) \quad \text{for all } x \in C.$$

(c) The set C is $\rho(\cdot)$-prox-regular if and only if the inequality in (b) of Theorem 3 holds with anyone of the normal cones $N^F(C;\cdot)$, $N^L(C;\cdot)$, $N^{Cl}(C;\cdot)$ in place of $N^P(C;\cdot)$.

PROOF. Suppose that C is $\rho(\cdot)$-prox-regular and fix any $x, x' \in C$. Take any sequences $(x_n)_n$ in C converging to x and $(v_n)_n$ converging weakly to v with $v_n \in N^p(C;x_n)$. Then, by (b) of Theorem 3 we have for $\beta := \sup_{n\in\mathbb{N}} \|v_n\| < +\infty$

$$\langle v_n, x' - x_n\rangle \le \frac{\|v_n\|}{2\rho(x_n)}\|x' - x_n\|^2 \le \frac{\beta}{2\rho(x_n)}\|x' - x_n\|^2,$$

and hence $\langle v, x' - x\rangle \le \frac{\beta}{2\rho(x)}\|x' - x\|^2$, which says that $v \in N^P(C;x)$. So, (a) is established.

Recall now that for any $x \in C$

$$N^P(C;x) \subset N^F(C;x) \subset N^L(C;x) \subset N^{Cl}(C;x). \tag{36}$$

Suppose that C is $\rho(\cdot)$-prox-regular. The arguments establishing (a) above show that $N^L(C;x) \subset N^P(C;x)$. This and (36) tell us that

$$N^P(C;x) = N^F(C;x) = N^L(C;x).$$

The convexity and closedness of $N^F(C;x)$ entails that $N^L(C;x)$ is closed and convex and hence it is also equal to $N^{Cl}(C;x)$ according to the equality $N^{Cl}(C;x) = \overline{\text{co}}N^L(C;x)$. We have then proved the equalities of assertion (b) concerning the normal cones.

Further, the equality $N^F(C;x) = N^{Cl}(C;x)$ ensures that $(N(^F(C;x))^0 = (N^{Cl}(C;x))^0$, thus $T^B(C;x) \subset T^{Cl}(C;x)$ because of the equality $(N^{Cl}(C;x))^0 = T(C;x)$ coming from the definition of the Clarke normal cone and because of the inclusion $T^B(C;x) \subset (N^F(C;x))^0$ (which is not difficult to check). Therefore the tangential regularity $T^B(C;x) = T^{Cl}(C;x)$ for all $x \in C$ is fulfilled because the inclusion $T^{Cl}(C;x) \subset T^B(C;x)$ always holds.

It remains to establish (c). By (36) it is obvious that the inequality in (b) of Theorem 3 holds with $N^P(C;\cdot)$ whenever it is so for $N^F(C;\cdot)$, $N^L(C;\cdot)$, or $N^{Cl}(C;\cdot)$, that is, the implication $\Leftarrow$ is true according to (b) of Theorem 3 again. Finally, the reverse implication is a direct consequence of (b) of the proposition. □

The above regularity result allows us to give a characterization of prox-regularity through a property of second order contact with the Bouligand tangent cone. Recall first that, for two extended real-valued convex functions $f, g : H \to \mathbb{R}\cup\{+\infty\}$ such that one is finite and continuous at a point of the effective domain of the other, one has the Fenchel duality result

$$\inf_{x\in H}[f(x) + g(x)] = -\min_{v\in H}[f^*(v) + g^*(-v)],$$

where f^* denotes the Fenchel conjugate of f given as

$$f^*(v) = \sup_{x\in X}[\langle v, x\rangle - f(x)] \quad \text{for all } v \in H.$$

For a fixed point $u \in H$ and a closed convex cone $K \subset H$, taking $f := \|\cdot - u\|$ and g as the indicator function ψ_K of K, it is easily seen that the latter formula reduces to

$$d(u, K) = \sup_{v \in K^0 \cap \mathbb{B}_H} \langle v, u \rangle. \tag{37}$$

PROPOSITION 8. The closed set C is $\rho(\cdot)$-prox-regular if and only if

$$d(x' - x, T^B(C; x)) \leq \frac{1}{2\rho(x)} \|x' - x\|^2 \quad \text{for all } x, x' \in C.$$

PROOF. Assume first the $\rho(\cdot)$-prox-regularity of C. Then Proposition 7 tells us that $T^{Cl}(C; \cdot) = T^B(C; \cdot)$. For $x, x' \in C$, taking as K the closed convex cone $T^{Cl}(C; x)$ in formula (37) yields

$$d(x' - x, T^B(C; x)) = d(x' - x, T^{Cl}(C; x)) = \sup_{v \in N^{Cl}(C;x) \cap \mathbb{B}_H} \langle v, x' - x \rangle.$$

Using the normal regularity $N^P(C; x) = N^{Cl}(C; x)$ coming from Proposition 7 and using (b) of Theorem 3 we obtain

$$d(x' - x, T^B(C; x)) \leq \frac{1}{2\rho(x)} \|x' - x\|^2.$$

Now assume that the condition of the proposition holds. Fix $x, x' \in C$ and $v \in N^P(C; x) \cap \mathbb{B}_H$. For $K = \operatorname{cl} \operatorname{co} T^B(C; x)$, it is easily verified that $\langle v, h \rangle \leq 0$ for all $h \in T^B(C; x)$ and hence $v \in K^0$. Therefore by formula (37) we have that

$$\langle v, x' - x \rangle \leq d(x' - x, K) \leq d(x' - x, T^B(C; x)) \leq \frac{1}{2\rho(x)} \|x' - x\|^2,$$

and this concludes the proof according to (b) of Theorem 3. □

The next proposition measures in some way the lack of convexity of prox-regular sets.

PROPOSITION 9. Assume that the closed set C of H is $\rho(\cdot)$-prox-regular. The following properties hold.
(i) Let $r > 0$ and let $u \in (\operatorname{co} C) \cap O_{\rho(\cdot)}(C)$. Then

$$d(u, C) = \|u - P_C(u)\| \leq \frac{1}{2\rho(P_C(u))} \sum_{i,j=1}^{n} t_i t_j \|x_i - x_j\|^2,$$

where $t_i \geq 0$, $\sum_{i=1}^{n} t_i = 1$, $x_i \in C$, and $u = \sum_{i=1}^{n} t_i x_i$.
In particular,
(ii) for any and $x, y \in C$ and $t \in [0, 1]$ such that $u_t := (1 - t)x + ty \in O_{\rho(\cdot)}(C)$ one has

$$d_C(u_t) \leq \frac{1}{2\rho(P_C(u_t))} \min\{t, 1 - t\} \|y - x\|^2.$$

Moreover,
(iii) there exists a positive $\varepsilon \leq 1$ and a continuous function $\alpha(\cdot)$ from $[0,\varepsilon]$ into $]0,+\infty]$ with $\alpha(0) = \rho(x)$ and such that for all $t \in [0,\varepsilon]$

$$d_C(u_t) \leq \frac{1}{2\alpha(t)} \min\{t, 1-t\}\|y-x\|^2.$$

PROOF. (i). By (g) in Theorem 3, $P_C(u)$ exists. Moreover, by (b) in the same theorem we have for each $i = 1, \ldots, n$

$$< u - P_C(u), x_i - P_C(u) > \leq \frac{1}{2\rho(P_C(u))} \|u - P_C(u)\| \, \|x_i - P_C(u)\|^2 ,$$

so that

$$< u - P_C(u), \sum_{i=1}^{n+1} t_i x_i - P_C(u) > \leq \frac{1}{2\rho(P_C(u))} \|u - P_C(u)\| \sum_{i=1}^{n+1} t_i \|x_i - P_C(u)\|^2 .$$

Recalling that $u = \sum_{i=1}^n t_i x_i$, we obtain

$$\|u - P_C(u)\| \leq \frac{1}{2\rho(P_C(u))} \sum_{i=1}^{n} t_i \|x_i - P_C(u)\|^2 . \tag{38}$$

Putting $I := \sum_{i=1}^n t_i \|x_i - u\|^2$, an elementary computation taking into account the condition $\sum_{i=1}^n t_i(u - x_i) = 0$ entails, for all $v \in \mathbb{R}^n$,

$$\sum_{i=1}^{n} t_i \|x_i - v\|^2 = \|u - v\|^2 + I. \tag{39}$$

Now we compute I. Taking $v = x_j$ in (39), we have

$$\sum_{i=1}^{n} t_i \|x_i - x_j\|^2 = \|x - x_j\|^2 + I.$$

Thus we obtain both

$$t_j \sum_{i=1}^{n} t_i \|x_i - x_j\|^2 = t_j \|u - x_j\|^2 + t_j I$$

and

$$\sum_{j=1}^{n} \sum_{i=1}^{n} t_j t_i \|x_i - x_j\|^2 = \sum_{j=1}^{n} t_j \|u - x_j\|^2 + \sum_{j=1}^{n} t_j I.$$

¿From $\sum_{j=1}^n t_j = 1$ and $I = \sum_{j=1}^n t_j \|u - x_j\|^2$, we see that

$$I = \frac{1}{2} \sum_{j=1}^{n} \sum_{i=1}^{n} t_j t_i \|x_i - x_j\|^2 .$$

Using this expression in (39) with $P_C(u)$ in place of v, we obtain

$$\sum_{i=1}^{n} t_i \|x_i - P_C(u)\|^2 = \|u - P_C(u)\|^2 + \frac{1}{2} \sum_{j=1}^{n} \sum_{i=1}^{n} t_j t_i \|x_i - x_j\|^2 .$$

Thus, recalling (38),

$$\|u - P_C(u)\| \le \frac{1}{2\rho(P_C(u))}\|u - P_C(u)\|^2 + \frac{\frac{1}{2\rho(P_C(u))}}{2}\sum_{j=1}^{n+1}\sum_{i=1}^{n} t_j t_i \|x_i - x_j\|^2 .$$

Since $\frac{1}{2\rho(P_C(u))}\|u - P_C(u)\| = \frac{1}{2\rho(P_C(u))}d(u, C) < \frac{1}{2}$, the proof of (i) is concluded. Observe that (ii) follows immediately from (i). To see (iii), note that since $u_t \to x \in O_{\rho(\cdot)}(C)$ as $t \downarrow 0$, the openness of $O_{\rho(\cdot)}(C)$ established in Proposition 4 ensures the existence of some positive $\varepsilon \le 1$ such that $u_t \in O_{\rho(\cdot)}(C)$ for all $t \in [0, \varepsilon]$. We may put $\alpha(t) := \rho(P_C(u_t))$ for all $t \in [0, \varepsilon]$ and we see that the function $\alpha(\cdot)$ is continuous thanks to the continuity of P_C on the open set $O_{\rho(\cdot)}(C)$ according to Theorem 3 and to Proposition 4, and thanks to the continuous dependence of u_t with respect to t. The inequality in (ii) then yields that the inequality in (iii) is verified with the function $\alpha(\cdot)$. □

REMARK 10. Condition (iii) in the above Proposition is actually equivalent to the prox-regularity of C, as it will be shown in (q) of Theorem 14.

Our next step is characterizing the $\rho(\cdot)$-prox-regularity of C by the Fréchet differentiability of its square distance function.

We start with the following proposition which has its own interest. Recall first that, for a convex continuous function f that is Gâteaux differentiable on an open set U, the equality $\langle \nabla^G f(x), h\rangle = \inf_{t>0} t^{-1}[f(x+th) - f(x)]$ ensures that the function $x \mapsto \langle \nabla^G f(x), h\rangle$ (for each $h \in H$) and its opposite are upper semicontinuous, and hence the mapping $\nabla^G f$ is norm-to-weak continuous on U.

PROPOSITION 11. For the closed set C and any nonempty open set U of H the following assertions are equivalent:

(i) The function d_C^2 is Fréchet regular on U, that is, $\partial^F(d_C^2)(x) = \partial^{Cl}(d_C^2)(x)$ for all $x \in U$.
(ii) d_C^2 is Fréchet differentiable on U.
(iii) d_C^2 is of class C^1 on U.
(iv) The mapping P_C is well defined on U and norm-to-norm continuous on U.
(v) The mapping P_C is well defined on U and norm-to-weak continuous on U.
(vi) d_C^2 is Gâteaux differentiable on U and $\mathrm{Proj}_C(u) \neq \emptyset$ for all $u \in U$.
(vii) d_C is Gâteaux differentiable on $U \setminus C$ and $\|\nabla^G d_c(u)\| = 1$ for all $u \in U \setminus C$.

If C is weakly closed relative to U, then one can add the following property to the list of equivalences:

(viii) The mapping P_C is well defined on U.

PROOF. The equivalence between (i) and (ii) comes from (a) and (e) in Lemma 5. Suppose now that (v) holds. For any $x \in U$ and any sequence $(x_n)_n$ of U converging to x in norm we have that $P_C(x_n) \xrightarrow[n]{w} P_C(x)$ and

$$\|x_n - P_C(x_n)\| = d_C(x_n) \xrightarrow[n]{} d_C(x) = \|x - P_C(x)\|,$$

and hence $P_C(x_n) \xrightarrow[n]{\|\ \|} P_C(x)$. So, $(v) \Rightarrow (iv)$, that is, $(v) \Leftrightarrow (iv)$ because the converse implication is obvious.

Suppose now that (ii) holds. We know by (e) in Lemma 5 that P_C is well defined on U and that $\nabla^F d_C^2(u) = 2(u - P_C(u))$ for all $u \in U$. Following an idea due to Asplund (see [**2**]), observe that

$$(40)\quad -d_C^2(x) = \sup_{y\in C}(-\|x-y\|^2) = -\|x\|^2 + \sup_{y\in C}(2\langle x,y\rangle - \|y\|^2) \quad \text{for all } x \in H$$

and that for $f(x) := \sup_{y\in C}(2\langle x,y\rangle - \|y\|^2)$ the function f is convex and continuous on H. Further, the Fréchet differentiability of d_C^2 on U entails that of f. The function f being convex continuous and Fréchet differentiable on U, we obtain that its derivative is norm-to-weak continuous on U according to the result recalled above. So, P_C is norm-to-weak continuous on U, and hence (v) holds. Moreover, by the above equivalence between (iv) and (v) we have the norm-to-norm continuity of P_C on U, which ensures the norm-to-norm continuity of $\nabla^F d_C^2$ on U. Thus, we have established the implications $(ii) \Rightarrow (iii)$ and $(ii) \Rightarrow (iv)$.

Since the implication $(iii) \Rightarrow (ii)$ is obvious, let us prove $(iv) \Rightarrow (iii)$. We know by (e) in Lemma 5 that at each point $u \in H$ where d_C^2 is Fréchet subdifferentiable we have $\partial^F(d_C^2)(u) = \{2(u - P_C(u))\}$. So, if (iv) holds (i.e., P_C is well defined and continuous on U), it follows from (18) that $\partial^{Cl}(d_C^2)(x) = \{2(x - P_C(x))\}$ for all $x \in U$. This says in particular that d_C^2 is Gâteaux differentiable on U with $\nabla^G(d_C^2)(x) = 2(x - P_C(x))$ for all $x \in U$. The norm-to-norm continuity of P_C again yields the existence of $\nabla^F(d_C^2)$ along with its norm-to-norm continuity on U. Consequently, the implication $(iv) \Rightarrow (iii)$ is true and the equivalence between the properties $(i) - (v)$ holds.

The property (ii) implies (vi) since (v) is true under (ii). Suppose (vi) and fix $u \in U$. Choosing any y in the nonempty set $\mathrm{Proj}_C(u)$ and taking any $h \in H$, the Gâteaux differentiability of d_C^2 at u tells us that for some $\varepsilon(t)$ converging to 0 as $t \to 0$ we have

$$(41)\quad \begin{aligned} t\langle \nabla^G d_C^2(u), h\rangle &= d_C^2(u+th) - d_C^2(u) + t\varepsilon(t) \\ &\le \|u+th-y\|^2 - \|u-y\|^2 + t\varepsilon(t) \\ &= 2t\langle u-y, h\rangle + t^2\|h\|^2 + t\varepsilon(t) \end{aligned}$$

and hence dividing by $t > 0$ and making $t \downarrow 0$ gives $\langle \nabla^G d_C^2(u), h\rangle \le 2\langle u-y, h\rangle$. This being true for all $h \in H$, it follows that $\nabla^G d_C^2(u) = 2(u-y)$ and hence $\mathrm{Proj}_C(u) = \{y\}$, that is, $P_C(u)$ exists and $\nabla^G d_C^2(u) = 2(u - P_C(u))$. Using the equality $f + d_C^2 = \|\ \|^2$ given by (40) we see that the convex function f is Gâteaux differentiable on U and $\nabla^G f(u) = 2P_C(u)$. The function f being convex and continuous on U, its Gâteaux derivative on the open set U is norm-to-weak continuous (see the result recalled before the proposition). Consequently, P_C is well defined and norm-to-weak continuous on U, that is, (v) is fulfilled. The equivalence between (ii) and (vi) is then established.

The equivalence $(ii) \Leftrightarrow (vii)$ is a direct consequence of Theorem 2.4 in [**64**].

Finally, under the weak closedness of C along with the existence of $P_C(u)$ for all $u \in U$, we have for any net $(u_i)_i$ in U converging strongly to $u \in U$

$$(42)\quad \|u_i - P_C(u_i)\| = d_C(u_i) \to d_C(u) = \|u - P_C(u)\|.$$

This tells us that the net $(P_C(u_i))_i$ is eventually bounded and hence its set of weak cluster points is nonempty. For any weak cluster point y of $(P_C(u_i))_i$ we have by the weak lower semicontinuity of $\| \ \|$ and by (42) that $\|u-y\| \le \|u-P_C(u)\|$. Since $y \in C$ according to the weak closedness of C, we obtain that $y = P_C(u)$. Consequently, P_C is norm-to-weak continuous on U. Therefore, $(viii)$ is equivalent to (v) whenever C is weakly closed. The proof of the proposition is then complete. □

The following crucial lemma (see [**106**]) will be also used in our characterization with the square distance function.

LEMMA 12. Assume that the distance function d_C to the closed set C is Fréchet differentiable on a neighborhood of a point $\bar{u} \notin C$. Then there exists some $\delta > 0$ such that whenever $u \in B(\bar{u}, \delta)$ and $P_C(u) = x$, there exists some $t > 0$ such that the point $u_t := u + t(u - x)$ also has $P_C(u_t) = x$.

PROOF. By Proposition 11, we may choose some $\varepsilon > 0$ such that $B(\bar{u}, 2\varepsilon) \subset H \setminus C$ and P_C is well-defined and continuous on $B(\bar{u}, 2\varepsilon)$, with d_C Fréchet differentiable on $B(\bar{u}, 2\varepsilon)$. Let $\sigma := \sup\{d_C(u) : u \in B(\bar{u}, \varepsilon)\}$. For all $u \in B(\bar{u}, \varepsilon)$ we have on one hand

$$d_C(u) \ge d_C(\bar{u}) - \|u - \bar{u}\| > 2\varepsilon - \varepsilon = \varepsilon,$$

hence $\varepsilon < d_C(u) \le \sigma$, and on the other hand for any positive $t < \varepsilon/\sigma$ and for $u_t := u + t(u - P_C(u))$

$$\begin{aligned} \|u_t - \bar{u}\| &= \|(u - \bar{u}) + t(u - P_C(u))\| \le \|u - \bar{u}\| + t\|u - P_C(u)\| \\ &\le \|u - \bar{u}\| + t d_C(u) < \varepsilon + (\varepsilon/\sigma)\sigma = 2\varepsilon, \end{aligned}$$

thus $u_t \in B(\bar{u}, 2\varepsilon)$. Fix positive numbers $\delta < \varepsilon$ and $s < \delta/\sigma$ (thus $s < 1$) such that

$$\|P_C(u_s) - P_C(u)\| < d_C(u) \quad \text{for all } u \in B(\bar{u}, \delta),$$

which is possible thanks to the continuity of d_C and P_C since $d_C(u) - \|P_C(u_s) - P_C(u)\| \to d_C(\bar{u}) > 0$ as $s \downarrow 0$ and $u \to \bar{u}$. This inequality assures us that for all $u \in B(\bar{u}, \delta)$ we have

$$\begin{aligned} d_C(u_s) &= \|u_s - P_C(u_s)\| = \|u + s(u - P_C(u)) - P_C(u_s)\| \\ &= \|(1 + s)(u - P_C(u_s)) + s(P_C(u_s) - P_C(u))\| \\ &\ge (1 + s)\|u - P_C(u_s)\| - s\|P_C(u_s) - P_C(u)\| \\ &> (1 + s)d_C(u) - s d_C(u) = d_C(u). \end{aligned}$$

Then for all $u \in B(\bar{u}, \delta)$ we have $d_C(u_s) > d_C(u)$ and also $\delta < d_C(u) < \delta/s$ because $\delta < \varepsilon$ and $s < \delta/\sigma$.

Take now any $u \in B(\bar{u}, \delta)$ and consider the closed set $D := \{w \in H : d_C(w) \ge d_C(u_s)\}$. As $u \notin D$, according to Lau's theorem (see [**80**]) there is a sequence $D \not\ni y_n \xrightarrow[n]{\| \|} u$ with $\mathrm{Proj}_D(y_n) \neq \emptyset$. Choosing $w_n \in \mathrm{Proj}_D(y_n)$ we have $d_C(w_n) = d_C(u_s)$ (because w_n is a boundary point of D). For all n large enough, $w_n \in B(\bar{u}, 2\delta)$ since $y_n \in B(\bar{u}, \delta)$ eventually and

$$\|y_n - w_n\| = d_D(y_n) \le \|y_n - u_s\| \xrightarrow[n]{} \|u - u_s\| = s\|u - P_C(u)\| = s d_C(u) < \delta. \tag{43}$$

This ensures that d_C is Fréchet differentiable at w_n and by (d) in Lemma 5

$$\nabla^F d_C(w_n) = (w_n - P_C(w_n))/d_C(w_n) \quad \text{and} \quad \|\nabla^F d_C(w_n)\| = 1.$$

Therefore, the half-space $E := \{v \in H : \langle -\nabla^F d_C(w_n), v\rangle \leq 0\}$ gives the Bouligand tangent cone to D at w_n, and since proximal normals must belong to the negative polar of this tangent cone, we obtain that $y_n - w_n \in E^0 = -[0, +\infty[\nabla^F d_C(w_n)$, that is, there exists some $\lambda_n > 0$ such that $y_n - w_n = -\lambda_n \nabla^F d_C(w_n)$, and hence

$$y_n - w_n = -\lambda_n(w_n - P_C(w_n))/d_C(w_n) \quad \text{and} \quad \lambda_n = \|y_n - w_n\|.$$

For n large enough, we have by (43) that $\lambda_n < \delta$ and hence

$$\lambda_n < \delta \leq d_C(u) < d_C(u_s) = d_C(w_n).$$

It follows that for $\alpha_n := \lambda_n/d_C(w_n)$ we have $\alpha_n \in]0,1[$ and $y_n = (1-\alpha_n)w_n + \alpha_n P_C(w_n)$. Hence, $P_C(y_n) = P_C(w_n)$ and

$$\lambda_n = \|y_n - w_n\| = d_C(w_n) - d_C(y_n) = d_C(u_s) - d_C(y_n).$$

Putting $t_n := \frac{\alpha_n}{1-\alpha_n} = \frac{d_C(u_s)-d_C(y_n)}{d_C(y_n)}$, we obtain $w_n = y_n + t_n(y_n - P_C(y_n))$. As $(t_n)_n$ converges to $t := (d_C(u_s) - d_C(u))/d_C(u) > 0$, we have $w_n \underset{n}{\longrightarrow} u_t$ and $u_t \in B(\bar{u}, 2\delta)$ by (43) and by the inclusion $u \in B(\bar{u}, \delta)$. So, by continuity of P_C over $B(\bar{u}, 2\delta)$ we get $P_C(w_n) \overset{\|\,\|}{\underset{n}{\longrightarrow}} P_C(u_t)$. But we also have $P_C(w_n) \overset{\|\,\|}{\underset{n}{\longrightarrow}} P_C(u)$ because $P_C(w_n) = P_C(y_n)$ and $y_n \underset{n}{\longrightarrow} u$. Finally, for this number t we have $P_C(u_t) = P_C(u)$ and hence the proof is complete. □

The next lemma follows from the preceeding one.

LEMMA 13. Let C be closed in H and let $\bar{x} \in C$. If the mapping P_C is well defined and norm-to-norm continuous on a neighborhood U of $\bar{x}$, then there exists some $\varepsilon > 0$ such that for all $x \in C \cap B(\bar{x}, \varepsilon)$ and all nonzero $v \in N^P(C; x)$ the equality $P_C(x + \varepsilon \frac{v}{\|v\|}) = x$ holds.

PROOF. Let $B(\bar{x}, \delta)$ be a ball on which P_C is well defined and norm-to-norm continuous. Take $\varepsilon < \delta/2$ and consider any nonzero $v \in N^P(C; x)$ with $\|x - \bar{x}\| < \varepsilon$. By definition of the proximal normal cone, there exists $\lambda > 0$ such that $P_C(x + \lambda v) = x$. Set $\lambda_s := \sup\{\lambda \leq \varepsilon : P_C(x + \lambda \frac{v}{\|v\|}) = x\}$. By the continuity of P_C on $B(\bar{x}, \delta)$ we have that $P_C(x + \lambda_s \frac{v}{\|v\|}) = x$. Suppose that $\lambda_s < \varepsilon$. As $x + \lambda_s \frac{v}{\|v\|}$ belongs to the open set $B(\bar{x}, \delta) \setminus C$ where d_C is Fréchet differentiable according to Proposition 11, by Lemma 12 there exists some $\eta > 0$ with $\lambda_s + \eta \leq \varepsilon$ such that $P_C(x + (\lambda_s + \eta)\frac{v}{\|v\|}) = x$. This gives a contradiction with the definition of λ_s. Hence $\lambda_s = \varepsilon$. □

The following theorem provides the characterization of the $\rho(\cdot)$-prox-regularity of C by the Fréchet differentiability of its square distance function, together with other characterizations. The statement and proof of the theorem make use of the sets $U_{\rho(\cdot)}(C)$ and $U^{\gamma}_{\rho(\cdot)}(C)$ in (28) and (29).

THEOREM 14. The closed set C of H is $\rho(\cdot)$-prox-regular if and only if anyone of the following assertions holds.

(f) For any positive $\gamma < 1$ the set $U^{\gamma}_{\rho(\cdot)}(C)$ is open and the mapping P_C is well-defined on $U^{\gamma}_{\rho(\cdot)}(C)$ and Lipschitz continuous there with $(1-\gamma)^{-1}$ as a Lipschitz constant, i.e.,

$$\|P_C(u_1) - P_C(u_2)\| \leq (1-\gamma)^{-1}\|u_1 - u_2\| \quad \text{for all } u_1, u_2 \in U^{\gamma}_{\rho(\cdot)}(C).$$

(g) The set $U_{\rho(\cdot)}(C)$ is open and coincides with $O_{\rho(\cdot)}(C)$, and the mapping P_C is well defined on $U_{\rho(\cdot)}(C)$ with for all $u_1, u_2 \in U_{\rho(\cdot)}(C)$

$$\|P_C(u_1) - P_C(u_2)\| \leq \left(1 - \frac{d_C(u_1)}{2\rho(P_C(u_1))} - \frac{d_C(u_2)}{2\rho(P_C(u_2))}\right)^{-1} \|u_1 - u_2\|.$$

(h) The set $U_{\rho(\cdot)}(C)$ is open and the mapping P_C is well defined on $U_{\rho(\cdot)}(C)$ and locally Lipschitz continuous on $U_{\rho(\cdot)}(C)$.
(i) The set $U_{\rho(\cdot)}(C)$ is open and the mapping P_C is well defined on $U_{\rho(\cdot)}(C)$ and norm-to-norm continuous on $U_{\rho(\cdot)}(C)$.
(j) The set $U_{\rho(\cdot)}(C)$ is open and the mapping P_C is well defined on $U_{\rho(\cdot)}(C)$ and norm-to-weak continuous on $U_{\rho(\cdot)}(C)$.
(k) The set $U_{\rho(\cdot)}(C)$ is open and d_C^2 is of class $C^{1,1}$ on $U_{\rho(\cdot)}(C)$ and $\nabla d_C^2(u) = 2(u - P_C(u))$ for all $u \in U_{\rho(\cdot)}(C)$.
(l) The set $U_{\rho(\cdot)}(C)$ is open and d_C^2 is of class C^1 on $U_{\rho(\cdot)}(C)$.
(m) The set $U_{\rho(\cdot)}(C)$ is open and d_C^2 is Fréchet differentiable on $U_{\rho(\cdot)}(C)$.
(n) The set $U_{\rho(\cdot)}(C)$ is open, Proj_C takes nonempty values on $U_{\rho(\cdot)}(C)$, and d_C^2 is Gâteaux differentiable on $U_{\rho(\cdot)}(C)$.
(o) The set $U_{\rho(\cdot)}(C)$ is open and d_C is Gâteaux differentiable on $U_{\text{rhod}}(C) \setminus C$ with $\nabla d_C(u) = (u - P_C(u))/d_C(u)$ for all $u \in U_{\rho(\cdot)}(C) \setminus C$.
(p) The set $U_{\rho(\cdot)}(C)$ is open and for any $u \in U_{\rho(\cdot)}(C) \setminus C$ and $x = P_C(u)$ one has

$$x = P_C\left(x + t\frac{u - x}{\|u - x\|}\right) \quad \text{for all } t \in [0, \rho(x)[.$$

(q) The set $U_{\rho(\cdot)}(C)$ is open and for all $x, y \in C$ there exist a positive $\varepsilon \leq 1$ and a continuous function $\alpha : [0, \varepsilon] \to]0, +\infty]$ with $\alpha(0) = \rho(x)$ and such that for all $t \in [0, \varepsilon]$

$$d_C(u_t) \leq \frac{1}{2\alpha(t)} \min\{t, 1-t\}\|y - x\|^2,$$

where $u_t := (1-t)x + ty$.

If C is weakly closed (which holds whenever H is finite dimensional), then one can add the following property to the list of equivalences:

(r) The set $U_{\rho(\cdot)}(C)$ is open and P_C is well defined on it.

PROOF. Theorem 3 and Proposition 4 assure us that the $\rho(\cdot)$-prox-regularity of C entails (f) and (g). On the other hand (f) obviously implies (h) and (h) implies (i). To see that (g) also implies (i), observe first that the assumption $U^{\gamma}_{\rho(\cdot)}(C) = O^{\gamma}_{\rho(\cdot)}(C)$ and (22) ensure that for all $u \in U^{\gamma}_{\rho(\cdot)}(C)$ we have $d_C(u) = \|u - P_C(u)\| < \gamma\rho(P_C(u))$. So, fixing any $\bar{u} \in U_{\rho(\cdot)}(C)$, the inequality of (g) yields for all $u \in U_{\rho(\cdot)}(C)$

$$\|P_C(u) - P_C(\bar{u})\| \le \frac{1}{2}\left(1 - \frac{d_C(\bar{u})}{\rho(P_C(\bar{u}))}\right)^{-1} \|u - \bar{u}\|,$$

which entails the continuity of P_C at $\bar{u}$ according to the openness assumption of $U_{\rho(\cdot)}(C)$, that is, the implication $(g) \Rightarrow (i)$ is true. Further, according to Proposition 11 and Lemma 5 the assertions (i), (j), (k), (l), (m), (n), and (o) are pairwise equivalent.

Let us show $(m) \Rightarrow (p)$. By Proposition 11 P_C is well defined on $U_{\rho(\cdot)}(C)$ and continuous there. Take any $u \in U_{\rho(\cdot)}(C) \setminus C$ and $x = P_C(u)$ (hence $d_C(u) < \rho(x)$), and take also any $t \in [d_C(u), \rho(x)[$. According to the Fréchet differentiability of d_C^2 on the open neighborhood $U_{\rho(\cdot)}(C)$ of u, we have by Lemma 12 the existence of some $\tau_0 > 0$ such that $u_\tau \in U_{\rho(\cdot)}(C)$ and $P_C(u_\tau) = x$ for all nonnegative $\tau < \tau_0$, where $u_\tau := u + \tau(u-x)/\|u-x\|$. The set of all $\tau \in [0, t - d_C(u)]$ such that $u_\tau \in U_{\rho(\cdot)}(C)$ and $P_C(u_\tau) = x$ is then nonempty. Let s be its supremum and $(\tau_n)_n$ be a sequence of that set converging to s. For each n we have $d_C(u_{\tau_n}) = \tau_n + d_C(u)$ and $\|u_{\tau_n} - x\| = d_C(u_{\tau_n})$. Combining this with $u_{\tau_n} \underset{n}{\longrightarrow} u_s$, we obtain $\|u_s - x\| = d_C(u_s)$ and $d_C(u_s) = s + d_C(u) \le t < \rho(x)$, and hence $u_s \in U_{\rho(\cdot)}(C)$ and $x = P_C(u_s)$. The function d_C being Fréchet differentiable on the open set $U_{\rho(\cdot)}(C)$, Lemma 12 assures us that $s = t - d_C(u)$ (otherwise, we would get a contradiction). Observing that

$$u_s = x + (s + d_C(u))(u-x)/\|u-x\| = x + t(u-x)/\|u-x\|,$$

we obtain that $x = P_C(x + t(u-x)/\|u-x\|)$ for all $t \in [d_C(u), \rho(x)[$. The same equality being easily seen to be still true for all $t \in [0, d_C(u)]$, it follows that it holds for all $t \in [0, \rho(x)[$, that is, (p) is fulfilled.

Let us prove that (p) entails the $\rho(\cdot)$-prox-regularity of C. Assume then (p) and fix any $x \in C$ and $v \in N^P(C;x) \cap \mathbb{B}_H$. By Definition 1 it is enough to show that $x = P_C(x + tv)$ for all positive $t < \rho(x)$. This is obviously true if $v = 0$. Suppose that v is nonzero and put $w := v/\|v\|$. By (13) there exists some $s > 0$ such that $x = P_C(x + sw)$. We may take $s > 0$ small enough that $x + sw \in U_{\rho(\cdot)}(C)$ and hence $u := x + sw \in U_{\rho(\cdot)}(C) \setminus C$. Then by (p) we have $x = P_C(x + t(u-x)/\|u-x\|)$ for all $t \in [0, \rho(x)[$, and hence $x = P_C(x + tv)$ for all positive $t < \rho(x)$. This means that C is $\rho(\cdot)$-prox-regular.

The ρ-prox-regularity of C ensuring (q) by Proposition 9, let us prove that (q) entails (b) in Theorem 3. Fix any $x, y \in C$ and any $v \in N^P(C;x) \cap \mathbb{B}_H$. Take ε and $\alpha(\cdot)$ as given by (q). By the first equality in (13) we know that $v \in \partial^P d_C(x)$ and hence by (9) there exists some $\sigma > 0$ and some positive $\varepsilon_0 \le \varepsilon$ such that for all $t \in]0, \varepsilon_0]$

$$\langle v, t(y-x)\rangle = \langle v, u_t - x\rangle \le d_C(u_t) + \sigma\|u_t - x\|^2 = d_C(u_t) + \sigma t^2\|y-x\|^2.$$

Therefore, for all $t \in]0, \varepsilon_0]$ we have according to the assumption (q)

$$\langle v, y-x\rangle \le \frac{1}{2\alpha(t)}\|y-x\|^2 + \sigma t\|y-x\|^2,$$

and making $t \downarrow 0$ we obtain $\langle v, y-x\rangle \le \frac{1}{2\rho(x)}\|y-x\|^2$ because $\lim_{t\downarrow 0} \alpha(t) = \rho(x)$ by the assumption (q).

Finally, under the weak closedness of C the equivalence between (r) and (j) follows from Proposition 11. The proof of the theorem in then complete. □

Instead of the differentiability of the distance function d_C, its Fréchet or proximal subdifferentiability can be used to characterize the $\rho(\cdot)$-prox-regularity of C.

PROPOSITION 15. The closed C of H is $\rho(\cdot)$-prox-regular if and only if one of the two following properties holds.

(i) $U_{\rho(\cdot)}(C)$ is open and $\partial^P d_C(u) \neq \emptyset$ for all $u \in U_{\rho(\cdot)}(C)$.
(ii) $U_{\rho(\cdot)}(C)$ is open and $\partial^F d_C(u) \neq \emptyset$ for all $u \in U_{\rho(\cdot)}(C)$.

PROOF. The $\rho(\cdot)$-prox-regularity of C ensures by (k) in Theorem 14 the openness of $U_{\rho(\cdot)}(C)$ along with the $C^{1,1}$-property of d_C on $U_{\rho(\cdot)}(C)$, and the latter is easily seen to entail in particular the ∂^P-subdifferentiability of d_C on $U_{\rho(\cdot)}(C)$. So, the $\rho(\cdot)$-prox-regularity of C implies (i). On the other hand, the implication $(i) \Rightarrow (ii)$ is obvious because $\partial^P f(\cdot) \subset \partial^F f(\cdot)$ for any function f. Finally, the fact that (ii) entails the $\rho(\cdot)$-prox-regularity of C is a direct consequence of (a) and (e) in Lemma 5 and of (m) in Theorem 14. □

3.2. Uniform prox-regularity. Let us consider now the case when $\rho(\cdot)$ is a constant function, that is, $\rho(x) = r$ for all $x \in C$, where $r \in]0, +\infty]$. We will write $O_r(C)$, $U_r(C)$, $O_r^\gamma(C)$, $U_r^\gamma(C)$ for the sets corresponding to the constant function associated with the value $r > 0$.

For any closed set C and for the constant r it is not difficult to see that

$$\limsup_{\substack{\|x-u\| \to d_C(u) \\ x \in C}} \frac{\|x-u\|}{\rho(x)} = \frac{d_C(u)}{r} = \liminf_{\substack{\mathrm{Dom}\,\mathrm{Proj}_C \ni u' \to u \\ x \in \mathrm{Proj}_C(u')}} \frac{\|x-u\|}{\rho(x)},$$

and this obviously assures us that

$$O_r(C) = \{u \in H : d_C(u) < r\} = U_r(C) \tag{44}$$

and

$$O_r^\gamma(C) = \{u \in H : d_C(u) < \gamma r\} = U_r^\gamma(C), \tag{45}$$

where $0 < \gamma < 1$. The middle sets in (44) and (45) being trivially open, the sets $U_r(C)$ and $U_r^\gamma(C)$, for any positive $\gamma < 1$, are then always open in H for any closed set C (at the opposite of $U_{\rho(\cdot)}(C)$ and $U_{\rho(\cdot)}^\gamma(C)$ for which we had to require the $\rho(\cdot)$-prox-regularity for their openness).

THEOREM 16. Let C be a closed subset of H and $r \in]0, +\infty]$. Then the following assertions are equivalent.

(a) The set C is r-prox-regular.
(b) For any $x, x' \in C$ and $v \in N^P(C; x)$ one has

$$\langle v, x' - x \rangle \leq \frac{1}{2r} \|v\| \, \|x' - x\|^2.$$

(c) For any $x \in C$, any $v \in N^P(C;x) \cap \mathbb{B}_H$, and any nonnegative real number $t < r$ one has $x = P_C(x + tv)$.

(d) For any $x_i \in C$, $v_i \in N^P(C;x_i)$ with $i = 1, 2$ one has

$$\langle v_1 - v_2, x_1 - x_2 \rangle \geq -\frac{1}{2}\left(\frac{\|v_1\|}{r} + \frac{\|v_2\|}{r}\right)\|x_1 - x_2\|^2.$$

(d') For any $x_i \in C$, $v_i \in N^P(C;x_i) \cap \mathbb{B}_H$ with $i = 1, 2$ one has

$$\langle v_1 - v_2, x_1 - x_2 \rangle \geq -\frac{1}{r}\|x_1 - x_2\|^2,$$

that is, the set-valued mapping $N^P(C;\cdot) \cap \mathbb{B}_H$ is $1/r$-hypomonotone.

(e) For any $u_i \in H$ and for any $y_i \in (I + N^P(C;\cdot))^{-1}(u_i)$ with $i = 1, 2$, one has

$$\langle y_1 - y_2, u_1 - u_2 \rangle \geq \left(1 - \frac{\|u_1 - y_1\|}{2r} - \frac{\|u_2 - y_2\|}{2r}\right)\|y_1 - y_2\|^2.$$

(f) For any positive $\gamma < 1$, for $u_i \in U_r^\gamma(C)$, and for $y_i \in (I + \gamma r\mathbb{B}_H \cap N^P(C;\cdot))^{-1}(u_i)$ with $i = 1, 2$, one has

$$\langle y_1 - y_2, u_1 - u_2 \rangle \geq (1 - \gamma)\|y_1 - y_2\|^2,$$

that is, the set-valued operator $\left(I + \gamma r\mathbb{B}_H \cap N^P(C;\cdot)\right)^{-1}$ is $(1 - \gamma)$-cocoercive on the set $U_r^\gamma(C)$.

(g) For any positive $\gamma < 1$ the mapping P_C is well-defined on $U_r^\gamma(C)$ and Lipschitz continuous on $U_r^\gamma(C)$ with $(1 - \gamma)^{-1}$ as a Lipschitz constant, i.e.,

$$\|P_C(u_1) - P_C(u_2)\| \leq (1 - \gamma)^{-1}\|u_1 - u_2\| \quad \text{for all } u_1, u_2 \in U_r^\gamma(C).$$

(g') For any positive $\gamma < 1$ the mapping P_C is well defined on $U_r^\gamma(C)$ and

$$P_C(u) = (I + \gamma r\mathbb{B}_H \cap N^P(C;\cdot))^{-1}(u) \quad \text{for all } u \in U_r^\gamma(C).$$

(g") For any positive $\gamma < 1$ and any $u \in U_r^\gamma(C)$ one has

$$\mathrm{Proj}_C(u) = (I + \gamma r\mathbb{B}_H \cap N^P(C;\cdot))^{-1}(u).$$

(h) The mapping P_C is well defined on $U_r(C)$ with for all $u_1, u_2 \in U_r(C)$

$$\|P_C(u_1) - P_C(u_2)\| \leq \left(1 - \frac{d_C(u_1)}{2r} - \frac{d_C(u_2)}{2r}\right)^{-1}\|u_1 - u_2\|.$$

(i) The mapping P_C is well defined on $U_r(C)$ and locally Lipschitz continuous there.

(j) The mapping P_C is well defined on $U_r(C)$ and norm-to-norm continuous there.

(k) The mapping P_C is well defined on $U_r(C)$ and norm-to-weak continuous there.

(l) The function d_C^2 is of class $C^{1,1}$ on $U_r(C)$ and $\nabla d_C^2(u) = 2(u - P_C(u))$ for all $u \in U_r(C)$.
(m) The function d_C^2 is of class C^1 on $U_r(C)$.
(n) The function d_C^2 is Fréchet differentiable on $U_r(C)$.
(o) The set-valued mapping Proj_C takes nonempty values on $U_r(C)$, and d_C^2 is Gâteaux differentiable on $U_r(C)$.
(p) The function d_C is Gâteaux differentiable on $U_r(C) \setminus C$ and $\nabla d_C(u) = (u - P_C(u))/d_C(u)$for all $u \in U_r(C) \setminus C$.
(q) For any $u \in U_r(C) \setminus C$ and $x = P_C(u)$ one has

$$x = P_C\left(x + t\frac{u-x}{\|u-x\|}\right) \quad \text{for all } t \in [0, r[.$$

(r) For all $x, y \in C$ and for all $t \in [0,1]$ such that $u_t \in U_r(C)$ one has

$$d_C(u_t) \le \frac{1}{2r}\min\{t, 1-t\}\|y - x\|^2,$$

where $u_t := (1-t)x + ty$.
(s) $\partial^P d_C(u) \neq \emptyset$ for all $u \in U_r(C)$.
(t) $\partial^F d_C(u) \neq \emptyset$ for all $u \in U_r(C)$.
If C is weakly closed, then one can add the following to the list of equivalences:
(u) P_C is well defined on $U_r(C)$.

PROOF. By Theorem 3 all the properties $(a) - (d)$ are equivalent, and by the comments just before the statement of the theorem and by Theorem 14 and Proposition 15 they are also equivalent to anyone of (g), $(h) - (q)$, (s), and (t). The equivalence of anyone of those properties with (r) of the theorem comes from (q) in Theorem 14 and from (i) in Proposition 9.

The r-prox-regularity of C implies (e) by the implication $(a) \Rightarrow (e)$ in Theorem 3 and (e) obviously implies (f). The property (d) trivially entails (d') and (d') implies (f) because, if $r < +\infty$ for $u_i \in U_r^\gamma(C)$ and $y_i \in (I + \gamma r\mathbb{B}_H \cap N^P(C;\cdot))^{-1}(u_i)$ we have $\|(u_i - y_i)/(\gamma r)\| \le 1$ and hence by (d')

$$(\gamma r)^{-1}\langle (u_1 - y_1) - (u_2 - y_2), y_1 - y_2\rangle \ge -r^{-1}\|y_1 - y_2\|^2,$$

which gives (f) (note that the case $r = +\infty$ is obvious). The arguments in the implication $(e) \Rightarrow (f)$ of the proof of Theorem 3 show that (f) implies (g) and (g'). Then (d), (d'), (e), (f) are each one equivalent to (a) because we already proved that (g') and (a) are equivalent. Taking into account the above implication $(f) \Rightarrow (g')$ and the trivial implication $(g') \Rightarrow (g'')$, it remains to show $(g'') \Rightarrow (b)$. Suppose (g'') and fix any $x \in C$ and $v \in N^P(C;x) \cap \mathbb{B}_H$. Consider any positive $t < r$ and take $t' \in]t, r[$. For the positive number $\gamma := t'/r < 1$, we have on one hand $x + tv \in U_r^\gamma(C)$ according to (45), and on the other hand $x + tv \in (I + \gamma r\mathbb{B}_H \cap N^P(C;\cdot))(x)$, i.e., $x \in (I + \gamma r\mathbb{B}_H \cap N^P(C;\cdot))^{-1}(x + tv)$. It follows from (g'') that $x \in \mathrm{Proj}_C(x + tv)$ and this means that (a) holds and completes the proof. □

Taking $r = +\infty$ yields the following characterizations of convexity of closed subsets of H.

COROLLARY 17. For the closed set C of H, the following properties are equivalent.

(a) C is convex.
(b) For any $x_i \in C$, $v_i \in N^P(C; x_i)$ with $i = 1, 2$ one has

$$\langle v_1 - v_2, x_1 - x_2 \rangle \geq 0,$$

that is, the set-valued mapping $N^P(C; \cdot)$ is monotone.
(c) The mapping P_C is well defined on H and 1-Lipschitz continuous there.
(d) The mapping P_C is well defined on H and norm-to-norm continuous there.
(e) The mapping P_C is well defined on H and norm-to-weak continuous there.
(f) The function d_C^2 is of class $C^{1,1}$ on H.
(g) The function d_C^2 is of class C^1 on H.
(h) The function d_C^2 is Fréchet differentiable on H.
(i) The set-valued mapping Proj_C takes nonempty values on H, and d_C^2 is Gâteaux differentiable on H.
(j) The function d_C is Gâteaux differentiable on $H \setminus C$ with $\|\nabla^G d_C(u)\| = 1$ for all $u \in H \setminus C$.
(k) $\partial^P d_C(u) \neq \emptyset$ for all $u \in H$.
(l) $\partial^F d_C(u) \neq \emptyset$ for all $u \in H$.
If C is weakly closed, then one can add the following to the list of equivalences:
(m) P_C is well defined on H.

PROOF. According to Theorem 16 all the properties $(b) - (m)$ of the corollary are equivalent to the r-prox-regularity of C for $r = +\infty$. By (r) of Theorem 16, the above properties are also equivalent to $d_C((1-t)x + ty) \leq 0$ for all $x, y \in C$, which corresponds to the convexity of the closed set C. The corollary then holds. □

The r-prox-regularity can also be characterized by the convexity of some perturbation of the square of the distance function.

PROPOSITION 18. The closed C is r-prox-regular if and only if for any positive $\gamma < 1$ the function $d_C^2 + \frac{\gamma}{1-\gamma}\|\ \|^2$ is convex on any open convex subset of $U_r^\gamma(C)$.

PROOF. Suppose that C is r-prox-regular. For any positive $\gamma < 1$ and for $\sigma := \gamma/(1-\gamma)$ the function $f := d_C^2 + \sigma\|\ \|^2$ is C^1 on the open set $U_r^\gamma(C)$ according to (m) in Theorem 16 and for all $u_i \in U_r^\gamma(C)$ with $i = 1, 2$ we have by (d) in Lemma 5

$$\begin{aligned} &\langle \nabla^F f(u_1) - \nabla^F f(u_2), u_1 - u_2 \rangle \\ &\quad = 2\langle (u_1 - P_C(u_1) + \sigma u_1) - (u_2 - P_C(u_2) + \sigma u_2), u_1 - u_2 \rangle \\ &\quad = 2(1+\sigma)\|u_1 - u_2\|^2 - 2\langle P_C(u_1) - P_C(u_2), u_1 - u_2 \rangle \\ &\quad \geq 2\left(1 + \sigma - (1-\gamma)^{-1}\right)\|u_1 - u_2\|^2 = 0, \end{aligned} \tag{46}$$

the last inequality being due to (g) of Theorem 16. This ensures that f is convex on any open convex subset of $U_r^\gamma(C)$.

Let us show the converse implication. Fix any $u \in U_r(C)$ and choose some positive $\gamma < 1$ such that $u \in U_r^\gamma(C)$. The set $U_r^\gamma(C)$ being open, there exists some open ball $B(u,\varepsilon) \subset U_r^{\gamma}(C)$. Our assumption tells us that the function $f := d_C^2 + \sigma\| \;\|^2$ is convex on $B(u,\varepsilon)$. This convexity property of the function f along with its continuity assures us that $\partial^F f(u) \neq \emptyset$, which yields that $\partial^F d_C^2(u) \neq \emptyset$. Therefore, the function d_C^2 is ∂^F-subdifferentiable on $U_r(C)$ and hence by (t) of Theorem 16 the set C is r-prox-regular. □

The next theorem provides a geometric characterization of r-prox-regularity. It uses the set of r-exterior points

$$E_r(C) := \{u \in H : d_C(u) \geq r\}$$

for any real number $r > 0$, along with the next lemma. In a similar way, we define the set of (exact) r-distance points to C by

$$D_r(C) := \{u \in H : d_C(u) = r\}.$$

LEMMA 19. Assume that the closed set C is r-prox-regular. Then for any positive real number $t \leq r$ and any $u \in U_t(C) \setminus C$ one has

$$d(u,C) + d(u,E_t(C)) = t.$$

PROOF. Fix any positive real number $t \leq r$ and any $u \in U_t(C) \setminus C$, and put $x := P_C(u)$ according to (i) of Theorem 16. Then for $u' := x + t(u-x)/\|u-x\|$ one has $x \in \mathrm{Proj}_C(u')$ by (q) of Theorem 16 and hence $d_C(u') = t$. Consequently, $u' \in E_t(C)$ and

$$d(u,C) + d(u,E_t(C)) \leq \|u-x\| + \|u-u'\| = \|u'-x\| = t. \tag{47}$$

On the other hand, for any $y \in E_t(C)$ one has

$$\|y-u\| \geq \|y - P_C(u)\| - \|u - P_C(u)\| \geq d_C(y) - d_C(u) \geq t - d_C(u),$$

which gives $d(u,E_t(C)) \geq t - d_C(u)$. Combining this inequality with (47) yields $d(u,C) + d(u,E_t(C)) = t$. □

Before proving the characterization of r-prox-regularity through the set of r-exterior points, we also need to recall the following theorem. Recall first that a function $f : H \to \mathbb{R}$ has a Gâteaux directional derivative at a point u in the full direction $h \in H$ provided that the limit $\lim_{t\to 0} t^{-1}[f(u+th) - f(u)]$ exists and is finite.

THEOREM 20. Let u be a point outside the closed set C such that $\mathrm{Proj}_C(u) \neq \emptyset$. If d_C has a Gâteaux directional derivative at u in the full direction $u - p$ for some $p \in \mathrm{Proj}_C(u)$, then d_C is Fréchet differentiable at u.

PROOF. See [**64, 15**]. □

The following is a geometric characterization of r-prox-regularity. Observe that it admits also a control theoretic interpretation: C is r-prox-regular if and only if every point of $U_r(C) \setminus C$ is *optimal* for the following minimum time problem: minimize the time taken to steer $u \in U_r(C) \setminus C$ to C subject to the (constant) dynamics $\dot{x} \in \mathbb{B}_H$. Optimal points (see Definition 2.24, p. 119 in [**6**]) are precisely those points which are *crossed* by optimal trajectories.

THEOREM 21. Assume that r is a positive real constant. Then, the closed set C is r-prox-regular if and only if

$$d(u,C) + d(u,E_r(C)) = r \quad \text{for all } u \in U_r(C) \setminus C.$$

PROOF. Lemma 19 assures us that the equality of the theorem holds whenever C is r-prox-regular. Suppose now that equality and consider any $u \in U_r(C) \setminus C$ for which $\mathrm{Proj}_C(u)$ is a singleton, say $P_C(u) =: x$. Then, for any $y \in]x,u[$ one has $P_C(y) = x$. This yields for any $\tau \in]-1, \|u-y\|/\|y-x\|[$

$$\tau^{-1}[d_C(y+\tau(y-x)) - d_C(y)] = \tau^{-1}[\|y+\tau(y-x)-x\| - \|y-x\|] = \|y-x\|,$$

which entails that d_C has a Gâteaux directional derivative at y in the full direction $y-x$ and hence by Theorem 20 the function d_C is Fréchet differentiable at y. By the equality assumption of the theorem, the function $d(\cdot, E_r(C))$ is also Fréchet differentiable at y. We then deduce by Lemma 5 that $P_{E_r(C)}(y) =: u'$ exists. Using successively the equality $d_C(u') = r$ and the equality assumption of the theorem (note that $y \in U_r(C) \setminus C$) we obtain

$$r \le \|x-u'\| \le \|y-x\| + \|y-u'\| = d(y,C) + d(y,E_r(C)) = r,$$

so $y \in]x,u'[$ and $x \in \mathrm{Proj}_C(u')$, the second inclusion being due to the fact that $\|x-u'\| = r = d_C(u')$. By (q) of Theorem 16 we conclude that C is r-prox-regular. □

3.3. Local prox-regularity. Our next step is considering the local concept of prox-regularity at a point of C.

DEFINITION 22. Let $r \in]0,+\infty]$ and $\delta \in]0,+\infty]$. The closed set C is said to be *r-prox-regular at a point $\bar{x} \in C$ with δ-latitude*, if for any $x \in C \cap B(\bar{x},\delta)$ and $v \in N^P(C;x)$ one has

$$x \in \mathrm{Proj}_C(x+tv) \quad \text{for all positive real number } t \le r.$$

We say that C is *r-prox-regular* (resp. *prox-regular*) at $\bar{x}$ when it so for some $\delta > 0$ (resp. for some $r > 0$ and $\delta > 0$).

Obviously, for $\delta = +\infty$ the above property for some $\bar{x} \in C$ corresponds to the r-prox-regularity of C.

The prox-regularity at all points of C is linked with the $\rho(\cdot)$-prox-regularity through the following proposition.

PROPOSITION 23. The closed set C is prox-regular at any of its points if and only if it is $\rho(\cdot)$-prox-regular for some continuous function $\rho(\cdot)$ from C into $]0,+\infty]$.

PROOF. Suppose that C is $\rho(\cdot)$-prox-regular for some continuous function $\rho(\cdot)$ from C into $]0,+\infty]$ and fix any $\bar{x} \in C$. Taking any positive number $r < \rho(\bar{x})$, the continuity of $\rho(\cdot)$ yields some positive δ such that $r < \rho(x)$ for all $x \in C \cap B(\bar{x},\delta)$. For any positive $t < r$, any $x \in C \cap B(\bar{x},\delta)$, and any $v \in N^P(C;x) \cap \mathbb{B}_H$ we then have $t < \rho(x)$ and hence by Definition 1 we obtain $x \in \mathrm{Proj}_C(x+tv)$. This assures us that C is prox-regular at $\bar{x}$.

Suppose now that C is prox-regular at any of its points, that is, for each $x \in C$ we have the nonemptiness of the set of positive r for which there exists some positive δ such that for all $x' \in C \cap B(x,\delta)$, $v \in N^P(C;x') \cap \mathbb{B}_H$, and $t \in]0,r[$ the inclusion $x' \in \mathrm{Proj}_C(x'+tv)$ holds. Denote by $s(x)$ the supremum of that set. It is easily checked that the function $s(\cdot)$ is lsc on C relative to C with $s(x) > 0$ for all $x \in C$, and hence by the paracompactness of the metric space C we may choose some continuous function $\rho(\cdot)$ on C with $0 < \rho(x) < s(x)$ for all $x \in C$. For each $x \in C$, the inequality $\rho(x) < s(x)$ assures us the existence of some r_x with $\rho(x) < r_x < s(x)$ and some $\delta_x > 0$ such that for all $x' \in C \cap B(x,\delta_x)$, $v \in N^P(C;x') \cap \mathbb{B}_H$, and $t \in]0,r_x[$, we have $x' \in \mathrm{Proj}_C(x'+tv)$. Consequently, for any $x \in C$ and any $v \in N^P(C;x) \cap \mathbb{B}_H$ we have $x \in \mathrm{Proj}_C(x+tv)$ for all positive $t \le \rho(x)$, that is, C is $\rho(\cdot)$-prox-regular according to Definition 1. □

The local r-prox-regularity is also related to the global r-prox-regularity as follows.

PROPOSITION 24. For a positive r the closed set C is r-prox-regular if and only if it is r-prox-regular at any of its points.

PROOF. The r-prox-regularity of C obviously ensures the local property. Suppose that C is r-prox-regular at any of its points. Fix any $x \in C$ and $v \in N^P(C;x) \cap \mathbb{B}_H$. The r-prox-regularity of C at x gives by Definition 22 some $\delta > 0$ such that for any $x' \in C \cap B(x,\delta)$ and any $v' \in N^P(C;x') \cap \mathbb{B}_H$ one has $x' \in \mathrm{Proj}_C(x'+tv')$ for all positive real number $t \le r$, and hence in particular for $x' = x$ and $v' = v$ we obtain $x \in \mathrm{Proj}_C(x+tv)$ for all positive real number $t \le r$. The set C is then r-prox-regular according to Definition 1. □

The local concept of prox-regularity enjoys properties similar to the global $\rho(\cdot)$-prox-regularity. A detailed study appears in [**106**], to which we refer for the proof of the following Theorem.

THEOREM 25. Let C be a closed set of H and let $\bar{x} \in C$. The following properties are equivalent.

(a) C is r-prox-regular at $\bar{x}$ with δ-latitude for some $r, \delta > 0$.
(b) For any $x \in C \cap B(\bar{x},\delta)$ and $v \in N^P(C;x)$ one has

$$\langle v, x' - x\rangle \le \frac{1}{2r}\|v\|\,\|x'-x\|^2 \quad \text{for all } x' \in C.$$

(c) The mapping P_C is well defined on a suitable neighborhood of $\bar{x}$ (depending only on r and δ) and locally Lipschitz continuous there.

Moreover, the following properties hold.

(d) If C is prox-regular at $\bar{x}$, then the normal cone $N^P(C;\cdot)$ is strongly $\times$ weakly sequentially closed at $\bar{x}$.
(e) If C is r-prox-regular with δ-latitude (resp. prox-regular) at $\bar{x}$, then for all x in $U_{r,\delta}(\bar{x},C)$ (resp. in some neighborhood of $\bar{x}$ relative to C) the normal regularity

$$N^P(C;x) = N^F(C;x) = N^L(C;x) = N^{Cl}(C;x)$$

is fulfilled along with the tangential regularity

$$T^B(C;x) = T^{Cl}(C;x).$$

3.4. Directional derivability of the metric projection. We conclude the section on general properties of prox-regular sets by showing that the metric projection on a prox-regular set C is directionally derivable at boundary points of C. The result can also be stated as derivability in a conical sense, see [**121**].

THEOREM 26. Assume that the closed set C of H is prox-regular at $\bar{x} \in C$. Then there exists some open neighborhood U of $\bar{x}$ such that for any $x \in C \cap U$ and any $y \in H$ one has

$$\lim_{t \downarrow 0} \frac{1}{t}[P(C; x+ty) - x] = P(T^B(C;x); y)$$

and the directional derivative $d'_C(x;y) := \lim_{t\downarrow 0} \frac{1}{t}[d_C(x+ty) - d_C(x)]$ of d_C at x in the direction y exists and

$$d'_C(x;y) = d(y, T^B(C;x)),$$

which translates the fact that $d'_C(x;\cdot)$ is a conical derivative.

PROOF. Assume that the closed set C is prox-regular at $\bar{x} \in C$ and fix any $y \in H$. By Theorem 25 there exist $\varepsilon, r > 0$ with $\varepsilon < 1/2$ such that P_C is a single-valued Lipschitz continuous mapping on $B(\bar{x}, \varepsilon)$ with $\beta \geq 0$ as a Lipschitz constant there and such that

$$\langle v' - v, x - x' \rangle \leq \frac{1}{r}\|x' - x\|^2 \tag{48}$$

for all $x, x' \in C \cap B(\bar{x}, \varepsilon)$ and all $v \in N^P(C;x)$, $v' \in N^P(C;x')$ with $\|v\| \leq 1$ and $\|v'\| \leq 1$. Fix $x \in C \cap B(\bar{x},\varepsilon)$ and $y \in X$. Choose $t_0 > 0$ such that for all $t \in]0, t_0]$ we have $x + ty \in B(\bar{x}, \varepsilon)$ and $P_C(x+ty) \in B(\bar{x},\varepsilon)$. For any $t \in]0,t_0]$ putting $h'_t := \frac{1}{t}[P_C(x+ty) - x]$, we see that

$$y - h'_t = \frac{1}{t}[x + ty - P_C(x+ty)]$$

and hence $y - h'_t \in N^P(C, P_C(x+ty))$.

Take now any sequence $(t_n)_n$ of $]0,t_0]$ converging to 0 and put $h''_n := h'_{t_n}$. The sequence $(y - h''_n)_n$ being bounded (because of the Lipschitz property of P_C), it has a subsequence $(y - h''_{\sigma(n)})_n$ converging weakly to $y - h$ for some vector $h \in H$. As $N^P(C;x) = N^L(C;x)$ (see Theorem 25), we have $y - h \in N^P(C;x)$. Put $h_n := h''_{\sigma(n)}$ for all $n \in \mathbb{N}$. By (48) there exists some constant $\gamma > 0$ independent of n such that for $\lambda_n := t_{\sigma(n)}$

$$\langle (y - h_n) - (y - h), x - P_C(x + \lambda_n y)\rangle \leq \frac{\gamma}{r}\|P_C(x+\lambda_n y) - x\|^2.$$

The latter inequality is equivalent to

$$\langle h - h_n, -\lambda_n h_n \rangle \leq \frac{\gamma}{r}\|P_C(x+\lambda_n y) - x\|^2,$$

that is,

$$\|h - h_n\|^2 - \langle h, h - h_n\rangle \leq \frac{\gamma}{r}\frac{1}{\lambda_n}\|[P_C(x+\lambda_n y) - x]\|^2.$$

Taking the Lipschitz property of P_C into account, we obtain

$$\|h - h_n\|^2 \le \langle h, h - h_n \rangle + \frac{\beta^2 \lambda_n \gamma}{r} \|y\|^2.$$

So, the sequence $(h_n)_n$ converges strongly to h and by definition of h_n we have $h \in T^B(C; x)$ (because $x + \lambda_n h_n = P_C(x + \lambda_n y) \in C$ for all n).

Take now any $z \in T^B(C; x)$. Since $T^B(C; x) = T^{Cl}(C; x)$ according to Theorem 25 again, there is a sequence $(z_n)_n$ converging to z such that $x + \lambda_n z_n \in C$ for all n. This entails that for all n

$$\|x + \lambda_n y - (x + \lambda_n z_n)\| \ge \|x + \lambda_n y - P_C(x + \lambda_n y)\|,$$

which is equivalent to

$$\|y - z_n\| \ge \|y + \frac{1}{\lambda_n}[x - P_C(x + \lambda_n y)]\|.$$

Then, passing to the limit with $n \to \infty$, we get $\|y - z\| \ge \|y - h\|$ for all z in the closed convex cone $T^B(C; x) = T^{Cl}(C; x)$, which means that $h = P(T^B(C; x); y)$. So, for any sequence of positive numbers $(t_n)_n$ converging to 0 there exists a subsequence $(t_{\sigma(n)})_n$ such that

$$\lim_{n \to \infty} \frac{1}{t_{\sigma(n)}} [P_C(x + t_{\sigma(n)} y) - x] = P(T^B(C; x); y).$$

Consequently

$$\lim_{t \downarrow 0} \frac{1}{t} [P_C(x + ty) - x] = P(T^B(C; x); y), \tag{49}$$

which is the desired equality of the theorem.

Concerning the equality relative to the directional derivative of the distance function d_C, observe first that for any positive number t small enough

$$\begin{aligned} \frac{1}{t}[d_C(x + ty) - d_C(x)] = \frac{1}{t} d_C(x + ty) &= \frac{1}{t}\|P_C(x + ty) - (x + ty)\| \\ &= \|\frac{1}{t}[P_C(x + ty) - x] - y\|. \end{aligned}$$

Then, taking (49) into account and passing to the limit when $t \downarrow 0$ give that the directional derivative $d'_C(x; y)$ exists and

$$d'_C(x; y) = \|P(T^B(C; x); y) - y\| = d(y, T^B(C; x)),$$

which completes the proof. □

As a consequence we compute the second order directional derivatives of the square distance at boundary points.

COROLLARY 27. Assume that the closed set C of H is prox-regular at $\bar{x} \in C$. Then there exists some open neighborhood U of $\bar{x}$ such that for any $x \in C \cap U$ and any $y \in H$ one has

$$\lim_{t \downarrow 0} \frac{1}{t} [\nabla d_C^2(x + ty) - \nabla d_C^2(x)] = 2(y - P(T^B(C; x); y)).$$

PROOF. It is an immediate consequence of the above Theorem and of the explicit computation of the gradient of $d_C^2(\cdot)$ (see (l) in Theorem 14). □

Another property concerning the metric projection to the tangent cone is ensured by the sequential strong × weak closedness property in Theorem 25.

PROPOSITION 28. Assume that the closed set C is prox-regular at $\bar{x} \in C$ and that $v \in T^B(C;\bar{x})$. Then

$$\lim_{C \ni x \to \bar{x}} P(T^B(C;x);v) = v.$$

PROOF. For $x \in C$ in some neighborhood U of $\bar{x}$ we know that the closed cone $T^B(C;x)$ is convex and hence $v(x) := P(T^B(C;x);v)$ exists. Take any sequence $(x_n)_n$ of $C \cap U$ converging to $\bar{x}$ and put $v_n := v(x_n)$. Obviously the sequence $(v_n)_n$ is bounded because $\|v - v_n\| \le \|v\|$. Fix any weakly convergent subsequence $(v_n)_{n \in K}$ tending to some $w \in H$. Observe that by (e) of Theorem 25

$$v - v_n \in (T^B(C;x_n))^0 = N^P(C;x_n).$$

The sequential strong × weak closedness property at $\bar{x}$ of $N^P(C;\cdot)$ (see Theorem 25) then says that $v - w \in N^P(C;\bar{x})$. Consequently, by (e) of Theorem 25 again $\langle v - w, v\rangle \le 0$ and hence $\|v\| \le \|w\|$. This entails $\|v\| \le \liminf_{n\to\infty} \|v_n\|$, and combining this with the equality

$$\|v - v_n\|^2 + \|v_n\|^2 = \|v\|^2$$

(coming from the convexity of the cone $T^B(C;x_n)$) we obtain $\lim_{n\to\infty} \|v - v_n\| = 0$. Therefore $\lim_{C \ni x \to \bar{x}} v(x) = v$. □

REMARK 29. It is clear from the proof that the result still holds with the Clarke tangent cone $T^{Cl}(C;\cdot)$ in place $T^B(C;\cdot)$ for any closed set C whose Clarke normal cone $N^{Cl}(C;\cdot)$ is closed at $\bar{x}$.

4. Operations on prox-regular sets

This section is related to the study of the preservation of prox-regularity under certain operations on sets. It is essentially concerned with the intersection of finitely many sets, the inverse image and the direct image. The convergence of prox-regular sets will be also discussed.

We start with the following example showing that the intersection of finitely many prox-regular sets may fail to be prox-regular. Consider in $\mathbb{R}^2$ endowed with its Euclidian norm the sets C_1 and C_2 in Figure 1.

The set C_1 is all the axis. The points $(x_i)_i$ form a sequence of C_1 converging to some $x \in C_1$. The set C_2 is the hatched surface delimited on one side by the closed arc $\widehat{x_1 x}$ of the half circle with radius $R = \|x_1 - x_2\|/2$ and on the other side by the arcs of the circles (*all with the same radius* $r < R$) $\widehat{x_i x_{i+1}}$, $i \in \mathbb{N}$, so that x belongs also to C_2. It is easily seen that while C_1 and C_2 are both prox-regular at all their points, their intersection $C_1 \cap C_2$ is not prox-regular at the point x.

To provide general sufficient conditions under which the prox-regularity of intersection or inverse image is preserved, we have to introduce the concept of normal

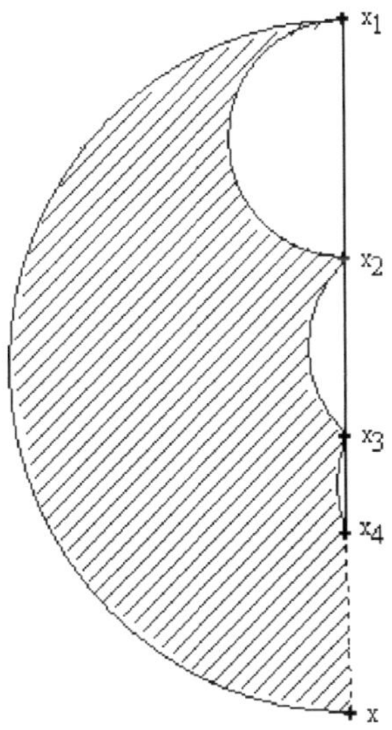

FIGURE 1.

cone property for intersection of finitely many sets or for inverse image set. We say that a finite family of closed sets $(C_k)_{k=1}^m$ of H has the *normal cone intersection property* near a point $\bar{x} \in \bigcap_{k=1}^m C_k$ with respect to a normal cone $\mathcal{N}(\cdot;\cdot)$ if there exists some $\gamma > 0$ and some neighborhood U of $\bar{x}$ such that for all $x \in U \cap C_1 \cap \cdots \cap C_m$ one has

$$\mathcal{N}\left(\bigcap_{k=1}^m C_k; x\right) \cap \mathbb{B} \subset \mathcal{N}(C_1;x) \cap \gamma\mathbb{B} + \cdots + \mathcal{N}(C_m;x) \cap \gamma\mathbb{B}. \tag{50}$$

Let now $F : H \to Y$ be a mapping from H into another Hilbert space Y and let D be a subset of Y and $\bar{x} \in F^{-1}(D)$. In the same way, we say that the inverse image set $F^{-1}(D)$ has the *normal cone inverse image property* at $\bar{x} \in F^{-1}(D)$ with respect to the normal cone $\mathcal{N}(\cdot;\cdot)$ if there exists some $\gamma > 0$ and some neighborhood U of $\bar{x}$ such that for all $x \in U \cap F^{-1}(D)$

$$\mathcal{N}(F^{-1}(D);x) \cap \mathbb{B}_H \subset DF(x)^* \left(\mathcal{N}(D;F(x)) \cap \gamma\mathbb{B}_Y\right). \tag{51}$$

More amenable conditions are provided by the notion of calmness. We say that the m sets $(C_k)_{k=1}^m$ of H are in a *metrically calm position* at the point $\bar{x} \in \bigcap_{k=1}^m C_k$ provided there exist a constant $\gamma > 0$ and a neighborhood U of $\bar{x}$ such that

$$d\left(x, \bigcap_{k=1}^m C_k\right) \le \gamma\left(d(x,C_1) + \cdots + d(x,C_m)\right) \quad \text{for all } x \in U. \tag{52}$$

Analogously, we say that the set-valued inverse of the mapping F above is *metrically calm* at $\bar{x}$ relatively to the set D if there exist a constant $\gamma > 0$ and a neighborhood U of $\bar{x}$ such that

$$d(x, F^{-1}(D)) \le \gamma d(F(x), D) \quad \text{for all } x \in U. \tag{53}$$

THEOREM 30. Let $(C_k)_{k=1}^m$ be a finite family of closed sets of H and let D be a closed set of Y.

(a) If all the sets C_k are prox-regular at a point $\bar{x}$ of their intersection and if they have the normal cone intersection property near $\bar{x}$ with respect to the Fréchet normal cone, then their intersection set $\bigcap_{k=1}^m C_k$ is prox-regular at $\bar{x}$.

(b) If a mapping $F : X \to Y$ is of class $C^{1,1}$ around a point $\bar{x} \in F^{-1}(D)$ and if the inverse image set $F^{-1}(D)$ has the normal cone inverse image property at $\bar{x}$ with respect to the Fréchet normal cone, then the inverse image set $F^{-1}(D)$ is prox-regular at $\bar{x}$.

PROOF. Assume that each set C_k is prox-regular at $\bar{x}$ and put $C := \bigcap_{k=1}^m C_k$. By Theorem 25 it is easily seen that there exist some real number $r > 0$ and some open neighborhood U of $\bar{x}$ such that all the sets C_k are prox-regular at all points of $C \cap U$ and such that for each $k = 1, \dots, m$ we have for all $x \in C_k \cap U$ and $w \in N^P(C_k; x) \cap \mathbb{B}_H$

$$\langle w, x' - x \rangle \le \frac{1}{2r} \|x' - x\|^2 \quad \text{for all } x' \in C_k \cap U. \tag{54}$$

Restricting the neighborhood U if necessary, we may suppose that for some $\gamma > 0$ the normal cone intersection property (50) also holds for points in $U \cap C_1 \cap \cdots \cap C_m$. Fix any $x \in U \cap C$ and $v \in N^P(C; x)$. By (50) there exists for each $i = 1, \dots, m$ some $w_i \in N^F(C_i; x)$ with $\|v_i\| \le \gamma$ such that $v = w_1 + \cdots + w_m$. Since $N^F(C_i; x) = N^P(C_i; x)$ (see Theorem 25), by (54) we obtain

$$\langle v, x' - x \rangle \le \frac{m\gamma}{2r} \|x' - x\|^2 \quad \text{for all } x' \in C \cap U,$$

which ensures that C is prox-regular at $\bar{x}$, that is, (a) holds.

Let us now establish (b). Fix some open convex neighborhood U of $\bar{x}$ over which (51) holds and over which the mapping F as well as its derivative $DF(\cdot)$ are Lipschitz continuous with Lipschitz constants K and K_1 respectively. For $S := F^{-1}(D)$ fix any $x \in U \cap S$ and $v \in N^P(S; x)$ with $\|v\| \le 1$. Fix by Theorem 25 some open neighborhood V of $F(\bar{x})$ and some constant $r > 0$ such that

$$\langle z, y' - y \rangle \le \frac{1}{2r} \|y' - y\|^2 \quad \text{for all } y, y' \in V \cap D, \text{ and } z \in N^P(D; y) \cap \mathbb{B}_Y. \tag{55}$$

Restricting the open convex neighborhood U of $\bar{x}$ if necessary, we may suppose that $F(U) \subset V$. By the normal cone property (51) with $\gamma > 0$, there is some $w \in N^F(D; F(x)) = N^P(D; F(x))$ with $\|w\| \le \gamma$ such that $v = DF(x)^*(w)$ and hence according to (55)

$$\langle w, y' - F(x) \rangle \le \frac{\gamma}{2r} \|y' - F(x)\|^2 \quad \text{for all } y' \in V \cap D. \tag{56}$$

Fix any $x' \in S \cap U$ and write

$$F(x') - F(x) = DF(x)(x' - x) + \int_0^1 \left(DF(x + t(x' - x)) - DF(x)\right)(x' - x)\, dt$$

and then

$$\langle v, x'-x\rangle = \langle w, F(x')-F(x)\rangle - \int_0^1 \langle w, (DF(x+t(x'-x)) - DF(x))\,(x'-x)\rangle\, dt.$$

Using (56) we obtain

$$\langle v, x'-x\rangle \le \frac{\gamma}{2r}\|F(x')-F(x)\|^2 + \gamma K_1\|x'-x\|^2 \le \frac{\gamma}{2}(\frac{K^2}{r}+K_1)\|x'-x\|^2.$$

The latter being true for all $x, x' \in S\cap U$ and $v \in N^P(S;x)\cap\mathbb{B}_H$, we conclude that the set S is prox-regular at $\bar{x}$. □

The cases with the presence of metric calmness may be deduced.

THEOREM 31. Let $(C_k)_{k=1}^m$ be a finite family of closed sets of H and let D be a closed set of Y.

(a) If all sets C_k are prox-regular at a point $\bar{x}$ in all the sets C_k and if the sets C_k are in a metrically calm position at $\bar{x}$, then their intersection set $\bigcap_{k=1}^{m} C_k$ is prox-regular at $\bar{x}$.

(b) If a mapping $F : X \to Y$ is of class $C^{1,1}$ around a point $\bar{x} \in F^{-1}(D)$ and metrically calm at $\bar{x}$ relatively to D and if D is prox-regular at $F(\bar{x})$, then the set $F^{-1}(D)$ is prox-regular at $\bar{x}$.

PROOF. Put $C := C_1\cap\cdots\cap C_m$ and fix some open neighborhood U of $\bar{x}$ such that all the sets C_k are prox-regular at all points of $C\cap U$ and such that the calmness inequality property (52) holds. Fix any $x \in C\cap U$ and take any $v \in N^F(C;x)$ with $\|v\| \le 1$. We have $v \in \partial^F d_C(x)$ since one knows that $\partial^F d_C(x) = N^F(C;x)\cap\mathbb{B}_H$ (see (7)). The calmness inequality (52) and the definition of Fréchet subgradient easily yields that $v \in \gamma\,\partial^F(d_{C_1}+\cdots+d_{C_m})(x)$ and hence in particular $v \in \gamma\,\partial^L(d_{C_1}+\cdots+d_{C_m})(x)$. The functions d_{C_k} being Lipschitz continuous, the formula of the Mordukhovich subdifferential of a finite sum of locally Lipschitz continuous functions (see the sum formula (3)) assures us that there exist $v_k \in \partial^L d_{C_k}(x)$ such that $\gamma^{-1}v = v_1+\cdots+v_m$. Since

$$\partial^L d_{C_k}(x) \subset N^L(C_k;x)\cap\mathbb{B}_H = N^F(C_k;x)\cap\mathbb{B}_H$$

(the inclusion coming from (8) and the equality from Theorem 25), we have $v_k \in N^F(C_k;x)\cap\mathbb{B}_H$ and hence we deduce that the normal cone intersection property (50) with respect to the Fréchet normal cone is satisfied near $\bar{x}$ for the sets $C_1,\ldots,C_m$. So, the assertion (a) follows from (a) in Theorem 30.

Let us now prove (b). Fix some open neighborhood U of $\bar{x}$ over which (53) holds and over which the mapping F as well as its derivative $DF(\cdot)$ are Lipschitz. For $S := F^{-1}(D)$ fix any $x \in U\cap S$ and $v \in N^F(S;x)$ with $\|v\| \le 1$. As above we have $v \in \partial^F d_S(x)$ and this entails by (53) that $v \in \gamma\,\partial^F(d_D\circ F)(x)$ and hence $v \in \gamma\,\partial^L(d_D\circ F)(x)$. The subdifferential chain rule (see the inclusion formula (4)) gives some $w \in \gamma\,\partial^L d_D(F(x))$ such that $v = DF(x)^*(w)$. Since

$$\partial^L d_D(F(x)) \subset N^L(D;F(x))\cap\mathbb{B}_Y = N^F(D;F(x))\cap\mathbb{B}_Y,$$

we obtain $w \in N^F(D;F(x))\cap\gamma\mathbb{B}_Y$, and this assures us that the normal cone property (51) holds. This gives the assertion (b) through (b) of Theorem 30. □

For the convenience of the reader, in the proof of Theorem 30 we made the choice to prove first the prox-regularity of the intersection and then to give the additional arguments yielding to the prox-regularity of the inverse image. However the case of the intersection can also be derived from that of the inverse image. Indeed putting $Y = H^m$, $F(x) = (x, \dots, x)$ for all $x \in H$, and $D = C_1 \times \cdots \times C_m$ we see that $F^{-1}(D) = \bigcap_{k=1}^{m} C_k$ and then some direct arguments and computation allow us to obtain through (b) the result of (a).

An analysis of the proofs of Theorem 30 and 31 reveals that a uniform version also holds. We state it only under the uniform calmness property in the case of the intersection of finitely many sets and leave the case of the inverse image to the reader.

THEOREM 32. Assume that all the closed sets C_k, $k = 1, \dots, m$, are r-prox-regular and that they are in a metrically calm position at any point of their intersection with the same modulus γ of metric calmness in (52). Then the set $\bigcap_{k=1}^{m} C_k$ is r'-prox-regular with $r' := r/(m\gamma)$.

Our next step now is to derive some useful corollaries. Let us begin with a set in the form $M := h^{-1}(0)$ with appropriate assumptions on h.

COROLLARY 33. Let $h : H \to Y$ be a mapping of class $C^{1,1}$ near $\bar{x} \in M := \{x \in H : h(x) = 0\}$ and such that $Dh(\bar{x})$ is onto. Then M is prox-regular at $\bar{x}$.

PROOF. The set M is obviously closed near $\bar{x}$. Since $Dh(\bar{x})$ is onto, the open mapping theorem gives us some real number $c > 0$ and some open neighborhood U of $\bar{x}$ such that $c\mathbb{B}_Y \subset Dh(x)(\mathbb{B}_H)$ for all $x \in U$. Restricting the neighborhood U if necessary, we know by the ontoness of $Dh(\bar{x})$ again (see, e.g.,[**41**]) that $N^F(M; x) = N^{Cl}(M; x) = Dh(x)^*(Y)$ for all $x \in U \cap M$. So, for any fixed $x \in U \cap M$ and $v \in N^F(M; x)$ with $\|v\| \leq 1$ we have $v = Dh(x)^*(w)$ for some $w \in Y$ and

$$\|v\| = \sup_{b \in \mathbb{B}_H} \langle b, Dh(x)^*(w) \rangle = \sup_{b \in \mathbb{B}_H} \langle w, Dh(x)(b) \rangle \geq \sup_{b' \in \mathbb{B}_Y} \langle w, cb' \rangle = c\|w\|,$$

and this ensures $\|w\| \leq 1/c$. Consequently, observing that the Fréchet normal cone to the singleton $\{0\}$ of Y is equal to $Y^* = Y$, we see that the normal cone property (51) holds with the prox-regular set $\{0\}$ of Y. The assertion (b) in Theorem 30 then tells us that M is prox-regular at $\bar{x}$. □

In the literature there are conditions with normal cones easy to handle ensuring the metric calmness inequalities (52) and (53). For example such a condition for (52) is known (see, e.g., [**75**, p. 548–549]) under the name of general position condition for intersection of finitely many sets. In our context of Hilbert space, the concept can be translated by saying that the closed sets $C_1, \dots, C_m$ are (*relative to a normal cone* $\mathcal{N}(\cdot;\cdot)$) in *sequential general position* at $\bar{x} \in \bigcap_{k=1}^{m} C_k$ whenever for any sequence of tuples $(x_{1,n}, \dots, x_{m,n}, v_{1,n}, \dots, v_{m,n})_n$ such that $x_{k,n} \in C_k$, $x_{k,n} \to \bar{x}$, $v_{k,n} \in \mathcal{N}(C_k; x_{k,n}) \cap \mathbb{B}_H$ and such that $\| \sum_{k=1}^{m} v_{k,n} \| \to 0$, one has $\|v_{k,n}\| \to 0$ for any $k = 1, \dots, m$.

Another condition which is much easier to handle involves as above a property with normal cones but at the fixed point $\bar{x}$. It probably appears for the first time as one of the assumptions of Theorem 4.10 of Federer's seminal paper [**63**] for sets of $\mathbb{R}^n$ which are not submanifolds. The sets C_k, $k = 1, \dots, m$, are (*relative to the normal cone* $\mathcal{N}(\cdot;\cdot)$) in *pointbased general position* at $\bar{x}$ provided the equality $v_1 + \cdots + v_m = 0$ with $v_k \in \mathcal{N}(C_k;\bar{x})$ entails $v_1 = \cdots = v_m = 0$.

Recall now that a closed set of X is *compactly epi-Lipschitzian* at $\bar{x} \in C$ (a concept due to J. M. Borwein and H. M. Strójwas [**19, 20, 21**]) if there exist a compact set Q, a neighborhood V of zero, a neighborhood U of $\bar{x}$ in X, and a positive real number ε such that

$$C \cap U + tV \subset C + tQ \quad \text{for all } t \in]0, \varepsilon].$$

Of course any closed subset of X is compactly epi-Lipschitzian at any of its points whenever the space X is finite dimensional.

COROLLARY 34. Let $(C_k)_{k=1}^m$ be a finite family of closed sets of X and $\bar{x} \in \bigcap_{k=1}^{m} C_k$. Then the following properties hold.

(a) If the sets C_k are in sequential general position at $\bar{x}$ with respect to the Fréchet normal cone and if each set C_k is prox-regular at $\bar{x}$, then the intersection $\bigcap_{k=1}^{m} C_k$ is a prox-regular at $\bar{x}$.
(b) If all the sets C_k except at most one are compactly epi-Lipschitzian at $\bar{x}$, then the sequential general position in (a) may be replaced by the corresponding pointbased general position at $\bar{x}$.
(c) If the space X is finite dimensional, again the sequential general position in (a) may be replaced by the corresponding pointbased general position at $\bar{x}$.

PROOF. (a) Since the Hilbert space X is an Asplund space, the sequential general position is known to entail the metric calmness property (52) (and even more) according to Proposition 6 in p. 548 and Theorem 2 in p. 545 of [**74**] (for example). Then the conclusion of (*a*) follows from Theorem 31.
(b) First observe that the prox-regularity of the sets C_k at $\bar{x}$ ensures that their Fréchet normal cones coincide near $\bar{x}$ with their limiting normal cones according to Theorem 25. The result then comes from the fact that the assumption of compactly epi-Lipschitzian property combined with the pointbased general position at $\bar{x}$ entails (see, e.g., [**78**, Theorem 3.4]) that the intersection is metrically regular and hence metrically calm at $\bar{x}$.
(c) This is a direct consequence of (*b*). □

The result in (c) was previously established in (5) of Theorem 4.10 in [**63**]. We state another uscful corollary.

COROLLARY 35. Let C be a closed set of H which is prox-regular at $\bar{x} \in C$ and let $h : H \to \mathbb{R}^m$ be a mapping of class $C^{1,1}$ near $\bar{x} \in M := \{x \in H : h(x) = 0\}$ and such that $Dh(\bar{x})$ is onto. Assume that the only vector $\lambda = (\lambda_1, \dots, \lambda_m)$ in $\mathbb{R}^m$ such that

$$\sum_{i=1}^{m} \lambda_i \nabla h_i(\bar{x}) \in N^F(C;\bar{x})$$

is the null vector $\lambda = (0, \dots, 0)$. Then $C \cap M$ is prox-regular at $\bar{x}$.

PROOF. Observe first that Corollary 33 tells us that M is prox-regular at $\bar{x}$. The ontoness property of $Dh(\bar{x})$ along with the continuity of Dh near $\bar{x}$ assures us that there exists some neighborhood U of $\bar{x}$ for which on one hand for some $c > 0$ we have $c\mathbb{B} \subset Dh(x)(\mathbb{B}_H)$ for all $x \in U$ (through the open mapping theorem), and on the other hand (see, e.g., [**41**])

$$\begin{aligned} N^F(M;x) &= N^{Cl}(M;x) \\ &= \{\lambda \circ Dh(x) : \lambda \in L(\mathbb{R}^m, \mathbb{R})\} \\ &= \left\{\sum_{i=1}^{m} \lambda_i \nabla h_i(x) : \lambda \in \mathbb{R}^m\right\}, \end{aligned}$$

where $L(\mathbb{R}^m, \mathbb{R})$ denotes the space of linear functions from $\mathbb{R}^m$ into $\mathbb{R}$. Take $(x_n, y_n)_n$ in $C \times M$ converging to $(\bar{x}, \bar{x})$ such that

$$(57) \quad v_n \in N^F(C;x_n) \cap \mathbb{B}_H, \quad w_n \in N^F(M;y_n) \cap \mathbb{B}_H, \quad \text{and} \quad \|v_n + w_n\| \to 0.$$

We may suppose that $x_n, y_n \in U$ for all integers n. Then for all n we have $w_n = \lambda_n \circ Dh(y_n)$ and hence

$$1 \geq \|w_n\| = \sup_{b \in \mathbb{B}_H} |\langle \lambda_n, Dh(y_n)(b)\rangle| \geq \sup_{u \in \mathbb{B}} |\langle \lambda_n, cu\rangle| = c\|\lambda_n\|,$$

the inequality being due to the inclusion $c\mathbb{B} \subset Dh(x_n)(\mathbb{B}_H)$. Consequently, we have $\|\lambda_n\| \leq 1/c$ for all integers n. Fix any convergent subsequence $(\lambda_{s(n)})_n$ with limit λ. We have $w_{s(n)} = \lambda_{s(n)} \circ Dh(x_{s(n)}) \to \lambda \circ Dh(\bar{x})$ in norm, and hence $v_{s(n)} \to -\lambda \circ Dh(\bar{x})$ in norm according to the convergence property in (57). Since $N^L(C;\bar{x}) = N^F(C;\bar{x})$ by the prox-regularity of C at $\bar{x}$, the first inclusion in (57) entails $\lambda \circ Dh(\bar{x}) \in -N^F(C;\bar{x})$. The qualification condition of the corollary then gives $\lambda = 0$. This being true for any convergent subsequence of the bounded sequence $(\lambda_n)_n$ in $\mathbb{R}^m$, all the sequence $(\lambda_n)_n$ converges to 0 in $\mathbb{R}^m$, and hence $\|w_n\| \to 0$. Combining this with the convergence in (57), we also obtain $\|v_n\| \to 0$. Therefore, the sets C and M are in sequential general position at $\bar{x}$ with respect to the Fréchet normal cone, and Corollary 34 says that $C \cap M$ is prox-regular at $\bar{x}$. □

The case of a real-valued function h (that is, $m = 1$) is of particular importance.

COROLLARY 36. Let C be a closed set of H which is prox-regular at $\bar{x} \in C$ and $h : H \to \mathbb{R}$ be a real-valued function of class $C^{1,1}$ near $\bar{x} \in M := \{x \in H : h(x) = 0\}$. If

$$\nabla h(\bar{x}) \notin -N^F(C;\bar{x}) \cup (-N^F(C;\bar{x})),$$

then $C \cap M$ is prox-regular at $\bar{x}$.

PROOF. Since $\nabla h(\bar{x}) \neq 0$, the continuous linear functional $Dh(\bar{x})$ from H into $\mathbb{R}$ is onto. So, the corollary is a direct consequence of Corollary 35. □

The next result is concerned with direct images of prox-regular sets. It is based on Theorem 4.18 in [**63**].

PROPOSITION 37. Let $C \subset H$ be r-prox-regular, let $0 < r' < r$, and let $f : O_{r'}(C) \to H$ be a one to one $\mathcal{C}^1$-function such that f^{-1}, and Df are Lipschitz continuous with constants L, M, respectively, and $\sup\{\|Df(x)\| : x \in C\} \leq N$. Set

$$r_1 = \min\left\{\frac{r}{M}, L^{-2}\left(M + \frac{N}{r}\right)^{-1}\right\}.$$

Then

$$f(C) \text{ is } r_1\text{-prox-regular.}$$

PROOF. Take $x, x' \in C$ such that $\|f(x') - f(x)\| \leq 2r'/L$. Being $T^B(C;x)$ convex and closed (see (b) in Proposition 7), recalling Proposition 8, there exists $u \in T^B(C;x)$ such that

$$\|x' - x - u\| \leq \frac{\|x' - x\|^2}{2r}.$$

Then it is easy to see that $Df(x)u \in T^B(f(C); f(x))$ and

$$\|Df(x)(x' - x) - Df(x)u\| \leq \frac{N}{2r}\|x' - x\|^2.$$

Since $\|x' - x\| \leq 2r'$, recalling (b) in Proposition 9 we obtain, for all $0 \leq \lambda \leq 1$,

$$d(\lambda x + (1 - \lambda)x'; C) \leq r',$$

and so, by Taylor's formula

$$\|f(x') - f(x) - Df(x)(x' - x)\| \leq \frac{M}{2}\|x' - x\|^2.$$

Therefore,

$$\begin{aligned} d(f(x') - f(x), T(f(C); f(x))) &\leq \|f(x') - f(x) - Df(x)u\| \\ &\leq \|f(x') - f(x) - Df(x)(x' - x)\| \\ &\quad + \|Df(x)(x' - x) - Df(x)u\| \\ &\leq \left(\frac{M}{2} + \frac{N}{2r}\right)\|x' - x\|^2 \\ &\leq L^2\left(\frac{M}{2} + \frac{N}{2r}\right)\|f(x') - f(x)\|^2. \end{aligned}$$

Recalling Proposition 8, the proof is complete. □

As a final result, we prove a theorem on the Hausdorff convergence of r_k-prox-regular sets, with $r_k \to r \in (0, +\infty]$. Together with its corollary on the convergence of functions with prox-regular epigraph, it may be seen as an extension to prox-regular sets of Attouch theorem (see, e.g., Theorem 12.35 in [**109**] and [**3, 15**]) on the Hausdorff convergence of convex sets and on the epigraphical convergence of convex functions. In other words, uniform prox-regularity together with Hausdorff convergence implies convergence of normal cones. For simplicity we restrict ourselves to the case of a prox-regularity parameter r_k independent of x.

THEOREM 38. Let $\{r_k : k \in \mathbb{N}\} \subset (0, +\infty]$ be such that $r_k \to r \in (0, +\infty]$. Let $\{C_k : k \in \mathbb{N}\}$ be a sequence of r_k-prox-regular subsets of H, Hausdorff converging to C. Then

(a) C is r-prox-regular;
(b) $P_{C_k}(\cdot)$ converges to $P_C(\cdot)$ uniformly on $B[C, r']$ for every $0 < r' < r$.

PROOF. First of all, observe that, by Hausdorff convergence, $d(\cdot, C_k)$ converges to $d(\cdot, C)$ uniformly in H. Then, fix $0 < r' < s < r$ and let k be so large that $B[C, r'] \subset U_s(C_k)$. Then all P_{C_k} are well defined and Lipschitz continuous with constant $r/(r-s)$ on $B[C, r']$. Recalling (l) in Theorem 16, $f_k(\cdot) := d(\cdot, C_k)^2$ is of class $\mathcal{C}^{1,1}$ on $B[C, r']$ and $Df_k(x) = 2(x - P_{C_k}(x))$. Let now $u \in H$, $\|u\| = 1$. Then, for all h with $0 < h < r - s$, for all k, and $x \in B[C, r']$ it holds

$$\left| \frac{f_k(x + hu) - f_k(x) - 2h\langle u, x - P_{C_k}(x)\rangle}{h} \right|$$
$$\leq \frac{2}{h} \int_0^h |t - \langle P_{C_k}(x + tu) - P_{C_k}(x), u\rangle| \, dt$$
$$\leq \frac{2}{h} \int_0^h t\Big(1 + \frac{r}{r-s}\Big) \, dt = \frac{2r - s}{r - s} h,$$

so that the difference ratios $(f_k(x+hu) - f_k(x))/h$ converge to $\langle Df_k(x), u\rangle$ uniformly with respect to $x \in B[C, r']$, $u \in H$, $\|u\| = 1$, and k as $h \to 0$. Set now $f(x) = d(x, C)^2$ and consider, for $u \in H$, $\|u\| = 1$, $0 < h < r - s$, the difference ratios

$$\frac{f_k(x + hu) - f_k(x)}{h}. \tag{58}$$

For all k large enough, (58) tends to $\langle Df_k(x), u\rangle$ for $h \to 0$, while for all $h > 0$ (58) converges to $(f(x+hu) - f(x))/h$ as $k \to \infty$ and all limits are uniform with respect to the rest of variables. By exchanging the limits one can therefore see that

$$\lim_{h \to 0} \frac{f(x + hu) - f(x)}{h} = \lim_{k \to \infty} \langle Df_k(x), u\rangle$$

and the limit is uniform with respect to x and u. Consequently, the sequence $(Df_k(x))_k$ being bounded, we see that the left member defines a continuous linear mapping in u, that is f is Gâteaux differentiable at x. Denoting by $D^G f(x)$ the Gâteaux derivative, we have

$$\lim_{k \to \infty} \sup_{x \in B[C, r'], \|u\| = 1} \langle Df_k(x) - D^G f(x), u\rangle = 0,$$

i.e.,

$$\lim_{k \to \infty} \sup_{x \in B[C, r']} \left\| Df_k(x) - D^G f(x) \right\| = 0.$$

Therefore $D^G f$ is continuous on $U_{r'}(C)$. Recalling (m) in Theorem 16 and (e) in Lemma 5, C is r-prox-regular and $P_{C_k}(\cdot)$ converges to $P_C(\cdot)$ uniformly in $B[C, r']$. □

The result can be extended to the Attouch-Wets or bounded Hausdorff convergence (see [**3, 15**]) and this extension has as an immediate consequence the following result.

COROLLARY 39. Let $\Omega \subset H$ be open and let $\{r_k\} \subset (0, +\infty]$ be such that $r_k \to r \in (0, +\infty]$. Let, for $k \in \mathbb{N}$, $f_k : \bar{\Omega} \to \mathbb{R} \cup \{+\infty\}$ be lower semicontinuous with r_k-prox-regular epigraph and let $f : \bar{\Omega} \to \mathbb{R} \cup \{+\infty\}$. Assume that $\mathrm{epi}(f_k)$ Attouch-Wets or bounded Hausdorff converges to $\mathrm{epi}(f)$. Then $\mathrm{epi}(f)$ is r-prox-regular.

5. Finite dimensional case

5.1. Introduction. We recall that sets with positive reach were introduced in [**63**], as a concept which could unify on one hand convexity and on the other hand $\mathcal{C}^2$-smoothness. The aim was establishing a polynomial expansion for the volume of a tubular neighborhood of a given set (also called parallel set), which generalized on one hand Steiner's Formula for convex sets and on the other hand Weyl's Tube Formula for n-dimensional submanifolds of $\mathbb{R}^n$ with $\mathcal{C}^2$-smooth boundary. More precisely, Federer proved for an r_0-prox-regular set $C \subset \mathbb{R}^n$ and any Borel set A of $\mathbb{R}^n$ and $0 \leq r < r_0$ that

$$\mathcal{L}^n(\{x \in \mathbb{R}^n : d(x, C) \leq r \text{ and } P_C(x) \in A\}) = \sum_{k=0}^{n} r^{n-k} \alpha(n-k) \Phi_k(C, A).$$

Therein, $\mathcal{L}^n$ denotes the n-dimensional Lebesgue measure in $\mathbb{R}^n$ and $\alpha(j)$ the $\mathcal{L}^n$-measure of the closed unit ball of $\mathbb{R}^j$. Of course, all $\Phi_k(C, \cdot)$ are countably additive measures, defining the curvatures measures considered in the title of [**63**]. The purpose was achieved through a detailed study of the newly introduced sets, together with a delicate approximation process. The interplay of nonsmooth analysis (not yet existing as an independent field) and of geometric measure theory (to some extent at its beginning) was strong and fruitful. We refer to the recent survey by Thäle [**111**] for a discussion of the curvature measures introduced in [**63**] and of more recent developments on the subject. In this section we concentrate on some aspects closer to nonsmooth analysis, proving some properties of $\rho(\cdot)$-prox-regular sets which are typically finite dimensional. We remark that positive reach is the same as r-prox-regularity, for some $r > 0$. Since the space is finite dimensional, there is not much loss of generality in concentrating in local properties.

5.2. Preliminaries. Let E be a Borel subset of $\mathbb{R}^n$ and $0 \leq k \leq n$, $k \in \mathbb{R}$. We denote by $\mathcal{L}^n(E)$ its Lebesgue measure, and by $\mathcal{H}^k(E)$ its k-dimensional Hausdorff measure, which is defined as follows.

DEFINITION 40. Let $E \subset \mathbb{R}^n$, and let $0 \leq k \leq n$, $k \in \mathbb{R}$ and $\delta > 0$. We set

$$H^k_\delta(E) = \inf\Big\{\sum_{i=1}^{\infty} \mathrm{diam}(U_i)^k : \bigcup_{i=1}^{\infty} U_i \supset E,\ \mathrm{diam}(U_i) < \delta\Big\}$$

and we define the (outer) k-dimensional Hausdorff measure of E as the possibly infinite number

$$H^k(E) := \sup_{\delta > 0} H^k_\delta(E) = \lim_{\delta \to 0^+} H^k_\delta(E).$$

We recall the following well known property of Hausdorff measures (see [**1**, pp. 72–80]), which follows directly from the definition:

PROPOSITION 41. If $f : \mathbb{R}^n \to \mathbb{R}^m$ is a Lipschitz continuous function with Lipschitz constant L, then for every $E \subset \mathbb{R}^n$

$$\mathcal{H}^k(f(E)) \leq L^k \mathcal{H}^k(E). \tag{59}$$

Through the Hausdorff measure, the concept of rectifiability of sets can be introduced as follows.

DEFINITION 42. Let $E \subset \mathbb{R}^n$ be a $\mathcal{H}^k$-measurable set with $0 \leq k \leq n$, $k \in \mathbb{N}$. We say that E is *countably $\mathcal{H}^k$-rectifiable* if there exist countably many Lipschitz continuous mappings $f_i : \mathbb{R}^k \to \mathbb{R}^n$ such that

$$\mathcal{H}^k\left(E \setminus \bigcup_{i=1}^{+\infty} f_i(\mathbb{R}^k)\right) = 0.$$

We say that E is *$\mathcal{H}^k$-rectifiable* if E is countably $\mathcal{H}^k$-rectifiable and $\mathcal{H}^k(E) < +\infty$.

Other concepts from Geometric Measure Theory will be used in our analysis for the finite dimensional case. In particular, we recall (see [**1**, pp. 117, 143]):

DEFINITION 43. Let $\Omega \subset \mathbb{R}^n$ be open.

(1) A function $u \in L^1(\Omega)$ is said to be of bounded variation in Ω ($u \in BV(\Omega)$) if its distributional derivative is representable by a vector-valued Radon measure in Ω, i.e., if

$$\int_\Omega u \frac{\partial \varphi}{\partial x_i}\, dx = -\int_\Omega \varphi \, dD_i u \text{ for all } \varphi \in C_c^\infty(\Omega),\ i = 1, \ldots, n,$$

for some Radon measure $Du = (D_1 u, \ldots, D_n u)$ (each $D_i u$ being a finite real-valued Radon measure). In general, the total variation of a vector-valued measure m is denoted by $|m|$ (and $\|m\|$ represents the norm of m, i.e;, $\|m\| := |m|(\Omega)$).

(2) We say that a $\mathcal{L}^n$-measurable set $E \subset \mathbb{R}^n$ has finite perimeter in Ω if its characteristic function 1_E has bounded variation in Ω. The real number

$$P(E, \Omega) := |D1_E|(\Omega)$$

is called the perimeter of E in Ω.

In addition to the notions of proximal and Fréchet normal vectors considered in the previous section, we recall now the concept of normal in De Giorgi's sense.

Let $E \subset \mathbb{R}^n$ be a set of finite perimeter in Ω; we call *reduced boundary of E in Ω* the set $\mathrm{bd}^* E$ of all points $x \in \Omega \cap \mathrm{supp}(|D1_E|)$ such that

$$\nu_E(x) := \lim_{r \to 0^+} \frac{D1_E(B(x,r))}{|D1_E(B(x,r))|} = \frac{dD1_E}{d|D1_E|}(x)$$

exists in $\mathbb{R}^n$ and satisfies $\|\nu_E(x)\| = 1$. For $x \in \mathrm{bd}^* E$, the vector $-\nu_E(x)$ is called the De Giorgi outer normal to E at x.

5.3. Further regularity results for sets. The first result is quite easy (see [**63**, Remark 4.15 (1)]).

PROPOSITION 44. Let $C \subset \mathbb{R}^n$ be closed and let $u \in \mathbb{R}^n$. Let $r > 0$ be given and assume that C is r-prox-regular. Then for all $0 < r' < r$ the set

$$C[u, r'] := \{x \in C \,:\, \|x - u\| \leq r'\}$$

is contractible in itself.

PROOF. Assume that $C[u, r']$ is nonempty. Then, for every $x \in C$ and $t \in [0, 1]$, $d_C((1-t)x + tu) \leq r'$, so that, recalling (g) in Theorem 3, the projection $P_C((1-t)x + tu)$ is well defined. Therefore, the homotopy

$$h(x, t) = P_C\left((1-t)x + tu\right), \quad x \in C[u, r'], \quad t \in [0, 1],$$

performs the task, since $(1-t)x + tu \in U_r(C)$ for all $t \in [0, 1]$ and $x \in C[u, r']$ because

$$d_C((1-t)x + tu) \leq d_C(x) + t\,\|x - u\| = t\,\|x - u\| \leq r' < r. \qquad \square$$

The second result essentially affirms that a prox-regular set can be partitioned according to the dimension of the normal cone, with an estimate on the Hausdorff measure of the corresponding sets.

PROPOSITION 45. Let $C \subset \mathbb{R}^n$ be $\rho(\cdot)$-prox-regular for a suitable continuous $\rho(\cdot)$. Then for all $k = 0, 1, \ldots, n$ the set

$$C^{(k)} = \{x \in C \,:\, \dim N(C; x) \geq n - k\}$$

is countably $\mathcal{H}^k$-rectifiable.

PROOF. There is no loss of generality in assuming that C is r-prox-regular for some $r > 0$.
Observe that $C^{(n)} = C$. Fix $k < n$ and let Π_k be a countable dense set of affine k-dimensional subspaces of $\mathbb{R}^n$. Suppose $0 < r' < r$. Then

$$C^{(k)} \subseteq \bigcup_{\sigma \in \Pi_k} P_C\left(\{u \in \sigma \,:\, d(u, C) \leq r'\}\right),$$

since for every $x \in C^{(k)}$ the normal cone $N(C; x)$ meets at least one $\sigma \in \Pi_k$. The result then follows by recalling the Lipschitz continuity of P_C in $O_r(C)$. $\square$

By combining Proposition 45 for $k = n - 2$ and Proposition 7 we obtain immediately the following

COROLLARY 46. Let $C \subset \mathbb{R}^n$ be $\rho(\cdot)$-prox-regular for a suitable continuous $\rho(\cdot)$. Then there exists a $\mathcal{H}^{n-2}$-countably rectifiable set $K \subset \mathrm{bdry}C$ such that $N(C; \cdot) \cap \mathbb{B}$ is single-valued and continuous on $\mathrm{bdry}C \setminus K$.

We now state without proof a result saying that a $\rho(\cdot)$-prox-regular set has (locally) finite perimeter in $\mathbb{R}^n$ and at $\mathcal{H}^{n-1}$-a.e. point of the reduced boundary the De Giorgi external normal coincides with the unit proximal normal, which therefore is unique.

PROPOSITION 47 (see §4 in [**47**]). Let $C \subset \mathbb{R}^n$ be closed and let $r > 0$ be given. Assume that C is r-prox-regular. Then C has locally finite perimeter in $\mathbb{R}^n$. Moreover, for every $x \in \mathrm{bd}^* C$ there exists v_x, $\|v_x\| = 1$, such that

$$N^P(C; x) = \mathbb{R}^+ v_x. \tag{60}$$

Furthermore, for $\mathcal{H}^{n-1}$-a.e. $x \in \mathrm{bd}^* C$,

$$v_x = -\nu_C(x). \tag{61}$$

5.4. Regularity results for functions. This section is devoted to some regularity properties of lower semicontinuous functions whose epigraph is $\rho(\cdot)$-prox-regular. A simple example, yet displaying most of the common features of this class of functions, is the epigraph of $\sqrt{|x|}$. In particular, such functions need not to be locally Lipschitz. Since semiconvex functions are characterized precisely by the joint properties of being locally Lipschitz and of having $\rho(\cdot)$-prox-regular epigraph for a suitable ρ (see Section 8), dropping the Lipschitz requirements appears to be a natural generalization which preserves the types of singularities which occur in the (semi)convex setting. The main result states that for almost every point x of the domain there exists a neighborhood of x where the function f is semiconvex. As a consequence, f is a.e. differentiable and admits a second order Taylor expansion along a.e. point of the domain.

THEOREM 48 (see Theorem 5.1 in [**47**]). Let $\Omega \subset \mathbb{R}^n$ be open, and let $f : \overline{\Omega} \to \mathbb{R} \cup \{+\infty\}$ be proper, lower semicontinuous, and such that $\mathrm{epi} f$ is $\rho(\cdot)$-prox-regular for a suitable continuous ρ (where $\overline{\Omega}$ denotes the closure of Ω). Then there exists a sequence of sets $\Omega_i \subseteq \Omega$ such that each Ω_i is compact in $\mathrm{dom} f$ and

(1) the union of Ω_i covers $\mathcal{L}^n$-almost all $\mathrm{dom} f$, i.e.,

$$\mathcal{L}^n\left(\mathrm{dom} f \setminus \bigcup_i \Omega_i\right) = 0; \tag{62}$$

(2) for all $x \in \bigcup_i \Omega_i$ there exist $\delta = \delta(x) > 0$, $L = L(x) > 0$ such that

$$\begin{gathered} f \text{ is Lipschitz continuous on } B(x,\delta) \text{ with constant } L, \\ \text{and hence semiconvex on } B(x,\delta); \end{gathered} \tag{63}$$

(3) for all $x \in \bigcup_i \Omega_i$

$$f \text{ is strictly Fréchet differentiable at } x. \tag{64}$$

The idea of the proof is the following. By Corollary 46, for $\mathcal{H}^n$-a.e. point (x, α) in the boundary of the epigraph of f the normal cone to $\mathrm{epi} f$ has dimension one. The point of the proof is showing that for $\mathcal{L}^n$-a.e. x this normal cone is not horizontal, and hence the Clarke subdifferential is nonempty and is a singleton. This is achieved by exploiting the property that the boundary of a suitable neighborhood of the epigraph is a $\mathcal{C}^{1,1}$-manifold, and therefore the projection into $\mathbb{R}^n$ of the points in it with horizontal normal has $\mathcal{L}^n$-measure zero. The complete proof can be found in [**47**, Section 5]. Here we present only the main arguments.

PROOF. We denote $x \in \mathbb{R}^{n+1}$ as $x = (y, \xi) = (y_1, \dots, y_n, \xi)$, with $y \in \mathbb{R}^n$ and $\xi \in \mathbb{R}$, and set $\pi_{\mathbb{R}^n}(y, \xi) = y$. Moreover $\{e_i\}$ will denote the canonical basis in $\mathbb{R}^{n+1}$.

Set $C = \text{epi} f$ and let $R > 0$ be sufficiently large. Set $\varphi_R = \max\{\frac{1}{2\rho(x)} : x \in C \cap \overline{B}(0, R)\}$ and take $0 < \rho_R < R$ such that $4\varphi_R \rho_R < 1$. Set, for $\lambda \in [0, 1]$,

$$C^{R,\lambda} = \left\{x + \lambda \rho_R v : x \in C \cap B[0, R],\ v \in N_C^P(x),\ \|v\| = 1\right\}.$$

Set also

$$C^R - \bigcup_{\lambda \in [0,1]} C^{R,\lambda},$$

and

$$\tilde{C}^R = C^{R,1}.$$

Observe that, by (l) in Theorem 14, $d(\cdot, C)$ is of class $\mathcal{C}^{1,1}$ in a neighborhood of $\tilde{C}^R$, and $\tilde{C}^R$ is a $\mathcal{C}^{1,1}$-manifold. Moreover, by (g) in Theorem 3,

(65) $$P_C \text{ is Lipschitz continuous with rank 2 on } \tilde{C}^R.$$

In view of Sard's lemma, it is not difficult to see (see, however, Claim 1 in the proof of Theorem 5.1 in [**47**]) that

(66) $$\mathcal{L}^n\left(\bigcup_{R>0}\left\{y : (y, \xi) \in \tilde{C}^R, \frac{\partial}{\partial \xi} d((y, \xi), C) = 0\right\}\right) = 0.$$

Consider now the set

$$C_0 = \left\{(x, \xi) \in C : N^P(C; (x, \xi)) \subseteq \mathbb{R}v,\ \|v\| = 1,\ \text{and } \langle v, e_{n+1}\rangle = 0\right\}.$$

We have □

CLAIM 1. $\mathcal{L}^n(\pi_{\mathbb{R}^n}(C_0)) = 0$.

PROOF OF CLAIM 1. By definition of C_0, for all $(x, \xi) \in C_0$ there exist $R = R(x) > 0$, $\lambda = \lambda(x) > 0$ and $w = w(x) \in \mathbb{R}^n$, $\|w\| = 1$, such that

(1) $(x, f(x)) \in B(0, R)$,
(2) $(w, 0) \in N^P(C; (x, f(x)))$,
(3) $(x + \lambda w, f(x)) := (y, f(x)) \in E_0^R$,

where $E_0^R = \left\{y : (y, \xi) \in \tilde{C}^R, \frac{\partial}{\partial \xi} d((y, \xi), C) = 0\right\}$.

Fix $\varepsilon > 0$. By (66) we can cover $\pi_{\mathbb{R}}^n(E_0^R)$ by countably many balls $B_i = B(z_i, r_i)$ such that

$$\sum_i \mathcal{H}^{n-1}(\text{bdry} B_i) < \varepsilon.$$

Define the cylinders

$$C_i = \{(z, \xi) : z \in \text{bdry} B_i,\ |\xi| \le R\}.$$

Observe that there is no loss of generality in taking the r_i so small that $4\varphi_R(\rho_R + 2r_i) < 1$. Therefore, recalling Corollary 4.15 (2) in [**43**], there exists $i \in \mathbb{N}$, $\sigma_x \geq 0$ such that $y \in B_i$, $(y + \sigma_x w, f(x)) \in C_i \cap \tilde{C}_R$, and

$$P_C(y + \sigma_x w, f(x)) = \{(x, f(x))\}.$$

So, for all $(x, \xi) \in C_0$ there exists $R > 0$ such that

$$\pi_{\mathbb{R}^n}(x, \xi) = \pi_{\mathbb{R}^n}(x, f(x)) \in (\pi_{\mathbb{R}^n} \circ P_C)\left(\bigcup_i C_i \cap \tilde{C}_R\right).$$

Recalling (g) in Theorem 3, $\pi_{\mathbb{R}^n} \circ P_C$ is Lipschitz continuous with constant 2 on each $C_i \cap \tilde{C}_R$. Therefore

$$\mathcal{L}^n\left((\pi_{\mathbb{R}^n} \circ P_C)\left(\bigcup_i C_i \cap \tilde{C}_R\right)\right) \leq 2^n \mathcal{H}^n\left(\bigcup_i C_i\right) \leq 2^{n+1}\varepsilon R.$$

Hence

$$\mathcal{L}^n\left(\pi_{\mathcal{R}^n}(C_0 \cap B(0, R))\right) = 0.$$

Claim 1 then follows immediately. □

It is not difficult to prove the following:

Claim 2. $\mathcal{L}^n\left(\mathrm{dom} f \cap \mathrm{bdry}\,\mathrm{dom} f\right) = 0.$

To conclude the proof of the theorem, assume that $\mathcal{L}^n(\mathrm{dom} f \setminus \mathrm{bdry}\,\mathrm{dom} f) > 0$ (otherwise there is nothing to prove). For all $\varepsilon > 0$ let

$$C_\varepsilon = \{(y, f(y)) : N^P(C; (y, f(y))) \subseteq \mathbb{R}v_y,\ \|v_y\| = 1 \text{ and } |\langle v_y, e_{n+1}\rangle| < \varepsilon\}.$$

Recalling Claim 1,

$$\mathcal{L}^n\left(\bigcap_{\varepsilon > 0} \pi_{\mathbb{R}^n}(C_\varepsilon)\right) = 0. \tag{67}$$

By Claim 2 and Corollary 46, let F_ε be an open subset of $\mathbb{R}^n$ such that

$$\left\{y \in \mathrm{dom} f : \dim N^P(C; (y, f(y))) > 1\right\} \cup (\mathrm{dom} f \cap \mathrm{bdry}\,\mathrm{dom} f) \cup (\pi_{\mathbb{R}^n}(C_\varepsilon)) \subset F_\varepsilon$$

and

$$\mathcal{L}^n(F_\varepsilon) < \mathcal{L}^n(\pi_{\mathbb{R}^n}(C_\varepsilon)) + \varepsilon.$$

Set, for $h \in \mathbb{N}$,

$$\Omega_h = \{y \in \mathrm{dom} f : d_{\mathrm{bdry}\Omega}(y) \geq 1/h,\ \|y\| \leq h\} \setminus F_{1/h}.$$

For h large enough, Ω_h is a nonempty subset of Ω, which is contained in the interior of $\mathrm{dom} f$ and is compact in $\mathrm{dom} f$. By construction we have

$$\mathcal{L}^n\left(\mathrm{dom} f \setminus \bigcup_{h \in \mathbb{N}} \Omega_h\right) = 0.$$

We claim that f is strictly Fréchet differentiable at all $y \in \bigcup_h \Omega_h$. Indeed, for all $y \in \bigcup_h \Omega_h$, say $y \in \Omega_{\overline{h}}$, the cone $N^P(C;(y,f(y)))$ is a half line, $N^P(C;(y,f(y))) = \mathbb{R}^+ v_y$, $\|v_y\| = 1$, and $|\langle v_y, e_{n+1}\rangle| \geq \frac{1}{h}$. Therefore the proximal subgradient $\partial_P f(y)$ is nonempty and is actually a singleton, say $\partial_P f(y) = \zeta_y$. The differentiability of f at x, then, will follow from the next claim:

CLAIM 3. Let $x \in \operatorname{int}\operatorname{dom} f$ be such that $\partial_P f(x)$ is a singleton. Then f is (strictly) Fréchet differentiable at x.

PROOF OF CLAIM 3. By assumption there exists $v_x \in \mathbb{R}^n$, $\|v_x\| = 1$, such that both $N^P(C;(x,f(x))) = \mathbb{R}^+ v_x$ and

$$\langle v_x, e_{n+1}\rangle \neq 0. \tag{68}$$

We observe first that f is continuous at x. For, should a sequence $y_n \to x$, $y_n \in \operatorname{int}\operatorname{dom} f$, exist such that $\lim f(y_n) := \overline{\xi} > f(x)$, the segment joining $(x,f(x))$ with $(x,\overline{\xi})$ would be in the boundary of $C = \operatorname{epi} f$. Then, all $v \in N^P(C;(x,\alpha))$ for all $f(x) < \alpha \leq \overline{\xi}$ ought to satisfy $\langle v, e_{n+1}\rangle = 0$ by the external sphere condition. By the regularity of the epigraph (recall Proposition 7) and the fact that $N^P(C;(x,f(x))) = \mathbb{R}^+ v_x$, we would obtain $\langle v_x, e_{n+1}\rangle = 0$, a contradiction with (68). Therefore, again by the regularity of the epigraph, there exist $\delta = \delta(x) > 0$ and $\eta = \eta(x) > 0$ such that $B(x,\delta) \subset \Omega$ and

$$\|y - x\| < \delta \text{ implies } N^P(C;(y,f(y))) \cap \{w : \|w\| = 1\} \subseteq \{w : |\langle w, e_{n+1}\rangle| \geq \eta\}.$$

This shows that the proximal subgradient of f is nonempty and uniformly bounded in $B(x,\delta)$. By Theorem 7.3, p. 52 in [**42**], f is Lipschitz continuous in $B(x,\delta)$. By the regularity of the epigraph, $\partial_C f(x) = \partial_P f(x) = \{\zeta_x\}$. Thus, by Proposition 2.2.4 in [**41**], f is strictly Fréchet differentiable at x, and Claim 3 is proved. □

5.5. Gradient formulas. The Clarke generalized gradient $\partial^{Cl} f$ of a Lipschitz continuous function f between finite dimensional spaces admits a well known representation as the convex hull of limits of gradients, namely

$$\partial^{Cl} f(x) = \operatorname{co}\{\lim Df(x_i) : x_i \to x, Df(x_i) \text{ exists}\}. \tag{69}$$

If f is not necessary Lipschitz continuous, such formula makes no sense, even if f is a.e. differentiable as the functions which are being considered here, since the set of limits may be empty. The example $f(x) = \sqrt{|x|}$, however, suggests that by interchanging the limit operation with the convex combinations one may obtain a representation formula for ∂f of a nature similar to (69). For instance, for $x = 0$, $\partial f(x) \ni 0 = \lim_{n\to\infty}(f'(1/n) + f'(-1/n))/2$. This conjecture is actually true, and we present here the statement (see, however, [**48, 50**]). We state first a result which requires that the normal cone at a certain point be pointed. In this case a representation which is entirely analogous to (69), both for sets and for subgradients, holds. The result is not surprising, since in this case an epi-Lipschitz condition can be invoked (see [**109**, p.386]).

THEOREM 49. Let $C \subset \mathbb{R}^n$ be $\rho(\cdot)$-prox-regular for a suitable continuous ρ. Fix $x \in \operatorname{bdry} C$ and assume that $N^P(C;x)$ is pointed. Set

$$C_1 = \{x \in \operatorname{bdry} C : \dim N^P(C;x) = 1\}.$$

Then

$$N^P(C;x) = \overline{\text{co}}\left\{\mathbb{R}_+ v \,:\, v = \lim v_k, v_k \in N^P(C;x_k), x_k \to x, x_k \in C_1\right\}. \tag{70}$$

Observe that the proximal subgradient can be represented by taking vectors with last coordinate $= -1$ in (70), with $C = \text{epi} f$.

Theorem 49 is based on the two following facts (see [**48**, 4.6]): 1) since $N^P(C;x)$ is pointed, it is the closed convex hull of its exposed rays (see [**108**, Corollary 18.7.1]); 2) if $\bar\zeta$, $\|\bar\zeta\| = 1$, belongs to an exposed ray of $N^P(C;x)$ and $\{x_k\} \subset C$ is such that $x_k \to x$ and $\zeta_k \in N^P(C;x_k)$, $\zeta_k \to \zeta$ with $\|\zeta\| = 1$, then $\zeta = \bar\zeta$.

In the general case, the representation formulas read as follows:

THEOREM 50. The following statements hold true.

(a) Let $\rho(\cdot)$ be given, and let $C \subset \mathbb{R}^n$ be $\rho(\cdot)$-prox-regular. Let x belong to the boundary of C . Then

$$N^P(C;x) \subseteq \mathbb{R}_+ \bigcap_{\delta>0} \overline{\text{co}}\{v \in N(C;y) \,:\, \|v\| = 1, y \in C_1 \cap B(x,\delta)\}; \tag{71}$$

(b) Let $\Omega \subseteq \mathbb{R}^n$ be open and let $f : \Omega \to \mathbb{R}$ be continuous with $\rho(\cdot)$-prox-regular epigraph, for a suitable continuous ρ. Let $x \in \Omega$. Then

$$\partial_P f(x) \subseteq \bigcap_{\delta>0} \overline{\text{co}}\{Df(y) \,:\, y \in \text{dom} Df \cap B(x,\delta)\}. \tag{72}$$

Thanks to Example 7.2 in [**27**], the inclusions in both (71) and (72) cannot be expected to be equalities.

The **proof of Theorem 50** requires some steps, which will be sketched.

Step 1. The first point is showing that a vector in the boundary of the normal cone to C at x can be approximated by normal vectors at points of a Lipschitz curve in C issuing from x, with a given tangent vector at x.

Let C be $\rho(\cdot)$-prox-regular, and let $x_0 \in C$ and p_0 be such that

$$p_0 \in \text{bdry} N^P(C;x_0).$$

Take $q \in \mathbb{R}^n \setminus \{0\}$ such that:

$$\langle q, p - p_0\rangle \le 0, \quad \forall p \in N^P(C;x_0). \tag{73}$$

Then there exist $\sigma > 0$ and a Lipschitz arc $x_q : [0,\sigma] \to \text{bdry} C$ such that:

$$x_q(0) = x_0, \tag{74}$$

$$\lim_{s\to 0^+} \frac{x_q(s) - x_0}{s} = q, \tag{75}$$

$$p(s) := p_0 + q - \frac{x_q(s) - x_0}{s} \in N^P(C;x_q(s)) \text{ for all } s \in [0,\sigma]. \tag{76}$$

Moreover, σ can be chosen independently of q and the limit in (75) is uniform with respect to q, provided q belongs to a fixed compact set.

PROOF. Set $\varphi(\cdot) = 1/(2\rho(\cdot))$, and let $r > 0$ be so small that $4r\varphi(y) \le 1$ for all $y \in B(x_0, r)$. Set $\sigma = r/(2\|p_0 + q\|)$ and observe that σ can be taken independently of q if q belongs to a fixed compact set. Then, recalling (i) in Theorem 3, for all $s \in [0, \sigma]$ the projection $x_q(s) := P_C(x_0 + s(p_0 + q))$ is unique and is Lipschitz with respect to s with constant $2\|p_0 + q\|$. Observe that $p_0 + q \notin N^P(C; x_0)$, because $q \neq 0$ is normal to $N^P(C; x_0)$ at p_0. Consequently, $x_q(s) \neq x_0$ for all $s \in [0, \sigma]$.

By construction, $x_0 + s(p_0 + q) - x_q(s) \in N^P(C; x_q(s))$, whence $p(s) \in N^P(C; x_q(s))$. Moreover $\|p(s)\| \le 2\|p_0 + q\|$ for all $s \in [0, \sigma]$. By definition of $p(s)$ and (d) in Theorem 3, we have

$$\begin{aligned} \langle p(s) - p_0, p(s) - (p_0 + q)\rangle &= \langle p(s) - p_0, \frac{x_0 - x_q(s)}{s}\rangle \\ &\le \frac{1}{s}(\varphi(x_0)\|p_0\| + \varphi(x_q(s))\|p(s)\|)\|x_q(s) - x_0\|^2 . \end{aligned} \tag{77}$$

Let $s_k \to 0$ be such that $p(s_k) \to \bar{p}$ and observe that $\bar{p} \in N^P(C; x_0)$ by graph-closedness (recall Proposition 7). Then by passing to the limit in (77) and recalling that $x_q(\cdot)$ is Lipschitz continuous, we obtain:

$$\|\bar{p} - p_0\|^2 - \langle q, \bar{p} - p_0\rangle \le 0.$$

By (73) we obtain that $\bar{p} = p_0$ and (75) follows. □

Step 2. The second point is showing that the curves constructed in the previous step cover enough of the boundary of C around x. Here and in the following, relbdry$K$ denotes the relative boundary of a convex set K.

Let $p_o \in$ relbdry$N^P(C; x)$, $p_o \neq 0$, and let dim$N^P(C; x) = d$, $0 < d \le n$. Consider the number σ as in the statement of Step 1, chosen uniformly with respect to q, $\|q\| \le 1$, such that (73) holds, and the curves x_q constructed in Step 1. Then for every $0 < \tau \le \sigma$ the set

$$\mathcal{X}^{\tau}_{p_o} := \{x_q(s) : s \in [0, \tau], \|q\| = 1, q \text{ satisfies } (73)\}$$

has positive $\mathcal{H}^{n-d+1}$-measure.

PROOF. Since $p_o \in$ relbdry$N^P(C; x)$, there exists $q_o \in$ span$N^P(C; x)$ with norm 1 and satisfying (73). If $d < n$, we choose also orthonormal vectors $q_1, \dots, q_{n-d}$ belonging to $(\text{span}N^P(C; x))^{\perp}$ and set

$$\begin{aligned} \mathcal{U} &= \text{co}\{q_o, \dots, q_{n-d}\} \\ \mathcal{U}' &= \text{co}\{0, \tau\mathcal{U}\}, \end{aligned}$$

for a fixed $0 < \tau \le \sigma$. Assume for simplicity that $x = 0$ and set, for a fixed $0 < \tau \le \sigma$,

$$\begin{aligned} V &= \text{co}\{0, \tau(q_o + p_o), \dots, \tau(q_{n-d} + p_o)\} \\ &= \{t(q + p_o) : q \in \mathcal{U}, t \in [0, \tau]\}. \end{aligned}$$

Note that both $\mathcal{U}'$ and V have positive $\mathcal{H}^{n-d+1}$-measure, and, for all $s \in]0, 1[$,

$$P_C(sV) \subset \{x_q(t) : q \in \mathcal{U}, t \in [0, \tau]\} = \mathcal{X}^{\tau}_{p_o}. \tag{78}$$

Set now $\mathcal{V}$ to be the $n-d+1$-dimensional space generated by $\mathcal{U}'$, and consider, for $s\in]0,\tau[$, the map $\psi_s:\mathcal{U}'\to\mathcal{V}$ defined as

$$\psi_s(v) := s^{-1}\left(P_{\mathcal{U}'}\circ P_C\right)\left(s(v+p_0)\right).$$

Observe that $\psi_s(\mathcal{U}') = s^{-1}\left(P_{\mathcal{U}'}\circ P_C\right)(sV)$, and for all $v\in\mathcal{U}'$ we have

$$\begin{aligned}\|\psi_s(v)-v\| &= s^{-1}\left\|\left(P_{\mathcal{U}'}\circ P_C\right)\left(s(v+p_0)\right)-sv\right\|\\ &\qquad \text{(since } sv\in\mathcal{U}' \text{ and } P_{\mathcal{U}'} \text{ is Lipschitz continuous with rank 1)}\\ &\le s^{-1}\left\|P_C(s(v+p_0))-sv\right\|.\end{aligned}$$

Therefore, recalling (75) we have that $\psi_s(v)\to v$ uniformly with respect to $v\in\mathcal{U}'$, as $s\to 0^+$. Therefore, by dominated convergence in $\mathcal{V}$,

$$\mathcal{H}^{n-d+1}(\psi_s(\mathcal{U}'))\to\mathcal{H}^{n-d+1}(\mathcal{U}')>0 \tag{79}$$

as $s\to 0^+$.

Recalling the definition of ψ_s and Proposition 41, for all s sufficiently small we have

$$\mathcal{H}^{n-d+1}\left(P_C(sV)\right)\ge s^{n-d+1}\mathcal{H}^{n-d+1}\left(\psi_s(\mathcal{U}')\right).$$

Therefore, noting that $r_1V\subseteq r_2V$ if $r_1\le r_2$,

$$\mathcal{H}^{n-d+1}(P_C(sV))>0 \qquad \text{for all } s\in]0,1[. \tag{80}$$

Recalling (78) and (79), the proof is complete. □

Step 3. The last step is an induction process which will only be sketched (see [**50**]).

Indeed, if the dimension of $N^P(C;x)$ is 1, the proof is easy. If the dimension d is larger than 1, then by combining Step 1 and Step 2 with Proposition 45 it is possible to approximate a vector in the boundary of $N^P(C;x)$ with normal vectors at points where the normal cone has dimension smaller than d, and so the induction can proceed. The completion of the proof is easy if $N^P(C;x)$ is not a half subspace, since in this case it is the convex hull of its relative boundary. Otherwise it is more delicate, and this is exactly the point where the interchange of the limit operation with the convex hull plays a role.

6. The sweeping process

The sweeping, or Moreau, process was introduced in the Seventies by J. J. Moreau, and extensively studied by himself and other authors (see, e.g., [**92**], [**93**], [**117**], [**34**]). A convex closed moving set $C(t)$ is considered, which drags a point that is constrained both to belong to $C(t)$ for all t, and to have the opposite of its velocity "normal" to the boundary of this set. Since the boundary is not necessarily smooth, the mathematical formulation of the sweeping process requires a differential inclusion, involving the normal cone $N(C(t);\cdot)$ to $C(t)$ in the sense of Convex Analysis. In other words, the problem

$$\left\{\begin{array}{rcl} -\dot{x}(t) &\in& N(C(t);x(t))\\ x(0) &=& x_0\in C(0)\\ x(t) &\in& C(t) \qquad \forall t\end{array}\right. \tag{81}$$

is considered. Under the further assumption that $C(t)$ be Lipschitz continuous in t, Moreau proves existence and uniqueness of a solution to (81) (see [**92**]), which is Lipschitz continuous with the same constant as $C(\cdot)$. There are two (equivalent) arguments to prove existence: a penalization technique, through Yosida approximations, and the "catching up algorithm". In both approaches a Cauchy sequence of approximate solutions is constructed, which converges *strongly* in $W^{1,2}$ to a solution. Results of Convex Analysis – such as the uniqueness of the projection and the differentiability of the distance from $C(t)$ – are fundamental tools. In this section the sweeping process is generalized to the case of $\rho(\cdot)$-prox-regular moving sets. Thanks to the properties implied by $\rho(\cdot)$-prox-regularity, the main features of the convex theory remain true: existence and uniqueness of trajectories hold, and the catching up algorithm produces a sequence which is Cauchy for the sup norm and converges weakly in $W^{1,2}$ to the solution. The cases of a merely continuous moving set, of a set moving with bounded variation, and of a set depending on both time and state, are a little more complicated. The continuous case will be described in Section 6.3 below and references will be given for the other cases.

Throughout the section, $C(t)$, $t \in [0,T]$, will denote a ρ-prox-regular moving subset of H, which is always supposed to be a possibly infinite dimensional Hilbert space. For simplicity we will suppose ρ to be constant.

6.1. The catching up algorithm. The catching up algorithm is developed here for r-prox-regular moving sets.

THEOREM 51. Let $C(\cdot)$ be a Hausdorff continuous multifunction from the interval $[0,T]$ into the closed nonempty subsets of H and let $\omega(\cdot)$ be its modulus of continuity. Let $r > 0$ be given and assume that $C(t)$ is r-prox-regular for all $t \in [0,T]$. Let $u_0 \in C(0)$. Then for each $n \in \mathbb{N}$ there exist functions $u_n : [0,T] \to H$, continuous and piecewise affine, and $\tau_n : [0,T] \to [0,T]$, piecewise constant, such that, for all $t \in]0,T[$,

(a) $u_n(\tau_n(t)) \in C(\tau_n(t))$;
(b) $-u_n'(t) \in N(C(\tau_n(t)); u_n(\tau_n(t)))$ (provided $t \neq \tau_n(t)$);
(c) $\|u_n(t) - u_n(\tau_n(t))\| \leq \omega(1/2^n)$;
(d) $d(u_n(t), C(t)) \leq 2\omega(1/2^n)$;
(e) $|\tau_n(t) - t| \leq 1/2^n$.

PROOF. We define, for all $n = 1, 2, \ldots$,

$$t_n^i = \frac{iT}{2^n}, \quad i = 0, \ldots, 2^n, \qquad I_n^i = [t_n^i, t_n^{i+1}], \quad i = 0, \ldots, 2^n - 1.$$

We have

$$\sup_{s,t \in I_n^i} H(C(s), C(t)) \leq \omega(1/2^n); \tag{82}$$

moreover, for all n large enough, the inequality

$$4\varphi_0 \omega(1/2^n) < 1 \tag{83}$$

holds, where

$$\varphi_0 = 1/(2r).$$

Set $u_n^0 = u_0 \in C(0)$. Since $\varphi_0 d(u_0, C(t_n^1)) < 1/4$, by (i) in Theorem 3, the projection $u_n^1 := P_{C(t_n^1)}(u_n^0)$ is a singleton. Analogously, $4\varphi_0 d(u_n^1, C(t_n^2) < 1$, so that $u_n^2 := P_{C(t_n^2)}(u_n^1)$ is well defined. With the same argument we can define recursively

$$u_n^{i+1} := P_{C(t_n^{i+1})}(u_n^i), \quad i = 1, \dots 2^n - 1. \tag{84}$$

By affine interpolation we define a corresponding sequence of approximate solutions $u_n : [0, T] \to H$; for $t \in I_n^i$ the explicit formula is

$$u_n(t) = u_n^i + \frac{t - t_n^i}{t_n^{i+1} - t_n^i}(u_n^{i+1} - u_n^i).$$

For simplicity of notation, we set also, for $t \in I_n^i$,

$$\tau_n(t) = t_n^{i+1}.$$

Observe that, for all $t \in]t_n^i, t_n^{i+1}[$, $u_n(\tau_n(t)) \in C(\tau_n(t))$, and

$$-u_n'(t) = \frac{u_n^i - u_n^{i+1}}{t_n^{i+1} - t_n^i} \in N(C(\tau_n(t)); u_n(\tau_n(t))). \tag{85}$$

Moreover,

$$\begin{aligned} \|u_n(t) - u_n(\tau_n(t))\| &\le \|u_n^{i+1} - u_n^i\| = d(u_n^i, C(t_n^{i+1})) \\ &\le H(C(t_n^i), C(t_n^{i+1})) \le \omega(1/2^n). \end{aligned} \tag{86}$$

Therefore,

$$d(u_n(t), C(t)) \le \|u_n(t) - u_n(\tau_n(t))\| + H(C(\tau_n(t)), C(t)) \le 2\omega(1/2^n). \tag{87}$$ □

REMARK 52. Assume that there exist $z \in H$ and $r' > 0$ such that $B(z, r') \subset C(t)$ for all t and that $\varphi_0(\|u_0 - z\| + r')^2 < r'$ (where $\varphi_0 = 1/(2r)$.) Then, in the above discretization procedure, without loss of generality we may assume that

$$\varphi_0(\|u_0 - z\| + r' + \omega(1/2^n))^2 < r' \tag{88}$$

for all n large enough.

The following technical lemma, applied to the discretized solutions constructed in the above theorem, will be crucial in establishing an *a priori* estimate on the sequence of approximate solutions.

LEMMA 53. Let $r > 0$ be given, together with r-prox-regular sets $C_1, \dots, C_n \subset H$. Assume that $2e(C_i, C_{i+1}) < r$, $i = 1, \dots, n-1$, and that there exist $z \in H$, $r' > 0$ such that $B[z, r'] \subset C_i$, $i = 1, \dots, n$. Let $u_0 \in H$ be such that $2d(u_0, C_1) < r$, and set $u_i = P_{C_i}(u_{i-1})$, $i = 1, \dots, n$. Set $C_0 = \{u_0\}$ and $d = \max\{e(C_i, C_{i+1}) : i = 0, \dots, n-1\} + \|u_0 - z\| + r'$, and assume that

$$d^2 < 2rr'. \tag{89}$$

Then (set $\varphi_0 = 1/(2r)$)
a) $\|u_{i+1} - z\| \leq \|u_i - z\|$, $\quad i = 0, \ldots, n-1$;
b) $\sum_{i=1}^n \|u_i - u_{i-1}\| \leq \max\left\{\frac{\|u_0 - z\|^2 - (r')^2}{2(r' - \varphi_0 d^2)}, 0\right\}$.

PROOF. We suppose $z = 0$. We prove first that

$$\|u_1 - u_0\| \leq \frac{\|u_0\|^2 - \|u_1\|^2}{2(r' - \varphi_0 d^2)}, \tag{90}$$

which implies a) for $i = 0$, by (89). To this aim, set $x' = r'(u_0 - u_1)/\|u_0 - u_1\|$ and observe that $x' \in C_1$, and $\|u_1 - x'\| \leq \|u_1 - u_0\| + \|u_0\| + \|x'\| \leq d$. Then, by applying (b) in Theorem 3 with $C = C_1$, $x = u_1$ and $v = u_0 - u_1$, one obtains

$$\begin{aligned}\|u_0\|^2 &= \|u_1\|^2 + \|u_0 - u_1\|^2 + 2\langle x', u_0 - u_1\rangle + 2\langle u_1 - x', u_0 - u_1\rangle \\ &\geq \|u_1\|^2 + 2r'\|u_1 - u_0\| - 2\varphi_0 \|u_1 - u_0\| \|u_1 - x'\|^2 \\ &\geq \|u_1\|^2 + 2\|u_1 - u_0\|(r' - \varphi_0 d^2),\end{aligned}$$

from which (90) follows. Next, we show by induction that

$$\|u_{i+1} - u_i\| \leq \frac{\|u_i\|^2 - \|u_{i+1}\|^2}{2(r' - \varphi_0 d^2)} \tag{91}$$

for $i = 1, \ldots, n-1$. Indeed, set $x' = r'(u_i - u_{i+1})/\|u_i - u_{i+1}\|$ and observe that, by the inductive assumption, $\|u_{i+1} - x'\| \leq \|u_i - u_{i+1}\| + \|u_i\| + r' \leq e(C_i, C_{i+1}) + \|u_0\| + r' \leq d$. Then, the same argument as for $i = 0$ yields (91). By summing all the estimates, we obtain

$$\sum_{i=1}^n \|u_i - u_{i-1}\| \leq \frac{\|u_0\|^2 - \|u_n\|^2}{2(r - \varphi_0 d^2)}.$$

By the same argument of Valadier [**117**, p. 3.9] we deduce the inequality b) from the one above. □

6.2. Lipschitz continuous moving sets. The section is devoted to proving the following existence and uniqueness result.

THEOREM 54. Let $C(\cdot)$ be Lipschitz continuous with constant L, let $r > 0$ be given and assume that for all $t \in [0, T]$ the set $C(t)$ is r-prox-regular.

Then the problem (81) admits one and only one L-Lipschitz solution.

PROOF. *Uniqueness.* Let u_1, u_2 be two L-Lipschitz solutions of (81) on $[0, T]$ and set $s(t) = \|u_2(t) - u_1(t)\|^2$. Recalling (d) in Theorem 3 we obtain, for a.e. $t \in [0, T]$,

$$\frac{d}{dt}s(t) = 2\langle u_2(t) - u_1(t), \dot{u}_2(t) - \dot{u}_1(t)\rangle \leq \frac{L}{r}s(t).$$

We conclude the proof by applying Gronwall's lemma.
Existence. We prove first that the sequence of approximate solutions constructed in Theorem 51 is Cauchy in $\mathcal{C}^0([0, T]; H)$. The convergence to a solution will then follow by the (strong×weak)-closedness of the graph of $N(C(t); \cdot)$ for each t.

Let $\{u_n(\cdot)\}$ be the sequence constructed in Theorem 51, with $\omega(r) = Lr$. Observe, in particular, that each u_n is L-Lipschitz. By r-prox-regularity, for all n, a.e. $t \in [0,T]$, and all $y \in C(\tau_n(t))$ we have

$$\langle -\dot{u}_n(t), y - u_n(\tau_n(t))\rangle \le \frac{L}{2r} \|y - u_n(\tau_n(t))\|^2 .$$

By setting $y_n^m(t) = P_{C(\tau_n(t))}(u_m(\tau_m(t)))$ one obtains in particular, for all m, n,

$$\langle -\dot{u}_n(t), y_n^m(t) - u_n(\tau_n(t))\rangle \le \frac{L}{2r} \|y_n^m(t) - u_n(\tau_n(t))\|^2 ,$$

from which it follows, for a suitable constant K independent of m, n,

$$\langle -\dot{u}_n(t), u_m(\tau_m(t)) - u_n(\tau_n(t))\rangle \le \frac{L}{2r} \|u_m(\tau_m(t)) - u_n(\tau_n(t))\|^2 + K\Big(\frac{1}{2^m} + \frac{1}{2^n}\Big).$$

By changing the order of m and n in the above inequality and adding the corresponding inequalities we deduce

$$\begin{aligned}&\langle \dot{u}_n(t) - \dot{u}_m(t), u_n(\tau_n(t)) - u_m(\tau_m(t))\rangle \\ &\qquad \le \frac{L}{r} \|u_m(\tau_m(t)) - u_n(\tau_n(t))\|^2 + K\Big(\frac{1}{2^m} + \frac{1}{2^n}\Big),\end{aligned}$$

or, using again the Lipschitz continuity of the $u_n(\cdot)$,

$$\langle \dot{u}_n(t) - \dot{u}_m(t), u_n(t) - u_m(t)\rangle \le \frac{2L}{r} \|u_m(t) - u_n(t)\|^2 + K'\Big(\frac{1}{2^m} + \frac{1}{2^n}\Big),$$

for a suitable constant K' independent of m, n. Set now $r_{mn}(t) = \|u_n(t) - u_m(t)\|^2$. The above calculations yield

$$\dot{r}_{mn}(t) \le \frac{4L}{r} r_{mn}(t) + 2K'\Big(\frac{1}{2^m} + \frac{1}{2^n}\Big),$$

and from Gronwall's inequality we obtain, for a suitable constant A independent of m, n,

$$r_{mn}(t) \le A\Big(\frac{1}{2^m} + \frac{1}{2^n}\Big),$$

for all $t \in [0,T]$, i.e., $\{u_n\}$ is Cauchy in $\mathcal{C}^0([0,T]; H)$. By taking a subsequence we may assume that there exists $u \in W^{1,2}([0,T]; H)$ such that $u_n \to u$ uniformly in $[0,T]$ and $\dot{u}_n \to \dot{u}$ weakly in $L^2([0,T]; H)$. By standard arguments (see, e.g., [**116**, p. 17]) we also have

$$\dot{u}(t) \in \bigcap_{n=1}^{\infty} \operatorname{cl} \operatorname{co} \bigcup_{k \ge n} \{\dot{u}_k(t)\} \quad \text{for a.e. } t \in [0,T],$$

together with

$$u(t) \in C(t) \quad \text{for all } t \in [0,T].$$

It remains to show that $-\dot{u}(t) \in N(C(t); u(t))$ for a.e. $t \in [0,T]$. To this aim, observe that by arguing as in the convergence step, we can find a sequence of numbers $\eta_n \downarrow 0$

such that for all n, a.e. t, and all $a \in C(t) \cap (u(t) + 2L\mathbb{B})$ one has

$$\langle -\dot{u}_n(t), a - u(t)\rangle \le \frac{L}{2r}\|a - u(t)\|^2 + \eta_n.$$

Putting $\varepsilon_n := \sup_{k \ge n} \eta_k$, we see that ε_n decreases to 0, and, for a.e. t and all n, we have

$$-\mathrm{cl\,co} \bigcup_{k \ge n} \{\dot{u}_k(t)\} \subset \Big\{ v \in H : \langle v, a - u(t)\rangle \le \frac{L}{2r}\|a - u(t)\|^2 + \varepsilon_n \quad \forall a \in C(t) \cap (u(t) + 2L\mathbb{B}) \Big\},$$

so that for a.e. $t \in [0, T]$

$$\langle -\dot{u}(t), a - u(t)\rangle \le \frac{L}{2r}\|a - u(t)\|^2 \quad \forall a \in C(t) \cap (u(t) + 2L\mathbb{B}).$$

But this is enough to conclude that $-\dot{u}(t) \in N(C(t); u(t))$. □

REMARK 55. It is possible to consider the problem (81) also without assuming the ρ-prox-regularity of the moving set, but only its closedness. In this case Clarke normal cone needs to be considered. However, due to the lack of (hypo)-monotonicity, no uniqueness is to be expected. Instead, existence of L-Lipschitz solutions holds, under a suitable compactness assumption, which implies both the nonemptiness of the metric projection and the convergence of the approximate solutions. The convergence argument uses also a joint (with respect to both the time t and the state x) measurability result for Clarke normal cone to moving closed sets. Details can be found in [**44**].

Another approach to the existence of solutions to (81) is based on the use of the distance function. Indeed it is proved in [**113**] that, under the r-prox-regularity and the L-Lipschitz property of the moving set $C(t)$, a mapping $u(\cdot)$ is a solution of the constrained differential inclusion (81) if and only if it is a solution of the unconstrained differential inclusion

$$\begin{cases} -\dot{x}(t) \in L\,\partial d_{C(t)}(x(t)) \\ x(0) = x_0 \in C(0). \end{cases} \tag{92}$$

Although the differential inclusion (92) is not subject to the constraint $x(t) \in C(t)$, it is shown that any of its solutions verifies this constraint. Without the prox-regularity assumption of $C(t)$, it can also be proved, e.g., in finite dimensions, the existence of a certain solution of (92) which is also a solution of (81) taking $N(C(t); \cdot)$ as the Clarke normal cone. Other results in this direction have been also established in [**69**].

6.3. Continuous moving sets. Now the case of a BV solution of (81) with a continuous moving set $C(\cdot)$ will be considered. The gain of regularity is due to the nonempty interior assumption for $C(\cdot)$, together with a compatibility condition between r and the diameter of the ball contained in $C(t)$ for all t, which is automatically satisfied when $C(t)$ is convex. The section ends with a counterexample showing that the compatibility condition cannot be removed.

Following [**93**], by a BV solution we mean a continuous selection from $C(\cdot)$, such that $u(0) = u_0$ and such that

$$-\frac{du}{|du|}(t) \in N_{C(t)}(u(t)) \qquad |du| - \text{a.e.}, \tag{93}$$

where $|du|$ denotes the total variation of the differential measure du.
With $\text{Var}(u; 0, T)$ we mean the total variation of the function $u : [0, T] \to H$.

THEOREM 56. Let $C : [0, T] \to H$ be a set-valued map, $u_0 \in C(0)$ and $r > 0$. Assume that

1. C is Hausdorff continuous;
2. $C(t)$ is r-prox-regular for all t;
3. there exist $z \in H$ and $r' > 0$ such that $B(z, r) \subset C(t)$ for all t;
4. the compatibility condition

$$(\|u_0 - z\| + r')^2 < 2rr'$$

holds.

Then the problem (81) admits a unique continuous BV solution $u : [0, T] \to H$.

PROOF. First we observe that it is easy to state an integral formulation of the problem (for the proof, see, e.g., [**51**, Proposition 3.1]): under our assumptions, a continuous BV selection u from C satisfies (93) if there exists a positive measure μ on $[0, T]$, with $\|du\| << \mu$, and $c \in [0, +\infty)$ such that, for every continuous selection ϕ from C the inequalities

$$\int_s^t \langle u - \phi, du \rangle \le c \int_{[s,t]} \|u - \phi\|^2 \, d\mu, \qquad s, t \in [0, T],\ s < t \tag{94}$$

hold.

Let now $\{u_n\}$ be the sequence of approximate solutions constructed in Theorem 51 and satisfying (88). The existence proof is divided into some steps.

Estimate on the total variation of u_n. Since u_n is piecewise affine, it is clear that

$$\text{Var}(u_n; 0, T) = \sum_{i=0}^{2^n - 1} \left\| u_n(t_n^{i+1}) - u_n(t_n^i) \right\| = \sum_{i=0}^{2^n - 1} \left\| u_n^{i+1} - u_n^i \right\|.$$

By (83), Lemma 53 above can be applied, yielding

$$\text{Var}(u_n; 0, T) \le \max \left\{ \frac{\|u_0\|^2 - (r')^2}{2(r' - \varphi_0 d^2)}, 0 \right\} =: K, \tag{95}$$

where $d = \|u_0\| + r' + \max\{\omega(1/2^n)\}$.

The sequence $\{u_n\}$ is Cauchy. This section is devoted to the proof of the following claim:
The sequence of approximate solutions $\{u_n\}$ satisfies the estimate

$$\|u_m - u_n\|_\infty^2 \leq 12K\omega(1/2^n)(1 + 6\omega(1/2^n)\varphi_0)e^{8\varphi_0 K} \qquad (m > n), \tag{96}$$

hence it is Cauchy for the uniform convergence on $[0, T]$.

By (85) and the assumption 2, for Lebesgue-a.e. t, if $y \in C(\tau_n(t))$ then:

$$\langle u_n(\tau_n(t)) - y, u_n'(t)\rangle \leq \varphi_0 \|u_n'(t)\| \, \|y - u_n(\tau_n(t))\|^2. \tag{97}$$

By (86) and (82), for all $m, n \in \mathbb{N}$ there exists $y = y(m) \in C(\tau_n(t))$ such that

$$\begin{aligned}\|u_m(t) - y\| &= d(u_m(t), C(\tau_n(t))) \\ &\leq \|u_m(t) - u_m(\tau_m(t))\| + d(u_m(\tau_m(t)), C(\tau_n(t))) \\ &\leq \omega(1/2^m) + H(C(\tau_m(t)), C(\tau_n(t))) \leq 2\omega(1/2^m) + \omega(1/2^n) \\ &< 2\left(\omega(1/2^m) + \omega(1/2^n)\right) =: a_{n,m}.\end{aligned}$$

Hence (97) implies that for Lebesgue-a.e. t

$$\begin{aligned}\langle u_n(t) - u_m(t), u_n'(t)\rangle &\leq \varphi_0 \|u_n'(t)\| \, \|y - u_n(\tau_n(t))\|^2 + a_{n,m} \|u_n'(t)\| \\ &\leq \|u_n'(t)\| \left(a_{n,m} + \varphi_0(\|u_n(t) - u_m(t)\| + a_{n,m})^2\right) \\ &\leq \left(A_{n,m} + 2\varphi_0 \|u_n(t) - u_m(t)\|^2\right) \|u_n'(t)\|,\end{aligned} \tag{98}$$

where $A_{n,m} := a_{n,m} + 2a_{n,m}^2\varphi_0$. Similarly, one obtains for a.e. t

$$\langle u_m(t) - u_n(t), u_m'(t)\rangle \leq (A_{n,m} + 2\varphi_0 \|u_m(t) - u_n(t)\|^2) \|u_m'(t)\|. \tag{99}$$

Integrating and summing (98) and (99), we conclude that

$$\begin{aligned}&\int_0^t \langle u_n(s) - u_m(s), u_n'(s) - u_m'(s)\rangle \, ds \\ &\qquad \leq \int_0^t (A_{n,m} + 2\varphi_0 \|u_m(s) - u_n(s)\|^2)(\|u_m'(s)\| + \|u_n'(s)\|) \, ds.\end{aligned}$$

Since $u_n(0) = u_m(0)$, we obtain from (95) and the preceding inequality

$$\|u_n(t) - u_m(t)\|^2 \leq 4A_{n,m}K + 4\varphi_0 \int_0^t \|u_n(s) - u_m(s)\|^2 (\|u_n'(s)\| + \|u_m'(s)\|) \, ds.$$

Hence, by Gronwall's Lemma:

$$\begin{aligned}\|u_n(t) - u_m(t)\|^2 &\leq 4A_{n,m}K\exp\left(4\varphi_0 \int_0^t (\|u_n'(s)\| + \|u_m'(s)\|) \, ds\right) \\ &\leq 4A_{n,m}K\exp(8\varphi_0 K),\end{aligned}$$

which concludes the proof.

Convergence. By the above claim, the sequence $\{u_n\}$ converges uniformly to a continuous function $u : [0, T] \to H$, which, by (87), is a selection from $C(t)$. Moreover, we deduce from (95) that

$$\mathrm{Var}(u; 0, T) \leq K.$$

In order to prove that u is a solution, we only need to show that the integral inequality (94) holds. To this aim, let ϕ be a continuous selection from C and $s < t$. By (84), φ_0-prox-regularity implies that

$$\int_s^t \langle u_n(\tau_n(\sigma)) - \phi(\tau_n(\sigma)), u_n'(\sigma)\rangle \, d\sigma \leq \varphi_0 \int_s^t \|u_n(\tau_n(\sigma)) - \phi(\tau_n(\sigma))\|^2 \, \|u_n'(\sigma)\| \, d\sigma.$$

Using the notation of differential measures this can be rewritten as

$$\int_s^t \langle u_n \circ \tau_n - \phi \circ \tau_n, du_n\rangle \leq \varphi_0 \int_s^t \|u_n \circ \tau_n - \phi \circ \tau_n\|^2 \, |du_n|. \tag{100}$$

By using Helly-Banach compactness theorem we know that du_n converges to du, in the weak* topology of measures. Moreover, we can suppose that $\|du_n\|$ converges in the weak* topology to a nonnegative finite measure μ, such that $\mu \geq \|du\|$ in the ordering of real measures. To see this inequality, observe that for every continuous nonnegative function f one has $\int f \, d\mu = \lim_{n\to\infty} \int f |du_n| \geq \lim_{n\to\infty} \left\|\int f \, du_n\right\| = \left\|\int f \, du\right\|$. By the uniform convergence of $\{u_n\}$, the continuity of ϕ and since $0 \leq \tau_n(t) - t \leq T/2^n$, it follows that $u_n \circ \tau_n - \phi \circ \tau_n \to u - \phi$ uniformly on $[0, T]$. Therefore (100) implies that

$$\int_s^t \langle u - \phi, du\rangle \leq \varphi_0 \int_{[s,t]} \|u - \phi\|^2 \, d\mu,$$

hence concluding the proof.

The uniqueness part is similar to the Lipschitz case, so we omit the proof (see, e.g., [**51**, Theorem 4.4]). □

EXAMPLE 57. The compatibility condition 4. in the statement of Theorem 56 cannot be dropped.

Indeed, let, for $t > 0$, $\Theta(t)$ be the counterclockwise rotation in $\mathbb{R}^2$ around the origin by the angle π/t, so that $\Theta(1/2)$ is the identity in $\mathbb{R}^2$, and let $S(t) = \Theta(t)\{(x, y) : -1 < x < 1, |y| < t\}$. Let $C(0) = \{(x, y) : x^2 + y^2 \leq 1\}$ and, for $0 < t \leq 1/2$, let

$$C(t) = C(0) \setminus S(t).$$

Observe that C is Hausdorff continuous and, for $t > 0$, $C(t)$ is t-prox-regular. Thus there is a lack of uniform prox-regularity with respect to t.

The problem (93) with the initial condition $u(0) = u_0 = (0, 0)$ cannot have a continuous BV solution. In fact, let by contradiction u be such a solution. Observe that, for $t > 0$, $C(t)$ is the union of two disjoint convex sets, $C_1(t)$ and $C_2(t)$, and by continuity $u(t)$ must belong to only one of them for all t, say to $C_1(t)$. Therefore, for $t > 0$, $u(t)$ is a solution of the classical sweeping process $u'(t) \in -N(C_1(t); u(t))$. We want to estimate the variation of u on the interval $[0, 1/2]$ by taking the sum of its variations on intervals of the type $I_k = [1/(2(k + 1)), 1/(2k)]$. To this aim, observe that, at most after half a turn of the rotation, $u(t)$ is caught by bdryC_1, and so it is forced to rotate as well. Therefore, the variation of u on I_k is not smaller than the length of the arc of radius $1/(2(k+1))$ (i.e., the minimal norm of points in $C_1(t)$, $t \in I_k$) and angle π; namely,

$$\mathrm{Var}\left(u_n; \frac{1}{2(k+1)}, \frac{1}{2k}\right) \geq \frac{\pi}{2(k+1)}.$$

Thus u cannot have bounded variation on $[0, 1/2]$.

6.4. Sets moving with bounded variation. In this subsection, the set $C(t)$ is not required to move in a continuous way, but instead it is assumed that it has a right continuous bounded variation. This means that its variation function $\mathrm{var}_C(\cdot)$ is finite and right continuous on $[0, T]$, where $\mathrm{var}_C(t)$ is the supremum over the set of all subdivisions of $[0, t]$ of the numbers

$$\sum H(C(t_i), C(t_{i-1}))$$

with $0 = t_0 < t_1 < \cdots < t_k = t$. Let us denote by μ the Radon measure associated with the function $\mathrm{var}_C(\cdot)$. So, for all $0 \leq s \leq t \leq T$

$$H(C(t), C(s)) \leq \mu(]s, t]).$$

Then a mapping $u : [0, T] \to H$ is a solution of the differential inclusion

$$\begin{cases} -dx & \in \; N(C(t); x(t)) \\ x(0) & = \; x_0 \in C(0) \\ x(t) & \in \; C(t) \qquad \forall t \end{cases}$$

if it is a right continuous selection from $C(\cdot)$ with bounded variation on $[0, T]$ and such that

$$-\frac{du}{d\mu}(t) \in N(C(t); u(t)) \quad \mu - \text{a.e.}$$

With the above assumptions and the r-prox-regularity of $C(t)$, an existence theorem is established in [**62**]. Uniqueness is also proved under an additional hypothesis.

6.5. Time and state depending moving sets. For the case when the moving set depends both on the time and on the state, i.e. $C(t, x)$, we refer the reader to [**37**] and [**40**].

7. Other applications

7.1. Minimum time problems. The main motivation for the study of functions with $\rho(\cdot)$-prox-regular epigraph or hypograph was their possible application to the regularity theory of minimum time functions. The present section is devoted to a brief overview of this topic.

Minimum time problems are widely studied from several viewpoints (see, e. g., [**6**, Chapter IV], [**23**], [**33**] and references therein). In particular, it is well known that in general the minimum time function T is not everywhere differentiable. It is also well known that suitable controllability conditions imply its Hölder continuity, but this fact does not contain information on differentiability. In a 1995 paper (see now Chapter 8 in the book [**33**]), Cannarsa and Sinestrari found a connection between the dynamics and the target which actually implies the semiconcavity (or semiconvexity if a different set of assumptions is taken) of T. Regularity properties of semiconcave functions were also thoroughly studied (see again [**33**] and references therein), thus providing a rich set of information on the structure of T. The link between semiconcave functions and r-prox-regular sets, through the hypograph, is explained in Proposition 66 below: a locally Lipschitz function is semiconcave if and only if there exists $r > 0$ such that its hypograph is r-prox-regular. Of course an entirely symmetric characterization for semiconvex functions can be expressed using the epigraph. The starting point of the study of functions with $\rho(\cdot)$-prox-regular hypo/epigraph was the observation that the minimum time function is Lipschitz continuous if and only if a strong controllability assumption is satisfied (see [**33**, Chapter 8]), while in general T is only Hölder. A natural question therefore was trying to understand whether the structure of the minimum time function remains unchanged if the controllability assumptions are weakened, i.e., whether the hypograph/epigraph of T are $\rho(\cdot)$-prox-regular if T is supposed to be only continuous. The answer is – essentially – positive and it is contained in [**49, 53**]. Here the results are only stated, for the sake of brevity.

Consider the controlled initial value problem

$$\begin{cases} \dot{y}(t) &= f(y(t), u(t)) \quad a.e., \\ u(t) &\in \mathcal{U} \qquad\qquad\quad a.e., \\ y(0) &= x, \end{cases} \tag{101}$$

where the function $f : \mathbb{R}^n \times \mathcal{U} \longrightarrow \mathbb{R}^n$ and the compact control set $\mathcal{U} \subset \mathbb{R}^n$ are given. The control function $u(\cdot)$ is measurable. For all $t > 0$, we denote by $\mathcal{U}_{ad}$ the set of admissible controls, i.e., the measurable functions $u : \mathbb{R} \to \mathbb{R}^n$, such that $u(t) \in \mathcal{U}$ a.e. We assume $f(\cdot, \cdot)$ to be Lipschitz with respect to the first variable, and for any $u(\cdot) \in \mathcal{U}_{ad}$, the unique Carathéodory solution of (101) is denoted by $y^{x,u}(\cdot)$.

Suppose we are now given a closed nonempty set $\mathcal{S} \subset \mathbb{R}^n$, which is called the target. For fixed $x \in \mathbb{R}^n \setminus \mathcal{S}$, we denote

$$\theta(x, u) := \min \{t \geq 0 \,:\, y^{x,u}(t) \in \mathcal{S}\}.$$

Then, $\theta(x, u) \in]0, \infty]$, and we note that $\theta(x, u)$ is the time taken for the trajectory $y^{x,u}(\cdot)$ to reach $\mathcal{S}$. The *minimum time* $T(x)$ to reach $\mathcal{S}$ from x is defined by

$$T(x) := \inf \{\theta(x, u) \,:\, u(\cdot) \in \mathcal{U}_{ad}\}.$$

As an example, consider the following minimum time function, arising in a classical linear problem: the rocket railroad car (see, e.g., [**6**, p. 242]).

Set $R_- = \{(x_1, x_2) : x_1 > -\frac{1}{2}x_2|x_2|\}$, $R_+ = \{(x_1, x_2) : x_1 < -\frac{1}{2}x_2|x_2|\}$ and

$$T(x_1, x_2) = \begin{cases} x_2 + 2\sqrt{\frac{x_2^2}{2} + x_1}, & (x_1, x_2) \in \overline{R}_- \\ -x_2 + 2\sqrt{\frac{x_2^2}{2} - x_1}, & (x_1, x_2) \in \overline{R}_+. \end{cases}$$

The function T is actually the minimum time to reach the origin for the control system:

$$\begin{cases} \dot{\xi}_1 &= \xi_2, & \xi_1(0) &= x_1 \\ \dot{\xi}_2 &\in [-1, 1], & \xi_2(0) &= x_2. \end{cases}$$

It is easily seen that T is not Lipschitz continuous and that its epigraph is r-prox-regular, for a suitable r. Here the (second order) Kalman rank controllability condition is satisfied. Several other examples can be given.

The main results are essentially as follows (more general versions can be found in [**49, 53**]).

THEOREM 58. Consider the problem of reaching the target $\mathcal{S}$ by a solution of the controlled dynamics (101) and assume that the minimum function T is everywhere finite and continuous. The following results hold true.

1) (linear case) Let $f(x, u) = Ax + Bu$ be linear in both x and u and let $\mathcal{S}$ be convex. Then there exists a continuous $\rho(\cdot)$ such that the epigraph of T is $\rho(\cdot)$-prox-regular in $\mathbb{R}^n \setminus \mathcal{S}$.
2) (nonlinear case) Let f be of class $\mathcal{C}^2$ in both x and u and let the complement of the target be r-prox-regular for a suitable r. Assume furthermore that for all $x \notin \mathcal{S}$ the normal cone to the hypograph of T $N(\text{hypo}(T); (x, T(x))$ be pointed. Then there exists a continuous $\rho(\cdot)$ such that the hypograph of T is $\rho(\cdot)$-prox-regular in $\mathbb{R}^n \setminus \mathcal{S}$.

Consequently, in both cases T satisfies the properties appearing in the statement of Theorem 48.

The proof of both results uses an explicit computation of the proximal subdifferential (resp. superdifferential) of T. The nonlinear result is no longer true if the pointedness assumption is dropped (see [**52**, Example 2]).

7.2. Hamilton–Jacobi equations. In this section, Ω will be an open subset of $\mathbb{R}^n$ and $H : \Omega \times \mathbb{R} \times \mathbb{R}^n \to \mathbb{R}$ will be a continuous function. We are interested in the following general nonlinear first order equation:

$$H(x, u, Du) = 0, \quad x \in \Omega \tag{102}$$

in the unknown $u : \Omega \to \mathbb{R}$. It is well known that a semiconcave function u which satisfies (102) a.e. is also a viscosity solution (see, e.g., [**6**, Corollary 5.2 (i)] or [**33**, Proposition 5.3.1]). If one focuses the attention on noncoercive Hamiltonians (i.e., does not require that $\lim_{\|p\| \to +\infty} H(x, u, p) = +\infty$), then solutions which are not necessarily locally Lipschitz continuous naturally appear. We present here conditions implying that a continuous a.e. solution of (102) with $\rho(\cdot)$-prox-regular hypograph is also a viscosity solution.

THEOREM 59. Let Ω be an open subset of $\mathbb{R}^n$, $u : \Omega \to \mathbb{R}$ be a continuous function with $\rho(\cdot)$-prox-regular hypograph for a suitable $\rho(\cdot)$. Let $H : \Omega \times \mathbb{R} \times \mathbb{R}^n \to \mathbb{R}$ and $\alpha : \Omega \to \mathbb{R}$ be continuous and such that $H(x, u, \cdot)$ is convex for all $(x, u) \in \Omega \times \mathbb{R}$. Suppose furthermore that

$$H(x, u(x), Du(x)) = \alpha(x) \quad \text{for a.e. } x \in \Omega.$$

Then, if either

(a) $(x, u) \mapsto H(x, u, p)$ is continuous uniformly with respect to $p \in \{Du(x) : x \in \text{dom} Du\}$,

or

(b) $H(x, u, \lambda p) = \lambda H(x, u, p)$ for all $(x, u, p) \in \Omega \times \mathbb{R} \times \mathbb{R}^n$ and all $\lambda \geq 0$, $u(\cdot)$ is a viscosity solution of the equation $H(x, u, Du) = \alpha$.

The proof can be found in [**50**] and is a consequence of the representation of the generalized gradient given in (72).

7.3. Continuous selections and parametrizations. For concepts and methods of set-valued analysis we refer to [**4, 5, 116**].

The famous Michael's selection theorem (see, e.g., [**87**]) states that if F is a lower semicontinuous set-valued function with closed convex values from a paracompact topological space X to a Banach space E, then there exists a continuous function $f : X \to E$ such that $f(x) \in F(x)$ for all $x \in X$, i.e., a continuous selection from F. If F is not convex-valued, continuous selections may not exist in general (see, e.g., Examples 1 and 2, pp. 68, 69, in [**4**]), although several nonconvex results are available.

If H is finite dimensional and G is a Lipschitz continuous compact convex valued map from H into itself, there exists a Lipschitz continuous map $g : H \times \mathbb{B} \to H$ such that $G(x) = g(x, \mathbb{B})$ for all $x \in H$ (see [**5**, Theorem 9.6.2] and references therein). Moreover, the Lipschitz constant λ of g with respect to the first variable x is proportional to the Lipschitz constant L of G, while the Lipschitz constant Λ of g with respect to the second variable p is proportional to the maximal norm of the elements of G. This type of result is called a *parametrization* of G. If G is not convex-valued, parametrizations may not exist, as well as if G is convex valued but H is infinite dimensional (see however the nonconvex parametrization result contained in [**76**]).

In this section we present two simple results, which are actually corollaries of the convex theorems stated above, where the values of the set-valued maps are assumed to be r-prox-regular.

Let $C \subset H$ be closed. We recall that if the diameter of C, diam C, is such that

$$s := \frac{n}{2(n+1)} \left(\frac{\operatorname{diam} C}{r} \right)^2 < 1,$$

then the metric projection P_C from $\operatorname{co} C$ into C is a Lipschitz continuous single valued map, with $\operatorname{Lip}(P_C) = 1/(1-s)$ (see Proposition 9 and (g) in Theorem 3). Actually in this case C is a retract of its convex hull, with a special Lipschitz continuous retraction, namely the metric projection (see also [**68**]). We are therefore ready to prove the following

PROPOSITION 60. Let $H = \mathbb{R}^n$ and let $r > 0$ be given. Let F be a Lipschitz continuous set valued map from H into H with compact r-prox-regular values and let $0 \leq \eta < 1$ be such that

$$\frac{n}{2(n+1)} (\operatorname{diam} F(x))^2 \leq r^2 \eta \qquad \forall x \in H. \tag{103}$$

Then there exists a Lipschitz continuous single valued map $f : H \times \mathbb{B} \to H$ such that $F(x) = f(x, \mathbb{B})$ for all $x \in H$. Moreover, the Lipschitz constant λ of f with respect to the first variable x is proportional to the Lipschitz constant L of F, while the Lipschitz constant Λ of f with respect to the second variable p is proportional to the maximal norm of the elements of F.

PROOF. Apply the parametrization result in [**5**] to the Lipschitz continuous set-valued map $\operatorname{co} G$, finding a Lipschitz continuous map $\tilde{f}$ such that $\operatorname{co} F(x) = \tilde{f}(x, \mathbb{B})$ for all $x \in H$. The proof is then concluded by setting, for all $x \in H, p \in \mathbb{B}$,

$$f(x,p) = P_{F(x)}(\tilde{f}(x,p)). \qquad \square$$

With a similar argument one can also prove the following

PROPOSITION 61. *Let X be a paracompact topological space and let $r > 0$ be given. Let F be a Hausdorff continuous set-valued map from X into the closed r-prox-regular subsets of H. Let $0 \leq \eta < 1$ be such that (103) holds for all $x \in X$. Then there exists a continuous selection $f : X \to H$ from F.*

A more precise result, where the Hausdorff continuity assumption on F is "parametrized with respect to r" and becomes lower semicontinuity for $r = +\infty$ (in order to generalize Michael's selection theorem), can be found in [**46**]. It is based on a kind of "geodesic convexity" which is satisfied by a certain class of $\rho(\cdot)$-prox-regular sets.

7.4. Isoperimetric inequalities. The well known isoperimetric inequality (we make reference to the original paper by De Giorgi [**60**]) states that if C is a Borel set in $\mathbb{R}^n$, $n \geq 2$, with finite measure $|C|$ and finite perimeter $P(C)$, the ball with the same volume has smaller perimeter, i.e.,

$$n\omega_n^{\frac{1}{n}} |C|^{\frac{n-1}{n}} \leq P(C). \tag{104}$$

Moreover equality holds in (104) if and only if C is a ball.

Several authors have given *quantitative* versions of the isoperimetric inequality, in the sense that they have improved (104) by introducing a further term which is zero if and only if the inequality is an equality. This term is a kind of measure of the "distance" of C from being a sphere. In [**66**], (104) is refined to become

$$n\omega_n^{\frac{1}{n}} |C|^{\frac{n-1}{n}} (1 + d(C)) \leq P(C),$$

$d(C)$ being the *spherical deviation* of C, which is actually the Hausdorff distance between a suitable normalization of C and the unit ball. The result in [**66**] requires some assumptions on C. In particular, C needs to be "nearly spherical", in the sense that both its Hausdorff distance from the unit ball and the norm of the gradient of a suitable representation of the boundary must be small. The main result in [**66**] applies to the class of compact *convex* sets with nonempty interior having "isoperimetric deficiency" (the isoperimetric deficiency of a set C is defined by

$$\Delta(C) = \frac{P(C)}{n\omega_n} \left(\frac{|C|}{\omega_n} \right)^{-\frac{n-1}{n}} - 1,$$

where ω_n denotes the n-dimensional Lebesgue measure of the unit ball $\mathcal{B}$ in $\mathbb{R}^n$) small enough.

The paper [**52**], which contains also references to the work of other authors on the subject, deals mainly with results on the line of [**66**] for a class of $\rho(\cdot)$-prox-regular sets. The results are based on several geometric estimates on a compact $\rho(\cdot)$-prox-regular set C with nonempty interior, which give simple conditions guaranteeing that C is "nearly spherical". This permits to apply the general result of [**66**], proving a quantitative version of (104) of the same type of [**66**]. Following [**66**], we deal with a closed set C with the property that there exist $0 < s < R$ such that

$$B(0,s) \subset C \subset B(0,R). \tag{105}$$

The result is obtained through suitable modifications of some arguments of convex analysis. If C is convex, the main result of [**66**] is recovered. More precisely, some compatibility conditions involving only ρ, s, and R are assumed. Such conditions vanish if the parameter $\rho(\cdot)$ is identically $+\infty$, i.e., if C is convex, and the estimates become the convex ones. We ask, essentially, $\rho(\cdot)$ to be large enough, and ensure in turn that

(1) C satisfies an internal cone condition with uniform width (which implies that an *a priori* estimate on the gradient of a suitable parametrization of the boundary can be derived from an L^∞ estimate of the same parametrization);
(2) the barycentre of C belongs to C;
(3) the convex hull of C is contained in a neighborhood of C where the distance to C is smooth;
(4) the isoperimetric deficiency of C and of its convex hull can be estimated each one from the other.

For the sake of brevity we present only the shorter geometric result (the result on the barycentre) and the statement of the main result. In this section the Lebesgue n-dimensional measure of a Borel set K will be denoted by $|K|$.

PROPOSITION 62. Let C be r-prox-regular and such that $|C| > 0$. Then C has nonempty interior. Assume now that C satisfies (105) and set $\varphi_0 = 1/(2r)$. Then, if furthermore

$$\varphi_0(R-s) < 1 \quad \text{and} \quad \frac{n-1}{n}\frac{\omega_{n-1}}{\omega_n}\left(\left(\frac{R}{s}\right)^2 - 1\right)^{\frac{n+3}{2}} \frac{\varphi_0^2 s^2}{(1+\varphi_0 s)^2 - \varphi_0^2 R^2} < 1, \tag{106}$$

the barycentre $b(C)$ of C belongs to the interior of C.

PROOF. To prove the first statement, assume by contradiction that $\mathrm{int}C = \emptyset$. Then a.e. point of C is on one hand a point with density 1, on the other hand a boundary point. But, thanks to the very definition of r-prox-regularity, every point in $\mathrm{bdry}C$ must have density not larger than $1/2$, a contradiction.

Assume now that C satisfies (105) with $b = 0$ and let by contradiction $b(C) \notin \mathrm{int}C$. Choose $y \in P_C(b(C)) \subset \mathrm{bdry}C$. Then, since by assumption $B(0,s) \subseteq C$, by r-prox-regularity we have

$$B\left(y + \frac{v}{2\varphi_0}, \frac{1}{2\varphi_0}\right) \cap B(0,s) = \emptyset, \tag{107}$$

where $v = (b(C) - y)/\|b(C) - y\|$ if $b(C) \notin C$, and v is any unit vector in $N^P(C; y)$ if $b(C) \in \text{bdry}C$.

Let P be the intersection of the ball $B\left(y + \frac{v}{2\varphi_0}, \frac{1}{2\varphi_0}\right)$ with the segment between 0 and $y + \frac{v}{2\varphi_0}$. Take the hyperplane H passing through the point P and perpendicular to the vector $y + \frac{v}{2\varphi_0}$. Recalling (107), this hyperplane splits the space into the half space H_1, which contains $B(0, s)$, and H_2, which contains $B(y + \frac{v}{2\varphi_0}, \frac{1}{2\varphi_0})$.

Putting $r_1 = \|P\|$ and $v_0 = \frac{y+\frac{v}{2\varphi_0}}{\left\|y+\frac{v}{2\varphi_0}\right\|}$, one can easily check that

$$\begin{aligned} H &= r_1 v_0 + \{x \in \mathbb{R}^n | < x, v_0 >= 0\}, \\ H_1 &= \{x \in \mathbb{R}^n | < x, v_0 >< r_1\}, \\ H_2 &- \{x \in \mathbb{R}^n | < x, v_0 >\geq r_1\}. \end{aligned}$$

Dividing the set C into three parts, $C_1 = B(0, s)$, $C_2 = (H_1 \cap C) \setminus B(0, s)$, and $C_3 = H_2 \cap C$, we obtain

$$< b(C), v_0 >= \frac{1}{|C|}\left(\int_{C_1} < x, v_0 > dx + \int_{C_2} < x, v_0 > dx + \int_{C_3} < x, v_0 > dx\right).$$

Observing that $\int_{C_1} < x, v_0 > dx = 0$ and $\int_{C_2} < x, v_0 > dx \leq r_1|C_2|$, we obtain also

$$(108) \qquad < b(C), v_0 >\leq \frac{1}{|C|}\left(r_1|C_2| + \int_{C_3} < x, v_0 > dx\right).$$

Since $C \subseteq B(0, R)$, we see that $C \cap H_2 \subseteq (B(0, R) \cap H_2) \setminus B(y + \frac{v}{2\varphi_0}, \frac{1}{2\varphi_0})$. Now observe that the set $(B(0, R) \cap H_2) \setminus B(y + \frac{v}{2\varphi_0}, \frac{1}{2\varphi_0})$ is contained in the set C obtained by making a 2π-rotation around the axis through 0 and P of the triangle PQ_1Q_2. Observe that the base length of this triangle is $\sqrt{R^2 - r_1^2}$, while its height is $h := \frac{\varphi_0(R^2 - r_1^2)}{\sqrt{(1+\varphi_0 r_1)^2 - \varphi_0^2 R^2}}$. Observe also that C is contained in H_2.

Thus,

$$|C \cap H_2| < |C| = \frac{(n-1)\omega_{n-1}}{n}\left(R^2 - r_1^2\right)^{\frac{n+1}{2}} \frac{\varphi_0}{\sqrt{(1 + \varphi_0 r_1)^2 - \varphi_0^2 R^2}}.$$

Moreover, $\|x - \pi_H(x)\| \leq h$ for all $x \in C$, whence $< x, v_0 >\leq r_1 + h$. Therefore

$$\int_{C_3} < x, v_0 > dx \leq (r_1 + h)|H_2 \cap C| \leq r_1|H_2 \cap C| + h|C|.$$

Combining with (108) we obtain

$$< b(C), v_0 >\leq \frac{1}{|C|}\left(r_1|C_2| + r_1|C_3| + h|C|\right).$$

Thus

$$\begin{aligned} &< b(C), v_0 > \\ &\leq \frac{1}{|C|}\left(r_1|C_2| + r_1|C_3| + \frac{(n-1)\omega_{n-1}}{n}\left(R^2 - r_1^2\right)^{\frac{n+3}{2}} \frac{\varphi_0^2}{(1 + \varphi_0 r_1)^2 - \varphi_0^2 R^2}\right). \end{aligned}$$

By construction $s \le r_1$. Using (106) we therefore obtain

$$\frac{(n-1)\omega_{n-1}}{n}\left(R^2-r_1^2\right)^{\frac{n+3}{2}}\frac{\varphi_0^2}{(1+\varphi_0 r_1)^2-\varphi_0^2R^2}<\omega_n s^{n+1}\le r_1|C_1|.$$

Consequently,

$$< b(C), v_0 >< \frac{1}{|C|}(r_1|C_2|+r_1|C_3|+r_1|C_1|)=r_1,$$

which is absurd because $b(C)$ belongs to H_2 by construction. The proof is concluded. □

THEOREM 63. Let $r>0$ be given and let $C\subset\mathbb{R}^n$ be a compact r-prox-regular set with $|C|>0$. Then there exist $d>0$, $\sigma>0$ and a continuous strictly increasing function $f:[0,+\infty)\longrightarrow[0,+\infty)$ such that, if $d(C)<d$ and $\Delta(C)<\sigma$, the estimate

$$d(C)\le f(\Delta(C)) \tag{109}$$

holds true. Furthermore, σ and f can be explicitly computed and depend only on the space dimension n, while d depends only $n, r, |C|$.

7.5. Infinite dimensional minimum time problems with constant dynamics. Let $C\subset H$ be a nonempty closed set. The distance $d(u,C)$ from $u\notin C$ to C is obviously the minimum time to reach C from u subject to the constant dynamics $\dot{y}\in\mathbb{B}$. Since $\rho(\cdot)$-prox-regularity is characterized by the property that $d(\cdot,C)$ be of class $\mathcal{C}^{1,1}$ around C and - equivalently - by the well posedness of the optimal control (which is actually $P_C(u)-u$), it is natural to investigate whether similar properties hold for a more general (constant) convex dynamics F, or for a more general target. This idea has been pursued in [**54**], where, in particular, sufficient conditions for the well posedness of the optimal control are given for an r-prox-regular target in terms of r and the strict convexity of F. More generally, in [**67**] the well posedness of the optimal control is established through local conditions on both the target and the dynamics, which include $\rho(\cdot)$-prox-regularity as a particular case. In [**67**], $\rho(\cdot)$-prox-regularity is interpreted as a condition involving a lower bound for a kind of (negative) curvature.

8. Miscellaneous results and examples

We begin with some examples of prox-regular sets. The space H, unless otherwise stated, will be supposed to be possibly infinite dimensional.

EXAMPLE 64. 1) Closed convex sets are $+\infty$-prox-regular, since all proximal normals are realized by balls of arbitrarily large radius. The converse statement follows from Corollary 17.
2) Closed sets with $\mathcal{C}^{1,1}$-boundary are $\frac{1}{L}$-prox-regular, where L is the Lipschitz constant of both the parametrization of the boundary and of its gradient. Indeed, without loss of generality one can express locally C as the epigraph of a $\mathcal{C}^{1,1}$-function f from a subspace of codimension one of H into $\mathbb{R}$. If we represent the points of

bdry$C$ as $(x, f(x)) =: X$, a normal vector can be written as $V = (Df(x), -1)$. Taking $Y = (y, \beta)$, with $\beta \geq f(y)$, recalling Taylor's formula we obtain

$$\begin{aligned}\langle V, Y - X\rangle &\leq \langle Df(x), y - x\rangle - (f(y) - f(x))\\ &= \int_0^1 \langle Df(x) - Df(x + t(y - x)), y - x\rangle\, dt\\ &\leq \frac{L}{2}\|y - x\|^2\\ &\leq \frac{L}{2}\|V\|\,\|Y - X\|^2.\end{aligned}$$

Recalling (b) in Theorem 3, the proof is complete.
3) The set

$$C = \{x = (x_1, \ldots, x_n) \in \mathbb{R}^n \,:\, \|x\| \geq 1, \max|x_i| \leq 1\}$$

is 1-prox-regular.
4) (see [**29**, Proposition 1.9]) Let $\Omega \subset \mathbb{R}^n$ be an open bounded domain, and c, s be positive numbers and set

$$K_c = \left\{u \in H_0^1(\Omega) \,:\, \int_\Omega \|\nabla u\|^2 \leq c^2\right\}, \quad M_s = \left\{u \in L^2(\Omega) \,:\, \int_\Omega \|u\|^2 = s^2\right\}.$$

Then $M_\rho \cap K_c$ is $\rho(\cdot)$-prox-regular in $L^2(\Omega)$, provided c^2/s^2 is not an eigenvalue of $-\Delta$ in $H_0^1(\Omega)$.
Indeed, K_c is convex and closed in $L^2(\Omega)$ and M_s is the 0-level set of the $\mathcal{C}^{1,1}$ function $h(u) = \|u\|_{L^2}^2 - s^2$, whose gradient is

$$Dh(u) = 2u.$$

Let $u \in K_c$. We now show that

$$N(K_c; u) = \begin{cases}\{0\} & \text{if } \int_\Omega \|\nabla u\|^2 < c^2 \text{ or } \Delta u \notin L^2(\Omega)\\ \{-\lambda\Delta u \,:\, \lambda \geq 0\} & \text{if } \Delta u \in L^2(\Omega).\end{cases}$$

Indeed, let $w \in L^2(\Omega)$ be a nonzero normal vector to K_c at u and let $\bar{w} \in H_0^1(\Omega)$ be such that $-\Delta\bar{w} = w$ in Ω in the sense of distributions. Then, for all $v \in K_c$ we have

$$\begin{aligned}0 \geq \int_\Omega w(v - u) &= \int_\Omega -\Delta\bar{w}(v - x)\\ &= \int_\Omega \langle\nabla\bar{w}, \nabla v - \nabla u\rangle.\end{aligned}$$

By taking $v(x) = c\bar{w}(x)/\|\nabla\bar{w}\|_{L^2}$ one therefore obtains

$$c\|\nabla\bar{w}\|_{L^2} \leq \int_\Omega \langle\nabla\bar{w}, \nabla u\rangle \leq \|\nabla\bar{w}\|_{L^2}\|\nabla u\|_{L^2},$$

from which it follows that $\|\nabla u\|_{L^2} = c$ and that $\nabla\bar{w} = \alpha\nabla u$ for some $\alpha > 0$, i.e., $\bar{w} = \alpha u$. Hence $Dh(u) \in N(K_c; u) \cup (-N(K_c; u))$ if and only if $\Delta u \in L^2(\Omega)$,

$\int_\Omega \|\nabla u(x)\|^2 = c^2$, and $2u = -\lambda\Delta u$ for some $\lambda \neq 0$, that is, $u = -\gamma\Delta u$ for some $\gamma \neq 0$. Fixing $u \in K_c \cap M_s$ we may then write

$$c^2 = \int_\Omega \langle \nabla u, \nabla u \rangle = -\int_\Omega u\Delta u = \frac{1}{\gamma}\int_\Omega \|u\|^2 .$$

So $\gamma > 0$ and $c^2 = s^2/\gamma$ since $u \in M_s$. This yields

$$-\Delta u = \frac{c^2}{s^2} u,$$

which is impossible according to our assumptions. Therefore

$$Dh(u) \notin N(K_c; u) \cup (-N(K_c; u)),$$

and the result follows by applying Corollary 36.

A rather important class of prox-regular sets in finite dimensions is provided by hypographs of semiconcave functions. Such mappings are widely used both in control theory and in partial differential equations, and the book [**33**] contains a general treatment and several applications of this concept. Here we switch to the symmetric concept of semiconvexity and give a well known and important characterization of semiconvex functions through their epigraph.

DEFINITION 65. Let $U \subseteq \mathbb{R}^n$ be open and $f : U \to \mathbb{R}$ be a lower semicontinuous and locally bounded from above function. We say that f is locally semiconvex on U if for every $\bar{x} \in U$ there exist constants $\delta, \sigma > 0$ such that, for every $x, y \in B(\bar{x}, \delta)$ and for every $\zeta \in \partial_P f(x)$ the inequality

$$f(y) \geq f(x) + \langle \zeta, y - x \rangle - \sigma \|y - x\|^2 \tag{110}$$

holds.

Of course, every convex function is semiconvex, and every $\mathcal{C}^2$-function is locally semiconvex. Recalling Theorem 5.1 in [**43**], local semiconvexity is equivalent to (a localized version of) the so called *lower-$\mathcal{C}^2$ property*. Other equivalent definitions of semiconvexity can be found in [**33**]. Observe that, recalling (10), it is not difficult to see that a locally semiconvex function is necessarily locally Lipschitz continuous. Indeed, let by contradiction $\{x_n\} \subset B(\bar{x}, \delta)$ and $\zeta_n \in \partial_P f(x_n)$ be such that $x_n \to x \in B(\bar{x}, \delta)$ and $\|\zeta_n\| \to +\infty$. Pick $s > 0$ such that $B(x, 2s) \subset B(\bar{x}, \delta)$ and set $y_n = x_n + s\zeta_n / \|\zeta_n\|$. Then, for all n,

$$\begin{aligned} f(y_n) - f(x_n) &\geq \langle \zeta_n, y_n - x_n \rangle - \sigma \|y_n - x_n\|^2 \\ &= \|\zeta_n\| s - \sigma s^2 \to +\infty. \end{aligned}$$

On the other hand, the left hand side of the first inequality is bounded from above by definition of semiconvexity, and this is a contradiction.

The following result (see Theorem 5.2 in [**43**] or Theorem 2.3 in [**65**]) provides the characterization of semiconvex functions through the epigraph.

PROPOSITION 66. Let $U \subset \mathbb{R}^n$ be open, convex and bounded. Let $f : U \to \mathbb{R}$ be locally Lipschitz continuous. The following assertions are equivalent:

(a) f is locally semiconvex on U;
(b) for every $\bar{x} \in U$ there exist $\delta, r > 0$ such that the epigraph of the restriction of f to $B[\bar{x}, \delta]$, $\mathrm{epi}_{\bar{x},\delta} f$, is r-prox-regular.

PROOF. Let (a) hold and take $\bar{x} \in U$. Let $\delta > 0$ and $\sigma > 0$ be given by Definition 65 and let $x, y \in B[\bar{x}, \delta]$, and $\zeta \in \partial_P f(x)$, so that (110) holds. This in turn implies, for all $\alpha \geq f(y)$,

$$\alpha \geq f(x) + \langle \zeta, y - x \rangle - \sigma \left(\|y - x\|^2 + (\alpha - f(x))^2 \right).$$

The above inequality can be rewritten as

$$\langle (\zeta, -1), (y, \alpha) - (x, f(x)) \rangle \leq \sigma \left\| (y, \alpha) - (x, f(x)) \right\|^2,$$

which gives the result, since – f being locally Lipschitz continuous – the normal cone to epi$f$ at $(x, f(x))$ is generated by vectors of the form $(\zeta, -1)$, $\zeta \in \partial f(x)$. Conversely, let $r > 0$ be such that for all $x, y \in B[\bar{x}, \delta]$, $\zeta \in \partial_P f(x)$, and $\alpha \geq f(y)$ the inequality

$$\langle (\zeta, -1), (y, \alpha) - (x, f(x)) \rangle \leq \frac{1}{2r} \left\| (\zeta, -1) \right\| \left\| (y, \alpha) - (x, f(x)) \right\|^2 \tag{111}$$

holds. Let L be the Lipschitz constant of f on $B[\bar{x}, \delta]$. Then (111) implies, taking $\alpha = f(y)$,

$$\langle \zeta, y - x \rangle - (f(y) - f(x)) \leq \frac{\sqrt{L^2 + 1}}{2r} \left(\|y - x\|^2 + |f(y) - f(x)|^2 \right).$$

Since $|f(y) - f(x)| \leq L \|y - x\|$, the result follows immediately. □

The last part of the section contains two examples dealing with a comparison between prox-regularity and two similar conditions: we show that they are not equivalent to prox-regularity. The first one concerns the equality of Clarke and proximal normal cones, while the second one concerns a weaker external sphere condition.

EXAMPLE 67. A set which is not $\rho(\cdot)$-prox-regular for any ρ, but is such that at every point the Clarke and the proximal normal cone coincide (see Example 7.2 in [**45**]; Remark 4.16 in [**43**] contains comparisons with further not equivalent conditions).

Consider the two curves $\gamma_1 = \{(x, \sqrt{x}), x \geq 0\}$ and $\gamma_2 = \{(x, 2\sqrt{x}), x \geq 0\}$, and the decreasing sequences

$$x'_n = \frac{1}{n^2} - \frac{2}{n^4}, \quad x''_n = \frac{1}{n^2} - \frac{1}{n^4}, \quad n \geq 2.$$

Observe that $x''_{n+1} < x'_n$ for all n and the points

$$P_n = \left(x'_n, \frac{1}{n} \right), \quad Q_n = \left(x''_n, \frac{1}{n} \right)$$

belong to the open region between γ_1 and γ_2. Taking into account that the segments $[x'_n, x''_n]$ are disjoint we define a $\mathcal{C}^2$ function $\psi :]0, +\infty[\to \mathbb{R}^+$ such that $\psi(x) = 1/n$ for $x \in [x'_n, x''_n]$ and $\sqrt{x} < \psi(x) < 2\sqrt{x}$ for all $x > 0$. Clearly, ψ can be continuously extended to the whole of $\mathbb{R}^+$ by setting $\psi(0) = 0$. Set $C = \{(x, y) : \psi(x) \leq y \leq 2\sqrt{x},\ x \geq 0\}$. Then every normal vector to C is proximal (at $(0, 0)$ by construction, at other points by $\mathcal{C}^2$-smoothness of ψ and γ_2). Observe that any neighborhood of $(0, 0)$ must contain points whose projection onto C is not a singleton. Therefore, C is not $\rho(\cdot)$-prox-regular for any ρ.

The following condition, called *external sphere condition*, appears in connection with some kind of regularity of the distance function. In fact, let C be a closed set. If $\overline{\mathbb{R}^n \setminus C}$ has the external sphere condition, then the distance function $d(\cdot, C)$ is semiconcave up to the boundary of C (see [**33**, Proposition 2.2.4]).

DEFINITION 68. Let $C \subset \mathbb{R}^n$ be closed and let $r > 0$ be given. We say that C satisfies an external sphere condition of radius r if for every $x \in \text{bdry}C$ there exists $v \in N^P(C; x)$, $v \neq 0$, which is realized by an r-ball.

Of course every r-prox-regular set satisfies an external sphere condition of radius r, since *every* proximal normal is realized by an r-ball.

EXAMPLE 69. A set which is the closure of an open set and satisfies an external sphere condition with uniform radius, but is not $\rho(\cdot)$-prox-regular (see [**85**], Example 3.2.1).

Denote by $B_r[x, y]$ the closed ball of radius r centered at $(x, y) \in \mathbb{R}^2$. Let, for $k = 1, 2, \ldots$,

$$C_k^1 := B_1\left[\frac{1}{2k+1}, 1\right] \cup B_1\left[\frac{1}{2k}, 1\right] \cup \left\{\left[\frac{1}{2k+2}, \frac{1}{2k+1}\right] \times [0, 1]\right\}$$

$$C_k^2 := B_1\left[\frac{1}{2k+1}, -1\right] \cup B_1\left[\frac{1}{2k}, -1\right] \cup \left\{\left[\frac{1}{2k+2}, \frac{1}{2k+1}\right] \times [-1, 0]\right\},$$

and

$$S := \bigcup_{k=1}^{\infty} (C_k^1 \cup C_k^2) \cup B_1[0, 1] \cup B_1[0, -1].$$

Observe that S is closed and satisfies an interior sphere condition of radius 1, i.e., the complement C of the interior of S satisfies an external sphere condition. We claim that C is not prox-regular. In fact, the points

$$P_k := \left(\frac{1}{2}\left(\frac{1}{2k+2} + \frac{1}{2k+1}\right), 0\right)$$

belong to the interior of S, but have two metric projections into the boundary of C, namely the points

$$Q_k := \left(\frac{1}{2k+2}, 0\right), \quad R_k := \left(\frac{1}{2k+1}, 0\right).$$

Since $\|P_k - Q_k\| = \|P_k - R_k\| \to 0$, S cannot be $\rho(\cdot)$-prox-regular for any ρ.

The papers [**99, 100**] contain various results and examples on the relationships between the external sphere condition with uniform radius $r > 0$ and prox-regularity. In particular, the authors prove that a closed epi-Lipschitz set satisfying the external sphere condition of Definition 68 is prox-regular.

9. Comments

As it was mentioned in the Introduction, prox-regularity was studied independently by several authors under different names. In this section we give a brief account of that, together with some further comments on the results appearing in the present note.

Federer's name (see the 1959 paper [**63**]) of *sets with positive reach* refers essentially to the fact that all proximal normals are realized by a ball with locally uniform radius. The *reach* of a boundary point x in a closed set C is the largest radius r with the property that for all $x' \in \mathbb{R}^n$ such that $\|x' - x\| < r$ the metric projection onto C of x' is a singleton.

The 1985 paper [**59**] (see also the survey by Marino and Tosques in [**39**]) is one of the main works of a line of research which aimed to extend the theory of evolution equations of maximal monotone type. In that paper functions whose Fréchet subgradient satisfies a hypomonotonicity property similar to (d) in Theorem 3 play a major role. Similarly, φ-convex functions are introduced as those whose Fréchet subgradient satisfies a property similar to (b) in Theorem 3. The name φ-convex (actually p-convex) sets is due to A. Canino [**29**] (paper appeared in 1988), and comes from applying the φ-convexity property to indicator functions of closed sets. In [**29**] (see also [**30, 31, 32**]), the existence of geodesics on infinite dimensional prox-regular sets is established and some properties of such sets are studied, including the existence and uniqueness of the metric projection (see (f) in Theorem 3). Finding sufficient conditions such that the epigraph of a φ-convex function be prox-regular is actually an open problem: $f(x) = -\sqrt{|x|}$, $x \in \mathbb{R}$, is φ-convex but its epigraph is obviously not prox-regular, while the converse is true (see [**47**, Theorem 3.2]).

The analysis of prox-regularity begun with the 1996 paper [**103**] of Poliquin and Rockafellar, where prox-regular functions are defined and studied largely in view of an analysis of second order concepts and properties. We also refer to [**104, 81**] for other studies on prox-regular functions and to [**101, 102, 82, 84**] for the previous similar class of pln functions. The name of *prox-regular function* comes from the fact that the Moreau envelope

$$e_\lambda f(x) := \inf_{y \in H} \left[f(y) + \frac{1}{\lambda} \|x - y\|^2 \right]$$

of such a function has the property of being $\mathcal{C}^{1,1}$, and the Moreau proximal mapping $\mathrm{Prox}_f(x)$ given by the argmin of the latter function is single-valued and locally Lipschitz continuous (see [**103, 109, 12, 13, 14, 8**]). When f is the indicator function of a set C, the proximal mapping is the projection mapping and the local and global properties of such sets have been studied in Hilbert space by Poliquin, Rockafellar and Thibault [**106**] (2000). In the present survey we choose to follow this denomination, as it seems to us the most fit to a nonsmooth analyst reader.

Previously to [**106**], a paper by Shapiro and another one by Clarke, Stern and Wolenski appeared. Shapiro considered in his 1994 paper [**110**] the closed sets C of a Hilbert space making a second order contact with its tangent cone (see Proposition 8) around a point x_0. The given name therein is $O(2)$-convexity. Existence and uniqueness of the metric projection P_C in a neighborhood of x_0 is established, together with some results on the directional differentiability of P_C. The paper was in fact motivated by those results. Clarke, Stern and Wolenski in their 1996 paper [**43**] defined a closed set C of a Hilbert space to be proximally smooth when the distance function d_C is continuously differentiable on an open tube around C, and they proved several results for such sets. Their aim was both characterizing proximally smooth sets and, in finite dimensions, locally Lipschitz continuous functions whose epigraph is proximally smooth.

The works by Colombo, Marigonda and Wolenski [**47, 48, 50**] are finite dimensional and are motivated by the applications of prox-regularity to differentiability and structural properties of the minimum time function in control theory contained in [**49, 52**].

Górniewicz (see Chapter VI in [**68**]) studies topological properties of solution sets to a class of constrained differential inclusions. Actually the constraint is required to satisfy property (g) in Theorem 16, and the given name is *proximate retract*. This emphasizes the fact that a prox-regular set is a neighborhood retract with a special retraction, namely the metric projection.

Vial [**120**] in 1983 introduced and studied the concept of *weak convexity* in finite dimensions. Given $R > 0$ and points x_0, x_1 such that $\|x_1 - x_0\| \leq 2R$ the set

$$D_R(x_0, x_1) = \bigcap_{a:\{x_0,x_1\}\subset \bar{B}(a,R)} \bar{B}(a, R)$$

is considered and a set C is called weakly convex with constant R if for any two points $x_0, x_1 \in C$ such that $0 < \|x_1 - x_0\| < 2R$ the property

$$C \cap \operatorname{int} D_R(x_0, x_1) \neq \emptyset \tag{112}$$

holds. Balls realizing proximal normals are called *supporting balls* and some properties of prox-regular sets are proved, including local connectedness. Observe that condition (112) makes perfect sense in every Banach space. In particular, in their 2007 paper [**76**] the authors study weakly convex sets with constant R and show that if the space is both uniformly convex and uniformly smooth, then such sets are R-prox-regular. This paper contains also several references to the Russian literature, where concepts related to prox-regularity do appear. Further characterizations of prox-regularity in Banach spaces appear in [**14, 15**].

Several papers are devoted to normal and tangential regularity concepts more general than prox-regularity. We mention here [**25, 45, 99, 7, 58, 77, 97**] and the references in those papers, together with the recent finite dimensional work [**98**], where the author shows that the hypograph of a continuous function satisfying the external sphere condition of Definition 68 is, outside a *closed* set of measure zero, prox-regular.

Characterizations and properties of $\rho(\cdot)$-prox-regular sets. In the finite dimensional setting, Federer [**63**, 1958] started, as definition, with the uniqueness of the metric projection of any point in a neighborhood of C, and he proved properties (c) and (g) of Theorem 16 for such sets. In the Hilbert context Canino [**29**] took for definition that the indicator of C is φ-convex in the sense of Degiovanni, Marino and Tosques

[**59**]. Concerning the set $O_{\rho(\cdot)}(C)$ described in (22) she showed for such sets that property (g) of Theorem 3 and property (h) of Theorem 16 (see Proposition 2.9, p. 123), the tangential regularity in (b) of Proposition 7 (see Proposition 2.14, p. 131). Property (h) of Theorem 16 has been independently proved by Shapiro ([**110**, Theorem 2.2, p. 132]), Clarke, Stern and Wolenski (see [**43**, Theorem 4.8, p. 134]), Poliquin, Rockafellar and Thibault (see [**106**, Theorems 1.3, p. 5234]). In [**106**] it is shown that Property (h) is in fact a characterization of prox-regularity. An alternative proof appears in the paper [**45**] by Colombo and Goncharov; in [**110**], an inequality like the one in Proposition 8 is used as definition (of $O(2)$-convex sets) and in [**43**] the authors started with the C^1 property of d_C^2 on the r-open enlargement of C. The fact that the second order contact property in Proposition 8 characterizes the prox-regularity has been shown in finite dimensions by Federer (see [**63**, Theorem 4.18, p. 449]) and in the Hilbert setting by Poliquin, Rockafellar and Thibault (see [**106**, Proposition 3.6, p. 5243]). Clarke, Stern and Wolenski [**43**] proved that each one of properties (b), (m), and (o) in Theorem 16 characterizes the r-prox-regularity. The other characterizations in Theorem 16 (as stated) appear in Poliquin, Rockafellar and Thibault [**106**] except (h), (p), and (r). The characterization (p) is due to Bernard, Thibault and Zlateva [**14**] (in the context of uniformly convex Banach space). The implications $(a) \Rightarrow (q)$ in Theorem 14 and $(a) \Rightarrow (r)$ in Theorem 16 first appeared in Canino [**29**, Proposition]. The fact that (q) in Theorem 14 ensures the $\rho(\cdot)$-prox-regularity is new. Another proof of (r) in Theorem 16 can be found in Colombo-Goncahrov [**45**, (ii) in Theorem 6.3, p. 213]. The fact that each one of properties (f)-(p) of Theorem 14 ensure the $\rho(\cdot)$-prox-regularity is new in that $\rho(\cdot)$-context.

Our proofs of characterizations of prox-regular sets, see Theorems 3 and 14, Proposition 11, Lemma 12, follow for a large part the approach and arguments in [**106**].

The case when $\rho(\cdot)$ is constant. The characterization in Proposition 18 is due to Poliquin, Rockafellar and Thibault (see [**106**, Corollary 4.4, p. 5247]). The equality in Theorem 21 has been seen to hold by Federer [**63**, Corollary 4.9, p. 441] in finite dimensions. Theorem 21 has been proved by Clarke, Stern and Wolenski [**43**, Theorem 4.1 (c), p. 129] under the additional assumption that $\mathrm{Proj}_C(u)$ is nonempty. The full generality of the theorem was established by Bernard, Thibault and Zlateva (see [**14**, Theorem 5.10, p. 557]). In Corollary 17, the characterization (m) of convexity is due to Bunt [**28**] and Motzkin [**96**] in finite dimensions and to Klee [**79**] in Hilbert space (for a survey see [**73**]); characterizations (c)-(e) in the same corollary are due to Asplund [**2**] and (b) is due to Poliquin [**101**] in finite dimensions and Correa-Jofre-Thibault [**55**] in infinite dimensions. Poliquin, Rockafellar and Thibault [**106**] introduced the concept of prox-regularity around a fixed point and provided a complete study of the concept in Hilbert space. We refer to Bernard, Thibault and Zlateva [**14, 15**] for the study in uniformly convex spaces.

Directional derivability of the metric projection. Theorem 26 has been first stated and proved by Zarantonello [**121**] for convex sets of Hilbert spaces. The result has been extended to prox-regular sets independently by Canino [**29**, Proposition 2.12, p. 128] and Shapiro [**110**, Theorem 3.1, p. 135] (see also [**16**]). Their proofs are different from ours (which is in the line of [**15**, Proposition 8.1] where the directional derivability property has been generalized to uniformly convex spaces). Proposition 28 comes from Canino [**32**, Proposition 1.13].

Operations on prox-regular sets. Proposition 37 and Theorem 38 have been first established by Federer [**63**, Theorem 4.19, p. 449, and Theorem 5.9, p. 457]. Theorem 30 and Corollaries 33 and 35 are new and (a) of Corollary 34 as well. The other statements of Corollary 34 along with Theorem 31 are due to Bernard, Thibault and Zlateva [**15**]. Theorem 38 has its analogues in [**63, 15**] and extend them to the case of varying thickness r_k. We also refer to [**105**] for calculus on prox-regular functions in finite dimensions, to [**71, 107**] for parametrized families, and to [**17**] for integration. The study of spectral prox-regular sets can be found in [**57**].
Finite dimensional regularity results for sets and functions. Section 5 is based on [**47, 48, 50**], to which we refer for the complete proofs. We emphasize here that the important topic of curvature measures and their applications to integral geometry is not developed here. We refer to [**111**] for a survey including recent developments.
Sweeping process. Although the literature on the sweeping process driven by a moving prox-regular set is rather rich (see, e.g., [**26, 113, 114, 40, 24, 36, 61, 69, 86**]) and nowadays covers several issues originally established for its convex version, we prefer to focus on some basic facts such as the catching up algorithm and its convergence, and to sets moving with bounded variation. The results we present here are based on [**44, 51**] and [**113, 62**].
Other applications. Most applications of prox-regularity in Section 7 are only sketched both for the sake of brevity and because they sometimes require techniques outside nonsmooth analysis. Precise references concerning Section 7 are given in the text. For algorithmic applications of prox-regularity, we refer to [**56, 70, 72, 83**] and the references therein.

Bibliography

[1] L. Ambrosio, N. Fusco and D. Pallara, *Functions of Bounded Variation and Free Discontinuity Problems*, Oxford Science Publications, Clarendon, Oxford (2000).
[2] E. Asplund, *Čebyšev sets in Hilbert spaces*, Trans. Amer. Math. Soc. 144 (1969), 235–240.
[3] H. Attouch and R.J.-B. Wets, *Isometries for the Legendre-Fenchel transform*, Trans. Amer. Math. Soc. 296 (1986), 33–60.
[4] J.P. Aubin and A.Cellina, *Differential Inclusions. Set-Valued Maps and Viability Theory*, Springer, Berlin (1984).
[5] J.P. Aubin and H. Frankowska, *Set-Valued Analysis*, Birkhäuser, Boston (1990).
[6] M. Bardi and I. Capuzzo-Dolcetta, *Optimal Control and Viscosity Solutions of Hamilton–Jacobi–Bellman Equations*, Birkhäuser, Boston (1997).
[7] D. Aussel, A. Daniilidis and L. Thibault, *Subsmooth sets: functional characterizations and related concepts*, Trans. Amer. Math. Soc. 357 (2005), 1275–1301.
[8] M. Bacák, J.M. Borwein, A. Eberhard and B.S. Mordukhovich *Infimal convolution and Lipschitzian properties of subdifferentials for prox-regular functions in Hilbert spaces*, to appear.
[9] H. Benabdellah, *Existence of solutions to nonconvex sweeping process*, J. Differential Equations 164 (2000), 286–295.
[10] H. Benabdellah, C. Castaing, A. Salvadori and A. Syam, *Nonconvex sweeping processes*, J. Appl. Anal. 2 (1996), 217–240.
[11] F. Bernard and L. Thibault, *Prox-regularity of functions and sets in Banach spaces*, Set-valued Anal. 12 (2004), 25–47.
[12] F. Bernard and L. Thibault, *Prox-regular functions in Hilbert spaces*, J. Math. Anal. Appl. 303 (2005), 1–14.
[13] F. Bernard and L. Thibault, *Uniform prox-regularity of functions in Hilbert spaces*, Nonlinear Analysis 60 (2005), 187–207.
[14] F. Bernard, L. Thibault and N. Zlateva, *Characterization of prox-regular sets in uniformly convex Banach spaces*, J. Convex Anal. 13 (2006), 525–559.

[15] F. Bernard, L. Thibault and N. Zlateva, *Prox-regular sets in uniformly convex Banach space: Tangential and various other properties*, to appear in Trans. Amer. Math. Soc.

[16] J.F. Bonnans and A.S. Shapiro, *Perturbation analysis of optimization problems*, Springer series in operations research (2000).

[17] S. Boralugada and R.A. Poliquin, *Local integration of prox-regular functions in in Hilbert spaces*, J. Convex Anal. 13 (2006), 27–36.

[18] J.M. Borwein and J.R. Giles, *The proximal normal formula in Banach spaces*, Trans. Amer. Math. Soc. 302 (1987), 371–381.

[19] J.M. Borwein and H. Strójwas, *Tangential approximations*, Nonlinear Anal. 9 (1985), 1347–1366.

[20] J.M. Borwein and H. Strójwas, *Proximal analysis and boundaries of closed sets in Banach space. I: Theory*, Canad. J. Math. 38 (1986), 431–452.

[21] J.M. Borwein and H. Strójwas, *Proximal analysis and boundaries of closed sets in Banach space. II: Applications*, Canad. J. Math. 39 (1987), 428–472.

[22] J.M. Borwein and Qiji J. Zhu, *Techniques of variational analysis*, CMS Books in Mathematics, Springer-Verlag, New-York (2005).

[23] U. Boscain and B. Piccoli, *Optimal syntheses for control systems on 2-D manifolds*, Springer, Berlin (2004).

[24] M. Bounkhel and D-L. Azzam, *Existence results on the second-order nonconvex sweeping processes with perturbations*, Set-Valued Anal. 12 (2004), 291–318.

[25] M. Bounkhel and L. Thibault, *On various notions of regularity of sets in nonsmooth analysis*, Nonlinear Anal. 48 (2002), 223–246.

[26] M; Bounkhel and L. Thibault, *Nonconvex sweeping process and prox-regularity in Hilbert space*, J. Nonlinear Convex Anal. 6 (2005), 359–374.

[27] J.V. Burke, A.S. Lewis and M.L. Overton, *Approximating subdifferentials by random sampling of gradients*, Math. Oper. Res. 27 (2002), 567–584.

[28] L.N.H. Bunt, *Bijdrage tot de theorie de convexe puntverzamelingen*, Thesis, Univ. of Groningen, Amsterdam (1934).

[29] A. Canino, *On p-convex sets and geodesics*, J. Differential Equations 75 (1988), 118–157.

[30] A. Canino, *Existence of a closed geodesic on p-convex sets*, Ann. Inst. Henri Poincaré 5 (1988), 501–518.

[31] A. Canino, *Local properties of geodesics on p-convex sets*, Ann. Mat. Pura Appl. 159 (1991), 17–44.

[32] A. Canino, *Periodic solutions of quadratic Lagrangian systems on p-convex sets*, Ann. Fac. Sci. Toulouse Math. 12 (1991), 37–60.

[33] P. Cannarsa and C. Sinestrari, *Semiconcave functions, Hamilton–Jacobi Equations, and Optimal Control*, Birkhäuser, Boston (2004).

[34] C. Castaing, *A new class of evolution equations in a Hilbert space*, in Multifunctions and Integrands (ed. G. Salinetti), Lecture Notes in Mathematics 1091, Springer, Berlin (1984).

[35] C. Castaing, T.X. Duc Ha and M. Valadier, *Evolution equations governed by sweeping process*, Set-Valued Analysis 1 (1993), 109–139.

[36] C. Castaing, A.G. Ibrahim and M. Yarou, *Existence problems in second order evolution inclusions: discretization and variational approach*, Taiwanese J. Math. 12 (2008), 1433–1475.

[37] C. Castaing, A.G. Ibrahim and M. Yarou, *Some contributions to nonconvex sweeping process*, J. Nonlinear Convex Anal. 10 (2009), 1–20.

[38] C. Castaing and M. Valadier, *Convex analysis and measurable multifunctions*, Lectures Notes in Mathematics, Vol. 580 springer-Verlag, Berlin (1977).

[39] A. Cellina (ed.), *Methods of nonconvex analysis*, Lecture Notes in Mathematics, Springer (Berlin) 1990.

[40] N. Chemetov and M. D. P. Monteiro Marques, *Non-convex quasi-variational differential inclusions*, Set-Valued Anal. 15 (2007), 209–221.

[41] F.H. Clarke, *Optimization and nonsmooth analysis*, Wiley Intersciences, New York (1983).

[42] F.H. Clarke, Yu.S. Ledyaev, R.J. Stern and P.R. Wolenski, *Nonsmooth analysis and control theory*, Springer, Berlin (1998).

[43] F.H. Clarke, R.L. Stern and P.R. Wolenski, *Proximal smoothness and the lower C^2 property*, J. Convex Analysis 2 (1995), 117–144.

[44] G. Colombo and V. Goncharov, *The sweeping processes without convexity*, Set-Valued Anal. 7 (1999), 357–374.

[45] G. Colombo and V. Goncharov, *Variational inequalities and regularity properties of closed sets in Hilbert spaces*, J. Convex Anal. 8 (2001), 197–221.

[46] G. Colombo and V. Goncharov, *Continuous selections via geodesics*, Topological Methods in Nonlin. Anal. 18 (2001), 171–182.

[47] G. Colombo and A. Marigonda, *Differentiability properties for a class of non-convex functions*, Calc. Var. 25 (2006), 1–31.

[48] G. Colombo and A. Marigonda, *Singularities for a class of non-convex sets and functions, and viscosity solutions of some Hamilton–Jacobi equations*, J. Convex Anal. 15 (2008), 105–129.

[49] G. Colombo, A. Marigonda and P. R. Wolenski, *Some new regularity properties for the minimal time function*, SIAM J. Control 44 (2006), 2285–2299.

[50] G. Colombo, A. Marigonda and P. R. Wolenski, *A representation formula of the generalized gradient for a class of non-Lipschitz functions*, in preparation.

[51] G. Colombo and M. D.P. Monteiro Marques, *Sweeping by a continuous φ-convex set*, J. Differ. Equations 187 (2003), 46–72.

[52] G. Colombo and Nguyen T.K., *Quantitative isoperimetric inequalities for a class of nonconvex sets*, Calc. Var., in print.

[53] G. Colombo and Nguyen T.K., *On the structure of the minimum time function*, SIAM J. Control, in print

[54] G. Colombo and P.R. Wolenski, *Variational analysis for a class of minimal time functions in Hilbert spaces*, J. Convex Analysis 11 (2004), 335–361.

[55] R. Correa, A. Jofre and L. Thibault, *Subdifferential monotonicity as characterization of convex functions*, Numer. Funct. Anal. Optimiz. 15 (1994), 531–536.

[56] A. Daniilidis, W.L. Hare and J. Malick, *Geometrical interpretation of the predictor-corrector type algorithms in structured optimization problems*, Optimization 55 (2006), 481–503.

[57] A. Daniilidis, A.S. Lewis, J. Malick and H. Sendov, *Prox-regularity of spectral functions and spectral sets*, J. Convex Anal. 15 (2008), 547–560.

[58] A. Daniilidis and L. Thibault, *Subsmooth sets and metrically subsmooth sets and functions in Banach space*, to appear.

[59] M. Degiovanni, A. Marino and M. Tosques, *Evolution equations with lack of convexity*, Nonlinear Anal. 9 (1985), 1401–1443.

[60] E. De Giorgi, *Sulla proprietà isoperimetrica dell'ipersfera, nella classe degli insiemi aventi frontiera orientata di misura finita*, (Italian) Atti Accad. Naz. Lincei. Mem. Cl. Sci. Fis. Mat. Nat., Sez. I 8 (1958), 33–44; also (in English) in De Giorgi, *Selected papers*, L. Ambrosio, G. Dal Maso, M. Forti, M. Miranda, S. Spagnolo (Eds.), Springer, Berlin (2006).

[61] J.F. Edmond and L. Thibault, *Relaxation of an optimal control problem involving a perturbed sweeping process*, Math. Program. ser B 104 (2005), 347–373.

[62] J.F. Edmond and L. Thibault, *BV solution of nonconvex sweeping process with perturbation*, J. Differential Equations 226 (2006), 135–179.

[63] H. Federer, *Curvature measures*, Trans. Amer. Math. Soc. 93 (1959), 418–491.

[64] S. Fitzpatrick, *Differentiation of real-valued functions and continuity of metric projections*, Proc. Amer. Math. Soc. 91 (1984), 544–548.

[65] J.H.G. Fu, *Tubular neighborhoods in Euclidean spaces* Duke Math. J. 52 (1985), 1025–1046.

[66] B. Fuglede, *Stability in the isoperimetric problem for convex or nearly spherical domains in $\mathbb{R}^n$*, Trans. Am. Math. Soc. 314 (1989), 619–638.

[67] V.V. Goncharov and F. Pereira, *Neighbourhood retractions of nonconvex sets in a Hilbert space via sublinear functionals*, J. Convex Anal., 18 (2011), no. 1.

[68] L. Górniewicz, *Topological Fixed Point Theory for Multivalued Mappings*, Kluwer Academic Publishers, Dordrecht (1999).

[69] T. Haddad, A. Jourani and L. Thibault, *Reduction of sweeping process to unconstrained differential inclusion*, Pac. J. Optim. 4 (2008), 493–512.

[70] W.L. Hare and A.S. Lewis, *Identifying active constraints via partial smoothness and prox-regularity*, J. Convex Anal. 11 (2004), 251–266.

[71] W.L. Hare and R.A. Poliquin, *Prox-regularity and stability of the proximal mapping*, J. Convex Anal. 14 (2007), 589–606.

[72] W.L. Hare and C. Sagastizábal, *Computing proximal points of nonconvex functions*, Math. Program. 116 (2009), Ser. B, 221–258.

[73] J.-B. Hiriart-Urruty, *Ensembles de Tchebychev vs. ensembles convexes: l'état de la situation vu par l'analyse non lisse*, Ann. Sci. Math. Que. 22 (1998), 47–62.

[74] A.D. Ioffe, *Proximal analysis and approximate subdifferentials*, J. London Math. Soc. 41 (1990), 175–192.

[75] A.D. Ioffe, *Metric regularity and subdifferential calculus*, Russian Math. Surveys 55 (2000), 501–558.

[76] G.E. Ivanov and M.V. Balashov, *Lipschitz parameterizations of multivalued mappings with weakly convex values*, Izv. Ross. Akad. Nauk Ser. Mat. 71 (2007), 47–68 (in Russian; translation in Izv. Math. 71 (2007), 1123–1143).

[77] A. Jourani, *Weak regularity of functions and sets in Asplund spaces*, Nonlinear Analysis 45 (2006), 660–676.

[78] A. Jourani and L. Thibault, *Metric regularity for strongly compactly Lipschitzian mappings*, Nonlinear Analysis 24 (1995), 229–240.

[79] V. Klee, *Convexity of Chebyshev sets*, Math. Annalen 142 (1961), 292–304.

[80] K.S. Lau, *Almost Chebyshev subsets in reflexive Banach spaces*, Indiana Univ. Math. J. 27 (1978), 791–795.

[81] A. Levi, R.A. Poliquin and R.T. Rockafellar, *Stability of locally optimal solutions*. SIAM J. Optim. 10 (2000), 580–604.

[82] A. Levi, R.A. Poliquin and L. Thibault, *A partial extension of Attouch's theorem and its applications to second-order differentiation*, Trans. Amer. Math. Soc. 347 (1995), 1269–1294.

[83] A.S. Lewis, D.R. Luke and J. Malick, *Local linear convergence for alternating and averaged nonconvex projections*, Found. Comput. Math. 9 (2009), 485–513.

[84] S. Marcellin and L. Thibault, *Evolution problems associated with primal lower nice functions*, J. Convex Anal. 13 (2006), 385–421.

[85] A. Marigonda, *Differentiability Properties for a Class of Non-Lipschitz Functions and Applications*, Ph.D. Thesis, University of Padova, 2006.

[86] B. Maury and J. Venel, *A mathematical framework for a crowd motion model*, C. R. Math. Acad. Sci. Paris 346 (2008), no. 23–24, 1245–1250.

[87] E. Michael, *Selected selection theorems*, Amer. Math. Monthly 63 (1956), 233–238.

[88] M. D. P. Monteiro Marques, *Differential Inclusions in Nonsmooth Mechanical Problems-Shocks and Dry Friction*, Birkhäuser, Boston (1993).

[89] J.J. Moreau, *Rafle par un convexe variable I*. Sém. Anal. Convexe Montpellier, Exposé 15, (1971).

[90] J.J. Moreau, *Rafle par un convexe variable II*. Sém. Anal. Convexe Montpellier, Exposé 3, (1972).

[91] J.J. Moreau, *Multi-applications à rétraction finie*, Ann. Scuola Norm. Sup. Pisa 1 (1974), 169–203.

[92] J. J. Moreau, *On unilateral constraints, friction and plasticity*, in New Variational Techniques in Mathematical Physics, Proceedings from CIME (Capriz and Stampacchia eds.), Cremonese, Rome (1974), 173–322.

[93] J.J. Moreau, *Evolution problem associated with a moving convex set in a Hilbert space*, J. Differential Equations 26 (1977), 347–374.

[94] M.D.P. Monteiro Marques, *Differential inclusions in Nonsmooth Mechanical Problems, Shocks and Dry Friction*. Birkhäuser, Basel, (1993).

[95] B.S. Mordukhovich, *Variational analysis and generalized differentiation, I: Basic theory; II: Applications*, Vol. 330 and 331, Springer, Berlin, (2006).

[96] T.S. Motzkin, *Sur quelques propriétés caractéristiques des ensembles convexes*, Att. R. Acad. Lincei Rend. 21 (1935), 562–567.

[97] H.V. Ngai and J.-P. Penot, *Approximately convex sets*, J. Nonlinear Convex Anal. 8 (2007), 337–355.

[98] T.K. Nguyen, *Hypographs satisfying an external sphere condition and the regularity of the minimum time function*, J. Math. Anal. Appl. 372 (2010), 611–628.

[99] C. Nour, R. J. Stern and J. Takche, *Proximal smoothness and the exterior sphere condition*, J. Convex Anal. 16 (2009), 501–514.

[100] C. Nour, R. J. Stern and J. Takche, *The ϑ-exterior sphere condition, φ-convexity, and local semiconcavity*, submitted.

[101] R.A. Poliquin, *Integration of subdifferentials of nonconvex functions*, Nonlinear Anal. 17 (1991), 385–398.

[102] R.A. Poliquin, *An extension of Attouch's theorem and its application to second-order epi-differentiation of convexly composite functions*, Trans. Amer. Math. Soc. 332 (1992), 861–874.

[103] R.A. Poliquin and R.T. Rockafellar, *Prox-regular functions in variational analysis*, Trans. Amer. math. Soc. 348 (1996), 1805–1838.

[104] R.A. Poliquin and R.T. Rockafellar, *Generalized Hessian properties of regularized non-smooth functions*, SIAM J. Optimization 6 (1996), 1121–1137.

[105] R.A. Poliquin and R.T. Rockafellar, *A calculus of prox-regularity*, J. Convex Anal. 17 (2010), to appear.

[106] R.A. Poliquin, R.T. Rockafellar and L. Thibault, *Local differentiability of distance functions*, Trans. Amer. Math. Soc. 352 (2000), 5231–5249.

[107] S. Robinson, *Aspects of the projector on prox-regular sets* Variational analysis and applications, 963–973, Nonconvex Optim. Appl., 79, Springer, New York, (2005).

[108] R. T. Rockafellar, *Convex Analysis*, Princeton Univesity Press, Princeton (1972).

[109] R.T. Rockafellar and R.J.-B. Wets, *Variational analysis*, Springer, Berlin (1998).

[110] A. Shapiro, *Existence and differentiability of metric projections in Hilbert spaces*, SIAM J. Optimization 4 (1994), 130–141.

[111] Ch. Thäle, *50 Years Sets with Positive Reach – A Survey*, Surveys in Mathematics and its Applications, Vol. 3 (2008), pp. 123–165.

[112] L. Thibault, *On subdifferentials of optimal value functions*, SIAM J. Control Optim. 29 (1991), 1019–1036.

[113] L. Thibault, *Sweeping process with regular and nonregular sets*, J. Differential Equations 193 (2003), 1–26.

[114] L. Thibault, *Regularization of nonconvex sweeping process in Hilbert space*, Set-Valued Anal. 16 (2008), 319–333.

[115] L. Thibault and D. Zagrodny, *Integration of subdifferentials of lower semicontinuous functions on Banach spaces*, J. Math. Anal. Appl. 189 (1995), 33–58.

[116] A. A. Tolstonogov, *Differential Inclusions in a Banach Space*, Kluwer, Dordrecht (2000).

[117] M. Valadier, *Quelques résultats de base concernant le processus de rafle*, Sémin. Anal. Conv. Montpellier (1988), exp; 3 (30 pages).

[118] M. Valadier, *Quelques problèmes d'entrainement unilatéral en dimension finie*, Sémin. Anal. Conv. Montpellier (1988), exp. 8 (21 pages).

[119] J.-P. Vial, *Strong convexity of sets and functions* J. Math. Econom. 9 (1982), 187–205.

[120] J.-P. Vial, *Strong and weak convexity of sets and functions*, Math. Oper. Res. 8 (1983), 231–259.

[121] E.H. Zarantonello, *Projections on convex sets in Hilbert space and spectral theory I and II*, in E.H. Zarantonello, ed., Contributions to Nonlinear Functional Analysis, Academic Press, New York (1971), pp 237–424.

Giovanni Colombo, Dipartimento di Matematica Pura ed Applicata, Università di Padova, via Trieste 63, 35121 Padova (Italy)
E-mail address: colombo@math.unipd.it

Lionel Thibault, Département de Mathématiques, Université de Montpellier II, Place Eugène Bataillon, 34095 Montpellier CEDEX 5
E-mail address: thibault@math.univ-montp2.fr

Multiplicity of solutions for nonlinear elliptic equations with combined nonlinearities

Leszek Gasiński* and S. Papageorgiou

ABSTRACT. We consider a nonlinear elliptic problem with concave terms (i.e. $(p-1)$-sublinear term) and a Carathéodory perturbation, which is assumed to be $(p-1)$-superlinear or $(p-1)$-linear near $\pm\infty$. We show that in each case the problem has five nonlinear smooth solutions: four of constant sign (two positive and two negative) and the fifth nodal (sign changing). The four constant sign solutions are obtained for equations driven by general nonlinear nonhomogeneous differential operators. The nodal solutions are produced for the particular case of the p-Laplacian. In the semilinear case, we produce a sixth nontrivial smooth solution. Our approach combines critical point and Morse theories and also uses truncation techniques and the method of upper-lower solutions.

CONTENTS

2000 AMS Subject Classification: 35J65, 58E05

Key words. Concave term, $(p-1)$-superlinear function, $(p-1)$-linear function, multiple solutions, constant sign solutions, nodal solutions, Morse theory, extremal solutions, upper and lower solutions, truncations, nonlinear regularity.

*This research has been partially supported by the Ministry of Science and Higher Education of Poland under Grant no. N201 542438.

1. Introduction

Let $\Omega \subseteq \mathbb{R}^N$ be a bounded domain with a C^2-boundary $\partial\Omega$. In this paper we study the following nonlinear Dirichlet problem:

$$\left\{ \begin{array}{l} -\operatorname{div} a\,(z, \nabla u(z)) = \beta(z)\big|u(z)\big|^{q-2}u(z) + f(z, u(z)) \text{ in } \Omega, \\ u\big|_{\partial\Omega} = 0. \end{array} \right. \tag{1.1}$$

Here $a\colon \overline{\Omega} \times \mathbb{R}^N \longrightarrow \mathbb{R}^N$ is a continuous map with $a(z, \cdot)$ strictly monotone and satisfying certain other regularity conditions (see hypotheses $H(a)$ in Section 3). As a special case of the differential operator $\operatorname{div} a\,(z, \nabla u(z))$, we recover the p-Laplacian $\operatorname{div}\big(\|\nabla u\|^{p-2}\nabla u\big)$, $1 < p < +\infty$. Also $\beta \in L^\infty(\Omega)_+ \setminus \{0\}$ is a parameter of the problem and $1 < q < p$. Therefore the function $\zeta \longmapsto \beta(z)|\zeta|^{q-2}\zeta$ is a "concave" term (i.e., a $(p-1)$-sublinear nonlinearity). The perturbation $f(z, \zeta)$ is a Carathéodory function. Concerning this perturbation, we consider two distinct cases. In the first, we assume that the nonlinearity $f(z, \cdot)$ exhibits a $(p-1)$-superlinear growth near infinity. However, we do not assume the usual in such cases Ambrosetti-Rabinowitz condition. Instead, we employ a weaker, more general condition. A very particular case of our setting is when the right hand side has the form:

$$\lambda|\zeta|^{q-2}\zeta + |\zeta|^{r-2}\zeta,$$

with $\lambda > 0$, $1 < q < p < r < p^*$. Here

$$p^* = \begin{cases} \dfrac{Np}{N-p} & \text{if } N > p, \\ +\infty & \text{if } N \leqslant p \end{cases}$$

is the critical Sobolev exponent. This particular case was investigated by mamy authors. The first work on problems with combined effects of "concave" (sublinear) and "convex" (superlinear) terms, is that of Ambrosetti-Brézis-Cerami [**3**] for semilinear problems (i.e. $p = 2$) driven by the Dirichlet Laplacian. They show that the problem exhibits certain bifurcation phenomena and that there exists $\lambda^* > 0$, such that for all $\lambda \in (0, \lambda^*)$ the problem admits at least two positive solutions. Their result was extended to problems driven by the radial p-Laplacian by Ambrosetti-García Azorero-Peral [**4**] and for the general p-Laplacian by García Azorero-Manfredi-Peral Alonso [**23**] and by Guo-Zhang [**30**] (in the latter, there is the restriction $p \geqslant 2$). We should also mention the semilinear work of Li-Wu-Zhou [**36**], who have a nonlinearity of the form:

$$h(z)\zeta^q + f(z, \zeta),$$

with $h \in L^\infty(\Omega) \setminus \{0\}$, $0 < q < 1$ and $f \in C(\overline{\Omega} \times \mathbb{R})$ superlinear in $\zeta \in \mathbb{R}$ and satisfying a condition more general than the Ambrosetti-Rabinowitz condition, analogous to the one employed here. They show that for all h with small L^∞-norm, the problem has two positive solutions.

The second case that we investigate here, is when $f(z, \cdot)$ exhibits a $(p-1)$-linear growth near $\pm\infty$ (asymptotically "linear" problems). This situation was studied by Li-Wu-Zhou [**36**] (semilinear problems, i.e., $p = 2$) and by Guo-Zhang [**30**] (nonlinear problems, i.e., driven by the p-Laplacian with $p \geqslant 2$). In both papers, they do not allow for resonance with respect to the principal eigenvalue $\lambda_1 > 0$ asymptotically at $\pm\infty$ and their multiplicity results produce fewer solutions. We should also mention the works of Perera [**45**] and de Paiva-Massa [**17**], where $h(z) = -\lambda$ with $\lambda > 0$ and with $f(z, \zeta) = f(\zeta) \in C^1(\overline{\Omega})$ satisfying more restrictive conditions.

In both cases under various restrictions on the maps $a(z, y)$ and $f(z, \zeta)$, we show for $\beta \in L^\infty(\Omega)_+$ with small L^∞-norm, the problem has at least five nontrivial smooth solutions: four of constant sign (two positive and two negative) and the fifth nodal (sign-changing). To obtain the nodal solution we need to restrict ourselves to the case of the p-Laplacian. In the semilinear case with the Laplace differential operator, we obtain additional solutions and prove multiplicity results guaranteeing the existence of six nontrivial smooth solutions of the equations. However, we are unable to determine the sign of the sixth solution.

Recently there have been papers dealing with p-Laplacian equations and proving multiplicity results with precise sign information. We mention the works of Aizicovici-Papageorgiou-Staicu [**2**], Bartsch-Liu-Weth [**7**], Carl-Motreanu [**9**], Carl-Perera [**10**], Filippakis-Kristaly-Papageorgiou [**21**], Gasiński-Papageorgiou [**26**], Motreanu-Motreanu-Papageorgiou [**40, 41, 42**], Papageorgiou-Rocha-Staicu [**44**], Zhang-Chen-Li [**51**], Zhang-Li [**52**]. However, none of the aforementioned works deals with the combined effects of different nonlinearities ("combined nonlinearities").

Our tools are drawn from variational analysis, in particular critical point theory, from the theory of nonlinear operators of monotone type and from Morse theory. We also use truncation techniques and the method of upper-lower solutions. In the next section, for the convenience of the reader, we recall some of these tools.

2. Mathematical background

Let us start with recalling some notions and facts from critical point theory. So, let X be a Banach space and let X^* be its topological dual. By $\langle \cdot, \cdot \rangle$ we denote the duality brackets for the pair (X^*, X). Let $\varphi \in C^1(X)$. A point $x_0 \in X$ is called a *critical point of* φ, if $\varphi'(x_0) = 0$. A number $c \in \mathbb{R}$ is said to be a *critical value of* φ, if there exists a critical point $x_0 \in X$, such that $\varphi(x_0) = c$.

We say that φ satisfies the *Palais-Smale condition at the level* $c \in \mathbb{R}$ (the *PS_c-condition* for short), if every sequence $\{x_n\}_{n\geqslant 1} \subseteq X$, such that

$$\varphi(x_n) \longrightarrow c \quad \text{in } \mathbb{R}$$

and

$$\varphi'(x_n) \longrightarrow 0 \quad \text{in } X^*,$$

has a strongly convergent subsequence. If this is true at every level $c \in \mathbb{R}$, then we say that φ satisfies the *Palais-Smale condition* (the *PS-condition* for short).

More generally, we say that φ satisfies the *Cerami condition at the level* $c \in \mathbb{R}$ (the *C_c-condition* for short), if every sequence $\{x_n\}_{n\geqslant 1} \subseteq X$, such that

$$\varphi(x_n) \longrightarrow c \quad \text{in } \mathbb{R}$$

and

$$(1 + \|x_n\|)\varphi'(x_n) \longrightarrow 0 \quad \text{in } X^*,$$

has a strongly convergent subsequence. If this is true at every level $c \in \mathbb{R}$, then we say that φ satisfies the *Cerami condition* (the *C-condition* for short).

We have the following general minimax characterization of certain critical values of a C^1-functional, known in the literature as the "mountain pass theorem".

THEOREM 2.1. *If X is a Banach space, $\varphi \in C^1(X)$, $x_0, x_1 \in X$, $\varrho > 0$, $\|x_1 - x_0\| > \varrho$,*

$$\max\{\varphi(x_0), \varphi(x_1)\} \leqslant \inf\{\varphi(x) : \|x - x_0\| = \varrho\} = \eta,$$
$$c = \inf_{\gamma \in \Gamma} \max_{0 \leqslant t \leqslant 1} \varphi(\gamma(t)),$$

where

$$\Gamma = \{\gamma \in C([0,1]; X) : \gamma(0) = x_0, \ \gamma(1) = x_1\}$$

and φ satisfies the C_c-condition, then $c \geqslant \eta$ and c is a critical value of φ. Moreover, if $c = \eta$, then there exists a critical point x of φ with $\varphi(x) = c$ and $\|x - x_0\| = \varrho$.

For every $c \in \mathbb{R}$, we set:

$$\varphi^c = \{x \in X : \varphi(x) \leqslant c\},$$
$$K^\varphi = \{x \in X : \varphi'(x) = 0\},$$
$$K_c^\varphi = \{x \in K^\varphi : \varphi(x) = c\}.$$

Another basic tool in our variational approach is the so called second deformation theorem (see e.g., Gasiński-Papageorgiou [**25**, p. 628]).

THEOREM 2.2. *If $\varphi \in C^1(X)$, $a \in \mathbb{R}$, $a < b \leqslant +\infty$, φ satisfies the C_c-condition for every $c \in [a, b)$, φ has no critical values in (a, b) and $\varphi^{-1}(a)$ contains at most a finite number of critical points of φ, then there exists a continuous map $h\colon [0,1] \times (\varphi^b \setminus K_b^\varphi) \longrightarrow \varphi^b$, such that*

(a) *$h(0, u) = u$ for every $u \in \varphi^b \setminus K_b^\varphi$;*
(b) *$h(1, u) \in \varphi^a$ for all $u \in \varphi^b \setminus K_b^\varphi$;*
(c) *$h(t, u) = u$ for all $t \in [0, 1]$ and all $u \in \varphi^a$;*
(d) *$\varphi(h(t, u)) \leqslant \varphi(h(s, u))$ for all $0 \leqslant s \leqslant t \leqslant 1$ and all $u \in \varphi^b \setminus K_b^\varphi$.*

In particular this theorem implies that φ^a is a strong deformation retract of $\varphi^b \setminus K_b^\varphi$.

Now let us recall some basic things from Morse theory (critical groups) that we will need in the study of problem (1.1). The main references are the books of Chang [**11**] and Mawhin-Willem [**37**]. So, let (Y_1, Y_2) is a topological pair with $Y_2 \subseteq Y_1 \subseteq X$. For every integer $k \geqslant 0$, by $H_k(Y_1, Y_2)$ we denote the k-th relative singular homology group for the pair (Y_1, Y_2) with integer coefficients. Then the critical groups of φ at an isolated critical point $x_0 \in X$ with $c = \varphi(x_0)$ are defined by

$$C_k(\varphi, x_0) = H_k(\varphi^c \cap U, \ \varphi^c \cap U \setminus \{x_0\}) \quad \forall k \geqslant 0,$$

where U is a neighbourhood of x_0, such that $K^\varphi \cap \varphi^c \cap U = \{x_0\}$. The excision property of singular homology, implies that the above definition of critical groups is independent of the particular choice of the neighbourhood U.

Suppose that $\varphi \in C^1(X)$ satisfies the C-condition and $\inf \varphi(K^\varphi) > -\infty$. Let

$$c < \inf \varphi(K^\varphi).$$

Then the critical groups of φ at infinity are defined by

$$C_k(\varphi, \infty) = H_k(X, \varphi^c) \quad \forall k \geqslant 0$$

(see Bartsch-Li [**6**]). Recall that the deformation theorem is still valid under the C-condition (see e.g., Bartolo-Benci-Fortunato [**5**] or Gasiński-Papageorgiou [**25**]). So, invoking the deformation theorem, it follows that the above definition is independent of the particular choice of the level $c < \inf \varphi(K^\varphi)$.

Suppose that K^φ is a finite set and define

$$M(t,x) = \sum_{k \geqslant 0} \operatorname{rank} C_k(\varphi, x) t^k \quad \forall x \in K^\varphi$$

$$P(t,\infty) = \sum_{k \geqslant 0} \operatorname{rank} C_k(\varphi, \infty) t^k.$$

The Morse relation says that

$$\sum_{x \in K} M(t,x) = P(t,\infty) + (1+t)Q(t), \tag{2.1}$$

where

$$Q(t) = \sum_{k \geqslant 0} \beta_k t^k$$

is a formal series in t with nonnegative integer coefficients.

A nonlinear map $A\colon X \longrightarrow X^*$ is of type $(S)_+$, if for every sequence $\{x_n\}_{n\geqslant 1} \subseteq X$, such that

$$x_n \longrightarrow x \text{ weakly in } X \quad \text{and} \quad \limsup_{n \to +\infty} \langle A(x_n), x_n - x \rangle \leqslant 0,$$

we have

$$x_n \longrightarrow x \quad \text{in } X.$$

This notion is useful when checking the PS or C-condition.

Recall that, if $G\colon \mathbb{R}^N \longrightarrow \mathbb{R}$ is a C^1 convex function, then

$$(\nabla_y G(y),\ h-u)_{\mathbb{R}^N} \leqslant G(h) - G(y) \quad \forall h \in \mathbb{R}^N$$

(see Gasiński-Papageorgiou [**25**]).

In our use of truncation techniques, we introduce a mild nonsmoothness in the energy functional. In order to deal with this, we employ the nonsmooth critical point theory (see Gasiński-Papageorgiou [**24**]). So, let us briefly recall some definitions and facts from this theory, which we will need later. Let $\varphi\colon X \longrightarrow \mathbb{R}$ be a locally

Lipschitz function. The generalized directional derivative of φ at $x \in X$ in the direction $h \in X$, is defined by

$$\varphi^0(x;h) = \limsup_{\substack{x' \to x \\ \lambda \searrow 0}} \frac{\varphi(x'+\lambda h) - \varphi(x')}{\lambda}.$$

It is easy to see that the function $h \longmapsto \varphi^0(x;h)$ is sublinear and Lipschitz continuous. So, from the Hahn-Banach theorem it follows that $\varphi^0(x;\cdot)$ is the support function of a nonempty, convex and w^*-compact set, defined by

$$\partial\varphi(x) = \{x^* \in X^* : \langle x^*, h\rangle \leqslant \varphi^0(x;h) \text{ for all } h \in X\}.$$

The multifunction $X \ni x \longmapsto \partial\varphi(x) \subseteq X^*$ is called the generalized (or Clarke) subdifferential of φ. If $\varphi \in C^1(X)$, then φ is locally Lipschitz and $\partial\varphi(x) = \{\varphi'(x)\}$. Also, if $\varphi\colon X \longrightarrow \mathbb{R}$ is continuous and convex, then φ is locally Lipschitz and the generalized subdifferential of φ coincides with the subdifferential in the sense of convex analysis, given by

$$\partial_c\varphi(x) = \{x^* \in X^* : \langle x^*, h\rangle \leqslant \varphi(x+h) - \varphi(x) \text{ for all } h \in X\}.$$

Let $\varphi\colon X \longrightarrow \mathbb{R}$ be a locally Lipschitz function. Extending the notion of the C_c-condition to locally Lipschitz functionals $\varphi\colon X \longrightarrow \mathbb{R}$, we require that for every sequence $\{x_n\}_{n\geqslant 1} \subseteq X$, such that

$$\varphi(x_n) \longrightarrow c \quad \text{in } \mathbb{R}$$

and

$$(1 + \|x_n\|)m^\varphi(x_n) \longrightarrow 0,$$

where

$$m^\varphi(x) = \inf\{\|x^*\|_* : x^* \in \partial\varphi(x)\},$$

has a strongly convergent subsequence. We say that φ satisfies the *C-condition*, if it satisfies the C_c-condition at every level $c \in \mathbb{R}$.

Also, we say that a point $x \in X$ is a *critical point of* φ, if $0 \in \partial\varphi(x)$. If $x \in X$ is a *critical point of* φ, then the value $c = \varphi(x)$ is called a critical value of φ. It is easy to check that if $x \in X$ is a local extremum of φ (i.e., a local minimum or a local maximum), then $x \in X$ is a critical point of φ. Theorem 2.1 is still valid if instead of $\varphi \in C^1(X)$ we have a locally Lipschitz function and the C_c-condition is interpreted in the above sense.

In the analysis of problem (1.1) in addition to the Sobolev space $W_0^{1,p}(\Omega)$, we will use the space

$$C_0^1(\overline{\Omega}) = \{u \in C^1(\overline{\Omega}) : u|_{\partial\Omega} = 0\}.$$

We know that this is an ordered Banach space with positive cone

$$C_+ = \{u \in C_0^1(\overline{\Omega}) : u(z) \geqslant 0 \text{ for all } z \in \overline{\Omega}\}.$$

This cone has a nonempty interior, given by

$$\operatorname{int} C_+ = \left\{ u \in C_+ : \ u(z) > 0 \text{ for all } z \in \Omega, \ \frac{\partial u}{\partial n}(z) < 0 \text{ for all } z \in \partial\Omega \right\}.$$

Here $n(\cdot)$ denotes the outward unit normal on $\partial\Omega$. Concerning ordered Banach space, the following simple fact will be useful in our subsequent considerations.

LEMMA 2.3. *If X is an ordered Banach space, C is the positive (order) cone of X, $x_0 \in \operatorname{int} C$, then for every $y \in X$, there exists $t = t(y) > 0$, such that $tx_0 - y \in \operatorname{int} C$.*

PROOF. Since $x_0 \in \operatorname{int} C$, we can find $\delta > 0$, such that

$$\overline{B}_\delta(x_0) = \left\{ x \in X : \ \|x - x_0\| \leqslant \delta \right\} \subseteq \operatorname{int} C.$$

Let $y \in X \setminus \{0\}$ (if $y = 0$, then the lemma is clearly true for all $t > 0$). We have

$$x_0 - \delta \frac{y}{\|y\|} \in \operatorname{int} C.$$

Choosing $t = \frac{\|y\|}{\delta}$, we have $tx_0 - y \in \operatorname{int} C$. □

We will also need some basic facts about the spectrum of the negative p-Laplacian with Dirichlet boundary conditions. So, we consider the following nonlinear eigenvalue problems

$$\begin{cases} -\Delta_p u(z) = \lambda |u(z)|^{p-2} u(z) \text{ in } \Omega, \\ u|_{\partial\Omega} = 0, \end{cases} \tag{2.2}$$

where

$$\Delta_p u = \operatorname{div}\left(\|\nabla u\|^{p-2} \nabla u \right).$$

Every number $\lambda \in \mathbb{R}$ for which problem (2.2) has a nontrivial solution is said to be an *eigenvalue* of $\left(-\Delta_p, \ W_0^{1,p}(\Omega)\right)$. The smallest eigenvalue λ_1 is positive, isolated, simple (i.e., if u, v are eigenfunctions corresponding to λ_1, then $u = \xi v$ for some $\xi \in \mathbb{R}$) and admits the following variational characterization

$$\lambda_1 = \inf \left\{ \frac{\|\nabla u\|_p^p}{\|u\|_p^p} : \ u \in W_0^{1,p}(\Omega), \ u \neq 0 \right\}. \tag{2.3}$$

The infimum is realized on the corresponding one-dimensional eigenspace. In what follows by $\widehat{u}_1$ we denote the L^p-normalized (i.e., $\|\widehat{u}_1\|_p = 1$) eigenfunction corresponding to $\lambda_1 > 0$. The nonlinear regularity theory (see e.g., Gasiński-Papageorgiou [**25**, pp. 737–738]) implies that $\widehat{u}_1 \in C_0^1(\overline{\Omega})$. Moreover, from (2.2) it is clear that $\widehat{u}_1$ does not change sign and so we may assume that $\widehat{u}_1 \in C_+$. In fact the nonlinear maximum principle of Vázquez [**49**] implies that $\widehat{u}_1 \in \operatorname{int} C_+$.

Since the p-Laplacian operator is $(p-1)$-homogeneous, the Ljusternik-Schnirelmann theory provides an increasing sequence $\{\lambda_k\}_{k \geqslant 1} \subseteq \mathbb{R}_+$ of eigenvalues, such that $\lambda_k \longrightarrow +\infty$ as $k \to +\infty$. These eigenvalues are known as the LS-eigenvalues of $\left(-\Delta_p, \ W_0^{1,p}(\Omega)\right)$. If $p = 2$ (linear eigenvalue problem), then the LS-eigenvalues are all the eigenvalues of $\left(-\Delta,\ H_0^1(\Omega)\right)$. If $p \neq 2$ (nonlinear eigenvalue problem), we do not know if this is the case. However, since $\lambda_1 > 0$ is isolated, we can define

$$\lambda_2^* = \inf\left\{\lambda:\ \lambda \text{ is an eigenvalue of } \left(-\Delta_p,\ W_0^{1,p}(\Omega)\right),\ \lambda > \lambda_1\right\} > \lambda_1.$$

Because the set $\sigma(p)$ of eigenvalues of $\left(-\Delta_p,\ W_0^{1,p}(\Omega)\right)$ is closed, λ_2^* is the second eigenvalue of $\left(-\Delta_p,\ W_0^{1,p}(\Omega)\right)$. We have

$$\lambda_2^* = \lambda_2,$$

i.e., the second eigenvalue and the second LS-eigenvalue coincide. Then λ_2 admits a minimax characterization provided by the Ljusternik-Schnirelman theory. However, for our purposes this is not convenient. Instead, we will use an alternative one due to Cuesta-de Figueiredo-Gossez [**14**]. Let us introduce the following sets

$$\begin{aligned}
\partial B_1^{L^p} &= \left\{u \in L^p(\Omega):\ \|u\|_p = 1\right\};\\
M &= W_0^{1,p}(\Omega) \cap \partial B_1^{L^p};\\
\Gamma_0 &= \left\{\gamma_0 \in C\left([-1,1], M\right):\ \gamma_0(-1) = -\widehat{u}_1,\ \gamma_0(1) = \widehat{u}_1\right\}.
\end{aligned}$$

PROPOSITION 2.4. $\lambda_2 = \inf\limits_{\gamma_0 \in \Gamma_0} \max\limits_{-1 \leqslant t \leqslant 1} \left\|\nabla \gamma_0(t)\right\|_p^p.$

A similar characterization for the second eigenvalue of the negative p-Laplacian with Neumann boundary conditions can be found in Aizicovici-Papageorgiou-Staicu [**1**].

As we already mentioned, for the linear eigenvalue problem $p = 2$, we have

$$\sigma(p) = \{\lambda_k\}_{k \geqslant 1}.$$

For every integer $k \geqslant 1$, let $E(\lambda_k)$ be the eigenspace corresponding to the eigenvalue λ_k. We have that

$$E(\lambda_k) \subseteq C_0^1(\overline{\Omega})$$

and it has the unique continuation property. This means that, if $u \in E(\lambda_k)$ vanishes on a set of positive measure, then $u(z) = 0$ for all $z \in \overline{\Omega}$. For any integer $m \geqslant 1$, we set

$$\overline{H}_m = \bigoplus_{k=1}^{m} E(\lambda_k) \quad \text{and} \quad \widehat{H}_m = \overline{H}_m^{\perp} = \overline{\bigoplus_{k \geqslant m+1} E(\lambda_k)}.$$

Then the eigenvalues $\{\lambda_k\}_{k \geqslant 1}$ admit the following variational characterizations

$$\lambda_1 = \min\left\{\frac{\|\nabla u\|_2}{\|u\|_2^2}:\ u \in H_0^1(\Omega),\ u \neq 0\right\} \tag{2.4}$$

$$\begin{aligned}
\lambda_m &= \max\left\{\frac{\|\nabla \overline{u}\|_2}{\|\overline{u}\|_2^2}:\ \overline{u} \in \overline{H}_m,\ \overline{u} \neq 0\right\}\\
&= \min\left\{\frac{\|\nabla \widehat{u}\|_2}{\|\widehat{u}\|_2^2}:\ \widehat{u} \in \widehat{H}_{m-1},\ \widehat{u} \neq 0\right\} \quad \forall m \geqslant 2.
\end{aligned} \tag{2.5}$$

The minimum in (2.4) is attained on $E(\lambda_1) = \mathbb{R}\widehat{u}_1$, while the maximum and minimum in (2.5) are realized on $E(\lambda_m)$. We will need the following simple inequalities concerning the component spaces in the orthogonal decomposition

$$H_0^1(\Omega) = \overline{H}_m \oplus \widehat{H}_m.$$

LEMMA 2.5. *(a) If $m \geqslant 0$ is an integer, $\eta \in L^\infty(\Omega)_+$, $\eta(z) \leqslant \lambda_{m+1}$ for almost all $z \in \Omega$ and $\eta \neq \lambda_{m+1}$, then there exists $\xi_0 > 0$, such that*

$$\|\nabla u\|_2^2 - \int_\Omega \eta u^2\, dz \geqslant \xi_0 \|\nabla u\|_2^2 \quad \forall u \in \widehat{H}_m.$$

(b) If $m \geqslant 1$ is an integer, $\eta \in L^\infty(\Omega)_+$, $\eta(z) \geqslant \lambda_m$ for almost all $z \in \Omega$ and $\eta \neq \lambda_m$, then there exists $\widehat{\xi}_0 > 0$, such that

$$\|\nabla u\|_2^2 - \int_\Omega \eta u^2\, dz \leqslant -\widehat{\xi}_0 \|\nabla u\|_2^2 \quad \forall u \in \overline{H}_m.$$

PROOF. (a) Let

$$\psi_0(u) = \|\nabla u\|_2^2 - \int_\Omega \eta u^2\, dz.$$

From the variational characterization of λ_{m+1} (see (2.4) and (2.5)), we have that $\psi_0 \geqslant 0$. We proceed by contradiction. So, suppose that the lemma is not true. Then we can find a sequence $\{u_n\}_{n\geqslant 1} \subseteq \widehat{H}_m$, such that

$$\|\nabla u_n\|_2 = 1 \quad \forall n \geqslant 1$$

and

$$\psi_0(u_n) \longrightarrow 0^+ \quad \text{as } n \to +\infty.$$

We may assume that

$$\begin{aligned} u_n &\longrightarrow u \quad && \text{weakly in } H_0^1(\Omega), \\ u_n &\longrightarrow u \quad && \text{in } L^2(\Omega). \end{aligned}$$

So, in the limit as $n \to +\infty$, we obtain

$$\|\nabla u\|_2^2 \leqslant \int_\Omega \eta u^2\, dz \leqslant \lambda_{m+1} \|u\|_2^2, \tag{2.6}$$

so using also (2.4) and (2.5), we have

$$\|\nabla u\|_2^2 = \lambda_{m+1} \|u\|_2^2, \tag{2.7}$$

Since $u \in \widehat{H}_m$, from (2.7), it follows that $u \in E(\lambda_{m+1})$. If $u = 0$, then

$$u_n \longrightarrow 0 \quad \text{in } H_0^1(\Omega),$$

a contradiction to the fact that $\|\nabla u_n\| = 1$ for all $n \geqslant 1$. Therefore, we have $u \in E(\lambda_m)\setminus\{0\}$ and so $u(z) \neq 0$ for almost all $z \in \Omega$ (unique continuation property). Then from the first inequality in (2.6) and the hypothesis on η, we obtain

$$\|\nabla u\|_2^2 < \lambda_{m+1}\|u\|_2^2,$$

which contradicts the fact that $u \in E(\lambda_{m+1})$.

(b) Let

$$\widehat{\psi}_0(u) = \int_\Omega \eta u^2\, dz - \|\nabla u\|_2^2.$$

Again from the variational characterization of λ_{m+1} (see (2.4) and (2.5)), we have that $\widehat{\psi}_0 \geqslant 0$. Arguing indirectly, suppose that we can find a sequence $\{u_n\}_{n\geqslant 1} \subseteq \overline{H}_m$, such that

$$\|\nabla u_n\|_2 = 1 \quad \forall n \geqslant 1$$

and

$$\widehat{\psi}_0(u_n) \longrightarrow 0^+ \quad \text{as } n \to +\infty.$$

Because of the finite dimensionality of the component space $\overline{H}_m$, we may assume that

$$u_n \longrightarrow u \quad \text{in } H_0^1(\Omega),$$

for some $u \in \overline{H}_m$, so

$$\|\nabla u\|_2^2 = \int_\Omega \eta u^2\, dz \geqslant \lambda_m\|u\|_2^2$$

and thus

$$\|\nabla u\|_2^2 = \int_\Omega \eta u^2\, dz \geqslant \lambda_m\|u\|_2^2 \tag{2.8}$$

and $u \in E(\lambda_m)$. As before $u \neq 0$ and from the unique continuation property and the hypothesis on η, we reach a contradiction. □

Let us recall the definitions of upper and lower solutions for problem (1.1).

DEFINITION 2.6. *(a) We say that $\overline{u} \in W^{1,p}(\Omega)$ is an upper solution for problem (1.1), if*

$$\begin{cases} \displaystyle\int_\Omega (a(z,\nabla\overline{u}),\nabla h)_{\mathbb{R}^N}\, dz \geqslant \int_\Omega \beta|\overline{u}|^{q-2}\overline{u}h\, dz + \int_\Omega f(z,\overline{u})h\, dz \\ \qquad\qquad\qquad\qquad\qquad\qquad \forall h \in C_c^1(\overline{\Omega}),\ h \geqslant 0, \\ \overline{u}|_{\partial\Omega} \geqslant 0. \end{cases}$$

We say that $\overline{u}$ is a strict upper solution, if it is an upper solution for (1.1) but not a solution.
(b) We say that $\underline{u} \in W^{1,p}(\Omega)$ is a lower solution for problem (1.1), if

$$\begin{cases} \int_\Omega (a(z, \nabla \underline{u}), \nabla h)_{\mathbb{R}^N} \, dz \leqslant \int_\Omega \beta |\underline{u}|^{q-2} \underline{u} h \, dz + \int_\Omega f(z, \underline{u}) h \, dz \\ \qquad \qquad \forall h \in C_c^\infty(\overline{\Omega}), \ h \geqslant 0, \\ \underline{u}|_{\partial\Omega} \leqslant 0. \end{cases}$$

We say that $\underline{u}$ is a strict lower solution, if it is a lower solution for (1.1) but not a solution.

Finally, in what follows $\|u\| = \|\nabla u\|_p$ for all $u \in W_0^{1,p}(\Omega)$ and by $|\cdot|_N$ we denote the Lebesgue measure on $\mathbb{R}^N$. Moreover, for every $\zeta \in \mathbb{R}$, we set $\zeta^\pm = \max\{\pm\zeta, 0\}$ and if X is an ordered Banach space with positive cone $C \subseteq X$, then we write $u \leqslant v$ if and only if $v - u \in C$.

3. Superlinear perturbations: Solutions of constant sign

The hypotheses on the map $a(z, y)$ are the following.

$\underline{H(a)}$ $a(z, y) = h(z, \|y\|)\, y$ for all $(z, y) \in \overline{\Omega} \times \mathbb{R}^N$, with $h(z, r) > 0$ for all $(z, r) \in \overline{\Omega} \times (0, +\infty)$ and

(i): $a \in C\left(\overline{\Omega} \times \mathbb{R}^N; \mathbb{R}^N\right) \cap C^1\left(\overline{\Omega} \times (\mathbb{R}^N \setminus \{0\}); \mathbb{R}^N\right)$;

(ii): for every $z \in \overline{\Omega}$ and every $y \in \mathbb{R}^N \setminus \{0\}$, we have

$$\left\|\nabla_y a(z, y)\right\| \leqslant c_1 \|y\|^{p-2},$$

for some $c_1 > 0$ and $1 < p < +\infty$;

(iii): for every $z \in \overline{\Omega}$ and every $y \in \mathbb{R}^N \setminus \{0\}$, we have

$$(\nabla_y a(z, y)\xi, \ \xi)_{\mathbb{R}^N} \geqslant c_0 \|y\|^{p-2} \|\xi\|^2,$$

for some $c_0 > 0$, all $\xi \in \mathbb{R}^N$.

REMARK 3.1. *Set $g(z, r) = h(z, r) r$ for all $(z, r) \in \overline{\Omega} \times \mathbb{R}_+$ ($\mathbb{R}_+ = [0, +\infty)$). Hypotheses $H(a)$ imply the one dimensional estimate*

$$c_0 r^{p-2} \leqslant g_r'(z, r) \leqslant c_1 r^{p-2} \quad \forall z \in \overline{\Omega}, \ r > 0,$$

so

the map $r \longmapsto g(z, r)$ is strictly increasing for all $z \in \overline{\Omega}$.

Let

$$G_0(z, r) = \int_0^r g(z, s) \, ds \quad \forall (z, r) \in \overline{\Omega} \times \mathbb{R}_+.$$

We have

$$\frac{\partial}{\partial r} G_0(z,r) = g(z,r) \quad \forall z \in \overline{\Omega},\ r \geqslant 0.$$

Therefore for all $z \in \overline{\Omega}$ *the function* $r \longmapsto G_0(z,r)$ *is strictly convex and strictly increasing on* $\mathbb{R}_+$. *We set*

$$G(z,y) = G_0(z,\|y\|) \quad \forall (z,y) \in \overline{\Omega} \times \mathbb{R}^N.$$

Then for all $z \in \overline{\Omega}$, *the function* $y \longmapsto G(z,y)$ *is strictly convex. Moreover, for all* $(z,y) \in \overline{\Omega} \times (\mathbb{R}^N \setminus \{0\})$, *we have*

$$\nabla_y G(z,y) = (G_0)'_r(z,\|y\|)\frac{y}{\|y\|} = g(z,\|y\|)\frac{y}{\|y\|} = h(z,\|y\|y) = a(z,y). \tag{3.1}$$

These observations lead to the following lemma.

LEMMA 3.2. *If hypotheses* $H(a)$ *hold, then*
(a) *for all* $z \in \overline{\Omega}$, *the map* $y \longmapsto a(z,y)$ *is maximal monotone and strictly monotone;*
(b) *for all* $(z,y) \in \overline{\Omega} \times \mathbb{R}^N$, *we have*

$$\big\|a(z,y)\big\| \leqslant \frac{c_1}{p-1}\|y\|^{p-1};$$

(c) *for all* $(z,y) \in \overline{\Omega} \times \mathbb{R}^N$, *we have*

$$\big(a(z,y),\ y\big)_{\mathbb{R}^N} \geqslant \frac{c_0}{p-1}\|y\|^p.$$

PROOF. (a) Since the map $y \longmapsto G(z,\cdot)$ is continuous and strictly convex, because of (3.1), it follows that for all $z \in \overline{\Omega}$, the map $y \longmapsto a(z,y)$ is maximal monotone and strictly monotone.

(b) Since $a(z,0) = 0$, for all $(z,y) \in \overline{\Omega} \times \mathbb{R}^N$, we have

$$a(z,y) = \int_0^1 \frac{d}{dt} a(z,ty)\,dt,$$

so, by the chain rule, we have

$$a(z,y) = \int_0^1 \nabla_y a(z,ty) y\,dt.$$

Using also hypothesis $H(a)(ii)$, we obtain

$$\big\|a(y,z)\big\| \leqslant \int_0^1 \big\|\nabla_y a(z,ty)\big\|\|y\|\,dt \leqslant c_1\|y\|^{p-1}\int_0^1 t^{p-2}\,dt = \frac{c_1}{p-1}\|y\|^{p-1}.$$

(c) Similarly, using the chain rule and hypothesis $H(a)(iii)$, for all $(z,y) \in \overline{\Omega} \times \mathbb{R}^N$, we have

$$\begin{aligned}(a(z,y),\, y)_{\mathbb{R}^N} &= \int_0^1 \frac{d}{dt}(a(z,ty),\, y)_{\mathbb{R}^N}\, dt \\ &= \int_0^1 (\nabla_y a(z,ty)y,\, y)_{\mathbb{R}^N}\, dt \\ &\geqslant c_0\|y\|^p \int_0^1 t^{p-2} dt \\ &= \frac{c_0}{p-1}\|y\|^p.\end{aligned}$$

From this lemma it follows that □

COROLLARY 3.3. *If hypotheses $H(a)$ hold, then for all $(z,y) \in \overline{\Omega} \times \mathbb{R}^N$, we have*

$$\frac{c_0}{p(p-1)}\|y\|^p \leqslant G(z,y) \leqslant \frac{c_1}{p(p-1)}\|y\|^p.$$

EXAMPLE 3.4. *Let $\vartheta \in C^1(\overline{\Omega})$, $\vartheta(z) > 0$ for all $z \in \overline{\Omega}$ and $K \in C^1(\overline{\Omega}; \mathbb{R}^{N\times N})$ with $K(z) > 0$ for all $z \in \overline{\Omega}$ (i.e., the $N \times N$-matrix $K(z)$ is strictly positive). Then for $1 < p < +\infty$, the following maps satisfy hypotheses $H(a)$:*

- $a_1(z,y) = a_1(y) = \|y\|^{p-2}y;$
- $a_2(z,y) = \vartheta(z)\|y\|^{p-2}y;$
- $a_3(z,y) = \vartheta(z)\left(1 + \|y\|^2\right)^{\frac{p-2}{2}} y;$
- $a_4(z,y) = \vartheta(z)\left(\|y\|^{p-2}y + \frac{\|y\|^{p-2}y}{1+\|y\|^p}\right);$
- $a_5(z,y) = K(z)y$ *(here $p = 2$).*

Note that $a_1(y)$ corresponds to the p-Laplacian differential operator and in this case

$$G_1(y) = \frac{1}{p}\|y\|^p \quad \forall y \in \mathbb{R}^N.$$

In the sequel by $\langle\cdot,\cdot\rangle$ we denote the duality brackets for the pair of spaces $\left(W^{-1,p'}(\Omega), W_0^{1,p}(\Omega)\right)$ $(\frac{1}{p} + \frac{1}{p'} = 1)$. Let $V\colon W_0^{1,p}(\Omega) \longrightarrow W^{-1,p'}(\Omega)$ be the nonlinear map, defined by

$$\langle V(u), h\rangle = \int_\Omega (a(z,\nabla u), \nabla h)_{\mathbb{R}^N} \quad \forall u,h \in W_0^{1,p}(\Omega).$$

The next proposition establishes the properties of V.

PROPOSITION 3.5. *If hypotheses $H(a)$ hold, then V is maximal monotone, strictly monotone, coercive and of type $(S)_+$.*

PROOF. From hypotheses $H(a)$ and Lemma 3.2, it follows that V is demicontinuous, strictly monotone, hence it is maximal monotone (see Gasiński-Papageorgiou [**25**, p. 309]). Also from Lemma 3.2(c), we have

$$\langle V(u), u\rangle = \int_V (a(z,\nabla u),\, \nabla u)_{\mathbb{R}^N}\, dz \geqslant \frac{c_0}{p-1}\|\nabla u\|_p^p,$$

so V is coercive.

It remains to show that V is of type $(S)_+$. To this end, we consider a sequence $\{u_n\}_{n\geqslant 1} \subseteq W_0^{1,p}(\Omega)$, such that

$$u_n \longrightarrow u \quad \text{weakly in } W_0^{1,p}(\Omega) \quad \text{and} \quad \limsup_{n\to+\infty}\langle V(u_n),\ u_n - u\rangle \leqslant 0. \tag{3.2}$$

Then the monotonicity of V implies that

$$\lim_{n\to+\infty}\langle V(u_n) - V(u),\ u_n - u\rangle = 0. \tag{3.3}$$

Let

$$\xi_n(z) = \big(a(z,\nabla u_n(z)) - a(z,\nabla u(z)),\ \nabla u_n(z) - \nabla u(z)\big)_{\mathbb{R}^N}.$$

Note that $\xi_n \geqslant 0$ and from (3.3) it follows that

$$\xi_n \longrightarrow 0 \quad \text{in } L^1(\Omega).$$

So, by passing to a suitable sequence if necessary, we may assume that

$$\xi_n(z) \longrightarrow 0 \quad \text{for almost all } z \in \Omega \tag{3.4}$$

and

$$0 \leqslant \xi_n(z) \leqslant \eta(z) \quad \text{for almost all } z \in \Omega, \tag{3.5}$$

with $\eta \in L^1(\Omega)_+$. From the definition of ξ_n and using Lemma 3.2(b)-(c) and (3.5), for almost all $z \in \Omega$ and all $n \geqslant 1$, we have

$$\begin{aligned} \eta(z) \ &\geqslant\ \frac{c_0}{p-1}\big(\|\nabla u_n(z)\|^p + \|\nabla u(z)\|^p\big) - \frac{c_1}{p-1}\|\nabla u_n(z)\|^{p-1}\|\nabla u(z)\| \\ &\quad -\frac{c_1}{p-1}\|\nabla u(z)\|^{p-1}\|\nabla u_n(z)\|. \end{aligned} \tag{3.6}$$

From (3.6), we see that for all $z \in \Omega\setminus E$, $|E|_N = 0$, we have that the sequence $\{\nabla u_n(z)\}_{n\geqslant 1} \subseteq \mathbb{R}^N$ is bounded. So, for at least a subsequence (depending in general on $z \in \Omega \setminus E$), we can say that

$$\nabla u_n(z) \longrightarrow \vartheta(z) \quad \text{in } \mathbb{R}^N.$$

Because of (3.4), in the limit $n \to +\infty$, we obtain

$$\big(a(z,\vartheta(z)) - a(z,\nabla u(z)),\ \vartheta(z) - \nabla u(z)\big)_{\mathbb{R}^N} = 0 \quad \forall z \in \Omega\setminus E,$$

so, due to the strict convexity of $a(z,\cdot)$, we have

$$\vartheta(z) = \nabla u(z) \quad \forall z \in \Omega \setminus E.$$

Therefore we have

$$\nabla u_n(z) \longrightarrow \nabla u(z) \quad \text{for almost all } z \in \Omega.$$

From (3.6) it is clear that the sequence $\{\|\nabla u_n(\cdot)\|^p\}_{n\geqslant 1} \subseteq L^1(\Omega)$ is uniformly integrable. So, invoking Vitali's theorem (see e.g., Gasiński-Papageorgiou [**25**, p. 901]), we have that

$$\|\nabla u_n\|_p \longrightarrow \|\nabla u\|_p.$$

This together with (3.2) and the fact that the Lebesgue space $L^p(\Omega;\mathbb{R}^N)$ is uniformly convex, imply that

$$\nabla u_n \longrightarrow \nabla u \quad \text{in } L^p(\Omega;\mathbb{R}^N),$$

hence

$$u_n \longrightarrow u \quad \text{in } W_0^{1,p}(\Omega).$$

This proves that V is of type $(S)_+$. □

The next result links local minimizers in the spaces $C_0^1(\overline{\Omega})$ and $W_0^{1,p}(\Omega)$ for a large class of C^1-functionals. The first such result was proved by Brézis-Nirenberg [**8**] for equations involving the Laplacian. Their result was extended to the p-Laplacian by García Azorero-Manfredi-Peral Alonso [**23**] and Guo-Zhang [**30**] (in the latter under the restriction that $p \geqslant 2$). Extensions to the Neumann case can be found in Motreanu-Motreanu-Papageorgiou [**42, 43**] and to nonsmooth problems by Iannizzotto-Papageorgiou [**31**] and Kyritsi-Papageorgiou [**32**] (in the latter under the restriction that $p \geq 2$).

So let $f_0\colon \Omega\times\mathbb{R} \longrightarrow \mathbb{R}$ be a Carathéodory function with subcritical growth in $\zeta\in\mathbb{R}$, i.e.,

$$|f_0(z,\zeta)| \leqslant a(z) + c|\zeta|^{r-1} \quad \text{for almost all } z\in\Omega, \text{ all } \zeta\in\mathbb{R},$$

with $a \in L^\infty(\Omega)_+$, $c>0$, $1<r<p^*$.

Let

$$F_0(z,\zeta) = \int_0^\zeta f_0(z,s)\,ds$$

and consider the C^1-functional $\varphi_0\colon W_0^{1,p}(\Omega) \longrightarrow \mathbb{R}$, defined by

$$\varphi_0(u) = \int_\Omega G(z,\nabla u)\,dz - \int_\Omega F(z,u)\,dz \quad \forall u \in W_0^{1,p}(\Omega).$$

PROPOSITION 3.6. *If $u_0 \in W_0^{1,p}(\Omega)$ is a local $C_0^1(\overline{\Omega})$-minimizer of φ_0, i.e., there exists $r_1>0$, such that*

$$\varphi_0(u_0) \leqslant \varphi_0(u_0+h) \quad \forall h \in C_0^1(\overline{\Omega}),\ \|h\|_{C_0^1(\overline{\Omega})} \leqslant r_1,$$

then $u_0 \in C_0^1(\overline{\Omega})$ and it is also a local $W_0^{1,p}(\Omega)$-minimizer of φ_0, i.e., there exists $r_2>0$, such that

$$\varphi_0(u_0) \leqslant \varphi_0(u_0+h) \quad \forall h \in W_0^{1,p}(\Omega),\ \|h\| \leqslant r_2.$$

PROOF. Let $h \in C_0^1(\overline{\Omega})$ and $\lambda \neq 0$ with $|\lambda| > 0$ small. Then by hypothesis, we have

$$\varphi_0(u_0) \leqslant \varphi_0(u_0 + \lambda h),$$

so

$$0 \leqslant \langle \varphi_0'(u_0), h \rangle. \tag{3.7}$$

Since $h \in C_0^1(\overline{\Omega})$ is arbitrary and $C_0^1(\overline{\Omega})$ is dense in $W_0^{1,p}(\Omega)$, from (3.7), it follows that

$$\varphi_0'(u_0) = 0,$$

so

$$V(u_0) = N_0(u_0),$$

where

$$N_0(u)(\cdot) = f_0(\cdot, u(\cdot)) \quad \forall u \in W_0^{1,p}(\Omega)$$

and thus

$$\begin{cases} -\operatorname{div} a(z, \nabla u_0(z)) = f_0(z, u_0(z)) \text{ in } \Omega, \\ u|_{\partial\Omega} = 0. \end{cases} \tag{3.8}$$

Theorem 7.1 of Ladyzhenskaya-Uraltseva [**33**, p. 286] and (3.8) imply that $u_0 \in L^\infty(\Omega)$. Then invoking Theorem 1 of Lieberman [**34**] (see also Tolksdorf [**48**]), we have $u_0 \in C_0^{1,\gamma}(\overline{\Omega})$ for some $\gamma \in (0, 1)$.

Next we show that u_0 is also a local $W_0^{1,p}(\Omega)$-minimizer of φ_0. We argue by contradiction. So, suppose that u_0 is not a local $W_0^{1,p}(\Omega)$-minimizer of φ_0. Exploiting the compactness of the embedding $W_0^{1,p}(\Omega) \subseteq L^r(\Omega)$, we can easily verify that φ_0 is sequentially weakly lower semicontinuous. Also for every $\varepsilon > 0$ the set

$$\overline{B}_\varepsilon = \{u \in W_0^{1,p}(\Omega) : \ \|u\| \leqslant \varepsilon\}$$

is weakly compact. Hence, by the Weierstrass theorem, for every $\varepsilon > 0$ small enough, we can find $v_\varepsilon \in \overline{B}_\varepsilon$, such that

$$\varphi_0(u_0 + v_\varepsilon) = \inf\{\varphi_0(u_0 + v) : v \in \overline{B}_\varepsilon\} < \varphi_0(u_0).$$

Then by the Lagrange multiplier rule (see, e.g., Clarke [**12**, p. 228]), we can find $\lambda_\varepsilon \leqslant 0$, such that

$$\varphi_0'(u_0 + v_\varepsilon) = \lambda_\varepsilon A(v_\varepsilon),$$

so

$$V(u_0 + v_\varepsilon) = N(u_0 + v_\varepsilon) + \lambda_\varepsilon A(v_\varepsilon)$$

and thus

$$-\operatorname{div} a(z, \nabla(u_0 + v_\varepsilon)(z)) = f_0(z, (u_0 + v_\varepsilon)(z)) - \lambda_\varepsilon \Delta_p v_\varepsilon(z) \text{ in } \Omega. \tag{3.9}$$

We consider two distinct cases depending on the values of the Lagrange multipliers λ_ε, $\varepsilon \in (0, 1]$ small enough.

Case I. $|\lambda_\varepsilon| \leqslant 1$ for all $\varepsilon \in (0,1]$ small enough.

We set $y_\varepsilon = u_0 + v_\varepsilon$. Then (3.9) becomes

$$-\operatorname{div} a\left(z, \nabla y_\varepsilon(z)\right) - |\lambda_\varepsilon| \Delta_p (y_\varepsilon - u_0)(z) = f_0\left(z, y_\varepsilon(z)\right) \text{ in } \Omega. \tag{3.10}$$

Subtracting (3.8) from (3.10), we obtain

$$\begin{aligned} &-\operatorname{div}\Big[a\left(z,\nabla y_\varepsilon(z)\right) - a\left(z, \nabla u_0(z)\right) + |\lambda_\varepsilon| \big\|\nabla(y_\varepsilon - u_0)(z)\big\|^{p-2} \nabla(y_\varepsilon - u_0)(z)\Big] \\ &\quad = f_0\left(z, y_\varepsilon(z)\right) - f_0\left(z, u_0(z)\right) \text{ in } \Omega. \end{aligned} \tag{3.11}$$

We set

$$L_\varepsilon(z,y) = a(z,y) - a\left(z, \nabla u_0(z)\right) + |\lambda_\varepsilon| \|y - \nabla u_0(z)\|^{p-2}\left(y - \nabla u_0(z)\right).$$

We have already established earlier that $u_0 \in C_0^1(\overline{\Omega})$. Hence $L_\varepsilon \in C(\overline{\Omega} \times \mathbb{R}^N; \mathbb{R}^N)$. Using L_ε, equation (3.11) becomes

$$-\operatorname{div} L_\varepsilon\left(z, \nabla y_\varepsilon(z)\right) = f_0\left(z, y_\varepsilon(z)\right) - f_0\left(z, u_0(z)\right) \text{ in } \Omega. \tag{3.12}$$

Hypotheses $H(a)$ and Lemma 2.1 of Damascelli [**15**] imply that
- if $1 < p < 2$, then

$$\left(a(z,y) - a(z,y'),\ y - y'\right)_{\mathbb{R}^N} \geqslant \frac{c_2}{(\|y\| + \|y'\|)^{2-p}} \|y - y'\|^2, \tag{3.13}$$

for some $c_2 > 0$ and all $y, y' \in \mathbb{R}^N$.
- if $p \geqslant 2$, then

$$\left(a(z,y) - a(z,y'),\ y - y'\right)_{\mathbb{R}^N} \geqslant c_2 \|y - y'\|^p, \tag{3.14}$$

for some $c_2 > 0$ and all $y, y' \in \mathbb{R}^N$.

First we assume that $p \geqslant 2$. Then setting $\xi = \|\nabla u_0\|_\infty$ (recall $u_0 \in C_0^1(\overline{\Omega})$), we have

$$\begin{aligned} \left(L_\varepsilon(z,y),\ y\right)_{\mathbb{R}^N} &= \left(a(z,y) - a(z, \nabla u_0(z)),\ y - \nabla u_0(z)\right)_{\mathbb{R}^N} \\ &\quad + \left(a(z,y) - a(z,\nabla u_0(z)),\ \nabla u_0(z)\right)_{\mathbb{R}^N} \\ &\quad + |\lambda_\varepsilon| \|y - \nabla u_0(z)\|^{p-2} \left(y - \nabla u_0(z),\ y\right)_{\mathbb{R}^N} \\ &\geqslant c_2 \|y - \nabla u_0(z)\|^{p-2} - \xi \big\|a(z,y) - a(z, \nabla u_0(z))\big\| \\ &\quad + |\lambda_\varepsilon| \|y - \nabla u_0(z)\|^{p-2} \left(\|y\|^2 - \xi\|y\|\right) \\ &\geqslant c_3 \|y\|^p - c_4, \end{aligned} \tag{3.15}$$

for some $c_3, c_4 > 0$. Here we have used (3.14), inequality (2.1) of Damascelli [**15**] and Young's inequality with $\varepsilon > 0$.

Next suppose that $1 < p < 2$. Then, using (3.13), we have

$$(L_\varepsilon(z,y),\ y)_{\mathbb{R}^N} \geqslant \frac{c_2}{(\|y\|+\xi)^{2-p}}\big\|y-\nabla u_0(z)\big\|^2 - \xi\big\|a(z,y)-a(z,\nabla u_0(z))\big\| \tag{3.16}$$
$$+ |\lambda_\varepsilon|\big\|y-\nabla u_0(z)\big\|^{p-2}\left(\|y\|^2-\xi\|y\|\right).$$

Suppose that $\big\|y-\nabla u_0(z)\big\| \geqslant 1$. Since $0 < 2-p < 1$, we have

$$(\|y\|+\xi)^{2-p} \leqslant \|y\|^{2-p} + \xi^{2-p}$$

and

$$\|y\| \leqslant \big\|y-\nabla u_0(z)\big\| + \xi,$$

so

$$\|y\|^{2-p} \leqslant \big\|y-\nabla u_0(z)\big\|^{2-p} + \xi^{2-p}.$$

Therefore and since $\|y-\nabla u_0(z)\| \geqslant 1$, we have

$$\begin{aligned}(\|y\|+\xi)^{2-p} &\leqslant \big\|y-\nabla u_0(z)\big|^{2-p} + 2\xi^{2-p}\\ &\leqslant (1+2\xi^{2-p})\|y-\nabla u_0(z)\|^{2-p}.\end{aligned}$$

Then, from (3.16) and using the facts that $1 < p < 2$ and $\|y-\nabla u_0(z)\| \geqslant 1$, we have

$$\begin{aligned}&(L_\varepsilon(z,y),\ y)_{\mathbb{R}^N}\\ &\quad\geqslant \frac{c_2}{1+2\xi^{2-p}}\big\|y-\nabla u_0(z)\big\|^p - \xi\big\|a(z,y)-a\left(z,\nabla u_0(z)\right)\big\| - |\lambda_\varepsilon|\xi\|y\|\\ &\quad\geqslant c_3\|y\|^p - c_4.\end{aligned} \tag{3.17}$$

Here we have used inequality (2.5) of Damascelli [**15**] and Young's inequality with $\varepsilon > 0$.

Now suppose that $\|y-\nabla u_0(z)\| < 1$. Then $\|y\| \leqslant 1+\xi$ and so from (3.16), we have

$$\begin{aligned}(L_\varepsilon(z,y),\ y)_{\mathbb{R}^N} &\geqslant \frac{c_2}{(1+\xi)^{2-p}}\big\|y-\nabla u_0(z)\big\|^2 - \xi\big\|a(z,y)-a(z,\nabla u_0(z))\big\|\\ &\quad - |\lambda_\varepsilon|\big\|y-\nabla u_0(z)\big\|^{p-2}\left(\|y\|-\xi\right)|y\|\\ &\geqslant c_3\|y\|^p - c_4\end{aligned} \tag{3.18}$$

(since $1 < p < 2$). From (3.15), (3.17) and (3.18), we conclude that

$$(L_\varepsilon(z,y),\ y)_{\mathbb{R}^N} \geqslant c_3\|y\|^p - c_4 \quad \forall (z,y) \in \overline{\Omega}\times\mathbb{R}^N. \tag{3.19}$$

Because of (3.7), we can apply Theorem 7.1 of Ladyzhenskaya-Uraltseva [**33**, p. 286] and infer that

$$y_\varepsilon \in L^\infty \quad \text{and} \quad \|y_\varepsilon\|_\infty \leqslant M_0 \quad \text{for all } \varepsilon \in (0,1] \text{ small enough,} \tag{3.20}$$

for some $M_0 > 0$. Using hypothesis $H(a)(iii)$ and recalling that $p+1 > 2$, for every $z \in \overline{\Omega}$, $y \in \mathbb{R}^N \setminus \{0\}$ and $\xi \in \mathbb{R}^N$, we have

$$\begin{aligned}(\nabla_y L_\varepsilon(z,y)\xi,\ \xi)_{\mathbb{R}^N} &= (\nabla_y a(z,y)\xi,\ \xi)_{\mathbb{R}^N} + |\lambda_\varepsilon| \|y - \nabla u_0(z)\|^{p-2} \\ &\quad \times \left(\left(\mathrm{id} + \frac{(p-2)(y-\nabla u_0(z), y - \nabla u_0(z))_{\mathbb{R}^N}}{\|y-\nabla u_0(z)\|^2}\right)\xi, \xi\right)_{\mathbb{R}^N} \\ &\geqslant c_0 \|y\|^{p-2}\|\xi\|^2. \end{aligned} \tag{3.21}$$

Also, let $z, z' \in \Omega$. Then for every $y \in \mathbb{R}^N$, we have

$$\begin{aligned}&\big\|L_\varepsilon(z,y) - L_\varepsilon(z',y)\big\| \\ &\quad\leqslant \|a(z,y) - a(z',y)\| + \big\|a(z,\nabla u_0(z)) - a(z',\nabla u_0(z'))\big\| \\ &\qquad + |\lambda_\varepsilon| \Big| \big\|y - \nabla u_0(z)\big\|^{p-2}(y-\nabla u_0(z)) - \big\|y - \nabla u_0(z')\big\|^{p-2}(y - \nabla u_0(z'))\Big| \\ &\quad\leqslant c_5 \|z-z'\|^\beta, \end{aligned} \tag{3.22}$$

for some $c_5 > 0$ and $\beta \in (0,\gamma)$ (recall that $u_0 \in C_0^{1,\gamma}(\overline{\Omega})$). Because of (3.20), (3.21) and (3.22), we can apply Theorem 1 of Lieberman [**34**] (see also Theorem 3 of Tolksdorf [**48**]) and obtain $\alpha \in (0,1)$ and $M_1 > 0$, such that

$$y_\varepsilon \in C_0^{1,\alpha}(\overline{\Omega}) \quad \text{and} \quad \|y_\varepsilon\|_{C_0^{1,\alpha}(\overline{\Omega})} \leqslant M_1 \quad \text{for all } \varepsilon \in (0,1], \text{ small enough.} \tag{3.23}$$

Case II. $|\lambda_\varepsilon| > 1$ for all $\varepsilon \in (0,1]$ small enough.

In this case, we set

$$S_\varepsilon(z,y) = \frac{1}{|\lambda_\varepsilon|}\big[a(z, \nabla u_0(z) + y) - a(z, \nabla u_0(z))\big] + \|y\|^{p-2}y.$$

We have

$$\begin{aligned}(S_\varepsilon(z,y),\ y)_{\mathbb{R}^N} &= \frac{1}{|\lambda_\varepsilon|}\left(a(z,\nabla u_0(z)+y) - a(z,\nabla u_0(z)),\ y\right)_{\mathbb{R}^N} + \|y\|^p \\ &\geqslant \|y\|^p. \end{aligned} \tag{3.24}$$

Subtracting (3.8) from (3.9) and dividing with $|\lambda_\varepsilon|$, we obtain

$$-\mathrm{div}\, S_\varepsilon(z, \nabla v_\varepsilon(z)) = \frac{1}{|\lambda_\varepsilon|}\Big[f_0(z,(u_0+v_\varepsilon)(z)) - f(z,u_0(z))\Big] \text{ in } \Omega. \tag{3.25}$$

Because of (3.24), from (3.25) and Theorem 7.1 of Ladyzhenskaya-Uraltseva [**33**, p. 286], we infer that

$$v_\varepsilon \in L^\infty(\Omega) \quad \text{and} \quad \|v_\varepsilon\|_\infty \leqslant M_2 \quad \text{for all } \varepsilon \in (0,1], \text{ small enough,} \tag{3.26}$$

for some $M_2 > 0$. Using hypothesis $H(a)(iii)$, for every $z \in \overline{\Omega}$, every $y \in \mathbb{R}^N \setminus \{0\}$ and every $\xi \in \mathbb{R}^N$, we have

$$\begin{aligned}(\nabla_y S_\varepsilon(z,y)\xi,\ \xi)_{\mathbb{R}^N} &= \frac{1}{|\lambda_\varepsilon|}(\nabla_y a(z, \nabla u_0(z)+y)\xi,\ \xi)_{\mathbb{R}^N} \\ &\quad + \|y\|^{p-2}\left(\left(\mathrm{id} + (p-2)\frac{(y,\ y)_{\mathbb{R}^N}}{\|y\|^2}\right)\xi,\ \xi\right)_{\mathbb{R}^N} \\ &\geqslant \|y\|^{p-2}\,(1-(2-p))\,\|\xi\|^2 \\ &= c_6\|y\|^{p-2}\|\xi\|^2,\end{aligned} \tag{3.27}$$

with $c_6 = p-1 > 0$. Moreover, if $z, z' \in \Omega$, then for every $y \in \mathbb{R}^N$, we have

$$\begin{aligned}\big\|S_\varepsilon(z,y) - S_\varepsilon(z',y)\big\| &\leqslant \big\|a(z, \nabla u_0(z)+y) - a(z', \nabla u_0(z')+y)\big\| \\ &\quad + \big\|a(z, \nabla u_0(z)) - a(z', \nabla u_0(z'))\big\| \\ &\leqslant c_7\|z-z'\|^{\beta'},\end{aligned} \tag{3.28}$$

for some $c_7 > 0$ and $\beta' \in (0,\gamma)$ (as before recall that $u_0 \in C_0^{1,\gamma}(\overline{\Omega})$).

Again from (3.26), (3.27), (3.28) and Theorem 1 of Lieberman [**34**] (see also Theorem 3 of Tolksdorf [**48**]), we infer that there is $\delta \in (0,1)$ and $M_3 > 0$, such that

$$v_\varepsilon \in C_0^{1,\delta}(\overline{\Omega}) \quad \text{and} \quad \|v_\varepsilon\|_{C_0^{1,\delta}(\overline{\Omega})} \leqslant M_3 \quad \text{for all } \varepsilon \in (0,1], \text{ small enough.} \tag{3.29}$$

Now let $\varepsilon_n \searrow 0$ and let us set

$$y_n = y_{\varepsilon_n} \quad \text{(Case I)} \quad \text{and} \quad v_n = v_{\varepsilon_n} \quad \text{(Case II)} \quad \forall n \geqslant 1.$$

Because of (3.23) (respectively (3.29)) and recalling that for every $\sigma \in (0,1)$, the embedding $C_0^{1,\sigma}(\overline{\Omega}) \subseteq C_0^1(\overline{\Omega})$ is compact, we see that, passing to a subsequence if necessary, we may assume

$$\begin{aligned} y_n &\longrightarrow \widehat{y} \qquad \text{in } C_0^1(\overline{\Omega}), \\ v_n &\longrightarrow \widehat{v} \qquad \text{in } C_0^1(\overline{\Omega}). \end{aligned}$$

Recall that $y_n = u_0 + v_n$ for all $n \geqslant 1$ and $v_n \longrightarrow 0$ in $W_0^{1,p}(\Omega)$. therefore

$$\widehat{y} = u_0 \quad \text{and} \quad \widehat{v} = 0.$$

Hence, we can find $n_0 \geqslant 1$, such that

$$\|v_n\|_{C_0^1(\overline{\Omega})} \leqslant r_1 \quad \forall n \geqslant n_0.$$

This then implies that

$$\varphi_0(u_0) \leqslant \varphi_0(u_0 + v_n) \quad \forall n \geqslant n_0,$$

which contradicts the choice of the sequence $\{v_n\}_{n\geqslant 1}$. This concludes the proof. $\square$

The hypotheses on the perturbation f are the following.

$\underline{H(f)_1}$ $f\colon \Omega \times \mathbb{R} \longrightarrow \mathbb{R}$ is a function, such that

(i): for all $\zeta \in \mathbb{R}$, the function $z \longmapsto f(z,\zeta)$ is measurable;

(ii): for almost all $z \in \Omega$, the function $\zeta \longmapsto f(z,\zeta)$ is continuous and we have $f(z,0) = 0$;

(iii): for almost all $z \in \Omega$ and all $\zeta \in \mathbb{R}$, we have

$$\big|f(z,\zeta)\big| \leqslant a(z) + c|\zeta|^{r-1},$$

with $a \in L^\infty$, $c > 0$ and $p < r < p^*$;

(iv): if $F(z,\zeta) = \int_0^\zeta f(z,s)\,ds$, then

$$\lim_{|\zeta|\to+\infty} \frac{F(z,\zeta)}{|\zeta|^p} = +\infty,$$

uniformly for almost all $z \in \Omega$ and there exist $\beta_0 > 0$ and $\mu > q$, with $\mu \in \Big((r-p)\max\big\{\frac{N}{p},1\big\},\ p^*\Big)$, such that

$$\liminf_{|\zeta\to+\infty|} \frac{f(z,\zeta)\zeta - pF(z,\zeta)}{|\zeta|^\mu} \geqslant \beta_0$$

uniformly for almost all $z \in \Omega$;

(v): there exist $a_- < 0 < a_+$ and $\beta^*, \xi^* > 0$, such that

$$\beta^* a_+^{q-1} + f(z,a_+) \leqslant 0 \leqslant \beta^*|a_-|^{q-2}a_- + f(z,a_-) \text{ for almost all } z \in \Omega.$$

and the function $\zeta \longmapsto f(z,\zeta) + \xi^*|\zeta|^{p-2}\zeta$ is nondecreasing on $[a_-,a_+]$ for almost all $z \in \Omega$;

(vi): there exists $\eta \in L^\infty(\Omega)_+$, $\eta(z) \geqslant \beta(z)$ for almost all $z \in \Omega$, $\eta \neq \beta$ and

$$\liminf_{\zeta\to 0} \frac{qF(z,\zeta)}{|\zeta|^q} \geqslant -\eta(z)$$

uniformly for almost all $z \in \Omega$.

REMARK 3.7. *Hypothesis $H(f)_1(iv)$ implies that the potential $F(z,\cdot)$ is p-superlinear near $\pm\infty$. This is the case if the perturbation $f(z,\cdot)$ is $(p-1)$-superlinear near $\pm\infty$. Note however that we do not use the Ambrosetti-Rabinowitz condition, common when dealing with "superlinear problems". We recall that the Ambrosetti-Rabinowitz condition says that there exist $M_4 > 0$ and $\tau > p$, such that*

$$f(z,\zeta)\zeta \geqslant \tau F(z,\zeta) > 0 \quad \text{for almost all } z \in \Omega, \text{ all } |\zeta| \geqslant M_4. \tag{3.30}$$

An integration of (3.30) gives

$$F(z,\zeta) \geqslant c_8|\zeta|^\tau \quad \text{for almost all } z \in \Omega, \text{ all } |\zeta| \geqslant M_4,$$

for some $c_8 > 0$. This shows that $F(z,\cdot)$ is p-superlinear near $\pm\infty$. Here, instead of (3.30), we employ hypothesis $H(f)_1(v)$ which is weaker. Similar conditions can be found in Costa-Magalhães [**13**]*, Fei* [**20**] *(for Hamiltonian systems) and Li-Wu-Zhou* [**36**] *(where $p = 2$).*

EXAMPLE 3.8. *Consider the following nonlinearity (for the sake of simplicity we drop the z-dependence):*

$$f(\zeta) = \begin{cases} |\zeta|^{q-2}\zeta - \xi_1|\zeta|^{\tau-2}\zeta & \text{if } |\zeta| \leqslant 1, \\ |\zeta|^{p-2}\zeta\left(\ln|\zeta| + \dfrac{1}{p}\right) - \xi_1\dfrac{\zeta}{|\zeta|} & \text{if } |\zeta| > 1, \end{cases}$$

where $\tau > q$ and $\xi_1 > 1$. In this case we have

$$F(\zeta) = \begin{cases} \dfrac{1}{q}|\zeta|^q - \dfrac{\xi_1}{\tau}|\zeta|^\tau & \text{if } |\zeta| \leqslant 1, \\ \frac{1}{p}|\zeta|^p\ln|\zeta| + \xi_1|\zeta| & \text{if } |\zeta| > 1. \end{cases}$$

This nonlinearity satisfies hypotheses $H(f)_1$ (take $a_- = -1 < 1 = a_+$ and $\beta^* > 0$ suitably small, while $r = p + \varepsilon$ with $\varepsilon > 0$ small, such that $N\varepsilon < p^2$ and $\mu = p$). This function does not satisfy Ambrosetti-Rabinowitz condition.

We start by producing two nontrivial smooth solutions of constant sign. So, let $\varphi\colon W_0^{1,p}(\Omega) \longrightarrow \mathbb{R}$ be the energy (Euler) functional for problem (1.1), defined by

$$\varphi(u) = \int_\Omega G(z,\nabla u)\,dz - \frac{1}{q}\int_\Omega \beta|u|^q\,dz - \int_\Omega F(z,u)\,dz \quad \forall u \in W_0^{1,p}(\Omega).$$

Evidently $\varphi \in C^1\left(W_0^{1,p}(\Omega)\right)$.

PROPOSITION 3.9. *If hypotheses $H(a)$ and $H(f)_1$ hold and $\|\beta\|_\infty \leqslant \beta^*$, then problem (1.1) has two solutions $u_0 \in \operatorname{int} C_+$ and $v_0 \in -\operatorname{int} C_+$ which are local minimizers of the functional φ and $a_- < v_0(z) < u_0(z) < a_+$ for all $z \in \overline{\Omega}$.*

PROOF. Consider the following truncation of the reaction term

$$\widehat{f}_+(z,\zeta) = \begin{cases} 0 & \text{if } \zeta < 0, \\ \beta(z)\zeta^{q-1} + f(z,\zeta) & \text{if } \zeta \in [0,a_+], \\ \beta(z)a_+^{q-1} + f(z,a_+) & \text{if } a_+ < \zeta. \end{cases} \tag{3.31}$$

We set

$$\widehat{F}_+(z,\zeta) = \int_0^\zeta \widehat{f}_+(z,s)\,ds$$

and introduce the C^1-functional $\widehat{\varphi}_+\colon W_0^{1,p}(\Omega) \longrightarrow \mathbb{R}$, defined by

$$\widehat{\varphi}_+(u) = \int_\Omega G(z,\nabla u)\,dz - \int_\Omega \widehat{F}_+(z,u)\,dz \quad \forall u \in W_0^{1,p}(\Omega).$$

Clearly $\widehat{\varphi} \in C^1\left(W_0^{1,p}(\Omega)\right)$ and in addition $\widehat{\varphi}_+$ is coercive (see (3.31)) and sequentially lower semicontinuous. So, by the Weierstrass theorem, we can find $u_0 \in$

$W_0^{1,p}(\Omega)$, such that

$$\widehat{m}_+ = \inf \widehat{\varphi}_+ = \widehat{\varphi}_+(u_0). \tag{3.32}$$

By virtue of hypothesis $H(f)_1(vi)$, for a given $\varepsilon > 0$, we can find $\delta = \delta(\varepsilon) \in (0, a_+)$, such that

$$\frac{1}{p}\left(-\eta(z) - \varepsilon\right)|\zeta|^q \leqslant F(z,\zeta) \quad \text{for almost all } z \in \Omega, \text{ all } |\zeta| \leqslant \delta. \tag{3.33}$$

Let $\widehat{u}_1$ be the L^p-normalized principal eigenfunction of $\left(-\Delta_p,\ W_0^{1,p}(\Omega)\right)$ (we know that $\widehat{u}_1 \in \operatorname{int} C_+$; see Section 2). We choose $t > 0$ small enough, such that

$$t\widehat{u}_1 \in [0,\delta] \quad \forall z \in \overline{\Omega}.$$

Then, since $\delta < a_+$ (see (3.31)) and using Corollary 3.3 and (3.33), we have

$$\begin{aligned}
\widehat{\varphi}_+(t\widehat{u}_1) &= \int_\Omega G(z, t\nabla \widehat{u}_1)\,dz - \int_\Omega \widehat{F}(z, t\widehat{u}_1)\,dz \\
&= \int_\Omega G(z, t\nabla \widehat{u}_1)\,dz - \frac{t^q}{q}\int_\Omega \beta \widehat{u}_1^q\,dz - \int_\Omega F(z, t\widehat{u}_1)\,dz \\
&\leqslant \frac{c_1}{p(p-1)} t^p \|\nabla \widehat{u}_1\|_p^p - \frac{t^q}{q}\int_\Omega (\beta - \eta)\widehat{u}_1^q\,dz + \frac{\varepsilon}{q} t^q \|\widehat{u}_1\|_q^q \\
&= \frac{c_1}{p(p-1)} t^p \lambda_1 - \frac{t^q}{q}\left[\int_\Omega (\beta - \eta)\widehat{u}_1^q\,dz + \varepsilon \|\widehat{u}_1\|_q^q\right].
\end{aligned} \tag{3.34}$$

But $\beta(z) - \eta(z) \leqslant 0$ for almost all $z \in \Omega$ and the inequality is strict on a set of positive measure. Also $\widehat{u}_1(z) > 0$ for all $z \in \Omega$ (since $\widehat{u}_1 \in \operatorname{int} C_+$). Hence

$$\sigma_1 = \int_\Omega (\beta - \eta)\widehat{u}_1^q\,dz < 0.$$

Choosing $\varepsilon, t > 0$ small enough and since $q < p$, from (3.34), it follows that

$$\widehat{\varphi}_+(t\widehat{u}_1) < 0,$$

so

$$\widehat{m}_+ = \widehat{\varphi}_+(u_0) < 0 = \widehat{\varphi}_+(0)$$

(see (3.32)) and finally

$$u_0 \neq 0. \tag{3.35}$$

From (3.32), we have

$$\widehat{\varphi}_+'(u_0) = 0,$$

so

$$V(u_0) = \widehat{N}_+(u_0), \tag{3.36}$$

where

$$\widehat{N}_+(u)(\cdot) = \widehat{f}_+(\cdot, u(\cdot)) \quad \forall u \in W_0^{1,p}(\Omega).$$

On (3.36) we act with $-u_0^- \in W_0^{1,p}(\Omega)$ and obtain

$$\langle V(u_0), -u_0^- \rangle = \int_\Omega \big(a(z, \nabla u_0), -\nabla u_0^-\big)_{\mathbb{R}^N} \, dz = 0$$

(see (3.31)), so using also Lemma 3.2(c), we have

$$\frac{c_0}{p-1} \|\nabla u_0^-\|_p^p \leqslant 0$$

and thus

$$u_0^- = 0, \quad \text{i.e., } u_0 \geqslant 0$$

(see (3.35)). Also, on (3.36) we act with $(u_0 - a_+)^+ \in W_0^{1,p}(\Omega)$. Then

$$\langle V(u_0),\ (u_0 - a_+)^+ \rangle = \int_\Omega \Big(\beta a_+^{q-1} + f(z, a_+)\Big) (u_0 - a_+)^+ dz$$

(see (3.31)), so

$$\int_{\{u_0 > a_+\}} \big(a(z, \nabla u_0\ \nabla u_0\big)_{\mathbb{R}^N} \, dz = \int_{\{u_0 > a_+\}} \Big(\beta a_+^{q-1} + f(z, a_+)\Big) (u_0 - a_+) \, dz.$$

Thus, using also Lemma 3.2(c) and hypothesis $H(f)_1(v)$, we have

$$\int_\Omega \frac{c_0}{p-1} \big\|\nabla(u_0 - a_+)\big\|^p \, dz \leqslant 0$$

and so

$$(u_0 - a_+)^+ = 0 \quad \text{i.e., } u_0 \leqslant a_+.$$

Therefore we have

$$0 \leqslant u_0 \leqslant a_+$$

and so by virtue of (3.31), equation (3.36) becomes

$$V(u_0) = \beta |u_0|^{q-2} u_0 + N(u_0),$$

with

$$N(u)(\cdot) = f(\cdot, u(\cdot)) \quad \forall u \in W_0^{1,p}(\Omega),$$

so

$$\begin{cases} -\operatorname{div} a\,(z, \nabla u_0(z)) = \beta(z) u_0(z)^{p-1} + f\,(z, u_0(z)) \text{ in } \Omega, \\ u|_{\partial\Omega} = 0, \end{cases}$$

Thus, recalling that $\|\beta\|_\infty \leqslant \beta^*$ and using hypothesis $H(f)_1(v)$, we have

$$
\begin{aligned}
-\operatorname{div} a\,(z, \nabla u_0(z)) + \xi^* u_0(z)^{p-1} &= \beta(z) u_0(z)^{p-1} + f\,(z, u_0(z)) + \xi^* u_0(z)^{p-1} \\
&\leqslant \beta^* a_+^{p-1} + f(z, a_+) + \xi^* a_+^{p-1} \\
&\leqslant \xi^* a_+^{p-1} \quad \text{for almost all } z \in \Omega \qquad (3.37)
\end{aligned}
$$

and so

$$
-\operatorname{div} a\,(z, \nabla u_0(z)) + \xi^* \left(u_0(z)^{p-1} - a_+^{p-1}\right) \leqslant 0 \quad \text{for almost all } z \in \Omega.
$$

Invoking Theorem 1 of Serrin [**47**] (the tangency principle; see also Pucci-Serrin [**46**, p. 35]), we obtain

$$
u_0(z) < a_+ \quad \text{for all } z \in \overline{\Omega}. \tag{3.38}
$$

Moreover, from (3.37) and hypothesis $H(f)_1(v)$, we have

$$
-\operatorname{div} a\,(z, \nabla u_0(z)) \leqslant \xi^* u_0(z)^{p-1} \quad \text{for almost all } z \in \Omega.
$$

Invoking Theorem 3 of Montenegro [**38**] (see also Theorem 1.2 of Zhang [**50**]), we obtain

$$
u_0 \in \operatorname{int} C_+. \tag{3.39}
$$

From (3.38) and (3.39), we infer that we can find $\varrho > 0$ small enough, such that

$$
0 < u(z) < a_+ \quad \forall z \in \overline{\Omega},\ u \in \overline{B}_\varrho^{C_0^1(\overline{\Omega})}(u_0),
$$

where

$$
\overline{B}_\varrho^{C_0^1(\overline{\Omega})}(u_0) = \left\{u \in C_0^1(\overline{\Omega}) :\ \|u - u_0\|_{C_0^1(\overline{\Omega})} \leqslant \varrho\right\}.
$$

Note that

$$
\widehat{\varphi}_+\Big|_{\overline{B}_\varrho^{C_0^1(\overline{\Omega})}(u_0)} = \varphi\Big|_{\overline{B}_\varrho^{C_0^1(\overline{\Omega})}(u_0)}.
$$

So, it follows that u_0 is a local $C_0^1(\overline{\Omega})$-minimizer of φ. Invoking Proposition 3.6, we have that u_0 is a local $W_0^{1,p}(\Omega)$-minimizer of φ.

Similarly, we consider the following truncation of the reaction

$$
\widehat{f}_-(z, \zeta) = \begin{cases} \beta(z)|a_-|^{q-2}a_- + f(z, a_-) & \text{if } \zeta < a_-, \\ \beta(z)|\zeta|^{q-2}\zeta + f(z, \zeta) & \text{if } \zeta \in [a_-, 0], \\ 0 & \text{if } 0 < \zeta. \end{cases} \tag{3.40}
$$

We set

$$
\widehat{F}_-(z, \zeta) = \int_0^\zeta \widehat{f}_-(z, s)\, ds
$$

and introduce the C^1-functional $\widehat{\varphi}_-: W_0^{1,p}(\Omega) \longrightarrow \mathbb{R}$, defined by

$$\widehat{\varphi}_-(u) = \int_\Omega G(z, \nabla u)\, dz - \int_\Omega \widehat{F}_-(z,u)\, dz \quad \forall u \in W_0^{1,p}(\Omega).$$

Working as above, using this time $\widehat{\varphi}_-$ and (3.40), we obtain $v_0 \in -\operatorname{int} C_+$, a second constant sign smooth solution of (1.1) which is a local minimizer of φ. □

Next using variational methods, we produce two more nontrivial, constant sign smooth solutions of (1.1). To do this we need to strengthen the hypotheses on the map $a(z,y)$ as follows.

$\underline{H(a)'}$ $a(z,y) = h(z, \|y\|)\, y$ for all $(z,y) \in \overline{\Omega} \times \mathbb{R}^N$, with $h(z,r) > 0$ for all $(z,r) \in \overline{\Omega} \times (0,+\infty)$ and

(i)-(iii): the same as $H(a)(i) - (iii)$;

(iv): $G(z,0) = 0$ for all $z \in \overline{\Omega}$ and there exists $\widehat{\sigma} \in L^1(\Omega)$, such that

$$\widehat{\sigma}(z) \leqslant pG(z,y) - (a(z,y),\ y)_{\mathbb{R}^N} \qquad \forall (z,y) \in \overline{\Omega} \times \mathbb{R}^N.$$

Remark 3.10. *All the maps $a_k(z,y)$, $k = 1,2,3,4,5$, given in Example 3.4 satisfy this extra condition $H(a)'(iv)$.*

Recall that a set $S \subseteq W^{1,p}(\Omega)$ is said to be downward (respectively upward) directed, if for any $u, v \in S$ we can find $y \in S$, such that $y \leqslant u$, $y \leqslant v$ (respectively $u \leqslant y$, $v \leqslant y$). The next lemma can be proved as in Gasiński-Papageorgiou [**26**], with minor modifications.

Lemma 3.11. ***(a)*** *The set of lower solutions for problem (1.1) is upward directed. In fact, if $\underline{u}_1, \underline{u}_2 \in W^{1,p}(\Omega)$ are two lower solutions for problem (1.1), then so is $\underline{u} = \max\{\underline{u}_1, \underline{u}_2\} \in W^{1,p}(\Omega)$;*
(b) *The set of upper solutions for problem (1.1) is downward directed. In fact, if $\overline{u}_1, \overline{u}_2 \in W^{1,p}(\Omega)$ are two upper solutions for problem (1.1), then so is $\overline{u} = \min\{\overline{u}_1, \overline{u}_2\} \in W^{1,p}(\Omega)$;*

Using this lemma we will produce extremal constant sign solutions in the ordered intervals:

$$[0, a_+] = \left\{ u \in W_0^{1,p}(\Omega) :\ 0 \leqslant u(z) \leqslant a_+ \text{ for almost all } z \in \Omega \right\},$$
$$[a_-, 0] = \left\{ u \in W_0^{1,p}(\Omega) :\ a_- \leqslant u(z) \leqslant 0 \text{ for almost all } z \in \Omega \right\}.$$

Proposition 3.12. *If hypotheses $H(a)'$ and $H(f)_1$ hold, then problem (1.1) has a biggest solution in $[0, a_+]$ and a smallest solution in $[a_-, 0]$.*

Proof. Let S_+ be the set of solutions of problem (1.1) in $[0, a_+]$. From Proposition 3.9, we know that $S_+ \neq \emptyset$.

First we show that S_+ is upward directed. To this end let $u_1, u_2 \in S_+$. Evidently u_1, u_2 are also lower solutions for problem (1.1). So, by virtue of Lemma 3.11(a), $\underline{u} = \max\{u_1, u_2\} \in W_0^{1,p}(\Omega)$ is a lower solution too and $\underline{u} \leqslant a_+$. Note that $\overline{u} = a_+$ is an upper solution for problem (1.1) (see hypothesis $H(f)_1(v)$). Hence truncating the reaction term at $\{\underline{u}(z), a_+\}$ and proceeding as before using the direct method (see also Gasiński-Papageorgiou [**24**]), we obtain a solution $\widehat{u}_0 \in [\underline{u}, a_+]$ of (1.1). Then

$$u_1 \leqslant \widehat{u}_0, \quad u_2 \leqslant \widehat{u}_0 \quad \text{and} \quad \widehat{u}_0 \in S_+.$$

This proves that S_+ is upward directed.

Next we show that S_+ has a maximal element. To this end, let $C \subseteq S_+$ be a chain, i.e., a totally ordered subset of S_+. According to Dunford-Schwartz [**19**, p. 336], we can find a sequence $\{u_n\}_{n\geqslant 1} \subseteq C$, such that

$$\sup C = \sup_{n\geqslant 1} u_n.$$

Because C is totally ordered, we may assume that the sequence $\{u_n\}_{n\geqslant 1}$ is increasing. Since $\{u_n\}_{n\geqslant 1} \subseteq S_+$, we have

$$V(u_n) = \beta u_n^{q-1} + N(u_n) \quad \forall n \geqslant 1, \tag{3.41}$$

so

$$\frac{c_0}{p-1}\|\nabla u_n\|_p^p \leqslant a_+^{q-1}\|\beta\|_1 + \int_\Omega f(z, u_n)u_n\, dz \quad \forall n \geqslant 1$$

and thus the sequence $\{u_n\}_{n\geqslant 1} \subseteq W_0^{1,p}(\Omega)$ is bounded (see also hypothesis $H(f)_1(iii)$).

Therefore, passing to a subsequence if necessary, we may assume that

$$u_n \longrightarrow u \quad \text{weakly in } W_0^{1,p}(\Omega), \tag{3.42}$$

$$u_n \longrightarrow u \quad \text{in } L^r(\Omega). \tag{3.43}$$

On (3.41) we act with $u_n - u \in W_0^{1,p}(\Omega)$ and pass to the limit as $n \to +\infty$. Then, using (3.42), we have

$$\lim_{n\to+\infty} \langle V(u_n), u_n - u\rangle = 0$$

and so, using Proposition 3.5, we see that

$$u_n \longrightarrow u \quad \text{in } W_0^{1,p}(\Omega).$$

So, if in (3.41) we pass to the limit as $n \to +\infty$, then

$$V(u) = \beta|u|^{q-2}u + N(u) \quad \text{and} \quad u = \sup C,\ u \in S_+.$$

Invoking the Kuratowski-Zorn lemma, we infer that S_+ has a maximal element u^*. We show that $u^* \in S_+$ is the biggest element of S_+. Indeed, let $u \in S_+$. Then, since S_+ is upward directed, we can find $\widehat{u} \in S_+$, such that $\max\{u^*, u\} \leqslant \widehat{u}$. The maximality of u^* implies that $\widehat{u} = u^*$.

Similarly for S_-, the set of solutions of problem (1.1) in $[a_-, 0]$, using this time Lemma 3.11(b), we obtain $v_* \in S_-$, the smallest solution of problem (1.1) in $[a_-, 0]$. □

Now we are ready to produce two more nontrivial, smooth constant sign solutions. In addition to $u_0 \in \operatorname{int} C_+$ and $v_0 \in -\operatorname{int} C_+$ obtained in Proposition 3.9.

THEOREM 3.13. *If hypotheses $H(a)'$ and $H(f)_1$ hold and $\|\beta\|_\infty \leqslant \beta^*$, then problem (1.1) has at least four smooth constant sign solutions*

$$u_0, \widehat{u} \in \mathit{int} C_+, \ u_0 \leqslant \widehat{u}, \ u_0 \neq \widehat{u} \text{ and } u_0(z) < a_+ \text{ for all } z \in \overline{\Omega}$$

and

$$v_0, \widehat{v} \in -\mathit{int} C_+, \ \widehat{v} \leqslant v_0, \ \widehat{v} \neq v_0 \text{ and } a_- < v_0(z) \text{ for all } z \in \overline{\Omega}.$$

PROOF. From Proposition 3.9 we already have two smooth constant sign solutions

$$u_0 \in \operatorname{int} C_+, \quad \text{and} \quad v_0 \in -\operatorname{int} C_+$$

with

$$a_- < v_0(z) < 0 < u_0(z) < a_+ \quad \text{for all } z \in \overline{\Omega}.$$

First we will produce a second positive smooth solution distinct from u_0. For this purpose we introduce the following truncation of the reaction term:

$$\widehat{f}_0^+(z,\zeta) = \begin{cases} \beta(z)u_0(z)^{q-1} + f(z,u_0(z)) & \text{if } \zeta \leqslant u_0(z), \\ \beta(z)\zeta^{q-1} + f(z,\zeta) & \text{if } u_0(z) < \zeta. \end{cases} \tag{3.44}$$

This is a Carathéodory function. Let

$$\widehat{F}_0^+(z,\zeta) = \int_0^\zeta \widehat{f}_0^+(z,s)\,ds$$

and consider the C^1-functional $\widehat{\varphi}_0^+ : W_0^{1,p}(\Omega) \longrightarrow \mathbb{R}$, defined by

$$\widehat{\varphi}_0^+(u) = \int_\Omega G(z,\nabla u)\,dz - \int_\Omega \widehat{F}_0^+(z,u)\,dz \quad \forall u \in W_0^{1,p}(\Omega).$$

CLAIM 1. $\widehat{\varphi}_0^+$ satisfies the C-condition.

We consider a sequence $\{u_n\}_{n\geqslant 1} \subseteq W^{1,p}(\Omega)$, such that

$$\left|\widehat{\varphi}_0^+(u_n)\right| \leqslant M_4 \quad \forall n \geqslant 1, \tag{3.45}$$

for some $M_4 > 0$ and

$$(1+\|u_n\|)(\widehat{\varphi}_0^+)'(u_n) \longrightarrow 0 \quad \text{in } W^{-1,p'}(\Omega). \tag{3.46}$$

From (3.45), we have

$$\left|\langle(\widehat{\varphi}_0^+)'(u_n),\ h)\rangle\right| \leqslant \frac{\varepsilon_n}{1+\|u_n\|}\|h\| \quad \forall h \in W_0^{1,p}(\Omega),$$

with $\varepsilon_n \searrow 0$, so

$$\left|\langle V(u_n),\ h\rangle - \int_\Omega \widehat{N}_0^+(u_n)h\,dz\right| \leqslant \frac{\varepsilon_n}{1+\|u_n\|}\|h\| \quad \forall n \geqslant 1, \tag{3.47}$$

where

$$\widehat{N}_0^+(u)(\cdot) = \widehat{f}_0^+(\cdot,u(\cdot)) \quad \forall u \in W_0^{1,p}(\Omega).$$

In (3.47), we choose $h = -u_n^- \in W_0^{1,p}(\Omega)$. Then, from (3.44), we have

$$\frac{c_0}{p-1}\|\nabla u_n^-\|_p^p \leqslant c_8\|\nabla u_n^-\|_p \quad \forall n \geqslant 1$$

for some $c_8 > 0$. Thus the sequence $\{u_n^-\}_{n\geqslant 1} \subseteq W_0^{1,p}(\Omega)$ is bounded.

Next, in (3.47) we choose $h = u_n^+ \in W_0^{1,p}(\Omega)$. Then, for all $n \geqslant 1$, we have

$$\begin{aligned} &-\int_\Omega \big(a(z,\nabla u_n),\ \nabla u_n^+\big)_{\mathbb{R}^n}\,dz + \int_{\{u_n>u_0\}} \beta(u_n^+)^q\,dz \\ &+\int_{\{u_n\leqslant u_0\}} f(z,u_0)u_n^+\,dz + \int_{\{u_n>u_0\}} f(z,u_n)u_n^+\,dz \leqslant \varepsilon_n. \end{aligned} \tag{3.48}$$

Also from (3.45) and the boundedness of the sequence $\{u_n^-\}_{n\geqslant 1} \subseteq W_0^{1,p}(\Omega)$, we have

$$\begin{aligned} &\int_\Omega pG(z,\ \nabla u_n^+)\,dz - \int_{\{u_n>u_0\}} \frac{p}{q}\beta(u_n^+)^q\,dz - \int_{\{u_n\leqslant u_0\}} pf(z,u_0)u_n^+\,dz \\ &-\int_{\{u_n>u_0\}} p\big(F(z,u_n^+) - F(z,u_0)\big)\,dz \quad \leqslant M_5 \quad \forall n \geqslant 1, \end{aligned} \tag{3.49}$$

for some $M_5 > 0$. We add (3.48) and (3.49) and we obtain

$$\begin{aligned} &\int_\Omega \Big[pG(z,\ \nabla u_n^+) - \big(a(z,\nabla u_n),\ \nabla u_n^+\big)_{\mathbb{R}^n}\Big]dz \\ &\quad + \int_{\{u_n>u_0\}} \big[f(z,u_n)u_n^+ - pF(z,u_n^+)\big]dz \\ &\quad \leqslant M_6 + \left(\frac{p}{q}-1\right)\int_\Omega \beta(u_n^+)^q\,dz \quad \forall n \geqslant 1, \end{aligned}$$

for some $M_6 > 0$ and using hypotheses $H(a)'(iv)$ and $H(f)_1(iii)$, we have

$$\int_\Omega \big[f(z,u_n)u_n^+ - pF(z,u_n^+)\big]dz \leqslant M_7 + \left(\frac{p}{q}-1\right)\int_\Omega \beta(u_n)^q\,dz \quad \forall n \geqslant 1, \tag{3.50}$$

for some $M_7 > 0$.

By virtue of hypothesis $H(f)_1(iv)$ we can find $\beta_1 \in (0,\beta_0)$ and $M_8 \geqslant \|u_0\|_\infty > 0$, such that

$$\beta_1\zeta^\mu \leqslant f(z,\zeta)\zeta - pF(z,\zeta) \quad \text{for almost all } z \in \Omega, \text{ all } \zeta \geqslant M_8. \tag{3.51}$$

On the other hand, hypothesis $H(f)_1(iii)$ implies that

$$\big|f(z,\zeta)\zeta - pF(z,\zeta)\big| \leqslant c_9 \quad \text{for almost all } z \in \Omega, \text{ all } \zeta \in [0,M_8], \tag{3.52}$$

for some $c_9 > 0$. Combining (3.51) and (3.52), we infer that

$$\beta_1\zeta^\mu - c_{10} \leqslant f(z,\zeta)\zeta - pF(z,\zeta) \quad \text{for almost all } z \in \Omega, \text{ all } \zeta \geqslant 0, \tag{3.53}$$

for some $c_{10} > 0$.

We return to (3.50) and use (3.53). Then

$$\beta_1 \int_\Omega (u_n^+)^\mu \, dz \leqslant M_9 + \left(\frac{p}{q} - 1\right) \int_\Omega \beta (u_n^+)^q \, dz \quad \forall n \geqslant 1,$$

for some $M_9 > 0$ and so the sequence $\{u_n^+\}_{n\geqslant 1} \subseteq L^\mu(\Omega)$ is bounded (since $\mu > q$; see hypothesis $H(f)_1(iv)$).

It is clear from hypothesis $H(f)_1(iv)$ that we can always assume that $\mu \leqslant r < p^*$. Then we can find $t \in [0,1)$, such that $\frac{1}{r} = \frac{1-t}{\mu} + \frac{t}{p^*}$. Invoking the interpolation inequality (see e.g., Gasiński-Papageorgiou [**25**, p. 905]), we have

$$\|u_n^+\|_r \leqslant \|u_n^+\|_\mu^{1-t} \|u_n^+\|_{p^*}^t \quad \forall n \geqslant 1,$$

so, using also the boundedness of the sequence $\{u_n^+\}_{n\geqslant 1} \subseteq L^\mu(\Omega)$, we have

$$\|u_n^+\|_r^r \leqslant M_{10} \|u_n^+\|_{p^*}^{tr} \quad \forall n \geqslant 1, \tag{3.54}$$

for some $M_{10} > 0$.

From (3.47) with $h = u_n^+ \in W_0^{1,p}(\Omega)$, we have

$$\left| \int_\Omega \left(a(z, \nabla u_n),\ \nabla u_n^+\right)_{\mathbb{R}^N} dz - \int_\Omega \widehat{f}_0^+(z, u_n) u_n^+ \, dz \right| \leqslant \varepsilon_n \quad \forall n \geqslant 1, \tag{3.55}$$

From hypothesis $H(f)_1(iii)$ and (3.44), we see that

$$\left|\widehat{f}_0^+(z,\zeta)\right| \leqslant c_{11}(1 + |\zeta|^{r-1}) \quad \text{for almost all } z \in \Omega, \text{ all } \zeta \in \mathbb{R},$$

for some $c_{11} > 0$. Using this growth condition in (3.55), if $p \neq N$, we obtain

$$\begin{aligned} \frac{c_0}{p-1} \|\nabla u_n^+\|_p^p &\leqslant M_{11} + c_{12}\left(\|u_n^+\| + \|u_n^+\|_r^r\right) \\ &\leqslant M_{11} + c_{13}\left(\|u_n^+\| + \|u_n^+\|^{tr}\right) \quad \forall n \geqslant 1, \end{aligned} \tag{3.56}$$

for some $M_{11}, c_{12}, c_{13} > 0$. A simple calculation shows that

$$\left[\ \mu > (r-p) \max\left\{\frac{N}{p}, 1\right\}\ \right] \iff [tr < p].$$

So, from (3.56), it follows that the sequence $\{u_n^+\}_{n\geqslant 1} \subseteq W_0^{1,p}(\Omega)$ is bounded, so from the boundedness of the sequence $\{u_n^-\}_{n\geqslant 1} \subseteq W_0^{1,p}(\Omega)$, also the sequence $\{u_n\}_{n\geqslant 1} \subseteq W_0^{1,p}(\Omega)$.

Similarly, if $p = N$, by replacing in the above argument p^* by $r > 0$ large enough, we obtain the boundedness of the sequence $\{u_n\}_{n\geqslant 1} \subseteq W_0^{1,p}(\Omega)$.

Passing to a subsequence if necessary, we may assume that

$$u_n \longrightarrow u \text{ weakly in } W_0^{1,p}(\Omega), \tag{3.57}$$

$$u_n \longrightarrow u \text{ in } L^r(\Omega). \tag{3.58}$$

In (3.47), we choose $h = u_n - u \in W_0^{1,p}(\Omega)$ and pass to the limit as $n \to +\infty$. Using (3.58), we obtain

$$\lim_{n\to+\infty} \langle V(u_n),\ u_n - u\rangle = 0,$$

so, from Proposition 3.5, we have

$$u_n \longrightarrow u \quad \text{in } W_0^{1,p}(\Omega).$$

This proves Claim 1.

CLAIM 2. $u_0 \in \operatorname{int} C_+$ is a local minimizer of $\widehat{\varphi}_0^+$.

We introduce the following truncation of the nonlinearity

$$\overline{f}_0^+(z,\zeta) = \begin{cases} \widehat{f}_0^+(z,\zeta) & \text{if } \zeta \leqslant a_+, \\ \widehat{f}_0^+(z,a_+) & \text{if } a_+ < \zeta. \end{cases} \tag{3.59}$$

This is a Carathéodory function. We set

$$\overline{F}_0^+(z,\zeta) = \int_0^\zeta \overline{f}_0^+(z,s)\,ds$$

and consider the C^1-functional $\overline{\varphi}_0^+ \colon W_0^{1,p}(\Omega) \longrightarrow \mathbb{R}$, defined by

$$\overline{\varphi}_0^+(u) = \int_\Omega G(z,\nabla u)\,dz - \int_\Omega \overline{F}_0^+(z,u)\,dz \quad \forall u \in W^{1,p}(\Omega).$$

Evidently $\overline{\varphi}_0^+$ is coercive and sequentially weakly lower semicontinuous. Hence by virtue of the Weierstrass theorem, we can find $\widehat{u}_0 \in W_0^{1,p}(\Omega)$, such that

$$\overline{m}_0^+ = \inf \overline{\varphi}_0^+ = \overline{\varphi}_0^+(\widehat{u}_0),$$

so

$$(\overline{\varphi}_0^+)'(\widehat{u}_0) = V(\widehat{u}_0) - \overline{N}_0^+(\widehat{u}_0) = 0, \tag{3.60}$$

where

$$\overline{N}_0^+(u)(\cdot) = \overline{f}_0^+(\cdot,u(\cdot)) \quad \forall u \in W_0^{1,p}(\Omega).$$

Also we have

$$V(u_0) = \beta u_0^{q-1} + N(u_0). \tag{3.61}$$

Subtracting (3.61) from (3.60), we obtain

$$V(\widehat{u}_0) - V(u_0) = \overline{N}_0^+(u_0) - N(u_0) - \beta u_0^{q-1}. \tag{3.62}$$

On (3.62) we act with the test function $(u_0 - \widehat{u}_0)^+ \in W_0^{1,p}(\Omega)$. Since $u_0 \leqslant a_+$ (see Proposition 3.5), from (3.59), we have

$$\big\langle V(\widehat{u}_0) - V(u_0),\ (u_0 - \widehat{u}_0)^+ \big\rangle = 0,$$

so

$$\int_{\{u_0 > \widehat{u}_0\}} \big(a(z,\nabla\widehat{u}_0) - a(z,\nabla u_0),\ \nabla u_0 - \nabla\widehat{u}_0\big)_{\mathbb{R}^N}\,dz = 0 \tag{3.63}$$

By virtue of the strict monotonicity of $a(z,\cdot)$ (see Lemma 3.2(a)), from (3.63), we infer that

$$\left|\{u_0 > \widehat{u}_0\}\right|_N = 0, \quad \text{i.e., } u_0 \leqslant \widehat{u}_0. \tag{3.64}$$

On the other hand, if on (3.60) we act with $(\widehat{u}_0 - a_+) \in W_0^{1,p}(\Omega)$, using Lemma 3.2(c), (3.59) and hypothesis $H(f)_1(v)$, we obtain

$$\begin{aligned}\frac{c_0}{p-1}\int_\Omega \left\|\nabla(\widehat{u}-a_+)^+\right\|^p dz &\leqslant \int_{\{\widehat{u}_0>a_+\}} \left(a(z,\nabla\widehat{u}_0,\ \nabla\widehat{u}_0\right)_{\mathbb{R}^N} dz \\ &= \int_\Omega \Big(\beta a_+^{q-1} + f(z,a_+)\Big)(\widehat{u}-a_+)\,dz \leqslant 0.\end{aligned}$$

Then $\widehat{u}_0 \leqslant a_+$ and moreover, using Theorem 1 of Serrin [**47**] and hypothesis $H(f)_1(v)$ as before, we have

$$\widehat{u}_0(z) < a_+ \quad \forall z \in \overline{\Omega}. \tag{3.65}$$

Because of (3.44), (3.59), (3.64) and (3.65), equation (3.60) becomes

$$V(\widehat{u}_0) = \beta\widehat{u}_0^{q-1} + N(\widehat{u}_0)$$

and so $\widehat{u}_0$ is a solution of problem (1.1).

By virtue of Proposition 3.12, we may assume that $u_0 \in \operatorname{int} C_+$ is the biggest solution of (1.1) in $[0,a_+]$. Then from (3.64), (3.65) and the fact that $\widehat{u}_0$ is a solution of problem (1.1), we infer that $\widehat{u}_0 = u_0$. So, we conclude that u_0 is a local minimizer of $\widehat{\varphi}_0^+$ (see (3.65)). This proves Claim 2.

Without any loss of generality, we may assume that $u_0 \in \operatorname{int} C_+$ is an isolated critical point of $\widehat{\varphi}_0^+$. Indeed, otherwise we have a whole sequence $\{u_n\}_{n\geqslant 1}$ of distinct critical points of $\widehat{\varphi}_0^+$, such that $u_n \longrightarrow u_0$ in $W_0^{1,p}(\Omega)$. As above we verify that $u_0 \leqslant u_n$ for all $n \geqslant 1$ and from (3.42), it follows that u_n are solutions of (1.1) for $n \geqslant 1$. Hence we have produced a whole sequence of distinct solutions in $\operatorname{int} C_+$ (nonlinear regularity theory) and we are done. So, we can assume that u_0 is an isolated critical point of $\widehat{\varphi}_0^+$. Arguing as in the proof of Proposition 6 of Motreanu-Motreanu-Papageorgiou [**39**], we can find $\varrho > 0$ small enough, such that

$$\widehat{\varphi}^+(u_0) < \inf\left\{\widehat{\varphi}_0^+(u) : \ \|u-u_0\| = \varrho\right\} = \widehat{\eta}_0^+. \tag{3.66}$$

Since $\widehat{u}_1 \in \operatorname{int} C_+$, using Lemma 2.3, we can find $t_1 > 0$, such that $u_0 \leqslant t\widehat{u}_1$ for all $t > t_1$. Hence, using also Corollary 3.3 and hypothesis $H(f)_1(iii)$, we have

$$\begin{aligned}\widehat{\varphi}_0^+(t\widehat{u}_1) &= \int_\Omega G(z,t\nabla\widehat{u}_1)\,dz - \int_\Omega \widehat{F}_0^+(z,t\widehat{u}_1)\,dz \\ &\leqslant \int_\Omega G(z,t\nabla\widehat{u}_1)\,dz - \int_\Omega \big(F(z,t\widehat{u}_1) - F(z,u_0) + f(z,u_0)u_0\big)\,dz \\ &\leqslant \frac{c_1t^p}{p(p-1)}\|\nabla\widehat{u}_1\|_p^p - \int_\Omega F(z,t\widehat{u}_1) + M_{12},\end{aligned} \tag{3.67}$$

for some $M_{12} > 0$. By virtue of hypothesis $H(f)_1(iv)$, for a given $\xi > 0$, we can find $M_{13} = M_{13}(\xi) > 0$, such that

$$F(z,\zeta) \geqslant \frac{\xi}{p}|\zeta|^p \quad \text{for almost all } z \in \Omega, \text{ all } |\zeta| \geqslant M_{13}. \tag{3.68}$$

Then, using (3.68) and hypothesis $H(f)_1(iii)$, we have

$$\begin{aligned} \int_\Omega F(z,t\widehat{u}_1)\,dz &= \int_{\{t\widehat{u}_1 \geqslant M_{13}\}} F(z,t\widehat{u}_1)dz + \int_{\{0\leqslant t\widehat{u}_1 < M_{13}\}} F(z,t\widehat{u}_1)dz \\ &\geqslant \frac{\xi}{p}t^p \int_{\{t\widehat{u}_1 \geqslant M_{13}\}} \widehat{u}_1^p\,dz - M_{14}, \end{aligned} \tag{3.69}$$

for some $M_{14} > 0$. Using (3.69) in (3.67) and recalling that $\|\nabla \widehat{u}_1\|_p^p = \lambda_1 \|\widehat{u}_1\|_p^p$, we obtain

$$\widehat{\varphi}_0^+(t\widehat{u}_1) \leqslant \frac{t^p}{p}\left(\frac{c_1\lambda_1}{p-1} - \xi\right) \int_{\{t\widehat{u}_1 \geqslant M_{13}\}} \widehat{u}_1^p\,dz + M_{15}, \tag{3.70}$$

for some $M_{15} > 0$. Note that

$$\chi_{\{t\widehat{u}_1 \geq M_{13}\}}(z) \longrightarrow \chi_\Omega(z) \quad \text{for almost all } z \in \Omega.$$

Hence

$$\int_{\{t\widehat{u}_1 \geqslant M_{13}\}} \widehat{u}_1^p\,dz \longrightarrow \|\widehat{u}_1\|_p^p \quad \text{as } t \to +\infty.$$

Since $\xi > 0$ was arbitrary, choosing $\xi > \frac{c_1\lambda_1}{p-1}$, from (3.70) we infer that

$$\widehat{\varphi}_0^+(t\widehat{u}_1) \longrightarrow -\infty \quad \text{as } t \to +\infty.$$

Therefore, using also (3.66), for $t > 0$ large enough, we have

$$\widehat{\varphi}_0^+(t\widehat{u}_1) \leqslant \widehat{\varphi}_0^+(u_0) < \widehat{\eta}_0^+ \quad \text{and} \quad \|t\widehat{u}_1\| > \varrho. \tag{3.71}$$

Because of Claim 1 and (3.66) and (3.71), we see that we can apply the mountain pass theorem (see Theorem 2.1) and obtain $\widehat{u} \in W_0^{1,p}(\Omega)$, such that

$$\widehat{\varphi}_0^+(u_0) < \widehat{\eta}_0^+ \leqslant \widehat{\varphi}_0^+(\widehat{u}) \tag{3.72}$$

and

$$(\widehat{\varphi}_0^+)'(\widehat{u}) = 0. \tag{3.73}$$

From (3.72), it follows that $\widehat{u} \neq u_0$. From (3.73), we have

$$V(\widehat{u}) = \widehat{N}_0^+(\widehat{u}). \tag{3.74}$$

Subtracting (3.61) from (3.74) and acting on the resulting equation with $(u_0 - \widehat{u})^+ \in W_0^{1,p}(\Omega)$, as before we obtain that

$$u_0 \leqslant \widehat{u}, \quad u_0 \neq \widehat{u}.$$

Hence (3.74) becomes

$$V(\widehat{u}) = \beta\widehat{u}^{q-1} + N(\widehat{u}),$$

so $\widehat{u}$ is a solution of (1.1) and $\widehat{u} \in \operatorname{int} C_+$ (nonlinear regularity theory).

If we consider the following truncation of the reaction term:

$$\widehat{f_0^-}(z,\zeta) = \begin{cases} \beta(z)|v_0(z)|^{q-2}v_0(z) + f(z,v_0(z)) & \text{if} \quad \zeta \leqslant v_0(z), \\ \beta(z)\zeta^{q-2}\zeta + f(z,\zeta) & \text{if} \quad v_0(z) < \zeta \end{cases}$$

we set

$$\widehat{F_0^-}(z,\zeta) = \int_0^\zeta \widehat{f_0^-}(z,s)\,ds$$

and work with the C^1-functional $\widehat{\varphi_0^-} : W_0^{1,p}(\Omega) \longrightarrow \mathbb{R}$, defined by

$$\widehat{\varphi_0^-}(u) = \int_\Omega G(z,\nabla u)\,dz - \int_\Omega \widehat{F_0^-}(z,u)\,dz \quad \forall u \in W_0^{1,p}(\Omega),$$

as above we obtain a second negative solution

$$\widehat{v} \in -\operatorname{int} C_+, \quad \widehat{v} \leqslant v_0, \quad \widehat{v} \neq v_0. \qquad \square$$

We can have another such multiplicity result producing four smooth solutions of constant sign (two positive and two negative), by modifying the behaviour of f near zero, keeping the $(p-1)$-superlinear growth near $\pm\infty$.

More precisely, the new hypotheses on the nonlinearity f are the following.

$\underline{H(f)_2}$ $f\colon \Omega \times \mathbb{R} \longrightarrow \mathbb{R}$ is a function, such that

(i)-(iv): the same as $H(f)_1(i)-(iv)$;

(v): we have

$$\lim_{\zeta\to 0} \frac{f(z,\zeta}{|\zeta|^{p-2}\zeta} = 0$$

uniformly for almost all $z \in \Omega$;

(vi): we have

$$\beta(z)|\zeta|^q + f(z,\zeta)\zeta \geqslant 0 \quad \text{for almost all } z \in \Omega, \text{ all } \zeta \in \mathbb{R}.$$

REMARK 3.14. *Note that now hypothesis $H(f)_2(v)$ dictates a strict $(p-1)$-superlinear growth near zero. In the previous case with $H(f)_1(vi)$ we had the possibility of $(p-1)$-sublinear or $(p-1)$-linear or $(p-1)$-superlinear growth for $f(z,\cdot)$ near zero. However, now instead of hypothesis $H(f)_1(v)$, we have the global sign condition on the reaction (see hypothesis $H(f)_2(vi)$). Hypotheses $H(f)_2$ allow us to recover the results of* [**23**] *and* [**30**].

EXAMPLE 3.15. *Let*

$$f(\zeta) = |\zeta|^{r-2}\zeta,$$

with $p < r < p^*$. *Then this nonlinearity satisfies hypotheses* $H(f)_2$ *but not* $H(f)_1$. *This is the case considered by García Azorero-Manfredi-Peral Alonso* [**23**] *and Guo-Zhang* [**30**], *with* $\beta = \lambda > 0$ *and the p-Laplacian as differential operator (also in* [**30**] *we assume that* $p > 2$*). Another function which satisfies hypotheses* $H(f)_2$ *(but not* $H(f)_1$*) is*

$$f(\zeta) = |\zeta|^{p-2}\zeta\Big(\ln(1+|\zeta|) + \frac{1}{p}\frac{|\zeta|}{1+|\zeta|}\Big).$$

In this case

$$F(\zeta) = \frac{1}{p}|\zeta|^p \ln(1+|\zeta|).$$

Note that this f *does not satisfy the Ambrosetti-Rabinowitz condition. So, our work here significantly generalizes García Azorero-Manfredi-Peral Alonso* [**23**] *and Guo-Zhang* [**30**].

We consider the positive and negative truncations of the reaction term

$$\widetilde{f}_+(z,\zeta) = \begin{cases} 0 & \text{if } \ \zeta \leqslant 0, \\ \beta(z)\zeta^{q-1} + f(z,\zeta) & \text{if } \ \zeta > 0, \end{cases} \tag{3.75}$$

$$\widetilde{f}_-(z,\zeta) = \begin{cases} \beta(z)\zeta^{q-2}\zeta + f(z,\zeta) & \text{if } \ \zeta < 0, \\ 0 & \text{if } \ \zeta \geqslant 0. \end{cases} \tag{3.76}$$

We set

$$\widetilde{F}_+(z,\zeta) = \int_0^\zeta \widetilde{f}_+(z,s)\,ds, \quad \widetilde{F}_-(z,\zeta) = \int_0^\zeta \widetilde{f}_-(z,s)\,ds$$

and define the C^1-functionals $\widetilde{\varphi}_\pm \colon W_0^{1,p}(\Omega) \longrightarrow \mathbb{R}$ by

$$\widetilde{\varphi}_\pm = \int_\Omega G(z,\nabla u)\,dz - \int_\Omega \widetilde{F}_\pm(z,u)\,dz \quad \forall u \in W_0^{1,p}(\Omega).$$

Reasoning as in Claim 1 in the proof of Theorem 3.13, we obtain the following result.

PROPOSITION 3.16. *If hypotheses* $H(a)'$ *and* $H(f)_2$ *hold, then the functionals* $\widetilde{\varphi}_\pm$ *satisfy the* C*-condition.*

The next proposition is a crucial step in the direction of satisfying the mountain pass geometry in our new setting.

PROPOSITION 3.17. *If hypotheses* $H(a)'$ *and* $H(f)_2$ *hold, then there exists* $\beta_0^* > 0$, *such that, if* $\|\beta\|_\infty < \beta_0^*$, *then we can find* $\varrho = \varrho(\|\beta\|_\infty) > 0$ *for which we have*

$$\inf\big\{\widetilde{\varphi}_\pm(u) : \|u\| = \varrho\big\} \geqslant \eta > 0.$$

PROOF. We do the proof for $\widetilde{\varphi}_+$, the proof for $\widetilde{\varphi}_-$ being similar.

By virtue of hypotheses $H(f)_2(iii)$ and (v), for a given $\varepsilon > 0$ we can find $c_{14} = c_{14}(\varepsilon) > 0$, such that

$$F(z,\zeta) \leqslant \varepsilon|\zeta|^p + c_{14}|\zeta|^r \quad \text{for almost all } z \in \Omega, \text{ all } \zeta \in \mathbb{R}. \tag{3.77}$$

Then, using (3.75), (3.77) and Corollary 3.3, for all $u \in W_0^{1,p}(\Omega)$, we have

$$\begin{aligned}\widetilde{\varphi}_+(u) &= \int_\Omega G(z,\nabla u)\,dz - \int_\Omega \widetilde{F}_+(z,u)\,dz \\ &\geqslant \int_\Omega G(z,\nabla u)\,dz - \frac{1}{q}\int_\Omega \beta(u^+)^q\,dz - \varepsilon\|u\|^p - c_{15}\|u\|^r \\ &\geqslant \left(\frac{c_0}{p(p-1)} - \varepsilon\right)\|u\|^p - c_{16}\|\beta\|_\infty\|u\|^q - c_{15}\|u\|^r,\end{aligned}$$

for some $c_{15} > 0$. We choose $\varepsilon \in \left(0, \frac{c_0}{p(p-1)}\right)$. Then

$$\widetilde{\varphi}_+(u) \geqslant \left(c_{17} - c_{16}\|\beta\|_\infty\|u\|^{q-p} - c_{15}\|u\|^{r-p}\right)\|u\|^p, \tag{3.78}$$

for $c_{17} = \frac{c_0}{p(p-1)} - \varepsilon > 0$. Consider the function

$$\widetilde{\sigma}(t) = c_{16}\|\beta\|_\infty t^{q-p} + c_{15}t^{r-p}, \quad t > 0. \tag{3.79}$$

Evidently $\widetilde{\sigma}(t)$ is continuous and $\widetilde{\sigma} \longrightarrow +\infty$ as $t \searrow 0$ and $t \to +\infty$ due to the fact that $q < p < r$. Therefore, we can find $t_0 \in (0,+\infty)$, such that

$$\widetilde{\sigma}(t_0) = \inf\left\{\widetilde{\sigma}(t) : t \geqslant 0\right\},$$

so

$$\widetilde{\sigma}'(t_0) = c_{16}\|\beta\|_\infty(q-p)t_0^{q-p-1} + c_{15}(r-p)t_0^{r-p-1} = 0$$

and thus

$$t_0 = \left(\frac{c_{16}\|\beta\|_\infty(p-q)}{c_{15}(r-p)}\right)^{\frac{1}{r-q}}. \tag{3.80}$$

We consider $\widetilde{\sigma}(t_0)$. Then it is clear from (3.79) and (3.80) that we can find β_0^*, such that

$$\text{if } \|\beta\|_\infty < \beta_0^*, \quad \text{then } \widetilde{\sigma}(t_0) < c_{17}.$$

Then from (3.78), we infer that

$$\widetilde{\varphi}_+(u) \geqslant \eta > 0 \quad \forall u \in W_0^{1,p}(\Omega), \|u\|_\infty = \varrho(\|\beta\|_\infty) = t_0.$$

Similarly for the functional $\widetilde{\varphi}_-$. □

PROPOSITION 3.18. *If hypotheses $H(a)'$ and $H(f)_2$ hold and $y \in C_+ \setminus \{0\}$ with $\|y\|_p = 1$, then $\widetilde{\varphi}_\pm(ty) \longrightarrow -\infty$ as $t \to \pm\infty$.*

PROOF. By virtue of hypothesis $H(f)_2(iv)$, for a given $\xi > 0$, we can find $M_{16} = M_{16}(\xi) > 0$, such that

$$F(z,\zeta) \geqslant \frac{\xi}{p}|\zeta|^p \quad \text{for almost all } z \in \Omega, \text{ all } |\zeta| \geqslant M_{16}.$$

This, combined with hypothesis $H(f)_2(iii)$, imply that

$$F(z,\zeta) \geqslant \frac{\xi}{p}|\zeta|^p - c_{18} \quad \text{for almost all } z \in \Omega, \text{ all } \zeta| \in \mathbb{R}, \tag{3.81}$$

for some $c_{18} > 0$. Then, using (3.81), for $t > 0$, we have

$$\begin{aligned} \widetilde{\varphi}_+(tu) &= \frac{t^p}{p}\|\nabla u\|_p^p - \int_\Omega \widetilde{F}_+(z,tu)\,dz \\ &\leqslant \frac{t^p}{p}\left(\|\nabla u\|_p^p - \xi\right) + c_{18}|\Omega|_N. \end{aligned} \tag{3.82}$$

Since $\xi > 0$ is arbitrary and $u \in W_0^{1,p}(\Omega)$ with $\|u\|_p = 1$ fixed, from (3.82), we conclude that

$$\widetilde{\varphi}_+(tu) \longrightarrow -\infty \quad \text{as } t \to +\infty.$$

Similarly we show that

$$\widetilde{\varphi}_-(tu) \longrightarrow -\infty \quad \text{as } t \to -\infty.$$ □

Now we are ready for the multiplicity result on solutions of constant sign under hypotheses $H(f)_2$.

THEOREM 3.19. *If hypotheses $H(a)'$ and $H(f)_2$ hold, then there exists $\beta_0^* > 0$, such that, if $\|\beta\|_\infty < \beta_0^*$, then problem (1.1) has four smooth solutions of constant sign*

$$u_0, \widehat{u} \in int C_+, \; u_0 \neq \widehat{u} \quad \text{and} \quad v_0, \widehat{v} \in -int C_+, \; v_0 \neq \widehat{v}.$$

PROOF. Propositions 3.16, 3.17 and 3.18 permit the application of the mountain pass theorem (see Theorem 2.1). So, we obtain $u_0 \in W_0^{1,p}(\Omega)$, such that

$$\widetilde{\varphi}_+(0) = 0 < \eta \leqslant \widetilde{\varphi}_+(u_0) \tag{3.83}$$

and

$$(\widetilde{\varphi}_+)'(u_0) = 0. \tag{3.84}$$

From (3.83) we see that $u_0 \neq 0$. From (3.84), we have

$$V(u_0) = \beta|u_0|^{q-2}u_0 + \widetilde{N}_+(u_0), \tag{3.85}$$

with

$$\widetilde{N}_+(u)(\cdot) = \widetilde{f}_+(\cdot, u(\cdot)) \quad \forall u \in W_0^{1,p}(\Omega).$$

If on (3.85) we act with $-u_0^- \in W_0^{1,p}(\Omega)$, then using (3.76), we obtain that $u_0 \geqslant 0$. So, (3.85) becomes

$$V(u_0) = \beta u_0^{q-1} + N(u_0)$$

(see (3.76)) and so u_0 is a solution of problem (1.1) and $u_0 \in C_+ \setminus \{0\}$ (nonlinear regularity principle).

Using hypothesis $H(f)_2(vi)$, we have

$$-\operatorname{div} a\left(z, \nabla u_0(z)\right) = \beta(z)\left|u_0(z)\right|^{q-2} u_0(z)^{q-1} + f\left(z, u_0(z)\right) \geqslant 0 \quad \text{in } \Omega.$$

Invoking Theorem 3 of Montenegro [**38**] (see also Theorem 1.2 of Zhang [**50**]), we infer that $u \in \operatorname{int} C_+$.

Claim. $-\infty < \inf\left\{\widetilde{\varphi}_+(u) : \|u\| \leqslant \varrho\right\} < 0$ (here $\varrho > 0$ is as in Proposition 3.17).

Because $\widetilde{\varphi}_+$ maps bounded sets into bounded sets, we have

$$-\infty < \inf\left\{\widetilde{\varphi}_+(u) : \|u\| \leqslant \varrho\right\}.$$

By virtue of hypothesis $H(f)_2(v)$, for a given $\varepsilon > 0$, we can find $\delta = \delta(\varepsilon) > 0$, such that

$$F(z, \zeta) \geqslant -\frac{\varepsilon}{p}|\zeta|^p \quad \text{for almost all } z \in \Omega, \text{ all } |\zeta| \leqslant \delta. \tag{3.86}$$

We can find $\xi \in (0,1)$ small enough, such that

$$\xi \widehat{u}_1(z) \in [0, \delta] \quad \forall z \in \overline{\Omega}$$

and

$$\|t\widehat{u}_1\| \leqslant \varrho.$$

Then, using (3.75), Corollary 3.3, (3.86) and the facts that $\|\nabla \widehat{u}_1\|_p^p = \lambda_1 \|\widehat{u}_1\|_p^p$ and $\|\widehat{u}_1\|_p = 1$, we have

$$\begin{aligned}
\widetilde{\varphi}_+(\xi \widehat{u}_1) &= \int_\Omega G(z, \xi \widehat{u}_1)\, dz - \frac{\xi^q}{q} \int_\Omega \beta \widehat{u}_1^q\, dz - \int_\Omega F(z, \xi \widehat{u}_1)\, dz \\
&\leqslant \frac{c_1}{p(p-1)} \xi^p \lambda_1 - \frac{\xi^q}{q} \int_\Omega \beta \widehat{u}_1^q\, dz + \frac{\varepsilon \xi^p}{p} \\
&= \frac{\xi^p}{p}\left(\frac{c_1}{p-1}\lambda_1 + \varepsilon\right) - \frac{\xi^q}{q} \int_\Omega \beta \widehat{u}_1^q\, dz.
\end{aligned}$$

Since $q < p$, by choosing $\xi \in (0,1)$ even smaller if necessary, we have

$$\widetilde{\varphi}_+(\xi \widehat{u}_1) < 0,$$

so

$$\inf\left\{\widetilde{\varphi}_+(u) : \|u\| \leqslant \varrho\right\} < 0.$$

This proves the Claim.

Let

$$\vartheta = \inf_{\partial B_\varrho} \widetilde{\varphi}_+ - \inf_{\overline{B}_\varrho} \widetilde{\varphi}_+ > 0$$

(see Proposition 3.17 and the Claim). Let $\varepsilon \in (0, \vartheta)$. We can apply the Ekeland variational principle (see e.g., Gasiński-Papageorgiou [**25**, p. 579]) and obtain $u_\varepsilon \in \overline{B}_\varrho$, such that

$$\widetilde{\varphi}_+(u_\varepsilon) \leqslant \inf_{\overline{B}_\varrho} \widetilde{\varphi}_+ + \varepsilon \tag{3.87}$$

and

$$\widetilde{\varphi}_+(u_\varepsilon) \leqslant \widetilde{\varphi}_+(y) + \varepsilon \|y - u_\varepsilon\| \quad \forall u \in \overline{B}_\varrho. \tag{3.88}$$

From (3.87) and since $\varepsilon \in (0, \vartheta)$, we have

$$\widetilde{\varphi}_+(u_\varepsilon) < \inf_{\overline{B}_\varrho} \widetilde{\varphi}_+ + \vartheta = \inf_{\partial B_\varrho} \widetilde{\varphi}_+$$

(recall the choice of ϑ) and so

$$\|u_\varepsilon\| < \varrho. \tag{3.89}$$

Consider the functional

$$\overline{B}_\varrho \ni y \longmapsto \psi_+^\varepsilon(y) = \widetilde{\varphi}_+(y) + \varepsilon \|y - u_\varepsilon\|.$$

Evidently ψ_+^ε is locally Lipschitz and from (3.88) we have that u_ε is a minimizer of $\widetilde{\varphi}_+$. Hence, using (3.89), we have

$$0 \in \partial \widetilde{\varphi}_+(u_\varepsilon)$$

(see also Clarke [**12**]), so

$$0 \leqslant (\psi_+^\varepsilon)^0(u_\varepsilon; h) \quad \forall h \in W_0^{1,p}(\Omega)$$

and thus

$$-\varepsilon \|h\| \leqslant \langle \widetilde{\varphi}_+'(u_\varepsilon),\ h \rangle \quad \forall h \in W_0^{1,p}(\Omega),$$

so finally we get

$$\|\widetilde{\varphi}_+'(u_\varepsilon)\| = \varepsilon. \tag{3.90}$$

Let $\varepsilon \searrow 0$ and let us set

$$u_n = u_{\varepsilon_n} \quad \forall n \geq 1.$$

Then $\{u_n\}_{n \geqslant 1} \subseteq \overline{B}_\varrho$ and

$$\widetilde{\varphi}_+'(u_n) \longrightarrow 0 \quad \text{in } W^{-1,p'}(\Omega)$$

(see (3.90)) and

$$\widetilde{\varphi}_+(u_n) \longrightarrow \inf_{\overline{B}_\varrho} \widetilde{\varphi}_+$$

(see (3.87)). So, by virtue of Proposition 3.17, passing to a subsequence if necessary, we may assume that

$$u_n \longrightarrow \widehat{u} \quad \text{in } W_0^{1,p}(\Omega).$$

Hence

$$\widetilde{\varphi}_+(\widehat{u}) = \inf_{\overline{B}_\varrho} \widetilde{\varphi}_+ < 0 = \widetilde{\varphi}_+(0).$$

Therefore $\widehat{u} \neq 0$ is a local minimizer of $\widetilde{\varphi}_+$ and so

$$\widetilde{\varphi}'_+(\widehat{u}) = V(\widehat{u}) - \beta|\widehat{u}|^{q-2}\widehat{u} - \widetilde{N}_+(\widehat{u}).$$

As before, we infer that $\widehat{u} \in \operatorname{int} C_+$ and solve problem (1.1). Moreover, we have

$$\widetilde{\varphi}_+(\widehat{u}) < 0 < \eta \leqslant \widetilde{\varphi}_+(u_0),$$

hence $\widehat{u} \neq u_0$.

Similarly working with $\widetilde{\varphi}_-$ we obtain two negative smooth solutions $v_0, \widehat{v} \in -\operatorname{int} C_+$, $v_0 \neq \widehat{v}$. □

So, in this section under two distinct sets of hypotheses on the nonlinearity f (both assuming $(p-1)$-superlinearity for $f(z,\cdot)$ near $\pm\infty$ without using Ambrosetti-Rabinowitz condition), we produced four distinct nontrivial smooth solutions of constant sign (two positive and two negative). In the next section we focus our attention on the existence of nodal (sign-changing) solutions.

4. Superlinear perturbations: Nodal solutions

First we investigate the existence of nodal solutions under a stronger version of hypotheses $H(f)_1$. Recall that hypothesis $H(f)_1(vi)$ allowed for a rather general behaviour of $f(z,\cdot)$ near zero. More precisely under hypothesis $H(f)_1(vi)$, the nonlinearity $f(z,\cdot)$ can be $(p-1)$-superlinear, $(p-1)$-linear or $(p-1)$-sublinear near zero. Now to produce a nodal solution in the superlinear case, we need to restrict further the growth near zero, which has to be $(p-1)$-linear and we need a sign condition. More precisely the new stronger version of hypotheses $H(f)_1$ on the nonlinearity f are the following

$H(f)_3$ $f\colon \Omega \times \mathbb{R} \longrightarrow \mathbb{R}$ is a function, such that

(i)-(v): the same as $H(f)_1(i) - (v)$;

(vi): there exists $\widehat{\eta} \in L^\infty(\Omega)_+$, such that

$$\lambda_2 < \liminf_{\zeta \to 0} \frac{f(z,\zeta)}{|\zeta|^{p-2}\zeta} \leqslant \limsup_{\zeta \to 0} \frac{f(z,\zeta)}{|\zeta|^{p-2}\zeta} \leqslant \widehat{\eta}(z),$$

uniformly for almost all $z \in \Omega$ and

$$\beta(z)|\zeta|^q + f(z,\zeta)\zeta \geqslant 0 \quad \text{for almost all } z \in \Omega, \text{ all } \zeta \in [a_-, a_+].$$

REMARK 4.1. *The sign condition introduced in hypothesis $H(f)_3(vi)$, forces that*

$$\beta(z)a_+^{p-1} + f(z,a_+) = 0 = \beta(z)|a_-|^{q-2}a_- + f(z,a_-) \quad \textit{for almost all } z \in \Omega$$

(see hypothesis $H(f)_3(v)$). Note that now the growth of $f(z,\cdot)$ near zero is strictly $(p-1)$-linear.

EXAMPLE 4.2. *Consider the function f, defined by*

$$f(\zeta) = \begin{cases} \eta|\zeta|^{p-2}\zeta - \dfrac{\eta r}{p}|\zeta|^{r-2}\zeta & \textit{if } |\zeta| \leqslant 1, \\ \xi|\zeta|^{p-2}\zeta\left(\ln|\zeta| + \dfrac{1}{p}\right) - \vartheta\dfrac{\zeta}{|\zeta|} & \textit{if } |\zeta| > 1, \end{cases}$$

with $\eta > \lambda_2$, $r < p$, $\vartheta > 0$ and $\frac{\vartheta-\eta}{\eta} > \frac{r}{p} > 1$. In this case

$$F(\zeta) = \begin{cases} \dfrac{\eta}{p}|\zeta|^p - |\zeta|^r & \textit{if } |\zeta| \leqslant 1, \\ \dfrac{\xi}{p}|\zeta|^p \ln|\zeta| - \vartheta|\zeta| + \vartheta & \textit{if } |\zeta| > 1. \end{cases}$$

Of course in this example, the z-independence of f together with hypothesis $H(f)_3(v)$ imply that the parameter $\beta(z)$ is z-independent too. So, assuming $\beta(z) \equiv \lambda > 0$, we have that $a_+ = -a_-$ and $\lambda + \eta a_+^{p-q} = \eta\frac{r}{p}a_+^{r-q}$.

Another restriction that we have to make, is that we replace the general differential operator $\operatorname{div} a(z,\nabla u)$ by the p-Laplacian. So, the problem under consideration is the following:

$$\begin{cases} -\Delta_p u(z) = \beta(z)\big|u(z)\big|^{q-2}u(z) + f(z,u(z)) \text{ in } \Omega, \\ u\big|_{\partial\Omega} = 0, \ 1 < q < p. \end{cases} \tag{4.1}$$

The strategy that we employ to produce nodal solutions originates in the semilinear work (i.e., $p = 2$) of Dancer-Du [**16**]. Roughly speaking, we produce a smallest positive solution $u_+ \in \operatorname{int} C_+$ and a biggest negative solution $v_- \in -\operatorname{int} C_+$. Then we focus on the order interval $[v_-, u_+]$. Clearly any nontrivial solution of the problem located in this order interval is nodal.

PROPOSITION 4.3. *If hypotheses $H(f)_3(i)$, (ii), (iii) and (vi) hold and*

$$\operatorname*{ess\,inf}_{\Omega} \beta > 0,$$

then problem (4.1) has a lower solution $\underline{u} \in \operatorname{int} C_+$ and an upper solution $\overline{v} \in -\operatorname{int} C_+$.

PROOF. Note that hypothesis $H(f)_3(vi)$ imply that there exists $\delta_0 \in (0,1)$, such that

$$f(z,\zeta)\zeta \geqslant 0 \quad \text{for almost all } z \in \Omega, \text{ all } |\zeta| \leqslant \delta_0 \tag{4.2}$$

(local sign condition). Let $\xi \in (0,1)$ be small enough, such that

$$\xi\widehat{u}_1(z) \in [0,\delta_0] \quad \forall z \in \overline{\Omega} \tag{4.3}$$

and

$$\lambda_1\xi^{p-1}\widehat{u}_1(z)^{p-1} < \overline{\beta}_0\xi^{q-1}\widehat{u}_1(z)^{q-1} \quad \forall z \in \Omega \tag{4.4}$$

(recall that $q < p$), where $\overline{\beta}_0 = \operatorname{ess\,inf} \beta > 0$. We set

$$\underline{u} = \xi\widehat{u}_1 \in \operatorname{int} C_+.$$

Then, using (4.3), (4.4) and (4.2), we have

$$\begin{aligned} -\Delta_p\underline{u}(z) &= -\Delta_p(\xi\widehat{u}_1)(z) = \lambda_1\xi^{p-1}\widehat{u}_1(z)^{p-1} \\ &< \overline{\beta}_0\xi^{q-1}\widehat{u}_1(z)^{q-1} \\ &\leqslant \beta(z)\overline{u}(z)^{q-1} + f(z,\underline{u}(z)) \quad \text{for almost all } z \in \Omega. \end{aligned}$$

So, $\underline{u} \in \operatorname{int} C_+$ is a lower solution for problem (4.1).

In a similar fashion, we produce $\overline{v} \in -\operatorname{int} C_+$, an upper solution for problem (4.1). □

Next reasoning similar as in the proof of Proposition 3.12, we will produce extremal solutions of constant sign for problem (4.1).

PROPOSITION 4.4. *If hypotheses $H(f)_3$ hold and*

$$\operatorname*{ess\,inf}_{\Omega} \beta > 0,$$

then problem (4.1) has a smallest positive solution $u_+ \in \operatorname{int} C_+$ and a biggest negative solution $v_- \in -\operatorname{int} C_+$.

PROOF. Let $\widehat{S}_+$ be the set of solutions u of problem (4.1) satisfying $\underline{u} \leqslant u$ with $\underline{u} \in \operatorname{int} C_+$ being the lower solution obtained in Proposition 4.3. Let $u_0 \in \operatorname{int} C_+$ be the solution obtained in Proposition 3.9. Recall (see the proof of Proposition 4.3) that $\underline{u} = \xi\widehat{u}_1$ with $\xi \in (0,1)$ small enough. By making ξ even smaller if necessary, we can have $\underline{u} \leqslant u_0$ (see Lemma 2.3). Hence $\widehat{S}_+ \neq \emptyset$.

We show that $\widehat{S}_+$ is downward directed. To this end, let $u_1, u_2 \in \widehat{S}_+$. Then both are upper solutions for (4.1) and so by Lemma 3.11(b), $\overline{u} = \min\{u_1, u_2\} \in W_0^{1,p}(\Omega)$ is an upper solution too and $\underline{u} \leqslant \overline{u}$. We consider the truncation:

$$\overline{f}_+(z,\zeta) = \begin{cases} \beta(z)\underline{u}(z)^{q-1} + f(z,\underline{u}(z)) & \text{if } \zeta < \underline{u}(z), \\ \beta(z)\zeta^{q-1} + f(z,\zeta) & \text{if } \underline{u}(z) \leqslant \zeta \leqslant \overline{u}(z), \\ \beta(z)\overline{u}(z)^{q-1} + f(z,\overline{u}(z)) & \text{if } \overline{u}(z) < \zeta. \end{cases} \tag{4.5}$$

We set

$$\overline{F}_+(z,\zeta) = \int_0^\zeta \overline{f}_+(z,s)\,ds$$

and introduce the C^1-functional $\overline{\varphi}_+ : W_0^{1,p}(\Omega) \longrightarrow \mathbb{R}$, defined by

$$\overline{\varphi}_+(u) = \frac{1}{p}\|\nabla u\|_p^p - \int_\Omega \overline{F}_+(z, u(z))\, dz.$$

Functional $\overline{\varphi}_+$ is coercive and sequentially weakly lower semicontinuous. So, we can find $\widehat{u}_0 \in W_0^{1,p}(\Omega)$, such that

$$\overline{\varphi}_+(\widehat{u}_0) = \inf\{\overline{\varphi}_+(u) \colon u \in W_0^{1,p}(\Omega)\},$$

so

$$\overline{\varphi}_+'(\widehat{u}_0) = 0$$

and thus

$$A(\widehat{u}_0) = \overline{N}_+(\widehat{u}_0), \tag{4.6}$$

where $A\colon W_0^{1,p}(\Omega) \longrightarrow W^{-1,p'}(\Omega)$ is the nonlinear map, defined by

$$\langle A(u), y\rangle = \int_\Omega \|\nabla u\|^{p-2}(\nabla u, \nabla y)_{\mathbb{R}^N}\, dz \quad \forall u, y \in W_0^{1,p}(\Omega)$$

and

$$\overline{N}_+(u)(\cdot) = \overline{f}_+(\cdot, u(\cdot)) \quad \forall u \in W_0^{1,p}(\Omega).$$

Acting on (4.6) with $(\underline{u} - \widehat{u}_0)^+ \in W_0^{1,p}(\Omega)$, using (4.5), Proposition 4.3 and Definition 2.6(b), we obtain

$$\begin{aligned}
\langle A(\widehat{u}_0),\ (\underline{u} - \widehat{u}_0)^+\rangle &= \int_{\{\underline{u}>\widehat{u}_0\}} \overline{f}_+(z, \widehat{u}_0)(\underline{u} - \widehat{u}_0)\, dz \\
&= \int_{\{\underline{u}>\widehat{u}_0\}} \beta \underline{u}^{q-1}(\underline{u} - \widehat{u}_0)\, dz \int_{\{\underline{u}>\widehat{u}_0\}} f(z, \underline{u})(\underline{u} - \widehat{u}_0)\, dz \\
&\geqslant \langle A(\underline{u}),\ (\underline{u} - u_0)^+\rangle,
\end{aligned} \tag{4.7}$$

so, using (4.7), we have

$$\int_{\{\underline{u}>\widehat{u}_0\}} \big(\|\nabla \widehat{u}_0\|^{p-2}\nabla \widehat{u}_0 - \|\nabla \underline{u}\|^{p-2}\nabla \underline{u},\ \nabla \underline{u} - \nabla \widehat{u}_0\big)_{\mathbb{R}^N}\, dz \geqslant 0.$$

Thus

$$\big|\{\underline{u} > u_0\}\big|_N = 0 \quad \text{i.e., } \underline{u} \leqslant u_0.$$

In a similar fashion by acting on (4.6) with $(\widehat{u}_0 - \overline{u}) \in W_0^{1,p}(\Omega)$, we obtain $\widehat{u}_0 \leqslant \overline{u}$. So, (4.6) becomes

$$A(\widehat{u}_0) = \beta \widehat{u}_0^{q-1} + N(\widehat{u}_0),$$

so $\widehat{u}_0 \in \operatorname{int} C_+$ is a solution of (4.1), $\underline{u} \leqslant \widehat{u}_0$, i.e., $\widehat{u}_0 \in \widehat{S}_+$.

This proves that the set $\widehat{S}_+$ is downward directed.

We consider a chain $C \subseteq \widehat{S}_+$. As in the proof of Proposition 3.12, we can find a sequence $\{u_n\}_{n\geqslant 1} \subseteq C$ which is decreasing (since C is totally ordered), such that $\inf C = \inf\limits_{n\geqslant 1} u_n$. We have

$$A(u_n) = \beta u_n^{q-1} + N(u_n) \quad \forall n \geqslant 1, \tag{4.8}$$

so

$$\|\nabla u_n\|_p^p \leqslant c_{19}\left(1 + \|u_1\|_q^q\right) \quad \forall n \geqslant 1,$$

for some $c_{19} > 0$ (recall that $0 \leqslant u_n \leqslant u_1$ for $n \geqslant 1$), thus the sequence $\{u_n\}_{n\geqslant 1} \subseteq W_0^{1,p}(\Omega)$ is bounded.

Passing to a subsequence if necessary, we may assume that

$$u_n \longrightarrow u \quad \text{weakly in } W_0^{1,p}(\Omega), \tag{4.9}$$

$$u_n \longrightarrow u \quad \text{in } L^r(\Omega). \tag{4.10}$$

Using (4.11) and acting on (4.8) with $u_n - u \in W^{1,p}(\Omega)$ and passing to the limit as $n \to +\infty$ and obtain

$$\lim_{n\to+\infty} \langle A(u_n), u_n - u\rangle = 0,$$

so using Proposition 3.5, we have

$$u_n \longrightarrow u \quad \text{in } W_0^{1,p}(\Omega). \tag{4.11}$$

Therefore, passing to the limit in (4.11) as $n \to +\infty$ and using the fact that $\underline{u} \leqslant u_n$ for $n \geqslant 1$, we have

$$A(u) = \beta u^{q-1} + N(u), \quad \underline{u} \leqslant u,$$

so $u \in \widehat{S}_+$ and $u = \inf C$. The Kuratowski-Zorn lemma implies that the set $\widehat{S}_+$ has a minimal element $u_* \in \widehat{S}_+$. Since the set $\widehat{S}_+$ is downward directed we conclude that u_* is the smallest solution of (4.1) in the set

$$T_+ = \left\{u \in W_0^{1,p}(\Omega) : \underline{u} \leqslant u\right\}.$$

Using this fact, we will produce the extremal positive solution for problem (4.1). So let $\xi_n \searrow 0$ and let us set

$$\underline{u}_n = \xi_n \widehat{u}_+ \quad \forall n \geqslant 1.$$

Let

$$T_+^n = \left\{u \in W_0^{1,p}(\Omega) : \underline{u}_n \leqslant u\right\}.$$

From the previous part of the proof we know that problem (4.1) has a smallest solution $u_*^n \in T_+^n$ in T_+^n. Clearly the sequence $\{u_*^n\}_{n\geqslant 1} \subseteq W_0^{1,p}(\Omega)$ is bounded and so we may assume that

$$u_*^n \longrightarrow u_+ \quad \text{weakly in } W_0^{1,p}(\Omega), \tag{4.12}$$

$$u_*^n \longrightarrow u_+ \quad \text{in } L^r(\Omega). \tag{4.13}$$

We have

$$A(u_*^n) = \beta(u_*^n)^{q-1} + N(u_*^n) \quad \forall n \geqslant 1. \tag{4.14}$$

Acting with $u_n - u_+$ on (4.14), using (4.12) and passing to the limit as $n \to +\infty$, we obtain

$$\lim_{n\to+\infty} \langle A(u_*^n),\ u_*^n - u_+ \rangle = 0,$$

so

$$u_*^n \longrightarrow u_+ \quad \text{in } W_0^{1,p},$$

so, using also (4.14), we have

$$A(u_+) = \beta u_+^{q-1} + N(u_+)$$

and thus u_+ is a solution of (4.1) and $u_+ \in C_+$ (nonlinear regularity principle).

We need to show that $u_+ \neq 0$ in which case by virtue of the nonlinear maximum principle of Vázquez [**49**], we have $u_+ \in \operatorname{int} C_+$.

To show the nontriviality of u_+ we proceed as follows. We consider the following auxiliary Dirichlet problem:

$$\begin{cases} -\Delta_p u(z) = \beta(z)\left|u(z)\right|^{q-1} \text{ in } \Omega, \\ u\big|_{\partial\Omega} = 0,\ u \geqslant 0. \end{cases} \tag{4.15}$$

Observing that the energy (Euler) functional of (4.15), defined by

$$\psi_+(u) = \frac{1}{p}\|\nabla u\|_p^p - \frac{1}{q}\int_\Omega \beta(u^+)^q\,dz \quad \forall u \in W_0^{1,p}(\Omega)$$

is coercive (since $q < p$) and sequentially weakly lower semicontinuous, we can find $\underline{u} \in W_0^{1,p}(\Omega)$, such that

$$\psi_+(\underline{u}) = \inf\left\{\psi_+(u) :\ u \in W_0^{1,p}(\Omega)\right\}. \tag{4.16}$$

Exploiting the fact that $q < p$, we can see that for $\xi \in (0,1)$ small enough, we have $\psi_+(\xi\widehat{u}_1) < 0$. Hence

$$\psi_+(\underline{u}) < 0 = \psi_+(0), \quad \text{i.e., } \underline{u} \neq 0.$$

From (4.16), we have

$$\psi_+'(\underline{u}) = 0,$$

so

$$A(\underline{u}) = \beta(\underline{u}^+)^{q-1}.$$

Acting with $-\underline{u}^- \in W_0^{1,p}(\Omega)$, we obtain $\underline{u} \geqslant 0$ and so

$$A(\underline{u}) = \beta\underline{u}^{q-1}, \quad \underline{u} \geqslant 0, \quad \underline{u} \neq 0,$$

so $\underline{u} \in \operatorname{int} C_+$ (nonlinear regularity and Vázquez [**49**]) and solves (4.15).

Arguing by contradiction, suppose that $u_+ = 0$. Then

$$u_*^n \longrightarrow u_+ = 0 \quad \text{in } W_0^{1,p}(\Omega).$$

Recall that for all $n \geqslant 1$, we have

$$\begin{cases} -\Delta_p u_*^n(z) = \beta(z) u_*^n(z)^{q-1} + f\left(z, u_*^n(z)\right) \text{ in } \Omega, \\ u_*^n|_{\partial\Omega} = 0. \end{cases}$$

Since $u_*^n \leqslant u_*^1$ for all $n \geqslant 1$, invoking Theorem 1 of Lieberman [**34**], we can find $M_{16} > 0$ and $\gamma \in (0,1)$, such that

$$u_*^n \in C_0^{1,\gamma}(\overline{\Omega}) \quad \text{and} \quad \|u_*^n\|_{C_0^{1,\gamma}(\overline{\Omega})} \leqslant M_{16} \quad \forall n \geqslant 1. \tag{4.17}$$

Recalling that the embedding $C_0^{1,\gamma}(\overline{\Omega}) \subseteq C_0^1(\overline{\Omega})$ is compact, from (4.17) we infer that

$$u_*^n \longrightarrow u_+ = 0 \quad \text{in } C_0^1(\overline{\Omega}).$$

So, for $n \geqslant 1$ large enough, we have

$$\|u_*^n\|_{C_0^1(\overline{\Omega})} \leqslant \delta_0 \tag{4.18}$$

(see (4.2)). Since $u_*^n \in \operatorname{int} C_+$, by virtue of Lemma 2.3, we can find $\xi_n > 0$, such that

$$\xi_n \underline{u} \leqslant u_*^n \quad \forall n \geqslant 1.$$

Let $\xi_n > 0$ be the biggest such positive number for which this is true. Suppose that $\xi_n \in (0,1)$ for all $n \geqslant 1$. Then, using (4.18), the facts that $\xi_n \in (0,1)$, $q < p$, for $n \geqslant 1$ large enough, we have

$$\begin{aligned} -\Delta_p u_*^n(z) &= \beta(z) u_*^n(z)^{q-1} + f\left(z, u_*^n(z)\right) \\ &\geqslant \beta(z) u_*^n(z)^{q-1} \\ &> \beta(z) \xi_n^{p-1} \underline{u}(z)^{q-1} \\ &= -\Delta_p(\xi_n \underline{u})(z) \quad \text{for almost all } z \in \Omega. \end{aligned} \tag{4.19}$$

From (4.19), using Proposition 2.2 of Guedda-Véron [**29**], we obtain

$$u_*^n - \xi_n \underline{u} \in \operatorname{int} C_+ \quad \forall n \geqslant 1, \text{ large enough},$$

which contradicts the maximality of ξ_n. Therefore $\xi_n \geqslant 1$ and so

$$\underline{u} \leqslant u_*^n \quad \forall n \geqslant 1, \text{ large enough},$$

hence

$$\underline{u} \leqslant u_+ = 0,$$

a contradiction. This proves that $u_+ \neq 0$ and so $u_+ \in \operatorname{int} C_+$ (see hypothesis $H(f)_3(vi)$) and the nonlinear maximum principle of Vázquez [**49**]. Evidently $u_+ \in \operatorname{int} C_+$ is the smallest positive solution of (4.1).

In a similar fashion using the upper solution $\overline{v} \in -\operatorname{int} C_+$ (see Proposition 4.3) and the set $\widehat{S}_-$ of solutions v of (4.1), such that $v \leqslant \overline{v}$, we obtain $v_- \in -\operatorname{int} C_+$, the biggest negative solution of (4.1). □

Using these extremal solutions of constant sign for problem (4.1), we will produce a nodal solution, the fifth nontrivial smooth solution of (4.1).

THEOREM 4.5. *If hypotheses $H(f)_3$ hold and $\beta \in L^\infty(\Omega)_+$ satisfies*

$$\operatorname{ess\,sup} \beta > 0 \quad \text{and} \quad \|\beta\|_\infty \leqslant \beta^*,$$

then problem (4.1) has at least five nontrivial smooth solutions:

$$\begin{gathered} u_0, \widehat{u} \in \operatorname{int} C_+, \quad u_0 \leqslant \widehat{u}, \quad u_0 \neq \widehat{u}, \quad u_0(z) < a_+ \quad \forall z \in \overline{\Omega}, \\ v_0, \widehat{v} \in \operatorname{int} C_+, \quad \widehat{v} \leqslant v_0, \quad \widehat{v} \neq v_0, \quad a_- < v_0(z) \quad \forall z \in \overline{\Omega}, \end{gathered}$$

and $y_0 \in C_0^1(\overline{\Omega}) \setminus \{0\}$ nodal.

PROOF. Theorem 3.13 gives us the four constant sign smooth solutions

$$\begin{gathered} u_0, \widehat{u} \in \operatorname{int} C_+, \ u_0 \leqslant \widehat{u}, \ u_0 \neq \widehat{u}, \\ v_0, \widehat{v} \in -\operatorname{int} C_+, \ \widehat{v} \leqslant v_0, \ \widehat{v} \neq v_0, \\ a_- < v_0(z) < u_0 < a_+ \quad \forall z \in \overline{\Omega}. \end{gathered}$$

Next we will produce the nodal solution. Let $u_+ \in \operatorname{int} C_+$ and $v_- \in -\operatorname{int} C_+$, be the two extremal constant sign solutions obtained in Proposition 4.4. Using them, we introduce the following truncations of the reaction term:

$$\widehat{g}_+(z,\zeta) = \begin{cases} 0 & \text{if } \zeta < 0, \\ \beta(z)\zeta^{q-1} + f(z,\zeta) & \text{if } \zeta \in [0, u_+(z)], \\ \beta(z)u_+(z)^{q-1} + f(z, u_+(z)) & \text{if } u_+(z) < \zeta, \end{cases}$$

$$\widehat{g}_-(z,\zeta) = \begin{cases} \beta(z)|v_-(z)|^{q-2}v_-(z) + f(z, v_-(z)) & \text{if } \zeta < v_-(z), \\ \beta(z)|\zeta|^{q-2}\zeta + f(z,\zeta) & \text{if } \zeta \in [v_-(z), 0], \\ 0 & \text{if } 0 < \zeta, \end{cases}$$

$$\widehat{g}_0(z,\zeta) = \begin{cases} \beta(z)|v_-(z)|^{q-2}v_-(z) + f(z, v_-(z)) & \text{if } \zeta < v_-(z), \\ \beta(z)|\zeta|^{q-2}\zeta + f(z,\zeta) & \text{if } \zeta \in [v_-(z), u_+(z)], \\ \beta(z)u_+(z)^{q-1} + f(z, u_+(z)) & \text{if } u_+(z) < \zeta. \end{cases}$$

All three are Carathéodory function. We set

$$\widehat{G}_\pm(z,\zeta) = \int_0^\zeta \widehat{g}_\pm(z,s)\,ds \quad \text{and} \quad \widehat{G}_0(z,\zeta) = \int_0^\zeta \widehat{g}_0(z,s)\,ds$$

and introduce the corresponding C^1 energy functionals $\widehat{\varphi}_\pm, \widehat{\varphi}_0 \colon W_0^{1,p}(\Omega) \longrightarrow \mathbb{R}$, defined by

$$\widehat{\varphi}_\pm(u) = \frac{1}{p}\|\nabla u\|_p^p - \int_\Omega \widehat{G}_\pm(z,u)\,dz \quad \forall u \in W_0^{1,p}(\Omega)$$
$$\widehat{\varphi}_0(u) = \frac{1}{p}\|\nabla u\|_p^p - \int_\Omega \widehat{G}_0(z,u)\,dz \quad \forall u \in W_0^{1,p}(\Omega).$$

In what follows, we set

$$[v_-, u_+] = \big\{y \in W_0^{1,p}(\Omega) :\ v_-(z) \leqslant y(z) \leqslant u_+(z) \quad \text{for almost all } z \in \Omega\big\}.$$

CLAIM 1. The critical points of $\widehat{\varphi}_+$ are $\{0, u_+\}$, the critical points of $\widehat{\varphi}_-$ are $\{v_-, 0\}$, and the critical points of $\widehat{\varphi}_0$ are located in the order interval $[v_-, u_+]$.

We do the proof for $\widehat{\varphi}_+$, the proofs for $\widehat{\varphi}_-$ and $\widehat{\varphi}_0$ being similar. So, suppose that $u \in W_0^{1,p}(\Omega)$ is a critical point of $\widehat{\varphi}_+$. Then

$$V(u) = \beta|u|^{q-2}u + \widehat{N}_+(u), \tag{4.20}$$

with

$$\widehat{N}_+(y)(\cdot) = \widehat{g}_+(\cdot, y(\cdot)) \quad \forall y \in W_0^{1,p}(\Omega).$$

Acting on (4.20) with $-u^- \in W_0^{1,p}(\Omega)$ we obtain that $u \geqslant 0$. Also, acting on (4.20) with $(u-u_+)^+ \in W_0^{1,p}(\Omega)$ and Proposition 4.4, we obtain

$$\begin{aligned}\big\langle V(u),\ (u-u_+)^+\big\rangle &= \int_\Omega \beta u_+^{q-1}(u-u_+)^+ dz + \int_\Omega f(z,u_+)(u-u_+)^+ dz \\ &= \big\langle V(u_+),\ (u-u_+)^+\big\rangle,\end{aligned}$$

so

$$\int_{\{u>u_+\}} \big(\|\nabla u\|^{p-2}\nabla u - \|\nabla u_+\|^{p-2}\nabla u_+,\ \nabla u - \nabla u_+\big)_{\mathbb{R}^N}\, dz = 0$$

and so

$$\big|\{u > u_+\}\big|_N = 0, \quad \text{i.e., } u \leqslant u_+.$$

Hence $0 \leqslant u \leqslant u_+$. If $u \neq 0$, then $u = u_+$ due to the extremality of u_+.

Similarly we show that the critical points of $\widehat{\varphi}_-$ are $\{0, v_-\}$ and that the critical points of $\widehat{\varphi}_0$ are located in the order interval $[v_-, u_+]$. This proves Claim 1.

CLAIM 2. Both u_+ and v_- are local minimizers of $\widehat{\varphi}_0$.

Clearly the functional $\widehat{\varphi}_+$ is coercive and sequentially weakly lower semicontinuous. So, we can find $\widetilde{u} \in W_0^{1,p}(\Omega)$, such that

$$\widehat{m}_+ = \inf \widehat{\varphi}_+ = \widehat{\varphi}(\widetilde{u}).$$

By virtue of hypothesis $H(f)_3(vi)$, we can find $\delta > 0$, such that

$$\frac{1}{p}\lambda_2\zeta^p \leqslant F(z,\zeta) \quad \text{for almost all } z \in \Omega, \text{ all } \zeta \in [0,\delta]. \tag{4.21}$$

Since $\widehat{u}_1, u_+ \in \operatorname{int} C_+$, we can find $\xi > 0$ small enough, such that

$$0 \leqslant \xi \widehat{u}_1 \leqslant \delta \quad \forall z \in \overline{\Omega}$$

and

$$\xi \widehat{u}_1 \leqslant u_+$$

(see Lemma 2.3). Then, recalling that $\|\nabla u\|_p^p = \lambda_1 \|u\|_p^p$ and $\|u\|_p = 1$, we have

$$\widehat{\varphi}_+(\xi \widehat{u}_1) \leqslant \frac{1}{p}\xi^p \lambda_1 - \frac{1}{p}\xi^p \lambda_2 = \frac{1}{p}\xi^p(\lambda_1 - \lambda_2) < 0.$$

Therefore

$$\widehat{m}_+ = \widehat{\varphi}_+(\widetilde{u}) < 0 = \widehat{\varphi}_+(0),$$

hence $\widetilde{u} \neq 0$. Since $\widetilde{u}$ is a critical point of $\widehat{\varphi}_+$, we must have $\widetilde{u} = u_+$ (see Claim 1). Since $u_+ \in \operatorname{int} C_+$, we can find $r > 0$ small enough, such that, if

$$\overline{B}_r^{C_0^1(\overline{\Omega})}(u_+) = \Big\{u \in C_0^1(\overline{\Omega}) : \ \|u - u_+\| \leqslant r\Big\},$$

then

$$\widehat{\varphi}_+\Big|_{\overline{B}_r^{C_0^1(\overline{\Omega})}(u_+)} = \widehat{\varphi}_0\Big|_{\overline{B}_r^{C_0^1(\overline{\Omega})}(u_+)}.$$

This means that u_+ is a local $C_0^1(\overline{\Omega})$-minimizer of $\widehat{\varphi}_0$, hence by Proposition 3.6 (see also García Azorero-Manfredi-Peral Alonso [**23**]), u_+ is a local $W_0^{1,p}(\Omega)$-minimizer of $\widehat{\varphi}_0$. Similarly for v_- using this time the functional $\widehat{\varphi}_-$. This proves Claim 2.

Without any loss of generality we assume that

$$\widehat{\varphi}_0(v_-) \leqslant \widehat{\varphi}_0(u_+) < 0. \tag{4.22}$$

Moreover, as before (see (3.66)), we can find $\varrho > 0$ small enough, such that

$$\varrho < \|u_+ - v_-\| \quad \text{and} \quad \widehat{\varphi}_0(u_+) < \overline{\eta}_0 = \inf\big\{\widehat{\varphi}(u) : \ \|u - u_+\| = \varrho\big\}. \tag{4.23}$$

The functional $\widehat{\varphi}_0$ being coercive, it satisfies the PS-condition. This fact together with (4.22) and (4.23) permit the application of the mountain pass theorem (see Theorem 2.1). Therefore, there exists $y_0 \in W_0^{1,p}(\Omega)$, such that

$$\widehat{\varphi}_0(v_-) \leqslant \widehat{\varphi}_0(u_+) < \overline{\eta}_0 \leqslant \widehat{\varphi}_0(y_0) = \inf_{\overline{\gamma} \in \overline{\Gamma}} \max_{1 \leqslant t \leqslant 1} \widehat{\varphi}_0\left(\overline{\gamma}(t)\right), \tag{4.24}$$

where

$$\overline{\Gamma} = \left\{\overline{\gamma} \in C\left([-1,1]; W_0^{1,p}(\Omega)\right) : \ \overline{\gamma}(-1) = v_-, \ \overline{\gamma}(1) = u_+\right\}$$

and

$$\widehat{\varphi}_0'(u_0) = 0, \tag{4.25}$$

hence $u_0 \in [v_-, u_+]$ (see Claim 1).

From (4.25), it follows that y_0 is a solution of (4.1) and $y_0 \in C_0^1(\overline{\Omega})$ (nonlinear regularity theory). From (4.24) we see that $y_0 \notin \{u_+, v_-\}$. Since $y_0 \in [v_-, u_+]$ (see (4.25)), if we show that $y_0 \neq 0$, then y_0 must be nodal (recall the extremality of the solutions u_+, v_-; see Proposition 4.4).

To show the nontriviality of y_0, we will use the minimax expression in (4.24). According to that expression, to establish the nontriviality of the solution u_0, it is enough to produce a path $\overline{\gamma}^* \in \overline{\Gamma}^*$, such that $\widehat{\varphi}_0|_{\overline{\gamma}^*} < 0$.

To this end, by virtue of hypotheses $H(f)_3(vi)$, we can find $\delta, \delta_0 > 0$, such that

$$\frac{\mu}{p}(\lambda_2 + \delta)|\zeta|^p \leqslant F(z, \zeta) \quad \text{for almost all } z \in \Omega, \text{ all } |\zeta| \leqslant \delta_0. \tag{4.26}$$

Let

$$\begin{aligned} \partial L_1^{L^p} &= \{u \in L^p(\Omega) : \|u\|_p = 1\} \\ M &= W_0^{1,p}(\Omega) \cap \partial B_1^{L^p} \\ M_c &= C_0^1(\overline{\Omega}) \cap \partial B_1^{L^p}. \end{aligned}$$

We endow M with the relative $W_0^{1,p}(\Omega)$-topology and M_c with the relative $C_0^1(\overline{\Omega})$-topology.

Then M_c is dense in M and $C([-1,1]; M_c)$ is dense in $C([-1,1]; M)$. We introduce

$$\begin{aligned} \Gamma_0 &= \{\gamma_0 \in C([-1,1]; M) : \gamma_0(-1) = -\widehat{u}_1, \ \gamma_0(1) = \widehat{u}_1\}, \\ \Gamma_0^c &= \{\gamma_0 \in C([-1,1]; M_c) : \gamma_0(-1) = -\widehat{u}_1, \ \gamma_0(1) = \widehat{u}_1\}. \end{aligned}$$

Evidently Γ_0^c is dense in Γ_0. By virtue of Proposition 2.4, we can find $\widehat{\gamma}_0 \in \Gamma_0^c$, such that

$$\max_{-1 \leqslant t \leqslant 1} \left\|\nabla \widehat{\gamma}_0(t)\right\|_p^p < \lambda_2 + \delta. \tag{4.27}$$

Since $\widehat{\gamma}_0 \in \Gamma_0^c$ and $u_+ \in \operatorname{int} C_+$, $v_- \in -\operatorname{int} C_+$, we can find $\xi > 0$ small enough, such that

$$v_-(z) \leqslant \xi \widehat{\gamma}_0(t)(z) \leqslant u_+(z) \quad \forall z \in \overline{\Omega}, \ t \in [-1,1] \tag{4.28}$$

and

$$\xi\left|\widehat{\gamma}_0(t)(z)\right| \leqslant \delta_0 \quad \forall z \in \overline{\Omega}, \ t \in [-1,1]. \tag{4.29}$$

Then, using (4.27), (4.26) and (4.28), for all $t \in [-1,1]$, we have

$$\begin{aligned} \widehat{\varphi}_0(\varepsilon \widehat{\gamma}_0(t)) &= \frac{\xi^p}{p}\left\|\nabla \widehat{\gamma}_0(t)\right\|_p^p - \frac{\xi^q}{q}\int_\Omega \beta\left|\widehat{\gamma}_0(t)\right|^q dz - \int_\Omega F(z, \xi\widehat{\gamma}_0(t)(z))\, dz \\ &< \frac{\xi^p}{p}(\lambda_2 + \delta) - \int_\Omega F(z, \xi\widehat{\gamma}_0(t)(z))\, dz \\ &< \frac{\xi^p}{p}(\lambda_2 + \delta - \lambda_2 - \delta) = 0. \end{aligned}$$

Therefore, if we set $\gamma_0 = \xi\widehat{\gamma}_0$, then

$$\widehat{\varphi}_0\Big|_{\gamma_0} < 0. \tag{4.30}$$

Next we produce a continuous path which joins $\varepsilon\widehat{u}_1$ and u_+ and along which $\widehat{\varphi}_0$ is strictly negative. For this purpose, let

$$a = \widehat{m}_+ = \inf \widehat{\varphi}_+ = \widehat{\varphi}_+(u_+) < 0 = \widehat{\varphi}_+(0) = b.$$

The functional $\widehat{\varphi}_+$ being coercive, it satisfies the PS-condition. Also, it does not have critical values in (a, b) and $\widehat{\varphi}_+^{-1}(a) = \{u_+\}$. So, we can apply the second deformation theorem (see Theorem 2.2) and obtain a deformation

$$h_+ \colon [0,1] \times \left(\widehat{\varphi}_+^b \setminus \{0\}\right) \longrightarrow \widehat{\varphi}_+^b,$$

such that

$$h_+(t, \cdot)\Big|_{K_a} = id\Big|_{K_a} \quad \forall t \in [0,1]$$

(in our case $K_a = \{u_+\}$) and

$$h_+\left(1, \widehat{\varphi}_+^b \setminus \{0\}\right) \subseteq \widehat{\varphi}_+^a. \tag{4.31}$$

and

$$\widehat{\varphi}_+\left(h_+(t,u)\right) \leqslant \widehat{\varphi}_+\left(h_+(s,u)\right) \quad \forall 0 \leqslant s \leqslant t \leqslant 1,\ u \in \widehat{\varphi}_+^b \setminus \{0\}. \tag{4.32}$$

Recall that u_+ is the minimizer of $\widehat{\varphi}_+$ (see the proof of Claim 2). Hence

$$\widehat{\varphi}_+^a = \{u_+\}.$$

Note that $\xi\widehat{u}_1 \in \widehat{\varphi}_+^b$ (see (4.28) and (4.30)). So, we can define

$$\gamma_+(t) = h_+(t, \xi\widehat{u}_1) \quad \forall t \in [0,1].$$

Clearly γ_+ is continuous. Moreover, we have:

- $\gamma_+(0) = h_+(0, \xi\widehat{u}_1) = \xi\widehat{u}_1$ (h_+ being a deformation);
- $\gamma_+(1) = h_+(1, \xi\widehat{u}_1) = u_+$ (see (4.31) and recall that $\widehat{\varphi}_+^a = \{u_1\}$).

Therefore γ_+ is a continuous path which joins $\varepsilon\widehat{u}_1$ and u_+. In addition, from (4.30) and (4.32), we have

$$\widehat{\varphi}_+\Big|_{\gamma_+} < 0.$$

Note that hypothesis $H(f)_3(vi)$ (the sign condition on the reaction term) implies that

$$\widehat{\varphi}_0 \leqslant \widehat{\varphi}_+.$$

Therefore

$$\widehat{\varphi}_0\Big|_{\gamma_+} < 0. \tag{4.33}$$

In a similar fashion we produce a continuous path γ_- which joins v_- and $-\varepsilon\widehat{u}_1$ and such that

$$\widehat{\varphi}_0\Big|_{\gamma_-} < 0. \tag{4.34}$$

We concatenate the path $\gamma_-, \gamma_0, \gamma_+$ and produce a continuous path $\overline{\gamma}^*$ which joins v_- and u_+, i.e.,

$$\overline{\gamma}^* \in \overline{\Gamma}.$$

Moreover, from (4.30), (4.33) and(4.34), we have

$$\widehat{\varphi}_0\Big|_{\overline{\gamma}^*} < 0,$$

hence y_0 is nontrivial and so $y_0 \in C_0^1(\overline{\Omega})$ is a nodal solution of (4.1). □

REMARK 4.6. *It remains an open problem if we can have Theorem 4.5 when the p-Laplacian is replaced by the more general nonhomogeneous differential operator* $\operatorname{div} a(z, \nabla u)$.

What about nodal solutions under hypotheses $H(f)_2$? We recall that hypotheses $H(f)_2$ dictate a $(p-1)$-superlinear growth for $f(z,\cdot)$ both near $\pm\infty$ and near zero. So, it implies different requirements on $f(z,\cdot)$ than $H(f)_3$. So, we need to change our approach. In this case too, we need to pass to the p-Laplacian problem (4.1) and strengthen hypotheses $H(f)_2$.

$\underline{H(f)_4}$ $f\colon \Omega \times \mathbb{R} \longrightarrow \mathbb{R}$ is a function, such that

(i)-(v): the same as hypotheses $H(f)_2(i)-(v)$;

(vi): for almost all $z \in \Omega$, the function $\zeta \longmapsto f(z,\zeta)$ is nondecreasing.

REMARK 4.7. *Evidently hypothesis $H(f)_4(iv)$ implies that*

$$f(z,\zeta)\zeta \geqslant 0 \quad \textit{for almost all } z \in \Omega, \ \textit{all } \zeta \in \mathbb{R}$$

and of course

$$\beta(z)(|\zeta|^q + f(z,\zeta)\zeta \geqslant 0 \quad \textit{for almost all } z \in \Omega, \ \textit{all } \zeta \in \mathbb{R}.$$

EXAMPLE 4.8. *The function*

$$f(\zeta) = |\zeta|^{p-2}\zeta\Big(\ln(1+|\zeta|) + \frac{\zeta}{p(1+|\zeta|)}\Big)$$

satisfies hypotheses $H(f)_4$ (but not the Ambrosetti-Rabinowitz condition and $H(f)_3$).

Now our proof instead of using Proposition 2.4, will employ arguments from Morse theory (critical groups).

Thanks to hypothesis $H(f)_4(vi)$ and reasoning as in the previous case (see the proofs of Proposition 4.3 and 4.4), we obtain the following proposition.

PROPOSITION 4.9. *If hypotheses $H(f)_4$ hold and*

$$\operatorname*{ess\,inf}_{\Omega} \beta > 0,$$

then problem (4.1) has a smallest positive solution $u_+ \in \operatorname{int} C_+$ and a biggest negative solution $v_- \in -\operatorname{int} C_+$.

Using these extremal constant sign solutions, as above (see the proof of Theorem 4.5), we truncate at extremal solutions $\{v_-, u_+\}$ and introduce the C^1-functional $\widehat{\varphi}\colon W_0^{1,p}(\Omega) \longrightarrow \mathbb{R}$, defined by

$$\widehat{\varphi}_0(u) = \frac{1}{p}\|\nabla u\|_p^p - \int_\Omega \widehat{G}_0(z,u)\,dz \quad \forall u \in W_0^{1,p}(\Omega).$$

PROPOSITION 4.10. *If hypotheses $H(f)_4$ hold and*

$$\operatorname*{ess\,inf}_{\Omega} \beta > 0,$$

then

$$H_k(\widehat{\varphi}_0, 0) = 0 \quad \forall k \geqslant 0.$$

PROOF. Let

$$\widehat{f}_0(z,\zeta) = \begin{cases} f\left(z, v_-(z)\right) & \text{if } \zeta < v_-(z), \\ f(z,\zeta) & \text{if } v_-(z) \leqslant \zeta \leqslant u_+(z), \\ f\left(z, u_+(z)\right) & \text{if } u_+(z) < \zeta \end{cases}$$

and

$$\widehat{F}_0(z,\zeta) = \int_0^\zeta \widehat{f}_0(z,s)\,ds.$$

Note that, from hypotheses $H(f)_4(vi)$, we have

$$0 \leqslant \widehat{f}_0(z,\zeta)\zeta \leqslant f(z,\zeta)\zeta \quad \text{for almost all } z \in \Omega, \text{ all } \zeta \in \mathbb{R},$$

so, using also hypotheses $H(f)_4(v)$, we have

$$\lim_{\zeta \to 0} \frac{\widehat{f}_0(z,\zeta)}{|\zeta|^{p-2}\zeta} = 0 \quad \text{uniformly for almost all } z \in \Omega. \tag{4.35}$$

From (4.35) we see that for a given $\varepsilon > 0$, we can find $c_\varepsilon > 0$, such that

$$\widehat{F}_0(z,\zeta) \geqslant -\frac{\varepsilon}{p}|\zeta|^p - c_\varepsilon|\zeta|^r \quad \text{for almost all } z \in \Omega, \text{ all } \zeta \in \mathbb{R}. \tag{4.36}$$

Let $u \in C_0^1(\overline{\Omega})$ and $t > 0$. Evidently for $t > 0$ small enough, we will have

$$v_- \leqslant tu \leqslant u_+ \tag{4.37}$$

(see Lemma 2.3). So, from (4.36) and (4.37), we have

$$\begin{aligned}\widehat{\varphi}_0(tu) &= \frac{t^p}{p}\|\nabla u\|_p^p - \frac{t^q}{q}\int_\Omega \beta|u|^q\,dz + \frac{\varepsilon t^p}{p\lambda_1}\|\nabla u\|_p^p + c_\varepsilon t^r\|u\|_q^q \\ &= \frac{t^p}{p}\Big(1+\frac{\varepsilon}{\lambda_1}\Big)\|\nabla u\|_p^p - \frac{t^q}{q}\int_\Omega \beta|u|^q\,dz + c_\varepsilon t^r\|u\|_q^q \qquad (4.38)\end{aligned}$$

(see also (3.33)). Since the embedding $C_0^1(\overline{\Omega}) \subseteq W_0^{1,p}(\Omega)$ is dense, we infer that (4.38) is true for $t>0$ and $u \in W_0^{1,p}(\Omega)$. Because $q<p<r$, we see that for each $u \in W_0^{1,p}(\Omega)\setminus\{0\}$, we can find $t^* = t^*(u) \in (0,1)$ small enough, such that

$$\widehat{\varphi}_0(tu) < 0 \quad \forall t \in (0,t^*). \qquad (4.39)$$

Let $u \in W_0^{1,p}(\Omega)\setminus\{0\}$, such that $\widehat{\varphi}_0(u)=0$. Then, since $\widehat{\varphi}_0(u)=0$ and using hypotheses $H(f)_4(iv)$ and (v), we have

$$\begin{aligned}\frac{d}{dt}\widehat{\varphi}_0(tu)\Big|_{t=1} &= \Big(1-\frac{q}{p}\Big)\|\nabla u\|_p^p + \int_\Omega \Big(q\widehat{F}_0(z,u) - \widehat{f}_0(z,u)u\Big)\,dz \\ &\geqslant \Big(1-\frac{q}{p}\Big)\|\nabla u\|_p^p - \int_\Omega \widehat{f}_0(z,u)u\,dz \\ &\geqslant \Big(1-\frac{q}{p}\Big)\|\nabla u\|_p^p - \int_\Omega f(z,u)u\,dz \\ &\geqslant c_{20}\|\nabla u\|_p^p - c_{21}\|\nabla u\|_p^r, \qquad (4.40)\end{aligned}$$

for some $c_{20}, c_{21} > 0$. Since $q<p<r$, from (4.40), it follows that we can find $\varrho \in (0,1)$, such that

$$\frac{d}{dt}\widehat{\varphi}_0(tu)\Big|_{t=1} > 0 \quad \forall u \in W_0^{1,p}(\Omega), \text{ with } 1 < \|u\| \leqslant \varrho,\ \varphi(u)=0. \qquad (4.41)$$

Now let $u \in \widehat{\varphi}_0^0 \cap \big(\overline{B}_\varrho \setminus \{0\}\big)$.

Claim 1. $\widehat{\varphi}_0(tu) \leqslant 0$ for all $t \in [0,1]$.

We argue indirectly. So, suppose that we can find $\widehat{t} \in (0,1)$, such that

$$\widehat{\varphi}_0(\widehat{t}u) > 0.$$

Since $u \in \widehat{\varphi}_0^0$ and $\widehat{\varphi}_0$ is continuous, we can define

$$t_1 = \min\big\{t \in (\widehat{t},1] : \widehat{\varphi}_0(tu) = 0\big\} > \widehat{t} > 0.$$

Evidently $t_1 > 0$ is the first time instant in $(\widehat{t},1]$ that the function $t \longmapsto \widehat{\varphi}_0(tu)$ assumes the zero value. We have

$$\widehat{\varphi}_0(tu) > \quad \forall t \in [\widehat{t}, t_1). \qquad (4.42)$$

Let us set $y = t_1 u$. Then

$$0 < \|y\| < \|u\| \leqslant \varrho \quad \text{and} \quad \widehat{\varphi}_0(y) = 0.$$

So, by virtue of (4.41), we have

$$\frac{d}{dt}\widehat{\varphi}_0(ty)\bigg|_{t=1} > 0. \tag{4.43}$$

Also, using (4.42), we have

$$\widehat{\varphi}_0(t_1 u) = 0 < \widehat{\varphi}_0(tu) \quad \forall t \in [\widehat{t}, t_1),$$

so

$$\frac{d}{dt}\widehat{\varphi}_0(ty)\bigg|_{t=1} > t_1 \frac{d}{dt}\widehat{\varphi}_0(tu)\bigg|_{t=t_1} = t_1 \lim_{t \to t_1^-} \frac{\widehat{\varphi}_0(tu)}{t - t_1} \leqslant 0. \tag{4.44}$$

Combining (4.43) and (4.44), we reach a contradiction. This proves Claim 1.

By choosing $\varrho > 0$ even smaller if necessary, we may assume that the origin is the only critical point of $\widehat{\varphi}_0$ (otherwise we have a whole sequence of nontrivial critical points of $\widehat{\varphi}_0$ converging to the origin and so we have a whole sequence of distinct nodal solutions of (4.1) (see Claim 1 in the proof of Theorem 4.5) and we are done). Consider the homotopy

$$\widehat{h}_0 \colon [0,1] \times \left(\widehat{\varphi}_0^0 \cap \overline{B}_\varrho\right) \longrightarrow \widehat{\varphi}_0^0 \cap \overline{B}_\varrho,$$

defined by

$$\widehat{h}_0(t,u) = (1-t)u \quad \forall (t,u) \in [0,1] \times \left(\widehat{\varphi}_0^0 \cap \overline{B}_\varrho\right).$$

Claim 1 guarantees that this homotopy is well-defined. Hence the set $\widehat{\varphi}_0^0 \cap \overline{B}_\varrho$ is contractible.

Next let $u \in \overline{B}_\varrho$ with $\widehat{\varphi}_0(u) > 0$.

Claim 2. There exists unique $t_u \in (0,1)$, such that $\widehat{\varphi}_0\left(t_u u\right) = 0$.

The continuity of the function $t \longmapsto \widehat{\varphi}_0(tu)$ and the fact that $\widehat{\varphi}_0(u) > 0$ together with (4.39), imply that we can find $t_u \in (0,1)$, such that

$$\widehat{\varphi}_0\left(t_u u\right) = 0.$$

We need to show that $t_u \in (0,1)$ is unique. We argue by contradiction. So, suppose that we can find

$$0 < t_u^1 < t_u^2 < 1,$$

such that

$$\widehat{\varphi}_0(t_u^1 u) = \widehat{\varphi}_0(t_u^2 u) = 0.$$

Claim 1 implies that

$$\widehat{\varphi}_0\left(t\, t_u^2 u\right) \leqslant 0 \quad \forall t \in [0,1].$$

Hence $\frac{t_u^1}{t_u^2} \in (0,1)$ is a maximizer of the function $t \longmapsto \widehat{\varphi}_0\left(t\, t_u^2, u\right)$ and so

$$\frac{d}{dt}\widehat{\varphi}_0\left(t\, t_u^1 u\right)\Big|_{t=1} = \frac{t_u^1}{t_u^2}\frac{d}{dt}\widehat{\varphi}_0\left(t\, t_u^1 u\right)\Big|_{t=\frac{t_u^1}{t_u^2}} = 0. \tag{4.45}$$

Note that $u \in \overline{B}_\varrho \setminus \{0\}$ and $\widehat{\varphi}_0\left(t_u^1 u\right) = 0$. Comparing (4.41) and (4.45), we reach a contradiction. This proves Claim 2.

Claims 1 and 2 imply that, if $u \in \overline{B}_\varrho \setminus \{0\}$ with $\widehat{\varphi}_0(u) > 0$, then

$$\widehat{\varphi}_0(tu) < 0 \quad \text{for } t \in (0, t_u)\,, \tag{4.46}$$

$$\widehat{\varphi}_0(tu) > 0 \quad \text{for } t \in (t_u, 1)\,. \tag{4.47}$$

We define the map $\vartheta\colon \overline{B}_\varrho \setminus \{0\} \longrightarrow [0,1]$ by

$$\vartheta(u) = \begin{cases} 1 & \text{if} \quad u \in \widehat{\varphi}_0^0 \cap \left(\overline{B}_\varrho \setminus \{0\}\right), \\ t(u) & \text{if} \quad u \in \overline{B}_\varrho \setminus \{0\},\ \widehat{\varphi}_0(u) > 0. \end{cases} \tag{4.48}$$

Claims 1 and 2 guarantee that ϑ is well defined.

Claim 3. ϑ is continuous.

It is clear from (4.48) that it is enough to check the continuity of ϑ at those $u \in \overline{B}_\varrho \setminus \{0\}$, such that $\widehat{\varphi}_0(u) = 0$. So, suppose that

$$u_n \longrightarrow u, \quad \text{with } \widehat{\varphi}_0(u_n) > 0 \quad \forall n \geqslant 1$$

and $\{u_n\}_{n\geqslant 1} \subseteq \overline{B}_\varrho \setminus \{0\}$, $u \in \overline{B}_\varrho \setminus \{0\}$. Proceeding indirectly and passing to a suitable subsequence if necessary, we may assume that

$$t_{u_n} \leqslant t_0 < 1 \quad \forall n \geqslant 1.$$

From (4.46), (4.47), we have

$$\widehat{\varphi}_0(tu_n) > 0 \quad \forall t \in (t_0, 1],\ n \geqslant 1,$$

so

$$\widehat{\varphi}_0(tu) \geqslant 0 \quad \forall t \in (t_0, 1].$$

Using Claim 1 and recalling that $\widehat{\varphi}_0(u) = 0$, we have

$$\widehat{\varphi}_0(tu_n) = 0 \quad \forall t \in (t_0, 1],$$

so

$$\frac{d}{dt}\widehat{\varphi}_0(tu)\Big|_{t=1} = 0,$$

which contradicts (4.41). This proves Claim 3.

Now let $\overline{\eta}\colon \overline{B}_\varrho \setminus \{0\} \longrightarrow \widehat{\varphi}_0^0 \cap (\overline{B}_\varrho \setminus \{0\})$ be defined by

$$\overline{\eta}(u) = \begin{cases} u & \text{if } u \in \widehat{\varphi}_0^0 \cap (\overline{B}_\varrho \setminus \{0\}), \\ \vartheta(u)u & \text{if } u \in \overline{B}_\varrho \setminus \{0\},\ \widehat{\varphi}_0(u) > 0. \end{cases} \tag{4.49}$$

Claim 2 and (4.48) imply that $\overline{\eta}$ is well defined and Claim 3 guarantees the continuity of $\overline{\eta}$. Then from (4.49), it follows that $\widehat{\varphi}_0^0 \cap (\overline{B}_\varrho \setminus \{0\})$ is a retract of $\overline{B}_\varrho \setminus \{0\}$. We know that $\overline{B}_\varrho \setminus \{0\}$ is contractible in itself. Hence $\widehat{\varphi}_0^0 \cap (\overline{B}_\varrho \setminus \{0\})$ is contractible in itself (see Dugundji [**18**, p. 325]).

From the definition of critical groups at an isolated point (see Section 2), we have

$$C_k(\widehat{\varphi}_0, 0) = H_k\left(\widehat{\varphi}_0^0 \cap \overline{B}_\varrho,\ \widehat{\varphi}_0^0 \cap (\overline{B}_\varrho \setminus \{0\})\right) \quad \forall k \geqslant 0. \tag{4.50}$$

The contractibility of the set $\widehat{\varphi}_0^0 \cap \overline{B}_\varrho$ and $\widehat{\varphi}_0^0 \cap (\overline{B}_\varrho \setminus \{0\})$ implies that

$$H_k\left(\widehat{\varphi}_0^0 \cap \overline{B}_\varrho,\ \widehat{\varphi}_0^0 \cap (\overline{B}_\varrho \setminus \{0\})\right) = 0 \quad \forall k \geqslant 0$$

(see Granas-Dugundji [**28**, p. 389]), so, from (4.50), we have

$$C_k(\widehat{\varphi}_0, 0) = 0 \quad \forall k \geqslant 0.$$

□

Now we are ready for the multiplicity result.

THEOREM 4.11. *If hypotheses $H(f)_4$ hold, then there exists $\beta_0^* > 0$, such that for every $\beta \in L^\infty(\Omega)_+$ with*

$$\operatorname*{ess\,inf}_{\Omega} \beta > 0 \quad \text{and} \quad \|\beta\|_\infty < \beta_0^*,$$

problem (4.1) has at least five nontrivial smooth solutions:

$$u_0, \widehat{u} \in \operatorname{int} C_+,\ u_0 \neq \widehat{u}, \quad v_0, \widehat{v} \in -\operatorname{int} C_+,\ v_0 \neq \widehat{v}, \quad \text{and} \quad y_0 \in C_0^1(\overline{\Omega}) \text{ nodal.}$$

PROOF. From Theorem 3.19, we already have four smooth constant sign solutions:

$$u_0, \widehat{u} \in \operatorname{int} C_+,\ u_0 \neq \widehat{u}, \quad v_0, \widehat{v} \in -\operatorname{int} C_+,\ v_0 \neq \widehat{v}.$$

We consider the functional $\widehat{\varphi}_0 \in C^1\left(W_0^{1,p}(\Omega)\right)$ (see the proof of Theorem 4.5). Suppose that $\{0, u_+, v_-\}$ are the only critical points of $\widehat{\varphi}_0$. From Claim 2 in the proof of Theorem 4.5, we have that $\{u_+, v_-\}$ are local minimizers of $\widehat{\varphi}_0$. Hence

$$C_k(\widehat{\varphi}_0, u_+) = C_k(\widehat{\varphi}, v_-) = \delta_{k,0}\mathbb{Z} \quad \forall k \geqslant 0 \tag{4.51}$$

(see Mawhin-Willem [**37**, p. 175]). Also from Proposition 4.10, we have

$$C_k(\widehat{\varphi}_0, 0) = 0 \quad \forall k \geqslant 0 \tag{4.52}$$

We know that $\widehat{\varphi}_0$ is coercive, hence it is bounded below. So, we can find $\lambda < \inf \widehat{\varphi}_0$. Then $\widehat{\varphi}_0^\lambda = \emptyset$. From the definition of critical groups at infinity (see Section 2) and since $\widehat{\varphi}_0^\lambda = 0$, we have

$$C_k(\widehat{\varphi}_0, \infty) = H_k\left(W_0^{1,p}(\Omega),\ \widehat{\varphi}_0^\lambda\right) = H_k\left(W_0^{1,p}(\Omega)\right) = \delta_{k,0}\mathbb{Z} \quad \forall k \geqslant 0. \tag{4.53}$$

From (4.51), (4.52), (4.53) and the Morse relation (2.1) with $t = -1$, we have

$$2(-1)^0 = (-1)^0,$$

a contradiction. This means that $\widehat{\varphi}_0$ has a fourth critical point $y_0 \notin \{0, u_+, v_-\}$. From Claim 1 in the proof of Theorem 4.5, we know that $y_0 \in [v_-, u_+]$, so y_0 is a solution of (4.1), which is nodal (due to the extremality of u_+, v_-) and $y_0 \in C_0^1(\overline{\Omega})$ (nonlinear regularity theory). □

REMARK 4.12. *A particular case of Theorem 4.11 is when*

$$f(z,\zeta) = f(\zeta) = |\zeta|^{r-2}\zeta,$$

with $p < r < p^*$. This case was studied by García Azorero-Manfredi-Peral Alonso [**23**] *and Guo-Zhang* [**30**] *(with $p > 2$), who obtained two positive solutions. The question of existence of nodal solutions was not addressed in the aforementioned works.*

It is natural to ask what happens in Theorem 4.11 when $\|\beta\|_\infty = \beta_0^*$. The next proposition shows that we can still obtain two nontrivial smooth solutions of constant sign.

PROPOSITION 4.13. *If hypotheses $H(f)_4$ hold and $\beta_0^* > 0$ is as in Theorem 4.11, then for every $\beta \in L^\infty(\Omega)_+$, with*

$$\operatorname*{ess\,inf}_{\Omega} \beta > 0 \quad \text{and} \quad \|\beta\|_\infty = \beta_0^*,$$

problem (4.1) has at least two smooth constant sign solutions

$$u_0 \in \operatorname{int} C_+ \quad \text{and} \quad v_0 \in -\operatorname{int} C_+.$$

PROOF. Let $\{\beta\}_{n\geqslant 1} \subseteq L^\infty(\Omega) \setminus \{0\}$ be an increasing sequence with

$$\operatorname*{ess\,inf}_{\Omega} \beta_n > 0, \quad \|\beta_n\|_\infty < \beta_0^* = \|\beta\|_\infty$$

and

$$\beta_n \longrightarrow \beta \quad \text{in } L^\infty(\Omega).$$

From the proof of Theorem 3.19, we know that we can find $u_n \in \operatorname{int} C_+$, such that

$$\begin{cases} -\Delta_p u_n(z) = \beta_n(z) u_n(z)^{q-1} + f(z, u_n(z)) \text{ in } \Omega, \\ u_n|_{\partial\Omega} = 0, \quad \|u_n\| \leqslant \varrho\beta_0^*, \quad n \geqslant 1. \end{cases} \tag{4.54}$$

From (4.54) and Theorems 6.2.6 and 6.2.7 of Gasiński-Papageorgiou [**25**, pp. 737–738], we can find $\gamma \in (0,1)$ and $M_{17} > 0$, such that

$$u_n \in C_0^{1,\gamma}(\overline{\Omega}) \quad \text{and} \quad \|u_n\|_{C_0^1(\overline{\Omega})} \leqslant M_{17} \quad \forall n \geqslant 1.$$

Exploiting the compactness of the embedding $C_0^{1,\gamma}(\overline{\Omega}) \subseteq C_0^1(\overline{\Omega})$ and by passing to a subsequence if necessary, we may assume that

$$u_n \longrightarrow u_0 \quad \text{in } C_0^1(\overline{\Omega}) \quad \text{as } n \to +\infty. \tag{4.55}$$

Let $\underline{u}_n \in \operatorname{int} C_+$ be a solution of

$$\begin{cases} -\Delta_p u(z) = \beta_n(z) u^+(z)^{q-1} \text{ in } \Omega, \\ u|_{\partial\Omega} = 0. \end{cases}$$

As above via nonlinear regularity theory, we can have

$$\underline{u}_n \longrightarrow \underline{u} \quad \text{in } C_0^1(\overline{\Omega}) \quad \text{as } n \to +\infty. \tag{4.56}$$

Recall that $\underline{u}_n$ is a minimizer of

$$\psi_+^n(u) = \frac{1}{p}\|\nabla u\|_p^p - \frac{1}{q}\int_\Omega \beta_n (u^+)^q \, dz \quad \forall u \in W_0^{1,p}(\Omega) \quad \forall n \geqslant 1.$$

Also, let $\psi_+ \colon W_0^{1,p}(\Omega) \longrightarrow \mathbb{R}$ be defined by

$$\psi_+(u) = \frac{1}{p}\|\nabla u\|_p^p - \frac{1}{q}\int_\Omega \beta (u^+)^q \, dz \quad \forall u \in W_0^{1,p}(\Omega).$$

Note that $\{\psi_+^n\}_{n \geqslant 1}$ is a decreasing sequence and

$$\psi_+^n(u) \longrightarrow \psi_+(u) \quad \forall u \in W_0^{1,p}(\Omega).$$

Then due to (4.55), we have that $\underline{u} \in C_+ \cap \overline{B}_\varrho$ is a minimizer of the limit functional ψ_+ and because

$$\inf_{\overline{B}_\varrho} \psi_+ < 0 = \psi_+(0),$$

we have that $\underline{u} \neq 0$ and so $\underline{u} \in \operatorname{int} C_+$ (see Vázquez [**49**]). Recall that $\underline{u}_n \leqslant u_n$, hence $\underline{u} \leqslant u_0$ and so $u_0 \in \operatorname{int} C_+$ and solves (4.1).

Similarly for the negative solutions, considering this time the auxiliary Dirichlet problem

$$\begin{cases} -\Delta_p v(z) = \beta_n(z)|v(z)|^{q-2} v(z) \quad \text{in } \Omega, \\ v|_{\partial\Omega} = 0, \quad v \leqslant 0. \end{cases} \tag{4.57}$$

Then the approximating negative solutions $\{v_n\}_{n \geqslant 1}$ satisfy $v_n \leqslant \underline{v}_n$ with $\underline{v}_n \in -\operatorname{int} C_+$ being a solution of (4.57) obtained be the direct method. ☐

5. Asymptotically linear perturbations

In this section we investigate problem (1.1) when the perturbation f of the concave term is asymptotically $(p-1)$-linear. So, the hypotheses on f are the following:

$\underline{H(f)_5}$ $f\colon \Omega \times \mathbb{R} \longrightarrow \mathbb{R}$ is a function, such that

(i) for all $\zeta \in \mathbb{R}$, the function $z \longmapsto f(z,\zeta)$ is measurable;

(ii) for almost all $z \in \Omega$, the function $\zeta \longmapsto f(z,\zeta)$ is continuous and we have $f(z,0)=0$;

(iii) for almost all $z \in \Omega$ and all $\zeta \in \mathbb{R}$, we have

$$|f(z,\zeta)| \leqslant a(z) + c|\zeta|^{p-1},$$

with $a \in L^\infty(\Omega)_+$, $c>0$;

(iv) we have

$$\lim_{|\zeta|\to+\infty} \frac{pF(z,\zeta)}{|\zeta|^p} = \frac{c_0}{p-1}\lambda_1 \quad \text{uniformly for almost all } z \in \Omega$$

and

$$\limsup_{|\zeta|\to+\infty} \big(f(z,\zeta)\zeta - pF(z,\zeta)\big) = -\infty \quad \text{uniformly for almost all } z \in \Omega;$$

(v) there exist $a_- < 0 < a_+$ and $\xi^* > 0$, such that

$$\beta(z)a_+^{q-1} + f(z,a_+) \leqslant 0 \leqslant \beta(z)a_-^{q-1} + f(z,a_-) \quad \text{for almost all } z \in \Omega$$

and also the function

$$\zeta \longmapsto \xi^*|\zeta|^{p-2}\zeta + \beta(z)|\zeta|^{q-2}\zeta + f(z,\zeta)$$

is nondecreasing on $[a_-,a_+]$;

(vi) there exists $\eta \in L^\infty(\Omega)_+$ with $\eta(z) \geqslant \beta(z)$ for almost all $z \in \Omega$, $\eta \neq \beta$ and

$$\liminf_{\zeta\to 0} \frac{qF(z,\zeta)}{|\zeta|^q} \geqslant -\eta(z) \quad \text{uniformly for almost all } z \in \Omega.$$

REMARK 5.1. *Hypothesis $H(f)_5(iv)$ is a resonance condition. Indeed, if the differential operator is the p-Laplacian (i.e., $a(y) = \|y\|^{p-2}y$, $y \in \mathbb{R}$), then*

$$\nabla_y a(y) = (p-1)\|y\|^{p-2}$$

and so

$$(\nabla_y a(y)\xi,\ \xi)_{\mathbb{R}^N} = (p-1)\|y\|^{p-2}\|\xi\|^2 \quad \forall \xi \in \mathbb{R}^N$$

(compare with hypothesis $H(a)(iii)$). Hence the first part of hypothesis $H(f)_5(iv)$ becomes

$$\lim_{|\zeta|\to+\infty} \frac{pF(z,\zeta)}{|\zeta|^p} = \lambda_1 \quad \textit{uniformly for almost all } z \in \Omega,$$

which is a resonance condition at infinity with respect to the principal eigenvalue $\lambda_1 > 0$. *The fact that the problem is asymptotically resonant at* $\pm\infty$ *necessitates the second asymptotic condition in hypothesis* $H(f)_5(iv)$. *Hypothesis* $H(f)_5(vi)$ *implies that at zero we can also have a resonance. These hypotheses complement the ones used by Gasiński-Papageorgiou* [**27**], *who have the p-Laplacian, concave terms and asymptotically at* $\pm\infty$ *the quotient* $\frac{pF(z,\zeta)}{|\zeta|^p}$ *is in the spectral interval* (λ_1, λ_2) *with no resonance allowed. Resonant problems with the p-Laplacian and no concave terms, were studied by Gasiński-Papageorgiou* [**24**]

EXAMPLE 5.2. *The following function satisfies hypotheses* $H(f)_5$. *As before, for the sake of simplicity, we drop the z-dependence:*

$$f(\zeta) = \begin{cases} c|\zeta|^{\tau-2}\zeta - \widehat{c}|\zeta|^{r-2}\zeta & \text{if } |\zeta| \leqslant 1, \\ \dfrac{c_0\lambda_1}{p-1}|\zeta|^{p-2}\zeta - |\zeta|^{\vartheta-2}\zeta - \dfrac{c_0\lambda_1}{p-1} + c - \widehat{c} - 1 & \text{if } |\zeta| > 1, \end{cases}$$

with $\widehat{c} > c > 0$, $q < \tau \leqslant p < r$, $\vartheta < p$. *The z-independence of f implies that* $\beta(z) = \lambda > 0$.

First we look for smooth constant sign solutions.

As before (see the proof of Proposition 3.9) by truncating the reaction at the pairs $\{0, a_+\}$ and $\{0, a_-\}$, using the direct method and hypothesis $H(f)_5(vi)$, we obtain the following proposition.

PROPOSITION 5.3. *If hypotheses* $H(a)$ *and* $H(f)_5$ *hold and* $\beta \in L^\infty(\Omega)_+ \setminus \{0\}$, *then problem (1.1) has two smooth constant sign solutions*

$$u_0 \in int C_+, \quad v_0 \in -int C_+, \quad a_- < v_0(z) < u_0(z) < a_+ \quad \forall z \in \overline{\Omega},$$

both local minimizers of φ.

Next using truncation techniques and the nonsmooth mountain pass theorem (the truncations introduce a very mild nonsmoothness in the energy functional), we produce two more smooth solutions of constant sigh. To this end we will need the following stronger version of hypotheses on a:
$\underline{H(a)''}$ $a(z,y) = h(z, \|y\|)\, y$ for all $(z,y) \in \overline{\Omega} \times \mathbb{R}^N$, with $h(z,r) > 0$ for all $(z,r) \in \overline{\Omega} \times (0, +\infty)$ and

(i)-(iii) the same as $H(a)(i) - (iii)$;

(iv) $G(z,0) = 0$ for all $z \in \overline{\Omega}$ and there exist functions $\widehat{\sigma}, \overline{\sigma} \in L^1(\Omega)$, such that

$$\widehat{\sigma}(z) \leqslant pG(z,y) - (a(z,y),\ y)_{\mathbb{R}^N} \leqslant \overline{\sigma}(z) \quad \forall (z,y) \in \Omega \times \mathbb{R}^N.$$

REMARK 5.4. *In the Examples 3.4, all maps* $a_1, \ldots, a_5$ *satisfy* $H(a)'$, *while* a_1, a_2, a_3 *and* a_5 *satisfy* $H(a)''$.

THEOREM 5.5. *If hypotheses* $H(a)''$ *and* $H(f)_5$ *hold and* $\beta \in L^\infty(\Omega)_+ \setminus \{0\}$, *then problem (1.1) has at least four nontrivial smooth constant sign solutions*

$$u_0, \widehat{u} \in int C_+, \ u_0 \neq \widehat{u}, \ u_0 \leqslant \widehat{u}, \quad v_0, \widehat{v} \in -int C_+, \ v_0 \neq \widehat{v}, \ \widehat{v} \leqslant v_0.$$

PROOF. From Proposition 5.3 we already have two solutions:

$$u_0 \in \operatorname{int} C_+ \quad \text{and} \quad v_0 \in -\operatorname{int} C_+.$$

Let

$$\xi_+ = \frac{1}{q}\int_\Omega \beta u_0^q \, dz + \int_\Omega F(z,u_0)\, dz$$

and consider the locally Lipschitz functional $\psi_+ \colon W_0^{1,p}(\Omega) \longrightarrow \mathbb{R}$, defined by

$$\begin{aligned} \psi_+(u) = {} & \int_\Omega G\left(z, \nabla(u_0+u)\right) dz - \int_\Omega \widetilde{F}_+\left(z, u_0+u^+\right) dz \\ & + \int_\Omega \left(\beta u_0^{q-1} + f(z,u_0)\right) u^- \, dz + \xi_+, \end{aligned} \tag{5.1}$$

where

$$\widetilde{F}_+(z,\zeta) = \int_0^\zeta \widetilde{f}_+(z,s)\, ds,$$

with $\widetilde{f}_+$ given by (3.75). Note that

$$G\left(z, \nabla(u_0+u)\right) = G\left(z, \nabla(u_0+u^+)\right) + G\left(z, \nabla(u_0-u^-)\right) - G(z, \nabla u_0). \tag{5.2}$$

Using (5.2) in (5.1), we obtain

$$\begin{aligned} \psi_+(u) = {} & \varphi(u_0+u^+) - \varphi(u_0) + \int_\Omega G\left(z, \nabla(u_0-u^-)\right) dz \\ & + \int_\Omega \left(\beta u_0^{q-1} + f(z,u_0)\right) u^- \, dz. \end{aligned} \tag{5.3}$$

CLAIM 4. We can find $\varrho > 0$ small enough, such that

$$\sigma_+ = \inf\left\{\psi_+(u) : \ \|u\| = \varrho\right\} \geqslant \psi_+(0). \tag{5.4}$$

Since u_0 is a local minimizer of φ (see Proposition 5.3), we can find $l > 0$ small enough, such that

$$\mu_\varrho = \inf\left\{\varphi(u) : \ u \in \overline{B}_l(u_0) \setminus B_\varrho(u_0)\right\} > \varphi(u_0) \tag{5.5}$$

(see Gasiński-Papageorgiou [**26**, proof of Theorem 3.4]).

We fix $\varrho \in (0,l)$ and let $\|u\| = \varrho$. Then

$$\|u^+\| \geqslant \frac{\varrho}{2} \quad \text{or} \quad \|u^-\| \geqslant \frac{\varrho}{2}.$$

First suppose that $\|u^+\| \geqslant \frac{\varrho}{2}$. Then from (5.3) and (5.5), recalling that $V(u_0) = \beta u_0^{q-1} + N(u_0)$, using the fact $\widehat{\mu}_{\frac{\varrho}{2}} = \mu_{\frac{\varrho}{2}} - \varphi(u_0) > 0$ and the continuity of the map

$u \longmapsto \int_\Omega G(z, \nabla u)\,dz$, we have

$$\begin{aligned}\psi_+(u) &\geqslant \mu_{\frac{\varrho}{2}} - \varphi(u_0) + \int_\Omega G\left(z, \nabla(u_0 - u^-)\right)\,dz + \left\langle V(u_0), u^- \right\rangle \\ &= \widehat{\mu}_{\frac{\varrho}{2}} + \int_\Omega \Big(G\left(z, \nabla(u_0 - u^-)\right) + \left(a(z, \nabla u_0), \nabla u^-\right)_{\mathbb{R}^N}\Big)\,dz \\ &= \widehat{\mu}_{\frac{\varrho}{2}} + \int_\Omega G(z, \nabla u_0)\,dz \\ &> \int_\Omega G(z, \nabla u_0)\,dz = \psi_+(0). \end{aligned} \tag{5.6}$$

Next suppose that $\|u^-\| \geqslant \frac{\varrho}{2}$. Then from (5.3) and since u_0 is a minimizer of φ on $\overline{B}_l(u_0)$ (choosing $l > 0$ smaller, if necessary) and since $u_0 \in \operatorname{int} C_+$ solves (1.1), we have

$$\begin{aligned}\psi_+(u) &\geqslant \int_\Omega G\left(z, \nabla(u_0 - u^-)\right)\,dz + \int_\Omega \Big(\beta u_0^{q-1} + f(z, u_0)\Big)\,u^-\,dz \\ &= \int_\Omega G\left(z, \nabla(u_0 - u^-)\right)\,dz + \left\langle V(u_0), u^- \right\rangle \\ &= \int_\Omega \Big(G\left(z, \nabla(u_0 - u^-)\right) + \left(a(z, \nabla u_0), \nabla u^-\right)_{\mathbb{R}^n}\Big)\,dz \\ &\geqslant \int_\Omega G(z, \nabla u_0)\,dz = \psi_0(0). \end{aligned} \tag{5.7}$$

From (5.6) and (5.7), we conclude that (5.4) holds. This proves Claim 1.

CLAIM 5. ψ_+ satisfies the C-condition.

Let $\{u_n\}_{n\geqslant 1} \subseteq W_0^{1,p}(\Omega)$ be a sequence, such that

$$\left|\psi_+(u_n)\right| \leqslant M_{18} \quad \forall n \geqslant 1, \tag{5.8}$$

for some $M_{18} > 0$ and

$$(1 + \|u_n\|)m_{\psi_+}(u_n) \longrightarrow 0 \quad \text{as } n \to \infty \tag{5.9}$$

(see Section 2). We can find $u_n^* \in \partial\psi_+(u_n)$, such that $m_{\psi_+}(u_n) = \|u_n^*\|$ and

$$u_n^* = V(u_0 + u_n) - \beta(u_0 + u_n^+)^{q-1} - N(u_0 + u_n^+) + v_n^* \quad \forall n \geqslant 1, \tag{5.10}$$

with $v_n^* = h_0 \partial i_-(u_n)$, where

$$h_0(z) = \beta(z)u_0(z)^{q-1} + f\left(z, u_n(z)\right)$$

and $i_- : \mathbb{R} \longrightarrow \mathbb{R}$ is the function, defined by $i_-(u) = u^-$. From the nonsmooth change rule (see Clarke [**12**, p. 45]), we have

$$\partial i_-(\zeta) = \begin{cases} -1 & \text{if } u < 0, \\ [-1, 0] & \text{if } u = 0, \\ 0 & \text{if } u > 0. \end{cases} \tag{5.11}$$

Also recall that

$$V(u_0+u_n) = V(u_0+u_n^+) + V(u_0-u_n^-) - V(u_0). \tag{5.12}$$

From (5.9), using (5.10) and (5.12), for all $n \geqslant 1$ and all $h \in W_0^{1,p}(\Omega)$ we have

$$\begin{aligned} &\Big|\big\langle V(u_0+u_n^+) + V(u_0-u_n^-) - V(u_0),\ h\big\rangle - \int_\Omega \beta(u_0+u_n^+)^{q-1} h\,dz \\ &\quad - \int_\Omega f(z,u_0+u_n^+)h\,dz + \int_\Omega v_n^* h\,dz\Big| \leqslant \frac{\varepsilon_n}{1+\|u_n\|}\|h\|, \end{aligned} \tag{5.13}$$

with $\varepsilon_n \searrow 0^+$. In (5.13), we choose $h = u_n^- \in W_0^{1,p}(\Omega)$. Using Lemma 3.2, we have

$$\begin{aligned} &\big\langle V(u_0+u_n^+), -u_n^-\big\rangle + \big\langle V(u_0-u_n^-), -u_n^-\big\rangle + \big\langle V(u_0), u_n^-\big\rangle \\ &\quad = \big\langle V(u_0-u_n^-), -u_n^-\big\rangle \\ &\quad = \big\langle V(u_0-u_n^-), u_0-u_n^-\big\rangle - \big\langle V(u_0-u_n^-), u_0\big\rangle \\ &\quad \geqslant \frac{c_0}{p-1}\|\nabla(u_0-u_n^-)\|_p^p - \frac{c_0}{p-1}\|\nabla(u_0-u_n^-)\|_p^{p-1}\|u_0\|. \end{aligned} \tag{5.14}$$

Also, using hypothesis $H(f)_5(v)$, we have

$$\begin{aligned} &-\int_\Omega \big(\beta(u_0+u_n^+)^{q-1} + f(z,u_0+u_n^+)\big)(-u_n^-)\,dz \\ &\quad = \int_\Omega \Big(\beta u_0^{q-1} + f(z,u_0)\Big)u_n^-\,dz \geqslant 0, \end{aligned} \tag{5.15}$$

and using hypothesis $H(f)_5(v)$ and (5.11), we have

$$\int_\Omega v_n^*(-u_n^-)\,dz = \int_\Omega h_0 u_n^-\,dz \geqslant 0. \tag{5.16}$$

Returning to (5.13) and using (5.14), (5.15) and (5.16), we obtain

$$\frac{c_0}{p-1}\|\nabla(u_0-u_n^-)\|_p^p \leqslant c_{22}\Big(1+\|\nabla(u_0-u_n^-)\|_p^{p-1}\Big) \quad \forall n \geqslant 1,$$

for some $c_{22} > 0$ and so the sequence $\{u_0-u_n^-\}_{n\geqslant 1} \subseteq W_0^{1,p}(\Omega)$ is bounded and thus the sequence $\{u_n^-\}_{n\geqslant 1} \subseteq W_0^{1,p}(\Omega)$ is bounded.

Suppose that the sequence $\{u_n^+\}_{n\geqslant 1} \subseteq W_0^{1,p}(\Omega)$ is not bounded. By passing to a suitable subsequence if necessary, we may assume that

$$\|u_n^+\| \longrightarrow +\infty \quad \text{as } n \to +\infty,$$

hence

$$\|u_0+u_n^+\| \longrightarrow 0 \quad \text{as } n \to +\infty.$$

We set

$$v_n = \frac{u_0 + u_n^+}{\|u_0 + u_n^+\|} \quad \forall n \geqslant 1.$$

Evidently

$$v_n \geqslant 0 \quad \text{and} \quad \|v_n\| = 1 \quad \forall n \geqslant 1.$$

So, we may assume that

$$\begin{aligned} v_n &\longrightarrow v \quad \text{weakly in } W_0^{1,p}(\Omega), \\ v_n &\longrightarrow v \quad \text{in } L^p(\Omega) \end{aligned}$$

and $v \geqslant 0$. By virtue of hypothesis $H(f)_5(iv)$, for a given $\varepsilon > 0$, we can find $M_{19} = M_{19}(\varepsilon) > 0$, such that for almost all $z \in \Omega$ and all $|\zeta| \geqslant M_{19}$, we have

$$\frac{1}{p}\Big(\frac{c_0}{p-1}\lambda_1 - \varepsilon\Big)|\zeta|^p \leqslant F(z,\zeta) \leqslant \frac{1}{p}\Big(\frac{c_0}{p-1}\lambda_1 + \varepsilon\Big)|\zeta|^p. \tag{5.17}$$

From (5.8), since $h_0 \geqslant 0$, $\xi_+ > 0$ and using Corollary 3.3, we have

$$\begin{aligned} \frac{M_{18}}{\|u_0 + u_n^+\|^p} &= \frac{1}{\|u_0 + u_n^+\|^p}\int_\Omega G\left(z, \nabla(u_0 + u_n^+)\right) dz - \frac{1}{q\|u_0 + u_n^+\|^{p-q}}\int_\Omega \beta y_n^q \, dz \\ &\quad - \int_\Omega \frac{F(z, u_0 + u_n^+)}{\|u_0 + u_n^+\|^p} dz + \int_\Omega \frac{h_0}{\|u_0 + u_n^+\|^p} u_n^- \, dz + \frac{\xi_+}{\|u_0 + u_n^+\|^p} \\ &\geqslant \frac{c_0}{p(p-1)}\|\nabla y_n\|^p - \frac{1}{q\|u_0 + u_n^+\|^{p-q}}\int_\Omega \beta y_n^q \, dz \\ &\quad - \int_\Omega \frac{F(z, u_0 + u_n^+)}{\|u_0 + u_n^+\|^p} dz. \end{aligned} \tag{5.18}$$

From (5.17) and using hypothesis $H(f)_5(iii)$, we have

$$\begin{aligned} &\int_\Omega \frac{F(z, u_0 + u_n^+)}{\|u_0 + u_n^+\|^p} dz \\ &\quad = \int_{\{u_0 + u_n^+ \geqslant M_{19}\}} \frac{F(z, u_0 + u_n^+)}{\|u_0 + u_n^+\|^p} dz + \int_{\{0 \leqslant u_0 + u_n^+ < M_{19}\}} \frac{F(z, u_0 + u_n^+)}{\|u_0 + u_n^+\|^p} dz \\ &\quad \leqslant \frac{1}{p}\Big(\frac{c_0}{p-1}\lambda_1 + \varepsilon\Big)\|y_n\|_p^p + \frac{c_{23}}{\|u_0 + u_n^+\|^p} \quad \forall n \geqslant 1, \end{aligned}$$

for some $c_{23} > 0$. Using this estimate in (5.18) and since $q < p$, we have

$$\frac{c_0}{p(p-1)}\left(\|\nabla y_n\|_p^p - \lambda_1\|y_n\|_p^p\right) - \frac{\varepsilon}{p}\|y_n\|_p^p \leqslant \varepsilon_n',$$

with $\varepsilon_n' \searrow 0$. Passing to the limit as $n \to +\infty$ and recalling that $\varepsilon > 0$ was arbitrary, we obtain

$$\|\nabla y\|_p^p \leqslant \lambda_1\|y\|_p^p,$$

so

$$y = 0 \quad \text{or} \quad y = \lambda \widehat{u}_1,$$

for some $\lambda > 0$ (see (2.3)).

If $y = 0$, then

$$y_n \longrightarrow 0 \quad \text{in } W_0^{1,p}(\Omega),$$

a contradiction to the fact that $\|y_n\| = 1$ for all $n \geqslant 1$.

Hence $y = \lambda \widehat{u}_1 \in \operatorname{int} C_+$. This implies that

$$(u_0 + u_n^+)(z) \longrightarrow +\infty \quad \text{for almost all } z \in \Omega.$$

By virtue of hypothesis $H(f)_5(iv)$, for almost all $z \in \Omega$, we have

$$f\left(z, (u_0 + u_n^+)(z)\right)(u_0 + u_n^+)(z) - pF\left(z, (u_0 + u_n^+)(z)\right) \longrightarrow -\infty.$$

This, by Fatou's lemma (see hypothesis $H(f)_5(iv)$) implies that

$$\int_\Omega \left(f(z, (u_0 + u_n^+))(u_0 + u_n^+) - pF(z, u_0 + u_n^+)\right) dz \longrightarrow -\infty. \tag{5.19}$$

On the other hand, exploiting the fact that the sequence $\{u_n^-\}_{n \geqslant 1} \subseteq W_0^{1,p}(\Omega)$ is bounded, we have

$$\begin{aligned} p\psi_+(u_n) = & \int_\Omega pG\left(z, \nabla(u_0 + u_n)\right) dz - \frac{p}{q} \int_\Omega \beta(u_0 + u_n^+)^q \, dz \\ & - \int_\Omega pF(z, u_0 + u_n^+) \, dz + \int_\Omega p\left(\beta u_0^{q-1} + f(z, u_0)\right) u_n^- \, dz + p\xi_+ \\ \leqslant & \int_\Omega pG\left(z, \nabla(u_0 + u_n^+)\right) dz - \frac{p}{q} \int_\Omega \beta(u_0 + u_n^+)^q \, dz \\ & - \int_\Omega pF(z, u_0 + u_n^+) \, dz + c_{24} \quad \forall n \geqslant 1, \end{aligned} \tag{5.20}$$

with $c_{24} > 0$. Also, using (5.10) and (5.12), we have

$$\begin{aligned} -\langle u_n^*, u_0 + u_n^+ \rangle = & -\left\langle V(u_0 + u_n^+), u_0 + u_n^+ \right\rangle - \left\langle V(u_0 - u_n^-), u_0 + u_n^+ \right\rangle \\ & + \left\langle V(u_0), u_0 + u_n^+ \right\rangle + \int_\Omega \beta(u_0 + u_n^+)^q \, dz \\ & + \int_\Omega f(z, u_0 + u_n^+)(u_0 + u_n^+) \, dz - \int_\Omega v_n^*(u_0 + u_n^+) \, dz \\ \leqslant & -\int_\Omega \left(a\left(z, \nabla(u_0 + u_n^+)\right), \ \nabla(u_0 + u_n^+)\right)_{\mathbb{R}^n} dz \\ & + \int_\Omega \beta(u_0 + u_n^+)^q \, dz \ \int_\Omega f(z, u_0 + u_n^+)(u_0 + u_n^+) \, dz \\ & - \int_\Omega v_n^*(u_0 + u_n^+) \, dz + c_{25} \quad \forall n \geqslant 1, \end{aligned} \tag{5.21}$$

for some $c_{25} > 0$. Adding (5.20) and (5.21) and using hypothesis $H(a)''(iv)$, (5.11) and the fact that $q < p$, we obtain

$$p\psi_+(u_n) - \langle u_n^*, u_0 + u_n^+\rangle \leqslant \int_\Omega \big(f(z, u_0 + u_n^+)(u_0 + u_n^+) - pF(z, u_0 + u_n^+)\big)\, dz + c_{26} \quad \forall n \geqslant 1, \tag{5.22}$$

for some $c_{26} > 0$. From (5.8) and (5.9), we have

$$-c_{27} \leqslant p\psi_+(u_n) - \langle u_n^*,\ u_0 + u_n^*\rangle \quad \forall n \geqslant 1,$$

for some $c_{27} > 0$, so using (5.22), we have

$$-c_{28} \leqslant \int_\Omega \big(f(z, u_0 + u_n^+)(u_0 + u_n^+) - pF(z, u_0 + u_n^+)\big)\, dz \quad \forall n \geqslant 1, \tag{5.23}$$

for $c_{28} = c_{27} + c_{26} > 0$.

Comparing (5.19) and (5.23), we reach a contradiction. Therefore the sequence $\{u_n^+\}_{n\geqslant 1} \subseteq W_0^{1,p}(\Omega)$ is bounded which together with the boundedness of $\{u_n^-\}_{n\geqslant 1} \subseteq W_0^{1,p}(\Omega)$ implies that the sequence $\{u_n\}_{n\geqslant 1} \subseteq W_0^{1,p}(\Omega)$ is bounded. Therefore, passing to a subsequence if necessary, we may assume that

$$u_n \longrightarrow u \quad \text{weakly in } W_0^{1,p}(\Omega), \tag{5.24}$$

$$u_n \longrightarrow u \quad \text{in } L^p(\Omega). \tag{5.25}$$

From (5.13) with $h = u_n - u \in W_0^{1,p}(\Omega)$ and (5.24), we obtain

$$\lim_{n\to+\infty} \langle V(u_n), u_n - u\rangle = 0,$$

so, using also Proposition 3.5, we have

$$u_n \longrightarrow u \quad \text{in } W_0^{1,p}(\Omega).$$

this proves Claim 2.

CLAIM 6. There exists $u \in W_0^{1,p}(\Omega)$ with $\|u\| > \varrho$ ($\varrho > 0$ as in Claim 1), such that

$$\psi_+(u) < \psi_+(0) \leqslant \sigma_+.$$

Hypothesis $H(f)_5(iv)$ implies that for a given $\xi > 0$, we can find $M_{20} = M_{20}(\xi) > 0$, such that

$$f(z,\zeta)\zeta - pF(z,\zeta) \leqslant -\xi \quad \text{for almost all } z \in \Omega, \text{ all } |\zeta| \geqslant M_{20}. \tag{5.26}$$

We have

$$\begin{aligned}\frac{d}{d\zeta}\left(\frac{F(z,\zeta)}{\zeta^p}\right) &= \frac{\zeta^p f(z,\zeta) - p\zeta^{p-1}F(z,\zeta)}{\zeta^{2p}} \\ &= \frac{f(z,\zeta)\zeta - pF(z,\zeta)}{\zeta^{p+1}} \\ &\leqslant -\frac{\xi}{\zeta^{p+1}} \quad \text{for almost all } z \in \Omega, \text{ all } \zeta \geqslant M_{20},\end{aligned}$$

so

$$\frac{F(z,v)}{v^p} - \frac{F(z,y)}{y^p} \leqslant \frac{\xi}{p}\left(\frac{1}{v^p} - \frac{1}{y^p}\right) \quad \text{for almost all } z \in \Omega, \text{ all } v \geqslant y \geqslant M_{20}.$$

Let $y \to +\infty$. Using hypothesis $H(f)_5(iv)$, we obtain

$$\frac{c_0\lambda_1}{p-1} - \frac{F(z,y)}{y^p} \leqslant -\frac{\xi}{py^p},$$

so

$$\frac{c_0\lambda_1}{p-1}y^p - F(z,y) \leqslant -\xi \quad \text{for almost all } z \in \Omega, \text{ all } y \geqslant M_{20}. \tag{5.27}$$

Let $t > 0$. Then using Corollary 3.3, we have

$$\begin{aligned} p\varphi(tu_1) &\leqslant \frac{c_1 t^p}{p(p-1)}\|\nabla \widehat{u}_1\|_p^p - \frac{t^q}{q}\int_\Omega \beta\widehat{u}_1\,dz - \int_\Omega pF(z,t\widehat{u}_1)\,dz \\ &\leqslant \int_\Omega \left(\frac{c_1\lambda_1}{p(p-1)}(t\widehat{u}_1)^p - pF(z,t\widehat{u}_1)\right)dz \\ &= \int_{\{t\widehat{u}_1 \geqslant M_{20}\}} \left(\frac{c_1\lambda_1}{p(p-1)}(t\widehat{u}_1)^p - pF(z,t\widehat{u}_1)\right)dz \\ &\quad + \int_{\{0 \leqslant t\widehat{u}_1 < M_{20}\}} \left(\frac{c_1\lambda_1}{p(p-1)}(t\widehat{u}_1)^p - pF(z,t\widehat{u}_1)\right)dz \\ &\leqslant -\xi\big|\{t\widehat{u}_1 \geqslant M_{20}\}\big|_N + c_{29}\big|\{0 \leqslant t\widehat{u}_1 < M_{20}\}\big|_N, \end{aligned} \tag{5.28}$$

for some $c_{29} > 0$. Note that

$$\big|\{0 \leqslant t\widehat{u}_1 < M_{20}\}\big|_N \longrightarrow 0 \quad \text{as } t \to +\infty$$

(recall that $\widehat{u}_1 \in \operatorname{int} C_+$).

So, from (5.28) and since $\xi > 0$ was arbitrary, it follows that

$$\varphi(t\widehat{u}_1) \longrightarrow -\infty \quad \text{as } t \to +\infty. \tag{5.29}$$

Since $\widehat{u}_1 \in \operatorname{int} C_+$, using Lemma 2.3, we can find $t > 0$ large enough, such that

$$t\widehat{u}_1 - u_0 \in \operatorname{int} C_+.$$

Then from (5.3), we have

$$\begin{aligned} \psi_+(t\widehat{u}_1 - u_0) &= \varphi(t\widehat{u}_1) - \varphi(u_0) + \int_\Omega G(z,\nabla u_0)\,dz \\ &= \varphi(t\widehat{u}_1) - \varphi(u_0) + \psi_+(0), \end{aligned}$$

so, using (5.29), we have

$$\psi_+(t\widehat{u}_1 - u_0) \longrightarrow -\infty \quad \text{as } t \to +\infty.$$

This proves Claim 3.

Claims 1, 2 and 3 permit the application of the mountain pass theorem (see Gasiński-Papageorgiou [**24**, p. 140]) and so we obtain $w_0 \in W_0^{1,p}(\Omega) \setminus \{0\}$, such that

$$0 \in \partial\psi_+(w_0),$$

so

$$V(u_0 + w_0) = \beta(u_0 + w_0^+)^{q-1} + N(u_0 + w_0^+) - v_0^*, \tag{5.30}$$

where $v_0^* \in h_0 \partial i_-(w_0)$. Also recall that

$$V(u_0) = \beta u_0^{q-1} + N(u_0). \tag{5.31}$$

Subtracting (5.31) from (5.30), we obtain

$$V(u_0+w_0) - V(u_0) = \beta\left((u_0 + w_0^+)^{q-1} - u_0^{q-1}\right) + N(u_0+w_0^+) - N(u_0) - v_0^*. \tag{5.32}$$

On (5.32), we act with $-w_0^- \in W_0^{1,p}(\Omega)$ and obtain

$$\left\langle V(u_0 + w_0^-) - V(u_0),\ -w_0^- \right\rangle \leqslant 0$$

(see (5.11)), so, due to strict monotonicity of V, we have $w_0^- = 0$, hence

$$w_0 \geqslant 0, \quad w_0 \neq 0.$$

From (5.30) and if $\widehat{u} = u_0 + w_0 \geqslant 0$, we have

$$\begin{cases} -\operatorname{div} a\left(z, \nabla\widehat{u}(z)\right) = \beta(z)\widehat{u}(z)^{q-1} + f\left(z, \widehat{u}(z)\right) - v_0^*(z) & \text{in } \Omega, \\ \widehat{u}|_{\partial\Omega} = 0. \end{cases}$$

By Stampacchia's theorem (see e.g., [**25**] [**25**, p. 195]), we have that

$$\nabla\widehat{u}(z) = \nabla u_0(z) \quad \text{for almost all } z \in w_0^{-1}(0).$$

Since

$$-\operatorname{div} a\left(z, \nabla u_0(z)\right) = \beta(z) u_0(z)^{q-1} + f\left(z, u_0(z)\right) \quad \text{in } z \in \Omega$$

(see (5.31)), we obtain

$$v_0^*(z) = 0 \quad \text{for almost all } z \in w_0^{-1}(0).$$

Recall that $w_0 \geqslant 0$. Then from (5.11), we conclude that $v_0^* = 0$ and so $\widehat{u} \in W_0^{1,p}(\Omega) \setminus \{0\}$, $\widehat{u} \geqslant u_0$, $\widehat{u} \neq u_0$ is a solution of (1.1). As before nonlinear regularity implies that $\widehat{u} \in \operatorname{int} C_+$.

Similarly, working with

$$\psi_-(y) = \int_\Omega G\left(z, \nabla(v_0 + y)\right) dz - \frac{1}{q}\int_\Omega \widetilde{F}_-(z, v_0 - y^-)\, dz + \int_\Omega \left(\beta|v_0|^{q-2}v_0 + f(z, v_0)\right) y^+\, dz + \xi_-,$$

where

$$\xi_- = \frac{1}{q}\int_\Omega \beta|v_0|^q\,dz + \int_\Omega F(z,v_0)\,dz,$$

we obtain a second negative smooth solution

$$\widehat{v} \in -\mathrm{int}\,C_+, \quad \widehat{v} \leqslant v_0, \quad \widehat{v} \neq v_0. \qquad \square$$

As before, to produce a nodal solution we need to strengthen the hypotheses on f near zero and consider the p-Laplacian (i.e., problem (4.1)).
$\underline{H(f)_6}$ $f\colon \Omega \times \mathbb{R} \longrightarrow \mathbb{R}$ is a function, such that

(i)-(v) the same as hypotheses $H(f)_5(i)-(v)$;
(vi) there exists $\widehat{\eta} \in L^\infty(\Omega)_+$, such that

$$\lambda_2 < \liminf_{|\zeta|\to 0} \frac{f(z,\zeta)}{|\zeta|^{p-2}\zeta} \leqslant \limsup_{|\zeta|\to 0} \frac{f(z,\zeta)}{|\zeta|^{p-2}\zeta} \leqslant \widehat{\eta}(z),$$

uniformly for almost all $z \in \Omega$ and

$$\beta(z)|\zeta|^q + f(z,\zeta)\zeta \geqslant 0 \quad \text{for almost all } z \in \Omega, \text{ all } \zeta \in [a_-, a_+].$$

REMARK 5.6. *Again the sign condition in $H(f)_6(vi)$ implies that in $H(f)_6(v)$, we have*

$$\beta(z)a_+^{q-1} + f(z,a_+) = 0 = \beta(z)|a_-|^{q-2}a_- + f(z,a_-) \quad \text{for almost all } z \in \Omega.$$

EXAMPLE 5.7. *The following function satisfies hypotheses $H(f)_6$, when $\beta(z) = \lambda > 0$.*

$$f(\zeta) = \begin{cases} x|\zeta|^{p-2}\zeta - \widehat{c}|\zeta|^{r-2}\zeta & \text{if } |\zeta| \leqslant 1, \\ \lambda_1|\zeta|^{p-2}\zeta - \xi|\zeta|^{\vartheta-2}\zeta & \text{if } |\zeta| > 1 \end{cases}$$

where $\widehat{c} > c > \lambda_2$, $\xi = \lambda_1 + \widehat{c} - c > 0$ and $\vartheta < p < r$.

Then as before (see Proposition 4.4), we can have extremal solutions of constant sign $u_+ \in \mathrm{int}\,C_+$, $v_- \in -\mathrm{int}\,C_+$ and reasoning as in Theorem 4.5, we obtain

THEOREM 5.8. *If hypotheses $H(f)_6$ hold and $\beta \in L^\infty(\Omega)_+ \setminus \{0\}$, then problem (4.1) has at least five nontrivial smooth solutions:*

$$u_0, \widehat{u} \in int C_+, \quad u_0 \leqslant \widehat{u}, \quad u_0 \neq \widehat{u}, \quad u_0(z) < a_+ \quad \forall z \in \overline{\Omega},$$

$$v_0, \widehat{v} \in -int C_+, \quad \widehat{v} \leqslant v_0, \quad \widehat{v} \neq v_0, \quad a_- < v_0(z) \quad \forall z \in \overline{\Omega},$$

and $y_0 \in C_0^1(\overline{\Omega}) \setminus \{0\}$ nodal.

6. Semilinear problems

In this section we consider the semilinear problem (i.e., $p = 2$). So, the problem under consideration is the following:

$$\left\{ \begin{array}{l} -\Delta u(z) = \beta(z)\left|u(z)\right|^{q-2}u(z) + f(z, u(z)) \text{ in } \Omega, \\ u|_{\partial\Omega} = 0, \ 1 < q < 2, \ N \geqslant 2. \end{array} \right. \tag{6.1}$$

Using Morse theory, we will produce additional nontrivial smooth solutions.

First we consider the case of a superlinear perturbation f:

$\underline{H(f)_7}$ $f\colon \Omega \times \mathbb{R} \longrightarrow \mathbb{R}$ is a function, such that

(i) for all $\zeta \in \mathbb{R}$, the function $z \longmapsto f(z,\zeta)$ is measurable;

(ii) for almost all $z \in \Omega$, the function $\zeta \longmapsto f(z,\zeta)$ is C^1 and $f(z,0) = 0$;

(iii) for almost all $z \in \Omega$ and all $\zeta \in \mathbb{R}$, we have

$$\left|f'_\zeta(z,\zeta)\right| \leqslant a(z) + c|\zeta|^{r-2},$$

with $a \in L^\infty(\Omega)_+$, $c > 0$ and $2 < r < 2^*$;

(iv) we have

$$\lim_{|\zeta|\to+\infty} \frac{F(z,\zeta)}{|\zeta|^2} = +\infty \quad \text{uniformly for almost all } z \in \Omega$$

and there exist $\mu \in \left((r-2)\frac{N}{2}, 2^*\right)$, $\mu > q$ and $\beta_0 > 0$, such that

$$\liminf_{|\zeta|\to+\infty} \frac{f(z,\zeta)\zeta - 2F(z,\zeta)}{|\zeta|^\mu} \geqslant \beta_0 \quad \text{uniformly for almost all } z \in \Omega;$$

(v) we have

$$\lim_{\zeta\to 0} \frac{f(z,\zeta)}{\zeta} = 0 \quad \text{uniformly for almost all } z \in \Omega;$$

(vi) for almost all $z \in \Omega$, the map $\zeta \longmapsto f(z,\zeta)$ is nondecreasing;

EXAMPLE 6.1. *The function $f(\zeta) = \zeta \ln(1+\zeta^2)$ satisfies hypotheses $H(f)_7$ but not the Ambrosetti-Rabinowitz condition.*

The energy functional $\varphi\colon H^1_0(\Omega) \longrightarrow \mathbb{R}$ of problem (1.1) is given by

$$\varphi(u) = \frac{1}{2}\|\nabla u\|_2^2 - \frac{1}{q}\int_\Omega \beta|u|^q\,dz - \int_\Omega F(z,u)\,dz \quad \forall u \in H^1_0(\Omega).$$

Evidently $\varphi \in C^2\left(H^1_0(\Omega) \setminus \{0\}\right) \cap C^{2-0}\left(H^1_0(\Omega)\right)$.

PROPOSITION 6.2. *If hypotheses $H(f)_7$ hold, then*

$$C_k(\varphi,\infty) = 0 \quad \forall k \geqslant 0.$$

PROOF. By virtue of hypotheses $H(f)_7(iii)$ and (iv), for a given $\xi > 0$, we can find $c_{30} = c_{30}(\xi) > 0$, such that

$$F(z,\zeta) \geqslant \frac{\xi}{2}|\zeta|^2 - c_{30} \quad \text{for almost all } z \in \Omega, \text{ all } \zeta \in \mathbb{R}. \tag{6.2}$$

Let $u \in \partial B_1 = \{u \in H_0^1(\Omega) : \|u\| = 1\}$ and $t > 0$. Then, using (6.2) and the fact that $\|u\| = 1$, we have

$$
\begin{aligned}
\varphi(tu) &= \frac{t^2}{2}\|\nabla u\|_2^2 - \frac{t^q}{q}\int_\Omega \beta u^q\,dz - \int_\Omega F(z,tu)\,dz \\
&\leqslant \frac{t^2}{2} - \frac{\xi t^2}{2}\|u\|_2^2 + c_{30}|\Omega|_N \\
&= \frac{t^2}{2}\left(1 - \xi\|u\|_2^2\right) + c_{30}|\Omega|_N.
\end{aligned} \tag{6.3}
$$

We choose $\xi > \frac{1}{\|u\|_2^2}$. Then from (6.3), we see that

$$
\varphi(tu) \longrightarrow -\infty \quad \text{as } t \to +\infty. \tag{6.4}
$$

Hypothesis $H(f)_7$ implies that there exist $\beta_1 \in (0, \beta_0)$ and $M_{21} > 0$, such that

$$
f(z,\zeta)\zeta - 2F(z,\zeta) \geqslant \beta_1|\zeta|^\mu \quad \text{for almost all } z \in \Omega, \text{ all } |\zeta| \geqslant M_{21}. \tag{6.5}
$$

Then, using (6.5) and hypothesis $H(f)_7(iii)$, for $u \in H_0^1(\Omega)$, we have

(6.6)

$$
\begin{aligned}
&\int_\Omega (2F(z,u) - f(z,u)u)\,dz \\
&= \int_{\{|u|\geqslant M_{21}\}} (2F(z,u) - f(z,u)u)\,dz + \int_{\{|u|<M_{21}\}} (2F(z,u) - f(z,u)u)\,dz \\
&\leqslant -\beta_1 \int_{\{|u|\geqslant M_{21}\}} |u|^\mu\,dz + c_{31},
\end{aligned}
$$

for some $c_{31} > 0$. Let $c_{32} = c_{31} + 1 > 0$ and choose $\xi < -c_{32}$. By virtue of (6.4), for $t > 0$ large and for $u \in \partial B_1$, we have

$$
\varphi(tu) = \frac{t^2}{2}\|\nabla u\|_2^2 - \frac{t^q}{q}\int_\Omega \beta|u|^q\,dz - \int_\Omega F(z,tu)\,dz \leqslant \xi. \tag{6.7}
$$

Then

$$
\begin{aligned}
\frac{d}{dt}\varphi(tu) &= \langle \varphi'(tu), u\rangle \\
&= t\|\nabla u\|_2^2 - t^{q-1}\int_\Omega \beta|u|^q\,dz - \int_\Omega f(z,tu)u\,dz \\
&= \frac{1}{t}\left(t^2\|\nabla u\|_2^2 - t^q\int_\Omega \beta|u|^q\,dz - \int_\Omega f(z,tu)tu\,dz\right) \\
&\leqslant \frac{1}{t}\Bigg(t^2\|\nabla u\|_2^2 - t^q\int_\Omega \beta|u|^q\,dz - \int_\Omega 2F(z,tu)\,dz \\
&\qquad -\beta_1 t^\mu \int_{\{|tu|\geqslant M_{21}\}} |u|^\mu\,dz + c_{31}\Bigg) \\
&\leqslant \frac{1}{t}\left(\xi + c_{31} + \frac{t^q(2-q)}{q}\int_\Omega \beta|u|^q\,dz - \beta_1 t^\mu \int_{\{|tu|\geqslant M_{21}\}} |u|^\mu\,dz\right) \\
&\leqslant \frac{1}{t}\left(+ \frac{t^q(2-q)}{q}\int_\Omega \beta|u|^q\,dz - \beta_1 t^\mu \int_{\{|tu|\geqslant M_{21}\}} |u|^\mu\,dz\right)
\end{aligned}
$$

(since $\xi < -c_{31}$). Since $\mu > q$ (see hypothesis $H(f)_7(iv)$), choosing $t > 0$ even larger, we see that

$$\frac{d}{dt}\varphi(tu) < 0.$$

So, there exists unique $\tau(u) > 0$, such that

$$\varphi(\tau(u)u) = \xi \quad \forall u \in \partial B_1.$$

The implicit function theorem guarantees that $\tau \in C(\partial B_1)$. For $u \neq 0$, we set

$$\widehat{\tau}(u) = \frac{1}{\|u\|}\tau\left(\frac{u}{\|u\|}\right).$$

Evidently $\widehat{\tau} \in C^1\left(H_0^1(\Omega) \setminus \{0\}\right)$ and $\varphi(\widehat{\tau}(u)u) = \xi$. Moreover, $\varphi(u) = \xi$ implies that $\widehat{\tau}(u) = 1$. We set

$$\widehat{\tau}_0(u) = \begin{cases} 1 & \text{if } \varphi(u) \leqslant \xi, \\ \widehat{\tau}(u) & \text{if } \varphi(u) > \xi. \end{cases} \tag{6.8}$$

Evidently $\widehat{\tau}_0 \in C\left(H_0^1(\Omega) \setminus \{0\}\right)$. Let $h\colon [0,1] \times \left(H_0^1(\Omega) \setminus \{0\}\right) \longrightarrow H_0^1(\Omega) \setminus \{0\}$ be defined by

$$h(t,u) = (1-t)u + t\widehat{\tau}_0(u)u.$$

This map is continuous and we have

$$h(0,u) = u, \quad h(1,u) \in \varphi^{\xi} \quad \forall u \in H_0^1(\Omega) \setminus \{0\}$$

and

$$h(1,u) = u \quad \forall t \in [0,1], \text{ all } u \in \varphi^{\xi}.$$

This means that the sublevel set φ^{ξ} is a strong deformation retract of $H_0^1(\Omega) \setminus \{0\}$. Also using the radial retraction we see that ∂B_1 is a retract of $H_0^1(\Omega) \setminus \{0\}$. We assume that the set of critical points of φ is finite (otherwise we have an infinity of solutions and so we are done). By choosing ξ even smaller if necessary, we can have that $\xi < \inf \varphi(K^{\varphi})$ (K^{φ} being the critical set of φ). Then

$$\varphi^{\xi} \text{ and } \partial B_1 \text{ are homotopy equivalent}$$

(see Dugundji [**18**, pp. 325, 365]), so

$$H_k\left(H_0^1(\Omega), \varphi^{\xi}\right) = H_k\left(H_0^1(\Omega), \partial B_1\right) \quad \forall k \geqslant 0$$

and thus

$$C_k(\varphi, \infty) = 0 \quad \forall k \geqslant 0$$

(since ∂B_1 is contractible, see Granas-Dugundji [**28**, p. 389]). ☐

THEOREM 6.3. *If hypotheses $H(f)_7$ hold, then there exists $\beta_0^* > 0$, such that for every $\beta \in L^\infty(\Omega)_+$ with*

$$ess\,inf \beta > 0 \quad \text{and} \quad \|\beta\|_\infty < \beta_0^*,$$

problem (6.1) has at least six nontrivial smooth solutions:

$$\begin{aligned} &u_0, \widehat{u} \in int\, C_+, \quad \widehat{u} - u_0 \in int\, C_+, \\ &v_0, \widehat{v} \in -int\, C_+, \quad v_0 - \widehat{v} \in int\, C_+, \\ &y_0 \in C_0^1(\overline{\Omega}) \text{ nodal and } w_0 \in C_0^1(\overline{\Omega}). \end{aligned}$$

PROOF. From Theorem 4.11 we already have five nontrivial smooth solutions:

$$u_0, \widehat{u} \in \operatorname{int} C_+,\ u_0 \neq \widehat{u}, \quad v_0, \widehat{v} \in -\operatorname{int} C_+,\ v_0 \neq \widehat{v}, \quad \text{and} \quad y_0 \in C_0^1(\overline{\Omega}) \text{ nodal}.$$

Note that hypothesis $H(f)_7(ii)$ implies the existence of $\xi^* > 0$, such that for almost all $z \in \Omega$, the function

$$\zeta \longmapsto \xi^* |\zeta|^{p-2}\zeta + \beta(z)|\zeta|^{q-2}\zeta + f(z, \zeta)$$

is nondecreasing on $[-\gamma_0, \gamma_0]$, with

$$\gamma_0 = \max\left\{\|\widehat{u}_\infty\|, \|\widehat{v}\|_\infty\right\}.$$

Then

$$\Delta(\widehat{u} - u_0)(z) \leqslant \xi^*(\widehat{u} - u_0)(z) \quad \text{for almost all } z \in \Omega,$$

so

$$\widehat{u} - u_0 \in \operatorname{int} C_+$$

(see Vázquez [**49**]).

Similarly, we show that $v_0 - \widehat{v} \in \operatorname{int} C_+$.

Recall that $\varphi \in C^2\left(H_0^1(\Omega) \setminus \{0\}\right) \cap C^{2-0}\left(H_0^1(\Omega)\right)$ and $\widehat{u} \in \operatorname{int} C_+$, $\widehat{v} \in -\operatorname{int} C_+$ and $y_0 \in C_0^1(\overline{\Omega}) \setminus \{0\}$ are critical points of mountain pass type. Hence

$$C_k(\varphi, \widehat{u}) = C_k(\varphi, \widehat{v}) = C_k(\varphi, y_0) = \delta_{k,1}\mathbb{Z} \quad \forall k \geqslant 0 \tag{6.9}$$

(see Mawhin-Willem [**37**]). Also $u_0 \in \operatorname{int} C_+$, $v_0 \in -\operatorname{int} C_+$ are local minimizers of φ. Hence

$$C_k(\varphi, u_0) = C_k(\varphi, v_0)\delta_{k,0}\mathbb{Z} \quad \forall k \geqslant 0. \tag{6.10}$$

Let $\widehat{\varphi}_0 \in C^{2-0}\left(H_0^1(\Omega)\right)$ be the functional obtained by truncating the reaction at the extremal solutions u_+ and v_- (see the proof of Theorem 4.5). Then for $\varrho > 0$ small enough, we have

$$\varphi\Big|_{\overline{B}_\varrho^{C_0^1(\overline{\Omega})}} = \widehat{\varphi}_0\Big|_{\overline{B}_\varrho^{C_0^1(\overline{\Omega})}}. \tag{6.11}$$

Also from Liu-Wu [**35**], we know that

$$(6.12)\quad C_k(\varphi,0)=C_k\left(\varphi|_{C_0^1(\overline{\Omega})},0\right) \quad\text{and}\quad C_k(\widehat{\varphi}_0,0)=C_k\left(\widehat{\varphi}_0|_{C_0^1(\overline{\Omega})},0\right) \quad \forall k\geqslant 0.$$

From (6.11) and (6.12), it follows that

$$C_k(\varphi,0)=C_k(\widehat{\varphi}_0,0)\quad \forall k\geqslant 0,$$

so using also Proposition 4.10, we have

$$(6.13)\qquad C_k(\varphi,0)=0\quad \forall k\geqslant 0.$$

Finally, from Proposition 6.2, we have

$$(6.14)\qquad C_k(\varphi,\infty)=0\quad \forall k\geqslant 0.$$

Suppose that $\{0,u_0,\widehat{u},v_0,\widehat{v},y_0\}$ are the only critical points of φ. From the Morse relation (2.1) with $t=-1$ and (6.9), (6.10), (6.13) and (6.14), we have

$$(-1)^1=0,$$

a contradiction. This means that φ has one more critical point

$$w_0\ \notin\ \{0,u_0,\widehat{u},v_0,\widehat{v},y_0\}.$$

Then w_0 is a solution of (6.1) and $w_0\in C_0^1(\overline{\Omega})$ (regularity theory). □

Next we investigate the case of an asymptotically linear perturbation. The hypotheses on f are the following:
$\underline{H(f)_8}$ $f\colon\Omega\times\mathbb{R}\longrightarrow\mathbb{R}$ is a function, such that
(i)-(iii) are the same as hypotheses $H(f)_7(i)-(iii)$;
(iv) there exist integer $m\geqslant 2$ and functions $\eta,\widehat{\eta}\in L^\infty(\Omega)_+$, such that

$$\lambda_m\leqslant\eta(z)\leqslant\widehat{\eta}(z)\leqslant\lambda_{m+1}\quad\text{for almost all } z\in\Omega,$$

with $\eta\neq\lambda_m$, $\widehat{\eta}\neq\lambda_{m+1}$ and

$$\eta(z)\leqslant\liminf_{|\zeta|\to+\infty}\frac{f(z,\zeta)}{\zeta}\leqslant\limsup_{|\zeta|\to+\infty}\frac{f(z,\zeta)}{\zeta}\leqslant\widehat{\eta}(z),$$

uniformly for almost all $z\in\Omega$;
(v) there exist $a_-<0<a_+$, such that

$$\beta(z)a_+^{q-1}+f(z,a_+)\leqslant 0\leqslant\beta(z)|a_-|^{q-2}a_-+f(z,a_-),$$

for almost all $z\in\Omega$ and

$$\beta(z)|\zeta|^q+f(z,\zeta)\zeta\geqslant 0\quad\text{for almost all } z\in\Omega,\ \text{all } \zeta\in[a_-,a_+];$$

(vi) there exists $\eta_0\in L^\infty(\Omega)_+$, such that

$$\lambda_2<\liminf_{\zeta\to 0}\frac{f(z,\zeta)}{\zeta}\leqslant\limsup_{\zeta\to 0}\frac{f(z,\zeta)}{\zeta}\leqslant\eta_0(z),$$

uniformly for almost all $z\in\Omega$.

EXAMPLE 6.4. *The following function satisfies hypotheses $H(f)_8$:*

$$f(\zeta) = \begin{cases} c(\zeta - |\zeta|\zeta) & \text{if } |\zeta| \leqslant 1, \\ \vartheta\zeta + \dfrac{\sigma}{\zeta} - (\vartheta + \sigma) & \text{if } |\zeta| > 1, \end{cases}$$

with $\beta(z) \equiv \lambda > 0$, where $c > \lambda_2$, $\vartheta, \sigma > 0$ and $\vartheta = \sigma - 1$.

THEOREM 6.5. *If hypotheses $H(f)_8$ hold and $\beta \in L^\infty(\Omega)_+ \setminus \{0\}$, then problem (6.1) has at least six nontrivial smooth solutions:*

$$\begin{array}{lll} u_0, \widehat{u} \in int C_+, & \widehat{u} - u_0 \in int C_+, & u_0(z) < a_+ \quad \forall z \in \overline{\Omega} \\ v_0, \widehat{v} \in -int C_+, & v_0 - \widehat{v} \in int C_+, & a_- < v_0(z) \quad \forall z \in \overline{\Omega} \\ y_0 \in C_0^1(\overline{\Omega}) \text{ nodal and } w_0 \in C_0^1(\overline{\Omega}) \setminus \{0\}. & & \end{array}$$

PROOF. Truncating at the pairs $\{0, a_+\}$ and $\{a_-, 0\}$ and using the direct method, we obtain two solutions

$$u_0 \in \operatorname{int} C_+, \quad u_0(z) \leqslant a_+ \quad \forall z \in \overline{\Omega}$$

and

$$v_0 \in -\operatorname{int} C_+, \quad a_- \leqslant v_0(z) \quad \forall z \in \overline{\Omega}.$$

Hypothesis $H(f)_8(ii)$ guarantees the existence of a $\xi^* > 0$, such that for almost all $z \in \Omega$, the map

$$\zeta \longmapsto \xi^*\zeta + \beta(z)|\zeta|^{q-2}\zeta + f(z, \zeta)$$

is nondecreasing on $[a_-, a_+]$. Then, using hypothesis $H(f)_8(v)$, we have

$$\xi^* u_0(z) + \beta(z) u_0(z)^{q-1} + f(z, u_0(z)) \leqslant \xi^* a_+ \quad \text{for almost all } z \in \Omega. \tag{6.15}$$

Therefore,

$$\begin{aligned} \Delta(a_+ - u_0)(z) &= -\Delta u_0(z) = \beta(z) u_0(z)^{q-1} + f(z, u_0(z)) \\ &\leqslant \xi^*(a_+ - u_0)(z) \quad \text{for almost all } z \in \Omega, \end{aligned}$$

so

$$u_0(z) < a_+ \quad \forall z \in \overline{\Omega}$$

(see Vázquez [**49**]). Similarly, we show that

$$a_- < v_0(z) \quad \forall z \in \overline{\Omega}.$$

Hence using Proposition 3.6, we have that u_0, v_0 are local minimizers of φ. Thus

$$C_k(\varphi, u_0) = C_k(\varphi, v_0) = \delta_{k,0}\mathbb{Z} \quad \forall k \geqslant 0. \tag{6.16}$$

As before, let

$$\xi_+ = \frac{1}{q}\int_\Omega \beta u_0^q\,dz + \int_\Omega F(z,u_0)\,dz,$$
$$\xi_- = \frac{1}{q}\int_\Omega \beta |v_0|^q\,dz + \int_\Omega F(z,v_0)\,dz$$

and consider the functionals $\psi_\pm\colon H_0^1(\Omega) \longrightarrow \mathbb{R}$, defined by

$$\begin{aligned}\psi_+(u) &= \frac{1}{2}\|\nabla(u_0+u)\|_2^2 - \int_\Omega \widetilde{F}_+(z,u_0+u^+)\,dz \\ &\quad + \int_\Omega \Big(\beta u_0^{q-1} + f(z,u_0)\Big)\,u^-\,dz + \xi_+ \quad \forall u \in H_0^1(\Omega), \\ \psi_-(u) &= \frac{1}{2}\|\nabla(v_0+u)\|_2^2 - \int_\Omega \widetilde{F}_-(z,v_0+u^-)\,dz \\ &\quad + \int_\Omega \big(\beta|v_0|^{q-2}v_0 + f(z,v_0)\big)\,u^-\,dz + \xi_- \quad \forall u \in H_0^1(\Omega).\end{aligned}$$

Here

$$\widetilde{F}_\pm(z,\zeta) = \int_0^\zeta \widetilde{f}_\pm(z,s)\,ds,$$

with $\widetilde{f}_\pm(z,\zeta)$ given by (3.75) and (3.76).

Note that $\psi_\pm \in C^2\big(H_0^1(\Omega)\setminus\{0\}\big) \cap C^{1-0}\big(H_0^1(\Omega)\big)$. As in the proof of Theorem 5.5, via the nonsmooth mountain pass theorem, we obtain two more smooth constant sign solutions:

$$\widehat{u} \in \operatorname{int} C_+, \quad \widehat{u} \neq u_0, \quad u_0 \leqslant \widehat{u}$$

and

$$\widehat{v} \in -\operatorname{int} C_+, \quad \widehat{v} \neq v_0, \quad \widehat{v} \leqslant v_0.$$

As in the proof of Theorem 6.3, we show that

$$\widehat{u} - u_0 \in \operatorname{int} C_+ \quad \text{and} \quad v_0 - \widehat{v} \in \operatorname{int} C_+.$$

Recall (see the proof of Theorem 5.5), that

$$\widehat{u} = u_0 + \widetilde{u}, \quad \widehat{v} = v_0 - \widetilde{v},$$

with $\widetilde{u} \in \operatorname{int} C_+$, $\widetilde{v} \in -\operatorname{int} C_+$ being critical points of ψ_+ and ψ_- respectively of mountain pass type. Hence

$$C_k(\psi_+,\widetilde{u}) = C_k(\psi_-,\widetilde{v}) = \delta_{k,1}\mathbb{Z} \quad \forall k \geqslant 0. \tag{6.17}$$

From (5.3) and since $\widetilde{u} \in \operatorname{int} C_+$, we can find $\varrho > 0$ small, such that

$$\psi_+(u) = \varphi(u_0+u) - \varphi(u_0) - \frac{1}{p}\|\nabla u_0\|_p^p \quad \forall u \in \overline{B}_\varrho^{C_0^1(\overline{\Omega})}(\widetilde{u}).$$

As before, from Liu-Wu [**35**], we infer that

$$\begin{aligned} C_k(\psi_+, \widetilde{u}) &= C_k\left(\varphi(u_0 + \cdot) - \varphi(u_0) - \frac{1}{p}\|\nabla u_0\|_p^p,\ \widetilde{u}\right) \\ &= C_k(\varphi, u_0 + \widetilde{u}) = C_k(\varphi, \widehat{u}) \qquad \forall k \geqslant 0, \end{aligned}$$

so, using also (6.17), we have

$$C_k(\varphi, \widehat{u}) = \delta_{k,1}\mathbb{Z} \quad \forall k \geqslant 0. \tag{6.18}$$

Similarly, we show that

$$C_k(\varphi, \widehat{v}) = \delta_{k,1}\mathbb{Z} \quad \forall k \geqslant 0. \tag{6.19}$$

Next, truncating the extremal solutions of constant sign, we produce a nodal solution $y_0 \in C_0^1(\overline{\Omega}) \setminus \{0\}$, for which as above we have

$$u_+ - y_0 \in \operatorname{int} C_+, \quad y - v_- \in \operatorname{int} C_+,$$

where $u_+ \in \operatorname{int} C_+$ and $v_- \in -\operatorname{int} C_+$ are two extremal constant sign solutions. Since y_0 is a critical point of mountain pass type, a new appeal to Liu-Wu [**35**], implies that

$$C_k(\varphi, y_0) = \delta_{k,1}\mathbb{Z} \quad \forall k \geqslant 0. \tag{6.20}$$

Moreover, from Proposition 4.10, we have

$$C_k(\varphi, 0) = 0 \quad \forall k \geqslant 0 \tag{6.21}$$

(the effect of the concave term). Using hypothesis $H(f)_8(iv)$ and Lemma 2.5, we show that

$$\varphi\big|_{\widehat{H}_m} \text{ is coercive and } \varphi\big|_{\overline{H}_m} \text{ is anticoercive.}$$

Invoking Proposition 3.8 of Bartsch-Li [**6**], we infer that

$$C_{d_m}(\varphi, \infty) \neq 0,$$

where $d_m = \dim \overline{H}_m \geqslant 2$. This means that there is a critical point w_0 of φ, such that

$$C_{d_m}(\varphi, w_0) \neq 0. \tag{6.22}$$

Comparing (6.22) with (6.16), (6.18), (6.19), (6.20), (6.21), we infer that

$$w_0 \notin \{0, u_0, v_0, \widehat{u}, \widehat{v}, y_0\}.$$

Therefore w_0 is the sixth nontrivial solution of (6.1) and $w_0 \in C_0^1(\Omega) \setminus \{0\}$ by regularity theory. $\square$

Bibliography

[1] Aizicovici, S., Papageorgiou, N.S., Staicu, V., *The spectrum and an index formula for the Neumann p-Laplacian and multiple solutions for problems with crossing nonlinearity*, Discrete Contin. Dyn. Syst., **25** (2009), 431–456.

[2] Aizicovici, S., Papageorgiou, N.S., Staicu, V., *Existence of multiple solutions with precise sign information for superlinear Neumann problems*, Ann. Mat. Pura Appl., **188** (2009), 679–719.

[3] Ambrosetti, A., Brézis, H., Cerami, G., *Combined effects of concave and convex nonlinearities in some elliptic problems*, J. Funct. Anal., **122** (1994), 519–543.

[4] Ambrosetti, A., García Azorero, J., Peral, I., *Multiplicity results for some nonlinear elliptic equations*, J. Funct. Anal., **137** (1996), 219–242.

[5] Bartolo, P., Benci, V., Fortunato, D., *Abstract critical point theorems and applications to some nonlinear problems with "strong" resonance at infinity*, Nonlinear Anal., **7** (1983), 981–1012.

[6] Bartsch, T., Li, S.-J., *Critical point theory for asymptotically quadratic functionals and applications to problems with resonance*, Nonlinear Anal., **28** (1997), 419–441.

[7] Bartsch, T., Liu, Z., Weth, T., *Nodal solutions of a p-Laplacian equations*, Proc. London Math. Soc., **91** (2005), 129–152.

[8] Brézis, H., Nirenberg, L., *H^1 versus C^1 local minimizers*, C. R. Acad. Sci. Paris Sér. I Math., **317** (1993), 465–472.

[9] Carl, S., Motreanu, D., *Constant-sign and sign-changing solutions for nonlinear eigenvalue problems*, Nonlinear Anal., **68** (2008), 2668–2676.

[10] Carl, S., Perera, K., *Sign-changing and multiple solutions for the p-Laplacian*, Abstr. Appl. Anal., **7** (2002), 613–625.

[11] Chang, K.-C., *Infinite-Dimensional Morse Theory and Multiple Solution Problems*, Progress in Nonlinear Differential Equations and Their Applications, Vol. 6, Birkhäuser Verlag, Boston, MA, 1993.

[12] Clarke, F.H., *Optimization and Nonsmooth Analysis*, Wiley, New York, 1983.

[13] Costa, D.G., Magalhães, C.A., *Existence results for perturbations of the p-Laplacian*, Nonlinear Anal., **24** (1995), 409–418.

[14] Cuesta, M., de Figueiredo, D., Gossez, J.-P., *The beginning of the Fučik spectrum for the p-Laplacian*, J. Differential Equations, **159** (1999), 212–238.

[15] Damascelli, L., *Comparison theorems for some quasilinear degenerate elliptic operators and applications to symmetry and monotonicity results*, Ann. Inst. H. Poincaré Anal. Non Linéaire, **15** (1998), 493–516.

[16] Dancer, N., Du., Y., *On sign-changing solutions of certain semilinear elliptic problems*, Appl. Anal., **56** (1995), 193–206.

[17] de Paiva, F.O., Massa, E., *Multiple solutions for some elliptic equations with a nonlinearity concave at the origin*, Nonlinear Anal., **66** (2007), 2940–2946.

[18] Dugundji, J., *Topology*, Allyn and Bacon Inc., Boston, 1966.

[19] Dunford, N., Schwartz, J.T., *Linear Operators. I. General Theory*, Pure and Applied Mathematics, Vol. 7, Wiley, New York, 1958.

[20] Fei, G., *On periodic solutions of superquadratic Hamiltonian systems*, Electron. J. Differential Equations, **8** (2002), 1–12.

[21] Filippakis, M., Kristaly, A., Papageorgiou, N.S., *Existence of five nonzero solutions with exact sign for a p-Laplacian equation*, Discrete Contin. Dyn. Syst., **24** (2009), 405–440.

[22] Filippakis, M., Papageorgiou, N.S., *Solvability of nonlinear variational-hemivariational inequalities*, J. Math. Anal. Appl., **311** (2005), 162–181.

[23] García Azorero, J., Manfredi, J., Peral Alonso, I., *Sobolev versus Hölder local minimizers and global multiplicity for some quasilinear elliptic equations*, Commun. Contemp. Math., **2** (2000), 385–404.

[24] Gasiński, L., Papageorgiou, N.S., *Nonsmooth Critical Point Theory and Nonlinear Boundary Value Problems*, Chapman and Hall/CRC Press, Boca Raton, FL, 2005.

[25] Gasiński, L., Papageorgiou, N.S., *Nonlinear Analysis*, Chapman and Hall/CRC Press, Boca Raton, FL, 2006.

[26] Gasiński, L., Papageorgiou, N.S., *Nodal and multiple constant sign solutions for resonant p-Laplacian equations with a nonsmooth potential*, Nonlinear Anal., **71** (2009), 5747–5772.

[27] Gasiński, L., Papageorgiou, N.S., *Multiple solutions for nonlinear Dirichlet problems with concave terms*, submitted.

[28] Granas, A., Dugundji, J., *Fixed Point Theory*, Springer, New York, 2003.
[29] Guedda, M., Véron, L., *Quasilinear elliptic equations involving critical Sobolev exponents*, Nonlinear Anal., **13** (1989), 879–902.
[30] Guo, Z., Zhang, Z., *$W^{1,p}$ versus C^1 local minimizers and multiplicity results for quasilinear elliptic equations*, J. Math. Anal. Appl., **286** (2003), 32–50.
[31] Iannizzotto, A., Papageorgiou, N.S., *Existence of three nontrivial solutions for nonlinear Neumann hemivariational inequalities*, Nonlinear Anal., **70** (2009), 3285–3297.
[32] Kyritsi, S.Th., Papageorgiou, N.S., *Multiple solutions of constant sign for nonlinear nonsmooth eigenvalue problems near resonance*, Calc. Var. Partial Differential Equations, **20** (2004), 1–24.
[33] Ladyzhenskaya, O.A., Uraltseva, N., *Linear and Quasilinear Elliptic Equations*, Mathematics in Science and Engineering, Vol. 46, Academic Press, New York, 1968.
[34] Lieberman, G.M., *Boundary regularity for solutions of degenerate elliptic equations*, Nonlinear Anal., **12** (1988), 1203–1219.
[35] Liu, J.-Q., Wu, S., *Calculating critical groups of solutions for elliptic problems with jumping nonlinearity*, Nonlinear Anal., **49** (2002), 779–797.
[36] Li, S.-J., Wu, S., Zhou, H.-S., *Solutions to semilinear elliptic problems with combined nonlinearities*, J. Differential Equations, **185** (2002), 200–224.
[37] Mawhin, J., Willem, M., *Critical Point Theory and Hamiltonian Systems*, Applied Mathematical Sciences, Vol. 74, Springer-Verlag, New York, 1989.
[38] Montenegro, M., *Strong maximum principles for supersolutions of quasilinear elliptic equations*, Nonlinear Anal., **37** (1999), 431–448.
[39] Motreanu, D., Motreanu, V.V., Papageorgiou, N.S., *A degree theoretic approach for multiple solutions of constant sign for nonlinear elliptic equations*, Manuscripta Math., **124** (2007), 507–531.
[40] Motreanu, D., Motreanu, V.V., Papageorgiou, N.S., *Multiple nontrivial solutions for nonlinear eigenvalue problems*, Proc. Amer. Math. Soc., **135** (2007), 3649–3658.
[41] Motreanu, D., Motreanu, V.V., Papageorgiou, N.S., *A unified approach for multiple constant sign and nodal solutions*, Adv. Differential Equations, **121** (2007), 1363–1392.
[42] Motreanu, D., Motreanu, V.V., Papageorgiou, N.S., *Nonlinear Neumann problems near resonance*, Indiana Univ. Math. J., **58** (2009), 1257–1280.
[43] Motreanu, D., Motreanu, V.V., Papageorgiou N.S., *Multiple consatant sign and nodal solutions for nonlinear Neumann eigenvalue problems*, Ann. Scuola Norm. Sup. Pisa.
[44] Papageorgiou, N.S., Rocha, E.M., Staicu, V., *Multiplicity theorem for superlinear elliptic problems*, Calc. Var. Partial Differential Equations, **33** (2008), 199–230.
[45] Perera, K., *Multiplicity result for some elliptic problems with concave nonlinearity*, J. Differential Equations, **140** (1997), 133–141.
[46] Pucci, P., Serrin, J., *The Maximum Principle*, Birkhäuser Verlag, Basel, 2007.
[47] Serrin, J., *On the strong maximum principle for quasilinear second order differential inequalities*, J. Funct. Anal., **5** (1970), 184–193.
[48] Tolksdorf, P., *Regularity for a more general class of quasilinear elliptic equations*, J. Differential Equations, **51** (1984), 126–150.
[49] Vázquez, J.L., *A strong maximum principle for some quasilinear elliptic equations*, Appl. Math. Optim., **12** (1984), 191–202.
[50] Zhang, Q., *A strong maximum principle for differential equations with nonstandard $p(x)$-growth condition*, J. Math. Anal. Appl., **312** (2005), 24–32.
[51] Zhang, Z., Chen, J.-Q., Li, S.-J., *Construction of pseudo-gradient vector field and sign-changing multiple solutions involving p-Laplacian*, J. Differential Equations, **201** (2004), 287–303.
[52] Zhang, Z., Li, S.-J., *On sign-changing and multiple solutions of the p-Laplacian*, J. Funct. Anal., **197** (2003), 447–468.

Leszek Gasiński, Jagiellonian University, Institute of Computer Science, ul. Łojasiewicza 6, 30-348 Kraków, Poland
E-mail address: `Leszek.Gasinski@ii.uj.edu.pl`

S. Papageorgiou, National Technical University, Department of Mathematics, Zografou Campus, Athens 15780, Greece
E-mail address: `npapg@math.ntua.gr`

Study of some semilinear elliptic problems on $\mathbb{R}^N$ via variational methods

Alexandru Kristály* and Nikolaos S. Papageorgiou

Abstract. By using various variational methods, we present some multiplicity results for elliptic problems defined on $\mathbb{R}^N$ which involve either sublinear or oscillatory nonlinear terms. Some stability properties of solutions are also established.

Contents

1. Introduction and motivation

Let $f : \mathbb{R}^N \times \mathbb{R} \to \mathbb{R}$ be a continuous function, $V : \mathbb{R}^N \to \mathbb{R}$ be a positive potential, and consider the problem

$$-\Delta u + V(x)u = f(x,u), \quad x \in \mathbb{R}^N, \ u \in H^1(\mathbb{R}^N). \tag{1}$$

The study of this problem is motivated by mathematical physics; it is well-known that certain kinds of solitary waves in nonlinear Klein-Gordon or Schrödinger equations are solutions of (1), see Rabinowitz [**27**], Strauss [**35**]. Due to its importance, many papers are concerned with the existence and multiplicity of solutions of (1); without seek of completeness, we refer the reader to the works of Bartsch-Wang [**6, 7**], Strauss [**35**], Willem [**39**]. The aforementioned papers have two common features: the nonlinearity $s \mapsto f(x,s)$ is *subcritical* and *superlinear at infinity*. To be

*The research of A. Kristály was supported by the János Bolyai Research Scholarship of the Hungarian Academy of Sciences (Budapest, Hungary) and by PN II IDEI_527 of CNCSIS (Bucharest, Romania).

more precise, most of the authors use the well-known Ambrosetti-Rabinowitz type condition:

(AR) There is a $\eta > 2$ such that

$$0 < \eta F(x,s) \le f(x,s)s \quad \text{for each } x \in \mathbb{R}^N, \ \ s \in \mathbb{R} \setminus \{0\},$$

where $F(x,s) = \int_0^s f(x,t)dt$.

As we know, this condition implies the *superlinearity* of the function $s \mapsto f(x,s)$ at infinity, i.e., there exist some numbers $C > 0, s_0 > 0$ such that

$$|f(x,s)| \ge C|s|^{\eta-1} \quad \text{for each } x \in \mathbb{R}^N, \ \ |s| \ge s_0.$$

The objective of the present paper is to guarantee existence/multiplicity of solutions for equations of type (1) whenever the nonlinear term f does not fulfill any Ambrosetti-Rabinowitz type condition. Namely, the following two cases will be considered:

a) f has a *sublinear growth at infinity*;

b) f has a suitable *oscillatory behaviour* at the origin or at infinity.

Let us briefly discuss them.

1.1. Sublinear problems. We assume that

($F1$) There exist $Q \in L^1(\mathbb{R}^N) \cap L^\infty(\mathbb{R}^N)$, $Q \not\equiv 0$, and $q \in (0,1)$ such that

$$|f(x,s)| \le Q(x)|s|^q \quad \text{for each } (x,s) \in \mathbb{R}^N \times \mathbb{R}.$$

In such a case the energy functional associated with (1) is coercive and bounded from below; thus the existence of at least *one* solution is always expected. However, one may happen that (1) has only the trivial solution, even if the nonlinearity fulfills (F1). Indeed, if we consider for instance $V = 1$, $f(x,s) = \lambda Q(x)\sin^2 s$ with $Q \not\equiv 0$ as above, and $0 < \lambda < (2\|Q\|_{L^\infty})^{-1}$, then (1) has only the trivial solution. Thus, this fact motivates the study of an *eigenvalue problem* rather than problem (1). On account of this statement, we shall investigate the following general eigenvalue problem

$$(\mathrm{P}_{\lambda,\mu}) \qquad -\Delta u + V(x)u = \lambda(f(x,u) + \mu g(x,u)), \quad x \in \mathbb{R}^N, \ \ u \in H^1(\mathbb{R}^N),$$

where the functions $f, g : \mathbb{R}^N \times \mathbb{R} \to \mathbb{R}$ are continuous functions verifying the above sublinearity hypothesis, while $\lambda, \mu \in \mathbb{R}$ are some parameters. Roughly speaking - under further assumptions on V, f and g, which will be specified later - our main results can be formulated as follows:

> If we fix $\mu \in \mathbb{R}$ (resp. $\lambda \in \mathbb{R}$) of certain range, we guarantee the existence of a non-degenerate interval $\Lambda_\mu \subset \mathbb{R}$ (resp. $\Pi_\lambda \subset \mathbb{R}$) such that $(\mathrm{P}_{\lambda,\mu})$ has at least two nontrivial, weak solutions whenever $\lambda \in \Lambda_\mu$ (resp. $\mu \in \Pi_\lambda$).

Before stating our results precisely, we mention that the potential $V : \mathbb{R}^N \to \mathbb{R}$ also has an important role concerning the existence and behaviour of solutions of $(\mathrm{P}_{\lambda,\mu})$. When $V(x)$ is a positive constant, or V is radially symmetric, it is natural to look for radially symmetric solutions of $(\mathrm{P}_{\lambda,\mu})$, see e.g., [**35, 39**]. Motivated by the work of Rabinowitz [**27**] (where $V \in \mathcal{C}(\mathbb{R}^N, \mathbb{R})$, $\inf_{\mathbb{R}^N} V > 0$, and $V(x) \to +\infty$ as $|x| \to +\infty$), Bartsch-Wang [**7**] considered a new class of potentials, namely:

(BW) $V \in \mathcal{C}(\mathbb{R}^N, \mathbb{R})$ satisfies $\inf_{\mathbb{R}^N} V > 0$, and for any $M > 0$,

$$m(\{x \in \mathbb{R}^N : V(x) \leq M\}) < +\infty,$$

where m denotes the Lebesgue measure in $\mathbb{R}^N$.

Under (BW), Bartsch-Wang [**7**] proved the existence of infinitely many solutions of $(\mathrm{P}_{\lambda,\mu})$ (for any fixed $\lambda > 0$) when f is subcritical, odd and verifies (AR). Furtado-Maia-Silva [**12**] studied $(\mathrm{P}_{\lambda,\mu})$ in the case when F (defined in (AR)) has some sort of resonance with a local nonquadraticity condition at infinity, while the potential V verifies (BW). Gazzola-Rădulescu [**13**] studied $(\mathrm{P}_{\lambda,\mu})$ when V verifies (BW), f is not necessarily continuous and satisfies an appropriate non-smooth (AR) condition. We assume that the potential V satisfies:

$(V1)$ $V \in L^\infty_{loc}(\mathbb{R}^N)$, $\quad \mathrm{essinf}_{\mathbb{R}^N} V > 0$; and

$(V2)$ One of the following conditions is satisfied:

$(V2A)$ $N \geq 2$ and $V : \mathbb{R}^N \to \mathbb{R}$ is radially symmetric;

$(V2B)$ For any $M > 0$ and any $r > 0$ there holds:

$$m(\{x \in B(y,r) : V(x) \leq M\}) \to 0 \quad \text{as } |y| \to +\infty,$$

where $B(y,r)$ denotes the open ball in $\mathbb{R}^N$ with center y and radius $r > 0$.

Note that hypotheses $(V1) - (V2B)$ are weaker conditions than (BW), see Bartsch-Pankov-Wang [**5**]. Requiring $(V1) - (V2B)$, Bartsch-Liu-Weth [**4**] proved recently the existence of three solutions of $(\mathrm{P}_{\lambda,\mu})$ for any fixed $\lambda > 0$, with f subcritical and verifying (AR).

Problems involving only sublinear terms (in certain papers this type of problem is referred to one involving *pure concave nonlinearities*) have been also studied by several authors, see for instance Yu [**40**], Chabrowski-do Ó [**9**]. Recently, Ricceri [**29**], [**30**] established a new variational principle, ensuring the existence of at least *three* distinct critical points of a coercive functional acting on a reflexive Banach space. By means of Ricceri's principle, several nonlinear elliptic eigenvalue problems have been treated (involving only concave nonlinearities), see for instance Ricceri [**30**], [**31**], Motreanu-Marano [**21**] (on bounded domains); Kristály [**18**], [**19**] (on unbounded domains). By using a version of the Mountain Pass theorem, Marano-Motreanu [**22**] established a counterpart of the main result of [**29**]. Other multiplicity results for elliptic eigenvalue problems involving pure concave nonlinearities on bounded domains can be found in Perera [**26**] (*three* or *four* nontrivial solutions), as well as in Ambrosetti-Badiale [**2**] and Wang [**38**] where *infinitely many* solutions are obtained, assuming the oddness of the involved nonlinearities.

1.2. Elliptic problems with oscillatory terms. Having infinitely many solutions for a given equation, after a 'small' perturbation of it, one expects to find still many solutions for the perturbed equation; moreover, once the perturbation tends to zero, the number of solutions for the perturbed equation should tend to infinity. Such phenomenon is well-known in the case of the equation $\sin s = c$ with $c \in (-1, 1)$ fixed, and its perturbation $\sin s = c + \varepsilon s$, $s \in \mathbb{R}$; the perturbed equation has more and more solutions as $|\varepsilon|$ decreases to 0. To the best of our knowledge, this natural phenomenon has been first exploited in an abstract framework by Krasnosel'skii [**15**]. More precisely, by using topological methods, Krasnosel'skii asserts the existence of more and more critical points of an even C^1-class functional perturbed by a non-even term tending to zero, the critical points of the perturbed functional being the solutions for the studied equation.

Later on, Krasnosel'skii's idea served for further developments; in order to describe them, we consider the equation

$$(\mathrm{E}_\varepsilon) \qquad -\triangle u + V(x)u = f(x,u) + \varepsilon g(x,u) \ \text{ in } \ \Omega,$$

where $\Omega \subseteq \mathbb{R}^N$ is an open domain, $V : \Omega \to \mathbb{R}$ is a measurable function, while $f, g : \Omega \times \mathbb{R} \to \mathbb{R}$ are Carathéodory functions. Subject to certain boundary condition, we assume the unperturbed equation

$$(\mathrm{E}_0) \qquad -\triangle u + V(x)u = f(x,u) \ \text{ in } \ \Omega,$$

has *infinitely many* distinct solutions. Then, the main question is:

(q) Fixing $k \in \mathbb{N}$, can one find a number $\varepsilon_k > 0$ such that the perturbed equation (E_ε) has at least k distinct solutions whenever $\varepsilon \in [-\varepsilon_k, \varepsilon_k]$?

Two different classes of results are available in the literature answering affirmatively question (q), both for *bounded* domains subjected to zero Dirichlet boundary condition, and $V \equiv 0$:

A. *Perturbation of symmetric problems.* Assume $f(x,s) = -f(x,-s)$ for every $(x,s) \in \Omega \times \mathbb{R}$. It is well-known that if the energy functional has the Mountain Pass Geometry, problem (E_0) has infinitely many solutions, due to the symmetric version of the Mountain Pass theorem, see Ambrosetti-Rabinowitz [**1**]. Furthermore, question (q) was fully answered by Li and Liu [**20**] for arbitrarily continuous nonlinearity g, following the topological approach developed by Degiovanni-Lancelotti [**10**] and Degiovanni-Rădulescu [**11**];
B. *Perturbation of oscillatory problems.* Assume $f(x,\cdot)$ oscillates near the origin or at infinity, uniformly with respect to $x \in \Omega$. Special kinds of oscillations produce infinitely many solutions for (E_0), as shown by Omari-Zanolin [**24**], and Saint Raymond [**34**]. Concerning the perturbed problem, Anello-Cordaro [**3**] answered question (q), by using the abstract variational principle of Ricceri [**33**].

Here present a third, *direct method* for answering question (q) whenever the nonlinear term $f(x,\cdot)$ belongs to a wide class of oscillatory functions. Although our method works in various frameworks (for instance, the domain Ω is bounded, and the studied problem is subject to Dirichlet, Neumann or more general boundary conditions), we illustrate this approach by treating the problem

$$(\mathrm{P}_\varepsilon) \qquad \begin{cases} -\triangle u + u = Q(x)[f(u) + \varepsilon g(u)], & x \in \mathbb{R}^N,\ N \geq 2, \\ u \geq 0 \\ u(x) \to 0 \ \text{ as } \ |x| \to \infty, \end{cases}$$

where $f : [0,\infty) \to \mathbb{R}$ is a continuous nonlinearity which oscillates near the origin or at infinity, see hypotheses (f_1^0) and (f_2^0), or (f_1^∞) and (f_2^∞) in Section 3; see also [**17**]. On the nonlinear term $g : [0, \infty) \to \mathbb{R}$ we assume only its *continuity* and that $g(0) = 0$. Actually, we prove that: (a) the unperturbed problem (P_0), i.e. $\varepsilon = 0$ in (P_ε), has infinitely many distinct solutions; (b) the number of distinct solutions for (P_ε) becomes greater and greater whenever $|\varepsilon|$ is smaller and smaller. In fact, our method surprisingly shows that (a) and (b) are *equivalent* in the sense that they are deducible from each other.

1.3. Plan of the paper. In Section 2 we give the precise form of our main results concerning *sublinear problems*; Subsection 2.1 contains some auxiliary result while in Subsections 2.2 and 2.3 we prove our results. In Section 3 we are dealing with elliptic problems which involve oscillatory terms, answering question (q) in various cases in the context of problem (P_ε).

2. Elliptic problems on $\mathbb{R}^N$ involving sublinear terms

In this section we assume the assumptions from Subsection 1.1 are fulfilled. In view of $(V1)$, we introduce the Hilbert space

$$E = \{u \in H^1(\mathbb{R}^N) : \int_{\mathbb{R}^N} V(x)u^2 < +\infty\}$$

which will be endowed with the inner product

$$\langle u, v\rangle_E = \int_{\mathbb{R}^N} (\nabla u \nabla v + V(x)uv) \quad u, v \in E,$$

and with the induced norm $\|\cdot\|_E$. Note that solutions of $(\mathrm{P}_{\lambda,\mu})$ are being sought in E which can be continuously embedded into $L^p(\mathbb{R}^N)$ whenever $2 \le p < 2^*$. Here, 2^* denotes the critical Sobolev exponent, i.e., $2^* = 2N/(N-2)$ for $N \ge 3$ and $2^* = +\infty$ for $N = 1, 2$.

Let $f, g : \mathbb{R}^N \times \mathbb{R} \to \mathbb{R}$ be two continuous functions and let $F(x, s) = \int_0^s f(x, t)dt$ and $G(x, s) = \int_0^s g(x, t)dt$, respectively. We assume:

$(FG1)$ There exist $Q \in L^1(\mathbb{R}^N) \cap L^\infty(\mathbb{R}^N)$, $Q \not\equiv 0$, and $q \in (0, 1)$ such that

$$\max\{|f(x, s)|, |g(x, s)|\} \le Q(x)|s|^q \quad \text{for each } (x, s) \in \mathbb{R}^N \times \mathbb{R}.$$

$(FG2)$ $\lim_{s\to 0} \frac{f(x,s)}{s} = \lim_{s\to 0} \frac{g(x,s)}{s} = 0$ uniformly for each $x \in \mathbb{R}^N$.

$(FG3)$ $f(\cdot, s)$ and $g(\cdot, s)$ are radially symmetric functions for each $s \in \mathbb{R}$.

$(F4)$ There exist $R_0 > 0$ and $s_0 \in \mathbb{R}$ such that $\min_{|x|\le R_0} F(x, s_0) > 0$.

We introduce the functions $\mathcal{F}, \mathcal{G} : E \to \mathbb{R}$, defined by

$$\mathcal{F}(u) = \int_{\mathbb{R}^N} F(x, u(x))dx, \quad \mathcal{G}(u) = \int_{\mathbb{R}^N} G(x, u(x))dx.$$

A standard argument, which is based on the facts that $Q \in L^1(\mathbb{R}^N) \cap L^\infty(\mathbb{R}^N)$, f, g satisfy $(FG1)$, and the embedding $E \hookrightarrow L^p(\mathbb{R}^N)$ is continuous $(2 \le p < 2^*)$, show that the functionals $\mathcal{F}, \mathcal{G} : E \to \mathbb{R}$ are well defined, are of class $\mathcal{C}^1$, and

$$\mathcal{F}'(u)(v) = \int_{\mathbb{R}^N} f(x, u(x))v(x)dx, \quad \mathcal{G}'(u)(v) = \int_{\mathbb{R}^N} g(x, u(x))v(x)dx \tag{2}$$

for each $u, v \in E$ (see, for instance [**39**, Lemma 3.10]). Finally, let $\mathcal{H}_{\lambda,\mu} : E \to \mathbb{R}$ be the energy functional associated to $(\mathrm{P}_{\lambda,\mu})(\mathrm{P}_{\lambda,\mu})$, defined by

$$\mathcal{H}_{\lambda,\mu}(u) = \frac{1}{2}\|u\|_E^2 - \lambda\mathcal{F}(u) - \lambda\mu\mathcal{G}(u).$$

Our results can be stated as follows.

THEOREM 2.1. *Let $f, g : \mathbb{R}^N \times \mathbb{R} \to \mathbb{R}$ be two continuous functions and $V : \mathbb{R}^N \to \mathbb{R}$ be a potential which satisfy $(FG1) - (FG2)$, $(F4)$, and $(V1) - (V2)$, respectively. Assume moreover that $(FG3)$ is verified whenever $(V2A)$ holds. Then, there exists $\mu_0 > 0$ such that to every $\mu \in [-\mu_0, \mu_0]$ it corresponds a nonempty open interval $\Lambda_\mu \subset (0, \infty)$ and a number $\sigma_\mu > 0$ for which $(\mathrm{P}_{\lambda,\mu})$ has at least two distinct, nontrivial weak solutions $v_{\lambda\mu}$ and $w_{\lambda\mu}$ with the property that*

$$\max\{\|v_{\lambda\mu}\|_E, \|w_{\lambda\mu}\|_E\} \leq \sigma_\mu$$

whenever $\lambda \in \Lambda_\mu$. Moreover, $v_{\lambda\mu}$ and $w_{\lambda\mu}$ are radially symmetric whenever $(V2A)$ holds.

From the point of view of the eigenvalues, the counterpart of Theorem 2.1 is the following:

THEOREM 2.2. *Under the assumptions of Theorem 2.1, there exists $\lambda_0 > 0$ such that to every $\lambda \in (\lambda_0, \infty)$ it corresponds a nonempty open interval $\Pi_\lambda \subset \mathbb{R}$ and a number $\sigma_\lambda > 0$ for which $(\mathrm{P}_{\lambda,\mu})$ has at least two distinct, nontrivial weak solutions $v_{\lambda\mu}$ and $w_{\lambda\mu}$ with $\mathcal{H}_{\lambda,\mu}(v_{\lambda\mu}) < 0 < \mathcal{H}_{\lambda,\mu}(w_{\lambda\mu})$ and $\max\{\|v_{\lambda\mu}\|_E, \|w_{\lambda\mu}\|_E\} \leq \sigma_\lambda$ whenever $\mu \in \Pi_\lambda$. Moreover, $v_{\lambda\mu}$ and $w_{\lambda\mu}$ are radially symmetric whenever $(V2A)$ holds.*

Although the two theorems above are completely independent, as a simple by-product of Theorem 2.2 we obtain the following result whose conclusion partially goes back to Theorem 2.1.

THEOREM 2.3. *Under the assumptions of Theorem 2.1, there exists $\overline{\mu} > 0$ such that for every $\mu \in [-\overline{\mu}, \overline{\mu}]$ the set*

$$\{\lambda > 0 : (\mathrm{P}_{\lambda,\mu}) \textit{ has at least two distinct, nontrivial weak solutions}\}$$

contains an interval.

REMARK 2.4. (a) If we omit $(F4)$ in the above results, then we may assume $f = g = 0$, thus $(\mathrm{P}_{\lambda,\mu})$ has only the trivial solution.

(b) The above theorems do not work in general for every $\mu \in \mathbb{R}$. Indeed, if one consider a function f which verifies the hypotheses of our results, and we set $g = -f$, then for $\mu = 1$, the problem $(\mathrm{P}_{\lambda,\mu})$ has only the trivial solution.

(c) The above theorems do not work in general for every $\lambda \in \mathbb{R}$. To see this, consider the nonlinearities $f(x, s) = Q(x)h(s)$ and $g(x, s) = Q(x)k(s)$ which fulfill the hypotheses of the above theorems where $Q : \mathbb{R}^N \to \mathbb{R}$ is from $(FG1)$, while h and k are Lipschitz continuous functions with Lipschitz constants $L_h > 0$ and $L_k > 0$, respectively. For instance, one can consider $Q(x) = (1 + |x|^\alpha)^{-\beta}$ with

$\alpha, \beta > 0$ such that $\alpha\beta > N \geq 3$, and $h(s) = \sin^2 s$, $k(s) = \arctan^2 s$. Fix $\mu \in \mathbb{R}$ arbitrarily. Then, if $0 < \lambda < (\|Q\|_{L^\infty})^{-1}(L_h + |\mu| L_k)^{-1}$, the problem $(\mathrm{P}_{\lambda,\mu})$ has only the zero solution. Indeed, let us observe the critical points of $\mathcal{H}_{\lambda,\mu}$ are precisely the fixed points of the operator $A_{\lambda,\mu} = \lambda(\mathcal{F}' + \mu\mathcal{G}')$. Since $A_{\lambda,\mu}$ is a contraction for λ's specified above, then $A_{\lambda,\mu}$ has a unique fixed point. Since $A_{\lambda,\mu}(0) = 0$ (due to $(FG1)$), then 0 will be the unique solution of $(\mathrm{P}_{\lambda,\mu})$.

Assumption $(FG2)$ means that both $f(x,\cdot)$ and $g(x,\cdot)$ are superlinear at the origin. One may ask what happens if such assumption does not hold. For simplicity, we consider a special case of $(\mathrm{P}_{\lambda,\mu})$, namely, when $f(x,s) = Q(x)f(s)$ and $\mu = 0$. Therefore, the functional $\mathcal{F} : E \to \mathbb{R}$ has the form

$$\mathcal{F}(u) = \int_{\mathbb{R}^N} Q(x)F(u(x))dx, \quad u \in E. \tag{3}$$

We assume that

- $(f1)$ There exist $c > 0$ and $q \in (0,1)$ such that $|f(s)| \leq c|s|^q$ for every $s \in \mathbb{R}$,
- $(Q1)$ $Q \in L^1(\mathbb{R}^N) \cap L^\infty(\mathbb{R}^N)$, and
- $(Q2)$ $Q \geq 0$, and $\sup_{R>0} \operatorname{essinf}_{|x|\leq R} Q(x) > 0$.

THEOREM 2.5. *Let $f : \mathbb{R} \to \mathbb{R}$ be a nonzero, non-decreasing continuous function such that $(f1)$ is fulfilled. Assume that V, $Q : \mathbb{R}^N \to \mathbb{R}$ satisfy $(V1) - (V2)$ and $(Q1) - (Q2)$, and Q is radially symmetric whenever $(V2A)$ holds.*

Then for each number $\zeta > 0$ there exist a number $\lambda > 0$ and $w \in \mathcal{F}^{-1}([0,\zeta[) \cap C_0^\infty(\mathbb{R}^N)$ such that the problem

$$-\triangle u + V(x)u = \lambda Q(x)f(u+w), \quad x \in \mathbb{R}^N \tag{P_λ^w}$$

has at least three distinct weak solutions.

2.1. Auxiliary results. We suppose that assumptions of Theorem 2.1 are fulfilled.

LEMMA 2.6. *Let $p \in (2, 2^*)$. For each $\varepsilon > 0$ there is $c(\varepsilon) > 0$ such that*

- (i) $\max\{|f(x,s)|, |g(x,s)|\} \leq \varepsilon|s| + c(\varepsilon)|s|^{p-1}$ *for every* $(x,s) \in \mathbb{R}^N \times \mathbb{R}$;
- (ii) $\max\{|F(x,s)|, |G(x,s)|\} \leq \varepsilon s^2 + c(\varepsilon)|s|^p$ *for every* $(x,s) \in \mathbb{R}^N \times \mathbb{R}$.

PROOF. Taking into account $(FG1)$, $(FG2)$ and the facts that $Q \in L^\infty(\mathbb{R}^N)$, and $q + 1 < 2 < p$, the claims easily follow. □

Due to this result, one can show in a standard way (see for instance [**39**, Lemma 3.10]) that the energy functional $\mathcal{H}_{\lambda,\mu} : E \to \mathbb{R}$ is well-defined and of class $\mathcal{C}^1$. Moreover,

$$\mathcal{H}'_{\lambda,\mu}(u)(v) = \langle u, v\rangle_E - \lambda \int_{\mathbb{R}^N} f(x,u)v - \lambda\mu \int_{\mathbb{R}^N} g(x,u)v, \quad \forall u, v \in E,$$

thus the critical points of $\mathcal{H}_{\lambda,\mu}$ are exactly the weak solutions of $(\mathrm{P}_{\lambda,\mu})$.

The embedding $H^1(\mathbb{R}^N) \hookrightarrow L^p(\mathbb{R}^N)$ is not compact for any $p \in [2, 2^*]$. However, when $(V1)-(V2B)$ hold, the embedding $E \hookrightarrow L^p(\mathbb{R}^N)$ is compact when $2 \leq p < 2^*$, cf. Bartsch-Pankov-Wang [**5**]. On the other hand, when $(V1) - (V2A)$ hold, in general, the space E cannot be compactly embedded into $L^p(\mathbb{R}^N)$. In this case, introducing the subspace of radially symmetric functions of E, i.e.

$$E_{\text{rad}} = \{u \in E : u(gx) = u(x) \quad \text{for each} \quad g \in O(N), \quad \text{a.e. } x \in \mathbb{R}^N\},$$

the embedding $E_{\text{rad}} \hookrightarrow L^p(\mathbb{R}^N)$ is compact whenever $N \geq 2$ and $2 < p < 2^*$, cf. Strauss [**35**]. Taking into account $(FG3)$, the functional $\mathcal{H}_{\lambda,\mu}$ is $O(N)$-invariant, thus the critical points of the functional $\mathcal{H}_{\lambda,\mu}$ restricted to the space E_{rad}, i.e. $\mathcal{H}^r_{\lambda,\mu} = \mathcal{H}_{\lambda,\mu}|_{E_{\text{rad}}}$, are critical points of $\mathcal{H}_{\lambda,\mu}$, cf. Palais [**25**]. In order to consider simultaneously the two cases, we introduce some new notations. Let

$$X = \begin{cases} E_{\text{rad}}, & \text{if}(V2A)\text{holds}, \\ E, & \text{if}(V2B)\text{holds}, \end{cases} \quad \text{and} \quad \mathcal{R}_{\lambda,\mu} = \begin{cases} \mathcal{H}^r_{\lambda,\mu}, & \text{if}(V2A)\text{holds}, \\ \mathcal{H}_{\lambda,\mu}, & \text{if}(V2B)holds. \end{cases}$$

On account of these notations, it is sufficient to find critical points of $\mathcal{R}_{\lambda,\mu}$ on X. We further denote by $\langle\cdot,\cdot\rangle_X$, $\|\cdot\|_X$, $\mathcal{F}_X$, and $\mathcal{G}_X$, the restriction of $\langle\cdot,\cdot\rangle_E$, $\|\cdot\|_E$, $\mathcal{F}$, and $\mathcal{G}$ to the space X, respectively.

In the sequel, we denote by $S_p > 0$ the best Sobolev embedding constant of $X \hookrightarrow L^p(\mathbb{R}^N)$, $p \in [2, 2^*)$. The usual norm on $L^p(\mathbb{R}^N)$ is $\|\cdot\|_p$, $p \in [1, \infty]$.

LEMMA 2.7. *There exists $u_0 \in X$ such that $\mathcal{F}_X(u_0) > 0$.*

PROOF. Let $R_0 > 0$ and $s_0 \in \mathbb{R}$ from $(F4)$ and fix $\varepsilon \in (0, R_0/2)$. Fix also a radially symmetric function $u_\varepsilon \in C^\infty(\mathbb{R}^N)$ such that $u_\varepsilon(x) = 0$ for $|x| \geq R_0$, $u_\varepsilon(x) = s_0$ for $|x| \leq R_0 - \varepsilon$, and $\|u_\varepsilon\|_\infty \leq |s_0|$. Due to $(V1)$, u_ε belongs to X in both cases, i.e. either $(V2A)$ or $(V2B)$ hold. Denoting by $A_0 = \min_{|x|\leq R_0} F(x, s_0) > 0$ and by $\text{Vol}(B_r)$ the volume of the ball $B(0,r)$, we have

$$\begin{aligned} \mathcal{F}_X(u_\varepsilon) &= \int_{|x|\leq R_0-\varepsilon} F(x, u_\varepsilon(x)) + \int_{|x|\geq R_0} F(x, u_\varepsilon(x)) + \int_{R_0-\varepsilon<|x|<R_0} F(x, u_\varepsilon(x)) \\ &\geq A_0 \,\text{Vol}(B_{R_0/2}) - \int_{R_0-\varepsilon<|x|<R_0} |F(x, u_\varepsilon(x))| \\ &\geq A_0 \,\text{Vol}(B_{R_0/2}) - \max_{|x|\in[R_0/2,R_0],|s|\leq|s_0|} |F(x,s)|(\text{Vol}(B_{R_0}) - \text{Vol}(B_{R_0-\varepsilon})). \end{aligned}$$

If $\varepsilon \to 0^+$, the last term of the above expression becomes as small as we want. Thus, setting $\varepsilon = \varepsilon_0$ small enough, $u_0 = u_{\varepsilon_0} \in X$ verifies the requirement. □

LEMMA 2.8. *Let $\lambda, \mu \in \mathbb{R}$ be arbitrary fixed. Then every bounded sequence $\{u_n\} \subset X$ such that $\|\mathcal{R}'_{\lambda,\mu}(u_n)\|_{X^*} \to 0$, contains a strongly convergent subsequence.*

PROOF. Taking a subsequence if necessary, we may assume that $\{u_n\}$ converges to u, weakly in X, and strongly in $L^p(\mathbb{R}^N)$ for some $p \in (2, 2^*)$ arbitrarily fixed. Therefore,

$$\begin{aligned} \|u - u_n\|_X^2 &= \langle u, u - u_n\rangle_X + \mathcal{R}'_{\lambda,\mu}(u_n)(u_n - u) \\ &\quad - \lambda \int_{\mathbb{R}^N} f(x, u_n)(u - u_n) - \lambda\mu \int_{\mathbb{R}^N} g(x, u_n)(u - u_n) \end{aligned}$$

$$\leq \langle u, u - u_n \rangle_X + \|\mathcal{R}'_{\lambda,\mu}(u_n)\|_{X^*} \|u_n - u\|_X + |\lambda|(1 + |\mu|)\|Q\|_{p/(p-q-1)} \|u_n\|_p^q \|u - u_n\|_p.$$

Thus, $\|u - u_n\|_X \to 0$ as $n \to \infty$. □

LEMMA 2.9. *For every $\lambda, \mu \in \mathbb{R}$, $\mathcal{R}_{\lambda,\mu}$ is coercive and bounded from below on X. In particular, $\mathcal{R}_{\lambda,\mu}$ satisfies the Palais-Smale condition.*

PROOF. By $(FG1)$ we have for every $u \in X$ that

$$\mathcal{R}_{\lambda,\mu}(u) \geq \frac{1}{2}\|u\|_X^2 - |\lambda|(1 + |\mu|)\|Q\|_{2/(1-q)} S_2^{q+1} \|u\|_X^{q+1}. \tag{4}$$

Since $q < 1$, the first assertion holds. Now, consider a sequence $\{u_n\} \subset X$ such that $\{\mathcal{R}_{\lambda,\mu}(u_n)\}$ is bounded and $\|\mathcal{R}'_{\lambda,\mu}(u_n)\|_{X^*} \to 0$ as $n \to \infty$. In particular, $\{u_n\} \subset X$ is bounded. On account of Lemma 2.8, $\mathcal{R}_{\lambda,\mu}$ satisfies the Palais-Smale condition. □

2.2. Proof of Theorem 2.1.

LEMMA 2.10. *For every $\mu \in \mathbb{R}$,*

$$\lim_{\rho \to 0^+} \frac{\sup\{\mathcal{F}_X(u) + \mu\mathcal{G}_X(u) : \|u\|_X < \sqrt{2\rho}\}}{\rho} = 0.$$

PROOF. Fix arbitrarily $\varepsilon > 0$ and $p \in (2, 2^*)$. Due to Lemma 2.6, for every $u \in X$, one has

$$\mathcal{F}_X(u) + \mu\mathcal{G}_X(u) \leq (1 + |\mu|)(\varepsilon S_2^2 \|u\|_X^2 + c(\varepsilon) S_p^p \|u\|_X^p).$$

Therefore, for every $\rho > 0$,

$$0 \leq \frac{\sup\{\mathcal{F}_X(u) + \mu\mathcal{G}_X(u) : \|u\|_X < \sqrt{2\rho}\}}{\rho} \leq (1 + |\mu|)(2\varepsilon S_2^2 + c(\varepsilon) 2^{\frac{p}{2}} S_p^p \rho^{\frac{p}{2}-1}).$$

When $\rho \to 0^+$, the right hand side of the above inequality tends to 0, due to the arbitrariness of $\varepsilon > 0$, which concludes the proof. □

LEMMA 2.11. *For every $\lambda, \mu \in \mathbb{R}$, the functional $\mathcal{R}_{\lambda,\mu}$ is sequentially weakly lower semicontinuous.*

PROOF. Since the function $u \mapsto \|u\|_X^2$ is sequentially weakly lower semicontinuous, it is enough to prove that $\mathcal{F}_X + \mu\mathcal{G}_X$ is sequentially weakly continuous for every $\mu \in \mathbb{R}$. Let us assume the contrary, i.e., let $\{u_n\}$ be a sequence in X which converges weakly to $u \in X$ and the sequence $\{(\mathcal{F}_X + \mu\mathcal{G}_X)(u_n)\}$ does not converge to $(\mathcal{F}_X + \mu\mathcal{G}_X)(u)$ as $n \to \infty$. Therefore, there exist $\varepsilon_0 > 0$ and a subsequence of $\{u_n\}$, denoted again by $\{u_n\}$, such that

$$0 < \varepsilon_0 \leq |(\mathcal{F}_X + \mu\mathcal{G}_X)(u_n) - (\mathcal{F}_X + \mu\mathcal{G}_X)(u)|$$

for every $n \in \mathbb{N}$, and $u_n \to u$ strongly in $L^p(\mathbb{R}^N)$ for some $p \in (2, 2^*)$. (Note that the embedding $X \hookrightarrow L^p(\mathbb{R}^N)$ is compact.) By the mean value theorem, (2) and $(FG1)$, for some $\theta_n \in (0,1)$, the above inequality gives

$$\begin{aligned} 0 < \varepsilon_0 &\leq |(\mathcal{F}'_X + +\mu\mathcal{G}'_X)(u + \theta_n(u_n - u))(u_n - u)| \\ &\leq (1+|\mu|)\int_{\mathbb{R}^N} Q(x)|u + \theta_n(u_n - u)|^q |u_n - u| \\ &\leq (1+|\mu|)\|Q\|_{p/(p-q-1)}(\|u\|_p + \|u_n - u\|_p)^q \|u_n - u\|_p. \end{aligned}$$

The last term tends to 0, which is a contradiction. □

The main ingredient in the proof of Theorem 2.1 is a Ricceri-type critical point theorem, see Ricceri [**29, 30**]. Here, we recall a refinement of this result, established by Bonanno [**8**].

THEOREM 2.12. *Let Y be a separable and reflexive real Banach space, and let $\Phi, J : Y \to \mathbb{R}$ be two continuously Gâteaux differentiable functionals. Assume that there exists $x_0 \in Y$ such that $\Phi(x_0) = J(x_0) = 0$ and $\Phi(x) \geq 0$ for every $x \in Y$ and that there exists $x_1 \in Y$, $\rho > 0$ such that*

(i) *$\rho < \Phi(x_1)$ and $\sup_{\Phi(x)<\rho} J(x) < \rho\frac{J(x_1)}{\Phi(x_1)}$. Further, put*

$$\bar{a} = \frac{\zeta\rho}{\rho\frac{J(x_1)}{\Phi(x_1)} - \sup_{\Phi(x)<\rho} J(x)},$$

with $\zeta > 1$, assume that the functional $\Phi - \lambda J$ is sequentially weakly lower semicontinuous, satisfies the Palais-Smale condition; and

(ii) *$\lim_{\|x\|\to+\infty}(\Phi(x) - \lambda J(x)) = +\infty$, for every $\lambda \in [0, \bar{a}]$.*

Then there is an open interval $\Lambda \subset [0, \bar{a}]$ and a number $\sigma > 0$ such that for each $\lambda \in \Lambda$, the equation $\Phi'(x) - \lambda J'(x) = 0$ admits at least three distinct solutions in Y having norm less than σ.

PROOF OF THEOREM 2.1. Let $u_0 \in X$ the element from Lemma 2.7, and fix

$$\mu_0 = \frac{\mathcal{F}_X(u_0)}{|\mathcal{G}_X(u_0)| + 1}.$$

Now, we apply Theorem 2.12 by choosing $Y = X$, $\Phi = \frac{1}{2}\|\cdot\|_X^2$, and $J = J_\mu = \mathcal{F}_X + \mu\mathcal{G}_X$ for $\mu \in [-\mu_0, \mu_0]$. Note that $\mathcal{R}_{\lambda,\mu} = \Phi - \lambda J_\mu$.

A simple calculation shows that for every $\mu \in [-\mu_0, \mu_0]$ we have

$$J_\mu(u_0) = \mathcal{F}_X(u_0) + \mu\mathcal{G}_X(u_0) \geq \mu_0 > 0. \tag{5}$$

On account of the above inequality and Lemma 2.10, for every $\mu \in [-\mu_0, \mu_0]$ one can choose $\rho_\mu > 0$ so small that

$$\rho_\mu < \min\{1, \frac{\|u_0\|_X^2}{2}\}; \tag{6}$$

$$\frac{\sup\{J_\mu(u) : \|u\|_X < \sqrt{2\rho_\mu}\}}{\rho_\mu} < \frac{J_\mu(u_0)}{\|u_0\|_X^2}. \tag{7}$$

Now, choosing $x_1 = u_0$, $x_0 = 0$, $\zeta = 1 + \rho_\mu$ and

$$\overline{a} = \overline{a}_\mu = \frac{1+\rho_\mu}{2J_\mu(u_0)\|u_0\|_X^{-2} - \sup\{J_\mu(u) : \|u\|_X < \sqrt{2\rho_\mu}\}\rho_\mu^{-1}},$$

all the assumptions of Theorem 2.12 are verified, cf. Lemmas 2.11 and 2.9, respectively.

Then there is an open interval $\Lambda_\mu \subset [0, \overline{a}_\mu]$ and a number $\sigma_\mu > 0$ such that for any $\lambda \in \Lambda_\mu$, the functional $\mathcal{R}_{\lambda,\mu} = \Phi - \lambda J_\mu$ admits at least three distinct critical points $u^i_{\lambda,\mu} \in X$ ($i \in \{1,2,3\}$), having norm less than σ_μ, concluding the proof of Theorem 2.1. □

REMARK 2.13. On account of (6), (7) and (5), for every $\mu \in [-\mu_0, \mu_0]$ one has

$$\overline{a}_\mu < \frac{2\|u_0\|_X^2}{J_\mu(u_0)} \leq \frac{2\|u_0\|_X^2}{\mu_0} = \frac{2\|u_0\|_X^2}{\mathcal{F}_X(u_0)}(1 + |\mathcal{G}_X(u_0)|).$$

Since the right hand side does not depend on $\mu \in \mathbb{R}$, we have a uniform estimation of Λ_μ, i.e., for every $\mu \in [-\mu_0, \mu_0]$,

$$\Lambda_\mu \subset [0, \frac{2\|u_0\|_X^2}{\mathcal{F}_X(u_0)}(1 + |\mathcal{G}_X(u_0)|)].$$

2.3. Proof of Theorems 2.2 and 2.3. Let us define

$$c_\mathcal{G} = \int_{\mathbb{R}^N} |G(x, u_0(x))|dx \quad \text{and} \quad \lambda_0 = \frac{\|u_0\|_X^2}{2\mathcal{F}_X(u_0)},$$

where $u_0 \in X$ is from Lemma 2.7. For every $\lambda > \lambda_0$ let also

$$\mu^*_\lambda = \frac{1}{1+c_\mathcal{G}}(1 - \frac{\lambda_0}{\lambda})\mathcal{F}_X(u_0). \tag{8}$$

LEMMA 2.14. *Let $\lambda > \lambda_0$ and $|\mu| < \mu^*_\lambda$. Then $\inf_{u\in X} \mathcal{R}_{\lambda,\mu}(u) < 0$.*

PROOF. It is sufficient to show that $\mathcal{R}_{\lambda,\mu}(u_0) < 0$ whenever $\lambda > \lambda_0$ and $|\mu| < \mu^*_\lambda$. Due to the choice of λ_0 and μ^*_λ, one has

$$\begin{aligned}\mathcal{R}_{\lambda,\mu}(u_0) &= \frac{1}{2}\|u_0\|_X^2 - \lambda\mathcal{F}_X(u_0) - \lambda\mu\mathcal{G}_X(u_0)\\ &\leq (\lambda_0 - \lambda)\mathcal{F}_X(u_0) + \lambda|\mu|c_\mathcal{G}\\ &= -\lambda(1 + c_\mathcal{G})\mu^*_\lambda + \lambda|\mu|c_\mathcal{G} < 0.\end{aligned}$$ □

LEMMA 2.15. *For every $\lambda > \lambda_0$ and $\mu \in \mathbb{R}$ which complies with $|\mu| < \mu^*_\lambda$, the functional $\mathcal{R}_{\lambda,\mu}$ has the Mountain Pass geometry.*

PROOF. Fix $p \in (2, 2^*)$ arbitrary. By Lemma 2.6 one has for every $\varepsilon > 0$ that there exists $c(\varepsilon) > 0$ such that

$$\max\{|\mathcal{F}_X(u)|, |\mathcal{G}_X(u)|\} \leq \varepsilon S_2^2\|u\|_X^2 + c(\varepsilon)S_p^p\|u\|_X^p \quad \text{for every } u \in X.$$

Thus, for every $u \in X$ one has

$$\begin{aligned}\mathcal{R}_{\lambda,\mu}(u) &\geq \frac{1}{2}\|u\|_X^2 - \lambda|\mathcal{F}_X(u)| - \lambda|\mu||\mathcal{G}_X(u)| \\ &\geq \Big(\frac{1}{2} - \lambda(1+|\mu|)\varepsilon S_2^2\Big)\|u\|_X^2 - \lambda(1+|\mu|)c(\varepsilon)S_p^p\|u\|_X^p.\end{aligned}$$

Let $\varepsilon = (4\lambda(1+|\mu|)S_2^2)^{-1}$. Then

$$\mathcal{R}_{\lambda,\mu}(u) \geq \Big(\frac{1}{4} - \lambda(1+|\mu|)c(\lambda,\mu)S_p^p\rho^{p-2}\Big)\rho^2 \equiv \eta(\lambda,\mu) > 0$$

if $\|u\|_X = \rho < \min\{(4\lambda(1+|\mu|)c(\lambda,\mu)S_p^p)^{\frac{1}{2-p}}, \|u_0\|_X\}$. Moreover, by construction, $\rho < \|u_0\|_X$ and by the proof of Lemma 2.14 we have $\mathcal{R}_{\lambda,\mu}(u_0) < 0$. Thus, $\mathcal{R}_{\lambda,\mu}$ has indeed the Mountain Pass geometry. □

Proof of Theorem 2.2. Fix $\lambda > \lambda_0$ and $\mu \in (-\mu_\lambda^*, \mu_\lambda^*) \equiv \Pi_\lambda$. Lemma 2.9 ensures in particular that there exists an element $v_{\lambda\mu} \in X$ such that $\mathcal{R}_{\lambda,\mu}(v_{\lambda\mu}) = \inf_{u\in X)} \mathcal{R}_{\lambda,\mu}(u)$. By using Lemma 2.14, $\mathcal{R}_{\lambda,\mu}(v_{\lambda\mu}) < 0$.

On the other hand, by Lemma 2.15, we find $w_{\lambda\mu} \in X$ such that $\mathcal{R}'_{\lambda,\mu}(w_{\lambda\mu}) = 0$ and $\mathcal{R}_{\lambda,\mu}(w_{\lambda\mu}) \geq \eta(\lambda,\mu) > 0$ (see for instance [**28**, Theorem 2.2]). The mountain pass level $\mathcal{R}_{\lambda,\mu}(w_{\lambda\mu})$ is characterized as

$$\mathcal{R}_{\lambda,\mu}(w_{\lambda\mu}) = \inf_{g\in\Gamma}\max_{t\in[0,1]} \mathcal{R}_{\lambda,\mu}(g(t)), \tag{9}$$

where $\Gamma = \{g \in C([0,1];X) : g(0) = 0,\ g(1) = u_0\}$. Let $g_0 : [0,1] \to X$, defined by $g_0(t) = tu_0$. Since $g_0 \in \Gamma$, by using (9), one has for every $\mu \in \Pi_\lambda$

$$\begin{aligned}\mathcal{R}_{\lambda,\mu}(w_{\lambda\mu}) &\leq \max_{t\in[0,1]} \mathcal{R}_{\lambda,\mu}(tu_0) \\ &\leq \frac{1}{2}\|u_0\|_X^2 + \lambda \max_{t\in[0,1]} (|\mathcal{F}_X(tu_0)| + \mu_\lambda^*|\mathcal{G}_X(tu_0)|) \equiv C_\lambda.\end{aligned}$$

By using (4), for every $\mu \in \Pi_\lambda$ we have

$$\|w_{\lambda\mu}\|_X^2 \leq 2\lambda(1+\mu_\lambda^*)\|Q\|_{2/(1-q)}S_2^{q+1}\|w_{\lambda\mu}\|_X^{q+1} + 2C_\lambda.$$

Since $q+1 < 2$ we have at once that there exits a number $\sigma_\lambda^1 > 0$ such that $\|w_{\lambda\mu}\|_X \leq \sigma_\lambda^1$ for every $\mu \in \Pi_\lambda$. Moreover, since $\mathcal{R}_{\lambda,\mu}(v_{\lambda\mu}) < 0$ for every $\mu \in \Pi_\lambda$, a similar argument as above (put formally $C_\lambda = 0$) shows the existence of $\sigma_\lambda^2 > 0$ such that $\|v_{\lambda\mu}\|_X \leq \sigma_\lambda^2$ for every $\mu \in \Pi_\lambda$. Thus, letting $\sigma_\lambda = \max\{\sigma_\lambda^1, \sigma_\lambda^2\}$, the proof is completed. □

Remark 2.16. On account of (8), for every $\lambda > \lambda_0$ one has

$$\mu_\lambda^* < \frac{\mathcal{F}_X(u_0)}{1+c_\mathcal{G}}.$$

Since the right hand side does not depend on $\lambda \in \mathbb{R}$, we have a uniform estimation of Π_λ, i.e., for every $\lambda > \lambda_0$,

$$\Pi_\lambda \subset \Big[-\frac{\mathcal{F}_X(u_0)}{1+c_\mathcal{G}}, \frac{\mathcal{F}_X(u_0)}{1+c_\mathcal{G}}\Big].$$

Proof of Theorem 2.3. Let us fix $\overline{\lambda} > \lambda_0$, $\gamma \in (0, \overline{\lambda} - \lambda_0)$ and define

$$\overline{\mu} = \mu^*_{\overline{\lambda}+\gamma} \frac{\overline{\lambda} - \lambda_0 - \gamma}{\overline{\lambda} - \lambda_0 + \gamma},$$

where $\mu^*_{(\cdot)}$ is defined in (8).

Now, fix $\mu \in \mathbb{R}$ such that $|\mu| \leq \overline{\mu}$. It is easy to verify that the inequality $\overline{\mu} < \mu^*_{\lambda}$ holds for every $\lambda \in (\overline{\lambda} - \gamma, \overline{\lambda} + \gamma)$; thus, for every $\lambda \in (\overline{\lambda} - \gamma, \overline{\lambda} + \gamma)$ one has $\mu \in (-\mu^*_{\lambda}, \mu^*_{\lambda}) = \Pi_\lambda$. By applying Theorem 2.2 for each $\lambda \in (\overline{\lambda} - \gamma, \overline{\lambda} + \gamma)$, we conclude that $(\mathrm{P}_{\lambda,\mu})$ has at least two distinct, nontrivial, weak solutions. Consequently, the interval $(\overline{\lambda} - \gamma, \overline{\lambda} + \gamma)$ is included in the set defined in the statement of Theorem 2.3, which completes the proof. □

2.4. Proof of Theorem 2.5. We assume that hypotheses of Theorem 2.5 are fulfilled. First, we prove some lemmas.

Lemma 2.17. $\inf_X \mathcal{F}_X = 0$; $\sup_X \mathcal{F}_X = +\infty$.

Proof. Since $f(0) = 0$ and f is non-decreasing, it is easy to prove that the function $F(s) = \int_0^s f(t)dt$ is non-negative for any $s \in \mathbb{R}$. Since $\mathcal{F}_X(0) = 0$, and the potential Q is non-negative, see $(Q2)$, the first assertion of the lemma holds. Furthermore, since f is nonzero and non-decreasing, then $\sup_{\mathbb{R}} F = +\infty$. Let $\{a_n\} \subset \mathbb{R}$ be a sequence such that $F(a_n) \to +\infty$ as $n \to +\infty$. By $(Q2)$, one can fix $R_0 > 0$ such that $b_{R_0} = \operatorname{essinf}_{|x| \leq R_0} Q(x) > 0$. Let $\{u_n\} \subset X$ be such that $u_n(x) = a_n$ for any $x \in B(0, R_0)$. Since Q and F are non-negative functions, one has

$$\sup_X \mathcal{F}_X \geq \mathcal{F}_X(u_n) \geq b_{R_0} F(a_n) R_0^N \omega_N \to +\infty$$

as $n \to +\infty$, as claimed. Here, ω_N denotes the volume of the N-dimensional unit ball. □

Lemma 2.18. *Let $\zeta > 0$ and $\gamma \in \mathbb{R}$ be two fixed numbers. Then there exist $u = u(\zeta, \gamma) \in X$ and $r = r(\zeta, \gamma) > 0$ such that $\mathcal{F}_X(u) = \zeta$, and $u(x) = \gamma$ for any $x \in B(0, r)$.*

Proof. Fix $\overline{\zeta} \in]\zeta, +\infty[$, and $R_0 > 0$ such that $b_{R_0} = \operatorname{essinf}_{|x| \leq R_0} Q(x) > 0$ (cf. $(Q2)$). Fix $r > 0$ so small that

$$2r < R_0; \tag{10}$$

$$\|Q\|_{L^\infty} \max_{[-|\gamma|, |\gamma|]} F \omega_N (2r)^N < \zeta; \tag{11}$$

$$\overline{\zeta}(1 - (2r/R_0)^N) > \zeta. \tag{12}$$

Let $u_0 \in X$ be such that $u_0(x) = \gamma$ for any $x \in B(0, r)$, $u_0(x) = 0$ for any $x \notin B(0, 2r)$, and $\|u_0\|_{L^\infty} \leq |\gamma|$. Then, due to (11) one has

$$\begin{aligned}
\mathcal{F}_X(u_0) &= \int_{\mathbb{R}^N} Q(x) F(u_0(x)) dx \\
&\leq \|Q\|_{L^\infty} F(\gamma) \omega_N r^N + \|Q\|_{L^\infty} \max_{[-|\gamma|, |\gamma|]} F \omega_N (2^N - 1) r^N \\
&< \zeta.
\end{aligned}$$

On the other hand, since $\inf_{\mathbb{R}} F = 0$ and $\sup_{\mathbb{R}} F = +\infty$, there exists $\overline{\xi} \in \mathbb{R}$ such that

$$F(\overline{\xi}) = \overline{\zeta}(b_{R_0}\omega_N R_0^N)^{-1}. \tag{13}$$

According to (10), we may define $u_1 \in X$ such that $u_1(x) = \gamma$ for any $x \in B(0,r)$, and $u_1(x) = \overline{\xi}$ for any $x \in B(0,R_0) \setminus B(0,2r)$. Since the functions Q and F are non-negative, by (12) and (13) we have

$$\mathcal{F}_X(u_1) \geq F(\overline{\xi})\mathrm{essinf}_{2r\leq|x|\leq R_0} Q(x)\omega_N(R_0^N - (2r)^N) > \zeta.$$

Define the set

$$S_r^\gamma = \{u \in X : u(x) = \gamma \text{ for each } x \in B(0,r)\}.$$

Taking into account the above constructions, we have two elements $u_0, u_1 \in S_r^\gamma$ such that $\mathcal{F}_X(u_0) < \zeta < \mathcal{F}_X(u_1)$. Since the function $X \ni u \mapsto \mathcal{F}_X(u)$ is continuous and the set S_r^γ is connected (being convex), then there exists $u \in S_r^\gamma$ such that $\mathcal{F}_X(u) = \zeta$. □

LEMMA 2.19. *Let $\zeta > 0$ be a fixed number. Then the set $\mathcal{F}_X^{-1}([\zeta, +\infty[)$ is not convex.*

PROOF. Since f is non-decreasing, then $\mathcal{F}_X'$ is monotone, cf. (2) and ($Q2$). Therefore, $\mathcal{F}_X$ is a convex function and the level set $\mathcal{F}_X^{-1}(]-\infty, \zeta])$ is convex. If we assume that $\mathcal{F}_X^{-1}([\zeta, +\infty[)$ is also convex, then $\mathcal{F}_X^{-1}(\zeta) = \mathcal{F}_X^{-1}(]-\infty,\zeta]) \cap \mathcal{F}_X^{-1}([\zeta, +\infty[)$ will be convex as well. Thus, in order to get the conclusion of the lemma, it is enough to prove that $\mathcal{F}_X^{-1}(\zeta)$ is not convex.

To this end, recall that $f(0) = 0$ and f is a nonzero function, i.e., there exists $\gamma \in \mathbb{R}$ such that $f(\gamma) \neq f(0) = 0$. By Lemma 2.18, there exist $u_0, u_1 \in X$ and $r > 0$ such that $u_0(x) = 0$ and $u_1(x) = \gamma$ for any $x \in B(0,r)$, and $u_0, u_1 \in \mathcal{F}_X^{-1}(\zeta)$. Arguing by contradiction, suppose that

$$\mathcal{F}_X(tu_0 + (1-t)u_1) = \zeta \text{ for each } t \in [0,1]. \tag{14}$$

After a differentiation in (14) in rapport of t and by using (2) one has

$$\int_{\mathbb{R}^N} Q(x)f(tu_0(x) + (1-t)u_1(x))(u_0(x) - u_1(x))dx = 0$$

for any $t \in [0,1]$. Choosing in particular $t = 0$ and $t = 1$ in the above relation, one has

$$\int_{\mathbb{R}^N} Q(x)[f(u_0(x)) - f(u_1(x))](u_0(x) - u_1(x))dx = 0.$$

Since the potential Q is non-negative and f is non-decreasing, we obtain

$$Q(x)[f(u_0(x)) - f(u_1(x))](u_0(x) - u_1(x)) = 0 \text{ for a.e. } x \in \mathbb{R}^N.$$

On the other hand, hypothesis ($Q2$) asserts the existence of a number $R_0 > 0$ such that $b_{R_0} = \mathrm{essinf}_{|x|\leq R_0} Q(x) > 0$. Now, applying the last relation on the ball

$B(0, \min\{R_0, r\})$ and exploring the choice of u_0 and u_1, respectively, we are led to $f(\gamma)\gamma = 0$, which contradicts the choice of the number γ. Thus, (14) is false, i.e., the set $\mathcal{F}_X^{-1}(\zeta)$ is not convex. □

Now, we are in the position to prove Theorem 2.5. Before to do this, we recall another recent critical point theorem of Ricceri [**32**].

THEOREM 2.20. *Let Y be a real Hilbert space and $J : Y \to \mathbb{R}$ a continuous Gâteaux differentiable, nonconstant functional, with compact derivative, such that*

$$\limsup_{\|x\|\to+\infty} \frac{J(x)}{\|x\|^2} \leq 0. \tag{15}$$

Then, for each $r \in]\inf_Y J, \sup_Y J[$ for which the set $J^{-1}([r, +\infty])$ is not convex and for every set $S \subseteq Y$ dense in Y, there exist $x_0 \in J^{-1}(]-\infty, r[) \cap S$ and $\lambda > 0$ such that the equation

$$x = \lambda J'(x) + x_0$$

has at least three distinct solutions.

PROOF OF THEOREM 2.5. We apply Theorem 2.20 by choosing $Y = X$, $J = \mathcal{F}_X$, and $S = C_0^\infty(\mathbb{R}^N) \cap X$. As in the previous section, we have that $\mathcal{F}_X$ is of class C^1, and $\mathcal{F}'_X$ is compact. Indeed, here we used only $(f1)$ and the fact that $Q \in L^1(\mathbb{R}^N) \cap L^\infty(\mathbb{R}^N)$. Lemma 2.18 implies in particular that $\mathcal{F}_X$ is not a constant functional. By using $(f1)$, one has

$$\mathcal{F}_X(u) = \int_{\mathbb{R}^N} Q(x)F(u(x))dx \leq c\kappa_2^{q+1}\|Q\|_{L^{2/(1-q)}}\|u\|_X^{q+1}.$$

Thus, the inequality (15) is clearly verified, since $q < 1$. Taking into account Lemmas 2.17 and 2.19, for every $\zeta \in]0, +\infty[$ there exist $w \in \mathcal{F}_X^{-1}(]-\infty, \zeta[) \cap C_0^\infty(\mathbb{R}^N) = \mathcal{F}_X^{-1}([0, \zeta[) \cap C_0^\infty(\mathbb{R}^N)$ and $\lambda > 0$ such that the equation

$$v = \lambda \mathcal{F}'_X(v) + w, \tag{16}$$

has three distinct solutions, say $v_i \in X$, $i \in \{1, 2, 3\}$. Due to (16) and (2), the elements v_i are weak solutions of the equation

$$-\Delta v + V(x)v = \lambda Q(x)f(v) - \Delta w + V(x)w, \quad x \in \mathbb{R}^N.$$

Therefore, the elements $u_i = v_i - w$ are weak solutions of (P_λ^w). This concludes the proof. □

3. Elliptic problems on $\mathbb{R}^N$ involving oscillatory terms

Throughout this section we assume

(Q) $Q : \mathbb{R}^N \to \mathbb{R}$ is a positive, continuous, radially symmetric potential such that $Q \in L^p(\mathbb{R}^N)$ for every $p \in [1, 2]$.

The aim of this section is to study problem

$$(\mathrm{P}_\varepsilon)\qquad \begin{cases} -\Delta u + u = Q(x)[f(u) + \varepsilon g(u)], & x \in \mathbb{R}^N,\ N \geq 2,\\ u \geq 0 \\ u(x) \to 0 \ \text{ as } \ |x| \to \infty, \end{cases}$$

where $f, g : [0, \infty) \to \mathbb{R}$ are a continuous nonlinearities, with $g(0) = 0$.

In order to formulate our results, we recall some notations. The Hilbert space $H^1(\mathbb{R}^N)$ is endowed with its usual inner product and norm,

$$\langle u, v\rangle_{H^1} = \int_{\mathbb{R}^N} (\nabla u \nabla v + uv)dx, \quad \|u\|_{H^1} = \sqrt{\langle u, u\rangle_{H^1}}, \quad u, v \in H^1(\mathbb{R}^N).$$

The space $L^q(\mathbb{R}^N)$ is endowed with its usual norm $\|\cdot\|_{L^q}$, $q \in [1, \infty]$.

Let $f \in C([0, \infty), \mathbb{R})$ and $F(s) = \int_0^s f(t)dt$, $s \geq 0$. We assume

(f_1^0) $-\infty < \liminf_{s\to 0^+} \frac{F(s)}{s^2} \leq \limsup_{s\to 0^+} \frac{F(s)}{s^2} = +\infty$;

(f_2^0) There exists a sequence $\{s_i\}_i \subset (0, \infty)$ converging to 0 such that $f(s_i) < 0$ for every $i \in \mathbb{N}$.

REMARK 3.1. (a) Hypotheses (f_1^0) and (f_2^0) imply an oscillatory behaviour of f near the origin.

(b) Let $\alpha, \beta, \gamma \in \mathbb{R}$ such that $0 < \alpha < 1 < \alpha + \beta$, and $\gamma \in (0, 1)$. Then, the function $f : [0, \infty) \to \mathbb{R}$ defined by $f(0) = 0$ and $f(s) = s^\alpha(\gamma + \sin s^{-\beta})$, $s > 0$, verifies (f_1^0) and (f_2^0), respectively.

The first result deals with the unperturbed problem (P_0):

THEOREM 3.2. *Assume* (Q) *and let* $f \in C([0, \infty), \mathbb{R})$ *satisfying* (f_1^0) *and* (f_2^0). *Then there exists a sequence* $\{u_i^0\}_i \subset H^1(\mathbb{R}^N)$ *of distinct, radially symmetric weak solutions of* (P_0) *such that*

$$\lim_{i\to\infty} \|u_i^0\|_{L^\infty} = \lim_{i\to\infty} \|u_i^0\|_{H^1} = 0. \tag{17}$$

Keeping in mind Theorem 3.2, we expect an affirmative answer to question (q) for the perturbed problem (P_ε). This is indeed the case:

THEOREM 3.3. *Assume* (Q), *let* $f \in C([0, \infty), \mathbb{R})$ *satisfying* (f_1^0) *and* (f_2^0), *and let* $g \in C([0, \infty), \mathbb{R})$ *with* $g(0) = 0$. *Then, for every* $k \in \mathbb{N}$, *there exists* $\varepsilon_k^0 > 0$ *such that* (P_ε) *has at least* k *distinct, radially symmetric weak solutions in* $H^1(\mathbb{R}^N)$ *whenever* $\varepsilon \in [-\varepsilon_k^0, \varepsilon_k^0]$. *Moreover, if the (first* k*) solutions are denoted by* $u_{i,\varepsilon}^0 \in H^1(\mathbb{R}^N)$, $i = \overline{1,k}$, *then*

$$\|u_{i,\varepsilon}^0\|_{L^\infty} < \frac{1}{i} \ \text{ and } \ \|u_{i,\varepsilon}^0\|_{H^1} < \frac{1}{i} \ \text{ for any } i = \overline{1,k};\ \varepsilon \in [-\varepsilon_k^0, \varepsilon_k^0]. \tag{18}$$

REMARK 3.4. Note that (17) and (18) are in a perfect concordance.

In the sequel, we will state the counterparts of Theorems 3.2 and 3.3 whenever f oscillates at infinity. We assume

(f_1^∞) $-\infty < \liminf_{s\to\infty} \frac{F(s)}{s^2} \leq \limsup_{s\to\infty} \frac{F(s)}{s^2} = +\infty$;

(f_2^∞) There exists a sequence $\{s_i\}_i \subset (0,\infty)$ converging to $+\infty$ such that $f(s_i) < 0$ for every $i \in \mathbb{N}$.

REMARK 3.5. (a) Hypotheses (f_1^∞) and (f_2^∞) imply an oscillatory behaviour of f at infinity.

(b) Let $\alpha, \beta, \gamma \in \mathbb{R}$ such that $1 < \alpha$, $|\alpha - \beta| < 1$, and $\gamma \in (0,1)$. Then, the function $f : [0,\infty) \to \mathbb{R}$ defined by $f(s) = s^\alpha(\gamma + \sin s^\beta)$ verifies the hypotheses (f_1^∞) and (f_2^∞), respectively.

Concerning problem (P_0), we have the counterpart of Theorem 3.2:

THEOREM 3.6. *Assume* (Q) *and let* $f \in C([0,\infty),\mathbb{R})$ *satisfying* (f_1^∞), (f_2^∞) *and* $f(0) = 0$. *Then there exists a sequence* $\{u_i^\infty\}_i \subset H^1(\mathbb{R}^N)$ *of radially symmetric weak solutions of* (P_0) *such that*

$$\lim_{i\to\infty} \|u_i^\infty\|_{L^\infty} = \infty. \tag{19}$$

REMARK 3.7. Note that beside of (f_1^∞) and (f_2^∞), no further growth condition is assumed on the nonlinear term at infinity. Actually, this is the reason why we are not able to give H^1-estimates for the solutions obtained in Theorem 3.6. However, if we assume that f has a *half-subcritical* growth at infinity, i.e., there exists $q \in (1, 2^*/2)$ and $c > 0$ such that

$$|f(s)| \le c(1 + s^{q-1}) \quad \text{for all } s \in [0,\infty), \tag{20}$$

then, we have

$$\lim_{i\to\infty} \|u_i^\infty\|_{H^1} = \infty, \tag{19'}$$

Here, the number 2^* is the usual critical exponent. Let us observe that relation (20) and the right side of (f_1^∞) imply $2 < q$. Thus, (19′) is possible for the low-dimensions $N = 2, 3$, since $2 < 2^*/2$ must hold. Another way to guarantee (19′) is to complete hypothesis (Q) by allowing for instance $Q \in L^\infty(\mathbb{R}^N)$ and (20) with $q \in (2, 2^*)$.

Throughout Theorem 3.6, another affirmative answer to (q) can be done:

THEOREM 3.8. *Assume* (Q), *let* $f \in C([0,\infty),\mathbb{R})$ *satisfying* (f_1^∞), (f_2^∞) *with* $f(0) = 0$, *and let* $g \in C([0,\infty),\mathbb{R})$ *with* $g(0) = 0$. *Then, for every* $k \in \mathbb{N}$, *there exists* $\varepsilon_k^\infty > 0$ *such that* (P_ε) *has at least* k *distinct, radially symmetric weak solutions in* $H^1(\mathbb{R}^N)$ *whenever* $\varepsilon \in [-\varepsilon_k^\infty, \varepsilon_k^\infty]$. *Moreover, if the (first* k*) solutions are denoted by* $u_{i,\varepsilon}^\infty \in H^1(\mathbb{R}^N)$, $i = \overline{1,k}$, *then*

$$\|u_{i,\varepsilon}^\infty\|_{L^\infty} > i - 1 \quad \textit{for any } i = \overline{1,k};\ \varepsilon \in [-\varepsilon_k^\infty, \varepsilon_k^\infty]. \tag{21}$$

REMARK 3.9. Relations (19) and (21) are also in concordance. Moreover, if both functions f and g verify (20) with $q \in (2, 2^*/2)$, then beside of (21), we also have

$$\|u_{i,\varepsilon}^\infty\|_{H^1} > i - 1 \quad \text{for any } i = \overline{1,k};\ \varepsilon \in [-\varepsilon_k^\infty, \varepsilon_k^\infty]. \tag{21'}$$

As we already pointed out, the method developed in the present paper is applicable in more general settings; not only the type of the domain Ω can vary with various boundary conditions, but also equations involving the p-Laplacian can be considered. We emphasize that existence of infinitely many solutions for elliptic problems in $\mathbb{R}^N$ involving the p-Laplacian and an oscillatory term has been already studied by Kristály [**16**]. However, in those papers the assumption $p > N \geq 2$ was essential, due to a Morrey-type embedding, and only the 'unperturbed' case was considered. Consequently, the unperturbed problem (P_0) in the present paper may be considered as a natural completion of [**16**] from the point of view of the parameter p and the space dimension N. Finally, we mention that elliptic problems involving decaying or unbounded terms can be also treated by this method, exploiting recent embedding results of Su-Wang-Willem [**36**].

3.1. Preliminaries. The proof of our theorems formulated above rely heavily in the principle of symmetric criticality for Szulkin-type functionals, which we state in the sequel for the sake of completeness. For further details, see the paper of Kobayashi-Ôtani [**14**].

3.1.1. *Principle of symmetric criticality for Szulkin-type functionals.* Let X be a real Banach space and X^* its dual. Let $E : X \to \mathbb{R}$ be a functional of class C^1 and let $\zeta : X \to \mathbb{R} \cup \{+\infty\}$ be a proper (i.e. $\not\equiv +\infty$), convex, lower semicontinuous function. Then, $I = E + \zeta$ is a *Szulkin-type functional*, see [**37**]. An element $u \in X$ is called a *critical point of* $I = E + \zeta$ if

$$E'(u)(v-u) + \zeta(v) - \zeta(u) \geq 0 \quad \text{for all } v \in X, \tag{22}$$

or equivalently,

$$0 \in E'(u) + \partial\zeta(u) \quad \text{in} \quad X^*,$$

where $\partial\zeta(u)$ stands for the subdifferential of the convex functional ζ at $u \in X$.

PROPOSITION 3.10. [**37**, p. 80] *Every local minimum point of* $I = E + \zeta$ *is a critical point of* I *in the sense of* (22).

Let G be a topological group acting linearly on X. We say that G acts *continuously on* X if the map $(g, u) \mapsto gu$ from $G \times X$ into X is continuous. A set M is called *G-invariant* if $gM = \{gu : u \in M\} \subseteq M$ for every $g \in G$. A function h on X is called *G-invariant* if $h(gu) = h(u)$ for every $u \in X$ and $g \in G$. The linear subspace of G-symmetric points of X is defined by

$$\Sigma = \mathrm{Fix}_G(X) = \{u \in X : gu = u \quad \text{for all } g \in G\}.$$

A special form of [**14**, Theorem 3.16] is the following result, known as the *principle of symmetric criticality for Szulkin functionals.*

THEOREM 3.11. *Let X be a reflexive Banach space and let $I = E + \zeta : X \to \mathbb{R} \cup \{+\infty\}$ be a Szulkin-type functional on X. If a compact group G acts linearly and continuously on X, and the functionals E and ζ are G-invariant, then the principle of symmetric criticality holds, i.e., fixing $u \in \Sigma$, we have*

$$0 \in (E|_\Sigma)'(u) + \partial(\zeta|_\Sigma)(u) \ \textit{in}\ \Sigma^* \implies 0 \in E'(u) + \partial\zeta(u) \ \textit{in}\ X^*.$$

3.1.2. *A key existence result.* Due to the fact that problems (P_0) and (P_ε) will be treated simultaneously, in this subsection we consider the generic problem

$$(\mathrm{P}_h) \qquad \begin{cases} -\Delta u + u = Q(x)h(u), & x \in \mathbb{R}^N,\ N \geq 2, \\ u(x) \to 0 \ \text{ as } \ |x| \to \infty, \end{cases}$$

and beside of hypothesis (Q), we assume that

(h_1) $h : [0, \infty) \to \mathbb{R}$ is a continuous, bounded function such that $h(0) = 0$;

(h_2) there are $0 < a < b$ such that $h(s) \leq 0$ for all $s \in [a, b]$.

Due to (h_1), we may extend h continuously to the whole $\mathbb{R}$, putting $h(s) = 0$ for all $s \leq 0$.

We introduce the energy functional $\mathcal{H}_h : H^1(\mathbb{R}^N) \to \mathbb{R}$ associated with problem (P_h), defined by

$$\mathcal{H}_h(u) = \frac{1}{2}\|u\|_{H^1}^2 - \int_{\mathbb{R}^N} Q(x)H(u(x))dx, \quad u \in H^1(\mathbb{R}^N),$$

where $H(s) = \int_0^s h(t)dt$, $s \in \mathbb{R}$. One can easily show that $\mathcal{H}_h$ is well-defined; indeed, by the mean value theorem, we have

$$(23) \qquad \int_{\mathbb{R}^N} Q(x)|H(u(x))|dx \leq M_h \|Q\|_{L^2}\|u\|_{L^2} < \infty, \quad u \in H^1(\mathbb{R}^N),$$

where $M_h = \sup_{s\in\mathbb{R}} |h(s)|$. Moreover, standard arguments show that $\mathcal{H}_h$ is of class C^1 on $H^1(\mathbb{R}^N)$.

Now, we denote by $H^1_{\mathrm{rad}}(\mathbb{R}^N)$ the radial functions in $H^1(\mathbb{R}^N)$, and let

$$\mathcal{R}_h = \mathcal{H}_h|_{H^1_{\mathrm{rad}}(\mathbb{R}^N)},$$

i.e., the restriction of $\mathcal{H}_h$ to $H^1_{\mathrm{rad}}(\mathbb{R}^N)$. Finally, considering the number $b \in \mathbb{R}$ from (h_2), we introduce

$$W^b = \{u \in H^1(\mathbb{R}^N) : \|u\|_{L^\infty} \leq b\} \ \text{ and } \ W^b_{\mathrm{rad}} = W^b \cap H^1_{\mathrm{rad}}(\mathbb{R}^N).$$

The main result of this section can be stated as follows.

THEOREM 3.12. *Assume that (h_1), (h_2) and (Q) hold. Then*

(i) *the functional $\mathcal{R}_h$ is bounded from below on W^b_{rad} and its infimum is attained at $u_h \in W^b_{\mathrm{rad}}$;*

(ii) *$u_h(x) \in [0, a]$ for a.e. $x \in \mathbb{R}^N$;*

(iii) *u_h is a weak solution of (P_h).*

PROOF. (i) Actually, $\mathcal{R}_h$ is bounded from below on the whole $H^1_{\mathrm{rad}}(\mathbb{R}^N)$. Indeed, due to (23), for all $u \in H^1_{\mathrm{rad}}(\mathbb{R}^N)$ we have

$$\begin{aligned} \mathcal{R}_h(u) &= \frac{1}{2}\|u\|_{H^1}^2 - \int_{\mathbb{R}^N} Q(x)H(u)dx \geq \frac{1}{2}\|u\|_{H^1}^2 - M_h\|Q\|_{L^2}\|u\|_{H^1} \\ &\geq -\frac{1}{2}M_h^2\|Q\|_{L^2}^2. \end{aligned}$$

Now, we prove that $\mathcal{R}_h$ attains its infimum on W^b_{rad}. Note that W^b_{rad} is convex and closed in $H^1_{\mathrm{rad}}(\mathbb{R}^N)$, thus weakly closed. Due to the boundedness from below of $\mathcal{R}_h$ on W^b_{rad}, it is enough to prove that $\mathcal{R}_h$ is sequentially weakly lower semicontinuous. The latter fact follows at once if we prove that $u \mapsto \int_{\mathbb{R}^N} Q(x)H(u)dx$, $u \in H^1_{\text{rad}}(\mathbb{R}^N)$, is sequentially weakly continuous. We argue by contradiction; let $\{u_i\}_i \subset H^1_{\text{rad}}(\mathbb{R}^N)$ be a sequence which converges weakly to $u \in H^1_{\text{rad}}(\mathbb{R}^N)$ but, up to a subsequence, one can find a number $\varepsilon_0 > 0$ such that

$$0 < \varepsilon_0 \leq \left| \int_{\mathbb{R}^N} Q(x)H(u_i)dx - \int_{\mathbb{R}^N} Q(x)H(u)dx \right| \text{ for all } i \in \mathbb{N},$$

and u_i converges strongly to u in $L^q(\mathbb{R}^N)$, for some $q \in (2, 2^*)$. Here, we employed the fact that $H^1_{\text{rad}}(\mathbb{R}^N)$ is compactly embedded into $L^q(\mathbb{R}^N)$ for all $q \in (2, 2^*)$. Using the mean value theorem and Hölder inequality, from the above inequality we deduce that

$$0 < \varepsilon_0 \leq M_h \int_{\mathbb{R}^N} Q(x)|u_i - u| \leq M_h \|Q\|_{L^{q/(q-1)}} \|u_i - u\|_{L^q}.$$

But the right hand side tends to 0 as $i \to \infty$, contradicting $\varepsilon_0 > 0$. This proves (i); let $u_h \in W^b_{\text{rad}}$ be a minimum point of $\mathcal{R}_h$ over W^b_{rad}.

(ii) Let $A = \{x \in \mathbb{R}^N : u_h(x) \notin [0, a]\}$ and suppose that meas$(A) > 0$. Define the function $\gamma : \mathbb{R} \to \mathbb{R}$ by $\gamma(s) = \min(s_+, a)$, where $s_+ = \max(s, 0)$. Now, set $w = \gamma \circ u_h$. Since γ is a Lipschitz function and $\gamma(0) = 0$, the theorem of Marcus-Mizel [**23**] shows that $w \in H^1(\mathbb{R}^N)$. In addition, w is radial, since $u_h \in H^1_{\text{rad}}(\mathbb{R}^N)$. Thus $w \in H^1_{\text{rad}}(\mathbb{R}^N)$. Moreover, by definition, $0 \leq w(x) \leq a$ for a.e. $\mathbb{R}^N$.

We introduce the sets

$$A_1 = \{x \in A : \ u_h(x) < 0\} \quad \text{and} \quad A_2 = \{x \in A : \ u_h(x) > a\}.$$

Thus, $A = A_1 \cup A_2$, and we have that $w(x) = u_h(x)$ for all $x \in \mathbb{R}^N \setminus A$, $w(x) = 0$ for all $x \in A_1$, and $w(x) = a$ for all $x \in A_2$. Moreover, we have

$$\begin{aligned}\mathcal{R}_h(w) - \mathcal{R}_h(u_h) &= \frac{1}{2}\left[\|w\|^2_{H^1} - \|u_h\|^2_{H^1}\right] - \int_{\mathbb{R}^N} Q(x)[H(w) - H(u_h)] \\ &= -\frac{1}{2}\int_A |\nabla u_h|^2 + \frac{1}{2}\int_A [w^2 - u_h^2] - \int_A Q(x)[H(w) - H(u_h)].\end{aligned}$$

Note that

$$\int_A [w^2 - u_h^2] = -\int_{A_1} u_h^2 + \int_{A_2} [a^2 - u_h^2] \leq 0.$$

Due to the fact that $h(s) = 0$ for all $s \leq 0$, one has

$$\int_{A_1} Q(x)[H(w) - H(u_h)] = 0.$$

By the mean value theorem, for a.e. $x \in A_2$, there exists $\theta(x) \in [a, u_h(x)] \subseteq [a, b]$ such that

$$H(w(x)) - H(u_h(x)) = H(a) - H(u_h(x)) = h(\theta(x))(a - u_h(x)).$$

Thus, on account of (h_2), one has

$$\int_{A_2} Q(x)[H(w) - H(u_h)] \geq 0.$$

Consequently, every term of the expression $\mathcal{R}_h(w) - \mathcal{R}_h(u_h)$ is non-positive. On the other hand, since $w \in W^b_{\text{rad}}$, then $\mathcal{R}_h(w) \geq \mathcal{R}_h(u_h) = \inf_{W^b_{\text{rad}}} \mathcal{R}_h$. So, every term in $\mathcal{R}_h(w) - \mathcal{R}_h(u_h)$ should be zero. In particular,

$$\int_{A_1} u_h^2 = \int_{A_2} [a^2 - u_h^2] = 0,$$

which imply that meas(A) should be 0, contradicting our assumption.

(iii) We divide this part into two steps.

Step 1. $\mathcal{H}_h'(u_h)(w - u_h) \geq 0$ *for every* $w \in W^b$.
Standard argument shows that W^b is closed and convex in $H^1(\mathbb{R}^N)$. Let ζ_{W^b} be the indicator function of the set W^b (i.e., $\zeta_{W^b}(u) = 0$ if $u \in W^b$, and $\zeta_{W^b}(u) = +\infty$, otherwise). Since $\mathcal{H}_h$ is of class C^1 on $H^1(\mathbb{R}^N)$, and ζ_{W^b} is convex, lower semicontinuous and proper (i.e., $\not\equiv +\infty$), we may define the Szulkin-type functional $I : H^1(\mathbb{R}^N) \to \mathbb{R} \cup \{+\infty\}$ by $I = \mathcal{H}_h + \zeta_{W^b}$. Since $W^b_{\text{rad}} = W^b \cap H^1_{\text{rad}}(\mathbb{R}^N)$, the restriction of ζ_{W^b} to $H^1_{\text{rad}}(\mathbb{R}^N)$ is precisely the indicator function $\zeta_{W^b_{\text{rad}}}$ of the set W^b_{rad}. Recall that u_h is a local minimum point of $\mathcal{R}_h$ relative to W^b_{rad} (see (i)), thus a local minimum point of the Szulkin-type functional $\tilde{I} : H^1_{\text{rad}}(\mathbb{R}^N) \to \mathbb{R} \cup \{+\infty\}$, defined by $\tilde{I} = \mathcal{R}_h + \zeta_{W^b_{\text{rad}}}$. Due to Proposition 3.10, u_h is a critical point of $\tilde{I}$, i.e.,

$$0 \in \mathcal{R}_h'(u_h) + \partial\zeta_{W^b_{\text{rad}}}(u_h) \quad \text{in} \quad (H^1_{\text{rad}}(\mathbb{R}^N))^*. \tag{24}$$

On the other hand, we introduce the action of the orthogonal group $G = O(N)$ on $H^1(\mathbb{R}^N)$ by

$$(gu)(x) = u(g^{-1}x), \quad \text{for all } g \in O(N),\ u \in H^1(\mathbb{R}^N),\ x \in \mathbb{R}^N.$$

Clearly, this action is linear and continuous on $H^1(\mathbb{R}^N)$. Since the potential $Q : \mathbb{R}^N \to \mathbb{R}$ is radial, one can easily check that the functional $\mathcal{H}_h$ is $O(N)$-invariant. Moreover, due to the fact that the set W^b is $O(N)$-invariant, the functional ζ_{W^b} is $O(N)$-invariant as well. The set $H^1_{\text{rad}}(\mathbb{R}^N)$ is exactly the subspace of $O(N)$-symmetric points of $H^1(\mathbb{R}^N)$. Therefore, on account of (24) and Theorem 3.11, we obtain

$$0 \in \mathcal{H}_h'(u_h) + \partial\zeta_{W^b}(u_h) \quad \text{in} \quad (H^1(\mathbb{R}^N))^*.$$

Consequently, for every $w \in H^1(\mathbb{R}^N)$, we have

$$\mathcal{H}_h'(u_h)(w - u_h) + \zeta_{W^b}(w) - \zeta_{W^b}(u_h) \geq 0,$$

which implies our claim.

Step 2. (Proof concluded) u_h *is a weak solution of* (P_h).
First of all, by a Strauss-type inequality (see for instance Willem [**39**, p. 76]), we have that $u_h(x) \to 0$ as $|x| \to \infty$. It remains to prove that

$$\int_{\mathbb{R}^N} \nabla u_h \nabla v + \int_{\mathbb{R}^N} u_h v - \int_{\mathbb{R}^N} Q(x) h(u_h) v = 0 \quad \text{for all } v \in H^1(\mathbb{R}^N). \tag{25}$$

By Step 1, we have

$$\int_{\mathbb{R}^N} \nabla u_h \nabla (w - u_h) + \int_{\mathbb{R}^N} u_h (w - u_h) - \int_{\mathbb{R}^N} Q(x) h(u_h)(w - u_h) \geq 0, \quad \forall\, w \in W^b.$$

Let us define the function $\gamma(s) = \operatorname{sgn}(s) \min(|s|, b)$, and fix $\varepsilon > 0$ and $v \in H^1(\mathbb{R}^N)$ arbitrarily. Since γ is Lipschitz continuous and $\gamma(0) = 0$, the element $w_\gamma = \gamma \circ (u_h + \varepsilon v)$ belongs to $H^1(\mathbb{R}^N)$, see Marcus-Mizel [**23**]. The explicit expression of the truncation function w_γ is

$$w_\gamma(x) = \begin{cases} -b, & \text{if } \ x \in \{u_h + \varepsilon v < -b\} \\ u_h(x) + \varepsilon v(x), & \text{if } \ x \in \{-b \leq u_h + \varepsilon v < b\} \\ b, & \text{if } \ x \in \{b \leq u_h + \varepsilon v\}. \end{cases}$$

Therefore, $w_\gamma \in W^b$. Taking $w = w_\gamma$ as a test function in the previous inequality, we obtain

$$\begin{aligned} 0 \leq & - \int_{\{u_h + \varepsilon v < -b\}} \left[|\nabla u_h|^2 + u_h(b + u_h) - Q(x) h(u_h)(b + u_h) \right] \\ & + \varepsilon \int_{\{-b \leq u_h + \varepsilon v < b\}} [\nabla u_h \nabla v + u_h v - Q(x) h(u_h) v] \\ & - \int_{\{b \leq u_h + \varepsilon v\}} [|\nabla u_h|^2 - u_h(b - u_h) + Q(x) h(u_h)(b - u_h)]. \end{aligned}$$

After a suitable rearrangement of the terms in this inequality, we obtain that

$$\begin{aligned} 0 \leq \varepsilon & \int_{\mathbb{R}^N} \nabla u_h \nabla v + \varepsilon \int_{\mathbb{R}^N} u_h v - \varepsilon \int_{\mathbb{R}^N} Q(x) h(u_h) v \\ & - \int_{\{u_h + \varepsilon v < -b\}} |\nabla u_h|^2 - \int_{\{b \leq u_h + \varepsilon v\}} |\nabla u_h|^2 \\ & + \int_{\{u_h + \varepsilon v < -b\}} [Q(x) h(u_h) - u_h](b + u_h + \varepsilon v) \\ & + \int_{\{b \leq u_h + \varepsilon v\}} [Q(x) h(u_h) - u_h](-b + u_h + \varepsilon v) \\ & - \varepsilon \int_{\{u_h + \varepsilon v < -b\}} \nabla u_h \nabla v - \varepsilon \int_{\{b \leq u_h + \varepsilon v\}} \nabla u_h \nabla v. \end{aligned}$$

Recalling the notation $M_h = \sup_{s \in \mathbb{R}} |h(s)| < \infty$, and taking into account that $u_h(x) \in [0, a] \subset [-b, b]$ for a.e. $x \in \mathbb{R}^N$, we have

$$\int_{\{u_h + \varepsilon v < -b\}} [Q(x) h(u_h) - u_h](b + u_h + \varepsilon v) \leq -\varepsilon \int_{\{u_h + \varepsilon v < -b\}} [M_h Q(x) + u_h(x)] v(x) dx$$

and

$$\int_{\{b\leq u_h+\varepsilon v\}} [Q(x)h(u_h) - u_h](-b + u_h + \varepsilon v) \leq \varepsilon M_h \int_{\{b\leq u_h+\varepsilon v\}} Q(x)v(x)dx.$$

Using the above estimates and dividing by $\varepsilon > 0$, we obtain

$$\begin{aligned} 0 \leq & \int_{\mathbb{R}^N} \nabla u_h \nabla v + \int_{\mathbb{R}^N} u_h v - \int_{\mathbb{R}^N} Q(x)h(u_h)v \\ & - \int_{\{u_h+\varepsilon v<-b\}} [\nabla u_h \nabla v + u_h v + M_h Q(x)v] \\ & - \int_{\{b\leq u_h+\varepsilon v\}} [\nabla u_h \nabla v - M_h Q(x)v]. \end{aligned}$$

Now, letting $\varepsilon \to 0^+$, and taking into account (ii), that is, $0 \leq u_h(x) \leq a$ for a.e. $x \in \mathbb{R}^N$, we have

$$m(\{u_h + \varepsilon v < -b\}) \to 0 \quad \text{and} \quad m(\{b \leq u_h + \varepsilon v\}) \to 0,$$

respectively. Consequently, the above inequality reduces to

$$0\leq \int_{\mathbb{R}^N} \nabla u_h \nabla v + \int_{\mathbb{R}^N} u_h v - \int_{\mathbb{R}^N} Q(x)h(u_h)v.$$

Putting $(-v)$ instead of v, we arrive to (25), i.e., u_h is a weak solution of (P_h). This ends the proof. □

3.1.3. *A special function from* $H^1_{\mathrm{rad}}(\mathbb{R}^N)$. We conclude this section by constructing a special function which will be useful in the proof of our theorems. In the sequel, since no confusion may arise, B_c will denote the closed N-dimensional ball with radius $c > 0$ and center 0.

Fix $\rho > 0$. For any $s > 0$ we introduce the function

$$w_s(x) = \begin{cases} 0, & \text{if } x \in \mathbb{R}^N \setminus B_\rho; \\ s, & \text{if } x \in B_{\rho/2}; \\ \frac{2s}{\rho}(\rho - |x|), & \text{if } x \in B_\rho \setminus B_{\rho/2}. \end{cases} \tag{26}$$

It is clear that $w_s \in H^1_{\mathrm{rad}}(\mathbb{R}^N)$ and

$$\|w_s\|^2_{H^1} \leq K(\rho)s^2 \tag{27}$$

where $K(\rho) = (4 + \rho^2)\rho^{N-2}\omega_N$.

3.2. Proof of Theorems 3.2 and 3.3. Due to (f_2^0) and to the continuity of f and g, we may fix the positive sequences $\{a_i\}_i$, $\{b_i\}_i$, and $\{\varepsilon_i\}_i$ such that $\lim_{i\to\infty} a_i = \lim_{i\to\infty} b_i = 0$, and for all $i \in \mathbb{N}$,

$$b_{i+1} < a_i < s_i < b_i < 1; \tag{28}$$

$$f(s) + \varepsilon g(s) \leq 0 \quad \text{for all } s \in [a_i, b_i] \quad \text{and} \quad \varepsilon \in [-\varepsilon_i, \varepsilon_i]. \tag{29}$$

For every $i \in \mathbb{N}$, we define the truncation functions $f_i, g_i : [0,\infty) \to \mathbb{R}$ by

$$f_i(s) = f(\min(s, b_i)) \quad \text{and} \quad g_i(s) = g(\min(s, b_i)). \tag{30}$$

By (f_1^0) and (f_2^0) we have $f(0) = 0$. Since $f_i(0) = g_i(0) = 0$, we may extend continuously the functions f_i and g_i to the whole real line, taking 0 for negative arguments. For every $s \in \mathbb{R}$ and $i \in \mathbb{N}$, let $F_i(s) = \int_0^s f_i(t)dt$ and $G_i(s) = \int_0^s g_i(t)dt$.

For every $i \in \mathbb{N}$ and $\varepsilon \in [-\varepsilon_i, \varepsilon_i]$ the function $h_{i,\varepsilon}^0 : [0,\infty) \to \mathbb{R}$ defined by $h_{i,\varepsilon}^0 = f_i + \varepsilon g_i$ is continuous, bounded, and $h_{i,\varepsilon}^0(0) = 0$. On account of relations (29) and (30), we have $h_{i,\varepsilon}^0(s) \le 0$ for all $s \in [a_i, b_i]$. Thus, we may apply Theorem 3.12 to the function $h_{i,\varepsilon}^0$ obtaining that for every $i \in \mathbb{N}$ and $\varepsilon \in [-\varepsilon_i, \varepsilon_i]$, the problem

$$(\mathrm{P}_{i,\varepsilon}^0) \qquad \begin{cases} -\triangle u + u = Q(x) h_{i,\varepsilon}^0(u), & x \in \mathbb{R}^N, \\ u(x) \to 0 \text{ as } |x| \to \infty, \end{cases}$$

has a radially symmetric, weak solution $u_{i,\varepsilon}^0 \in H^1(\mathbb{R}^N)$ such that

$$u_{i,\varepsilon}^0 \in [0, a_i] \text{ for a.e. } x \in \mathbb{R}^N; \tag{31}$$

$$u_{i,\varepsilon}^0 \text{ is the infimum of the functional } \mathcal{R}_i^\varepsilon \text{ on } W_{\mathrm{rad}}^{b_i}, \tag{32}$$

where

$$\mathcal{R}_i^\varepsilon(u) = \frac{1}{2}\|u\|_{H^1}^2 - \int_{\mathbb{R}^N} Q(x)[F_i(u) + \varepsilon G_i(u)], \quad u \in H_{\mathrm{rad}}^1(\mathbb{R}^N). \tag{33}$$

Due to (30) and (31), $u_{i,\varepsilon}^0$ is a weak solution not only for $(\mathrm{P}_{i,\varepsilon}^0)$ but also for our problem (P_ε). Consequently, it remains to prove that:

- (I_0) there are infinitely many distinct elements in the sequence $\{u_{i,0}^0\}_i$ verifying (17), see Theorem 3.2;
- (II_0) for every $k \in \mathbb{N}$, there are at least k distinct elements $u_{i,\varepsilon}^0$ verifying (18) when ε belongs to a certain interval around the origin, see Theorem 3.3.

3.2.1. *Proof of* (I_0); *Theorem 3.2 concluded.* For abbreviation, take $u_i^0 = u_{i,0}^0$ and $\mathcal{R}_i = \mathcal{R}_i^0$ for every $i \in \mathbb{N}$. We first prove that

$$\mathcal{R}_i(u_i^0) < 0 \quad \text{for all } i \in \mathbb{N}; \tag{34}$$

$$\lim_{i \to \infty} \mathcal{R}_i(u_i^0) = 0. \tag{35}$$

The left side of (f_1^0) implies the existence of $l_0 > 0$ and $\delta \in (0, b_1)$ such that

$$F(s) \ge -l_0 s^2 \quad \text{for all } s \in (0, \delta). \tag{36}$$

Let $L_0 > 0$ be large enough such that

$$\frac{1}{2}K(\rho) + l_0\|Q\|_{L^1} < L_0(\rho/2)^N \omega_N \min_{B_{\rho/2}} Q, \tag{37}$$

where $\rho > 0$ and $K(\rho)$ come from (27). Taking into account the right side of (f_1^0), there is a sequence $\{\tilde{s}_i\}_i \subset (0, \delta)$ such that $\tilde{s}_i \le a_i$ and $F(\tilde{s}_i) > L_0 \tilde{s}_i^2$ for all $i \in \mathbb{N}$.

Let $i \in \mathbb{N}$ fixed and $w_{\tilde{s}_i} \in H^1_{\rm rad}(\mathbb{R}^N)$ be the function from (26) corresponding to the value $\tilde{s}_i > 0$. Then $w_{\tilde{s}_i} \in W^{b_i}_{\rm rad}$, and on account of (27) and (36) one has

$$\begin{aligned}\mathcal{R}_i(w_{\tilde{s}_i}) &= \frac{1}{2}\|w_{\tilde{s}_i}\|^2_{H^1} - \int_{\mathbb{R}^N} Q(x)F_i(w_{\tilde{s}_i}(x))dx \\ &= \frac{1}{2}\|w_{\tilde{s}_i}\|^2_{H^1} - F(\tilde{s}_i)\int_{B_{\rho/2}} Q(x)dx - \int_{B_\rho \setminus B_{\rho/2}} Q(x)F(w_{\tilde{s}_i}(x))dx \\ &\leq \left[\frac{1}{2}K(\rho) - L_0(\rho/2)^N\omega_N \min_{B_{\rho/2}} Q + l_0\|Q\|_{L^1}\right]\tilde{s}_i^2.\end{aligned}$$

Consequently, using (37), we obtain that

$$\mathcal{R}_i(u_i^0) = \min_{W^{b_i}_{\rm rad}} \mathcal{R}_i \leq \mathcal{R}_i(w_{\tilde{s}_i}) < 0 \tag{38}$$

which proves in particular (34). Now, let us prove (35). For every $i \in \mathbb{N}$, by using the mean value theorem, (28), (30) and (31), we have

$$\mathcal{R}_i(u_i^0) \geq -\int_{\mathbb{R}^N} Q(x)F_i(u_i^0(x))dx \geq -\|Q\|_{L^1} \max_{s\in[0,1]} |f(s)|a_i.$$

Taking into account that $\lim_{i\to\infty} a_i = 0$, the above inequality and (38) leads to (35).

Due to (30) and (31), we observe that

$$\mathcal{R}_i(u_i^0) = \mathcal{R}_1(u_i^0) \text{ for all } i \in \mathbb{N}.$$

Combining this relation with (34) and (35), we see that the sequence $\{u_i^0\}_i$ contains infinitely many distinct elements.

It remains to prove relation (17). The first limit easily follows by (31), i.e. $\|u_i^0\|_{L^\infty} \leq a_i$ for all $i \in \mathbb{N}$, combined with $\lim_{i\to\infty} a_i = 0$. For the second limit, we use (38), (28), (30) and (31), obtaining for all $i \in \mathbb{N}$ that

$$\frac{1}{2}\|u_i^0\|^2_{H^1} < \|Q\|_{L^1} \max_{s\in[0,1]} |f(s)|a_i$$

which concludes the proof of Theorem 3.2.

3.2.2. *Proof of* (II$_0$)*; Theorem 3.3 concluded.* Let $\{\theta_i\}_i$ be a sequence with negative terms such that $\lim_{i\to\infty} \theta_i = 0$. By (35) and (38), we clear have that $\lim_{i\to\infty} \mathcal{R}_i(w_{\tilde{s}_i}) = 0$. Thus, up to a subsequence, we may assume that the sequence $\{(\theta_i, \mathcal{R}_i(u_i^0), \mathcal{R}_i(w_{\tilde{s}_i}), a_i)\}_i \subset \mathbb{R}^4$ which converges to $0_{\mathbb{R}^4}$, has the property that for all $i \in \mathbb{N}$,

$$\theta_i < \mathcal{R}_i(u_i^0) \leq \mathcal{R}_i(w_{\tilde{s}_i}) < \theta_{i+1}; \tag{39}$$

$$a_i < \min\left(\frac{1}{i}, \frac{1}{2i^2\|Q\|_{L^1}[\max_{[0,1]}|f| + \max_{[0,1]}|g| + 1]}\right). \tag{40}$$

Let us denote

$$\varepsilon_i' = \frac{\theta_{i+1} - \mathcal{R}_i(w_{\tilde{s}_i})}{\|Q\|_{L^1}[\max_{[0,1]}|g|+1]} \quad \text{and} \quad \varepsilon_i'' = \frac{\mathcal{R}_i(u_i^0) - \theta_i}{\|Q\|_{L^1}[\max_{[0,1]}|g|+1]}, \quad i \in \mathbb{N}.$$

Fix $k \in \mathbb{N}$. On account of (39),

$$\varepsilon_k^0 = \min(1, \varepsilon_1, \ldots, \varepsilon_k, \varepsilon_1', \ldots, \varepsilon_k', \varepsilon_1'', \ldots, \varepsilon_k'') > 0.$$

Then, for every $i \in \{1, \ldots, k\}$ and $\varepsilon \in [-\varepsilon_k^0, \varepsilon_k^0]$ we have

$$\begin{aligned}
\mathcal{R}_i^\varepsilon(u_{i,\varepsilon}^0) &\leq \mathcal{R}_i^\varepsilon(w_{\tilde{s}_i}) && \text{(see (32))}\\
&= \mathcal{R}_i(w_{\tilde{s}_i}) - \varepsilon \int_{\mathbb{R}^N} Q(x)G_i(w_{\tilde{s}_i})\\
&< \theta_{i+1}, && \text{(see the choice of } \varepsilon_i' \text{ and (28))}
\end{aligned}$$

and taking into account that $u_{i,\varepsilon}^0$ belongs to $W_{\mathrm{rad}}^{b_i}$, and u_i^0 is the minimum point of $\mathcal{R}_i$ over the set $W_{\mathrm{rad}}^{b_i}$, see relation (32) for $\varepsilon = 0$, we have

$$\begin{aligned}
\mathcal{R}_i^\varepsilon(u_{i,\varepsilon}^0) &= \mathcal{R}_i(u_{i,\varepsilon}^0) - \varepsilon \int_{\mathbb{R}^N} Q(x)G_i(u_{i,\varepsilon}^0)\\
&\geq \mathcal{R}_i(u_i^0) - \varepsilon \int_{\mathbb{R}^N} Q(x)G_i(u_{i,\varepsilon}^0)\\
&> \theta_i. && \text{(see the choice of } \varepsilon_i'' \text{ and (28))}
\end{aligned}$$

In conclusion, for every $i \in \{1, \ldots, k\}$ and $\varepsilon \in [-\varepsilon_k^0, \varepsilon_k^0]$ we have

$$\theta_i < \mathcal{R}_i^\varepsilon(u_{i,\varepsilon}^0) < \theta_{i+1},$$

thus

$$\mathcal{R}_1^\varepsilon(u_{1,\varepsilon}^0) < \ldots < \mathcal{R}_k^\varepsilon(u_{k,\varepsilon}^0).$$

But, $u_{i,\varepsilon}^0 \in W_{\mathrm{rad}}^{b_1}$ for every $i \in \{1, \ldots, k\}$, so $\mathcal{R}_i^\varepsilon(u_{i,\varepsilon}^0) = \mathcal{R}_1^\varepsilon(u_{i,\varepsilon}^0)$, see relation (30). Therefore, from above, we obtain that for every $\varepsilon \in [-\varepsilon_k^0, \varepsilon_k^0]$,

$$\mathcal{R}_1^\varepsilon(u_{1,\varepsilon}^0) < \ldots < \mathcal{R}_1^\varepsilon(u_{k,\varepsilon}^0).$$

In particular, this fact shows that the elements $u_{1,\varepsilon}^0, \ldots, u_{k,\varepsilon}^0$ are distinct whenever $\varepsilon \in [-\varepsilon_k^0, \varepsilon_k^0]$.

It remains to prove relation (18). The first relation directly follows by (31) and (40). To chech the second limit, we observe that for every $i \in \{1, \ldots, k\}$ and $\varepsilon \in [-\varepsilon_k^0, \varepsilon_k^0]$,

$$\mathcal{R}_1^\varepsilon(u_{i,\varepsilon}^0) = \mathcal{R}_i^\varepsilon(u_{i,\varepsilon}^0) < \theta_{i+1} < 0.$$

Consequently, for every $i \in \{1, \dots, k\}$ and $\varepsilon \in [-\varepsilon_k^0, \varepsilon_k^0]$, by a mean value theorem we obtain

$$\begin{aligned} \frac{1}{2}\|u_{i,\varepsilon}^0\|_{H^1}^2 &< \int_{\mathbb{R}^N} Q(x)[F_i(u_{i,\varepsilon}^0) + \varepsilon G_i(u_{i,\varepsilon}^0)] \\ &\le \|Q\|_{L^1}[\max_{[0,1]}|f| + \max_{[0,1]}|g|]a_i \qquad \text{(see (28), (31) and } \varepsilon_k^0 \le 1) \\ &< \frac{1}{2i^2}, \qquad \text{(see (40))} \end{aligned}$$

which concludes the proof of Theorem 3.3.

3.3. Proof of Theorems 3.6 and 3.8. The left side of (f_1^∞) implies the existence of $l_\infty > 0$ and $\delta > 0$ such that

$$F(s) \ge -l_\infty s^2 \quad \text{for all } s > \delta. \tag{41}$$

Fix a number $L_\infty > 0$ large enough such that

$$\frac{1}{2}K(\rho) + l_\infty\|Q\|_{L^1} < L_\infty(\rho/2)^N \omega_N \min_{B_{\rho/2}} Q, \tag{42}$$

where $\rho > 0$ and $K(\rho)$ are from (27). The right side of (f_1^∞) implies the existence of a sequence $\{\tilde{s}_i\}_i \subset (0, \infty)$ such that $\lim_{i\to\infty} \tilde{s}_i = \infty$, and

$$F(\tilde{s}_i) > L_\infty \tilde{s}_i^2 \quad \text{for all } i \in \mathbb{N}. \tag{43}$$

Since $\lim_{i\to\infty} s_i = \infty$, see (f_2^∞), we may fix a subsequence $\{s_{m_i}\}_i$ of $\{s_i\}_i$ such that $\tilde{s}_i \le s_{m_i}$ for all $i \in \mathbb{N}$. Due to the continuity of f and g, we may fix the positive sequences $\{a_i\}_i$, $\{b_i\}_i$, and $\{\varepsilon_i\}_i$ such that $\lim_{i\to\infty} a_i = \lim_{i\to\infty} b_i = +\infty$, and for all $i \in \mathbb{N}$,

$$a_i < s_{m_i} < b_i < a_{i+1}; \tag{44}$$

$$f(s) + \varepsilon g(s) \le 0 \quad \text{for all } s \in [a_i, b_i] \text{ and } \varepsilon \in [-\varepsilon_i, \varepsilon_i]. \tag{45}$$

In the same way as we did in (30), let us define the truncation functions $f_i, g_i : [0, \infty) \to \mathbb{R}$ by

$$f_i(s) = f(\min(s, b_i)) \quad \text{and} \quad g_i(s) = g(\min(s, b_i)). \tag{46}$$

Since $f_i(0) = g_i(0) = 0$, we may extend continuously the functions f_i and g_i to the whole real line, taking 0 for negative arguments. For every $s \in \mathbb{R}$ and $i \in \mathbb{N}$, let $F_i(s) = \int_0^s f_i(t)dt$ and $G_i(s) = \int_0^s g_i(t)dt$.

For every $i \in \mathbb{N}$ fixed and $\varepsilon \in [-\varepsilon_i, \varepsilon_i]$ the function $h_{i,\varepsilon}^\infty : [0, \infty) \to \mathbb{R}$ defined by $h_{i,\varepsilon}^\infty = f_i + \varepsilon g_i$ is continuous, bounded, and $h_{i,\varepsilon}^\infty(0) = 0$. On account of relations (45) and (46), one has $h_{i,\varepsilon}^\infty(s) \le 0$ for all $s \in [a_i, b_i]$. Consequently, we may apply

Theorem 3.12 to the function $h^\infty_{i,\varepsilon}$ obtaining that for every $i \in \mathbb{N}$ and $\varepsilon \in [-\varepsilon_i, \varepsilon_i]$, the problem

$$(\mathrm{P}^\infty_{i,\varepsilon}) \qquad \begin{cases} -\Delta u + u = Q(x)h^\infty_{i,\varepsilon}(u), & x \in \mathbb{R}^N, \\ u(x) \to 0 \ \text{ as } \ |x| \to \infty, \end{cases}$$

has a radially symmetric, weak solution $u^\infty_{i,\varepsilon} \in H^1(\mathbb{R}^N)$ such that

$$u^\infty_{i,\varepsilon} \in [0, a_i] \quad \text{for a.e. } x \in \mathbb{R}^N; \tag{47}$$

$$u^\infty_{i,\varepsilon} \text{ is the infimum of the functional } \mathcal{R}^\varepsilon_i \text{ on } W^{b_i}_{\text{rad}}, \tag{48}$$

where $\mathcal{R}^\varepsilon_i$ is defined exactly as in (33). Due to (46) and (47), $u^\infty_{i,\varepsilon}$ is a weak solution not only for $(\mathrm{P}^\infty_{i,\varepsilon})$ but also for the initial problem (P_ε). Consequently, we have to prove that:

- (I_∞) there are infinitely many distinct elements in the sequence $\{u^\infty_{i,0}\}_i$ verifying (19), see Theorem 3.6;
- (II_∞) for every $k \in \mathbb{N}$, there are at least k distinct elements $u^\infty_{i,\varepsilon}$ verifying (21) when ε belongs to a certain interval around the origin, see Theorem 3.8.

3.3.1. *Proof of* (I_∞)*; Theorem 3.6 concluded.* Let $u^\infty_i = u^\infty_{i,0}$ and $\mathcal{R}_i = \mathcal{R}^0_i$ for every $i \in \mathbb{N}$. We prove that

$$\lim_{i\to\infty} \mathcal{R}_i(u^\infty_i) = -\infty. \tag{49}$$

Let $i \in \mathbb{N}$ be fixed and $w_{\tilde{s}_i} \in H^1_{\text{rad}}(\mathbb{R}^N)$ be the function from (26) corresponding to the value $\tilde{s}_i > 0$. Then $w_{\tilde{s}_i} \in W^{b_i}_{\text{rad}}$, and on account of (27), (41) and (43), one has

$$\begin{aligned} \mathcal{R}_i(w_{\tilde{s}_i}) &= \frac{1}{2}\|w_{\tilde{s}_i}\|^2_{H^1} - \int_{\mathbb{R}^N} Q(x)F_i(w_{\tilde{s}_i}(x))dx \\ &= \frac{1}{2}\|w_{\tilde{s}_i}\|^2_{H^1} - F(\tilde{s}_i)\int_{B_\rho/2} Q(x)dx - \int_{(B_\rho\setminus B_{\rho/2})\cap\{w_{\tilde{s}_i}>\delta\}} Q(x)F(w_{\tilde{s}_i}(x))dx \\ &\quad - \int_{(B_\rho\setminus B_{\rho/2})\cap\{w_{\tilde{s}_i}\le\delta\}} Q(x)F(w_{\tilde{s}_i}(x))dx \\ &\le \left[\frac{1}{2}K(\rho) - L_\infty(\rho/2)^N\omega_N \min_{B_{\rho/2}} Q + l_\infty\|Q\|_{L^1}\right]\tilde{s}_i^2 + \|Q\|_{L^1}\max_{s\in[0,\delta]}|F(s)|. \end{aligned}$$

Using the fact that $\lim_{i\to\infty}\tilde{s}_i = \infty$ and (42), we have that $\lim_{i\to\infty}\mathcal{R}_i(w_{\tilde{s}_i}) = -\infty$. But, $\mathcal{R}_i(u^\infty_i) \le \mathcal{R}_i(w_{\tilde{s}_i})$ for all $i \in \mathbb{N}$, which implies (49).

Now, let us assume that in the sequence $\{u^\infty_i\}_i$ there are only finitely many distinct elements, say $\{u^\infty_1, \dots, u^\infty_{i_0}\}$ for some $i_0 \in \mathbb{N}$. Consequently, the sequence $\{\mathcal{R}_i(u^\infty_i)\}_i$ reduces mostly to the finite set $\{\mathcal{R}_{i_0}(u^\infty_1), \dots, \mathcal{R}_{i_0}(u^\infty_{i_0})\}$, which contradicts relation (49).

It remains to prove (19). Arguing by contradiction assume there exists a subsequence $\{u^\infty_{k_i}\}_i$ of $\{u^\infty_i\}_i$ such that for all $i \in \mathbb{N}$, we have $\|u^\infty_{k_i}\|_{L^\infty} \le M$ for some $M > 0$. In particular, $\{u^\infty_{k_i}\}_i \subset W^{b_l}_{\text{rad}}$ for some $l \in \mathbb{N}$. Thus, for every $k_i \ge l$, we have

$$\begin{aligned}\mathcal{R}_l(u_l^\infty) &= \min_{W_{\text{rad}}^{b_l}} \mathcal{R}_l = \min_{W_{\text{rad}}^{b_l}} \mathcal{R}_{k_i} \\ &\geq \min_{W_{\text{rad}}^{b_{k_i}}} \mathcal{R}_{k_i} = \mathcal{R}_{k_i}(u_{k_i}^\infty) \\ &\geq \min_{W_{\text{rad}}^{b_l}} \mathcal{R}_{k_i} && \text{(cf. hypothesis, } u_{k_i}^\infty \in W_{\text{rad}}^{b_l}) \\ &= \mathcal{R}_l(u_l^\infty).\end{aligned}$$

As a consequence,

$$\mathcal{R}_{k_i}(u_{k_i}^\infty) = \mathcal{R}_l(u_l^\infty) \quad \text{for all } i \in \mathbb{N}. \tag{50}$$

But, the sequence $\{\mathcal{R}_i(u_i^\infty)\}_i$ is non-increasing; indeed, due to (48) and (46), for all $i \in \mathbb{N}$, one has

$$\mathcal{R}_{i+1}(u_{i+1}^\infty) = \min_{W_{\text{rad}}^{b_{i+1}}} \mathcal{R}_{i+1} \leq \min_{W_{\text{rad}}^{b_i}} \mathcal{R}_{i+1} = \min_{W_{\text{rad}}^{b_i}} \mathcal{R}_i = \mathcal{R}_i(u_i^\infty).$$

Combining this latter fact with (50), one can find a number $i_0 \in \mathbb{N}$ such that $\mathcal{R}_i(u_i^\infty) = \mathcal{R}_l(u_l^\infty)$ for all $i \geq i_0$. This fact contradicts (49) which concludes the proof of Theorem 3.6.

Proof of (19′) *from Remark 3.7.* Assume that (20) holds for f with $q \in (2, 2^*/2)$. By contradiction, we assume that there exists a subsequence $\{u_{k_i}^\infty\}_i$ of $\{u_i^\infty\}_i$ such that for all $i \in \mathbb{N}$, we have $\|u_{k_i}^\infty\|_{H^1} \leq \tilde{M}$ for some $\tilde{M} > 0$. Now, let us fix $\alpha \in [2q, 2^*)$. On account of (20) and the mean value theorem, we have

$$\begin{aligned}\left| \int_{\mathbb{R}^N} Q(x) F_{k_i}(u_{k_i}^\infty(x)) dx \right| &\leq c \left(\|Q\|_{L^2} \|u_{k_i}^\infty\|_{L^2} + \|Q\|_{L^{\alpha/(\alpha-q)}} \|u_{k_i}^\infty\|_{L^\alpha}^q \right) \\ &\leq c \left(\|Q\|_{L^2} \tilde{M} + \|Q\|_{L^{\alpha/(\alpha-q)}} C_\alpha^q \tilde{M}^q \right) < \infty,\end{aligned}$$

where $C_\alpha > 0$ is the Sobolev embedding constant in $H^1(\mathbb{R}^N) \hookrightarrow L^\alpha(\mathbb{R}^N)$. Consequently, the sequence $\{\mathcal{R}_{k_i}(u_{k_i}^\infty)\}_i$ is bounded. Since the sequence $\{\mathcal{R}_i(u_i^\infty)\}_i$ is non-increasing, it will be bounded as well, which contradicts (49).

3.3.2. *Proof of* (II_∞); *Theorem 3.8 concluded.* Let $\{\theta_i\}_i$ be a sequence with negative terms such that $\lim_{i\to\infty} \theta_i = -\infty$. On account of the proof of Theorem 3.6, we may assume that the sequence $\{(\theta_i, \mathcal{R}_i(u_i^\infty), \mathcal{R}_i(w_{\tilde{s}_i}), a_i)\}_i \subset \mathbb{R}^4$ which converges to $(-\infty, -\infty, -\infty, \infty)$, has the property that for all $i \in \mathbb{N}$,

$$\theta_{i+1} < \mathcal{R}_i(u_i^\infty) \leq \mathcal{R}_i(w_{\tilde{s}_i}) < \theta_i; \tag{51}$$

$$a_i \geq i. \tag{52}$$

Let us denote

$$\varepsilon_i' = \frac{\theta_i - \mathcal{R}_i(w_{\tilde{s}_i})}{\|Q\|_{L^1}[\max_{[0,b_i]}|g| + 1]b_i} \quad \text{and} \quad \varepsilon_i'' = \frac{\mathcal{R}_i(u_i^\infty) - \theta_{i+1}}{\|Q\|_{L^1}[\max_{[0,b_i]}|g| + 1]b_i}, \quad i \in \mathbb{N}.$$

Fix $k \in \mathbb{N}$. Due to (51), we have

$$\varepsilon_k^\infty = \min(1, \varepsilon_1, \ldots, \varepsilon_k, \varepsilon_1', \ldots, \varepsilon_k', \varepsilon_1'', \ldots, \varepsilon_k'') > 0.$$

Then, for every $i \in \{1, \ldots, k\}$ and $\varepsilon \in [-\varepsilon_k^\infty, \varepsilon_k^\infty]$ we have

$$\begin{aligned} \mathcal{R}_i^\varepsilon(u_{i,\varepsilon}^\infty) &\leq \mathcal{R}_i^\varepsilon(w_{\tilde{s}_i}) && \text{see (48))} \\ &= \mathcal{R}_i(w_{\tilde{s}_i}) - \varepsilon \int_{\mathbb{R}^N} Q(x) G_i(w_{\tilde{s}_i}) \\ &< \theta_i, && \text{(see the choice of } \varepsilon_i',\ \tilde{s}_i \leq s_{m_i}, \text{ and (44))} \end{aligned}$$

and since $u_{i,\varepsilon}^\infty$ belongs to $W_{\rm rad}^{b_i}$, and u_i^∞ is the minimum point of $\mathcal{R}_i$ on the set $W_{\rm rad}^{b_i}$, see relation (48) for $\varepsilon = 0$, we have

$$\begin{aligned} \mathcal{R}_i^\varepsilon(u_{i,\varepsilon}^\infty) &= \mathcal{R}_i(u_{i,\varepsilon}^\infty) - \varepsilon \int_{\mathbb{R}^N} Q(x) G_i(u_{i,\varepsilon}^\infty) \\ &\geq \mathcal{R}_i(u_i^\infty) - \varepsilon \int_{\mathbb{R}^N} Q(x) G_i(u_{i,\varepsilon}^\infty) \\ &> \theta_{i+1}. && \text{(see the choice of } \varepsilon_i'',\ \tilde{s}_i \leq s_{m_i}, \text{ and (44))} \end{aligned}$$

Thus, for every $i \in \{1, \ldots, k\}$ and $\varepsilon \in [-\varepsilon_k^\infty, \varepsilon_k^\infty]$ we have

$$\theta_{i+1} < \mathcal{R}_i^\varepsilon(u_{i,\varepsilon}^\infty) < \theta_i. \tag{53}$$

In particular,

$$\mathcal{R}_k^\varepsilon(u_{k,\varepsilon}^\infty) < \ldots < \mathcal{R}_1^\varepsilon(u_{1,\varepsilon}^\infty) < 0. \tag{54}$$

By construction, $u_{i,\varepsilon}^\infty \in W_{\rm rad}^{b_k}$ for every $i \in \{1, \ldots, k\}$, see (44); thus, $\mathcal{R}_i^\varepsilon(u_{i,\varepsilon}^\infty) = \mathcal{R}_k^\varepsilon(u_{i,\varepsilon}^\infty)$, see relation (46). Therefore, (54) implies that for every $\varepsilon \in [-\varepsilon_k^\infty, \varepsilon_k^\infty]$,

$$\mathcal{R}_k^\varepsilon(u_{k,\varepsilon}^\infty) < \ldots < \mathcal{R}_k^\varepsilon(u_{1,\varepsilon}^\infty) < 0.$$

In particular, the elements $u_{1,\varepsilon}^\infty, \ldots, u_{k,\varepsilon}^\infty$ are distinct whenever $\varepsilon \in [-\varepsilon_k^\infty, \varepsilon_k^\infty]$.

Now, we prove relation (21). Fix $\varepsilon \in [-\varepsilon_k^\infty, \varepsilon_k^\infty]$. First, since $\mathcal{R}_1^\varepsilon(u_{1,\varepsilon}^\infty) < 0 = \mathcal{R}_1^\varepsilon(0)$, then $\|u_{1,\varepsilon}^\infty\|_{L^\infty} > 0$ which proves (21) for $i = 1$. We further prove that

$$\|u_{i,\varepsilon}^\infty\|_{L^\infty} > a_{i-1} \quad \text{for all } i \in \{2, \ldots, k\}. \tag{55}$$

Let us assume that there exists an element $i_0 \in \{2, \ldots, k\}$ such that $\|u_{i_0,\varepsilon}^\infty\|_{L^\infty} \leq a_{i_0-1}$. Since $a_{i_0-1} < b_{i_0-1}$, then $u_{i_0,\varepsilon}^\infty \in W_{\rm rad}^{b_{i_0-1}}$. Thus, on account of (48) and (46), we have

$$\mathcal{R}_{i_0-1}^\varepsilon(u_{i_0-1,\varepsilon}^\infty) = \min_{W_{\rm rad}^{b_{i_0-1}}} \mathcal{R}_{i_0-1}^\varepsilon \leq \mathcal{R}_{i_0-1}^\varepsilon(u_{i_0,\varepsilon}^\infty) = \mathcal{R}_{i_0}^\varepsilon(u_{i_0,\varepsilon}^\infty),$$

which contradicts (54). Thus, (55) holds true which can be combined with (52), obtaining relation (21). This ends the proof of Theorem 3.8.

Proof of (21′) *from Remark 3.9.* Assume that both f and g verify (20) with $q \in (2, 2^*/2)$. Fix $\alpha \in [2q, 2^*)$. We may assume that the sequence $\{\theta_i\}_i$ from (51) fulfills

$$\theta_i < -2c\|Q\|_{L^2}(i-1) - 2c\|Q\|_{L^{\alpha/(\alpha-q)}}C_\alpha^q(i-1)^q \quad \text{for all } i \in \mathbb{N} \tag{56}$$

where $c > 0$ comes from (20). Let us fix $\varepsilon \in [-\varepsilon_k^\infty, \varepsilon_k^\infty]$. We assume that $\|u_{i_0,\varepsilon}^\infty\|_{H^1} \leq i_0 - 1$ for some $i_0 \in \{1, \ldots, k\}$. Then, we have

$$\begin{aligned}
\frac{1}{2}\|u_{i_0,\varepsilon}^\infty\|_{H^1}^2 &= \mathcal{R}_{i_0}^\varepsilon(u_{i_0,\varepsilon}^\infty) + \int_{\mathbb{R}^N} Q(x)\left[F_{i_0}(u_{i_0,\varepsilon}^\infty) + \varepsilon G_{i_0}(u_{i_0,\varepsilon}^\infty)\right] \\
&< \theta_{i_0} + c(1+|\varepsilon|)[\|Q\|_{L^2}\|u_{i_0,\varepsilon}^\infty\|_{H^1} + \|Q\|_{L^{\alpha/(\alpha-q)}}C_\alpha^q\|u_{i_0,\varepsilon}^\infty\|_{H^1}^q] \\
\text{(see (53))} & \\
&\leq \theta_{i_0} + 2c\big[\|Q\|_{L^2}(i_0-1) + \|Q\|_{L^{\alpha/(\alpha-q)}}C_\alpha^q(i_0-1)^q\big], \qquad (\varepsilon_k^\infty \leq 1) \\
&< 0, \qquad \text{(see (56))}
\end{aligned}$$

contradiction. Therefore, relation (21′) is proved.

References

[1] A. Ambrosetti and P. H. Rabinowitz, Dual variational methods in critical point theory and applications. J. Functional Analysis **14** (1973), 349–381.

[2] A. Ambrosetti and M. Badiale, The dual variational priniciple and elliptic problems with discontinous nonlinearities, J. Math. Anal. Appl. **140** (1989), 363-373.

[3] G. Anello and G. Cordaro, Perturbation from Dirichlet problem involving oscillating nonlinearities. J. Differential Equations **234** (2007), no. 1, 80–90.

[4] T. Bartsch, Z. Liu and T. Weth, Sign changing solutions of superlinear Schrödinger equations, Comm. Partial Differential Equations, **29** (2004), 25-42.

[5] T. Bartsch, A. Pankov and Z.-Q. Wang, Nonlinear Schrödinger equations with steep potential well, Comm. Contemp. Math. **4** (2001), 549-569.

[6] T. Bartsch and Z.-Q. Wang, Multiple positive solutions for a nonlinear Schrödinger equation, Z. Angew. Math. Phys. **51** (2000), 366–384.

[7] T. Bartsch and Z.-Q. Wang, Existence and multiplicity results for some superlinear elliptic problems in $\mathbb{R}^N$, Comm. Partial Differential Equations, **20** (1995), 1725-1741.

[8] G. Bonanno, Some remarks on a three critical points theorem, Nonlinear Analysis TMA, **54** (2003), 651-665.

[9] J. Chabrowski and J. do Ó, On semilinear elliptic equation involving concave and convex nonlinearities, Math. Nachr. **233-234** (2002), 55-76.

[10] M. Degiovanni and S. Lancelotti, Perturbations of even nonsmooth functionals. Differential Integral Equations **8** (1995), no. 5, 981–992.

[11] M. Degiovanni and V. Rădulescu, Perturbation of nonsmooth symmetric nonlinear eigenvalue problems. C. R. Acad. Sci. Paris Sér. I, **329** (1999), 281–286.

[12] M. F. Furtado, L. A. Maia and E. A. B. Silva, On a double resonant problem in $\mathbb{R}^N$, Differential Integral Equations, **15** (2002), 1335-1344.

[13] F. Gazzola and V. Rădulescu, A nonsmooth critical point theory approach to some nonlinear elliptic equations in $\mathbb{R}^N$, Differential Integral Equations, **13** (2000), 47-60.

[14] J. Kobayashi and M. Ôtani, The principle of symmetric criticality for non-differentiable mappings. J. Funct. Anal. **214** (2004), no. 2, 428–449.

[15] M. A. Krasnosel'skii, Topological methods in the theory of nonlinear integral equations, Translated by A. H. Armstrong; translation edited by J. Burlak. A Pergamon Press Book, Macmillan, New York, 1964.

[16] A. Kristály, Infinitely many solutions for a differential inclusion problem in $\mathbb{R}^N$. J. Differential Equations **220** (2006) 511-530.

[17] A. Kristály, Detection of arbitrarily many solutions for perturbed elliptic problems involving oscillatory terms. J. Differential Equations **245** (2008), no. 12, 3849-3868.

[18] A. Kristály, Multiple solutions of a sublinear Schrdinger equation. NoDEA Nonlinear Differential Equations Appl. **14** (2007), no. 3-4, 291–301.
[19] A. Kristály, A double eigenvalue problem for Schrdinger equations involving sublinear nonlinearities at infinity. Electron. J. Differential Equations 2007, No. 42, 11 pp.
[20] S. Li and Z. Liu, Perturbations from symmetric elliptic boundary value problems. J. Differential Equations **185** (2002), no. 1, 271–280.
[21] S. Marano and D. Motreanu, On a three critical point theorem for non-differentiable functions and applications to nonlinear boundary value problems, Nonlinear Anal. T.M.A. **48** (2002), 37-52.
[22] S. Marano and D. Motreanu, Existence of two nontrivial solutions for a class of elliptic eigenvalue problems, Arch. Math. (Basel), **75** (2000), 53-58.
[23] M. Marcus and V. J. Mizel: Absolute continuity on tracks and mappings of Sobolev spaces. Arch. Rational Mech. Anal. **45** (1972) 294–320.
[24] P. Omari and F. Zanolin, Infinitely many solutions of a quasilinear elliptic problem with an oscillatory potential. Comm. Partial Differential Equations **21** (1996) 721–733.
[25] R. S. Palais, The principle of symmetric criticality, Comm. Math. Phys. **69** (1979), 19-30.
[26] K. Perera, Multiplicity results for some elliptic problems with concave nonlinearities, J. Differential Equations, **140** (1997), 133-141.
[27] P. H. Rabinowitz, On a class of nonlinear Schrödinger equations, Z. Angew. Math. Phys. **43** (1992), 270-291.
[28] P. H. Rabinowitz, Minimax Methods in Critical Point Theory with Applications to Differential Equations, CBMS Reg. Conf. Ser. in Math. , Amer. Math. Soc., Providence, 1986.
[29] B. Ricceri, On a three critical points theorem, Arch. Math. (Basel) **75** (2000), 220-226.
[30] B, Ricceri, Existence of three solutions for a class of elliptic eigenvalue problems, Math. Comput. Modelling, **32** (2000), 1485-1494.
[31] B. Ricceri, Three solutions for a Neumann problem, Top. Meth. Nonl. Anal. **20** (2002), 275-281.
[32] B. Ricceri, *A general multiplicity theorem for certain nonlinear equations in Hilbert spaces*, Proc. Amer. Math. Soc., **133** (2005), 3255-3261.
[33] B. Ricceri, A general variational principle and some of its applications. J. Comput. Appl. Math. **113** (2000) 401–410.
[34] J. Saint Raymond, On the multiplicity of the solutions of the equations $-\Delta u = \lambda f(u)$. J. Differential Equations **180** (2002) 65–88.
[35] W. A. Strauss, Existence of solitary waves in higher dimensions, Comm. Math. Phys. **55** (1977) 149-162.
[36] J. Su, Z.-Q. Wang and M. Willem, Weighted Sobolev embedding with unbounded and decaying radial potentials. J. Differential Equations **238** (2007), no. 1, 201–219.
[37] A. Szulkin, Minimax principles for lower semicontinuous functions and applications to nonlinear boundary value problems. Ann. Inst. H. Poincaré Anal. Non Linéaire **3** (1986), no. 2, 77–109.
[38] Z.-Q. Wang, Nonlinear boundary value problems with concave nonlinearities near the origin, NoDEA Nonlinear Differ. Equ. Appl. **8** (2001), 15-33.
[39] M. Willem, Minimax Theorems, Birkhäuser, Boston, 1995.
[40] L. S. Yu, Nonlinear p-Laplacian problems on unbounded domains, Proc. Amer. Math. Soc. **115** (1992), 1037-1045.

Alexandru Kristály Babeş-Bolyai University, Department of Economics, 400591 Cluj-Napoca, Romania
E-mail address: alexandrukristaly@yahoo.com

Nikolaos S. Papageorgiou National Technical University, Department of Mathematics, 15780 Zografou Campus, Athens, Greece
E-mail address: npapg@math.ntua.gr

Equations and inequalities in Orlicz–Sobolev spaces: Selected topics

Vy Khoi Le and Klaus Schmitt

Contents

This chapter consists of a collection of topics from some of our work on equations and inequalities in Orlicz–Sobolev spaces which formulate boundary value problems, whose principal operators have fast, slow, or general growths.

We are interested in equations or variational and hemivariational inequalities related to the equation

$$-\text{div}(a(|\nabla u|)\nabla u) + F(x, u) = 0 \text{ in } \Omega, \tag{1}$$

where $\Omega \subset \mathbb{R}^N$ is a bounded domain with sufficiently smooth boundary $\partial\Omega$ and $a : \mathbb{R} \to [0, \infty)$ is an even function with properties to be specified later. Let

$$\phi : \mathbb{R} \to \mathbb{R}, \ \phi(s) := a(s)s$$

in (1). Then, if $\phi(s) = |s|^{p-1}s$ $(p > 1)$, problem (1) is one containing the p-Laplacian. This has been extensively studied recently and a great variety of existence results are available. These results are usually obtained using variational methods, monotone operator methods or fixed point and degree theory arguments in the Sobolev space $W_0^{1,p}(\Omega)$. If, on the other hand, we assume that ϕ is an odd nondecreasing function such that

$$\phi(0) = 0, \quad \phi(t) > 0, \quad t > 0,$$
$$\lim_{t\to\infty} \phi(t) = \infty,$$

and ϕ is right continuous, then a Sobolev space setting for the problem is not appropriate and very general results are sparse.

The first general existence results using the theory of monotone type operators in Orlicz-Sobolev spaces were obtained in the pioneering papers [**20, 21**] and [**32**]. Subsequently, the case where ϕ is a general (nonhomogeneous) N-function has attracted much of attention. Further applications of the monotone operator approach to other problems, including those with measures or L^1 functions, are presented in e.g. [**2, 3, 6, 7, 22**] and their references. Topological and variational approaches for related problems were given recently in [**13, 17, 18, 19, 24, 25, 26, 27, 30, 33, 34, 36, 39, 44, 45, 47, 49, 50, 53, 56, 59, 60, 61, 62**] and their references. The sub- and supersolution approach to such problems in Orlicz–Sobolev spaces was presented in [**43, 51**]. Some applications of problems in Orlicz–Sobolev spaces to nonlinear elasticity and in a sandpile problem are given in [**48**] and [**51**].

In the following sections, we present some methods which we employed recently to study such problems, namely, topological, variational, and sub-supersolution approaches. Our presentation here is based on the papers [**56**], [**30**], [**52**] and [**43**]. We shall see that a natural way of formulating the boundary value problem (1) and its related problems will be a variational inequality formulation in suitable Orlicz-Sobolev spaces. Topological, variational, or sub-supersolution arguments are next developed for such variational inequalities. In section 1, topological arguments are employed to study equations and inequalities with fast growths for which several existence theorems for solutions and nonnegative solutions are established. In sections 2 and 3, we use a variational approach to investigate eigenvalue problems in Orlicz–Sobolev spaces and variational hemivariational inequalities with slowly growing principal operators. Equations and inequalities with general growths are considered in section 4 where we use a sub-supersolution method to study the existence and some properties of solutions in the noncoercive case.

1. Quasilinear elliptic equations and inequalities with rapidly growing coefficients

1.1. Introduction. In this section, to somewhat simplify the notation we consider the boundary value problem

$$\begin{cases} -\mathrm{div}(A(|\nabla u|^2)\nabla u) + F(x,u) = 0, & \text{in } \Omega \\ u = 0, \quad \text{on } \partial\Omega, \end{cases} \tag{2}$$

that is, $a(s) = A(s^2)$ in equation (1).

We note here that the problems to be studied, when formulated as operator equations, lead to the use of the topological degree for multivalued maps (cf. [**16**], [**58**]). To prepare for the main discussion and existence theorems in subsections 1.4 and 1.5, we first present some basic concepts and results related to Orlicz and Orlicz–Sobolev spaces and a variational inequality formulation of (2).

1.2. Preliminaries on Orlicz and Orlicz-Sobolev spaces. Let us put

$$\Phi(t) = \int_0^t \phi(s)ds, \quad t \in \mathbb{R}.$$

Then Φ is a Young (or N-) function (cf. [**1**], [**38**], [**41**]). Also, following these references, we denote by $\overline{\Phi}$ the conjugate Young function of Φ, i.e.,

$$\overline{\Phi}(t) = \sup\{ts - \Phi(s) : s \in \mathbb{R}\},$$

and by Φ^* the Sobolev conjugate of Φ, i.e.,

$$(\Phi^*)^{-1}(t) = \int_0^t \frac{\Phi^{-1}(s)ds}{s^{\frac{N+1}{N}}} ds, \tag{3}$$

provided that

$$\int_1^\infty \frac{\Phi^{-1}(s)ds}{s^{\frac{N+1}{N}}} ds = \infty. \tag{4}$$

Let Φ be a Young function. The Orlicz class $\tilde{L}_\Phi := \tilde{L}_\Phi(\Omega)$ is the set of all (equivalence classes) of measurable functions u defined on Ω such that

$$\int_\Omega \Phi(|u(x)|)dx < \infty.$$

The Orlicz space $L_\Phi := L_\Phi(\Omega)$ is the linear hull of $\tilde{L}_\Phi$, i.e., the set of all measurable functions u on Ω such that

$$\int_\Omega \Phi\left(\frac{|u(x)|}{k}\right) dx < \infty, \quad \text{for some } k > 0.$$

Then L_Φ is a Banach space when equipped with the norm (the Luxemburg norm)

$$\|u\|_\Phi = \inf\left\{k > 0 : \int_\Omega \Phi\left(\frac{|u|}{k}\right) dx \le 1\right\}$$

or the equivalent norm (the Orlicz norm)

$$\|u\|_{(\Phi)} = \sup\left\{\left|\int_\Omega uvdx\right| : v \in \tilde{L}_{\overline{\Phi}}, \int_\Omega \overline{\Phi}(|v|)dx \le 1\right\}.$$

If Φ_1 and Φ_2 are two Young functions, one writes

$$\Phi_1 \preceq \Phi_2,$$

provided there exist constants t_0 and k such that

$$\Phi_1(t) \le \Phi_2(kt), \quad t \ge t_0,$$

and one says that Φ_1 and Φ_2 are equivalent whenever

$$\Phi_1 \preceq \Phi_2 \quad \text{and} \quad \Phi_2 \preceq \Phi_1.$$

If Φ_1 and Φ_2 are equivalent, then they determine the same Orlicz space. Also, it is the case that the imbedding

$$L_{\Phi_2} \hookrightarrow L_{\Phi_1},$$

is continuous, whenever $\Phi_1 \le \Phi_2$.

It is clear that

$$L^\infty \subset L_\Phi \subset L^1,$$

where we have employed the convention that $L^p := L^p(\Omega)$, $1 \le p \le \infty$, denote the usual Lebesgue spaces. The closure of L^∞ in L_Φ is denoted by E_Φ, which is a separable Banach space. The Orlicz-Sobolev space $W^1 L_\Phi := W^1 L_\Phi(\Omega)$ is the set of all $u \in L_\Phi$ such that the distributional derivatives $\partial_i u = \dfrac{\partial u}{\partial x_i}$, $i = 1, \cdots, N$, are also in L_Φ. This is a Banach space with respect to the norm

$$\|u\|_{1,\Phi} = \|u\|_{W^1 L_\Phi} = \|u\|_\Phi + \sum_{i=1}^N \|\partial_i u\|_\Phi.$$

It is known (cf. [**38**], [**66**]) that L_Φ is the dual space of $E_{\overline{\Phi}}$, i.e.,

$$L_\Phi = (E_{\overline{\Phi}})^*, \text{ and } L_{\overline{\Phi}} = (E_\Phi)^*.$$

The space $W^1 E_\Phi$ is defined similarly. The spaces $W^1 L_\Phi$ and $W^1 E_\Phi$ can be identified with closed subspaces of the products $\prod_{i=0}^N L_\Phi$ and $\prod_{i=0}^N E_\Phi$, respectively. It is the case that

$$\prod_{i=0}^N L_\Phi = \left(\prod_{i=0}^N E_{\overline{\Phi}}\right)^*$$

and if we denote by $\sigma = \sigma(\prod L_\Phi, \prod E_{\overline{\Phi}})$ the weak* topology in $\prod L_\Phi$ and also the restriction of σ to the closed subspace $W^1 L_\Phi$, then $W^1 L_\Phi$ is closed under weak* convergence of $\prod L_\Phi$. Since $\prod E_{\overline{\Phi}}$ is separable, we have the following useful properties of $W^1 L_\Phi$:

- If $\{u_n\}$ is a bounded sequence in $W^1 L_\Phi$ (with respect to $\|\cdot\|_{1,\Phi}$), then $\{u_n\}$ has a subsequence which converges with respect to the topology σ to some $u \in W^1 L_\Phi$, i.e., a bounded set in $W^1 L_\Phi$ is relatively sequentially compact with respect to the weak* topology σ.

We denote by $W_0^1 L_\Phi$ the closure of $C_0^\infty(\Omega)$ with respect to the weak* topology σ.

We also mention a notion of relative growth of Young functions which will play a role in our considerations (cf. [**1**], [**38**], [**41**]). A Young function Φ_1 is said to grow essentially more slowly than another Young function Φ_2, abbreviated by

$$\Phi_1 \ll \Phi_2,$$

if

$$\lim_{t\to\infty} \frac{\Phi_1(t)}{\Phi_2(kt)} = 0, \quad \forall k > 0.$$

With the hypotheses imposed on Ω the following important imbedding results hold:

- The imbedding

$$W_0^1 L_\Phi \hookrightarrow L_{\Phi^*} \tag{5}$$

is continuous.

- If $\Psi \ll \Phi^*$, then the imbedding

$$W_0^1 L_\Phi \hookrightarrow L_\Psi \tag{6}$$

is compact. Hence, since $\Phi \ll \Phi^*$ (cf. Lemma 4.14, [**32**]), the imbedding

$$W_0^1 L_\Phi \hookrightarrow L_\Phi$$

is compact.

If it is the case that the growth of the function Φ is restricted so that Φ satisfies a Δ_2 condition (at infinity), then and only then is it the case that

$$L_\Phi = \tilde{L}_\Phi = E_\Phi,$$

with corresponding equalities for associated Orlicz-Sobolev spaces (cf. [**1**], [**38**], [**41**]).

REMARK 1. (a) In the imbeddings (5) and (6) above, we assume implicitly that (4) holds, so that the function Φ^* defined by (3) is a Young function. However, if

$$\int_1^\infty \frac{\Phi^{-1}(s)ds}{s^{\frac{N+1}{N}}} ds < \infty, \tag{7}$$

then the imbedding $W_0^1 L_\Phi \hookrightarrow C(\overline{\Omega})$ is continuous and thus (6) holds for all L_Ψ (i.e. the imbedding (6) is compact) (cf. [**1**], [**41**]).

(b) In the sequel, we consider equations and inequalities under the assumption that (4) holds and as a consequence Φ^* is a Young function. However, as noted in (a), all results presented here are valid in the (easier) case where (7) holds.

1.3. Variational inequality formulation. Now, we formulate and extend (2) as a variational inequality in a suitable Orlicz-Sobolev space. We note that variational inequalities in Orlicz-Sobolev spaces have been studied in [**35**] by monotone operator and Galerkin approximation methods. We consider here inequalities in these spaces but of a different nature and from a different viewpoint. As will be seen in the sequel, the (nonsmooth) convex functional we are interested in is the potential functional of the principal operator $-\text{div}(A(|\nabla u|^2)\nabla u)$, which is intrinsic to (2). In [**35**], the convex set K presents an extrinsic constraint to the problem.

Multiplying both sides of (2) by $v \in C_0^\infty(\Omega)$ and integrating by parts (provided these integrations may be performed), we see that the weak form of (2) is

$$\int_\Omega A(|\nabla u|^2)\nabla u \cdot \nabla v dx + \int_\Omega F(x,u)v dx = 0. \tag{8}$$

A natural choice of the space of test functions v is, of course, $W_0^1 L_\Phi$. However, the mapping $u \mapsto L(u)$, where

$$\langle L(u), v\rangle = \int_\Omega A(|\nabla u|^2)\nabla u \cdot \nabla v dx, \quad v \in W_0^1 L_\Phi,$$

is not necessarily defined on the whole space. The existence of solutions of (8), in that form, were studied in [**20**] and [**32**] by using methods extending solution methods for equations that contain coercive pseudo-monotone operators of Leray-Lions types. However, it is somewhat difficult to verify that the operator in (8) satisfies the growth conditions assumed in [**20**] and [**32**]. We are here also concerned with the existence of solutions of (2) (or (8)) which are either nonnegative or nonpositive; again, the methods in [**20**] or [**32**] do not seem suitable for such an investigation. We therefore propose here a different approach by formulating (8) as a variational inequality.

Consider the functional

$$J : W_0^1 L_\Phi \to \mathbb{R} \cup \{\infty\},\ J(u) := \int_\Omega \Phi(|\nabla u|)dx.$$

Since $\dfrac{\partial \Phi}{\partial \xi_i}(|\xi|) = A(|\xi|^2)\xi_i,\ i = 1, \dots, N$, we have, at least formally,

$$\langle J'(u), v\rangle = \int_\Omega \sum_{i=1}^N A(|\nabla u|^2)\,\partial_i u \partial_i v dx = \int_\Omega A(|\nabla u|^2)\nabla u \cdot \nabla v dx.$$

Let us now assume that F satisfies the growth condition

(9) $$|F(x, s)| \le B(x) + |\Psi_0'(s)|,\ s \in \mathbb{R},\ x \in \Omega,$$

where Ψ_0 is a differentiable Young function such that Ψ_0 is strictly convex and satisfies

(10) $$\Psi_0 \ll \Phi^*,$$

and

(11) $$B \in L_{\overline{\Psi}_0}.$$

We then may, for $u \in W^1 L_\Phi$, define $f(u) \in (W^1 L_\Phi)^*$ by

$$\langle f(u), v\rangle := \int_\Omega F(x, u)v dx,\ \forall v \in W^1 L_\Phi.$$

Some properties of F and f are presented in the following lemmas whose proofs are given in [**56**].

LEMMA 2. *The mapping*

$$f : W^1 L_\Phi \to (W^1 L_\Phi)^*$$

is continuous.

In many situations, it is more convenient to work in an Orlicz space which lies between L_{Φ^*} and L_{Ψ_0}. We choose a Young function Ψ such that

$$\Psi_0 \ll \Psi \ll \Phi^*. \tag{12}$$

We can replace, because of this ordering, Φ^* by Ψ in the proof of Lemma 2 and obtain:

LEMMA 3. *If $u \in L_\Psi$, then $\psi_0(u) \in L_{\overline{\Psi}_0}$ and*

$$F(\cdot, u) \in L_{\overline{\Psi}_0} \subset L_{\overline{\Psi}}. \tag{13}$$

Moreover,

$$\|F(\cdot, u)\|_{\overline{\Psi}} \leq C\|F(\cdot, u)\|_{\overline{\Psi}_0} \leq \|B\|_{\overline{\Psi}_0} + \|\psi_0(|u|)\|_{\overline{\Psi}_0}.$$

Furthermore, we have the following result:

LEMMA 4. *The mapping*

$$f : u \mapsto f(u) = F(\cdot, u)$$

is continuous and bounded from L_Ψ to $L_{\overline{\Psi}}$.

One may now formulate (8) (at least formally) as the equation:

$$J'(u) + f(u) = 0. \tag{14}$$

However, J is not differentiable in general (J is not even defined on the whole space $W_0^1 L_\Phi$, since J assumes, in general, finite values only on a convex, nondense subset of $W_0^1 L_\Phi$). On the other hand, since J is convex and lower semicontinuous (as will be proved later), we replace (14) by the inclusion

$$0 \in \partial J(u) + f(u), \tag{15}$$

where ∂J is the subdifferential of J; this, in turn, is equivalent to the variational inequality

$$\begin{cases} J(v) - J(u) + \langle f(u), v - u\rangle \geq 0, \ \forall v \in W_0^1 L_\Phi \\ u \in W_0^1 L_\Phi. \end{cases} \tag{16}$$

The advantage of this formulation is that solutions of (15) are included in the effective domain of the functional J,

$$D(J) = \{u \in W_0^1 L_\Phi : J(u) = \int_\Omega \Phi(|\nabla u|)dx < \infty\}.$$

Such a conclusion is not immediate in the work [**20**], [**32**]. We therefore may consider (16) as the variational inequality formulation of (8) (and hence (2)).

1.4. Solutions of the variational inequality. We now proceed to establish the existence of solutions solutions of (16) and more general inequalities. We first provide some properties of the functional J (we again refer to [**56**] for complete proofs).

LEMMA 5. *The functional J is convex and lower semicontinuous on W^1L_Φ. If Φ is strictly convex, then J is strictly convex on $W_0^1L_\Phi$.*

In what follows, we consider the following variational inequality associated with the boundary value problem (16):

$$\begin{cases} J(v) - J(u) + \langle f(u), v-u\rangle \geq 0, \ \forall v \in K \\ u \in K, \end{cases} \tag{17}$$

where K is a convex subset of $W_0^1L_\Phi$, closed with respect to the weak* topology and $0 \in K$ (in the case $K = W_0^1L_\Phi$, (17) reduces to (16)).

We consider the auxiliary problem:

$$\begin{cases} J(v) - J(u) + \langle f, v-u\rangle \geq 0, \ \forall v \in K \\ u \in K, \end{cases} \tag{18}$$

with $f \in L_{\overline{\Psi}}$, $\langle f, v\rangle = \int_\Omega fv dx$, $v \in W_0^1L_\Phi$.

We may rewrite (18) as

$$\begin{cases} J(v) + \langle f, v\rangle \geq J(u) + \langle f, u\rangle, \ \forall v \in K \\ u \in K, \end{cases}$$

and see that u solves (18) if and only if u is a minimizer of the problem

$$\min_{v\in K}[J(v) + \langle f, v\rangle]. \tag{19}$$

We will show that (19) has a solution.

To accomplish this we shall, in what follows in this section, make the additional assumption:

- $\overline{\Phi}$ satisfies a Δ_2 condition at infinity (cf. [**38**]), which has as a consequence that $L_{\overline{\Phi}} = E_{\overline{\Phi}}$.

It follows from [**32**] that in the space $W_0^1L_\Phi$ a Poincaré inequality holds and consequently that $\||\nabla u|\|_\Phi$ furnishes an equivalent norm for $W_0^1L_\Phi$. Thus for $u \in W_0^1L_\Phi$ we shall henceforth use

$$\|u\| := \||\nabla u|\|_\Phi.$$

We have the following lemma:

LEMMA 6. *The functional J is sequentially lower semicontinuous with respect to the weak* topology of $W_0^1L_\Phi$ and is coercive in the sense that*

$$\frac{J(u)}{\|u\|} \to \infty, \quad \textit{as} \ \|u\| \to \infty. \tag{20}$$

From this lemma, we obtain the following preliminary result:

THEOREM 7. *For each* $f \in L_{\overline{\Psi}} (\subset (W_0^1 L_\Phi)^*)$, *the set* U_f *of solutions of (19) (and thus of (18)) is nonempty, convex, and bounded in* $W_0^1 L_\Phi$, *and thus precompact in* L_Ψ. *The solution set is a singleton, whenever* Φ *is strictly convex.*

PROOF. For $v(\neq 0) \in W_0^1 L_\Phi$,

$$I(v) := J(v) + \langle f, v \rangle = \|v\| \left[\frac{J(v)}{\|v\|} + \left\langle f, \frac{v}{\|v\|} \right\rangle \right].$$

Since $\left\langle f, \frac{v}{\|v\|} \right\rangle$ is bounded and $\frac{J(v)}{\|v\|} \to \infty$, as $\|v\| \to \infty$, we obtain that

$$I(v) \to \infty, \quad \text{as } \|v\| \to \infty. \tag{21}$$

In particular, I is bounded from below on K. Let us choose a minimizing sequence $\{v_n\}$ of I on K, i.e., $v_n \in K$, $\forall n$ and $I(v_n) \to \inf_K I$. Since (21) holds, $\{v_n\}$ is bounded in $W_0^1 L_\Phi$. By passing to a subsequence, if necessary, we have

$$v_n \overset{*}{\rightharpoonup} u \ \text{ in } W_0^1 L_\Phi. \tag{22}$$

Since K is weak*-closed, $u \in K$, and Lemma 6 implies that $J(u) \le \liminf J(v_n)$.

Now, since the imbedding $W_0^1 L_\Phi \hookrightarrow L_\Psi$ is compact and $\{v_n\}$ is bounded in $W_0^1 L_\Phi$, $\{v_n\}$ is relatively compact in L_Ψ. By passing to a subsequence, we have $v_n \to v$ in L_Ψ. In particular, $v_n \to v$ in L^1. Using (22) and arguments as in Lemma 6, we conclude that $v_n \rightharpoonup u$ in L^1-weak. Hence, $v = u$ and $v_n \to u$ in L_Ψ. This proves

$$\langle f, v_n \rangle = \int_\Omega f v_n dx \to \int_\Omega f u dx = \langle f, u \rangle.$$

Thus,

$$I(u) = J(u) + \langle f, u \rangle \le \liminf [J(v_n) + \langle f, v_n \rangle] = \liminf I(v_n).$$

This shows that

$$I(u) = \min\{I(v) : v \in K\}$$

and that $U_f \neq \emptyset$. Since J is convex and $\langle f, \cdot \rangle$ is linear, the functional I is convex and it easily follows that U_f is convex. If $\{u_n\} \subset U_f$ and $u_n \overset{*}{\rightharpoonup} u$ in $W_0^1 L_\Phi$, then $u \in K$ and thus U_f is weak* closed in $W_0^1 L_\Phi$.

Since $I(u) = \inf_{v \in K} I(v) < \infty$, $\forall u \in U_f$, it follows from (21) that U_f is bounded in $W_0^1 L_\Phi$. Since the mapping

$$W_0^1 L_\Phi \hookrightarrow L_\Psi \tag{23}$$

is compact (cf. [**1**], [**20**]), U_f is relatively compact in L_Ψ. We next prove that U_f is closed in L_Ψ. Thus assume $\{u_n\} \subset U_f$ and

$$u_n \to u \ \text{ in } L_\Psi. \tag{24}$$

As observed previously, $\{u_n\}$ is bounded in $W_0^1 L_\Phi$ and by passing to a subsequence, if necessary, we may assume that $u_n \overset{*}{\rightharpoonup} \overline{u}$ in $W_0^1 L_\Phi$, with $\overline{u} \in U_f$. It follows that

$u_n \rightharpoonup \overline{u}$ in L^1. On the other hand, since $L_\Psi \hookrightarrow L^1$, (24) implies that $u_n \to u$ in L^1, and therefore $u = \overline{u} \in U_f$. This shows that U_f is closed and compact in L_Ψ.

If Φ is strictly convex, then Lemma 5 shows that J is strictly convex in $W_0^1 L_\Phi$. Since $\langle f, \cdot \rangle$ is linear, it follows that I is also strictly convex, from this immediately follows that U_f contains exactly one element. □

Now, we let

$$\mathcal{K} = \mathcal{K}(L_\Psi) = \{E \subset L_\Psi : E \neq \emptyset,\ E \text{ is convex and compact}\}(\subset 2^{L_\Psi} \setminus \{\emptyset\}).$$

Then $U_f \in \mathcal{K}(L_\Psi)$, $\forall f \in L_{\overline{\Psi}}$, and the mapping

$$P : L_{\overline{\Psi}} \to 2^{L_\Psi} \setminus \{\emptyset\}, \quad f \mapsto P(f) = U_f,$$

has the property that

$$P : L_{\overline{\Psi}} \to \mathcal{K}(L_\Psi).$$

We therefore have that u is a solution of (17) if and only if $u \in L_\Psi$ and

$$u \in U_{f(u)} = P[f(u)],$$

and we conclude that the variational inequality (17) is equivalent to the inclusion in L_Ψ

$$u \in L_\Psi : u \in P[f(u)] \ \ (\text{i.e., } 0 \in u - P[f(u)]). \tag{25}$$

We thus consider the multivalued compact vector field

$$u \mapsto u - P[f(u)].$$

(Note that it follows from the proof of Lemma 2 that $u \mapsto f(u)$ is a continuous mapping from L_Ψ to $L_{\overline{\Psi}}$.)

In order to verify the existence of solutions of (25) we shall use properties of the topological degree of set-valued compact vector fields. To be able to do this we need to establish some preparatory results.

LEMMA 8. *If S is bounded in $L_{\overline{\Psi}}$, then*

$$P(S) = \bigcup_{f \in S} P(f)$$

is bounded in $W_0^1 L_\Phi$.

PROOF. Assume there exists a sequence $\{w_n\} \subset P(S)$, $w_n \in P(f_n)$, $f_n \in S$, such that $\|w_n\| \to \infty$. We have

$$J(w_n) + \langle f_n, w_n \rangle \le J(\varphi) + \langle f_n, \varphi \rangle,$$

where $\varphi \in K \cap D(J)$ is fixed. Since

$$|\langle f_n, \varphi \rangle| \le \|f_n\|_{\overline{\Psi}} \|\varphi\|_\Phi \le c_1,$$

we have

$$J(w_n) + \langle f_n, w_n \rangle \le c_2,\ \forall n.$$

Hence,

$$\frac{J(w_n)}{\|w_n\|} + \left\langle f_n, \frac{w_n}{\|w_n\|} \right\rangle \le \frac{c_2}{\|w_n\|}, \ \forall n.$$

Because

$$J(w_n)/\|w_n\| \to \infty$$

(cf. Lemma 6) and since the sequences $\{\langle f_n, w_n, \|w_n\|\rangle\}$, $\{c_2/\|w_n\|\}$ are bounded (as $n \to \infty$), we obtain a contradiction. Hence, $P(S)$ is bounded in $W_0^1 L_\Phi$. □

LEMMA 9. *The mapping $P \circ f : L_\Psi \to 2^{L_\Psi} \setminus \{\emptyset\}$ is upper semicontinuous.*

PROOF. We prove (cf. [**16**]) that if E is closed in L_Ψ, then

$$(P \circ f)^{-1}(E) := \{u \in L_\Psi : (P \circ f)(u) \cap E \neq \emptyset\}$$

is also closed in L_Ψ (Proposition 24.1, [**16**]). Let

$$\{u_n\} \subset (P \circ f)^{-1}(E), \quad u_n \to u \text{ in } L_\Psi.$$

We show that $u \in (P \circ f)^{-1}(E)$, i.e., $P[f(u)] \cap E \neq \emptyset$. For each n, choose $w_n \in P[f(u_n)] \cap E$. Now, since $\{u_n\}$ is bounded in L_Ψ, Lemma 4 shows that $\{f(u_n)\}$ is bounded in $L_{\overline{\Psi}}$. From Lemma 8 it follows that $\{P[f(u_n)]\}$ is bounded in $W_0^1 L_\Phi$. In particular, $\{w_n\}$ is bounded in $W_0^1 L_\Phi$. Because of the compact imbedding $W_0^1 L_\Phi \hookrightarrow L_\Psi$, $\{w_n\}$ is relatively compact in L_Ψ. By passing to a subsequence, if necessary, we may assume that for some $w \in L_\Psi$,

$$w_n \to w \ \text{ in } L_\Psi. \tag{26}$$

Since E is closed in L_Ψ it follows that $w \in E$. On the other hand, we may also assume that

$$w_n \overset{*}{\rightharpoonup} w \ \text{ in } W_0^1 L_\Phi. \tag{27}$$

Hence $w \in K$. For $v \in K (\subset L_\Psi)$,

$$J(w_n) + \langle f(u_n), w_n\rangle \le J(v) + \langle f(u_n), v\rangle.$$

It follows from Lemma 4 that $f(u_n) \to f(u)$ in $L_{\overline{\Psi}}$, and thus, $\langle f(u_n), w_n\rangle \to \langle f(u), w\rangle$ (cf. (26)) and $\langle f(u_n), v\rangle \to \langle f(u), v\rangle$. Therefore Lemma 6 and (27) imply that

$$J(w) + \langle f(u), w\rangle \le J(v) + \langle f(u), v\rangle,$$

for all $v \in K$, and hence $w \in P[f(u)] \cap E$. □

LEMMA 10. *The mapping $P \circ f$ is compact.*

PROOF. Suppose S is bounded in L_Ψ. Lemma 4 implies that $f(S)$ is bounded in $L_{\overline{\Psi}}$ and Lemma 8 implies that $P[f(S)]$ is bounded in $W_0^1 L_\Phi$. Due to the compact imbedding $W_0^1 L_\Phi \hookrightarrow L_\Psi$, this set is relatively compact in L_Ψ. $\square$

It follows from Lemmas 9 and 10 that the degree

$$\mathrm{d}(I - P \circ f, O, a)$$

is well defined for all bounded open subsets O of L_Ψ, all $a \in L_\Psi$, provided $a \notin (I - P \circ f)(\partial O)$ (cf. [**16**], [**58**]). This implies our first general result, whose proof is a direct application of properties of degrees of set-valued compact perturbations of the identity (Theorem 7.3.6, [**58**]).

THEOREM 11. *If there exists a bounded open set U of L_Ψ such that*

$$0 \notin u - P[f(u)], \ \forall u \in \partial U,$$

and

$$\mathrm{d}(I - P \circ f, U, 0) \neq 0, \tag{28}$$

then the inclusion (25), and equivalently the variational inequality (17), has a solution.

1.5. Some particular cases. We next consider some particular situations in which (28) holds and we have existence of solutions.

1.5.1. *Sublinear (with respect to Φ) perturbations of signed nonlinearities.* We assume that

$$F(x,s) = G(x,s) + H(x,s), \quad \text{for a.e. } x \in \Omega, \ s \in \mathbb{R}, \tag{29}$$

where G, H satisfy the growth condition (9)-(10)-(11) and, moreover, H satisfies the sign condition

$$H(x,s)s \geq 0, \ \forall s \in \mathbb{R}, \ \text{a.e. } x \in \Omega. \tag{30}$$

Furthermore it will be required that G obeys the growth condition

$$|G(x,s)| \leq A_1(x) + |\Psi_1'(s)|, \quad \text{for a.e. } x \in \Omega, \ s \in \mathbb{R}, \tag{31}$$

where $A_1 \in L_{\overline{\Psi}}$ and Ψ_1 is a differentiable Young function such that

$$\Psi_1 \ll \Phi, \tag{32}$$

and Ψ is a Young function such that

$$\Psi_1 \ll \Psi, \quad \Psi_0 \ll \Psi, \quad \Psi \ll \Phi^*. \tag{33}$$

Note that if G satisfies (31)–(32), then it also satisfies (9)–(10)–(11).

We shall show that under these conditions, the degree

$$\mathrm{d}(I - P \circ f, U, 0) = 1,$$

$U = B_R(0)$ a ball (in L_Ψ) centered at the origin with $R > 0$, sufficiently large.

To see that Theorem 11 may be applied, we choose Ψ as given above and put $U = B_R(0) = \{u \in L_\Psi : \|u\|_\Psi < R\}$. Consider the family of operators:

$$\{I - \lambda P \circ f : 0 \leq \lambda \leq 1\} \tag{34}$$

on $\overline{U}$. We first show that the set

$$\bigcup \{\lambda P[f(u)] : u \in \overline{U},\ 0 \leq \lambda \leq 1\}$$

is relatively compact in L_Ψ.

In fact, as shown previously, the set $\{f(u) : u \in \overline{U}\}$ is bounded in $L_{\overline{\Psi}}$ (cf. Lemma 4). Then Theorem 7 implies that $\bigcup \{P[f(u)] : u \in \overline{U}\}$ is bounded in $W_0^1 L_\Phi$. Since $0 \leq \lambda \leq 1$, we immediately have that the set

$$\bigcup \{\lambda P[f(u)] : u \in \overline{U}, 0 \leq \lambda \leq 1\}$$

is also bounded in $W_0^1 L_\Phi$. Because $\Psi \ll \Phi^*$, this set is relatively compact in L_Ψ.

We next show that for $R > 0$, sufficiently large,

$$0 \notin \bigcup \{u - \lambda P[f(u)] : u \in \partial U,\ 0 \leq \lambda \leq 1\}. \tag{35}$$

In fact, assume $u (\neq 0) \in L_\Psi$ and $0 < \lambda \leq 1$ are such that

$$0 \in u - \lambda P[f(u)], \tag{36}$$

i.e., $u \in \lambda P[f(u)]$, and thus $\dfrac{u}{\lambda} \in P[f(u)]$. Using the definition of P and f, this means that

$$J(v) - J\left(\frac{u}{\lambda}\right) + \left\langle f(u), v - \frac{u}{\lambda}\right\rangle \geq 0, \quad \forall v \in K,$$

i.e.,

$$J\left(\frac{u}{\lambda}\right) + \int_\Omega F(x,u)\left(\frac{u}{\lambda}\right) dx \leq J(v) + \int_\Omega F(x,u) v dx.$$

Choosing $v = 0 \in K$ and using (29), one gets

$$\begin{aligned} 0 &\geq J\left(\frac{u}{\lambda}\right) + \frac{1}{\lambda}\int_\Omega G(x,u) u dx + \frac{1}{\lambda}\int_\Omega H(x,u) u dx \\ &\geq J\left(\frac{u}{\lambda}\right) + \frac{1}{\lambda}\int_\Omega G(x,u) u dx \ \ \text{(by (30))}. \end{aligned}$$

Since $J(0) = 0$, and J is convex, it follows that $J(u) \leq \lambda J\left(\dfrac{u}{\lambda}\right)$. Hence,

$$-\frac{1}{\lambda}\int_\Omega G(x,u) u dx \geq J\left(\frac{u}{\lambda}\right) \geq \frac{1}{\lambda} J(u),$$

and thus

$$\begin{aligned} J(u) &\le \int_\Omega |G(x,u)|\,|u|dx \\ &\le \int_\Omega A_1|u|dx + \int_\Omega |\Psi_1'(u)|\,|u|dx. \end{aligned}$$

Therefore,

$$\int_\Omega \Phi(|\nabla u|)dx \le \int_\Omega A_1|u|dx + \int_\Omega \Psi_1'(|u|)|u|dx. \tag{37}$$

Now, using the arguments as in Lemma 4 (cf. (13)), we have

$$\int_\Omega \Psi_1'(|u|)|u|dx \le \int_\Omega \Psi_1(2|u|)dx. \tag{38}$$

Combining (37)–(38) and Hölder's inequality, we obtain

$$\int_\Omega \Phi(|\nabla u|)dx \le \|A_1\|_{\overline{\Psi}}\|u\|_\Psi + \int_\Omega \Psi_1(2|u|)dx. \tag{39}$$

Also, using Lemma 5.7 of [**32**], we obtain constants $c > 0$ and $k > 0$ such that

$$c\int_\Omega \Phi(|\nabla u|)dx \ge \int_\Omega \Phi(k|u|)dx. \tag{40}$$

It follows from (32) that there exists $M > 0$ such that

$$\frac{\Psi_1(2|s|)}{\Phi(k|s|)} < \frac{1}{2c}, \quad \text{for all } s \in \mathbb{R}, |s| \ge M.$$

Hence (39) implies

$$\begin{aligned} \int_\Omega \Phi(|\nabla u|)dx &\le \|A_1\|_{\overline{\Psi}}\|u\|_\Psi + \left(\int_{\{|u|<M\}} + \int_{\{|u|\ge M\}}\right)\Psi_1(2|u|)dx \\ &\le \|A_1\|_{\overline{\Psi}}\|u\|_\Psi + \int_{\{|u|<M\}} \Psi_1(2M)dx + \frac{1}{2c}\int_{\{|u|\ge M\}} \Phi(k|u|)dx \\ &\le \|A_1\|_{\overline{\Psi}}\|u\|_\Psi + |\Omega|\Psi_1(2M) + \frac{1}{2c}\int_\Omega \Phi(k|u|)dx \\ &\le \|A_1\|_{\overline{\Psi}}\|u\|_\Psi + |\Omega|\Psi_1(2M) + \frac{1}{2}\int_\Omega \Phi(|\nabla u|)dx, \end{aligned}$$

and hence

$$\frac{1}{2}\int_\Omega \Phi(|\nabla u|)dx \le \|A_1\|_{\overline{\Psi}}\|u\|_\Psi + |\Omega|\Psi_1(2M). \tag{41}$$

On the other hand, since

$$\frac{\int_\Omega \Phi(|\nabla u|)dx}{\|u\|} \to \infty, \quad \text{as } \|u\|_\Phi \to \infty,$$

(see Lemma 6) there exists $R_1 > 0$ such that for all u, $\|u\| \geq R_1$, we have

$$\int_\Omega \Phi(|\nabla u|)dx \geq \left(\frac{3\|A_1\|_{\overline{\Psi}}}{\mu}\right)\|u\|, \tag{42}$$

where μ is the best imbedding constant for the imbedding

$$W_0^1 L_\Phi \hookrightarrow L_\Psi,$$

i.e.

$$\mu\|u\|_\Psi \leq \|u\|, \ \forall u \in W_0^1 L_\Phi.$$

Hence, for $\|u\|_\Psi \geq \frac{R_1}{\mu}$,

$$\frac{1}{2}\|A_1\|_{\overline{\Psi}}\|u\|_\Psi \leq |\Omega|\Psi_1(2M),$$

and therefore

$$\|u\|_\Psi \leq \frac{2|\Omega|\Psi_1(2M)}{\|A_1\|_{\overline{\Psi}}}.$$

Thus, if u satisfies (36), then

$$\|u\|_\Psi \leq \max\left\{\frac{R_1}{\mu}, \frac{2|\Omega|\Psi_1(2M)}{\|A_1\|_{\overline{\Psi}}}\right\}. \tag{43}$$

Hence, if we choose $R > \max\left\{\frac{R_1}{\mu}, \frac{2|\Omega|\Psi_1(2M)}{\|A_1\|_{\overline{\Psi}}}\right\}$, then (35) holds.

Consequently, the family (34) is an admissible homotopy family and it follows that

$$\mathrm{d}(I - P \circ f, U, 0) = \mathrm{d}(I, U, 0) = 1,$$

and therefore (25) and thus (17) have a solution.

The above calculations prove the following result:

THEOREM 12. *Assume that F is of the form (29), where G and H satisfy (30)–(32). Then (25) (or equivalently (17)) has a solution.*

We also note that the above calculations show (see (37) and (38)) that if

$$\int_\Omega \Phi(|\nabla u|)dx > \int_\Omega A_1|u|dx + \int_\Omega \Psi_1(2|u|)dx,$$

for $u \in W_0^1 L_\Phi, \|u\|_\Psi = r > 0$, we may choose $U = B_r(0)$ and obtain a similar result.

1.5.2. *The example above revisited.* Let us assume in the above example that G satisfies (31) with $\Psi_1 = \Psi$. Then the calculations leading up to (39) yield

$$\int_\Omega \Phi(|\nabla u|)dx \leq \|A_1\|_{\overline{\Psi}}\|u\|_\Psi + \int_\Omega \Psi(2|u|)dx. \tag{44}$$

Hence, if it is the case that

$$\alpha = \liminf_{R\to\infty} \left(\inf_{\|u\|_\Psi = R} \frac{\int_\Omega \Phi(|\nabla u|)dx}{\int_\Omega \Psi(2|u|)dx} \right) > 1, \tag{45}$$

then, using (44), the imbedding $W_0^1 L_\Phi \hookrightarrow L_\Psi$ and the fact that

$$\frac{\int_\Omega \Psi(|u|)dx}{\|u\|_\Psi} \to \infty, \text{ as } \|u\|_\Psi \to \infty,$$

we obtain again an a priori bound on $\|u\|_\Psi$ and hence a result similar to Theorem 12.

1.5.3. *Existence of nonnegative solutions.* If, in the previous section, it is the case that $F(x,0) = 0$, a.e in Ω, and $0 \in K$, then the results above do not guarantee the existence of nontrivial solutions of the variational inequality under consideration, or for that matter, if a nontrivial solution exists, it cannot be asserted that it be of one sign in Ω. In this section, we are concerned with the existence of solutions which are of one sign and we specifically restrict ourselves to the existence of nonnegative solutions of (25) and (17).

The result to follow is motivated by the considerations in [**15**] and [**55**].

A concept that will be needed below is the concept of *recession cone* (cf. [**54**]) of the convex set K, it is defined as follows:

$$rcK = \bigcap_{t>0} tK$$

and we note that since K is assumed convex and weak* closed, so will be the recession cone rcK.

Below, it will be more convenient to use, instead of the solution mapping P, defined above, the mapping Q defined as follows:

$$Q(f) = P(-f) = U_{-f}, \ f \in (W_0^1 L_\Phi)^*.$$

Hence, $u \in Q(f)$ if and only if

$$\begin{cases} J(v) - J(u) \geq \langle f, v - u\rangle, \ \forall v \in K \\ u \in K, \end{cases}$$

and (25) is equivalent to the inclusion

$$u \in L_\Psi : u \in Q[-f(u)], \ (\text{i.e., } 0 \in u - Q[-f(u)]).$$

It follows from its definition, that Q has all properties of P as stated in Lemmas 8, 9, 10, and Theorem 11. We assume in this section that F satisfies:

$$-A_1(x) - \Psi_1'(s) \leq F(x,s) \leq G(x,s), \text{ for a.e. } x \in \Omega, \quad \text{all } s \in \mathbb{R}^+. \tag{46}$$

Here, $A_1 \in L_{\overline{\Psi}}$, Ψ_1 is an Young function such that $\Psi_1 \ll \Phi$. G is a Carathéodory function with growth condition

$$0 \leq G(x,s) \leq B(x) + |\Psi'(s)|, \quad x \in \Omega, \ s \geq 0, \tag{47}$$

where Ψ is a differentiable Young function such that $\Psi \ll \Phi^*$, and $B \in L_{\overline{\Psi}}$,

$$G(x, 0) = 0, \quad \text{a.e. } x \in \Omega, \tag{48}$$

and

$$G(x, s) \text{ is nondecreasing with respect to } s \text{ for all } x \in \Omega. \tag{49}$$

We denote by

$$W^+ := (W_0^1 L_\Phi)^+ = \{u \in W_0^1 L_\Phi : u \geq 0, \quad \text{a.e. on } \Omega\}$$

the cone of all nonnegative functions of $W_0^1 L_\Phi$.

We shall prove the following result concerning the existence of nonnegative solutions of (25).

THEOREM 13. *Assume that (46)–(49) are satisfied and that*

$$W^+ \subset rcK. \tag{50}$$

Then (25) has a nonnegative solution, i.e., a solution in $K \cap W^+$.

PROOF. Let I and Q be as above. Extend $G(\cdot, s)$ to $\mathbb{R}$ by defining

$$G(x, s) = -G(x, -s) \text{ for } s \leq 0.$$

Then G is still a Carathéodory function and (47) becomes

$$G(x, s) \leq B(x) + |\Psi'(s)|, \quad x \in \Omega, \quad s \in \mathbb{R}. \tag{51}$$

We first prove that for each $w \in L_\Psi$, there exists a solution u in L_Ψ of the following inclusion

$$u \in Q[-G(\cdot, u) + w]. \tag{52}$$

Using the definition of Q, this inclusion can be written in the equivalent variational inequality form:

$$\begin{cases} J(v) - J(u) \geq \int_\Omega [-G(x, u) + w](v - u)dx, \ \forall v \in K, \\ u \in K. \end{cases} \tag{53}$$

or

$$\begin{cases} J(v) - J(u) + \int_\Omega G(x, u)(v - u)dx \geq \int_\Omega w(v - u)dx, \ \forall v \in K \\ u \in K. \end{cases} \tag{54}$$

The extension of G, (48), and (49), imply that $G(x, s)$ is nondecreasing with respect to $s \in \mathbb{R}$. Hence, the operator

$$u \mapsto S(u), \ S : L_\Psi \to (W_0^1 L_\Phi)^*$$

defined by the second integral,

$$\langle S(u), v\rangle := \int_\Omega G(x,u)v dx,$$

is monotone on L_Ψ. Since $G(\cdot,u)$ is of the same sign as u, S is nonnegative. It follows from (51) that $u \mapsto G(\cdot,u)$ is a continuous, bounded mapping from L_Ψ to $L_{\overline{\Psi}}$ and therefore that S is continuous and bounded. Using Lemma 6, we see that

$$\frac{1}{\|u\|}\left[J(u) + \int_\Omega G(x,u)u dx\right] \to \infty, \quad \text{as } \|u\| \to \infty. \tag{55}$$

By classical existence results for variational inequalities, it follows that for each $w \in L_\Psi$ the set U_w of solutions of (54) (and thus (53), (52)) is nonempty. Define M from $L_{\overline{\Psi}}$ to 2^{L_Ψ} by $M(w) = U_w$. We check that M is upper semicontinuous and compact. Let $\{w_n\}$ be a sequence in $L_{\overline{\Psi}}$ and let $u_n \in M(w_n)$, i.e.,

$$u_n \in Q[-G(\cdot,u_n) + w_n]. \tag{56}$$

If $\{w_n\}$ is bounded in $L_{\overline{\Psi}}$, then $\{u_n\}$ is bounded in $W_0^1 L_\Phi$, by the coercivity property (55). It follows that $\{u_n\}$ is relatively compact in $L_\Psi(\Omega)$. To prove the upper semicontinuity of M, we let $w_n \to w$ in $L_{\overline{\Psi}}$.

Since $\{w_n\}$ is bounded, $\{u_n\}$ is relatively compact and we can choose a subsequence $\{u_\eta\} \subset \{u_n\}$ such that

$$u_\eta \to u \text{ in } L_\Psi.$$

It follows from Lemma 4 that

$$-G(\cdot,u_\eta) + w_\eta \to -G(\cdot,u) + w \text{ in } L_{\overline{\Psi}},$$

hence, using the arguments in Lemma 9, we see the upper semicontinuity of M.

We next write the inclusion $u \in Q[f(u)]$ in the form

$$\begin{aligned} u &\in Q[-G(\cdot,u) - f(u) + G(\cdot,u)] \\ &\in Q\{-G(\cdot,u) + [G(\cdot,u) - F(\cdot,u)]\}, \end{aligned}$$

or, using the definition of M,

$$u \in M[G(\cdot,u) - F(\cdot,u)] := N(u). \tag{57}$$

We show that (57) has a solution in the positive cone of L_Ψ:

$$L_\Psi^+ := \{u \in L_\Psi : u \geq 0, \quad \text{a.e. in } \Omega\}.$$

To do this we first verify that $N[L_\Psi^+(\Omega)] \subset L_\Psi^+(\Omega)$. Let $w \in L_\Psi^+(\Omega)$ and

$$u \in N(w) = M[\tilde{Q}(w)],$$

with

$$\tilde{Q}(w) = G(\cdot,w) - F(\cdot,w) \geq 0, \text{ a.e. on } \Omega, \tag{58}$$

as follows from (46) and the fact that $w(x) \geq 0$, for a.e. $x \in \Omega$. The proof that $u \geq 0$, a.e. in Ω is based on a maximum principle argument, applied to Orlicz–Sobolev functions. By the definition of M, one has

$$u \in Q[-G(\cdot, u) + \tilde{Q}(w)],$$

or equivalently,

$$\begin{cases} J(v) - J(u) + \int_\Omega G(x,u)(v-u)dx \geq \int_\Omega \tilde{Q}(w)(v-u)dx, \ \forall v \in K \\ u \in K. \end{cases}$$

Choose $\varphi \in W_0^1 L_\Phi$ with $\varphi \leq 0$, a.e. in Ω. Since $-\varphi \in rcK$, (50) implies that $v = u - \varphi \in K$. Substituting $v = u - \varphi$ in the above inequality, one obtains

$$J(u-\varphi) - J(u) + \int_\Omega G(x,u)\varphi dx \leq \int_\Omega \tilde{Q}(w)\varphi dx \leq 0, \tag{59}$$

(cf. (58)). Let $\varphi = u^- = \min\{u, 0\}$. We show that $u^- \in W_0^1 L_\Phi$ and

$$\nabla u^-(x) = \begin{cases} \nabla u(x), & \text{if } u(x) \leq 0 \\ 0, & \text{if } u(x) > 0. \end{cases} \tag{60}$$

In fact, since $u \in W^{1,1}$, it follows that $u^- \in W^{1,1}$ and (60) holds. Then for each $k > 0$,

$$\int_\Omega \Phi(ku^-)dx = \int_{\{u\leq 0\}} \Phi(ku)dx \leq \int_\Omega \Phi(ku)dx,$$
$$\int_\Omega \Phi(k|\nabla u^-|)dx = \int_{\{u\leq 0\}} \Phi(k|\nabla u|)dx \leq \int_\Omega \Phi(k|\nabla u|)dx,$$

and thus $u^-, |\nabla u^-| \in L_\Phi$, which, of course, implies that $u^- \in W^1 L_\Phi(\Omega)$. Moreover, using trace theory for Orlicz–Sobolev functions (cf. [**41**]),

$$u^-|_{\partial\Omega} = \min\{u|_{\partial\Omega}, 0\} = 0,$$

and therefore $u^- \in W_0^1 L_\Phi$.

On the other hand, one has

$$G(x,u(x))u^-(x) = \begin{cases} 0, & \text{if } u(x) \geq 0 \\ G(x,u(x))u(x), & \text{if } u(x) \leq 0 \end{cases}$$
$$\geq 0, \quad \text{for a.e. } x \in \Omega.$$

Hence $\int_\Omega G(x,u)u^- dx \geq 0$ and it follows from (59) that $J(u-u^-) \geq J(u)$. Hence,

$$\int_{\{u\geq 0\}} \Phi(|\nabla u|)dx \geq \int_\Omega \Phi(|\nabla u|)dx,$$

i.e.,

$$\int_{\{u\leq 0\}} \Phi(|\nabla u|)dx \leq 0.$$

Consequently, $\nabla u = \nabla u^- = 0$, a.e. on the set $\{x \in \Omega : u(x) \leq 0\}$ and it follows from (60), that $\nabla u^- = 0$, a.e. on Ω. Since $u^- \in W_0^1 L_\Phi$, one obtains that $u^- = 0$ and therefore $u \geq 0$, a.e. on Ω. This shows that $u \in L_\Psi^+$ and that L_Ψ^+ is invariant under N.

Since $\tilde{Q}$ is continuous and M is upper semicontinuous, $N = M \circ \tilde{Q}$ is upper semicontinuous. In what follows, we consider the restriction of N to $L_\Psi^+(\Omega)$, which we simply still denote by N. Since $G - F$ is a continuous, bounded operator from L_Ψ to $L_{\overline{\Psi}}$ and M is compact from $L_{\overline{\Psi}}$ to 2^{L_Ψ}, we see that N is compact from L_Ψ into its subsets and from L_Ψ^+ into its subsets. It follows from [**23**] that the fixed point index of N with respect to the positive cone L_Ψ^+, $\text{ind}(N, B_r(0) \cap L_\Psi^+)$, is well defined, provided

$$0 \notin u - N(u),\ \forall u \in \partial B_r(0) \cap L_\Psi^+.$$

To show that

$$\text{ind}(N, B_r(0) \cap L_\Psi^+) = 1, \tag{61}$$

for R sufficiently large, we consider the family

$$\{u - \lambda N(u) : \lambda \in [0,1]\}$$

of compact perturbations of the identity and prove that

$$0 \notin u - \lambda N(u),\ \forall u \in \partial B_r(0) \cap L_\Psi^+,\ \forall \lambda \in [0,1]. \tag{62}$$

This clearly holds for $\lambda = 0$. Assume $0 < \lambda \leq 1$ and $u \in L_\Psi^+$ satisfies $u \in \lambda N(u)$, then $\frac{u}{\lambda} \in M[G(\cdot,u) - F(\cdot,u)]$, or equivalently,

$$\frac{u}{\lambda} \in Q\left[-G\left(\cdot, \frac{u}{\lambda}\right) + G(\cdot,u) - F(x,u)\right],$$

which in variational inequality form reads as:

$$\begin{cases} J(v) - J\left(\frac{u}{\lambda}\right) \geq \int_\Omega \left[-G\left(x, \frac{u}{\lambda}\right) + G(x,u) - F(x,u)\right]\left(v - \frac{u}{\lambda}\right) dx,\ \forall v \in K \\ u \in K. \end{cases}$$

Choosing $v = 0 \in K$ in this inequality, one obtains

$$J\left(\frac{u}{\lambda}\right) \leq \frac{1}{\lambda}\int_\Omega \left[G(x,u) - G\left(x, \frac{u}{\lambda}\right)\right] u dx - \frac{u}{\lambda}\int_\Omega F(x,u) u dx. \tag{63}$$

Since Φ is convex, it follows that

$$J\left(\frac{u}{\lambda}\right) = \int_\Omega \Phi\left(\frac{|\nabla u|}{\lambda}\right) dx \geq \frac{1}{\lambda}\int_\Omega \Phi(|\nabla u|)dx.$$

Since $u \in L_\Psi^+(\Omega)$ and $G(\cdot, u)$ is nondecreasing for $u \geq 0$, we have

$$G(\cdot, u) - G\left(\cdot, \frac{u}{\lambda}\right) \leq 0,$$

and therefore

$$\int_\Omega \left[G(x,u) - G\left(x, \frac{u}{\lambda}\right)\right] u dx \leq 0.$$

It follows from (63) that

$$\frac{1}{\lambda}\int_\Omega \Phi(|\nabla u|)dx \leq -\frac{1}{\lambda}\int_\Omega F(x,u)udx.$$

Therefore,

$$\begin{aligned}\int_\Omega \Phi(|\nabla u|)dx &\leq -\int_\Omega F(x,u)udx \\ &\leq \int_\Omega A_1(x)udx + \int_\Omega \Psi_1'(u)udx.\end{aligned}$$

Using the arguments as in the proof of Theorem 12, we conclude from this estimate that $\|u\|_\Psi \leq R_0$, for some fixed, positive constant R_0 depending only on Φ, A_1, Ψ_1, and Ω. Therefore,

$$0 \notin u - N(u), \quad \forall u \in L_\Psi^+, \ \|u\|_\Psi = R,$$

for all $R > 0$ sufficiently large.

Using the homotopy invariance property of the fixed point index for multivalued compact vector fields (cf. [**23**]), we obtain

$$\begin{aligned}\operatorname{ind}(N, B_R(0) \cap L_\Psi^+) &= \operatorname{ind}(0, B_R(0) \cap L_\Psi^+) \\ &= 1 \ \ (\text{since } 0 \in B_R(0) \cap L_\Psi^+).\end{aligned}$$

It follows that N has a fixed point in $B_R(0) \cap L_\Psi^+$, i.e., (25) has a nonnegative solution. □

REMARK 14. As in subsections 1.5.1, 1.5.2, the above existence result still holds if the Young function Ψ_1 in (46) satisfies only the subcriticality condition $\Psi_1 \ll \Phi^*$, provided a condition such as (45) is satisfied.

2. Principal eigenvalues for quasilinear elliptic differential operators

2.1. Introduction. We consider in this section eigenvalue problems for quasilinear elliptic partial differential equations which involve nonhomogeneous differential operators on general domains. We reduce the problem of the existence of eigenvalues of the nonlinear elliptic equation to the question of the existence of a minimum of a coercive extended real valued functional, defined in an Orlicz-Sobolev space, which is subject to a constraint. The existence of the eigenvalue then will follow by an application of general properties of subdifferentials and arguments as those presented by Kubrusly [**40**] and of Liusternik's theorem (applied in a suitable subspace) on Lagrange multipliers ([**69**], [**72**]). This will yield what will be called a principal eigenvalue (for a given level). The existence of nonprincipal eigenvalues will follow from arguments based on Liusternik-Schnirelman theory similar to those used in [**4**], [**14**] and [**70**] (see e.g. [**71**],[**28**], [**29**], [**67**], [**68**], and [**53**] for some developments in this direction). This section may be considered to be an eigenvalue theory for rapidly growing operators as considered in [**20**], [**32**], and [**33**].

The eigenvalue problem we shall consider is the following:

$$(E)\qquad \begin{cases} -\operatorname{div}(A(|\nabla u|^2)\nabla u) = \lambda\psi(u) & \text{in } \Omega \\ u = 0 & \text{on } \partial\Omega, \end{cases}$$

where Ω is a bounded domain in $\mathbb{R}^N$ with a Lipschitz continuous boundary.

We shall show the existence of values of λ such that problem (E) (either in the form stated or in a weaker sense as observed below) has nontrivial positive (negative) solutions. Assume that $\phi(s) = sA(s^2)$ and $\psi(s) = sB(s^2)$ are odd increasing homeomorphisms of $\mathbb{R}$ onto $\mathbb{R}$ and consider the Young functions $\Phi(s) = \int_0^s \phi(t)\,dt$, $\Psi(s) = \int_0^s \psi(t)\,dt$. If it is the case that $\Phi = \Psi$ (or more generally that Φ is equivalent to Ψ, see [**38**]) and Φ satisfies a Δ_2 condition, we also conclude that the set of such λ is bounded away from 0 by, say, λ_1, and for any $\lambda < \lambda_1$ problem (E) has no non-trivial solutions.

Next, as an application of our results, we will prove the existence of solutions for the problem

$$(N)\begin{cases} -\operatorname{div}(A(|\nabla u|^2)\nabla u) = p(x,u) & \text{in } \Omega \\ u = 0 & \text{on } \partial\Omega, \end{cases}$$

where $p : \Omega \times \mathbb{R} \to \mathbb{R}$ satisfies Carathéodory conditions and a growth condition of the form

$$|p(x,u)| \le a\phi(u) + b(x) \tag{64}$$

where $a \in \mathbb{R}$ is such that $a < \lambda_1$ and $b \in L_{\overline{\Phi}}$. If both Φ and $\overline{\Phi}$ satisfy a Δ_2 condition at infinity, then more detailed information on the eigenvalues will become available.

We shall use a variational approach to establish the existence of eigenvalues for problem (E). This will be accomplished by minimizing a coercive weak* lower semicontinuous convex functional on a suitable weak* closed set. The existence of so-called principal eigenvalues will then follow using results about Lagrange multipliers for the derivatives of the functionals restricted to suitable subspaces. Once the N-function Φ satisfies a Δ_2 condition we are able to prove that the principal eigenvalues are bounded below and a Fredholm like result will be verified for problem (N). Related works on eigenvalue problems in Orlicz–Sobolev spaces can be found in [**34, 53, 60, 64, 65, 71**] and the references therein.

2.2. Existence of eigenvalues. In this section we shall consider the existence of eigenvalues for problem (E), i.e. the existence of values of λ such that problem

(E) will have positive (or negative) solutions. To establish the existence of such eigenvalues, we shall minimize an appropriate functional subject to a constraint in the weak* closed subspace $W_0^1 L_\Phi$ of the Orlicz-Sobolev space $W^1 L_\Phi$. We define functionals $f, g : W_0^1 L_\Phi \to \mathbb{R} \cup \{\infty\}$ by

$$f(u) = \int_\Omega \Phi(|\nabla u|)\, dx, \tag{65}$$

and

$$g(u) = \int_\Omega \Psi(|u|)\, dx, \tag{66}$$

and consider the minimization problem:

$$\min_{\{u \in W_0^1 L_\Phi : g(u) = \mu\}} f(u), \tag{67}$$

where μ is a given positive number. Note that both functionals are convex.

We shall denote by

$$M_\mu = \{u \in W_0^1 L_\Phi : g(u) = \mu\}$$

and

$$\alpha = \inf_{u \in M_\mu} f(u).$$

2.2.1. *Preliminaries.* In order to prove the existence of eigenvalues of our problem we shall need a general result about Lagrange multipliers which is due to Kubrusly [**40**]. The theorem is stated below and applied later (in the space $U = W_0^1 L_\Phi$). We refer to the paper [**40**] for its proof. An alternative proof of Kubrusly's result together with other related arguments can also be found in [**42**].

THEOREM 15 (Theorem 2, [**40**]). *Let U be a normed linear space and $F, G : U \to \mathbb{R}$ be two Fréchet differentiable functionals whose gradients are $F', G' : U \to U^*$. Let $P : U \to \mathbb{R} \cup \{\infty\}$ be a proper, convex functional with effective domain $D(P)$.*

If $F + P$ achieves its minimum restricted to the nonempty set

$$\partial G_R = \{u \in U : G(u) = R\}, \quad R > 0,$$

at an element $u \in \partial G_R$ satisfying

$$\begin{cases} \{v \in U : \langle G'(u), v - u\rangle > 0\} \cap D(P) \neq \emptyset, \\ \{v \in U : \langle G'(u), v - u\rangle < 0\} \cap D(P) \neq \emptyset, \end{cases} \tag{68}$$

then there exists $\lambda \in \mathbb{R}$ such that u is a solution of the variational inequality:

$$\langle F'(u) - \lambda G'(u), v - u\rangle + P(v) - P(u) \geq 0, \quad \forall v \in U. \tag{69}$$

Some properties of the functionals and sets involved in our eigenvalue problem are given in the following results, whose proofs can be found in [**30**].

LEMMA 16. *The functional f is coercive.*

LEMMA 17. *The functional f is weak* lower semicontinuous.*

LEMMA 18. *The set M_μ is sequentially weak* closed.*

REMARK. As with this lemma, we can show that g is strongly continuous and compact, thus g is in fact completely continuous on $W_0^1 L_\Phi$.

2.2.2. *The minimization problem.* The following theorem provides the existence of solutions of the minimization problem (67).

THEOREM 19. *The minimization problem (67) has a solution.*

PROOF. It follows from the properties of the functional g that $M_\mu \neq \emptyset$. We choose $\{u_n\} \subset M_\mu$ such that

$$f(u_n) \to \alpha.$$

The coercivity of f therefore implies that $\{u_n\}$ is bounded in $W_0^1 L_\Phi$ which is in the dual of a separable Banach space. Hence $\{u_n\}$ has a weak* convergent subsequence converging to say $u \in W_0^1 L_\Phi$ and $u \in M_\mu$ since M_μ is weak* closed. By the weak* lower semicontinuity of f, $f(u) = \lim f(u_n) = \alpha$. u is therefore a solution of our constrained minimization problem.

As f is a proper functional (i.e. it assumes finite values) it follows, of course, that $\alpha < +\infty$. □

2.2.3. *Differentiability properties.* Let us assume that the functionals f and g are as above. Let us show that f (with a condition on $\overline{\Phi}$) and g are of class C^1 on $W_0^1 E_\Phi$ and $W_0^1 L_\Phi$, respectively.

LEMMA 20. *Assume that $\overline{\Phi}$ satisfies a Δ_2 condition. Then the functional $f : W_0^1 E_\Phi \to \mathbb{R}$ is C^1.*

PROOF. We observe that for $u, h \in W_0^1 E_\Phi$ and $t > 0$

$$\frac{1}{t}\left(f(u+th) - f(u)\right) = \int_\Omega \frac{1}{t}\int_{|\nabla u|}^{|\nabla u + t\nabla h|} sA(s^2)dsdx.$$

On the other hand, as $t \to 0$,

$$|\nabla u + t\nabla h| \to |\nabla u|$$

in L_Φ and hence in L^1 (see [**41**]) and thus via a subsequence may be assumed to converge almost everywhere. An easy calculation shows that

$$\left|\frac{1}{t}\int_{|\nabla u|}^{|\nabla u + t\nabla h|} sA(s^2)ds\right| \le \phi(|\nabla u| + |\nabla h|)|\nabla h|$$

with $\phi(|\nabla u| + |\nabla h|) \in L_{\overline{\Phi}}$ and $|\nabla h| \in L_\Phi$ and hence $\phi(|\nabla u| + |\nabla h|)|\nabla h| \in L^1$, because of a form of Hölder's inequality in Orlicz spaces ([**1**]). Thus, by the dominated convergence theorem

$$\begin{aligned}\lim_{t\to 0}\frac{1}{t}\left(f(u+th) - f(u)\right) &= \int_\Omega |\nabla u| A(|\nabla u|^2)\frac{\nabla u\cdot\nabla h}{|\nabla u|}dx\\ &= \int_\Omega A(|\nabla u|^2)\nabla u\cdot\nabla h dx\\ &= f'(u,h).\end{aligned}$$

We note that this Gâteaux derivative is linear with respect to h. We also have that f' is continuous. Indeed if we let $\{u_n\} \subset W_0^1 E_\Phi$ be a sequence such that $u_n \to u$ and consider

$$|\langle f'(u_n) - f'(u), h\rangle| = \left|\int_\Omega (A(|\nabla u_n|^2)\nabla u_n - A(|\nabla u|^2)\nabla u)\cdot\nabla h dx\right|, \tag{70}$$

then a lengthy exercise using Egoroff's theorem shows that there exists a sequence $\{\delta_n\} \subset \mathbb{R}_+$ with $\delta_n \to 0$ such that

$$\begin{aligned}&\sup_{\|h\|_{1,\Phi}\le 1}\left|\int_\Omega (A(|\nabla u_n|^2)\nabla u_n - A(|\nabla u|^2)\nabla u)\cdot\nabla h dx\right|\\ &\quad\le \|\phi(|\nabla u_n|) - \phi(|\nabla u|)\|_{\overline{\Phi}} + \delta_n.\end{aligned}$$

Since

$$\|\phi(|\nabla u_n|) - \phi(|\nabla u|)\|_{\overline{\Phi}} \to 0$$

as follows from Theorem 12, page 83 of [**66**] (it is here that we have used for the first time that $\overline{\Phi}$ satisfies a Δ_2 condition), we obtain that

$$||f'(u_n) - f'(u)||_{(W_0^1 E_\Phi)^*} \to 0$$

and thus the lemma is proved. □

We remark here, that if Φ also satisfies a Δ_2 condition, then the effective domain of f is the whole space $W_0^1 L_\Phi$ and f is C^1 there.

The verification that g is C^1 on $W_0^1 L_\Phi$ is somewhat analogous and proceeds using reasoning similar to the above (that $\overline{\Phi}$ satisfies a Δ_2 condition is not needed) and some of the reasoning employed in the proof of Lemma 18. In fact, we have the following result, whose full proof is again given in [**30**].

LEMMA 21. *The functional* $g : W_0^1 L_\Phi \to \mathbb{R}$ *is of class* C^1 *and*

$$\langle g'(u), v\rangle = \int_\Omega \psi(u) v dx.$$

REMARK. We here point out that in the nonreflexive case $W_0^1 E_\Phi$ is not necessarily weak* closed in $W^1 L_\Phi$.

2.2.4. *The main theorems.* We now consider the eigenvalue problem (E) in a general setting, without imposing constraints on the growth of Φ.

In the general situation, the functional f is not necessarily differentiable (and even may assume infinite values) on the whole space. Since f is convex (and lower semicontinuous) on $W_0^1 L_\Phi$, we can replace the derivative of f as considered in Theorem 23 by the subdifferential ∂f of f, and use a nonsmooth version of the Liusternik theorem. Another reason for considering the general case is that, unlike the case when Φ satisfies a polynomial growth condition, the derivative of f does not, in general, act on $W_0^1 L_\Phi$ continuously. Hence, we cannot apply Liusternik's theorem directly.

Before proving the result, we formulate (E) as an eigenvalue problem for an inclusion, or equivalently, for a variational inequality. Note that in the smooth case, (E) is formulated (at least formally) as the equation:

$$f'(u) - \lambda g'(u) = 0.$$

Replacing $f'(u)$ by the subdifferential $\partial f(u)$, this equation is extended as the inclusion

$$(E_1) \qquad \partial f(u) - \lambda g'(u) \ni 0.$$

The equivalent variational inequality form of this inclusion (cf. [**72**]) is

$$(E_2) \qquad \begin{cases} f(v) - f(u) - \lambda \langle g'(u), v-u \rangle \geq 0, \ \forall v \in W_0^1 L_\Phi \\ u \in W_0^1 L_\Phi. \end{cases}$$

Again, λ is called an eigenvalue of (E_2) if this variational inequality has a solution $u \neq 0$.

THEOREM 22. *Suppose that Φ and Ψ, are N-functions as given above, and assume Ψ grows essentially more slowly than Φ^*. Then for every constant $\mu > 0$, there exists $\lambda > 0$ such that (E_2) (and equivalently (E_1)) has a positive (negative) solution u such that $g(u) = \mu$.*

PROOF. We consider the constrained minimization problem (67). As checked in Theorem 19, for each $\mu > 0$, (67) has a positive (negative) solution u. To establish the existence of the Lagrange multiplier λ, we need the nonsmooth version of Liusternik's theorem established in [**40**]. In this case, (67) is just the minimization stated in Kubrusly's theorem and (69) is the same as our variational inequality (E_2).

It follows from the results in Section 2.2.1 (Lemma 21) that $G = g$ is differentiable on $W_0^1 L_\Phi(\Omega)$ and $G' = g'$ is given by

$$\langle g'(u), v \rangle = \int_\Omega \psi(u) v, \quad \forall v \in W_0^1 L_\Phi.$$

Let u be a solution of (67) (whose existence is proved in Theorem 19). We only need to check the nondegeneracy condition (68). First, it is clear that $u \neq 0$; hence $\psi(u) \neq 0$ in Ω (since ψ is strictly increasing and $\psi(0) = 0$). Choosing $v = 0$, it follows that $v \in D(f)$ $(f(0) = 0)$ and

$$\langle g'(u), v-u \rangle = -\int_\Omega \psi(u) u < 0,$$

(again, we use the strict monotonicity of ψ and the fact that $\psi(u)$ and u have the same sign). This verifies the second condition of (68).

To check the first condition, we note that since $\psi(u) \neq 0$ on Ω, we can choose a function $v_0 \in C_0^1(\overline{\Omega})$ such that

$$\int_\Omega \psi(u) v_0 > 0.$$

Since $\int_\Omega \psi(u) u < \infty$, there exists t sufficiently large such that

$$\int_\Omega \psi(u)(tv_0) > \int_\Omega \psi(u) u.$$

Choosing $v = tv_0$, we have $\langle g'(u), v - u\rangle > 0$. Moreover, $v \in C_0^1(\overline{\Omega}) \subset W_0^1 E_\Phi \subset D(f)$, and we have verified the second condition of (68). The existence of the Lagrange multiplier λ now follows from Kubrusly's theorem. Choosing $v = 0$ in (E_2), we have $-f(u) + \lambda\langle g'(u), u\rangle \geq 0$. Hence,

$$\lambda \geq f(u)\left(\int_\Omega \psi(u) u\right)^{-1} > 0.$$

THEOREM 23. *Suppose that Φ and Ψ, are N-functions as given above, and assume Ψ grows essentially more slowly than Φ^*. Further assume that Φ and $\overline{\Phi}$ satisfy a Δ_2 condition. Then for every constant $\mu > 0$ there exists $\lambda > 0$ such that (E) has a positive (negative) solution u such that $g(u) = \mu$.*

If $\Phi = \Psi$ and Φ satisfies a global Δ_2 condition, then the set of such eigenvalues is bounded below.

PROOF. The proof proceeds as follows. We set

$$M_\mu = \{u \in W_0^1 L_\Phi : g(u) = \mu\},$$

and let

$$\alpha = \inf_{u \in M_\mu} f(u).$$

We note that for every $\mu \geq 0$ $M_\mu \neq \emptyset$ (this will follows from the continuity of g).

Since in this case the functionals are differentiable, we may apply what has been established above and find that

$$\int_\Omega A(|\nabla u_\mu|^2)\nabla u_\mu \cdot \nabla v dx - \lambda_\mu \int_\Omega B(u_\mu^2) u_\mu v dx = 0, \quad \forall v \in W_0^1 E_\Phi, \tag{71}$$

i.e. u_μ is a weak solution of (E). Choosing $v = u_\mu$ in (71), we obtain that $\lambda_\mu > 0$. Also since $f(u_\mu) = f(|u_\mu|)$ and $g(u_\mu) = g(|u_\mu|)$, we may assume that u_μ is one-signed in Ω.

Let us next consider the special case that $\Phi = \Psi$, and let λ_μ be as above. It follows from (71) that

$$\lambda_\mu = \frac{\displaystyle\int_\Omega A(|\nabla u_\mu|^2)|\nabla u_\mu|^2 dx}{\displaystyle\int_\Omega A(u_\mu^2) u_\mu^2 dx}. \tag{72}$$

Since

$$\Phi(t) \le t\phi(t) \le \Phi(2t), \quad t \ge 0,$$

we obtain that

$$\frac{\int_\Omega \Phi(|\nabla u_\mu|)dx}{\int_\Omega \Phi(2|u_\mu|)dx} \le \lambda_\mu \le \frac{\int_\Omega \Phi(2|\nabla u_\mu|)dx}{\int_\Omega \Phi(|u_\mu|)dx} \tag{73}$$

and using the Poincaré inequality in (73) we obtain

$$\lambda_\mu \ge \frac{\int_\Omega \Phi(|\nabla u_\mu|)}{\int_\Omega \Phi(2d|\nabla u_\mu|)}. \tag{74}$$

Hence if $2d \le 1$ (which is the case for domains Ω with small diameter) we find that

$$\lambda_\mu \ge 1.$$

Otherwise, if Φ satisfies a Δ_2 condition, then

$$\lambda_\mu \ge \frac{1}{\alpha(2d)}.$$

Since the lower bounds so obtained are independent of μ, we find that

$$\lambda_1 = \inf\{\lambda_\mu : \ \mu \in \mathbb{R}\backslash\{0\}\} > 0. \tag{75}$$

In general, however, $\lambda_1 \ge 0$.

These calculations also show that problem (E) has no non-trivial solutions for $\lambda < \lambda_1$.

As an example we use the $N-$function

$$\Phi(s) = e^{s^2} - 1$$

whose Hölder's conjugate function $\overline{\Phi}$ satisfies a Δ_2 condition (see [**38**]) but Φ does not. Our result implies the existence of principal eigenvalues (in the sense made precise above) for the problem

$$\begin{cases} -\mathrm{div}(e^{|\nabla u|^2}\nabla u) = \lambda e^{u^2}u & \text{in } \Omega \\ u = 0 & \text{on } \partial\Omega. \end{cases}$$

2.3. An existence result. In this section we will show existence of solutions for the problem (N) in the introduction. Here we shall assume that

$$\tilde{\lambda}_1 = \inf_{u \in W_0^1 L_\Phi \backslash \{0\}} \frac{\int_\Omega \Phi(|\nabla u|)dx}{\int_\Omega \Phi(|u|)dx}$$

is positive (which is the case when Φ satisfies a Δ_2 condition, as was verified above).

We have the following theorem:

THEOREM 24. *Suppose that $\overline{\Phi}$ satisfies a Δ_2 condition. Then problem (N) has a solution provided that p satisfies the condition (64), with $a < \tilde{\lambda}_1$.*

PROOF. Consider the functional $G : W_0^1 L_\Phi \to \mathbb{R} \cup \{\infty\}$ defined by

$$G(u) := \int_\Omega \Phi(|\nabla u|)\, dx - \int_\Omega P(u)dx, \tag{76}$$

where

$$P(u)(x) = \int_0^{u(x)} p(x,s)ds.$$

We have already shown that the functional

$$u \mapsto \int_\Omega \Phi(|\nabla u|)\, dx$$

is weak* lower semicontinuous. On the other hand, if $u_n \rightharpoonup^* u$ then $\{u_n\}$ is bounded and hence precompact in L_Φ. The Niemitskii operator P maps L_Φ continuously to $L_{\overline{\Phi}}$ and thus for $u \in W_0^1 L_\Phi$, $P(u) \in \tilde{L}_{\overline{\Phi}}$. Hence each subsequence of $\{u_n\}$ has a subsequence with

$$\int_\Omega P(u_n)dx \to \int_\Omega P(u)dx,$$

and therefore G is weak* lower semicontinuous.

We next show that G is coercive also. In this regard we observe that for any $u \in W_0^1 L_\Phi$,

$$G(u) \geq \int_\Omega \Phi(|\nabla u|)\, dx - a\int_\Omega \Phi(|u|)\, dx - \int_\Omega budx.$$

Hence

$$G(u) \geq \left(1 - \frac{a}{\tilde{\lambda}_1}\right)\int_\Omega \Phi(|\nabla u|) - 2\|b\|_{\overline{\Phi}}\|u\|_{1,\Phi}.$$

It follows from Lemma 3.13 of [**32**] that (here it is needed that $\overline{\Phi}$ satisfies a Δ_2 condition)

$$\frac{\displaystyle\int_\Omega \Phi(|\nabla u|)\, dx}{\|u\|_{1,\Phi}} \to \infty \quad \text{as} \quad \|u\|_{1,\Phi} \to \infty,$$

and hence that the functional G is coercive. Thus G assumes its global minimum at say u, i.e. u satisfies

$$G(u) = \min_{v \in W_0^1 L_\Phi} G(v).$$

On the other hand, again, $G(u) < \infty$, and since $|\int_\Omega P(u)dx| < \infty$, we must have that $\int_\Omega \Phi(|\nabla u|)dx < \infty$.

The second term of the functional G is C^1 and therefore, by using again Kubrusly's theorem, we obtain that u must satisfy the variational inequality

$$\begin{cases} \int_\Omega \Phi(|\nabla v|)dx - \int_\Omega \Phi(|\nabla u|)dx - \int_\Omega p(x,u)(v-u)dx \geq 0, \forall v \in W_0^1 L_\Phi \\ u \in W_0^1 L_\Phi, \end{cases}$$

or since $\overline{\Phi}$ satisfies a Δ_2 condition, $\langle G'(u), h\rangle = 0$ for all $h \in W_0^1 L_\Phi$, i.e. $u \in W_0^1 L_\Phi$ satisfies

$$\int_\Omega A(|\nabla u|^2)\nabla u \cdot \nabla h dx - \int_\Omega p(x,u) h dx = 0,$$

implying that u is a weak solution of problem (N). □

3. Nontrivial solutions of variational-hemivariational inequalities with slowly growing principal parts

3.1. Introduction. This section is about a variational-hemivariational inequality arising from the following inclusion:

$$-\text{div}(a(|\nabla u|)\nabla u) + \partial_u G(x,u) \ni 0 \quad \text{in } \Omega, \tag{77}$$

with boundary condition

$$u = 0 \quad \text{on } \partial\Omega. \tag{78}$$

We are interested here in the case where the function $a(t)t$ has slow growth and $G(x,u)$ is a Carathéodory function that is locally Lipschitz in u and the lower order term $\partial_u G(x,u)$ in (77) is the generalized gradient of G with respect to u (cf. [**12**]). As above, the principal (higher order) part of the equation is represented by the function $\phi : t \mapsto a(t)t$, $t \in \mathbb{R}$, which is supposed to be increasing, continuous, odd, and vanishing at 0.

Let Φ denote the antiderivative of ϕ, $\Phi(t) = \int_0^t \phi(s)ds$ $(t \in \mathbb{R})$. The classical case $\Phi(t) = t^2$ corresponds to the semilinear Laplace inclusion. We are concerned here with the situation where Φ is growing very slowly, that is, $\Phi(t) = o(t^p)$ as $t \to \infty$ for all $p > 1$. Since the Hölder conjugate $\overline{\Phi}$ of Φ does not satisfy a Δ_2 condition (see Section 1.2 for more details on $\overline{\Phi}$ and Δ_2 condition), the functional $u \mapsto \int_\Omega \Phi(|\nabla u|)\,dx$, does not belong to class C^1. Moreover, the integral given by the lower order term is not differentiable in general. Therefore, problem (77)–(78) is formulated, in the weak form, not as a variational equation but naturally as a variational-hemivariational inequality in an Orlicz–Sobolev space.

To study the existence of nontrivial solutions, we shall use a version of the Mountain Pass theorem for variational-hemivariational inequalities. Note that in the case both Φ and $\overline{\Phi}$ satisfy Δ_2 conditions, we could prove a compactness condition for equations in $W_0^1 L_\Phi$, which implies the Palais–Smale (PS) condition (cf. [**44**]). However, there has not been proved a similar result in Orlicz–Sobolev spaces when either Φ or $\overline{\Phi}$ fails to satisfy this condition. We could in fact prove the existence and boundedness of Palais–Smale sequences $\{u_n\}$ of the variational-hemivariational inequality associated with problem (77)–(78). However, the convergence of the integrals $\{\int_\Omega \Phi(|\nabla u_n|)dx\}$ is, in our case, not strong enough to allow us to conclude the strong convergence of a subsequence of $\{u_n\}$ in Orlicz–Sobolev spaces. Therefore, we need here a version of the Mountain Pass theorem for variational-hemivariational inequalities without the (PS) condition.

In this section, we establish a general linking theorem for variational - hemivariational inequalities without the (PS) condition whose corollary, a Mountain Pass theorem for variational-hemivariational inequalities, will be needed for our investigation of problem (77)–(78). This abstract version of Mountain Pass theorem is next applied to prove the existence of nontrivial solutions of the variational-hemivariational inequality that formulates (77)-(78). We also observe that the arguments in our case also apply to inequalities in which the principal operators have "not very fast" growth.

3.2. Problem setting. The inclusion (77)-(78) can be formulated (in the weak form) as the inequality

$$\langle J'(u), v\rangle + \int_\Omega G^o(x, u; v)\, dx \geq 0,$$

for all test functions v, where J is the potential functional associated with the principal part:

$$J(u) = \int_\Omega \Phi(|\nabla u|)\, dx, \tag{79}$$

J' is the Gâteaux derivative of J, $\langle \cdot, \cdot \rangle$ is the pairing between the space of test functions and its dual, and $G^o(x, u; v)$ stands for the generalized directional derivative of G (with respect to u) in the direction of v. Since the growth of the principal term is represented by Φ, we choose the function space for the solutions and test functions as the first-order Orlicz–Sobolev space $W_0^1 L_\Phi$. In this space, we write the above inequality as the following variational-hemivariational inequality:

$$\begin{cases} J(v) - J(u) + \displaystyle\int_\Omega G^o(x, u; v - u)\, dx \geq 0, \ \forall v \in W_0^1 L_\Phi \\ u \in W_0^1 L_\Phi. \end{cases} \tag{80}$$

We are interested here in problems with principal terms given by a Young function Φ growing more slowly than any power $t^{1+\varepsilon}$ ($\varepsilon > 0$), that is

$$\Phi(t) \ll t^{1+\varepsilon}, \quad \forall \varepsilon > 0.$$

Typical examples of such Young functions are

$$\Phi(t) = \int_0^{|t|} [\ln(1+s)]^\beta\, ds \ (t \in \mathbb{R}) \tag{81}$$

($\Phi(t) = (|t| + 1)\ln(|t| + 1) - |t|$, when $\beta = 1$), or

$$\Phi(t) = |t|[\ln(|t| + 1)]^\beta \ (t \in \mathbb{R}), \tag{82}$$

(with β being a fixed positive constant). It is easy to check that in both cases

$$\lim_{t\to\infty} \frac{t\Phi'(t)}{\Phi(t)} = 1. \tag{83}$$

For such functions, their conjugates $\overline{\Phi}$ do not satisfy a Δ_2 condition. We refer to [**38**] (or [**41, 1**]) for basic properties of Δ_2 condition. We assume in this section that Φ satisfies the growth condition determined by (83).

3.3. Linking and Mountain Pass theorems. We shall need here a version of the Mountain Pass theorem for variational-hemivariational inequalities without the Palais–Smale condition. Since this compactness condition is not imposed, we obtain, instead of the existence of critical points of the associated functionals, only that of (PS) sequences in the sense of (89) and (90) below. Note that in the problem we are interested in here, the functional J is convex and finite everywhere, hence locally Lipschitz on the underlying Banach space where J is defined. We state the theorem for the general case of sums of convex, lower semicontinuous and locally Lipschitz functionals, due to its own interest and applicability in other situations as well.

Furthermore, we shall first establish a more general linking theorem which contains the needed Mountain Pass theorem as a particular case. Let us start with the definition of linking that we are interested in.

DEFINITION 25. (Definition 3.3, [**63**]) Let S be a nonempty closed subset of a Banach space X and let Q be a compact topological submanifold of X with nonempty boundary ∂Q (in the sense of manifolds with boundary). We say that S and Q link if the next properties hold:

$$S \cap \partial Q = \emptyset, \tag{84}$$

and

$$f(Q) \cap S \neq \emptyset \tag{85}$$

whenever $f \in \Gamma$, where

$$\Gamma = \{f \in C(Q, X) : f|_{\partial Q} = id_{\partial Q}\}. \tag{86}$$

We are now ready to state the following general minimax theorem with the above type of linking for functionals which are sums of convex and locally Lipschitz ones.

THEOREM 26. *Let the functional $I : X \to \mathbb{R} \cup \{+\infty\}$ on the Banach space $(X, \|\cdot\|)$ satisfy the following assumption:*
(H) *$I = P + \psi$, where $P : X \to \mathbb{R}$ is a locally Lipschitz functional and $\psi : X \to \mathbb{R} \cup \{+\infty\}$ is convex, proper, and lower semicontinuous.*

Let the closed sets S and Q link in X in the sense of Definition 25. Assume

$$\sup_{x\in Q} I(x) \in \mathbb{R},\ a := \sup_{x\in \partial Q} I(x) < b := \inf_{x\in S} I(x). \tag{87}$$

Then the number

$$c = \inf_{f\in\Gamma} \sup_{x\in Q} I(f(x)), \tag{88}$$

where Γ is given by (86), satisfies the following property: There exist sequences $\{u_n\}$ in X and $\{\epsilon_n\}$ in $(0, +\infty)$ such that

$$\varepsilon_n \downarrow 0,\ I(u_n) \to c \ \text{ as } n \to \infty, \tag{89}$$

and

$$P^o(u_n; v - u_n) + \psi(v) - \psi(u_n) \geq -\varepsilon_n \|v - u_n\|, \quad \forall v \in X, \ \forall n \in \mathbb{N}. \tag{90}$$

We refer to [**52**] for a detailed proof of this theorem. As a particular case, we have the following Mountain Pass theorem for variational-hemivariational inequalities without assuming the (PS) condition, which is formulated in the setting of hypothesis (H) above. This abstract result will be used in our variational approach for studying problem (77)–(78).

COROLLARY 27. *Let* $(X, \|\cdot\|)$ *be a Banach space and* $I : X \to \mathbb{R} \cup \{+\infty\}$ *satisfy assumption* (H). *Suppose that*

(i) $I(0) = 0$ *and there exist* $\beta, \rho > 0$ *such that*

$$I(u) \geq \beta, \quad \forall u \in X, \ \|u\| = \rho, \tag{91}$$

(ii) There exists $e \in X$ *such that*

$$\|e\| > \rho \quad \text{and} \quad I(e) \leq 0. \tag{92}$$

Let

$$c = \inf_{f \in \Gamma} \sup_{t \in [0,1]} I(f(t)) \ (\geq \beta), \tag{93}$$

where

$$\Gamma = \{f \in C([0,1], X) : f(0) = 0, f(1) = e\}. \tag{94}$$

Then, there exist sequences $\{u_n\} \subset X$ *and* $\{\varepsilon_n\} \subset (0, +\infty)$ *that satisfy (89) and (90).*

3.4. Existence of nontrivial solutions via Mountain Pass theorem. Let us consider now the necessary assumptions on G and Φ. First, assume that $G : \Omega \times \mathbb{R} \to \mathbb{R}$ is a Carathéodory function such that $G(x, 0) = 0$ and $G(x, \cdot) : \mathbb{R} \to \mathbb{R}$ is locally Lipschitz for almost all $x \in \Omega$. Furthermore, the generalized gradient of $G(x, \cdot)$ has the following growth condition:

$$|\xi| \leq a_1 + a_2 |s|^{\alpha - 1}, \tag{95}$$

for a.e. $x \in \Omega$, all $s \in \mathbb{R}$, all $\xi \in \partial_s G(x, s)$, where $1 < \alpha \leq \frac{N}{N-1}$ (subcritical condition). From the imbeddings between Orlicz and Orlicz–Sobolev spaces in Section 3.2 with $\Psi = \Phi_\alpha$ ($\Phi_\alpha(t) = t^\alpha$, $\forall t \geq 0$), we have the compact (hence continuous) embedding

$$W^1 L_\Phi \hookrightarrow L^\alpha(\Omega). \tag{96}$$

On the other hand, we refer to [**12**] for the basic concepts and results about the nonsmooth analysis of locally Lipschitz functionals.

Next, let us assume the following behavior of $G(x, t)$ when t is small:

$$\liminf_{t \to 0} \frac{G(x, t)}{\Phi(t)} > -\Lambda \text{ uniformly for almost all } x \in \Omega, \tag{97}$$

where

$$\Lambda = \inf_{u \in W_0^1 L_\Phi \setminus \{0\}} \frac{\int_\Omega \Phi(|\nabla u|)\, dx}{\int_\Omega \Phi(|u|)\, dx}.$$

Note that $\Lambda > 0$ (see e.g. [**32**]) and from its definition, we have

$$\int_\Omega \Phi(|\nabla u|)\, dx \geq \Lambda \int_\Omega \Phi(|u|)\, dx, \ \forall u \in W_0^1 L_\Phi. \tag{98}$$

We also suppose that there exist $t_1 > 0$ and $\gamma > 1$ such that

$$\sup_{x \in \Omega} G(x, t_1) < 0 \tag{99}$$

and

$$G^o(x, t; t) \leq \gamma G(x, t) \ \text{ for a.e. } x \in \Omega, \text{ all } t \text{ with } |t| \geq t_1. \tag{100}$$

It follows from this assumption that there exist $a_3, a_4 > 0$ such that

$$G(x, t) \leq -a_3 t^\gamma + a_4 t \ \text{ for a.e. } x \in \Omega, \text{ all } t \geq 0. \tag{101}$$

In fact, for $x \in \Omega$, $s \in \mathbb{R}$, and $t > 0$, we have the following formula

$$\partial_t[G(x,t)t^{-\gamma}] = \gamma t^{-1-\gamma}[\gamma^{-1} t \partial_t G(x,t) - G(x,t)].$$

Thus, for $t > t_1$, from Lebourg's theorem (cf. [**57**]), there exists $\tilde{t} \in (t_1, t)$ such that

$$\begin{aligned}
\left(\frac{t}{t_1}\right)^{-\gamma} G(x,t) - G(x,t_1) &\in \partial_\tau \left[\left(\frac{\tau}{t_1}\right)^{-\gamma} G(x,\tau) \right]_{\tau = \tilde{t}} (t - t_1) \\
&= t_1^\gamma \partial_\tau \left[\tau^{-\gamma} G(x,\tau)\right]_{\tau = \tilde{t}} (t - t_1) \\
&= \gamma t_1^\gamma \tilde{t}^{-1-\gamma} \left[\gamma^{-1} \tilde{t} \partial_t G(x, \tilde{t}) - G(x, \tilde{t})\right] (t - t_1).
\end{aligned}$$

Hence, from assumption (100),

$$\begin{aligned}
\left(\frac{t}{t_1}\right)^{-\gamma} G(x,t) - G(x,t_1) &\leq \gamma t_1^\gamma \tilde{t}^{-1-\gamma} [\gamma^{-1} G^o(x, \tilde{t}; \tilde{t}) - G(x, \tilde{t})](t - t_1) \\
&\leq 0.
\end{aligned}$$

Therefore,

$$G(x,t) \leq \frac{G(x,t_1)}{t_1^\gamma} t^\gamma \leq -a_3 t^\gamma \ \text{ for a.e. } x \in \Omega, \quad \text{all } t \geq t_1, \tag{102}$$

where $a_3 = -t_1^{-\gamma} \sup_{x \in \Omega} G(x, t_1) > 0$.

On the other hand, it follows from (95) and Lebourg's theorem that for a.e. $x \in \Omega$, all $t \geq 0$,

$$\begin{aligned}
|G(x,t)| &= |G(x,t) - G(x,0)| \\
&= |\xi t| \ \text{ for some } \xi \in \partial_s G(x,s), \ 0 \leq s \leq t, \\
&\leq t(a_1 + a_2 s^{\alpha - 1}) \\
&\leq (a_1 + a_2 t^{\alpha - 1}) t.
\end{aligned}$$

Thus,

$$|G(x,t)| \le (a_1 + a_2 t_1^{\alpha-1})t, \tag{103}$$

for a.e. $x \in \Omega$, all $t \in [0,t_1]$. Choosing $a_4 = a_1 + a_2 t_1^{\alpha-1} + a_3 t_1^{\gamma-1} (> 0)$, we see that

$$|G(x,t)| \le a_4 t - a_3 t^\gamma \ \text{ for a.e. } x \in \Omega, \quad \text{all } t \in [0,t_1].$$

Combining this estimate with (102), we obtain (101).

Concerning Φ, we assume that

$$\Phi \text{ satisfies a } \Delta_2 \text{ condition (on } \mathbb{R}) \tag{104}$$

and

$$k \ge \sup_{t>0} \frac{\Phi(2t)}{\Phi(t)} (< \infty). \tag{105}$$

Moreover, suppose that

$$k_1 = \frac{\ln k}{\ln 2} < \alpha \tag{106}$$

(or equivalently, $k < 2^\alpha$).

The following theorem is our main existence results for nontrivial solutions of the inequality (80) (or the inclusion (77)–(78)).

THEOREM 28. *Suppose Φ is a Young function satisfying a Δ_2 condition and the growth condition (83). Assume G satisfies (95), (99), and (100). Let k be given by (105) and assume that (97) and (106) hold.*

Under these assumptions, the variational-hemivariational inequality (80) (with J given by (79)) has a nontrivial solution.

The proof of this result is an application of the Mountain Pass theorem (Corollary 27) stated in Section 3.2. We first recall the following lemma concerning an estimate of $\|u\|_\Phi$.

LEMMA 29 (Lemma 3, [**47**]). *Assume Φ is a Young function that satisfies a Δ_2 condition on $\mathbb{R}$, that is*

$$\Phi(2t) \le k\Phi(t), \ \forall t > 0, \tag{107}$$

for some $k > 1$. Then, for each $R > 0$, there is $c > 0$ such that for all u with $\|u\|_\Phi \le R$ or $\int_\Omega \Phi(u)\,dx \le R$, we have

$$\int_\Omega \Phi(u)\,dx \ge c\|u\|_\Phi^{\ln k/\ln 2}. \tag{108}$$

PROOF OF THEOREM 28. We consider the Banach space $X = W_0^1 L_\Phi$ with norm $\|\cdot\| = \|\cdot\|_{W_0^1 L_\Phi}$. Let us define the functional $I : W_0^1 L_\Phi \to \mathbb{R}$ by $I = J + P$, where J is given by (79) and

$$P(u) = \int_\Omega G(x,u)\,dx, \quad u \in W_0^1 L_\Phi. \tag{109}$$

It follows from (95) and Lebourg's theorem (see also (103)) that G has the growth

$$|G(x,s)| \le a_5 + a_6|s|^\alpha \text{ for a.e. } x \in \Omega, \text{ all } s \in \mathbb{R}. \tag{110}$$

Therefore, from the locally Lipschitz property of $G(x,\cdot)$, we see that P is locally Lipschitz on $L^\alpha(\Omega)$ and thus on $W_0^1 L_\Phi$, because of the embedding (96). We show that under the assumptions of Theorem 28, all the assumption of the Mountain Pass theorem in Section 3.3 (Corollary 27) are fulfilled.

Let us check the first condition (91) in Corollary 27. From (97), there are $s_1 \in (0,\infty)$ and $\varepsilon_1 \in (0,\Lambda)$ such that

$$G(x,s) \ge (-\Lambda + \varepsilon_1)\Phi(s) \tag{111}$$

for a.e. $x \in \Omega$, all s with $|s| < s_1$. From (110), there exists $a_7 > 0$ such that

$$|G(x,s)| \le a_7|s|^\alpha \tag{112}$$

for a.e. $x \in \Omega$, all s with $|s| \ge s_1$. Combining (111) and (112), we have

$$G(x,s) \ge (-\Lambda + \varepsilon_1)\Phi(s) - a_7|s|^\alpha, \tag{113}$$

for a.e. $x \in \Omega$, all $s \in \mathbb{R}$. Therefore,

$$\begin{aligned}\int_\Omega G(x,u)\,dx &\ge (-\Lambda + \varepsilon_1)\int_\Omega \Phi(u)\,dx - a_7\int_\Omega |u|^\alpha\,dx \\ &\ge \frac{-\Lambda+\varepsilon_1}{\Lambda}\int_\Omega \Phi(|\nabla u|)\,dx - a_7\int_\Omega |u|^\alpha\,dx,\end{aligned} \tag{114}$$

for all $u \in W_0^1 L_\Phi$. For simplicity, if there is no confusion, we shall in the sequel use C to denote a generic positive constant. By means of the continuous embedding from $W_0^1 L_\Phi$ into $L^\alpha(\Omega)$ and Lemma 29 with $R = 1$, we see that there is a constant $C > 0$ such that

$$\int_\Omega |u|^\alpha\,dx \le C\|u\|^\alpha_{W_0^1 L_\Phi} \le C\left[\int_\Omega \Phi(|\nabla u|)\,dx\right]^{\frac{\ln 2}{\ln k}\alpha}, \tag{115}$$

for all $u \in W_0^1 L_\Phi$ with $\|u\|_{W_0^1 L_\Phi} = r < 1$. Combining (114) with (115) yields the following estimate for $I(u)$:

$$\begin{aligned}I(u) &\ge \int_\Omega \Phi(|\nabla u|)\,dx - \frac{\Lambda - \epsilon_1}{\Lambda}\int_\Omega \Phi(|\nabla u|)\,dx - a_7\int_\Omega |u|^\alpha\,dx \\ &\ge \frac{\epsilon_1}{\Lambda}\int_\Omega \Phi(|\nabla u|)\,dx - a_7 C\left[\int_\Omega \Phi(|\nabla u|)\,dx\right]^{\frac{\ln 2}{\ln k}\alpha} \\ &= \left\{\frac{\epsilon_1}{\Lambda} - a_7 C\left[\int_\Omega \Phi(|\nabla u|)\,dx\right]^{\frac{\ln 2}{\ln k}\alpha - 1}\right\}\int_\Omega \Phi(|\nabla u|)\,dx.\end{aligned} \tag{116}$$

Since

$$\lim_{\|u\|\to 0}\left[\int_\Omega \Phi(|\nabla u|)\,dx\right]^{\frac{\ln 2}{\ln k}\alpha - 1} = 0$$

(note that $\dfrac{\ln 2}{\ln k}\alpha - 1 > 0$), we have, by choosing $r > 0$, sufficiently small, the following estimate:

$$a_7 C \left[\int_\Omega \Phi(|\nabla u|)\,dx\right]^{\frac{\ln 2}{\ln k}\alpha - 1} \le \frac{\epsilon_1}{2\Lambda},$$

for all $u \in W_0^1 L_\Phi$ with $\|u\| = r$. From (116) and (108) (with u replaced by $|\nabla u|$) ,

$$I(u) \ge \frac{\epsilon_1}{2\Lambda}\int_\Omega \Phi(|\nabla u|)\,dx \ge \frac{\epsilon_1}{2\Lambda} C \|u\|_{W_0^1 L_\Phi}^{\frac{\ln k}{\ln 2}} = \frac{C\epsilon_1}{2\Lambda} r^{\frac{\ln k}{\ln 2}} > 0,$$

for all $u \in W_0^1 L_\Phi$ with $\|u\| = r$. We have checked (91), so (i) of Corollary 27 holds here.

Now, let us check condition (ii) in Corollary 27. Let us fix a number $\gamma_0 \in (1, \gamma)$ with $\gamma > 1$ given in (100). From (83), there exists $T_0 > 0$ such that

$$\frac{t\Phi'(t)}{\Phi(t)} \le \gamma_0, \quad \forall t \ge T_0. \tag{117}$$

Hence,

$$\ln\left(\frac{\Phi(t)}{\Phi(T_0)}\right) = \int_{T_0}^{t} \frac{\Phi'(s)}{\Phi(s)}\,ds \le \int_{T_0}^{t} \frac{\gamma_0}{s}\,ds = \ln\left(\frac{t^{\gamma_0}}{T_0^{\gamma_0}}\right), \quad \forall t \ge T_0, \tag{118}$$

implying that

$$\Phi(t) \le \frac{\Phi(T_0)}{T_0^{\gamma_0}} t^{\gamma_0}, \quad \forall t \ge T_0.$$

Therefore, for some constants $C_1, C_2 > 0$ we have

$$\Phi(t) \le C_1 t^{\gamma_0} + C_2, \quad \forall t \ge 0. \tag{119}$$

Let us fix $\phi_0 \in C_0^1(\Omega)$ such that $\phi_0 \ge 0$ on Ω and $\phi_0 \ne 0$. For $\lambda > 0$, let $u = u_\lambda = \lambda\phi_0 (\ge 0)$. The following estimates are a consequence of (119) and (101)

(120)

$$\begin{aligned} I(u) &\le \int_\Omega (C_1|\nabla u|^{\gamma_0} + C_2)\,dx + \int_\Omega (-a_3 u^\gamma + a_4 u)\,dx \\ &= -\lambda^\gamma \left(a_3 \int_\Omega \phi_0^\gamma\,dx - C_1\lambda^{\gamma_0-\gamma}\int_\Omega |\nabla\phi_0|^{\gamma_0}\,dx - C_2|\Omega|\lambda^{-\gamma} - a_4\lambda^{-\gamma}\int_\Omega \phi_0\,dx\right). \end{aligned}$$

As $\lambda \to \infty$, $\lambda^{\gamma_0-\gamma}, \lambda^{-\gamma} \to 0$ and the number in the parentheses tends to $a_3 \int_\Omega \phi_0^\gamma$ $dx (> 0)$.

Hence, the right hand side of (120) tends to $-\infty$ as $\lambda \to \infty$. For $\lambda > 0$ sufficiently large, $I(u_\lambda) < 0$ and u_λ is outside the ball centered at 0 with radius r.

We have checked both conditions in Corollary 27. By that result, there exist a sequence $\{u_n\}$ in $W_0^1 L_\Phi$ and a sequence $\{\varepsilon_n\}$ in $(0, \infty)$ such that $\varepsilon_n \downarrow 0$, $I(u_n) \to c$ (c is given in (93)) and for every $n \in \mathbb{N}$,

$$\int_\Omega \Phi(|\nabla v|)\,dx - \int_\Omega \Phi(|\nabla u_n|)\,dx + P^o(u_n; v - u_n) \ge -\varepsilon_n \|v - u_n\|,$$

for all $v \in W_0^1 L_\Phi$. From Aubin–Clarke's theorem (cf. [**12**]), we have that

$$P^o(u;v) \le \int_\Omega G^o(x,u(x);v(x))dx, \quad \forall u,\ v \in W_0^1 L_\Phi.$$

Hence,

$$\begin{aligned}\int_\Omega \Phi(|\nabla v|)\,dx - \int_\Omega \Phi(|\nabla u_n|)\,dx + \int_\Omega G^o(x,u_n(x);v(x)-u_n(x))\,dx \\ \ge -\varepsilon_n\|v-u_n\|, \forall v \in W_0^1 L_\Phi.\end{aligned} \tag{121}$$

We show that the sequence $\{u_n\}$ is bounded in $W_0^1 L_\Phi$. In fact, because $\gamma > 1$, we can choose $\gamma_0 \in (1,\gamma)$ sufficiently close to 1 such that $2^{\gamma_0} < \gamma + 1$. By using calculations as in (118), we have for all $t \ge T_0$,

$$\ln\left(\frac{\Phi(2t)}{\Phi(t)}\right) = \int_t^{2t} \frac{\Phi'(s)}{\Phi(s)}ds \le \int_t^{2t} \frac{\gamma_0}{s}ds = \gamma_0 \ln 2.$$

Hence, $\Phi(2t) \le 2^{\gamma_0}\Phi(t)$, $\forall t \ge T_0$, and thus

$$\Phi(2t) \le 2^{\gamma_0}\Phi(t) + C_3, \quad \forall t \in \mathbb{R}, \tag{122}$$

with $C_3 = \sup\{\Phi(2t) : 0 \le t \le T_0\} \in (0,\infty)$. Hence,

$$\int_\Omega \Phi(2|\nabla u|)\,dx \le 2^{\gamma_0}\int_\Omega \Phi(|\nabla u|)\,dx + C_3|\Omega|, \quad \forall u \in W_0^1 L_\Phi. \tag{123}$$

Letting $v = 2u_n$ in (121) and using (123), one gets

$$\begin{aligned}(2^{\gamma_0}-1)\int_\Omega \Phi(|\nabla u_n|)\,dx + C_3|\Omega| &\ge \int_\Omega \Phi(2|\nabla u_n|)\,dx - \int_\Omega \Phi(|\nabla u_n|)\,dx \\ &\ge -\int_\Omega G^o(x,u_n(x);u_n(x))\,dx - \varepsilon_n\|u_n\|.\end{aligned} \tag{124}$$

Without loss of generality, we may only consider n such that $\|u_n\| > 1$. It follows from Theorem 9.5, Chapter 2, [**38**], that

$$\int_\Omega \Phi(|\nabla u_n|)\,dx \ge \||\nabla u_n|\|_\Phi = \|u_n\|. \tag{125}$$

From (100) and (95), there exists a constant $a_8 > 0$ such that

$$G^o(x,t;t) \le \gamma G(x,t) + a_8, \quad \forall t \in \mathbb{R}. \tag{126}$$

From (124)–(126),

$$\begin{aligned}(2^{\gamma_0}-1)\int_\Omega \Phi(|\nabla u_n|)\,dx + c_3|\Omega| \\ \ge -\gamma\int_\Omega G(x,u_n)\,dx - a_8|\Omega| - \varepsilon_n\int_\Omega \Phi(|\nabla u_n|)\,dx.\end{aligned} \tag{127}$$

On the other hand, since $I(u_n) \to c$,

$$\int_\Omega \Phi(|\nabla u_n|)\,dx + \int_\Omega G(x,u_n)\,dx = c + \delta_n, \tag{128}$$

with $\delta_n \to 0$ as $n \to \infty$.

It follows from (127) and (128) that

$$(2^{\gamma_0} - 1 + \varepsilon_n)\int_\Omega \Phi(|\nabla u_n|)\,dx \ge \gamma \int_\Omega \Phi(|\nabla u_n|)\,dx - \gamma(c+\delta_n) - (a_8 + c_3)|\Omega|.$$

It follows that

$$(\gamma + 1 - 2^{\gamma_0} - \varepsilon_n)\int_\Omega \Phi(|\nabla u_n|)\,dx \le \gamma(c + \sup \delta_n) + (a_8 + c_3)|\Omega| < \infty, \tag{129}$$

for all $n \in \mathbb{N}$. On the other hand, by the choice of γ_0, because $\varepsilon_n \to 0$ as $n \to \infty$, we have $\gamma + 1 - 2^{\gamma_0} - \varepsilon_n \ge \varepsilon_0 > 0$, for all n large. Estimate (129) shows that the sequence $\left\{\int_\Omega \Phi(|\nabla u_n|)\,dx\right\}$ is bounded, implying the boundedness of $\{u_n\}$ in $W_0^1 L_\Phi$ (see (125)).

Since $\{u_n\}$ is bounded in $W_0^1 L_\Phi$, by passing to a subsequence, if necessary, we can assume that

$$u_n \rightharpoonup^* u \quad \text{in} \quad W_0^1 L_\Phi. \tag{130}$$

This implies that $\int_\Omega u_n \phi\,dx \to \int_\Omega u\phi\,dx$ and $\int_\Omega \partial_i u_n \phi\,dx \to \int_\Omega \partial_i u\phi\,dx$, for all $\phi \in E_{\overline{\Phi}}$, and thus for all $\phi \in L^\infty(\Omega)$, i.e.,

$$\nabla u_n \rightharpoonup \nabla u \quad \text{in } [L^1(\Omega)]^N (\text{-weak}). \tag{131}$$

Since the function $\xi \mapsto \Phi(|\xi|)$ is convex, continuous on $\mathbb{R}^N$ and $\Phi(|\xi|) \ge 0$, $\forall \xi \in \mathbb{R}^N$, we have

$$\int_\Omega \Phi(|\nabla u|)\,dx \le \liminf \int_\Omega \Phi(|\nabla u_n|)\,dx, \tag{132}$$

(cf. e.g. [**5**]). Put $\Phi_1^*(t) = |t|^{N/(N-1)}$, $t \in \mathbb{R}$. (We assume here that $N > 1$, thus $p^* = \frac{N}{N-1}$ is the Sobolev conjugate exponent of $p = 1$ and Φ_1^* is a Young function; trivial modifications are needed for the case $N = 1$.) Straightforward calculations show that $\Phi_1^* \ll \Phi^*$ (Φ^* is the Sobolev conjugate of Φ). Consequently, the embedding $W_0^1 L_\Phi \hookrightarrow L_{\Phi_1^*} (= L^{N/(N-1)}(\Omega))$ is compact. It follows that $u_n \to u$ in $L^{N/(N-1)}(\Omega)$ and thus in $L^\alpha(\Omega)$. From the growth condition (95) and Fatou's lemma, we see that

$$\limsup \int_\Omega G^o(x, u_n(x); v(x) - u_n(x))\,dx \le \int_\Omega G^o(x, u(x); v(x) - u(x))\,dx. \tag{133}$$

Also, since the sequence $\{u_n\}$ is bounded in $W_0^1 L_\Phi$, we have

$$\varepsilon_n \|v - u_n\| \to 0. \tag{134}$$

Letting $n \to \infty$ in (121) and noting (132), (133), and (134), we have that u is a solution of (80). Note that $u \ne 0$. In fact, suppose by contradiction that $u = 0$.

Letting $v = 0$ in (121), one gets

$$\int_\Omega \Phi(|\nabla u_n|)\,dx \le \int_\Omega G^o(x, u_n; -u_n)\,dx + \varepsilon_n \|u_n\|, \quad \forall n \in \mathbb{N}.$$

From (133), we have $\lim_{n\to\infty} \int_\Omega \Phi(|\nabla u_n|)\,dx = 0$. Also, $G(\cdot, u_n) \to G(\cdot, 0) = 0$ in $L^1(\Omega)$ and thus $\int_\Omega G(x, u_n)\,dx \to 0$. Hence,

$$I(u_n) = \int_\Omega \Phi(|\nabla u_n|)\,dx + \int_\Omega G(x, u_n)\,dx \to 0$$

as $n \to \infty$. This contradicts (89) and (93) and completes our proof. □

Let us conclude this section with some further remarks. Related to the assumptions on Φ, we note that with Φ given by (81) or (82), since the function $t\Phi'(t)/\Phi(t)$ is decreasing on $(0, \infty)$, we have

$$\sup_{t>0} \frac{t\Phi'(t)}{\Phi(t)} = \lim_{t\to 0} \frac{t\Phi'(t)}{\Phi(t)} = 1 + \beta = k_1. \tag{135}$$

Hence, one has (104) and (105) with

$$k = 2^{1+\beta}. \tag{136}$$

If Φ is given by (81) or (82), then (106) holds if

$$1 + \beta < \alpha. \tag{137}$$

Using similar arguments as in [**47**], we can extend the above results and assumptions to variational-hemivariational inequalities that contains locally Lipschitz lower order terms and principal terms defined by Young functions Φ with "not very fast" growth (i.e. when Φ satisfies a Δ_2 condition, but $\overline{\Phi}$ may or may not satisfy this condition). Examples of such functions are

$$\Phi(t) = \int_0^{|t|} s^{p-1}[\ln(1+s)]^\beta\,ds,$$

or

$$\Phi(t) = |t|^p[\ln(1+|t|)]^\beta, \quad t \in \mathbb{R},$$

with $\beta > 0, p \ge 1$. Since these functions Φ are not equivalent to any power functions t^p $(p \in [1, \infty])$, the regular setting in ordinary Sobolev spaces is not suitable for such inequalities.

4. Existence results and properties of solutions in quasilinear variational inequalities with general growth

4.1. Introduction - problem setting. In this section, we study the following boundary value problem

$$\begin{cases} -\operatorname{div}(a(|\nabla u|)\nabla u) = f(x, u) & \text{in } \Omega \\ u = 0 & \text{on } \partial\Omega. \end{cases} \tag{138}$$

Our main interest here is that the principal operator, given by the function $\phi(t) = a(t)t$, can have general growth (either very fast or very slow or any growth in between) in both coercive and noncoercive cases. This fact is reflected in the assumption that the anti-derivative Φ of ϕ,

$$\Phi(t) = \int_0^t \phi(s)dx, \quad t \in \mathbb{R}, \tag{139}$$

is an arbitrary N-function (cf. [**1, 38**]). In this section, we study problem (138) with ϕ having general growth. Note that in a related work (cf. [**51**]), we proposed a sub-supersolution approach to investigate an equation related to a sandpile problem where the principal operator had very fast (exponential) growth.

A weak formulation of (138) is the following variational equation:

$$\int_\Omega a(|\nabla u|)\nabla u \cdot \nabla v dx = \int_\Omega f(x,u)v dx, \tag{140}$$

for all v in some appropriate function space. A natural choice for this function space is the first order Orlicz–Sobolev space $W_0^1 L_\Phi$. However, the operator

$$v \mapsto \int_\Omega a(|\nabla u|)\nabla u \cdot \nabla v dx, \quad v \in W_0^1 L_\Phi,$$

is not defined for every $u \in W_0^1 L_\Phi$. Therefore, one approach is to require that u belong to a subset $D(\Delta_\Phi)$ of $W_0^1 L_\Phi$, usually defined by

$$D(\Delta_\Phi) = \{u \in W_0^1 L_\Phi : \phi(|\nabla u|) \in L_{\overline{\Phi}}\},$$

where $\overline{\Phi}$ is the Hölder conjugate function of Φ, defined in Section 4.2 (cf. e.g. [**32, 34, 64**]). Another approach, which we follow here, is based on the observation that the functional

$$J : u \mapsto \int_\Omega \Phi(|\nabla u|)dx,$$

is convex from $W_0^1 L_\Phi$ into $[0, +\infty]$, and (at least formally or when u and v are sufficiently smooth)

$$\langle J'(u), v\rangle = \int_\Omega a(|\nabla u|)\nabla u \cdot \nabla v dx, \tag{141}$$

(J' is the Fréchet derivative of J and $\langle \cdot, \cdot \rangle$ stands for the dual pairing between $W_0^1 L_\Phi$ and its dual). However, since J is not differentiable (not even finite) on the entire space $W_0^1 L_\Phi$, a natural extension of (140) in view of (141) is the inclusion

$$\tilde{f}(u) \in \partial J(u), \tag{142}$$

where $\tilde{f}(u) \in (W_0^1 L_\Phi)^*$ is the Niemitskii functional associated with $f(x,u)$, $\langle \tilde{f}(u), v\rangle = \int_\Omega f(x,u)v dx$ for $u, v \in W_0^1 L_\Phi$, and $\partial J(u)$ is the subdifferential of J at u in the sense of convex analysis, $\partial J(u) = \{l \in (W_0^1 L_\Phi)^* : \langle l, v-u\rangle \le J(v) - J(u), \forall v \in W_0^1(\Omega)\}$. Therefore, an alternate (weak) formulation of (138) without the restriction to $D(\Delta_\Phi)$ is the following variational inequality

$$(143)\quad \begin{cases} \int_\Omega \Phi(|\nabla v|)dx - \int_\Omega \Phi(|\nabla u|)dx \geq \int_\Omega f(x,u)(v-u)dx, \ \forall v \in W_0^1 L_\Phi \\ u \in W_0^1 L_\Phi. \end{cases}$$

In the case there are constraints (such as the presence of obstacles or unilateral conditions) the problem is formulated as the following more general variational inequality:

$$(144)\quad \begin{cases} \int_\Omega \Phi(|\nabla v|)dx - \int_\Omega \Phi(|\nabla u|)dx \geq \int_\Omega f(x,u)(v-u)dx, \ \forall v \in K \\ u \in K, \end{cases}$$

where K is the set of admissible functions reflecting the imposed constraints. In what follows, we always assume that K is convex and (sequentially) closed with respect to the weak* topology in $W_0^1 L_\Phi$. For this variational inequality to be properly defined, we assume that

$$(145)\quad K \cap D_0 \neq \emptyset,$$

where D_0 is the effective domain of J in $W_0^1 L_\Phi$,

$$D_0 = \left\{ u \in W_0^1 L_\Phi : \int_\Omega \Phi(|\nabla u|)dx < \infty \right\}.$$

In this section, we consider the existence and certain properties of solutions of the inequality (144) in the noncoercive case. The existence results in the coercive case, although expected and are a preparation for noncoercive problems, seem new and have their own interest. Note that a problem is coercive when the principal term "dominates," or has faster growth than, the lower order term in certain sense (see Theorem 30 and conditions (150) and (151) below). When such a condition is not fulfilled the problem is generally called noncoercive. We define here appropriate concepts of subsolutions and supersolutions for (144) and prove, in the noncoercive case, the existence of solutions between sub- and supersolutions. We also investigate some properties of solution sets between sub- and supersolutions such as their compactness (in certain topologies), their directedness, and the existence of extremal solutions.

4.2. Existence of solutions in coercive problems. Let $f : \Omega \times \mathbb{R} \to \mathbb{R}$ be a Carathéodory function with the growth condition:

$$(146)\quad |f(x,u)| \leq h(x) + c_1 \overline{\Psi}^{-1} \circ \Psi(c_2 u) \ \text{ for a.e. } \ x \in \Omega, \text{ all } u \in \mathbb{R},$$

where $c_1, c_2 \geq 0$ and Ψ is an N-function such that

$$(147)\quad \Psi \ll \Phi^*,$$

and $h \in L_{\overline{\Psi}}$.

It follows from (146) that $f(\cdot, u) \in L_{\overline{\Psi}}$ whenever $u \in L_\Psi$. Moreover, the Niemitskii operator $u \mapsto f(\cdot, u)$ is bounded and continuous from L_Ψ into $L_{\overline{\Psi}}$. Therefore,

$$f(\cdot, u)v \in L^1(\Omega), \quad \forall u, \ v \in L_\Psi. \tag{148}$$

Because of (147), the embedding

$$W^1 L_\Phi \hookrightarrow E_\Psi \tag{149}$$

is compact. Hence, (148) holds for all $u, v \in W^1 L_\Phi$. Moreover, the mapping $\tilde{f} : W^1 L_\Phi \to (W^1 L_\Phi)^*$, defined by

$$\langle \tilde{f}(u), v \rangle = \int_\Omega f(x, u) v dx, \quad \forall u, \ v \in W^1 L_\Phi,$$

is completely continuous.

We first need an abstract existence theorems for the inequality (144) in the coercive case, whose detailed proof is presented in [**43**].

THEOREM 30. *(a) If $0 \in K$ and there exists $R > 0$ such that $K \cap \overline{B_R(0)} \neq \emptyset$ and*

$$\int_\Omega \Phi(|\nabla u|) dx - \int_\Omega f(x, u) u dx > 0, \tag{150}$$

for all $u \in K$ such that $\|u\| (= \|u\|_{W_0^1 L_\Phi}) = R$. Then (144) has a solution.

(b) Assume (145) and the following coercivity assumption:

$$\lim_{\|u\|_{W_0^1 L_\Phi} \to \infty} \left[\int_\Omega \Phi(|\nabla u|) dx - \int_\Omega f(x, u)(u - u_0) dx \right] = \infty, \tag{151}$$

for some $u_0 \in K \cap D_0$, then (144) has a solution.

4.3. Noncoercive case - Sub-supersolution method. In the case the coercivity condition (150) or (151) is not fulfilled, (144) may not have solutions. However, if subsolutions and/or supersolutions, in certain suitable sense, exist, then we can relax the coercivity condition, yet still have the solvability of (144). Furthermore, some extra properties such as enclosure or comparison of solutions can also be obtained.

Let us begin with the concepts of sub- and supersolutions for the variational inequality (144). Similar concepts were proposed in [**51**] for a variational inequality whose principal term were given by a specific N-function with very fast growth.

DEFINITION 31. A function $\underline{u} \in W^1 L_\Phi$ is called a subsolution of (144) if there exists an N-function Ψ satisfying

$$\Psi \ll \Phi^*, \tag{152}$$

such that

$$\int_\Omega \Phi(|\nabla \underline{u}|) dx < \infty, \quad f(\cdot, \underline{u}) \in L_{\overline{\Psi}}, \tag{153}$$

and

$$\int_\Omega \Phi(|\nabla v|)dx - \int_\Omega \Phi(|\nabla \underline{u}|)dx \geq \int_\Omega f(x,\underline{u})(v-\underline{u})dx, \quad \forall v \in \underline{u} \wedge K. \tag{154}$$

Similarly, a function $\overline{u} \in W^1 L_\Phi$ is a supersolution of (144) if

$$\int_\Omega \Phi(|\nabla \overline{u}|)dx < \infty, \quad f(\cdot, \overline{u}) \in L_{\overline{\Psi}}, \tag{155}$$

and

$$\int_\Omega \Phi(|\nabla v|)dx - \int_\Omega \Phi(|\nabla \overline{u}|)dx \geq \int_\Omega f(x,\overline{u})(v-\overline{u})dx, \quad \forall v \in \overline{u} \vee K. \tag{156}$$

As in [**46**], for measurable functions u, v and sets A, B of measurable functions on Ω, we use the notation $u \wedge v = \min\{u,v\}$, $u \vee v = \max\{u,v\}$, $A^*B = \{u^*v : u \in A, v \in B\}$, and $u * A = \{u\}^* A$ with $* \in \{\wedge, \vee\}$. The main result of this section is the following existence theorem for (144) when sub- and supersolutions exist. Note that if both sub- and supersolution exist then the N-functions Ψ in conditions (153) and (155) may be different. However, by replacing both functions by a common function preceeding both (cf. [**1, 38**]) and still satisfies (152), we can assume for simplicity of notation that both the sub- and supersolution satisfy (153) and (155) with the same function Ψ.

THEOREM 32. *If there is a pair of sub- and supersolution of (144) such that*

$$\underline{u} \vee K \subset K, \quad \overline{u} \wedge K \subset K, \tag{157}$$

$$\underline{u} \leq \overline{u} \text{ a.e. on } \Omega, \tag{158}$$

and f has the following growth condition between $\underline{u}$ and $\overline{u}$: For some $h \in L_{\overline{\Psi}}$,

$$|f(x,u)| \leq h(x) \text{ for a.e. } x \in \Omega, \text{ all } u \in [\underline{u}(x), \overline{u}(x)]. \tag{159}$$

Then, there exists a solution u of (144) such that

$$\underline{u} \leq u \leq \overline{u} \text{ a.e. on } \Omega. \tag{160}$$

PROOF. We use the usual truncation-regularization method. A key of this combined method is the choice of an appropriate truncation function. This function needs to be chosen substantially different from earlier proofs in [**9, 10, 11**] simply because we work in Orlicz–Sobolev rather than regular Sobolev spaces. The following truncation function is crucial for our calculations in the sequel. For $x \in \Omega$, $u \in \mathbb{R}$, let us put

$$b(x,u) = \begin{cases} -1 & \text{if } u - \underline{u}(x) < -1 \\ u - \underline{u}(x) & \text{if } -1 \leq u - \underline{u}(x) < 0 \\ 0 & \text{if } \underline{u}(x) \leq u \leq \overline{u}(x) \\ u - \overline{u}(x) & \text{if } 0 < u - \overline{u}(x) \leq 1 \\ 1 & \text{if } u - \overline{u}(x) > 1. \end{cases} \tag{161}$$

One can check that b is a Carathéodory function with the following properties:

$$-1 \le b(x,u) \le 1, \text{ for a.e. } x \in \Omega, \text{ all } u \in \mathbb{R}, \tag{162}$$

$$\begin{cases} b(x,u) < 0 & \text{if } u < \underline{u}(x) \\ b(x,u) > 0 & \text{if } u > \overline{u}(x). \end{cases} \tag{163}$$

We also define, for $x \in \Omega$, $u \in \mathbb{R}$,

$$\begin{aligned} T(x,u) &= [u \vee \underline{u}(x)] \wedge \overline{u}(x) \\ &= \begin{cases} \underline{u}(x) & \text{if } u < \underline{u}(x) \\ u & \text{if } \underline{u}(x) \le u \le \overline{u}(x) \\ \overline{u}(x) & \text{if } u > \overline{u}(x), \end{cases} \end{aligned} \tag{164}$$

and

$$\begin{aligned} \tilde{f}(x,u) &= f(x, T(x,u)) \\ &= \begin{cases} f(x, \underline{u}(x)) & \text{if } u < \underline{u}(x) \\ f(x,u) & \text{if } \underline{u}(x) \le u \le \overline{u}(x) \\ f(x, \overline{u}(x)) & \text{if } u > \overline{u}(x). \end{cases} \end{aligned} \tag{165}$$

It is clear that T and $\tilde{f}$ are Carathéodory functions on $\Omega \times \mathbb{R}$. It follows from (159) that

$$|\tilde{f}(x,u)| \le h(x) \text{ for a.e. } x \in \Omega, \text{ all } u \in \mathbb{R}. \tag{166}$$

We consider the following auxiliary variational inequality:

$$\begin{cases} \int_\Omega \Phi(|\nabla v|)dx - \int_\Omega \Phi(|\nabla u|)dx + \int_\Omega b(x,u)(v-u)dx \\ \ge \int_\Omega \tilde{f}(x,u)(v-u)dx, \ \forall v \in K \\ u \in K. \end{cases} \tag{167}$$

Let us first verify that (167) has solutions in K. Put $g(x,u) = \tilde{f}(x,u) - b(x,u)$ $(x \in \Omega, u \in \mathbb{R})$. Then g is a Carathéodory function and, from (162) and (166), it follows that

$$|g(x,u)| \le h + 1 \text{ for a.e. } x \in \Omega, \text{ all } u \in \mathbb{R}, \tag{168}$$

with $h + 1 \in L_{\overline{\Psi}}$. In particular, g satisfies the growth condition (146). Moreover, (167) is simply (144) with f replaced by g. Let u_0 be any (fixed) function in $K \cap D_0$. We check that (167) satisfies the coercivity condition (151), i.e.,

$$\lim_{\|u\|_{W_0^1 L_\Phi} \to \infty} \left[\int_\Omega \Phi(|\nabla u|)dx - \int_\Omega g(x,u)(u-u_0)dx \right] = +\infty. \tag{169}$$

In fact, since $\Psi \ll \Phi^*$, we can choose an N-function M such that

$$M \ll \Phi, \tag{170}$$

and

$$\Psi \ll M^*. \tag{171}$$

From (171), we have $W_0^1 E_M \hookrightarrow W_0^1 L_M \hookrightarrow E_\Psi$ where the first embedding is continuous and the second is compact. Since $h+1 \in L_{\overline{\Psi}}$, we see from these imbeddings that

$$\eta : u \mapsto \int_\Omega (h+1) u dx, \tag{172}$$

is a bounded linear functional on $W_0^1 E_M$, that is, $\eta \in (W_0^1 E_M)^* = W^{-1} L_{\overline{M}}$ from the classical complementary system among Orlicz–Sobolev spaces:

$$\begin{pmatrix} W_0^1 L_M & W^{-1} L_{\overline{M}} \\ W_0^1 E_M & W^{-1} E_{\overline{M}} \end{pmatrix}$$

(cf. e.g. [**32**]). Hence, there exist $f_0, f_1, \ldots, f_N \in L_{\overline{M}}$ such that

$$\langle \eta, u \rangle_{W_0^1 E_M, W^{-1} L_{\overline{M}}} = \int_\Omega \left(f_0 u + \sum_{j=1}^N f_j \partial_j u \right) dx, \quad \forall u \in W_0^1 E_M. \tag{173}$$

From (170), one has $L_\Phi \hookrightarrow E_M$ and thus $W_0^1 L_\Phi \hookrightarrow W_0^1 E_M$. Thus, from (172) and (173),

$$\int_\Omega (h+1) u dx = \int_\Omega \left(f_0 u + \sum_{j=1}^N f_j \partial_j u \right) dx, \quad \forall u \in W_0^1 L_\Phi. \tag{174}$$

Moreover, we obtain from (170) that $\overline{\Phi} \ll \overline{M}$ and thus $L_{\overline{M}}$ is continuously embedded into $E_{\overline{\Phi}}$. Therefore, $f_j \in E_{\overline{\Phi}}$ for all $j \in \{0, 1, \ldots, N\}$ and, in particular,

$$\int_\Omega \overline{\Phi}(C f_j) dx < \infty, \quad \forall C \in \mathbb{R}, \ \forall j \in \{0, 1, \ldots, N\}. \tag{175}$$

On the other hand, from Poincaré's inequality in Orlicz–Sobolev spaces (cf. [**32**]), there are $C_0, K_0 > 0$ such that

$$\int_\Omega \Phi(C_0 u) dx \le K_0 \int_\Omega \Phi(|\nabla u|) dx, \quad \forall u \in W_0^1 L_\Phi. \tag{176}$$

In the following estimates, we repeatedly use Young's inequality

$$|xy| \le \varepsilon \Phi(x) + \varepsilon^{-1} \overline{\Phi}(y), \quad \forall \ x, \ y \in \mathbb{R}, \ \forall \varepsilon > 0.$$

Let ε be a fixed positive constant to be determined later. We have

$$\begin{aligned}\int_\Omega |f_0 u| dx &\le \varepsilon \int_\Omega \Phi(C_0 u) dx + \frac{1}{\varepsilon}\int_\Omega \overline{\Phi}\left(\frac{1}{C_0} f_0\right) dx \\ &\le \varepsilon K_0 \int_\Omega \Phi(|\nabla u|) dx + \frac{1}{\varepsilon}\int_\Omega \overline{\Phi}\left(\frac{1}{C_0} f_0\right) dx.\end{aligned} \tag{177}$$

For $j = 1, \dots, N$,

$$\begin{aligned}\int_\Omega |f_j \partial_j u| dx &\le \varepsilon \int_\Omega \Phi(|\partial_j u|) dx + \frac{1}{\varepsilon}\int_\Omega \overline{\Phi}(f_j) dx \\ &\le \varepsilon \int_\Omega \Phi(|\nabla u|) dx + \frac{1}{\varepsilon}\int_\Omega \overline{\Phi}(f_j) dx.\end{aligned} \tag{178}$$

Hence, from (174), (177), (178), and (168), noting that $|u| \in W_0^1 L_\Phi$ whenever $u \in W_0^1 L_\Phi$ and $\int_\Omega (h+1)|u_0| dx < \infty$, we obtain

$$\begin{aligned}&\int_\Omega \Phi(|\nabla u|) dx - \int_\Omega g(x,u)(u-u_0) dx \\ &\ge \int_\Omega \Phi(|\nabla u|) dx - \int_\Omega (h+1)|u| dx - \int_\Omega (h+1)|u_0| dx \\ &\ge \int_\Omega \Phi(|\nabla u|) dx - \int_\Omega \left(|f_0||u| dx + \sum_{j=1}^N |f_j||\partial_j u| \right) dx - \int_\Omega (h+1)|u_0| dx \\ &\ge (1 - \varepsilon K_0 - \varepsilon N) \int_\Omega \Phi(|\nabla u|) dx \\ &\quad - \frac{1}{\varepsilon}\int_\Omega \left[\overline{\Phi}\left(\frac{1}{C_0} f_0\right) + \sum_{j=1}^N \overline{\Phi}(f_j) \right] dx - \int_\Omega (h+1)|u_0| dx.\end{aligned}$$

Choosing $\varepsilon \in (0, [2(K_0+N)]^{-1})$ and putting

$$C_1 = \frac{1}{\varepsilon}\int_\Omega \left[\overline{\Phi}\left(\frac{1}{C_0} f_0\right) + \sum_{j=1}^N \overline{\Phi}(f_j) \right] dx + \int_\Omega (h+1)|u_0| dx \in [0, \infty),$$

(see (175)), we have

$$\int_\Omega \Phi(|\nabla u|) dx - \int_\Omega g(x,u)(u-u_0) dx \ge \frac{1}{2}\int_\Omega \Phi(|\nabla u|) dx - C_1. \tag{179}$$

Since a subset W of $W_0^1 L_\Phi$ is bounded (in the $\|\cdot\|_{W_0^1 L_\Phi}$-norm) if and only if $\sup\left\{\int_\Omega \Phi(|\nabla u|) dx : u \in W\right\} < \infty$, we see that $\int_\Omega \Phi(|\nabla u|) dx \to \infty$ as $\|u\|_{W_0^1 L_\Phi} \to \infty$. Hence, (179) proves (169).

In view of Theorem 30, we see that (167) has solutions in K. Let u be any solution of (167). In this next step, we show that

$$\underline{u}(x) \le u(x) \le \overline{u}(x) \quad \text{for a.e. } x \in \Omega, \tag{180}$$

and u is in fact a solution of (144).

Let us verify that $u \geq \underline{u}$ a.e. in Ω, the other inequality in (180) is proved in a similar way. First, we note that

$$\int_\Omega \Phi(|\nabla(u \wedge v)|)dx + \int_\Omega \Phi(|\nabla(u \vee v)|)dx = \int_\Omega \Phi(|\nabla u|)dx + \int_\Omega \Phi(|\nabla v|)dx, \tag{181}$$

for all $u, v \in W^1L_\Phi$. In fact, from Stampacchia's theorem (cf. [**31**]), we have

$$\nabla(u \wedge v) = \begin{cases} \nabla u & \text{a.e. on } \{x \in \Omega : u(x) < v(x)\} \\ \nabla u = \nabla v & \text{a.e. on } \{x \in \Omega : u(x) = v(x)\} \\ \nabla v & \text{a.e. on } \{x \in \Omega : u(x) > v(x)\}, \end{cases}$$

and a similar result for $\nabla(u \vee v)$. Hence,

$$\begin{aligned} &\int_\Omega \Phi(|\nabla(u \wedge v)|)dx + \int_\Omega \Phi(|\nabla(u \vee v)|)dx \\ &= \left[\int_{\{x\in\Omega:u(x)<v(x)\}} \Phi(|\nabla u|)dx + \int_{\{x\in\Omega:u(x)\geq v(x)\}} \Phi(|\nabla v|)dx\right] \\ &\quad + \left[\int_{\{x\in\Omega:u(x)<v(x)\}} \Phi(|\nabla v|)dx + \int_{\{x\in\Omega:u(x)\geq v(x)\}} \Phi(|\nabla u|)dx\right] \\ &= \left(\int_{\{x\in\Omega:u(x)<v(x)\}} + \int_{\{x\in\Omega:u(x)\geq v(x)\}}\right) [\Phi(|\nabla u|) + \Phi(|\nabla v|)]dx \\ &= \int_\Omega \Phi(|\nabla u|)dx + \int_\Omega \Phi(|\nabla v|)dx. \end{aligned}$$

Letting $v = \underline{u} \wedge u \in \underline{u} \wedge K$ in (154) gives

$$\begin{aligned} \int_\Omega \Phi(|\nabla(\underline{u} \wedge u)|)dx - \int_\Omega \Phi(|\nabla \underline{u}|)dx &\geq \int_\Omega f(x,\underline{u})(\underline{u} \wedge u - \underline{u})dx \\ &= -\int_\Omega f(x,\underline{u})(\underline{u} - u)^+dx. \end{aligned} \tag{182}$$

From (157), we have $\underline{u} \vee u \in K$. Letting $v = \underline{u} \vee u$ in (167) and noting that $\underline{u} \vee u - u = (\underline{u} - u)^+$, one gets

$$\begin{aligned} &\int_\Omega \Phi(|\nabla(\underline{u} \vee u)|)dx - \int_\Omega \Phi(|\nabla u|)dx + \int_\Omega b(x,u)(\underline{u} - u)^+dx \\ &\quad \geq \int_\Omega \tilde{f}(x,u)(\underline{u} - u)^+dx. \end{aligned} \tag{183}$$

Adding (182) with (183) and taking into account (181), we obtain

$$\begin{aligned} \int_\Omega b(x,u)(\underline{u} - u)^+dx &\geq \int_\Omega [\tilde{f}(x,u) - f(x,\underline{u})](\underline{u} - u)^+dx \\ &= \int_{\{x\in\Omega:\underline{u}(x)\geq u(x)\}} [f(x,T(x,u(x))) - f(x,\underline{u}(x))][\underline{u}(x) - u(x)]dx. \end{aligned}$$

If $\underline{u}(x) > u(x)$ then by (164), $T(x, u(x)) = \underline{u}(x)$ and thus $f(x, T(x, u(x))) - f(x, \underline{u}(x)) = 0$. It follows that

$$\begin{aligned} 0 &\le \int_\Omega b(x, u(x))[\underline{u}(x) - u(x)]^+ dx \\ &= \int_{\{x\in\Omega:\underline{u}(x)>u(x)\}} b(x, u(x))[\underline{u}(x) - u(x)]dx. \end{aligned}$$

However, if $\underline{u}(x) > u(x)$, then by (163), $b(x, u(x)) < 0$ and thus

$$b(x, u(x))[\underline{u}(x) - u(x)] < 0.$$

This shows that the set $\{x \in \Omega : \underline{u}(x) > u(x)\}$ must have measure 0, i.e., $u(x) \ge \underline{u}(x)$ for a.e. $x \in \Omega$. Following a similar proof for $\overline{u}$, we obtain (180).

Finally, it follows from (180) and (161)-(164)-(165) that $b(x, u(x)) = 0$ and $\tilde{f}(x, u(x)) = f(x, u(x))$ for a.e. $x \in \Omega$. (167) therefore reduces to (144) and our proof is complete. □

REMARK 33. (a) If $K = W_0^1 L_\Phi$ then (157) holds if $\underline{u} \le 0$ on $\partial\Omega$, $\overline{u} \ge 0$ on $\partial\Omega$, and we get the familiar boundary conditions for sub- and supersolutions in equations (cf. e.g. [**37**] for one of the pioneering works in which such concepts, for equations in variational (weak) form, were introduced).

(b) If instead of the existence of both sub- and supersolutions, we assume only that a subsolution $\underline{u}$ of (144) exists and that f has the following one-sided growth condition above $\underline{u}$:

$$|f(x, u)| \le h_1(x) + d_1 \overline{\Psi_1}^{-1} \Psi_1(d_2 u) \quad \text{(for a.e. } x \in \Omega, \text{ all } u \in [\underline{u}(x), \infty)),$$

where Ψ_1 is an N-function such that $\Psi_1 \ll \Phi$ and $h_1 \in L_{\overline{\Psi_1}}$. The proof of this result is similar to that of Theorem 32. One also has an analogous existence result when a supersolution exists and f satisfies a growth condition below that supersolution. Therefore, in the case the principal operators are given by N-functions, the results here extend those in [**46**] to variational inequalities with non-polynomial growths.

(c) Theorem 32 can be extended to the existence and enclosure between several subsolutions and supersolutions. We have the following existence-enclosure theorem.

THEOREM 34. *Assume $\underline{u}_1, \dots, \underline{u}_k$ are subsolutions and $\overline{u}_1, \dots, \overline{u}_m$ are supersolutions of (144) such that*

$$\underline{u}_i \vee K \subset K, \quad \overline{u}_j \wedge K \subset K, \qquad \forall i \in \{1, \dots, k\}, \quad j \in \{1, \dots, m\}, \tag{184}$$

$$\underline{u} := \max\{\underline{u}_1, \dots, \underline{u}_k\} \le \overline{u} := \min\{\overline{u}_1, \dots, \overline{u}_m\}, \tag{185}$$

and f satisfies the growth condition (159) for a.e. $x \in \Omega$, all $u \in [\underline{u}_0, \overline{u}_0]$, where

$$\underline{u}_0 := \min\{\underline{u}_1, \dots, \underline{u}_k\}, \quad \overline{u}_0 := \max\{\overline{u}_1, \dots, \overline{u}_m\}. \tag{186}$$

Then, (144) has a solution u satisfying (160) with $\underline{u}$ and $\overline{u}$ given by (185).

The proof of Theorem 34 follows the same lines as those in Theorem 32 and is omitted. The only noteworthy change in the proof is about the function $\tilde{f}(x, u)$

which is now given by

$$\tilde{f}(x,u) = f(x,T(x,u)) + \sum_{i=1}^{k} |f(x,T_{i0}(x,u)) - f(x,T(x,u))| \\ - \sum_{j=1}^{m} |f(x,T_{0j}(x,u)) - f(x,T(x,u))|, \tag{187}$$

where $T(x,u)$ is defined in (164) (with $\underline{u}$ and $\overline{u}$ now given by (185)) and $T_{ij}(x,u) = [u \vee \underline{u}_i(x)] \wedge \overline{u}_j(x) (= [u \wedge \overline{u}_j(x)] \vee \underline{u}_i(x))$ for $i \in \{0,1,\dots,k\}$, $j \in \{0,1,\dots,m\}$.

4.4. Existence of extremal solutions. As in the case of equations and inequalities in regular Sobolev spaces (cf. [**11**] and the references therein), we can obtain not only the existence of solutions but also other properties of the solution set of (144) between sub- and supersolutions. In fact, we have the following result.

THEOREM 35. *Assume the conditions in Theorem 32, that is, there exist a subsolution $\underline{u}$ and a supersolution $\overline{u}$ of (144) such that (157), (158), and (159) are satisfied. Let*

$$S = \{u \in K : \underline{u} \le u \le \overline{u} \text{ a.e. on } \Omega \text{ and } u \text{ is a solution of (144)}\}. \tag{188}$$

Then,

(a) *S is (sequentially) compact with respect to the weak* topology of $W_0^1 L_\Phi$ and is, in particular, a bounded subset of $W_0^1 L_\Phi$.*

(b) *If*

$$S \wedge K, \quad S \vee K \subset K, \tag{189}$$

then (144) has a least solution and a greatest solution between $\underline{u}$ and $\overline{u}$, i.e., there exist $u_*, u^* \in S$ such that $u_* \le u \le u^*$ a.e. in Ω for all $u \in S$.

PROOF. (a) If $u \in S$ then $|u| \le |\underline{u}| + |\overline{u}|$ a.e. on Ω. Since $\underline{u}, \overline{u} \in E_\Psi$, we have

$$\int_\Omega \Psi(u)dx \le \int_\Omega \Psi(|\underline{u}| + |\overline{u}|)dx \ (< \infty),$$

and

$$\|u\|_{L_\Psi} \le \| |\underline{u}| + |\overline{u}| \|_{L_\Psi} \le \|\underline{u}\|_{L_\Psi} + \|\overline{u}\|_{L_\Psi}, \quad \forall u \in S.$$

Therefore, S is a bounded set in E_Ψ (with respect to the $\|\cdot\|_{L_\Psi}$-norm).

Let v_0 be a fixed function in $K \cap D_0$. For $u \in S$, we have from (144),

$$\int_\Omega \Phi(|\nabla v_0|)dx - \int_\Omega \Phi(|\nabla u|)dx \ge \int_\Omega f(x,u)(v_0 - u)dx.$$

Thus, from (159),

$$\begin{aligned}\int_\Omega \Phi(|\nabla u|)dx &\le \int_\Omega \Phi(|\nabla v_0|)dx + \int_\Omega |f(x,u)|(|u| + |v_0|)dx \\ &\le \int_\Omega \Phi(|\nabla v_0|)dx + \|h\|_{L_{\overline{\Psi}}}(\|v_0\|_{L_\Psi} + \|\underline{u}\|_{L_\Psi} + \|\overline{u}\|_{L_\Psi})(< \infty).\end{aligned}$$

Since this holds for all $u \in S$, we obtain the boundedness of S in $W_0^1 L_\Phi$. Let us prove that S is sequentially compact with respect to the weak* topology of $W_0^1 L_\Phi$. From the above discussions, S is relatively (sequentially) compact with respect to that topology. We only need to check now that S is sequentially closed in that weak* topology. Assume $\{u_n\}$ is a sequence in S and $u_n \rightharpoonup^* u$ (in the weak* topology) in $W_0^1 L_\Phi$. Because of the compact embedding $W_0^1 L_\Phi \hookrightarrow E_\Phi$, we have $u_n \to u$ in E_Ψ. From the growth condition (159) and usual convergence arguments based on Lebesgue dominated convergence theorem, we have $f(\cdot, u_n) \to f(\cdot, u)$ in $L_{\overline{\Psi}}$. Hence,

$$\int_\Omega f(x, u_n)(v - u_n)dx \to \int_\Omega f(x, u)(v - u)dx, \quad \forall v \in K. \tag{190}$$

From the lower semicontinuity of the functional $J : u \mapsto \int_\Omega \Phi(|\nabla u|)dx$ with respect to the weak* topology in $W_0^1 L_\Phi$, we have

$$\int_\Omega \Phi(|\nabla u|)dx \leq \liminf \int_\Omega \Phi(|\nabla u_n|)dx. \tag{191}$$

For each $n \in \mathbb{N}$,

$$\int_\Omega \Phi(|\nabla v|)dx - \int_\Omega \Phi(|\nabla u_n|)dx \geq \int_\Omega f(x, u_n)(v - u_n)dx, \quad \forall v \in K.$$

Passing to the limit in this inequality and using (190)–(191), we see that u is a solution of (144). On the other hand, by passing to a subsequence if necessary, we have $u_n(x) \to u(x)$ for a.e. $x \in \Omega$. Then, $\underline{u} \leq u \leq \overline{u}$ a.e. in Ω and thus $u \in S$. We have shown that S is sequentially weak* closed in $W_0^1 L_\Phi$ and therefore sequentially compact with respect to that topology in $W_0^1 L_\Phi$.

(b) With our preparations given above, the proof of (b) is just an extension of the arguments for equations and inequalities in regular Sobolev spaces (cf. e.g. [**8, 9, 10, 11**]) and is presented here for the sake of completeness. First, note that under assumption (189), any solution $u \in S$ is both a subsolution and a supersolution in the sense of Definition 31. In fact, we have from (159) that $f(\cdot, u) \in L_{\overline{\Psi}}$. Since $K \cap D_0 \neq \emptyset$, all solutions $u \in S$ must satisfy $\int_\Omega \Phi(|\nabla u|)dx < \infty$. Furthermore, since $u \wedge K \subset K$, (154) follows from (144). Hence, u is a subsolution of (144). The proof for supersolution is similar.

In the next step, we note that since $S \subset E_\Psi$ and the Banach space $(E_\Psi, \|\cdot\|_{L_\Psi})$ is separable, S is separable with respect to the L_Ψ-topology and there exists a sequence $\{w_n\}$ in S such that the set $\{w_n : n \in \mathbb{N}\}$ is dense in S (with the L_Ψ topology). We construct another sequence $\{u_n\}$ in S inductively as follows. Put $u_1 = w_1$. Assume $u_1, \ldots, u_n$ are constructed such that

$$u_{j-1} \leq u_j \text{ a.e. on } \Omega, \quad \forall j \in \{2, \ldots, n\}, \tag{192}$$

and

$$u_j \geq w_j \text{ a.e. on } \Omega, \quad \forall j \in \{1, \ldots, n\}. \tag{193}$$

The next function u_{n+1} is constructed as follows. Since $u_n, w_{n+1} \in S$, they are also subsolutions of (144). Moreover,

$$\underline{u} \le \min\{u_n, w_{n+1}\} \le \max\{u_n, w_{n+1}\} \le \overline{u} \text{ a.e. on } \Omega,$$

condition (184) and the growth condition (159) (with (186)) in Theorem 34 are satisfied with the subsolutions u_n, w_{n+1} and the supersolution $\overline{u}$. Invoking that theorem, we see that there exists a solution u of (144) such that $\max\{u_n, w_{n+1}\} \le u \le \overline{u}$. Letting u_{n+1} be any of such solutions, we immediately have $u_{n+1} \in S$ and furthermore,

$$u_n \le u_{n+1} \quad \text{and} \quad w_{n+1} \le u_{n+1} \text{ a.e. on } \Omega.$$

We see that properties (192) and (193) are now extended up to $n+1$.

As a consequence, we obtain an increasing sequence $\{u_n\}$ in S such that

$$u_n \ge w_n \text{ a.e. on } \Omega, \quad \forall n \in \mathbb{N}. \tag{194}$$

For $x \in \Omega$, let

$$u^*(x) = \sup\{u_n(x) : n \in \mathbb{N}\} = \lim_{n\to\infty} u_n(x). \tag{195}$$

Then u^* is measurable on Ω and $\underline{u} \le u^* \le \overline{u}$ a.e. on Ω. In particular, $u \in L_\Psi$. Moreover, from part (a), the set $\{u_n : n \in \mathbb{N}\}$ is a bounded and thus weak* (sequentially) relatively compact in $W_0^1 L_\Phi$. Together with (195), this shows that

$$u_n \rightharpoonup^* u^*(\text{weak}^*) \text{ in } W_0^1 L_\Phi. \tag{196}$$

In particular, $u^* \in W_0^1 L_\Phi$. It also follows from (196) and the weak* sequential compactness of S in part (a) that $u^* \in S$. Let us verify that u^* is moreover the greatest element of S in the (almost everywhere) pointwise ordering. In fact, let v be any function in S. From the density of $\{w_n : n \in \mathbb{N}\}$, we can extract a subsequence $\{w_{n_k}\}$ of $\{w_n\}$ such that $w_{n_k} \to v$ in L_Ψ. By passing to a subsequence if needed, we also have $w_{n_k}(x) \to v(x)$ for a.e. $x \in \Omega$. Combining this limit with (194) and (195), we have

$$v(x) = \lim_{k\to\infty} w_{n_k}(x) \le \lim_{k\to\infty} u_{n_k}(x) = u^*(x), \quad \text{for a.e. } x \in \Omega.$$

We proved that $v \le u^*$ a.e. on Ω for any $v \in S$. The existence of the least element of S is proved in the same way. □

4.5. Some particular cases. In this section, we consider some particular cases of the above discussions. Let K_0 be the (sequential) closure of $K \cap W_0^1 E_\Phi$ with respect to the weak* topology of $W_0^1 L_\Phi$. Since K is weak* closed, $K_0 \subset K$. In the following result, it is stated that in some particular cases of K, the variational inequality (144) has the usual form of operational equality or inequality in Orlicz–Sobolev spaces as presented in [**32, 34, 35, 64**].

THEOREM 36. (a) *Assume* $0 \in K$. *If* u *is a solution of (144) then*

$$\phi(|\nabla u|) \in L_{\overline{\Phi}}, \tag{197}$$

and

$$\int_\Omega a(|\nabla u|)\nabla u\cdot\nabla(v-u)dx \geq \int_\Omega f(x,u)(v-u)dx, \quad \forall v\in K_0. \tag{198}$$

(b) *If $K\cap W_0^1L_\Phi$ is weak* dense in K (that is $K_0=K$), in particular, if Φ satisfies a Δ_2 condition, then any solution u of (144) is also a solution of the variational inequality:*

$$\int_\Omega a(|\nabla u|)\nabla u\cdot\nabla(v-u)dx \geq \int_\Omega f(x,u)(v-u)dx, \quad \forall v\in K. \tag{199}$$

(c) *If $K=W_0^1L_\Phi$, then u satisfies the following variational equality (cf. (140))*

$$\int_\Omega a(|\nabla u|)\nabla u\cdot\nabla v dx = \int_\Omega f(x,u)v dx, \quad \forall v\in W_0^1L_\Phi,$$

i.e., equation (138) in the weak sense.

PROOF. The proof of (a) relies on an application of the usual Mean value theorem as presented in [**34**] (see also [**64**]). Let u be a solution of (144). Since $0\in K$, we have $\lambda u\in K$, $\forall\lambda\in[0,1]$. Since $\nabla g(\xi)=a(|\xi|)\xi$ for $g(\xi)=\Phi(|\xi|)$, $\xi\in\mathbb{R}^N$, letting $v=\lambda u$ in (144) ($\lambda\in(0,1)$) and using the Mean value theorem for g, we have

$$\theta=\theta(\lambda,x,u)\in(\lambda,1) \tag{200}$$

such that

$$\begin{aligned}\int_\Omega \Phi(\lambda|\nabla u|)dx - \int_\Omega \Phi(|\nabla u|)dx &= \int_\Omega a(\theta|\nabla u|)\theta\nabla u\cdot[(\lambda-1)\nabla u]dx\\ &\geq (\lambda-1)\int_\Omega f(x,u)udx.\end{aligned}$$

Hence,

$$\int_\Omega a(\theta|\nabla u|)\theta|\nabla u|^2dx \leq \int_\Omega f(x,u)udx.$$

From (200) and the continuity of a, we have

$$\lim_{\lambda\to1} a(\theta(\lambda,x,u)|\nabla u(x)|) = a(|\nabla u(x)|) \text{ for a.e. } x\in\Omega.$$

Therefore, by Fatou's lemma,

$$\int_\Omega \phi(|\nabla u|)|\nabla u|dx = \int_\Omega a(|\nabla u|)|\nabla u|^2dx \leq \int_\Omega f(x,u)udx \ (<\infty).$$

Since $\overline{\Phi}(\phi(|\nabla u|))\leq\phi(|\nabla u|)|\nabla u|$, we have $\int_\Omega \overline{\Phi}(\phi(|\nabla u|))dx<\infty$, implying (197).

For $v\in K\cap W_0^1E_\Phi$ and $\lambda\in(0,1)$, letting $w=\lambda v+(1-\lambda)u\in K$ in place of v in (144) and using again the Mean value theorem, we see that for each $x\in\Omega$, there is $\theta = \theta(x, \lambda, u, v) \in (0,1)$ such that

$$\begin{aligned}
&\int_\Omega a(|\lambda\theta\nabla v + (1-\lambda\theta)\nabla u|)[\lambda\theta\nabla v + (1-\lambda\theta)\nabla u]\cdot\nabla(v-u)dx \\
&\quad = \frac{1}{\lambda}\left[\int_\Omega \Phi(|\nabla(\lambda v + (1-\lambda)u)|)dx - \int_\Omega \Phi(|\nabla u|)dx\right] \\
&\quad \geq \int_\Omega f(x,u)(v-u)dx.
\end{aligned} \tag{201}$$

As above, we have

$$a(|\lambda\theta\nabla v + (1-\lambda\theta)\nabla u|)[\lambda\theta\nabla v + (1-\lambda\theta)\nabla u] \to a(|\nabla u|)\nabla u, \tag{202}$$

a.e. on Ω as $\lambda \to 0^+$, and

$$\begin{aligned}
a(|\lambda\theta\nabla v + (1-\lambda\theta)\nabla u|)|\lambda\theta\nabla v + (1-\lambda\theta)\nabla u| &= \phi(|\lambda\theta\nabla v + (1-\lambda\theta)\nabla u|) \\
&\leq \phi(|\nabla v|) + \phi(|\nabla u|) \ \text{ a.e. on } \Omega.
\end{aligned} \tag{203}$$

Moreover, since $v \in W_0^1 E_\Phi$,

$$\phi(|\nabla v|) \in L_{\overline{\Phi}}, \tag{204}$$

and

$$|\nabla v|, |\nabla u| \in L_\Phi. \tag{205}$$

Letting $\lambda \to 0^+$ in (201), by Lebesgue's dominated convergence theorem, which is applicable due to (197) and (202)–(205), we see that v satisfies inequality (198). Since $K \cap W_0^1 E_\Phi$ is $(W_0^1 L_\Phi)$ weak* dense in K_0, (198) also holds for $v \in K_0$.

The claims in (b) and (c) follow directly from (a) and the weak* density of $W_0^1 E_\Phi$ in $W_0^1 L_\Phi$. □

As another property of solutions of (144) in the case both Φ and its Hölder conjugate $\overline{\Phi}$ satisfy Δ_2 conditions, we have the following stronger property of S compared to that in Theorem 35(a).

THEOREM 37. *Assume Φ and $\overline{\Phi}$ satisfy Δ_2 conditions and let $\underline{u}$ and $\overline{u}$ be sub- and supersolutions of (144) that satisfy conditions (157)–(159). Then the set S of solutions of (144) between $\underline{u}$ and $\overline{u}$ is compact in $W_0^1 L_\Phi$.*

PROOF. As in the proof of Theorem 35 (a) and from the property of S established there, to prove the compactness of S in $W_0^1 L_\Phi$, one only needs to verify that any sequence $\{u_n\} \subset S$ such that $u_n \rightharpoonup^* u$ in S, we must have $u_n \to u$ strongly in $W_0^1 L_\Phi$. This is a direct consequence of Theorem 4 in [**44**]. In fact, from Theorem 36, we see that u is a solution of (199) and also

$$\int_\Omega a(|\nabla u_n|)\nabla u_n \cdot \nabla(v-u_n)dx \geq \int_\Omega f(x,u_n)(v-u_n)dx, \ \forall v \in K, \forall n \in \mathbb{N}. \tag{206}$$

Letting $v = u_n$ in (199),$v = u$ in (206), and adding the inequalities thus obtained, we get

$$(0 \leq) \int_\Omega [a(|\nabla u_n|)\nabla u_n - a(|\nabla u|)\nabla u] \cdot (\nabla u_n - \nabla u)dx$$
$$\leq \int_\Omega [f(x, u_n) - f(x, u)](u_n - u)dx, \quad \forall n \in \mathbb{N}.$$

From (190), we have $\int_\Omega [f(x, u_n) - f(x, u)](u_n - u)dx \to 0$ as $n \to \infty$. Hence,

$$\lim_{n\to\infty} \int_\Omega [a(|\nabla u_n|)\nabla u_n - a(|\nabla u|)\nabla u] \cdot (\nabla u_n - \nabla u)dx = 0.$$

It follows from Theorem 4, [**44**], that $u_n \to u$ in $W_0^1 L_\Phi$, which completes our proof. □

Bibliography

[1] R. Adams, *Sobolev spaces*, Academic Press, New York, 1975.

[2] L. Aharouch, A. Benkirane, and M. Rhoudaf, *Strongly nonlinear elliptic variational unilateral problems in Orlicz space*, Abstr. Appl. Anal. (2006), Art. ID 46867, 20.

[3] ———, *Existence results for some unilateral problems without sign condition with obstacle free in Orlicz spaces*, Nonlinear Anal. **68** (2008), 2362–2380.

[4] H. Amann, *Liusternik–Schnirelman theory and non-linear eigenvalue problems*, Math. Ann. **199** (1972), 55–72.

[5] J. Ball, J. Currie, and P. Olver, *Null Lagrangians, weak continuity, and variational problems of arbitrary order*, J. Funct. Anal. **41** (1981), 135–174.

[6] A. Benkirane and A. Elmahi, *An existence theorem for a strongly nonlinear elliptic problem in Orlicz spaces*, Nonlinear Anal. **36** (1999), 11–24.

[7] A. Benkirane and A. Youssfi, *Existence of bounded solutions for a class of strongly nonlinear elliptic equations in Orlicz-Sobolev spaces*, Aust. J. Math. Anal. Appl. **5** (2008), Art. 7, 26.

[8] S. Carl, V. K. Le, and D. Motreanu, *The sub-supersolution method and extremal solutions for quasilinear hemivariational inequalities*, Differential Integral Equations **17** (2004), no. 1-2, 165–178.

[9] ———, *Existence and comparison principles for general quasilinear variational-hemivariational inequalities*, J. Math. Anal. Appl. **302** (2005), no. 1, 65–83.

[10] ———, *Existence, comparison, and compactness results for quasilinear variational-hemivariational inequalities*, Int. J. Math. Math. Sci. (2005), no. 3, 401–417.

[11] ———, *Nonsmooth variational problems and their inequalities. comparison principles and applications*, Springer Monographs in Mathematics, Springer, New York, 2007.

[12] F. H. Clarke, *Optimization and nonsmooth analysis*, SIAM, Philadelphia, 1990.

[13] P. Clément, M. García-Huidobro, R. Manásevich, and K. Schmitt, *Mountain pass type solutions for quasilinear elliptic equations*, Calc. Var. and PDE **11** (2000), 33–62.

[14] C. Coffman, *Lyusternik–Schnirelman theory and eigenvalue problems for monotone potential operators*, J. Funct. Anal. **14** (1973), 237–252.

[15] H. Dang and K. Schmitt, *Existence of nonnegative solutions of semilinear elliptic equations with subcritical exponents*, Topol. Methods Nonlinear Anal. **5** (1995), 125–140.

[16] K. Deimling, *Nonlinear functional analysis*, Springer, Berlin, 1985.

[17] G. Dinca and P. Matei, *Variational and topological methods for operator equations involving duality mappings on Orlicz-Sobolev spaces*, Electron. J. Differential Equations (2007), No. 93, 47 pp. (electronic).

[18] ———, *Multiple solutions for operator equations involving duality mappings on Orlicz-Sobolev spaces*, Differential Integral Equations **21** (2008), 891–916.

[19] ———, *Multiple solutions to operator equations involving duality mappings on Orlicz-Sobolev spaces via the mountain pass theorem*, Rev. Roumaine Math. Pures Appl. **53** (2008), 419–437.

[20] T. Donaldson, *Nonlinear elliptic boundary value problems in Orlicz–Sobolev spaces*, J. Diff. Equations **10** (1971), 507–528.

[21] T. Donaldson and N. Trudinger, *Orlicz–Sobolev spaces and imbedding theorems*, J. Functional Analysis **8** (1971), 52–75.

[22] A. Elmahi and D. Meskine, *Elliptic inequalities with lower order terms and L^1 data in Orlicz spaces*, J. Math. Anal. Appl. **328** (2007), 1417–1434.

[23] P. M. Fitzpatrick and W. V. Petryshyn, *Fixed point theorems and the fixed point index for multivalued mappings incones*, J. London Math. Soc. **12** (1976), 75–85.

[24] N. Fukagai, M. Ito, and K. Narukawa, *Positive solutions of quasilinear elliptic equations with critical Orlicz-Sobolev nonlinearity on $\mathbf{R}^N$*, Funkcial. Ekvac. **49** (2006), 235–267.

[25] ———, *Quasilinear elliptic equations with slowly growing principal part and critical Orlicz-Sobolev nonlinear term*, Proc. Roy. Soc. Edinburgh Sect. A **139** (2009), 73–106.

[26] N. Fukagai and K. Narukawa, *Multiple positive solutions of nonlinear eigenvalue problems associated to a class of p-Laplacian like operators*, Commun. Contemp. Math. **5** (2003), 737–759.

[27] ———, *On the existence of multiple positive solutions of quasilinear elliptic eigenvalue problems*, Ann. Mat. Pura Appl. (4) **186** (2007), 539–564.

[28] M. García-Huidobro, R. Manásevich, and K. Schmitt, *On principal eigenvalues of p-Laplacian like operators*, J. Differential Equations **130** (1996), 235–246.

[29] ———, *Some bifurcation results for a class of p-Laplacian like operators*, Diff. Int. Equations **10** (1997), 51–66.

[30] M. García-Huidobro, V. K. Le, R. Manásevich, and K. Schmitt, *On pricipal eigenvalues for quasilinear elliptic differential operators: An Orlicz–Sobolev space setting*, Nonl. Diff. Eq. Appl. **6** (1999), 207–225.

[31] D. Gilbarg and N. Trudinger, *Elliptic partial differential equations of second order*, Springer, Berlin, 1983.

[32] J. P. Gossez, *Nonlinear elliptic boundary value problems for equations with rapidly or slowly increasing coefficients*, Trans. Amer. Math. Soc. **190** (1974), 163–205.

[33] ———, *A strongly nonlinear elliptic problem in Orlicz–Sobolev spaces*, Proc. Sympos. Pure Math. **45** (1986), 455–462.

[34] J. P. Gossez and R. Manásevich, *On a nonlinear eigenvalue problem in Orlicz-Sobolev spaces*, Proc. Roy. Soc. Edinburgh Sect. A **132** (2002), 891–909.

[35] J. P. Gossez and V. Mustonen, *Variational inequalities in Orlicz–Sobolev spaces*, Nonlinear Analysis **11** (1987), 379–392.

[36] N. Halidias and V. K. Le, *Multiple solutions for quasilinear elliptic Neumann problems in Orlicz-Sobolev spaces*, Bound. Value Probl. (2005), no. 3, 299–306.

[37] P. Hess, *On the solvability of nonlinear elliptic boundary value problems*, Indiana Univ. Math. J. **25** (1976), 461–466.

[38] M. A. Krasnosels'kii and J. Rutic'kii, *Convex functions and Orlicz spaces*, Noordhoff, Groningen, 1961.

[39] A. Kristály, M. Mihăilescu, and V. Rădulescu, *Two non-trivial solutions for a non-homogeneous Neumann problem: an Orlicz-Sobolev space setting*, Proc. Roy. Soc. Edinburgh Sect. A **139** (2009), 367–379.

[40] R. S. Kubrusly, *Variational methods for nonlinear eigenvalue inequalities*, Differential Integral Equations **3** (1990), 923–932.

[41] A. Kufner, O. John, and S. Fučic, *Function spaces*, Noordhoff, Leyden, 1977.

[42] A. Le, V. Le, and K. Schmitt, *Lagrange multiplier rule for nonlinear inequalities*, Vietnam J. Math. **32** (2004), 43–47.

[43] V. K. Le, *Some existence results and properties of solutions in quasilinear variational inequalities with general growths*, Differ. Equ. Dyn. Syst., to appear.

[44] ———, *A global bifurcation result for quasilinear elliptic equations in Orlicz-Sobolev spaces*, Topol. Methods Nonlinear Anal. **15** (2000), 301–327.

[45] ———, *A bifurcation result for equations with anisotropic p-Laplace-like operators*, Abstr. Appl. Anal. **6** (2001), 229–251.

[46] ———, *Subsolution-supersolution method in variational inequalities*, Nonlinear Analysis **45** (2001), 775–800.

[47] ———, *Nontrivial solutions of mountain pass type of quasilinear equations with slowly growing principal parts*, Differential Integral Equations **15** (2002), 839–862.

[48] ———, *Noncoercive problems in nonlinear elasticity with stored energies having fast or slow growth: Orlicz-Sobolev space setting*, J. Optim. Theory Appl. **119** (2003), 83–103.
[49] ———, *Generic existence result for an eigenvalue problem with rapidly growing principal operator*, ESAIM Control Optim. Calc. Var. **10** (2004), 677–691.
[50] ———, *Some existence and bifurcation results for quasilinear elliptic equations with slowly growing principal operators*, Houston J. Math. **32** (2006), 921–943 (electronic).
[51] ———, *Sub-supersolutions in a variational inequality related to a sandpile problem*, ANZIAM J. **48** (2006), 179–197.
[52] V. K. Le and D. Motreanu, *On nontrivial solutions of variational-hemivariational inequalities with slowly growing principal parts*, Z. Anal. Anwend. **28** (2009), 277–293.
[53] V. K. Le, D. Motreanu, and V. V. Motreanu, *On a nonsmooth eigenvalue problem in Orlicz–Sobolev spaces*, Applicable Analysis, to appear.
[54] V. K. Le and K. Schmitt, *Global bifurcation in variational inequalities: Applications to obstacle and unilateral problems*, Springer, Appl. Math. Sci., Vol. 123, New York, 1997.
[55] ———, *On boundary value problems for degenerate quasilinear elliptic equations and inequalities*, J. Diff. Eq. **144** (1998), 170–218.
[56] ———, *Quasilinear elliptic equations and inequalities with rapidly growing coefficients*, J. London Math. Soc. **62** (2000), 852–872.
[57] G. Lebourg, *Valeur moyenne pour gradient généralisé*, C. R. Acad. Sci. Paris **281** (1975), 795–797.
[58] N. G. Lloyd, *Degree theory*, Cambridge University Press, 1978.
[59] M. Mihăilescu and V. Rădulescu, *Existence and multiplicity of solutions for quasilinear nonhomogeneous problems: an Orlicz-Sobolev space setting*, J. Math. Anal. Appl. **330** (2007), 416–432.
[60] ———, *Eigenvalue problems associated with nonhomogeneous differential operators in Orlicz-Sobolev spaces*, Anal. Appl. (Singap.) **6** (2008), 83–98.
[61] ———, *Neumann problems associated to nonhomogeneous differential operators in Orlicz-Sobolev spaces*, Ann. Inst. Fourier (Grenoble) **58** (2008), 2087–2111.
[62] ———, *Nonhomogeneous Neumann problems in Orlicz-Sobolev spaces*, C. R. Math. Acad. Sci. Paris **346** (2008), 401–406.
[63] D. Motreanu and P. D. Panagiotopoulos, *Minimax theorems and qualitative properties of the solutions of hemivariational inequalities*, Nonconvex Optimization and its Applications, vol. 29, Kluwer Academic Publishers, Dordrecht, 1999.
[64] V. Mustonen and M. Tienari, *An eigenvalue problem for generalized Laplacian in Orlicz-Sobolev spaces*, Proc. Roy. Soc. Edinburgh Sect. A **129** (1999), 153–163.
[65] ———, *Remarks on inhomogeneous elliptic eigenvalue problems*, Partial differential equations, Lecture Notes in Pure and Appl. Math., 229 (New York), Dekker, 2002, pp. 259–265.
[66] M. N. Rao and Z. D. Ren, *Theory of Orlicz spaces*, Marcel Dekker, New York, 1985.
[67] K. Schmitt, *On boundary value problems for quasilinear elliptic equations*, Fields Institute Communications **21** (1999), 419–427.
[68] K. Schmitt and I. Sim, *Bifurcation problems associated with generalized Laplacians*, Adv. Diff. Equations **9** (2004), 797–828.
[69] M. Struwe, *Variational methods*, 2nd ed., Springer, Berlin, 1991.
[70] A. Szulkin, *Ljusternik–Schnirelmann theory on C^1-manifolds*, Ann. Inst. Henri Poincaré **5** (1988), 119–139.
[71] M. Tienari, *Ljusternik-Schnirelmann theorem for the generalized Laplacian*, J. Differential Equations **161** (2000), 174–190.
[72] E. Zeidler, *Nonlinear functional analysis and its applications, vol.1: Fixed-point theorems*, Springer, Berlin, 1986.

Vy Khoi Le, Department of Mathematics and Statistics, Missouri University of Science and Technology, Rolla, Missouri 65409
E-mail address: vy@mst.edu

Klaus Schmitt, Department of Mathematics, University of Utah, 155 South 1400 East, Salt Lake City, Utah 84112
E-mail address: schmitt@math.utah.edu

Non-smooth critical point theory

Roberto Livrea and Salvatore Angelo Marano

ABSTRACT. Extensions of main existence, multiplicity, and structure results in critical point theory to non-smooth functions given by a locally Lipschitz continuous term plus a convex, proper, lower semi-continuous function are presented and discussed. The problem of weakening the Palais-Smale conditions is also treated. Some abstract consequences as well as applications to elliptic hemivariational or variational-hemivariational inequalities are then pointed out.

CONTENTS

1. Introduction

The critical point theory for C^1 functions in a real Banach space is by now well established and excellent monographs devoted to various aspects of it are already available [**R1, St, Gh2, S, J**]. In particular, starting with the famous Mountain Pass Theorem (briefly, MPT) by Ambrosetti-Rabinowitz [**R1**, Theorem 2.2], a number of existence, multiplicity, or structure results have been obtained. Let us mention here

2000 *Mathematics Subject Classification.* 58E05, 49J35, 49J52.

Key words and phrases. Critical point, Mountain Pass Theorem, critical set, non-smooth function, weak Palais-Smale condition, splitting, variational-hemivariational inequality, elliptic equation with discontinuous nonlinearity.

(a) some dual versions of the MPT; vide [**R1**, Theorem 3.2] and, as regards the more general case of linking subsets, [**D**, Theorem 2.2].
(b) the Saddle Point Theorem [**R1**, Theorem 4.6].
(c) the Generalized MPT [**R1**, Theorem 5.3].
(d) the results where the strict inequality occurring in the MPT is weakened to allow also equality; see e.g. [**D**, Theorem 2.1].
(e) Benci-Rabinowitz's critical point theorem [**R1**, Theorem 5.29], which unifies (b) and (c).
(f) the Brézis-Nirenberg critical point theorem in presence of splitting [**BN**, Theorem 4].
(g) Pucci-Serrin's and Ghoussoub-Preiss' results on the structure of the critical set; vide [**PS3**] and [**GhPr**], respectively.
(h) the Ghoussoub min-max principle [**Gh1**], which can be exploited to get more general versions of (b)–(d), (f), and (g); cf. [**Gh2**].

Applications of the above-mentioned results - chiefly to differential equations - are countless, and their number is always increasing.

In 1981, through techniques of non-smooth analysis previously introduced by Clarke [**Cl**], K.-C. Chang, in the seminal paper [**Ch**], developed a critical point theory for locally Lipschitz continuous functions f in a real Banach space X, extending both the MPT and some of its variants to this more general framework. Let

$$f^0(x;z) := \limsup_{w\to x, t\to 0^+} \frac{f(w+tz)-f(w)}{t}, \qquad x,z\in X.$$

Obviously, if $f\in C^1(X,\mathbb{R})$ then $f^0(x;z) = \langle f'(x), z\rangle$. We say that $x\in X$ is a critical point of f when

$$f^0(x;z)\geq 0 \quad \forall z\in X, \tag{1}$$

which can also be written as $0\in\partial f(x)$, where

$$\partial f(x) := \{x^*\in X^* : \langle x^*, z\rangle \leq f^0(x;z) \forall z\in X\}.$$

The problem of finding a point $x\in X$ such that (1) holds is often called a *hemivariational inequality* and reveals extremely useful in studying elliptic equations with discontinuous nonlinearities. 'The interest in extending the theory of nonlinear PDEs to PDEs with discontinuous nonlinearities has been increasing daily. The reasons are twofold:

(1) Many free boundary problems may be reduced to BVPs for PDEs with discontinuous nonlinearities.
(2) Sometimes, in dealing with a PDE problem, it is beneficial to put the original PDE into a large category, for instance, PDEs with discontinuous nonlinearities.

Simply speaking, a PDE with discontinuous nonlinearity is as follows:

$$-\Delta u = \phi(u) \quad \text{in } \Omega, \quad u=0 \quad \text{on } \partial\Omega,$$

where ϕ is only a locally bounded measurable function. [...] Since $\phi(t)$ may be discontinuous, the corresponding functional

$$f(u) = \frac{1}{2}\int_\Omega |\nabla u(x)|^2 dx + \int_\Omega dx \int_0^{u(x)} \phi(t)dt$$

may be non-differentiable; variational methods for non-differentiable functions are needed'(from [**Ch**, Introduction]). We refer to [**C, J, BM0, GP, MR**] and the items therein for a complete account of the theory.

Later on, in 1986, A. Szulkin, in [**Su**], treated the case $f := \Phi + \psi$, where $\Phi : X \to \mathbb{R}$ belongs to $C^1(X, \mathbb{R})$ while $\psi : X \to \mathbb{R} \cup \{+\infty\}$ is convex, proper, and lower semi-continuous. A point $x \in X$ is called a critical point of f if

$$\langle \Phi'(x), z - x\rangle + \psi(z) - \psi(x) \geq 0 \quad \forall z \in X. \tag{2}$$

Inequalities of this type arise in many problems of Physics [**DL**] and examples given in [**Su**] include variational inequalities as well as variational equations with single- and multi-valued operators. Accordingly, a standard but meaningful case of ψ is $\psi := I_K$, with I_K being the indicator function of some nonempty, convex, closed set $K \subseteq X$.

After the papers of Chang [**Ch**] and Szulkin [**Su**], a critical point theory that applies to functionals $f : X \to \mathbb{R} \cup \{+\infty\}$ fulfilling the structural hypothesis

(H_f) *$f(x) := \Phi(x) + \psi(x)$ for all $x \in X$, where $\Phi : X \to \mathbb{R}$ is locally Lipschitz continuous while $\psi : X \to \mathbb{R} \cup \{+\infty\}$ is convex, proper, and lower semi-continuous,*

has been worked out by Motreanu-Panagiotopoulos [**MP**] in 1998. The notion of critical point is similar to that of Szulkin, but $\Phi^0(x; z - x)$ takes now the place of $\langle \Phi'(x), z - x\rangle$. Hence, (2) becomes

$$\Phi^0(x; z - x) + \psi(z) - \psi(x) \geq 0 \quad \forall z \in X. \tag{3}$$

'Such a mathematical model - called a *variational-hemivariational inequality* - can be used to formulate various problems in Mechanics involving both monotone and non-monotone possibly multivalued boundary conditions, whose complete description can be found in the books of Panagiotopoulos [**P**] and Naniewicz-Panagiotopoulos [**NaPa**]' (from [**MP**, p. 59]).

The main purpose of this chapter is to provide a short account of recent results, as well as their applications, in critical point theory for functions f that comply with (H_f). Section 2 contains basic definitions and preliminary results. The next one is devoted to non-smooth versions of classical critical point theorems under the usual Mountain Pass Geometry. Roughly speaking, Section 4 extends (h) to Motreanu-Panagiotopoulos' framework. The nature of min-max generated critical points is investigated in Section 6, while Section 7 deals with possible extensions of (f). A special mention deserves Section 5, where the key results of Section 4 are established under a weaker Palais-Smale hypothesis, that contains the so-called non-smooth Cerami condition.

Finally, it is worth noting that there is also a critical point theory for merely continuous functionals defined in metric spaces, which will not be presented here. Main results are due to Degiovanni and his co-authors; see [**DeG, Co**].

2. Elements of non-smooth analysis

Let $(X, \|\cdot\|)$ be a real Banach space. If U is a subset of X, we write $\mathrm{int}(U)$ for the interior of U, $\bar{U}$ for the closure of U, and ∂U for the boundary of U. Moreover, when $x \in X$ and $r > 0$, we define $B(x,r) := \{z \in X : \|z - x\| < r\}$, $B_r := B(0,r)$, as well as

$$d(x,U) := \inf_{z\in U} \|x - z\|, \quad N_r(U) := \{z \in X : d(z,U) < r\}.$$

The symbol X^* indicates the dual space of X, while $\langle \cdot, \cdot \rangle$ stands for the duality pairing between X^* and X. A function $\Phi : X \to \mathbb{R}$ is called coercive provided

$$\lim_{\|x\|\to+\infty} \Phi(x) = +\infty.$$

If to every $x \in X$ there correspond a neighborhood V_x of x and a constant $L_x \geq 0$ such that

$$|\Phi(z) - \Phi(w)| \leq L_x \|z - w\| \quad \forall z, w \in X,$$

then we say that f is locally Lipschitz continuous. In this case $\Phi^0(x; z)$ stands for the generalized directional derivative of Φ at the point $x \in X$ along the direction $z \in X$, namely

$$\Phi^0(x; z) := \limsup_{w\to x, t\to 0^+} \frac{\Phi(w + tz) - \Phi(w)}{t}.$$

It is known [**Cl**, Proposition 2.1.1] that Φ^0 is upper semi-continuous on $X \times X$. The generalized gradient of the function Φ in x, denoted by $\partial\Phi(x)$, is the set

$$\partial\Phi(x) := \{x^* \in X^* : \langle x^*, z\rangle \leq \Phi^0(x; z) \quad \forall z \in X\}.$$

Proposition 2.1.2 of [**Cl**] ensures that $\partial\Phi(x)$ is nonempty, convex, in addition to weak* compact, and that

$$\Phi^0(x; z) = \max\{\langle x^*, z\rangle : x^* \in \partial\Phi(x)\}.$$

Hence, it makes sense to put

$$m_\Phi(x) := \min\{\|x^*\|_{X^*} : x^* \in \partial\Phi(x)\}.$$

EXAMPLE 1. Let $j \in L^\infty_{\mathrm{loc}}(\mathbb{R})$. Put, for every $\delta > 0$ and $t \in \mathbb{R}$,

$$\underline{j}_\delta(t) := \mathrm{essinf}\{j(\tau) : |\tau - t| < \delta\}, \quad \overline{j}_\delta(t) := \mathrm{esssup}\{j(\tau) : |\tau - t| < \delta\},$$

$$\underline{j}(t) := \lim_{\delta\to 0^+} \underline{j}_\delta(t), \quad \overline{j}(t) := \lim_{\delta\to 0^+} \overline{j}_\delta(t).$$

Is is evident that $\underline{j}(t) \leq \overline{j}(t)$. Moreover, if

$$J(\xi) := \int_0^\xi j(t)dt, \quad \xi \in \mathbb{R},$$

then, on account of [**MR**, Proposition 1.7], one has

$$\partial J(\xi) = [\underline{j}(\xi), \overline{j}(\xi)].$$

Given $x, z \in X$, the symbol $[x, z]$ denotes the line segment joining x to z, i.e.,

$$[x, z] := \{x + t(z - x) : t \in [0, 1]\}.$$

The following result [**Cl**, Theorem 2.3.7] will be repeatedly used in the sequel.

THEOREM 2. *If U is an open subset of X, $x, z \in U$, while $\Phi : U \to \mathbb{R}$ is locally Lipschitz continuous in U then there exist $w \in [x, z]$ and $w^* \in \partial\Phi(w)$ such that $\Phi(z) - \Phi(x) = \langle w^*, z - x\rangle$.*

Now, let $\psi : X \to \mathbb{R} \cup \{+\infty\}$ be convex, proper, and lower semi-continuous. The function ψ is continuous on $\text{int}(D_\psi)$, where, as usual, $D_\psi := \{x \in X : \psi(x) < +\infty\}$; see for instance [**De**, Exercise 1, p. 296]. If $\partial\psi(x)$ indicates the sub-differential of ψ at the point $x \in X$, $D_{\partial\psi} := \{x \in X : \partial\psi(x) \neq \emptyset\}$, and X is reflexive then Theorem 23.5 in the same reference gives

$$\text{int}(D_\psi) = \text{int}(D_{\partial\psi}). \tag{4}$$

Recall also that

$$\psi(z) - \psi(x) \geq \langle z^*, z - x\rangle \quad \forall z \in D_\psi, z^* \in \partial\psi(x),$$

provided $x \in D_{\partial\psi}$. Moreover, by [**MM1**, Proposition 1.1], one has:

LEMMA 3. *Suppose X is reflexive and $x \in \text{int}(D_\psi)$. Then for every $x_n \to x$ in X and every $z_n^* \in \partial\psi(x_n)$, $n \in \mathbb{N}$, there exist a subsequence $\{z_{k_n}^*\}$ besides a point $z^* \in \partial\psi(x)$ such that $z_{k_n}^* \rightharpoonup z^*$ in X^*.*

Finally, let f be a function on X complying with the structural hypothesis

(H_f) *$f(x) := \Phi(x) + \psi(x)$ for all $x \in X$, where $\Phi : X \to \mathbb{R}$ is locally Lipschitz continuous while $\psi : X \to \mathbb{R} \cup \{+\infty\}$ is convex, proper, and lower semi-continuous.*

We say that $x \in X$ is a critical point of f when

$$\Phi^0(x; z - x) + \psi(z) - \psi(x) \geq 0 \quad \forall z \in X. \tag{5}$$

Given a real number a, write

$$K_a(f) := \{x \in X : f(x) = a, \ \ x \text{ is a critical point of } f\},$$

besides $f^a := \{x \in X : f(x) \geq a\}$ and $f_a := \{x \in X : f(x) \leq a\}$.

Sometimes, the following more restrictive assumption will be employed.

(H'_f) *f satisfies (H_f). Moreover, ψ is continuous on any nonempty compact set $A \subseteq X$ such that $\sup_{x \in A} \psi(x) < +\infty$.*

Although less general than (H_f), this condition still works in all the most important concrete situations.

If S is a closed subset of X, the Palais-Smale hypothesis around the set S at the level $c \in \mathbb{R}$ takes the form

$(\text{PS})_{f,S,c}$ *Every sequence $\{x_n\} \subseteq X$ such that $d(x_n, S) \to 0$, $f(x_n) \to c$, and*

$$\Phi^0(x_n; z - x_n) + \psi(z) - \psi(x_n) \geq -\varepsilon_n \|z - x_n\| \quad \text{for all } n \in \mathbb{N}, \ \ z \in X, \tag{6}$$

where $\varepsilon_n \to 0^+$, possesses a convergent subsequence.

When $S := X$ we simply write $(\mathrm{PS})_{f,c}$ in place of $(\mathrm{PS})_{f,X,c}$. The symbol $(\mathrm{PS})_f$ indicates $(\mathrm{PS})_{f,c}$ for all $c \in \mathbb{R}$.

Condition $(\mathrm{PS})_{f,c}$ coincides with the one exploited by Motreanu and Panagiotopoulos [**MP**, Definition 3.2]. If $\psi \equiv 0$, then it reduces to the Palais-Smale assumption introduced in [**Ch**], which can be easily verified via Proposition 3.1 of [**Su**]. Finally, for $\Phi \in C^1(X, \mathbb{R})$ and $\psi \equiv 0$, $(\mathrm{PS})_{f,S,c}$ includes the Palais-Smale condition adopted by Ghoussoub [**Gh1**, p. 440].

Likewise the C^1-case, there is a meaningful relation between $(\mathrm{PS})_f$ and coerciveness, as the next result shows. For the proof we refer the reader to [**MoMoPa**].

THEOREM 4. *Suppose the function f satisfies* (H_f) *and* $(\mathrm{PS})_f$. *If, moreover,*

$$\liminf_{\|x\| \to +\infty} f(x) > -\infty$$

then f is coercive.

The following definition of linking sets will be used in the sequel. For a deep treatment see [**S**].

DEFINITION 5. Let Q be a compact set in X and let Q_0 be a nonempty closed subset of Q. Put

$$\Gamma := \{\gamma \in C^0(Q, X) : \gamma|_{Q_0} = \mathrm{id}|_{Q_0}\}.$$

We say that the pair (Q, Q_0) links with S provided $Q_0 \cap S = \emptyset$ and for every $\gamma \in \Gamma$ one has $\gamma(Q) \cap S \neq \emptyset$. When Q is a topological manifold in X having a nonempty boundary $Q_0 := \partial Q$ and (Q, Q_0) links with S then we simply say that Q and S link.

Finally, the next version [**E1**, pp. 444, 456] of the famous variational principle of Ekeland will be also employed.

THEOREM 6. *Let (Z, δ) be a complete metric space and let $I : Z \to \mathbb{R} \cup \{+\infty\}$ be a proper, lower semi-continuous, bounded below function. Then to each $\varepsilon, r > 0$ and $\bar{z} \in Z$ satisfying $I(\bar{z}) \le \inf_{z \in Z} I(z) + \varepsilon$ there corresponds a point $z_0 \in Z$ such that*

$$I(z_0) \le I(\bar{z}), \quad \delta(z_0, \bar{z}) \le \frac{1}{r}, \quad I(z) - I(z_0) \ge -\varepsilon r \delta(z, z_0) \quad \forall z \in Z.$$

3. Critical points of non-smooth functions

From now on, $(X, \|\cdot\|)$ denotes a real Banach space while $f : X \to \mathbb{R} \cup \{+\infty\}$. A first elementary but meaningful result is the following [**MM0**, Proposition 2.1].

THEOREM 7. *Let f satisfy* (H_f). *Then each local minimum is also a critical point of f.*

PROOF. Suppose x is a local minimum of f and fix $z \in X$. Using the convexity of ψ yields

$$0 \le f((1-t)x + tz) - f(x) \le \Phi(x + t(z - x)) - \Phi(x) + t(\psi(z) - \psi(x))$$

for all small $t > 0$. Dividing by t and letting $t \to 0^+$ we infer (5). $\square$

Moreover, one has

THEOREM 8. *Assume f is bounded below and fulfils* $(\mathrm{PS})_f$ *in addition to* (H_f). *Then each minimizing sequence for f possesses a convergent subsequence. So, in particular, f attains its minimum at some $x_0 \in X$.*

PROOF. Let $\{x_n\} \subseteq X$ fulfil $\lim_{n\to+\infty} f(x_n) = \inf_{x\in X} f(x)$. Passing to a subsequence if necessary, we may suppose $f(x_n) \leq \inf_{x\in X} f(x) + 1/n^2$, $n \in \mathbb{N}$. By Theorem 6, for every $n \in \mathbb{N}$ there exists a point $z_n \in X$ enjoying the following properties:

$$f(z_n) \leq f(x_n); \tag{7}$$

$$\|z_n - x_n\| \leq \frac{1}{n}; \tag{8}$$

$$f(z) - f(z_n) \geq -\frac{1}{n}\|z - z_n\| \quad \forall z \in X. \tag{9}$$

Through (7) we obtain that $\{f(z_n)\}$ is bounded, while (9) leads to

$$\Phi^0(z_n; x - z_n) + \psi(x) - \psi(z_n) \geq -\frac{1}{n}\|x - z_n\| \quad \forall x \in X. \tag{10}$$

Indeed, if $x \in X$ and $z := z_n + t(x - z_n)$, with $t \in]0,1[$, then from (9) besides (H_f) it follows

$$\Phi(z_n + t(x - z_n)) - \Phi(z_n) + t\,[\psi(x) - \psi(z_n)] \geq -\frac{t}{n}\|x - z_n\|.$$

Dividing by t and letting $t \to 0^+$ we achieve (10). At this point, condition $(\mathrm{PS})_f$ forces $z_n \to x_0$ for suitable $x_0 \in X$, where a subsequence is considered when necessary, and thus, by (8), also $x_n \to x_0$. □

The following deformation lemma is a special case of Theorem 3.1 in [**MP**]. It can also be established by easily adapting the proof of [**Su**, Proposition 2.3].

PROPOSITION 9. *Suppose f fulfils* (H_f) *and* $(\mathrm{PS})_{f,c}$ *for some $c \in \mathbb{R}$ with $K_c(f) = \emptyset$. Let $\bar{\varepsilon} > 0$. Then there exists $\varepsilon \in]0, \bar{\varepsilon}[$ such that to every compact set $A \subseteq X$ satisfying*

$$c \leq \sup_{x\in A} f(x) \leq c + \varepsilon$$

there corresponds a closed neighborhood W of A as well as a continuous function $\alpha : W \times [0, \bar{s}] \to X$, $\bar{s} > 0$, having the properties:

- (p_1) $\|x - \alpha(x, s)\| \leq s \quad \forall (x, s) \in W \times [0, \bar{s}]$.
- (p_2) $f(\alpha(x, s)) - f(x) \leq Ms \quad \forall (x, s) \in W \times [0, \bar{s}]$, *where the constant $M > 0$ does not depend on (x, s).*
- (p_3) $f(\alpha(x, s)) - f(x) \leq -2\varepsilon s \quad \forall (x, s) \in W \times [0, \bar{s}]$ *with $f(x) \geq c - \varepsilon$.*
- (p_4) $\sup_{x\in A} f(\alpha(x, s)) - \sup_{x\in A} f(x) \leq -2\varepsilon s \quad \forall s \in [0, \bar{s}]$.

Through the same arguments made in the proof of [**Su**, Theorem 3.2], but applying Proposition 9 in place of [**Su**, Proposition 2.3], we obtain the result below, which unifies [**Su**, Theorem 3.2] and [**Ch**, Theorem 3.4].

THEOREM 10. *Let the function* f *satisfy* (H_f) *and* $(\mathrm{PS})_f$. *If* $f(0) = 0$ *and, moreover,*

(i_1) *there exist* $a, \rho > 0$ *such that* $f|_{\partial B_\rho} \geq a$,
(i_2) $f(e) \leq 0$ *for some* $e \notin B_\rho$,

then f *has a critical value* $c \geq a$ *that may be characterized by*

$$c = \inf_{\gamma \in \Gamma} \sup_{t \in [0,1]} f(\gamma(t)),$$

where $\Gamma := \{\gamma \in C^0([0,1], X) : \gamma(0) = 0, \gamma(1) = e\}$.

A much more general result, where the strict inequality

$$f(e) < \inf_{x \in \partial B_\rho} f(x)$$

is weakened to allow also equality, will be presented in Section 4.

Finally, a useful sufficient condition for $(\mathrm{PS})_f$ holds true is the following.

THEOREM 11. *Let* X *be reflexive and let* f *fulfil* (H_f). *Assume that:*

(i_3) *There exist a real Banach space* $\tilde{X}$ *and a locally Lipschitz continuous function* $\tilde{\Phi} : \tilde{X} \to \mathbb{R}$ *such that* X *compactly embeds in* $\tilde{X}$ *and* $\tilde{\Phi}|_X = \Phi$.
(i_4) f *is coercive.*
(i_5) *Each sequence* $\{x_n\} \subseteq D_\psi$ *complying with* $x_n \rightharpoonup x$ *in* X *and*

$$\limsup_{n \to +\infty} \psi(x_n) \leq \psi(x)$$

possesses a strongly convergent subsequence.

Then f *satisfies condition* $(\mathrm{PS})_f$.

PROOF. Pick $c \in \mathbb{R}$, $\{x_n\} \subseteq X$ such that $f(x_n) \to c$ and (6) holds for some $\varepsilon_n \to 0^+$. By (i_4) the sequence $\{x_n\}$ is bounded. Therefore, $x_n \rightharpoonup x$ and $x_n \to x$ in $\tilde{X}$, where a subsequence is considered when necessary. Setting $z := x$ in (6) produces

$$\tilde{\Phi}^0(x_n; x - x_n) + \psi(x) - \psi(x_n) \geq -\varepsilon_n \|x - x_n\|, \quad n \in \mathbb{N}.$$

Because of the upper semi-continuity of $\tilde{\Phi}^0$ this implies $\limsup_{n \to +\infty} \psi(x_n) \leq \psi(x)$. Now the conclusion follows at once from (i_5). □

4. The limit case

Once observed that the 'geometry' of the MPT, also in the more general framework of linking sets, requires that a strict inequality is satisfied, a natural question arises in order to weaken this condition, i.e., whether the conclusions of the obtained critical points theorems remain true when standard strict inequalities are weakened to be equalities.

Complete and satisfactory results in this direction have been first obtained in the C^1 case by Pucci-Serrin [**PS1**, Theorem 1], Rabinowitz [**R2**, Theorem 2.13], Ghoussoub-Preiss [**GhPr**, Theorem 1.bis], and, in the framework of linking sets, Du [**D**, Theorem 2.1]. Later on, Motreanu and Varga studied the same question for locally Lipschitz continuous functionals [**MV**, Theorem 2.1]. The more general case of functions having the structure (H_f) was treated first in [**MM1**], where a deformation result for non-smooth functionals which can be expressed as a sum of a locally Lipschitz continuous function and a concave, proper, upper semi-continuous function (see Theorem 15 below) is presented. It represents the main tool in order to obtain a general min-max principle for functions of (H_f) type.

More precisely, let $g : X \to [-\infty, +\infty[$ be a function defined on a real reflexive Banach space $(X, \|\cdot\|)$ such that

(h_g) *$g = \Psi + \beta$ where $\Psi : X \to \mathbb{R}$ is locally Lipschitz and $\beta : X \to [-\infty, +\infty[$ is concave, proper (i. e., $\neq -\infty$), and upper semi-continuous.*

Put

$$D_\beta = \{x \in X : \beta(x) > -\infty\}$$

and

$$\partial\beta(x) = -\partial(-\beta)(x), x \in X,$$

where $\partial(-\beta)$ means the sub-differential of $-\beta$ in the sense of convex analysis.

The function g is said to fulfil the Palais-Smale condition around a set $B \subseteq X$ at a level $c \in \mathbb{R}$ if:

$(\mathrm{ps})_{g,B,c}$ *Each sequence $\{x_n\} \subseteq X$ such that $d(x_n, B) \to 0$, $g(x_n) \to c$, and*

$$\Psi^0(x_n; x_n - x) + \beta(x_n) - \beta(x) \geq -\varepsilon_n \|x_n - x\| \quad \text{for all } n \in \mathbb{N},\ x \in X,$$

where $\varepsilon_n \to 0^+$, possesses a convergent subsequence.

The following assumption will be posited in the sequel.

(g) *A and B are two nonempty closed subsets of X such that*

$$A \cap B = \emptyset, \quad A \subseteq g^c, \quad B \subseteq g_c, \quad K_c(g) \cap B = \emptyset.$$

Moreover, there exists $\varepsilon_0 > 0$ satisfying $N_{\varepsilon_0}(B) \subseteq \mathrm{int}(D_\beta)$.

The next three lemmas represent preliminary steps to establish the announced deformation result.

LEMMA 12. [**MM1**, Lemma 2.2] *Let (h_g), (g) and $(\mathrm{ps})_{g,B,c}$ be fulfilled. Then there exist $\varepsilon_1 \in]0, \varepsilon_0[$, $\sigma > 0$ such that for every $x \in N_{\varepsilon_1}(B) \cap g^{c-\varepsilon_1} \cap g_{c+\varepsilon_1}, x^* \in \partial\Psi(x), z^* \in \partial\beta(x)$ one has $\|x^* + z^*\| \geq \sigma$.*

LEMMA 13. [**MM1**, Lemma 2.3] *Suppose the function g satisfies (h_g), (g) and $(\mathrm{ps})_{g,B,c}$, while ε_1, σ are as in Lemma 12. Then to every $x \in N_{\varepsilon_1}(B) \cap g^{c-\varepsilon_1} \cap g_{c+\varepsilon_1}$ there corresponds a point $\xi_x \in X$ such that*

$$\|\xi_x\| = 1, \langle x^* + z^*, \xi_x \rangle \geq \sigma \quad \forall x^* \in \partial\Psi(x), \ z^* \in \partial\beta(x).$$

LEMMA 14. [**MM1**, Lemma 2.4] *Let (h_g), (g) and $(\mathrm{ps})_{g,B,c}$ be fulfilled and let ε_1, σ be as in Lemma 12. Then for every $x \in N_{\varepsilon_1}(B) \cap g^{c-\varepsilon_1} \cap g_{c+\varepsilon_1}$ there exists $\delta_x > 0$ such that*

$$\langle x^* + z^*, \xi_x \rangle > \frac{\sigma}{2} \quad \forall x^* \in \partial\Psi(x'), \ z^* \in \partial\beta(x''), \ x', \ x'' \ B(x, \delta_x),$$

where ξ_x is given in Lemma 13.

The following result extends [**D**, Lemma 1.1] and [**MV**, Theorem 1.1] to the more general framework of functional complying with (h_g).

THEOREM 15. [**MM1**, Theorem 2.5] *Assume the function g satisfies* (h_g), (g), $(\mathrm{ps})_{g,B,c}$ *and the set* $N_{\varepsilon_1}(B) \cap g^{c-\varepsilon_1} \cap g_{c+\varepsilon_1}$, *with* ε_1 *as in Lemma 12, is closed. Then there exist* $\varepsilon > 0$ *and a homeomorphism* $\eta : X \to X$ *having the following properties:*

(a) $\eta(x) = x$ *for every* $x \in A$,
(b) $\eta(B) \subseteq g_{c-\varepsilon}$.

REMARK 16. For the proof we refer the reader to [**MM1**]. However, it may be of interest to emphasize that the homeomorphism η is obtained through the unique solution of a suitable Cauchy problem in a Banach space, whose right hand-side is carefully constructed using a particular locally Lipschitz continuous vector field, which exists thanks to Lemmas 12-14.

Here is our first general critical point result [**MM1**, Theorem 3.1].

THEOREM 17. *Suppose Q and S link, in the sense of Definition 5, while the function f satisfies the following assumptions in addition to* (H_f).

(f_1) $\sup_{x\in Q} f(x) < \infty$.
(f_2) $\partial Q \subseteq f_a$ *and* $S \subseteq f^a$ *for some* $a \in \mathbb{R}$.
(f_3) *Setting*

$$c = \inf_{\gamma\in\Gamma} \sup_{z\in\gamma(Q)} f(z)$$

either $(\mathrm{PS})_{f,c}$ *or* $(\mathrm{PS})_{f,S,c}$ *holds according to whether* $c > a$ *or* $c = a$. *Further, there exists* $\varepsilon_0 > 0$ *such that*

$(\mathrm{f}_{3.1})$ $N_{\varepsilon_0}(S) \subseteq \mathrm{int}(D_\psi)$, *and*
$(\mathrm{f}_{3.2})$ *the set* $N_\delta(S) \cap f^{c-\delta} \cap f_{c+\delta}$, $\delta \in]0, \varepsilon_0[$, *is closed.*

Then one has

(j_1) $c \geq a$,
(j_2) $K_c(f) \setminus \partial Q \neq \emptyset$, *and*
(j_3) $K_c(f) \cap S \neq \emptyset$ *if* $c = a$.

A dual version of Theorem 17 is the following; see [**MoMo**, Theorem 2].

THEOREM 18. *Let Q and S link and let f fulfil* (f_1) *besides* (H_f). *Define*

$$a := \inf_{x\in S} f(x), \quad b := \sup_{\gamma^*\in\Gamma^*} \inf_{z\in\gamma^*(Q)} f(z),$$

where $\Gamma^* := \{\gamma^* \in C^0(X,X) : \gamma^*$ *is a homeomorphism and* $\gamma^*|_{\partial Q} = \mathrm{id}|_{\partial Q}\}$. *If* $\partial Q \subseteq f_a$, *condition* $(\mathrm{PS})_{f,S,a}$ *holds,* $a = b$, *and, moreover, there exists* $\varepsilon_0 > 0$ *such that* $N_{\varepsilon_0}(S) \subseteq \mathrm{int}(D_\psi)$ *and the set* $N_\delta(S) \cap f^{a-\delta} \cap f_{a+\delta}$, $\delta \in]0, \varepsilon_0[$, *is closed, then* $K_a(f) \cap S \neq \emptyset$.

Another general min-max principle for non-smooth functionals can be achieved via a completely different approach. In particular, by adapting the duality and perturbation techniques developed in [**Gh1**] for C^1 functions and using the new structural hypothesis (H'_f), Theorem 1 of [**Gh1**] can be generalized to non-differentiable functions. Although less general than (H_f), condition (H'_f) still works in all the most important concrete situations. For example, think to case when ψ is the indicator function of some convex closed subset of X.

The deformation result below extends [**Gh1**, Lemma 1] to the framework of functions satisfying (H_f). It will be essential in formulating our non-smooth min-max principle.

THEOREM 19. [**LM1**, Theorem 2.2] *Suppose* (H_f) *is fulfilled,* $\varepsilon > 0$*, while* B, C *are two nonempty closed sets in* X*. If* C *is compact,* $B \cap C = \emptyset$*,* $C \subseteq D_\psi$*, and, moreover,*

(a_1) *to each* $x \in C$ *there corresponds a point* $\xi_x \in X$ *such that*

$$\Phi^0(x; \xi_x - x) + \psi(\xi_x) - \psi(x) < -\varepsilon\|\xi_x - x\|,$$

then for every $k > 1$ *there exist* $t_0 \in (0,1]$*,* $\alpha \in C^0([0,1]\times X, X)$*, and* $\varphi \in C^0(X, \mathbb{R}_0^+)$ *with the following properties:*

(i_1) $\alpha(t, D_\psi) \subseteq D_\psi$ $\forall t \in [0, t_0)$ *and* $\alpha(t,x) = x$ $\forall (t,x) \in [0,t_0) \times B$.
(i_2) $\|\alpha(t,x) - x\| \le kt$ $\forall (t,x) \in [0,t_0) \times X$.
(i_3) $f(\alpha(t,x)) - f(x) \le -\varepsilon\varphi(x)t$ $\forall (t,x) \in [0,t_0) \times D_\psi$.
(i_4) $\varphi(x) = 1$ $\forall x \in C$.

PROOF. Pick $x \in C$. By (a_1) and the upper semi-continuity of the function $(z,w) \mapsto \Phi^0(z;\xi_x - w) + \psi(\xi_x) - \psi(w) + \varepsilon\|\xi_x - w\|$ on $X \times X$, we can find a positive number $\delta_x < \|\xi_x - x\|$ such that

$$\Phi^0(z;\xi_x - w) + \psi(\xi_x) - \psi(w) < -\varepsilon\|\xi_x - w\| \quad \forall z, w \in B(x,\delta_x). \tag{11}$$

Moreover, δ_x can be chosen small enough such that Φ is Lipschitz continuous in $B(x,\delta_x)$. Define

$$V_x := B(x, \delta_x/4), \quad U_x := N_{\delta_x/4}(V_x) = B(x, \delta_x/2).$$

The family $\mathcal{B} = \{V_x : x \in C\}$ represents an open covering of C. Since this set is compact, $\mathcal{B}$ possesses a finite sub-covering $\{V_{x_j} : j = 1,2,\dots,m\}$ to which we can associate a continuous partition of unity $\{\chi_j : j = 1,2,\dots,m\}$, with $\sum_{j=1}^m \chi_j(x) = 1$ for every $x \in C$ and $\sum_{j=1}^m \chi_j(x) \le 1$ for all $x \in X$. To simplify notation, write

$$\xi_j := \xi_{x_j}, \quad \delta_j := \delta_{x_j}, \quad V_j := V_{x_j}, \quad U_j := U_{x_j}.$$

Let $l : X \to [0,1]$ be a continuous function such that

$$l(x) = 1 \quad \forall x \in C, \quad l(x) = 0 \quad \forall x \in X \setminus \cup_{j=1}^m V_j \cup B.$$

Put $\delta_0 = \frac{1}{2}\min_{1\le j\le m}\delta_j$, fix $k > 1$, and choose a positive number t_0 with $t_0 \le 1$ such that

$$t_0 < \min\left\{\frac{\delta_0}{1+k^2}, \min_{1\le j\le m} d(\xi_j, U_j)\right\}. \tag{12}$$

Starting with $\alpha_0(t,x)=x$, for every $(t,x)\in[0,1]\times X$, we define by induction on j, $1\le j\le m$, the functions $\alpha_j:[0,1]\times X\to X$ as follows:

$$\alpha_j(t,x):=\begin{cases}\alpha_{j-1}(t,x)+tl(x)\chi_j(x)\dfrac{\xi_j-\alpha_{j-1}(t,x)}{\|\xi_j-\alpha_{j-1}(t,x)\|} & \text{if}\alpha_{j-1}(t,x)\in U_j,\\ \\ \alpha_{j-1}(t,x) & \text{otherwise.}\end{cases}$$

Observe that, since $\xi_j\notin B(x_j,\delta_j)$, α_j is well-defined. We claim that for any $t\in[0,t_0)$ one has:

$$\alpha_j(t,D_\psi)\subseteq D_\psi; \tag{13}$$

$$\|\alpha_j(t,x)-\alpha_{j-1}(t,x)\|\le kl(x)\chi_j(x)t \quad \forall x\in X; \tag{14}$$

$$f(\alpha_j(t,x))-f(\alpha_{j-1}(t,x))\le -\varepsilon l(x)\chi_j(x)t \quad \forall x\in D_\psi. \tag{15}$$

Indeed, fix $j=1$. If $x=\alpha_0(t,x)\notin U_1$ then (13) and (14) are obvious. Since $x\notin V_1$, we get $\chi_1(x)=0$, which yields (15). Suppose now $x=\alpha_0(t,x)\in U_1$. In this case, (13) immediately follows from $\xi_1\in D_\psi$ and the convexity of D_ψ while (14) is trivial. In order to verify (15), by Theorem 2, there exist suitable $z\in[\alpha_0(t,x),\alpha_1(t,x)]$ and $z^*\in\partial\Phi(z)$ such that

$$\begin{aligned}f(\alpha_1(t,x))-f(\alpha_0(t,x))&=\langle z^*,\alpha_1(t,x)-\alpha_0(t,x)\rangle+\psi(\alpha_1(t,x))-\psi(\alpha_0(t,x))\\ &\le\tau\left[\Phi^0(z;\xi_1-\alpha_0(t,x))+\psi(\xi_1)-\psi(\alpha_0(t,x))\right],\end{aligned} \tag{16}$$

where

$$\tau:=\frac{tl(x)\chi_1(x)}{\|\xi_1-\alpha_0(t,x)\|}<1.$$

On account of (12) and (14) one has

$$\begin{aligned}\|z-x_1\|&\le\|z-\alpha_0(t,x)\|+\|\alpha_0(t,x)-x_1\|\\ &<\|\alpha_1(t,x)-\alpha_0(t,x)\|+\frac{\delta_1}{2}\le kl(x)\chi_1(x)t+\frac{\delta_1}{2}<\delta_1.\end{aligned}$$

Therefore, by (11),

$$\Phi^0(z;\xi_1-\alpha_0(t,x))+\psi(\xi_1)-\psi(\alpha_0(t,x))<-\varepsilon\|\xi_1-\alpha_0(t,x)\|. \tag{17}$$

Combining (16) and (17) yields (15). Assume now (13)–(15) hold up to $j-1$; we will prove them for $\alpha_j(t,x)$. The same reasoning made before gives (13) and (14). Let us next observe that

$$\chi_j(x)=0 \quad \text{as soon as} \quad \alpha_{j-1}(t,x)\notin U_j. \tag{18}$$

Indeed, if $\chi_j(x)\ne0$ for some point x fulfilling $\alpha_{j-1}(t,x)\notin U_j$ then $x\in V_j$. So, thanks to (14),

$$\begin{aligned}\|\alpha_{j-1}(t,x)-x_j\|&\le\|\alpha_{j-1}(t,x)-x\|+\|x-x_j\|\\ &<\sum_{h=1}^{j-1}\|\alpha_h(t,x)-\alpha_{h-1}(t,x)\|+\frac{\delta_j}{4}\le kl(x)t\sum_{h=1}^{j-1}\chi_h(x)+\frac{\delta_j}{4}<\frac{\delta_j}{2}\end{aligned}$$

that is $\alpha_{j-1}(t,x) \in B(x_j, \delta_j/2)$ against the condition $\alpha_{j-1}(t,x) \notin U_j$. Therefore, if $\alpha_{j-1}(t,x) \notin U_j$, inequality (15) immediately follows from (18), while if $\alpha_{j-1}(t,x) \in U_j$ it can be established arguing exactly as in the case $j = 1$, with

$$\tau = \frac{tl(x)\chi_j(x)}{\|\xi_j - \alpha_{j-1}(t,x)\|} < \frac{d(\xi_j, B(x_j, \delta_j/2))}{\|\xi_j - \alpha_{j-1}(t,x)\|} \leq 1.$$

Suitable arguments relying on (18) then imply that each α_j is continuous.

Define $\alpha(t,x) = \alpha_m(t,x)$ and $\varphi(x) = l(x)\sum_{j=1}^m \chi_j(x)$, $(t,x) \in [0,1] \times X$. Since $x \in B$ implies $l(x) = 0$, by (14) we get

$$\alpha(t,x) = \alpha_{m-1}(t,x) = \ldots = \alpha_0(t,x) = x,$$

and (i_1) is proved. Assertions (i_2) and (i_3) are consequences of the properties of l and χ_j $(j = 1, 2, \ldots, m)$ as well as (14) and (15), respectively. Finally, since $l_{|C} \equiv 1$ one has $\varphi(x) = l(x)\sum_{j=1}^m \chi_j(x) = 1$ for all $x \in C$, which shows (i_4). □

The notion of homotopy-stable family due to Ghoussoub, and recalled in Definition 20 below, will be useful in the sequel, since it allow to state critical point results requiring very general Mountain Pass Geometries.

DEFINITION 20. Let $B \subseteq X$ be closed and let $\mathcal{F}$ be a class of compact sets in X. We say that $\mathcal{F}$ is a *homotopy-stable family with extended boundary* B when for every $A \in \mathcal{F}$ and every $\eta \in C^0([0,1] \times X, X)$ such that $\eta(t,x) = x$ for all $(t,x) \in (\{0\} \times X) \cup ([0,1] \times B)$ one has $\eta(\{1\} \times A) \in \mathcal{F}$.

Some meaningful situations are special cases of this notion.

EXAMPLE 21. If Q denotes a compact set in X, Q_0 is a nonempty closed subset of Q, γ_0 belongs to $C^0(Q_0, X)$, and $\Gamma = \{\gamma \in C^0(Q,X) : \gamma_{|Q_0} = \gamma_0\}$, then the family $\mathcal{F} = \{\gamma(Q) : \gamma \in \Gamma\}$ enjoys the above-mentioned properties with $B = \gamma_0(Q_0)$. In particular, it holds when Q is a compact topological manifold in X having nonempty boundary Q_0, while $\gamma_0 = \mathrm{id}_{|Q_0}$.

We will make use of the following assumptions:

(a_2) *$\mathcal{F}$ is a homotopy-stable family with extended boundary B, the function f fulfills condition (H'_f), and*

$$c := \inf_{A \in \mathcal{F}} \sup_{x \in A} f(x) < +\infty. \tag{19}$$

(a_3) *There exists a closed subset F of X such that*

$$(F \cap A) \setminus B \neq \emptyset \quad \forall A \in \mathcal{F} \quad \text{and} \quad \sup_{x \in B} f(x) \leq \inf_{x \in F} f(x).$$

By (a_2) and (a_3) one has

$$\inf_{x \in F} f(x) \leq c. \tag{20}$$

THEOREM 22. *Let (a_2)–(a_3) be satisfied. Then, to every sequence $\{A_n\} \subseteq \mathcal{F}$ such that $\lim_{n\to+\infty} \sup_{x \in A_n} f(x) = c$ there corresponds a sequence $\{x_n\} \subseteq X \setminus B$ fulfilling the following properties:*

(i_5) $\lim_{n\to+\infty} f(x_n) = c$.

(i_6) $\Phi^0(x_n; v - x_n) + \psi(v) - \psi(x_n) \geq -\varepsilon_n \|v - x_n\|$ $\forall n \in \mathbb{N}, v \in X$, *where* $\varepsilon_n \to 0^+$.

(i_7) $\lim_{n\to+\infty} d(x_n, F) = 0$ *provided* $\inf_{x\in F} f(x) = c$.

(i_8) $\lim_{n\to+\infty} d(x_n, A_n) = 0$.

PROOF. First we consider the case

$$\inf_{x\in F} f(x) = c. \tag{21}$$

Fix $\varepsilon > 0$ and let $A_\varepsilon \in \mathcal{F}$ be such that

$$c \leq \sup_{x\in A_\varepsilon} f(x) < c + \frac{\varepsilon^2}{8}. \tag{22}$$

We claim that there exists $x_\varepsilon \in X$ such that

$$c - \varepsilon^2/8 \leq f(x_\varepsilon) \leq c + \frac{5\varepsilon^2}{4}, \tag{23}$$

$$\Phi^0(x_\varepsilon; v - x_\varepsilon) + \psi(v) - \psi(x_\varepsilon) \geq -5\varepsilon\|v - x_\varepsilon\| \ \forall v \in X, \tag{24}$$

$$d(x_\varepsilon, F) \leq \frac{3\varepsilon}{2}, \tag{25}$$

$$d(x_\varepsilon, A_\varepsilon) \leq \frac{\varepsilon}{2}, \tag{26}$$

which provides a sequence $\{x_n\} \subseteq X \setminus B$ enjoying properties (i_5)–(i_8). To simplify notation, put

$$A'_\varepsilon := \{1\} \times A_\varepsilon, \quad F_\varepsilon := N_\varepsilon(F), \quad G_\varepsilon := (\{0\} \times X) \cup ([0,1] \times ((A_\varepsilon \setminus F_\varepsilon) \cup B)),$$

and let $\mathcal{L}$ be the space of all $\eta \in C^0([0,1] \times X, X)$ such that

$$\eta(t,x) = x \quad \forall (t,x) \in G_\varepsilon, \qquad \sup_{(t,x)\in[0,1]\times X} \|\eta(t,x) - x\| < +\infty.$$

An easy computation shows that $\mathcal{L}$, equipped with the metric ρ of uniform convergence, is complete. Moreover, since $(\{0\} \times X) \cup ([0,1] \times B) \subseteq G_\varepsilon$ it results in

$$\eta(A'_\varepsilon) \in \mathcal{F} \quad \forall \eta \in \mathcal{L}. \tag{27}$$

Define now, for every $x \in X$,

$$f_1(x) = \max\left\{0, \varepsilon^2 - \varepsilon d(x,F)\right\}, \quad f_2(x) = \min\left\{\varepsilon^2/8, \varepsilon d(x, (A_\varepsilon \setminus F_\varepsilon) \cup B)\right\},$$
$$g(x) = f(x) + f_1(x) + f_2(x).$$

Moreover, let $I : \mathcal{L} \to \mathbb{R} \cup \{+\infty\}$ be the function defined by

$$I(\eta) = \sup_{z\in\eta(A'_\varepsilon)} g(z), \quad \forall \eta \in \mathcal{L},$$

which clearly is lower semi-continuous. Combining (27) with assumption (a_3) one has

$$(\eta(A'_\varepsilon) \cap F) \setminus B \neq \emptyset.$$

Hence, for each $\eta \in \mathcal{L}$ we get

$$\begin{aligned} I(\eta) &\geq \sup_{z\in\eta(A'_\varepsilon)\cap F} g(z) \geq \sup_{z\in\eta(A'_\varepsilon)\cap F} (f(z)+f_1(z)) \\ &= \sup_{z\in\eta(A'_\varepsilon)\cap F} f(z)+\varepsilon^2 \geq c+\varepsilon^2, \end{aligned}$$

that is

(28) $$\inf_{\eta\in\mathcal{L}} I(\eta) \geq c+\varepsilon^2.$$

Therefore, bearing in mind the above inequality and (22), putting $\overline{\eta}(t,x)=x$ for each $(t,x)\in[0,1]\times X$, one has

(29) $$I(\overline{\eta}) < \inf_{\eta\in\mathcal{L}} I(\eta)+\frac{\varepsilon^2}{4}.$$

By Theorem 6, there exists $\eta_0 \in \mathcal{L}$ such that

(30) $$I(\eta_0) \leq I(\overline{\eta}),$$

(31) $$\rho(\overline{\eta},\eta_0) \leq \frac{\varepsilon}{2},$$

(32) $$I(\eta) \geq I(\eta_0) - \frac{\varepsilon}{2}\rho(\eta,\eta_0) \quad \forall \eta\in\mathcal{L}.$$

From (30) one has

$$\sup_{z\in\eta_0(A'_\varepsilon)} \psi(z) \leq I(\overline{\eta}) - \min_{z\in\eta_0(A'_\varepsilon)} \Phi(z) < +\infty.$$

Thus, by (H'_f), the function ψ turns out to be continuous on $\eta_0(A'_\varepsilon)$ as well as $x\mapsto g(\eta_0(1,x))$, $\forall x\in A_\varepsilon$. So, the set

$$C := \left\{ z\in\eta_0(A'_\varepsilon) : g(z) = \max_{w\in\eta_0(A'_\varepsilon)} g(w) \right\}$$

is nonempty and compact. Now, we write $B' = (A_\varepsilon \setminus F_\varepsilon)\cup B$ and verify that

(33) $$B'\cap C = \emptyset.$$

To this end, first we show that there exists $z_0 \in (\eta_0(A'_\varepsilon)\cap F)\setminus B$ such that

(34) $$f(z_0) = \max_{x\in\eta_0(A'_\varepsilon)\cap F} f(x).$$

Let $\widehat{z}\in\eta_0(A'_\varepsilon)\cap F$ be such that $f(\widehat{z}) = \max_{z\in\eta_0(A'_\varepsilon)\cap F} f(z)$. If $\widehat{z}\notin B$ then we can choose $z_0=\widehat{z}$. Otherwise, by (a_3), there exists $z_0 \in (\eta_0(A'_\varepsilon)\cap F)\setminus B$ such that

$$\max_{z\in\eta_0(A'_\varepsilon)\cap F} f(z) = f(\widehat{z}) \leq \sup_{x\in B} f(x) \leq \inf_{x\in F} f(x) \leq f(z_0) \leq \max_{z\in\eta_0(A'_\varepsilon)\cap F} f(z)$$

and (34) holds. Moreover,

(35) $$\max_{z\in\eta_0(A'_\varepsilon)} g(z) \geq f(z_0)+f_1(z_0)+f_2(z_0) \geq c+\varepsilon^2+f_2(z_0),$$

with $z_0 \notin B \cup \overline{(A_\varepsilon \setminus F_\varepsilon)}$ by the fact that $(\eta_0(A'_\varepsilon) \cap F) \cap \overline{(A_\varepsilon \setminus F_\varepsilon)} = \emptyset$. Hence, $f_2(z_0) > 0$. On the other hand, $f_{1|X\setminus F_\varepsilon} \equiv f_{2|A_\varepsilon \setminus F_\varepsilon} \equiv 0$ implies

$$\sup_{z\in A_\varepsilon \setminus F_\varepsilon} g(z) \le \sup_{z\in A_\varepsilon} f(z) < c + \frac{\varepsilon^2}{8} \tag{36}$$

and, taking into account that $\sup_{z\in B} f(z) \le c$, we get

$$\sup_{z\in B\cap F_\varepsilon} g(z) = \sup_{z\in B\cap F_\varepsilon} (f+f_1)(z) \le \sup_{z\in B} f(z) + \varepsilon^2 \le c + \varepsilon^2. \tag{37}$$

From (35), (36) and (37) one has

$$\sup_{z\in B'} g(z) = \max\left\{ \sup_{z\in A_\varepsilon\setminus F_\varepsilon} g(z),\ \sup_{z\in B\cap F_\varepsilon} g(z),\ \sup_{z\in B\setminus F_\varepsilon} g(z) \right\} \le \max_{z\in\eta_0(A'_\varepsilon)} g(z) - f_2(z_0),$$

which clearly ensures that $B' \cap C = \emptyset$.

Now, we prove that there exists $x_\varepsilon \in C$ satisfying (24). Suppose not, for $k \in]1,2[$ we can apply Theorem 19 to B' and C with 5ε in place of ε. Therefore, there exist $t_0 \in]0,1]$, $\alpha \in C^0([0,1]\times X, X)$, and $\varphi \in C^0(X, \mathbb{R}_0^+)$ satisfying (i_1)–(i_4). Pick $\lambda \in [0, t_1]$, where $t_1 \in]0, t_0]$, and define

$$\eta_\lambda(t,x) := \alpha(\lambda t, \eta_0(t,x)) \quad \forall (t,x) \in [0,1]\times X.$$

We have $\eta_\lambda \in \mathcal{L}$ and $\rho(\eta_0, \eta_\lambda) \le \lambda k$ for each λ. Hence, by (32), one achieves

$$I(\eta_\lambda) \ge I(\eta_0) - \frac{\lambda k \varepsilon}{2}, \tag{38}$$

while (i_3) furnishes

$$\sup_{z\in\eta_\lambda(A'_\varepsilon)} f(z) \le \sup_{z\in\eta_0(A'_\varepsilon)} f(z) - 5\varepsilon\lambda \min_{z\in\eta_0(A'_\varepsilon)} \varphi(z) \le I(\eta_0) - 5\varepsilon\lambda \min_{z\in\eta_0(A'_\varepsilon)} \varphi(z) < +\infty.$$

Hence, by (H'_f) the function $x \mapsto \psi(\eta_\lambda(1,x))$ is continuous in A_ε as well as $x \mapsto g(\eta_\lambda(1,x))$. So, there exists $x_\lambda \in A_\varepsilon$ such that

$$g(\eta_\lambda(1,x_\lambda)) = I(\eta_\lambda).$$

Then, for each $x \in A_\varepsilon$ one has

$$g(\eta_\lambda(1,x_\lambda)) \ge g(\eta_0(1,x)) - \frac{\lambda\varepsilon k}{2}. \tag{39}$$

At this point we can show that

$$\begin{aligned} -\frac{\lambda\varepsilon k}{2} &\le g(\eta_\lambda(1,x_\lambda)) - g(\eta_0(1,x)) \le -5\lambda\varepsilon\varphi(\eta_0(1,x_\lambda)) + 2\varepsilon\rho(\eta_\lambda,\eta_0) \\ &\le -5\lambda\varepsilon\varphi(\eta_0(1,x_\lambda)) + 2\varepsilon\lambda k, \end{aligned}$$

from which it follows

$$\varphi(\eta_0(1,x_\lambda)) \le \frac{k}{2} \quad \forall \lambda \in [0,t_1]. \tag{40}$$

If $\widehat{x} \in A_\varepsilon$ is a cluster point of $\{x_\lambda : \lambda \in [0, t_1]\}$ as $\lambda \to 0^+$, bearing in mind that the function $(\lambda, x) \mapsto g(\alpha(\lambda, \eta_0(1, x)))$ is continuous in $[0, t_1] \times A_\varepsilon$, by (39) for every $x \in A_\varepsilon$ one has

$$g(\eta_0(1, \widehat{x})) \geq g(\eta_0(1, x)),$$

which ensures $\eta_0(1, \widehat{x}) \in C$ and $\varphi(\eta_0(1, \widehat{x})) = 1$ from (i_4). On the other hand, (40) implies $\varphi(\eta_0(1, \widehat{x})) < 1$. This contradiction shows that there exists $x_\varepsilon \in C$ satisfying (24). Put $x_\varepsilon := \eta_0(1, \widetilde{x})$, with $\widetilde{x} \in A_\varepsilon$. Clearly $\widetilde{x} \in F_\varepsilon$, because otherwise $x_\varepsilon = \widetilde{x} \in A_\varepsilon \setminus F_\varepsilon \subseteq B'$, which contradicts (33). Consequently, on account of (31),

$$d(x_\varepsilon, F) \leq \|\eta_0(1, \widetilde{x}) - \bar{\eta}(1, \widetilde{x})\| + \varepsilon \leq \rho(\eta_0, \bar{\eta}) + \varepsilon \leq \frac{3\varepsilon}{2},$$
$$d(x_\varepsilon, A_\varepsilon) \leq \|\eta_0(1, \widetilde{x}) - \bar{\eta}(1, \widetilde{x})\| \leq \rho(\eta_0, \bar{\eta}) \leq \frac{\varepsilon}{2}.$$

This shows (25) and (26). Thus, the conclusion is achieved once we verify (23). Bearing in mind the choice of x_ε, (30), (22), and the properties of f_1 and f_2, one has

$$f(x_\varepsilon) \leq g(x_\varepsilon) = I(\eta_0) \leq I(\bar{\eta}) = \sup_{x \in A_\varepsilon} g(x) < c + \frac{\varepsilon^2}{8} + \varepsilon^2 + \frac{\varepsilon^2}{8} = c + \frac{5\varepsilon^2}{4}.$$

Through (28) we then get

$$f(x_\varepsilon) = I(\eta_0) - f_1(\varepsilon) - f_2(\varepsilon) \geq \inf_{\eta \in \mathcal{L}} I(\eta) - \varepsilon^2 - \frac{\varepsilon^2}{8} \geq c - \frac{\varepsilon^2}{8}.$$

The above inequalities imply (23) and complete the proof under condition (21). Let us now consider the case

$$\inf_{x \in F} f(x) < c. \tag{41}$$

Fix $\varepsilon > 0$ and choose $A_\varepsilon \in \mathcal{F}$ fulfilling

$$c \leq \sup_{x \in A_\varepsilon} f(x) < c + \frac{\varepsilon^2}{4}. \tag{42}$$

We claim the existence of $x_\varepsilon \in X \setminus B$ such that

$$c \leq f(x_\varepsilon) \leq c + \frac{\varepsilon^2}{4}, \tag{43}$$
$$\Phi^0(x_n; v - x_\varepsilon) + \Psi(v) - \Psi(x_\varepsilon) \geq -\varepsilon\|v - x_\varepsilon\| \quad \forall v \in X, \tag{44}$$
$$d(x_\varepsilon, A_\varepsilon) \leq \frac{\varepsilon}{2}, \tag{45}$$

which provides a sequence $\{x_n\}$ in $X \setminus B$ enjoying properties (i_5), (i_6), and (i_8). To achieve this we can exploit the reasoning made in the preceding case (21), replacing the function g with f, the space $\mathcal{L}$ with the space of all $\eta \in C^0([0,1] \times X, X)$ such that

$$\eta(t, x) = x \ \forall (t, x) \in (\{0\} \times X) \cup ([0,1] \times B), \qquad \sup_{[0,1] \times X} \|\eta(t, x) - x\| < +\infty,$$

and the set B' with B. □

The result just proved furnishes a general procedure to construct Palais-Smale sequences. As a straightforward, but meaningful, consequence we obtain the next Theorem 23 which represents an extension of Theorem 1.bis and Corollary 2 in [**Gh1**].

THEOREM 23. *Let* (a_2) *and* (a_3) *be fulfilled. Suppose that either* $(\mathrm{PS})_{f,c}$ *or* $(\mathrm{PS})_{f,F,c}$, *holds according to whether* $\inf_{x\in F} f(x) < c$ *or* $\inf_{x\in F} f(x) = c$. *Then* $K_c(f) \neq \emptyset$. *If, moreover,* $\inf_{x\in F} f(x) = c$, *then* $K_c(f) \cap F \neq \emptyset$.

PROOF. Theorem 22 provides a sequence $\{x_n\} \subset X \setminus B$ with properties (i_5)–(i_8). Thus, the Palais-Smale condition allows us to assume $x_n \to x$ in X, where a subsequence is considered if necessary. The conclusion follows at once from (i_5)–(i_7). □

In view of the great generality of $\mathcal{F}$, B, and F, Theorem 23 furnishes more refined versions of several known results. Theorem 24 below includes Theorem 17 as well as Theorem 3.2 of [**MP**] with (H_f) replaced by (H'_f), Theorems 1 and 2 in [**Ra**] (vide also [**C**, Theorem 7.3.1]), and Theorem (1.bis) of [**GhPr**].

THEOREM 24. *Let* Q, Q_0, γ_0, Γ *be as in Example 21 and let* f *satisfy the following assumptions, in addition to* (H'_f).

- (a_4) $\sup_{x\in Q} f(\gamma(x)) < +\infty$ *for some* $\gamma \in \Gamma$.
- (a_5) *There exists a closed subset* F *of* X *such that* $(\gamma(Q) \cap F) \setminus \gamma_0(Q_0) \neq \emptyset$ $\forall \gamma \in \Gamma$ *and, moreover,* $\sup_{x\in Q_0} f(\gamma_0(x)) \leq \inf_{x\in F} f(x)$.
- (a_6) *Setting*

$$c := \inf_{\gamma\in\Gamma} \sup_{x\in Q} f(\gamma(x)), \tag{46}$$

either $(\mathrm{PS})_{f,c}$ *or* $(\mathrm{PS})_{f,F,c}$ *is fulfilled, according to whether* $\inf_{x\in F} f(x) < c$ *or* $\inf_{x\in F} f(x) = c$.

Then the conclusion of Theorem 23 holds true.

Appropriate versions of the Saddle Point Theorem and of the Generalized Mountain Pass Theorem can also be drawn from Theorem 24.

THEOREM 25. [**CMM**, Theorem 2.2] *Let* $X := V \oplus E$, *where* $V \neq \{0\}$ *is finite dimensional, and let* f *satisfy* (H'_f). *Assume there exists an* $r > 0$ *such that*

- (a_7) $\sup_{x\in \bar{B}_r\cap V} \psi(x) < +\infty$,
- (a_8) $\sup_{x\in \partial B_r\cap V} f(x) \leq \inf_{x\in E} f(x)$, *and*
- (a_9) *either* $(\mathrm{PS})_{f,c}$ *or* $(\mathrm{PS})_{f,E,c}$ *holds true, according to whether* $\inf_{x\in E} f(x) < c$ *or* $\inf_{x\in E} f(x) = c$, *where* c *is given by (46) written for* $Q := \bar{B}_r \cap V$, $Q_0 := \partial B_r \cap V$.

Then $K_c(f) \neq \emptyset$. *If, moreover,* $\inf_{x\in E} f(x) = c$ *then* $K_c(f) \cap E \neq \emptyset$.

PROOF. Apply Theorem 24 with $F = E$. □

THEOREM 26. [**CMM**, Theorem 2.3] *Suppose* $X := V \oplus E$, *where* V *is finite dimensional,* (H'_f) *holds, and there are* $r > 0$, $e \in \partial B_1 \cap E$, $\rho \in]0, r[$ *such that*

- (a_{10}) $\sup_{x\in Q} \psi(x) < +\infty$,

(a_{11}) $\sup_{x\in Q_0} f(x) \leq \inf_{x\in F} f(x)$,

and (a_6) *is fulfilled for* $Q := (\bar{B}_r \cap V) \oplus [0, re]$, Q_0 *the boundary of* Q *relative to* $V \oplus \operatorname{span}\{e\}$, $F := \partial B_\rho \cap E$.
Then the conclusion of Theorem 24 remains true.

In particular, when X is reflexive and

$$\Phi(x) := \Phi_1(x) + \Phi_2(x), \quad x \in X, \tag{47}$$

with $\Phi_1, \Phi_2 : X \to \mathbb{R}$ locally Lipschitz continuous, Theorem 26 produces the following

THEOREM 27. [**CMM**, Theorem 3.1] *Let* X, f, Q, Q_0, S *be as in Theorem 26 and let (47),* (a_{10}) – (a_{11}) *be satisfied. Assume that:*

(h_1) *If* $\{x_n\} \subseteq D_\psi$, $x_n \rightharpoonup x$ *in* X, *and there exists a* $\{\xi_n^*\} \subseteq X^*$ *fulfilling*

$$\xi_n^* \in \partial\Phi_1(x_n) \forall n \in \mathbb{N}, \quad \limsup_{n\to+\infty} \langle \xi_n^*, x_n - x\rangle \leq 0$$

then $\{x_n\}$ *has a strongly convergent subsequence.*

(h_2) *Every sequence* $\{x_n\} \subseteq D_\psi$ *weakly converging to some point* x *in* X *admits a subsequence* $\{x_{k_n}\}$ *such that* $\limsup_{n\to+\infty} \Phi_2^0(x_{k_n}; x - x_{k_n}) \leq 0$.

Suppose further that either D_ψ *is bounded or* $\inf_{x\in S} f(x) = c$, *with* c *given by* (46). *Then* $K_c(f) \neq \emptyset$. *If, moreover,* $\inf_{x\in S} f(x) = c$ *then* $K_c(f) \cap S \neq \emptyset$.

Finally, we want to point out that the techniques adopted in Theorems 22 and 23 can also be exploited when f is symmetric. In particular, if f satisfies condition (H'_f) and, in addition, Φ and ψ are both even, then it is possible to show that the deformation constructed in Theorem 22 is odd. Hence, Theorem 23 can be restated in the symmetric case, where, moreover, $\mathcal{F}$ is replaced by an arbitrary union of homotopy-stable families whose members are symmetric. Details of these arguments, together with some other interesting multiplicity results, can be found in [**CLM**].

5. Weakening the Palais-Smale condition

Through this section, $(X, \|\cdot\|)$ denotes a real reflexive Banach space while $f : X \to \mathbb{R} \cup \{+\infty\}$ is a function fulfilling the structural hypothesis (H_f). Let $h : [0, +\infty[\to [0, +\infty[$ be a continuous function enjoying the following property:

$$\int_0^{+\infty} \frac{1}{1 + h(\xi)} d\xi = +\infty. \tag{48}$$

We say that f satisfies a *weak Palais-Smale condition at the level* $c \in \mathbb{R}$ when for some h as above one has:

$(PS)_c^h$ *Every sequence* $\{x_n\} \subseteq X$ *such that* $f(x_n) \to c$ *and there exist* $y_n^* \in \partial\Phi(x_n)$, $z_n^* \in \partial\psi(x_n)$, $n \in \mathbb{N}$, *satisfying* $(1 + h(\|x_n\|))\|y_n^* + z_n^*\|_{X^*} \to 0$ *possesses a convergent subsequence.*

REMARK 28. For $\psi \equiv 0$, namely in the locally Lipschitz continuous setting, $(\mathrm{PS})_c^h$ reduces to the weak Palais-Smale hypothesis of [**LM2**]. So, in particular, if $h \equiv 0$ then we obtain Chang's compactness condition [**Ch**, Definition 2] while setting $h(\xi) := \xi$, $\xi \in [0, +\infty[$, produces a non-smooth version, previously adopted in [**KP**], of the so-called Cerami assumption [**J**, Section 13.1].

REMARK 29. Under the further request that h be nondecreasing, the following compactness hypothesis at the level $c \in \mathbb{R}$ is already known [**MoMoPa**, Definition 1.3]:

$(\mathrm{ps})_c^h$ *Every sequence* $\{x_n\} \subseteq X$ *such that* $f(x_n) \to c$ *and*

$$\Phi^0(x_n; z - x_n) + \psi(z) - \psi(x_n) \geq -\frac{\varepsilon_n}{1 + h(\|x_n\|)}\|z - x_n\| \quad \forall n \in \mathbb{N}, z \in X,$$

where $\varepsilon_n \to 0^+$, *has a convergent subsequence.*

Obviously, $(\mathrm{ps})_c^h$ includes $(\mathrm{PS})_c^h$ as a special case. This is true because if $x \in D_{\partial\psi}$ and $z \in D_\psi$ then

$$\Phi^0(x; z - x) + \psi(z) - \psi(x) \geq \langle y^* + z^*, z - x\rangle \geq -\|y^* + z^*\|_{X^*}\|z - x\| \tag{49}$$

for all $y^* \in \partial\Phi(x)$, $z^* \in \partial\psi(x)$. However, we do not need here that h be monotone. Moreover, condition $(\mathrm{ps})_c^h$ has been introduced in [**MoMoPa**] to treat different topics.

Given $x, z \in X$, write $\mathcal{P}(x, z)$ for the family of all piecewise C^1 paths $p : [0, 1] \to X$ such that $p(0) = x$ and $p(1) = z$. Moreover, put

$$l_h(p) := \int_0^1 \frac{\|p'(t)\|}{1 + h(\|p(t)\|)} dt, \quad p \in \mathcal{P}(x, z),$$

as well as

$$\delta_h(x, z) := \inf\{l_h(p) : p \in \mathcal{P}(x, z)\}. \tag{50}$$

For $h(\xi) := \xi$, $\xi \in [0, +\infty[$, the function $\delta_h : X \times X \to \mathbb{R}$ defined by (50) coincides with the geodesic distance introduced in [**E**, p. 138]. Exploiting (48) and the arguments of [**E**, p. 138] (cf. besides [**Co**, Section 4], where a more general situation is treated) yields the following basic properties of δ_h.

- (p_1) $\delta_h(x, z) \leq \|x - z\|$ *for all* $x, z \in X$.
- (p_2) *If* U *is a nonempty bounded subset of* X *then there exists a constant* $c_U > 0$ *such that*
$$\delta_h(x, z) \geq c_U\|x - z\| \quad \forall x, z \in U.$$
- (p_3) δ_h *turns out to be a distance on* X *and the metric topology derived from* δ_h *coincides with the norm topology.*
- (p_4) δ_h*-bounded and norm-bounded sets in* X *are the same.*

Through (p_1), (p_2), and (p_4) one easily verifies that the metric space (X, δ_h) is complete.

LEMMA 30. *Let* (H_f) *be satisfied, let* C *be a nonempty subset of* $\mathrm{int}(D_\psi)$, *and let* $\sigma > 0$ *be such that*

$$(1 + h(\|x\|))\|y^* + z^*\|_{X^*} \geq \sigma \quad \forall x \in C,\ y^* \in \partial\Phi(x),\ z^* \in \partial\psi(x). \tag{51}$$

Then to every $x \in C$ there corresponds a $\xi_x \in X$ such that $\|\xi_x\| = 1$ and

$$(1 + h(\|x\|))\langle y^* + z^*, \xi_x\rangle \le -\sigma \quad \forall y^* \in \partial\Phi(x),\ z^* \in \partial\psi(x). \tag{52}$$

PROOF. Fix $x \in C$. By (4) one has $\partial\psi(x) \neq \emptyset$. Hence, $\partial\Phi(x) + \partial\psi(x)$ turns out to be nonempty and convex. We claim that it is also closed. Indeed, pick $\{y_n^*\} \subseteq \partial\Phi(x)$, $\{z_n^*\} \subseteq \partial\psi(x)$ complying with $y_n^* + z_n^* \to w^*$ in X^*. Since X is reflexive, Proposition 2.1.2 in [**Cl**] yields a $y^* \in \partial\Phi(x)$ such that $y_n^* \rightharpoonup y^*$, where a subsequence is considered when necessary. Thanks to Lemma 3, the choice of $\{z_n^*\}$ forces $w^* - y^* \in \partial\psi(x)$, from which the assertion follows.

Next, observe that $0 \notin \partial\Phi(x) + \partial\psi(x)$. Through Corollary III.20 [**B**] we thus obtain $u^* \in \partial\Phi(x)$, $v^* \in \partial\psi(x)$ satisfying

$$B(0, \delta^*) \cap (\partial\Phi(x) + \partial\psi(x)) = \emptyset, \quad \text{where} \quad \delta^* := \|u^* + v^*\|_{X^*} > 0.$$

Now, the Hahn-Banach Theorem [**B**, Theorem I.6] provides a point $\hat{\xi}_x \in X$ with the properties $\|\hat{\xi}_x\| = 1$ and, whenever $x^* \in \partial\Phi(x)$, $z^* \in \partial\psi(x)$,

$$\langle y^* + z^*, \hat{\xi}_x\rangle \ge \langle w^*, \hat{\xi}_x\rangle \quad \forall w^* \in B(0, \delta^*).$$

Since

$$\|u^* + v^*\|_{X^*} = \|u^* + v^*\|_{X^*}\|\hat{\xi}_x\| = \max\{\langle w^*, \hat{\xi}_x\rangle : \ w^* \in \overline{B(0, \delta^*)}\},$$

the above inequality leads to

$$\langle y^* + z^*, \hat{\xi}_x\rangle \ge \|u^* + v^*\|_{X^*}.$$

On account of (51) this implies

$$(1 + h(\|x\|))\langle y^* + z^*, -\hat{\xi}_x\rangle \le -(1 + h(\|x\|))\|u^* + v^*\|_{X^*} \le -\sigma$$

for all $y^* \in \partial\Phi(x)$, $z^* \in \partial\psi(x)$. The conclusion is evidently achieved once we choose $\xi_x := -\hat{\xi}_x$. □

LEMMA 31. *Under the assumptions of Lemma 30, for every $x \in C$ there exists a $\delta_x > 0$ such that*

$$(1 + h(\|w\|))\langle y^* + z^*, \xi_x\rangle < -\frac{\sigma}{2} \quad \forall y^* \in \partial\Phi(u),\ z^* \in \partial\psi(v),\ u, v, w \in B(x, \delta_x).$$

PROOF. If the conclusion were false then we could find $x \in C$, $\{u_n\}$, $\{v_n\}$, $\{w_n\} \subseteq X$, and $\{y_n^*\}, \{z_n^*\} \subseteq X^*$ fulfilling the following conditions:

$$u_n \to x, \quad y_n^* \in \partial\Phi(u_n) \quad \forall n \in \mathbb{N}; \tag{53}$$

$$v_n \to x, \quad z_n^* \in \partial\psi(v_n) \quad \forall n \in \mathbb{N}; \tag{54}$$

$$w_n \to x, \quad (1 + h(\|w_n\|))\langle y_n^* + z_n^*, \xi_x\rangle \ge -\frac{\sigma}{2} \quad \forall n \in \mathbb{N}. \tag{55}$$

Due to the reflexivity of X and (53), Proposition 2.1.2 in [**Cl**] yields a $y^* \in X^*$ such that, taking a subsequence if necessary, $y_n^* \rightharpoonup y^*$ in X^*, while Proposition 2.1.5 of the same reference forces $y^* \in \partial\Phi(x)$. Since $x \in C \subseteq \text{int}(D_\psi) = \text{int}(D_{\partial\psi})$, combining (54) with Lemma 3 we obtain, up to subsequences, $z_n^* \rightharpoonup z^*$ for some $z^* \in \partial\psi(x)$. Now, from (55) it follows, as $n \to +\infty$,

$$(1 + h(\|x\|))\langle y^* + z^*, \xi_x \rangle \geq -\frac{\sigma}{2}$$

where $y^* \in \partial\Phi(x)$ and $z^* \in \partial\psi(x)$. However, this contradicts (52). □

REMARK 32. The proofs of Lemmas 30 and 31 above are chiefly patterned after those of Lemmas 2.3 and 2.4, respectively, in [**MM1**].

The following assumptions will be posited in the sequel.

(b_1) $f : X \to \mathbb{R} \cup \{+\infty\}$ *satisfies* (H'_f).

(b_2) $\mathcal{F}$ *denotes a homotopy-stable family with extended boundary* B *and*

$$c := \inf_{A \in \mathcal{F}} \sup_{x \in A} f(x) < +\infty. \tag{56}$$

(b_3) *There exists a closed subset* F *of* X *such that*

$$(A \cap F) \setminus B \neq \emptyset \quad \forall A \in \mathcal{F} \tag{57}$$

as well as

$$\sup_{x \in B} f(x) \leq \inf_{x \in F} f(x). \tag{58}$$

(b_4) *For suitable* $r, \mu > 0$ *one has* $N_r(f^{-1}(]-\infty, c+\mu])) \subseteq D_\psi$.

(b_5) $h : [0, +\infty[\to [0, +\infty[$ *is a continuous function fulfilling (48), while* δ_h *indicates the metric defined by (50).*

Combining (b_2) with (b_3) immediately yields

$$\inf_{x \in F} f(x) \leq c. \tag{59}$$

THEOREM 33. *Let* (b_1)–(b_5) *be satisfied. Then to every sequence* $\{A_n\} \subseteq \mathcal{F}$ *such that* $\sup_{x \in A_n} f(x) \to c$ *there corresponds a sequence* $\{x_n\} \subseteq X \setminus B$ *with the following properties:*

(i_1) $f(x_n) \to c$.

(i_2) *For each* $n \in \mathbb{N}$ *we can find* $y_n^* \in \partial\Phi(x_n)$, $z_n^* \in \partial\psi(x_n)$ *such that*

$$\lim_{n \to +\infty} (1 + h(\|x_n\|))\|y_n^* + z_n^*\|_{X^*} = 0.$$

(i_3) $\delta_h(x_n, F) \to 0$ *provided* $\inf_{x \in F} f(x) = c$.

(i_4) $\delta_h(x_n, A_n) \to 0$.

PROOF. On account of (59) we first consider the case

$$\inf_{x \in F} f(x) = c. \tag{60}$$

Pick $\varepsilon > 0$ so small that $\frac{\varepsilon^2}{8} \leq \mu$,

$$\{x \in X : \delta_h(x, f^{-1}(]-\infty, c+\mu])) \leq \varepsilon/2\} \subseteq N_r(f^{-1}(]-\infty, c+\mu])),$$

and choose $A_\varepsilon \in \mathcal{F}$ fulfilling

$$c \le \sup_{x \in A_\varepsilon} f(x) < c + \frac{\varepsilon^2}{8}. \tag{61}$$

We shall prove the existence of points $x_\varepsilon \in X \setminus B$, $y_\varepsilon^* \in \partial\Phi(x_\varepsilon)$, and $z_\varepsilon^* \in \partial\psi(x_\varepsilon)$ such that

$$c - \frac{\varepsilon^2}{8} \le f(x_\varepsilon) \le c + \frac{5\varepsilon^2}{4}, \tag{62}$$

$$(1 + h(\|x_\varepsilon\|))\|y_\varepsilon^* + z_\varepsilon^*\|_{X^*} < 6\varepsilon, \tag{63}$$

$$\delta_h(x_\varepsilon, F) \le \frac{3\varepsilon}{2}, \tag{64}$$

$$\delta_h(x_\varepsilon, A_\varepsilon) \le \frac{\varepsilon}{2}, \tag{65}$$

which obviously provides a sequence $\{x_n\} \subseteq X \setminus B$ enjoying properties (i_1)–(i_4). To shorten notation, write

$$A'_\varepsilon := \{1\} \times A_\varepsilon, \quad F_\varepsilon := \{x \in X : \ \delta_h(x, F) < \varepsilon\},$$
$$G_\varepsilon := (\{0\} \times X) \cup ([0,1] \times ((A_\varepsilon \setminus F_\varepsilon) \cup B)).$$

We denote by $\mathcal{L}$ the space of all $\eta \in C^0([0,1] \times X, X)$ satisfying

$$\eta(t,x) = x \quad \forall (t,x) \in G_\varepsilon, \qquad \sup_{(t,x)\in[0,1]\times X} \delta_h(\eta(t,x), x) < +\infty.$$

Since $(\{0\} \times X) \cup ([0,1] \times B) \subseteq G_\varepsilon$, one clearly has

$$\eta(A'_\varepsilon) \in \mathcal{F} \quad \forall \eta \in \mathcal{L}. \tag{66}$$

A simple computation then shows that $\mathcal{L}$, equipped with the uniform distance

$$\rho(\eta_1, \eta_2) := \sup_{(t,x)\in[0,1]\times X} \delta_h(\eta_1(t,x), \eta_2(t,x)), \quad \eta_1, \eta_2 \in \mathcal{L},$$

is complete. Define, for every $x \in X$,

$$f_1(x) := \max\{0, \varepsilon^2 - \varepsilon\delta_h(x, F)\}, \quad f_2(x) := \min\{\varepsilon^2/8, \varepsilon\delta_h(x, (A_\varepsilon \setminus F_\varepsilon) \cup B)\},$$
$$g(x) := f(x) + f_1(x) + f_2(x).$$

Moreover, set

$$I(\eta) := \sup_{z \in \eta(A'_\varepsilon)} g(z), \quad \eta \in \mathcal{L}.$$

The function $I : \mathcal{L} \to \mathbb{R} \cup \{+\infty\}$ is evidently lower semi-continuous. Gathering (66) and (57) together yields

$$(\eta(A'_\varepsilon) \cap F) \setminus B \ne \emptyset \quad \forall \eta \in \mathcal{L}.$$

Consequently, on account of (60),

$$I(\eta) \ge \sup_{z \in \eta(A'_\varepsilon)\cap F} g(z) \ge \sup_{z \in \eta(A'_\varepsilon)\cap F} f(z) + \varepsilon^2 \ge c + \varepsilon^2, \quad \eta \in \mathcal{L},$$

which means

$$\inf_{\eta\in\mathcal{L}} I(\eta) \geq c+\varepsilon^2. \tag{67}$$

Let $\bar{\eta}(t,x) := x$ for all $(t,x)\in[0,1]\times X$. Since, by (61) and (67),

$$I(\bar{\eta}) \leq \sup_{x\in A_\varepsilon} f(x)+\varepsilon^2+\frac{\varepsilon^2}{8} < c+\varepsilon^2+\frac{\varepsilon^2}{4} \leq \inf_{\eta\in\mathcal{L}} I(\eta)+\frac{\varepsilon^2}{4}, \tag{68}$$

Theorem 6 can be applied. Hence, there exists an $\eta_0\in\mathcal{L}$ such that

$$I(\eta_0) \leq I(\bar{\eta}), \tag{69}$$

$$\rho(\eta_0,\bar{\eta}) \leq \frac{\varepsilon}{2}, \tag{70}$$

$$(\eta) \geq I(\eta_0) - \frac{\varepsilon}{2}\rho(\eta,\eta_0) \quad \forall \eta\in\mathcal{L}. \tag{71}$$

Inequality (69) immediately yields

$$\sup_{z\in\eta_0(A'_\varepsilon)} \psi(z) \leq I(\bar{\eta}) - \min_{z\in\eta_0(A'_\varepsilon)} \Phi(z) < +\infty.$$

Consequently, due to (b_1), the function $z\mapsto\psi(z)$ is continuous in $\eta_0(A'_\varepsilon)$. Since the same holds for $z\mapsto g(z)$, $z\in\eta_0(A'_\varepsilon)$, the set

$$C := \left\{ w\in\eta_0(A'_\varepsilon) : g(w) = I(\eta_0) = \sup_{z\in\eta_0(A'_\varepsilon)} g(z) \right\}$$

turns out nonempty and compact. We claim that for some $z_0\in(\eta_0(A'_\varepsilon)\cap F)\setminus B$ one has

$$f(z_0) = \max_{z\in\eta_0(A'_\varepsilon)\cap F} f(z). \tag{72}$$

In fact, let $\hat{z}\in\eta_0(A'_\varepsilon)\cap F$ satisfy $f(\hat{z}) = \max_{z\in\eta_0(A'_\varepsilon)\cap F} f(z)$. If $\hat{z}\notin B$, then the assertion is true with $z_0 := \hat{z}$. Otherwise, exploiting (57), pick any $z_0\in(\eta_0(A'_\varepsilon)\cap F)\setminus B$. Thanks to (58) we obtain

$$\max_{z\in\eta_0(A'_\varepsilon)\cap F} f(z) = f(\hat{z}) \leq \sup_{x\in B} f(x) \leq \inf_{x\in F} f(x) \leq f(z_0) \leq \max_{z\in\eta_0(A'_\varepsilon)\cap F} f(z),$$

and (72) holds. The point z_0 does not lie in $(A_\varepsilon\setminus F_\varepsilon)\cup B$. So, $f_2(z_0)>0$. Because of (60) this implies

$$\max_{z\in\eta_0(A'_\varepsilon)} g(z) \geq f(z_0)+f_1(z_0)+f_2(z_0) \geq c+\varepsilon^2+f_2(z_0) > c+\varepsilon^2. \tag{73}$$

Through (61), (58), and (60) we then achieve

$$\sup_{x\in A_\varepsilon\setminus F_\varepsilon} g(x) \leq \sup_{x\in A_\varepsilon} f(x) < c+\frac{\varepsilon^2}{8}, \tag{74}$$

$$\sup_{x\in B\cap F_\varepsilon} g(x) \leq \sup_{x\in B} f(x)+\varepsilon^2 \leq \inf_{x\in F} f(x)+\varepsilon^2 = c+\varepsilon^2, \tag{75}$$

$$\sup_{x\in B\setminus F_\varepsilon} g(x) = \sup_{x\in B\setminus F_\varepsilon} f(x) \leq \sup_{x\in B} f(x) \leq \inf_{x\in F} f(x) = c. \tag{76}$$

Let $B' := (A_\varepsilon \setminus F_\varepsilon) \cup B$. From (73)–(76) it results in

$$\sup_{x\in B'} g(x) \le c + \varepsilon^2 < \max_{z\in\eta_0(A'_\varepsilon)} g(z),$$

which clearly forces

$$B' \cap C = \emptyset. \tag{77}$$

We shall next prove that:

(78) There exist $x_\varepsilon \in C$, $y_\varepsilon^* \in \partial\Phi(x_\varepsilon)$, and $z_\varepsilon^*\psi(x_\varepsilon)$ satisfying (63).

To this end, let us first note that, by construction, (70), (61), the choice of ε, and (b_4), one has

$$\begin{aligned} C \subseteq \eta_0(A'_\varepsilon) &\subseteq \{x \in X : \delta_h(x, A_\varepsilon) \le \varepsilon/2\} \\ &\subseteq \{x \in X : \delta_h(x, f^{-1}(]-\infty, c+\mu])) \le \varepsilon/2\} \\ &\subseteq N_r(f^{-1}(]-\infty, c+\mu])) \subseteq \operatorname{int}(D_\psi). \end{aligned} \tag{79}$$

So, if (78) were false then all the hypotheses of Lemma 31 would be satisfied with $\sigma := 6\varepsilon$. Thus, for each $x \in C$ we could find a $\delta_x > 0$, besides a $\xi_x \in X$, as in that result. Let $r_x \in]0, \delta_x[$ be small enough to have also Φ Lipschitz continuous on $B(x, r_x)$,

$$B' \cap B(x, r_x) = \emptyset \tag{80}$$

by (77), as well as

$$h(\|w\|) < h(\|x\|) + 1 \quad \forall w \in B(x, r_x). \tag{81}$$

Put $V_x := B(x, r_x)$. The family $\mathcal{B} := \{V_x : \ x \in C\}$ represents an open covering of C. Since C is compact, $\mathcal{B}$ possesses a finite sub-covering $\{V_{x_j} : \ j = 1, 2, \ldots, m\}$ to which we can associate a continuous partition of unity $\{\chi_j : \ j = 1, 2, \ldots, m\}$. For simplicity of notation, write $\xi_j := \xi_{x_j}$, $r_j := r_{x_j}$, besides $V_j := V_{x_j}$. Moreover, define, for every $x \in X$,

$$v(x) := \begin{cases} \sum_{j=1}^m \chi_j(x)\xi_j & \text{if} x \in \cup_{j=1}^m V_j, \\ 0 & \text{otherwise}, \end{cases} \qquad u(x) := (1 + h(\|x\|))v(x).$$

Observe that $\|v(x)\| \le 1$ for all $x \in X$ because $\|\xi_j\| = 1$, $j = 1, 2, \ldots, m$. Hence, the function $u : X \to X$ is continuous,

$$\|u(x)\| \le 1 + h(\|x\|) \quad \forall x \in X, \tag{82}$$

and, on account of (80),

$$u(x) = 0 \quad \text{in} \quad B'. \tag{83}$$

Through (81)–(82) we then obtain

$$\|u(x)\| \le 2 + M, \quad x \in X, \tag{84}$$

where $M := \max_{x\in C} h(\|x\|)$. Now, set, provided $n \in \mathbb{N}$,

$$\eta_n(t,x) := \eta_0(t,x) + \frac{t}{n} u(\eta_0(t,x)) \quad \forall (t,x) \in [0,1] \times X. \tag{85}$$

From $\eta_0 \in \mathcal{L}$, (83), (84), and (70) it follows $\eta_n \in \mathcal{L}$. Further,

$$\begin{aligned}\rho(\eta_n, \eta_0) &= \sup_{(t,x)\in[0,1]\times X} \delta_h(\eta_n(t,x), \eta_0(t,x)) \\ &\leq \frac{1}{n} \sup_{(t,x)\in[0,1]\times X} \int_0^1 \frac{\|u(\eta_0(t,x))\|}{1 + h(\|\eta_0(t,x) + \frac{t\tau}{n} u(\eta_0(t,x))\|)} d\tau.\end{aligned}$$

This inequality, combined with (71), leads to

$$I(\eta_n) - I(\eta_0) \geq -\frac{\varepsilon}{2n} \sup_{(t,x)\in[0,1]\times X} \int_0^1 \frac{\|u(\eta_0(t,x))\|}{1 + h(\|\eta_0(t,x) + \frac{t\tau}{n} u(\eta_0(t,x))\|)} d\tau$$

for all $n \in \mathbb{N}$. Letting $n \to +\infty$ and using (82) we thus get

$$\liminf_{n\to+\infty} \frac{I(\eta_n) - I(\eta_0)}{1/n} \geq -\frac{\varepsilon}{2} > -\varepsilon. \tag{86}$$

In view of (79), (84), and (85) one has $\eta_n(A'_\varepsilon) \subseteq \text{int}(D_\psi)$ for any sufficiently large n. So, the function $x \mapsto \psi(\eta_n(1,x))$ is continuous on A_ε; vide (b_1). Since the same holds for $x \mapsto g(\eta_n(1,x))$, $x \in A_\varepsilon$, there exists a point $x_n \in A_\varepsilon$ such that $I(\eta_n) = g(\eta_n(1,x_n))$. The compactness of A_ε forces, along a subsequence when necessary, $x_n \to x_0$ for some $x_0 \in A_\varepsilon$ as well as

$$\eta_0(1,x_0) \in C,$$

because g is continuous on $\text{int}(D_\psi)$ and thus

$$g(\eta_0(1,x)) = \lim_{n\to+\infty} g(\eta_n(1,x)) \leq \lim_{n\to+\infty} g(\eta_n(1,x_n)) = g(\eta_0(1,x_0)) \quad \forall x \in A_\varepsilon.$$

Now, Theorem 2 gives $x'_n \in [\eta_0(1,x_n), \eta_n(1,x_n)]$, $y^*_n \in \partial\Phi(x'_n)$ such that

$$\Phi(\eta_n(1,x_n)) - \Phi(\eta_0(1,x_n)) = \langle y^*_n, \eta_n(1,x_n) - \eta_0(1,x_n)\rangle. \tag{87}$$

Since ψ is convex, we can find a $z^*_n \in \partial\psi(\eta_n(1,x_n))$ fulfilling

$$\psi(\eta_0(1,x_n)) - \psi(\eta_n(1,x_n)) \geq \langle z^*_n, \eta_0(1,x_n) - \eta_n(1,x_n)\rangle. \tag{88}$$

From (87)–(88) it easily follows

$$f(\eta_n(1,x_n)) - f(\eta_0(1,x_n)) \leq \frac{1}{n}\langle y^*_n + z^*_n, u(\eta_0(1,x_n))\rangle, \tag{89}$$

with $y^*_n \in \partial\Phi(x'_n)$, $z^*_n \in \partial\psi(\eta_n(1,x_n))$.

Let $J_0 := \{j : \ \eta_0(1,x_0) \in V_j\}$ and let $J_n := \{j \in J_0 : \ \eta_0(1,x_n), \eta_n(1,x_n) \in V_j\}$. Obviously, $J_0 \neq \emptyset$. Moreover, $J_n = J_0$ for any n large enough, as $\eta_0(1,x_n) \to$

$\eta_0(1,x_0)$ and $\eta_n(1,x_n) \to \eta_0(1,x_0)$ by (84)–(85). Therefore, $x'_n \in V_j$ $(j \in J_n)$ too. On account of Lemma 31 we thus get

$$\begin{aligned}\langle y_n^* + z_n^*, u(\eta_0(1,x_n))\rangle &= \sum_{j\in J_n} \langle y_n^* + z_n^*, \xi_j\rangle \chi_j(\eta_0(1,x_n))(1 + h(\|\eta_0(1,x_n)\|)) \\ &< -3\varepsilon \sum_{j\in J_0} \chi_j(\eta_0(1,x_n)). \end{aligned} \tag{90}$$

Gathering (89)–(90) together yields

$$f(\eta_n(1,x_n)) - f(\eta_0(1,x_n)) < -\frac{3\varepsilon}{n} \sum_{j\in J_0} \chi_j(\eta_0(1,x_n)). \tag{91}$$

Now, it is evident that

$$\begin{aligned} f_i(\eta_n(1,x_n)) - f_i(\eta_0(1,x_n)) &\le \varepsilon\delta_h(\eta_n(1,x_n), \eta_0(1,x_n)) \\ &\le \frac{\varepsilon}{n} \int_0^1 \frac{\|u(\eta_0(1,x_n))\|}{1 + h(\|\eta_0(1,x_n) + \frac{\tau}{n} u(\eta_0(1,x_n))\|)} d\tau \quad i = 1,2, \quad n \in \mathbb{N}. \end{aligned} \tag{92}$$

Through (91) and (92) we thus achieve

$$\begin{aligned} &I(\eta_n) - I(\eta_0) \le g(\eta_n(1,x_n)) - g(\eta_0(1,x_n)) \\ &\le \frac{1}{n}\left(-3\varepsilon \sum_{j\in J_0} \chi_j(\eta_0(1,x_n)) + 2\varepsilon \int_0^1 \frac{\|u(\eta_0(1,x_n))\|}{1 + h(\|\eta_0(1,x_n) + \frac{\tau}{n} u(\eta_0(1,x_n))\|)} d\tau \right), \end{aligned}$$

which implies

$$\limsup_{n\to+\infty} \frac{I(\eta_n) - I(\eta_0)}{1/n} \le -\varepsilon.$$

However, this contradicts (86). Hence, Claim (78) holds true.

Choose $x_\varepsilon \in C$, $y_\varepsilon^* \in \partial\Phi(x_\varepsilon)$, $z_\varepsilon^* \in \partial\psi(x_\varepsilon)$ complying with (78). By (77) one has $x_\varepsilon \notin B$, i.e., (63) is fulfilled. Observe that $x_\varepsilon = \eta_0(1,\hat{x})$ for some $\hat{x} \in A_\varepsilon$ and that $\hat{x} \in F_\varepsilon$, because otherwise $x_\varepsilon = \hat{x} \in A_\varepsilon \setminus F_\varepsilon \subseteq B'$ as $\eta_0 \in \mathcal{L}$, against (77). Consequently, due to (70),

$$\begin{aligned} \delta_h(x_\varepsilon, F) &\le \delta_h(\eta_0(1,\hat{x}), \bar{\eta}(1,\hat{x})) + \delta_h(\hat{x}, F) \le \frac{3\varepsilon}{2}, \\ \delta_h(x_\varepsilon, A_\varepsilon) &\le \delta_h(\eta_0(1,\hat{x}), \bar{\eta}(1,\hat{x})) \le \frac{\varepsilon}{2}. \end{aligned}$$

This shows (64)–(65). The conclusion is thus achieved once we verify (62). Bearing in mind the choice of x_ε, (69), (61), besides the properties of f_1 and f_2, one achieves

$$f(x_\varepsilon) \le g(x_\varepsilon) = I(\eta_0) \le I(\bar{\eta}) = \sup_{x\in A_\varepsilon} g(x) < c + \frac{\varepsilon^2}{8} + \varepsilon^2 + \frac{\varepsilon^2}{8} = c + \frac{5\varepsilon^2}{4}.$$

Through (67) we then get

$$f(x_\varepsilon) = I(\eta_0) - f_1(x_\varepsilon) - f_2(x_\varepsilon) \ge \inf_{\eta\in\mathcal{L}} I(\eta) - \varepsilon^2 - \frac{\varepsilon^2}{8} \ge c - \frac{\varepsilon^2}{8}.$$

Combining the two inequalities above yields (62) and completes the proof when (60) holds.

Let us now come to the case

$$\inf_{x\in F} f(x) < c. \tag{93}$$

Pick $\varepsilon > 0$ so small that $\frac{\varepsilon^2}{4} \leq \mu$,

$$\{x \in X : \delta_h(x, f^{-1}(]-\infty, c+\mu])) \leq \varepsilon/2\} \subseteq N_r(f^{-1}(]-\infty, c+\mu])),$$

and choose $A_\varepsilon \in \mathcal{F}$ fulfilling

$$c \leq \sup_{x\in A_\varepsilon} f(x) < c + \frac{\varepsilon^2}{4}. \tag{94}$$

We shall seek points $x_\varepsilon \in X \setminus B$, $y_\varepsilon^* \in \partial\Phi(x_\varepsilon)$, $z_\varepsilon^* \in \partial\psi(x_\varepsilon)$ such that

$$c \leq f(x_\varepsilon) < c + \frac{\varepsilon^2}{4}, \tag{95}$$

$$(1 + h(\|x_\varepsilon\|))\|y_\varepsilon^* + z_\varepsilon^*\|_{X^*} < 6\varepsilon, \tag{96}$$

$$\delta_h(x_\varepsilon, A_\varepsilon) \leq \frac{\varepsilon}{2}, \tag{97}$$

which obviously provides a sequence $\{x_n\} \subseteq X \setminus B$ enjoying properties (i_1), (i_2), and (i_4). Denote by $\mathcal{L}$ the space of all $\eta \in C^0([0,1] \times X, X)$ such that

$$\eta(t,x) = x \quad \forall (t,x) \in (\{0\} \times X) \cup ([0,1] \times B), \qquad \sup_{(t,x)\in[0,1]\times X} \delta_h(\eta(t,x), x)) < +\infty.$$

One clearly has $\eta(A'_\varepsilon) \in \mathcal{F}$ for every $\eta \in \mathcal{L}$. Moreover $\mathcal{L}$, when equipped with the uniform distance ρ is complete. If

$$I(\eta) := \sup_{z\in\eta(A'_\varepsilon)} f(z), \quad \eta \in \mathcal{L},$$

then $\inf_{\eta\in\mathcal{L}} I(\eta) \geq c$ by (56) while the function $I : \mathcal{L} \to \mathbb{R} \cup \{+\infty\}$ is evidently lower semi-continuous, bounded from below, and, due to (94), satisfies (68). Through Theorem 6 we thus get an η_0 in $\mathcal{L}$ complying with (69)–(71). Observe that, because of (69) and (94), one has $\sup_{z\in\eta_0(A'_\varepsilon)} f(z) < +\infty$. Hence, owing to ($b_1$), the function $x \mapsto f(\eta_0(1,x))$ turns out continuous on A_ε. This implies that the set

$$C := \left\{ w \in \eta_0(A'_\varepsilon) : \; f(w) = I(\eta_0) = \sup_{z\in\eta_0(A'_\varepsilon)} f(z) \right\}.$$

is nonempty, compact, and such that

$$f(x) \geq \inf_{\eta\in\mathcal{L}} I(\eta) \geq c \quad \forall x \in C. \tag{98}$$

Exploiting (58), (93), and (98) yields

$$\sup_{x\in B} f(x) \leq \inf_{x\in F} f(x) < c \leq \inf_{x\in C} f(x),$$

namely $B \cap C = \emptyset$. Now, the same reasoning adopted before gives points $x_\varepsilon \in C$, $y_\varepsilon^* \in \partial\Phi(x_\varepsilon)$, $z_\varepsilon^* \in \partial\psi(x_\varepsilon)$ fulfilling (96). Since x_ε lies in C, from (98), (69), and (94) we immediately infer (95). Indeed,

$$c \leq \inf_{x \in C} f(x) \leq f(x_\varepsilon) = I(\eta_0) \leq I(\bar{\eta}) = \sup_{x \in A_\varepsilon} f(x) < c + \frac{\varepsilon^2}{4}.$$

Finally, (97) can be verified exactly as in the preceding case. □

REMARK 34. By (b_1), condition (56) is equivalent to the claim that, for some $A \in \mathcal{F}$, the function ψ is continuous on A.

REMARK 35. Theorem 33 reduces to [**LM2**, Theorem 3.1] whenever $\psi \equiv 0$. Through (49) we then realize that it includes Theorem 22 (for $h \equiv 0$) as a special case.

The following critical point result represents an almost direct but meaningful consequence of Theorem 33.

THEOREM 36. *Suppose* (b_1)–(b_5) *and* $(\mathrm{PS})_c^h$, *with* c *given by* (56), *hold true. Then* $K_c(f) \neq \emptyset$. *If, moreover,* $\inf_{x \in F} f(x) = c$ *then* $K_c(f) \cap F \neq \emptyset$.

PROOF. Let $\{x_n\} \subseteq X \setminus B$ satisfy (i_1)–(i_4) of Theorem 33. Due to $(\mathrm{PS})_c^h$ we may assume that $x_n \to x$ in X, where a subsequence is considered when necessary. Since, taking into account (49),

$$\Phi^0(x_n; z - x_n) + \psi(z) - \psi(x_n) \geq -\|y_n^* + z_n^*\|_{X^*} \|z - x_n\|$$

for all $z \in X$, $n \in \mathbb{N}$, $y_n^* \in \partial\Phi(x_n)$, $z_n^* \in \partial\psi(x_n)$, the upper semi-continuity of Φ^0, besides (i_2), yield

$$\Phi^0(x; z - x) + \psi(z) - \psi(x) \geq 0 \quad \forall z \in X.$$

On the other hand, by (i_1) and (b_4),

$$x \in f^{-1}(]-\infty, c]) \subseteq f^{-1}(]-\infty, c + \mu]) \subseteq \mathrm{int}(D_\psi),$$

i.e., f turns out to be continuous at x. Consequently, (i_1) gives $f(x) = c$, which means $x \in K_c(f)$. Now, suppose $\inf_{x \in F} f(x) = c$. Because of (p_3) the set F is δ_h-closed. Thanks to (i_3) one thus has $x \in F$ and, a fortiori, $K_c(f) \cap F \neq \emptyset$. □

Making suitable choices of $\mathcal{F}$, B, and F, more refined versions of several recent results can easily be drawn from Theorem 36. By way of example, we find the result below, which generalizes Theorem 3.3 in [**LM2**]; cf. also Theorem 24. Keep the same notation introduced in Example 21.

THEOREM 37. *Let* (b_1) *and* (b_5) *be fulfilled. Assume that:*

(b_6) $\sup_{x \in Q} f(\gamma(x)) < +\infty$ *for some* $\gamma \in \Gamma$.

(b_7) *There exists a closed subset* F *of* X *such that* $(\gamma(Q) \cap F) \setminus \gamma_0(Q_0) \neq \emptyset$ *for all* $\gamma \in \Gamma$ *and, moreover,* $\sup_{x \in Q_0} f(\gamma_0(x)) \leq \inf_{x \in F} f(x)$.

(b_8) *Setting* $c := \inf_{\gamma \in \Gamma} \sup_{x \in Q} f(\gamma(x))$, *both* $(\mathrm{PS})_c^h$ *and* (b_4) *holds.*

Then the conclusion of Theorem 36 is true.

REMARK 38. When $\inf_{x\in F} f(x) = c$ condition $(\mathrm{PS})_c^h$ that appears in Theorems 36 and 37 can be replaced by the following weaker one, as an elementary argument shows.

$(\mathrm{PS})_{F,c}^h$ *Every sequence* $\{x_n\} \subseteq X$ *such that* $\delta_h(x_n, F) \to 0$, $f(x_n) \to c$, *and there exist* $y_n^* \in \partial\Phi(x_n)$, $z_n^* \in \partial\psi(x_n)$, $n \in \mathbb{N}$, *satisfying*

$$\lim_{n\to+\infty} (1 + h(\|x_n\|))\|y_n^* + z_n^*\|_{X^*} = 0$$

possesses a convergent subsequence.

REMARK 39. Theorems 36 and 37 might be exploited to get further information on the critical set $K_c(f)$, as already made in [**LM1, BoM**] for $h \equiv 0$ or $\psi \equiv 0$, respectively.

REMARK 40. For a concrete functional f, which fulfils the above-mentioned weak Palais-Smale assumption on the Sobolev space $X : H_0^1(\Omega)$, we refer the reader to [**MM2**, Section 3].

6. On the structure of the critical set

This section is devoted to a qualitative study of the the critical set $K_c(f)$. Following the ideas developed in Pucci-Serrin [**PS2, PS3**] and Ghoussoub [**Gh2**] for C^1 functions, it is possible to combine the results obtained before with appropriate topological arguments, thus providing classes of non-smooth functions whose critical sets exhibit nonlocal minima or, more precisely, saddle points. Keep the same notation as Section 4.

THEOREM 41. [**LM1**, Theorem 4.1] *Suppose that*

(a_{12}) $\mathcal{F}$ *is a homotopy-stable family with extended boundary* B, *whose members are path-wise connected and contain* B.
(a_{13}) f *fulfils conditions* (H'_f), *(19), and* $(\mathrm{PS})_{f,c}$, *while the set* f^c *is closed.*
(a_{14}) $\sup_{x\in B} f(x) < c$.

Then $K_c(f)$ *possesses a nonlocal minimum point.*

PROOF. Put $F = \partial f^c$. For every $A \in \mathcal{F}$ one has $A \cap f^c \neq \emptyset$ as well as

$$(A \cap F) \setminus B \neq \emptyset. \tag{99}$$

Indeed, (19) forces $\sup_{x\in A} f(x) \geq c$. Hence, the first assertion is obvious when $\sup_{x\in A} f(x) > c$. Otherwise it follows from (H'_f), being $\sup_{x\in A} f(x)$ a maximum. To verify (99) we first observe that $(A \cap F) \setminus B = A \cap F$, by the fact that ($a_{14}$) implies $B \subseteq X \setminus f^c$. If $A \cap F = \emptyset$ for some $A \in \mathcal{F}$, then $A \cap \mathrm{int}(f^c) \neq \emptyset$ and $A \subseteq \mathrm{int}(f^c) \cup (X \setminus f^c)$. Since A is path-wise connected, one has $A \subseteq f^c$, so that $A \cap B = \emptyset$, against (a_{12}), and (99) holds. We next have

$$c \leq \inf_{x\in f^c} f(x) \leq \inf_{x\in F} f(x) \leq \inf_{x\in A\cap F} f(x) \leq \sup_{x\in A} f(x) \;\; \forall A \in \mathcal{F},$$

namely $c = \inf_{x \in F} f(x)$. Hence, in view of (a_{14}), assumption (a_3) is satisfied. Moreover (a_2) follows from (a_{12}) – (a_{13}). Therefore we can apply Theorem 23, and $K_c(f) \cap F \neq \emptyset$. Pick $x \in K_c(f) \cap F$. By the choice of F it follows that for every $\delta > 0$ there exists $x_\delta \in B(x, \delta)$ fulfilling $f(x_\delta) < c = f(x)$, that is x is a nonlocal minimum point. □

If $x_0, \ x_1 \in X$, define

$$c := \inf_{\gamma \in \Gamma} \sup_{x \in [x_0, x_1]} f(\gamma(x)),$$

where $\Gamma = \{\gamma \in C^0([x_0, x_1], X) : \gamma(x_i) = x_i, i = 0, 1\}$. A nonempty closed subset F of X is said to separate x_0 and x_1 when x_0 and x_1 do not lie in the same connected component of $X \setminus F$; i.e.,

$$\gamma([x_0, x_1]) \cap F \setminus \{x_0, x_1\} \neq \emptyset \quad \forall \gamma \in \Gamma. \tag{100}$$

Recall that any compact set of an infinite-dimensional Banach space cannot separate two points in its complement.

A critical point $x \in X$ is called a saddle point for f provided that to every $\delta > 0$ there correspond $x', x'' \in B(x, \delta)$ such that $f(x') < f(x) < f(x'')$.

THEOREM 42. [**LM1**, Theorem 4.3] *Suppose X is infinite dimensional, the function f satisfies (H'_f), and $x_0, x_1 \in D_\psi$. If*

(a_{15}) *condition $(PS)_{f,c}$ holds, while the set f^c is closed, and*

(a_{16}) $\max\{f(x_0), \ f(x_1)\} < c$,

then $K_c(f)$ possesses a saddle point.

PROOF. Put

$$\begin{aligned} M_c &:= \{x \in K_c(f) : \ x \text{ is a local minimum of } f\}, \\ P_c &:= \{x \in K_c(f) : \ x \text{ is a proper local maximum for } f\}, \\ S_c &:= \{x \in K_c(f) : \ x \text{ is a saddle point of } f\}. \end{aligned}$$

Clearly,

$$K_c(f) = M_c \cup P_c \cup S_c. \tag{101}$$

Moreover, for each $x \in M_c$ there exists a $\delta_x > 0$ satisfying $B(x, \delta_x) \subseteq f^c$. If $N = \cup_{x \in M_c} B(x, \delta_x)$, one has

$$M_c \subseteq N \subseteq \overline{N} \subseteq f^c. \tag{102}$$

Thanks to $(PS)_{f,c}$ the set $K_c(f)$ is compact. Since the same holds regarding $\overline{M}_c$ while $x_0, x_1 \notin \overline{M}_c$ in view of (a_{16}), we may assume that $x_0, x_1 \notin \overline{N}$. Therefore, the set

$$F = \partial N \cup (f^c \setminus N) \tag{103}$$

is nonempty closed and does not contain x_0 and x_1. Reasoning as in the proof of Theorem 41 we infer

$$\partial f^c \cap \gamma([x_0, x_1]) \setminus \{x_0, x_1\} \neq \emptyset \quad \forall \gamma \in \Gamma.$$

This clearly forces (100), because $\partial f^c \subseteq f^c \setminus N \subseteq F$. We next have

$$c \leq \inf_{x\in F} f(x) \leq \inf_{x\in\gamma([x_0,x_1])\cap F} f(x) \leq \sup_{x\in\gamma([x_0,x_1])} f(x) \ \forall\gamma \in \Gamma,$$

that is $c = \inf_{x\in F} f(x)$. Thus, Theorem 24 can be applied with $Q = [x_0, x_1]$, $Q_0 = \{x_0, x_1\}$, and $\gamma_0 = \mathrm{id}_{|Q_0}$, so that $K_c(f) \cap F \neq \emptyset$. Suppose now, contrary to our claim, that $S_c = \emptyset$. From (101)–(103) it follows $P_c \cap F \neq \emptyset$. Exploiting Theorem 1 in [**K**, Chapter VIII, Section 57, III] provides a closed connected set $\hat{F} \subseteq F$ such that

$$\gamma([x_0, x_1]) \cap \hat{F} \setminus \{x_0, x_1\} \neq \emptyset \quad \forall\gamma \in \Gamma. \tag{104}$$

We now proceed exactly as before to achieve $P_c \cap \hat{F} \neq \emptyset$. The set $P_c \cap \hat{F}$ is relatively open in $\hat{F}$ whereas $P_c \cap \hat{F} = K_c(f) \cap \hat{F}$ is closed. The connectedness of $\hat{F}$ implies $P_c \cap \hat{F} = \hat{F}$. Therefore, in view of (104), $\hat{F}$ turns out to be a compact set separating x_0 and x_1, which is impossible because X is infinite dimensional. □

REMARK 43. The above proof was inspired by those of Theorem 2 and Corollary 3 in [**F**]. Moreover, Theorem 42 goes back to [**PS2**, Theorem 8] in the C^1 case, while a locally Lipschitz continuous variant of it can be found in [**BM**].

Keep the same notation as Section 5. The next result extends Corollary 4.14 of [**PS1**] to locally Lipschitz functions satisfying a weak compactness condition.

THEOREM 44. *Let* (b_1), (b_3), *and* $(\mathrm{PS})_c^h$, *with* c *as in (56), be satisfied. Assume also that:*

(b_9) *The members of* $\mathcal{F}$ *are path-wise connected and contain* B.
(b_{10}) $\sup_{x\in B} f(x) < c$.

Then $K_c(f)$ *possesses a nonlocal minimum point.*

PROOF. Choose $F := \partial f^c$. For every $A \in \mathcal{F}$ one has $A \cap f^c \neq \emptyset$, besides (99). Indeed, (59) implies $\max_{x\in A} f(x) \geq c$. Thus, the first claim is obvious. To verify (99) we first observe that $(A \cap F) \setminus B = A \cap F$, because $B \subseteq X \setminus f^c$ on account of (b_{10}). If $A \cap F = \emptyset$ for some $A \in \mathcal{F}$ then $A \cap \mathrm{int}(f^c) \neq \emptyset$ and $A \subseteq \mathrm{int}(f^c) \cup (X \setminus f^c)$. Since A is path-wise connected we get $A \subseteq f^c$, which implies $A \cap B = \emptyset$. However, this contradicts (b_9). We next have

$$c \leq \inf_{x\in f^c} f(x) \leq \inf_{x\in F} f(x) \leq \inf_{x\in A\cap F} f(x) \leq \max_{x\in A} f(x) \quad \forall A \in \mathcal{F},$$

namely, $c = \inf_{x\in F} f(x)$. Hence, by (b_{10}), inequality (58) holds, and hypothesis (b_2) is satisfied. Consequently, thanks to Theorem 36, $K_c(f) \cap F \neq \emptyset$. Pick $x \in K_c(f) \cap F$. Bearing in mind the choice of F, for every $\delta > 0$ there exists an $x_\delta \in B(x, \delta)$ such that $f(x_\delta) < c = f(x)$, i.e., x is not a local minimum for f. □

THEOREM 45. *Suppose* (b_1), (b_3), *and* $(\mathrm{PS})_a^h$, $a \in \mathbb{R}$, *hold true. If* $x_0, x_1 \in X$ *are two local minima of* f *then it has at least three critical points.*

PROOF. Obviously, x_0 and x_1 turn out to be critical points for f; cf. Theorem 7 . It is not restrictive to assume that $f(x_1) \leq f(x_0) = 0$. Let $r > 0$ satisfy

$$r < \|x_1 - x_0\|, \qquad \inf_{x\in \overline{B}(x_0,r)} f(x) \geq f(x_0) = 0. \tag{105}$$

Define $Q := [x_0, x_1]$, $Q_0 := \{x_0, x_1\}$, $\gamma_0 := \mathrm{id}|_{Q_0}$, $\Gamma := \{\gamma \in C^0(Q, X) : \gamma|_{Q_0} = \gamma_0\}$, $\mathcal{F} := \{\gamma(Q) : \gamma \in \Gamma\}$, and $B := \{x_0, x_1\}$. Hypothesis (b_2) holds, because $\mathcal{F}$ is a homotopy stable family with extended boundary B; see Example 21. Setting $F := \partial B(x_0, r)$ one evidently has $(A \cap F) \setminus B \neq \emptyset$ for all $A \in \mathcal{F}$ as well as

$$\sup_{x \in B} f(x) = \max\{f(x_0), f(x_1)\} = 0 \leq \inf_{x \in F} f(x), \tag{106}$$

which yields (b_3). Since $(\mathrm{PS})_c^h$, with

$$c := \inf_{\gamma \in \Gamma} \max_{x \in \gamma([x_0, x_1])} f(x),$$

is true too, all the assumptions of Theorem 36 are satisfied. So, $K_c(f) \neq \emptyset$ and $K_c(f) \cap F \neq \emptyset$ provided $\inf_{x \in F} f(x) = c$. On account of (106) the conclusion is obvious when $\inf_{x \in F} f(x) < c$. Otherwise, it easily follows from $K_c(f) \cap F \neq \emptyset$, the choice of F, and (105). □

Theorem 36 extends [**Gh2**, Corollary 4.14] while Theorem 45 represents a non-smooth variant of the famous [**PS1**, Corollary 1]. Other structure results can be found in [**BM**].

Combining [**BC**, Theorem 3.1] with Theorem 45 yields the multiplicity result below. We shall assume that:

(b_{11}) *X is reflexive. Moreover, there exists a Banach space $\tilde{X}$ such that X compactly embeds in $\tilde{X}$.*

(b_{12}) *$\Phi, \Psi : \tilde{X} \to \mathbb{R}$ turn out to be locally Lipschitz continuous, $\Phi(0) = \Psi(0) = 0$, and $\Phi|_X$ is coercive.*

(b_{13}) *For some $r > 0$, $x_1 \in X$ one has $r < \Phi(x_1)$ as well as*

$$\sup_{z \in \Phi_r} \Psi(z) < r \frac{\Psi(x_1)}{\Phi(x_1)}.$$

To shorten notation, define

$$f_\lambda := \Phi - \lambda \Psi, \quad \lambda \in \Lambda_r, \tag{107}$$

where

$$\Lambda_r := \left] \frac{\Phi(x_1)}{\Psi(x_1)}, \frac{r}{\sup_{z \in \Phi_r} \Psi(z)} \right[.$$

(b_{14}) *f_λ fulfils $(\mathrm{PS})_a^h$, $a \in \mathbb{R}$, and is coercive for all $\lambda \in \Lambda_r$.*

THEOREM 46. *Let (b_3) and (b_{11})–(b_{14}) be satisfied. Then the function f_λ given by (107) possesses at least three critical points (two local minima and a nonlocal minimum point) provided $\lambda \in \Lambda_r$.*

The proof of this result is analogous to that of [**BoM**, Theorem 2.6]. So, we omit it.

7. Critical points in presence of splitting

Theorem 24 can be fruitfully exploited to obtain multiple critical points. In fact, for example, it represents the main tool in establishing a non-smooth version of the famous Brézis-Nirenberg critical point theorem [**BN**, Theorem 5] in presence of splitting. If X is reflexive and $X = V \oplus E$, with $0 < \dim(V) < +\infty$, $\dim(E) > 0$, then the result below holds.

THEOREM 47. [**LMM**, Theorem 3.1] *Assume that*

(f_1) *f turns out to be bounded below and fulfils* $(\mathrm{PS})_f$ *besides* (H'_f),

(f_2) *$x_0 \in X$ is a global minimum point of the function f,*

(f_3) *the set $X \setminus f^a$ is open for some $a > 0$,*

(f_4) *there is an $r \in]0, \|x_0\|/2[$ such that $f_{|\bar{B}_r \cap E} \geq 0$, $f_{|\bar{B}_r \cap V} \leq 0$, and $f_{|\partial B_r \cap V} < 0$.*

If, moreover, $\inf_{x \in X} f(x) < f(0)$ and $f(0) = 0$ then the function f possesses at least two nontrivial critical points.

Before establishing Theorem 47 we shall prove some preliminary results.

PROPOSITION 48. *Let f be bounded below and fulfil* $(\mathrm{PS})_f$ *in addition to* (H_f). *Assume the global minimum point x_0 is unique. Then, for every $\rho_0 > 0$ there exists a $\rho > 0$ such that*

$$U_\rho := \{x \in X : f(x) < f(x_0) + \rho\} \subseteq B(x_0, \rho_0).$$

PROOF. Arguing by contradiction one can find a $\rho_0 > 0$ and sequence $\{x_n\}$ in X such that

$$f(x_n) < f(x_0) + \frac{1}{n^2}, \quad \|x_n - x_0\| \geq \rho_0 \quad \forall n \in \mathbb{N}.$$

Now, Theorem 6 provides a point $z_n \in X$ satisfying (7)–(9). Set $z := z_n + t(x - z_n)$, with $x \in X$ and $t \in]0, 1[$. As in the proof of Theorem 8, inequality (9), besides the convexity of ψ, lead to (10). Thus, by condition $(\mathrm{PS})_f$, there exists a subsequence $\{z_{k_n}\}$ of $\{z_n\}$ strongly converging to some $z \in X$. Since $f(z_{k_n}) \to f(x_0)$ in view of (7), we have $f(z) = f(x_0)$, which forces $z = x_0$ taking into account the uniqueness of x_0. On the other hand, $x_{k_n} \to x_0$ due to (8). However, this is impossible because $\|x_{k_n} - x_0\| \geq \rho_0$ for all $n \in \mathbb{N}$, and the conclusion follows. □

PROPOSITION 49. *Suppose* (f_1)–(f_2) *hold true, in addition to*

(f_5) *$x_0 \neq 0$. Moreover, x_0 and eventually 0 are the only critical points for f.*

(f_6) *There exist two disjoint open neighborhoods U_0 and N_0 of x_0 and 0, respectively, and a constant $b > \inf_{x \in X} f(x)$, satisfying $f_b \setminus (U_0 \cup N_0) \subseteq D_{\partial\psi}$.*

Then there exists $\sigma > 0$ such that for every $x \in f_b \setminus (U_0 \cup N_0)$, $x^* \in \partial\Phi(x)$, $z^* \in \partial\psi(x)$ one has $\|x^* + z^*\|_{X^*} \geq \sigma$.

PROOF. Arguing by contradiction one could construct three sequences $\{x_n\} \subseteq X$, $\{x_n^*\}, \{z_n^*\} \subseteq X^*$ with the following properties:

$$x_n \in f_b \setminus (U_0 \cup N_0), \quad n \in \mathbb{N}; \tag{108}$$

$$x_n^* \in \partial\Phi(x_n) \text{ and } z_n^* \in \partial\psi(x_n) \quad \forall n \in \mathbb{N}; \tag{109}$$

$$\|x_n^* + z_n^*\|_{X^*} \to 0 \text{ as } n \to +\infty. \tag{110}$$

From (109) we obtain easily

$$\begin{aligned} \Phi^0(x_n; x - x_n) + \psi(x) - \psi(x_n) &\geq \langle x_n^*, x - x_n \rangle + \langle z_n^*, x - x_n \rangle \\ &\geq -\|x_n^* + z_n^*\|_{X^*} \|x - x_n\| \end{aligned}$$

for all $n \in \mathbb{N}$, $x \in X$. Thanks to $(\mathrm{PS})_f$, setting $\varepsilon_n := \|x_n^* + z_n^*\|_{X^*}$ and using (110) produces $x_n \to \bar{x}$ in X, where a subsequence is considered when necessary.

Moreover, by (108), $\bar{x} \in f_b \setminus (U_0 \cup N_0)$. Since Φ^0 and $-\psi$ are upper semi-continuous, this forces both $\bar{x} \in D_\psi$ and

$$\Phi^0(\bar{x}; x - \bar{x}) + \psi(x) - \psi(\bar{x}) \geq 0 \quad \forall x \in X,$$

namely $\bar{x}$ turns out a critical point of f different from x_0 and 0, against hypothesis (f_5). □

PROPOSITION 50. *Let* (f_1)–(f_2), (f_5)–(f_6) *be satisfied and let* σ *be as in Proposition 49. Assume that:*

(f_7) *If* $\{x_n\} \subseteq f_b \setminus (U_0 \cup N_0)$, $x_n \to x$ *in* X, *and* $x_n^* \in \partial\psi(x_n)$ *for all* $n \in \mathbb{N}$, *then to each* $z \in X$ *there corresponds an* $x^* \in \partial\psi(x)$ *such that* $\langle x^*, z\rangle \leq \limsup_{n\to+\infty} \langle x_n^*, z\rangle$.

Then there exists a locally Lipschitz continuous function $F : f_b \setminus (U_0 \cup N_0) \to X$ *such that, for every* $x \in f_b \setminus (U_0 \cup N_0)$, $\|F(x)\| \leq 1$ *and*

$$\langle x^* + z^*, F(x)\rangle > \frac{\sigma}{2} \quad \forall x^* \in \partial\Phi(x),\ z^* \in \partial\psi(x). \tag{111}$$

PROOF. From now on, W denotes the set $f_b \setminus (U_0 \cup N_0)$. Pick $x \in W$. We first claim that the infimum

$$\delta(x) := \inf\{\|x^* + z^*\|_{X^*} :\ x^* \in \partial\Phi(x),\ z^* \in \partial\psi(x)\} \tag{112}$$

is attained. To show this, fix $\{x_n^*\} \subseteq \partial\Phi(x)$ and $\{z_n^*\} \subseteq \partial\psi(x)$ fulfilling

$$\lim_{n\to+\infty} \|x_n^* + z_n^*\|_{X^*} = \delta(x). \tag{113}$$

Since X is reflexive while $\partial\Phi(x)$ is weak* compact, we can find an $\bar{x}^* \in \partial\Phi(x)$ such that, along a subsequence if necessary, $x_n^* \rightharpoonup \bar{x}^*$. By (113) the sequence $\{z_n^*\}$ turns out bounded. So, as before, $z_n^* \rightharpoonup \bar{z}^*$ for some $\bar{z}^* \in \partial\psi(x)$. One clearly has

$$\|\bar{x}^* + \bar{z}^*\|_{X^*} \leq \liminf_{n\to+\infty} \|x_n^* + z_n^*\|_{X^*},$$

which implies $\|\bar{x}^* + \bar{z}^*\|_{X^*} = \delta(x)$.

Proposition 49 ensures that $\delta(x) \geq \sigma > 0$. Hence, $B_{\delta(x)}$ is nonempty and, on account of (112),

$$B_{\delta(x)} \cap (\partial\Phi(x) + \partial\psi(x)) = \emptyset.$$

Now, the Hahn-Banach Theorem [**B**, Theorem I.6] provides a point $\xi_x \in X$ with the properties $\|\xi_x\| = 1$ and, whenever $x^* \in \partial\Phi(x)$, $z^* \in \partial\psi(x)$,

$$\langle x^* + z^*, \xi_x\rangle \geq \langle w^*, \xi_x\rangle \quad \forall w^* \in B_{\delta(x)}. \tag{114}$$

Since

$$\|x^* + z^*\|_{X^*} = \|x^* + z^*\|_{X^*}\|\xi_x\| = \max\left\{\langle w^*, \xi_x\rangle : w^* \in \overline{B}_{\delta(x)}\right\},$$

inequality (114) and Proposition 49 lead to

$$\langle x^* + z^*, \xi_x\rangle \geq \|x^* + z^*\|_{X^*} \geq \sigma \tag{115}$$

for all $x^* \in \partial\Phi(x)$, $z^* \in \partial\psi(x)$.

We next show that to each $x \in W$ there corresponds a neighbourhood V_x of x such that as soon as $v \in V_x \cap W$ one has

$$\langle x^* + z^*, \xi_x \rangle > \frac{\sigma}{2} \ \forall x^* \in \partial\Phi(v), \ \ z^* \in \partial\psi(v). \tag{116}$$

Indeed, if the assertion were false then one could find $x \in W$, $\{x_n\} \subseteq X$, and $\{x_n^*\}, \{z_n^*\} \subseteq X^*$ satisfying the following conditions:

$$x_n \to x, \ \ x_n^* \in \partial\Phi(x_n), \ \ z_n^* \in \partial\psi(x_n), \quad n \in \mathbb{N}, \tag{117}$$

$$\langle x_n^* + z_n^*, \xi_x \rangle \leq \frac{\sigma}{2} \quad \forall n \in \mathbb{N}. \tag{118}$$

Due to the reflexivity of X and (117), Proposition 2.1.2 in [**Cl**] yields an $x^* \in X^*$ such that $x_n^* \rightharpoonup x^*$ in X^*, where a subsequence is considered when necessary, while Proposition 2.1.5 of the same reference forces $x^* \in \partial\Phi(x)$. From (118) we thus get

$$\limsup_{n \to +\infty} \langle z_n^*, \xi_x \rangle \leq \frac{\sigma}{2} - \langle x^*, \xi_x \rangle.$$

Now, exploiting (f_7) provides a point $z^* \in \partial\psi(x)$ such that

$$\langle z^*, \xi_x \rangle \leq \frac{\sigma}{2} - \langle x^*, \xi_x \rangle,$$

which contradicts (115).

The family $\mathcal{V} := \{V_x : x \in W\}$ represents an open covering of W. Since, by [**Du**, Theorem VIII.2.4], this set is para-compact, $\mathcal{V}$ possesses an open locally finite refinement $\{V_i : i \in I\}$. Moreover, on account of (116), to each $i \in I$ there corresponds a $\xi_i \in X$ fulfilling $\|\xi_i\| = 1$ as well as, whenever $x \in V_i \cap W$,

$$\langle x^* + z^*, \xi_i \rangle > \frac{\sigma}{2} \quad \forall x^* \in \partial\Phi(x), \ \ z^* \in \partial\psi(x). \tag{119}$$

Let $\{\rho_i : i \in I\}$ be a partition of unity subordinated to $\{V_i : i \in I\}$ such that each ρ_i turns out locally Lipschitz continuous; for a possible construction we refer to [**MM1**, p. 145]. Define

$$F(x) := \sum_{i \in I} \rho_i(x)\xi_i, \quad x \in W. \tag{120}$$

The function F is evidently locally Lipschitz continuous and one has $\|F(x)\| \leq 1$ because $\sum_{i \in I} \rho_i(x) \leq 1$ in X. Exploiting (119) we then see at once that

$$\langle x^* + z^*, F(x) \rangle > \frac{\sigma}{2} \quad \forall x^* \in \partial\Phi(x), \ \ z^* \in \partial\psi(x),$$

which completes the proof. □

Proof of Theorem 47. One clearly has $x_0 \neq 0$ because

$$f(x_0) = \inf_{x \in X} f(x) < f(0).$$

Arguing by contradiction, assume (f_5) holds true. It is not restrictive to write (f_4) for $r = 1$. Let us first note that

$$f(x_0) < \inf_{x \in \overline{B}_1 \cap V} f(x). \tag{121}$$

Indeed, $\overline{B}_1 \cap V$ turns out a compact subset of $\{x \in X : f(x) < a\}$ while, due to (f_3), f is locally Lipschitz continuous on this set. Thus, one can find an $x_1 \in \overline{B}_1 \cap V$ fulfilling $f(x_1) = \inf_{x \in \overline{B}_1 \cap V} f(x)$. If $f(x_1) = f(x_0)$ then x_1 would be a global minimum point for f. Since $\|x_1\| \leq 1 < \|x_0\|$ and $f(x_0) < f(0)$, we would get $x_1 \in K(f) \setminus \{0, x_0\}$, which contradicts ($f_5$).

Now, from $f(x_0) < 0 < a$ it easily follows

$$f(x) < 0 \ \text{ in } \ \overline{B(x_0, \rho_0)} \tag{122}$$

for some $\rho_0 \in]0, \frac{\|x_0\|}{2}[$. Let $\rho > 0$ be as in Proposition 48. On account of (121), we may assume that

$$2\rho < \min\{\rho_0, 1\} \quad \text{and} \quad f(x_0) + \rho < \inf_{x \in \overline{B}_1 \cap V} f(x). \tag{123}$$

Moreover, by decreasing ρ when necessary, hypothesis (f_4) leads to

$$\sup_{x \in \partial B_1 \cap V} f(x) < -2\rho L < 0, \tag{124}$$

where L denotes a Lipschitz constant for f on a suitable closed ball centered at 0, which contains $\overline{B}_{2\rho}$. Pick any $b \in]0, a[$. Through (f_4) and (123) we obtain

$$\partial B_1 \cap V \subseteq \{x \in X : f(x) < b\} \setminus \overline{U_\rho \cup B_{2\rho}}. \tag{125}$$

Observe next that (f_6) is satisfied with $U_0 := U_\rho$, $N_0 := B_{2\rho}$ because, in view of Proposition 48,

$$U_\rho \cap B_{2\rho} \subseteq B(x_0, \rho_0) \cap B_{2\rho} = \emptyset$$

while, due to the choice of b besides (f_3),

$$f_b \subseteq \{x \in X : f(x) < a\} \subseteq \operatorname{int}(D_\psi) = \operatorname{int}(D_{\partial\psi}). \tag{126}$$

Exploiting this inclusion we also see that (f_7) holds true. Indeed, since the set $\{x \in X : f(x) < a\}$ is open, ψ turns out locally Lipschitz continuous on a neighbourhood of f_b. If $\{x_n\} \subseteq f_b \setminus (U_0 \cup N_0)$, $x_n \to x$ in X, $x_n^* \in \partial\psi(x_n)$ for all $n \in \mathbb{N}$, and $z \in X$ then, by [**Ch**, Proposition 6], there exists a sequence $\{w_n^*\} \subseteq \partial\psi(x)$ fulfilling

$$\lim_{n \to +\infty} \langle x_n^* - z_n^*, z \rangle = 0. \tag{127}$$

We may suppose $w_n^* \rightharpoonup x^*$ for some $x^* \in \partial\psi(x)$, where a subsequence is considered when necessary. Consequently, owing to (127),

$$\langle x^*, z \rangle = \lim_{n \to +\infty} \langle w_n^*, z \rangle = \lim_{n \to +\infty} \langle w_n^* - x_n^* + x_n^*, z \rangle \leq 0 + \limsup_{n \to +\infty} \langle x_n^*, z \rangle,$$

as desired. At this point, Proposition 50 can be applied, and we get a locally Lipschitz continuous function $F : f_b \setminus (U_0 \cup N_0) \to X$ enjoying property (111) besides $\|F(x)\| \le 1$. In particular, (111) evidently forces $F(x) \neq 0$ for all $x \in f_b \setminus (U_0 \cup N_0)$.

Fix $z \in \partial B_1 \cap V$. On account of (125) it makes sense to consider the Cauchy problem

$$\begin{cases} \dfrac{d\eta_z(t)}{dt} = -F(\eta_z(t)), \\ \eta_z(0) = z. \end{cases} \tag{128}$$

By the basic existence-uniqueness theorem for ordinary differential equations in Banach spaces it possesses a unique local solution η_z. Let T_z be the maximum of $\{T \in]0, +\infty] : \eta_z$ is defined on $[0, T[\}$. We claim that $T_z < +\infty$. In fact, since f turns out locally Lipschitz continuous on a neighbourhood of f_b, Proposition 9 in [**Ch**] yields

$$\frac{d}{dt} f(\eta_z(t)) \le \max_{x^* \in \partial\Phi(\eta_z(t)),\ z^* \in \partial\Psi(\eta_z(t))} \langle x^* + z^*, -F(\eta_z(t))\rangle$$

for almost every $t \in [0, T_z[$. Thanks to (111) we thus have

$$\frac{d}{dt} f(\eta_z(t)) \le -\frac{\sigma}{2}. \tag{129}$$

Integrating over $[0, t]$, $t \in]0, T_z[$, provides

$$f(\eta_z(t)) - f(z) \le -\frac{\sigma}{2} t, \tag{130}$$

which clearly leads to

$$T_z \le \frac{2}{\sigma} \left(f(z) - \inf_{x \in X} f(x) \right) < +\infty.$$

Observe next that

$$\eta_z(t) = z - \int_0^t F(\eta_z(\tau)) d\tau \quad \forall t \in [0, T_z[.$$

Consequently, due to the boundedness of F, $\eta_z(t)$ converges as $t \to T_z$. Setting

$$w_z := \lim_{t \to T_z} \eta_z(t) \tag{131}$$

it results in $w_z \in \partial(f_b \setminus (U_0 \cup N_0))$, because $[0, T_z[$ is maximal. By (130) and (125), the point w_z cannot belong to the boundary of f_b. Therefore, $w_z \in \partial(U_0 \cup N_0)$. If $w_z \in \partial N_0$ then $\|w_z\| = 2\rho$. Using (130) again, (131), besides (124), one has

$$f(w_z) < f(z) < -2\rho L. \tag{132}$$

Since f is Lipschitz continuous on $\overline{B}_{2\rho}$, we also obtain

$$f(w_z) = f(w_z) - f(0) \ge -L\|w_z\| = -2\rho L,$$

which contradicts (132). Hence,

$$w_z \in \partial U_0 \quad \forall z \in \partial B_1 \cap V. \tag{133}$$

Now, pick an $e \in \partial B_1 \cap E$ and define

$$Q := \left((\overline{B}_1 \cap V) \oplus [0, e]\right) \cap \overline{B}_1. \tag{134}$$

The boundary Q_0 of Q relative to $V \oplus \operatorname{span}\{e\}$ is given by

$$Q_0 = \{e\} \cup (\overline{B}_1 \cap V) \cup (\partial B_1 \cap (\{\mu z : \mu \in]0,1], z \in \partial B_1 \cap V\} \oplus]0, e])).$$

Write $\gamma_0(e) := x_0$, $\gamma_0(x) := x$ for all $x \in \overline{B}_1 \cap V$, as well as

$$\gamma_0(x) := \begin{cases} \eta_z(2\lambda T_z) & \text{if } 0 < \lambda < \dfrac{1}{2}, \\ w_z & \text{if } \lambda = \dfrac{1}{2}, \\ (2\lambda - 1)x_0 + (2 - 2\lambda)w_z & \text{if } \dfrac{1}{2} < \lambda \leq 1 \end{cases} \tag{135}$$

provided $x := \mu z + \lambda e$, with $\lambda, \mu \in]0, 1]$, $z \in \partial B_1 \cap V$, and $\|x\| = 1$. A simple computation ensures that $\gamma_0 : Q_0 \to X$ turns out continuous. Moreover,

$$f(\gamma_0(x)) \leq 0 \quad \forall x \in Q_0. \tag{136}$$

Indeed, we evidently have $f(\gamma_0(e)) = f(x_0) < 0$ while, in view of (f_7), $f(\gamma_0(x)) = f(x) \leq 0$ for any $x \in \overline{B}_1 \cap V$. Put $x := \mu z + \lambda e$, where $\lambda, \mu \in]0, 1]$, $z \in \partial B_1 \cap V$. If $\lambda < \frac{1}{2}$ then, thanks to (129),

$$f(\gamma_0(x)) = f(\eta_z(2\lambda T_z)) \leq f(\eta_z(0)) = f(z) \leq 0.$$

The same reasoning yields (136) for $\lambda = \frac{1}{2}$. So, suppose $\lambda > \frac{1}{2}$. Because of (133), besides Proposition 48, it results in

$$\|(2\lambda - 1)x_0 + (2 - 2\lambda)w_z - x_0\| \leq \|x_0 - w_z\| \leq \rho_0. \tag{137}$$

From (122) we thus achieve

$$f(\gamma_0(x)) = f((2\lambda - 1)x_0 + (2 - 2\lambda)w_z) \leq 0,$$

and (136) is proved.

Let us next verify that

$$\|\gamma_0(x)\| \geq 2\rho \quad \forall x \in \partial B_1 \cap Q. \tag{138}$$

When $x := e$ or $x \in \overline{B}_1 \cap V$, this inequality is an immediate consequence of the choice of ρ_0 and (123). Pick $x := \mu z + \lambda e$, with $\|x\| = 1$, $\lambda, \mu \in]0, 1]$, $z \in \overline{B}_1 \cap V$. Since $\eta_z(t)$, $t \in]0, T_z[$, does not belong to N_0, (138) holds true for $0 < \lambda < \frac{1}{2}$. If $\frac{1}{2} \leq \lambda \leq 1$ then exploiting (137) besides (123) we infer

$$\gamma_0(x) \in \overline{B(x_0, \rho_0)} \subseteq X \setminus B_{2\rho},$$

namely $\|\gamma_0(x)\| \geq 2\rho$, as desired.

Now, define

$$\Gamma := \left\{\gamma \in C^0(Q, X) : \gamma|_{Q_0} = \gamma_0\right\}, \quad c := \inf_{\gamma\in\Gamma} \sup_{x\in Q} f(\gamma(x)),$$

in addition to $F := \partial B_\rho \cap E$. Gathering (136), ($f_4$), the inequality $\rho < 1$ together one has

$$\sup_{x\in Q_0} f(\gamma_0(x)) \le 0 \le \inf_{x\in F} f(x).$$

Through (138) and [**BN**, Lemma 3] we then get $(\gamma(Q) \cap F) \setminus \gamma_0(Q_0) \neq \emptyset$ for all $\gamma \in \Gamma$. Hence, assumption (a_5) of Theorem 24 is satisfied. To verify (a_4), observe at first that the set $\operatorname{conv}(\gamma_0(Q_0))$ turns out compact, because so is Q_0, while from (136), (126) it follows $\operatorname{conv}(\gamma_0(Q_0)) \subseteq \operatorname{int}(D_\psi)$. Thus, by the Generalized Theorem of Tietze [**Bo**, p. 77], there exists a $\hat\gamma \in \Gamma$ such that $\hat\gamma(Q) \subseteq \operatorname{conv}(\gamma_0(Q_0))$. Since f is continuous on $\operatorname{int}(D_\psi)$, this implies $\sup_{x\in Q} f(\hat\gamma(x)) < +\infty$, i.e., hypothesis ($a_4$) holds true too. Therefore, thanks to Theorem 24, $K_c(f) \neq \emptyset$. One clearly has $\inf_{x\in F} f(x) \le c$. If $\inf_{x\in F} f(x) < c$ then the function f possesses a critical point different from x_0 and 0. Otherwise, $K_c(f) \cap F \neq \emptyset$, which always leads to the same conclusion. However, this contradicts condition (f_5). □

REMARK 51. Hypothesis (f_4) is obviously satisfied in the meaningful special case when

(f_4') *for some* $r > 0$ *one has* $f_{|\bar B_r \cap E} \ge 0$ *as well as* $f_{|\bar B_r \cap V\setminus\{0\}} < 0$,

namely 0 turns out to be a local minimum of $f|_E$ and a proper local maximum for $f|_V$. If $\dim(V) \ge 2$ then (f_4) may be replaced by the more general condition:

(f_4'') *There is an* $r \in]0, \|x_0\|/2[$ *such that*

$$f|_{\bar B_r\cap E} \ge 0, \quad f|_{\bar B_r \cap V} \le 0, \quad f|_{\bar B_r\cap V\setminus\{0\}} \not\equiv 0.$$

8. Examples and applications

We conclude the chapter devoting this section to discuss with possible applications of some of the theoretical results previously established to suitable classes of hemivariational or variational-hemivariational inequalities.

Let Ω a nonempty bounded domain of the real Euclidean N-space $(\mathbb{R}^N, |\cdot|)$, $N \ge 3$, having a smooth boundary $\partial\Omega$. Denote by $|\Omega|$ the Lebesgue measure of Ω and by $H_0^1(\Omega)$ the closure of $C_0^\infty(\Omega)$ with respect to the norm

$$\|u\| := \left(\int_\Omega |\nabla u(x)|^2 dx\right)^{1/2}.$$

Let 2^* be the critical exponent for the Sobolev embedding $H_0^1(\Omega) \hookrightarrow L^p(\Omega)$. Recall that $2^* = \frac{2N}{N-2}$, if $p \in [1, 2^*]$ then there exists a positive constant c_p such that

$$\|u\|_{L^p(\Omega)} \le c_p \|u\| \tag{139}$$

for all $u \in H_0^1(\Omega)$ and, in particular, the embedding is compact whenever $p \in [1, 2^*[$; see [**R1**, Proposition B.7]. Let $\{\lambda_n\}$ be the sequence of eigenvalues of the operator

$-\Delta$ in $H_0^1(\Omega)$, with $0 < \lambda_1 < \lambda_2 \le \ldots \le \lambda_n \le \ldots$, and let $\{\varphi_n\}$ be a corresponding sequence of eigenfunctions normalized as follows:

$$\begin{aligned} &\|\varphi_n\|^2 = 1 = \lambda_n \|\varphi_n\|^2_{L^2(\Omega)}, \quad n \in \mathbb{N}, \\ &\textstyle\int_\Omega \nabla\varphi_m(x) \cdot \nabla\varphi_n(x)dx = \int_\Omega \varphi_m(x)\varphi_n(x)dx = 0 \quad \text{provided } m \ne n. \end{aligned} \tag{140}$$

We shall adopt the same notation introduced at the end of Section 4. Fix $k \in \mathbb{N}$ and $\lambda \in [\lambda_k, \lambda_{k+1}]$. Given $\Phi_2 : H_0^1(\Omega) \to \mathbb{R}$ locally Lipschitz continuous and $\psi : H_0^1(\Omega) \to \mathbb{R} \cup \{+\infty\}$ convex, proper, besides lower semi-continuous, indicate with $(\mathrm{P}_{\lambda,\Phi_2,\psi})$ the elliptic variational-hemivariational inequality problem:
Find $u \in D_\psi$ such that

$$-\int_\Omega \nabla u(x)\cdot\nabla(v-u)(x)dx + \lambda\int_\Omega u(x)(v(x)-u(x))dx \le \Phi_2^0(u; v-u) + \psi(v) - \psi(u)$$

for all $v \in D_\psi$.
A meaningful special case occurs when

$$\Phi_2(u) = \int_\Omega J(x, u(x))dx, \quad u \in H_0^1(\Omega), \tag{141}$$

where

$$J(x,\xi) := \int_0^\xi -j(x,t)dt, \quad (x,\xi) \in \Omega \times \mathbb{R}, \tag{142}$$

and the function $j : \Omega \times \mathbb{R} \to \mathbb{R}$ satisfies the conditions

(j_1) *j is measurable with respect to each variable separately,*
(j_2) *there exist $a_1 > 0$, $p \in [1, 2^*]$ such that*

$$|j(x,t)| \le a_1\left(1 + |t|^{p-1}\right) \quad \forall (x,t) \in \Omega \times \mathbb{R}.$$

Indeed, from (j_1)–(j_2) it follows that J turns out well defined, $J(\cdot,\xi)$ is measurable, while $J(x,\cdot)$ is locally Lipschitz continuous. Hence, its generalized directional derivative J_x^0 with respect to the variable ξ can be considered. Moreover, on account of formula (9) at p. 84 in [**Cl**], any solution u to $(\mathrm{P}_{\lambda,\Phi_2,\psi})$ also fulfils the inequality

$$\begin{aligned} &-\int_\Omega \nabla u(x)\cdot\nabla(v-u)(x)dx + \lambda\int_\Omega u(x)(v(x)-u(x))dx \\ &\le \int_\Omega J_x^0(u(x); v(x)-u(x))dx + \psi(v) - \psi(u) \end{aligned} \tag{143}$$

for all $v \in D_\psi$. When $j \in C^0(\Omega \times \mathbb{R})$ and $\psi \equiv 0$ then the function $u \in H_0^1(\Omega)$ turns out a weak solution to the Dirichlet problem

$$-\Delta u = \lambda u + j(x,u) \text{ in } \Omega, \qquad u = 0 \text{ on } \partial\Omega.$$

Now, put

$$X := H_0^1(\Omega), \quad V := \operatorname{span}\{\varphi_1, \varphi_2, \ldots, \varphi_k\}, \quad E := V^\perp,$$

and

$$Q(r) := (\overline{B}_r \cap V) \oplus [0, r\varphi_{k+1}], \quad r > 0.$$

Hence, each $u \in Q(r)$ can be uniquely written as $u = \bar{u} + t\varphi_{k+1}$, with $\bar{u} \in \overline{B}_r \cap V$, $t \in [0, r]$.

THEOREM 52. *Let* (h_2) *be satisfied and let* $r > 0$ *be such that*

$$\sup_{u \in Q(r)} \psi(u) < +\infty,$$

$$\Phi_2(\bar{u}) + \psi(\bar{u}) \leq \frac{1}{2}\left(\frac{\lambda}{\lambda_k} - 1\right)\|\bar{u}\|^2 \quad \forall \bar{u} \in \overline{B}_r \cap V, \tag{144}$$

$$\Phi_2(u) + \psi(u) \leq \frac{1}{2}\left(\frac{\lambda}{\lambda_k} - 1\right)\|\bar{u}\|^2 - \frac{1}{2}\left(1 - \frac{\lambda}{\lambda_{k+1}}\right)t^2 \quad \forall u \in Q(r) \setminus B_r. \tag{145}$$

Assume that:

(h_3) *There exists a* $\rho \in]0, r[$ *fulfilling*

$$\inf_{\tilde{u} \in \partial B_\rho \cap E} \frac{1}{\rho^2}\left(\Phi_2(\tilde{u}) + \psi(\tilde{u})\right) \geq -\frac{1}{2}\left(1 - \frac{\lambda}{\lambda_{k+1}}\right).$$

(h_4) *Either* D_ψ *is bounded or* (145) *holds for all* $u \in Q(r)$.

Then ($P_{\lambda,\Phi_2,\psi}$) *admits at least one nontrivial solution* $u \in H_0^1(\Omega)$. *When the second assertion of* (h_4) *is true we also have* $u \in E$, $\|u\| = \rho$, *as well as*

$$\frac{\lambda}{2}\|u\|^2_{L^2(\Omega)} - \Phi_2(u) - \psi(u) = \frac{\rho^2}{2}.$$

PROOF. The conclusion comes from Theorem 27. To see this, choose $Q := Q(r)$, $S := \partial B_\rho \cap E$, and denote by Q_0 the boundary of Q relative to W, with

$$W := \operatorname{span}\{\varphi_1, \varphi_2, \ldots, \varphi_k, \varphi_{k+1}\}. \tag{146}$$

Moreover, for every $u \in X$, define

$$\Phi_1(u) := \frac{1}{2}\left(\|u\|^2 - \lambda\|u\|^2_{L^2(\Omega)}\right), \quad \Phi(u) := \Phi_1(u) + \Phi_2(u),$$
$$f(u) := \Phi(u) + \psi(u).$$

Let us verify (a_{11}). Every $u \in Q_0$ can be written as $u = \sum_{i=1}^{k+1} t_i\varphi_i$, where $t_1, \ldots, t_{k+1} \in \mathbb{R}$. Hence, exploiting (144), (145), besides (140), yields either

$$f(u) = \frac{1}{2}\sum_{i=1}^{k}\left(1 - \frac{\lambda}{\lambda_i}\right)t_i^2 + \Phi_2(\bar{u}) + \psi(\bar{u}) \leq \frac{1}{2}\left(1 - \frac{\lambda}{\lambda_k}\right)\|\bar{u}\|^2 + \Phi_2(\bar{u}) + \psi(\bar{u}) \leq 0,$$

or

$$\begin{aligned} f(u) &= \frac{1}{2}\sum_{i=1}^{k}\left(1 - \frac{\lambda}{\lambda_i}\right)t_i^2 + \frac{1}{2}\left(1 - \frac{\lambda}{\lambda_{k+1}}\right)t_{k+1}^2 + \Phi_2(u) + \psi(u) \\ &\leq \frac{1}{2}\left(1 - \frac{\lambda}{\lambda_k}\right)\|\bar{u}\|^2 + \frac{1}{2}\left(1 - \frac{\lambda}{\lambda_{k+1}}\right)t_{k+1}^2 + \Phi_2(u) + \psi(u) \leq 0, \end{aligned}$$

according to whether $t_{k+1} = 0$ or $t_{k+1} > 0$. Consequently, $\sup_{u\in Q_0} f(u) \le 0$. Now, for every $\tilde{u} \in S$ one has $\tilde{u} = \sum_{i=k+1}^{\infty} t_i \varphi_i$ for suitable $t_i \in \mathbb{R}$ and $\|\tilde{u}\| = \rho$. At this point (140) and (h_3) furnish

$$f(\tilde{u}) = \frac{1}{2}\sum_{i=k+1}^{\infty}\left(1-\frac{\lambda}{\lambda_i}\right)t_i^2 + \Phi_2(\tilde{u}) + \psi(\tilde{u}) \ge \frac{1}{2}\left(1-\frac{\lambda}{\lambda_{k+1}}\right)\rho^2 + \Phi_2(\tilde{u}) + \psi(\tilde{u}) \ge 0.$$

Therefore, $\inf_{\tilde{u}\in S} f(\tilde{u}) \ge 0$, and ($a_{11}$) is true.
We next show (h_1). To this end, observe at first that by [**Cl**, Proposition 2.2.4] one has $\partial\Phi_1(u) = \{\Phi_1'(u)\}$, $u \in X$, because $\Phi_1 \in C^1(X, \mathbb{R})$. Fix a sequence $\{u_n\} \subseteq D_\psi$ with $u_n \rightharpoonup u$ in X and

$$\limsup_{n\to+\infty} \langle \Phi_1'(u_n), u_n - u\rangle \le 0. \tag{147}$$

Up to subsequences, we may suppose $u_n \to u$ in $L^2(\Omega)$. Since

$$\begin{aligned}\langle \Phi_1'(u_n), u_n - u\rangle &= \langle \Phi_1'(u), u_n - u\rangle + \langle \Phi_1'(u_n) - \Phi_1'(u), u_n - u\rangle \\ &= \langle \Phi_1'(u), u_n - u\rangle + \|u_n - u\|^2 - \lambda\|u_n - u\|^2_{L^2(\Omega)} \quad \forall n \in \mathbb{N},\end{aligned}$$

from (147) it easily follows $\limsup_{n\to+\infty} \|u_n - u\|^2 \le 0$, namely $u_n \to u$ in X. Finally, if (145) is true for all $u \in Q$ then $\sup_{u\in Q} f(u) \le 0$. Bearing in mind that

$$0 \le \inf_{\tilde{u}\in S} f(\tilde{u}) \le c \le \sup_{u\in Q} f(u) \le 0,$$

as (Q, Q_0) links with S while $\mathrm{id}|_Q \in \Gamma$, we get

$$\inf_{\tilde{u}\in S} f(\tilde{u}) = c = 0. \tag{148}$$

Now, on account of (h_4), Theorem 27 can be applied. Thus, there exists a point $u \in D_\psi$ satisfying

$$\Phi^0(u; v-u) + \psi(v) - \psi(u) \ge 0 \quad \forall v \in D_\psi \tag{149}$$

besides $f(u) = c$. Due to the smoothness of Φ_1, inequality (149) becomes

$$-\int_\Omega \nabla u(x)\cdot\nabla(v-u)(x)dx + \lambda\int_\Omega u(x)(v(x)-u(x))dx \le \Phi_2^0(u; v-u) + \psi(v) - \psi(u),$$

$v \in D_\psi$, i.e. u turns out a solution to ($P_{\lambda,\Phi_2,\psi}$). When (145) holds on the whole Q, by (148) we also have $u \in S$ in addition to $f(u) = 0$. Hence, $u \in E$, $\|u\| = \rho$, and

$$\frac{\lambda}{2}\|u\|^2_{L^2(\Omega)} - \Phi_2(u) - \psi(u) = \frac{\rho^2}{2}.$$

To complete the proof, observe that the solution u we find is nontrivial. In fact, (144) implies $f(0) \le 0$. Hence, if $c > 0$ the assertion immediately comes from $f(u) = c$. Otherwise, it is a consequence of $\|u\| = \rho > 0$. □

The above result contains Theorem 3 of [**MoMo**] as a special case. Making suitable choices of Φ_2 and ψ, other existence results can be easily achieved through Theorem 52. By way of example, we present the result below, where Φ_2 is defined in (141) while ψ denotes the indicator function of some closed ball centered at the origin. It essentially coincides with Theorem 3.3 in [**MR**].

THEOREM 53. *Suppose the function* $j : \Omega \times \mathbb{R} \to \mathbb{R}$ *fulfils* (j_1), (j_2) *with* $p \in]2, 2^*[$, *and, moreover,*

(j_3) *for some* $a_2 > 0$, $q \in]2, p]$ *one has* $J(x, \xi) \le \min\{0, a_2(1 - |\xi|^q)\}$ *in* $\Omega \times \mathbb{R}$,
(j_4) *there exists a* $\rho > 0$ *such that*

$$\inf_{\tilde{u} \in \partial B_\rho \cap E} \int_\Omega J(x, \tilde{u}(x))dx \ge -\frac{1}{2}\left(1 - \frac{\lambda}{\lambda_{k+1}}\right),$$

where J *is given by* (142). *Then to any sufficiently large* $R > 0$ *there corresponds a function* $u \in \overline{B}_R \setminus \{0\} \subseteq H_0^1(\Omega)$ *satisfying*

$$-\int_\Omega \nabla u(x) \cdot \nabla (v - u)(x)dx + \lambda \int_\Omega u(x)(v(x) - u(x))dx \le \int_\Omega J_x^0(u(x); v(x) - u(x))dx$$

for all $v \in \overline{B}_R$.

PROOF. The set W defined in (146) is a finite dimensional subspace of $L^q(\Omega)$ because $q \le p < 2^*$. Hence, one can find an $a_3 > 0$ such that $\|u\|_{L^q(\Omega)} \ge a_3\|u\|$, $u \in W$. From (j_3) we obtain

$$\int_\Omega J(x, u(x))dx \le a_2|\Omega| - a_2 a_3^q \|u\|^q \quad \forall u \in W.$$

The condition $q > 2$ leads to

$$a_2|\Omega| - a_2 a_3^q \|u\|^q \le -\frac{1}{2}\left(1 - \frac{\lambda}{\lambda_{k+1}}\right)\|u\|^2$$

provided $\|u\| \ge r$ and $r > \rho$ is large enough. Therefore,

$$\Phi_2(u) \le -\frac{1}{2}\left(1 - \frac{\lambda}{\lambda_{k+1}}\right)\|u\|^2 \le \frac{1}{2}\left(\frac{\lambda}{\lambda_k} - 1\right)\|\bar{u}\|^2 - \frac{1}{2}\left(1 - \frac{\lambda}{\lambda_{k+1}}\right)t^2$$

for all $u \in Q(r) \setminus B_r$, $u = \bar{u} + t\varphi_{k+1}$. This means that (145) holds true with $\psi : X \to \mathbb{R} \cup \{+\infty\}$ defined by

$$\psi(u) := 0 \quad \text{when } u \in \overline{B}_{r\sqrt{2}}, \qquad \psi(u) := +\infty \quad \text{otherwise.} \tag{150}$$

Again from (j_3) one has

$$\Phi_2(\bar{u}) \le 0 \le \frac{1}{2}\left(\frac{\lambda}{\lambda_k} - 1\right)\|\bar{u}\|^2 \quad \forall \bar{u} \in \overline{B}_r \cap V,$$

namely (144) is satisfied. Since (h_3) immediately comes from (j_4) while (h_4) is obvious, we only need to verify (h_2). Let $\{u_n\} \subseteq D_\psi$ be such that $u_n \rightharpoonup u$ in X. On account of the condition $p < 2^*$ we may suppose, along a subsequence when necessary, $u_n(x) \to u(x)$ and $|u_n(x)| \leq w(x)$, $n \in \mathbb{N}$, for almost every $x \in \Omega$, where $w \in L^p(\Omega)$. Consequently, by (j_2),

$$\begin{aligned}|J_x^0(u_n(x); u(x) - u_n(x))| &\leq a_1 \left(1 + |u_n(x)|^{p-1}\right) |u(x) - u_n(x)| \\ &\leq a_1 \left(1 + w(x)^{p-1}\right) (|u(x)| + w(x)) \quad \text{a.e. in } \Omega, \quad \forall n \in \mathbb{N}.\end{aligned}$$

Exploiting formula (9) at p. 84 in [**Cl**], Fatou's lemma, and the upper semicontinuity of J_x^0 finally yields

$$\begin{aligned}\limsup_{n\to+\infty} \Phi_2^0(u_n; u - u_n) &\leq \limsup_{n\to+\infty} \int_\Omega J_x^0(u_n(x); u(x) - u_n(x))dx \\ &\leq \int_\Omega J_x^0(u(x); 0)dx = 0,\end{aligned}$$

as desired. At this point Theorem 52 can be applied to get a nontrivial solution $u \in \overline{B}_R$, with $R := r\sqrt{2}$, of $(\mathrm{P}_{\lambda,\Phi_2,\psi})$ and, a fortiori, of (143) written for ψ defined in (150). □

Given a function $a \in L^\infty(\Omega)$, consider the eigenvalue problem

(151) $$\begin{cases} -\Delta u + a(x)u = \lambda u & \text{in } \Omega, \\ u = 0 & \text{on } \partial\Omega. \end{cases}$$

It is well known [**GT**, Section 8.12] that (151) possesses a sequence $\{\lambda_n\}$ of eigenvalues fulfilling $\lambda_1 < \lambda_2 \leq \cdots \leq \lambda_n \leq \cdots$ (the number of times an eigenvalue appears in the sequence equals its multiplicity) and, moreover, that (vide [**A**, p. 14])

(152) $$\lambda_1 > \operatorname{essinf}_{x\in\Omega} a(x).$$

Let $\{\phi_n\}$ be a corresponding sequence of eigenfunctions normalized as follows:

(153) $$\int_\Omega [|\nabla\phi_n(x)|^2 + a(x)\phi_n(x)^2]dx = \lambda_n \int_\Omega \phi_n(x)^2 dx = \lambda_n$$

for every $n \in \mathbb{N}$;

(154) $$\int_\Omega [\nabla\phi_m(x) \cdot \nabla\phi_n(x) + a(x)\phi_m(x)\phi_n(x)]dx = \int_\Omega \phi_m(x)\phi_n(x)dx = 0$$

provided $m, n \in \mathbb{N}$ and $m \neq n$.
For simplicity, we suppose that

(155) $$\lambda_s < 0 < \lambda_{s+1} \quad \text{for some } s \in \mathbb{N}.$$

Let $g : \mathbb{R} \to \mathbb{R}$ be a function satisfying the conditions

(g_1) *g is measurable,*
(g_2) *there exist $a_1 > 0$, $p \in]2, 2^*[$ such that $|g(t)| \leq a_1(1 + |t|^{p-1})$ $\forall t \in \mathbb{R}$.*

Hence, the functions $G : \mathbb{R} \to \mathbb{R}$ and $\mathcal{G} : H_0^1(\Omega) \to \mathbb{R}$ defined by putting

$$G(\xi) := \int_0^{\xi} -g(t)dt \quad \forall \xi \in \mathbb{R}, \quad \mathcal{G}(u) := \int_{\Omega} G(u(x))dx \quad \forall u \in H_0^1(\Omega),$$

respectively, turn out locally Lipschitz continuous besides well defined. So we can consider their generalized directional derivatives G^0 and $\mathcal{G}^0$. On account of formula (9) at p. 84 in [**Cl**] one has

$$\mathcal{G}^0(u; v) \leq \int_{\Omega} G^0(u(x); v(x))dx, \quad u, v \in H_0^1(\Omega). \tag{156}$$

For our application, we will further assume

(g_3) $\lim_{t \to 0} \frac{g(t)}{t} = 0$,
(g_4) $\limsup_{|t| \to +\infty} \frac{g(t)}{t} < 0$, *and*
(g_5) *there exists a* $\xi_0 \in \mathbb{R}$ *such that* $G(\xi_0) < 0$.

Hypothesis (g_4) assures the existence of two positive constants β, γ satisfying

$$g(t) \geq -\beta t - \gamma \quad \forall t \leq 0, \quad g(t) \leq -\beta t + \gamma \quad \forall t \geq 0. \tag{157}$$

Now, let $\lambda, \mu > 0$. Define

$$r_{\lambda,\mu} := \frac{\lambda\gamma}{c_1} + \sqrt{\left(\frac{\lambda\gamma}{c_1}\right)^2 + 2\mu}, \tag{158}$$

with c_1 as in (139) written for $p = 1$. A set $K_\lambda \subseteq H_0^1(\Omega)$ is called of type (K_λ^g) provided

(K_λ^g) K_λ *turns out convex closed in* $H_0^1(\Omega)$. *Moreover, there exists a* $\mu > 0$ *such that* $\overline{B}_{r_{\lambda,\mu}} \subset K_\lambda$.

Given $\lambda > 0$ and K_λ satisfying (K_λ^g), denote by (P_λ) the elliptic variational-hemivariational inequality problem:

Find $u \in K_\lambda$ *such that*

$$-\int_{\Omega} \nabla u(x) \cdot \nabla(v-u)(x)dx - \int_{\Omega} a(x)u(x)(v-u)(x)dx \leq \lambda \mathcal{G}^0(u; v-u)$$

for all $v \in K_\lambda$.

Due to (156), any solution u of (P_λ) also fulfils the inequality

$$-\int_{\Omega} \nabla u(x) \cdot \nabla(v-u)(x)dx - \int_{\Omega} a(x)u(x)(v-u)(x)dx$$
$$\leq \lambda \int_{\Omega} G^0(u(x); (v-u)(x))dx \quad \forall v \in K_\lambda.$$

When g is continuous, while $K_\lambda := H_0^1(\Omega)$, the function $u \in H_0^1(\Omega)$ turns out a weak solution to the Dirichlet problem

$$-\Delta u + a(x)u = \lambda g(u) \text{ in } \Omega, \quad u = 0 \text{ on } \partial\Omega,$$

which has been previously investigated in [**BN**] under more restrictive conditions.

THEOREM 54. *Suppose* (g_1)–(g_5) *hold true. Then, for every* λ *sufficiently large, problem* (P_λ) *possesses at least two nontrivial solutions.*

PROOF. Write $X := H_0^1(\Omega)$ and define, whenever $u \in X$,

$$\Phi(u) := \frac{1}{2}\int_\Omega \left(|\nabla u(x)|^2 + a(x)u^2(x)\right) dx + \lambda \mathcal{G}(u)$$

as well as

$$\psi(u) := \begin{cases} 0 & \text{if } u \in K_\lambda, \\ +\infty & \text{otherwise,} \end{cases} \qquad f(u) := \Phi(u) + \psi(u),$$

where $\lambda > 0$ while $K_\lambda \subseteq H_0^1(\Omega)$ is of type (K_λ^g). Owing to (g_1), (g_2) the function $\Phi : X \to \mathbb{R}$ turns out locally Lipschitz continuous. Consequently, f satisfies condition (H_f'). We claim that

$$\text{(159)} \qquad f \text{ is bounded below and coercive for any } \lambda > -\frac{\alpha}{\beta},$$

with $\alpha := \operatorname{essinf}_{x\in\Omega} a(x)$. Fix $\lambda > -\frac{\alpha}{\beta}$. If $u \in X$ then from (157) it follows

$$\int_{\Omega(u(x)\geq 0)} dx \int_0^{u(x)} g(t)dt \leq \int_{\Omega(u(x)\geq 0)} \left(-\frac{\beta}{2}u^2(x) + \gamma u(x)\right) dx$$

besides

$$\begin{aligned}\int_{\Omega(u(x)\leq 0)} dx \int_0^{u(x)} g(t)dt &\leq \int_{\Omega(u(x)\leq 0)} \int_{u(x)}^0 (\beta t + \gamma)dt \\ &= \int_{\Omega(u(x)\leq 0)} \left(-\frac{\beta}{2}u^2(x) - \gamma u(x)\right) dx.\end{aligned}$$

Gathering these inequalities together yields

$$\int_\Omega dx \int_0^{u(x)} g(t)dt \leq -\frac{\beta}{2}\|u\|^2_{L^2(\Omega)} + \gamma\|u\|_{L^1(\Omega)},$$

which clearly means

$$\text{(160)} \qquad \mathcal{G}(u) \geq \frac{\beta}{2}\|u\|^2_{L^2(\Omega)} - \gamma\|u\|_{L^1(\Omega)} \quad \forall u \in X.$$

Now, through (139) and (160) we obtain

$$\begin{aligned}f(u) \geq \Phi(u) &\geq \frac{1}{2}\|u\|^2 + \frac{1}{2}(\alpha + \lambda\beta)\|u\|^2_{L^2(\Omega)} - \lambda\gamma\|u\|_{L^1(\Omega)} \\ &\geq \frac{1}{2}\|u\|^2 + \frac{1}{2}(\alpha + \lambda\beta)\|u\|^2_{L^2(\Omega)} - \frac{\lambda\gamma}{c_1}\|u\|,\end{aligned}$$

i.e., due to the choice of λ,

$$\text{(161)} \qquad f(u) \geq \frac{1}{2}\|u\|^2 - \frac{\lambda\gamma}{c_1}\|u\|, \quad u \in X.$$

Therefore, (159) holds true. Let us next show that the function f satisfies condition $(\mathrm{PS})_f$ provided $\lambda > -\frac{\alpha}{\beta}$. So, pick a sequence $\{u_n\} \subseteq X$ such that $\{f(u_n)\}$ turns out bounded and

$$\Phi^0(u_n; v - u_n) + \psi(v) - \psi(u_n) \geq -\varepsilon_n \|v - u_n\| \tag{162}$$

for all $n \in \mathbb{N}$, $v \in X$, where $\varepsilon_n \to 0^+$. By (162) one evidently has $\{u_n\} \subseteq K_\lambda$. Since f is coercive, the sequence $\{u_n\}$ turns out bounded. Thus, passing to a subsequence if necessary, we may suppose both $u_n \rightharpoonup u$ in X and $u_n \to u$ in $L^2(\Omega)$. The point u belongs to K_λ because this set is weakly closed. Exploiting (162) with $v := u$ we then get

$$\begin{aligned} &\int_\Omega \nabla u_n(x) \cdot \nabla(u - u_n)(x) dx + \int_\Omega a(x) u_n(x)(u - u_n)(x) dx \\ &\quad + \lambda \mathcal{G}^0(u_n; u - u_n) \geq -\varepsilon_n \|u - u_n\| \quad \forall n \in \mathbb{N}. \end{aligned} \tag{163}$$

From $u_n \to u$ in $L^2(\Omega)$ it follows

$$\lim_{n \to +\infty} \int_\Omega a(x) u_n(x)(u - u_n)(x) dx = 0. \tag{164}$$

The upper semi-continuity of $\mathcal{G}^0$ on $L^2(\Omega) \times L^2(\Omega)$ forces

$$\limsup_{n \to +\infty} \mathcal{G}^0(u_n; u - u_n) \leq \mathcal{G}^0(u; 0) = 0. \tag{165}$$

Taking account of (164)–(165) besides the weak convergence of $\{u_n\}$ to u and letting $n \to +\infty$ in (163) yields

$$\limsup_{n \to \infty} \int_\Omega |\nabla u_n(x)|^2 dx \leq \int_\Omega |\nabla u(x)|^2 dx,$$

namely, by [**B**, Proposition III.30], $u_n \to u$ in X. Hence, hypothesis (f_1) in Theorem 47 is fulfilled.
Through (g_5) we can construct an $u_0 \in \overline{B}_{r_{\lambda,\mu}}$ such that $\mathcal{G}(u_0) < 0$. Therefore, $\inf_{u \in X} f(u) < 0$ provided

$$\lambda > -\frac{1}{2\mathcal{G}(u_0)} \int_\Omega [|\nabla u_0(x)|^2 + a(x) u_0(x)^2] dx,$$

while $f(0) = \lambda \mathcal{G}(0) = 0$.

Our next objective is to verify (f_3) of the same result. Since K_λ turns out of type (K^g_λ), the set

$$\{u \in X : f(u) < \mu\} \quad \text{is open.} \tag{166}$$

Indeed, inequality (161) ensures that

$$\{u \in X : \Phi(u) < \mu\} \subseteq B_{r_{\lambda,\mu}} \subseteq K_\lambda.$$

Consequently,

$$\{u \in X : f(u) < \mu\} = \{u \in K_\lambda : \Phi(u) < \mu\} = \{u \in X : \Phi(u) < \mu\}$$

which leads to (166).
Finally, reasoning as in [**BM**, p. 137] we obtain

$$\lim_{u \to 0} \frac{\mathcal{G}(u)}{\|u\|^2} = 0 \tag{167}$$

while to any $\varepsilon > 0$ there corresponds a $\delta \in]0, 1[$ such that

$$\mathcal{G}(u) \geq -\|u\|^2 \left(\frac{\varepsilon}{2} c_2^2 + \frac{2a_1 c_p^p}{\delta^p} \|u\|^{p-2} \right) \quad \forall u \in X, \tag{168}$$

with c_2, c_p given by (139). Write $V := \text{span}\{\phi_1, \ldots, \phi_s\}$ and $E := V^\perp$, where the orthogonal complement is taken in $H_0^1(\Omega)$. One clearly has $X = V \oplus E$, $\dim(E) > 0$, besides $0 < \dim(V) < +\infty$. Moreover, if $u \in V$ then $u = \sum_{i=1}^s t_i \phi_i$ for some $t_1, \ldots, t_s \in \mathbb{R}$. A simple computation shows that

$$\|u\|^2 \leq (\lambda_s - \alpha) \|u\|^2_{L^2(\Omega)}, \quad u \in V, \tag{169}$$

with $\lambda_s - \alpha > \lambda_1 - \alpha > 0$ because of (152). Thanks to (K_λ^g) and (153)–(155) we get

$$f(u) = \Phi(u) = \frac{1}{2} \sum_{i=1}^s t_i^2 \lambda_i + \lambda \mathcal{G}(u) \leq \frac{1}{2} \lambda_s \|u\|^2_{L^2(\Omega)} + \lambda \mathcal{G}(u)$$

whenever $\|u\| \leq r_{\lambda,\mu}$. By (167), the above inequality, and (169), for every $\sigma > 0$ there exists a $\rho \in]0, r_{\lambda,\mu}[$ satisfying

$$f(u) \leq \left[\frac{\lambda_s}{2} + \lambda\sigma(\lambda_s - \alpha) \right] \|u\|^2_{L^2(\Omega)} \qquad \forall u \in \overline{B}_\rho \cap V.$$

At this point, choose $\sigma > 0$ so small that $\frac{\lambda_s}{2} + \lambda\sigma(\lambda_s - \alpha) < 0$, which is possible on account of (155). Bearing in mind (139) we have

$$f(u) < 0 \quad \forall u \in \overline{B}_R \cap V \setminus \{0\}, \tag{170}$$

where $R := \frac{\rho}{c_2}$. Let us next prove that

$$\int_\Omega \left(|\nabla u(x)|^2 + a(x) u(x)^2 \right) dx \geq \theta \|u\|^2 \quad \text{in } E \tag{171}$$

for a suitable constant $\theta > 0$. Indeed, if the assertion were false then there would exist a sequence $\{u_n\} \subseteq E$ enjoying the properties

$$\|u_n\| = 1, \quad n \in \mathbb{N}, \tag{172}$$

$$\int_\Omega \left(|\nabla u_n(x)|^2 + a(x) u_n(x)^2 \right) dx < \frac{1}{n} \quad \forall n \in \mathbb{N}. \tag{173}$$

Passing to a subsequence when necessary, we may suppose $u_n \rightharpoonup u$ in $H_0^1(\Omega)$ as well as $u_n \to u$ in $L^2(\Omega)$, with $u \in E$. Thus, letting $n \to +\infty$ in (173) yields

$$\int_\Omega \left(|\nabla u(x)|^2 + a(x)u(x)^2\right) dx \le 0. \tag{174}$$

From $u \in E$ it follows $u = \sum_{i=s+1}^{+\infty} t_i\phi_i$, where $t_i \in \mathbb{R}$, $i \ge s+1$. Through (153)–(155) we obtain

$$\lambda_{s+1}\|u\|_{L^2(\Omega)}^2 \le \int_\Omega \left(|\nabla u(x)|^2 + a(x)u(x)^2\right) dx. \tag{175}$$

Gathering (174) and (175) together leads to $u = 0$. By (173) this forces $u_n \to 0$ in $H_0^1(\Omega)$, against (172). Combining (171) with (168) provides

$$f(u) \ge \|u\|^2 \left[\frac{\theta}{2} - \lambda\left(\frac{\varepsilon}{2}c_2^2 + \frac{2a_1c_p^p}{\delta^p}\|u\|^{p-2}\right)\right] \quad \forall u \in E.$$

Pick $\varepsilon > 0$ and $r \in]0, R[$ such that

$$\frac{\theta}{2} - \lambda\left(\frac{\varepsilon}{2}c_2^2 + \frac{2a_1c_p^p}{\delta^p}r^{p-2}\right) > 0.$$

Then, thanks to (176) we have

$$f(u) \ge 0 \quad \forall u \in \overline{B}_r \cap E. \tag{176}$$

Finally, taking account of Remark 51, (170) and (176) immediately yield condition (f_4) of Theorem 47.

We are now in a position to apply this result. So, there exist at least two points $u_1, u_2 \in X \setminus \{0\}$ such that

$$\Phi^0(u_i; v - u_i) + \psi(v) - \psi(u_i) \ge 0$$

for all $v \in X$, $i = 1, 2$. The choice of ψ gives both $u_i \in K_\lambda$ and $\Phi^0(u_i; v - u_i) \ge 0$, $v \in K_\lambda$, $i = 1, 2$, namely u_1, u_2 turn out nontrivial solutions to problem (P_λ), which completes the proof. □

REMARK 55. Reading the above arguments we realize that the conclusion of Theorem 54 holds true as soon as

$$\lambda > \max\left\{-\frac{\alpha}{\beta}, \frac{c_1}{\gamma}\|u_0\|, -\frac{1}{2\mathcal{G}(u_0)}\int_\Omega [|\nabla u_0(x)|^2 + a(x)u_0(x)^2]dx\right\},$$

where $\alpha := \operatorname{essinf}_{x\in\Omega} a(x)$, β and γ are given by (157), c_1 comes from (139) written for $p = 1$, while $u_0 \in H_0^1(\Omega)$ fulfils $\mathcal{G}(u_0) < 0$.

Finally, Theorem 47 provides also the existence of multiple weak solutions to the hemivariational inequality problem, say (P), with p-Laplacian

$$\begin{cases} -\Delta_p u \in \partial J(u) & \text{in}\Omega, \\ \dfrac{\partial u}{\partial n_p} = 0 & \text{on}\partial\Omega, \end{cases}$$

where $2 \le p < +\infty$, $\Delta_p u := \mathrm{div}(|\nabla u|^{p-2}\nabla u)$, the potential $J : \mathbb{R} \to \mathbb{R}$ is locally Lipschitz continuous, while $\frac{\partial u}{\partial n_p} = |\nabla u|^{p-2}\nabla u \cdot n$, with $n(x)$ being the outward unit normal vector to $\partial\Omega$ at the point $x \in \partial\Omega$.

Let $p' := p/(p-1)$. A function $u \in W^{1,p}(\Omega)$ is called a weak solution to (P) provided there exists $v \in L^{p'}(\Omega)$ such that $v(x) \in \partial J(u(x))$ almost everywhere in Ω and

$$\int_\Omega |\nabla u(x)|^{p-2}\nabla u(x) \cdot \nabla w(x) dx = \int_\Omega v(x)w(x)dx \quad \forall w \in W^{1,p}(\Omega),$$

namely u turns out to be a solution of the hemivariational inequality

$$\int_\Omega |\nabla u(x)|^{p-2}\nabla u(x) \cdot \nabla w(x) dx \le \int_\Omega J^0(u(x); w(x))dx, \quad w \in W^{1,p}(\Omega).$$

The following result is established in [**MMB**]; see [**MMB**, Theorem 3.1]. Further existence results concerning problem (P) may be found in [**MaP**].

THEOREM 56. *Assume that:*

(J_1) *J is locally Lipschitz continuous and $J(0) = 0$.*

(J_2) *There exists a constant $a_1 > 0$ such that for every $\xi \in \mathbb{R}$ one has*

$$|y| \le a_1(1 + |\xi|^{p-1}) \quad \forall y \in \partial J(\xi).$$

(J_3) $\limsup_{|\xi|\to 0} \frac{J(\xi)}{|\xi|^p} = 0$.

(J_4) $\limsup_{|\xi|\to +\infty} \frac{J(\xi)}{|\xi|^p} < 0$.

(J_5) *There are $\mu > p$, $M > 0$ such that $\mu J(\xi) \le -J^0(\xi; -\xi)$ for any $|\xi| \ge M$.*

(J_6) *There exists a $\rho > 0$ such that $J(\xi) > 0$ provided $0 < |\xi| < \rho$.*

Then (P) possesses at least two nontrivial weak solutions $u_1, u_2 \in W^{1,p}(\Omega)$.

A meaningful special case of Theorem 56 occurs when

$$J(\xi) := \int_0^\xi j(t)dt \quad \forall \xi \in \mathbb{R}, \tag{177}$$

where the function $j : \mathbb{R} \to \mathbb{R}$ satisfies the conditions

(J_1') *j is measurable,*

(J_2') *there exists a constant $a_1 > 0$ such that*

$$|j(t)| \le a_1(1 + |t|^{p-1}), \quad t \in \mathbb{R}.$$

Indeed, under (J_1')–(J_2'), J turns out well defined and locally Lipschitz continuous. Moreover, one has

$$\partial J(\xi) = [\underline{j}(\xi), \overline{j}(\xi)], \quad \xi \in \mathbb{R},$$

with $\underline{j}$ and $\overline{j}$ being as in Example 1. Now, through Theorem 56 we directly infer the following

THEOREM 57. *Let $j : \mathbb{R} \to \mathbb{R}$ fulfil (J_1')–(J_2') and let J be given by (177). Suppose (J_3)–(J_6) hold true. Then the problem*

$$\begin{cases} \underline{j}(u) \le -\Delta_p u \le \overline{j}(u) & \text{in } \Omega, \\ \dfrac{\partial u}{\partial n_p} = 0 & \text{on } \partial\Omega \end{cases}$$

admits at least two nontrivial weak solutions $u_1, u_2 \in W^{1,p}(\Omega)$.

REMARK 58. It is worth pointing out that the inequalities $\underline{j}(u) \le -\Delta_p u \le \overline{j}(u)$ reduce to the equation $\Delta_p u = j(u)$ whenever $j \in C^0(\mathbb{R})$.

A related situation is treated by the result below, patterned after [**Ma**, Theorem 3.1]; cf. also [**MM0**, Theorem 4.2].

LEMMA 59. [**MMB**, Lemma 2.1] *Let* $u \in W^{1,p}(\Omega)$ *and let* $E \subseteq \mathbb{R}$ *be such that* $|E| = 0$. *Then* $\Delta_p u(x) = 0$ *at almost all* $x \in u^{-1}(E)$.

For $j : \mathbb{R} \to \mathbb{R}$ we write

$$D_j := \{t \in \mathbb{R} : \ j \text{ is discontinuous at } t\}$$

and, besides (J_2'), assume that

(J_1'') $|D_j| = 0$,

(J_7) *the conditions* $t \in D_j$, $\underline{j}(t) \le 0 \le \overline{j}(t)$ *imply* $j(t) = 0$.

Since(J_1'') forces (J_1'), the function J defined in (177) turns out well defined and locally Lipschitz continuous.

THEOREM 60. *If* (J_1''), (J_2'), *and* (J_3)–(J_7) *are satisfied then the problem*

$$\begin{cases} -\Delta_p u = j(u) & \text{in } \Omega, \\ \dfrac{\partial u}{\partial n_p} = 0 & \text{on } \partial\Omega \end{cases}$$

possesses at least two nontrivial weak solutions $u_1, u_2 \in W^{1,p}(\Omega)$.

PROOF. The preceding result yields $u_i \in W^{1,p}(\Omega) \setminus \{0\}$, $i = 1, 2$, such that $\frac{\partial u}{\partial n_p} = 0$ on $\partial\Omega$ as well as

$$\underline{j}(u_i(x)) \le -\Delta_p u_i(x) \le \overline{j}(u_i(x)) \quad \text{a.e. in } \Omega. \tag{178}$$

Thanks to Lemma 59, from (J_1'') it follows $\Delta_p u_i(x) = 0$ for almost all $x \in u_i^{-1}(D_j)$. Through (178) and (J_7) we thus have

$$-\Delta_p u_i(x) = 0 = j(u_i(x)) \quad \text{a.e. in } u_i^{-1}(D_j).$$

Finally, since at almost all $x \in \Omega \setminus u_i^{-1}(D_j)$ expression (178) evidently becomes $-\Delta_p u_i(x) = j(u_i(x))$, the conclusion is achieved. □

Bibliography

[ABT] S. Adly, G. Buttazzo and M. Théra, *Critical points for nonsmooth energy functions and applications*, Nonlinear Anal. **32** (1998), 711–718.

[A] H. Amann, *Nonlinear operators in ordered Banach spaces and some applications to nonlinear boundary value problems,* in: Lecture Notes in Math. 543, Springer-Verlag, 1976, 1–55.

[BM0] G. Barletta and S.A. Marano, *Some remarks on the critical point theory for locally Lipschitz functions*, Glasgow Math. J. **45** (2003), 131–141.

[BM] G. Barletta and S. A. Marano, *The structure of the critical set in the mountain-pass theorem for nondifferentiable functions*, Differential Integral Equations **16** (2003), 1001–1012.

[BC] G. Bonanno and P. Candito, *Non-differentiable functionals and applications to elliptic problems with discontinuous nonlinearities*, J. Differential Equations **224** (2008), 3031–3959.

[BoM] G. Bonanno and S.A. Marano, *On the structure of the critical set of non-differentiable functions with a weak compactness condition*, Appl. Anal., to appear.

[B] H. Brézis, *Analyse Fonctionelle–Théorie et Applications*, Masson, Paris, 1983

[BN] H. Brézis and L. Nirenberg, *Remarks on finding critical points*, Comm. Pure. Appl. Math. **44** (1991), 939–963.

[Bo] K. Borsuk, *Theory of Retracts*, PWN, Warsaw, 1967.

[CLM] P. Candito, R. Livrea and D. Motreanu, *$\mathbb{Z}_2$-symmetric critical point theorems for non-differentiable functions*, Glasgow Math. J. **50** (2008), 447–466.

[CMM] P. Candito, *S. A. Marano and D. Motreanu, Critical poits for a class of non-differentiable functions and applications*, Discrete Contin. Dyn. Syst. **13** (2005), 175–194.

[C] J. Chabrowski, *Variational Methods for Potential Operator Equations*, de Gruyter Ser. Nonlinear Anal. Appl. **24**, de Gruyter, Berlin, 1997.

[Ch] K.-C. Chang, *Variational methods for nondifferentiable functions and their applications to partial differential equations*, J. Math. Anal. Appl. **80** (1981), 102–129.

[Cl] F.H. Clarke, *Optimization and Nonsmooth Analysis*, Classics Appl. Math. **5**, SIAM, Philadelphia, 1990.

[Co] J.-N. Corvellec, *Quantitative deformation theorems and critical point theory*, Pacific J. Math. **187** (1999), 263–279.

[DeG] M. Degiovanni, *Nonsmooth critical point theory and applications*, Nonlinear Anal. **30** (1997), 89–99.

[De] K. Deimling, *Nonlinear Functional Analysis*, Springer, Berlin, 1985.

[D] Y. Du, *A deformation lemma and some critical point theorems*, Bull. Austral. Math. Soc. **43** (1991), 161-168.

[Du] J. Dugundji, *Topology*, Allyn and Bacon, Boston, 1966.

[DL] C. Duvaut and J.-L. Lions, *Inequlities in Mechanics and Physics*, Springer, Berlin, 1976.

[E] I. Ekeland, *Convexity Methods in Hamiltonian Mechanics*, Ergeb. Math. Grenzgeb. (3) **19**, Springer, Berlin, 1990.

[E1] I. Ekeland, *Nonconvex minimization problems*, Bull. Amer. Math. Soc. (N.S.) **1** (1979), 443–474.

[F] G. Fang, *The structure of the critical set in the general mountain pass principle*, Ann. Fac. Sci. Toulouse Math. **3** (1994), 345–362.

[GP] L. Gasiński and N.S. Papageorgiou, *Nonsmooth Critical Point Theory and Nonlinear Boundary Value Problems*, Ser. Math. Anal. Appl. **8**, Chapman and Hall/CRC Press, Boca Raton, 2005.

[Gh1] N. Ghoussoub, *A min-max principle with a relaxed boundary condition*, Proc. Amer. Math. Soc. **117** (1993), 439–447.

[Gh2] N. Ghoussoub, *Duality and Perturbation Methods in Critical Point Theory*, Cambridge Tracts in Math. **107**, Cambridge Univ. Press, Cambridge, 1993.

[GhPr] N. Ghoussoub and D. Preiss, *A general mountain pass principle for locating and classifyng critical points*, Ann. Inst. H. Poincaré Anal. Non Linéaire **6** (1989), 321–330.

[GT] D. Gilbarg and N.S. Trudinger, *Elliptic Partial Differential Equations of Second Order*, 2nd edn, Springer-Verlag, Berlin, 1983.

[J] Y. Jabri, *The Mountain Pass Theorem: Variants, Generalizations and some Applications*, Encyclopedia Math. Appl., Cambridge Univ. Press, Cambridge, 2003.

[KP] N.C. Kourogenis and N.S. Papageorgiou, *Nonsmooth critical point theory and nonlinear elliptic equations at resonance*, J. Austral. Math. Soc. Ser. A **69** (2000), 245–271.

[K] C. Kuratowski, *Topology*, Vol. II, Academic. Press, New York, 1968.

[LM1] R. Livrea and S.A. Marano, *Existence and classification of critical points for nondifferentiable functions*, Adv. Differential Equations **9** (2004), 961–978.

[LM2] R. Livrea and S.A. Marano, *A min-max principle for non-differentiable functions with a weak compactness condition*, Comm. Pure Appl. Anal. **8**, 1019–1029.

[LMM] R. Livrea, S. A. Marano and D. Motreanu, *Critical points or nondifferentiable functions in presence of splitting*, J. Differential Equations **226** (2006), 704–725.

[Ma] S.A. Marano, *Elliptic boundary-value problems with discontinuous nonlinearities*, Set-Valued Anal. **3** (1995), 167–180.

[MMB] S. A. Marano and G. Molica Bisci, *Multiplicity results for a Neumann problem with p-Laplacian and non-smooth potential*, Rend. Circ. Mat. Palermo (2) **55** (2006), 113–122.

[MM0] S.A. Marano and D. Motreanu, *On a three critical points theorem for non-differentiable functions and applications to nonlinear boundary value problems*, Nonlinear Anal. **48** (2002), 37–52.

[MM1] S. A. Marano and D. Motreanu, *A deformation theorem and some critical points results for non-differentiable functions*, Topol. Methods Nonlinear Anal. **22** (2003), 139–158.

[MM2] S.A. Marano and D. Motreanu, *Critical points of non-smooth functions with a weak compactness condition*, J. Math. Anal. Appl. **358** (2009), 189–201.

[MaP] S.A. Marano and N.S. Papageorgiou, *On a Neumann problem with p-Laplacian and non-smooth potential*, Differential Integral Equations **19** (2006), 1301–1320.

[MB] G. Molica Bisci, *Some remarks on a recent critical point result of nonsmooth analysis*, Matematiche (Catania), **64** (2009), 97–112.

[MoMo] D. Motreanu and V. V. Motreanu, *Nonsmooth variational problems in the limit case and duality*, J. Global Optim. **29** (2004), 439–453.

[MoMoPa] D. Motreanu, V. V. Motreanu and D. Pasca, *A version of Zhong's coercivity result for a general class of nonsmooth functionals*, Abst. Appl. Anal. **7** (2002), 601–612.

[MP] D. Motreanu and P. D. Panagiotopoulos, *Minimax Theorems and Qualitative Properties of the Solutions of Hemivariational Inequalities*, Nonconvex Optim. Appl. **29**, Kluwer, Dordrecht, 1998.

[MR] D. Motreanu and V. Radulescu, *Variational and Non-Variational Methods in Nonlinear Analysis and Boundary Value Problems*, Nonconvex Optim. Appl. **67**, Kluwer, Dordrecht, 2003.

[MV] D. Motreanu and C. Varga, *Some critical point results for locally Lipschitz functionals*, Comm. Appl. Nonlinear Anal. **4** (1997), 17-33.

[NaPa] Z. Naniewicz and P.D. Panagiotopoulos, *Mathematical Theory of Hemivariational Inequalities and Applications*, Marcel Dekker, New York, 1995.

[P] P. D. Panagiotopoulos, *Hemivariational Inequalities. Applications in Mechanics and Engeneering*, Springer, Berlin, 1993.

[PS1] P. Pucci and J. Serrin, *A mountain pass theorem*, J. Differential Equations **63** (1985), 142–149.

[PS2] P. Pucci and J. Serrin, *Extensions of the mountain pass theorem*, J. Funct. Anal. **59** (1984), 185–210.

[PS3] P. Pucci and J. Serrin, *The structure of the critiical set in the mountain pass theorem*, Trans. Amer. Math. Soc. **299** (1987), 115-132.

[R1] P. H. Rabinowitz, *Minimax Methods in Critical Point Theory with Applications to Differential Equations*, CBMS Reg. Conf. Ser. Math. **65**, Amer. Math. Soc., Providence, RI, 1986.

[R2] P. H. Rabinowitz, *Some aspects of critical point theory*, Proceedings of the 1982 Changchun Symposium on Differential Geometry and Differential Equation (s.S. Chern, R. Wang and M. Chi, eds.), Science Press, Bijing, 1986, pp. 185-232.

[Ra] V. Radulescu, *Mountain pass theorem for non-differentiable functions and applications*, Proc. Japan Acad. Ser. AMath. Sci. **69** (1993), 193–198.

[S] M. Schechter, *Linking Methods in Critical Point Theory*, Birkäuser, Boston, 1999.

[St] M. Struwe, *Variational Methods. Applications to Nonlinear Partial Differential Equations and Hamiltonian Systems*, Second Edition, Ergeb. Math. Grenzgeb **34**, Springer-Verlag, Berlin, 1996.

[Su] A. Szulkin, *Minimax principles for lower semicontinuous functions and applications to nonlinear boundary value problems*, Ann. Inst. Henri Poincaré **3** (1986), 77–109.

Roberto Livrea, Department P.A.U., University of Reggio Calabria, Salita Melissari, 89100 Reggio Calabria, Italy
E-mail address: roberto.livrea@unirc.it

Salvatore Angelo Marano, Department of Mathematics and Informatics, University of Catania, Viale A. Doria 6, 95125 Catania, Italy
E-mail address: marano@dmi.unict.it

Evolution hemivariational inequalities with applications

Stanisław Migórski*

ABSTRACT. We present a survey on mathematical modeling of nonconvex and nonsmooth problems arising in mechanics and engineering. The approach to such problems is based on the notion of hemivariational inequality and focuses on the following aspects. First we present the ideas leading to inequality problems encountered in mechanics and describe a model example. Then we give results on the existence of solutions to hemivariational inequalities of parabolic and hyperbolic types and associated evolution inclusions of first and second order. Finally we formulate and study three selected models of mechanics to which the theory is applicable. We also indicate issues which are not covered in the chapter.

CONTENTS

1. Introduction

Mathematical modeling of problems arising in mechanics is a fascinating field of research which combine analysis, differential equations, theoretical mechanics and

---

2000 *Mathematics Subject Classification.* Primary 34G20, 35A15, 35J85, 35K90, 35K85, 35L85, 35K55, 49J40, 49K40, 74Axx.

Key words and phrases. Hemivariational inequalities, subdifferential, parabolic, hyperbolic, multifunction, nonconvex, viscoelasticity, piezoelectricity, thermoviscoelasticity.

*The author was supported in part by the Ministry of Science and Higher Education of Poland under grant no. N201 027 32/1449.

numerical simulations. During the last twenty years variational methods based on the formulations with nonconvex energy functionals and in particular hemivariational inequalities play an important role in the understanding of many physical phenomena.

The hemivariational inequalities have been introduced in 1980–1981 by Panagiotopoulos in order to describe several important mechanical and engineering problems (cf. [**100**]–[**104**]). In physics and mechanics there are many formulations leading to variational inequalities. They appear when the constitutive laws and/or the boundary conditions are derived from the convex and nondifferentiable potentials. The variational inequalities have a precise physical meaning. They express the principles of virtual work and power introduced by Fourier in 1823. The prototypes of boundary value problems leading to variational inequalities are the Signorini-Fichera problem and the friction problem of elasticity. However, in mechanical problems one often meets nonmonotone relations between stress and strain or between the displacement (or velocity) and reactions. The lack of monotonicity results from the fact that the corresponding superpotentials are nonconvex. This means that the description of some phenomena needs variational formulations involving the nonconvex energy functionals. The hemivariational inequalities make use of the generalized gradient of Clarke–Rockafellar (cf. [**16**]) and they are natural generalizations of variational inequalities. There are many situations in mechanics experimentally verified when the constitutive laws or boundary conditions are represented by relations involving decreasing branches (for instance nonmonotone friction laws between a structure and a foundation, semipermeable problems, reaction-displacement laws in interface problems, debonding and delamination problems, etc.). On the other hand the lack of convexity causes new difficulties and the methods of multivalued analysis used so far seems to be insufficient. Though the first papers on hemivariational inequalities have had the engineering character, the hemivariational inequalities constitute now the basis for the effective treatment of many unilateral problems. The theory of hemivariational inequalities is today applicable to problems in mechanics and economics, and it is systematically developed from the mathematical point of view.

The purpose of this paper is primarily to discuss the issues related to evolution hemivariational inequalities to provide an orientation for non-experts and possibly to widen knowledge of experts dealing often with only a rather specific topic or model. Besides, the presentation gives an overview of models of various sorts. The particular models differ not only by the description of the physical situation but also by the ways they treat the phenomena and the mathematical methods being introduced or applied. The paper concerns with the following issues

(1) existence and uniqueness of solutions for evolution inclusions of first and second order,
(2) existence and uniqueness of solutions to hemivariational inequalities of parabolic and hyperbolic types,
(3) inverse problems for parabolic hemivariational inequality,
(4) variational sensitivity analysis for systems described by hyperbolic hemivariational inequality,
(5) vanishing acceleration approach to quasistatic hemivariational inequality,
(6) vanishing viscosity for hyperbolic hemivariational inequality,

(7) hyperbolic hemivariational inequality with impulses,
(8) applications of hemivariational inequalities in the modeling of contact problems in viscoelasticity, thermoviscoelasticity and piezoviscoelectricity.

In the stationary case the hemivariational inequalities can be formulated as the substationary point problems for the corresponding nondifferentiable, nonsmooth and nonconvex energy functionals. The formulations of mechanical problems as hemivariational inequalities allow to give positive answers to unsolved or partially unsolved problems, e.g. the delamination problem and the adhesive contact in cracks. There are numerous industrial applications of hemivariational inequalities including the area of contact mechanics and frictional modeling. The goal of this contribution is to give a survey on evolution hemivariational inequalities and illustrate the results by providing several applications. We have tried to keep the presentation as simple as possible and this makes it accessible also to the newcomer to this field. Due to a limited scope, the list of references certainly cannot be considered as complete, however, it is partly influenced by author's own activities.

Although we restrict ourselves to evolution problems, the subject of hemivariational inequalities is today very broad and can not be exhausted within the limits of our presentation. The material which is not covered in this chapter include the following topics.

Noncoercive problems. In recent years, many engineers and mathematicians have focused increasingly their attention on non-coercive evolution hemivariational inequalities. New existence theorems for a general class of non-coercive variational–hemivariational inequalities have been originally proved by Liu [**50, 51**]. This represented a breakthrough in this area. Noncoercive damping in dynamic hemivariational inequality have been also studied by Liu and Migórski [**54**].

Convergence of solutions and regularization. The asymptotic behavior of solutions to hemivariational inequalities have been studied by Migórski [**67**] for abstract parabolic problems, Liu et al. [**56**] for stationary boundary hemivariational inequality, including homogenization, and Liu and Migórski [**55**] by using the G-convergence of parabolic operators. Homogenization method in inverse problems for boundary hemivariational inequalities were studied by Migórski in [**70**]. For non-coercive evolution hemivariational inequalities, which are ill-posed in general, by means of the so-called Browder–Tikhonov regularization method, Liu [**52**] designed a strongly convergent approximation procedure for the ill-posed problems, which is one of a major accomplishment in the area. The extension of the Browder–Tikhonov regularization to hyperbolic hemivariational inequalities can be found in Xiao and Huang [**115**]. The result on the convergence of solutions to parabolic hemivariational inequalities is given by Liu and Migórski in [**55**].

Eigenvalue problems. By use of the theory of multivalued pseudomonotone operators, the notion of generalized gradient of Clarke and the property of the first eigenfunction, Liu [**53**] built a Landesman–Lazer theory in the nonsmooth framework of quasilinear parabolic hemivariational inequalities. This was one of the most important contributions to the theory of hemivariational inequalities.

Method of subsolution and supersolution. Using the method of sub- and supersolution, Carl, Le and Motreanu [**10, 12**] obtained the existence results for solutions for some quasilinear evolution hemivariational inequalities. Results on comparison and existence of extremal solutions can be found in Carl [**5**]–[**8**], Carl and Gilbert [**9**], and Carl et al. [**10, 11, 13**]. The evolution hemivariational inequalities have been studied for parabolic problems by Papageorgiou [**109**] who combined the method of lower and upper solutions with truncation and penalization techniques.

Optimal control problems. The results on existence of solutions to evolution inclusions and their control can be found in Migórski [**64, 65**]. Optimal shape design problems for a class of hemivariational inequalities were treated by Denkowski and Migórski [**19**]. Results on optimal control of a system of hemivariational inequalities modeling thermoviscoelastic contact problem of mechanics are delivered by Denkowski et al. [**26**] and control problems of contact models described by evolution hemivariational inequalities have been treated by Migórski and Ochal [**80**] for parabolic problems and Migórski [**77**] for hyperbolic ones. The sensitivity of optimal solutions to control problems for the systems described by stationary and evolution hemivariational inequalities under perturbations of state relations and of cost functionals have been investigated by Denkowski and Migórski [**20, 22, 23**]. Inverse and parameter identification problems for evolution hemivariational inequalities were investigated by Migórski in [**66, 68**]. The approximation of control problems for elliptic hemivariational inequalities have been considered by Miettinen and Haslinger [**58**], Panagiotopoulos and Haslinger [**106**], and the approximation of a parabolic hemivariational inclusion and optimal control problems have been studied by Debinska et al. [**18**].

Numerical treatment of hemivariational inequalities. The results can be found in Haslinger et al. [**43**], Miettinen et al. [**59**], Miettinen [**57**], Miettinen and Haslinger [**58**] and the references therein. Finite element approximation for parabolic problems can be found in Miettinen and Panagiotopoulos [**62, 63**].

Other topics. Recent applications to problems of economics can be found in Naniewicz [**95**]. The hemivariational inequalities modeling the problems of fluid flow have been studied by Migórski [**71**], and Migórski and Ochal [**87**] while antiperiodic solutions for hemivariational inequalities were considered by Park and Ha [**110**]. The existence results for evolution hemivariational inequalities can be found in Goeleven et al. [**36**], Miettinen and Panagiotopoulos [**61**] and Migórski [**69**] for parabolic problems, and Panagiotopoulos [**105**], Panagiotopoulos and Pop [**108**], Migórski [**73**]–[**75**], Gasinski [**33**], Gasinski and Smolka [**34**], Ochal [**99**] and Park [**111**] for hyperbolic problems. The semilinear problems for hyperbolic hemivariational inequalities with hysteresis operators were formulated and studied by Miettinen and Panagiotopoulos [**60**]. The recent results on the decay of energy of hyperbolic hemivariational inequalities are provided by Kalita in [**45**].

Recent books and monographs on mathematical theory of hemivariational inequalities include Carl et al. [**11**], Goeleven et al. [**37**], Haslinger et al. [**43**], Motreanu and Panagiotopoulos [**93**], Naniewicz and Panagiotopoulos [**96**], Panagiotopoulos [**100, 104**], and we refer the reader there for a wealth of additional information about these and related topics. The results on Mathematical Theory of Contact Mechanics can be found in several monographs, e.g. Eck et al. [**30**], Han and Sofonea [**38**], Shillor et al. [**112**], Sofonea et al. [**113**] and Sofonea and Matei [**114**].

2. Preliminaries

Given a reflexive Banach space Y, we denote by $\langle \cdot, \cdot \rangle$ the pairing between Y and its dual Y^*. We recall some definitions for a multivalued operator $T \colon Y \to 2^{Y^*}$ (see e.g. Zeidler [**116**] and Denkowski et al. [**28**]).

An operator T is said to be *pseudomonotone* if it satisfies

(a) for every $y \in Y$, Ty is a nonempty, convex and weakly compact set in Y^*;
(b) T is u.s.c. from every finite dimensional subspace of Y into Y^* endowed with the weak topology;
(c) if $y_n \to y$ weakly in Y, $y_n^* \in Ty_n$ and $\limsup \langle y_n^*, y_n - y\rangle \leq 0$, then for each $z \in Y$ there exists $y^*(z) \in Ty$ such that $\langle y^*(z), y - z\rangle \leq \liminf \langle y_n^*, y_n - z\rangle$.

Let $L\colon D(L) \subset Y \to Y^*$ be a linear densely defined maximal monotone operator. An operator T is said to be *pseudomonotone with respect to* $D(L)$ (shortly L*-pseudomonotone*) if and only if (a) and (b) hold and

(d) if $\{y_n\} \subset D(L)$ is such that $y_n \to y$ weakly in Y, $Ly_n \to Ly$ weakly in Y^*, $y_n^* \in Ty_n$, $y_n^* \to y^*$ weakly in Y^* and $\limsup \langle y_n^*, y_n\rangle \leq \langle y^*, y\rangle$, then $y^* \in Ty_n$ and $\langle y_n^*, y_n\rangle \to \langle y^*, y\rangle$.

An operator T is said to be *coercive* if there exists a function $c\colon \mathbb{R}^+ \to \mathbb{R}$ with $c(r) \to \infty$ as $r \to \infty$ such that $\langle y^*, y\rangle \geq c(\|y\|_Y)\,\|y\|_Y$ for every $(y, y^*) \in$ *Graph*(T).

A single-valued operator $T\colon Y \to Y^*$ is said to be *demicontinuous* if it is continuous from Y to Y^* endowed with weak topology. $T\colon Y \to Y^*$ is *pseudomonotone* if for each sequence $\{y_n\} \subseteq Y$ such that it converges weakly to $y_0 \in Y$ and $\limsup \langle Ty_n, y_n - y_0\rangle \leq 0$, we have $\langle Ty_0, y_0 - y\rangle \leq \liminf \langle Ty_n, y_n - y\rangle$ for all $y \in Y$.

The following surjectivity result (see Denkowski et al. [**27**]) holds for operators which are pseudomonotone with respect to $D(L)$.

PROPOSITION 1. If Y is a reflexive, strictly convex Banach space, $L\colon D(L) \subset Y \to Y^*$ is a linear densely defined maximal monotone operator and $T\colon Y \to 2^{Y^*} \setminus \{\emptyset\}$ is bounded, coercive and pseudomonotone with respect to $D(L)$, then $L + T$ is surjective.

Finally, we recall the definitions of the generalized directional derivative and the generalized gradient of Clarke for a locally Lipschitz function $h\colon E \to \mathbb{R}$, where E is a Banach space (see [**16, 17**]). The generalized directional derivative of h at x in the direction v, denoted by $h^0(x; v)$, is defined by

$$h^0(x; v) = \limsup_{y \to x,\ t \downarrow 0} \frac{h(y + tv) - h(y)}{t}. \tag{1}$$

The generalized gradient of h at x, denoted by $\partial h(x)$, is a subset of a dual space E^* given by

$$\partial h(x) = \{\, \zeta \in E^* \mid h^0(x; v) \geq \langle \zeta, v\rangle \text{ for all } v \in E \,\}. \tag{2}$$

At this place recall that a function $g\colon E \to \mathbb{R}$, where E is a Banach space, is said to be one-sided directional Gâteaux differentiable at $x \in E$, if there exists $g'(x; v)$ such that

$$\limsup_{t \downarrow 0} \frac{g(x + tv) - g(x)}{t} = g'(x; v) \qquad \text{for all } v \in E \tag{3}$$

and the functional $v \mapsto g'(x; v)$ is the one-sided directional Gâteaux derivative of g at x with respect to the direction v. For a convex function $g\colon E \to \mathbb{R}$, its subdifferential at the point $x \in X$ is defined by

$$\partial g(x) = \{\, \zeta \in E^* \mid g(z) - g(x) \geq \langle \zeta, z - x \rangle \text{ for all } z \in E \,\}. \tag{4}$$

The locally Lipschitz function h is called regular (in the sense of Clarke) at $x \in E$ if for all $v \in E$ the one-sided directional derivative $h'(x; v)$ (see (3)) exists and satisfies $h^0(x; v) = h'(x; v)$ for all $v \in E$.

3. Physical background

The aim of this section is to give a short explanation of the origins of the inequality problems encountered in mechanics. Our presentation is based on the works of Germain [**35**], Nayroles [**97**] and Panagiotopoulos [**100, 107**].

A mechanical system is a triplet $\Sigma = \{\, \mathcal{U}, \mathcal{F}, \langle \cdot, \cdot \rangle_{\mathcal{F} \times \mathcal{U}} \,\}$ which consists of

- a linear vector space $\mathcal{U}$ of virtual displacements (or velocities) which take place in a time interval,
- a linear vector space $\mathcal{F}$ of forces
- a duality pairing (bilinear form) $\langle \cdot, \cdot \rangle_{\mathcal{F} \times \mathcal{U}}$ between $\mathcal{F}$ and $\mathcal{U}$ with the properties
  - (i) for all $u \in \mathcal{U}$, $u \neq 0$, there is $f \in \mathcal{F}$ such that $\langle f, u \rangle_{\mathcal{F} \times \mathcal{U}} \neq 0$,
  - (ii) for every $f \in \mathcal{F}$, $f \neq 0$, there exists $u \in \mathcal{U}$ such that $\langle f, u \rangle_{\mathcal{F} \times \mathcal{U}} \neq 0$.

If $\mathcal{U}$ represents the space of displacements (for static problems), then the duality $\langle f, u \rangle_{\mathcal{F} \times \mathcal{U}}$ expresses the work of the force $f \in \mathcal{F}$ for the displacement $u \in \mathcal{U}$. If $\mathcal{U}$ is the space of virtual velocities (for dynamic problems), then $\langle f, u \rangle_{\mathcal{F} \times \mathcal{U}}$ is the virtual power produced by $f \in \mathcal{F}$ for the velocity $u \in \mathcal{U}$. The bilinear form $\langle \cdot, \cdot \rangle_{\mathcal{F} \times \mathcal{U}}$ is called the *virtual work* or the *virtual power*, respectively. Thus the force $f \in \mathcal{F}$ is a linear continuous functional on the space $\mathcal{U}$.

The motion of a body occupied by the mechanical system is governed by the following *principle* of virtual power, being the main postulate of mechanics

> At any time and for any field of virtual velocities, the virtual power of all internal and external forces impressed on the system is equal to the virtual power of the inertial (or d'Alembert) forces.

For static problems (cf. Panagiotopoulos [**107**]), the principle of virtual power takes a form known as principle of virtual work. The principle of virtual work is equivalent to the equilibrium equation (cf. (7) and (9) below). Examples of mechanical systems and more information on the principles of virtual work and power can be found in Panagiotopoulos [**100**], Germain [**35**] and Nayroles [**97**].

Let us now consider a mechanical system Σ and suppose that only certain subsets of $\mathcal{U}$ and $\mathcal{F}$ are admissible for the system. Then a multivalued mapping $A \colon \mathcal{U}_1 \to \mathcal{F}$ such that

$$f \in A(u) \quad \text{for all } u \in \mathcal{U}_1 \subset \mathcal{U} \tag{5}$$

introduces a *law* on system Σ (or we say that Σ is subjected to a law). The domain and the range of the map A are called the domain and the range of the mechanical system, respectively. For example, if $f = -\nabla W(u)$, where $W \colon \mathcal{U} \to \mathbb{R}$ is such that $\nabla W(u)$ exists for all $u \in \mathcal{U}$, then the single-valued law $A = -\nabla W$ is called a potential law and W is the potential. Other reasons that only subsets of $\mathcal{U}$ and

$\mathcal{F}$ are admissible for the system Σ are *constraints* (e.g. boundary conditions). A constraint on the system Σ is a multivalued mapping $B\colon \mathcal{U}_1 \subset \mathcal{U} \to \mathcal{F}$ such that

$$f \in B(u) \quad \text{for all } u \in \mathcal{U}_1. \tag{6}$$

In this case f is called the reaction on the constraint B. Obviously, in this abstract setting it is not always possible to make a clear distinction between a mechanical law and a constraint. From a mathematical viewpoint we do not make any distinction between a law and a constraint. For dynamic problems laws and constraints analogous to (5) and (6) are obtained if $\mathcal{U}$ represents the field of velocities.

Consider now the equilibrium of a mechanical system Σ subjected to certain forces f_i, laws and constraints. Given laws $A_i\colon \mathcal{U}_i \subset \mathcal{U} \to \mathcal{F}$,

$$f_i \in A_i(u) \quad \text{for all } u \in \mathcal{U}_i,\ i = 1, \dots, n$$

and constraints $B_j\colon \mathcal{U}_j \subset \mathcal{U} \to \mathcal{F}$,

$$f_j \in B_j(u) \quad \text{for all } u \in \mathcal{U}_j,\ j = n+1, \dots p,$$

from the principle of virtual work, we find that at the *position of equilibrium*

$$\sum_{i=1}^{n} f_i + \sum_{j=n+1}^{p} f_j = 0.$$

Therefore, at the position of equilibrium, we have

$$\text{find } u \in \bigcap_{i=1}^{p} \mathcal{U}_i \quad \text{such that} \quad 0 \in \sum_{i=1}^{n} A_i(u) + \sum_{j=n+1}^{p} B_j(u). \tag{7}$$

In order to move a step forward and introduce the notion of *generalized variational inequality* (and in particular variational inequality and hemivariational inequality), we now make a restriction concerning the form of the laws or constraints defined on the mechanical system Σ. In what follows we consider the static case.

So assume that the system is subjected to multivalued mechanical laws (or constraints) $A_i\colon \mathcal{U} \to \mathcal{F}$ such that

$$-f_i \in A_i(u) \quad \text{for } i = 1, \dots, n \tag{8}$$

where $A_i = \widehat{\partial}_i \Phi_i$ for all $i = 1, \dots, n$ and $\widehat{\partial}_i \Phi_i$ denotes the generalization of the classical gradient of functions $\Phi_i\colon \mathcal{U} \to \overline{\mathbb{R}} = \mathbb{R} \cup \{\pm\infty\}$ defined by

$$\widehat{\partial}_i \Phi_i(u) = \{\, f \in \mathcal{F} \mid D_i\Phi_i(u)(v) \geq \langle f, v \rangle_{\mathcal{F}\times\mathcal{U}} \quad \text{for all } v \in \mathcal{U} \,\}.$$

The derivative $D_i\Phi_i(u)(v)$ expresses physically the virtual work (or power) for the virtual displacement (or virtual velocity) v at the position u. Therefore the subdifferential laws (or constraints) (8) have now the following equivalent form

$$D_i\Phi_i(u)(v) \geq \langle -f_i, v \rangle_{\mathcal{F}\times\mathcal{U}} \quad \text{for all } v \in \mathcal{U},\ i = 1, \dots, n.$$

Due to the equilibrium equation

$$\sum_{i=1}^{n} f_i = 0, \tag{9}$$

from the relation $-f_i \in \widehat{\partial}_i \Phi_i(u)$, $i = 1, \dots, n$, we readily obtain the *generalized variational inequality*

$$\sum_{i=1}^{n} D_i \Phi_i(u)(v-u) \geq 0 \quad \text{for all } v \in \mathcal{U}. \tag{10}$$

At this point variational and hemivariational inequalities (cf. (12) and (11), respectively) can be considered as special cases of the inequality (10) by assuming further hypotheses on the functions Φ_i. We remark that there is not a unique theory for the generalized gradient $\widehat{\partial}_i \Phi_i$ and the directional derivative $D_i \Phi_i(u)(v)$. One of the possible approaches is given by Clarke [**16**] under hypothesis that Φ_i are locally Lipschitz functions. When Φ_i are convex functions, then we know that $\widehat{\partial}_i \Phi_i$ reduces to the subdifferential in the sense of convex analysis. We consider these two cases below.

Case I. Let us assume that for all $i = 1, \dots, n$, $\Phi_i \colon \mathcal{U} \to \mathbb{R}$ are *locally Lipschitz* functions on $\mathcal{U}$. Now we observe that in this case, we have

$$D_i \Phi_i(u)(v) = \Phi_i^0(u; v), \qquad \widehat{\partial}_i \Phi_i(u) = \partial \Phi_i(u)$$

for all u, $v \in \mathcal{U}$, $i = 1, \dots, n$, where $\Phi_i^0(u; v)$ and $\partial \Phi_i(u)$ denote the generalized directional derivative of Φ at $u \in X$ in the direction $v \in X$ and the generalized gradient of Clarke (given by (1) and (2), respectively). Now the inequality (10) takes the form

$$\sum_{i=1}^{n} \Phi_i^0(u; v-u) \geq 0 \quad \text{for all } v \in \mathcal{U}, \tag{11}$$

which is called *hemivariational inequality.*

Case II. Suppose that for all $i = 1, \dots, n$, $\Phi_i \colon \mathcal{U} \to \mathbb{R}$ are convex functions. It is well known (cf. Propositions 2.2.6 and 2.2.7 of Clarke [**16**]) that such functions are locally Lipschitz, one-sided directional Gâteaux differentiable, the Clarke gradient $\partial \Phi_i$ coincides with the subdifferential of convex analysis $\partial \Phi_i$ and the directional derivative $\Phi_i^0(u; v)$ equals to the Gâteaux directional derivative $\Phi'(u; v)$. Hence, in Case II, we have

$$D_i \Phi_i(u)(v) = \Phi_i'(u; v), \qquad \widehat{\partial}_i \Phi_i(u) = \partial \Phi_i(u)$$

for all u, $v \in \mathcal{U}$, $i = 1, \dots, n$ and the inequality (10) reduces to the classical *variational inequality* of the form

$$\sum_{i=1}^{n} (\Phi_i(v) - \Phi_i(u)) \geq 0 \quad \text{for all } v \in \mathcal{U}. \tag{12}$$

In this case it is also clear that the physical laws (or constraints) $A_i = \partial \Phi_i$, $i = 1, \dots, n$ are of monotone nature or more precisely $\partial \Phi_i \colon \mathcal{U} \to 2^{\mathcal{U}^*}$ are multivalued maximal monotone operators (cf. Proposition 2.2.9 of Clarke [**16**]). Moreover, if $\Phi_i : \mathcal{U} \to \mathbb{R}$ are locally Lipschitz functions such

- Φ_i are convex for $i = 1, \ldots, n' < n$,
- Φ_i are nonconvex for $i = n' + 1, \ldots, n$,

then the abstract inequality (10) gives *variational–hemivariational inequality* of the form:

$$\sum_{i=1}^{n'} (\Phi_i(v) - \Phi_i(u)) + \sum_{i=n'+1}^{n} \Phi_i^0(u; v-u) \geq 0 \quad \text{for all } v \in \mathcal{U}.$$

4. A model example

In this section we describe the destruction support problem which serves as an illustration of a possible formulation of a nonmonotone multivalued boundary condition. This formulation leads to an elliptic hemivariational inequality.

Let Ω be a bounded open subset of $\mathbb{R}^d$ ($d = 3$) occupied by a linear elastic body in its undeformed state. We suppose that the boundary $\Gamma = \partial\Omega$ of Ω consists of three open disjoint parts Γ_D, Γ_N and Γ_C, i.e. $\Gamma = \overline{\Gamma}_D \cup \overline{\Gamma}_N \cup \overline{\Gamma}_C$. A point $x \in \Omega$ is refereed to a fixed Cartesian coordinate system. For i, $j = 1, \ldots, d$, we use the standard notation

$\nu = \{\nu_i\}$	the outward unit normal vector to Γ;
$\sigma = \{\sigma_{ij}\}$	the stress tensor;
$\varepsilon = \{\varepsilon_{ij}\}$	the strain tensor;
$u = \{u_i\}$	the displacement vector;
$f_1 = \{f_{1i}\}$	the volume force vector;
$\sigma = \{\sigma_i = \sigma_{ij}\nu_j\}$	the boundary force (the stress vector on Γ).

We also decompose the stress vector $\sigma = \{\sigma_i\}$ on Γ and the displacement $u = \{u_i\}$ on Γ into the normal and tangential components

$$\sigma_\nu = (\sigma\nu) \cdot \nu, \quad \sigma_\tau = \sigma\nu - \sigma_\nu \nu,$$
$$u_\nu = u \cdot \nu, \quad u_\tau = u - u_\nu \nu.$$

Working in the framework of the small deformation theory (cf. [**98**]), we have in Ω the following pointwise relations

$$\sigma_{ij,j}(u) = -f_{1i} \quad \text{(the equilibrium equation)}, \tag{13}$$

$$\varepsilon_{ij}(u) = \frac{1}{2}(u_{i,j} + u_{j,i}) \quad \text{(the strain–displacement law)},$$

$$\sigma_{ij}(u) = a_{ijkl}\varepsilon_{kl}(u) \quad \text{(the constitutive equation, Hooke's law)}, \tag{14}$$

where $\sigma_{ij,j}(u) = \frac{\partial}{\partial x_j}\sigma_{ij}(u)$ and the elasticity tensor $\{a_{ijkl}\}$ is assumed to satisfy the ellipticity and symmetry properties $a_{ijkl}\varepsilon_{ij}\varepsilon_{kl} \geq \alpha\varepsilon_{ij}\varepsilon_{ij}$ a.e. in Ω, for all $\varepsilon_{ij} = \varepsilon_{ji} \in \mathbb{R}$ with $\alpha > 0$, $a_{ijkl} = a_{jikl} = a_{klij}$. In order to give a complete formulation of the problem we consider the following boundary conditions

on Γ_D : $u = \overline{u}$ (prescribed displacement),
on Γ_N : $\sigma = f_2$ (i.e. $\sigma_i = f_{2i}(x)$ are prescribed tractions),
on Γ_C : $\sigma_\tau = C_\tau$ (the tangential forces $C_\tau = C_\tau(x)$ are known),
σ_ν satisfies the idealized law

$$\begin{cases} \sigma_\nu = 0 & \text{if } u_\nu < 0; \\ \sigma_\nu + k_0\, u_\nu = 0 & \text{if } 0 \le u_\nu < \varepsilon; \\ -k_0\, \varepsilon \le \sigma_\nu \le 0 & \text{if } u_\nu = \varepsilon; \\ \sigma_\nu = 0 & \text{if } u_\nu > \varepsilon. \end{cases} \tag{15}$$

For simplicity in the Dirichlet condition on Γ_D we take $\overline{u} = 0$. The constant k_0 is supposed to be positive (Winkler's coefficient) and $\varepsilon > 0$. Conditions (15) describe unilateral contact between the body and the Winkler-type support: in the noncontact region $\sigma_\nu = 0$, in the contact region the support generates a reaction force which is proportional to its deformation $-\sigma_\nu \sim u_\nu$, the destruction of the support appears when the tractions reach the limited value (the maximal value of reactions that can be maintained by the support in $k_0\varepsilon$) and again $\sigma_\nu = 0$ in the region where the support is destroyed. Since the diagram $(u_\nu, -\sigma_\nu)$ on Γ_C is nonmonotone (cf. Figure 1), it is not possible to formulate the destruction support problem as a variational inequality.

In order to give the variational formulation of the above problem, we introduce the function $\beta\colon \mathbb{R} \to \mathbb{R}$ such that

$$\beta(t) = \begin{cases} 0 & \text{if } t < 0; \\ k_0 t & \text{if } t \in [0, \varepsilon); \\ 0 & \text{if } t \ge \varepsilon \end{cases}$$

and define the multivalued map $\widehat{\beta}\colon \mathbb{R} \to 2^{\mathbb{R}}$ which is obtained from β by filling in the jump at $t = \varepsilon$.

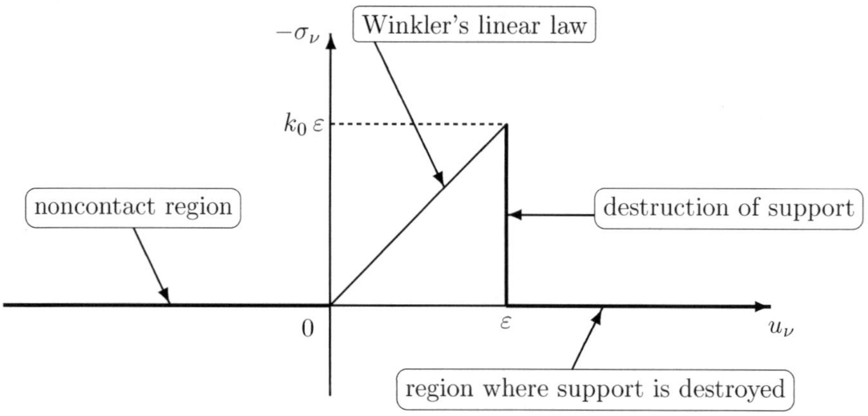

FIGURE 1. Nonmonotone diagram for the Winkler-type support.

Generally, given $\beta \in L^{\infty}_{loc}(\mathbb{R})$ we define $\widehat{\beta}$ as follows

$$\widehat{\beta}(\xi) = [\underline{\beta}(\xi), \overline{\beta}(\xi)] \subset \mathbb{R}, \tag{16}$$

where

$$\underline{\beta}(\xi) = \lim_{\delta \to 0^+} \operatorname*{ess\,inf}_{|t-\xi| \le \delta} \beta(t), \qquad \overline{\beta}(\xi) = \lim_{\delta \to 0^+} \operatorname*{ess\,sup}_{|t-\xi| \le \delta} \beta(t).$$

From a result of Chang [**14**], it is known that there exists a locally Lipschitz function $j \colon \mathbb{R} \to \mathbb{R}$ determined (up to an additive constant) by the relation $j(t) = \int_0^t \beta(s)\, ds$ and

$$\partial j(t) \subset \widehat{\beta}(t).$$

Additionaly, if $\lim_{t \to \xi^{\pm}} \beta(t)$ exist for every $\xi \in \mathbb{R}$, then we have

$$\partial j(t) = \widehat{\beta}(t) \quad \text{for } t \in \mathbb{R}.$$

Taking the above into consideration, the boundary condition (15) can be written as follows

$$-\sigma_\nu \in \widehat{\beta}(u_\nu) = \partial j(u_\nu) \quad \text{on } \Gamma_C, \tag{17}$$

where $j \colon \mathbb{R} \to \mathbb{R}$ is of the form

$$j(t) = \int_0^t \beta(s)\, ds = \begin{cases} 0 & \text{if } t < 0; \\ \dfrac{1}{2} k_0 t^2 & \text{if } 0 \le t \le \varepsilon; \\ \dfrac{1}{2} k_0 \varepsilon^2 & \text{if } t > \varepsilon. \end{cases}$$

Now let us introduce the space V of kinematically admissible displacements

$$V = \{\, v \in H^1(\Omega; \mathbb{R}^d) \mid v = 0 \text{ on } \Gamma_D \,\}.$$

Let $v \in V$. Multiplying the equilibrium equation (13) by $v - u$, integrating over Ω and applying the Green formula, we have

(18)

$$\sum_{i,j=1}^{d} \int_\Omega \sigma_{ij}(u) \frac{\partial (v_i - u_i)}{\partial x_j}\, dx = \sum_{i=1}^{d} \int_\Omega f_{1i}(v_i - u_i)\, dx + \sum_{i,j=1}^{d} \int_\Gamma \sigma_{ij}(u)(v_i - u_i)\nu_j\, d\Gamma.$$

Hence, taking into account the boundary conditions, symmetry of the stress tensor, we arrive at the problem

$$a(u, v - u) + \int_{\Gamma_C} (-\sigma_\nu (v_\nu - u_\nu))\, d\Gamma = \langle l, v - u \rangle \quad \text{for all } v \in V, \tag{19}$$

where

$$a(u,v) = \sum_{i,j=1}^{d} \int_{\Omega} \sigma_{ij}(u)\varepsilon_{ij}(v)\,dx,$$

$$\langle l, v\rangle = \sum_{i=1}^{d} \int_{\Omega} f_{1i} v_i\,dx + \sum_{i=1}^{d} \int_{\Gamma_N} f_{2i} v_i\,d\Gamma + \sum_{i=1}^{d} \int_{\Gamma_C} C_{\tau i} v_{\tau i}\,d\Gamma.$$

From (17) and the definition of the Clarke subdifferential, we have $j^0(u_\nu; v_\nu - u_\nu) \geq -\sigma_\nu(v_\nu - u_\nu)$ for all $v \in V$. Introducing this inequality into (19) we obtain the problem: find $u \in V$ such that

$$\text{(20)} \qquad a(u, v-u) + \int_{\Gamma_C} j^0(u_\nu; v_\nu - u_\nu)\,d\Gamma \geq \langle l, v-u\rangle \quad \text{for all } v \in V.$$

The above problem is called *hemivariational inequality.*

An analogous hemivariational inequality can be obtained when on Γ_C a pointwise boundary condition for the tangential force

$$-\sigma_\tau \in \partial j(u_\tau)$$

is assumed and the normal forces $\sigma_\nu = C_\nu$, $C_\nu = C_\nu(x)$ are prescribed. Such relation between u_τ and $-\sigma_\tau$ appears, for instance, in rock interface analysis in geomechanics and describes a nonmonotone friction law (see Panagiotopoulos [**100, 107**]).

We conclude this section with an observation on the variational formulation of a class of problems in which the constitutive laws are expressed by means of subdifferential relations. Consider the law

$$\text{(21)} \qquad \sigma \in \partial w(\varepsilon)$$

between the stress tensor σ and the strain tensor ε. Here ∂w denotes the Clarke subdifferential of a function $w\colon \mathbb{R}^6 \to \mathbb{R}$. As before the relation (21) is understood pointwise at every point in Ω (recall that $\Omega \subset \mathbb{R}^3$). Suppose that in the destruction support problem, the elastic body is not longer linear but the material law is of the form (21) (this means that in the model we replace now the Hooke law (14) by the nonlinear multivalued relation (21)). Thus

$$w^0(\varepsilon(u); \varepsilon(v) - \varepsilon(u)) \geq \sum_{i,j=1}^{3} \sigma_{ij}(u)(\varepsilon_{ij}(v) - \varepsilon_{ij}(u)) \quad \text{for all } v \in V.$$

Proceeding as before, from (18), we obtain the following hemivariational inequality: find $u \in V$ such that

$$\text{(22)} \quad \int_{\Omega} w^0(\varepsilon(u); \varepsilon(v) - \varepsilon(u))\,dx + \int_{\Gamma_C} j^0(u_\nu; v_\nu - u_\nu)\,d\Gamma \geq \langle l, v-u\rangle \quad \text{for all } v \in V.$$

We make a remark about two particular cases of relation (21). Namely, if w is defined by $w(\varepsilon) = \frac{1}{2}(A\varepsilon, \varepsilon)_{\mathbb{R}^6} = \frac{1}{2}\sum_{ijkh} a_{ijkh}\varepsilon_{ij}\varepsilon_{hk}$ with symmetric, coercive tensor $A = \{a_{ijkh}\}$, i, j, k, $h = 1, 2, 3$, then $\partial w = \nabla w$, hence (21) becomes the Hooke law $\sigma = A\varepsilon$ (cf. (14)) and (22) reduces to (20). If the function w in the relation (21) is continuously differentiable, then we get $\sigma_{ij} = \dfrac{\partial w(\varepsilon)}{\partial \varepsilon_{ij}}$ which corresponds to a nonlinear elastic material.

5. Evolution first order hemivariational inequalities

In this section we treat first order hemivariational inequalities. Before we pass to the formulations of results, we provide an important example which comes from the nonconvex problems for semipermeable media and leads to a scalar time dependent hemivariational inequality. Such problems were firstly considered in the convex setting by Duvaut and Lions [**29**], where semipermeability relations were assumed to be monotone and they led to variational inequalities. The nonmonotone semipermeability conditions are realized by various types of membranes, natural and artificial ones. These conditions arise in electrostatics, hydraulics and in the description of the flow of Bingham's fluids. Below we formulate an interior semipermeability problem in the language of heat conduction.

Let Ω be an open bounded subset of $\mathbb{R}^d$, ($d = 2, 3$ in applications), $0 < T < +\infty$ and $Q = \Omega \times (0, T)$. We consider the time-dependent heat equation

$$\frac{\partial u}{\partial t} - \Delta u = f \text{ in } Q. \tag{23}$$

We suppose that $f = f_1 + f_2$ with f_2 prescribed and

$$-f_1(x,t) \in \partial j(u(x,t)) \quad \text{a.e. in } Q. \tag{24}$$

On the boundary $\partial\Omega$ the temperature u satisfies the Dirichlet (homogeneous, for simplicity) boundary condition

$$u = 0 \quad \text{on } \partial\Omega \times (0, T) \tag{25}$$

and moreover, at $t = 0$ the temperature is given

$$u(x, 0) = u_0(x) \quad \text{in } \Omega. \tag{26}$$

The function $j \colon \mathbb{R} \to \mathbb{R}$ is assumed to be locally Lipschitz and ∂j denotes its Clarke's subdifferential in the second variable. In a more general case j can depend also in a measurable way on $x \in \Omega$.

It is known (cf. Duvaut and Lions [**29**], Panagiotopoulos [**101, 107**]) that (24) describes the behavior of a semipermeable membrane of finite thickness occupying the set Ω or the behaviour of temperature controller producing in Ω heat in order to regulate the temperature in the interior of Ω. For instance, on Figure 2(a), if the temperature $u < h_1$, then the region Ω supplies constant heat per unit volume, say c. When $u = h_1$ the heat is supplied (for constant temperature) until a value d is reached. Then the relation between supplied heat and temperature follows the parabola until h_2 is reached. At the temperature $u = h_2$ the heat is changed from value $-a$ to $-b$, and then the heat supply remains constant while the temperature u may increase.

Similarily the graph of Figure 2(b) corresponds to a temperature control problem when we regulate the temperature to deviate as little as possible from the interval $[h_1, h_2]$ (cf. Panagiotopoulos [**101**] and Naniewicz and Panagiotopoulos [**96**] for futher details). Assuming that the function j is generated by a L^∞_{loc} function β via the formula (17), we obtain the following parabolic problem for the Laplace operator with a discontinuous nonlinearity in the lower order term $\widehat{\beta}$ of the form

$$\begin{cases} u'(t) + Au(t) + \eta(t) = f(t) \quad \text{a.e. } t \in (0, T) \\ \eta(x,t) \in \widehat{\beta}(u(x,t)) \quad \text{a.e. } (x,t) \in Q \\ u(0) = u_0. \end{cases} \tag{27}$$

The problem (27) belongs to a class of the interior hemivariational inequality since the nonmonotone relation holds in the domain under consideration. In the case of nonmonotone boundary semipermeability conditions which are expressed by the Clarke subdifferential, we are lead to the the boundary hemivariational inequality. An example of the latter is given in the following problem

$$\begin{cases} \dfrac{\partial u}{\partial t} + A(t,u) + a_0(x,t,u) = f \quad \text{in } \Omega \times (0,T) \\ -\dfrac{\partial u}{\partial \nu_A} \in \partial j(x,t,u) \quad \text{on } \Gamma \times (0,T) \\ u = 0 \quad \text{on } (\partial\Omega \setminus \Gamma) \times (0,T) \\ u(0) = u_0 \quad \text{in } \Omega, \end{cases} \tag{28}$$

where Ω is a bounded domain in $\mathbb{R}^d$ with regular boundary, $\partial u / \partial \nu_A$ denotes the outward conormal derivative on Γ related to the operator A which is, for example, of the form

$$A(t,v) = -\sum_{i=1}^{N} \frac{\partial}{\partial x_i} a_i(x,t,v,Dv) + a_0(x,t,v).$$

In order to formulate the boundary parabolic hemivariational inequality and deliver existence result, we recall the following notation.

Let Ω be a bounded open subset of $\mathbb{R}^d$ with sufficiently regular boundary $\partial\Omega$ and let Γ be an open subset of $\partial\Omega$ with positive surface measure. We introduce

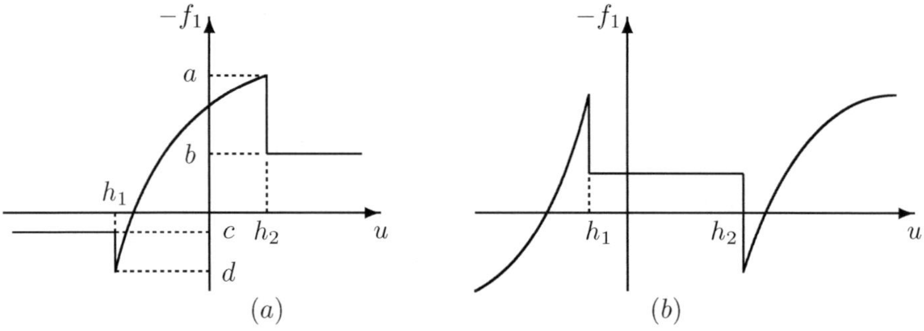

FIGURE 2. Semipermeability relations.

the following spaces $V = \{v \in H^1(\Omega) \mid \gamma v = 0 \text{ on } \partial\Omega \setminus \Gamma\}$ and $H = L^2(\Omega)$, where $\gamma\colon H^1(\Omega) \to L^2(\partial\Omega)$ denotes the trace operator. Identifying H with its dual, we have an evolution triple $V \subset H \subset V^*$ (cf. Zeidler [**116**], Denkowski et. al [**28**]) with dense, continuous and compact embeddings. We denote by $\langle \cdot, \cdot \rangle$ the duality of V and its dual V^* as well as the inner product on H, by $\|\cdot\|$, $|\cdot|$ and $\|\cdot\|_{V^*}$ the norms in V, H and V^*, respectively. In what follows we need the spaces $\mathcal{V} = L^2(0,T;V)$, $\mathcal{H} = L^2(0,T;H)$ and $\mathcal{W} = \{\, w \in \mathcal{V} \mid w' \in \mathcal{V}^* \,\}$, where the time derivative involved in the definition of $\mathcal{W}$ is understood in the sense of vector valued distributions. Equipped with the norm $\|v\|_{\mathcal{W}} = \|v\|_{\mathcal{V}} + \|v'\|_{\mathcal{V}^*}$ the space $\mathcal{W}$ becomes a separable reflexive Banach space. Clearly $\mathcal{W} \subset \mathcal{V} \subset \mathcal{H} \subset \mathcal{V}^*$. The duality for the pair $(\mathcal{V}, \mathcal{V}^*)$ is given by $\langle\langle f, v \rangle\rangle = \int_0^T \langle f(t), v(t) \rangle \, dt$. It is well known that every function in $\mathcal{W}$ is, after an eventual modification on a set of Lebesgue measure zero, continuous from $[0,T]$ into H and the embedding of $\mathcal{W}$ into $C([0,T];H)$ is continuous.

Of course the function space V for the problem (28) reflects the Dirichlet boundary condition. Using the Green formula we can formulate the parabolic hemivariational inequality corresponding to problem (28): find $u \in \mathcal{W}$ such that

$$
(29)\qquad \begin{cases} \langle u'(t) + A(t,u(t)) - f(t), v - u(t) \rangle + \displaystyle\int_\Gamma j^0(x,t,\gamma u(t); \gamma v - \gamma u(t)) \, d\Gamma \\ \quad \geq 0 \quad \text{for all } v \in V \text{and a.e. } t \in (0,T) \\ u(0) = u_0. \end{cases}
$$

The hypotheses on the data of (29) are the following.

$\underline{H(A)}$: $A\colon (0,T) \times V \to V^*$ is an operator such that

(i) $A(\cdot, v)$ is measurable on $(0,T)$ for all $v \in V$;
(ii) $A(t, \cdot)$ is pseudomonotone for a.e. $t \in (0,T)$;
(iii) $\|A(t,v)\|_{V^*} \leq a(t) + b\|v\|$ for all $v \in V$, a.e. $t \in (0,T)$ with $a \in L^2(0,T)$, $a \geq 0$, $b > 0$;
(iv) $\langle A(t,v), v \rangle \geq \alpha \|v\|^2$ for all $v \in V$, a.e. $t \in (0,T)$ with $\alpha > 0$.

$\underline{H(j)}$: $j\colon \Gamma \times (0,T) \times \mathbb{R} \to \mathbb{R}$ is a function such that

(i) $j(\cdot, t, \xi)$ is measurable for all $t \in (0,T)$, $\xi \in \mathbb{R}$ and $j(\cdot, t, 0) \in L^1(\Gamma)$;
(ii) $j(x,t,\cdot)$ is locally Lipschitz for all $t \in (0,T)$, $x \in \Gamma$;
(iii) $\eta \in \partial j(x,t,\xi) \implies |\eta| \leq c\,(1 + |\xi|)$ for all $t \in (0,T)$, $x \in \Gamma$ with $c > 0$;
(iv) $j^0(x,t,\xi;-\xi) \leq d\,(1 + |\xi|)$ for all $t \in (0,T)$, $x \in \Gamma$ with $d \geq 0$.

$\underline{(H_0)}$: $u_0 \in H$, $f \in \mathcal{V}^*$.

In the hypothesis $H(j)$ the symbol ∂j denotes the Clarke gradient of j with respect to the variable ξ and j^0 stands for the generalized directional derivative of j at the point ξ in the direction $-\xi$.

The following auxiliary result concerns the properties of the functional $J\colon (0,T) \times L^2(\Gamma) \to \mathbb{R}$ given by

$$
(30)\qquad J(t,v) = \int_\Gamma j(x,t,v(x)) \, d\Gamma, \qquad t \in (0,T), \quad v \in L^2(\Gamma).
$$

LEMMA 2. *Assume that $j\colon \Gamma \times (0,T) \times \mathbb{R} \to \mathbb{R}$ satisfies hypothesis $H(j)$. Then the functional J given by (30) is well defined and locally Lipschitz in the second* *variable (in fact, Lipschitz in the second variable on bounded subsets of $L^2(\Gamma)$), its generalized gradient satisfies the linear growth condition*

$$\zeta \in \partial J(t,v) \quad \Longrightarrow \quad ||\zeta||_{L^2(\Gamma)} \le c_1 \left(1 + ||v||_{L^2(\Gamma)}\right) \quad \text{with } c_1 > 0 \tag{31}$$

and for its generalized directional derivative we have

$$J^0(t,u;v) \le \int_\Gamma j^0(x,t,u(x);v(x))\, d\Gamma \quad \text{for } t \in (0,T), \quad u,v \in L^2(\Gamma) \tag{32}$$

and

$$J^0(t,u;-u) \le d_1 \left(1 + ||u||_{L^2(\Gamma)}\right) \quad \text{with } d_1 > 0. \tag{33}$$

Moreover, if additionally either $j(x,t,\cdot)$ or $-j(x,t,\cdot)$ is regular in the sense of Clarke, then $J(t,\cdot)$ or $-J(t,\cdot)$ is regular, respectively,

$$J^0(t,u;v) = \int_\Gamma j^0(x,t,u(x);v(x))\, d\Gamma \quad \text{for } t \in (0,T), \quad u,v \in L^2(\Gamma) \tag{34}$$

and

$$\partial(J \circ \gamma)(t,v) = \gamma^* \circ \partial J(t,\gamma v) \quad \text{for } t \in (0,T), \quad v \in V, \tag{35}$$

where $\gamma^* \colon L^2(\Gamma) \to V^*$ denotes the adjoint operator of γ given by

$$\gamma^* z(v) = \int_\Gamma z(x)\gamma v(x)\, d\Gamma \quad \text{for } v \in V \text{ and } z \in L^2(\Gamma).$$

For the proof, we refer to Lemma 2 of [**81**].

Example 3. Let the function $j \colon \mathbb{R}^d \to \mathbb{R}$ be given by $j(\xi) = \min\{f_1(\xi), f_2(\xi)\}$, $\xi \in \mathbb{R}^d$, where $f_i \colon \mathbb{R}^d \to \mathbb{R}$, $i = 1,2$ are convex and quadratic. Using Theorem 2.5.1 of Clarke [**16**], we know that $\partial j(\xi) \subset co\{\nabla f_1(\xi), \nabla f_2(\xi)\}$ and hence the subdifferential has at most a linear growth. Next, from Proposition 2.1.2 of Clarke [**16**], we have $j^0(\xi;\eta) = \max\{\xi^* \cdot \eta \mid \xi^* \in \partial j(\xi)\}$. Therefore

$$j^0(\xi;-\xi) = \max\{\xi^* \cdot (-\xi) \mid \xi^* = \lambda \nabla f_1(\xi) + (1-\lambda)\nabla f_2(\xi), \quad \lambda \in (0,1)\} \le 0,$$

because $\nabla f_i(\xi) \cdot \xi \ge 0$, $i = 1,2$. Hence the function j satisfies hypothesis $H(j)$(iii) and (iv).

In order to establish the existence result for problem (29), we will introduce an operator evolution inclusion. Consider the following auxiliary inequality problem: find $u \in \mathcal{W}$ such that

$$\begin{cases} \langle u'(t) + A(t,u(t)) - f(t), v\rangle + J^0(t,\gamma u(t);\gamma v) \ge 0 & \text{for all } v \in V, \text{a.e. } t \\ u(0) = u_0, \end{cases} \tag{36}$$

where the functional J is defined by (30).

REMARK 4. We notice that if hypothesis $H(j)$ holds, then every solution to problem (36) is also a solution to (29). This fact is an easy consequence of inequality (32) of Lemma 2. If additionally either $j(x,t,\cdot)$ or $-j(x,t,\cdot)$ is regular, then $u \in \mathcal{W}$ solves (29) if and only if $u \in \mathcal{W}$ solves (36). This equivalence of problems (29) and (36) follows from (34) of Lemma 2.

Let $Z = H^{\delta}(\Omega)$ with a fixed $\delta \in (\frac{1}{2}, 1)$. Denoting by $i\colon V \to Z$ the embedding injection and by $\overline{\gamma}\colon Z \to L^2(\Gamma)$ the trace operator, for all $v \in V$ we have $\gamma v = \overline{\gamma}(iv)$. In what follows we omit the notation of the embedding i and we simply write $\gamma v = \overline{\gamma} v$ for $v \in V$. So we have $V \subset Z \subset H \subset Z^* \subset V^*$ with all embeddings being compact. Also $\mathcal{W} \subset \mathcal{V} \subset \mathcal{Z} \subset \mathcal{H} \subset \mathcal{Z}^* \subset \mathcal{V}^*$, where $\mathcal{Z} = L^2(0,T;Z)$. Now we associate with problem (36) the following inclusion

$$(37) \qquad \begin{cases} u'(t) + A(t,u(t)) + \eta(t) = f(t) & \text{for a.e. } t \in (0,T) \\ \eta(t) \in \overline{\gamma}^* \left(\partial J(t, \overline{\gamma} u(t))\right) & \text{for a.e. } t \in (0,T) \\ u(0) = u_0. \end{cases}$$

REMARK 5. We note that if hypothesis $H(j)$ holds, then every solution to problem (37) is also a solution to (36). If additionally either $j(x,t,\cdot)$ or $-j(x,t,\cdot)$ is regular, then the problems (36) and (37) are equivalent.

5.1. Existence via the Galerkin method. In this section we describe the Galerkin method which can be used to prove the existence of solutions to the interior hemivariational inequality (27).

Let Ω be a bounded open subset of $\mathbb{R}^d$ with Lipschitz boundary $\partial\Omega$. Let V be a reflexive Banach space such that $V \subset H^1(\Omega)$ and (V, H, V^*) be an evolution triple of spaces with $H = L^2(\Omega)$. As before, we consider $\mathcal{W} = \{\, w \in \mathcal{V} \mid w' \in \mathcal{V}^* \,\}$, where $\mathcal{V} = L^2(0,T;V)$.

Consider the following hypotheses on the data.

$\underline{H(A)_1}$: $A\colon V \to V^*$ is a linear continuous operator which is coercive in the following sense $\langle Av, v\rangle \geq \alpha\|v\|^2 - \alpha_0|v|_H^2$ for all $v \in V$ with $\alpha > 0$ and $\alpha_0 \geq 0$.

$\underline{H(\beta)}$: $\beta \in L^{\infty}_{loc}(\mathbb{R})$ satisfies a growth condition $|\beta(r)| \leq c(1+|r|)$ a.e $r \in \mathbb{R}$ with $c > 0$.

In order to formulate the existence result, we define a regularization of β by

$$\beta^h(r) = \frac{1}{h}\int_{-\infty}^{+\infty} \beta(r-s)\rho\left(\frac{s}{h}\right) ds,$$

where $h \in (0,1)$, $\rho \in C_0^{\infty}((-1,1))$, $\rho \geq 0$ and $\int_{-1}^{1} \rho(s)\, ds = 1$. Let $\{V_h\}_h \subset V$ be a family of finite dimensional subspaces of V such that for every $v \in V$ there exists $\{v_h\}_h$, $v_h \in V_h$, $v_h \to v$ in V, as $h \to 0$. Subsequently, let $\{\varphi_{jh}\}_{j=1}^{n_h}$ be a linearly independent basis of V_h, $n_h = \dim V_h$ and $\{u_{0h}\}_h$ be a sequence approximating the initial condition such that $u_{0h} \in V_h$ for all $h \in (0,1)$ and $u_{0h} \to u_0$ in H as $h \to 0$. Putting

$$u_h(t) = \sum_{j=1}^{n_h} c_{jh}(t)\varphi_{jh}, \quad u_{0h} = \sum_{j=1}^{n_h} d_{jh}\varphi_{jh},$$

we formulate a regularized Galerkin problem for (27): find u_h such that

$$\begin{cases} \langle u_h'(t), v_h\rangle + \langle Au_h(t), v_h\rangle + \langle \beta^h(u_h(t)), v_h\rangle = \langle f(t), v_h\rangle \\ \qquad\qquad\qquad\qquad\qquad\qquad \text{for all } v_h \in V_h, \text{ a.e. } t \in (0,T) \\ u_h(0) = u_{0h}. \end{cases} \tag{38}$$

THEOREM 6. *Under hypotheses $H(A)_1$, $H(\beta)$ and (H_0), the problem (27) admits at least one solution $u \in \mathcal{W}$ with $\eta \in L^2(Q)$. Moreover, for all $h \in (0,1)$, there exists at least one solution u_h to the regularized Galerkin problem (38) and we can find a subsequence of $\{u_h\}$ which converges to some solution u of (27) weakly in $\mathcal{V}$, as $h \to 0$.*

The main difficulty in the proof of the above result is to show the compactness of the sequence of solutions u_h of the regularized Galerkin problem (38) in $L^2(Q)$. This have been done in [**57**] by exploiting the following compactness result.

LEMMA 7. *Let $\{u_h\}$ be a sequence bounded in $\mathcal{V} \cap L^\infty(0,T;L^1(\Omega))$. If $u_h(t) \to u(t)$ weakly in $L^1(\Omega)$ for a.e. $t \in (0,T)$, then for a some subsequence $u_h \to u$ in $L^2(Q)$.*

5.2. Existence via theory of pseudomonotone operators. In this section we will shortly indicate a method which allows to obtain an existence result for the boundary hemivariational inequality (29) by exploiting the theory of pseudomonotone operators (cf. Browder and Hess [**4**]). In the stationary case such approach have been used by Naniewicz and Panagiotopoulos in [**96**]. Having in mind Remarks 4 and 5, to show the existence of solutions to the hemivariational inequality (29), it is enough to establish existence of solutions to the inclusion (37).

THEOREM 8. *If hypotheses $H(A)$, $H(j)$ and (H_0) hold, then the inclusion (37) admits a solution.*

The proof consists of two parts. First we assume that the initial condition is more smooth, i.e. $u_0 \in V$ and in the second step we consider $u_0 \in H$.

For $u_0 \in V$, let $\mathcal{A}\colon \mathcal{V} \to \mathcal{V}^*$ be the Nemitsky (superposition) operator corresponding to A and defined by $(\mathcal{A}v)(t) = A(t, v(t) + u_0)$. Let $\mathcal{N}\colon \mathcal{V} \to 2^{\mathcal{V}^*}$ be the multivalued Nemitsky operator corresponding to $\overline{\gamma}^* \circ \partial J(t, \overline{\gamma}\,\cdot)$, i.e.

$$\mathcal{N}v = \{w \in \mathcal{Z}^* \mid w(t) \in \overline{\gamma}^*\left(\partial J(t, \overline{\gamma}(v(t) + u_0))\right) \quad \text{a.e. } t \in (0,T)\}.$$

In a consequence, from (37), we obtain the following inclusion

$$\begin{cases} z'(t) + \mathcal{A}z + \mathcal{N}z \ni f \\ z(0) = 0. \end{cases} \tag{39}$$

Note that $z \in \mathcal{W}$ is a solution to problem (37) if and only if $z - u_0 \in \mathcal{W}$ solves (39).

Let $L\colon D(L) \subset \mathcal{V} \to \mathcal{V}^*$ be the operator defined by $Lv = v'$ with $D(L) = \{w \in \mathcal{W} \mid w(0) = 0\}$. It is known (see e.g. [**116, 28**]) that L is densely defined maximal monotone operator. Let the operator $\mathcal{T}\colon \mathcal{V} \to 2^{\mathcal{V}^*}$ be given by $\mathcal{T}z = \mathcal{A}z + \mathcal{N}z$. So now problem (39) takes the form

$$\text{find } z \in D(L) \quad \text{such that} \quad Lz + \mathcal{T}z \ni f.$$

In order to show the existence of a solution to problem (39), we can prove that the operator $\mathcal{T}$ is bounded, coercive and L-pseudomonotone. Then, we apply the surjectivity result of Proposition 1. For a detailed proof we refer to Migórski and Ochal [**81**].

5.3. Inverse problem. In this section we consider the following nonlinear parabolic initial-boundary value problem with multivalued nonmonotone boundary conditions

$$\frac{\partial u}{\partial t} - \nabla(k(|\nabla u|^2)\nabla u) + q(x,t)u = f_1 \quad \text{in} \quad Q, \tag{40}$$

$$-k(|\nabla u|^2)\frac{\partial u}{\partial n} \in \partial j(x,t,u) \quad \text{on} \quad \Gamma_1 \times (0,T), \tag{41}$$

$$u = 0 \quad \text{on} \quad \Gamma_2 \times (0,T), \tag{42}$$

$$k(|\nabla u|^2)\frac{\partial u}{\partial n} = \varphi \quad \text{on} \quad \Gamma_3 \times (0,T), \tag{43}$$

$$u(0) = u_0 \quad \text{in} \quad \Omega, \tag{44}$$

where $Q = \Omega \times (0,T)$, Ω is a bounded open subset of $\mathbb{R}^d$ with a regular boundary $\partial\Omega = \Gamma = \overline{\Gamma}_1 \cup \overline{\Gamma}_2 \cup \overline{\Gamma}_3$, $\Gamma_k \cap \Gamma_l = \emptyset$ for $k \neq l$, k, $l = 1$, 2, 3, $m(\Gamma_2) > 0$, $0 < T < \infty$ and $\partial j(x,t,u)$ denotes the Clarke generalized gradient of a locally Lipschitz function $j(x,t,\cdot)\colon \mathbb{R} \to \mathbb{R}$ at a point $u \in \mathbb{R}$. We deal with the identification problem of determination of the unknown coefficient k from the additional information on the solution: the Dirichlet type boundary measured data, the data measured at the final time instant or the nonlocal measured data. The identification problem is formulated as an optimal control one where the space of controls consists of the class of admissible coefficients. For the coefficient k we assume the following conditions

$\underline{(A_1)}$: $k \in C([0,+\infty))$, $c_1 \le k(s) \le c_2$ for all $s \in [0,+\infty)$ with $0 < c_1 \le c_2$,

$\underline{(A_2)}$: $\sum_{i=1}^{d} \left(k(|\xi|^2)\xi_i - k(|\zeta|^2)\zeta_i\right)(\xi_i - \zeta_i) \ge c_3|\xi - \zeta|^2$ for all ξ, $\zeta \in \mathbb{R}^d$, where $c_3 > 0$.

Moreover, the regularity of the data is as follows

$\underline{(A_3)}$: $q \in L^\infty(Q)$, $q > 0$, $f_1 \in L^2(Q)$, $\varphi \in L^2(0,T;L^2(\Gamma_3))$, $u_0 \in L^2(\Omega)$.

We consider the inverse coefficient problem which consists in the determination of the unknown coefficient k in the problem (40)-(44) from various measured data: the Dirichlet type boundary measure data

$$u(x,t) = g(x,t), \quad (x,t) \in \Gamma_3 \times (0,T) \tag{45}$$

given on the part Γ_3 of the boundary Γ, the final measure data

$$u(x,T) = \psi(x), \quad x \in \Omega, \tag{46}$$

the nonlocal measure data

$$\int_\Omega u(x,t)dx = \Phi(t), \quad t \in (0,T]. \tag{47}$$

The hemivariational inequality (40)–(44) is referred to as the direct problem. The motivation to deal with the parabolic problem (40)-(44) comes from the semipermeability theory of nonsmooth mechanics. The problem of this type arise in electrostatics, hydraulics and flow through porous media (cf. [**107, 96**]). It is also met in certain heat conduction problems, e.g. the behavior of semipermeability membrane of finite thickness and temperature control problems (cf. [**29, 104**]). The solution of problem (40)–(44) represents the temperature. For some examples of subdifferential boundary condition (41), cf. [**96, 83**].

The form of the nonlinearity in the differential operator in (40) appears in a wide class of inverse problems (cf. [**39, 40, 42**]) as well as in the deformation theory of plasticity. In the latter the nonlinear boundary value problems related to the elliptic operator $\nabla(k(|\nabla \cdot|^2)\nabla \cdot)$ include nonlinear Lame system, elastoplastic bending of an incompressible hardening plate, and elastoplastic torsion of a bar. The function k describes the elastoplastic behavior of a deformable plate and usually is called the modulus of plasticity. It appears in the relationship $\mathcal{T} = k(\xi^2)\xi$ between the intensities of shift strain $\xi = (2\varepsilon_{ij}^D\varepsilon_{ij}^D)^{1/2}$ and tangential stress $\mathcal{T} = (2\sigma_{ij}^D\sigma_{ij}^D)^{1/2}$, $i,\ j = 1,\ldots,d$ (where $\varepsilon^D = \{\varepsilon_{ij}^D\}$ and $\sigma^D = \{\sigma_{ij}^D\}$ denote the deviators of the deformation and stress, respectively) and it is derived from the Hencky correlation $\sigma_{ij}^D = 2k(\xi^2)\varepsilon_{ij}^D$, cf. [**41, 42**]. We mention that the model (40)–(44) was studied in [**42**] with unilateral Signorini condition on $\Gamma_1 \times (0,T)$ which corresponds to a convex potential $j(x,t,\cdot)$.

We now give a variational formulation of (40)–(44) and present inverse problems. Let $V = \{v \in H^1(\Omega) \mid v = 0 \text{ on } \Gamma_2\}$. Identifying $H = L^2(\Omega)$ with its dual we have an evolution triple of spaces (V, H, V^*) with dense, continuous and compact embeddings. For every $v \in V$ we denote by γv its trace on Γ, where $\gamma\colon V \to H^{1/2}(\Gamma) \subset L^2(\Gamma)$ is the trace map. Let $Z = H^\delta(\Omega)$ with $\delta \in (1/2, 1)$. Denoting by $i\colon V \to Z$ the embedding injection and by $\overline{\gamma}\colon Z \to L^2(\Gamma)$ the trace operator, for all $v \in V$ we have $\gamma v = \overline{\gamma}(iv)$. For simplicity we omit the notation of the embedding and we write $\gamma v = \overline{\gamma} v$ for $v \in V$. Thus we have $V \subset Z \subset H \subset Z^* \subset V^*$ with all embeddings being compact. We also introduce the following spaces $\mathcal{V} = L^2(0,T;V)$, $\mathcal{H} = L^2(0,T;H)$, $\mathcal{V}^* = L^2(0,T;V^*)$ and $\mathcal{W} = \{w \in \mathcal{V} \mid w' \in \mathcal{V}^*\}$. We have $\mathcal{W} \subset \mathcal{V} \subset \mathcal{Z} \subset \mathcal{H} \subset \mathcal{Z}^* \subset \mathcal{V}^*$, where $\mathcal{Z} = L^2(0,T;Z)$ and $\mathcal{Z}^* = L^2(0,T;Z^*)$ denote its dual.

In what follows, we denote by $\mathcal{K}$ the class of admissible coefficients $k\colon [0,+\infty) \to \mathbb{R}$ which satisfy the conditions (A_1) and (A_2). For every $k \in \mathcal{K}$, we associate a nonlinear operator $A_k\colon (0,T) \times V \to V^*$ of the form

$$\langle A_k(t,u), v\rangle = \int_\Omega \big(k(|\nabla u|^2)\nabla u \cdot \nabla v + q(t)uv\big)\, dx \quad \text{for } u,\ v \in V. \tag{48}$$

Next, using the hypothesis (A_3), we define $f \in L^2(0,T,V^*)$ by

$$\langle f(t), v\rangle = \int_\Omega f_1(x,t)v(x)\, dx + \int_{\Gamma_3} \varphi(x,t)\gamma v(x)\, d\Gamma \quad \text{for } v \in V.$$

The weak formulation of (40)–(44) is given (cf. [**90**]) by the following dynamic hemivariational inequality: find $u \in \mathcal{W}$ such that

$$
(49) \qquad \begin{cases} \langle u'(t) + A_k(t, u(t)) - f(t), v\rangle + \displaystyle\int_{\Gamma_1} j^0(x, t, \gamma u(t); \gamma v)\, d\Gamma \geq 0 \\ \qquad\qquad\qquad\qquad\qquad\qquad \text{for all } v \in V \text{ and a.e. } t \in (0, T) \\ u(0) = u_0. \end{cases}
$$

The problem (49) is called the direct problem. We consider the coefficient identification problem which consists in the determination of the unknown coefficient k in the problem (49) from various measured data. In practice, due to measurement errors, exact equality in the equations (45), (46) and (47) is usually not achieved. For this reason the inverse problems are formulated as the optimal control one

$$
(50) \qquad \min\{\widehat{F}(u(k)) \mid k \in K_c\},
$$

where $K_c \subseteq \mathcal{K}$ is a set of admissible coefficients, $\widehat{F}\colon \mathcal{V} \cap C(0,T;H) \to \mathbb{R}$ is a prescribed cost functional and $u = u(k)$ represents the solution to the problem (49) corresponding to the coefficient k. The possible choice of the cost functional $\widehat{F}$ is one of the following

$$
F_1(u) = \int_0^T \int_{\Gamma_3} |\gamma u(x,t) - g(x,t)|^2 \, d\Gamma(x) dt,
$$

with $g \in L^2(0, T; L^2(\Gamma_3))$ being the Dirichlet type boundary measure data,

$$
F_2(u) = \int_\Omega |u(x,T) - \psi(x)|^2 \, dx,
$$

with $\psi \in L^2(\Omega)$ being the measure of the solution in final time, and

$$
F_3(u) = \int_0^T \left(\int_\Omega u(x,t)\, dx - \Phi(t) \right)^2 dt,
$$

where $\Phi \in L^2(0,T)$ measures the solution at time t.

We suppose that the coefficient k in the operator A_k is fixed and therefore we will write A in place of A_k. The problem under consideration reads as follows: find $u \in \mathcal{W}$ such that

$$
(51) \qquad \begin{cases} \langle u'(t) + A(t, u(t)) - f(t), v\rangle + \displaystyle\int_{\Gamma_1} j^0(x, t, \gamma u(t); \gamma v)\, d\Gamma \geq 0 \\ \qquad\qquad\qquad\qquad\qquad\qquad \text{for all } v \in V \text{ and a.e. } t \in (0, T) \\ u(0) = u_0. \end{cases}
$$

We need the following hypotheses on the data of (51).

$\underline{H(A)_2}$: $A\colon (0,T) \times V \to V^*$ is an operator such that

- (i) $A(\cdot, v)$ is measurable on $(0,T)$ for all $v \in V$;
- (ii) $A(t, \cdot)$ is pseudomonotone for a.e. $t \in (0,T)$;
- (iii) $\|A(t,v)\|_{V^*} \leq a(t) + b\|v\|$ a.e. t, for all $v \in V$ with $a \in L^2(0,T)$, $a \geq 0$, $b > 0$;

(iv) $\langle A(t,v),v\rangle \geq \alpha\|v\|^2 - \alpha_1\|v\|^\rho - \alpha_2(t)$ a.e. $t \in (0,T)$, for all $v \in V$ with $\alpha > 0$, $\alpha_1 \geq 0$, $\alpha_2 \in L^1(0,T)$, $\alpha_2 \geq 0$ and $0 < \rho < 2$;
(v) $A(t,\cdot)$ is strongly monotone, i.e. there exists $m_1 > 0$ such that $\langle A(t,u) - A(t,v), u - v\rangle \geq m_1\|u-v\|^2$ for a.e. $t \in (0,T)$ and for all $u, v \in V$.

$\underline{H(j)_1}$: $j\colon \Gamma_1 \times (0,T) \times \mathbb{R} \to \mathbb{R}$ is a function such that
(i) $j(\cdot,\cdot,r)$ is measurable on $\Gamma_1 \times (0,T)$, for all $r \in \mathbb{R}$ and $j(\cdot,t,0) \in L^1(\Gamma_1)$ for a.e. $t \in (0,T)$;
(ii) $j(x,t,\cdot)$ is locally Lipschitz for a.e. $(x,t) \in \Gamma_1 \times (0,T)$;
(iii) $|\partial j(x,t,r)| \leq c\,(1+|r|)$ for a.e. $(x,t) \in \Gamma_1 \times (0,T)$, all $r \in \mathbb{R}$ with $c > 0$;
(iv) $j^0(x,t,r;-r) \leq d\,(1+|r|)$ for a.e. $(x,t) \in \Gamma_1 \times (0,T)$, all $r \in \mathbb{R}$ with $d \geq 0$;
(v) either $j(x,t,\cdot)$ or $-j(x,t,\cdot)$ is regular for a.e. $(x,t) \in \Gamma_1 \times (0,T)$;
(vi) there exists $m_2 \geq 0$ such that $(s_1 - s_2)(r_1 - r_2) \geq -m_2|r_1 - r_2|^2$ for all $s_i \in \partial j(x,t,r_i)$, $r_i \in \mathbb{R}$, $i = 1, 2$ and a.e. $(x,t) \in \Gamma_1 \times (0,T)$.

$\underline{(H_1)}$: $m_1 \geq m_2\|\gamma\|^2$, where $\|\gamma\|$ is the norm in $\mathcal{L}(V, L^2(\Gamma))$ of the trace operator.

We introduce the functional $J\colon (0,T) \times L^2(\Gamma_1) \to \mathbb{R}$ given by

$$J(t,v) = \int_{\Gamma_1} j(x,t,v(x))\,d\Gamma \quad \text{for } t \in (0,T), \quad v \in L^2(\Gamma_1). \tag{52}$$

As before, it is convenient to associate with problem (51) the following evolution inclusion

$$\begin{cases} u'(t) + A(t,u(t)) + \overline{\gamma}^*\partial J(t,\overline{\gamma}u(t)) \ni f(t) & \text{for a.e. } t \in (0,T) \\ u(0) = u_0. \end{cases} \tag{53}$$

Lemma 9. *Under hypothesis $H(j)_1(i)$-(iii), every solution of the evolution inclusion (53) is also a solution to (51). If additionally $H(j)_1(v)$ holds, then $u \in \mathcal{W}$ solves (53) if and only if it solves (51).*

The following result is on solvability and unique solvability of the hemivariational inequality under consideration.

Theorem 10. *Under hypotheses $H(A)_2(i)$-(iv), $H(j)_1(i)$-(iv) and (H_0), problem (53) has a solution. Moreover, this solution is unique provided $H(A)_2(i)$, (iii)-(v), $H(j)_1(i)$-(iv), (vi), (H_0) and (H_1) hold.*

We pass to a result on the existence of solutions to the inverse problems (50).

Lemma 11. *Let $k \in \mathcal{K}$ and $q \in L^\infty(Q)$, $q \geq 0$. Then the operator $A_k\colon (0,T) \times V \to V^*$ defined by (48) satisfies conditions $H(A)_2$.*

We associate with the hemivariational inequality (49) the following evolution inclusion

$$\begin{cases} u'(t) + A_k(t,u(t)) + \overline{\gamma}^*\partial J(t,\overline{\gamma}u(t)) \ni f(t) & \text{for a.e. } t \in (0,T) \\ u(0) = u_0. \end{cases} \tag{54}$$

From Theorem 10, we infer the following

COROLLARY 12. Let $k \in \mathcal{K}$. Under hypotheses (A_3), $H(j)_1$ and $c_3 \geq m_2\|\gamma\|^2$, the inclusion (54) admits a unique solution.

The crucial role in the existence of solutions to the inverse problem plays the following result on the dependence of solution to the direct problem on the coefficient.

THEOREM 13. *Let the hypootheses $H(j)_1(i)$-(iv), (vi), (A_3) hold and $c_3 > m_2\|\gamma\|^2$. Suppose a sequence $\{k_n\} \subset \mathcal{K}$ of coefficients converges pointwise in $[0, +\infty)$ to a function $k \in \mathcal{K}$. Then the sequence $\{u_n\} = \{u(k_n)\}_{n\in\mathbb{N}}$ of (unique) solutions to (54) converges in $\mathcal{V} \cap C(0,T;H)$ to the solution $u = u(k)$ of (54).*

From the above theorem and Lemma 9, we easily deduce

COROLLARY 14. Let the hypotheses $H(j)_1$, (A_3) hold and $c_3 > m_2\|\gamma\|^2$. Assume a sequence $\{k_n\} \subset \mathcal{K}$ of coefficients converges pointwise in $[0, +\infty)$ to a function $k \in \mathcal{K}$. Then the sequence $\{u_n\} = \{u(k_n)\}_{n\in\mathbb{N}}$ of (unique) solutions to (49) converges in $\mathcal{V} \cap C(0,T;H)$ to the solution $u = u(k)$ of (49).

We are now in a position to state the main result on the existence of solutions to inverse coefficient problem (50). We admit the following hypotheses.

$\underline{H(K_c)}$: The family $K_c \subseteq \mathcal{K}$ of functions is equicontinuous, i.e. for a given $\varepsilon > 0$ we can find $\delta > 0$ such that for every $k \in K_c$, we have $|k(s_1) - k(s_2)| \leq \varepsilon$ for all s_1, $s_2 \in [0, +\infty)$ with $|s_1 - s_2| \leq \delta$.

$\underline{H(\widehat{F})}$: The cost functional $\widehat{F}$ is strongly lower semicontinuous on $\mathcal{V} \cap C(0,T;H)$.

THEOREM 15. *If hypotheses $H(j)_1$, (A_3), $c_3 > m_2\|\gamma\|^2$, $H(K_c)$ and $H(F)$ hold, then the problem (50) admits a solution.*

PROOF. Let $\{k_n\}_{n\in\mathbb{N}}$ be a minimizing sequence for the problem (50), that is, $k_n \in K_c$ and $\lim \widehat{F}(u(k_n)) = \inf\{\widehat{F}(u(k)) \mid k \in K_c\} = m$. Since K_c is an equicontinuous and equibounded, by the Ascoli-Arzela theorem, it follows that $\{k_n\}$ admits a uniformly convergent subsequence on any compact subset of $[0, +\infty)$. By a standard diagonal argument, there are a subsequence, still denoted by $\{k_n\}$ and a function $k \in K_c$ such that k_n converges to k, as $n \to \infty$ uniformly on every compact subset of $[0, +\infty)$. From Corollary 14 we deduce that the sequence $u_n = u(k_n)$ converges in $\mathcal{V} \cap C(0,T;H)$ to the solution $u = u(k)$ of (49). Finally, due to the lower semicontinuity of $\widehat{F}$, we have $m \leq \widehat{F}(u(k)) \leq \liminf_{n\to\infty} \widehat{F}(u(k_n)) = m$ which completes the proof. □

We can easily observe that the hypothesis $H(\widehat{F})$ is satisfied for the cost functionals F_i, $i = 1, 2, 3$. Thus, we conclude that the identification problems for these functionals have optimal solutions. Finally, we provide a simple example of a nonconvex potential j which satisfies the hypothesis $H(j)_1$.

EXAMPLE 16. Let $j\colon \mathbb{R} \to \mathbb{R}$ be defined by

$$j(r) = \begin{cases} 0, & \text{if } r < 0 \\ -r(r-2), & \text{if } 0 \leq r < 1 \\ 1, & \text{if } r \geq 1. \end{cases}$$

Then its subdifferential is given by

$$\partial j(r) = \begin{cases} 0, & \text{if } r < 0 \\ [0,2], & \text{if } r = 0 \\ -2r+2, & \text{if } 0 < r < 1 \\ 0, & \text{if } r \geq 1. \end{cases}$$

It is immediate that j satisfies $H(j)_1(i)$–(ii) and $H(j)_1(iii)$ with $c = 2$. Using the formula $j^0(r; -r) = \max\{\eta(-r) \mid \eta \in \partial j(r)\}$ for $r \in \mathbb{R}$, we see that $H(j)_1(iv)$ holds with $d = 2$. Next, we can easily check that $\eta_1 \leq \eta_2 - 2(r_1 - r_2)$ for all $r_1 < r_2$ and $\eta_i \in \partial j(r_i)$, $i = 1, 2$. This condition implies $H(j)_1(vi)$ with $m_2 = 2$. Finally, the function j can be represented as the difference of convex functions (i.e. it is of d.c. type), $j(r) = j_1(r) - j_2(r)$, where

$$j_1(r) = \begin{cases} r^2 - 2r + 2, & \text{if } r < 0 \\ 2, & \text{if } 0 \leq r < 1 \\ r^2 - 2r + 3, & \text{if } r \geq 1 \end{cases}$$

and $j_2(r) = r^2 - 2r + 2$ for $r \in \mathbb{R}$. Since j_1 and j_2 are convex functions, ∂j_i, $i = 1$, 2 have a sublinear growth with ∂j_2 being a singleton, we deduce that either j or $-j$ is regular with $\partial j(r) = \partial j_1(r) - \partial j_2(r)$. Hence $H(j)_1(v)$ is also satisfied.

6. Evolution second order hemivariational inequalities

In this section we study second order dynamic hemivariational inequalities. In order to emphasize the relationships between these inequalities and evolution multivalued equations, we begin with a more general result on existence and uniqueness of solutions to evolution inclusions. Then, by specifying a multifunction in the subdifferential form we obtain, in a particular case, existence and uniqueness of solutions to second order hemivariational inequalities.

6.1. Existence and uniqueness of solutions. In this part we study a class of second order nonlinear evolution inclusions. For this class we give a result on the existence and uniqueness of solutions to the Cauchy problem for the inclusion under investigation. The proof consists of two main parts. First we consider the Cauchy problem for a nonlinear inclusion without a Lipschitz time dependent elasticity operator. We prove the unique solvability of this problem using the surjectivity result for pseudomonotone multivalued operators. In the second part of the proof, we apply the Banach Contraction Principle to show that a suitable contraction operator has a unique fixed point which will be the solution of the problem under consideration.

We begin with the notation needed for the statement of the problem. Let V and Z be separable and reflexive Banach spaces with the duals V^* and Z^*, respectively. Let H denote a separable Hilbert space and we identify H with its dual. We suppose that $V \subset H \subset V^*$ and $Z \subset H \subset Z^*$ form evolution triples of spaces where all embedings are continuous, dense and compact (see e.g. Chapter 23.4 of [**116**], Chapter 3.4 of [**28**]) and that V is compactly embedded in Z. Let $\|\cdot\|$ and $|\cdot|$ denote the norms in V and H, respectively, and let $\langle \cdot, \cdot \rangle$ be the duality pairing between V^*

and V. We also introduce the following spaces $\mathcal{V} = L^2(0,T;V)$, $\mathcal{Z} = L^2(0,T;Z)$, $\hat{\mathcal{H}} = L^2(0,T;H)$, $\mathcal{Z}^* = L^2(0,T;Z^*)$, $\mathcal{V}^* = L^2(0,T;V^*)$ and $\mathcal{W} = \{v \in \mathcal{V} \mid v' \in \mathcal{V}^*\}$.

The nonlinear evolution inclusion under consideration is as follows.

PROBLEM. $\mathcal{P}$: find $u \in \mathcal{V}$ such that $u' \in \mathcal{W}$ and

$$\begin{cases} u''(t) + A(t,u'(t)) + B(t,u(t)) + F(t,u(t),u'(t)) \ni f(t) & \text{a.e. } t \in (0,T), \\ u(0) = u_0, \quad u'(0) = u_1, \end{cases}$$

where A, $B\colon (0,T)\times V \to V^*$ are nonlinear operators and $F\colon (0,T)\times V \times V \to 2^{Z^*}$ is a multivalued mapping.

Let us notice that the initial conditions in Problem $\mathcal{P}$ have sense in V and H since the embeddings $\{v \in \mathcal{V} \mid v' \in \mathcal{W}\} \subset C(0,T;V)$ and $\mathcal{W} \subset C(0,T;H)$ are continuous.

A solution to Problem $\mathcal{P}$ is understood as follows.

DEFINITION 17. A function $u \in \mathcal{V}$ is a solution of Problem $\mathcal{P}$ if and only if $u' \in \mathcal{W}$ and there exists $z \in \mathcal{Z}^*$ such that

$$\begin{cases} u''(t) + A(t,u'(t)) + B(t,u(t)) + z(t) = f(t) & \text{a.e. } t \in (0,T), \\ z(t) \in F(t,u(t),u'(t)) \quad \text{a.e. } t \in (0,T), \\ u(0) = u_0, \quad u'(0) = u_1. \end{cases}$$

We will need the following hypotheses on the data.

$\underline{H(A)_3}$: $A\colon (0,T) \times V \to V^*$ is an operator such that
- (i) $A(\cdot,v)$ is measurable on $(0,T)$ for all $v \in V$;
- (ii) $A(t,\cdot)$ is pseudomonotone for a.e. $t \in (0,T)$;
- (iii) $\|A(t,v)\|_{V^*} \le a(t) + b\|v\|$ for all $v \in V$, a.e. $t \in (0,T)$ with $a \in L^2(0,T)$, $a \ge 0$, $b > 0$;
- (iv) $\langle A(t,v), v\rangle \ge \alpha\|v\|^2$ for all $v \in V$, a.e. $t \in (0,T)$ with $\alpha > 0$.

$\underline{H(A)_4}$: The operator $A\colon (0,T) \times V \to V^*$ satisfies $H(A)_3(i)$, (iii), (iv) and
- (v) $A(t,\cdot)$ is hemicontinuous for a.e. $t \in (0,T)$;
- (vi) $A(t,\cdot)$ is strongly monotone for a.e. $t \in (0,T)$, i.e. there is a $m_1 > 0$ such that $\langle A(t,v) - A(t,u), v - u\rangle \ge m_1\|v-u\|^2$ for all u, $v \in V$, a.e. $t \in (0,T)$.

REMARK 18. The hypothesis $H(A)_4$ implies $H(A)_3$. Indeed, strong monotonicity clearly implies monotonicity which with hemicontinuity entails (cf. Proposition 27.6(a), p.586, of Zeidler [**116**]) pseudomonotonicity. We also recall (cf. Remark 1.1.13 of [**28**]) that for monotone operators, demicontinuity and hemicontinuity are equivalent notions.

$\underline{H(B)}$: $B\colon (0,T) \times V \to V^*$ is such that
- (i) $B(\cdot,v)$ is measurable on $(0,T)$ for all $v \in V$;
- (ii) $B(t,\cdot)$ is Lipschitz continuous for a.e. $t \in (0,T)$, i.e. $\|B(t,u) - B(t,v)\|_{V^*} \le L_B\|u-v\|$ for all u, $v \in V$, a.e. $t \in (0,T)$ with $L_B > 0$;
- (iii) $\|B(t,v)\|_{V^*} \le b_0(t) + b_1\|v\|$ for all $v \in V$, a.e. $t \in (0,T)$ with $b_0 \in L^2(0,T)$ and b_0, $b_1 \ge 0$.

REMARK 19. 1) If the condition $H(B)$(ii) holds and $B(\cdot,0) \in L^2(0,T;V^*)$, then $\|B(t,v)\|_{V^*} \leq b(t) + L_B\|v\|$ for all $v \in V$, a.e. $t \in (0,T)$, where $b(t) = \|B(t,0)\|_{V^*}$, $b \in L^2(0,T)$, $b \geq 0$. 2) If $B \in L^\infty(0,T;\mathcal{L}(V,V^*))$, the assumption $H(B)$(ii) holds.

$\underline{H(F)}$: The multifunction $F\colon (0,T) \times V \times V \to 2^{Z^*}$ has nonempty, convex, closed values and
- (i) $F(\cdot,u,v)$ is measurable on $(0,T)$ for all u, $v \in V$;
- (ii) $F(t,\cdot,\cdot)$ is upper semicontinuous from $V \times V$ into w-Z^* for a.e. $t \in (0,T)$, where $V \times V$ is endowed with $(Z \times Z)$-topology;
- (iii) $\|F(t,u,v)\|_{Z^*} \leq d_0(t) + d_1\|u\| + d_2\|v\|$ for all u, $v \in V$, a.e. $t \in (0,T)$ with $d_0 \in L^2(0,T)$ and d_0, d_1, $d_2 \geq 0$.

$\underline{H(F)_1}$: The multifunction $F\colon (0,T) \times V \times V \to 2^{Z^*}$ satisfies $H(F)$ and
- (iv) $\langle F(t,u_1,v_1) - F(t,u_2,v_2), v_1 - v_2\rangle_{Z^*\times Z} \geq -m_2\|v_1 - v_2\|^2 - m_3\|v_1 - v_2\|\|u_1 - u_2\|$ for all u_i, $v_i \in V$, $i = 1, 2$, a.e. $t \in (0,T)$ with m_2, $m_3 \geq 0$.

$\underline{(H_2)}$: $f \in \mathcal{V}^*$, $u_0 \in V$, $u_1 \in H$.

$\underline{(H_3)}$: $\alpha > 2\sqrt{3}c_e(d_1T + d_2)$, where $c_e > 0$ is the embedding constant of V into Z, i.e. $\|\cdot\|_Z \leq c_e\|\cdot\|$.

$\underline{(H_4)}$: $m_1 > m_2 + \dfrac{1}{\sqrt{2}}m_3T$.

REMARK 20. The conditions (H_3) and (H_4) give a restriction on the length of time interval T unless $d_1 = m_3 = 0$. This means that under (H_3) and (H_4), the existence and uniqueness results of Theorems 22 and 23 below are local and hold for a sufficiently small time interval. On the other hand, if the data satisfy (H_3) and (H_4) with $d_1 = m_3 = 0$, then these results are global in time. For example, we observe that if the multifunction $F(t,u,\cdot)$ is monotone for $u \in V$, a.e. $t \in (0,T)$, i.e. $\langle F(t,u,v_1) - F(t,u,v_2), v_1 - v_2\rangle_{Z^*\times Z} \geq 0$ for all u, $v_i \in V$, $i = 1, 2$, a.e. $t \in (0,T)$, then the hypothesis (H_4) clearly holds with $m_2 = m_3 = 0$ and every $m_1 > 0$.

The following observation concerns the existence of $\mathcal{Z}^*$ selections of the multifunction F which appears in Problem $\mathcal{P}$ (cf. Definition 17). It is known that a multifunction $\mathcal{F}\colon \Omega \times X \to 2^Y \setminus \{\emptyset\}$ which is measurable in $\omega \in \Omega$ and upper semicontinuous in $x \in X$ is not necessarily jointly measurable (here Ω is a measurable space, X, Y are metric spaces), cf. Example 7.2, Chapter 2 of [**44**]. In a consequence, the theorems on the existence of measurable selections of measurable multifunctions (cf. e.g. Chapter 4 of [**27**]) are not directly applicable in this case. Therefore it is not immediately clear that under the hypothesis $H(F)$ the multifunction $t \mapsto F(t,u(t),u'(t))$ has a measurable selection. The following lemma deals with this issue. We define a multifunction $G\colon W^{1,2}(0,T;V) \to 2^{\mathcal{Z}^*}$ by $G(u) = \{z \in \mathcal{Z}^* \mid z(t) \in F(t,u(t),u'(t)) \text{ a.e. on } (0,T)\}$.

LEMMA 21. *If $F\colon (0,T) \times V \times V \to 2^{Z^*}$ satisfies $H(F)$, then G has nonempty, weakly compact and convex values in $\mathcal{Z}^*$.*

First we provide a theorem on the unique solvability of the Cauchy problem for the evolution inclusion without the elasticity operator. This result will play a crucial role in the proof of the solvability of Problem $\mathcal{P}$. Consider the following problem.

PROBLEM. $\mathcal{Q}$: find $u \in \mathcal{V}$ such that $u' \in \mathcal{W}$ and

$$\begin{cases} u''(t) + A(t, u'(t)) + F(t, u(t), u'(t)) \ni f(t) \quad \text{a.e.} \quad t \in (0,T), \\ u(0) = u_0, \quad u'(0) = u_1. \end{cases}$$

THEOREM 22. *Under the hypotheses $H(A)_3$, $H(F)$, (H_2) and (H_3), Problem $\mathcal{Q}$ admits a solution. If $H(A)_4$, $H(F)_1$, (H_2), (H_3) and (H_4) hold, then the solution of Problem $\mathcal{Q}$ is unique.*

Next, we have the following existence result.

THEOREM 23. *Under the hypotheses $H(A)_4$, $H(B)$, $H(F)_1$, (H_2), (H_3), and (H_4), Problem $\mathcal{P}$ admits a unique solution.*

The idea of the proof is as follows. Let $\eta \in \mathcal{V}^*$. We consider the following auxiliary problem: find $u \in \mathcal{V}$ such that $u' \in \mathcal{W}$ and

$$\begin{cases} u''(t) + A(t, u'(t)) + F(t, u(t), u'(t)) \ni f(t) - \eta(t) \quad \text{a.e. } t \in (0,T), \\ u(0) = u_0, \quad u'(0) = u_1. \end{cases} \tag{55}$$

From Theorem 22, we know that for every $\eta \in \mathcal{V}^*$, the problem (55) has a unique solution $u_\eta \in \mathcal{V}$ such that $u'_\eta \in \mathcal{W}$. Then we consider the operator $\Lambda\colon \mathcal{V}^* \to \mathcal{V}^*$ defined by

$$(\Lambda\eta)(t) = B(t, u_\eta(t)) \quad \text{for } \eta \in \mathcal{V}^*, \text{ a.e. } t \in (0,T), \tag{56}$$

where $u_\eta \in \mathcal{V}$ is a unique solution to (55). We are able to show that the following estimate holds

$$\|(\Lambda\eta_1)(t) - (\Lambda\eta_2)(t)\|^2_{V^*} \le c \int_0^t \|\eta_1(s) - \eta_2(s)\|^2_{V^*}\, ds \tag{57}$$

for every η_1, $\eta_2 \in L^2(0,T;V^*)$, a.e. $t \in (0,T)$ with a constant $c > 0$. Hence, we deduce that Λ has a unique fixed point in $\mathcal{V}^*$, i.e. there exists a unique $\eta^* \in \mathcal{V}^*$ such that $\Lambda\eta^* = \eta^*$. Now have now all the ingredients to conclude the proof of the theorem.

Existence. Let $\eta^* \in \mathcal{V}^*$ be the fixed point of the operator Λ. We denote by u the solution of the problem (55) for $\eta = \eta^*$, i.e. $u = u_{\eta^*}$. The regularity of u follows from Theorem 22. Furthermore, since $\eta^* = \Lambda\eta^*$, we have

$$\eta^*(t) = B(t, u_{\eta^*}(t)) \quad \text{for a.e. } t \in (0,T).$$

Hence we conclude that u is a solution of Problem $\mathcal{P}$.

Uniqueness. The uniqueness of solutions of Problem $\mathcal{P}$ is a consequence of Theorem 22 and the uniqueness of the fixed point of Λ. This concludes the proof of the theorem.

In what follows we apply Theorem 23 to a problem in which the multivalued term F in Problem $\mathcal{P}$ is an operator of a subdifferential type. To this end, analogously as before, let Ω be an open bounded subset of $\mathbb{R}^d$ with Lipschitz continuous boundary Γ. We also assume that $\Gamma = \Gamma_D \cup \Gamma_N \cup \Gamma_C$ with mutually disjoint parts Γ_D,

Γ_N and Γ_C such that $m(\Gamma_D) > 0$. In order to define an operator F, we introduce first the functional $J\colon (0,T) \times L^2(\Gamma_C;\mathbb{R}^d)^2 \to \mathbb{R}$ given by

$$(58) \qquad J(t,u,z) = \int_{\Gamma_C} g(x,t,u(x),z(x))\, d\Gamma, \quad t \in (0,T),\ u,\ \ z \in L^2(\Gamma_C;\mathbb{R}^d).$$

We assume the following conditions on the integrand of J.

<u>$H(g)$</u> : $g\colon \Gamma_C \times (0,T) \times \mathbb{R}^d \times \mathbb{R}^d \to \mathbb{R}$ is a function such that

(i) $g(\cdot,\cdot,\xi,\eta)$ is measurable for all $\xi,\ \eta \in \mathbb{R}^d$ and $g(\cdot,\cdot,0,0) \in L^1(\Gamma_C \times (0,T))$;

(ii) $g(x,t,\cdot,\cdot)$ is locally Lipschitz on $\mathbb{R}^d \times \mathbb{R}^d$ for all $(x,t) \in \Gamma_C \times (0,T)$;

(iii) $\|\partial g(x,t,\xi,\eta)\|_{\mathbb{R}^d\times\mathbb{R}^d} \le c_g\,(1 + \|\xi\|_{\mathbb{R}^d} + \|\eta\|_{\mathbb{R}^d})$ for all $(x,t) \in \Gamma_C \times (0,T)$, $\xi,\ \eta \in \mathbb{R}^d$ with $c_g > 0$,
where ∂g denotes the Clarke subdifferential of g with respect to (ξ,η).

LEMMA 24. *Suppose that $H(g)$ holds. Then the functional J defined by (58) satisfies*

a) *$J(\cdot,u,z)$ is measurable for all $u,\ z \in L^2(\Gamma_C;\mathbb{R}^d)$ and $J(\cdot,0,0) \in L^1(0,T)$;*

b) *$J(t,\cdot,\cdot)$ is Lipschitz continuous on bounded subsets of $L^2(\Gamma_C;\mathbb{R}^d)^2$;*

c) *$\|\partial J(t,u,z)\|_{L^2(\Gamma_C;\mathbb{R}^d)^2} \le \sqrt{3}\,c_g\,\big(1 + \|u\|_{L^2(\Gamma_C;\mathbb{R}^d)} + \|z\|_{L^2(\Gamma_C;\mathbb{R}^d)}\big)$ for $t \in (0,T)$ and $u,\ z \in L^2(\Gamma_C;\mathbb{R}^d)$;*

d) *for $t \in (0,T)$, $u,\ z,\ v,\ w \in L^2(\Gamma_C;\mathbb{R}^d)$ we have*

$$J^0(t,u,z;v,w) \le \int_{\Gamma_C} g^0(x,t,u(x),z(x);v(x),w(x))\, d\Gamma,$$

where J^0 denotes the directional derivative of $J(t,\cdot,\cdot)$ at a point $(u,z) \in L^2(\Gamma_C;\mathbb{R}^d)^2$ in the direction $(v,w) \in L^2(\Gamma_C;\mathbb{R}^d)^2$ and g^0 stands for the directional derivative of $g(x,t,\cdot,\cdot)$.

For the proof of lemma, we refer to Migórski and Ochal [**83**].

Now we introduce the following spaces: V is a closed subspace of $H^1(\Omega;\mathbb{R}^d)$, $Z = H^\theta(\Omega;\mathbb{R}^d)$ with a fixed $\theta \in (1/2,1)$ and $H = L^2(\Omega;\mathbb{R}^d)$. Let $\gamma\colon Z \to L^2(\Gamma_C;\mathbb{R}^d)$ be the trace operator and let $\gamma^*\colon L^2(\Gamma_C;\mathbb{R}^d) \to Z^*$ stands for its adjoint. We define the multivalued mapping $F\colon (0,T) \times V \times V \to 2^{Z^*}$ by

$$(59) \qquad F(t,u,v) = S\,R^*\,\partial J(t,R(u,v)) \quad \text{for } u,v \in V,\ \ t \in (0,T),$$

where $S\colon Z^* \times Z^* \to Z^*$ and $R\colon Z \times Z \to L^2(\Gamma_C;\mathbb{R}^d)^2$ are given by $S(z_1,z_2) = z_1 + z_2$, $R(u,v) = (\gamma u, \gamma v)$. The operator $R^*\colon L^2(\Gamma_C;\mathbb{R}^d)^2 \to Z^* \times Z^*$ given by $R^*(z_1,z_2) = (\gamma^* z_1, \gamma^* z_2)$ is the adjoint of R and ∂J denotes the Clarke subdifferential of J defined by (58) with respect to (u,v).

LEMMA 25. *Under the hypotheses $H(g)$ the multivalued map F defined by (59) satisfies $H(F)$ with $d_0(t) = \sqrt{3}\,c_g\,\|\gamma\|$ and $d_1 = d_2 = \sqrt{3}\,c_e\,c_g\,\|\gamma\|^2$, where $\|\gamma\| = \|\gamma\|_{\mathcal{L}(Z;L^2(\Gamma_C;\mathbb{R}^d))}$.*

Now from Theorem 23 and Lemma 25 we obtain the following.

COROLLARY 26. Under the hypotheses $H(A)_3$, $H(B)$, $H(g)$, (H_2) and

$\underline{(H_5)}$: $$\alpha > 6\, c_e^2\, c_g (T+1) \|\gamma\|^2,$$

Problem $\mathcal{P}$ with F given by (59) admits a solution.

PROPOSITION 27. Let the multifunction F be defined by (59) with J of the form (58) with the integrand g satisfying $H(g)$. Then every solution u to Problem $\mathcal{P}$ is a solution to the following inequality problem: find $u \in \mathcal{V}$ with $u' \in \mathcal{W}$ such that

$$(60) \qquad \begin{cases} \langle u''(t) + A(t, u'(t)) + Bu(t) - f(t), v \rangle_{V^* \times V} \\ \qquad + \displaystyle\int_{\Gamma_C} g^0(x, t, u(t), u'(t); v(x), v(x))\, d\Gamma \geq 0 \\ \qquad\qquad\qquad\qquad \text{for all } v \in V, \ \text{a.e. } t \in (0, T) \\ u(0) = u_0, \ u'(0) = u_1. \end{cases}$$

From Corollary 26 and Proposition 27 we obtain the main result of this section.

THEOREM 28. *If the hypotheses $H(A)_3$, $H(B)$, $H(g)$, (H_2) and (H_5) hold, then the inequality problem (60) has at least one solution.*

Specifying the function g, we obtain a series of results for hyperbolic hemivariational inequalities with different types of subdifferential boundary conditions, i.e. the nonmonotone normal compliance condition, the simplified Coulomb friction law, the nonmonotone normal damped response condition, the viscous contact with Tresca's friction law, the viscous contact with power-law friction boundary conditions, the normal damped response with time dependent Tresca's friction, the normal damped response with power-law friction, the nonmonotone variants of the friction law of Coulomb as well as the sawtooth laws generated by nonconvex superpotentials. For details, we refer to [**83**].

We comment that a result on existence of solutions to hyperbolic hemivariational inequalities with multivalued damping operator can be found in [**24**]. For a result on the evolution inclusion with a memory term, being the generalization of Problem $\mathcal{P}$, we refer to Kulig [**47**].

6.2. Sensitivity analysis. In this section we consider the hemivariational inequality of the form

$$\begin{cases} u''(t) + A(u'(t)) + \partial J(u(t)) \ni f(t) & \text{a.e. in } (0,T) \times \Omega \\ u(0) = u^0, \ u'(0) = u^1 \text{ in } \Omega. \end{cases}$$

We provide a result on the sensitivity of the solution set for this problem. The perturbations appear in the multivalued operator A, the functional J, the right hand side f and initial conditions u^0 and u^1.

We recall the following definitions needed in the sequel. Let $\Omega \subset \mathbb{R}^N$ be an open bounded set with regular boundary. We fix p and q such that $1 < p < +\infty$ and $1/p + 1/q = 1$, and we admit:

DEFINITION 29. (cf. [**15**]) For fixed $m_i \in L^1(\Omega)$, $c_i > 0$, $i = 1, 2$, we denote

$$\mathcal{M}_\Omega(\mathbb{R}^N) = \{a \colon \Omega \times \mathbb{R}^N \to 2^{\mathbb{R}^N} \text{ such that (i) - (iii) below hold}\}$$

(i) $a(x,\xi)$ is maximal monotone with respect to ξ for all $x \in \Omega$;
(ii) a is $\mathcal{L}(\Omega) \otimes \mathcal{B}(\mathbb{R}^N) \otimes \mathcal{B}(\mathbb{R}^N)$ measurable
(i.e. $a^{-1}(C) \in \mathcal{L}(\Omega) \otimes \mathcal{B}(\mathbb{R}^N)$ for all closed set $C \subset \mathbb{R}^N$);
(iii) for every (ξ,η) with $\eta \in a(x,\xi)$ we have

$$|\eta|^q \leq m_1(x) + c_1(\eta,\xi)_{\mathbb{R}^N} \quad \text{and} \quad |\xi|^p \leq m_2(x) + c_2(\eta,\xi)_{\mathbb{R}^N}.$$

This class is large enough e.g. it contains the mappings of the form $a(x,\xi) = \partial_\xi \psi(x,\xi)$ for some $\psi\colon \Omega \times \mathbb{R}^N \to [0,+\infty)$ which is measurable in both variables, convex in ξ and satisfies $c_1 |\xi|^p \leq \psi(x,\xi) \leq c_2 |\xi|^p$.

Let $V = W_0^{1,p}(\Omega)$, $V_1 = L^p(\Omega)$, $H = L^2(\Omega)$, $V_1^* = L^q(\Omega)$, $V^* = W^{-1,q}(\Omega)$ and $\vec{V_1} = L^p(\Omega,\mathbb{R}^N) \simeq V_1^N$, $\vec{V_1}^* = L^q(\Omega,\mathbb{R}^N) \simeq V_1^{*N}$. Notice that for $p \geq 2$ we have

$$V \subset V_1 \subseteq H \subseteq V_1^* \subset V^*$$

and the first inclusion is a compact embedding. Let $Q = (0,T) \times \Omega$ and let us define the spaces $\mathcal{V} = L^p(0,T;V)$, $\mathcal{V}_1 = L^p(0,T;V_1)$, $\mathcal{H} = L^2(0,T;H) \simeq L^2(Q)$, $\mathcal{V}_1{}^* = L^q(0,T;V_1^*)$, $\mathcal{V}^* = L^q(0,T;V^*)$, $\vec{\mathcal{V}_1} = L^p(0,T;\vec{V_1})$, $\vec{\mathcal{V}_1}^* = L^q(0,T;\vec{V_1}^*)$ and

$$\mathcal{W}_{pq} = \{y \in \mathcal{V} \mid y' \in \mathcal{V}^*\}, \qquad \mathcal{Y} = \{y \in \mathcal{V} \mid y' \in \mathcal{W}_{pq}\}.$$

So $\mathcal{W}_{pq} \subset \mathcal{V} \subset \mathcal{V}_1 \subseteq \mathcal{H} \subseteq \mathcal{V}_1{}^* \subset \mathcal{V}^*$ and the embedding $\mathcal{W}_{pq}$ into $\mathcal{V}_1$ is compact.

Consider a sequence ($k \in \overline{\mathbb{N}} = \mathbb{N} \cup \{+\infty\}$) of problems

$$(HVI)_k \qquad \begin{cases} u''(t) + A_k(u'(t)) + \partial J_k(u(t)) \ni f_k(t) & \text{a.e. in } (0,T)\times\Omega \\ u(0) = u_k^0,\ u'(0) = u_k^1 \text{ in } \Omega. \end{cases}$$

We admit the hypothesis below which are satisfied uniformly in $k \in \overline{\mathbb{N}}$.

$\underline{H(A)_5}$: $A_k\colon V \to 2^{V^*} \setminus \emptyset$ correspond to functions $a_k \in \mathcal{M}_\Omega(\mathbb{R}^N)$ (i.e. $A_k v = -\mathrm{div} a_k(x,Dv) =: \{-\mathrm{div}\eta \in V^* : \eta(x) \in a(x,Dv(x))\}$) and $a_k \xrightarrow{G} a_\infty$;

$\underline{H(J)}$: $J_k\colon (0,T) \times V_1 \to \mathbb{R}$ satisfies the conditions
(i) $J_k(\cdot,z)$ is measurable for all $z \in V_1$;
(ii) $J_k(t,\cdot)$ is locally Lipschitz for a.e. $t \in (0,T)$;
(iii) $||\partial J_k(t,z)||_{V_1^*} \leq c_J(1 + ||z||_{V_1}^{2/q})$ for all $z \in V_1$, $t \in (0,T)$ with some $c_J > 0$;
(iv) $K(s - \mathcal{V}_1, w - \mathcal{V}_1^*)$–$\limsup Gr\partial J_k(t,\cdot) \subset Gr\partial J_\infty(t,\cdot)$ for a.e. t, where Gr denotes the graph of the corresponding multivalued operator.

$\underline{(H_6)}$: either $f_k,\ f_\infty \in \mathcal{V}^*$, $f_k \xrightarrow{s-\mathcal{V}^*} f_\infty$ or $f_k,\ f_\infty \in \mathcal{V}_1^*$, $f_k \xrightarrow{w-\mathcal{V}_1^*} f_\infty$
and
$y_k^0,\ y_\infty^0 \in V$, $y_k^0 \xrightarrow{w-V} y_\infty^0$, and $y_k^1,\ y_\infty^1 \in H$, $y_k^1 \xrightarrow{s-H} y_\infty^1$.

$\underline{(H_7)}$: if $p = 2$, then $\alpha > c_J\delta^2 T$, where $\alpha > 0$ is the coercivity constant for A_k and $\delta > 0$ is an embedding constant of V into V_1.

For the definition of G-convergence of multivalued operators, we refer to [**15**], and the notion and basic properties of the Kuratowski convergence of sets can be found in [**27**].

Now, the inclusion $(HVI)_k$ reads as follows

$$\begin{cases} \text{Find } y_k \in \mathcal{Y},\ \eta_k \in \vec{\mathcal{V}_1}^* \text{ and } \xi_k \in \mathcal{V}_1{}^* \text{ such that} \\ y_k''(t,x) - \mathrm{div}\eta_k(t,x) + \xi_k(t,x) = f_k(t,x) \quad \text{a.e. in } Q \\ \eta_k(t,x) \in a_k(x, Dy_k'(t,x)), \quad \xi_k(t,x) \in \partial J_k(t, y_k(t,x)) \quad \text{a.e. in } Q \\ y_k(0,x) = y_k^0(x), \quad y_k'(0,x) = y_k^1(x) \text{ in } \Omega. \end{cases}$$

For the proof of the following result we refer to Denkowski and Migórski [**23**].

THEOREM 30. *Under hypotheses* $H(A)_5$, $H(J)$, (H_6) *and* (H_7) *it holds:*

(j) *The solution set* $\mathcal{S}(A_k, J_k, f_k, y_k^0, y_k^1)$ *of* (HVI_k) *is nonempty for every* $k \in \overline{\mathbb{N}}$,

(jj) *For every sequence* $y_k \in \mathcal{S}(A_k, J_k, f_k, y_k^0, y_k^1)$ *there exists a subsequence* $\{y_{k_n}\}$ *such that* $y_{k_n} \to y_\infty$ *weakly in* $\mathcal{V}$, $y_{k_n}' \to y_\infty'$ *weakly in* $\mathcal{W}$, *as* $n \to \infty$ *and* $y_\infty \in \mathcal{S}(A_\infty, J_\infty, f_\infty, y_\infty^0, y_\infty^1)$.

6.3. Quasistatic hemivariational inequalities. In this section we are concerned with the asymptotic behavior of a sequence of solutions to second order hemivariational inequalities as the power of the acceleration forces vanishes. We formulate an abstract result and comment on its physical motivations and applications.

Let V and Z be reflexive, separable Banach spaces, and let H be a Hilbert space. Suppose that $V \subset Z \subset H \approx H^* \subset Z^* \subset V^*$, where H^*, Z^* and V^* denote dual spaces to H, Z and V, respectively, all embeddings are continuous and V is compactly embedded in Z.

Consider the following evolution hemivariational inequality of the form

$$\begin{cases} \varepsilon u''(t) + A(t)u'(t) + Bu(t) + M^* \partial J(t, Mu(t)) \ni f(t) \quad \text{a.e. } t \in (0,T) \\ u(0) = u_0, \quad \sqrt{\varepsilon} u'(0) = u_1, \end{cases} \tag{61}$$

where $\varepsilon > 0$. In addition to our previous notation, let X be a Banach space. We assume the following hypotheses.

$\underline{H(A)_6}$: $A \in L^\infty(0,T; \mathcal{L}(V,V^*))$ is an operator such that $A(t)$ is coercive, i.e. there is a constant $\alpha > 0$ such that for a.e. $t \in (0,T)$, $\langle A(t)v, v\rangle \geq \alpha \|v\|_V^p$ for all $v \in V$.

$\underline{H(B)_1}$: $B \in \mathcal{L}(V,V^*)$ is monotone (nonnegative) and symmetric.

$\underline{H(J)_1}$: $J\colon (0,T) \times X \to R$ is a function such that

(i) $J(\cdot, x)$ is measurable on $(0,T)$ for all $x \in X$;

(ii) $J(t,\cdot)$ is locally Lipschitz on X for a.e. $t \in (0,T)$;

(iii) $\|\partial J(t,x)\|_{X^*} \leq c(1 + \|x\|_X^{2/q})$ for all $x \in X$, a.e. $t \in (0,T)$ with $c > 0$.

$\underline{H(M)}$: $M \in \mathcal{L}(Z,X)$.

$\underline{(H_8)}$: if $p = 2$, then $\alpha > 2\,c\,T\,c_e^2\,\|M\|\,\max\{1, \|M\|\}$, where $c_e > 0$ is an embedding constant of V into Z and $\|M\| = \|M\|_{\mathcal{L}(Z,X)}$.

The crucial step in the proof of the existence result for (61) is to derive the following uniform estimate.

LEMMA 31. *Let $\varepsilon > 0$ be fixed. Assume hypotheses $H(A)_6$, $H(B)_1$, $H(J)_1$, $H(M)$ and (H_2), and let u be a solution to (61). If $p > 2$, then there exists a constant $C > 0$ independent of ε such that*

$$\|u\|_{C(0,T;V)} + \|u'\|_{\mathcal{V}} + \sqrt{\varepsilon}\|u'\|_{L^\infty(0,T;H)} + \varepsilon\|u''\|_{\mathcal{V}^*} \\ \leq C\left(1 + \|u_0\|_V^{2/q} + |u_1|_H^{2/q} + \|f\|_{\mathcal{V}^*}^{2/q}\right).$$

Moreover, this estimate still holds for $p = 2$ provided (H_8) is satisfied.

The following is the existence result for a hemivariational inequality (61).

THEOREM 32. *If hypotheses $H(A)_6$, $H(B)_1$, $H(J)_1$, $H(M)$, (H_2) and (H_8) hold, then for every fixed $\varepsilon > 0$ the problem (61) admits at least one solution.*

THEOREM 33. *Let the hypotheses $H(A)_6$, $H(B)_1$, $H(J)_1$, $H(M)$, (H_2) and (H_8) hold. For every ε fixed, let u_ε be a solution to (61) given by Theorem 32. Then there exists u fulfilling*

$$u \in L^\infty(0,T;V), \quad u' \in \mathcal{W}$$

such that the following convergences hold

$$\begin{aligned} &u_\varepsilon \to u \quad \text{weakly* in} \quad L^\infty(0,T;V), \\ &u_\varepsilon' \to u' \quad \text{weakly in} \quad \mathcal{V}, \\ &\sqrt{\varepsilon}u_\varepsilon' \to 0 \quad \text{weakly* in} \quad L^\infty(0,T;H), \\ &\varepsilon u_\varepsilon'' \to 0 \quad \text{weakly in} \quad \mathcal{V}^*, \end{aligned}$$

as $\varepsilon \to 0$. Moreover, the limit function u is a solution to the following problem

$$\begin{cases} A(t)u'(t) + Bu(t) + M^*\partial J(t, Mu(t)) \ni f(t) \quad \text{a.e. } t \in (0,T) \\ u(0) = u_0. \end{cases} \tag{62}$$

We recall that a function $u \in L^\infty(0,T;V)$ is called a solution to (62) if and only if $u' \in \mathcal{V}$ and there exists $\eta \in \mathcal{Z}^*$ such that

$$\begin{cases} A(t)u'(t) + Bu(t) + \eta(t) = f(t) \quad \text{a.e. } t \in (0,T) \\ \eta(t) \in M^*\partial J(t, Mu(t)) \quad \text{a.e. } t \in (0,T) \\ u(0) = u_0. \end{cases}$$

The study of inclusions of type (62) is motivated by several contact problems of solid mechanics. It is well known that the dynamic equations of motion, representing momentum conservation, that govern the evolution of the state of the body, are of the form $u_i'' - \sigma_{ij,j} = f_i$, where u is a displacement, σ is the stress tensor and f is the density (per unit volume) of applied forces. In many cases we are interested in situations in which the system configuration and the external forces and tractions evolve slowly in time in such a way that the accelerations in the system are rather small and negligible, so that the inertial (the second order time derivative) terms can be neglected. In such a way, we obtain the quasistatic approximation (equilibrium equations) for the equations of motion, $-\operatorname{Div}\sigma = f$, where Div is the divergence operator. In this approximation, at each time instant the system is in equilibrium, and the external forces are balanced by the internal stresses. Rigorous mathematical treatment of quasistatic problems is recent. The rapid growth of the theory of quasistatic contact models can be seen from several monographs and many papers dedicted to such phenomena, cf. Duvaut and Lions [**29**], Han and Sofonea [**38**], Shillor et al. [**112**], Sofonea et al. [**113**] and the references therein. When we assume that the process is slow and the accelerations are negligible, mathematically it means that the system changes character, from being of a hyperbolic type to an elliptic or a parabolic type. To our knowledge quasistatic hemivariational inequalities have not been studied in the literature so far and a recent existence result for such problems is provided in Migórski and Ochal [**89**]. We mention that quasistatic problems modeled by variational inequalities with strongly monotone and Lipschitz operators were considered in [**38, 112, 113**]. We also point out that the result of Theorem 33 can be applicable to several challenging and important quasistatic models of contact phenomena. We refer to [**89**] for the application to a viscoelastic contact model governed by a quasistatic hemivariational inequality. Finally, we remark that the question of uniqueness of solutions to both dynamical and quasistatic hemivariational inequality is left open. It would be also interesting to extend the result of Theorem 33 to a class of problems with nonlinear operators $A(t)$ and B.

6.4. Vanishing viscosity for hemivariational inequalities. In this section we present another result on the behavior of solutions when the viscosity term in a hemivariational inequality tends to zero. We start with the physical setting of a model which will be also needed in the next section.

We consider a linearly elastic body which occupies the reference configuration $\Omega \subset \mathbb{R}^d$, $d = 2, 3$. We suppose that Ω is a bounded domain with Lipschitz boundary Γ and Γ is divided into two disjoint measurable parts Γ_D and Γ_N such that $meas(\Gamma_D) > 0$. We put $Q = \Omega \times (0, T)$, where $(0, T)$ with $T < +\infty$ denotes a time interval. The body is clamped along Γ_D, so the displacement field vanishes there. We denote by ψ the density of surface tractions on Γ_N. We suppose that the nonmonotone skin effects (e.g. skin friction, adhesive forces etc.) appear in Ω (cf. [**107, 86**]). In order to describe such effects we assume that the volume forces φ consist of two parts: φ_2 is given and φ_1 is the reaction of the constraint introducing the skin effects. So we may write $\varphi = \varphi_1 + \varphi_2$, where φ_2 is the prescribed external loading and φ_1 is a possibly multivalued function of the displacement

$$-\varphi_1(x,t) \in \partial j(x,t,u(x,t)) \quad \text{in } \Omega$$

and ∂j is the Clarke generalized gradient of a given function $j(x,t,\cdot)$.

We denote by $u = (u_1, \ldots, u_d)$ the displacement vector, by $\sigma = (\sigma_{ij})$ the stress tensor and by $\varepsilon(u) = (\varepsilon_{ij}(u))$ the linearized (small) strain tensor, where $i, j = 1, \ldots, d$. We suppose the Hooke elastic constitutive relation $\sigma_{ij}(t) = C_{ijkl}\varepsilon_{kl}(u(t))$ in Q, where $\{C_{ijkl}\}$, i, j, k, $l = 1, \ldots, d$ is the elasticity tensor. Denoting by u_0 and u_1 the initial displacement and the initial velocity, respectively, the classical formulation of the problem is as follows: find a displacement field $u\colon Q \to \mathbb{R}^d$ such that

$$
(63) \qquad \begin{cases} u''(t) - \operatorname{div} \sigma(t) = \varphi(t) & \text{in } Q; \\ \sigma(t) = \mathcal{C}\varepsilon(u(t)) & \text{in } Q; \\ \varphi(t) = \varphi_1(t) + \varphi_2(t) & \text{in } Q; \\ u = 0 & \text{on } \Gamma_D \times (0,T); \\ \sigma(t)\nu = \psi(t) & \text{on } \Gamma_N \times (0,T); \\ -\varphi_1(t) \in \partial j(t, u(t)) & \text{in } Q; \\ u(0) = u_0, \quad u'(0) = u_1 & \text{in } \Omega. \end{cases}
$$

We set now the problem (63) in a variational form. Let $H = L^2(\Omega; \mathbb{R}^d)$, $H_1 = H^1(\Omega; \mathbb{R}^d)$ and $\mathcal{H} = L^2(\Omega; \mathcal{S}_d)$. We denote by V the closed subspace of H_1 defined by $V = \{ v \in H_1 \mid v = 0 \quad \text{on } \Gamma_D \}$. Identifying H with its dual, we have an evolution triple of spaces $V \subset H \subset V^*$ with dense, continuous and compact embeddings (cf. [**116, 28**]). We need the following spaces $\mathcal{V} = L^2(0,T;V)$, $\widehat{\mathcal{H}} = L^2(0,T;H)$ and $\mathcal{W} = \{ w \in \mathcal{V} \mid w' \in \mathcal{V}^* \}$, where the time derivative involved in the definition of $\mathcal{W}$ is understood in the sense of vector valued distributions.

We suppose the regularity for the body forces and surface tractions

$$
\varphi_2 \in L^2(0,T;H), \quad \psi \in L^2(0,T;H^{1/2}(\Gamma_N;\mathbb{R}^d))
$$

and we define $g \in \widehat{\mathcal{H}}^*$ by

$$
(64) \qquad \langle g(t), h \rangle_{H^* \times H} = \langle \varphi_2(t), h \rangle_H + \langle \psi(t), h \rangle_\Gamma
$$

for all $h \in H$ and a.e. $t \in (0,T)$. Let $v \in V$. Using the equation of motion in (63) and the Green formula (assuming the smoothness of the functions involved), we have

$$
\langle u''(t), v \rangle_{V^* \times V} + \langle \sigma(t), \varepsilon(v) \rangle_{\mathcal{H}} = \langle \varphi_1(t), v \rangle_{V^* \times V} + \langle \varphi_2(t), v \rangle_{V^* \times V} + \int_\Gamma \sigma(t)\nu \cdot v \, d\Gamma
$$

for $v \in V$ and a.e. $t \in (0,T)$. Taking into account the boundary conditions, we get

$$
\int_\Gamma \sigma(t)\nu \cdot v \, d\Gamma = \langle \psi_1(t), v \rangle_\Gamma .
$$

From the definition of the subdifferential and the multivalued relation in (63), we obtain

$$
\langle \varphi_1(t), v \rangle_{V^* \times V} \geq - \int_\Omega j^0(x, t, u(x,t); v(x)) \, dx
$$

for $v \in V$ and a.e. $t \in (0,T)$. Hence we arrive to the following variational formulation of (63) which is known as an interior hemivariational inequality: find $u \in C(0,T;V) \cap C^1(0,T;H)$ with $u'' \in \mathcal{V}^*$ such that

$$
(65) \qquad \begin{cases} \langle u''(t), v \rangle_{V^* \times V} + \langle \mathcal{C}\varepsilon(u(t)), \varepsilon(v) \rangle_{\mathcal{H}} + \displaystyle\int_\Omega j^0(x, t, u(t); v) \, dx \\ \qquad\qquad \geq \langle g(t), v \rangle_{H^* \times H} \quad \text{for all } v \in V, \text{ a.e. } t \in (0,T) \\ u(0) = u_0, \quad u'(0) = u_1. \end{cases}
$$

We need the following hypotheses on the data.

$\underline{H(\mathcal{C})}$: The elasticity operator $\mathcal{C}\colon \Omega \times \mathcal{S}_d \to \mathcal{S}_d$ is of the form $\mathcal{C}(x,\varepsilon) = \mathbb{E}(x)\varepsilon$ (the Hooke law) with a symmetric and coercive elasticity tensor $\mathbb{E}$, i.e. $\mathbb{E} = (C_{ijkl})$, $i,j,k,l = 1,\dots,d$ with $C_{ijkl} \in L^\infty(\Omega)$, $C_{ijkl} = C_{jikl} = C_{lkij}$ and $C_{ijkl}(x)\chi_{ij}\chi_{kl} \geq \alpha\chi_{ij}\chi_{ij}$ for all symmetric tensors $\chi = (\chi_{ij})$ and for a.e. $x \in \Omega$ with $\alpha > 0$.

$\underline{H(j)_2}$: $j\colon \Omega \times (0,T) \times \mathbb{R}^d \to \mathbb{R}$ is a function such that

(i) $j(\cdot,\cdot,\xi)$ is measurable for all $\xi \in \mathbb{R}^d$;

(ii) $j(x,t,\cdot)$ is locally Lipschitz for all $(x,t) \in Q$;

(iii) $\|\partial j(x,t,\xi)\|_{\mathbb{R}^d} \leq c\,(1 + \|\xi\|_{\mathbb{R}^d})$ for all $(x,t) \in Q$, $\xi \in \mathbb{R}^d$ with $c > 0$.

$\underline{H(g)_1}$: $\psi \in L^2(0,T;H^{1/2}(\Gamma_N;\mathbb{R}^d))$, $\varphi_2 \in L^2(0,T;H)$, $u_0 \in V$ and $u_1 \in H$.

We associate with the hemivariational inequality (65) a second order evolution inclusion. Consider the operator $A\colon V \to V^*$ and the functional $J\colon (0,T) \times H \to \mathbb{R}$ given by

(66) $$\langle Au, v\rangle_{V^*\times V} = (\mathcal{C}(x,\varepsilon(u)),\varepsilon(v))_{\mathcal{H}} \quad \text{for } u,v \in V \text{ and } t \in (0,T),$$

(67) $$J(t,v) = \int_\Omega j(x,t,v(x))\,dx \quad \text{for } v \in H \text{ and } t \in (0,T).$$

Their properties are given in the following two lemmata.

LEMMA 34. *Under the assumption $H(\mathcal{C})$ the operator A defined by (66) satisfies*

$\underline{H(A)_7}$: *$A\colon V \to V^*$ is a bounded, linear, coercive and symmetric operator, (i.e. $A \in \mathcal{L}(V,V^*)$, $\langle Av,v\rangle_{V^*\times V} \geq \alpha\|v\|^2$, $\langle Av,w\rangle_{V^*\times V} = \langle Aw,v\rangle_{V^*\times V}$ for all v, $w \in V$).*

LEMMA 35. *Under the assumption $H(j)_2$ the functional J defined by (67) satisfies*

$\underline{H(J)_2}$: *$J\colon (0,T) \times H \to \mathbb{R}$ is a functional such that*

(i) *$J(\cdot,v)$ is measurable for all $v \in H$;*

(ii) *$J(t,\cdot)$ is well defined and Lipschitz continuous on bounded subsets of H;*

(iii) *$|\partial J(t,u)| \leq \widetilde{c}\,(1 + |u|)$ for $t \in (0,T)$ and $u \in H$ with $\widetilde{c} > 0$;*

(iv) *for $t \in (0,T)$, u, $v \in H$, we have*

$$J^0(t,u;v) \leq \int_\Omega j^0(x,t,u(x);v(x))\,dx,$$

where $\partial J(t,u)$ denotes the Clarke subdifferential of $J(t,\cdot)$ at a point $u \in H$ and $J^0(t,u;v)$ stands for the directional derivative of $J(t,\cdot)$ at a point $u \in H$ in the direction $v \in H$.

We also observe that if g is defined by (64) and $H(g)_1$ is satisfied, then the following condition holds

$\underline{(H_9)}$: $g \in \widehat{\mathcal{H}}^*$, $\quad u_0 \in V$, $\quad u_1 \in H$.

We associate with the hemivariational inequality (65) the following evolution inclusion: find $u \in C(0,T;V) \cap C^1(0,T;H)$ such that $u'' \in \mathcal{V}^*$ and

$$(68)\qquad \begin{cases} u''(t) + Au(t) + \partial J(t, u(t)) \ni g(t) & \text{a.e. } t \in (0,T) \\ u(0) = u_0, \ u'(0) = u_1. \end{cases}$$

PROPOSITION 36. If $H(\mathcal{C})$, $H(J)_2$, (H_9) hold and the operator A is defined by (66), then every solution to the problem (68) is a solution to the hemivariational inequality (65).

In view of Proposition 36, in order to provide an existence result for (65), it is enough to establish the existence of solutions to (68). To this end we consider the following Cauchy problem for second order evolution inclusion

$$(69)\qquad \begin{cases} u''(t) + Au(t) \in F(t, u(t)) & \text{a.e. } t \in (0,T) \\ u(0) = u_0, \ u'(0) = u_1. \end{cases}$$

A function $u \in C(0,T;V) \cap C^1(0,T;H)$ is called a solution to the problem (69) if and only if $u'' \in \mathcal{V}^*$, $u''(t) + Au(t) = f(t)$ a.e. $t \in (0,T)$, $u(0) = u_0$, $u'(0) = u_1$ and $f(t) \in F(t, u(t))$ a.e. $t \in (0,T)$.

The following hypothesis is needed.

$\underline{H(F)_2}$: $F\colon (0,T) \times H \to \mathcal{P}_{fc}(H)$ is a multifunction such that
- (i) $F(\cdot, v)$ is measurable for all $v \in H$;
- (ii) $Gr\, F(t, \cdot)$ is sequentially closed in $H \times w\text{-}H$, a.e. $t \in (0,T)$;
- (iii) $|F(t,v)| \le a(t) + b|v|$ for a.e. $t \in (0,T)$, where $a \in L^2_+(0,T)$ and $b > 0$,

where $\mathcal{P}_{fc}(H)$ denotes the family of nonempty, closed and convex subsets of H and $w\text{-}H$ stands for the space H endowed with the weak topology.

THEOREM 37. *If the hypotheses $H(A)_7$, $H(F)_2$ hold, $u_0 \in V$ and $u_1 \in H$, then the problem (69) admits a solution.*

Before we give a proof of Theorem 37, we formulate auxiliary results. We will use the "parabolic regularization" method, cf. Lions [**49**] and Migórski and Ochal [**86**]. For every $\varepsilon > 0$, we consider the following regularized problem:

$$(70)\qquad \begin{cases} \text{find } u_\varepsilon \in \mathcal{V} \text{ such that } u'_\varepsilon \in \mathcal{W} \text{ and} \\ u''_\varepsilon(t) + \varepsilon A u'_\varepsilon(t) + A u_\varepsilon(t) \in F(t, u_\varepsilon(t)) & \text{a.e. } t \in (0,T) \\ u_\varepsilon(0) = u_0, \ u'_\varepsilon(0) = u_1. \end{cases}$$

It is well known that $\{v \in \mathcal{V} \mid v' \in \mathcal{V}\} \subset C(0,T;V)$ and $\mathcal{W} \subset C(0,T;H)$ continuously so the initial conditions in (70) make sense for $u_0 \in V$ and $u_1 \in H$.

We establish the a priori estimate for the solutions of (70).

PROPOSITION 38. Suppose the hypotheses of Theorem 37. If u_ε is a solution to (70), then for every $\varepsilon > 0$ we have

$$(71)\qquad \|u_\varepsilon\|_{C(0,T;V)} + \|u'_\varepsilon\|_{C(0,T;H)} + \|u''_\varepsilon\|_{\mathcal{V}^*} + \sqrt{\varepsilon}\|u'_\varepsilon\|_{\mathcal{V}} \le C\,(1 + \|u_0\| + |u_1|)$$

with a positive constant C.

Next, following the lines of the proof of Theorem 3.1 in Migórski [**65**], we obtain

PROPOSITION 39. Under the hypotheses of Theorem 37, for every fixed $\varepsilon > 0$, the problem (70) admits a solution.

Now we prove Theorem 37. Let u_ε be a solution to (70). We have

$$u''_\varepsilon + \varepsilon \mathcal{A} u'_\varepsilon + \mathcal{A} u_\varepsilon = f_\varepsilon, \tag{72}$$

$$f_\varepsilon(t) \in F(t, u_\varepsilon(t)) \quad \text{for a.e. } t \in (0,T), \tag{73}$$

$$u_\varepsilon(0) = u_0, \quad u'_\varepsilon(0) = u_1, \tag{74}$$

where $\mathcal{A}\colon \mathcal{V} \to \mathcal{V}^*$ is the Nemitsky operator corresponding to A, i.e. $(\mathcal{A}v)(t) = Av(t)$ for $v \in \mathcal{V}$. By Proposition 38 we know that for any $\varepsilon > 0$ we have

$$\|u_\varepsilon\|_{C(0,T;V)} + \|u'_\varepsilon\|_{C(0,T;H)} + \|u''_\varepsilon\|_{\mathcal{V}^*} + \sqrt{\varepsilon}\|u'_\varepsilon\|_{\mathcal{V}} \le C\left(1 + \|u_0\| + |u_1|\right),$$

where $C > 0$ is independent of ε. Passing to a subsequence, we may assume that as $\varepsilon \to 0$

$$u_\varepsilon \to u \quad \text{weakly-}*\text{-}L^\infty(0,T;V), \tag{75}$$

$$u'_\varepsilon \to u' \quad \text{weakly-}*\text{-}L^\infty(0,T;H), \tag{76}$$

$$u''_\varepsilon \to u'' \quad \text{weakly in } \mathcal{V}^*, \tag{77}$$

$$\varepsilon u'_\varepsilon \to 0 \quad \text{in } \mathcal{V}. \tag{78}$$

Since $\mathcal{A}$ is (by $H(A)_7$) linear and continuous, from (75), we obtain

$$\mathcal{A}u_\varepsilon \to \mathcal{A}u \quad \text{weakly in } \mathcal{V}^*, \tag{79}$$

while (78) gives

$$\|\varepsilon \mathcal{A} u'_\varepsilon\|_{\mathcal{V}^*} \le c\|\varepsilon u'_\varepsilon\|_{\mathcal{V}} \to 0$$

which implies

$$\varepsilon \mathcal{A} u'_\varepsilon \to 0 \quad \text{in } \mathcal{V}^*. \tag{80}$$

Using the compactness of the embedding $\mathcal{W} \subset \mathcal{H}$, by (75) and (76), we have $u_\varepsilon \to u$ in $\mathcal{H}$ and hence

$$u_\varepsilon(t) \to u(t) \quad \text{in } H \text{ for a.e. } t \in (0,T). \tag{81}$$

From the following estimate

$$\|f_\varepsilon\|_{\mathcal{H}} \le C\left(1 + \|u_0\| + |u_1|\right),$$

we may suppose that

$$f_\varepsilon \to f \quad \text{weakly in } \mathcal{H}. \tag{82}$$

Exploiting (77), (79), (80) and (82), from (72) we deduce $u'' + \mathcal{A}u = f$. Next using the convergences (81) and (82), and passing to the limit in (73), due to $H(F)$ and the Convergence Theorem of Aubin and Cellina [**1**], we have $f(t) \in F(t, u(t))$ for a.e. $t \in (0,T)$. Since the embeddings $\mathcal{W} \subset C(0,T;H)$ and $\{v \in L^\infty(0,T;H) \mid v' \in L^2(0,T;V^*)\} \subset C(0,T;V^*)$ are continuous, from (75)–(77), we deduce

$$u_\varepsilon(0) \to u(0) \quad \text{weakly in } H,$$
$$u'_\varepsilon(0) \to u'(0) \quad \text{weakly in } V^*$$

and (74) implies $u(0) = u_0$ and $u'(0) = u_1$.

From (75)–(77) we know that

$$u \in L^\infty(0,T;V), \quad u' \in L^\infty(0,T;H) \text{ and } u'' \in \mathcal{V}^*. \tag{83}$$

Hence, by using the embeddings $\{v \in \mathcal{V} \mid v' \in \mathcal{V}\} \subset C(0,T;V)$ and $\mathcal{W} \subset C(0,T;H)$, we get

$$u \in C(0,T;H), \quad u' \in C(0,T;V^*). \tag{84}$$

In order to conclude the proof that u is a solution to (69), we need to show that u is more regular, i.e.

$$u \in C(0,T;V) \cap C^1(0,T;H). \tag{85}$$

To this end we employ the following result: if X and Y are two Banach spaces, X is reflexive and the embedding $X \subset Y$ is dense and continuous, then

$$L^\infty(0,T;X) \cap C_s(0,T;Y) = C_s(0,T;X),$$

where $C_s(0,T;X) = \{u \in L^\infty(0,T;X) \mid t \mapsto \langle u^*, u(t)\rangle_{X^* \times X}$ is continuous on $[0,T]$, for all $u^* \in X^*\}$ is the space of scalarly continuous functions from $[0,T]$ to X. From this result, (83) and (84), we deduce

$$u \in C_s(0,T;V), \quad u' \in C_s(0,T;H). \tag{86}$$

Next, since $u''(t) + Au(t) = f(t)$ and $f(t) \in F(t, u(t))$ for a.e. $t \in (0,T)$, similarly as in the proof of Proposition 38, we obtain

$$\int_0^t \langle u''(s), u'(s)\rangle_{V^* \times V}\, ds + \int_0^t \langle Au(s), u'(s)\rangle_{V^* \times V}\, ds = \int_0^t \langle f(s), u'(s)\rangle_H\, ds$$

and

$$\frac{1}{2}|u'(t)|^2 - \frac{1}{2}|u_1|^2 + \frac{1}{2}\langle Au(t), u(t)\rangle_{V^* \times V} - \frac{1}{2}\langle Au_0, u_0\rangle_{V^* \times V} = \int_0^t \langle f(s), u'(s)\rangle_H\, ds$$

for all $t \in [0,T]$. We introduce the energy function $E\colon [0,T] \to \mathbb{R}$ by $E(t) = \frac{1}{2}|u'(t)|^2 + \frac{1}{2}\langle Au(t), u(t)\rangle_{V^* \times V}$. Thus for all $t \in [0,T]$ we have $E(t) - E(0) = \int_0^t \langle f(s), u'(s)\rangle_H\, ds$, which implies the continuity of $E(\cdot)$ on $[0,T]$.

Let t_n, $t \in [0,T]$, $t_n \to t$ as $n \to \infty$ and let

$$\xi_n = \frac{1}{2}|u'(t_n) - u'(t)|^2 + \frac{1}{2}\langle Au(t_n) - Au(t), u(t_n) - u(t)\rangle_{V^* \times V}.$$

The symmetry of A and the routine calculations yield

$$\xi_n = E(t_n) + E(t) - \langle Au(t), u(t_n)\rangle_{V^* \times V} - \langle u'(t_n), u'(t)\rangle_H.$$

Consequently the continuity of $E(\cdot)$ and (86) entails

$$\lim_n \xi_n = 2E(t) - \langle Au(t), u(t)\rangle_{V^* \times V} - |u'(t)|^2 = 0.$$

On the other hand

$$\frac{1}{2}|u'(t_n) - u'(t)|^2 + \frac{\alpha}{2}\|u(t_n) - u(t)\|^2 \leq \xi_n \to 0$$

which implies $u \in C(0,T;V)$ and $u' \in C(0,T;H)$. Hence (85) follows and completes the proof of the theorem. □

We conclude this section with the following

THEOREM 40. *Under the hypotheses $H(\mathcal{C})$, $H(j)_2$ and $H(g)_1$, the hemivariational inequality (65) admits a solution.*

PROOF. It is known that $H(A)_7$, $H(J)_2$ and (H_9) hold. We verify that the multifunction $F\colon (0,T) \times H \to \mathcal{P}_{fc}(H)$ given by $F(t,v) = -\partial J(t,v) + f(t)$ satisfies $H(F)_2$. The condition $H(F)_2(i)$ follows from $H(J)_2(i)$. Since the subdifferential has nonempty, w-compact and convex values, the multifunction F is $\mathcal{P}_{fc}(H)$-valued. It is also known (see Proposition 5.6.10 of [**28**]) that $\partial J(t,\cdot)$ is usc from H into w-H. Hence (cf. Proposition 4.1.14 of [**28**]) F satisfies $H(F)_2(ii)$. The condition $H(F)_2(iii)$ follows from $H(J)_2(iii)$.

We can apply Theorem 37 to deduce that the evolution inclusion (68) has a solution. Finally, using Proposition 36 we have that the hemivariational inequality (65) admits a solution. The proof is completed. □

6.5. Hyperbolic hemivariational inequalities with impulses. In this section we consider hyperbolic hemivariational inequalities with impulses. As we already have observed such problems are associated with second order evolution inclusions with impulses considered in the framework of an infinite dimensional evolution triple of spaces.

We start with the following notation. Given $I = [0,T]$, $0 < T < +\infty$, $0 \leq \tau_1 < \tau_2 \leq T$ and an evolution triple (V, H, V^*), we define

$$W(\tau_1, \tau_2) = \{v \in L^2(\tau_1, \tau_2; V) \mid \dot{v} \in L^2(\tau_1, \tau_2; V^*)\},$$

where the time derivative is understood in the sense of vector-valued distributions. Endowed with the norm $\|v\|_{W(\tau_1,\tau_2)} = \|v\|_{L^2(\tau_1,\tau_2;V)} + \|\dot{v}\|_{L^2(\tau_1,\tau_2;V^*)}$, the space $W(\tau_1,\tau_2)$ becomes a separable, reflexive Banach space. It is well known (cf. [**28**]) that the space $W(\tau_1,\tau_2)$ is embedded continuously in $C(\tau_1,\tau_2;H)$ (the space of continuous functions on $[\tau_1,\tau_2]$ with values in H), i.e. every element of $W(\tau_1,\tau_2)$, after a possible modification on a set of measure zero, has a unique continuous representative in $C(\tau_1, \tau_2; H)$. Moreover, since V is embedded compactly in H, then so does $W(\tau_1, \tau_2)$ into $L^2(\tau_1, \tau_2; H)$ (cf. [**28**]). We denote by $\|\cdot\|$, $|\cdot|$ and $\|\cdot\|_{V^*}$ the norms in V, H and V^*, respectively. The duality brackets for the pair (V, V^*) is denoted by $\langle \cdot, \cdot \rangle$.

Let $D = \{t_1, \dots, t_m\}$ be a finite set of points such that

$$0 = t_0 < t_1 < t_2 < \dots < t_m < t_{m+1} = T.$$

The elements of D are called impulsive points. We need the following space of piecewise continuous functions

$$\begin{aligned} PC(I; V) = \{v \colon I \to V \text{ such that } v \text{ is continuous at } t \in I \setminus D, v \text{ is left} \\ \text{continuous at } t \in D, \text{the right limits } v(t_i^+) \text{ exist for} \\ i = 1, \dots, m\}. \end{aligned}$$

Evidently, $PC(I; V)$ is a Banach space with norm $\|v\|_{PC(I;V)} = \sup_{t \in I} \|v(t)\|$. Analogously we define the space $PC(I; H)$ furnished with the supremum norm.

Let $\sigma_i = (t_i, t_{i+1})$ for $i = 0, 1, \dots, m$. We define

$$PW(I) = \{v \colon I \to V \text{ such that } v|_{\sigma_i} \in W(\sigma_i) \text{ for } i = 0, \dots, m\}$$

which becomes a Banach space with norm $\|v\|_{PW(I)} = \sum_{i=0}^{m} \|v|_{\sigma_i}\|_{W(\sigma_i)}$.

We also denote $PC^1(I; H) = \{v \in PC(I; H) \mid \dot{v} \in PC(I; H)\}$.

Now, we consider a mechanical contact problem involving a nonmonotone multivalued term. We formulate this problem as a hyperbolic hemivariational inequality with impulses and then we associate with the latter an impulsive evolution inclusion.

Let Ω be an open bounded subset of $\mathbb{R}^d$ with Lipschitz continuous boundary $\Gamma = \partial\Omega$ and $Q = \Omega \times (I \setminus D)$. We consider the following problem.

$$\frac{\partial^2 u}{\partial t^2}(t) - \operatorname{div} \sigma(t) = \varphi(t) \quad \text{in } Q \tag{87}$$

$$\sigma(t) = \mathcal{C}\varepsilon\Big(\frac{\partial u}{\partial t}(t)\Big) + \mathcal{G}\varepsilon(u(t)) \quad \text{in } Q \tag{88}$$

$$\varphi(x,t) = \varphi_1(x,t) + \varphi_2(x,t) \quad \text{in } Q \tag{89}$$

$$u = 0 \quad \text{on } \Gamma_D \times (I \setminus D) \tag{90}$$

$$\sigma(t)\nu = \psi(t) \quad \text{on } \Gamma_N \times (I \setminus D) \tag{91}$$

$$-\varphi_1(x,t) \in \partial j(x, t, u(x,t)) \quad \text{in } Q \tag{92}$$

$$u(0) = u_0, \quad \frac{\partial u}{\partial t}(0) = v_0 \quad \text{in } \Omega \tag{93}$$

$$u(x, t_i^+) \in u(x, t_i^-) + B_{\mathbb{R}}\Big(0, g_i\Big(|u(x, t_i^-)|, |\frac{\partial u}{\partial t}(x, t_i^-)|\Big)\Big) \tag{94}$$

$$\frac{\partial u}{\partial t}(x, t_i^+) \in \frac{\partial u}{\partial t}(x, t_i^-) + B_{\mathbb{R}}\Big(0, h_i\Big(|u(x, t_i^-)|, |\frac{\partial u}{\partial t}(x, t_i^-)|\Big)\Big), \quad i = 1, \dots, m. \tag{95}$$

The system (87)–(95) serves as a mathematical model for a contact problem in viscoelasticity with nonmonotone and nonconvex superpotential laws. More precisely, the reference configuration set $\Omega \subset \mathbb{R}^d$, $d = 2, 3$, is occupied by a viscoelastic body. We suppose that the boundary Γ is divided into two disjoint measurable parts Γ_D and Γ_N such that $meas(\Gamma_D) > 0$. The body is clamped along Γ_D, so the displacement field vanishes there. We denote by ψ the density of surface tractions on Γ_N. We suppose that the nonmonotone skin effects (e.g. skin friction, adhesive forces etc.) appear in Ω (cf. [**107, 85**]). In order to describe such effects we assume that the volume forces φ consist of two parts: φ_2 is given and φ_1 is the reaction of the constraint introducing the skin effects. So we may write $\varphi = \varphi_1 + \varphi_2$ (cf. (89)), where φ_2 is the prescribed external loading and φ_1 is a possibly multivalued function of the displacement satisfying (92), where ∂j is the Clarke generalized gradient of a given function $j(x, t, \cdot)$. The displacement field is denoted by $u = (u_1, \dots, u_d)$, σ stands for the stress tensor and ε denotes the linearized (small) strain tensor. The relation (87) is a dynamic equation of motion. We suppose the viscoelastic constitutive relationship of Kelvin-Voigt type (88), where $\mathcal{C}$ and $\mathcal{G}$ are given viscosity and elasticity functions, respectively. Conditions (90) and (91) represent the displacement and traction boundary conditions, respectively while u_0 and u_1 in (93) denote the initial displacement and the initial velocity, respectively. Conditions (94) and (95) are the impulse constraints relations. The classical problem with impulses is to find a displacement field $u\colon Q \to \mathbb{R}^d$ such that (87)–(95) hold.

In order to set the problem (87)–(95) in a variational form, we recall that $H = L^2(\Omega; \mathbb{R}^d)$, $\mathcal{H} = L^2(\Omega; \mathcal{S}_d)$ and V is the closed subspace of $H^1(\Omega; \mathbb{R}^d)$ defined by $V = \{v \in H^1(\Omega; \mathbb{R}^d) \mid v = 0 \text{ on } \Gamma_D\}$.

In the study of the problem (87)–(95) we need the following hypotheses.

<u>$H(\mathcal{C})_1$</u> : the viscosity operator $\mathcal{C}\colon Q \times \mathcal{S}_d \to \mathcal{S}_d$ satisfies the Carathéodory condition (i.e. $\mathcal{C}(\cdot, \cdot, \varepsilon)$ is measurable on Q for all $\varepsilon \in \mathcal{S}_d$ and $\mathcal{C}(x, t, \cdot)$ is continuous on $\mathcal{S}_d$ for a.e. $(x, t) \in Q$) and

(i) $\|\mathcal{C}(x, t, \varepsilon)\|_{\mathcal{S}_d} \le c_1 (b(x, t) + \|\varepsilon\|_{\mathcal{S}_d})$ for $\varepsilon \in \mathcal{S}_d$, a.e. $(x, t) \in Q$ with $b \in L^2_+(Q)$, $c_1 > 0$;

(ii) $(\mathcal{C}(x, t, \varepsilon_1) - \mathcal{C}(x, t, \varepsilon_2)) : (\varepsilon_1 - \varepsilon_2) \ge 0$ for all ε_1, $\varepsilon_2 \in \mathcal{S}_d$ and a.e. $(x, t) \in Q$;

(iii) $\mathcal{C}(x, t, \varepsilon) : \varepsilon \ge c_2 \|\varepsilon\|^2_{\mathcal{S}_d}$ for all $\varepsilon \in \mathcal{S}_d$ and a.e. $(x, t) \in Q$ with $c_2 > 0$.

<u>$H(\mathcal{G})$</u> : the elasticity operator $\mathcal{G}\colon \Omega \times \mathcal{S}_d \to \mathcal{S}_d$ is of the form $\mathcal{G}(x, \varepsilon) = \mathbb{E}(x)\varepsilon$ (Hooke's law) with a symmetric and positive elasticity tensor $\mathbb{E} \in L^\infty(\Omega)$, i.e. $\mathbb{E} = (g_{ijkl})$, $i, j, k, l = 1, \dots, d$ with $g_{ijkl} = g_{jikl} = g_{lkij}$ and $g_{ijkl}(x)\chi_{ij}\, \chi_{kl} \ge 0$ for a.e. $x \in \Omega$ and for all symmetric tensors $\chi = \{\chi_{ij}\}$.

<u>$H(g, h)$</u> : $g_i, h_i\colon \mathbb{R}_+ \times \mathbb{R}_+ \to \mathbb{R}_+$ are bounded and upper semicontinuous with respect to both variables, for $i = 1, \dots, m$.

The external loading, boundary tractions and the initial data have the following regularity.
$\underline{(H_{10})}$: $\psi \in L^2(I; H^{1/2}(\Gamma_N; \mathbb{R}^d))$, $\varphi_2 \in L^2(I; H)$, $u_0 \in V$ and $u_1 \in H$.

The variational formulation of the problem (87)–(95) reads as follows: find a displacement field $u\colon I \setminus D \to V$ such that $u \in PC(I; V)$, $\dot{u} \in PW(I) \cap PC(I; H)$ and

$$(96)\quad \begin{cases} \langle \ddot{u}(t), v\rangle + \langle \mathcal{C}\varepsilon(\dot{u}(t)) + \mathcal{G}\varepsilon(u(t)), \varepsilon(v)\rangle_{\mathcal{H}} + \int_\Omega j^0(x,t,u(x,t);v(x))\,dx \\ \qquad\qquad \geq \langle \widetilde{\psi}(t), v\rangle_{H^*\times H} \quad \text{for all } v \in V \text{ and a.e. } t \in I \setminus D \\ u(0) = u_0, \quad \dot{u}(0) = v_0, \\ u(t_i^+) \in u(t_i^-) + B_{\mathbb{R}}\left(0, g_i\left(|u(t_i^-)|, |\dot{u}(t_i^-)|\right)\right), \\ \dot{u}(t_i^+) \in \dot{u}(t_i^-) + B_{\mathbb{R}}\left(0, h_i\left(|u(t_i^-)|, |\dot{u}(t_i^-)|\right)\right) \quad \text{for } i = 1, \ldots, m, \end{cases}$$

where $\langle \widetilde{\psi}, v\rangle_{H^*\times H} = \langle \varphi_2(t), v\rangle_H + \langle \psi(t), v\rangle_\Gamma$.

We associate with the hemivariational inequality (96) an evolution impulsive inclusion. To this end, let $A\colon I \times V \to V^*$, $B\colon V \to V^*$ and $J\colon I \times H \to \mathbb{R}$ be defined by

$$(97)\qquad \langle A(t,u), v\rangle = \langle \mathcal{C}(x,t,\varepsilon(u)), \varepsilon(v)\rangle_{\mathcal{H}} \quad \text{for } u, v \in V \text{ and } t \in I,$$

$$(98)\qquad \langle Bu, v\rangle = \langle \mathcal{G}(x, \varepsilon(u)), \varepsilon(v)\rangle_{\mathcal{H}} \quad \text{for } u, v \in V,$$

$$(99)\qquad J(t,v) = \int_\Omega j(x,t,v(x))\,dx \quad \text{for } v \in H \text{ and } t \in I.$$

We define the multifunctions $G_i\colon V \times H \to 2^V$ and $H_i\colon V \times H \to 2^H$ by $G_i(u,v)(\cdot) = B_V\left(0, g_i\left(|u(\cdot)|, |v(\cdot)|\right)\right)$, $H_i(u,v)(\cdot) = B_H\left(0, h_i\left(|u(\cdot)|, |v(\cdot)|\right)\right)$ for $u \in V$, $v \in H$. We also observe that if (H_{10}) is satisfied, then $\widetilde{\psi} \in \mathcal{H}^*$, $u_0 \in V$ and $v_0 \in H$.

Consider the following evolution inclusion with impulses.

$$(100)\quad \begin{cases} \ddot{u}(t) + A(t, \dot{u}(t)) + Bu(t) + \partial J(t, u(t)) \ni \widetilde{\psi}(t) \quad \text{for } t \in I \setminus D \\ u(0) = u_0, \quad \dot{u}(0) = v_0 \\ u(t_i^+) \in u(t_i^-) + G_i(u(t_i^-), \dot{u}(t_i^-)) \\ \dot{u}(t_i^+) \in \dot{u}(t_i^-) + H_i(u(t_i^-), \dot{u}(t_i^-)) \quad \text{for } i = 1, \ldots, m. \end{cases}$$

The reason to introduce problem (100) is stated in the following lemma. For details we refer to Migórski and Ochal [**85**].

LEMMA 41. *Under the above hypotheses, every solution to problem (100) is a solution to the hemivariational inequality (96).*

The result on existence and properties of solutions to problem (100) can be found in Theorem 3 of [**85**]. Summing up, we obtain the following

THEOREM 42. *If $H(\mathcal{C})_1$, $H(\mathcal{G})$, $H(j)_2$, (H_{10}), $H(g,h)$ hold and $u_0 \in H_0^1(\Omega)$, $v_0 \in L^2(\Omega)$, then problem (87)–(95) has a solution $u \in PC(I; H_0^1(\Omega))$ with $\frac{\partial u}{\partial t} \in PW(I) \cap PC(I; L^2(\Omega))$ and the set of all solutions is compact in $PC^1(I; L^2(\Omega))$.*

We would like to recall that recently the impulsive differential equations and inclusions have been studied by several authors. This is due to the fact that many real phenomena and processes in mechanics, biology, physics, chemistry, biotechnology, etc. are characterized by the situation that at certain instants in time, the system parameters (e.g. displacement, velocities) undergo rapid changes. The duration of these changes is often neglected and it is assumed that the changes are represented by parameter jumps. One of natural tools for mathematical modeling and simulation of such phenomena is the theory of impulsive differential equations. This theory has started in the 1980s and today it covers various kind of problems which are motivated by numerous applications. To our knowledge the impulsive second order evolution inclusions with multivalued jump operators formulated above have been considered in the literature only recently by Migórski and Ochal [**85**]. We mention that the model with multivalued jump sizes may arise in a control problem where we want to control the jump sizes in order to achieve given objectives.

7. Applications of evolution hemivariational inequalities

The theory of hemivariational inequalities plays an important role in several nonlinear boundary value problems arising in mechanics, physics and engineering sciences. We concentrate on three selected applications: integrodifferential hemivariational inequalities in viscoelasticity, hemivariational inequalities in thermoviscoelasticity and bilateral contact problem with adhesion in piezoviscoelasticity. We formulate the corresponding mechanical problems, describe hemivariational inequalities and provide results on the existence of solutions to such problems.

7.1. Integrodifferential hemivariational inequalities of viscoelasticity. We begin with the description of the mechanical problem. A viscoelastic body occupies in its reference configuration a set $\Omega \subset \mathbb{R}^d$ ($d = 2, 3$ in applications). The body is acted upon by volume forces and surface tractions and, as a result, its state is evolving. We are interested in dynamic evolution process of the mechanical state of the body on the time interval $[0, T]$ with $T > 0$. The boundary Γ of Ω is assumed to be Lipschitz continuous and therefore the unit outward normal vector ν exists a.e. on Γ. Moreover, Γ is assumed to be divided into three mutually disjoint parts Γ_D, Γ_N and Γ_C such that $m(\Gamma_D) > 0$. We suppose that the body is clamped on Γ_D, so the displacement field vanishes there. Volume forces of density f_1 act in Ω and surface tractions of density f_2 are applied on Γ_N. The body may come in contact with an obstacle over the potential contact surface Γ_C.

We denote by $\mathcal{S}_d$ the linear space of second order symmetric tensors on $\mathbb{R}^d$, or equivalently, the space $\mathbb{R}_s^{d\times d}$ of symmetric matrices of order d and we use the notation $Q = \Omega \times (0, T)$. For simplicity we skip the dependence of various functions on the spatial variable $x \in \Omega \cup \Gamma$. Then, the frictional contact problem under consideration can be stated as follows: find the displacement field $u\colon Q \to \mathbb{R}^d$ and the stress tensor $\sigma\colon Q \to \mathcal{S}_d$ such that

$$u''(t) - \operatorname{div} \sigma(t) = f_1(t) \quad \text{in } Q \tag{101}$$

$$\sigma(t) = \mathcal{A}(t, \varepsilon(u'(t))) + \mathcal{B}\varepsilon(u(t)) + \int_0^t \mathcal{C}(t-s)\varepsilon(u(s))\, ds \quad \text{in } Q \tag{102}$$

$$u(t) = 0 \quad \text{on } \Gamma_D \times (0, T) \tag{103}$$

(104) $\sigma(t)\nu = f_2(t) \quad \text{on } \Gamma_N \times (0,T)$

(105) $-\sigma_\nu(t) \in \partial j_\nu(t, u'_\nu(t)), \quad -\sigma_\tau(t) \in \partial j_\tau(t, u'_\tau(t)) \quad \text{on } \Gamma_C \times (0,T)$

(106) $u(0) = u_0, \quad u'(0) = v_0 \quad \text{in } \Omega.$

The equation (101) is the equation of motion and div denotes the divergence operator for tensor valued functions; equation (102) represents the viscoelastic constitutive law which will be discussed below. Conditions (103) and (104) are the displacement and the traction boundary conditions, respectively. Conditions (105) represent the frictional contact condition in which j_ν and j_τ are given functions and the subscripts n and τ for σ and u' indicate normal and tangential components of tensors and vectors. Here the symbol ∂j denotes the Clarke subdifferential of j with respect to the last variable. Note also that the explicit dependence of the functions j_ν and j_τ with respect to the time variable allows to model situations when the frictional contact conditions depend on the temperature, which plays the role of a parameter, and which evolution in time is prescribed. Finally, conditions (106) represent the initial conditions where u_0 and v_0 denote the initial displacement and the initial velocity, respectively.

Concrete examples of frictional models which lead to subdifferential boundary conditions of the form (105) with the functions j_ν and j_τ satisfying assumptions $H(j_\nu)$ and $H(j_\tau)$ below can be found in [**83**]. We remark that these examples include the viscous contact and the contact with nonmonotone normal damped response, associated to a nonmonotone friction law, to Tresca's friction law or to a power-law friction. The examples of the constitutive law (102) and the construction of rheological models obtained by connecting springs and dashpots can be found in Chapter 6 of [**38**]. We note that when $\mathcal{C} = 0$ the constitutive law (102) reduces to a viscoelastic constitutive law with short memory $\sigma(t) = \mathcal{A}(t, \varepsilon(u'(t)) + \mathcal{B}\varepsilon(u(t))$ and, in the case when $\mathcal{A} = 0$, it reduces to a viscoelastic constitutive law with long memory $\sigma(t) = \mathcal{B}\varepsilon(u(t)) + \int_0^t \mathcal{C}(t-s)\varepsilon(u(s))\,ds$.

In order to provide the variational formulation of the frictional contact problem (101)–(106), we recall the following notation which will be used in this section. The inner products and the corresponding norms on $\mathbb{R}^d$ and $\mathcal{S}_d$ are given by

$$u \cdot v = u_i v_i, \qquad \|v\|_{\mathbb{R}^d} = (v \cdot v)^{1/2} \qquad \text{for all } u, v \in \mathbb{R}^d,$$
$$\sigma : \tau = \sigma_{ij}\tau_{ij}, \qquad \|\tau\|_{\mathcal{S}_d} = (\tau : \tau)^{1/2} \qquad \text{for all } \sigma, \tau \in \mathcal{S}_d.$$

Here and everywhere in this section the indices i, j, k, l run from 1 to d. Summation convention over repeated indices is used and the index that follows a comma indicates a partial derivative. Next, we introduce the spaces

$$H = L^2(\Omega;\mathbb{R}^d), \qquad \mathcal{H} = \{\tau = \{\tau_{ij}\} \mid \tau_{ij} = \tau_{ji} \in L^2(\Omega)\} = L^2(\Omega;\mathcal{S}_d),$$
$$H_1 = \{u \in H \mid \varepsilon(u) \in \mathcal{H}\} = H^1(\Omega;\mathbb{R}^d), \qquad \mathcal{H}_1 = \{\tau \in \mathcal{H} \mid \operatorname{div}\tau \in H\},$$

where $\varepsilon\colon H^1(\Omega;\mathbb{R}^d) \to L^2(\Omega;\mathcal{S}_d)$ and $\operatorname{div}\colon \mathcal{H}_1 \to L^2(\Omega;\mathbb{R}^d)$ denote the deformation and the divergence operators, respectively, given by

$$\varepsilon(u) = \{\varepsilon_{ij}(u)\}, \quad \varepsilon_{ij}(u) = \frac{1}{2}(u_{i,j} + u_{j,i}), \quad \operatorname{div}\sigma = \{\sigma_{ij,j}\}.$$

The spaces H, $\mathcal{H}$, H_1 and $\mathcal{H}_1$ are Hilbert spaces equipped with the inner products

$$\langle u, v\rangle_H = \int_\Omega u \cdot v \, dx, \qquad \langle \sigma, \tau\rangle_{\mathcal{H}} = \int_\Omega \sigma : \tau \, dx,$$
$$\langle u, v\rangle_{H_1} = \langle u, v\rangle_H + \langle \varepsilon(u), \varepsilon(v)\rangle_{\mathcal{H}}, \qquad \langle \sigma, \tau\rangle_{\mathcal{H}_1} = \langle \sigma, \tau\rangle_{\mathcal{H}} + \langle \operatorname{div} \sigma, \operatorname{div} \tau\rangle_H.$$

The associated norms in H, $\mathcal{H}$, H_1 and $\mathcal{H}_1$ are denoted by $\|\cdot\|_H$, $\|\cdot\|_{\mathcal{H}}$, $\|\cdot\|_{H_1}$ and $\|\cdot\|_{\mathcal{H}_1}$, respectively. We also recall that the following Green formula holds

$$(107) \qquad \langle \sigma, \varepsilon(v)\rangle_{\mathcal{H}} + \langle \operatorname{div} \sigma, v\rangle_H = \int_\Gamma \sigma\nu \cdot v \, d\Gamma(x) \quad \text{for } v \in H_1, \ \sigma \in \mathcal{H}_1.$$

In the study of problem (101)–(106) we consider the following assumptions on the viscosity operator $\mathcal{A}$, on the elasticity operator $\mathcal{B}$ and on the relaxation operator $\mathcal{C}$.

$\underline{H(\mathcal{A})}$: $\mathcal{A}\colon Q \times \mathcal{S}_d \to \mathcal{S}_d$ is such that
- (i) $\mathcal{A}(\cdot, \cdot, \varepsilon)$ is measurable on Q, for all $\varepsilon \in S_d$;
- (ii) $\mathcal{A}(x, t, \cdot)$ is continuous on $\mathcal{S}_d$ for a.e. $(x, t) \in Q$;
- (iii) $\|\mathcal{A}(x, t, \varepsilon)\|_{\mathcal{S}_d} \le c_1 \left(b(x, t) + \|\varepsilon\|_{\mathcal{S}_d}\right)$ for all $\varepsilon \in \mathcal{S}_d$, a.e. $(x, t) \in Q$ with $b \in L^2(Q)$, $b \ge 0$ and $c_1 > 0$;
- (iv) $(\mathcal{A}(x, t, \varepsilon_1) - \mathcal{A}(x, t, \varepsilon_2)) : (\varepsilon_1 - \varepsilon_2) \ge m_1 \|\varepsilon_1 - \varepsilon_2\|^2_{\mathcal{S}_d}$ for all $\varepsilon_1, \varepsilon_2 \in \mathcal{S}_d$, a.e. $(x, t) \in Q$ with $m_1 > 0$;
- (v) $\mathcal{A}(x, t, \varepsilon) : \varepsilon \ge c_2 \|\varepsilon\|^2_{\mathcal{S}_d}$ for all $\varepsilon \in \mathcal{S}_d$, a.e. $(x, t) \in Q$ with $c_2 > 0$.

$\underline{H(\mathcal{B})}$: $\mathcal{B}\colon \Omega \times \mathcal{S}_d \to \mathcal{S}_d$ is such that $\mathcal{B}(x, \varepsilon) = b(x)\varepsilon$ and
- (i) $b(x) = \{b_{ijkl}(x)\}$ with $b_{ijkl} = b_{jikl} = b_{lkij} \in L^\infty(\Omega)$;
- (ii) $b_{ijkl}(x)\varepsilon_{ij}\varepsilon_{kl} \ge 0$ for all $\varepsilon = \{\varepsilon_{ij}\} \in \mathcal{S}_d$, a.e. $x \in \Omega$.

$\underline{H(\mathcal{C})}$: $\mathcal{C}\colon Q \times \mathcal{S}_d \to \mathcal{S}_d$ is such that $\mathcal{C}(x, t, \varepsilon) = c(x, t)\varepsilon$ and $c(x, t) = \{c_{ijkl}(x, t)\}$ with $c_{ijkl} = c_{jikl} = c_{lkij} \in L^2(0, T; L^\infty(\Omega))$.

The contact and frictional potentials j_ν and j_τ satisfy the following hypotheses.

$\underline{H(j_\nu)}$: $j_\nu : \Gamma_C \times (0, T) \times \mathbb{R} \to \mathbb{R}$ satisfies
- (i) $j_\nu(\cdot, \cdot, r)$ is measurable for all $r \in \mathbb{R}$ and $j_\nu(\cdot, \cdot, 0) \in L^1(\Gamma_C \times (0, T))$;
- (ii) $j_\nu(x, t, \cdot)$ is locally Lipschitz for a.e. $(x, t) \in \Gamma_C \times (0, T)$;
- (iii) $|\partial j_\nu(x, t, r)| \le c_\nu (1 + |r|)$ for a.e. $(x, t) \in \Gamma_C \times (0, T)$, all $r \in \mathbb{R}$ with $c_\nu > 0$;
- (iv) $j^0_\nu(x, t, r; -r) \le d_\nu (1 + |r|)$ for all $r \in \mathbb{R}$, a.e. $(x, t) \in \Gamma_C \times (0, T)$ with $d_\nu \ge 0$;
- (v) $(\eta_1 - \eta_2)(r_1 - r_2) \ge -m_\nu |r_1 - r_2|^2$ for all $\eta_i \in \partial j_\nu(x, t, r_i)$, $r_i \in \mathbb{R}$, $i = 1, 2$, a.e. $(x, t) \in \Gamma_C \times (0, T)$ with $m_\nu \ge 0$.

$\underline{H(j_\tau)}$: $j_\tau : \Gamma_C \times (0, T) \times \mathbb{R}^d \to \mathbb{R}$ satisfies
- (i) $j_\tau(\cdot, \cdot, \xi)$ is measurable for all $\xi \in \mathbb{R}^d$ and $j_\tau(\cdot, \cdot, 0) \in L^1(\Gamma_C \times (0, T))$;
- (ii) $j_\tau(x, t, \cdot)$ is locally Lipschitz for a.e. $(x, t) \in \Gamma_C \times (0, T)$;
- (iii) $\|\partial j_\tau(x, t, \xi)\|_{\mathbb{R}^d} \le c_\tau (1 + \|\xi\|_{\mathbb{R}^d})$ for a.e. $(x, t) \in \Gamma_C \times (0, T)$, all $\xi \in \mathbb{R}^d$ with $c_\tau > 0$;
- (iv) $j^0_\tau(x, t, \xi; -\xi) \le d_\tau (1 + \|\xi\|_{\mathbb{R}^d})$ for all $\xi \in \mathbb{R}^d$, a.e. $(x, t) \in \Gamma_C \times (0, T)$ with $d_\tau \ge 0$;
- (v) $(\eta_1 - \eta_2, \xi_1 - \xi_2)_{\mathbb{R}^d} \ge -m_\tau \|\xi_1 - \xi_2\|^2_{\mathbb{R}^d}$ for all $\eta_i \in \partial j_\tau(x, t, \xi_i)$, $\xi_i \in \mathbb{R}^d$, $i = 1, 2$, a.e. $(x, t) \in \Gamma_C \times (0, T)$ with $m_\tau \ge 0$.

The volume force and traction densities satisfy

$\underline{H(f)}$: $f_1 \in L^2(0,T;H), \quad f_2 \in L^2(0,T;L^2(\Gamma_N;\mathbb{R}^d))$

and finally the initial data have the regularity

$\underline{H(0)}$: $u_0 \in V, \quad v_0 \in H.$

Here $V = \{v \in H^1(\Omega;\mathbb{R}^d) \mid v = 0 \text{ on } \Gamma_D\}$. On V we consider the inner product and the corresponding norm given by

$$\langle u,v\rangle_V = \langle \varepsilon(u), \varepsilon(v)\rangle_{\mathcal{H}}, \quad \|v\| = \|\varepsilon(v)\|_{\mathcal{H}} \text{ for } u,v \in V.$$

It follows from Korn's inequality that $\|\cdot\|_{H^1(\Omega;\mathbb{R}^d)}$ and $\|\cdot\|$ are the equivalent norms on V, and therefore $(V, \langle\cdot,\cdot\rangle_V)$ is a Hilbert space. As usual, we denote by V^* the dual space of V and by $\langle\cdot,\cdot\rangle$ the duality pairing of V and V^*. Let $\mathcal{V} = L^2(0,T;V)$ and $\mathcal{W} = \{w \in \mathcal{V} \mid w' \in \mathcal{V}^*\}$. We obtain $\mathcal{W} \subset \mathcal{V} \subset L^2(0,T;H) \subset \mathcal{V}^*$, where all the embeddings are continuous.

We now turn to the variational formulation of the contact problem (101)–(106). We introduce the operators $A\colon (0,T)\times V \to V^*$, $B\colon V \to V^*$ and $C\colon (0,T)\times V \to V^*$ defined by

$$\langle A(t,u),v\rangle = \langle \mathcal{A}(t,\varepsilon(u)), \varepsilon(v)\rangle_{\mathcal{H}} \quad \text{for } u,v \in V \text{ and } t \in (0,T), \tag{108}$$

$$\langle Bu,v\rangle = \langle \mathcal{B}\varepsilon(u), \varepsilon(v)\rangle_{\mathcal{H}} \quad \text{for } u,v \in V, \tag{109}$$

$$\langle C(t)u,v\rangle = \langle \mathcal{C}(t)\varepsilon(u), \varepsilon(v)\rangle_{\mathcal{H}} \quad \text{for } u,v \in V \text{ and } t \in (0,T) \tag{110}$$

and the function $f\colon (0,T) \to V^*$ given by

$$\langle f(t),v\rangle = \langle f_1(t),v\rangle_H + (f_2(t),v)_{L^2(\Gamma_N;\mathbb{R}^d)} \quad \text{for } u,v \in V, \text{ a.e. } t \in (0,T). \tag{111}$$

Assume that (u,σ) is a couple of regular functions which solve (101)–(106) and let $v \in V$ and $t \in (0,T)$. We use the equation of motion (101), the Green formula (107), take into account the boundary conditions (103) and (104) use the definition of the Clarke subdifferential, (105) and (111), the constitutive law (102), notation (108)–(110) and the initial conditions (106) to obtain the following variational formulation of the mechanical problem (101)–(106):

$$\begin{cases} \text{find } u\colon (0,T) \to V \quad \text{such that } u \in \mathcal{V},\ u' \in \mathcal{W} \text{ and} \\ \langle u''(t) + A(t,u'(t)) + Bu(t) + \displaystyle\int_0^t C(t-s)u(s)\,ds, v\rangle \\ \quad + \displaystyle\int_{\Gamma_C} \big(j_\nu^0(x,t,u_\nu'(x,t);v_\nu(x)) + j_\tau^0(x,t,u_\tau'(x,t);v_\tau(x))\big)\,d\Gamma \\ \qquad \geq \langle f(t),v\rangle \quad \text{for all } v \in V \text{ and a.e. } t \in (0,T) \\ u(0) = u_0, \quad u'(0) = v_0. \end{cases} \tag{112}$$

The unique solvability of the problem (112) is given by the following result.

THEOREM 43. *Assume that* $H(\mathcal{A})$, $H(\mathcal{B})$, $H(\mathcal{C})$, $H(j_\nu)$, $H(j_\tau)$, $H(f)$, $H(0)$ *hold and* $m_1 > (m_\nu + m_\tau)\|\gamma\|^2$. *Then problem* (112) *admits at least one solution. If, in addition,*

$$\begin{cases} \text{either } j_\nu(x,t,\cdot) \text{ or } -j_\nu(x,t,\cdot) \text{ is regular and} \\ \text{either } j_\tau(x,t,\cdot) \text{ or } -j_\tau(x,t,\cdot) \text{ is regular,} \end{cases} \tag{113}$$

then problem (112) *has a unique solution.*

The proof of Theorem 43 is based on a result for the following abstract evolutionary inclusion

$$
(114)\qquad \begin{cases} \text{find } u \in \mathcal{V} \text{ with } u' \in \mathcal{W} \text{ such that} \\ u''(t) + A(t, u'(t)) + Bu(t) + \displaystyle\int_0^t C(t-s)u(s)\,ds \\ \qquad\qquad + \gamma^* \partial J(t, \gamma u'(t)) \ni f(t) \quad \text{a.e. } t \in (0,T), \\ u(0) = u_0, \quad u'(0) = v_0. \end{cases}
$$

We consider the following hypotheses on the data.

$\underline{H(A)}$: $A\colon (0,T) \times V \to V^*$ is such that
(i) $A(\cdot, v)$ is measurable on $(0,T)$ for all $v \in V$;
(ii) $A(t,\cdot)$ is strongly monotone i.e. $\langle A(t,u) - A(t,v), u - v\rangle \geq m_1\|u - v\|^2$ for all $u, v \in V$, a.e. $t \in (0,T)$ with $m_1 > 0$;
(iii) $\|A(t,v)\|_{V^*} \leq a(t) + b\|v\|$ for all $v \in V$, a.e. $t \in (0,T)$ with $a \in L^2(0,T)$, $a \geq 0$, $b > 0$;
(iv) $\langle A(t,v), v\rangle \geq \alpha\|v\|^2$ for all $v \in V$, a.e. $t \in (0,T)$ with $\alpha > 0$.

$\underline{H(B)}$: $B\colon V \to V^*$ is a bounded, linear, monotone and symmetric operator, i.e. $B \in \mathcal{L}(V,V^*)$, $\langle Bv, v\rangle \geq 0$ for all $v \in V$, $\langle Bv, w\rangle = \langle Bw, v\rangle$ for all v, $w \in V$.

$\underline{H(C)}$: $C \in L^2(0,T;\mathcal{L}(V,V^*))$.

$\underline{H(J)}$: $J\colon (0,T) \times L^2(\Gamma_C;\mathbb{R}^d) \to \mathbb{R}$ is a functional such that
(i) $J(\cdot, v)$ is measurable for all $v \in L^2(\Gamma_C;\mathbb{R}^d)$ and $J(\cdot,0) \in L^1(0,T)$;
(ii) $J(t,\cdot)$ is locally Lipschitz for a.e. $t \in (0,T)$;
(iii) $\|\partial J(t,v)\|_{L^2(\Gamma_C;\mathbb{R}^d)} \leq c_0\left(1 + \|v\|_{L^2(\Gamma_C;\mathbb{R}^d)}\right)$ for all $v \in L^2(\Gamma_C;\mathbb{R}^d)$, a.e. $t \in (0,T)$ with $c_0 > 0$;
(iv) $J^0(t,v;-v) \leq d_0\left(1 + \|v\|_{L^2(\Gamma_C;\mathbb{R}^d)}\right)$ for all $v \in L^2(\Gamma_C;\mathbb{R}^d)$, a.e. $t \in (0,T)$ with $d_0 \geq 0$;
(v) $(z_1 - z_2, w_1 - w_2)_{L^2(\Gamma_C;\mathbb{R}^d)} \geq -m_2\|w_1 - w_2\|^2_{L^2(\Gamma_C;\mathbb{R}^d)}$ for all $z_i \in \partial J(t,w_i)$, $w_i \in L^2(\Gamma_C;\mathbb{R}^d)$, $i = 1,2$, a.e. $t \in (0,T)$ with $m_2 \geq 0$.

$\underline{(H_0)}$: $f \in \mathcal{V}^*$, $u_0 \in V$, $v_0 \in H$.

$\underline{(H_1)}$: $m_1 > m_2\,\|\gamma\|^2$, where $\|\gamma\| = \|\gamma\|_{\mathcal{L}(Z,L^2(\Gamma;\mathbb{R}^d))}$ and $Z = H^\delta(\Omega;\mathbb{R}^d)$, $\delta \in (1/2,1)$.

The unique solvability of the problem (114) is given by the following existence and uniqueness result (cf. [**91**]).

THEOREM 44. *Assume that $H(A)$, $H(B)$, $H(C)$, $H(J)$, (H_0) and (H_1) hold. Then the problem* (114) *admits a unique solution.*

Finally, we apply Theorem 44 in the study of a class of second order hemivariational inequalities. The problem we are interested in is formulated as follows

$$
(115)\qquad \begin{cases} \text{find } u \in \mathcal{V} \text{ with } u' \in \mathcal{W} \text{ such that} \\ \langle u''(t) + A(t, u'(t)) + Bu(t) + \displaystyle\int_0^t C(t-s)u(s)\,ds, v\rangle \\ \quad + \displaystyle\int_{\Gamma_C} j^0(x,t,u'(t);v)\,d\Gamma \geq \langle f(t), v\rangle \quad \text{for all } v \in V, \text{ a.e. } t \in (0,T), \\ u(0) = u_0, \quad u'(0) = v_0. \end{cases}
$$

In the problem (115) we consider the following hypothesis.

$\underline{H(j)}$: $j\colon \Gamma_C \times (0,T) \times \mathbb{R}^d \to \mathbb{R}$ is such that

(i) $j(\cdot,\cdot,\xi)$ is measurable for all $\xi \in \mathbb{R}$ and $j(\cdot,\cdot,0) \in L^1(\Gamma_C \times (0,T))$;

(ii) $j(x,t,\cdot)$ is locally Lipschitz a.e. $x \in \Gamma_C$, $t \in (0,T)$;

(iii) $|\partial j(x,t,\xi)| \le \widetilde{c}\,(1 + \|\xi\|_{\mathbb{R}^d})$ for all $\xi \in \mathbb{R}^d$, a.e. $x \in \Gamma_C$, $t \in (0,T)$ with $\widetilde{c} > 0$;

(iv) $j^0(x,t,\xi;-\xi) \le \widetilde{d}\,(1 + \|\xi\|_{\mathbb{R}^d})$ for all $\xi \in \mathbb{R}^d$, a.e. $x \in \Gamma_C$, $t \in (0,T)$ with $\widetilde{d} \ge 0$;

(v) $(\eta_1 - \eta_2, \xi_1 - \xi_2)_{\mathbb{R}^d} \ge -m_2\|\xi_1 - \xi_2\|^2_{\mathbb{R}^d}$ for all $\eta_i \in \partial j(x,t,\xi_i)$, $\xi_i \in \mathbb{R}^d$, $i = 1, 2$ with $m_2 \ge 0$,

where j^0 and ∂j denote the directional derivative and the Clarke generalized gradient of $j(x,t,\cdot)$, respectively.

From Theorem 44 we obtain the following existence result.

Corollary 45. Assume that $H(A)$, $H(B)$, $H(C)$, $H(j)$, (H_0) and (H_1) hold. Then the hemivariational inequality (115) has at least a solution. If, in addition, either $j(x,t,\cdot)$ or $-j(x,t,\cdot)$ is regular, then the hemivariational inequality (115) admits a unique solution.

Proof of Theorem 43. The proof is based on arguments similar to those used in [**83**] and therefore we skip the details. The main steps of the proof are the following.

a) Under the assumptions $H(\mathcal{A})$, $H(\mathcal{B})$ and $H(\mathcal{C})$, the operators A, B and C defined by (108), (109) and (110) satisfy hypotheses $H(A)$, $H(B)$ and $H(C)$, respectively.

b) Let $j\colon \Gamma_C \times (0,T) \times \mathbb{R}^d \to \mathbb{R}$ be the function defined by

$$j(x,t,\xi) = j_\nu(x,t,\xi_\nu) + j_\tau(x,t,\xi_\tau) \quad \text{a.e. } (x,t) \in \Gamma_C \times (0,T), \text{ all } \xi \in \mathbb{R}^d.$$

It can be proved that, under the assumptions $H(j_\nu)$ and $H(j_\tau)$, the function j satisfies $H(j)$ with $\widetilde{c} = \max\{c_\nu, c_\tau\}$, $\widetilde{d} = \max\{d_\nu, d_\tau\}$ and $m_2 = m_\nu + m_\tau$.

c) The assumptions $H(f)$ and $H(0)$ combined with (111) imply that (H_0) holds. It is clear that (H_1) also is satisfied.

The steps above allow us to apply Corollary 45 to obtain the existence of a solution to the problem (115). It is easy to observe that the regularity hypotheses on j_ν, j_τ or $-j_\nu$, $-j_\tau$ imply the regularity of j or $-j$, respectively. In this case, we deduce the uniqueness of a solution to (112). □

We introduce now the following definition.

Definition 46. A couple of functions $u\colon (0,T) \to V$ and $\sigma\colon (0,T) \to V$ which satisfies (112) and (102) is called a *weak solution* of the frictional contact problem (101)–(106).

We conclude with the following result, which represents a direct consequence of Theorem 43.

Corollary 47. Under the hypotheses of Theorem 43, the frictional contact problem (101)–(102) admits a weak solution which satisfies $u \in H^1(0,T;V)\cap$

$C(0,T;V)$, $u' \in C(0,T;H)$, $u'' \in L^2(0,T;V^*)$, $\sigma \in C(0,T;\mathcal{H})$, $\operatorname{div}\sigma \in L^2(0,T;V^*)$. If, in addition, (113) holds, then the weak solution is unique.

7.2. Hemivariational inequalities in thermoviscoelasticity. In this section we first describe the classical model of thermoviscoelasticity, then we present its variational formulation and provide a result on existence and uniqueness of solutions.

Let Ω be an open bounded domain in $\mathbb{R}^d$, $d = 2, 3$, with a Lipschitz continuous boundary $\Gamma = \partial\Omega$. Let Γ be divided into three mutually disjoint measurable parts Γ_D, Γ_N and Γ_C such that $m(\Gamma_D) > 0$. We consider a viscoelastic body, occupying volume Ω, which is supposed to be stress free and at a constant temperature, conveniently set as zero. We assume that the temperature changes accompanying the deformations are small and they do not produce any changes in the material parameters which are regarded temperature independent. We confine ourselves to the classical linear thermoviscoelasticity theory and study the evolution of the system state in a time interval $[0,T]$ with $T > 0$. We suppose that the body is clamped on Γ_D, the volume forces of density $f_1(x,t) = (f_1^1(x,t), \ldots, f_1^d(x,t))$ act in Ω and the surface tractions of density $f_2(x,t) = (f_2^1(x,t), \ldots, f_2^d(x,t))$ are applied on Γ_N. Moreover, the body is subjected to a heat source term per unit volume $g = g(x,t)$ and it comes in contact with a rigid foundation over the potential contact surface Γ_C. We also put $Q = \Omega \times (0,T)$.

We denote by $u(x,t) = (u_1(x,t), \ldots, u_d(x,t))$ the displacement vector, by $\sigma = \{\sigma_{ij}\}$ the stress tensor, by $\varepsilon(u) = \{\varepsilon_{ij}(u)\}$ the linearized (small) strain tensor, where $i, j = 1, \ldots, d$ and by $\theta = \theta(x,t)$ the temperature. We assume with no loss of generality that the material density and the specific heat at constant deformation are constants, both set equal to one. The system of equations of motion assuming small displacements and the law of conservation of energy takes the form

$$\begin{aligned} u_i'' - \partial_j \sigma_{ij} &= f_{1i} \quad \text{in } Q, \qquad i = 1, \ldots, d, \\ \theta' + \partial_i q_i &= -c_{ij}\partial_j u_i' + g \quad \text{in } Q. \end{aligned}$$

Further the heat flux vector $q = (q_1, \ldots, q_d)$ satisfies the Fourier law of heat conduction $q_i = -k_{ij}\partial_j\theta$ in Q. The behavior of the body is governed by the thermoviscoelastic constitutive equation of the Kelvin-Voigt type

$$\sigma_{ij} = a_{ijkl}\partial_l u_k' + b_{ijkl}\partial_l u_k - c_{ij}\theta \quad \text{in } Q, \quad i, j = 1, \ldots, d,$$

where $\{a_{ijkl}\}$ and $\{b_{ijkl}\}$, i, j, k, $l = 1, \ldots, d$ are the viscosity and elasticity tensors, respectively, $\{c_{ij}\}$, i, $j = 1, \ldots, d$ being the so-called coefficients of thermal expansion.

Our main interest lies in the contact and friction boundary conditions on the surface Γ_C. As concerns the contact condition we assume that the normal stress σ_ν and the normal velocity u_ν' satisfy the nonmonotone normal damped response condition of the form

$$-\sigma_\nu \in \partial j_\nu(x, t, u_\nu') \quad \text{on } \Gamma_C \times (0,T).$$

The friction relation is given by

$$-\sigma_\tau \in \partial j_\tau(x, t, u_\tau') \quad \text{on } \Gamma_C \times (0,T)$$

and describes the multivalued law between the tangential force σ_τ on Γ_C and the tangential velocity u'_τ, see also Migórski [**72**]. Moreover, we suppose that between the boundary temperature and the heat flux vector the following relation holds $q \cdot \nu \in \partial j(x,t,\theta)$ which we write as

$$-\frac{\partial \theta}{\partial \nu_K} \in \partial j(x,t,\theta) \quad \text{on } \Gamma_C \times (0,T),$$

where $\frac{\partial \theta}{\partial \nu_K} = k_{ij}\, \partial_j \theta\, \nu_i$. Here $j_\nu \colon \Gamma_C \times (0,T) \times \mathbb{R} \to \mathbb{R}$, $j_\tau \colon \Gamma_C \times (0,T) \times \mathbb{R}^d \to \mathbb{R}$ and $j \colon \Gamma_C \times (0,T) \times \mathbb{R} \to \mathbb{R}$ are locally Lipschitz functions in their last variables and ∂j_ν, ∂j_τ, ∂j represent their Clarke subdifferentials. Finally for the sake of simplicity we assume that $\theta = 0$ on $(\Gamma_D \cup \Gamma_N) \times (0,T)$.

Denoting by u_0, u_1 and θ_0 the initial displacement, the initial velocity and the initial temperature, respectively, the classical formulation of the problem can be stated as follows (P):

find a displacement field $u \colon Q \to \mathbb{R}^d$ and a temperature $\theta \colon Q \to \mathbb{R}$ such that

$$\begin{cases} u''_i - \partial_j \left(a_{ijkl} \partial_l u'_k \right) - \partial_j \left(b_{ijkl} \partial_l u_k \right) + \partial_j \left(c_{ij} \theta \right) = f_{1i} \quad \text{in } Q \\ \theta' - \partial_i \left(k_{ij} \partial_j \theta \right) + c_{ij} \partial_j u'_i = g \quad \text{in } Q \\ u = 0 \quad \text{on } \Gamma_D \times (0,T) \\ \sigma n = f_2 \quad \text{on } \Gamma_N \times (0,T) \\ -\sigma_\nu \in \partial j_\nu(x,t,u'_\nu) \quad \text{on } \Gamma_C \times (0,T) \\ -\sigma_\tau \in \partial j_\tau(x,t,u'_\tau) \quad \text{on } \Gamma_C \times (0,T) \\ -\dfrac{\partial \theta}{\partial \nu_K} \in \partial j(x,t,\theta) \quad \text{on } \Gamma_C \times (0,T) \\ \theta = 0 \quad \text{on } (\Gamma_D \cup \Gamma_N) \times (0,T) \\ u(0) = u_0, \quad u'(0) = u_1, \quad \theta(0) = \theta_0 \quad \text{in } \Omega. \end{cases}$$

We pass to the variational formulation of the above problem. We introduce the following spaces

$$E = \left\{ v \in H^1(\Omega;\mathbb{R}^d) \mid v = 0 \quad \text{on } \Gamma_D \right\},$$
$$V = \left\{ \eta \in H^1(\Omega) \mid \eta = 0 \quad \text{on } \Gamma_D \cup \Gamma_N \right\}.$$

On E we consider the inner product and the corresponding norm given by

$$(u,v)_E = \langle \varepsilon(u), \varepsilon(v) \rangle_{L^2(\Omega;\mathcal{S}_d)}, \qquad \|v\|_E = \|\varepsilon(v)\|_{L^2(\Omega;\mathcal{S}_d)}.$$

From the Korn inequality, it follows that $\|\cdot\|_{H^1(\Omega;\mathbb{R}^d)}$ and $\|\cdot\|_E$ are the equivalent norms on E. Let $H = L^2(\Omega;\mathbb{R}^d)$ and $Z = H^\delta(\Omega;\mathbb{R}^d)$ with a fixed $\delta \in (1/2,1)$. Denoting by $i \colon E \to Z$ the embedding injection and by $\gamma \colon Z \to L^2(\Gamma;\mathbb{R}^d)$ the trace operator, for all $v \in E$, we have $\gamma_0 v = \gamma(iv)$. For simplicity we omit the notation of the embedding and write $\gamma_0 v = \gamma v$ for $v \in E$. Moreover, by $\gamma^* \colon L^2(\Gamma;\mathbb{R}^d) \to Z^*$ we denote the adjoint operator to γ. Identifying H with its dual, we have the following evolution fivefold of spaces with dense, continuous and compact embeddings $E \subset Z \subset H \subset Z^* \subset E^*$.

We also introduce the following spaces of vector valued functions $\mathcal{E} = L^2(0,T;E)$, $\mathcal{Z} = L^2(0,T;Z)$, $\mathcal{H} = L^2(0,T;H)$ and $\mathbb{E} = \{v \in \mathcal{E} \mid v' \in \mathcal{E}^*\}$, where the time derivative is understood in the sense of vector valued distributions. Endowed with the norm $\|v\|_{\mathbb{E}} = \|v\|_{\mathcal{E}} + \|v'\|_{\mathcal{E}^*}$, the space $\mathbb{E}$ becomes a separable reflexive Banach space. We have $\mathbb{E} \subset \mathcal{E} \subset \mathcal{Z} \subset \mathcal{H} \subset \mathcal{Z}^* \subset \mathcal{E}^*$ with dense and continuous embeddings. The duality for the pair $(\mathcal{E}, \mathcal{E}^*)$ is denoted by $\langle w, z\rangle_{\mathcal{E}^* \times \mathcal{E}} = \int_0^T \langle w(s), z(s)\rangle_{E^* \times E}\, ds$. It is well known (see e.g. [**28**]) that the embeddings $\mathbb{E} \subset C(0,T;H)$ and $\{v \in \mathcal{E} \mid v' \in \mathbb{E}\} \subset C(0,T;E)$ are continuous and $\mathbb{E} \subset \mathcal{Z}$ is compact.

Similarly, we introduce $Y = H^\delta(\Omega)$ with the same $\delta \in (1/2, 1)$ and we have the evolution fivefold of spaces $V \subset Y \subset L^2(\Omega) \subset Y^* \subset V^*$ with dense, continuous and compact embeddings. Let $\mathcal{V} = L^2(0,T;V)$, $\mathcal{Y} = L^2(0,T;Y)$ and $\mathcal{W} = \{\eta \in \mathcal{V} \mid \eta' \in \mathcal{V}^*\}$. We obtain

$$\mathcal{W} \subset \mathcal{V} \subset \mathcal{Y} \subset L^2(0,T;L^2(\Omega)) \subset \mathcal{Y}^* \subset \mathcal{V}^*,$$

where all the embeddings are dense and continuous. Furthermore, we know that the embeddings $\mathcal{W} \subset C(0,T;L^2(\Omega))$ and $\{\eta \in \mathcal{V} \mid \eta' \in \mathcal{W}\} \subset C(0,T;V)$ are continuous and $\mathcal{W} \subset \mathcal{Y}$ is compact.

Similarly as before, we denote by $\gamma_s\colon Y \to L^2(\Gamma)$ the trace operator for scalar valued functions, by $\gamma_s^*\colon L^2(\Gamma) \to Y^*$ its adjoint and we write $\gamma_0^s v = \gamma_s v$ for $v \in V$. We assume the following regularity for the density of heat sources, the body forces and surface tractions: $g \in \mathcal{V}^*$, $f_1 \in L^2(0,T;E^*)$, $f_2 \in L^2(0,T;L^2(\Gamma_N;\mathbb{R}^d))$, and we define $f \in \mathcal{E}^*$ by

$$\langle f(t), v\rangle_{E^* \times E} = \langle f_1(t), v\rangle_{E^* \times E} + (f_2(t), \gamma_0 v)_{L^2(\Gamma_N;\mathbb{R}^d)} \tag{116}$$

for $v \in E$ and a.e. $t \in (0,T)$. The fact that $f \in \mathcal{E}^*$ follows easily from the continuity of the trace operator. To obtain the variational formulation of the problem, we use a standard technique multiplying the first equation in (P) by $v \in E$ and using the Green formula. Then taking into account the boundary condition on Γ_N we have

$$\begin{aligned}&\langle u''(t) + Au'(t) + Bu(t) + C_1\theta(t), v\rangle_{E^* \times E}\\ &\quad + \int_{\Gamma_C} \left(j_\nu^0(x,t,u_\nu'; v_\nu) + j_\tau^0(x,t,u_\tau'; v_\tau)\right) d\Gamma \geq \langle f(t), v\rangle_{E^* \times E}\end{aligned}$$

for all $v \in E$ and a.e. $t \in (0,T)$, where $A\colon E \to E^*$, $B\colon E \to E^*$ and $C_1\colon V \to E^*$ are given by

$$\begin{aligned}\langle Aw, v\rangle_{E^* \times E} &= \int_\Omega a_{ijkl} \frac{\partial w_k}{\partial x_l} \frac{\partial v_i}{\partial x_j}\, dx,\\ \langle Bw, v\rangle_{E^* \times E} &= \int_\Omega b_{ijkl} \frac{\partial w_k}{\partial x_l} \frac{\partial v_i}{\partial x_j}\, dx,\\ \langle C_1\theta, v\rangle_{E^* \times E} &= -\int_\Omega c_{ij} \theta \frac{\partial v_i}{\partial x_j}\, dx\end{aligned}$$

for all $w, v \in E$ and $\theta \in V$.

Next let $\eta \in V$. Multiplying the second equation in (P) by η, using again the Green formula and the heat flux boundary condition, we arrive at the following inequality

$$\langle \theta'(t) + C_2\theta(t) + C_3u'(t), \eta\rangle_{V^* \times V} + \int_{\Gamma_C} j^0(x,t,\theta(t); \eta)\, d\Gamma \geq \langle g(t), \eta\rangle_{V^* \times V}$$

for all $\eta \in V$ and a.e. $t \in (0,T)$, where the operators $C_2 \colon V \to V^*$ and $C_3 \colon E \to V^*$ are defined by

$$\langle C_2\theta, \eta\rangle_{V^*\times V} = \int_\Omega k_{ij} \frac{\partial \theta}{\partial x_i}\frac{\partial \eta}{\partial x_j}\, dx,$$

$$\langle C_3 w, \eta\rangle_{V^*\times V} = \int_\Omega c_{ij} \frac{\partial w_i}{\partial x_j}\eta\, dx$$

for all θ, $\eta \in V$ and $w \in E$. Hence we obtain the following variational formulation of the problem (P), which consists of two inequalities which are called hemivariational inequalities: find a displacement field $u \colon (0,T) \to E$ with $u' \colon (0,T) \to E$ and $u'' \colon (0,T) \to E^*$ and a temperature $\theta \colon (0,T) \to V$ with $\theta' \colon (0,T) \to V^*$ such that

$$(117)\quad \begin{cases} \langle u''(t) + Au'(t) + Bu(t) + C_1\theta(t), v\rangle_{E^*\times E} \\ \quad + \int_{\Gamma_C} \big(j_\nu^0(x,t,u_\nu'(t);v_\nu) + j_\tau^0(x,t,u_\tau'(t);v_\tau)\big)\, d\Gamma \ge \langle f(t), v\rangle_{E^*\times E} \\ \qquad\qquad\qquad\qquad \text{for } v \in E, \text{a.e. } t \\ \langle \theta'(t) + C_2\theta(t) + C_3 u'(t), \eta\rangle_{V^*\times V} \\ \quad + \int_{\Gamma_C} j^0(x,t,\theta(t);\eta)\, d\Gamma \ge \langle g(t), \eta\rangle_{V^*\times V} \quad \text{for } \eta \in V, \text{a.e. } t \\ u(0) = u_0, \quad u'(0) = u_1, \ \theta(0) = \theta_0. \end{cases}$$

Next we embed the inequality problem (117) into a more general class of problems for evolution inclusions for which we provide an existence result. We formulate a system of coupled evolution inclusions related to the system (117) of hemivariational inequalities. The hypotheses on the data of the problem (117) are as follows.

<u>$H(a)$</u> : $a_{ijkl} \in L^\infty(\Omega)$, $a_{ijkl} = a_{jikl} = a_{lkij}$, $a_{ijkl}(x)\chi_{ij}\chi_{kl} \ge \alpha_1\chi_{ij}\chi_{ij}$, for all $\chi = \{\chi_{ij}\} \in \mathcal{S}_d$, a.e. $x \in \Omega$ with $\alpha_1 > 0$.

<u>$H(b)$</u> : $b_{ijkl} \in L^\infty(\Omega)$, $b_{ijkl} = b_{jikl} = b_{lkij}$, $b_{ijkl}(x)\chi_{ij}\chi_{kl} \ge 0$, for all $\chi = \{\chi_{ij}\} \in \mathcal{S}_d$, a.e. $x \in \Omega$.

<u>$H(c)$</u> : $c_{ij} \in L^\infty(\Omega)$, $c_{ij} = c_{ji}$.

<u>$H(k)$</u> : $k_{ij} \in L^\infty(\Omega)$, $k_{ij} = k_{ji}$, $k_{ij}(x)\xi_i\xi_j \ge \alpha_2\xi_i\xi_j$ for all $\xi = \{\xi_i\} \in \mathbb{R}^d$, a.e. $x \in \Omega$ with $\alpha_2 > 0$.

<u>$H(j_\nu)$</u> : $j_\nu \colon \Gamma_C \times (0,T) \times \mathbb{R} \to \mathbb{R}$ is such that

(i) $j_\nu(\cdot,\cdot,r)$ is measurable for all $r \in \mathbb{R}$;
(ii) $j_\nu(x,t,\cdot)$ is locally Lipschitz for a.e. $(x,t) \in \Gamma_C \times (0,T)$;
(iii) $|\eta| \le c_\nu\,(1 + |r|)$ for all $\eta \in \partial j_\nu(x,t,r), r \in \mathbb{R}$, a.e. $(x,t) \in \Gamma_C \times (0,T)$ with $c_\nu > 0$;
(iv) $j_\nu^0(x,t,r;-r) \le d_\nu\,(1 + |r|)$ for all $r \in \mathbb{R}$, a.e. $(x,t) \in \Gamma_C \times (0,T)$ with $d_\nu \ge 0$.

<u>$H(j_\tau)$</u> : $j_\tau \colon \Gamma_C \times (0,T) \times \mathbb{R}^d \to \mathbb{R}$ is such that

(i) $j_\tau(\cdot,\cdot,\xi)$ is measurable for all $\xi \in \mathbb{R}^d$;
(ii) $j_\tau(x,t,\cdot)$ is locally Lipschitz for a.e. $(x,t) \in \Gamma_C \times (0,T)$;
(iii) $\|\eta\| \le c_\tau\,(1 + \|\xi\|)$ for all $\eta \in \partial j_\tau(x,t,\xi), \xi \in \mathbb{R}^d$, a.e. $(x,t) \in \Gamma_C \times (0,T)$
with $c_\tau > 0$;
(iv) $j_\tau^0(x,t,\xi;-\xi) \le d_\tau\,(1 + \|\xi\|)$ for all $\xi \in \mathbb{R}^d$, a.e. $(x,t) \in \Gamma_C \times (0,T)$ with $d_\tau \ge 0$.

$\underline{H(j)}$: $j\colon \Gamma_C \times (0,T) \times \mathbb{R} \to \mathbb{R}$ is such that
(i) $j(\cdot,\cdot,r)$ is measurable for all $r \in \mathbb{R}$;
(ii) $j(x,t,\cdot)$ is locally Lipschitz for a.e. $(x,t) \in \Gamma_C \times (0,T)$;
(iii) $|\eta| \le c\,(1+|r|)$ for all $\eta \in \partial j(x,t,r), r \in \mathbb{R}$, a.e. $(x,t) \in \Gamma_C \times (0,T)$ with $c > 0$;
(iv) $j^0(x,t,r;-r) \le d\,(1+|r|)$ for all $r \in \mathbb{R}$, a.e. $(x,t) \in \Gamma_C \times (0,T)$ with $d \ge 0$.

$\underline{H(f)}$: $f_1 \in L^2(0,T;E^*)$, $f_2 \in L^2(0,T;L^2(\Gamma_N;\mathbb{R}^d))$, $g \in \mathcal{V}^*$, $u_0 \in E$, $u_1 \in H$ and $\theta_0 \in L^2(\Omega)$.

In the hypotheses $H(j_\nu)$, $H(j_\tau)$ and $H(j)$ the Clarke subdifferentials are taken with respect to the last variables of j_ν, j_τ and j, respectively.

Let us consider the functionals $J_1\colon (0,T) \times L^2(\Gamma_C;\mathbb{R}^d) \to \mathbb{R}$ and $J_2\colon (0,T) \times L^2(\Gamma_C) \to \mathbb{R}$ defined by

$$J_1(t,v) = \int_{\Gamma_C} \Big(j_\nu(x,t,v_\nu(x)) + j_\tau(x,t,v_\tau(x)) \Big)\, d\Gamma, \quad v \in L^2(\Gamma_C;\mathbb{R}^d),\ t \in (0,T),$$

$$J_2(t,\theta) = \int_{\Gamma_C} j(x,t,\theta(x))\, d\Gamma, \quad \theta \in L^2(\Gamma_C), \quad t \in (0,T).$$

Under the assumptions $H(j_\nu)$, $H(j_\tau)$ and $H(j)$ from Lemmas 4 and 5 of Denkowski and Migórski [**21**] (cf. also [**22**]), we know that J_1 and J_2 are well defined, $J_1(t,\cdot)$, $J_2(t,\cdot)$ are Lipschitz on every bounded subsets and satisfy suitable growth and sign conditions. We now formulate the following system of evolution inclusions:

find $u \in \mathcal{E}$ with $u' \in \mathbb{E}$ and $\theta \in \mathcal{W}$ such that

$$(118) \qquad \begin{cases} u''(t) + Au'(t) + Bu(t) + C_1\theta(t) + \gamma^*\partial J_1(t,\gamma u'(t)) \ni f(t) \ \text{ a.e. } t \\ \theta'(t) + C_2\theta(t) + C_3 u'(t) + \gamma_s^*\partial J_2(t,\gamma_s\theta(t)) \ni g(t) \ \text{ a.e. } t \\ u(0) = u_0, \quad u'(0) = u_1, \ \theta(0) = \theta_0. \end{cases}$$

By Proposition 6 in [**21**], we have

PROPOSITION 48. Under hypotheses $H(j_\nu)$, $H(j_\tau)$ and $H(j)$, every solution to the problem (118) is a solution to the system (117) of hemivariational inequalities. The converse also holds provided $j_\nu(x,t,\cdot)$, $j_\tau(x,t,\cdot)$ and $j(x,t,\cdot)$ (or $-j_\nu(x,t,\cdot)$, $-j_\tau(x,t,\cdot)$ and $-j(x,t,\cdot)$) are regular in the sense of Clarke.

The following is the main existence result for the problem (118). For the proof we refer to Denkowski and Migórski [**21**].

THEOREM 49. *Under the hypotheses $H(a)$, $H(b)$, $H(c)$, $H(k)$, $H(j_\nu)$, $H(j_\tau)$, $H(j)$ and $H(f)$, the problem (118) has a solution.*

Under some additional assumptions on the subdifferential terms we can get (cf. Theorem 15 of Denkowski and Migórski [**21**]) the uniqueness of solutions to the problem (118).

$\underline{H(J)}$: $J_1\colon (0,T) \times L^2(\Gamma_C;\mathbb{R}^d) \to \mathbb{R}$ and $J_2\colon (0,T) \times L^2(\Gamma_C) \to \mathbb{R}$ satisfy the following relaxed monotonicity conditions

$$(z_1 - z_2, v_1 - v_2)_{L^2(\Gamma_C;\mathbb{R}^d)} \ge -m_1 \|v_1 - v_2\|^2_{L^2(\Gamma_C;\mathbb{R}^d)}$$

for all $z_i \in \partial J_1(t, v_i)$, $v_i \in L^2(\Gamma_C; \mathbb{R}^d)$, $i = 1, 2$ and a.e. t with $m_1 > 0$,

$$(\eta_1 - \eta_2, \theta_1 - \theta_2)_{L^2(\Gamma_C)} \geq -m_2 \|\theta_1 - \theta_2\|^2_{L^2(\Gamma_C)}$$

for all $\eta_i \in \partial J_2(t, \theta_i)$, $\theta_i \in L^2(\Gamma_C)$, $i = 1, 2$ and a.e. t with $m_2 > 0$.

$\underline{(H_1)}$: $\alpha_1 > m_1 \|\gamma\|^2$ and $\alpha_2 > m_2 \|\gamma_s\|^2$, where $\|\gamma\| = \|\gamma\|_{\mathcal{L}(E, L^2(\Gamma; \mathbb{R}^d))}$ and $\|\gamma_s\| = \|\gamma_s\|_{\mathcal{L}(V, L^2(\Gamma))}$.

For the examples of functionals which satisfy $H(J)$ we refer to Migórski [**72**] and Migórski and Ochal [**83**].

THEOREM 50. *Besides the hypotheses of Theorem 49, we assume additionally $H(J)$ and (H_1). Then the problem (118) admits a unique solution.*

From Proposition 48 and Theorem 50 we have

COROLLARY 51. If the hypotheses of Theorem 50 are satisfied, and j_ν, j_τ and j or $-j_\nu$, $-j_\tau$ and $-j$ are regular with respect to their third variables, then the system (117) of hemivariational inequalities has a unique solution $u \in \mathcal{E}$ with $u' \in \mathbb{E}$ and $\theta \in \mathcal{W}$.

7.3. Bilateral contact problem with adhesion in piezoviscoelasticity. In this section we describe the mechanical piezoelectricity problem with adhesion and we present its variational formulation.

We consider the following physical setting. A viscoelastic piezoelectric body occupies, in its undeformed state, a bounded open set Ω of $\mathbb{R}^d$, $d = 2, 3$ with a regular boundary Γ. The body is in adhesive contact with an insulator obstacle, the so-called foundation. The contact is supposed to be bilateral, so during the process there is no gap between the body and the foundation. We consider two partitions of Γ. A first partition is given by three disjoint measurable parts Γ_D, Γ_N and Γ_C such that $m(\Gamma_D) > 0$, and a second one consists of two disjoint measurable parts Γ_1 and Γ_2 such that $m(\Gamma_1) > 0$ and $\Gamma_C \subset \Gamma_2$. We suppose that the body is clamped on Γ_D, thus the displacement field vanishes there. Surface tractions of density g act on Γ_N, Γ_C is a contacting surface and f_0 denotes the density of body forces. For simplicity, we assume also free electric charges. Due to the adhesive contact, we suppose a nonmonotone possibly multivalued law between the shear and the tangential displacement. This law depends also on a bonding field, denoted by β, which describes the pointwise fractional density of active bonds on Γ_C.

Following [**31, 32**], the evolution of the bonding field is governed by an ordinary differential equation depending on the displacement and considered on contact surface. When $\beta = 1$ at a point of the contact part, the adhesion is complete and all the bonds are active, when $\beta = 0$ all bonds are inactive and there is no adhesion, when $0 < \beta < 1$ the adhesion is partial and a fracture β of the bonds is active. Let $Q = \Omega \times (0, T)$ with $(0, T)$ being a finite time interval.

The classical dynamical model for the process under consideration is as follows: find a displacement field $u \colon Q \to \mathbb{R}^d$, an electric potential $\varphi \colon Q \to \mathbb{R}$ and a bonding field $\beta\colon \Gamma_C \times (0,T) \to [0,1]$ such that

$$u''(t) - \operatorname{Div}\sigma(t) = f_0(t) \quad \text{in } Q \tag{119}$$

$$\operatorname{div} D(t) = 0 \quad \text{in } Q \tag{120}$$

$$\sigma(t) = \mathcal{A}\varepsilon(u(t)) + \mathcal{B}\varepsilon(u'(t)) - \mathcal{P}^\top E(\varphi(t)) \quad \text{in } Q \tag{121}$$

$$D(t) = \mathcal{D}E(\varphi(t)) + \mathcal{P}\varepsilon(u(t)) \quad \text{in } Q \tag{122}$$

$$u(t) = 0 \quad \text{on } \Gamma_D \times (0,T) \tag{123}$$

$$\sigma(t)\nu = g(t) \quad \text{on } \Gamma_N \times (0,T) \tag{124}$$

$$\varphi(t) = 0 \quad \text{on } \Gamma_1 \times (0,T) \tag{125}$$

$$D(t)\cdot\nu = 0 \quad \text{on } \Gamma_2 \times (0,T) \tag{126}$$

$$u_\nu(t) = 0 \quad \text{on } \Gamma_C \times (0,T) \tag{127}$$

$$-\sigma_\tau(t) \in \partial j(\beta(t), u_\tau(t)) \quad \text{on } \Gamma_C \times (0,T) \tag{128}$$

$$\beta'(t) = F(t, u(t), \beta(t)) \quad \text{on } \Gamma_C \times (0,T) \tag{129}$$

$$\beta(0) = \beta_0 \quad \text{on } \Gamma_C \tag{130}$$

$$u(0) = u_0, \quad u'(0) = u_1 \text{ in } \Omega. \tag{131}$$

We provide comments on the equations and conditions in the above model.

In (119)–(131), in order to simplify the notation, we do not indicate explicitely the dependence of various functions and tensors on the spatial variable $x \in \Omega \cup \Gamma$. Equations (119) and (120) are the equation of motion for the stress field and the equilibrium equation for the electric displacement field, respectively. Recall that Div is the divergence operator for tensor valued functions given by $\operatorname{Div}\sigma = (\sigma_{ij,j})$ and div stands for the divergence operator for vector valued functions, i.e. $\operatorname{div} D = (D_{i,i})$. The relations (121) and (122) represent the electroviscoelastic constitutive law of the material in which $\mathcal{A}$, $\mathcal{B}$, $\mathcal{D}$ and $\mathcal{P}$ are respectively the (fourth order) elasticity tensor, the (fourth order) viscosity tensor, the (second order) electric permittivity tensor and the (third order) piezoelectric tensor. The equation (121) describes the converse effect and (122) models the direct effect. Furthermore, $\sigma\colon Q \to \mathcal{S}_d$, $\sigma = (\sigma_{ij})$ and $D\colon Q \to \mathbb{R}^d$, $D = (D_i)$ denote the stress tensor and the electric displacement field, respectively and $\mathcal{P}^\top$ is the tensor transposed to $\mathcal{P}$. We use here the notation $p^\top$ to denote the transpose of the tensor p given $p\tau \cdot v = \tau \cdot p^\top v$ for $\tau \in \mathcal{S}_d$ and $v \in \mathbb{R}^d$. The decoupled state (purely viscoelastic and purely electric deformations) can be obtained by setting the piezoelectric tensor $\mathcal{P} = 0$. The elastic strain–displacement and electric field–potential relations are given by

$$\varepsilon(u) = \frac{1}{2}(\nabla u + (\nabla u)^\top) \quad \text{in } Q,$$
$$E(\varphi) - -\nabla\varphi \quad \text{in } Q,$$

where $\varepsilon(u) = (\varepsilon_{ij}(u))$ and $E(\varphi) = (E_i(\varphi))$ denote the linear strain tensor and the electric vector field, respectively. Here $u\colon Q \to \mathbb{R}^d$, $u = (u_i)$ and $\varphi\colon \Omega \to \mathbb{R}$ are the displacement vector field and the electric potential (scalar field), respectively. The equations (123) and (124) are the displacement and traction boundary conditions, respectively while (125) and (126) represent the electric boundary conditions. Condition (127) shows that the contact is bilateral and (128) is the subdifferential boundary condition with a nonconvex nonsmooth superpotential j. This relation says that the tangential traction σ_τ depends on the intensity of adhesion β and the tangential displacement u_τ. The evolution of the adhesion (bonding) field β is governed by the ordinary differential equation (129) on the contact surface and (130) represents a given initial bonding field (cf. [**32**]). The adhesive evolution rate function F depends on both the bonding field and the displacement and may change sign. This allows for rebonding after debonding took place, and it allows for possible cycles of debonding and rebonding (cf. [**112**] for examples of F). Finally, the initial values for the displacement and the velocity are prescribed in (131).

We now turn to the variational formulation of the problem (119)–(131). We introduce the spaces for the displacement and the electric potential:

$$V = \{ v \in H^1(\Omega;\mathbb{R}^d) \mid v = 0 \quad \text{on } \Gamma_D,\ v_\nu = 0 \quad \text{on } \Gamma_C \},$$
$$\Phi = \{ \varphi \in H^1(\Omega) \mid \varphi = 0 \quad \text{on } \Gamma_1 \},$$

which are closed subspaces of $H^1(\Omega;\mathbb{R}^d)$ and $H^1(\Omega)$, respectively. Similarly as in the previous chapter, let $H = L^2(\Omega;\mathbb{R}^d)$ and $Z = H^\delta(\Omega;\mathbb{R}^d)$ with $\delta \in (\frac{1}{2}, 1)$. We have $V \subset Z \subset H \subset Z^* \subset V^*$ with all embeddings being dense and compact, so (V, H, V^*) forms an evolution triple of spaces. We also need the space $\mathcal{Z} = L^2(0,T;Z)$. Then $\mathcal{W} \subset \mathcal{V} \subset \mathcal{Z} \subset \widehat{\mathcal{H}} \subset \mathcal{Z}^* \subset \mathcal{V}^*$ with $\mathcal{Z}^* = L^2(0,T;Z^*)$. Following the Poincaré inequality $\|\varphi\|_{H^1(\Omega)} \le c\|\nabla\varphi\|_H$ for $\varphi \in \Phi$ with $c > 0$ and introducing the inner product $\langle \varphi, \psi \rangle_\Phi = \int_\Omega \nabla\varphi \cdot \nabla\psi \, dx$ for φ, $\psi \in \Phi$, it follows that $\|\cdot\|_{H^1(\Omega)}$ and $\|\cdot\|_\Phi$ are equivalent norms on Φ. Therefore $(\Phi, \|\cdot\|_\Phi)$ is also a Hilbert space.

Assuming sufficient smoothness of the functions and tensors involved in the problem (119)–(131), multiplying (119) by $v \in V$ and using the Green formula, we have

$$\langle u''(t), v \rangle_{V^* \times V} + \langle \sigma(t), \varepsilon(v) \rangle_{\mathcal{H}} = \int_\Gamma \sigma(t)\nu \cdot v \, d\Gamma + \langle f_0(t), v \rangle_{V^* \times V}$$

for a.e. $t \in (0,T)$. Using the boundary conditions (124), (127) and (128), we get

$$\int_\Gamma \sigma(t)\nu \cdot v \, d\Gamma = \int_{\Gamma_N} g(t) \cdot v \, d\Gamma + \int_{\Gamma_C} \sigma_\tau(t) \cdot v_\tau \, d\Gamma,$$
$$-\sigma_\tau(t) \cdot \xi \le j^0(x, \beta(x,t), u_\tau(x,t); \xi) \ \text{ for a.e. } (x,t) \in \Gamma_C \times (0,T), \ \text{ all } \xi \in \mathbb{R}^d.$$

Hence

(132)
$$\langle u''(t), v \rangle_{V^* \times V} + \langle \sigma(t), \varepsilon(v) \rangle_{\mathcal{H}} + \int_{\Gamma_C} j^0(x, \beta(x,t), u_\tau(x,t); v_\tau(x)) \, d\Gamma \ge \langle f(t), v \rangle_{V^* \times V}$$

for a.e. $t \in (0,T)$ and all $v \in V$, where

$$\langle f(t), v \rangle_{V^* \times V} = \langle f_0(t), v \rangle_H + \langle g(t), v \rangle_{L^2(\Gamma_N;\mathbb{R}^d)} \quad \text{for a.e. } t \in (0,T), \text{all } v \in V.$$

Next, let $\psi \in \Phi$. From (120), we have

$$-\langle D(t), \nabla\psi \rangle_H = 0. \tag{133}$$

Inserting (121) into (132) and (122) into (133), we obtain the following variational formulation: find a displacement field $u\colon (0,T) \to V$, an electric potential

$\varphi\colon (0,T) \to \Phi$ and a bonding field $\beta\colon (0,T) \to L^2(\Gamma_C)$ such that

(134)
$$\begin{cases} \langle u''(t), v\rangle_{V^*\times V} + \langle \mathcal{A}\varepsilon(u(t)), \varepsilon(v)\rangle_{\mathcal{H}} + \langle \mathcal{B}\varepsilon(u'(t)), \varepsilon(v)\rangle_{\mathcal{H}} + \langle \mathcal{P}^\top \nabla\varphi(t), \varepsilon(v)\rangle_{\mathcal{H}} \\ \qquad + \int_{\Gamma_C} j^0(x, \beta(t), u_\tau(t); v_\tau)\, d\Gamma \geq \langle f(t), v\rangle_{V^*\times V} \quad \text{for a.e. } t, \text{ all } v \in V \\ \langle \mathcal{D}\nabla\varphi(t), \nabla\psi\rangle_H - \langle \mathcal{P}\varepsilon(u(t)), \nabla\psi\rangle_H = 0 \quad \text{for a.e. } t, \text{ all } \psi \in \Phi \\ \beta'(t) = F(t, u(t), \beta(t)) \quad \text{on } \Gamma_C, \text{ for a.e. } t \in (0,T) \\ \beta(0) = \beta_0 \quad \text{on } \Gamma_C, \quad u(0) = u_0, \quad u'(0) = u_1 \quad \text{in } \Omega. \end{cases}$$

We impose the following hypotheses.

$\underline{H(a)}$: The elasticity tensor field $a = (a_{ijkl})$ satisfies $a_{ijkl} = a_{klij} = a_{ijlk} \in L^\infty(\Omega)$ and $a_{ijkl}(x)\tau_{ij}\tau_{kl} \geq m_a \tau_{ij}\tau_{ij}$ for a.e. $x \in \Omega$, all $\tau = (\tau_{ij}) \in \mathcal{S}_d$ with $m_a > 0$.

$\underline{H(b)}$: The viscosity tensor field $b = (b_{ijkl})$ satisfies $b_{ijkl} = b_{klij} = b_{ijlk} \in L^\infty(\Omega)$ and $b_{ijkl}(x)\tau_{ij}\tau_{kl} \geq m_b \tau_{ij}\tau_{ij}$ for a.e. $x \in \Omega$, all $\tau = (\tau_{ij}) \in \mathcal{S}_d$ with $m_b > 0$.

$\underline{H(d)}$: The dielectric tensor field $d = (d_{ij})$ satisfies $d_{ij} = d_{ji} \in L^\infty(\Omega)$ and $d_{ij}(x)\xi_i\xi_j \geq m_d \xi_i\xi_i$ for a.e. $x \in \Omega$, all $\xi = (\xi_i) \in \mathbb{R}^d$ with $m_d > 0$.

$\underline{H(p)}$: The piezoelectric tensor field $p = (p_{ijk})$ satisfies $p_{ijk} = p_{ikj} \in L^\infty(\Omega)$.

$\underline{H(j)}$: The superpotential $j\colon \Gamma_C \times \mathbb{R} \times \mathbb{R}^d \to \mathbb{R}$ is a function such that
- (i) $j(\cdot, r, \xi)$ is measurable for all $(r, \xi) \in \mathbb{R} \times \mathbb{R}^d$, $j(\cdot, 0, 0) \in L^1(\Gamma_C)$;
- (ii) $j(x, r, \cdot)$ is locally Lipschitz for a.e. $x \in \Gamma_C$ and all $r \in \mathbb{R}$;
- (iii) $\|\partial j(x, r, \xi)\|_{\mathbb{R}^d} \leq c_1(1 + \|\xi\|_{\mathbb{R}^d})$ for $(x, r, \xi) \in \Gamma_C \times \mathbb{R} \times \mathbb{R}^d$ with $c_1 > 0$;
- (iv) either $j(x, r, \cdot)$ or $-j(x, r, \cdot)$ is regular for a.e. $x \in \Gamma_C$ and all $r \in \mathbb{R}$;
- (v) $\limsup j^0(x, r_n, \xi_n; \zeta) \leq j^0(x, r, \xi; \zeta)$ for all $r_n \to r$ in $\mathbb{R}$, $\xi_n \to \xi$ in $\mathbb{R}^d$, all $\zeta \in \mathbb{R}^d$, a.e. $x \in \Gamma_C$,

where the Clarke subdifferential and the generalized directional derivative are taken with respect to the last variable of j.

$\underline{(H_0)}$: $f_0 \in L^2(0,T; L^2(\Omega; \mathbb{R}^d))$, $g \in L^2(0,T; L^2(\Gamma_N; \mathbb{R}^d))$, $u_0 \in V$, $u_1 \in H$, $\beta_0 \in L^2(\Gamma_C)$ and $0 \leq \beta_0(x) \leq 1$ a.e. $x \in \Gamma_C$.

$\underline{H(F)}$: The adhesive evolution rate function $F\colon \Gamma_C \times (0,T) \times \mathbb{R}^d \times \mathbb{R} \to \mathbb{R}$ satisfies
- (i) $F(\cdot, \cdot, \xi, r)$ is measurable on $\Gamma_C \times (0,T)$ for all $(\xi, r) \in \mathbb{R}^d \times \mathbb{R}$ and $F(x, t, \cdot, \cdot)$ is continuous on $\mathbb{R}^d \times \mathbb{R}$ for a.e. $(x,t) \in \Gamma_C \times (0,T)$;
- (ii) $|F(x, t, \xi, r_1) - F(x, t, \xi, r_2)| \leq L_F |r_1 - r_2|$ for all $\xi \in \mathbb{R}^d$, $r_1, r_2 \in \mathbb{R}$ and a.e. $(x,t) \in \Gamma_C \times (0,T)$ with $L_F > 0$;
- (iii) $|F(x, t, \xi_1, r_1) - F(x, t, \xi_2, r_2)| \leq L_F (\|\xi_1 - \xi_2\|_{\mathbb{R}^d} + |r_1 - r_2|)$ for all $\xi_i \in \mathbb{R}^d$, $r_i \in [0,1]$, $i = 1, 2$ and a.e. $(x,t) \in \Gamma_C \times (0,T)$ with $L_F > 0$;
- (iv) $F(x, t, \xi, 0) = 0$, $F(x, t, \xi, r) \geq 0$ for $r \leq 0$ and $F(x, t, \xi, r) \leq 0$ for $r \geq 1$, for a.e. $(x,t) \in \Gamma_C \times (0,T)$, all $\xi \in \mathbb{R}^d$.

The concrete examples of functions j and F which satisfy the hypotheses $H(j)$ and $H(F)$ are given in [**88**]. For the bonding field we will use the set

$$\mathcal{Q} = \{\theta\colon [0,T] \to L^2(\Gamma_C) \mid 0 \leq \theta(t) \leq 1 \text{ for all } t \in [0,T], \text{ a.e. on } \Gamma_C\}.$$

We define the following bilinear forms $a\colon V\times V\to\mathbb{R}$, $b\colon V\times V\to\mathbb{R}$, $d\colon \Phi\times\Phi\to\mathbb{R}$, $p\colon V\times\Phi\to\mathbb{R}$, $p^\top:\Phi\times V\to\mathbb{R}$ by

$$a(u,v)=\int_\Omega a_{ijkl}(x)\frac{\partial u_k}{\partial x_l}\frac{\partial v_i}{\partial x_j}\,dx \quad \text{for } u,v\in V,$$
$$b(u,v)=\int_\Omega b_{ijkl}(x)\frac{\partial u_k}{\partial x_l}\frac{\partial v_i}{\partial x_j}\,dx \quad \text{for } u,v\in V,$$
$$d(\varphi,\psi)=\int_\Omega d_{ij}(x)\frac{\partial \varphi}{\partial x_i}\frac{\partial \psi}{\partial x_j}\,dx \quad \text{for } \varphi,\psi\in \Phi,$$
$$p(u,\varphi)=\int_\Omega p_{ijk}(x)\frac{\partial u_i}{\partial x_j}\frac{\partial \varphi}{\partial x_k}\,dx \quad \text{for } u\in V,\ \varphi\in\Phi,$$
$$p^\top(\varphi,u)=\int_\Omega p_{kij}(x)\frac{\partial \varphi}{\partial x_k}\frac{\partial u_i}{\partial x_j}\,dx \quad \text{for } \varphi\in\Phi,\ u\in V.$$

Then, we have

$$\begin{aligned} a(u,v)&=\langle\mathcal{A}\varepsilon(u),\varepsilon(v)\rangle_{\mathcal{H}} \quad \text{for } u,v\in V,\\ b(u,v)&=\langle\mathcal{B}\varepsilon(u),\varepsilon(v)\rangle_{\mathcal{H}} \quad \text{for } u,v\in V,\\ d(\varphi,\psi)&=\langle\mathcal{D}\nabla\varphi,\nabla\psi\rangle_{H} \quad \text{for } \varphi,\psi\in\Phi,\\ p(u,\varphi)&=\langle\mathcal{P}\varepsilon(u),\nabla\varphi\rangle_{H} \quad \text{for } u\in V,\ \varphi\in\Phi,\\ p^\top(\varphi,u)&=\langle\mathcal{P}^\top\nabla\varphi,\varepsilon(u)\rangle_{\mathcal{H}} \quad \text{for } \varphi\in\Phi, u\in V,\end{aligned}$$

where the elasticity operator $\mathcal{A}\colon\Omega\times\mathcal{S}_d\to\mathcal{S}_d$ is given by $\mathcal{A}(x,\varepsilon)=a(x)\varepsilon$, $a(x)=(a_{ijkl}(x))$, the viscosity operator $\mathcal{B}\colon\Omega\times\mathcal{S}_d\to\mathcal{S}_d$ is given by $\mathcal{B}(x,\varepsilon)=b(x)\varepsilon$, $b(x)=(b_{ijkl}(x))$, the electric permittivity operator $\mathcal{D}\colon\Omega\times\mathbb{R}^d\to\mathbb{R}^d$ is defined by $\mathcal{D}(x,\xi)=d(x)\xi$, $d(x)=(d_{ij}(x))$, the piezoelectric operator $\mathcal{P}\colon\Omega\times\mathcal{S}_d\to\mathbb{R}^d$ is given by $\mathcal{P}(x,\varepsilon)=p(x)\varepsilon$, $p(x)=(p_{ijk}(x))$, the transpose to $\mathcal{P}$, $\mathcal{P}^\top:\Omega\times\mathbb{R}^d\to\mathcal{S}_d$ is given by $\mathcal{P}^\top(x,\xi)=p^\top(x)\xi$, $p^\top(x)=(p^\top_{ijk}(x))=(p_{kij}(x))$.

Using the above notation, the problem (134) is formulated as follows: find $u\in\mathcal{V}$ with $u'\in\mathcal{W}$, $\varphi\in C(0,T;\Phi)$ and $\beta\in W^{1,\infty}(0,T;L^2(\Gamma_C))\cap\mathcal{Q}$ such that

$$(135)\quad \begin{cases} \langle u''(t),v\rangle_{V^*\times V}+a(u(t),v)+b(u'(t),v)+p^\top(\varphi(t),v)\\ \quad +\displaystyle\int_{\Gamma_C} j^0(x,\beta(t),u_\tau(t);v_\tau)\,d\Gamma\ge\langle f(t),v\rangle_{V^*\times V} \quad \text{for a.e. } t,\ \text{all } v\in V\\ d(\varphi(t),\psi)=p(u(t),\psi) \quad \text{a.e. } t\in(0,T) \text{ and all } \psi\in\Phi\\ \beta'(t)=F(t,u(t),\beta(t)) \quad \text{on } \Gamma_C,\ \text{for a.e. } t\in(0,T)\\ \beta(0)=\beta_0 \quad \text{on } \Gamma_C, \quad u(0)=u_0, \quad u'(0)=u_1 \quad \text{in } \Omega.\end{cases}$$

This problem is a system coupled with the hemivariational inequality for the displacement, the time dependent stationary equation for the electric potential and the ordinary differential equation for the bonding field.

From the Lax-Milgram lemma, we deduce

LEMMA 52. *Assume $H(d)$, $H(p)$ and let $z\in V$. Then there exists a unique element $\varphi_z\in\Phi$ such that*

$$d(\varphi_z,\psi)=p(z,\psi) \quad \textit{for all } \psi\in\Phi$$

and the map $R\colon V\ni z\mapsto\varphi_z\in\Phi$ is linear and continuous.

Let us observe that if u, $u' \in \mathcal{V}$ and $W^{1,2}(0,T;V) \subset C(0,T;V)$, then $u \in C(0,T;V)$. From Lemma 52, we obtain the following result for the time dependent variational equation.

COROLLARY 53. If $H(d)$, $H(p)$ hold and $u \in C(0,T;V)$, then the problem

$$\begin{cases} \text{find } \varphi \in C(0,T;\Phi) \text{ such that} \\ d(\varphi(t),\psi) = p(u(t),\psi) \quad \text{for a.e. } t \in (0,T), \text{ all } \psi \in \Phi \end{cases}$$

has a unique solution $\varphi \in C(0,T;\Phi)$ and $\|\varphi\|_{C(0,T;\Phi)} \le c\|u\|_{C(0,T;V)}$ with $c > 0$. For all $t \in [0,T]$, we have $\varphi(t) = Ru(t)$, where the operator R is defined in Lemma 52.

Subsequently, since for every $\varphi \in \Phi$, the linear form $v \mapsto p^\top(\varphi, v)$ is continuous on V, so there exists a unique element $P\varphi \in V^*$ such that $p^\top(\varphi,v) = \langle P\varphi, v\rangle_{V^*\times V}$ for all $v \in V$ and $P \in \mathcal{L}(\Phi, V^*)$. We observe that

$$p^\top(\varphi, v) = p(v,\varphi) \quad \text{for all } v \in V, \ \varphi \in \Phi. \tag{136}$$

Analogously, we introduce operators A, $B \in \mathcal{L}(V,V^*)$ by $\langle Au, v\rangle_{V^*\times V} = a(u,v)$ and

$$\langle Bu, v\rangle_{V^*\times V} = b(u,v) \tag{137}$$

for all u, $v \in V$, respectively.

We are now in a position to reformulate the system (135). Since for $u \in \mathcal{V}$ such that $u' \in \mathcal{W}$, the second equation in (135) is uniquely solvable (cf. Corollary 53), we have

$$p^\top(\varphi(t),v) = \langle P\varphi(t), v\rangle_{V^*\times V} = \langle PRu(t), v\rangle_{V^*\times V} \quad \text{for all } v \in V, \ \text{a.e. } t \in (0,T).$$

Define the operator $G\colon V \to V^*$ by

$$Gv = Av + PRv \quad \text{for all } v \in V. \tag{138}$$

LEMMA 54. *Under hypotheses $H(a)$, $H(d)$ and $H(p)$, the operator $G\colon V \to V^*$ defined by (138) is linear, bounded, symmetric and coercive.*

Then the problem (135) has the following form: find $u \in \mathcal{V}$ with $u' \in \mathcal{W}$ and $\beta \in W^{1,\infty}(0,T;L^2(\Gamma_C)) \cap \mathcal{Q}$ such that

$$\begin{cases} \langle u''(t) + Bu'(t) + Gu(t), v\rangle_{V^*\times V} + \displaystyle\int_{\Gamma_C} j^0(x,\beta(t),u_\tau(t);v_\tau)\,d\Gamma \\ \qquad\qquad \ge \langle f(t), v\rangle_{V^*\times V} \text{ for a.e. } t \in (0,T), \text{ all } v \in V \\ \beta'(t) = F(t,u(t),\beta(t)) \text{ on } \Gamma_C, \text{ for a.e. } t \in (0,T) \\ \beta(0) = \beta_0 \text{ on } \Gamma_C, \quad u(0) = u_0, \quad u'(0) = u_1 \quad \text{in } \Omega. \end{cases} \tag{139}$$

Now we consider the differential equation which governs the evolution of the bonding field. We solve it for the adhesion field under the assumption that the displacement is given and study the continuous dependence of the adhesion on the displacement.

Lemma 55. *Assume that $H(F)$ holds, $\beta_0 \in L^2(\Gamma_C)$ satisfies $0 \le \beta_0(x) \le 1$ a.e. on Γ_C and $z \in L^2(0,T;L^2(\Gamma_C;\mathbb{R}^d))$. Then there exists $\beta \in W^{1,\infty}(0,T;L^2(\Gamma_C)) \cap \mathcal{Q}$ a unique solution of the Cauchy problem*

$$\begin{cases} \beta'(x,t) = F(x,t,z(x,t),\beta(x,t)) & a.e.\ on\ \Gamma_C \times (0,T) \\ \beta(x,0) = \beta_0(x) & on\ \Gamma_C. \end{cases} \tag{140}$$

Moreover, given $z_i \in L^2(0,T;L^2(\Gamma_C;\mathbb{R}^d))$ and denoting by $\beta_i \in W^{1,\infty}(0,T;L^2(\Gamma_C)) \cap \mathcal{Q}$ the unique solutions corresponding to z_i, $i = 1, 2$, we have

$$\|\beta_1(t) - \beta_2(t)\|_{L^2(\Gamma_C)} \le c \int_0^t \|z_1(s) - z_2(s)\|_{L^2(\Gamma_C;\mathbb{R}^d)}\, ds \tag{141}$$

for a.e. $t \in (0,T)$ with $c > 0$.

Let us define the operator $S\colon L^2(0,T;L^2(\Gamma_C;\mathbb{R}^d)) \to W^{1,\infty}(0,T;L^2(\Gamma_C)) \cap \mathcal{Q}$ by $S(z) = \beta$, where $z \in L^2(0,T;L^2(\Gamma_C;\mathbb{R}^d))$ and $\beta \in W^{1,\infty}(0,T;L^2(\Gamma_C))$ is the unique solution to the Cauchy problem (140) corresponding to z.

Corollary 56. From Lemma 55, it follows that under the hypothesis $H(F)$ and $\beta_0 \in L^2(\Gamma_C)$ with $0 \le \beta_0(x) \le 1$ a.e. on Γ_C, the operator S is well defined and continuous from $L^2(0,T;L^2(\Gamma_C;\mathbb{R}^d))$ to $L^\infty(0,T;L^2(\Gamma_C))$. The latter follows from (141) and the Hölder inequality.

Having in mind the definition of the operator S, we know that under the hypotheses of Lemma 55, the problem (139) can be formulated as follows: find $u \in \mathcal{V}$ such that $u' \in \mathcal{W}$ and

$$\begin{cases} \langle u''(t) + Bu'(t) + Gu(t), v\rangle_{V^* \times V} + \displaystyle\int_{\Gamma_C} j^0(x, S(\gamma u)(t), u_\tau(t); v_\tau)\, d\Gamma \\ \qquad\qquad \ge \langle f(t), v\rangle_{V^* \times V} \quad \text{for a.e. } t, \text{ all } v \in V \\ u(0) = u_0, \quad u'(0) = u_1. \end{cases} \tag{142}$$

In this way we have eliminated the unknowns φ and β in the problem (135). In what follows, we find a solution to (142) and then we recover φ and β by using Corollary 53 and the operator S.

We can now state the main result of this section.

Theorem 57. *Assume that the hypotheses $H(a)$, $H(b)$, $H(d)$, $H(j)$, $H(p)$, (H_0) hold, $S\colon L^2(0,T;L^2(\Gamma_C;\mathbb{R}^d)) \to L^\infty(0,T;L^2(\Gamma_C))$ is a continuous operator and*

$$m_a > 2\, c_1 T c_e^2 \|\gamma\|^2 \tag{143}$$

where $c_e > 0$ is the embedding constant of V onto Z, m_a and c_1 are given by $H(a)$ and $H(j)(iii)$, respectively. Then the problem (142) admits a solution.

The proof of Theorem 57 can be found in [**88**]. We conclude this section with the following

COROLLARY 58. Under the hypotheses $H(a)$, $H(b)$, $H(d)$, $H(j)$, $H(p)$, (H_0), $H(F)$ and (143), the problem (135) has a solution (u, φ, β) such that $u \in \mathcal{V}$ with $u' \in \mathcal{W}$, $\varphi \in C(0,T;\Phi)$ and $\beta \in W^{1,\infty}(0,T;L^2(\Gamma_C)) \cap \mathcal{Q}$.

We conclude this section with further comments on results for contact problems with nonsmooth and nonconvex superpotentials.

For contact problems with slip dependent friction, we refer to Migórski and Ochal [**82**], Migórski [**79**], Kulig [**46**], and Li and Liu [**48**], and results on quasistatic frictional contact are provided in Motreanu and Sofonea [**94**]. The hemivariational inequalities modeling contact with wear and adhesion have been studied in Bartosz [**2, 3**], and viscoelastic problems were investigated by Denkowski et al. [**25**], Migórski and Ochal [**84**], and Migórski et al. [**92**]. The applications of the theory of hemivariational inequalities to problems of piezoelectricity can be found in Migórski [**76, 78**].

References

[1] J.P. Aubin and A. Cellina, *Differential Inclusions. Set-Valued Maps and Viability Theory*, Springer-Verlag, Berlin, New York, Tokyo (1984).

[2] K. Bartosz, Hemivariational inequality approach to the dynamic viscoelastic sliding contact problem with wear, *Nonlinear Analysis. Theory, Methods & Applications* **65** (2006), 546–566.

[3] K. Bartosz, Hemivariational inequalities modeling dynamic contact problems with adhesion, *Nonlinear Analysis. Theory, Methods & Applications* **71** (2009), 1747–1762.

[4] F.E. Browder and P. Hess, Nonlinear mappings of monotone type in Banach spaces, *Journal of Functional Analysis* **11** (1972), 251–294.

[5] S. Carl, Enclosure of solutions for quasilinear dynamic hemivariational inequalities, *Nonlinear World* **3** (1996), 281–298.

[6] S. Carl, Existence of extremal solutions of boundary hemivariational inequalities, *Journal of Differential Equations* **171** (2001), 370–396.

[7] S. Carl, Extremal solutions of parabolic hemivariational inequalities, *Nonlinear Analysis* **47** (2001), 5077–5088.

[8] S. Carl, Comparison results for a class of quasilinear evolutionary hemivariational inequalities, *Discrete Contin. Dyn. Syst.* 2007, Dynamical Systems and Differential Equations. suppl. 221–229.

[9] S. Carl and R.P. Gilbert, Extremal solutions of a class of dynamic hemivariational inequalities, *Journal of Inequalities and Applications* **7** (2002), 479–502.

[10] S. Carl, V. K. Le and D. Motreanu, Existence and comparison results for quasilinear evolution hemivariational inequalities, *Electronic Journal of Differential Equations* **57** (2004), 1–17.

[11] S. Carl, V. K. Le and D. Motreanu, *Nonsmooth Variational Problems and Their Inequalities*, Springer, New York (2007).

[12] S. Carl and D. Motreanu, Extremal solutions of quasilinear parabolic inclusions with generalized Clarkes gradient, *J. Differential Equations* **191** (2003), 206–233.

[13] S. Carl, Vy K. Le and D. Motreanu, Evolutionary variational–hemivariational inequalities: Existence and comparison results, *J. Math. Anal. Appl.* **345** (2008), 545–558.

[14] K. C. Chang, Variational methods for nondifferentiable functionals and applications to partial differential equations, *Journal of Mathematical Analysis and Applications* **80** (1981), 102–129.

[15] V. ChiadoPiat, G. Dal Maso and A. Defranceschi, *G*-convergence of monotone operators, *Annales de l'Institut Henri Poincaré* **7** (1990), 124–160.

[16] F. H. Clarke, *Optimization and Nonsmooth Analysis*, Wiley, Interscience, New York (1983).

[17] F.H. Clarke, Y. S. Ledyaev, R. J. Stern and P.R. Wolenski, *Nonsmooth Analysis and Control Theory*, Springer, New York (1998).

[18] A. Dębińska–Nagórska, A. Just and Z. Stempień, The optimal control problem and approximation of some parabolic hemivariational inclusion, ZAMM **88** (2008), 218–226.

[19] Z. Denkowski and S. Migórski, Optimal shape design problems for a class of systems described by hemivariational inequalities, *J. Global Optimization* **12** (1998), 37–59.

[20] Z. Denkowski and S. Migórski, Sensitivity of optimal solutions to control problems for systems described by hemivariational inequalities, *Control and Cybernetics* **33** (2004), 211–236.

[21] Z. Denkowski and S. Migórski, A system of evolution hemivariational inequalities modeling thermoviscoelastic frictional contact, *Nonlinear Analysis* **60** (2005), 1415–1441.

[22] Z. Denkowski and S. Migórski, Hemivariational inequalities in thermoviscoelasticity, *Nonlinear Analysis* **63** (2005), e87–e97.

[23] Z. Denkowski and S. Migórski, On Sensitivity of Optimal Solutions to Control Problems for Hyperbolic Hemivariational Inequalities, Chapter XII in the book "Control and Boundary Analysis," J. Cagnol and J.-P. Zolesio (eds), Marcel Dekker, 2005, 147–156.

[24] Z. Denkowski and S. Migórski, Existence of solutions to evolution second order hemivariational inequalities with multivalued damping, System modeling and optimization, 203–215, IFIP Int. Fed. Inf. Process., 166, Kluwer Acad. Publ., Boston, MA, 2005. Proceedings.

[25] Z. Denkowski, S. Migórski and A. Ochal, Existence and uniqueness to a dynamic bilateral frictional contact problem in viscoelasticity, *Acta Appl. Math.* **94** (2006), 251–276.

[26] Z. Denkowski, S. Migórski and A. Ochal, Optimal control for a class of mechanical thermoviscoelastic frictional contact problems, *Control and Cybernetics* **36** (2007), 611–632.

[27] Z. Denkowski, S. Migórski and N.S. Papageorgiou, *An Introduction to Nonlinear Analysis: Theory*, Kluwer Academic/Plenum, Boston, Dordrecht, London, New York (2003).

[28] Z. Denkowski, S. Migórski and N.S. Papageorgiou, *An Introduction to Nonlinear Analysis: Applications*, Kluwer Academic/Plenum, Boston, Dordrecht, London, New York (2003).

[29] G. Duvaut and J. L. Lions, *Les Inéquations en Mécanique et en Physique*, Dunod, Paris (1972).

[30] C. Eck, J. Jarušek and M. Krbeč, *Unilateral Contact Problems: Variational Methods and Existence Theorems*, Chapman/CRC Press, New York (2005).

[31] M. Frémond, Adhérence des solides, *J. Mécanique Théorique et Appliquée* **6** (1987), 383–407.

[32] M. Frémond, *Non-Smooth Thermomechanics.* Springer, Berlin (2002).

[33] L. Gasiński, Existence of solutions for hyperbolic hemivariational inequalities, *Journal of Mathematical Analysis and Applications* **276** (2002), 723–746.

[34] L. Gasiński and M. Smołka, An existence theorem for wave-type hyperbolic hemivariational inequalities, *Math. Nachr.* **242** (2002), 79–90.

[35] P. Germain, *Cours de Mécanique des milieux continus I*, Masson, Paris (1973).

[36] D. Goeleven, M. Miettinen and P.D. Panagiotopoulos, Dynamic hemivariational inequalities and their applications, *Journal of Optimization Theory and Applications* **103** (1999), 567–601.

[37] D. Goeleven, D. Motreanu, Y. Dumont and M. Rochdi, *Variational and Hemivariational Inequalities: Theory, Methods and Applications*, Kluwer Academic Publishers, Boston, Dordrecht, London (2003).

[38] W. Han and M. Sofonea, *Quasistatic Contact Problems in Viscoelasticity and Viscoplasticity*, American Mathematical Society/International Press, Providence, RI (2002).

[39] A. Hasanov, An inverse coefficient problem for an elasto-plastic medium, *SIAM J. Appl. Math.* **55** (1995), 1736–1752.

[40] A. Hasanov, Inverse coefficient problems for monotone potential operators, *Inverse Problems* **13** (1997), 1265–1278.

[41] A. Hasanov, Monotonicity of nonlinear boundary value problems related to deformation theory of plasticity, *Mathematical Problems in Engineering*, volume 2006, Article ID 58143, pages 1–18.

[42] A. Hasanov and Z. Liu, An inverse coefficient problem for a nonlinear parabolic variational inequality, *Appl. Math. Letters* **21** (2008), 563–570.

[43] J. Haslinger, M. Miettinen and P.D. Panagiotopoulos, *Finite Element Method for Hemivariational Inequalities. Theory, Methods and Applications,* Kluwer Academic Publishers, Boston, Dordrecht, London (1999).

[44] S. Hu and N. S. Papageorgiou, *Handbook of Multivalued Analysis, Volume I: Theory,* Kluwer, Dordrecht, The Netherlands (1997).

[45] P. Kalita, Decay of energy of a system described by hyperbolic hemivariational inequality, *in preparation.*

[46] A. Kulig, Hemivariational inequality approach to the dynamic viscoelastic contact problem with nonmonotone normal compliance and slip-dependent friction, *Nonlinear Analysis. Real World Applications* **9** (2008), 1741–1755.

[47] A. Kulig, Nonlinear Evolution Inclusions and Hemivariational Inequalities for Nonsmooth Problems in Contact Mechanics, PhD thesis, Jagiellonian University, Krakow, Poland, 2009.

[48] Y. Li and Z. Liu, Dynamic contact problem for viscoelastic piezoelectric materials with slip dependent friction, *Nonlinear Analysis. Theory, Methods & Applications* **71** (2009), 1414–1424.

[49] J.L. Lions, *Quelques méthodes de résolution des problémes aux limites non linéaires*, Dunod, Paris (1969).

[50] Z. Liu, Some convergence results for evolution hemivariational inequalities, *J. Global Optim.* **29** (2004), 85–95.

[51] Z. Liu, Existence results for evolution noncoercive hemivariational inequalities, *J. Optim. Theory Appl.* **120** (2004), 417–427.

[52] Z. Liu, Browder–Tikhonov regularization of non-coercive evolution hemivariational inequalities, *Inverse Problems* **21** (2005), 13–20.

[53] Z. Liu, Existence results for quasilinear parabolic hemivariational inequalities, *J. Diff. Eqns.* **244** (2008), 1395–1409.

[54] Z. Liu and S. Migórski, Noncoercive damping in dynamic hemivariational inequality with application to problem of piezoelectricity, *Discrete and Continuous Dynamical Systems Series B* **9** (2008), 129–143.

[55] Z. Liu and S. Migórski, On the convergence of solutions to parabolic hemivariational inequality, in preparation.

[56] Z. Liu, S. Migórski and A. Ochal, Homogenization of boundary hemivariational inequalities in linear elasticity, *J. Math. Anal. Appl.* **340** (2008), 1347–1361.

[57] M. Miettinen, A parabolic hemivariational inequality, *Nonlinear Analysis. Theory, Methods & Applications* **26** (1996), 725–734.

[58] M. Miettinen and J. Haslinger, Approximation of optimal control problems of hemivariational inequalities, *Numer. Funct. Anal. and Optimiz.* **13** (1992), 43–68.

[59] M. Miettinen, M. Mäkëla and J. Haslinger, On numerical solution of hemivariational inequalities by nonsmooth optimization methods, *J. Global Optimization* **6** (1995), 401–425.

[60] M. Miettinen and P.D. Panagiotopoulos, Hysteresis and hemivariational inequalities: semilinear case, *J. Global Optimization* **13** (1998) 269–298.

[61] M. Miettinen and P.D. Panagiotopoulos, On parabolic hemivariational inequalities and applications, *Nonlinear Analysis. Theory, Methods & Applications* **35** (1999), 885–915.

[62] M. Mietinen and P.D. Panagiotopoulos, Finite element approximation of parabolic hemivariational inequalities, *Numer. Funct. Anal. Optimiz.* **19** (1999), 565–585.

[63] M. Miettinen and P.D. Panagiotopoulos, On parabolic hemivariational inequalities and applications, *Nonlinear Analysis* **35** (1999) 885–915.

[64] S. Migórski, Existence, variational and optimal control problems for nonlinear second order evolution inclusions, *Dynamic Systems and Applications* **4** (1995), 513–528.

[65] S. Migórski, Existence and relaxation results for nonlinear second order evolution inclusions, *Discussiones Mathematicae. Differential Inclusions* **15** (1995), 129–148.

[66] S. Migórski, Identification of nonlinear heat transfer laws in problems modeled by hemivariational inequalities, Proceedings of International Symposium on Inverse Problems in Engineering Mechanics 1998 (ISIP'98), M. Tanaka and G. S. Dulikravich (Eds.). Elsevier Science B.V. (1998) 27–37.

[67] S. Migórski, Existence and convergence results for evolution hemivariational inequalities. *Topological Methods in Nonlinear Analysis* **16** (2000), 125–144.

[68] S. Migórski, Parameter identification for evolution hemivariational inequalities and applications, Proceedings of International Symposium on Inverse Problems in Engineering Mechanics 2001 (ISIP2001), M. Tanaka and G. S. Dulikravich (Eds.), Elsevier Science B.V. (2002), 211–218.

[69] S. Migórski, On existence of solutions for parabolic hemivariational inequalities, *J. Comp. Appl. Math.* **129** (2001), 77–87.

[70] S. Migórski, Homogenization technique in inverse problems for boundary hemivariational inequalities, *Inverse Problems in Engineering* **11** (2003), 229–242.

[71] S. Migórski, Hemivariational inequalities modeling viscous incompressible fluids, *Journal of Nonlinear and Convex Analysis* **5** (2004), 217–227.

[72] S. Migórski, Dynamic hemivariational inequality modeling viscoelastic contact problem with normal damped response and friction, *Applicable Analysis* **84** (2005), 669–699.

[73] S. Migórski, Boundary hemivariational inequalities of hyperbolic type and applications, *Journal of Global Optimization* **31** (2005), 505–533.

[74] S. Migórski, Hemivariational inequalities modeling dynamic viscoelastic contact problems with friction. Advances in analysis, 295–304, World Sci. Publ., Hackensack, NJ, 2005. Proceedings.

[75] S. Migórski, Evolution hemivariational inequality for a class of dynamic viscoelastic nonmonotone frictional contact problems, *Computers & Mathematics with Applications* **52** (2006), 677–698.

[76] S. Migórski, Hemivariational inequality for a frictional contact problem in elasto-piezoelectricity, *Discrete and Continuous Dynamical Systems B* **6** (2006), 1339–1356.

[77] S. Migórski, Optimal control of contact problems described by evolution hemivariational inequalities, Frontiers on Optimization and Control, Hou S.H., Yang X.M., Chen G.Y. (eds.), The Hong Kong Polytechnic University, Hong Kong, 2007, 97–123.

[78] S. Migórski, Dynamic Viscoelastic Piezoelectric Contact with Adhesion, *Institute of Physics Conference Series* **96** (2008), 1–6.

[79] S. Migórski, A class of hemivariational inequalities for electroelastic contact problems with slip dependent friction, *Discrete and Continuous Dynamical Systems, S* **1** (2008), 117–126.

[80] S. Migórski and A. Ochal, Optimal control of parabolic hemivariational inequalities, *Journal of Global Optimization* **17** (2000), 285–300.

[81] S. Migórski and A. Ochal, Boundary hemivariational inequality of parabolic type, *Nonlinear Analysis. Theory, Methods & Applications* **57** (2004), 579–596.

[82] S. Migórski and A. Ochal, Hemivariational inequality for viscoelastic contact problem with slip dependent friction, *Nonlinear Analysis. Theory, Methods & Applications* **61** (2005), 135–161.

[83] S. Migórski and A. Ochal, A unified approach to dynamic contact problems in viscoelasticity, *Journal of Elasticity* **83** (2006), 247–275.

[84] S. Migórski and A. Ochal, Existence of solutions for second order evolution inclusions with application to mechanical contact problems, *Optimization* **55** (2006), 101–120.

[85] S. Migórski and A. Ochal, Nonlinear impulsive evolution inclusions of second order, *Dynamic Systems and Applications* **16** (2007), 155–174.

[86] S. Migórski and A. Ochal, Vanishing viscosity for hemivariational inequality modeling dynamic problems in elasticity, *Nonlinear Analysis* **66** (2007), 1840–1852.

[87] S. Migórski and A. Ochal, Navier–Stokes problems modeled by evolution hemivariational inequalities, *Discrete and Continuous Dynamical Systems*, Supplement (2007), 731–740.

[88] S. Migórski and A. Ochal, Dynamic bilateral contact problem for viscoelastic piezoelectric materials with adhesion, *Nonlinear Analysis. Theory, Methods & Applications* **69** (2008) 495–509.

[89] S. Migórski and A. Ochal, Quasistatic hemivariational inequality via vanishing acceleration approach, *SIAM Journal on Mathematical Analysis* **41** (2009), 1415–1435.

[90] S. Migórski and A. Ochal, An inverse coefficient problem for a parabolic hemivariational inequality, *Applicable Analysis*, **89** (2010), 243–256.

[91] S. Migórski, A. Ochal and M. Sofonea, Integrodifferential hemivariational inequalities with applications to viscoelastic frictional contact, *Mathematical Models and Methods in Applied Sciences* **18** (2008), 271–290.

[92] S. Migórski, A. Ochal and M. Sofonea, Analysis of a dynamic elastic-viscoplastic contact problem with friction, *Discrete and Continuous Dynamical Systems, B* **10** (2008), 887–902.

[93] D. Motreanu and P.D. Panagiotopoulos, *Minimax Theorems and Qualitative Properties of the Solutions of Hemivariational Inequalities and Applications*, Kluwer Academic Publishers, Boston, Dordrecht, London (1999).

[94] D. Motreanu and M. Sofonea, Evolutionary variational inequalities arising in quasistatic frictional contact problems for elastic materials, *Abstract and Applied Analysis* **4** (1999), 255–279.

[95] Z. Naniewicz, Pseudo-monotonicity and economic equilibrium problem in reflexive Banach space, *Math. Oper. Res.* **32** (2007) 436–466.

[96] Z. Naniewicz and P. D. Panagiotopoulos, *Mathematical Theory of Hemivariational Inequalities and Applications*, Marcel Dekker, New York, Basel, Hong Kong (1995).
[97] B. Nayroles, Point de vue algébrique. Convexité et intégrandes convexes en Méchanique des solides, in: *New Variational Techniques in Mathematical Physics*, CIME Edizioni Cremonese, Roma 1974.
[98] J. Nečas and I. Hlaváček, *Mathematical Theory of Elastic anf Elasto-plastic Bodies: An Introduction*, Elsevier, Amsterdam, Oxford, New York (1981).
[99] A. Ochal, Existence results for evolution hemivariational inequalities of second order, *Nonlinear Analysis: Theory, Methods & Applications* **60** (2005), 1369–1391.
[100] P. D. Panagiotopoulos, *Inequality Problems in Mechanics and Applications. Convex and Nonconvex Energy Functions*, Birkhäuser, Basel (1985).
[101] P. D. Panagiotopoulos, Nonconvex problems of semipermeable media and related topics, *Z. Angew. Math. Mech.* **65** (1985) 29–36.
[102] P. D. Panagiotopoulos, Optimal control of systems governed by hemivariational inequalities. Necessary conditions, in: *International Series of Numerical Mathematics*, K. H. Hoffmann, J. Sprekels, Eds., vol. 95, Birkhäuser, Basel (1990).
[103] P. D. Panagiotopoulos, Coercive and semicoercive hemivariational inequalities, *Nonlinear Analysis* **16** (1991), 209–231.
[104] P. D. Panagiotopoulos, *Hemivariational Inequalities, Applications in Mechanics and Engineering*, Springer, Berlin (1993).
[105] P.D. Panagiotopoulos, Modelling of nonconvex nonsmooth energy problems: dynamic hemivariational inequalities with impact effects, *Journal of Computational and Applied Mathematics* **63** (1995), 123–138.
[106] P. D. Panagiotopoulos and J. Haslinger, Optimal control and identification of structures involving multivalued nonmonotonicities. Existence and approximation results, *Eur. J. Mech., A/Solids* **11** (1992), 425–445.
[107] P.D. Panagiotopoulos, Hemivariational inequalities and Fan-variational inequalities. new applications and results, *Atti sem. Mat. Fis. Univ. Modena*, **43** (1995), 159–191.
[108] P. D. Panagiotopoulos and G. Pop, On a type of hyperbolic variational-hemivariational inequalities, *Journal of Applied Analysis* **5** (1999), 95–112.
[109] N. S. Papageorgiou, On the existence of solutions for nonlinear parabolic problems with nonmonotone discontinuities, *Journal of Mathematical Analysis and Applications* **205** (1997), 434–453.
[110] J.Y. Park and T.G. Ha, Existence of antiperiodic solutions for hemivariational inequalities, *Nonlinear Analysis. Theory, Methods & Applications* **68** (2008), 747–767.
[111] S. H. Park, J. Y. Park and J. M. Jeong, Boundary stabilization of hyperbolic hemivariational inequalities, *Acta Appl. Math.* **104** (2008), 139–150.
[112] M. Shillor, M. Sofonea and J.J. Telega, *Models and Analysis of Quasistatic Contact*, Springer-Verlag, Berlin (2004).
[113] M. Sofonea, W. Han and M. Shillor, *Analysis and Approximation of Contact Problems with Adhesion or Damage*, Chapman & Hall/CRC, Boca Raton (2006).
[114] M. Sofonea and A. Matei, *Variational Inequalities with Applications. A study of Antiplane Frictional Contact Problems*, Advances in Mechanics and Mathematics, Springer (2009).
[115] Y. Xiao and N. Huang, Browder-Tikhonov regularization for a class of evolution second order hemivariational inequalities, *Journal of Global Optimization*, **45** (2009), 371–388.
[116] E. Zeidler, *Nonlinear Functional Analysis and Applications, II A/B*, Springer Verlag, New York (1990).

Stanisław Migórski, Faculty of Mathematics and Computer Science, Jagiellonian University, Institute of Computer Science, ul. Stanisława Łojasiewicza 6, 30348 Krakow, Poland

E-mail address: migorski@ii.uj.edu.pl

Morse theory and applications to variational problems

Kanishka Perera

Contents

1. Introduction

The purpose of this chapter is to introduce the reader to the topic of Morse theory, with applications to variational problems. General references are Milnor [**68**], Mawhin and Willem [**66**], Chang [**21**], and Benci [**12**]; see also Perera et al. [**83**] and Perera and Schechter [**84**].

We consider a real-valued function Φ defined on a real Banach space $(W, \|\cdot\|)$. We say that Φ is Fréchet differentiable at $u \in W$ if there is an element $\Phi'(u)$ of the dual space $(W^*, \|\cdot\|^*)$, called the Fréchet derivative of Φ at u, such that

$$\Phi(u+v) = \Phi(u) + (\Phi'(u), v) + \mathrm{o}(\|v\|) \quad \text{as} \quad v \to 0 \quad \text{in} \quad W,$$

where $(\cdot, \cdot)$ is the duality pairing. The functional Φ is continuously Fréchet differentiable on W, or belongs to the class $C^1(W, \mathbb{R})$, if Φ' is defined everywhere and the map $W \to W^*$, $u \mapsto \Phi'(u)$ is continuous. We assume that $\Phi \in C^1(W, \mathbb{R})$ for the rest of the chapter. We say that u is a critical point of Φ if $\Phi'(u) = 0$. A real number $c \in \Phi(W)$ is a critical value of Φ if there is a critical point u with $\Phi(u) = c$, otherwise it is a regular value. We use the notations

$$\Phi_a = \big\{u \in W : \Phi(u) \ge a\big\}, \quad \Phi^b = \big\{u \in W : \Phi(u) \le b\big\}, \quad \Phi_a^b = \Phi_a \cap \Phi^b,$$

$$K = \big\{u \in W : \Phi'(u) = 0\big\}, \quad \widetilde{W} = W \setminus K, \quad K_a^b = K \cap \Phi_a^b, \quad K^c = K_c^c$$

for the various superlevel, sublevel, critical, and regular sets of Φ.

We begin by recalling the compactness condition of Palais and Smale and its weaker variant given by Cerami in Section 2. Then we state the first and second deformation lemmas under the Cerami's condition in Section 3. In Section 4 we define the critical groups of an isolated critical point and summarize the basic results of Morse theory. These include the Morse inequalities, Morse lemma and its generalization splitting lemma, shifting theorem of Gromoll and Meyer, and the handle body theorem. Next we discuss the minimax principle in Section 5. Section 6 contains a discussion of homotopical linking, pairs of critical points with nontrivial critical groups produced by homological linking, and nonstandard geometries without a finite dimensional closed loop. We recall the notion of local linking and an alternative for a critical point produced by a local linking in Section 7. In Section 8 we present some recent results on critical groups associated with jumping nonlinearities. We conclude with a result on nontrivial critical groups associated with the p-Laplacian in Section 9.

2. Compactness conditions

It is usually necessary to assume some sort of a "compactness condition" when seeking critical points of a functional. The following condition was originally introduced by Palais and Smale [**74**]: Φ satisfies the Palais-Smale compactness condition at the level c, or $(\mathrm{PS})_c$ for short, if every sequence $(u_j) \subset W$ such that

$$\Phi(u_j) \to c, \quad \Phi'(u_j) \to 0,$$

called a $(\mathrm{PS})_c$ sequence, has a convergent subsequence; Φ satisfies (PS) if it satisfies $(\mathrm{PS})_c$ for every $c \in \mathbb{R}$, or equivalently, if every sequence such that $\Phi(u_j)$ is bounded and $\Phi'(u_j) \to 0$, called a (PS) sequence, has a convergent subsequence.

The following weaker version was introduced by Cerami [**18**]: Φ satisfies the Cerami condition at the level c, or $(\mathrm{C})_c$ for short, if every sequence such that

$$\Phi(u_j) \to c, \quad \big(1 + \|u_j\|\big)\, \Phi'(u_j) \to 0,$$

called a $(\mathrm{C})_c$ sequence, has a convergent subsequence; Φ satisfies (C) if it satisfies $(\mathrm{C})_c$ for every c, or equivalently, if every sequence such that $\Phi(u_j)$ is bounded and $\big(1 + \|u_j\|\big)\, \Phi'(u_j) \to 0$, called a (C) sequence, has a convergent subsequence. This condition is weaker since a $(\mathrm{C})_c$ (resp. (C)) sequence is clearly a $(\mathrm{PS})_c$ (resp. (PS)) sequence also.

The limit of a $(\mathrm{PS})_c$ (resp. (PS)) sequence is in K^c (resp. K) since Φ and Φ' are continuous. Since any sequence in K^c is a $(\mathrm{C})_c$ sequence, it follows that K^c is a compact set when $(\mathrm{C})_c$ holds.

3. Deformation lemmas

An essential tool for locating critical points is the deformation lemmas, which allow to lower sublevel sets of a functional, away from its critical set. The main ingredient in their proofs is a suitable negative pseudo-gradient flow, a notion due to Palais [**76**]: a pseudo-gradient vector field for Φ on $\widetilde{W}$ is a locally Lipschitz continuous mapping $V : \widetilde{W} \to W$ satisfying

$$\|V(u)\| \leq \|\Phi'(u)\|^*, \quad 2\,(\Phi'(u), V(u)) \geq \left(\|\Phi'(u)\|^*\right)^2 \quad \forall u \in \widetilde{W}.$$

Such a vector field exists, and may be chosen to be odd when Φ is even.

For $A \subset W$, let

$$N_\delta(A) = \left\{u \in W : \operatorname{dist}(u, A) \leq \delta\right\}$$

be the δ-neighborhood of A. The first deformation lemma provides a local deformation near a (possibly critical) level set of a functional.

Lemma 1 (First Deformation Lemma). *If $c \in \mathbb{R}$, C is a bounded set containing K^c, $\delta, k > 0$, and Φ satisfies* $(\mathrm{C})_c$, *then there are an $\varepsilon_0 > 0$ and, for each $\varepsilon \in (0, \varepsilon_0)$, a map $\eta \in C([0,1] \times W, W)$ satisfying*

(i) $\eta(0, \cdot) = id_W$,

(ii) *$\eta(t, \cdot)$ is a homeomorphism of W for all $t \in [0,1]$,*

(iii) *$\eta(t, \cdot)$ is the identity outside $A = \Phi^{c+2\varepsilon}_{c-2\varepsilon} \setminus N_{\delta/3}(C)$ for all $t \in [0,1]$,*

(iv) $\|\eta(t,u) - u\| \leq \left(1 + \|u\|\,\right) \delta/k \quad \forall (t,u) \in [0,1] \times W$,

(v) *$\Phi(\eta(\cdot, u))$ is nonincreasing for all $u \in W$,*

(vi) $\eta(1, \Phi^{c+\varepsilon} \setminus N_\delta(C)) \subset \Phi^{c-\varepsilon}$.

When Φ is even and C is symmetric, η may be chosen so that $\eta(t, \cdot)$ is odd for all $t \in [0,1]$.

First deformation lemma under the $(\mathrm{PS})_c$ condition is due to Palais [**75**]; see also Rabinowitz [**94**]. The proof under the $(\mathrm{C})_c$ condition was given by Cerami [**18**] and Bartolo, Benci, and Fortunato [**8**]. The particular version given here can be found in Perera et al. [**83**].

The second deformation lemma implies that the homotopy type of sublevel sets can change only when crossing a critical level.

Lemma 2 (Second Deformation Lemma). *If $-\infty < a < b \leq +\infty$ and Φ has only a finite number of critical points at the level a, has no critical values in (a,b), and satisfies $(\mathrm{C})_c$ for all $c \in [a,b] \cap \mathbb{R}$, then Φ^a is a deformation retract of $\Phi^b \setminus K^b$, i.e., there is a map $\eta \in C([0,1] \times (\Phi^b \setminus K^b), \Phi^b \setminus K^b)$, called a deformation retraction of $\Phi^b \setminus K^b$ onto Φ^a, satisfying*

(i) $\eta(0, \cdot) = id_{\Phi^b \setminus K^b}$,

(ii) $\eta(t, \cdot)|_{\Phi^a} = id_{\Phi^a} \quad \forall t \in [0,1]$,

(iii) $\eta(1, \Phi^b \setminus K^b) = \Phi^a$.

Second deformation lemma under the $(\mathrm{PS})_c$ condition is due to Rothe [**103**], Chang [**20**], and Wang [**115**]. The proof under the $(\mathrm{C})_c$ condition can be found in Bartsch and Li [**9**], Perera and Schechter [**90**], and Perera et al. [**83**].

4. Critical groups

In Morse theory the local behavior of Φ near an isolated critical point u is described by the sequence of critical groups

$$C_q(\Phi, u) = H_q(\Phi^c \cap U, \Phi^c \cap U \setminus \{u\}), \quad q \geq 0 \tag{1}$$

where $c = \Phi(u)$ is the corresponding critical value, U is a neighborhood of u containing no other critical points, and H denotes singular homology. They are independent of the choice of U by the excision property.

For example, if u is a local minimizer, $C_q(\Phi, u) = \delta_{q0}\, \mathcal{G}$ where δ is the Kronecker delta and $\mathcal{G}$ is the coefficient group. A critical point u with $C_1(\Phi, u) \neq 0$ is called a mountain pass point.

Let $-\infty < a < b \leq +\infty$ be regular values and assume that Φ has only isolated critical values $c_1 < c_2 < \cdots$ in (a, b), with a finite number of critical points at each level, and satisfies $(\mathrm{PS})_c$ for all $c \in [a, b] \cap \mathbb{R}$. Then the Morse type numbers of Φ with respect to the interval (a, b) are defined by

$$M_q(a, b) = \sum_i \operatorname{rank} H_q(\Phi^{a_{i+1}}, \Phi^{a_i}), \quad q \geq 0$$

where $a = a_1 < c_1 < a_2 < c_2 < \cdots$. They are independent of the a_i by the second deformation lemma, and are related to the critical groups by

$$M_q(a, b) = \sum_{u \in K_a^b} \operatorname{rank} C_q(\Phi, u).$$

Writing $\beta_j(a, b) = \operatorname{rank} H_j(\Phi^b, \Phi^a)$, we have

THEOREM 3 (Morse Inequalities). *If there is only a finite number of critical points in Φ_a^b, then*

$$\sum_{j=0}^{q} (-1)^{q-j} M_j \geq \sum_{j=0}^{q} (-1)^{q-j} \beta_j, \quad q \geq 0,$$

and

$$\sum_{j=0}^{\infty} (-1)^j M_j = \sum_{j=0}^{\infty} (-1)^j \beta_j$$

when the series converge.

Critical groups are invariant under homotopies that preserve the isolatedness of the critical point; see Rothe [**102**], Chang and Ghoussoub [**19**], and Corvellec and Hantoute [**23**].

THEOREM 4. *If Φ_t, $t \in [0, 1]$ is a family of C^1-functionals on W satisfying* (PS), *u is a critical point of each Φ_t, and there is a closed neighborhood U such that*

(i) *U contains no other critical points of Φ_t,*
(ii) *the map $[0, 1] \to C^1(U, \mathbb{R})$, $t \mapsto \Phi_t$ is continuous,*

then $C_*(\Phi_t, u)$ are independent of t.

When the critical values are bounded from below and Φ satisfies (C), the global behavior of Φ can be described by the critical groups at infinity introduced by Bartsch and Li [**9**]

$$C_q(\Phi, \infty) = H_q(W, \Phi^a), \quad q \geq 0$$

where a is less than all critical values. They are independent of a by the second deformation lemma and the homotopy invariance of the homology groups.

For example, if Φ is bounded from below, $C_q(\Phi, \infty) = \delta_{q0}\,\mathcal{G}$. If Φ is unbounded from below, $C_q(\Phi, \infty) = \widetilde{H}_{q-1}(\Phi^a)$ where $\widetilde{H}$ denotes the reduced groups.

THEOREM 5. *If $C_k(\Phi, \infty) \neq 0$ and Φ has only a finite number of critical points and satisfies* (C), *then Φ has a critical point u with $C_k(\Phi, u) \neq 0$.*

The second deformation lemma implies that $C_q(\Phi, \infty) = C_q(\Phi, 0)$ if $u = 0$ is the only critical point of Φ, so Φ has a nontrivial critical point if $C_q(\Phi, 0) \neq C_q(\Phi, \infty)$ for some q.

Now suppose that W is a Hilbert space $(H, (\cdot, \cdot))$ and $\Phi \in C^2(H, \mathbb{R})$. Then the Hessian $A = \Phi''(u)$ is a self-adjoint operator on H for each u. When u is a critical point the dimension of the negative space of A is called the Morse index of u and is denoted by $m(u)$, and $m^*(u) = m(u) + \dim \ker A$ is called the large Morse index. We say that u is nondegenerate if A is invertible. The Morse lemma describes the local behavior of the functional near a nondegenerate critical point.

LEMMA 6 (Morse Lemma). *If u is a nondegenerate critical point of Φ, then there is a local diffeomorphism ξ from a neighborhood U of u into H with $\xi(u) = 0$ such that*

$$\Phi(\xi^{-1}(v)) = \Phi(u) + \frac{1}{2}(Av, v), \quad v \in \xi(U).$$

Morse lemma in $\mathbb{R}^n$ was proved by Morse [**69**]. Palais [**75**], Schwartz [**110**], and Nirenberg [**72**] extended it to Hilbert spaces when Φ is C^3. Proof in the C^2 case is due to Kuiper [**47**] and Cambini [**16**].

A direct consequence of the Morse lemma is

THEOREM 7. *If u is a nondegenerate critical point of Φ, then*

$$C_q(\Phi, u) = \delta_{qm(u)}\,\mathcal{G}.$$

The handle body theorem describes the change in topology as the level sets pass through a critical level on which there are only nondegenerate critical points.

THEOREM 8 (Handle Body Theorem). *If c is an isolated critical value of Φ for which there are only a finite number of nondegenerate critical points u_i, $i = 1, \ldots, k$, with Morse indices $m_i = m(u_i)$, and Φ satisfies* (PS), *then there are an $\varepsilon > 0$ and homeomorphisms φ_i from the unit disk D^{m_i} in $\mathbb{R}^{m_i}$ into H such that*

$$\Phi^{c-\varepsilon} \cap \varphi_i(D^{m_i}) = \Phi^{-1}(c - \varepsilon) \cap \varphi_i(D^{m_i}) = \varphi_i(\partial D^{m_i})$$

and $\Phi^{c-\varepsilon} \cup \bigcup_{i=1}^k \varphi_i(D^{m_i})$ is a deformation retract of $\Phi^{c+\varepsilon}$.

The references for Theorems 3, 7, and 8 are Morse [**69**], Pitcher [**92**], Milnor [**68**], Rothe [**100, 101, 103**], Palais [**75**], Palais and Smale [**74**], Smale [**111**], Marino and Prodi [**64**], Schwartz [**110**], Mawhin and Willem [**66**], and Chang [**21**].

The splitting lemma generalizes the Morse lemma to degenerate critical points. Assume that the origin is an isolated degenerate critical point of Φ and 0 is an isolated point of the spectrum of $A = \Phi''(0)$. Let $N = \ker A$ and write $H = N \oplus N^\perp$, $u = v + w$.

LEMMA 9 (Splitting Lemma). *There are a ball $B \subset H$ centered at the origin, a local homeomorphism ξ from B into H with $\xi(0) = 0$, and a map $\eta \in C^1(B \cap N, N^\perp)$ such that*

$$\Phi(\xi(u)) = \frac{1}{2}(Aw, w) + \Phi(v + \eta(v)), \quad u \in B.$$

Splitting lemma when A is a compact perturbation of the identity was proved by Gromoll and Meyer [**42**] for $\Phi \in C^3$ and by Hofer [**44**] in the C^2 case. Mawhin and Willem [**65, 66**] extended it to the case where A is a Fredholm operator of index zero. The general version given here is due to Chang [**21**].

A consequence of the splitting lemma is

THEOREM 10 (Shifting Theorem). *We have*

$$C_q(\Phi, 0) = C_{q-m(0)}(\Phi|_{\mathcal{N}}, 0) \quad \forall q$$

where $\mathcal{N} = \xi(B \cap N)$ is the degenerate submanifold of Φ at 0.

Shifting theorem is due to Gromoll and Meyer [**42**]; see also Mawhin and Willem [**66**] and Chang [**21**].

Since $\dim \mathcal{N} = m^*(0) - m(0)$, shifting theorem gives us the following Morse index estimates when there is a nontrivial critical group.

COROLLARY 11. *If $C_k(\Phi, 0) \neq 0$, then*

$$m(0) \leq k \leq m^*(0).$$

It also enables us to compute the critical groups of a mountain pass point of nullity at most one.

THEOREM 12. *If u is a mountain pass point of Φ and $\dim \ker \Phi''(u) \leq 1$, then*

$$C_q(\Phi, u) = \delta_{q1}\, \mathcal{G}.$$

This result is due to Ambrosetti [**3, 4**] in the nondegenerate case and to Hofer [**44**] in the general case.

Shifting theorem also implies that all critical groups of a critical point with infinite Morse index are trivial, so the above theory is not suitable for studying strongly indefinite functionals. An infinite dimensional Morse theory particularly well suited to deal with such functionals was developed by Szulkin [**113**]; see also Kryszewski and Szulkin [**46**].

The following important perturbation result is due to Marino and Prodi [**63**]; see also Solimini [**112**].

THEOREM 13. *If some critical value of Φ has only a finite number of critical points u_i, $i = 1, \dots, k$ and $\Phi''(u_i)$ are Fredholm operators, then for any sufficiently small $\varepsilon > 0$ there is a C^2-functional Φ_ε on H such that*

(i) $\|\Phi_\varepsilon - \Phi\|_{C^2(H)} \leq \varepsilon$,

(ii) $\Phi_\varepsilon = \Phi$ *in* $H \setminus \bigcup_{i=1}^k B_\varepsilon(u_j)$,

(iii) Φ_ε *has only nondegenerate critical points in* $B_\varepsilon(u_j)$ *and their Morse indices are in* $[m(u_i), m^*(u_i)]$,

(iv) Φ *satisfies* (PS) $\implies \Phi_\varepsilon$ *satisfies* (PS).

5. Minimax principle

Minimax principle originated in the work of Ljusternik and Schnirelmann [**60**] and is a useful tool for finding critical points of a functional. Note that the first deformation lemma implies that if c is a regular value and Φ satisfies $(\mathrm{C})_c$, then the family $\mathcal{D}_{c,\,\varepsilon}$ of maps $\eta \in C([0,1] \times W, W)$ satisfying

(i) $\eta(0,\cdot) = id_W$,
(ii) $\eta(t,\cdot)$ is a homeomorphism of W for all $t \in [0,1]$,
(iii) $\eta(t,\cdot)$ is the identity outside $\Phi_{c-2\varepsilon}^{c+2\varepsilon}$ for all $t \in [0,1]$,
(iv) $\Phi(\eta(\cdot,u))$ is nonincreasing for all $u \in W$,
(v) $\eta(1,\Phi^{c+\varepsilon}) \subset \Phi^{c-\varepsilon}$

is nonempty for all sufficiently small $\varepsilon > 0$. We say that a family $\mathcal{F}$ of subsets of W is invariant under $\mathcal{D}_{c,\,\varepsilon}$ if

$$M \in \mathcal{F},\ \eta \in \mathcal{D}_{c,\,\varepsilon} \implies \eta(1,M) \in \mathcal{F}.$$

THEOREM 14 (Minimax Principle). *If $\mathcal{F}$ is a family of subsets of W,*

$$c := \inf_{M \in \mathcal{F}} \sup_{u \in M} \Phi(u)$$

is finite, $\mathcal{F}$ is invariant under $\mathcal{D}_{c,\,\varepsilon}$ for all sufficiently small $\varepsilon > 0$, and Φ satisfies $(\mathrm{C})_c$, then c is a critical value of Φ.

We say that a family Γ of continuous maps γ from a topological space X into W is invariant under $\mathcal{D}_{c,\,\varepsilon}$ if

$$\gamma \in \Gamma,\ \eta \in \mathcal{D}_{c,\,\varepsilon} \implies \eta(1,\cdot) \circ \gamma \in \Gamma.$$

Minimax principle is often applied in the following form, which follows by taking $\mathcal{F} = \{\gamma(X) : \gamma \in \Gamma\}$ in Theorem 14.

THEOREM 15. *If Γ is a family of continuous maps γ from a topological space X into W,*

$$c := \inf_{\gamma \in \Gamma} \sup_{u \in \gamma(X)} \Phi(u)$$

is finite, Γ is invariant under $\mathcal{D}_{c,\,\varepsilon}$ for all sufficiently small $\varepsilon > 0$, and Φ satisfies $(\mathrm{C})_c$, then c is a critical value of Φ.

Some references for Theorems 14 and 15 are Palais [**76**], Nirenberg [**73**], Rabinowitz [**97**] and Ghoussoub [**41**].

Minimax methods were introduced in Morse theory by Marino and Prodi [**64**]. The following result is due to Liu [**57**].

THEOREM 16. *If $\sigma \in H_k(\Phi^b, \Phi^a)$ is a nontrivial singular homology class where $-\infty < a < b \le +\infty$ are regular values,*

$$c := \inf_{z \in \sigma} \sup_{u \in |z|} \Phi(u)$$

where $|z|$ denotes the support of the singular chain z, Φ satisfies $(\mathrm{C})_c$, and K^c is a finite set, then there is a $u \in K^c$ with $C_k(\Phi,u) \neq 0$.

6. Linking

The notion of homotopical linking is useful for obtaining critical points via the minimax principle.

DEFINITION 17. Let A be a closed proper subset of a topological space X, $g \in C(A, W)$ such that $g(A)$ is closed, B a nonempty closed subset of W such that $\operatorname{dist}(g(A), B) > 0$, and

$$\Gamma = \big\{\gamma \in C(X, W) : \gamma(X) \text{ is closed}, \ \gamma|_A = g\big\}.$$

We say that (A, g) homotopically links B with respect to X if

$$\gamma(X) \cap B \neq \emptyset \quad \forall \gamma \in \Gamma.$$

When $g : A \subset W$ is the inclusion and $X = \big\{tu : u \in A,\ t \in [0, 1]\big\}$, we simply say that A homotopically links B.

Some standard examples of homotopical linking are the following.

EXAMPLE 18. If $u_0 \in W$, U is a bounded neighborhood of u_0, and $u_1 \notin \overline{U}$, then $A = \{u_0, u_1\}$ homotopically links $B = \partial U$.

EXAMPLE 19. If $W = W_1 \oplus W_2$, $u = u_1 + u_2$ is a direct sum decomposition of W with W_1 nontrivial and finite dimensional, then $A = \big\{u_1 \in W_1 : \|u_1\| = R\big\}$ homotopically links $B = W_2$ for any $R > 0$.

EXAMPLE 20. If $W = W_1 \oplus W_2$, $u = u_1 + u_2$ is a direct sum decomposition with W_1 finite dimensional and $v \in W_2$ with $\|v\| = 1$, then $A = \big\{u_1 \in W_1 : \|u_1\| \leq R\big\} \cup \big\{u = u_1 + tv : u_1 \in W_1, t \geq 0, \|u\| = R\big\}$ homotopically links $B = \big\{u_2 \in W_2 : \|u_2\| = r\big\}$ for any $0 < r < R$.

THEOREM 21. *If (A, g) homotopically links B with respect to X,*

$$c := \inf_{\gamma \in \Gamma} \sup_{u \in \gamma(X)} \Phi(u)$$

is finite, $a := \sup \Phi(g(A)) \leq \inf \Phi(B) =: b$, and Φ satisfies $(\mathrm{C})_c$, then $c \geq b$ and is a critical value of Φ. If $c = b$, then Φ has a critical point with critical value c on B.

Many authors have contributed to this result. The special cases that correspond to Examples 18, 19, and 20 are the well-known mountain pass lemma of Ambrosetti and Rabinowitz [**5**] and the saddle point and linking theorems of Rabinowitz [**96, 95**], respectively. See also Ahmad, Lazer, and Paul [**1**], Castro and Lazer [**17**], Benci and Rabinowitz [**13**], Ni [**71**], Chang [**21**], Qi [**93**], and Ghoussoub [**40**]. The version given here can be found in Perera et al. [**83**].

Morse index estimates for a critical point produced by a homotopical linking have been obtained by Lazer and Solimini [**50**], Solimini [**112**], Ghoussoub [**40**], Ramos and Sanchez [**98**], and others. However, the notion of homological linking introduced by Benci [**10, 11**] and Liu [**57**] is better suited for obtaining critical points with nontrivial critical groups.

DEFINITION 22. Let A and B be disjoint nonempty subsets of W. We say that A homologically links B in dimension k if the inclusion $i : A \subset W \setminus B$ induces a nontrivial homomorphism

$$i_* : \widetilde{H}_k(A) \to \widetilde{H}_k(W \setminus B).$$

In Examples 18, 19, and 20, A homologically links B in dimensions 0, $\dim W_1 - 1$, and $\dim W_1$, respectively.

THEOREM 23. *If A homologically links B in dimension k, $\Phi|_A \leq a < \Phi|_B$ where a is a regular value, and Φ has only a finite number of critical points in Φ_a and satisfies* $(\mathrm{C})_c$ *for all $c \geq a$, then Φ has a critical point u_1 with*

$$\Phi(u_1) > a, \quad C_{k+1}(\Phi, u_1) \neq 0.$$

This follows from Theorem 16. Indeed, since the composition $\widetilde{H}_k(A) \to \widetilde{H}_k(\Phi^a) \to \widetilde{H}_k(W \setminus B)$ induced by the inclusions $A \subset \Phi^a \subset W \setminus B$ is i_*, $\widetilde{H}_k(\Phi^a)$ is nontrivial, and since W is contractible, it then follows from the exact sequence of the pair (W, Φ^a) that $H_{k+1}(W, \Phi^a)$ is nontrivial.

Note that when W_1 is infinite dimensional in Examples 19 and 20 the set A is contractible and therefore does not link B homotopically or homologically. Schechter and Tintarev [**109**] introduced yet another notion of linking according to which A links B in those examples as long as W_1 or W_2 is finite dimensional; see also Ribarska, Tsachev, and Krastanov [**99**] and Schechter [**107, 108**]. Moreover, according to their definition of linking, if A and B are disjoint closed bounded subsets of W such that A links B and $W \setminus A$ is connected, then B links A. If, in addition, $a := \sup \Phi(A) \leq \inf \Phi(B) =: b$ and Φ is bounded on bounded sets and satisfies (C), this then yields a pair of critical points u_1 and u_2 with $\Phi(u_1) \geq b \geq a \geq \Phi(u_2)$. The following analogous result for homological linking was obtained in Perera [**77**], where it was shown that the second critical point also has a nontrivial critical group. We assume that Φ has only a finite number of critical points and satisfies (C) for the rest of this section.

THEOREM 24. *If A homologically links B in dimension k and B is bounded, $\Phi|_A \leq a < \Phi|_B$ where a is a regular value, and Φ is bounded from below on bounded sets, then Φ has two critical points u_1 and u_2 with*

$$\Phi(u_1) > a > \Phi(u_2), \quad C_{k+1}(\Phi, u_1) \neq 0, \quad C_k(\Phi, u_2) \neq 0.$$

COROLLARY 25. *Let $W = W_1 \oplus W_2$, $u = u_1 + u_2$ be a direct sum decomposition with* $\dim W_1 = k < \infty$. *If $\Phi \leq a$ on $\{u_1 \in W_1 : \|u_1\| \leq R\} \cup \{u = u_1 + tv : u_1 \in W_1, t \geq 0, \|u\| = R\}$ for some $R > 0$ and $v \in W_2$ with $\|v\| = 1$, $\Phi > a$ on $\{u_2 \in W_2 : \|u_2\| = r\}$ for some $0 < r < R$, where a is a regular value, and Φ is bounded from below on bounded sets, then Φ has two critical points u_1 and u_2 with*

$$\Phi(u_1) > a > \Phi(u_2), \quad C_{k+1}(\Phi, u_1) \neq 0, \quad C_k(\Phi, u_2) \neq 0.$$

It was also shown in Perera [**77**] that the assumptions that B is bounded and Φ is bounded from below on bounded sets can be relaxed as follows; see also Schechter [**105**].

THEOREM 26. *If A homologically links B in dimension k, $\Phi|_A \le a < \Phi|_B$ where a is a regular value, and Φ is bounded from below on a set $C \supset B$ such that the inclusion-induced homomorphism $\widetilde{H}_k(W \setminus C) \to \widetilde{H}_k(W \setminus B)$ is trivial, then Φ has two critical points u_1 and u_2 with*

$$\Phi(u_1) > a > \Phi(u_2), \quad C_{k+1}(\Phi, u_1) \neq 0, \ C_k(\Phi, u_2) \neq 0.$$

COROLLARY 27. *Let $W = W_1 \oplus W_2$, $u = u_1 + u_2$ be a direct sum decomposition with* $\dim W_1 = k < \infty$. *If $\Phi \le a$ on $\{u_1 \in W_1 : \|u_1\| = R\}$ for some $R > 0$, $\Phi > a$ on W_2, where a is a regular value, and Φ is bounded from below on $\{tv + u_2 : t \ge 0,\ u_2 \in W_2\}$ for some $v \in W_1 \setminus \{0\}$, then Φ has two critical points u_1 and u_2 with*

$$\Phi(u_1) > a > \Phi(u_2), \quad C_k(\Phi, u_1) \neq 0, \ C_{k-1}(\Phi, u_2) \neq 0.$$

The following theorem of Perera and Schechter [**89**] gives a critical point with a nontrivial critical group in a saddle point theorem with nonstandard geometrical assumptions that do not involve a finite dimensional closed loop; see also Perera and Schechter [**85**] and Lancelotti [**48**].

THEOREM 28. *Let $W = W_1 \oplus W_2$, $u = u_1 + u_2$ be a direct sum decomposition with* $\dim W_1 = k < \infty$. *If Φ is bounded from above on W_1 and from below on W_2, then Φ has a critical point u_1 with*

$$\inf \Phi(W_2) \le \Phi(u_1) \le \sup \Phi(W_1), \quad C_k(\Phi, u_1) \neq 0.$$

7. Local linking

In many applications Φ has the trivial critical point $u = 0$ and we are interested in finding others. The notion of local linking was introduced by Li and Liu [**58, 53**], who used it to obtain nontrivial critical points under various assumptions on the behavior of Φ at infinity; see also Brezis and Nirenberg [**14**] and Li and Willem [**54**].

DEFINITION 29. Assume that the origin is a critical point of Φ with $\Phi(0) = 0$. We say that Φ has a local linking near the origin if there is a direct sum decomposition $W = W_1 \oplus W_2$, $u = u_1 + u_2$ with W_1 finite dimensional such that

$$\begin{cases} \Phi(u_1) \le 0, & u_1 \in W_1,\ \|u_1\| \le r \\ \Phi(u_2) > 0, & u_2 \in W_2,\ 0 < \|u_2\| \le r \end{cases}$$

for sufficiently small $r > 0$.

Liu [**57**] showed that this yields a nontrivial critical group at the origin.

THEOREM 30. *If Φ has a local linking near the origin with* $\dim W_1 = k$ *and the origin is an isolated critical point, then $C_k(\Phi, 0) \neq 0$.*

The following alternative obtained in Perera [**78**] gives a nontrivial critical point with a nontrivial critical group produced by a local linking.

THEOREM 31. *If Φ has a local linking near the origin with* $\dim W_1 = k$, $H_k(\Phi^b, \Phi^a) = 0$ *where* $-\infty < a < 0 < b \leq +\infty$ *are regular values, and Φ has only a finite number of critical points in Φ_a^b and satisfies* $(\mathrm{C})_c$ *for all* $c \in [a,b] \cap \mathbb{R}$, *then Φ has a critical point $u_1 \neq 0$ with either*

$$a < \Phi(u_1) < 0, \quad C_{k-1}(\Phi, u_1) \neq 0$$

or

$$0 < \Phi(u_1) < b, \quad C_{k+1}(\Phi, u_1) \neq 0.$$

When Φ is bounded from below, taking $a < \inf \Phi(W)$ and $b = +\infty$ gives the following three critical points theorem; see also Krasnosel'skii [**45**], Chang [**20**], Li and Liu [**58**], and Liu [**57**].

COROLLARY 32. *If Φ has a local linking near the origin with* $\dim W_1 = k \geq 2$, *is bounded from below, has only a finite number of critical points, and satisfies* (C), *then Φ has a global minimizer $u_0 \neq 0$ with*

$$\Phi(u_0) < 0, \quad C_q(\Phi, u_0) = \delta_{q0}\, \mathcal{G}$$

and a critical point $u_1 \neq 0, u_0$ with either

$$\Phi(u_1) < 0, \quad C_{k-1}(\Phi, u_1) \neq 0$$

or

$$\Phi(u_1) > 0, \quad C_{k+1}(\Phi, u_1) \neq 0.$$

Theorems 30 and 31 and Corollary 32 also hold under the more general notion of homological local linking introduced in Perera [**79**].

8. Jumping nonlinearities

In this section we present some results on critical groups associated with jumping nonlinearities recently obtained in Perera and Schechter [**84**]; see also Dancer [**28, 29**] and Perera and Schechter [**86, 87, 88**].

Let H be a Hilbert space with the inner product $(\cdot,\cdot)$ and the associated norm $\|\cdot\|$. Recall that

(i) $f : H \to H'$, where H' is also a Hilbert space, is positive homogeneous if

$$f(su) = sf(u) \quad \forall u \in H, \ s \geq 0.$$

Taking $u = 0$ and $s = 0$ gives $f(0) = 0$.

(ii) $f : H \to H$ is monotone if

$$(f(u) - f(v), u - v) \geq 0 \quad \forall u, v \in H.$$

(iii) $f \in C(H, H)$ is a potential operator if there is a $F \in C^1(H, \mathbb{R})$, called a potential for f, such that

$$F'(u) = f(u) \quad \forall u \in H.$$

Replacing F with $F - F(0)$ gives a potential F with $F(0) = 0$.

Assume that there are positive homogeneous monotone potential operators $p, n \in C(H, H)$ such that

$$p(u) + n(u) = u, \quad (p(u), n(u)) = 0 \quad \forall u \in H.$$

We use the suggestive notation

$$u^+ = p(u), \quad u^- = -n(u),$$

so that

$$u = u^+ - u^-, \quad (u^+, u^-) = 0.$$

This implies

$$\|u\|^2 = \|u^+\|^2 + \|u^-\|^2,$$

in particular,

$$\|u^\pm\| \leq \|u\| .$$

Now let A be a self-adjoint operator on H with the spectrum $\sigma(A) \subset (0, \infty)$ and A^{-1} compact. Then $\sigma(A)$ consists of isolated eigenvalues λ_l, $l \geq 0$ of finite multiplicities satisfying

$$0 < \lambda_0 < \lambda_1 < \cdots < \lambda_l < \cdots .$$

Moreover,

$$D = D(A^{1/2})$$

is a Hilbert space with the inner product

$$(u, v)_D = (A^{1/2}u, A^{1/2}v) = (Au, v)$$

and the associated norm

$$\|u\|_D = \|A^{1/2}u\| = (Au, u)^{1/2} .$$

We have

$$\|u\|_D^2 = (Au, u) \geq \lambda_0 \, (u, u) = \lambda_0 \, \|u\|^2 \quad \forall u \in H,$$

so $D \hookrightarrow H$, and the embedding is compact since A^{-1} is a compact operator. Let E_l be the eigenspace of λ_l,

$$N_l = \bigoplus_{j=0}^{l} E_j, \qquad M_l = N_l^\perp \cap D.$$

Then

$$D = N_l \oplus M_l$$

is an orthogonal decomposition with respect to both $(\cdot,\cdot)$ and $(\cdot,\cdot)_D$. Moreover,

$$\|v\|_D^2 = (Av, v) \le \lambda_l\,(v, v) = \lambda_l\,\|v\|^2 \quad \forall v \in N_l, \tag{2}$$
$$\|w\|_D^2 = (Aw, w) \ge \lambda_{l+1}\,(w, w) = \lambda_{l+1}\,\|w\|^2 \quad \forall w \in M_l.$$

We assume that

$$w \in M_0 \setminus \{0\} \implies w^\pm \ne 0.$$

Consider the equation

$$Au = bu^+ - au^-, \quad u \in D \tag{3}$$

where $a, b \in \mathbb{R}$. We will call the right-hand side of (3) a jumping nonlinearity. The set $\Sigma(A)$ of points $(a, b) \in \mathbb{R}^2$ such that (3) has a nontrivial solution is called the Fučík spectrum of A. It is a closed subset of $\mathbb{R}^2$ (see [**84**, Proposition 4.4.3]). If $a = b = \lambda$, then (3) reduces to the eigenvalue problem

$$Au = \lambda u, \quad u \in D,$$

which has a nontrivial solution if and only if λ is one of the eigenvalues λ_l, so the points (λ_l, λ_l) are in $\Sigma(A)$.

Perhaps the best-known example is the semilinear elliptic boundary value problem

$$\begin{cases} -\Delta u = bu^+ - au^- & \text{in } \Omega \\ u = 0 & \text{on } \partial\Omega \end{cases} \tag{4}$$

where Ω is a bounded domain in $\mathbb{R}^n$, $n \ge 1$ and $u^\pm = \max\{\pm u, 0\}$ are the positive and negative parts of u, respectively. Here $H = L^2(\Omega)$, $D = H_0^1(\Omega)$ is the usual Sobolev space, and A is the inverse of the solution operator $S : H \to D$, $f \mapsto u = (-\Delta)^{-1} f$ of the problem

$$\begin{cases} -\Delta u = f(x) & \text{in } \Omega \\ u = 0 & \text{on } \partial\Omega. \end{cases}$$

Since the embedding $D \hookrightarrow H$ is compact, $A^{-1} = S$ is compact on H. The Fučík spectrum $\Sigma(-\Delta)$ of $-\Delta$ was introduced by Dancer [**26, 27**] and Fučík [**38**], who recognized its significance for the solvability of the problem

$$\begin{cases} -\Delta u = f(x, u) & \text{in } \Omega \\ u = 0 & \text{on } \partial\Omega \end{cases}$$

when f is a Carathéodory function on $\Omega \times \mathbb{R}$ satisfying

$$f(x, t) = bt^+ - at^- + \mathrm{o}(t) \quad \text{as } |t| \to \infty, \quad \text{uniformly a.e.}$$

In the ODE case $n = 1$, Fučík showed that $\Sigma(-d^2/dx^2)$ consists of a sequence of hyperbolic-like curves passing through the points (λ_l, λ_l), with one or two curves going through each point. In the PDE case $n \ge 2$ also, $\Sigma(-\Delta)$ consists, at least locally, of curves emanating from the points (λ_l, λ_l); see Gallouët and Kavian [**39**], Ruf [**104**], Lazer and McKenna [**49**], Lazer [**51**], Các [**15**], Magalhães [**61**], Cuesta and Gossez [**25**], de Figueiredo and Gossez [**31**], Schechter [**106**], and Margulies and Margulies [**62**]. In particular, it was shown in Schechter [**106**] that in the square $(\lambda_{l-1}, \lambda_{l+1})^2$, $\Sigma(-\Delta)$ contains two strictly decreasing curves, which may coincide, such that the points in the square that are either below the lower curve or above the upper curve are not in $\Sigma(-\Delta)$, while the points between them may or may not belong to $\Sigma(-\Delta)$ when they do not coincide.

When

$$(-u)^{\pm} = u^{\mp} \quad \forall u \in H, \tag{5}$$

as is the case in the above example, u solves (3) if and only if

$$A(-u) = a(-u)^+ - b(-u)^-$$

and hence $(a, b) \in \Sigma(A)$ if and only if $(b, a) \in \Sigma(A)$, so $\Sigma(A)$ is symmetric about the line $a = b$.

Equation (3) has a variational formulation. It is easy to see that A is a potential operator with the potential

$$\frac{1}{2}\,(Au, u) = \frac{1}{2}\,\|u\|_D^2\,.$$

The potentials of p and n are

$$\frac{1}{2}\,(p(u), u) = \frac{1}{2}\,(u^+, u^+ - u^-) = \frac{1}{2}\,\|u^+\|^2,$$

$$\frac{1}{2}\,(n(u), u) = \frac{1}{2}\,(-u^-, u^+ - u^-) = \frac{1}{2}\,\|u^-\|^2,$$

respectively (see [**84**, Proposition 4.3.2]). Let

$$I(u, a, b) = \|u\|_D^2 - a\,\|u^-\|^2 - b\,\|u^+\|^2, \quad u \in D.$$

Then $I(\cdot, a, b) \in C^1(D, \mathbb{R})$ with

$$I'(u) = 2(Au + au^- - bu^+),$$

so the critical points of I coincide with the solutions of (3). Thus, $(a, b) \in \Sigma(A)$ if and only if $I(\cdot, a, b)$ has a nontrivial critical point. When $(a, b) \notin \Sigma(A)$, every sequence $(u_j) \subset D$ such that $I'(u_j) \to 0$ converges to zero and hence I satisfies (PS) (see [**84**, Proposition 4.4.2]).

Now we describe the minimal and maximal curves of $\Sigma(A)$ in the square

$$Q_l = (\lambda_{l-1}, \lambda_{l+1})^2, \quad l \geq 1$$

constructed in Perera and Schechter [**84**]; see also Schechter [**106**]. When $(a, b) \in Q_l$, $I(v + y + w)$, $v + y + w \in N_{l-1} \oplus E_l \oplus M_l$ is strictly concave in v and strictly convex in w, i.e., if $v_1 \neq v_2 \in N_{l-1}$, $w \in M_{l-1}$,

$$I((1-t)\,v_1 + t\,v_2 + w) > (1-t)\,I(v_1 + w) + t\,I(v_2 + w) \quad \forall t \in (0, 1), \tag{6}$$

and if $v \in N_l$, $w_1 \neq w_2 \in M_l$,

$$I(v + (1-t)\, w_1 + t\, w_2) < (1-t)\, I(v + w_1) + t\, I(v + w_2) \quad \forall t \in (0,1) \tag{7}$$

(see [**84**, Proposition 4.6.1]).

Proposition 33 ([**84**, Proposition 4.7.1, Corollary 4.7.3, Proposition 4.7.4]). *Let* $(a,b) \in Q_l$.

(i) *There is a positive homogeneous map* $\theta(\cdot, a, b) \in C(M_{l-1}, N_{l-1})$ *such that* $v = \theta(w)$ *is the unique solution of*

$$I(v + w) = \sup_{v' \in N_{l-1}} I(v' + w), \quad w \in M_{l-1}.$$

Moreover,

$$I'(v + w) \perp N_{l-1} \iff v = \theta(w).$$

(ii) *There is a positive homogeneous map* $\tau(\cdot, a, b) \in C(N_l, M_l)$ *such that* $w = \tau(v)$ *is the unique solution of*

$$I(v + w) = \inf_{w' \in M_l} I(v + w'), \quad v \in N_l.$$

Moreover,

$$I'(v + w) \perp M_l \iff w = \tau(v).$$

Furthermore,

(i) θ *is continuous on* $M_{l-1} \times Q_l$, *and*

$$\theta(w, \lambda_l, \lambda_l) = 0 \quad \forall w \in M_{l-1},$$

(ii) τ *is continuous on* $N_l \times Q_l$, *and*

$$\tau(v, \lambda_l, \lambda_l) = 0 \quad \forall v \in N_l.$$

For $(a,b) \in Q_l$, let

$$\begin{aligned} \sigma(w, a, b) &= \theta(w, a, b) + w, \quad w \in M_{l-1}, \\ S_l(a,b) &= \sigma(M_{l-1}, a, b), \\ \zeta(v, a, b) &= v + \tau(v, a, b), \quad v \in N_l, \\ S^l(a,b) &= \zeta(N_l, a, b). \end{aligned}$$

Then S_l and S^l are topological manifolds modeled on M_{l-1} and N_l, respectively. Thus, S_l is infinite dimensional, while S^l is d_l-dimensional where

$$d_l = \dim N_l.$$

For $B \subset D$, set

$$\widetilde{B} = B \cap S$$

where

$$S = \{u \in D : \|u\|_D = 1\}$$

is the unit sphere in D. We say that B is a radial set if

$$B = \{su : u \in \widetilde{B},\ s \geq 0\}.$$

Since θ and τ are positive homogeneous, so are σ and ζ, and hence S_l and S^l are radial manifolds.

Let

$$K(a, b) = \{u \in D : I'(u, a, b) = 0\}$$

be the set of critical points of $I(\cdot, a, b)$. Since I' is positive homogeneous, K is a radial set. Moreover,

$$I(u) = \frac{1}{2}\,(I'(u), u)$$

(see [**84**, Proposition 4.3.2]) and hence

$$I(u) = 0 \quad \forall u \in K. \tag{8}$$

Since

$$D = N_{l-1} \oplus E_l \oplus M_l,$$

Proposition 33 implies

$$K = \{u \in S_l \cap S^l : I'(u) \perp E_l\}. \tag{9}$$

Together with (8), it also implies

$$K \subset \{u \in S_l \cap S^l : I(u) = 0\}. \tag{10}$$

Set

$$n_{l-1}(a, b) = \inf_{w \in \widetilde{M}_{l-1}} \sup_{v \in N_{l-1}} I(v + w, a, b),$$

$$m_l(a, b) = \sup_{v \in \widetilde{N}_l} \inf_{w \in M_l} I(v + w, a, b).$$

Since $I(u, a, b)$ is nonincreasing in a for fixed u and b, and in b for fixed u and a, $n_{l-1}(a, b)$ and $m_l(a, b)$ are nonincreasing in a for fixed b, and in b for fixed a. By Proposition 33,

$$n_{l-1}(a, b) = \inf_{w \in \widetilde{M}_{l-1}} I(\sigma(w, a, b), a, b),$$

$$m_l(a, b) = \sup_{v \in \widetilde{N}_l} I(\zeta(v, a, b), a, b).$$

PROPOSITION 34 ([**84**, Proposition 4.7.5, Lemma 4.7.6, Proposition 4.7.7]). *Let* $(a, b), (a', b') \in Q_l$.

(i) *Assume that* $n_{l-1}(a,b) = 0$. *Then*

$$I(u,a,b) \geq 0 \quad \forall u \in S_l(a,b),$$

$$K(a,b) = \{u \in S_l(a,b) : I(u,a,b) = 0\},$$

and $(a,b) \in \Sigma(A)$.

(a) *If* $a' \leq a$, $b' \leq b$, *and* $(a',b') \neq (a,b)$, *then* $n_{l-1}(a',b') > 0$,

$$I(u,a',b') > 0 \quad \forall u \in S_l(a',b') \setminus \{0\},$$

and $(a',b') \notin \Sigma(A)$.

(b) *If* $a' \geq a$, $b' \geq b$, *and* $(a',b') \neq (a,b)$, *then* $n_{l-1}(a',b') < 0$ *and there is a* $u \in S_l(a',b') \setminus \{0\}$ *such that*

$$I(u,a',b') < 0.$$

(ii) *Assume that* $m_l(a,b) = 0$. *Then*

$$I(u,a,b) \leq 0 \quad \forall u \in S^l(a,b),$$

$$K(a,b) = \{u \in S^l(a,b) : I(u,a,b) = 0\},$$

and $(a,b) \in \Sigma(A)$.

(a) *If* $a' \geq a$, $b' \geq b$, *and* $(a',b') \neq (a,b)$, *then* $m_l(a',b') < 0$,

$$I(u,a',b') < 0 \quad \forall u \in S^l(a',b') \setminus \{0\},$$

and $(a',b') \notin \Sigma(A)$.

(b) *If* $a' \leq a$, $b' \leq b$, *and* $(a',b') \neq (a,b)$, *then* $m_l(a',b') > 0$ *and there is a* $u \in S^l(a',b') \setminus \{0\}$ *such that*

$$I(u,a',b') > 0.$$

Furthermore,

(i) n_{l-1} *is continuous on* Q_l, *and*

$$n_{l-1}(\lambda_l,\lambda_l) = 0,$$

(ii) m_l *is continuous on* Q_l, *and*

$$m_l(\lambda_l,\lambda_l) = 0.$$

For $a \in (\lambda_{l-1},\lambda_{l+1})$, set

$$\nu_{l-1}(a) = \sup\{b \in (\lambda_{l-1},\lambda_{l+1}) : n_{l-1}(a,b) \geq 0\},$$

$$\mu_l(a) = \inf\{b \in (\lambda_{l-1},\lambda_{l+1}) : m_l(a,b) \leq 0\}.$$

Then

$$b = \nu_{l-1}(a) \iff n_{l-1}(a,b) = 0,$$

$$b = \mu_l(a) \iff m_l(a,b) = 0$$

(see [**84**, Lemma 4.7.8]).

THEOREM 35 ([**84**, Theorem 4.7.9]). *Let* $(a, b) \in Q_l$.

(i) *The function* ν_{l-1} *is continuous, strictly decreasing, and satisfies*
  (a) $\nu_{l-1}(\lambda_l) = \lambda_l$,
  (b) $b = \nu_{l-1}(a) \implies (a, b) \in \Sigma(A)$,
  (c) $b < \nu_{l-1}(a) \implies (a, b) \notin \Sigma(A)$.

(ii) *The function* μ_l *is continuous, strictly decreasing, and satisfies*
  (a) $\mu_l(\lambda_l) = \lambda_l$,
  (b) $b = \mu_l(a) \implies (a, b) \in \Sigma(A)$,
  (c) $b > \mu_l(a) \implies (a, b) \notin \Sigma(A)$.

(iii) $\nu_{l-1}(a) \leq \mu_l(a)$

Thus,

$$C_l : b = \nu_{l-1}(a), \qquad C^l : b = \mu_l(a)$$

are strictly decreasing curves in Q_l that belong to $\Sigma(A)$. They both pass through the point (λ_l, λ_l) and may coincide. The region

$$\mathrm{I}_l = \big\{(a, b) \in Q_l : b < \nu_{l-1}(a)\big\}$$

below the lower curve C_l and the region

$$\mathrm{I}^l = \big\{(a, b) \in Q_l : b > \mu_l(a)\big\}$$

above the upper curve C^l are free of $\Sigma(A)$. They are the minimal and maximal curves of $\Sigma(A)$ in Q_l in this sense. Points in the region

$$\mathrm{II}_l = \big\{(a, b) \in Q_l : \nu_{l-1}(a) < b < \mu_l(a)\big\}$$

between C_l and C^l, when it is nonempty, may or may not belong to $\Sigma(A)$.

For $(a, b) \in Q_l$, let

$$\mathcal{N}_l(a, b) = S_l(a, b) \cap S^l(a, b).$$

Since S_l and S^l are radial sets, so is $\mathcal{N}_l$. The next two propositions show that $\mathcal{N}_l$ is a topological manifold modeled on E_l and hence

$$\dim \mathcal{N}_l = d_l - d_{l-1}.$$

We will call it the null manifold of I.

PROPOSITION 36 ([**84**, Proposition 4.8.1, Lemma 4.8.3, Proposition 4.8.4]). *Let* $(a, b) \in Q_l$.

(i) *There is a positive homogeneous map* $\eta(\cdot, a, b) \in C(E_l, N_{l-1})$ *such that* $v = \eta(y)$ *is the unique solution of*

$$I(\zeta(v + y)) = \sup_{v' \in N_{l-1}} I(\zeta(v' + y)), \quad y \in E_l.$$

Moreover,

$$I'(\zeta(v + y)) \perp N_{l-1} \iff v = \eta(y).$$

(ii) *There is a positive homogeneous map* $\xi(\cdot, a, b) \in C(E_l, M_l)$ *such that* $w = \xi(y)$ *is the unique solution of*

$$I(\sigma(y+w)) = \inf_{w' \in M_l} I(\sigma(y+w')), \quad y \in E_l.$$

Moreover,

$$I'(\sigma(y+w)) \perp M_l \iff w = \xi(y).$$

(iii) *For all* $y \in E_l$,

$$\zeta(\eta(y)+y) = \sigma(y+\xi(y)),$$

i.e.,

$$\eta(y) = \theta(y+\xi(y)), \quad \xi(y) = \tau(\eta(y)+y).$$

Furthermore,

(i) η *is continuous on* $E_l \times Q_l$, *and*

$$\eta(y, \lambda_l, \lambda_l) = 0 \quad \forall y \in E_l,$$

(ii) ξ *is continuous on* $E_l \times Q_l$, *and*

$$\xi(y, \lambda_l, \lambda_l) = 0 \quad \forall y \in E_l.$$

Let

$$\varphi(y) = \zeta(\eta(y)+y) = \sigma(y+\xi(y)), \quad y \in E_l.$$

PROPOSITION 37 ([**84**, Proposition 4.8.5]). *Let* $(a,b) \in Q_l$.

(i) $\varphi(\cdot, a, b) \in C(E_l, D)$ *is a positive homogeneous map such that*

$$I(\varphi(y)) = \inf_{w \in M_l} \sup_{v \in N_{l-1}} I(v+y+w) = \sup_{v \in N_{l-1}} \inf_{w \in M_l} I(v+y+w), \quad y \in E_l$$

and

$$I'(\varphi(y)) \in E_l \quad \forall y \in E_l.$$

(ii) *If* $(a', b') \in Q_l$ *with* $a' \geq a$ *and* $b' \geq b$, *then*

$$I(\varphi(y, a', b'), a', b') \leq I(\varphi(y, a, b), a, b) \quad \forall y \in E_l.$$

(iii) φ *is continuous on* $E_l \times Q_l$.

(iv) $\varphi(y, \lambda_l, \lambda_l) = y \quad \forall y \in E_l$

(v) $\mathcal{N}_l(a,b) = \{\varphi(y,a,b) : y \in E_l\}$

(vi) $\mathcal{N}_l(\lambda_l, \lambda_l) = E_l$

By (9) and (10),

$$K = \{u \in \mathcal{N}_l : I'(u) \perp E_l\} \subset \{u \in \mathcal{N}_l : I(u) = 0\}. \tag{11}$$

The following theorem shows that the curves C_l and C^l are closely related to

$$\widetilde{I} = I|_{\mathcal{N}_l}.$$

THEOREM 38 ([**84**, Theorem 4.8.6]). *Let* $(a, b) \in Q_l$.

(i) *If* $b < \nu_{l-1}(a)$, *then*

$$\widetilde{I}(u, a, b) > 0 \quad \forall u \in \mathcal{N}_l(a, b) \setminus \{0\}.$$

(ii) *If* $b = \nu_{l-1}(a)$, *then*

$$\widetilde{I}(u, a, b) \geq 0 \quad \forall u \in \mathcal{N}_l(a, b),$$

$$K(a, b) = \big\{u \in \mathcal{N}_l(a, b) : \widetilde{I}(u, a, b) = 0\big\}.$$

(iii) *If* $\nu_{l-1}(a) < b < \mu_l(a)$, *then there are* $u_i \in \mathcal{N}_l(a, b) \setminus \{0\}$, $i = 1, 2$ *such that*

$$\widetilde{I}(u_1, a, b) < 0 < \widetilde{I}(u_2, a, b).$$

(iv) *If* $b = \mu_l(a)$, *then*

$$\widetilde{I}(u, a, b) \leq 0 \quad \forall u \in \mathcal{N}_l(a, b),$$

$$K(a, b) = \big\{u \in \mathcal{N}_l(a, b) : \widetilde{I}(u, a, b) = 0\big\}.$$

(v) *If* $b > \mu_l(a)$, *then*

$$\widetilde{I}(u, a, b) < 0 \quad \forall u \in \mathcal{N}_l(a, b) \setminus \{0\}.$$

By (11), solutions of (3) are in $\mathcal{N}_l$. The set $K(a, b)$ of solutions is all of $\mathcal{N}_l(a, b)$ exactly when $(a, b) \in Q_l$ is on both C_l and C^l (see [**84**, Theorem 4.8.7]). When λ_l is a simple eigenvalue, $\mathcal{N}_l$ is 1-dimensional and hence this implies that (a, b) is on exactly one of those curves if and only if

$$K(a, b) = \big\{t\, \varphi(y_0, a, b) : t \geq 0\big\}$$

for some $y_0 \in E_l \setminus \{0\}$ (see [**84**, Corollary 4.8.8]).

The following theorem gives a sufficient condition for the region II_l to be nonempty.

THEOREM 39 ([**84**, Theorem 4.9.1]). *If there are* $y_i \in E_l$, $i = 1, 2$ *such that*

$$\|y_1^+\| - \|y_1^-\| < 0 < \|y_2^+\| - \|y_2^-\|,$$

then there is a neighborhood $N \subset Q_l$ *of* (λ_l, λ_l) *such that every point* $(a, b) \in N \setminus \{(\lambda_l, \lambda_l)\}$ *with* $a + b = 2\lambda_l$ *is in* II_l.

When (5) holds, the conclusion of this theorem follows if there is a $y \in E_l$ such that

$$\|y^+\| \neq \|y^-\|.$$

Indeed,

$$\|(-y)^+\| - \|(-y)^-\| = -(\|y^+\| - \|y^-\|)$$

by (5) and hence $\|(\pm y)^+\| - \|(\pm y)^-\|$ have opposite signs. For problem (4), this result is due to Li et al. [**52**]. When λ_l is a simple eigenvalue, the region II_l is free of $\Sigma(A)$ (see [**84**, Theorem 4.10.1]). For problem (4), this is due to Gallouët and Kavian [**39**].

When $(a, b) \notin \Sigma(A)$, the origin is the only critical point of I and hence the critical groups $C_q(I, 0)$ are defined. We take the coefficient group to be the field $\mathbb{Z}_2$. For $B \subset D$, set

$$B^- = \{u \in B : I(u) < 0\}, \qquad B^+ = B \setminus B^-.$$

The following theorem gives our main results on $C_q(I, 0)$.

THEOREM 40 ([**84**, Theorem 4.11.2]). *Let* $(a, b) \in Q_l \setminus \Sigma(A)$.

(i) *If* $(a, b) \in \mathrm{I}_l$, *then*

$$C_q(I, 0) \approx \delta_{qd_{l-1}} \mathbb{Z}_2.$$

(ii) *If* $(a, b) \in \mathrm{I}^l$, *then*

$$C_q(I, 0) \approx \delta_{qd_l} \mathbb{Z}_2.$$

(iii) *If* $(a, b) \in \mathrm{II}_l$, *then*

$$C_q(I, 0) = 0, \quad q \le d_{l-1} \text{ or } q \ge d_l$$

and

$$C_q(I, 0) \approx \widetilde{H}_{q-d_{l-1}-1}(\widetilde{\mathcal{N}}_l^-), \quad d_{l-1} < q < d_l.$$

In particular, $C_q(I, 0) = 0$ *for all* q *when* λ_l *is simple.*

PROOF. Taking $U = D$ in (1) gives

$$C_q(I, 0) = H_q(I^0, I^0 \setminus \{0\}). \tag{12}$$

Since $(a, b) \notin \Sigma(A)$, I satisfies (PS), so I^0 is a deformation retract of D and $I^0 \setminus \{0\}$ deformation retracts to I^a for any $a < 0$ by the second deformation lemma, and hence

$$H_q(I^0, I^0 \setminus \{0\}) \approx H_q(D, I^a). \tag{13}$$

Since $\widetilde{H}_q(D) = 0$ for all q, the exact sequence

$$\cdots \longrightarrow \widetilde{H}_q(D) \longrightarrow H_q(D, I^a) \longrightarrow \widetilde{H}_{q-1}(I^a) \longrightarrow \widetilde{H}_{q-1}(D) \longrightarrow \cdots$$

of the pair (D, I^a) gives

$$H_q(D, I^a) \approx \widetilde{H}_{q-1}(I^a). \tag{14}$$

By [**84**, Remark 1.3.8], I^a is a deformation retract of D^- and hence

$$\widetilde{H}_{q-1}(I^a) \approx \widetilde{H}_{q-1}(D^-). \tag{15}$$

Writing $u \in D^-$ as $v + w \in N_l \oplus M_l$, let

$$\eta_1(u,t) = v + (1-t)\, w + t\, \tau(v), \quad (u,t) \in D^- \times [0,1].$$

We have

$$\begin{aligned} I(\eta_1(u,t)) &\le (1-t)\, I(v+w) + t\, I(v+\tau(v)) && \text{by (7)} \\ &\le I(u) && \text{by Proposition 33 (ii)} \\ &< 0, \end{aligned}$$

so η_1 is a deformation retraction of D^- onto S^{l-}. On the other hand,

$$\eta_2(u,t) = (1-t)\, u + t\, \pi(u), \quad (u,t) \in S^{l-} \times [0,1],$$

where

$$\pi : D \setminus \{0\} \to S, \quad u \mapsto \frac{u}{\|u\|_D}$$

is the radial projection onto S, is a deformation retraction of S^{l-} onto $\widetilde{S}^{l-}$ by the positive homogeneity of ζ and I. Thus,

$$\widetilde{H}_{q-1}(D^-) \approx \widetilde{H}_{q-1}(\widetilde{S}^{l-}). \tag{16}$$

Combining (12)–(16) gives

$$C_q(I,0) \approx \widetilde{H}_{q-1}(\widetilde{S}^{l-}). \tag{17}$$

If $(a,b) \in \mathrm{I}^l$, then $I < 0$ on $S^l \setminus \{0\}$ by (the proof of) Theorem 38 (v) and hence $\widetilde{S}^{l-} = \widetilde{S}^l$. Since the latter is homeomorphic to $\widetilde{N}_l$, (ii) follows. Since I_l and I^{l-1} are subsets of the same connected component of $\mathbb{R}^2 \setminus \Sigma(A)$, (i) follows from (ii).

Now let $(a,b) \in \mathrm{II}_l$. By Theorem 38 (iii), $\widetilde{S}^{l-}$ is a proper subset of $\widetilde{S}^l$ and hence $C_q(I,0) \approx \widetilde{H}_{q-1}(\widetilde{S}^{l-}) = 0$ for $q \ge d_l$. Since $\widetilde{H}_q(\widetilde{S}^l) = \delta_{q(d_l-1)}\, \mathbb{Z}_2$, the exact sequence

$$\cdots \longrightarrow \widetilde{H}_q(\widetilde{S}^l) \longrightarrow H_q(\widetilde{S}^l, \widetilde{S}^{l-}) \longrightarrow \widetilde{H}_{q-1}(\widetilde{S}^{l-}) \longrightarrow \widetilde{H}_{q-1}(\widetilde{S}^l) \longrightarrow \cdots$$

of the pair $(\widetilde{S}^l, \widetilde{S}^{l-})$ now gives

$$\widetilde{H}_{q-1}(\widetilde{S}^{l-}) \approx H_q(\widetilde{S}^l, \widetilde{S}^{l-}) / \delta_{q(d_l-1)}\, \mathbb{Z}_2. \tag{18}$$

By the Poincaré-Lefschetz duality theorem,

$$H_q(\widetilde{S}^l, \widetilde{S}^{l-}) \approx \check{H}^{d_l-1-q}(\widetilde{S}^{l+}) \tag{19}$$

where $\check{H}$ denotes Čech cohomology.

Writing $u \in \widetilde{S}^{l+}$ as $\zeta(v+y)$ with $v+y \in N_{l-1} \oplus E_l$, let

$$\eta_3(u,t) = \zeta((1-t)\,v + t\,\eta(y) + y), \quad (u,t) \in \widetilde{S}^{l+} \times [0,1].$$

We have

$$\begin{aligned} I(\eta_3(u,t)) &= \inf_{w \in M_l} I((1-t)\,v + t\,\eta(y) + y + w) \quad \text{by Proposition 33 (ii)} \\ &\geq \inf_{w \in M_l} [(1-t)\,I(v+y+w) + t\,I(\eta(y)+y+w)] \quad \text{by (6)} \\ &\geq (1-t)\,I(\zeta(v+y)) + t\,I(\zeta(\eta(y)+y)) \quad \text{by Proposition 33 (ii)} \\ &\geq I(u) \quad \text{by Proposition 36 (i)} \\ &\geq 0. \end{aligned}$$

If $\eta_3(u,t) = 0$, then $(1-t)\,v + t\,\eta(y) + y = 0$ and hence $y = 0$, so $u = \zeta(v)$. Since $u \neq 0$, then $v \neq 0$, so

$$\begin{aligned} I(u) &\leq I(v) \quad \text{by Proposition 33 (ii)} \\ &\leq \|v\|_D^2 - \underline{\lambda}\,\|v\|^2 \quad \text{where } \underline{\lambda} = \min\{a,b\} \\ &\leq -(\underline{\lambda} - \lambda_{l-1})\,\|v\|^2 \quad \text{by (2)} \\ &< 0, \end{aligned}$$

contrary to assumption. So $\eta_3(u,t) \neq 0$. Thus,

$$\eta_4 = \pi \circ \eta_3$$

is a deformation retraction of $\widetilde{S}^{l+}$ onto $\widetilde{\mathcal{N}}_l^+$, and hence

(20) $$\check{H}^{d_l-1-q}(\widetilde{S}^{l+}) \approx \check{H}^{d_l-1-q}(\widetilde{\mathcal{N}}_l^+).$$

Applying the Poincaré-Lefschetz duality theorem again gives

(21) $$\check{H}^{d_l-1-q}(\widetilde{\mathcal{N}}_l^+) \approx H_{q-d_{l-1}}(\widetilde{\mathcal{N}}_l, \widetilde{\mathcal{N}}_l^-).$$

By Theorem 38 (iii), $\widetilde{\mathcal{N}}_l^-$ is a proper subset of $\widetilde{\mathcal{N}}_l$ and hence $\widetilde{H}_{q-d_{l-1}-1}(\widetilde{\mathcal{N}}_l^-) = 0$ for $q \geq d_l$. Since $\widetilde{H}_{q-d_{l-1}-1}(\widetilde{\mathcal{N}}_l) = \delta_{qd_l}\,\mathbb{Z}_2$, the exact sequence

$$\begin{aligned} \cdots \longrightarrow \widetilde{H}_{q-d_{l-1}}(\widetilde{\mathcal{N}}_l) \longrightarrow H_{q-d_{l-1}}(\widetilde{\mathcal{N}}_l, \widetilde{\mathcal{N}}_l^-) \longrightarrow \widetilde{H}_{q-d_{l-1}-1}(\widetilde{\mathcal{N}}_l^-) \\ y \qquad \longrightarrow \widetilde{H}_{q-d_{l-1}-1}(\widetilde{\mathcal{N}}_l) \longrightarrow \cdots \end{aligned}$$

of the pair $(\widetilde{\mathcal{N}}_l, \widetilde{\mathcal{N}}_l^-)$ then gives

(22) $$H_{q-d_{l-1}}(\widetilde{\mathcal{N}}_l, \widetilde{\mathcal{N}}_l^-)/\delta_{q(d_l-1)}\,\mathbb{Z}_2 \approx \widetilde{H}_{q-d_{l-1}-1}(\widetilde{\mathcal{N}}_l^-).$$

Combining (17) – (22) gives $C_q(I,0) \approx \widetilde{H}_{q-d_{l-1}-1}(\widetilde{\mathcal{N}}_l^-)$, from which the rest of (iii) follows. □

For problem (4), this result is due to Dancer [**28, 29**] and Perera and Schechter [**86, 87, 88**]. It can be used, for example, to obtain nontrivial solutions of perturbed problems with nonlinearities that cross a curve of the Fučík spectrum, via a comparison of the critical groups at zero and infinity. Consider the operator equation

$$Au = f(u), \quad u \in D \tag{23}$$

where $f \in C(D, H)$ is a potential operator that maps bounded sets into bounded sets.

THEOREM 41 (see [**84**, Theorem 5.6.1]). *If*

$$f(u) = b_0 u^+ - a_0 u^- + o(\|u\|_D) \quad \text{as} \quad \|u\|_D \to 0,$$
$$f(u) = b u^+ - a u^- + o(\|u\|_D) \quad \text{as} \quad \|u\|_D \to \infty$$

for some (a_0, b_0) *and* (a, b) *in* $Q_l \setminus \Sigma(A)$ *that are on opposite sides of* C_l *or* C^l, *then* (23) *has a nontrivial solution.*

For problem (4), this was proved in Perera and Schechter [**87**]. It generalizes a well-known result of Amann and Zehnder [**2**] on the existence of nontrivial solutions for problems crossing an eigenvalue.

9. *p*-Laplacian

In this section we present a result on nontrivial critical groups associated with the p-Laplacian obtained in Perera [**80**]; see also Dancer and Perera [**30**] and Perera et al. [**83**].

Consider the nonlinear eigenvalue problem

$$\begin{cases} -\Delta_p\, u = \lambda\, |u|^{p-2}\, u & \text{in } \Omega \\ u = 0 & \text{on } \partial\Omega \end{cases} \tag{24}$$

where Ω is a bounded domain in $\mathbb{R}^n$, $n \geq 1$ and $\Delta_p\, u = \operatorname{div}\big(|\nabla u|^{p-2}\, \nabla u\big)$ is the p-Laplacian of u, $p \in (1, \infty)$. It is known that the first eigenvalue λ_1 is positive, simple, has an associated eigenfunction that is positive in Ω, and is isolated in the spectrum $\sigma(-\Delta_p)$; see Anane [**7**] and Lindqvist [**55, 56**]. So the second eigenvalue $\lambda_2 = \inf \sigma(-\Delta_p) \cap (\lambda_1, \infty)$ is also defined; see Anane and Tsouli [**6**]. In the ODE case $n = 1$, where Ω is an interval, the spectrum consists of a sequence of simple eigenvalues $\lambda_k \nearrow \infty$ and the eigenfunction associated with λ_k has exactly $k - 1$ interior zeroes; see Cuesta [**24**] or Drábek [**35**]. In the semilinear PDE case $n \geq 2$, $p = 2$ also $\sigma(-\Delta)$ consists of a sequence of eigenvalues $\lambda_k \nearrow \infty$, but in the quasilinear PDE case $n \geq 2$, $p \neq 2$ a complete description of the spectrum is not available.

Eigenvalues of (24) are the critical values of the C^1-functional

$$I(u) = \int_\Omega |\nabla u|^p, \quad u \in S = \big\{u \in W = W_0^{1,\,p}(\Omega) : \|u\|_{L^p(\Omega)} = 1\big\},$$

which satisfies (PS). Denote by $\mathcal{A}$ the class of closed symmetric subsets of S and by

$$\gamma^+(A) = \sup \left\{k \ge 1 : \exists \text{ an odd continuous map } S^{k-1} \to A\right\},$$

$$\gamma^-(A) = \inf \left\{k \ge 1 : \exists \text{ an odd continuous map } A \to S^{k-1}\right\}$$

the genus and the cogenus of $A \in \mathcal{A}$, respectively, where S^{k-1} is the unit sphere in $\mathbb{R}^k$. Then

$$\lambda_k^\pm = \inf_{\substack{A \in \mathcal{A} \\ \gamma^\pm(A) \ge k}} \sup_{u \in A} I(u), \quad k \ge 1$$

are two increasing and unbounded sequences of eigenvalues, but, in general, it is not known whether either sequence is a complete list. The sequence (λ_k^+) was introduced by Drábek and Robinson [**36**]; γ_k^- is also called the Krasnosel'skii genus [**45**].

Solutions of (24) are the critical points of the functional

$$I_\lambda(u) = \int_\Omega |\nabla u|^p - \lambda\, |u|^p, \quad u \in W_0^{1,\,p}(\Omega).$$

When $\lambda \notin \sigma(-\Delta_p)$, the origin is the only critical point of I_λ and hence the critical groups $C_q(I_\lambda, 0)$ are defined. Again we take the coefficient group to be $\mathbb{Z}_2$. The following theorem is our main result on them.

THEOREM 42 ([**80**, Proposition 1.1]). *The spectrum of $-\Delta_p$ contains a sequence of eigenvalues $\lambda_k \nearrow \infty$ such that $\lambda_k^- \le \lambda_k \le \lambda_k^+$ and*

$$\lambda \in (\lambda_k, \lambda_{k+1}) \setminus \sigma(-\Delta_p) \implies C_k(I_\lambda, 0) \ne 0.$$

Various applications of this sequence of eigenvalues can be found in Perera [**81, 82**], Liu and Li [**59**], Perera and Szulkin [**91**], Cingolani and Degiovanni [**22**], Guo and Liu [**43**], Degiovanni and Lancelotti [**33, 34**], Tanaka [**114**], Fang and Liu [**37**], Medeiros and Perera [**67**], Motreanu and Perera [**70**], and Degiovanni, Lancelotti, and Perera [**32**].

The eigenvalues λ_k are defined using the Yang index, whose definition and some properties we now recall. Yang [**116**] considered compact Hausdorff spaces with fixed-point-free continuous involutions and used the Čech homology theory, but for our purposes here it suffices to work with closed symmetric subsets of Banach spaces that do not contain the origin and singular homology groups.

Following [**116**], we first construct a special homology theory defined on the category of all pairs of closed symmetric subsets of Banach spaces that do not contain the origin and all continuous odd maps of such pairs. Let (X, A), $A \subset X$ be such a pair and $C(X, A)$ its singular chain complex with $\mathbb{Z}_2$-coefficients, and denote by $T_\#$ the chain map of $C(X, A)$ induced by the antipodal map $T(u) = -u$. We say that a q-chain c is symmetric if $T_\#(c) = c$, which holds if and only if $c = c' + T_\#(c')$ for some q-chain c'. The symmetric q-chains form a subgroup $C_q(X, A; T)$ of $C_q(X, A)$, and the boundary operator ∂_q maps $C_q(X, A; T)$ into

$C_{q-1}(X, A; T)$, so these subgroups form a subcomplex $C(X, A; T)$. We denote by

$$Z_q(X, A; T) = \{c \in C_q(X, A; T) : \partial_q c = 0\},$$
$$B_q(X, A; T) = \{\partial_{q+1} c : c \in C_{q+1}(X, A; T)\},$$
$$H_q(X, A; T) = Z_q(X, A; T)/B_q(X, A; T)$$

the corresponding cycles, boundaries, and homology groups. A continuous odd map $f : (X, A) \to (Y, B)$ of pairs as above induces a chain map $f_\# : C(X, A; T) \to C(Y, B; T)$ and hence homomorphisms

$$f_* : H_q(X, A; T) \to H_q(Y, B; T).$$

For example,

$$H_q(S^k; T) = \begin{cases} \mathbb{Z}_2, & 0 \le q \le k \\ 0, & q \ge k + 1 \end{cases}$$

(see [**116**, Example 1.8]).

Let X be as above, and define homomorphisms $\nu : Z_q(X; T) \to \mathbb{Z}_2$ inductively by

$$\nu(z) = \begin{cases} \mathrm{In}(c), & q = 0 \\ \nu(\partial c), & q > 0 \end{cases}$$

if $z = c + T_\#(c)$, where the index of a 0-chain $c = \sum_i n_i \, \sigma_i$ is defined by $\mathrm{In}(c) = \sum_i n_i$. As in [**116**], ν is well-defined and $\nu \, B_q(X; T) = 0$, so we can define the index homomorphism $\nu_* : H_q(X; T) \to \mathbb{Z}_2$ by $\nu_*([z]) = \nu(z)$. If F is a closed subset of X such that $F \cup T(F) = X$ and $A = F \cap T(F)$, then there is a homomorphism

$$\Delta : H_q(X; T) \to H_{q-1}(A; T)$$

such that $\nu_*(\Delta[z]) = \nu_*([z])$ (see [**116**, Proposition 2.8]). Taking $F = X$ we see that if $\nu_* \, H_k(X; T) = \mathbb{Z}_2$, then $\nu_* \, H_q(X; T) = \mathbb{Z}_2$ for $0 \le q \le k$. We define the Yang index of X by

$$i(X) = \inf \{k \ge -1 : \nu_* \, H_{k+1}(X; T) = 0\},$$

taking $\inf \emptyset = \infty$. Clearly, $\nu_* \, H_0(X; T) = \mathbb{Z}_2$ if $X \ne \emptyset$, so $i(X) = -1$ if and only if $X = \emptyset$. For example, $i(S^k) = k$ (see [**116**, Example 3.4]).

PROPOSITION 43 ([**116**, Proposition 2.4]). *If $f : X \to Y$ is as above, then $\nu_*(f_*([z])) = \nu_*([z])$ for $[z] \in H_q(X; T)$, and hence $i(X) \le i(Y)$. In particular, this inequality holds if $X \subset Y$.*

Thus, $k^+ - 1 \le i(X) \le k^- - 1$ if there are odd continuous maps $S^{k^+ - 1} \to X \to S^{k^- - 1}$, so

$$\gamma^+(X) \le i(X) + 1 \le \gamma^-(X). \tag{25}$$

PROPOSITION 44 ([**80**, Proposition 2.6]). *If* $i(X) = k \geq 0$, *then* $\widetilde{H}_k(X) \neq 0$.

PROOF. We have

$$\nu_* \, H_q(X;T) = \begin{cases} \mathbb{Z}_2, & 0 \leq q \leq k \\ 0, & q \geq k+1. \end{cases}$$

We show that if $[z] \in H_k(X;T)$ is such that $\nu_*([z]) \neq 0$, then $[z] \neq 0$ in $\widetilde{H}_k(X)$. Arguing indirectly, assume that $z \in B_k(X)$, say, $z = \partial c$. Since $z \in B_k(X;T)$, $T_\#(z) = z$. Let $c' = c + T_\#(c)$. Then $c' \in Z_{k+1}(X;T)$ since $\partial c' = z + T_\#(z) = 2z = 0$ mod 2, and $\nu_*([c']) = \nu(c') = \nu(\partial c) = \nu(z) \neq 0$, contradicting $\nu_* \, H_{k+1}(X;T) = 0$. □

LEMMA 45. *We have*

$$C_q(I_\lambda, 0) \approx \widetilde{H}_{q-1}(I^\lambda) \quad \forall q.$$

PROOF. Taking $U = \big\{u \in W : \|u\|_{L^p(\Omega)} \leq 1\big\}$ in (1) gives

$$C_q(I_\lambda, 0) = H_q(I_\lambda^0 \cap U, I_\lambda^0 \cap U \setminus \{0\}).$$

Since I_λ is positive homogeneous, $I_\lambda^0 \cap U$ radially contracts to the origin via

$$(I_\lambda^0 \cap U) \times [0,1] \to I_\lambda^0 \cap U, \quad (u,t) \mapsto (1-t)\, u$$

and $I_\lambda^0 \cap U \setminus \{0\}$ deformation retracts onto $I_\lambda^0 \cap S$ via

$$(I_\lambda^0 \cap U \setminus \{0\}) \times [0,1] \to I_\lambda^0 \cap U \setminus \{0\}, \quad (u,t) \mapsto (1-t)\, u + t\, u / \|u\|_{L^p(\Omega)},$$

so it follows from the exact sequence of the pair $(I_\lambda^0 \cap U, I_\lambda^0 \cap U \setminus \{0\})$ that

$$H_q(I_\lambda^0 \cap U, I_\lambda^0 \cap U \setminus \{0\}) \approx \widetilde{H}_{q-1}(I_\lambda^0 \cap S).$$

Since $I_\lambda|_S = I - \lambda$, $I_\lambda^0 \cap S = I^\lambda$. □

We are now ready to prove Theorem 42.

PROOF OF THEOREM 42. Set

$$\lambda_k = \inf_{\substack{A \in \mathcal{A} \\ i(A) \geq k-1}} \; \sup_{u \in A} I(u), \quad k \geq 1.$$

Then (λ_k) is an increasing sequence of critical points of I, and hence eigenvalues of $-\Delta_p$, by a standard deformation argument (see [**80**, Proposition 3.1]). By (25), $\lambda_k^- \leq \lambda_k \leq \lambda_k^+$, in particular, $\lambda_k \to \infty$.

Let $\lambda \in (\lambda_k, \lambda_{k+1}) \setminus \sigma(-\Delta_p)$. By Lemma 45, $C_k(I_\lambda, 0) \approx \widetilde{H}_{k-1}(I^\lambda)$, and $I^\lambda \in \mathcal{A}$ since I is even. Since $\lambda > \lambda_k$, there is an $A \in \mathcal{A}$ with $i(A) \geq k-1$ such that $I \leq \lambda$ on A. Then $A \subset I^\lambda$ and hence $i(I^\lambda) \geq i(A) \geq k-1$ by Proposition 43. On the other hand, $i(I^\lambda) \leq k-1$ since $I \leq \lambda < \lambda_{k+1}$ on I^λ. So $i(I^\lambda) = k-1$ and hence $\widetilde{H}_{k-1}(I^\lambda) \neq 0$ by Proposition 44. □

References

[1] S. Ahmad, A. C. Lazer and J. L. Paul. Elementary critical point theory and perturbations of elliptic boundary value problems at resonance. *Indiana Univ. Math. J.*, 25(10):933–944, 1976.

[2] H. Amann and E. Zehnder. Nontrivial solutions for a class of nonresonance problems and applications to nonlinear differential equations. *Ann. Scuola Norm. Sup. Pisa Cl. Sci. (4)*, 7(4):539–603, 1980.

[3] Antonio Ambrosetti. Differential equations with multiple solutions and nonlinear functional analysis. In *Equadiff 82 (Würzburg, 1982)*, volume 1017 of *Lecture Notes in Math.*, pages 10–37. Springer, Berlin, 1983.

[4] Antonio Ambrosetti. Elliptic equations with jumping nonlinearities. *J. Math. Phys. Sci.*, 18(1):1–12, 1984.

[5] Antonio Ambrosetti and Paul H. Rabinowitz. Dual variational methods in critical point theory and applications. *J. Functional Analysis*, 14:349–381, 1973.

[6] A. Anane and N. Tsouli. On the second eigenvalue of the p-Laplacian. In *Nonlinear partial differential equations (Fès, 1994)*, volume 343 of *Pitman Res. Notes Math. Ser.*, pages 1–9. Longman, Harlow, 1996.

[7] Aomar Anane. Simplicité et isolation de la première valeur propre du p-laplacien avec poids. *C. R. Acad. Sci. Paris Sér. I Math.*, 305(16):725–728, 1987.

[8] P. Bartolo, V. Benci and D. Fortunato. Abstract critical point theorems and applications to some nonlinear problems with "strong" resonance at infinity. *Nonlinear Anal.*, 7(9): 981–1012, 1983.

[9] Thomas Bartsch and Shujie Li. Critical point theory for asymptotically quadratic functionals and applications to problems with resonance. *Nonlinear Anal.*, 28(3):419–441, 1997.

[10] Vieri Benci. Some applications of the generalized Morse-Conley index. *Confer. Sem. Mat. Univ. Bari*, (218):32, 1987.

[11] Vieri Benci. A new approach to the Morse-Conley theory and some applications. *Ann. Mat. Pura Appl. (4)*, 158:231–305, 1991.

[12] Vieri Benci. Introduction to Morse theory: a new approach. In *Topological nonlinear analysis*, volume 15 of *Progr. Nonlinear Differential Equations Appl.*, pages 37–177. Birkhäuser Boston, Boston, MA, 1995.

[13] Vieri Benci and Paul H. Rabinowitz. Critical point theorems for indefinite functionals. *Invent. Math.*, 52(3):241–273, 1979.

[14] Haïm Brezis and Louis Nirenberg. Remarks on finding critical points. *Comm. Pure Appl. Math.*, 44(8–9):939–963, 1991.

[15] Nguyên Phuong Các. On nontrivial solutions of a Dirichlet problem whose jumping nonlinearity crosses a multiple eigenvalue. *J. Differential Equations*, 80(2):379–404, 1989.

[16] Alberto Cambini. Sul lemma di M. Morse. *Boll. Un. Mat. Ital. (4)*, 7:87–93, 1973.

[17] A. Castro and A. C. Lazer. Applications of a maximin principle. *Rev. Colombiana Mat.*, 10:141–149, 1976.

[18] Giovanna Cerami. An existence criterion for the critical points on unbounded manifolds. *Istit. Lombardo Accad. Sci. Lett. Rend. A*, 112(2):332–336 (1979), 1978.

[19] K. C. Chang and N. Ghoussoub. The Conley index and the critical groups via an extension of Gromoll-Meyer theory. *Topol. Methods Nonlinear Anal.*, 7(1):77–93, 1996.

[20] Kung Ching Chang. Solutions of asymptotically linear operator equations via Morse theory. *Comm. Pure Appl. Math.*, 34(5):693–712, 1981.

[21] Kung-ching Chang. *Infinite-dimensional Morse theory and multiple solution problems*, volume 6 of *Progress in Nonlinear Differential Equations and their Applications.* Birkhäuser Boston Inc., Boston, MA, 1993.

[22] Silvia Cingolani and Marco Degiovanni. Nontrivial solutions for p-Laplace equations with right-hand side having p-linear growth at infinity. *Comm. Partial Differential Equations*, 30(7–9):1191–1203, 2005.

[23] J.-N. Corvellec and A. Hantoute. Homotopical stability of isolated critical points of continuous functionals. *Set-Valued Anal.*, 10(2–3):143–164, 2002. Calculus of variations, nonsmooth analysis and related topics.

[24] M. Cuesta. On the Fučík spectrum of the Laplacian and the p-Laplacian. In *Proceedings of Seminar in Differential Equations*, pages 67–96, Kvilda, Czech Republic, May 29 – June 2,

2000. Centre of Applied Mathematics, Faculty of Applied Sciences, University of West Bohemia in Pilsen.

[25] Mabel Cuesta and Jean-Pierre Gossez. A variational approach to nonresonance with respect to the Fučik spectrum. *Nonlinear Anal.*, 19(5):487–500, 1992.

[26] E. N. Dancer. On the Dirichlet problem for weakly non-linear elliptic partial differential equations. *Proc. Roy. Soc. Edinburgh Sect. A*, 76(4):283–300, 1976/77.

[27] E. N. Dancer. Corrigendum: "On the Dirichlet problem for weakly nonlinear elliptic partial differential equations" [Proc. Roy. Soc. Edinburgh Sect. A **76** (1976/77), no. 4, 283–300; MR **58** #17506]. *Proc. Roy. Soc. Edinburgh Sect. A*, 89(1–2):15, 1981.

[28] E. N. Dancer. Remarks on jumping nonlinearities. In *Topics in nonlinear analysis*, volume 35 of *Progr. Nonlinear Differential Equations Appl.*, pages 101–116. Birkhäuser, Basel, 1999.

[29] E. Norman Dancer. Some results for jumping nonlinearities. *Topol. Methods Nonlinear Anal.*, 19(2):221–235, 2002.

[30] Norman Dancer and Kanishka Perera. Some remarks on the Fučík spectrum of the p-Laplacian and critical groups. *J. Math. Anal. Appl.*, 254(1):164–177, 2001.

[31] D. G. de Figueiredo and J.-P. Gossez. On the first curve of the Fučik spectrum of an elliptic operator. *Differential Integral Equations*, 7(5–6):1285–1302, 1994.

[32] M. Degiovanni, S. Lancelotti, and K. Perera. Nontrivial solutions of p-superlinear p-laplacian problems via a cohomological local splitting. Commun. Contemp. Math., 12(3):475–486, 2010.

[33] Marco Degiovanni and Sergio Lancelotti. Linking over cones and nontrivial solutions for p-Laplace equations with p-superlinear nonlinearity. 2007.

[34] Marco Degiovanni and Sergio Lancelotti. Linking solutions for p-Laplace equations with nonlinearity at critical growth. *J. Funct. Anal.*, 256(11):3643–3659, 2009.

[35] P. Drábek. *Solvability and bifurcations of nonlinear equations*, volume 264 of *Pitman Research Notes in Mathematics Series*. Longman Scientific & Technical, Harlow, 1992.

[36] Pavel Drábek and Stephen B. Robinson. Resonance problems for the p-Laplacian. *J. Funct. Anal.*, 169(1):189–200, 1999.

[37] Fei Fang and Shibo Liu. Nontrivial solutions of superlinear p-Laplacian equations. *J. Math. Anal. Appl.*, 351(1):138–146, 2009.

[38] Svatopluk Fučík. Boundary value problems with jumping nonlinearities. *Časopis Pěst. Mat.*, 101(1):69–87, 1976.

[39] Thierry Gallouët and Otared Kavian. Résultats d'existence et de non-existence pour certains problèmes demi-linéaires à l'infini. *Ann. Fac. Sci. Toulouse Math. (5)*, 3(3–4):201–246 (1982), 1981.

[40] N. Ghoussoub. Location, multiplicity and Morse indices of min-max critical points. *J. Reine Angew. Math.*, 417:27–76, 1991.

[41] Nassif Ghoussoub. *Duality and perturbation methods in critical point theory*, volume 107 of *Cambridge Tracts in Mathematics*. Cambridge University Press, Cambridge, 1993. With appendices by David Robinson.

[42] Detlef Gromoll and Wolfgang Meyer. On differentiable functions with isolated critical points. *Topology*, 8:361–369, 1969.

[43] Yuxia Guo and Jiaquan Liu. Solutions of p-sublinear p-Laplacian equation via Morse theory. *J. London Math. Soc. (2)*, 72(3):632–644, 2005.

[44] Helmut Hofer. The topological degree at a critical point of mountain-pass type. 45:501–509, 1986.

[45] M. A. Krasnosel'skii. *Topological methods in the theory of nonlinear integral equations*. Translated by A. H. Armstrong; translation edited by J. Burlak. A Pergamon Press Book. The Macmillan Co., New York, 1964.

[46] Wojciech Kryszewski and Andrzej Szulkin. An infinite-dimensional Morse theory with applications. *Trans. Amer. Math. Soc.*, 349(8):3181–3234, 1997.

[47] Nicolaas H. Kuiper. C^1-equivalence of functions near isolated critical points. pages 199–218. Ann. of Math. Studies, No. 69, 1972.

[48] Sergio Lancelotti. Existence of nontrivial solutions for semilinear problems with strictly differentiable nonlinearity. *Abstr. Appl. Anal.*, pages Art. ID 62458, 14, 2006.

[49] A. C. Lazer and P. J. McKenna. Critical point theory and boundary value problems with nonlinearities crossing multiple eigenvalues. II. *Comm. Partial Differential Equations*, 11(15):1653–1676, 1986.

[50] A. C. Lazer and S. Solimini. Nontrivial solutions of operator equations and Morse indices of critical points of min-max type. *Nonlinear Anal.*, 12(8):761–775, 1988.

[51] Alan Lazer. Introduction to multiplicity theory for boundary value problems with asymmetric nonlinearities. In *Partial differential equations (Rio de Janeiro, 1986)*, volume 1324 of *Lecture Notes in Math.*, pages 137–165. Springer, Berlin, 1988.

[52] Chong Li, Shujie Li and Zhaoli Liu. Existence of type (II) regions and convexity and concavity of potential functionals corresponding to jumping nonlinear problems. *Calc. Var. Partial Differential Equations*, 32(2):237–251, 2008.

[53] Shu Jie Li and J. Q. Liu. Morse theory and asymptotic linear Hamiltonian system. *J. Differential Equations*, 78(1):53–73, 1989.

[54] Shu Jie Li and Michel Willem. Applications of local linking to critical point theory. *J. Math. Anal. Appl.*, 189(1):6–32, 1995.

[55] Peter Lindqvist. On the equation div $(|\nabla u|^{p-2}\nabla u)+\lambda|u|^{p-2}u=0$. *Proc. Amer. Math. Soc.*, 109(1):157–164, 1990.

[56] Peter Lindqvist. Addendum: "On the equation div$(|\nabla u|^{p-2}\nabla u)+\lambda|u|^{p-2}u=0$" [Proc. Amer. Math. Soc. 109 (1990), no. 1, 157–164; MR 90h:35088]. *Proc. Amer. Math. Soc.*, 116(2):583–584, 1992.

[57] Jia Quan Liu. The Morse index of a saddle point. *Systems Sci. Math. Sci.*, 2(1):32–39, 1989.

[58] Jia Quan Liu and Shu Jie Li. An existence theorem for multiple critical points and its application. *Kexue Tongbao (Chinese)*, 29(17):1025–1027, 1984.

[59] Shibo Liu and Shujie Li. Existence of solutions for asymptotically 'linear' p-Laplacian equations. *Bull. London Math. Soc.*, 36(1):81–87, 2004.

[60] L. Ljusternik and L. Schnirelmann. *Methodes topologique dans les problémes variationnels.* Hermann and Cie, Paris, 1934.

[61] C. A. Magalhães. Semilinear elliptic problem with crossing of multiple eigenvalues. *Comm. Partial Differential Equations*, 15(9):1265–1292, 1990.

[62] Caryl Ann Margulies and William Margulies. An example of the Fučik spectrum. *Nonlinear Anal.*, 29(12):1373–1378, 1997.

[63] A. Marino and G. Prodi. Metodi perturbativi nella teoria di Morse. *Boll. Un. Mat. Ital. (4)*, 11(3, suppl.):1–32, 1975. Collection of articles dedicated to Giovanni Sansone on the occasion of his eighty-fifth birthday.

[64] Antonio Marino and Giovanni Prodi. La teoria di Morse per gli spazi di Hilbert. *Rend. Sem. Mat. Univ. Padova*, 41:43–68, 1968.

[65] J. Mawhin and M. Willem. On the generalized Morse lemma. *Bull. Soc. Math. Belg. Sér. B*, 37(2):23–29, 1985.

[66] Jean Mawhin and Michel Willem. *Critical point theory and Hamiltonian systems*, volume 74 of *Applied Mathematical Sciences*. Springer-Verlag, New York, 1989.

[67] Everaldo Medeiros and Kanishka Perera. Multiplicity of solutions for a quasilinear elliptic problem via the cohomological index. *Nonlinear Anal.*, 71(9):3654–3660, 2009.

[68] J. Milnor. Morse theory. pages vi+153, 1963.

[69] Marston Morse. Relations between the critical points of a real function of n independent variables. *Trans. Amer. Math. Soc.*, 27(3):345–396, 1925.

[70] Dumitru Motreanu and Kanishka Perera. Multiple nontrivial solutions of Neumann p-Laplacian systems. *Topol. Methods Nonlinear Anal.*, 34(1):41–48, 2009.

[71] Wei Ming Ni. Some minimax principles and their applications in nonlinear elliptic equations. *J. Analyse Math.*, 37:248–275, 1980.

[72] L. Nirenberg. Topics in nonlinear functional analysis. pages viii+259, 1974. With a chapter by E. Zehnder, Notes by R. A. Artino, Lecture Notes, 1973–1974.

[73] L. Nirenberg. Variational and topological methods in nonlinear problems. *Bull. Amer. Math. Soc. (N.S.)*, 4(3):267–302, 1981.

[74] R. S. Palais and S. Smale. A generalized Morse theory. *Bull. Amer. Math. Soc.*, 70:165–172, 1964.

[75] Richard S. Palais. Morse theory on Hilbert manifolds. *Topology*, 2:299–340, 1963.

[76] Richard S. Palais. Critical point theory and the minimax principle. In *Global Analysis (Proc. Sympos. Pure Math., Vol. XV, Berkeley, Calif, 1968)*, pages 185–212. Amer. Math. Soc., Providence, R.I., 1970.

[77] Kanishka Perera. Critical groups of pairs of critical points produced by linking subsets. *J. Differential Equations*, 140(1):142–160, 1997.

[78] Kanishka Perera. Critical groups of critical points produced by local linking with applications. *Abstr. Appl. Anal.*, 3(3–4):437–446, 1998.

[79] Kanishka Perera. Homological local linking. *Abstr. Appl. Anal.*, 3(1–2):181–189, 1998.

[80] Kanishka Perera. Nontrivial critical groups in p-Laplacian problems via the Yang index. *Topol. Methods Nonlinear Anal.*, 21(2):301–309, 2003.

[81] Kanishka Perera. Nontrivial solutions of p-superlinear p-Laplacian problems. *Appl. Anal.*, 82(9):883–888, 2003.

[82] Kanishka Perera. p-superlinear problems with jumping nonlinearities. In *Nonlinear analysis and applications: to V. Lakshmikantham on his 80th birthday. Vol. 1, 2*, pages 823–829. Kluwer Acad. Publ., Dordrecht, 2003.

[83] Kanishka Perera, Ravi P. Agarwal, and Donal O'Regan. *Morse-Theoretic Aspects of p-Laplacian Type Operators*. Mathematical Surveys and Monographs. American Mathematical Society, Providence, RI, 2010.

[84] Kanishka Perera and Martin Schechter. *Morse Theoretic Methods in Critical Point Theory*. Topics in Critical Point Theory. to appear.

[85] Kanishka Perera and Martin Schechter. Morse index estimates in saddle point theorems without a finite-dimensional closed loop. *Indiana Univ. Math. J.*, 47(3):1083–1095, 1998.

[86] Kanishka Perera and Martin Schechter. Type II regions between curves of the Fučik spectrum and critical groups. *Topol. Methods Nonlinear Anal.*, 12(2):227–243, 1998.

[87] Kanishka Perera and Martin Schechter. A generalization of the Amann-Zehnder theorem to nonresonance problems with jumping nonlinearities. *NoDEA Nonlinear Differential Equations Appl.*, 7(4):361–367, 2000.

[88] Kanishka Perera and Martin Schechter. The Fučík spectrum and critical groups. *Proc. Amer. Math. Soc.*, 129(8):2301–2308 (electronic), 2001.

[89] Kanishka Perera and Martin Schechter. Critical groups in saddle point theorems without a finite dimensional closed loop. *Math. Nachr.*, 243:156–164, 2002.

[90] Kanishka Perera and Martin Schechter. Double resonance problems with respect to the Fučík spectrum. *Indiana Univ. Math. J.*, 52(1):1–17, 2003.

[91] Kanishka Perera and Andrzej Szulkin. p-Laplacian problems where the nonlinearity crosses an eigenvalue. *Discrete Contin. Dyn. Syst.*, 13(3):743–753, 2005.

[92] Everett Pitcher. Inequalities of critical point theory. *Bull. Amer. Math. Soc.*, 64(1):1–30, 1958.

[93] Gui Jie Qi. Extension of Mountain Pass Lemma. *Kexue Tongbao (English Ed.)*, 32(12):798–801, 1987.

[94] P. H. Rabinowitz. Variational methods for nonlinear eigenvalue problems. pages 139–195, 1974.

[95] Paul H. Rabinowitz. Some critical point theorems and applications to semilinear elliptic partial differential equations. *Ann. Scuola Norm. Sup. Pisa Cl. Sci. (4)*, 5(1):215–223, 1978.

[96] Paul H. Rabinowitz. Some minimax theorems and applications to nonlinear partial differential equations. In *Nonlinear analysis (collection of papers in honor of Erich H. Rothe)*, pages 161–177. Academic Press, New York, 1978.

[97] Paul H. Rabinowitz. *Minimax methods in critical point theory with applications to differential equations*, volume 65 of *CBMS Regional Conference Series in Mathematics*. Published for the Conference Board of the Mathematical Sciences, Washington, DC, 1986.

[98] Miguel Ramos and Luis Sanchez. Homotopical linking and Morse index estimates in min-max theorems. *Manuscripta Math.*, 87(3):269–284, 1995.

[99] N. K. Ribarska, Ts. Y. Tsachev, and M. I. Krastanov. A saddle point theorem without a finite-dimensional closed loop. *C. R. Acad. Bulgare Sci.*, 51(11–12):13–16, 1998.

[100] E. H. Rothe. Some remarks on critical point theory in Hilbert space. In *Nonlinear Problems (Proc. Sympos., Madison, Wis., (1962)*, pages 233–256. Univ. of Wisconsin Press, Madison, Wis., 1963.

[101] E. H. Rothe. Some remarks on critical point theory in Hilbert space (continuation). *J. Math. Anal. Appl.*, 20:515–520, 1967.

[102] E. H. Rothe. On continuity and approximation questions concerning critical Morse groups in Hilbert space. In *Symposium on Infinite-Dimensional Topology (Louisiana State Univ., Baton Rouge, La., 1967)*, pages 275–295. Ann. of Math. Studies, No. 69. Princeton Univ. Press, Princeton, N. J., 1972.

[103] E. H. Rothe. Morse theory in Hilbert space. *Rocky Mountain J. Math.*, 3:251–274, 1973. Rocky Mountain Consortium Symposium on Nonlinear Eigenvalue Problems (Santa Fe, N.M., 1971).

[104] Bernhard Ruf. On nonlinear elliptic problems with jumping nonlinearities. *Ann. Mat. Pura Appl. (4)*, 128:133–151, 1981.

[105] Martin Schechter. Splitting subspaces and saddle points. *Appl. Anal.*, 49(1–2):33–48, 1993.

[106] Martin Schechter. The Fučík spectrum. *Indiana Univ. Math. J.*, 43(4):1139–1157, 1994.

[107] Martin Schechter. New linking theorems. *Rend. Sem. Mat. Univ. Padova*, 99:255–269, 1998.

[108] Martin Schechter. *Linking methods in critical point theory*. Birkhäuser Boston Inc., Boston, MA, 1999.

[109] Martin Schechter and Kyril Tintarev. Pairs of critical points produced by linking subsets with applications to semilinear elliptic problems. *Bull. Soc. Math. Belg. Sér. B*, 44(3): 249–261, 1992.

[110] J. T. Schwartz. Nonlinear functional analysis. pages vii+236, 1969. Notes by H. Fattorini, R. Nirenberg and H. Porta, with an additional chapter by Hermann Karcher, Notes on Mathematics and its Applications.

[111] S. Smale. Morse theory and a non-linear generalization of the Dirichlet problem. *Ann. of Math. (2)*, 80:382–396, 1964.

[112] Sergio Solimini. Morse index estimates in min-max theorems. *Manuscripta Math.*, 63(4): 421–453, 1989.

[113] Andrzej Szulkin. Cohomology and Morse theory for strongly indefinite functionals. *Math. Z.*, 209(3):375–418, 1992.

[114] Mieko Tanaka. On the existence of a non-trivial solution for the p-Laplacian equation with a jumping nonlinearity. *Tokyo J. Math.*, 31(2):333–341, 2008.

[115] Zhi Qiang Wang. A note on the second variation theorem. *Acta Math. Sinica*, 30(1):106–110, 1987.

[116] Chung-Tao Yang. On theorems of Borsuk-Ulam, Kakutani-Yamabe-Yujobô and Dyson. I. *Ann. of Math. (2)*, 60:262–282, 1954.

Kanishka Perera, Department of Mathematical Sciences, Florida Institute of Technology, Melbourne, FL 32901, USA

E-mail address: kperera@fit.edu

Quasiconvex optimization and its applications

Enkhbat Rentsen

ABSTRACT. We consider maximization and minimization of a quasiconvex function over an arbitrary set. Based on the global optimality conditions we propose some methods and algorithms for solving the above problems. Applications of quasiconvex functions are provided.

CONTENTS

Preface

The notion of convexity plays an important role in Economics and Engineering. In classical economic theory production and utility functions are assumed to be concave functions and as consequence, demand and cost functions are convex. In recent years, generalized convex functions such as quasiconvex functions are found to be useful in economics. The purpose of this chapter is to survey recent advances in quasiconvex optimization and its application.

Enkhbat Rentsen

1991 *Mathematics Subject Classification.* Primary 90C26, 93C05; Secondary 34H05, 34K35.

Key words and phrases. quasiconvex functions, maximization, minimization, global optimality conditions, optimal control.

Introduction

We consider two problems of maximizing and minimizing a quasiconvex function over an arbitrary set $D \subset R^n$:

$$f(x) \longrightarrow \max, \quad x \in D, \tag{1}$$

$$f(x) \longrightarrow \min, \quad x \in D. \tag{2}$$

Both problems (1) and (2) have many applications in economics and engineering. Economic analysis sometimes requires a maximization and sometimes a minimization problem.

Quasi-convexity was pioneered by Kantorovich [**45**], De Finetti [**15**], Fenchel [**26**], Arrow and Enthoven [**3**]. Due to their importance in applications, both convex and concave functions, along with their generalizations to quasiconvex and quasiconcave functions, are discussed frequently in Economics. There are many papers [**1, 17, 18, 19, 42, 50, 56, 63, 71**] devoted to optimization problems in microeconomic theory, most of which are concerned with convex minimization or, equivalently, concave maximization problems. It occurs in consumer theory however, where under reasonable assumptions, consumer's preferences can be represented by a quasiconcave utility function [**19**]. The quasiconcave case in microeconomics was studied in [**19**]. There is the work [**38**] for the discrete case related to economics. Under some changes of economic environment of consumer's preference or markets, we face sometimes quasiconvex maximization or, equivalently, quasiconcave minimization problems.

There is a relationship between problems (1.1) and (1.2) via their generalized duality under certain assumptions on f and D ([**57**]). For example, when f is quasiconvex satisfying $f(0) = \inf_{R^n \setminus \{0\}} f$ and D is a compact convex set containing 0 then

$$\sup_{x \in D} f(x) = -\inf_{v \notin D^o} f^H(v)$$

holds, where

$$f^H(v) = \begin{cases} -\inf\limits_{\langle v,x\rangle \le 1} f(x) & \text{if } v \neq 0 \\ -\sup\limits_{x \in R^n} f(x) & \text{if } v = 0 \end{cases}$$

is quasinconvex and D^o is the polar of D defined as $D^o = \{v \mid \langle v, x\rangle \le 1 \text{ for all } x \in D\}$ in [**57**].

Although nonconvex optimization problems play an important role in applications, it has been noted [**43**] that non-convexity creates great mathematical difficulties. There are a number of results [**3, 10, 11, 12, 13, 14, 15, 16, 17, 18, 19, 26, 43, 62**] on mathematical properties of quasiconvex functions, and quasiconvex (resp., quasiconcave) minimization (resp., maximization) problems. Convex maximization (equivalently, concave minimization) problem as a special case of problem (1.1), where f is convex and D is a polytope, is originated from Tuy [**69**]. There are many iterative algorithms [**9, 32, 33, 34, 35, 36, 37, 54, 55, 69, 70**] for solving the convex maximization problem by resorting to the cutting plane and/ or branch and bound techniques.

The global optimality condition for the problem of maximizing a convex function over an arbitrary set was first obtained by Strekalovsky [**66, 67**]. An attempt to construct an algorithm based on these conditions was done in [**21**]. There is another type of global optimality condition for the problem of maximizing a convex function over a convex set in [**30**] based on the so-called $\epsilon-$ subdifferentials. On the other hand, the problem of minimizing a quasiconvex function over a convex set is originated from Kantorovich [**45**], where an global optimality condition was obtained.

From a mathematical perspective, it is well-known that standard nonlinear programming techniques have not been successful for solving nonconvex optimization problems. Moreover, the general Lagrangean method provides only a stationary point without guarantee of local optimality. In addition to these deficiencies, classical methods do not yield conditions for global optimality. For these reasons, global optimization appears to require new methods different from the standard nonlinear programming techniques. We will focus mainly on the quasiconvex maximization and quasiconvex minization problems with a nonconvex feasible set. We consider problems (1) and (2), and present new global optimality conditions with applications in certain models of economics. Then we consider some algorithms for the convex case based on those conditions. We show also some connections between the global optimality conditions and the classical local conditions. We show also that the fractional programming problems such as

$$\frac{f_1(x)}{f_2(x)} \longrightarrow \max, \quad x \in D$$

and

$$\frac{f_1(x)}{f_2(x)} \longrightarrow \min, \quad x \in D,$$

under certain assumptions on f_1, f_2 and D, can be reduced to problems (1) and (2), respectively. Also, we consider nonconvex optimal control problem and develop methods and algorithms for solving them globally.

1. Quasiconvex maximization problem

1.1. Definition and properties of a quasiconvex function. We start this section by recalling the definition of a quasiconvex function and some of its well-known properties [**3, 14, 16**] which we use in the subsequent sections.

DEFINITION 1. A function $f : R^n \to R$ is said to be quasiconvex if

$$f(\alpha x + (1-\alpha)y) \leq \max\{f(x), f(y)\}$$

holds for all $x, y \in R^n$ and $\alpha \in [0,1]$.

DEFINITION 2. A function $f : R^n \to R$ is said strictly quasiconvex if the inequality

$$f(\alpha x + (1-\alpha)y) < \max\{f(x), f(y)\}$$

holds for all $x, y \in R^n, x \neq y$ and $\alpha \in (0,1)$.

It is clear that any convex function is a quasiconvex, but the converse is not necessarily true. If f is quasiconvex then $-f$ is called quasiconcave.

DEFINITION 3. A differentiable function $f : R^n \to R$ is said to be pseudoconvex if the inequality $\langle f'(x), y - x\rangle \geq 0$ for $x, y \in R^n$ implies that $f(x) \leq f(y)$.

Note again that any convex function is pseudoconvex, but the converse does not follow. It is also true that a pseudoconvex function is quasiconvex but not conversely.

LEMMA 4. *A function $f : R^n \to R$ is quasiconvex if and only if the set*

$$L_c(f) = \{x \in R^n \mid f(x) \leq c\}$$

is convex for all $c \in R$.

PROOF. Necessity. Suppose that $c \in R$ is an arbitrary number and $x, y \in L_c(f)$. By the definition of quasiconvexity, we have

$$f(\alpha x + (1-\alpha)y) \leq \max\{f(x), f(y)\} \leq c \quad \text{for all } \alpha \in [0,1],$$

which means that the set $L_c(f)$ is convex.

Sufficiency. Let $L_c(f)$ be a convex set for all $c \in R$. For arbitrary $x, y \in R^n$, define $c^o = \max\{f(x), f(y)\}$. Then $x \in L_{c^o}(f)$ and $y \in L_{c^o}(f)$. Consequently, $\alpha x + (1 - \alpha y) \in L_{c^o}(f)$, for any $\alpha \in [0,1]$. This completes the proof. □

LEMMA 5. *Let $f : R^n \to R$ be a quasiconvex and differentiable function. Then the inequality $f(x) \leq f(y)$ for $x, y \in R^n$ implies that*

$$\langle f'(y), x - y\rangle \leq 0,$$

where f' denotes the gradient and $\langle \cdot, \cdot \rangle$ denotes the scalar product of two vectors.

PROOF. Since f is quasiconvex,

$$f(\alpha x + (1-\alpha)y) \leq \max\{f(x), f(y)\} = f(y)$$

for all $\alpha \in [0,1]$ and $x, y \in R^n$ such that $f(x) \leq f(y)$. By Taylor's formula, there is a neighborhood of the point y on which:

$$f(y + \alpha(x-y)) - f(y) = \alpha\left(\langle f'(y), x - y\rangle + \frac{o(\alpha\|x-y\|)}{\alpha}\right) \leq 0, \quad \alpha > 0.$$

From the fact that $\dfrac{o(\alpha\|x-y\|)}{\alpha} \xrightarrow{\alpha \to o} 0$, we obtain $\langle f'(y), x - y\rangle \leq 0$ which completes the proof. □

1.2. Problem statement and global maximality condition. Consider the quasiconvex maximization problem,

$$f(x) \longrightarrow \max, \quad x \in D, \tag{3}$$

where $f : R^n \to R$ is a quasiconvex and differentiable function and $D \subset R^n$ is a nonempty arbitrary subset of R^n. Then an optimality condition for this problem is given by the following theorem:

THEOREM 6. **[22]** *Let z be a solution of problem* (3), *and let*

$$E_c(f) = \{y \in R^n \mid f(y) = c\}.$$

Then

$$\langle f'(y), x - y\rangle \leq 0 \quad \text{for all } y \in E_{f(z)}(f) \text{ and } x \in D. \tag{4}$$

If, in addition, $f'(y) \neq 0$ holds for all $y \in E_{f(z)}(f)$, then condition (4) *is sufficient for $z \in D$ being a solution to problem* (3).

PROOF. Necessity. Assume that z is a solution of problem (3) and let $y \in E_{f(z)}(f)$ and $x \in D$. Then we have $f(x) \leq f(y)$. Applying Lemma 5, we obtain $\langle f'(y), x - y\rangle \leq 0$.

Sufficiency. Suppose, on the contrary, that z is not a solution to problem (3); i.e., there exists an $u \in D$ such that $f(u) > f(z)$. The closed set $L_{f(z)}(f) = \{x \in R^n \mid f(x) \leq f(z)\}$ is convex by Lemma 4. Let y be the projection of u on $L_{f(z)}(f)$ such that

$$\|y - u\| = \min_{x \in L_{f(z)}(f)} \|x - u\|.$$

Clearly,

$$\|y - u\| > 0 \tag{5}$$

holds because $u \notin L_{f(z)}(f)$. Moreover, this y can be considered as a solution of the following convex minimization problem:

$$g(x) = \frac{1}{2}\|x - u\|^2 \longrightarrow \min, \quad x \in L_{f(z)}(f).$$

Applying the Lagrange method to this latter problem, we obtain the following optimality condition at the point y:

$$\begin{cases} \lambda_o \geq 0, \ \lambda \geq 0, \ \lambda_o + \lambda > 0 \\ \lambda_o g'(y) + \lambda f'(y) = 0 \\ \lambda(f(y) - f(z)) = 0 \end{cases} \tag{6}$$

If $\lambda_o = 0$, then (6) implies that $\lambda > 0, f(y) = f(z)$ and $f'(y) = 0$ which contradicts the assumption in the theorem. If $\lambda = 0$, then we have $\lambda_o > 0$ and $g'(y) = y - u = 0$, which also contradicts (5). So, without loss of generality, we can set $\lambda_o = 1$ and $\lambda > 0$ in (6) to obtain

$$y - u + \lambda f'(y) = 0, \quad \lambda > 0, \quad f(y) = f(z).$$

From this we conclude that $\langle f'(y), u - y\rangle > 0$, which contradicts (4). This last contradiction implies that the assumption that z is not a solution of problem (3) must be false. This completes the proof. □

REMARK 7. If $f(x)$ is convex in problem (3), then Strekalovsky [**66**] proved in 1987 the following theorem.

Theorem [**66**]. Suppose that D is an arbitrary set. Let a point $z \in D$ satisfy

$$-\infty < \inf_{R^n} f < f(z) < +\infty$$

and $L_{f(z)}(f) = \{x \in R^n \mid f(x) \leq f(z)\} \subset intdom\ f$ is compact. Then z is a global maximizer of problem (3) if and only if

$$\begin{aligned} \langle c, x - z\rangle &\leq 0 \quad \text{for all } x \in \overline{conv}\ D \text{ and } c \in \partial f(z), \\ \langle v, x - z\rangle &\leq 1 \quad \text{for all } x \in \overline{conv}\ D \text{ and } v \in S(f, z), \end{aligned}$$

where

$$\begin{aligned} S(f,z) &= \left\{ v \in R^n \,\middle|\, \begin{array}{l} \exists y \in R^n:\ y \neq z,\ y \in E_{f(z)}(f), \\ \exists \alpha > 0:\ \alpha v \in \partial f(y),\ \langle v, y - z\rangle = 1. \end{array} \right\}, \\ \partial f(z) &= \{c \in R^n \mid f(x) - f(z) \geq \langle c, x - z\rangle,\ x \in R^n\}, \\ intdom\ f &= \{x \in R^n \mid f(x) < +\infty\}, \end{aligned}$$

and $\overline{conv} D$ is the closed convex hull of D. If f is differentiable as well, then it can be easily shown that condition (4) for problem (3) is the same as the above condition. Thus Theorem 6 generalizes Strekalovsky's result to the quasiconvex differentiable case.

REMARK 8. There is another related result obtained by Hiriart-Urruty and Ledyaev in 1996 [**31**].

Theorem [**31**]. Suppose f is convex and D is closed convex in problem (3). Let a point $z \in D$ satisfy $-\infty \leq \inf_D f < f(z)$. Then z is a global maximizer of (3) if and only if

$$\partial f(x) \subset N(x|D) \quad \text{for all } x \in D \text{ and } x \in E_{f(z)}(f),$$

where $N(x|D)$ is the normal cone to D at x defined as:

$$N(x|D) = \{c \in R^n | \langle c, y - x\rangle \leq 0,\ y \in D\}.$$

REMARK 9. If D is convex, we obtain from (4) with $y = z$ the well-known local optimality condition [**61**]:

$$\langle f'(z), x - z\rangle \leq 0, \quad \forall x \in D.$$

Thus it is clear that the global optimality condition (4) is connected to the classical extremum theory. Note, however, that condition (4) does not require the convexity of D at all.

REMARK 10. Let us consider the variational inequality problem [**65**] of finding an $x \in D$ such that a vector valued function $g : R^n \to R$ becomes inward normal to the given closed convex subset $D \subset R^n$ at x, that is,

$$\langle g(x), y - x\rangle \geq 0, \quad \forall y \in D.$$

This problem can be solved, in some cases, by a componentwise projection method [**65**]. If D is closed and convex in problem (3), then its solution is given by a point $z \in D$ such that

$$\langle g(y), x - y\rangle \geq 0$$

holds for all $y \in E_{f(z)}(f) \cap D$ and $x \in D$ with $g(y) = -f'(y)$. This fact suggests some relationship between the quasiconvex maximization and the variational inequality problems.

REMARK 11. In order to conclude that a point z' in D is not a solution to problem (3), Theorem 6 tells that we need to find a pair $x, y \in R^n$ such that

$$\langle f'(y), x - y\rangle > 0, \quad f(y) = f(z'), \quad x \in D.$$

The following example illustrates the use of this property.
Example

$$f(x) = \frac{x_1^2 + x_2^2}{2x_1 + 4x_2 + 5},$$
$$D = \{x \in R^2 \,|\, 1 \leq x_1 \leq 3;\ 2 \leq x_2 \leq 5\}.$$

We can easily evaluate the gradient of f as:

$$f'(x) = \left(\frac{2x_1^2 + 8x_1x_2 + 10x_1 - 2x_2^2}{(2x_1 + 4x_2 + 5)^2}, \frac{4x_2^2 + 4x_1x_2 + 10x_2 - 4x_1^2}{(2x_1 + 4x_2 + 5)^2}\right).$$

Also, by using Lemma 4, we can show that f is quasiconvex on D. Now we want to check whether a point $x^o = (1, 2) \in D$ is globally optimal or not. Then consider a pair $u = (2, 2) \in D$ and $y = \left(\frac{1}{3}, \frac{2+2\sqrt{5}}{3}\right)$ satisfying $f(y) = f(x^o) = \frac{1}{3}$. We have $\langle f'(y), u - y\rangle > 0$ and it follows that x^o is not a global solution. In fact, we can show that the global solution is $x^* = (3, 5)$.

1.3. Preliminary results for maximization. First of all, we introduce the following definition.

DEFINITION 12. [**21**] A set D in R^n is called a simple set if the following conditions hold:

a) D is compact.
b) The problem of maximizing a linear function over D is solvable by a "simple" method.

We say that condition b) holds, for example, if it can be solved as a linear programming problem (i.e., if D is a convex polyhedron) or if an analytical form of the solution is explicitly given.

Consider the convex maximization problem as a particular case of problem (3):

$$f(x) \longrightarrow \max, \quad x \in D, \tag{7}$$

where $f : R^n \to R$ is convex and $D \subset R^n$ is a simple set. In addition, in problem (7) we assume that f is strongly convex and twice continuously differentiable. Then the global optimality condition of Theorem 6 can be stated without the assumption

$$f'(y) \neq 0 \quad \text{for all } y \in E_{f(z)}(f).$$

THEOREM 13. *A point $z \in D$ with $f'(z) \neq 0$ is a solution of problem* (7) *if and only if the following condition holds:*

$$\langle f'(y), x - y \rangle \leq 0 \quad \text{for all } y \in E_{f(z)}(f) \text{ and } x \in D,$$

where and $E_c(f) = \{y \in R^n | f(y) = c\}$.

The proof is completely analogous to the proof of Theorem 6.

Define the auxiliary function $\pi(y)$ as follows:

$$\pi(y) = \max_{x \in D} \langle f'(y), x - y \rangle, \quad y \in R^n. \tag{8}$$

For subsequent discussion, it is useful to derive some properties of $\pi(y)$.

LEMMA 14. *The function $\pi(y)$ is continuous on R^n.*

Denote by $D(y)$ the set:

$$D(y) = \{x \in D : \langle f'(y), x \rangle = \varphi(y)\}, \quad y \in R^n.$$

THEOREM 15. [**23**] *The directional derivative of $\pi(y)$ at any point $y \in R^n$ in any direction $h \in R^n$ of norm 1 exists, and is given by*

$$\frac{\partial \pi(y)}{\partial h} = \max_{x \in D(y)} \langle f''(y)h, x \rangle - \langle f''(y)y + f'(y), h \rangle.$$

LEMMA 16. *If the set D is strictly convex, then the function $\pi(y)$ becomes differentiable and the gradient is given by*

$$\pi'(y) = f''(y)(x(y) - y) - f'(y),$$

where

$$\langle f'(y), x(y) \rangle = \max_{x \in D} \langle f'(y), x \rangle.$$

1.4. Algorithm and its convergence. Before presenting an algorithm for solving problem (7), it is useful to restate Theorem 13 in a convenient way via the following function $\theta(z)$ defined for $z \in D$.

$$\theta(z) = \max_{y \in E_{f(z)}(f)} \pi(y),$$

where $\pi(y) = \max_{x \in D} \langle f'(y), x - y \rangle$. Since f is strongly convex, the set $E_{f(z)}(f)$ is compact [**46**]. Thus, clearly, $\theta(z) < +\infty$. Also, we can note that

$$\pi(y) \leq \theta(z) \quad \text{for all } y \in E_{f(z)}(f).$$

THEOREM 17. *Let $z \in D$ and $f'(z) \neq 0$. If $\theta(z) = 0$ then the point z is a solution to problem* (7).

PROOF. is immediate from the following inequalities:

$$\langle f'(y), x - y \rangle \leq \pi(y) = \max_{x \in D} \langle f'(y), x - y \rangle \leq \theta(z) = 0$$

which hold for all $x \in D$ and $y \in E_{f(z)}(f)$. □

Algorithm MAX

Input: A strongly convex function f and a simple set D.
Output: A solution x to problem (2.6); i.e., a global maximizer of f over D.
Step 1. Choose a point $x^0 \in D$ such that $f'(x^0) \neq 0$. Set $k := 0$.
Step 2. Solve the following problem

$$\pi(y) \longrightarrow \max, \quad y \in E_{f(x^k)}(f).$$

Let y^k be a solution of this problem, i.e.,

$$\pi(y^k) = \max_{y \in E_{f(x^k)}(f)} \max_{x \in D} \langle f'(y), x - y \rangle.$$

Let $\theta(x^k) := \pi(y^k)$, and let x^{k+1} be a solution satisfying

$$\pi(y^k) = \langle f'(y^k), x^{k+1} - y^k \rangle.$$

Step 3. If $\theta(x^k) = 0$ then output $x = x^k$ and terminate. Otherwise set $k := k + 1$ and return to Step 2. □

The convergence of this Algorithm is given by the following theorem.

THEOREM 18. [**21**] *Assume that $f : R^n \to R$ be strongly convex and continuously differentiable. Then the sequence $\{x^k\}$, $(k = 0, 1, \ldots)$ generated by Algorithm MAX is a maximizing sequence for problem* (7), *that is,*

$$\lim_{k \to \infty} f(x^k) = \max_{x \in D} f(x),$$

and every accumulation point of the sequence $\{x^k\}$ is a global maximizer of (7)).

2. Quasiconvex minimization problem

2.1. Problem statement and global minimality condition. Now we consider the following quasiconvex minimization problem:

$$f(x) \longrightarrow \min, \quad x \in D, \tag{9}$$

where $f : R^n \to R$ is a quasiconvex and continuously differentiable function and $D \subset R^n$ is a nonempty arbitrary subset. We can obtain the following global optimality condition, which is the counterpart of Theorem 6.

THEOREM 19. **[22]** *Let z be a solution of problem* (9). *Then*

$$\langle f'(x), x - y\rangle \geq 0 \quad \text{for all } y \in E_{f(z)}(f) \text{ and } x \in D, \tag{10}$$

where $E_c(f) = \{y \in R^n \mid f(y) = c\}$. *If, in addition,*

$$\lim_{\|x\|\to\infty} f(x) = +\infty \quad \text{and} \quad f'(x + \alpha f'(x)) \neq 0 \tag{11}$$

holds for all $x \in D$ and $\alpha \geq 0$, then condition (3.2) becomes sufficient.

PROOF. Necessity. Assume that z is a solution of problem (9). Let $x \in D$ and $y \in E_{f(z)}(f)$. Then we have $0 \geq f(z) - f(x) = f(y) - f(x)$, and Lemma 5 implies $\langle f'(x), x - y\rangle \geq 0$.

Sufficiency. Suppose, on the contrary, that z is not a solution of (9). Then there exists a $u \in D$ such that $f(u) < f(z)$. Construct a ray y_α for $\alpha > 0$ defined by

$$y_\alpha = u + \alpha f'(u).$$

We claim that $f(y_\alpha) > f(u)$ holds for all positive α. By Taylor's formula, we have

$$f(u + \alpha f'(u)) - f(u) = \alpha\left(\|f'(u)\|^2 + \frac{o(\alpha\|f'(u)\|)}{\alpha}\right)$$

for small $\alpha > 0$, where $\lim_{\alpha\to 0+} \frac{o(\alpha\|f'(u)\|)}{\alpha} = 0$. Therefore, there exists $\alpha_o > 0$ such that $f(y_\alpha) - f(u) > 0$ holds for all $\alpha \in (0, \alpha_o)$. Hence, by Lemma 5, we have $\langle f'(u + \alpha_o f'(u)), f'(u)\rangle \geq 0$ since $f'(u) \neq 0$ and $f'(u + \alpha_o f'(u)) \neq 0$ by the assumption. Note that for all $\gamma > 1$, we also have $f(u+\gamma\alpha_o f'(u)) > f(u+\alpha_o f'(u))$; for otherwise, we would have $f(u + \gamma\alpha_o f'(u)) \leq f(u + \alpha_o f'(u))$, and consequently, by Lemma 5, $\langle f'(u+\alpha_o f'(u)), \alpha_o(\gamma - 1)f'(u)\rangle \leq 0$, which would imply $\gamma \leq 1$ contradicting to the assumption that $\gamma > 1$. Moreover, we can show that $f(u + \gamma\alpha_o f'(u))$ is increasing in $\gamma > 0$. For if $f(u + \gamma'\alpha_o f'(u)) < f(u + \gamma\alpha_o f'(u))$ holds for some $\gamma' > \gamma$, then $\alpha_o(\gamma' - \gamma)\langle f'(u+\gamma\alpha_o f'(u)), f'(u)\rangle \leq 0$, which would contradict the fact that $\gamma' > \gamma$. These prove our claim $f(y_\alpha) > f(u)$ for all $\alpha > 0$.

Now it is obvious that the function $\varphi : R^+ \to R$ defined as

$$\varphi(\alpha) = f(y_\alpha)$$

is continuous on $[0, \infty)$. Also, by assumption (11), $\lim_{\alpha\to\infty} \varphi(\alpha) = +\infty$, and therefore, there exists an $\hat{\alpha}$ such that $\varphi(\hat{\alpha}) > f(z)$. Using the continuity of $\varphi(\alpha)$ and the inequalities $\varphi(\hat{\alpha}) > f(z) > f(u)$, there exists an $\bar{\alpha}$ such that

$$f(y + \bar{\alpha} f'(u)) = f(z),$$

which means that $y_{\bar{\alpha}} \in E_{f(z)}(f)$. On the other hand, we have $f'(u) = \frac{1}{\bar{\alpha}}(y_{\bar{\alpha}} - u)$. Thus we get

$$\langle f'(u), u - y_{\bar{\alpha}}\rangle = \frac{1}{\bar{\alpha}}\langle y_{\bar{\alpha}} - u, u - y_{\bar{\alpha}}\rangle = -\frac{1}{\bar{\alpha}}\|y_{\bar{\alpha}} - y\|^2 < 0,$$

which contradicts (10). This means that z must be a solution of (9). □

REMARK 20. If $f(x)$ is convex in problem (9), then the following theorem was obtained in Deitrich [**29**].

Theorem [**29**]. Let D be an arbitrary nonempty set and f be a lower-semicontinuous proper convex function. Suppose that the assumptions

$$D \subset M(\partial f),$$
$$-\infty \leq \inf_{R^n} f < f(x) \quad \text{for all } x \in D,$$
$$L_{f(v)}(f) = \{x \in R^n | f(x) \leq f(v)\} \subset intdom\ f \quad \text{for all } v \in D$$

are fulfilled. Then a point $z \in D$ is a global minimizer of problem (9) if and only if

$$\langle x^*, x - y \rangle \geq 0 \quad \text{for all } y \in E_{f(z)}(f), \quad x \in D \text{ and } x^* \in \partial f(x), \tag{12}$$

where

$$\partial f(x) = \{c \in R^n \mid f(v) - f(x) \geq \langle c, v - x \rangle, \ v \in R^n\},$$
$$intdom\ f = \{x \in R^n \mid f(x) < +\infty\},$$

and $M(\partial f)$ is the effective domain of the subdifferential mapping $\partial f : R^n \to 2^{R^n}$.

We can easily note that if f is differentiable as well then necessary condition (10) for problem (9) is the same as condition (12). Thus Theorem 19 is a generalization of this result to the quasiconvex differentiable case.

REMARK 21. If D is convex, then under some additional assumptions on f there is a global optimality condition by Kantorovich stated as follows.

Theorem [**45**]. If D is convex, then a point $z \in D$ is a solution of problem (9) if and only if there exists a $c \in R^n$ such that

$$\langle c, x \rangle \leq \langle c, z \rangle \quad \text{for all } x \in D,$$
$$\langle c, x \rangle \geq \langle c, z \rangle \quad \text{for all } x \in R^n \text{ with } f(x) \leq f(z).$$

REMARK 22. Note that by the assumption of Theorem 19, we have $f'(x) \neq 0$ for all $x \in D$. If f is strictly quasiconvex and D is convex, we do not need to use condition 10. Instead we have the following result which is immediate from the classical local optimality conditions [**72, 73**].

Theorem. Assume that f is strictly quasiconvex and differentiable, and D is convex in (9). Then a point $z \in D$ is the solution of problem (9) if and only if

$$\langle f'(z), x - z \rangle \geq 0 \quad \text{for all } x \in D.$$

Also, for the pseudoconvex case we have the following result.

Lemma. let $f : R^n \to R$ be pseudoconvex and $x_* \in D$. If $f'(x_*) = 0$, then x_* is a global minimizer of problem (9).

REMARK 23. There is a strong relationship between the quasiconvex minimization problem (9) and the variational inequality problem defined on a convex set [**20**].

For example, if f is continuously differentiable and convex, and D is convex, then the problem $\min_{x \in D} f(x) = f(x_*)$ and the variational problem

$$\langle f'(x), x - x_* \rangle \geq 0 \quad \text{for all } x \in D$$

are equivalent [**20**]. Therefore, condition (10) might be regarded as a generalization of this well known problem for the quasiconvex case with an arbitrary set D.

REMARK 24. In order to conclude that a point $z \in D$ is not a solution of (9) it is sufficient to have a pair $u, y \in R^n$ such that $\langle f'(u), u - y \rangle < 0, f(y) = f(z)$ and $u \in D$.

The use of Theorem 19 is illustrated in the following example.

Example. Consider the following problem:

$$f(x) = x_1^2 + x_2^2 \longrightarrow \min,$$

subject to

$$x \in D = D_1 \cup D_2,$$

where $D_1 = \{x \in R^2 : x_1 \leq -2,\ x_2 \geq 2\}, D_2 = \{x \in R^2 : \ x_1 x_2 \geq 1\}$. The Lagrange condition for this problem provides two stationary points $z^o = (1,1)$ and $z^1 = (-2,2)$. To see if z^1 is a global solution, consider points $u = (2, 0.6) \in D$ and $y = (2,2)$ satisfying $f(y) = f(z^1) = 8$. Evaluating $\langle f'(u), u - y \rangle$ we have $\langle f'(u), u - y \rangle < 0$. Consequently, the point z^1 is not a global solution. In fact, it is not difficult to see that z^0 is the global minimizer with $f(z^0) = 2$.

2.2. Preliminary results for minimization. Assume that $f : R^n \to R$ be a continuously differentiable function and $D \subset R^n$ be an arbitrary compact set. For the purpose of describing an algorithm for solving (9), we define the auxiliary functions $\alpha(y)$ and $\beta(y)$ for $y \in R^n$,

$$\alpha(y) = \min_{x \in D} \langle f'(x), y \rangle, \quad \beta(y) = \max_{x \in D} \langle f'(x), y \rangle.$$

LEMMA 25. *The above functions $\alpha(y)$ and $\beta(y)$ are continuous at point 0.*

Next we define another auxiliary functions $p(y)$:

$$p(y) = \min_{x \in D} \langle f'(x), x - y \rangle, \quad y \in R^n. \tag{13}$$

LEMMA 26. *Function $p(y)$ is continuous on R^n.*

2.3. An algorithm for the convex minimization problem. We consider the convex minimization problem as a special case of problem (9):

$$f(x) \longrightarrow \min, \quad x \in D, \tag{14}$$

where $f : R^n \to R$ is strongly convex and continuously differentiable, and D is an arbitrary compact set in R^n. In this case, then we can weaken condition (11) as shown in the next theorem.

THEOREM 27. [22] *Let z be a solution of problem* (14). *Then*

$$\langle f'(x), x - y\rangle \geq 0 \quad \text{for all } y \in E_{f(z)}(f) \text{ and } x \in D. \tag{15}$$

If, in addition,

$$\min_{x \in D} \|f'(x)\| > 0 \tag{16}$$

holds, then condition (15) *is also sufficient.*

Now using the function $p(y)$ of (13), we reformulate Theorem 27 in terms of function $\psi(z)$ defined as follows.

$$\psi(z) = \min_{y \in E_{f(z)}(f)} p(y), \quad z \in D.$$

THEOREM 28. *Assume that $f : R^n \to R$ is strongly convex and continuously differentiable, and D is a compact set in R^n. Let $\min_{x\in D} \|f'(x)\| > 0$. If $\psi(z) = 0$, then the point z is a solution to problem* (14).

PROOF. This is an obvious consequence of the following relations.

$$0 = \psi(z) \leq p(y) \leq \langle f'(x), x - y\rangle,$$

which are fulfilled for all $y \in E_{f(z)}(f)$ and $x \in D$. □

Now we are ready to present an algorithm for solving problem (14). We also suppose that one can efficiently solve the problem of computing $\min_{x\in D}\langle f'(x), x - y\rangle$ for any given $y \in R^n$. For example, this is possible in the following cases:

1. f is linear and D is a simple set defined as in Definition 12.
2. D is convex and f is such that $s''(x)$ is a nonnegative definite matrix, where $s''(x)$ is the matrix of second order derivatives of the function $s = \langle f'(x), x - y\rangle$ defined for $x \in D$ and $y \in R^n$.

Algorithm MIN

Input: A strongly convex function f and a compact set D.
Output: A solution x to the minimization problem (14).
Step 1. Choose a feasible point $x^0 \in D$. Set $k := 0$.
Step 2. Solve the following problem

$$p(y) \longrightarrow \min, \quad y \in E_{f(x^k)}(f).$$

Let y^k be a solution of this problem (i.e., $p(y^k) = \min_{x\in D}\langle f'(x), x - y^k\rangle = \min_{y \in E_{f(x^k)}(f)} p(y)$), and let x^{k+1} realizes $p(y^k)$ (i.e., $\psi(x^k) = p(y^k) = \langle f'(x^{k+1}), x^{k+1} - y^k\rangle$).
Step 3. If $\psi(x^k) = 0$ then output $x = x^k$ and terminate. Otherwise, let $k := k + 1$ and return Step 2. □

The convergence of this Algorithm is based on the following theorem.

THEOREM 29. [**22**]*Assume that $f : R^n \to R$ be strongly convex and continuously differentiable, and D is a compact set in R^n. Let $\min_{x\in D} \|f'(x)\| > 0$. Then the sequence $\{x^k,\ k = 0, 1, \ldots\}$ generated by Algorithm MIN is a minimizing sequence for problem* (14)*, that is,*

$$\lim_{k\to\infty} f(x^k) = \min_{x\in D} f(x),$$

and every accumulation point of the sequence $\{x^k\}$ is a global minimizer of (14).

REMARK 30. If D is convex, there are a number of well established convex minimization algorithms; e.g., [**4, 41, 46, 58, 72, 73, 75**]. Thus, the proposed algorithms mainly intended for the case with a nonconvex feasible set.

REMARK 31. In some case, computing $\psi(x^k)$, in Step 2 of Algorithm MIN, computing $\psi(x^k)$ may not be easier than the original problem. But if we can find a point $v \in D$ such that $\langle f'(v), v - y^k\rangle < 0$ during this computation, it implies $f(v) < f(x^k)$, improving the initial value of $f(x^k)$. Therefore this v may be used as x^{k+1} in Step 2. A modification of Algorithm MIN to incorporate this type of heuristic rule may turn out to be useful.

Like Algorithm MAX, note that the efficiency of Algorithm MIN depends on how we can solve successfully the following problem

$$p(y) \longrightarrow \min, \ \ y \in E_{f(x^k)}(f). \tag{17}$$

In general, we can again use in Algorithm MIN the numerical methods [**53, 74**] for such problem as (17) which find an approximate solution to the problem at each iteration k.

3. Nonconvex optimal control problems

3.1. Optimal control maximization problem. Consider the following optimal control problem:

$$J(u) = \phi(x(t_1)) \to \max \tag{18}$$

$$\begin{cases} \dot{x}(t) = A(t)x + B(t)u(t) + C(t) \\ x(t_0) = x^0 \end{cases} \tag{19}$$

$$u \in V = \left\{u \in L_2^{(r)}(T) \,|\, u(t) \in U, \, t \in T\right\}, \tag{20}$$

where $t \in T = [t_0, t_1]$, $-\infty < t_0 < t_1 < +\infty$, $u(t) = ((u_1(t), u_2(t), \ldots, u_r(t))^T$, $x(t) = (x_1(t), x_2(t), \ldots, x_n(t))^T$. Here $x^0 \in R^n$ is an initial state, t_0, t_1 and x^0 are given. Matrix functions $A(t)$, $B(t)$ and $C(t)$ are piecewise continuous on $[t_0, t_1]$ with dimensions $(n \times n)$, $(n \times r)$ and $(n \times 1)$ respectively. $U \subset R^r$ is a simple set, $\phi : R^n \to R$ is a strongly convex and differentiable function.

The problem (18)–(20) belongs to the class of nonconvex optimal control problem and application of Pontryagin's maximum principle [**40, 59**] can give only stationary processes $(x(\cdot), u(\cdot))$.

We will show how to apply Algorithm MAX to the solution of problem (18)–(20). It is well known that every admissible control $u \in V$ corresponds to a unique solution $x(t) = x(t, u)$ of the Cauchy problem (19) by the formula [**73, 72**]:

$$x(t,u) = F(t)x^0 + \int_{t_0}^{t} F(t)F^{-1}(\xi)\Big(B(\xi)u(\xi) + C(\xi)\Big)d\xi, \tag{21}$$

where $F(t)$ is a fundamental matrix with dimension $(n \times n)$ that satisfies the matrix equation

$$\dot{F} = A(t)F, \quad F(t_0) = E$$

on T with the unit matrix E. Note that a state $x(t, u)$ is an absolutely continuous vector-function of time t, which satisfies the system (19) almost everywhere in T [**72**]. Let us denote by

$$\tilde{D} = \tilde{D}(t_1) = \Big\{ y \in R^n \,|\, y = x(t_1, u), u \in V \Big\}$$

the reachable set of the control system (19)–(20). Under the above assumptions the set $\tilde{D} \subset R^n$ is convex and compact [**59, 73**]. Then we can present problem (18)–(20) in the following form:

$$\phi(x) \to \max, \quad x \in \tilde{D}. \tag{22}$$

Let a point x^* be a solution of problem (22), then the admissible control $u^* = u^*(t)$, $t \in T$ corresponding to x^* is the global optimal control solution to problem (18)–(20). Therefore, Algorithm MAX can also be used for solving the problem (18)–(20). As we have done before, write down the auxiliary function $\tilde{\pi}(y)$ for problem (22) as follows:

$$\tilde{\pi}(y) = \max_{x \in \tilde{D}} \langle \phi'(y), x - y \rangle, \quad y \in R^n. \tag{23}$$

Then we can formulate a theorem by analogy with Theorem 17 for corresponding assumptions.

THEOREM 32. *A control $u^* \in V$ is an optimal control of the problem* (18)–(20) *if and only if*

$$\max_{y:\phi(y)=\phi(x^*)} \tilde{\pi}(y) \le 0,$$

where $x^* = x^*(t_1, u^*) \in \tilde{D}(t_1)$.

PROOF. The proof is immediate from the proof of Theorem 17 since problems (18)–(20) and (22) are equivalent. □

Example:

$$\phi = x_1^2(3) + x_2^2(3) \to \max,$$

$$\begin{cases} \dot{x}_1 = x_2 \\ \dot{x}_2 = u \\ x_1(0) = 3 \\ x_2(0) = -4 \\ -2 \le u(t) \le 5, \ \ t \in T = [0, 3]. \end{cases}$$

We show that in this problem we have two controls $u^1(t) \equiv -2$ and $u^2(t) \equiv 5$ which satisfy the maximum principle. Indeed, first let us investigate the control u^1. Then

$$x_2(t) = x_2(0) + \int_0^t u(t)dt = -4 - 2t, \quad \forall t \in T,$$

$$x_1(t) = x_1(0) + \int_0^t x_2(t)dt = 3 + \int_0^t (-4 - 2t)dt = -t^2 - 4t + 3, \quad \forall t \in T$$

Write down the Hamiltonian function as follows:

$$H(\psi, u, t) = \psi_1 x_2 + \psi_2 u.$$

Solve the conjugate system

$$\begin{cases} \dot{\psi}_1 = -\dfrac{\partial H}{\partial x_1}, \\ \dot{\psi}_2 = -\dfrac{\partial H}{\partial x_2} = -\psi_1, \\ \psi_1(3) = -\dfrac{\partial \phi(x(t_1))}{\partial x_1} = 2x_1(3) \\ \psi_2(3) = -\dfrac{\partial \phi(x(t_1)))}{\partial x_2} = 2x_2(3) \end{cases}$$

Note that $x_1(3) = -18$ and $x_2(3) = -10$, and then

$$\psi_1(t) = -36, \quad \psi_2(t) = 36t - 128.$$

Admissible process (u^1, x_1^1, x_2^1) is optimal when

$$H(\psi^1(t), x^1(t), u^1(t), t) = \max_{v \in V} H(\psi^1(t), x^1(t), v, t), \quad t \in T$$

$$H(\psi^1(t), x^1(t), u^1(t), t) = -36(-4 - 2t) + (36t - 128)(-2) = 400.$$

On the other hand, we get

$$H(\psi^1(t), x^1(t), v, t) = 144 + 72t + (36t - 128)v.$$

Then

$$\max_{v \in V} H(\psi^1(t), x^1(t), v, t) = 400, \quad \forall t \in T.$$

Thus the control u^1satisfies the maximum principle. Analogously, we can show that u^2 also satisfies the maximum principle with $\phi(x(3), u^2) = \frac{1213}{4}$. In fact the control u^1 is the global optimal control for which the condition of Theorem 32 holds.

Now we first consider the following linearized optimal control problem

$$\langle \phi'(y), x \rangle \to \max, \quad x \in \tilde{D}. \tag{24}$$

In order to solve problem (24), we introduce a conjugate state

$$\psi(t,y) = \Big(\psi_1(t), \ldots, \psi_n(t)\Big)^T$$

as a solution of the following differential system:

$$\dot{\psi} = -A^T \psi, \quad \psi(t_1) = -\phi'(y) \text{ for every } y \in R^n. \tag{25}$$

This system has the unique piecewise differentiable solution $\psi(t) = \psi(t,y)$ defined on T. Let $u(t), t \in T$ be an admissible control, and $x(t)$ be its corresponding phase trajectory. It can be easily proved that the following relation holds almost everywhere on T.

$$\begin{aligned}
\frac{d}{dt}[\langle \psi(t), x(t)\rangle] &= \left\langle \frac{d\psi(t)}{dt}, x(t) \right\rangle + \left\langle \frac{dx(t)}{dt}, \psi(t) \right\rangle \\
&= -\langle \psi(t), Ax(t)\rangle + \langle A(t)x(t) + B(t)u(t) + C(t), \psi(t)\rangle \\
&= \langle B(t)u(t), \psi(t)\rangle + \langle C(t), \psi(t)\rangle.
\end{aligned}$$

Using this relation we can rewrite the functional of problem (24) in the following form:

$$\begin{aligned}
-\langle \phi(y), x(t_1, u)\rangle = \langle \psi(t_1), x(t_1)\rangle &= \int_{t_0}^{t_1} \frac{d}{dt}\langle \psi(t), x(t)\rangle dt + \langle \psi(t_0), x(t_0)\rangle \\
&= \int_{t_0}^{t_1} [\langle B(t)u(t), \psi(t)\rangle + \langle C(t), \psi(t)\rangle] dt + \langle \psi(t_0), x(t_0)\rangle.
\end{aligned}$$

It is known [**73**] that for an admissible control $z \in V$ to be optimal control in problem (24) it is necessary and sufficient that the condition

$$\langle \psi'(y), x(t_1, z)\rangle - \langle \phi'(y), x(t_1, u)\rangle \geq 0, \quad \forall u \in V$$

holds. Hence, taking into account the previous expressions, we obtain

$$\int_{t_0}^{t_1} \langle \psi(t), B(t)(u(t) - z(t))\rangle dt \geq 0, \quad \forall u \in V.$$

Using the local structure of the set V [**73**] we have

$$\langle \psi(t), B(t)(u(t) - z(t))\rangle dt \geq 0, \quad \forall u \in V$$

for almost all $t \in T$. Then a solution of problem (24) will be given by the following theorem.

THEOREM 33. [**73**] *Let $\psi(t) = \psi(t,y)$, $t \in T$ be a solution of conjugate system (25) for $y \in R^n$. Then for an admissible control $z(t) = z(t,y)$ to be optimal in problem (24) it is necessary and sufficient that the condition $\langle \psi(t,y), B(t)z(t,y)\rangle = \min_{u \in U} \langle \psi(t,y), B(t)u\rangle$ be fulfilled for almost every $t \in T$.*

Using this theorem we can easily show that the reachable set $\tilde{D} = \tilde{D}(t_1)$ is a simple set. Therefore, the value $\tilde{\pi}(\tilde{y})$ is calculated by the following scheme:

1. Solve problem (25) for a given $\tilde{y} \in R^n$. Let $\psi(t) = \psi(t, \tilde{y})$ be a solution to this problem.
2. Find the optimal control $z = z(t, \tilde{y})$ as a solution of the problem

$$\langle \psi(t), B(t)u \rangle \to \min, \quad u \in U$$

at each moment of $t \in T$.
3. Find a solution $\tilde{x} = x(t, z)$ of problem (19) for $u = z(t, \tilde{y})$.
4. Find $\tilde{x}(t_1) = \tilde{x}(t_1, z)$ by the formula (21) for $t = t_1$.
5. Calculate $\tilde{\pi}(\tilde{y})$ by the formula

$$\tilde{\pi}(\tilde{y}) = \max_{x \in \tilde{D}} \langle \phi'(\tilde{y}), x - \tilde{y} \rangle = \langle \phi'(\tilde{y}), \tilde{x}(t_1) - \tilde{y} \rangle.$$

Based on Algorithm MAX now we are ready to present an algorithm for solving problem (18)–(20).

Algorithm OPTMAX

Step 1. Let an arbitrary control $u^k \in V$ be given. Find $x^k = x(t_1, u^k)$ by solving the system (19) for $u = u^k$.

Step 2. Solve the problem $\tilde{\pi}(y) \to \max$, $\phi(y) = \phi(x^k)$. Let y_k be the solution of this problem and $z^k = z^k(t, y_k)$ be the a solution of the corresponding problem

$$\langle \phi'(y_k), x \rangle \to \max, \quad x \in \tilde{D}$$

or

$$\langle \psi^k(t), B(t)z^k \rangle = \min_{u \in U} \langle \psi^k(t), B(t)u \rangle, \quad t \in T,$$

where

$$\dot{\psi}^k(t) = -A^T(t)\psi^k(t), \quad \psi^k(t_1) = -\phi'(y_k)$$

Step 3. If $\tilde{\pi}(y_k) \leq 0$ then go to step 5. Otherwise, go to step 4.

Step 4. Set $u^{k+1} = z^k(t, y_k)$ and $k := k + 1$. Then go to step 1.

Step 5. Iteration is terminated. $u^* = u^k$ is a global optimal solution of problem (18)–(20).

Convergence of the algorithm is based on the following statement.

THEOREM 34. **[25]** *The sequence $\{u^k\} \subset V$ constructed by Algorithm OPTMAX is a maximizing sequence for problem* (18)–(20), *i.e.,*

$$\lim_{k \to \infty} J(u^k) = \max_{u \in V} J(u)$$

The list of some optimal control test problems solved by Algorithm OPTMAX is the following [**25**]:

$$(p1)\quad \begin{cases} x_1^2(1) + x_2^2(1) \to \max \\ \dot{x}_1 = u_1 \\ \dot{x}_2 = u_2 \\ x_1(0) = x_2(0) = 0, \quad u(t) \in U, \quad t \in [0,1] \\ U = \{u \in R^2 / -3 \le u_1 \le 4, \quad -2 \le u_2 \le 5\} \end{cases}$$

$$(p2)\quad \begin{cases} x_1^2(1) + x_2^2(1) \to \max \\ \dot{x}_1 = x_2 + u_1 \\ \dot{x}_2 = x_1 + u_2 \\ x_1(0) = x_2(0) = 0, \quad u(t) \in U, \quad t \in [0,1] \\ U = \{u \in R^2 / -3 \le u_i \le 4, \quad i = 1,2\} \end{cases}$$

$$(p3)\quad \begin{cases} x_1^2(t_1) + x_2^2(t_1) + x_3^2(t_1) \to \max \\ x_1 = -x_1 + x_2 \\ \dot{x}_2 = -x_1 - x_2 - 10u \\ \dot{x}_3 = x_2 \\ x_1(0) = x_2(0) = 0, \quad x_3(0) = -30, \quad T = [0, 2.8] \\ u(t) \in U = \{u \in R / -15 \le u \le 15\} \end{cases}$$

The results of the numerical experiments for these problems are shown in the following table.

Problems	Local value of ϕ	Global value of ϕ	Computing time (min:sec)
p1	13	41	0:51
p2	53.1448	94.4798	0:56
p3	149.055	263.519	0:6.78

3.2. Optimal control minimization problem. Consider the following optimal control problem:

$$J(u) = \phi(x(t_1)) \to \min, \tag{26}$$

$$\begin{cases} \dot{x}(t) = f(x,u,t) \\ x(t_0) = x^0 \end{cases} \tag{27}$$

$$u \in V = \left\{ u \in L_2^{(r)}(T) \mid u(t) \in U,\ t \in T \right\}, \tag{28}$$

where $t \in T = [t_0, t_1]$, $-\infty < t_0 < t_1 < +\infty$, $u(t) = ((u_1(t), u_2(t), \dots, u_r(t))^T$, $x(t) = ((x_1(t), x_2(t), \dots, x_n(t))^T$. Here $x^0 \in R^n$ is an initial state, t_0, t_1 and x^0 are given. $U \subset R^r$ is a compact set, $\phi : R^n \to R$ is a strongly convex and differentiable function. The vector function $f = (f_1, \dots, f_n)$ is continuous in its arguments (x, u, t) together with its partial derivatives with respect to x and satisfies Lipschitz's condition in x with the same constant L for all $u \in U, t \in T$:

$$\|f(x + \Delta x, u, t) - f(x, u, t)\| \le L \|\Delta x\|$$

In general, the problem (26)–(28) belongs also to the class of nonconvex optimal control problem and application of Pontryagin's maximum principle [**40, 59**] can give only stationary processes $(x(\cdot), u(\cdot))$.

We will show how to apply Algorithm MIN to the solution of problem (26)–(28). It is well known that every admissible control $u \in V$ corresponds to a unique solution $x(t) = x(t, u)$ of the Cauchy problem (27) [**73, 72**]. Note that a state $x(t, u)$ is an absolutely continuous vector-function of time t, which satisfies the system (27) almost everywhere in T [**72**]. Let us denote by

$$\tilde{D} = \tilde{D}(t_1) = \Big\{y \in R^n \,|\, y = x(t_1, u), u \in V\Big\}$$

the reachable set of the control system (27)–(28). Under the above assumptions the set $\tilde{D} \subset R^n$ is compact [**59, 73**]. Then we can reformulate problem (26)–(28) as follows:

$$\phi(x) \to \min, \quad x \in \tilde{D} \tag{29}$$

Let a point x^* be a solution of problem (29), then the admissible control $u^* = u^*(t)$, $t \in T$ corresponding to x^* is the global optimal control solution to problem (26)–(28). Therefore, Algorithm MIN can also be used for solving the problem (26)–(28). As we have done before, write down the auxiliary function $\tilde{p}(y)$ for problem (29) as follows:

$$\tilde{p}(y) = \min_{x \in \tilde{D}} \langle \phi'(x), x - y\rangle, \quad y \in R^n. \tag{30}$$

Then we can formulate a counterpart theorem of Theorem 32 in the following.

THEOREM 35. *Let* $\min_{x \in \tilde{D}} \|\phi'(x)\| > 0$, *then a control* $u^* \in V$ *is a global optimal control of problem* (26)–(28) *if and only if*

$$\min_{y:\phi(y)=\phi(x^*)} \tilde{p}(y) \geq 0,$$

where $x^* = x^*(t_1, u^*) \in \tilde{D}(t_1)$.

The proof is immediate from the proof of Theorem 32 since problems (26)–(28) and (29) are equivalent.

Algorithm OPTMIN

Step 1. Choose a feasible point $x^0 \in \tilde{D}$. Set $k := 0$.
Step 2. Solve the following problem

$$\tilde{p}(y) \longrightarrow \min, \ \ y \in E_{\phi(x^k)}(\phi).$$

Let y^k be a solution of this problem (i.e., $\tilde{p}(y^k) = \min_{x \in \tilde{D}} \langle \phi'(x), x - y^k\rangle = \min_{y \in E_{\phi(x^k)}(\phi)} \tilde{p}(y)$), and let x^{k+1} realizes $\tilde{p}(y^k)$ (i.e., $\tilde{\psi}(x^k) = \tilde{p}(y^k) = \langle \phi'(x^{k+1}), x^{k+1} - y^k\rangle$).

Step 3. If $\tilde{\psi}(x^k) = 0$ then output $x = x^k$ and terminate. Otherwise, let $k := k+1$ and return Step 2.

The convergence of this Algorithm is based on the following theorem.

THEOREM 36. [**23**] *The sequence* $\{u^k\} \subset V$ *generated by Algorithm OPTMIN is a minimizing sequence for problem* (26)–(28), *i.e.,*

$$\lim_{k\to\infty} J(u^k) = \min_{u\in V} J(u)$$

4. Applications

4.1. Optimization problems in microeconomic analysis. It is well known that, on the one hand, microeconomics is concerned with the behavior of individual economic units and their interactions. On the other hand, microeconomics may be thought of as a collection of techniques to solve problems that arise in relation to those economic units. Two important types of economic units typically considered are firms and consumers and two of the major analytic techniques employed in studying these are optimization and simultaneity.

The analysis of optimum choice from among a set of alternatives is a feature of many economic problems. Sometimes even the study of simultaneity is formulated as an optimization problem. When we solve such problems, we need to specify both the objectives or economic criteria with respect to which optimization takes place and the constraints, if any, that limit the range of options. For example, when we model the behavior of consumers, we often consider the constrained utility maximization problem. In this case, the objective is to secure the most desirable consumption opportunity and the constraint is the collection of possibilities determined by the prices of goods and the consumer's income. Thus optimization problems play important role in microeconomic analysis.

In this subsection, we will show that traditional microeconomic theory is very much related to optimization problems. Consider a maximization or a minimization problem of real valued function f over a given set D:

$$f(x) \longrightarrow \max, \quad x \in D \subset R^n, \tag{31}$$

where $f : D \longrightarrow R$ is the objective function, D is a subset of a finite dimensional Eucledian space R^n, and $x = (x_1, x_2, \ldots, x_n) \in D \subset R^n$. D is called the feasible set. Furthermore, f is referred to as an economic criteria function such as, for example, a utility function, a profit function, or a cost function. In economic terms, the feasible set D may be a commodity space, an input space, defined by a budget constraint, and so on. In each of the following examples we will specify the objective function and feasible set in appropriate economic terms.

There are many works [**1, 17, 18, 19, 42, 43, 50, 56, 63, 71**] devoted to optimization problems in Economics but most of them are concerned with only convex minimization or, equivalently, concave maximization problems.

EXAMPLE 37. [**19**] *The classical model of utility maximization.* The problem of preferences maximization can be written as:

$$(32)\qquad \begin{gathered} u(x) \longrightarrow \max, \quad x \in D, \\ D = \{x \in R^n \mid \sum_{i=1}^{n} p_i x_i \leq m, \;\; x_i \geq 0, \;\; i = 1, 2, \ldots, n\}, \end{gathered}$$

where $x = (x_1, x_2, \ldots, x_n)$ is a vector or basket of quantities of n goods, $u(x)$ a utility function representing the preferences of the individual, $p = (p_1, p_2, \ldots, p_n)$ a vector of market prices of goods $1, 2, \ldots, n$, and m income.

In problem (32) the consumer chooses the best basket by maximizing his utility function over the budget constrained set D. The following conditions are often imposed on the function u [**19**]:

a). u is continuous on $R^+ = \{x : \;\; x \geq 0\}$ and has continuous second order derivatives everywhere in the interior of R^+.
b). $\frac{\partial u(x)}{\partial x_i} > 0$ for $i = 1, 2, \ldots, n$ and all x in the interior of R^+.
c). u is strictly quasiconcave.

In an oligopsonistic market, the consumer influences the market price through the amount that he purchases. Thus, if the market for good i is oligopsonistic in this way, then p_i is a function of x_i, i.e $p_i = p_i(x_i)$. If all good markets are oligopsonistic, then the utility maximization problem is:

$$(33)\qquad \begin{gathered} u(x) \longrightarrow \max, \quad x \in D, \\ D = \{x \in R^n \mid \sum_{i=1}^{n} p_i(x_i) x_i \leq m, \;\; x_i \geq 0, \;\; i = 1, 2, \ldots, n\} \end{gathered}$$

Assuming that the utility function u is quasiconvex and differentiable on D, we obtain from the global optimality condition (4) that

$$\begin{cases} \displaystyle\sum_{i=1}^{n} \frac{\partial u(y)}{\partial x_i}(x_i - y_i) \leq 0 \\ u(y_1, y_2, \ldots, y_n) = u(z) \\ \displaystyle\sum_{i=1}^{n} p_i(x_i) x_i \leq m \\ x_i \geq 0, \;\; i = 1, 2, \ldots, n. \end{cases}$$

Now consider the classical utility maximization problem in an oligopsonistic market with a quasiconcave utility function. This can be equivalently presented as the following quasiconvex minimization problem:

$$-u(x) \longrightarrow \min, \quad x \in D.$$

Then global optimality conditions (10) for this problem are as follows:

$$\begin{cases} \sum_{i=1}^{n} \frac{\partial u(x)}{\partial x_i}(x_i - y_i) \leq 0 \\ u(y_1, y_2, \ldots, y_n) = u(z) \\ \sum_{i=1}^{n} p_i(x_i)x_i \leq m \\ x_i \geq 0, \quad i = 1, 2, \ldots, n. \end{cases}$$

If in this problem, we have $u(x) = \prod_{i=1}^{p} f_i$, where $f_i : R^n \to R$ is a finite, convex function for each $i = 1, 2, \ldots, p (p \geq 2)$, and $f_i(x) \geq 0$ for all $x \in D$ and $i = 1, 2, \ldots, p$, then utility maximization problem is represented as the multiplicative maximization problem. In general, the preference ordering theory [**18**] tells that utility functions exist but might be even discontinuous. The quasiconvex case better fits to the case of substitute goods [**71**].

EXAMPLE 38. [**19**] *Pareto optimality in consumption.* Let person k's utility function be

$$\mu_k = u^k(x_k), \quad k = 1, 2, \ldots, s,$$

where x_{ik} is an amount of good i held by person k, and $x_k = (x_{1k}, x_{2k}, \ldots, x_{nk})$. There are n commodities indexed by $i = 1, 2, \ldots, n$ and k persons identified as $k = 1, 2, \ldots, s$. Then, in order to find a point that is a Pareto optimal in consumption, we have to solve the following problem:

$$u^s(x_s) = u^s(x_{1s}, x_{2s}, \ldots, x_{ns}) \longrightarrow \max, \tag{34}$$

subject to constraints

$$u^k(x_{1k}, x_{2k}, \ldots, x_{nk}) = \bar{\mu}_k, \quad k = 1, 2, \ldots, s-1,$$

$$\sum_{k=1}^{s} x_{ik} = \bar{x}_i, \quad i = 1, 2, \ldots, n,$$

$$x_{ik} \geq 0, \quad i = 1, 2, \ldots, n; \quad k = 1, 2, \ldots, s,$$

where $\bar{x}_i$ is the total quantity of good i that can be distributed among the s persons.

EXAMPLE 39. [**19**] *Expenditure minimization.* Suppose that a consumer chooses baskets of goods to minimize his expenditure on goods given a particular, minimum level of utility u^0. This problem is:

$$p_1x_1 + p_2x_2 + \ldots p_nx_n \longrightarrow \min, \tag{35}$$

subject to constraint

$$u(x_1, x_2, \ldots, x_n) \geq u^0 x_1 \geq 0, \ldots, x_n \geq 0.$$

EXAMPLE 40. [**19**] *Transformation function.* The transformation function $x_n = t(x_1, \ldots, x_{n-1})$ is defined by first solving the following problem:

$$x_n = f^n(y_{n1}, y_{n2}, \ldots, y_{nm}) \longrightarrow \max, \tag{36}$$

subject to constraints

$$f^i(y_{i1}, y_{i2}, \dots, y_{im}) = \bar{x}_i, \quad i = 1, 2, \dots, n-1,$$
$$\sum_{i=1}^{n} y_{ij} \leq a_j, \quad j = 1, \dots, m,$$
$$y_{ij} \geq 0, \quad i = 1, 2, \dots, n; \; j = 1, 2, \dots, m,$$

where x_n is the quantity of output produced by industry n, the $\bar{x}_i$ are fixed quantities of outputs produced by industries $i = 1, \dots, n-1$, y_{ij} is the quantity of input j employed by industry i, f^i is industry $i's$ production function and a_j is the total quantity of input j available. Assume that each $f^i(y_i)$ is defined for all $y_i \geq 0$ such that

1). $f^i(0) = 0$ and $f^i(y_i) \geq 0$, for all $y_i \geq 0$.
2). $f^i(y_i)$ is continuous everywhere and twice continuously differentiable where $y_i > 0$.
3). The first order partial derivatives $\frac{\partial f^i(y_i)}{\partial y_{ij}} > 0$, for all $y_i > 0$ and $j = 1, 2, \dots, m$.
4). f^i is strictly quasiconcave.

Then the problem (36) under above conditions has a unique solution[**19**]. By deriving that solution for appropriate variations in $\bar{x}_1, \dots, \bar{x}_{n-1}$, the transformation function $x_n = t(x_1, \dots, x_{n-1})$ is obtained.

EXAMPLE 41. [**19**] *Welfare maximization.* One way to describe the maximization of social welfare is as follows:

$$W(\mu) = W(\mu_1, \mu_2, \dots, \mu_k) \longrightarrow \max, \tag{37}$$

subject to the constraint

$$\mu_k \leq \varphi(\mu_1, \mu_2, \dots, \mu_{k-1}),$$

where individuals' utility functions $\mu_k = u^k(x_{1k}, \dots, x_{nk})$ defined on commodity spaces, have the properties of continuity and differentiability a), differential increasingness b), and strict quasiconcavity c). The welfare function W satisfies the same conditions as u in a)-c) with respect to its variables $\mu_1, \mu_2, \dots, \mu_k$.

The grand utility possibility function φ is obtained as follows: First, the utility possibility function $\psi(\mu_1, \mu_2, \dots, \mu_{k-1}, x_1, x_2, \dots, x_n)$ describes possible distributions of utility obtainable from the various distributions of the market quantities $x_1, x_2, \dots, x_n$. Then φ is defined as:

$$\varphi(\mu_1, \mu_2, \dots, \mu_{k-1}) = \max_{x_n = t(x_1, x_2, \dots, x_{n-1})} \psi(\mu_1, \mu_2, \dots, \mu_{k-1}, x_1, x_2, \dots, x_n)$$

EXAMPLE 42. [**19**] *Competitive equilibrium.* Under certain conditions, economy-wide equilibrium under perfect competition can be found by maximizing the welfare function $W(\mu)$:

$$W = \sum_{k=1}^{s} \lambda_k u^k(x_{1k}, x_{2k}, \dots, x_{nk}) \longrightarrow \max, \tag{38}$$

subject to constraints

$$\sum_{k=1}^{s} x_{nk} = t\left(\sum_{k=1}^{s} x_{1k}, \sum_{k=1}^{s} x_{2k}, \dots, \sum_{k=1}^{s} x_{n-1k}\right)$$
$$x_{ik} \geq 0, \quad i = 1,2,\dots,n; \quad k = 1,2,\dots,s,$$

where $(\lambda_1, \lambda_2, \dots, \lambda_s)$ is a vector of certain positive constants (or weighs) for the individual utility functions.

EXAMPLE 43. [**19**] *The theory of the firm.* The theory of the firm postulates that the goal of the firm is to maximize its profit over input space $Y = \{y = (y_1, y_2, \dots, y_n) \mid y_j \geq 0, \ j = 1,2,\dots,n\}$.
The general maximization problem is:

$$\pi = R(y_1, y_2, \dots, y_n) - C(y_1, y_2, \dots, y_n) \longrightarrow \max, \quad y \in Y, \tag{39}$$

where π is the profit, R is the revenue obtained by selling the output produced by the inputs $y_1, \dots, y_n$, and C is the cost of those inputs.

Suppose the production function $f(y)$, the output price p, and input prices, $w = (w_1, w_2, \dots, w_n)$ are given. Then $R = pq = pf(y), C = \sum_{j=1}^{n} w_j y_j = \langle w, y \rangle$. In the long run, when all inputs are variable, the problem (39) becomes:

$$pf(y_1, y_2, \dots, y_n) - \sum_{j=1}^{n} w_j y_j \longrightarrow \max, \quad y \in Y, \tag{40}$$

In the short run, when the quantities of some inputs available for use in production cannot be varied, the profit maximization problem is:

$$pf(y_1, y_2, \dots, y_n) - \sum_{j=1}^{n} w_j y_j \longrightarrow \max, \quad y \in Y, \tag{41}$$

subject to constraints

$$y_j = y_j^o, \quad j = m+1, 2, \dots, n$$
$$y_j \geq 0, \quad j = 1,2,\dots,n.$$

where $m < n$.

EXAMPLE 44. [**19**] Profit Maximization for a Firm that is both a Monopolist and a Monopsonist in all of its input Markets.

As a monopolist, the price, p, of the firm's output depends on the quantity, x, it sells according to the inverse demand function $p = p(x)$. Thus revenue is defined as $R(x) = p(x)x$. As a monopsonist, the firm also influences the prices, w_j, of its inputs by varying the amounts it purchases of them:

$$w_j = w_j(y_j), \quad j = 1,2,\dots,n.$$

Since the cost of the j-th input is $w_j(y_j)y_j$, the problem of this firm is

$$p(x)x - \sum_{j=1}^{n} w_j(y_j)y_j \longrightarrow \max, \tag{42}$$

TABLE 1

Problem	Name	Convex	Nonconvex
(32)	Utility maximization	convex	—
(33)	Utility max.in Oligopsony	—	nonconvex
(34)	Pareto optimality in consumption	convex	—
(35)	Expenditure minimization	convex	—
(36)	Transformation function	convex	—
(37)	Welfare maximization	—	nonconvex
(38)	Competitive equilibrium	—	nonconvex
(39)	Theory of the firm	convex	—
(40)	Long run profit maximization	convex	—
(41)	Short run profit maximization	convex	—
(42)	Monopolist and monopsonist	—	nonconvex
(43)	Money capital and firm	—	nonconvex

subject to constraint

$$x \le f(y_1, y_2, \dots, y_n)$$
$$y_j \ge 0, \quad j = 1, 2, \dots, n,$$

Note that the production functions $f(y)$ in problems (40)–(42) are generally assumed to be continuously differentiable and strictly quasiconcave.

EXAMPLE 45. [**19**] *Money capital and the firm.* For given output price p and input price vector r, denote the average money capital requirement function by $C = C(p, r, x, y)$, where x is a quantity of output, and $y = (y_1, y_2, \dots, y_n)$ is a vector of input quantities. This function indicates the average money capital that the firm needs to support the production of output x with inputs y at prices p and r. The firm's rate of return earned on the money capital invested in it is given by:

$$\Phi(p, r, x, y) = \frac{px - \langle r, y\rangle}{C(p, r, x, y)}.$$

In this context, the profit maximization problem (actually a rate-of-return maximization problem) is formulated as follows:

$$\Phi(p, r, x, y) \longrightarrow \max, \tag{43}$$

subject to constraints $x \ge 0, y \ge 0$, where p and r are given.

Let us now classify problems (32)–(42) as convex or nonconvex under the assumption that all objective functions in these problems are concave.

It is apparent from table 1, that classical assumptions such as (a)–(b) or (1)–(4) make many microeconomic optimization problems convex. Also, with strictly quasiconcave objective functions, it has been shown in [**19**] that those optimization problems have unique global solutions.

TABLE 2

Years	Population (thousand)
2009	2677.8
2010	2720.8
2011	2732.6
2012	2759.3
2013	2785.4
2014	2811.0
2015	2836.1
2016	2860.7
2017	2884.7
2018	2908.3
2019	2931.3
2020	2953.7

4.2. Quasiconvexity in growth theory.

4.2.1. *Population growth.* Unlike the exponential growth, the logistic functions as solutions of the following differential equation

$$\begin{cases} L'(t) = mL(N - L) \\ L(0) = L_0. \end{cases} \tag{44}$$

are commonly used in modeling population growth providing steady growth.

The solution of (44) is

$$L = \frac{L_0 N}{L_0 + (N - L_0)^{-Nmt}}.$$

and belongs to the class of quasiconvex functions, where $L(t)$ is a population at time t, m is a given parameter and N is the maximum available population.

In practice, the parameters N and m are estimated from the nonlinear regression model by solving a global optimization problem based on population data. For example, we obtained the logistic function for Mongolian population data with $L_0 = 262.4$, $N = 3700$ and $m = 0.00004599$. Based on this model we predicted population growth up to 2020. Results are shown in the table 2.

4.2.2. *Classical economic growth theory.* Let $f(K, L)$ be a concave, differentiable homogeneous production function, s the savings (saving rate), $0 \le s \le 1$, and μ depreciation rate of capital.

Assume that the labor grows at exponential rate η which means that

$$L' = L_0 e^{\eta t}. \tag{45}$$

Define k as per capita capital function:

$$k(t) = \frac{K(t)}{L(t)}.$$

Then

$$k'(t) = \left(\frac{K(t)}{L(t)}\right)' = \frac{K'L - L'K}{L^2} = \frac{1}{L}\Big(K' - \frac{L'}{L}K\Big) = \frac{1}{L}\Big(K' - \eta K\Big).$$

If we substitute $f(K, L)$ into this equation, we have

$$\begin{aligned} k' &= \frac{1}{L}\Big(s(t)f(K,L) - \mu K - \eta K\Big) \\ &= s(t)f(\frac{K}{L}, 1) - \mu\frac{K(t)}{L(t)} - \eta\frac{K(t)}{L(t)} \\ &= s(t)\varphi(k) - (\mu + \eta)k, \end{aligned}$$

where $\varphi(k)$ is the Solow per capita production function. Then per capita capital accumulation equation is

$$\begin{cases} k' = s(t)\varphi(k) - (\mu + \eta)k \\ k(0) = k_0. \end{cases} \tag{46}$$

Economic equilibria condition is:

$$k' = s(t)\varphi(k) - (\mu + \eta)k = 0. \tag{47}$$

Now we consider per capita consumption function

$$c(t) = \frac{C(t)}{L(t)} = \frac{(1-s)f(K,L)}{L} = (1-s)\varphi(k) \tag{48}$$

Assume that s is a constant function.
Let us consider the per capita consumption maximization problem subject to economic equilibria. That is

$$c = (1-s)\varphi(k) \to \max, \tag{49}$$

$$s\varphi(k) - (\mu + \eta)k = 0. \tag{50}$$

Then problem (49)–(50) is equivalent to the following one dimensional problem:

$$\max_k \Big[\varphi(k) - (\mu + \eta)k\Big].$$

The solution k^* satisfies the equation

$$\varphi'(k^*) = (\mu + \eta). \tag{51}$$

So called the golden rule of level of accumulation is determined from (50) as:

$$s^* = \frac{\varphi'(k^*)k^*}{\varphi(k^*)}. \tag{52}$$

In order to further generalize the Solow growth model note that the main assumptions of the model were

- The production function is concave and homogeneous.
- Labor grows exponentially at rate η.
- The production f is autonomous, did not defend on time t.
- The saving function is constant.

4.2.3. *Economic growth with quasiconcave production functions.* Now we consider problem (49)–(50) for the case when

$$\begin{cases} \varphi'(k) = \gamma\varphi(M-\varphi) \\ \varphi(0) = \varphi_0 \\ L = L_0 e^{\eta t}. \end{cases} \tag{53}$$

The solution of (53) is

$$\varphi(k) = \frac{\varphi_0 M}{\varphi_0 + (M-\varphi_0)^{-M\gamma k}}.$$

Then problem (49)–(50) reduces to the problem:

$$\begin{cases} c = (1-s)\varphi(k) \to \max, \\ s\varphi(k) = (\mu+\eta)k. \end{cases} \tag{54}$$

Clearly, the function $\varphi(k)$ is quasiconcave and problem (54) is equivalent to one variable maximization problem

$$c = \varphi(k) - (\mu+\eta)k \to \max, \quad k > 0. \tag{55}$$

We find stationary points of the problem

$$c'(k) = \varphi'(k) - (\mu+\eta) = 0.$$

It can be easily calculated that

$$\varphi'(k) = \frac{M^2\gamma\varphi_0(M-\varphi_0)e^{-M\gamma k}}{\Big(\varphi_0 + (M-\varphi_0)e^{-M\gamma k}\Big)^2}.$$

If we substitute $z = e^{-M\gamma k}$ and $A = M^2\gamma\varphi_0(M-\varphi_0)$ then

$$\frac{Az}{\Big(\varphi_0 + (M-\varphi_0)z\Big)^2} = \mu + \eta.$$

This reduces to the following quadratic equation

$$(M-\varphi_0)^2(\mu+\eta)z^2 + \big[2\varphi_0(M-\varphi_0)(\mu+\eta) - A\big]z + (\mu+\eta)\varphi_0^2 = 0.$$

Denote by D:

$$D = \big[2\varphi_0(M-\varphi_0)(\mu+\eta) - A\big]^2 - 4(\mu+\eta)^2(M-\varphi_0)^2\varphi_0^2. \tag{56}$$

TABLE 3

φ_0	M	μ	η	γ	Solutions		
					k^*	c^*	s^*
1.5	10	0.01	0.023	0.2	4.06913	9.849	0.01345
0.3	5	0.2	0.014	0.5	2.7135	4.332	0.1182
4	13	0.043	0.032	0.7	0.89	12.924	0.005185

The solution z is found as

$$z^* = -\frac{\left[2\varphi_0(M-\varphi_0)(\mu+\eta) - A\right] + \sqrt{D}}{2(\mu+\eta)(M-\varphi_0)^2}. \tag{57}$$

Then

$$k^* = -\frac{1}{M\gamma} lnz^*.$$

It can be checked that $c''(k^*) = \varphi''(k^*) < 0$. This means that z^* is a global solution to the problem.

LEMMA 46. $0 < z^* < 1$.

The proof is obvious consequence of (56) and (57).

Problem (55) has been solved numerically for different parameters of $\varphi_0, M, \mu, \eta, \gamma$. We present the numerical results in the following table 3.

4.3. Quasiconvex maximization applications.

4.3.1. *Fractional programming.* Consider a fractional programming problem

$$f(x) = \frac{f_1(x)}{f_2(x)} \longrightarrow \max, \quad x \in D,$$

where f_1 is convex and differentiable on R^n, and f_2 is concave and differentiable on R^n. Suppose that f_1 and f_2 are defined positively on a ball B containing a subset $D \subset R^n$, that is,

$$f_1(x) > 0, \quad f_2(x) > 0 \quad \text{for all } x \in D \subset B.$$

We shall call this problem as the mixed fractional maximization problem. By Lemma 4, we can easily show that $f(x)$ is quasiconvex. Hence, we can apply condition (4) at a global solution z for the above problem as follows:

$$\sum_{i=1}^{n} \left(\frac{\partial f_1(y)}{\partial x_i} f_2(y) - \frac{\partial f_2(y)}{\partial x_i} f_1(y) \right) \frac{(x_i - y_i)}{f_2^2(y)} \le 0$$
$$\text{for all } y \in E_{f(z)}(f) \text{ and } x \in D.$$

Consider Money Capital and the Firm model [**19**].
For given output price p and input price vector r, denote the average money capital requirement function by $f_2 = f_2(x, y)$, where x is a quantity of output, and $y = (y_1, y_2, \dots, y_n)$ is a vector of input quantities. This function indicates the average

money capital that the firm needs to support the production of output x with inputs y at prices p and r. The firm's rate of return earned on the money capital invested in it is given by:

$$f(x,y) = \frac{f_1(x,y)}{f_2(x,y)},$$

where $f_1(x,y) = px - \langle r, y\rangle$ is regarded as profit earned by the firm and $f_1(x,y) > 0$, $f_2(x,y) > 0$. In this context, if f_2 is concave then the maximization problem (actually a rate-of-return maximization problem) is formulated as the mixed fractional maximization problem:

$$f(x,y) \longrightarrow \max$$

subject to

$$0 \leq x \leq \bar{x}, \quad 0 \leq y_i \leq \bar{y}_i, \quad i = 1, 2, \ldots, n,$$

where the maximum output $\bar{x}$ and the maximum input $\bar{y}_i$ are given. Optimality condition (2.2) for this problem is the same as in the previous example.

4.3.2. *Multiplicative programming.* The following special problem of the convex multiplicative programming problem [**8, 28, 48**] can be reduced to a quasiconcave minimization problem [**8, 28**]:

$$f(x) = \prod_{i=1}^{p} f_i \longrightarrow \min, \quad x \in D, \tag{58}$$

where $f_i : R^n \to R$ are finite, convex functions for $i = 1, 2, \ldots, p (p \geq 2)$, D is a compact convex set in R^n, and $f_i(x) \geq 0$ for all $x \in D$ and $j = 1, 2, \ldots, p$.

We can also have the global optimality condition for this problem for its equivalent quasiconvex maximization problem in the outcome space which is as follows [**28**]:

$$g(x) = -\prod_{i=1}^{p} x_i \longrightarrow \max, \quad x \in Y,$$

where the convex and compact set Y is constructed for a point in the feasible set D of problem (58). Then condition (4) for this problem becomes as follows.

$$\sum_{i=1}^{n} \prod_{j \neq i}^{p} y_j(x_i - y_i) \geq 0 \quad \text{for all } y \in E_{g(z)}(g) \quad \text{and} \quad x \in Y.$$

4.3.3. *D.C programming.* Consider the problem of maximizing the difference of two convex functions on a convex subset. We shall call this as D.C. maximization problem.

$$g(x) = g_1(x) - g_2(x) \longrightarrow \max \quad \text{subject to } h_i(x) \leq 0, \quad i = 1, 2, \ldots, m,$$

where g_1, g_2 and $h_i, i = 1, 2, \ldots, m$ are convex differentiable functions on R^n. This is equivalent to its counterpart D.C. minimization problem:

$$-g(x) = g_2(x) - g_1(x) \longrightarrow \min \quad \text{subject to } h_i(x) \leq 0, \quad i = 1, 2, \ldots, m.$$

It has been shown that this problem can be transformed into the equivalent quasiconvex maximization problem [**37**] as follows:

$$g_1(x) - x_{n+1} \longrightarrow \max \quad \text{subject to } g_2(x) \leq x_{n+1}, \quad h_i(x) \leq 0, \quad i = 1, 2, \ldots, m,$$

where $(x, x_{n+1}) \in R^n \times R$. Now we can apply condition (4) to this problem to obtain

$$\begin{cases} \displaystyle\sum_{i=1}^{n} \frac{\partial g_1(y)}{\partial x_i}(x_i - y_i) + y_{n+1} - x_{n+1} \leq 0 \\ g_1(y) - y_{n+1} = g_1(z) - z_{n+1} \\ g_2(x) \leq x_{n+1} \\ h_i(x) \leq 0, \ \ i = 1, 2, \ldots, m. \end{cases}$$

4.4. Quasiconvex minimization applications.

4.4.1. *Fractional programming.* Consider the fractional programming problem

$$f(x) = \frac{f_1(x)}{f_2(x)} \longrightarrow \min, \quad x \in D, \tag{59}$$

where f_1 is convex and differentiable on R^n, and f_2 is concave and differentiable on R^n. Suppose that f_1 and f_2 are defined positively on a ball B containing a subset $D \subset R^n$, that is,

$$f_1(x) > 0, \quad f_2(x) > 0 \text{ for all } x \in D \subset B.$$

We shall call this problem as the mixed fractional minimization problem. By Lemma 4, we can easily show that $f(x)$ is quasiconvex. Hence, the optimality condition (10) at a solution z of problem (59) is as follows:

$$\sum_{i=1}^{n} \left(\frac{\partial f_1(x)}{\partial x_i} f_2(x) - \frac{\partial f_2(x)}{\partial x_i} f_1(x) \right) \frac{(x_i - y_i)}{f_2^2(x)} \geq 0$$
$$\text{for all } y \in E_{f(z)}(f) \quad \text{and} \quad x \in D.$$

When f_1, f_2 and D were all convex, then problem (59) was studied in [**39, 64**].

4.4.2. *Average cost minimization.* It is well known that when the production is strictly concave, the cost function $C(x)$ is strictly convex. In economics, in the short run, the so-called average cost function $f(x) = \frac{C(x)}{x}$ is usually has to be minimized. It can be easily shown that the average cost function $f(x)$ is quasiconvex due to Lemma 4.

Then the average cost minimization problem is formulated as the following quasiconvex minimization problem:

$$f(x) = \frac{C(x)}{x} \longrightarrow \min, \quad x \geq 0.$$

4.4.3. *D.C. programming.* The so-called D.C. programming problem, where D.C. is an abbreviation for the difference of two convex functions [**37**], is an another example of a quasiconvex minimization problem:

$$\begin{aligned} g(x) = g_1(x) - g_2(x) \longrightarrow \min \\ \text{subject to } h_i(x) \le 0, \quad i = 1, 2, \ldots, m, \end{aligned} \tag{60}$$

where g_1, g_2 and $h_i i = 1, 2, \ldots, m$ are all finite convex functions on R^n.

Consider the D.C. programming problem (60). Assume additionally that the functions g_1 and g_2 are differentiable. This problem can be converted into an equivalent convex minimization problem with a nonconvex feasible set, which is sometimes called reverse convex constraint [**37**].

$$\begin{aligned} g_1(x) - x_{n+1} \longrightarrow \min, \\ \text{subject to: } g_2(x) \ge x_{n+1}, \quad h_i(x) \le 0, \; i = 1, 2, \ldots, m, \end{aligned}$$

where $(x, x_{n+1}) \in R^n \times R$. Now the condition (10) at a solution $z \in D$ becomes as follows.

$$\begin{cases} \sum_{i=1}^{n} \frac{\partial g_1(y)}{\partial x_i}(x_i - y_i) + y_{n+1} - x_{n+1} \ge 0 \\ g_1(y) - y_{n+1} = g_1(z) - z_{n+1} \\ g_2(x) \ge x_{n+1} \\ h_i(x) \le 0, \; i = 1, 2, \ldots, m. \end{cases}$$

Bibliography

[1] Akira Takayama, *Mathematical Economics*, Cambridge University Press, 1985.

[2] K.J. Arrow and L. Hurwics, *Reduction of Constrained Maxima to Saddle-Point Problems*, in J. Neyman(ed.), Proceedings of the Third Berkeley Symposium on Mathematical Statistics and Probability, Berkeley and Los Angeles, University of California Press, Vol.5,pp.1–20, 1956.

[3] K.J. Arrow and A.C. Enthoven, *Quasi-Concave Programming*, Econometrica, 29(4), pp.779–800, 1961.

[4] D.P. Bertsekas, *Constrained Optimization and Lagrangean Multiplier Methods*, Academic Press, New York, 1982.

[5] A. Chinchuluun, R. Enkhbat and P.M. Pardalos, *A Novel Approach for Nonconvex Optimal Control Problems*, Optimization, Vol.58, No.7, pp.781–789, 2009.

[6] A.Chinchuluun, P.M. Pardalos and R. Enkhbat, *Global Minimization Algorithms for Concave Quadratic Programming Problems*, Optimization, Taylor and Francis, Vol.54, No.6, pp.627–639, 2005.

[7] A. Chinchuluun, P.M. Pardalos and R. Enkhbat, *A Numerical Method for Concave Programming Problems*, in A. Rubinov and V. Jeyakumar (eds). Continuous Optimization: Current Trends and Modern Applications, pp.251–272, Springer, 2005.

[8] B. Jaumard, C. Meyer and H. Tuy, *Generalized Convex Multiplicative Programming via Quasiconcave Minimization*, Journal of Global Optimization, 10, pp.229–256, 1997.

[9] V.P. Bulatov, *The Embedding Methods in Extremum Problems*, Nauka, Novosivirsk, 1977.

[10] J.P. Crouzeix, *Duality between Direct and Indirect Utility Functions*, Journal of Mathematical Economics, 12(2), pp.149–165, 1983.

[11] J.P. Crouzeix and P.O. Lindberg, *Additively Decomposed Quasi-Convex Functions*, Mathematical Programming, 35(1), pp.42–57, 1986.

[12] G. Debreu, *Least Concave Utility Functions*, Journal of Mathematical Economics, 3(2), pp.121–129, 1976.

[13] G. Debreu and T.C. Koopmans, *Additively Decomposed Quasi-Convex Functions*, Mathematical Programming, 24(1), pp.1–38, 1982.
[14] W.E. Diewert, *Generalized Concavity in Optimization and Economics, ed. S. Schaible and Ziemba*, Academic Press, New-York, 1981.
[15] B. De Finetti, *Sulle Stratificazioni Convesse*, Annali Di Mathematica Pura ed Applicata, 4(30), pp.173–183, 1949.
[16] V.P. Demyanov and L.V. Vasiliev, *Nondifferentiable Optimization*, Nauka, Moscow, 1981.
[17] A.K. Dixit, *Optimization in Economic Theory*, Oxford University Press, 1976.
[18] Donald W. Katzner, *Static Demand Theory*, Macmillian, London, 1970.
[19] Donald W. Katzner, *Walrasian Microeconomics*, Addison-Wesley, New-York, 1988.
[20] I. Ekeland and R. Temam, *Convex Analysis and Variational Problems*, North-Holland, Amsterdam, 1976.
[21] R. Enkhbat, *An Algorithm for Maximizing a Convex Function over a Simple Set*, Journal of Global Optimization, 8, pp.379–391, 1996.
[22] R. Enkhbat and T. Ibaraki, *On the Maximization and Minimization of a Quasiconvex Function*, International Journal of Nonlinear and Convex Analysis, pp. 43–76, Japan, 2003
[23] R. Enkhbat, B. Barsbold and M. Kamada, *A Numerical Approach for Solving Some Convex Maximization Problems*, Journal of Global Optimization, Springer, Vol.35, pp.85–101, 2006.
[24] R. Enkhbat, *A Global Method for Some Class of Optimization and Control Problems*, International Journal of Mathematics and Mathematical Sciences, USA, Vol.23, No.9, pp.605–616, 2000.
[25] R. Enkhbat, *Theory, Methods and Applications of Concave Programming*, in book: Global Optimization: Theory, Methods and Applications 1–2, Lecture Notes in Decision Sciences, Vol.12(A), Global-Link Publisher, Hong Kong-London, pp.292–299, 2009
[26] W. Fenchel, *Convex Cones, Sets, and Functions*, Princeton University, Mimeo, 1953.
[27] I.V. Girsanov, *Lecture Notice on the Theory of Extremum*, Moscow State University, 1970.
[28] Harold P. Benson, *An Outcome Space Branch and Bound-Outer Approximation Algorithm for Convex Multiplicative Programming*, Journal of Global Optimization, 15, pp.315–342, 1999.
[29] Helmut Dietrich, *Global Optimization Conditions for Certain Nonconvex Minimization Problems*, Journal of Global Optimization, 5, pp.359–370, 1994.
[30] J.B. Hiriart-Urruty, *From Convex Optimization to Nonconvex Optimization*, Nonsmooth Optimization and Related Topics, Plenum, pp.219–239, 1989.
[31] J.B. Hiriart-Urruty and J.S. Ledyaev, *A Note on the Characterization of the Global Maxima of a (tangentially) Convex Function over a Convex Set*, Journal of Convex Analysis, 3(1), pp.55–61, 1996.
[32] R. Horst, *On the Global Minimization of a Concave Function: Introduction and Servey*, Operations Research Spectrum, 6, pp.195–200, 1984.
[33] R. Horst, *A General Class of Branch and Bound Methods in Global Optimization with some New Approaches for Concave Minimization*, Journal of Optimization Theory and Applications, 51, pp.271–291, 1986.
[34] R. Horst, *Outer Cut Methods in Global Optimization*, Springer- Verlag, Lecture Notes in Economics and Mathematical Systems, 304, pp.28–40, 1987.
[35] R. Horst, N.V. Thoai and H. Tuy, *Outer Approximation by Polyhedral Convex Sets*, Operations Research Spectrum, 9(3), pp.153–159, 1987.
[36] R. Horst, *A New Branch and Bound Approach for Concave Minimization Problems*, Springer-Verlag, Lecture Notes in Computer Science, 41, pp.330–337, 1987.
[37] R. Horst and H. Tuy, *Global Optimization(Deterministic Approaches)*, Springer, Berlin, 1990.
[38] T. Ibaraki and N. Katoh, *Resourse Allocation Problems: Algorithmic Approaches*, The MIT Press, Cambridge, London, 1988.
[39] T. Ibaraki, *Parametric Approaches to Practional Programs*, Mathematical Programming, 26, pp.345–362, 1983.
[40] A.D. Ioffe and V.M. Tihomirov, *Theory of Extremal Problems*, North-Holland, New York, 1979.
[41] S.L. Jacoby, J.S. Kowalik and J.T. Pizzo, *Iterative Methods for Nonlinear Optimization Problems*, Prentice-Hall, Inc., 1972.
[42] J.M. Henderson and R.E. Quandt, *Microeconomic theory: A Mathematical Approach*, McGraw-Hill, 1971.

[43] J. Eatwell, M. Milgate and P. Newman(editors), *The New Palgrave A Dictionary of Economics*, Vol.1-4, Macmillian, 1987.

[44] B. Kalantari and J.B. Rosen, *Algorithm for Large-Scale Global Minimization of Linearly Constrained Concave Quadratic Functions*, Mathematics of Operations Research, 12, pp.544–561, 1987.

[45] L.V. Kantorovich, *On an Effective Method for Solving some Classes of Extremum Problems*, Soviet Math. Doklady, 28(3), pp.212–215, 1940.

[46] V.G. Karmanov, *Mathematical Programming*, Mir Publisher, Moscow, 1989.

[47] K.J. Arrow, L. Hurwicz and H. Uzawa, *Studies in Linear and Non-Linear Programming*, Stanford University Press, 1958.

[48] H. Konno and T. Kuno, *Multiplicative Programming Problems in Horst and Pardalos, eds., Handbook of Global Optimization*, Kluwer Dordrecht, pp.369–405, 1995.

[49] H.W. Kuhn and A.W. Tucker, *Nonlinear Programming*, in J. Neyman(ed.), Proceedings of the Second Berkeley Symposium on Mathematical Statistics and Probability, Berkeley and Los Angeles, University of California Press, pp.481–492, 1951.

[50] Michael D. Intriligator, *Mathematical Optimization and Economic Theory*, Prentice-Hall, 1971.

[51] V.S. Michaelyevich, V.S. Gupal and V.I. Norkin, *Methods of Nonconvex Optimization*, Nauka, Moscow, 1987.

[52] K.M. Mital, *Optimization Methods in Operations Research and Systems Analysis*, Jonh Wiley and Sons, New-York, 1976.

[53] V.N. Nefedov, *Finding the Global Maximum of a Function of Several Variables on a Set Given by Inequality Constraints*, Journal of Numerical Mathematics and Mathematical Physics, 27(1), pp.35–51, 1987.

[54] P.M. Pardalos and J.B. Rosen, *Constrained Global Optimization: Algorithms and Applications*, Springer- Verlag, Lecture Notes in Computer, Science, 1987.

[55] P.M. Pardalos and J.B. Rosen, *Methods for Global Concave Minimization: A Bibliographic Survey*, SIAM Review, 28, pp.367–379,1986.

[56] Paul Madden, *Concavity and Optimization in Microeconomics*, Oxford University Press, 1986.

[57] P.T. Thach, *A Generalized Duality and Applications*, Journal of Global Optimization, 3, pp.311–324, 1993.

[58] V.V. Podinovskii and A.V. Timohov, *Introduction to Mathematical Programming*, Moscow University, 1988.

[59] L.S. Pontryagin, R.G. Boltyanski, R.V. Gamkrelidze and E.E. Mischenko, *Mathematical Theory of Optimal Processes*, Interscience Publ. Inc., New York, 1962.

[60] B.N. Pshenichnyi and Yu.M. Danilin, *Numerical Methods in Extremal Problems*, Nauka, Moscow, 1975.

[61] R.T. Rockafellar, *Convex Analysis*, Princeton University Press, Princeton, 1970.

[62] R.M. Roessler, *Eine Method zur Berechnung des Optimalen Produktions -Programms bei Konkaven Kosten*, Unternehmensforschung, 15, pp.103–111, 1971.

[63] R.E. Weintraub, *Mathematics for Economists*, Cambridge University Press, 1982.

[64] S. Schaible and T. Ibaraki, *Invited Review: Fractional programming*, European J.of Operational Research, 12, pp.325–338, 1983.

[65] S. M. Robinson, *Structural Methods in the Solution of Variational Inequalities*, in G.Dipillo and F.Giannessi(eds.), Nonlinear Optimization and Related Topics, Kluwer Academic Publishers, pp.369–380, 2000.

[66] A.S. Strekalovsky, *On the Global Extremum Problem*, Soviet Math.Doklady, 292(5), pp.1062–1066, 1987.

[67] A.S. Strekalovsky, *Global Optimality Conditions for Nonconvex Optimization*, Journal of Global Optimization, 12, pp.415–434, 1998.

[68] A.S. Strekalovsky and R. Enkhbat, *Global Maximum of Convex Functions on an Arbitrary Set*, Dep.in VINITI, Irkutsk, No.1063, pp.1–27, 1990.

[69] H. Tuy, *Concave Programming under Linear Constraints*, Soviet Math.Doklady, 159(1), pp.32–35, 1964.

[70] H. Tuy, *Normal Conical Algorithm for Concave Minimization over Polytopes*, Mathematical Programming,51, pp.229–245, 1991.

[71] H.R. Varian, *Microeconomic Analysis*, Norton, New-York, 1984.

[72] F.P. Vasiliev, *Numerical Methods for Solving Extremum Problems*, Nauka, Moscow, 1987.
[73] O.V. Vasiliev, *Optimization Methods*, World Federation Publishers, Atlanta, 1996.
[74] N.C. Vasiliev, *On the Finding the Global Minimum of a Quasiconcave Function*, Journal of Numerical Mathematics and Mathematical Physics, 23(2), pp.307–313, 1983.
[75] I.W. Zangwill, *Nonlinear Programming*, Prentice-Hall,1986.

Enkhbat Rentsen, Department of Mathematics and Computer, The National University of Mongolia

Current address: The School of Economic Studies, P.O.BOX 46/635, Ulaanbaatar, Mongolia 210646

E-mail address: renkhbat46@yahoo.com

Nonlinear eigenvalue problems

Biagio Ricceri

CONTENTS

Introduction

The aim of this chapter is to give a kind of synopsis of certain results on nonlinear eigenvalue problems that have been obtained by the author in the last decade.

In extreme synthesis, we can say that, given $k \in \mathbf{N}$ and a function $f : X \times I \to \mathbf{R}$, where X is a topological space and I a real interval, the common leading theme of most of such results is the existence of $\lambda \in I$ such that the function $f(\cdot, \lambda)$ has at least k, or even infinitely many, local minima in X.

A feature of these results is their applicability to nonlinear equations via variational methods. Actually, there is a broad literature on such applications: this is shown by the long list of references that we tried to make as complete and up-dated as possible.

After a first section of preliminaries, we dispose our theoretical results in five sections. Then, we give a few selected samples of their applications to Dirichlet and Neumann problems. In the final section, we make some comments on the presented results, providing, at the same time, a guide to the relevant literature.

1. Preliminaries

In this short section, we group a few definitions and notations that will be used throughout the sequel.

Let X be a real Banach space.

For a set $A \subseteq X$, we denote by $(\overline{A})_w$ its closure in the weak topology. We say that A is weakly connected if it is connected in the weak topology. The weakly connected components of A are its connected components in the weak topology.

We denote by $\mathscr{W}_X$ the class of all functionals $J : X \to \mathbf{R}$ possessing the following property: if $\{x_n\}$ is a sequence in X converging weakly to $x \in X$ and $\liminf_{n\to\infty} J(x_n) \leq J(x)$, then $\{x_n\}$ has a subsequence converging strongly to x.

For instance, if X is uniformly convex and $g : [0, +\infty[\to \mathbf{R}$ is a continuous, strictly increasing function, then, by a classical result, the functional $x \to g(\|x\|)$ belongs to the class $\mathscr{W}_X$.

If $J : X \to \mathbf{R}$ is a Gâteaux differentiable functional, the critical points of J are the zeros of its derivative, J'. If, in particular, X is a Hilbert space, we identify X with its dual. As a consequence, J' is thought as an operator from X into itself.

Moreover, J is said to satisfy the Palais-Smale condition if each sequence $\{x_n\}$ in X such that $\sup_{n\in\mathbf{N}} |J(x_n)| < +\infty$ and $\lim_{n\to+\infty} \|J'(x_n)\|_{X^*} = 0$ admits a strongly converging subsequence.

Also, when we say that the derivative of J admits a continuous inverse on X^*, we mean that there exists a continuous operator $B : X^* \to X$ such that $B(J'(x)) = x$ for all $x \in X$.

Further, we recall that an operator $A : X \to X^*$ is said to be a potential operator if there exists a a Gâteaux differentiable functional P on X (which is called a potential of A) such that $P' = A$.

Now, let (T, τ) be a topological space.

We denote by τ_s the topology on T whose members are the sets $A \subseteq T$ such that $T \setminus A$ is sequentially τ-closed. Clearly, τ_s is stronger than τ. We note that a function $f : T \to \mathbf{R}$ is sequentially τ-lower semicontinuous if and only if it is τ_s-lower semicontinuous.

Let S be another topological space and $F : T \to 2^S$ a multifunction. For $C \subseteq T$ and $D \subseteq S$, we put

$$F(C) = \bigcup_{t\in C} F(t)$$

and

$$F^-(D) = \{t \in T : F(t) \cap D \neq \emptyset\}.$$

F is said to be lower semicontinuous at $t \in T$ if, for every open set $\Omega \subseteq S$, the condition $t \in F^-(\Omega)$ implies that $t \in \operatorname{int}(F^-(\Omega))$.

Denote by $NLS(F(t))$ the set of all points $s \in F(t)$ for which there exist a neighbourhood Ω of s in S and a net $\{t_\alpha\}$ in $T \setminus F^-(\Omega)$ converging to t. It is clear that F is not lower semicontinuous at t if and only if the set $NLS(F(t))$ is non-empty.

If τ_1 is the topology on S, we will also write $NLS_{\tau_1}(F(t))$ to denote $NLS(F(t))$. Finally, F is said to be upper semicontinuous if, for every closed set $D \subseteq S$, the set $F^-(D)$ is closed in T.

2. A general variational principle

This section is devoted to the following general principle:

THEOREM 2.1. *Let X be a topological space, let $\sigma > 0$ and let $\Phi, \Psi : X \to \mathbf{R}$ be such that the function $\Phi + \lambda\Psi$ is sequentially lower semicontinuous for all $\lambda \geq \sigma$. Denote by $\hat{\tau}_\Psi$ the weakest topology on X for which Ψ is upper semicontinuous. Also, denote by I the set of all $\rho > \inf_X \Psi$ such that the set $\Psi^{-1}(]-\infty, \rho[)$ is contained in some sequentially compact subset of X. Assume that $I \neq \emptyset$. For each $\rho \in I$, denote by $\mathscr{F}_\rho$ the family of all sequentially compact subsets of X containing $\Psi^{-1}(]-\infty, \rho[)$, put*

$$\alpha(\rho) = \sup_{K \in \mathscr{F}_\rho} \inf_K \Phi$$

and assume that $\alpha(\rho) > -\infty$.

Then, the following conclusions hold:

(a) *For each $\rho \in I$ and each λ satisfying*

$$\lambda > \max\left\{\sigma, \inf_{x \in \Psi^{-1}(]-\infty,\rho[)} \frac{\Phi(x) - \alpha(\rho)}{\rho - \Psi(x)}\right\}$$

the restriction of the function $\Phi + \lambda\Psi$ to $\Psi^{-1}(]-\infty, \rho[)$ has a global minimum.

(b) *If*

$$\gamma := \liminf_{\rho \to (\sup I)^-} \inf_{x \in \Psi^{-1}(]-\infty,\rho[)} \frac{\Phi(x) - \alpha(\rho)}{\rho - \Psi(x)} < +\infty$$

then, for each $\lambda > \max\{\sigma, \gamma\}$, the following alternative holds: either $\Phi + \lambda\Psi$ has a global minimum, or there exists a sequence $\{x_n\}$ of $\hat{\tau}_\Psi$-local minima of $\Phi + \lambda\Psi$ such that $\Psi(x_n) < \sup I$ for each $n \in \mathbf{N}$ and

$$\lim_{n \to \infty} \Psi(x_n) = \sup I.$$

(c) *If Ψ is sequentially lower semicontinuous and*

$$\delta := \liminf_{\rho \to \left(\inf_X \Psi\right)^+} \inf_{x \in \Psi^{-1}(]-\infty,\rho[)} \frac{\Phi(x) - \alpha(\rho)}{\rho - \Psi(x)} < +\infty$$

then, for each $\lambda > \max\{\sigma, \delta\}$, there exists a sequence $\{x_n\}$ of $\hat{\tau}_\Psi$-local minima of $\Phi + \lambda\Psi$, with $\lim_{n\to\infty} \Psi(x_n) = \inf_X \Psi$, which converges to a global minimum of Ψ.

PROOF. Let us prove (a). Fix $\rho \in I$ and λ as in the conclusion. Observe that

$$\inf_{x \in \Psi^{-1}(]-\infty,\rho[)} \frac{\Phi(x) - \alpha(\rho)}{\rho - \Psi(x)} = \inf_{r < \alpha(\rho)} \inf_{x \in \Psi^{-1}(]-\infty,\rho[)} \frac{\Phi(x) - r}{\rho - \Psi(x)}.$$

So, there is $r_0 < \alpha(\rho)$ such that

$$\inf_{x\in\Psi^{-1}(]-\infty,\rho[)} \frac{\Phi(x)-r_0}{\rho-\Psi(x)} < \lambda.$$

Fix also $K \in \mathscr{F}_\rho$ so that $r_0 < \inf_K \Phi$. Consequently, we have

$$\inf_{x\in\Psi^{-1}(]-\infty,\rho[)} \frac{\Phi(x)-\inf_K \Phi}{\rho-\Psi(x)} < \lambda.$$

Therefore, there is $x_0 \in \Psi^{-1}(]-\infty,\rho[)$ such that

$$\frac{\Phi(x_0)-\inf_K \Phi}{\rho-\Psi(x_0)} < \lambda.$$

From this, we get

$$\Phi(x_0)+\lambda\Psi(x_0) < \lambda\rho + \inf_K \Phi. \tag{2.1}$$

Since $\lambda \geq \sigma$, the function $\Phi+\lambda\Psi$ is sequentially lower semicontinuous, and so, by sequential compactness, there exists $\hat{x} \in K$ such that

$$\Phi(\hat{x})+\lambda\Psi(\hat{x}) = \inf_{x\in K}(\Phi(x)+\lambda\Psi(x)). \tag{2.2}$$

To get the conclusion, it is enough to prove that $\Psi(\hat{x}) < \rho$. Arguing by contradiction, assume that $\Psi(\hat{x}) \geq \rho$. Then, taking (2.1) into account, we have

$$\Phi(x_0)+\lambda\Psi(x_0) < \lambda\Psi(\hat{x})+\Phi(\hat{x})$$

contradicting (2.2).

Let us prove (b). Let $\lambda > \max\{\sigma,\gamma\}$. Then, we can fix a sequence $\{\rho_n\}$ in $I\setminus\{\sup I\}$, with $\lim_{n\to\infty}\rho_n = \sup I$, such that

$$\inf_{x\in\Psi^{-1}(]-\infty,\rho_n[)} \frac{\Phi(x)-\alpha(\rho_n)}{\rho_n-\Psi(x)} < \lambda$$

for all $n \in \mathbf{N}$. Consequently, thanks to (a), for each $n \in \mathbf{N}$, there is $x_n \in \Psi^{-1}(]-\infty,\rho_n[)$ such that

$$\Phi(x_n)+\lambda\Psi(x_n) \leq \Phi(x)+\lambda\Psi(x) \tag{2.3}$$

for all $x \in \Psi^{-1}(]-\infty,\rho_n[)$. Now, if $\lim_{n\to\infty}\Psi(x_n) = \sup I$, we are done, since each set $\Psi^{-1}(]-\infty,\rho_n[)$ is $\hat{\tau}_\Psi$-open. Thus, suppose that $\liminf_{n\to\infty}\Psi(x_n) < \sup I$. Hence, there are an increasing sequence $\{n_k\}$ in $\mathbf{N}$ and a constant $c \in I$ such that

$$\Psi(x_{n_k}) < c$$

for all $k \in \mathbf{N}$. But the set $\Psi^{-1}(]-\infty,c[)$ is contained in some sequentially compact subset of X, and so there is a subsequence $\{x_{n_{k_r}}\}$ converging to some $x^* \in X$. Finally, fix $x \in X$. Since we definitively have $\rho_{n_{k_r}} > \Psi(x)$, taking into account the sequential lower semicontinuity of $\Phi + \lambda\Psi$, from (2.3) we get

$$\Phi(x^*) + \lambda\Psi(x^*) \leq \liminf_{r\to+\infty}(\Phi(x_{n_{k_r}}) + \lambda\Psi(x_{n_{k_r}})) \leq \Phi(x) + \lambda\Psi(x).$$

Hence, x^* is a global minimum of $\Phi + \lambda\Psi$.

Finally, ley us prove (c). Let $\lambda > \max\{\sigma, \delta\}$. Fix a sequence $\{\rho_n\}$ in I, with $\lim_{n\to\infty} \rho_n = \inf_X \Psi$, such that

$$\inf_{x\in\Psi^{-1}(]-\infty,\rho_n[)} \frac{\Phi(x) - \alpha(\rho_n)}{\rho_n - \Psi(x)} < \lambda$$

for all $n \in \mathbf{N}$. Thanks to (a), for each $n \in \mathbf{N}$, the restriction of $\Phi + \lambda\Psi$ to $\Psi^{-1}(]-\infty, \rho_n[)$ has a global minimum, say x_n. Hence, x_n is a $\hat{\tau}_\Psi$-local minimum of $\Phi + \lambda\Psi$. Finally, the sequence $\{x_n\}$ lies in $\Psi^{-1}(]-\infty, \max_{n\in\mathbf{N}} \rho_n[)$ (which is, in turn, contained in a sequentially compact subset of X), and so it admits a subsequence converging to a point which, by the sequential lower semicontinuity of Ψ, is a global minimum of Ψ. □

REMARK 2.2. Concerning the statement of Theorem 2.1, observe that when the space X is Hausdorff, the set $H = \cap\{K : K \in \mathscr{F}_\rho\}$ belongs to $\mathscr{F}_\rho$, and so we have

$$\alpha(\rho) = \inf_H \Phi.$$

We now derive from Theorem 2.1 more concrete results in reflexive Banach spaces.

THEOREM 2.3. *Let E be a reflexive real Banach space, X a closed, convex, unbounded subset of E, and $\Phi, \Psi : X \to \mathbf{R}$ two convex functionals, with Φ lower semicontinuous and Ψ continuous and satisfying $\lim_{x\in X,\|x\|\to+\infty} \Psi(x) = +\infty$. Put*

$$\lambda^* = \inf_{\rho>\inf_X \Psi} \inf_{x\in\Psi^{-1}(]-\infty,\rho[)} \frac{\Phi(x) - \inf_{\Psi^{-1}(]-\infty,\rho])} \Phi}{\rho - \Psi(x)}.$$

Then, for each $\lambda > \lambda^*$, the functional $\Phi+\lambda\Psi$ has a global minimum in X. Moreover, if $\lambda^* > 0$, for each $\mu < \lambda^*$, the functional $\Phi + \mu\Psi$ has no global minima in X.

PROOF. It is clear that we can apply Theorem 2.1 endowing X with the relativization of the weak topology. In particular, for each $\rho > \inf_X \Psi$, the set $\Psi^{-1}(]-\infty, \rho])$ is sequentially weakly compact owing to the reflexivity of E and to the coercivity of Ψ. Moreover, by the convexity of Ψ, the same set is equal to the closure of $\Psi^{-1}(]-\infty, \rho[)$. From this, it follows that $\Psi^{-1}(]-\infty, \rho])$ is the smallest sequentially weakly compact subset of X containing $\Psi^{-1}(]-\infty, \rho[)$. Hence, by Remark 2.2, we have

$$\alpha(\rho) = \inf_{\Psi^{-1}(]-\infty,\rho])} \Phi.$$

Now, let $\lambda > \lambda^*$ and choose $\rho > \inf_X \Psi$ so that

$$\lambda > \inf_{x\in\Psi^{-1}(]-\infty,\rho[)} \frac{\Phi(x) - \inf_{\Psi^{-1}(]-\infty,\rho])} \Phi}{\rho - \Psi(x)}.$$

Then, Theorem 2.1 ensures that the restriction of the functional $\Phi+\lambda\Psi$ to $\Psi^{-1}(]-\infty,\rho[)$ has a global minimum, say x_0. But, by assumption, Ψ is (strongly) continuous, and so the set $\Psi^{-1}(]-\infty,\rho[)$ is (strongly) open in X. In other words, x_0 is a local minimum for $\Phi+\lambda\Psi$ in the strong topology. Since $\Phi+\lambda\Psi$ is convex, x_0 is actually a global minimum for $\Phi+\lambda\Psi$ in X. Now, assume that $\lambda^*>0$. Let μ be such that the functional $\Phi+\mu\Psi$ has a global minimum in X, say x_1. We claim that $\mu\geq\lambda^*$ from which, of course, the second part of the conclusion follows. Indeed, fix $\rho>\Psi(x_1)$ and choose $x_2\in\Psi^{-1}(]-\infty,\rho])$ so that

$$\Phi(x_2)=\inf_{\Psi^{-1}(]-\infty,\rho])}\Phi.$$

Clearly, since $\lambda^*>0$, we have $\Psi(x_2)=\rho$. Hence, we get

$$\mu\geq\frac{\Phi(x_1)-\Phi(x_2)}{\Psi(x_2)-\Psi(x_1)}=\frac{\Phi(x_1)-\alpha(\rho)}{\rho-\Psi(x_1)}\geq\inf_{x\in\Psi^{-1}(]-\infty,\rho[)}\frac{\Phi(x)-\alpha(\rho)}{\rho-\Psi(x)}\geq\lambda^*,$$

as claimed. □

The next result is a particular version of Theorem 2.1 which is directly applicable to differential equations.

THEOREM 2.4. *Let X be a reflexive real Banach space, and let $\Phi,\Psi:X\to\mathbf{R}$ be two sequentially weakly lower semicontinuous and Gâteaux differentiable functionals. Assume also that Ψ is (strongly) continuous and satisfies $\lim_{\|x\|\to+\infty}\Psi(x)=+\infty$. For each $\rho>\inf_X\Psi$, put*

$$\varphi(\rho)=\inf_{x\in\Psi^{-1}(]-\infty,\rho[)}\frac{\Phi(x)-\inf_{\overline{(\Psi^{-1}(]-\infty,\rho[))}_w}\Phi}{\rho-\Psi(x)}.$$

Furthermore, set

$$\gamma=\liminf_{\rho\to+\infty}\varphi(\rho)$$

and

$$\delta=\liminf_{\rho\to\left(\inf_X\Psi\right)^+}\varphi(\rho).$$

Then, the following conclusions hold:

(a) *For each $\rho>\inf_X\Psi$ and each $\lambda>\varphi(\rho)$, the functional $\Phi+\lambda\Psi$ has a critical point which lies in $\Psi^{-1}(]-\infty,\rho[)$.*

(b) *For each* $\rho > \inf_X \Psi$ *and each* $\lambda > \varphi(\rho)$ *for which the functional* $\Phi + \lambda\Psi$ *satisfies the Palais-Smale condition, the following alternative holds: either the functional* $\Phi + \lambda\Psi$ *has a strict global minimum which lies in* $\Psi^{-1}(]-\infty, \rho[)$*, or the same functional has at least two critical points one of which lies in* $\Psi^{-1}(]-\infty, \rho[)$.

(c) *If* $\gamma < +\infty$*, then, for each* $\lambda > \gamma$*, the following alternative holds: either* $\Phi + \lambda\Psi$ *has a global minimum, or there exists a sequence* $\{x_n\}$ *of critical points of* $\Phi + \lambda\Psi$ *such that* $\lim_{n\to\infty} \Psi(x_n) = +\infty$.

(d) *If* $\delta < +\infty$*, then, for each* $\lambda > \delta$*, the following alternative holds: either there exists a global minimum of* Ψ *which is a local minimum of* $\Phi + \lambda\Psi$*, or there exists a sequence* $\{x_n\}$ *of pairwise distinct critical points of* $\Phi + \lambda\Psi$*, with* $\lim_{n\to\infty} \Psi(x_n) = \inf_X \Psi$*, which weakly converges to a global minimum of* Ψ.

PROOF. Endow X with the weak topology. As in the proof of Theorem 2.3, it is seen that $\overline{(\Psi^{-1}(]-\infty, \rho[))}_w$ is the smallest sequentially weakly compact subset of X containing $\Psi^{-1}(]-\infty, \rho[)$. So, by Remark 2.2, we have

$$\alpha(\rho) = \inf_{\overline{(\Psi^{-1}(]-\infty,\rho[))}_w} \Phi.$$

Now, (a) follows at once from Theorem 2.1 (part (a)) since any global minimum of the restriction of $\Phi + \lambda\Psi$ to $\Psi^{-1}(]-\infty, \rho[)$ is, by the continuity of Ψ, a local minimum of $\Phi + \lambda\Psi$ in the strong topology. Let $\rho > \inf_X \Psi$ and let $\lambda > \varphi(\rho)$ be such that the functional $\Phi + \lambda\Psi$ satisfies the Palais-Smale condition. Then, by (a), the functional $\Phi + \lambda\Psi$ has a local minimum, say x_0, which lies in $\Psi^{-1}(]-\infty, \rho[)$. Assume that there is no strict global minimum of $\Phi + \lambda\Psi$ lying in $\Psi^{-1}(]-\infty, \rho[)$. Consequently, there is some $x_1 \in X$, with $x_1 \neq x_0$, such that

$$\Phi(x_1) + \mu\Psi(x_1) \leq \Phi(x_0) + \lambda\Psi(x_0).$$

Next, choose $s \in]0, \|x_0 - x_1\|[$ so that

$$\Phi(x_0) + \lambda\Psi(x_0) \leq \Phi(x) + \lambda\Psi(x)$$

for all $x \in X$ satisfying $\|x - x_0\| \leq s$. Clearly, we have

$$\max\{\Phi(x_0) + \lambda\Psi(x_0), \Phi(x_1) + \lambda\Psi(x_1)\} \leq \inf_{\frac{s}{2}<\|x-x_0\|<s} (\Phi(x) + \lambda\Psi(x)).$$

As a consequence, we are allowed to apply Theorem 1 of [**156**] by virtue of which the functional $\Phi + \lambda\Psi$ has a critical point distinct from x_0 and x_1. Since, of course, x_0 is another critical point of $\Phi + \lambda\Psi$, the conclusion follows.

Assume $\gamma < +\infty$, and let $\lambda > \gamma$. Then, by Theorem 2.1 (part (b)), if $\Phi + \lambda\Psi$ has no global minima, there exists a sequence $\{x_n\}$ of local minima of $\Phi + \lambda\Psi$ in the strong topology such that $\lim_{n\to\infty} \Psi(x_n) = +\infty$, and so (c) follows.

Finally, assume $\delta < +\infty$, and let $\lambda > \delta$. Assume that no global minimum of Ψ is a local minimum of $\Phi + \lambda\Psi$. By Theorem 2.1 (part (c)), there exists a sequence $\{x_n\}$, of local minima of $\Phi + \lambda\Psi$ in the strong topology, with $\lim_{n\to\infty} \Psi(x_n) = \inf_X \Psi$, which weakly converges to a global minimum of Ψ. But then, since X equipped with the weak topology is Hausdorff, there is a subsequence whose terms are pairwise distinct, and we are done. □

It is also worth noticing the following corollary of Theorem 2.4.

THEOREM 2.5. *Let the assumptions of Theorem 2.4 be satisfied. Assume also that* $\delta < +\infty$.

Then, either the system

$$\begin{cases} \Phi'(x) = 0 \\ \Psi'(x) = 0 \end{cases}$$

has at least one solution, or, for each $\lambda > \delta$, *there exists a sequence of pairwise distinct critical points of* $\Phi + \lambda\Psi$ *which weakly converges to a global minimum of* Ψ. *In particular, the first case of the alternative occurs when there exists a Hausdorff vector topology on* X^* *with respect to which the operators* Φ' *and* Ψ' *are sequentially weakly continuous.*

PROOF. Assume that the system

$$\begin{cases} \Phi'(x) = 0 \\ \Psi'(x) = 0 \end{cases}$$

has no solutions. Let $\lambda > \delta$. Then, of course, no global minimum of Ψ can also be a local minimum of $\Phi + \lambda\Psi$. Consequently, by Theorem 2.4 (part (*d*)), there is a sequence $\{x_n\}$ of pairwise distinct critical points of $\Phi + \lambda\Psi$ which weakly converges to a global minimum of Ψ, say x^*. So, we have

$$\Phi'(x_n) + \lambda\Psi'(x_n) = 0 \tag{2.4}$$

for all $n \in \mathbf{N}$. Finally, observe that there is no Hausdorff vector topology on X^* with respect to which the operators Φ' and Ψ' are sequentially weakly continuous, since, otherwise, passing to the limit in (2.4), we would have

$$\Phi'(x^*) + \lambda\Psi'(x^*) = 0$$

from which

$$\begin{cases} \Phi'(x^*) = 0 \\ \Psi'(x^*) = 0, \end{cases}$$

against our assumption. □

We now point out another corollary of Theorem 2.4 concerning fixed points of potential operators in Hilbert spaces.

THEOREM 2.6. *Let* X *be a real Hilbert space, and let* $A : X \to X$ *be a potential operator, with sequentially weakly upper semicontinuous potential* P. *For each* $\rho > 0$, *put*

$$\varphi(\rho) = \inf_{\|x\|^2 < \rho} \frac{\sup_{\|y\|^2 \le \rho} P(y) - P(x)}{\rho - \|x\|^2}.$$

Furthermore, set

$$\gamma = \liminf_{\rho \to +\infty} \varphi(\rho)$$

and

$$\delta = \liminf_{\rho \to 0^+} \varphi(\rho).$$

Then, the following conclusions hold:

(a) *If there is $\rho > 0$ such that $\varphi(\rho) < \frac{1}{2}$, then the operator A has a fixed point whose norm is less than $\sqrt{\rho}$.*

(b) *If the functional $x \to \frac{1}{2}\|x\|^2 - P(x)$ satisfies the Palais-Smale condition and if there is $\rho > 0$ such that $\varphi(\rho) < \frac{1}{2}$, then the following alternative holds: either the functional $x \to \frac{1}{2}\|x\|^2 - P(x)$ has a strict global minimum whose norm is less than $\sqrt{\rho}$, or the operator A has at least two fixed points one of which has norm less than $\sqrt{\rho}$.*

(c) *If $\gamma < \frac{1}{2}$, then the following alternative holds: either the functional $x \to \frac{1}{2}\|x\|^2 - P(x)$ has a global minimum, or the set of all fixed points of A is unbounded.*

(d) *If $\delta < \frac{1}{2}$, then the following alternative holds: either 0 is a local minimum of the functional $x \to \frac{1}{2}\|x\|^2 - P(x)$, or there exists a sequence of pairwise distinct fixed points of A which weakly converges to 0.*

PROOF. Apply Theorem 2.4 taking $\Phi(x) = -P(x)$ and $\Psi(x) = \|x\|^2$ for all $x \in X$, and observe that $\Psi'(x) = 2x$. □

Here is a corollary of Theorem 2.6.

THEOREM 2.7. *Let X be a real Hilbert space, and let $A : X \to X$ be a potential operator, with sequentially weakly upper semicontinuous potential P satisfying*

$$\liminf_{r \to +\infty} \frac{\sup_{\|x\| \le r} P(x)}{r^2} < \frac{1}{2} < \limsup_{r \to +\infty} \frac{\sup_{\|x\| \le r} P(x)}{r^2}.$$

Then, the set of all fixed points of A is unbounded.

PROOF. We have

$$\begin{aligned}\liminf_{\rho \to +\infty} \inf_{\|x\|^2 < \rho} \frac{\sup_{\|y\|^2 \le \rho} P(y) - P(x)}{\rho - \|x\|^2} &\le \liminf_{\rho \to +\infty} \frac{\sup_{\|y\|^2 \le \rho} P(y) - P(0)}{\rho} \\ &= \liminf_{r \to +\infty} \frac{\sup_{\|x\| \le r} P(x)}{r^2} < \frac{1}{2}.\end{aligned}$$

On the other hand, since $\limsup_{r \to +\infty} \frac{\sup_{\|x\| \le r} P(x)}{r^2} > \frac{1}{2}$, there are a number $c > \frac{1}{2}$ and two sequences $\{r_n\}$, $\{x_n\}$, with $\lim_{n \to \infty} r_n = +\infty$ and $\|x_n\| \le r_n$ for all $n \in \mathbf{N}$, such that

$$P(x_n) \ge c r_n^2$$

for all $n \in \mathbf{N}$. Consequently, we have

$$P(x_n) - \frac{1}{2}\|x_n\|^2 \geq \left(c - \frac{1}{2}\right) r_n^2$$

for all $n \in \mathbf{N}$. From this, we then infer that the functional $x \to \frac{1}{2}\|x\|^2 - P(x)$ is unbounded below, and so it has no global minima. Now, the conclusion follows directly from (c) of Theorem 2.6. □

3. A development of the previous theory

In the present section, we essentially develop the theory built in Section 2 taking into account the number of connected components of the set $\overline{\Psi^{-1}(]-\infty,\rho[)}$.

If (X,τ) is a topological space, for any $\Psi : X \to]-\infty,+\infty]$, with $\mathrm{dom}(\Psi) := \{x \in X : \Psi(x) < +\infty\} \neq \emptyset$, we denote by τ_Ψ the smallest topology on X which contains both τ and the family of sets $\{\Psi^{-1}(]-\infty,r[)\}_{r\in\mathbf{R}}$.

Our main abstract result is as follows.

THEOREM 3.1. *Let (X,τ) be a Hausdorff topological space, and $\Phi,\Psi : X \to]-\infty,+\infty]$ two functions. Assume that there is $\rho > \inf_X \Psi$ such that the set $\overline{\Psi^{-1}(]-\infty,\rho[)}$ is compact and sequentially compact, has at least k connected components and each of them intersects the interior of* $\mathrm{dom}(\Phi)$. *Moreover, suppose that the function Φ is bounded below in $\overline{\Psi^{-1}(]-\infty,\rho[)}$ and that the function $\Psi+\lambda\Phi$ is sequentially lower semicontinuous for each $\lambda > 0$ small enough. Then, there exists $\lambda^* > 0$ such that, for each $\lambda \in]0,\lambda^*[$, the function $\Psi+\lambda\Phi$ has at least k τ_Ψ-local minima lying in* $\mathrm{dom}(\Phi)\cap\Psi^{-1}(]-\infty,\rho[)$.

PROOF. Denote by $\mathscr{C}$ the family of all connected components of $\overline{\Psi^{-1}(]-\infty,\rho[)}$. Note that these sets are closed in X since they are closed in $\overline{\Psi^{-1}(]-\infty,\rho[)}$ which is, in turn, closed in X. We now observe that there are k pairwise disjoint closed non-empty sets $C_1,\ldots,C_k$ such that

$$\overline{\Psi^{-1}(]-\infty,\rho[)} = \bigcup_{i=1}^{k} C_i.$$

We distinguish two cases. First, assume that $\mathscr{C}$ is finite. Let h be its cardinality. Let $B_1,..,B_h$ be the members of $\mathscr{C}$. Then, if we choose $C_i = B_i$ for $i = 1,\ldots,k-1$ and $C_k = \cup_{i=k}^{h} B_i$, we are clearly done. Now, assume that $\mathscr{C}$ is infinite. In this case, we prove our claim by induction. The claim is true, of course, if $k = 1$. Assume that it is true if $k = p$. So, we are assuming that there are p pairwise disjoint closed non-empty sets $D_1,\ldots,D_p$, such that

$$\overline{\Psi^{-1}(]-\infty,\rho[)} = \bigcup_{i=1}^{p} D_i.$$

Notice that at least one of the sets D_i must be disconnected, since, otherwise, we would have $\{D_1,\ldots,D_p\} = \mathscr{C}$, contrary to the assumption that $\mathscr{C}$ is infinite. Then, if D_{i^*} is disconnected, there are two disjoint closed non-empty sets E_1,E_2 such that $D_{i^*} = E_1 \cup E_2$. So, $D_1,\ldots,D_{i^*-1},D_{i^*+1},\ldots,D_p,E_1,E_2$ are $p+1$ pairwise disjoint closed non-empty sets whose union is $\overline{\Psi^{-1}(]-\infty,\rho[)}$. So, our claim is true for $k = p+1$, and hence, by induction, for any k.

Now, fix i $(1 \leq i \leq k)$. By compactness and Hausdorffness, it is clear that there exists an open set $A_i \subset X$ such that $C_i \subset A_i$ and $A_i \cap \cup_{j=1,j\neq i}^{k} C_j = \emptyset$. Furthermore, it is easily seen that, if we put

$$G_i = \{x \in A_i : \Psi(x) < \rho\},$$

we have

$$\overline{G_i} = C_i.$$

Clearly, C_i contains at least one member of $\mathscr{C}$, and hence, by assumption, int(dom $(\Phi)) \cap C_i \neq \emptyset$. This implies that $\mathrm{dom}(\Phi) \cap G_i \neq \emptyset$. Taken into account that, by assumption, $\inf_{C_i} \Phi$ is finite, put

$$\mu_i = \inf_{x \in \mathrm{dom}(\Phi) \cap G_i} \frac{\Phi(x) - \inf_{C_i} \Phi}{\rho - \Psi(x)}.$$

Let $\lambda' > 0$ be such that $\Psi + \lambda\Phi$ is sequentially lower semicontinuous for each $\lambda \in]0, \lambda']$. Fix $\mu > \max\{\mu_i, \frac{1}{\lambda'}\}$. Then, there exists $y \in \mathrm{dom}(\Phi) \cap G_i$ such that

$$\mu\rho > \mu\Psi(y) + \Phi(y) - \inf_{C_i} \Phi.$$

Moreover, since C_i is sequentially compact, there exists $x_i^* \in \mathrm{dom}(\Phi) \cap C_i$ such

$$\Phi(x_i^*) + \mu\Psi(x_i^*) \leq \Phi(x) + \mu\Psi(x)$$

for all $x \in C_i$. We claim that $x_i^* \in G_i$. Arguing by contradiction, assume that $\Psi(x_i^*) \geq \rho$. We then have

$$\Phi(x_i^*) + \mu\Psi(x_i^*) \geq \Phi(x_i^*) + \mu\rho > \Phi(x_i^*) + \Phi(y) + \mu\Psi(y) - \inf_{C_i} \Phi \geq \Phi(y) + \mu\Psi(y)$$

which is absurd. Now, let i vary. Put $\mu^* = \max\{\mu_1, \ldots, \mu_k, \frac{1}{\lambda'}\}$. Clearly, each set G_i is τ_Ψ-open, and hence each x_i^* is a τ_Ψ-local minimum of $\Phi + \mu\Psi$ for all $\mu > \mu^*$. Consequently, the points $x_1^*, \ldots, x_k^*$ satisfy the conclusion, taking $\lambda^* = \frac{1}{\mu^*}$, and the proof is complete. □

The next result provides a reasonable way to check the key assumption of Theorem 3.1.

PROPOSITION 3.2. *Let X be a Hausdorff topological space and $\Psi : X \to]-\infty, +\infty]$ a sequentially lower semicontinuous function. Assume that there is $r > \inf_X \Psi$ such that the set $\overline{\Psi^{-1}(]-\infty, r[)}$ is compact and first-countable. Moreover, assume that the set of all global minima of Ψ has at least k connected components.*

Then, there exists $\rho^* \in]\inf_X \Psi, r]$ such that, for each $\rho \in]\inf_X \Psi, \rho^*]$, the set $\overline{\Psi^{-1}(]-\infty, \rho[)}$ has at least k connected components.

PROOF. Arguing by contradiction, assume that there is a decreasing sequence $\{\rho_n\}$ in $]\inf_X \Psi, r[$, coverging to $\inf_X \Psi$, such that for each $n \in \mathbf{N}$, the set

$\overline{\Psi^{-1}(-\infty, \rho_n[)}$ has at most $k-1$ connected components. Clearly, we have

$$\Psi^{-1}(\inf_X \Psi) = \bigcap_{n=1}^{\infty} \overline{\Psi^{-1}(]-\infty, \rho_n[)}.$$

Reasoning as in the proof of Thoerem 3.1, we find k open and pairwise disjoint subsets of X, $\Omega_1, \ldots, \Omega_k$, such that $\Omega_i \cap \Psi^{-1}(\inf_X \Psi) \neq \emptyset$ for all $i = 1, \ldots, k$ and $\Psi^{-1}(\inf_X \Psi) \subseteq \bigcup_{i=1}^k \Omega_i$. Then, for each $n \in \mathbf{N}$, the set $\overline{\Psi^{-1}(]-\infty, \rho_n[)}$ cannot be contained in $\bigcup_{i=1}^k \Omega_i$, since, otherwise, it would have at least k connected components. So, $\{\overline{\Psi^{-1}(]-\infty, \rho_n[)} \cap (X \setminus \bigcup_{i=1}^k \Omega_i)\}$ is a non-increasing sequence of non-empty closed subsets of a compact one, and hence $\bigcap_{n=1}^{\infty} \overline{\Psi^{-1}(]-\infty, \rho_n[)} \cap (X \setminus \bigcup_{i=1}^k \Omega_i) \neq \emptyset$ which is absurd. □

So, putting Theorem 3.1 and Proposition 3.2 together, we clearly get

THEOREM 3.3. *Let (X, τ) be a Hausdorff topological space and $\Phi : X \to \mathbf{R}$, $\Psi : X \to]-\infty, +\infty]$ two functions. Assume that there is $r > \inf_X \Psi$ such that the set $\overline{\Psi^{-1}(]-\infty, r[)}$ is compact and first-countable. Moreover, suppose that the function Φ is bounded below in $\overline{\Psi^{-1}(]-\infty, r[)}$ and that the function $\Psi + \lambda\Phi$ is sequentially lower semicontinuous for each $\lambda \geq 0$ small enough. Finally, assume that the set of all global minima of Ψ has at least k connected components.*

Then, there exists $\lambda^* > 0$ such that, for each $\lambda \in]0, \lambda^*[$, the function $\Psi + \lambda\Phi$ has at least k τ_Ψ-local minima lying in $\Psi^{-1}(]-\infty, r[)$.

Arguing by contradiction, the use of Theorems 3.1 and 3.3 gives the following

THEOREM 3.4. *Let (X, τ) be a Hausdorff topological space and $\Psi : X \to]-\infty, +\infty]$ a sequentially lower semicontinuous function such that, for some $r > \inf_X \Psi$, the set $\overline{\Psi^{-1}(]-\infty, r[)}$ is compact and first-countable. Suppose that there are a function $\Phi : X \to \mathbf{R}$, bounded below in $\overline{\Psi^{-1}(]-\infty, r[)}$, and a sequence $\{\mu_n\}$ in $\mathbf{R}^+$ converging to 0 such that, for each $\lambda > 0$ small enough, the function $\Psi + \lambda\Phi$ is sequentially lower semicontinuous, and, for each $n \in \mathbf{N}$, the function $\Psi + \mu_n\Phi$ has at most k τ_Ψ-local minima lying in $\Psi^{-1}(]-\infty, r[)$.*

Then, for every $\rho \in]\inf_X \Psi, r]$, the sets $\overline{\Psi^{-1}(]-\infty, \rho[)}$ and $\Psi^{-1}(\inf_X \Psi)$ have at most k connected components. So, in particular, these sets are connected when $k = 1$.

REMARK 3.5. When $k = 1$, Theorem 3.4 ensures that, for every $\rho \in]\inf_X \Psi, r[$, the set $\Psi^{-1}(]-\infty, \rho])$ is connected. This follows from the equality

$$\Psi^{-1}(]-\infty, \rho]) = \bigcap_{\rho < s < r} \overline{\Psi^{-1}(]-\infty, s[)}$$

and from the fact that, for every $s \in]\inf_X \Psi, r]$, the set $\overline{\Psi^{-1}(]-\infty, s[)}$ is connected and compact.

An interesting consequence of Theorem 3.4 is the following two local minima result.

THEOREM 3.6. *Let (X, τ) be a Hausdorff topological space, and $\Phi : X \to \mathbf{R}$, $\Psi : X \to]-\infty, +\infty]$ two functions. Assume that there is $r > \inf_X \Psi$ such that the* *set* $\overline{\Psi^{-1}(]-\infty, r[)}$ *is compact and first-countable. Moreover, assume that there is a strict local minimum of* Ψ, *say* x_0, *such that* $\inf_X \Psi < \Psi(x_0) < r$. *Finally, suppose that the function* Φ *is bounded below in* $\overline{\Psi^{-1}(]-\infty, r[)}$ *and that the function* $\Psi + \lambda\Phi$ *is sequentially lower semicontinuous for each* $\lambda \geq 0$ *small enough.*

Then, there exists $\lambda^* > 0$ *such that, for each* $\lambda \in]0, \lambda^*[$, *the function* $\Psi + \lambda\Phi$ *has at least two* τ_Ψ*-local minima lying in* $\Psi^{-1}(]-\infty, r[)$.

PROOF. Arguing by contradiction, assume that the conclusion does not hold. Then, by Theorem 3.4 and Remark 3.5, the set $\Psi^{-1}(]-\infty, \Psi(x_0)])$ is connected. But this set contains x_0 as an isolated point (since x_0 is a strict local minimum of Ψ) and does not reduce to it (since $\inf_X \Psi < \Psi(x_0)$), against connectedness. □

With the aim to apply them to nonlinear differential equations, we now establish some consequences of the previous general results in the setting of reflexive real Banach spaces.

THEOREM 3.7. *Let* X *be a sequentially weakly closed subset of a reflexive real Banach space* E, *and* $\Phi : X \to \mathbf{R}$, $\Psi : X \to]-\infty, +\infty]$ *two functionals. Assume that there is* $\rho > \inf_X \Psi$ *such that the set* $(\overline{\Psi^{-1}(]-\infty, \rho[)})_w$ *is bounded and has at least* k *weakly connected components. Moreover, suppose that the functional* Φ *is bounded below in* $(\overline{\Psi^{-1}(]-\infty, \rho[)})_w$ *and that the functional* $\Psi + \lambda\Phi$ *is sequentially weakly lower semicontinuous for each* $\lambda > 0$ *small enough.*

Then, there exists $\lambda^* > 0$ *such that, for each* $\lambda \in]0, \lambda^*[$, *the functional* $\Psi + \lambda\Phi$ *has at least* k τ_Ψ*-local minima lying in* $\Psi^{-1}(]-\infty, \rho[)$, *where* τ *is the relative weak topology of* X.

PROOF. Apply Theorem 3.1, τ just being the relative weak topology. In particular, observe that, since E is reflexive, the weak closure of any bounded set is weakly compact. Moreover, it is sequentially weakly compact by the Eberlein-Smulyan theorem. □

Analogously, taking into account that any weakly compact set in a separable Banach space is metrizable with respect to the weak topology, from Theorem 3.3, we get

THEOREM 3.8. *Let* X *be a sequentially weakly closed subset of a reflexive and separable real Banach space* E, *and* $\Phi : X \to \mathbf{R}$, $\Psi : X \to]-\infty, +\infty]$ *two functionals. Assume that there is* $\rho > \inf_X \Psi$ *such that the set* $\Psi^{-1}(]-\infty, \rho[)$ *is bounded. Moreover, suppose that the functional* Φ *is bounded below in* $(\overline{\Psi^{-1}(]-\infty, \rho[)})_w$ *and that the functional* $\Psi + \lambda\Phi$ *is sequentially weakly lower semicontinuous for each* $\lambda \geq 0$ *small enough. Finally, assume that the set* $\Psi^{-1}(\inf_X \Psi)$ *has at least* k *weakly connected components.*

Then, the conclusion of Theorem 3.7 holds.

Arguing by contradiction, from Theorems 3.7 and 3.8 we then get

THEOREM 3.9. *Let* X *be a sequentially weakly closed subset of a reflexive and separable real Banach space, and* $\Psi : X \to]-\infty, +\infty]$ *a sequentially weakly lower semicontinuous functional such that, for some* $r > \inf_X \Psi$, *the set* $\Psi^{-1}(]-\infty, r[)$ *is bounded.*

Suppose that there are a functional $\Phi : X \to \mathbf{R}$*, bounded below in* $\overline{(\Psi^{-1}(]-\infty,r[))}_w$*, and a sequence* $\{\mu_n\}$ *in* $\mathbf{R}^+$ *converging to* 0 *such that, for each* $\lambda > 0$ *small enough, the functional* $\Psi + \lambda\Phi$ *is sequentially weakly lower semicontinuous, and, for each* $n \in \mathbf{N}$*, the functional* $\Psi + \mu_n\Phi$ *has at most* k τ_Ψ*-local minima lying in* $\Psi^{-1}(]-\infty,r[)$*, where* τ *is relative weak topology of* X

Then, for every $\rho \in]\inf_X \Psi, r]$*, the sets* $\overline{(\Psi^{-1}(]-\infty,\rho[))}_w$ *and* $\Psi^{-1}(\inf_X \Psi)$ *have at most k weakly connected components. So, in particular, these sets are weakly connected when* $k = 1$*.*

REMARK 3.10. In Theorem 3.9, for the part concerning $\overline{(\Psi^{-1}(]-\infty,\rho[))}_w$, the separability of the Banach space is superfluous.

The next result is an application of Theorems 3.7 and 3.8 to critical point theory.

THEOREM 3.11. *In addition to the assumptions of either Theorem 3.7 or Theorem 3.8, suppose that* $X = E$*, that the functionals* $\Psi, \Phi : X \to \mathbf{R}$ *are continuously Gâteaux differentiable, and that* $k \geq 2$*.*

Then, there exists $\lambda^* > 0$ *such that, for each* $\lambda \in]0, \lambda^*[$ *for which the functional* $\Psi + \lambda\Phi$ *satisfies the Palais-Smale condition, the same functional has at least* $k+1$ *critical points,* k *of which are lying in* $\Psi^{-1}(]-\infty,\rho[)$*.*

PROOF. By either Theorem 3.7 or Theorem 3.8, there exists $\lambda^* > 0$ such that, for each $\lambda \in]0, \lambda^*[$, the functional $\Psi + \lambda\Phi$ has at least k τ_Ψ-local minima lying in $\Psi^{-1}(]-\infty,\rho[)$, where τ is the weak topology of X. Note that Ψ, being C^1, is (norm) continuous. Consequently, the topology τ_Ψ is weaker than the strong topology, and so the above mentioned τ_Ψ-local minima of $\Psi + \lambda\Phi$ are local minima of this functional in the strong topology. Now, assuming that $\Psi + \lambda\Phi$ satisfies the Palais-Smale condition, the conclusion follows from Theorem (1.ter) of [**113**]. □

From Theorem 3.11, arguing by contradiction, we then obtain the following

THEOREM 3.12. *Let* X *be a reflexive and separable real Banach space and* $\Psi : X \to \mathbf{R}$ *a continuously Gâteaux differentiable and sequentially weakly lower semicontinuous functional such that, for some* $r > \inf_X \Psi$*, the set* $\Psi^{-1}(]-\infty,r[)$ *is bounded. Let* $k \in \mathbf{N}$ *with* $k \geq 2$*.*

Suppose that there are a continuously Gâteaux differentiable functional $\Phi : X \to \mathbf{R}$*, which is bounded below in* $\overline{(\Psi^{-1}(]-\infty,r[))}_w$*, and a sequence* $\{\mu_n\}$ *in* $\mathbf{R}^+$ *converging to* 0 *such that the functional* $\Psi + \lambda\Phi$ *is sequentially weakly lower semicontinuous for each* $\lambda > 0$ *small enough and, for each* $n \in \mathbf{N}$*, the functional* $\Psi + \mu_n\Phi$ *satisfies the Palais-Smale condition and has at most* k *critical points in* X*.*

Then, for every $\rho \in]\inf_X \Psi, r]$*, the sets* $\overline{(\Psi^{-1}(]-\infty,\rho[))}_w$ *and* $\Psi^{-1}(\inf_X \Psi)$ *have at most* $k-1$ *weakly connected components. So, in particular, these sets are weakly connected when* $k = 2$*.*

REMARK 3.13. When $k = 2$, Theorem 3.12 ensures that, for every $\rho \in]\inf_X \Psi, r[$, the set $\Psi^{-1}(]-\infty,\rho])$ is weakly connected (see Remark 3.5).

We now establish

THEOREM 3.14. *Let* X *be a reflexive and separable real Banach space, let* $\Psi : X \to \mathbf{R}$ *be a functional belonging to* $\mathscr{W}_X$ *and let* $J : X \to \mathbf{R}$ *be a sequentially weakly lower semicontinuous functional. Assume that*

$$\lim_{\|x\|\to+\infty} (\Psi(x) + J(x)) = +\infty. \tag{3.1}$$

Then, any strict local minimum of the functional $\Psi + J$ in the strong topology is so in the weak topology.

PROOF. So, let $x_0 \in X$. Assume that x_0 is not a strict local minimum of $\Psi + J$ in the weak topology. We want to prove that x_0 is not a strict local minimum of $\Psi + J$ in the strong topology. By (3.1), there is $\rho > 0$ such that

$$\Psi(x_0) + J(x_0) < \Psi(x) + J(x) \tag{3.2}$$

for all $x \in X$ satisfying $\|x\| > \rho$. Set

$$B = \{x \in X : \|x\| \leq \rho\} .$$

Since X is separable and reflexive, the set B, endowed with the relative weak topology, is metrizable. Let σ be a metric on B inducing such a topology. For each $n \in \mathbf{N}$, set

$$B_n = \left\{x \in B : \sigma(x, x_0) < \frac{1}{n}\right\}$$

and fix a weakly open set $V_n \subseteq X$ such that

$$B_n = V_n \cap B.$$

By assumption, for each $n \in \mathbf{N}$, there exists $x_n \in V_n \setminus \{x_0\}$ such that

$$\Psi(x_n) + J(x_n) \leq \Psi(x_0) + J(x_0). \tag{3.3}$$

Due to (3.2), we clearly have $x_n \in B_n$. Hence, the sequence $\{x_n\}$ weakly converges to x_0. From (3.3), recalling that J is sequentially weakly lower semicontinuous, we get

$$\begin{aligned}\liminf_{n\to\infty} \Psi(x_n) + J(x_0) &\leq \liminf_{n\to\infty} \Psi(x_n) + \liminf_{n\to\infty} J(x_n)\\ &\leq \liminf_{n\to\infty}(\Psi(x_n) + J(x_n)) \leq \Psi(x_0) + J(x_0)\end{aligned}$$

and so

$$\liminf_{n\to\infty} \Psi(x_n) \leq \Psi(x_0).$$

Then, since $\Psi \in \mathscr{W}_X$, the sequence $\{x_n\}$ has a subsequence strongly converging to x_0, and so, by (3.3), x_0 is not a strict local minimum of $\Psi + J$ in the strong topology of X. $\square$

Here is an application of Theorem 3.6.

THEOREM 3.15. *Let X be a uniformly convex and separable real Banach space, $g : [0,+\infty[\to \mathbf{R}$ a strictly increasing continuous function, and $J : X \to \mathbf{R}$ a sequentially weakly lower semicontinuous functional. For every $x \in X$, put*

$$\Psi(x) = g(\|x\|) + J(x).$$

Assume that the functional Ψ is coercive and has a strict, not global, local minimum, say x_0.

Then, for every $r > \Psi(x_0)$ and every functional $\Phi : X \to \mathbf{R}$ which is bounded below in $\overline{(\Psi^{-1}(]-\infty, r[))}_w$ and such that $\Psi + \lambda\Phi$ is sequentially weakly lower semicontinuous for each $\lambda > 0$ small enough, there exists $\lambda^* > 0$ such that, for each $\lambda \in]0, \lambda^*[$, the functional $\Psi + \lambda\Phi$ has at least two τ_Ψ-local minima lying in $\Psi^{-1}(]-\infty, r[)$, where τ is the weak topology of X.

PROOF. As we already observed, the functional $x \to g(\|x\|)$ belongs to $\mathscr{W}_X$. So, from Theorem 3.14, it follows that x_0 is a τ-strict local minimum of Ψ. Now, the conclusion follows directly from Theorem 3.6, taking as τ just the weak topology. □

We conclude this section with an application of Theorem 3.9 in the setting of Hilbert spaces.

THEOREM 3.16. *Let X be a separable real Hilbert space and $J : X \to \mathbf{R}$ a continuous, Gâteaux differentiable and sequentially weakly upper semicontinuous functional. For every $x \in X$, put*

$$\Psi(x) = \frac{1}{2}\|x\|^2 - J(x).$$

Assume that, for some $r > \inf_X \Psi$, the set $\Psi^{-1}(]-\infty, r[)$ is bounded. Moreover, suppose that the restriction of J' to $\Psi^{-1}(]-\infty, r[)$ is nonexpansive.

Then, for every $\rho \in]\inf_X \Psi, r[$, the sets $\Psi^{-1}(]-\infty, \rho])$ and $\Psi^{-1}(\inf_X \Psi)$ are weakly connected.

PROOF. Let us apply Theorem 3.9 taking as $\{\mu_n\}$ any sequence in $]0, +\infty[$ coverging to zero, and $\Phi(x) = \frac{1}{2}\|x\|^2$. Thus, Φ and Ψ are two continuous, Gâteaux differentiable and sequentially weakly lower semicontinuous functionals, and, for each $n \in \mathbf{N}$, the functional $\Psi + \mu_n\Phi$ admits at most one τ_Ψ-local minimum in $\Psi^{-1}(]-\infty, r[)$. This follows from the fact that the strong topology is stronger than τ_Ψ and that the retriction of $\frac{1}{1+\mu_n}J'$ to $\Psi^{-1}(]-\infty, r[)$ is a contraction, and so the equation $\Psi'(x) + \mu_n\Phi'(x) = 0$ has at most one solution in $\Psi^{-1}(]-\infty, r[)$. Hence, the hypotheses of Theorem 3.9 are satisfied, and the conclusion follows from it and from Remark 3.5. □

The following proposition is an useful complement to both Theorems 3.12 and 3.16.

PROPOSITION 3.17. *Let X be a real Hilbert space and $J : X \to \mathbf{R}$ a continuously Gâteaux differentiable functional whose derivative is compact. For every $x \in X$, put*

$$\Psi(x) = \frac{1}{2}\|x\|^2 - J(x).$$

Assume that, for some $r > \inf_X \Psi$, *the set* $\Psi^{-1}(]-\infty, r[)$ *is bounded.*

Then, the set $\Psi^{-1}(\inf_X \Psi)$ *is compact.*

PROOF. The functional Ψ is Gâteaux differentiable and its critical points are exactly the fixed points of J'. Let B be a closed ball in X containing $\Psi^{-1}(]-\infty, r[)$. Let $\{x_n\}$ be a sequence of fixed points of J' lying in B. Since this sequence is bounded and J' is compact, there is a subsequence $\{x_{n_k}\}$ such that $J'(x_{n_k})$ converges to some $z \in B$. Clearly, by continuity, z is a fixed point of J'. So, the set $\{x \in B : J'(x) = x\}$ is compact. Of course, it contains $\Psi^{-1}(\inf_X \Psi)$ which is closed since Ψ is continuous, and the conclusion follows. □

REMARK 3.18. Note that when we can apply Theorem 3.9 with $k = 1$ to a functional Ψ as in Proposition 3.17, then the set $\Psi^{-1}(\inf_X \Psi)$ is connected. This follows from the fact in any compact subset of a Banach space the relative strong and weak topologies coincide. This remark, in particular, applies to Theorem 3.12 (when $k = 2$) and to Theorem 3.16.

4. A further variational principle

The main result of this section reads as follows:

THEOREM 4.1. *Let* (X, τ) *be a topological space,* τ_1 *a topology on* X *weaker than* τ_s *and* $\Phi, \Psi : X \to \mathbf{R}$ *two functions such that, for each* $\lambda > 0$ *small enough, the function* $\Psi + \lambda\Phi$ *is* τ_1*-lower semicontinuous. Assume also that* $\inf_X \Psi > -\infty$ *and that there exists* $\rho > \inf_X \Psi$ *such that* $\Psi^{-1}(]-\infty, \rho])$ *is sequentially* τ*-compact,* $\inf_{\Psi^{-1}(]-\infty,\rho])} \Phi > -\infty$ *and there is no* τ_1*-local minimum of* Ψ *lying in* $\Psi^{-1}(\rho)$.

Then, for each $\delta > 0$, *there exists* $\lambda^* \in]0, \delta]$ *such that function* $\Psi + \lambda^*\Phi$ *has a* τ_1*-local minimum lying in* $\Psi^{-1}(]-\infty, \rho])$.

To prove Theorem 4.1, we need some other results that we are going to establish.

PROPOSITION 4.2. *Let* X, Λ *be two topological spaces and* $f : X \times \Lambda \to \mathbf{R}$ *a function such that, for each* $x \in X$, *the set* $\{\lambda \in \Lambda : f(x, \lambda) \geq 0\}$ *is closed. For each* $\lambda \in \Lambda$, *put*

$$F(\lambda) = \{x \in X : f(x, \lambda) \leq 0\}.$$

Assume that the multifunction F *is not lower semicontinuous at* $\lambda_0 \in \Lambda$.

Then, each point $x_0 \in NLS(F(\lambda_0))$ *is a local minimum of the function* $f(\cdot, \lambda_0)$ *such that* $f(x_0, \lambda_0) = 0$.

PROOF. Let $x_0 \in NLS(F(\lambda_0))$. So, there are a neighbourhood Ω of x_0 in X and a net $\{\mu_\alpha\}_{\alpha \in D}$ in $\Lambda \setminus F^-(\Omega)$ converging to λ_0. Thus, we have $f(x_0, \lambda_0) \leq 0$ as well as $f(x, \mu_\alpha) > 0$ for all $x \in \Omega$, $\alpha \in D$. Since each set $\{\lambda \in \Lambda : f(x, \lambda) \geq 0\}$ is closed, we then have

$$f(x_0, \lambda_0) = 0 \leq f(x, \lambda_0)$$

for all $x \in \Omega$, and the proof is complete. □

The next result is crucial.

THEOREM 4.3. *Let (X,τ) be a topological space, τ_1 a topology on X weaker than τ_s, I a real interval, and $F : I \to 2^X$ a multifunction, with τ_1-closed values, such that, for each $x \in X$, the set $I \setminus F^-(x)$ is a non-empty interval open in I. Finally, suppose that there is $\lambda_0 \in I$ such that $F(\lambda_0)$ is non-empty and sequentially τ-compact.*

Then, the multifunction F is not τ_1-lower semicontinuous at a point $\lambda^* \in I$ such that $NLS_{\tau_1}(F(\lambda^*)) \cap F(\lambda_0) \neq \emptyset$.

PROOF. For each $x \in X$, put

$$\Phi(x) = I \setminus F^-(x) .$$

Note that

$$\Phi^-(\lambda) = X \setminus F(\lambda)$$

for all $\lambda \in I$. Since τ_1 is weaker than τ_s, each set $X \setminus \Phi^-(\lambda)$ is sequentially τ-closed. Hence, by Proposition 5.5 in the next section, we can find a compact (non-degenerate) interval $[a,b] \subseteq I$, with $\lambda_0 \in [a,b]$, such that

$$\Phi(x) \cap [a,b] \neq \emptyset$$

for all $x \in X$. Therefore, each set $\Phi(x) \cap [a,b]$ is a non-empty interval open in $[a,b]$. For each $x \in X$, put

$$\alpha(x) = \inf(\Phi(x) \cap [a,b])$$

and

$$\beta(x) = \sup(\Phi(x) \cap [a,b]).$$

Let $x_0 \in X$ and $\epsilon \in]0, \beta(x_0) - \alpha(x_0)[$. Observe that the set $\Phi^-(\alpha(x_0)+\epsilon) \cap \Phi^-(\beta(x_0)-\epsilon)$ is a τ_1-neighbourhood of x_0, and for every point x belonging to it, one has

$$\alpha(x) < \alpha(x_0) + \epsilon$$

as well as

$$\beta(x_0) - \epsilon < \beta(x).$$

This shows that α is τ_1-upper semicontinuous and that β is τ_1-lower semicontinuous in X. Now, suppose that $\lambda_0 \in]a,b[$. Then, we have

$$F(\lambda_0) = \alpha^{-1}([\lambda_0, +\infty[) \cup \beta^{-1}(]-\infty, \lambda_0]).$$

Thus, since $F(\lambda_0)$ is non-empty, either $\alpha^{-1}([\lambda_0, +\infty[)$ or $\beta^{-1}(]-\infty, \lambda_0])$ is non-empty. For instance, suppose that $\beta^{-1}(]-\infty, \lambda_0])$ is non-empty. Since τ_1 is weaker than τ_s, the function β is τ_s-lower semicontinuous and so, as we observed above, sequentially τ-lower semicontinuous. Hence, as $F(\lambda_0)$ is sequentially τ-compact, there is $x^* \in X$ such that $\beta(x^*) = \inf_X \beta$. Since $\beta(x^*) \leq \lambda_0$, we have $\beta(x^*) \in]a,b[$. This implies, in particular, that $\beta(x^*)$ does not belong to $\Phi(x^*) \cap [a,b]$, since this set is open in $[a, b]$. Hence, we have $x^* \in F(\beta(x^*))$. Fix $c \in]\alpha(x^*), \beta(x^*)[$. Owing to the τ_1-upper semicontinuity of α, the set $\alpha^{-1}(]-\infty, c[)$ is a τ_1-neighbourhood of x^*. Now, let $x \in \alpha^{-1}(]-\infty, c[)$ and $\lambda \in]c, \beta(x^*)[$. Then, we have

$$\alpha(x) < c < \lambda < \beta(x^*) \leq \beta(x),$$

and so $\lambda \in \Phi(x)$, that is x does not belong to $F(\lambda)$. In other words, we have

$$F(\lambda) \cap \alpha^{-1}(]-\infty, c[) = \emptyset$$

for all $\lambda \in]c, \beta(x^*)[$. This shows that $x^* \in NLS_{\tau_1}(F(\beta(x^*))) \cap F(\lambda_0)$. In the case when $\alpha^{-1}([\lambda_0, +\infty[)$ is non-empty, α has at least one global maximum, say x'. Proceeding as before (with obvious changes), one realizes that $x' \in NLS_{\tau_1}(F(\alpha(x'))) \cap F(\lambda_0)$. Suppose now that $\lambda_0 = b$. If $\inf_X \beta < b$, since

$$\beta^{-1}(]-\infty, b[) \subseteq F(b)$$

and $F(b)$ is sequentially τ-compact, $\inf_X \beta$ is attained by β, and, as we have seen, each global minimum of β belongs to $NLS_{\tau_1}(F(\inf_X \beta)) \cap F(\lambda_0)$. If $\beta(x) = b$ for all $x \in X$, pick $x_0 \in F(b)$. Then, if $d \in]\alpha(x_0), b[$ and $\lambda \in]d, b[$, the set $\alpha^{-1}(]-\infty, d[)$ is a τ_1-neighbourhood of x_0 which does not meet $F(\lambda)$. So, $x_0 \in NLS_{\tau_1}(F(b))$. In other words, $F(b) = NLS_{\tau_1}(F(b))$. Finally, if $\lambda_0 = a$, reasoning as before (with obvious changes), one checks that $\sup_X \alpha$ is attained by α and that each global maximum of α belongs to $NLS_{\tau_1}(F(\sup_X \alpha)) \cap F(a)$. The proof is complete. □

REMARK 4.4. From the above proof, it is clear that if we assume $F(\lambda_0)$ to be τ-compact, then we get the same conclusion provided that τ_1 is weaker than τ.

A joint application of Proposition 4.2 and Theorem 4.3 gives

THEOREM 4.5. *Let (X, τ) be a topological space, τ_1 a topology on X weaker than τ_s, I a real interval and $f : X \times I \to \mathbf{R}$ a function satisfying the following conditions:*
(d_1) *for each $x \in X$, the set*

$$\{\lambda \in I : f(x, \lambda) \geq 0\}$$

is closed in I and the set

$$\{\lambda \in I : f(x, \lambda) > 0\}$$

is a non-empty interval open in I;
(d_2) *for each $\lambda \in I$, the set*

$$\{x \in X : f(x, \lambda) \leq 0\}$$

is τ_1-closed;
(d_3) *there is $\lambda_0 \in I$ such that the set*

$$\{x \in X : f(x, \lambda_0) \leq 0\}$$

is non-empty and sequentially τ-compact.

Then, there exists $\lambda^* \in I$ such that the function $f(\cdot, \lambda^*)$ has a τ_1-local minimum x^* such that $f(x^*, \lambda^*) = 0$ and $f(x^*, \lambda_0) \leq 0$.

PROOF. For each $\lambda \in I$, put

$$F(\lambda) = \{x \in X : f(x, \lambda) \leq 0\}.$$

By Theorem 4.3, the multifunction F is not τ_1-lower semicontinuous at a point $\lambda^* \in I$ such that $NLS_{\tau_1}(F(\lambda^*)) \cap F(\lambda_0) \neq \emptyset$. Any point of this set, by Proposition 4.2, satisfies the conclusion. □

Now, we can give the

PROOF OF THEOREM 4.1. Fix $\delta > 0$ so that $\Psi + \lambda\Phi$ is τ_1-lower semicontinuous for all $\lambda \in]0, \delta]$. Next, fix $r \in \mathbf{R}$ satisfying

$$r < \frac{\inf_X \Psi - \rho + \delta \inf_{\Psi^{-1}(]-\infty,\rho])} \Phi}{\delta}.$$

From this, it clearly follows that

$$\Psi(x) - \rho + \delta(\Phi(x) - r) > 0$$

for all $x \in \Psi^{-1}(] - \infty, \rho])$. Now, put

$$f(x, \lambda) = \Psi(x) - \rho + \lambda(\Phi(x) - r)$$

for all $(x, \lambda) \in X \times [0, \delta]$. At this point, we realize that Theorem 4.5 applies with such a function f. In particular, (d_1) is satisfied noticing that the set $\{\lambda \in [0, \delta] : f(x, \lambda) > 0\}$ contains 0 if $\Psi(x) > \rho$ and δ if $\Psi(x) \leq \rho$. Clearly, (d_3) is satisfied taking $\lambda_0 = 0$. So, the conclusion follows directly from that of Theorem 4.5. □

5. Some three critical points theorems

The aim of this section is to highlight the following three multiplicity results:

THEOREM 5.1. *Let X be a separable and reflexive real Banach space, $I \subseteq \mathbf{R}$ an interval, and $f : X \times I \to \mathbf{R}$ a function satisfying the following conditions:*

(α_1) *for each $x \in X$, the function $f(x, \cdot)$ is continuous and concave;*

(β_1) *for each $\lambda \in I$, the function $f(\cdot, \lambda)$ is sequentially weakly lower semicontinuous and Gâteaux differentiable, and $\lim_{\|x\| \to +\infty} f(x, \lambda) = +\infty$;*

(γ_1) *there exists a continuous concave function $h : I \to \mathbf{R}$ such that*

$$\sup_{\lambda \in I} \inf_{x \in X} (f(x, \lambda) + h(\lambda)) < \inf_{x \in X} \sup_{\lambda \in I} (f(x, \lambda) + h(\lambda)).$$

Then, there exist an open interval $J \subseteq I$ and a positive real number ρ, such that, for each $\lambda \in J$, the equation

$$f'_x(x, \lambda) = 0$$

has at least two solutions in X whose norms are less than ρ. If, in addition, the function f is (strongly) continuous in $X \times I$, and, for each $\lambda \in I$, the function $f(\cdot, \lambda)$ is C^1 and satisfies the Palais-Smale condition, then the above conclusion holds with "three" instead of "two".

THEOREM 5.2. *Let X be a reflexive real Banach space; $I \subseteq \mathbf{R}$ an interval; $\Psi : X \to \mathbf{R}$ a sequentially weakly lower semicontinuous C^1 functional, bounded on each bounded subset of X and whose derivative admits a continuous inverse on X^*; $J : X \to \mathbf{R}$ a C^1 functional with compact derivative. Assume that*

$$\lim_{\|x\|\to+\infty} (\Psi(x) - \lambda J(x)) = +\infty$$

for all $\lambda \in I$, and that there exists $\rho \in \mathbf{R}$ such that

$$\sup_{\lambda\in I} \inf_{x\in X} (\Psi(x) + \lambda(\rho - J(x))) < \inf_{x\in X} \sup_{\lambda\in I}(\Psi(x) + \lambda(\rho - J(x))).$$

Then, there exist a non-empty open set $A \subseteq I$ and a positive real number r with the following property: for every $\lambda \in A$ and every C^1 functional $\Gamma : X \to \mathbf{R}$ with compact derivative, there exists $\delta > 0$ such that, for each $\mu \in [0, \delta]$, the equation

$$\Psi'(x) = \lambda J'(x) + \mu\Gamma'(x)$$

has at least three solutions in X whose norms are less than r.

THEOREM 5.3. *Let X be a separable and reflexive real Banach space; $\Psi : X \to \mathbf{R}$ a coercive, sequentially weakly lower semicontinuous C^1 functional, belonging to $\mathscr{W}_X$, bounded on each bounded subset of X and whose derivative admits a continuous inverse on X^*; $J : X \to \mathbf{R}$ a C^1 functional with compact derivative. Assume that Ψ has a strict local minimum x_0 with $\Psi(x_0) = J(x_0) = 0$. Finally, setting*

$$\alpha = \max\left\{0, \limsup_{\|x\|\to+\infty} \frac{J(x)}{\Psi(x)}, \limsup_{x\to x_0} \frac{J(x)}{\Psi(x)}\right\},$$

$$\beta = \sup_{x\in\Psi^{-1}(]0,+\infty[)} \frac{J(x)}{\Psi(x)},$$

assume that $\alpha < \beta$.

Then, for each compact interval $[a, b] \subset \left]\frac{1}{\beta}, \frac{1}{\alpha}\right[$ (with the conventions $\frac{1}{0} = +\infty$, $\frac{1}{+\infty} = 0$) there exists $r > 0$ with the following property: for every $\lambda \in [a, b]$ and every C^1 functional $\Gamma : X \to \mathbf{R}$ with compact derivative, there exists $\delta > 0$ such that, for each $\mu \in [0, \delta]$, the equation

$$\Psi'(x) = \lambda J'(x) + \mu\Gamma'(x)$$

has at least three solutions whose norms are less than r.

To prove Theorem 5.1, we need the following result from [**182**]:

THEOREM 5.4. *Let X be a compact metric space, $I \subseteq \mathbf{R}$ a compact interval, and $f : X \times I \to \mathbf{R}$ a function satisfying the following conditions:*

(i) *for each $x \in X$, the function $f(x, \cdot)$ is continuous and concave;*
(ii) *for each $\lambda \in I$, the function $f(\cdot, \lambda)$ is lower semicontinuous;*
(iii) *one has*

$$a = \sup_{\lambda\in I} \inf_{x\in X} f(x, \lambda) < b = \inf_{x\in X} \sup_{\lambda\in I} f(x, \lambda).$$

Then, for each $c \in]a, b[$, *there exist a strictly increasing continuous function* $\varphi :]a, c[\to I$ *and an upper semicontinuous multifunction* $F :]a, c[\to 2^X$, *with non-empty closed values, such that, for every* $r \in]a, c[$, *one has*

$$\inf_{x \in F(r)} f(x, \varphi(r)) = r$$

and each $x \in F(r)$ *satisfying* $f(x, \varphi(r)) = r$ *is a local, non-absolute minimum of the function* $f(\cdot, \varphi(r))$. We also need

PROPOSITION 5.5. *Let* X, Y *be two topological spaces, and let* $F : X \to 2^Y$ *be a multifunction, with nonempty values, such that* $X \setminus F^-(y)$ *is sequentially closed for all* $y \in Y$ *and* $X \setminus F^-(y_0)$ *is sequentially compact for some* $y_0 \in Y$.

Then, for every non-decreasing sequence $\{Y_n\}$ *of subsets of* Y, *with* $\cup_{n \in \mathbf{N}} Y_n = Y$, *there exists* $n \in \mathbf{N}$ *such that* $F^-(Y_n) = X$.

PROOF. Arguing by contradiction, assume that, for every $n \in \mathbf{N}$, there is $x_n \in X$ such that

$$F(x_n) \cap Y_n = \emptyset. \tag{5.1}$$

Fix $\nu \in \mathbf{N}$ such that $y_0 \in Y_\nu$. Thus, for each $n > \nu$, one has $y_0 \notin F(x_n)$, that is $x_n \in X \setminus F^-(y_0)$. Consequently, since this latter set is sequentially compact, there is a subsequence of $\{x_n\}$, say $\{x_{n_k}\}$, converging to a point $x^* \in X \setminus F^-(y_0)$. Fix $y^* \in F(x^*)$ and choose $h \in \mathbf{N}$ such that $y^* \in Y_h$. Put $\mathbf{N}_1 = \{k \in \mathbf{N} : x_{n_k} \in F^-(y^*)\}$. Observe that $\mathbf{N}_1$ is infinite. Indeed, if $x_{n_k} \notin F^-(y^*)$ definitively, since, by assumption, the set $X \setminus F^-(y^*)$ is sequentially closed, we would have $x^* \notin F^-(y^*)$, that is $y^* \notin F(x^*)$, against our choice. Hence, we can take $k \in \mathbf{N}_1$ so that $n_k > h$. Therefore, $y^* \in F(x_{n_k}) \cap Y_{n_k}$, contradicting (5.1). □

PROOF OF THEOREM 5.1. Put

$$a = \sup_{\lambda \in I} \inf_{x \in X} (f(x, \lambda) + h(\lambda)),$$
$$b = \inf_{x \in X} \sup_{\lambda \in I} (f(x, \lambda) + h(\lambda))$$

and fix $c \in]a, b[$. Consider the multifunctions $F_0 : I \to 2^X$ and $G : X \to 2^I$ defined by

$$F_0(\lambda) = \{x \in X : f(x, \lambda) + h(\lambda) \leq c\}$$

and

$$G(x) = \{\lambda \in I : f(x, \lambda) + h(\lambda) > c\}.$$

Since $c < b$, one has $G(x) \neq \emptyset$ for all $x \in X$. Moreover, for each $\lambda \in I$, thanks to (β_1), the set $X \setminus G^-(\lambda)$ (equal to $F_0(\lambda)$) is bounded and sequentially weakly closed. Hence, it is sequentially weakly compact, due to reflexivity of X and to Eberlein-Smulyan's theorem. Consequently, if we fix any non-decreasing sequence $\{I_n\}$ of compact real intervals such that $I = \cup_{n \in \mathbf{N}} I_n$, by Proposition 5.5, there is $n \in \mathbf{N}$ such that

$$G(x) \cap I_n \neq \emptyset \tag{5.2}$$

for all $x \in X$. Now, observe that

$$S := \bigcup_{\lambda \in I_n} F_0(\lambda) \subseteq F_0(a_n) \cup F_0(b_n)$$

where $I_n = [a_n, b_n]$. This is due to the fact that, for each $x \in X$, the function $f(x, \cdot) + h(\cdot)$ is concave. Consequently, the set S is sequentially weakly compact. Since X is separable, S, endowed with the relativization of the weak topology, is metrizable. Now, put

$$a^* = \sup_{\lambda \in I_n} \inf_{x \in S} (f(x, \lambda) + h(\lambda))$$

and

$$b^* = \inf_{x \in S} \sup_{\lambda \in I_n} (f(x, \lambda) + h(\lambda)).$$

The function $\lambda \to \inf_{x \in S}(f(x, \lambda) + h(\lambda))$ is clearly upper semicontinuous, and so there is $\lambda_0 \in I_n$ such that

$$\inf_{x \in S} (f(x, \lambda_0) + h(\lambda_0)) = a^*.$$

On the other hand, we have

$$\inf_{x \in X} (f(x, \lambda_0) + h(\lambda_0)) \leq a < c,$$

and so

$$a^* < c.$$

Analogously, the function $x \to \sup_{\lambda \in I_n}(f(x, \lambda) + h(\lambda))$ is weakly lower semicontinuous in S, and so there is $x_0 \in S$ such that

$$\sup_{\lambda \in I_n} (f(x_0, \lambda)) + h(\lambda)) = b^*.$$

But, by (5.2), there is some $\lambda' \in I_n$ such that

$$f(x_0, \lambda') + h(\lambda') > c.$$

Hence

$$c < b^*.$$

At this point, we can apply Theorem 5.4 to the restriction of the function $(x, \lambda) \to f(x, \lambda) + h(\lambda)$ to $S \times I_n$ (S being endowed, of course, with the relativization of the weak topology). Therefore, there exist a strictly increasing continuous function $\varphi :]a^*, c[\to I_n$ and a weakly upper semicontinuous multifunction $F :]a^*, c[\to 2^S$, with non-empty weakly closed values, such that, for every $r \in]a^*, c[$, one has

$$\inf_{x \in F(r)} f(x, \varphi(r)) = r - h(\varphi(r))$$

and each $x \in F(r)$ satisfying $f(x, \varphi(r)) = r - h(\varphi(r))$ is a local, non-absolute minimum (in the weak topology) of the restriction of the function $f(\cdot, \varphi(r))$ to S. Now, for each $\lambda \in \varphi(]a^*, c[)$, put

$$P(\lambda) = \{x \in F(\varphi^{-1}(\lambda)) : f(x, \lambda) = \varphi^{-1}(\lambda) - h(\lambda)\}$$

and

$$Q(\lambda) = \left\{x \in X : f(x, \lambda) = \inf_{u \in X} f(u, \lambda)\right\}.$$

Of course, by (β_1), $P(\lambda)$ and $Q(\lambda)$ are non-empty, and $Q(\lambda) \subseteq S$. If $x \in P(\lambda)$, there exists a neighbourhood U of x in the weak topology such that

$$f(y, \lambda) \geq \varphi^{-1}(\lambda) - h(\lambda) \tag{5.3}$$

for all $y \in U \cap S$. But, if $y \notin S$, then $f(y, \mu) + h(\mu) > c$ for all $\mu \in I_n$. Thus, (5.3) actually holds for all $y \in U$. That is to say, x is a local, non-absolute minimum of the function $f(\cdot, \lambda)$ in the weak topology. At this point, to obtain the first part of the conclusion, it is enough to take $J = \varphi(]a^*, c[)$, and $\rho > \sup_{x \in S} \|x\|$. Indeed, since φ is continuous and increasing, $\varphi(]a^*, c[)$ is an open interval. Moreover, for each $\lambda \in \varphi(]a^*, c[)$, the points belonging to $P(\lambda) \cup Q(\lambda)$ (which are at least two since $P(\lambda) \cap Q(\lambda) = \emptyset$) are solutions of the equation $f'_x(x, \lambda) = 0$ whose norms are less than ρ.

Now, assume, in addition, that the function f is (strongly) continuous in $X \times I$ and that, for each $\lambda \in I$, the function $f(\cdot, \lambda)$ is C^1 and satisfies the Palais-Smale condition. By Lemma 5 of [**182**], conditions (α_1) and (β_1) imply that the function $f_{|S \times I_n}$ is weakly lower semicontinuous. By (α_1) again, the function $\lambda \to \inf_{x \in X} f(x, \lambda)$ is concave, and so it is continuous in the open interval $\varphi(]a^*, c[)$ (as well as $\lambda \to \varphi^{-1}(\lambda) - h(\lambda)$, of course). With this in mind, it is seen at once that the multifunctions P and Q are weakly upper semicontinuous. Since X is separable, each (strongly) open set $B \subseteq X$ is the union of a sequence of closed balls. Consequently, $P^-(B)$ and $Q^-(B)$ are $\mathscr{F}_\sigma$-sets in $\varphi(]a^*, c[)$. Then, by a result of Kuratowski and Ryll-Nardzewski ([**135**], p. 401), there esist two functions $\psi, \omega : \varphi(]a^*, c[) \to S$, of first Baire class with respect to the strong topology on X, such that $\psi(\lambda) \in P(\lambda)$ and $\omega(\lambda) \in Q(\lambda)$ for all $\lambda \in \varphi(]a^*, c[)$. Since $\varphi(]a^*, c[)$ is a Baire space, by a classical result ([**67**], Theorem 7.5), we can find a dense $\mathscr{G}_\delta$-set $C \subseteq \varphi(]a^*, c[)$ at each point of which both ψ and ω are continuous (with respect to the strong topology on X). Fix $\lambda^* \in C$. Then, since f is (strongly) continuous, there are a neighbourhood V of λ^* in $\varphi(]a^*, c[)$ and a number $\delta > 0$ such that, for each $\lambda \in V$, one has

$$\max\{\|\psi(\lambda) - \psi(\lambda^*)\|, \|\omega(\lambda) - \omega(\lambda^*)\|\} < \delta, \quad f(x, \lambda) < f(\psi(\lambda^*), \lambda^*) + 1$$

for all $x \in X$ satisfying $\|x - \psi(\lambda^*)\| < \delta$, and

$$f(x, \lambda) < f(\omega(\lambda^*), \lambda^*) + 1$$

for all $x \in X$ satisfying $\|x - \omega(\lambda^*)\| < \delta$. Now, for each $\lambda \in V$, denote by A_λ the set of all continuous functions $u : [0, 1] \to X$ such that $u(0) = \psi(\lambda)$ and $u(1) = \omega(\lambda)$. Also, put

$$\gamma(\lambda) = \inf_{u \in A_\lambda} \sup_{t \in [0,1]} f(u(t), \lambda).$$

Choose $v \in A_{\lambda^*}$ so that

$$\sup_{t\in[0,1]} f(v(t), \lambda^*) < \gamma(\lambda^*) + 1.$$

Since $v([0,1])$ is (strongly) compact and f is continuous, the function $\lambda \to \sup_{t\in[0,1]} f(v(t), \lambda)$ is continuous. So, there is an open interval $D \subseteq V$ containing λ^*, such that

$$\sup_{t\in[0,1]} f(v(t), \lambda) < \gamma(\lambda^*) + 1$$

for all $\lambda \in D$. Now, fix $\lambda \in D$, and define $u : [0,1] \to X$ by

$$u(t) = \begin{cases} \psi(\lambda) + 3t(\psi(\lambda^*) - \psi(\lambda)) & \text{if } t \in \left[0, \frac{1}{3}\right[\\ v(3t-1) & \text{if } t \in \left[\frac{1}{3}, \frac{2}{3}\right] \\ 3\omega(\lambda^*) - 2\omega(\lambda) + 3t(\omega(\lambda) - \omega(\lambda^*)) & \text{if } t \in \left]\frac{2}{3}, 1\right]. \end{cases}$$

Of course, $u \in A_\lambda$. Observe that

$$\sup_{t\in\left[0,\frac{1}{3}\right[} f(u(t), \lambda) < f(\psi(\lambda^*), \lambda^*) + 1 < \gamma(\lambda^*) + 2,$$

$$\sup_{t\in\left[\frac{1}{3},\frac{2}{3}\right]} f(u(t), \lambda) = \sup_{t\in[0,1]} f(v(t), \lambda) < \gamma(\lambda^*) + 1$$

and

$$\sup_{t\in\left]\frac{2}{3},1\right]} f(u(t), \lambda) < f(\omega(\lambda^*), \lambda^*) + 1 < \gamma(\lambda^*) + 2.$$

In other words, we have

$$\sup_{t\in[0,1]} f(u(t), \lambda) < \gamma(\lambda^*) + 2$$

and so

$$\gamma(\lambda) < \gamma(\lambda^*) + 2.$$

Now, we apply Theorem 1 of [**156**] to find a point $z_\lambda \in X$, different from $\psi(\lambda)$ and $\omega(\lambda)$, such that $f'_x(z_\lambda, \lambda) = 0$ and $f(z_\lambda, \lambda) = \gamma(\lambda)$. Finally, put $M = \max\{\gamma(\lambda^*) + 2, c - \inf_{\mu\in D} h(\mu)\}$, and consider the set

$$S_1 = \bigcup_{\lambda\in\overline{D}} \{x \in X : f(x, \lambda) \leq M\}.$$

For the same remark made on S, the set S_1 is sequentially weakly compact, and hence bounded. Now, to satisfy the conclusion, we can take $\rho > \sup_{x\in S_1} \|x\|$ and $J = D$. Indeed, for each $\lambda \in D$, the points $\psi(\lambda)$, $\omega(\lambda)$ and z_λ are three distinct solutions of the equation $f'_x(x, \lambda) = 0$ which lie in S_1. This concludes the proof. $\square$

We now are going to prove Theorem 5.2. We first need some preliminary facts and, in particular, a mini-max theorem.

It is well-known ([**119**], Theorem 7.4.4) that if X, Y are two topological space and $F : X \to 2^Y$ is a lower semicontinuous multifunction, with non-empty and connected values, then for every connected set $A \subseteq X$, the set $F(A)$ is connected.

We want to remark that such a result can be improved as follows:

PROPOSITION 5.6. *Let X, Y be two topological spaces, $F : X \to 2^Y$ a lower semicontinuous multifunction with non-empty values and $A \subseteq X$ a connected set. Assume that the set*

$$D := \{x \in A : F(x) \text{ is connected}\}$$

is dense in A.

Then, the set $F(A)$ is connected.

PROOF. Arguing by contradiction, assume that $F(A)$ is disconnected. So, there are two open sets $B, C \subseteq Y$ such that

$$F(A) \cap B \neq \emptyset, \quad F(A) \cap C \neq \emptyset, \quad F(A) \cap B \cap C = \emptyset, \quad F(A) \subseteq B \cup C.$$

From this, it clearly follows that

$$F^-(B) \cap A \neq \emptyset, \quad F^-(C) \cap A \neq \emptyset, \quad A \subseteq F^-(B) \cup F^-(C),$$

the latter equality holding since the values of F are non-empty. By lower semicontinuity, the sets $F^-(B)$ and $F^-(C)$ are open in X. As a consequence, since A is connected, we have

$$F^-(B) \cap F^-(C) \cap A \neq \emptyset.$$

But then, since D is dense in A, there exists $\hat{x} \in F^-(B) \cap F^-(C) \cap D$. At this point, since $F(\hat{x}) \cap B \cap C = \emptyset$ and $F(\hat{x}) \subseteq B \cup C$, it follows that the set $F(\hat{x})$ is disconnected, conflicting with the definition of D. □

From Theorem 7.3.12 of [**119**] and Proposition 5.6, we get, in particular, the following

PROPOSITION 5.7. *Let X, Y be two topological spaces, with X connected, and let $F : X \to 2^Y$ be a lower semicontinuous multifunction with non-empty values. Assume the set*

$$\{x \in X : F(x) \text{ is connected}\}$$

is dense in X.

Then, the set

$$\{(x, y) \in X \times Y : y \in F(x)\}$$

is connected.

For a generic set $S \subseteq X \times I$, for each $(x, \lambda) \in X \times I$, we set

$$S_x = \{\mu \in I : (x, \mu) \in S\}$$

and

$$S^\lambda = \{u \in X : (u, \lambda) \in S\}.$$

Let us recall the following result from [**157**]:

THEOREM 5.8. ([**157**], Theorem 2.3) *Let X be a topological space, $I \subseteq \mathbf{R}$ a compact interval and $S, T \subseteq X \times I$. Assume that S is connected and $S^\lambda \neq \emptyset$ for all $\lambda \in I$, while T_x is non-empty and connected for all $x \in X$, and T^λ is open for all $\lambda \in I$.*

Then, one has $S \cap T \neq \emptyset$.

We now prove the mini-max theorem which was mentioned above.

THEOREM 5.9. *Let X be a topological space, $I \subseteq \mathbf{R}$ an interval and $f : X \times I \to \mathbf{R}$. Assume that there exist a point $\lambda_0 \in I$, a set $D \subseteq I$, dense in I, and a number $\rho^* > \sup_I \inf_X f$, such that, for each $\rho \in]-\infty, \rho^*[$, the following conditions are satisfied:*
(a_1) for each $x \in X$, the set

$$\{\lambda \in I : f(x, \lambda) > \rho\}$$

is connected and the set

$$\{\lambda \in I : f(x, \lambda) < \rho\}$$

is open in I;
(b_1) for each $\lambda \in D$, the set

$$\{x \in X : f(x, \lambda) < \rho\}$$

is connected;
(c_1) the set

$$\{x \in X : f(x, \lambda) \leq \rho\}$$

is closed for each $\lambda \in I$ and compact for $\lambda = \lambda_0$.

Then, one has

$$\sup_{\lambda \in I} \inf_{x \in X} f(x, \lambda) = \inf_{x \in X} \sup_{\lambda \in I} f(x, \lambda).$$

PROOF. Fix a non-decreasing sequence $\{I_n\}$ of compact sub-intervals of I, with $I = \cup_{n \in \mathbf{N}} I_n$ and $\lambda_0 \in I_1$. Fix $n \in \mathbf{N}$. We claim that

$$\sup_{\lambda \in I_n} \inf_{x \in X} f(x, \lambda) = \inf_{x \in X} \sup_{\lambda \in I_n} f(x, \lambda). \tag{5.4}$$

Arguing by contradiction, assume that

$$\sup_{\lambda\in I_n}\inf_{x\in X} f(x,\lambda) < \inf_{x\in X}\sup_{\lambda\in I_n} f(x,\lambda).$$

Fix $n \in \mathbf{N}$ so that

$$\sup_{\lambda\in I_n}\inf_{x\in X} f(x,\lambda) < \rho < \max\left\{\rho^*, \inf_{x\in X}\sup_{\lambda\in I_n} f(x,\lambda)\right\}. \tag{5.5}$$

Put

$$S = \{(x,\lambda) \in X \times I_n : f(x,\lambda) < \rho\}$$

and

$$T = \{(x,\lambda) \in X \times I_n : f(x,\lambda) > \rho\}.$$

Observe that the multifunction $\lambda \to S^\lambda$ is lower semicontinuous (by (a_1)) and with non-empty values (by (5.5)). Then, in view of (b_1), Proposition 5.6 ensures that the set

$$\{(\lambda,x) \in I \times X : f(x,\lambda) < \rho\}$$

is connected. But this set is homeomorphic to S, and so S is connected. On the other hand, by (a_1), (c_1) and (5.8) again, T_x is non-empty and connected for all $x \in X$, while T^λ is open for all $\lambda \in I$. Consequently, by Theorem 5.5, we would have $S \cap T \neq \emptyset$ which is clearly false. So, (5.4) holds. Finally, let us prove the conclusion. Arguing by contradiction again, suppose that

$$\sup_{\lambda\in I}\inf_{x\in X} f(x,\lambda) < \inf_{x\in X}\sup_{\lambda\in I} f(x,\lambda).$$

Fix r satisfying

$$\sup_{\lambda\in I}\inf_{x\in X} f(x,\lambda) < r < \max\left\{\rho^*, \inf_{x\in X}\sup_{\lambda\in I} f(x,\lambda)\right\}.$$

For each $n \in \mathbf{N}$, put

$$C_n = \left\{x \in X : \sup_{\lambda\in I_n} f(x,\lambda) \leq r\right\}.$$

Note that $C_n \neq \emptyset$. Indeed, otherwise, we would have

$$r \leq \inf_{x\in X}\sup_{\lambda\in I_n} f(x,\lambda) = \sup_{\lambda\in I_n}\inf_{x\in X} f(x,\lambda) \leq \sup_{\lambda\in I}\inf_{x\in X} f(x,\lambda).$$

So, we have

$$C_n = \bigcap_{\lambda\in I_n}\{x \in X : f(x,\lambda) \leq r\}.$$

Consequently, $\{C_n\}$ is a non-increasing sequence of non-empty closed subsets of the compact set $\{x \in X : f(x, \lambda_0) \le r\}$. Therefore, one has $\cap_{n\in\mathbf{N}} C_n \neq \emptyset$. Let $x^* \in \cap_{n\in\mathbf{N}} C_n$. Then, one has

$$\sup_{\lambda\in I} f(x^*, \lambda) = \sup_{n\in\mathbf{N}} \sup_{\lambda\in I_n} f(x^*, \lambda) \le r$$

and so

$$\inf_{x\in X} \sup_{\lambda\in I} f(x, \lambda) \le r,$$

a contradiction. The proof is complete. □

REMARK 5.10. From the above proof, it clearly follows that, when I is compact, the compactness of the set $\{x \in X : f(x, \lambda_0) \le \rho\}$ becomes superfluous.

REMARK 5.11. Let X be a Hausdorff topological space and let $\Psi : X \to]-\infty, +\infty]$ be a function such that, for some $\rho^* > \inf_X \Psi$, the set $\Psi^{-1}(]-\infty, \rho])$ is compact for all $\rho < \rho^*$. Assume also that the set $\overline{\Psi^{-1}(]-\infty, \rho[)}$ is connected for all $\rho < \rho^*$. Then, the set $\Psi^{-1}(]-\infty, \rho[)$ is connected for all $\rho < \rho^*$.

To show this, let $r, s \in \mathbf{R}$, with $r < \sigma < \rho^*$. Since $\Psi^{-1}(]-\infty, \sigma])$ is closed, one has $\overline{\Psi^{-1}(]-\infty, \sigma[)} \subseteq \Psi^{-1}(]-\infty, \sigma])$ and hence

$$\Psi^{-1}(]-\infty, r]) = \bigcap_{r<\sigma<\rho^*} \overline{\Psi^{-1}(]-\infty, \sigma[)}.$$

Therefore, the set $\Psi^{-1}(]-\infty, r])$, as the intersection of a non-increasing sequence of compact and connected sets, is connected too. Finally, let $\rho \in]-\infty, \rho^*[$. Since

$$\Psi^{-1}(]-\infty, \rho[) = \bigcup_{r<\rho} \Psi^{-1}(]-\infty, r]),$$

it follows that the set $\Psi^{-1}(]-\infty, \rho[)$ is connected, as claimed.

A joint application of Theorems 5.9 and 3.9 (via Remark 5.11) gives

THEOREM 5.12. *Let X be a reflexive real Banach space, $I \subseteq \mathbf{R}$ an interval, and $\varphi : X \times I \to]-\infty, +\infty]$ a function such that $\varphi(x, \cdot)$ is quasi-concave in I for all $x \in X$, while $\varphi(\cdot, \lambda)$ is continuous, coercive and sequentially weakly lower semicontinuous in X for all $\lambda \in I$. Further, assume that*

$$\sup_{\lambda\in I} \inf_{x\in X} \varphi(x, \lambda) < \inf_{x\in X} \sup_{\lambda\in I} \varphi(x, \lambda). \tag{5.6}$$

Then, for each $\sigma > \sup_I \inf_X \varphi$, there exists a non-empty open set $D \subseteq I$ with the following property: for every $\lambda \in D$ and every sequentially weakly lower semicontinuous functional $\psi : X \to \mathbf{R}$, there exists $\gamma > 0$ such that, for each $\mu \in [0, \gamma]$, the functional $\varphi(\cdot, \lambda) + \mu\psi(\cdot)$ has at least two local minima lying in the set $\{x \in X : \varphi(x, \lambda) < \sigma\}$.

PROOF. Fix $\sigma > \sup_I \inf_X \varphi$. From Remark 5.11, it follows that if there is a set $A \subseteq I$, such that, for each $\rho < \sigma$ and each $\lambda \in A$, the set $\overline{(\{x \in X : \varphi(x, \lambda) < \rho\})_w}$ is weakly connected, then, for each $\rho < \sigma$ and each $\lambda \in A$ again, the set $\{x \in X : \varphi(x, \lambda) < \rho\}$ is weakly connected. Now, denote by $\tilde{D}$ the set of all $\lambda \in I$ for which there exist a sequentially weakly lower semicontinuous functional $\psi_\lambda : X \to \mathbf{R}$ and a sequence $\{\mu_n\}$ in $\mathbf{R}^+$ converging to 0 such that, for each $n \in \mathbf{N}$, the set $\{x \in X : \varphi(x, \lambda) < \sigma\}$ contains at most one local minimum of the functional $\varphi(\cdot, \lambda) + \mu_n \psi_\lambda(\cdot)$. Then, thanks to Theorem 3.9 (and Remark 3.10), the set $\overline{(\{x \in X : \varphi(x, \lambda) < \rho\})_w}$ is weakly connected for all $\lambda \in \tilde{D}$ and all $\rho \leq \sigma$. As a consequence, in view of (5.6) and Theorem 5.9, the set $\tilde{D}$ is not dense in I. At this point, it is enough to take $D = \operatorname{int}(I \setminus \tilde{D})$ to get the conclusion. □

We can now prove Theorem 5.2.

PROOF OF THEOREM 5.2. First, note that any C^1 functional on X with compact derivative is sequentially weakly continuous ([**196**], Corollary 41.9), and so, in particular, it is bounded on each bounded subset of X, due to the reflexivity of X. Consequently, the function $\varphi : X \times I \to \mathbf{R}$ defined by

$$\varphi(x, \lambda) = \Psi(x) + \lambda(\rho - J(x))$$

satisfies all the hypotheses of Theorem 5.12. Fix $\sigma > \sup_I \inf_X \varphi$ and consider a non-empty open set D with the property expressed in Theorem 5.12. Fix also a compact interval $[a, b] \subset D$. We have

$$\bigcup_{\lambda \in [a,b]} \{x \in X : \varphi(x, \lambda) < \sigma\}$$
$$\subseteq \{x \in X : \Psi(x) - aJ(x) < \sigma - a\rho\} \cup \{x \in X : \Psi(x) - bJ(x) < \sigma - b\rho\}.$$

Note that the set on the right-hand side is bounded, due to the coercivity assumption. Consequently, there is some $\eta > 0$ such that

$$\bigcup_{\lambda \in [a,b]} \{x \in X : \varphi(x, \lambda) < \sigma\} \subseteq B_\eta \tag{5.7}$$

where $B_\eta = \{x \in X : \|x\| < \eta\}$. Now, set

$$\hat{c} = \sup_{B_\eta} \Phi + \max\{|a|, |b|\} \sup_{B_\eta} |J|$$

and fix $r > \eta$ so that

$$\bigcup_{\lambda \in [a,b]} \{x \in X : \Psi(x) - \lambda J(x) \leq \hat{c} + 2\} \subseteq B_r. \tag{5.8}$$

Now, let $\Gamma : X \to \mathbf{R}$ be any C^1 functional with compact derivative. Next, choose a bounded C^1 function $g : \mathbf{R} \to \mathbf{R}$ such that

$$g(t) = t$$

for all $t \in [-\sup_{B_r} |\Gamma|, \sup_{B_r} |\Gamma|]$. Put

$$\tilde{\Gamma}(x) = g(\Gamma(x))$$

for all $x \in X$. So, $\tilde{\Gamma}$ is a bounded C^1 functional on X such that

$$\tilde{\Gamma}(x) = \Gamma(x) \tag{5.9}$$

for all $x \in B_r$. For any set $E \subset X$, we have

$$\tilde{\Gamma}'(E) = g'(\Gamma(E))\Gamma'(E)$$

and hence it is clear that the derivative of $\tilde{\Gamma}$ is compact. Now, fix $\lambda \in [a, b]$. Then, taking (5.7) into account, there exists $\gamma > 0$ such that, for each $\mu \in [0, \gamma]$, the functional $\Psi - \lambda J - \mu\tilde{\Gamma}$ has two local minima, say x_0, x_1, lying in B_η. Further, set

$$\delta = \min\left\{\gamma, \frac{1}{\sup_{\mathbf{R}} |g|}\right\}$$

and fix $\mu \in [0, \delta]$. Note that functional $\Psi - \lambda J - \mu\tilde{\Gamma}$ is coercive. Then, taking the assumptions on Ψ into account, by a classical result ([**196**], Example 38.25), the same functional satisfies the Palais-Smale condition. Consequently, if we denote by $\mathscr{S}$ the set of all continuous functions $u : [0, 1] \to X$ such that $u(0) = x_0, u(1) = x_1$, and set

$$c_{\lambda,\mu} = \inf_{u \in \mathscr{S}} \sup_{t \in [0,1]} (\Psi(u(t)) - \lambda J(u(t)) - \mu\tilde{\Gamma}(u(t))),$$

by Theorem 1 of [**156**], there exists $x_2 \in X$ such that

$$\Psi'(x_2) - \lambda J'(x_2) - \mu\tilde{\Gamma}'(x_2) = 0$$

and

$$\Psi(x_2) - \lambda J(x_2) - \mu\tilde{\Gamma}(x_2) = c_{\lambda,\mu}.$$

Now, observe that

$$\begin{aligned} c_{\lambda,\mu} &\leq \sup_{t \in [0,1]} (\Psi(x_0 + t(x_1 - x_0)) - \lambda J(x_0 + t(x_1 - x_0)) - \mu\tilde{\Gamma}(x_0 + t(x_1 - x_0))) \\ &\leq \sup_{B_\eta} \Psi + \max\{|a|, |b|\} \sup_{B_\eta} |J| + \delta \sup_{\mathbf{R}} |g| \leq \hat{c} + 1. \end{aligned}$$

Consequently, we clearly have

$$\Psi(x_2) - \lambda J(x_2) \leq \hat{c} + 2.$$

By (5.8), it then follows that $x_2 \in B_r$ and so, by (5.9), one has $\tilde{\Gamma}'(x_i) = \Gamma'(x_i)$ for $i = 0, 1, 2$. In conclusion, x_0, x_1 and x_2 are three solutions of the equation

$$\Psi'(x) = \lambda J'(x) + \mu\Gamma'(x)$$

lying in B_r, and the proof is complete. □

Finally, let us prove Theorem 5.3. We will derive it from the following more general result:

THEOREM 5.13. *Let X be a separable and reflexive real Banach space; $I \subseteq \mathbf{R}$ an interval; $\Psi : X \to \mathbf{R}$ a sequentially weakly lower semicontinuous C^1 functional, belonging to $\mathscr{W}_X$, bounded on each bounded subset of X and whose derivative admits a continuous inverse on X^*; $J : X \to \mathbf{R}$ a C^1 functional with compact derivative. Assume that, for each $\lambda \in I$, the functional $\Psi - \lambda J$ is coercive and has a strict local, not global minimum, say $\hat{x}_\lambda$.*

Then, for each compact interval $[a,b] \subseteq I$ for which $\sup_{\lambda\in[a,b]}(\Psi(\hat{x}_\lambda) - \lambda J(\hat{x}_\lambda)) < +\infty$, there exists $r > 0$ with the following property: for every $\lambda \in [a,b]$ and every C^1 functional $\Gamma : X \to \mathbf{R}$ with compact derivative, there exists $\delta > 0$ such that, for each $\mu \in [0,\delta]$, the equation

$$\Psi'(x) = \lambda J'(x) + \mu\Gamma'(x)$$

has at least three solutions whose norms are less than r.

PROOF. The proof follows the same pattern as that of Theorem 5.2, and so we only sketch it. Fix a compact interval $[a,b] \subseteq I$ such that $\sup_{\lambda\in[a,b]}(\Psi(\hat{x}_\lambda) - \lambda J(\hat{x}_\lambda)) < +\infty$. Fix also $\sigma > \sup_{\lambda\in[a,b]}(\Psi(\hat{x}_\lambda) - \lambda J(\hat{x}_\lambda))$. Now, let $\eta > 0$ be such that

$$\bigcup_{\lambda\in[a,b]} \{x \in X : \Psi(x) - \lambda J(x) < \sigma\} \subseteq B_\eta. \tag{5.10}$$

Set

$$\hat{c} = \sup_{B_\eta} \Psi + \max\{|a|,|b|\} \sup_{B_\eta} |J|$$

and fix $r > \eta$ so that

$$\bigcup_{\lambda\in[a,b]} \{x \in X : \Psi(x) - \lambda J(x) \leq \hat{c} + 2\} \subseteq B_r.$$

Let $\Gamma : X \to \mathbf{R}$ be any C^1 functional with compact derivative. Next, choose a bounded C^1 function $g : \mathbf{R} \to \mathbf{R}$ such that

$$g(t) = t$$

for all $t \in [-\sup_{B_r} |\Gamma|, \sup_{B_r} |\Gamma|]$. Put

$$\tilde{\Gamma}(x) = g(\Gamma(x))$$

for all $x \in X$. Now, fix $\lambda \in [a,b]$. Clearly, we are allowed to apply Theorem 3.15 to the functionals $\Psi - \lambda J$ and $-\tilde{\Gamma}$, taking as τ the weak topology of X. Therefore, taking into account (5.10) and the fact that the topology $\tau_{(\Psi-\lambda J)}$ is weaker than the strong one, there exists $\gamma > 0$ such that, for each $\mu \in [0,\gamma]$, the functional $\Psi - \lambda J - \mu\tilde{\Gamma}$ has two local minima, say x_0, x_1, lying in B_η. Finally, set

$$\delta = \min\left\{\gamma, \frac{1}{\sup_{\mathbf{R}} |g|}\right\}$$

and fix $\mu \in [0,\delta]$. From now on, the proof runs exactly as that of Theorem 5.2. $\square$

PROOF OF THEOREM 5.3. Fix $\lambda \in \left]\frac{1}{\beta}, \frac{1}{\alpha}\right[$. Then, since $\limsup_{x \to x_0} \frac{J(x)}{\Psi(x)} < \frac{1}{\lambda}$ and $\Psi(x) > 0$ for each $x \neq x_0$ in some neighbourhood U of x_0, there is a neighbourhood $V \subseteq U$ of x_0 such that $\Psi(x) - \lambda J(x) > 0$ for all $x \in V \setminus \{x_0\}$. Hence, x_0 is a strict local minimum of $\Psi - \lambda J$. On the other hand, since $\frac{1}{\lambda} < \beta$, there is some $x_1 \in X$, with $\min\{\Psi(x_1), J(x_1)\} > 0$, such that $\Psi(x_1) - \lambda J(x_1) < 0$. So, x_0 is not a global minimum of $\Psi - \lambda J$. Finally, since $\limsup_{\|x\| \to +\infty} \frac{J(x)}{\Psi(x)} < \frac{1}{\lambda}$, we have

$$\liminf_{\|x\| \to +\infty} \left(1 - \lambda \frac{J(x)}{\Psi(x)}\right) > 0$$

and so, since Ψ is coercive,

$$\lim_{\|x\| \to +\infty} (\Psi(x) - \lambda J(x)) = \lim_{\|x\| \to +\infty} \Psi(x) \left(1 - \lambda \frac{J(x)}{\Psi(x)}\right) = +\infty.$$

That is to say, $\Psi - \lambda J$ is coercive. Therefore, all the assumptions of Theorem 5.13 are satisfied with $I = \left]\frac{1}{\beta}, \frac{1}{\alpha}\right[$ and $\hat{x}_\lambda = x_0$, and the conclusion follows. □

6. Eigenvalue estimates for a class of potential operators

The aim of this chapter is to highlight Theorem 6.1 below which gives a uniform estimate from below for the eigenvalues of a certain class of potential operators in Hilbert spaces.

For any real Hilbert space X, denote by $\mathscr{A}_X$ the set of all C^1 functionals $I : X \to \mathbf{R}$ such that 0 is a global maximum of I and I' is Lipschitzian with Lipschitz constant less than 1. Set

$$\gamma_X = \inf_{I \in \mathscr{A}_X} \inf\{\lambda > 0 : x = \lambda I'(x) \text{ for some } x \neq 0\}.$$

We have:

THEOREM 6.1. *For any real Hilbert space $(X, \langle \cdot, \cdot \rangle)$, with $X \neq \{0\}$, one has*

$$\gamma_X = 3.$$

We first prove

PROPOSITION 6.2. *One has*

$$\gamma_{\mathbf{R}} = 3.$$

PROOF. Let $I_0 \in \mathscr{A}_{\mathbf{R}}$ and let $L < 1$ be the Lipschitz constant of I_0'. Set

$$I = I_0 - I_0(0).$$

Fix $\lambda \in]0, 3]$. Let us prove that 0 is the only solution of the equation

$$x = \lambda I'(x).$$

Arguing by contradiction, assume that

$$x_0 = \lambda I'(x_0)$$

for some $x_0 \neq 0$. It is not restrictive to assume that $x_0 > 0$ (otherwise, we would work with $I'(-x)$). Consider now the function $g : \mathbf{R} \to \mathbf{R}$ defined by

$$g(x) = \begin{cases} -\dfrac{x^2}{2} & \text{if } x < \dfrac{x_0}{3} \\ \dfrac{x^2}{2} - \dfrac{2x_0x}{3} + \dfrac{x_0^2}{9} & \text{if } \dfrac{x_0}{3} \leq x \leq x_0 \\ -\dfrac{x^2}{2} + \dfrac{4x_0x}{3} - \dfrac{8x_0^2}{9} & \text{if } x_0 > x. \end{cases}$$

Clearly, $g \in C^1(\mathbf{R})$. Let $x > 0$ with $x \neq x_0$. Let us prove that

$$g'(x) < I'(x).$$

We distinguish two cases. If $0 < x \leq \frac{x_0}{3}$, We have

$$g'(x) = -x < -Lx \leq I'(x).$$

If $x > \frac{x_0}{3}$, We have

$$\begin{aligned} g'(x) &= \frac{x_0}{3} - |x - x_0| < \frac{x_0}{3} - L|x - x_0| \\ &= \frac{\lambda I'(x_0)}{3} - L|x - x_0| \leq I'(x_0) - L|x - x_0| \leq I'(x). \end{aligned}$$

So, in particular, we get

$$I\left(\frac{4x_0}{3}\right) = \int_0^{\frac{4x_0}{3}} I'(x)dx > \int_0^{\frac{4x_0}{3}} g'(x)dx = g\left(\frac{4x_0}{3}\right) = 0$$

which contradicts the fact that the function I is non-positive, since 0 is a global maximum of I_0. From what we have just proven, it clearly follows that

$$3 \leq \gamma_{\mathbf{R}}.$$

Now, fix any $\mu > 1$. Continue to consider the function g defined above (for a fixed $x_0 > 0$). Clearly, the function $\frac{1}{\mu}g$ belongs to $\mathscr{A}_{\mathbf{R}}$ and

$$x_0 = 3\mu \frac{g(x_0)}{\mu}.$$

Of course, from this we infer that

$$\gamma_{\mathbf{R}} \leq 3\mu$$

and the conclusion clearly follows. □

PROOF OF THEOREM 6.1. First, let us prove that

$$\gamma_X \leq 3. \tag{6.1}$$

To this end, fix any $\varphi \in \mathscr{A}_{\mathbf{R}}$ and any $\lambda > 0$ such that

$$t = \lambda\varphi'(t)$$

for some $t \neq 0$. Fix also $u \in X$, with $\|u\| = 1$, and consider the functional I defined by putting

$$I(x) = \varphi(\langle u, x\rangle)$$

for all $x \in X$. It is readily seen that $I \in \mathscr{A}_X$. In particular, note that

$$I'(x) = \varphi'(\langle u, x\rangle)u.$$

Finally, set

$$\hat{x} = \lambda\varphi(t)u.$$

Of course, $\hat{x} \neq 0$. Since

$$\langle u, \hat{x}\rangle = \lambda\varphi'(t)$$

we also have

$$\langle u, \hat{x}\rangle = t$$

and so

$$\hat{x} = \lambda I'(\hat{x}).$$

From this, it clearly follows that

$$\gamma_X \leq \gamma_{\mathbf{R}}$$

and so (6.1) follows now from Proposition 6.2.

Now, let us prove that

$$3 \leq \gamma_X. \tag{6.2}$$

To this end, fix $I \in \mathscr{A}_X$, $\lambda > 0$ and $x \in X \setminus \{0\}$ such that

$$x = \lambda I'(x). \tag{6.3}$$

Then, consider the function $\varphi : \mathbf{R} \to \mathbf{R}$ defined by

$$\varphi(t) = I\left(\frac{tx}{\|x\|}\right)$$

for all $t \in \mathbf{R}$. Clearly, 0 is a global maximum for φ. Moreover, $\varphi \in C^1(\mathbf{R})$ and one has

$$\varphi'(t) = \left\langle I'\left(\frac{tx}{\|x\|}\right), \frac{x}{\|x\|}\right\rangle.$$

Therefore, if L is the Lipschitz constant of I', for each $t, s \in \mathbf{R}$, we have

$$|\varphi'(t) - \varphi'(s)| = \left|\left\langle I'\left(\frac{tx}{\|x\|}\right) - I'\left(\frac{sx}{\|x\|}\right), \frac{x}{\|x\|}\right\rangle\right|$$
$$\leq \left\|I'\left(\frac{tx}{\|x\|}\right) - I'\left(\frac{sx}{\|x\|}\right)\right\| \leq L|t-s|.$$

This shows that φ' is a contraction, and so $\varphi \in \mathscr{A}_{\mathbf{R}}$. Now, from (6.3), we get

$$\|x\| = \lambda\left\langle I'(x), \frac{x}{\|x\|}\right\rangle$$

that is

$$\|x\| = \lambda\varphi'(\|x\|).$$

From this, we infer that

$$\gamma_{\mathbf{R}} \leq \gamma_X.$$

So (6.2) follows from Proposition 6.2, and the proof is complete. □

7. Applications

In this chapter, we present some applications of results obtained in the previous chapters.

Let $\Omega \subset \mathbf{R}^n$ be a bounded domain with smooth boundary and let $p > 1$.

The space $W^{1,p}(\Omega)$ is considered with the norm

$$\|u\|_1 = \left(\int_\Omega (|\nabla u(x)|^p + |u(x)|^p)dx\right)^{\frac{1}{p}}$$

while $W_0^{1,p}(\Omega)$ is considered with the norm

$$\|u\|_2 = \left(\int_\Omega |\nabla u(x)|^p dx\right)^{\frac{1}{p}}.$$

If $p > n$, $W^{1,p}(\Omega)$ is compactly embedded in $C^0(\overline{\Omega})$, and so the constant

$$c^* := \sup_{u \in W^{1,p}(\Omega)\setminus\{0\}} \frac{\max_{\overline{\Omega}}|u|}{\|u\|_1} \tag{7.1}$$

is finite.

If $n \geq p$, we denote by $\mathscr{A}$ the class of all Carathéodory functions $\varphi : \Omega \times \mathbf{R} \to \mathbf{R}$ such that

$$\sup_{(x,\xi)\in\Omega\times\mathbf{R}} \frac{|\varphi(x,\xi)|}{1+|\xi|^q} < +\infty,$$

where $0 < q < \frac{pn-n+p}{n-p}$ if $p < n$ and $0 < q < +\infty$ if $p = n$. While, when $n < p$, we denote by $\mathscr{A}$ the class of all Carathéodory functions $\varphi : \Omega \times \mathbf{R} \to \mathbf{R}$ such that, for each $r > 0$, the function $x \to \sup_{|\xi|\leq r} |\varphi(x,\xi)|$ belongs to $L^1(\Omega)$.

Given $\varphi \in \mathscr{A}$, consider the Neumann problem

$$(\mathrm{NP}_\varphi) \qquad \begin{cases} -\mathrm{div}(|\nabla u|^{p-2}\nabla u) + |u|^{p-2}u = \varphi(x,u) & \text{in } \Omega \\ \dfrac{\partial u}{\partial \nu} = 0 & \text{on } \partial\Omega, \end{cases}$$

where ν is the outer unit normal to $\partial\Omega$.

Consider also the Dirichlet problem

$$(\mathrm{DP}_\varphi) \qquad \begin{cases} -\mathrm{div}(|\nabla u|^{p-2}\nabla u) = \varphi(x,u) & \text{in } \Omega \\ u = 0 & \text{on } \partial\Omega. \end{cases}$$

Let us recall that a weak solution of (NP_φ) is any $u \in W^{1,p}(\Omega)$ such that

$$\int_\Omega |\nabla u(x)|^{p-2}\nabla u(x)\nabla v(x)dx + \int_\Omega |u(x)|^{p-2}u(x)v(x)dx - \int_\Omega \varphi(x,u(x))v(x)dx = 0$$

for all $v \in W^{1,p}(\Omega)$, while a weak solution of (DP_φ) is any $u \in W_0^{1,p}(\Omega)$ such that

$$\int_\Omega |\nabla u(x)|^{p-2}\nabla u(x)\nabla v(x)dx - \int_\Omega \varphi(x,u(x))v(x)dx = 0$$

for all $v \in W_0^{1,p}(\Omega)$.

The functionals $\Psi_1, J_\varphi : W^{1,p}(\Omega) \to \mathbf{R}$ and $\Psi_2 : W_0^{1,p}(\Omega) \to \mathbf{R}$ defined by

$$\Psi_1(u) = \frac{1}{p}\|u\|_1,$$

$$J_\varphi(u) = \int_\Omega \left(\int_0^{u(x)} \varphi(x,\xi)d\xi \right) dx,$$

$$\Psi_2(u) = \frac{1}{p}\|u\|_2$$

are C^1 with derivatives given by

$$\Psi_1'(u)(v) = \int_\Omega |\nabla u(x)|^{p-2}\nabla u(x)\nabla v(x)dx + \int_\Omega |u(x)|^{p-2}u(x)v(x)dx$$

$$J_\varphi'(u)(v) = \int_\Omega \varphi(x,u(x))v(x)dx$$

for all $u, v \in W^{1,p}(\Omega)$ and

$$\Psi_2'(u)(v) = \int_\Omega |\nabla u(x)|^{p-2}\nabla u(x)\nabla v(x)dx$$

for all $u, v \in W_0^{1,p}(\Omega)$. Consequently, the weak solutions of problem (NP_φ) (resp. (DP_φ)) are exactly the critical points in $W^{1,p}(\Omega)$ (resp. $W_0^{1,p}(\Omega)$) of the functional $\Psi_1 - J_\varphi$ (resp. $\Psi_2 - J_\varphi$). Moreover, J_φ' is compact, while Ψ_1' (resp. Ψ_2') is a homeomorphism between $W^{1,p}(\Omega)$ (resp. $W_0^{1,p}(\Omega)$) and its dual. Clearly, $\Psi_1 \in \mathscr{W}_{W^{1,p}(\Omega)}$ (resp. $\Psi_2 \in \mathscr{W}_{W_0^{1,p}(\Omega)}$) since $W^{1,p}(\Omega)$ (resp. $W_0^{1,p}(\Omega)$) is uniformly convex.

We begin with three applications of Theorem 2.4.

THEOREM 7.1. *Let $p = 2$ and let $f \in \mathscr{A}$. Assume that there are two constants $r \geq 0$ and $c > 2$ such that*

$$0 < c \int_0^\xi f(x,t)dt \leq \xi f(x,\xi)$$

for all $(x,\xi) \in \Omega \times \mathbf{R}$ with $|\xi| \geq r$, $\xi \neq 0$.

Then, for each $\rho > 0$ and each λ satisfying

$$\lambda > \inf_{u \in B_\rho} \frac{\sup_{B_\rho} J_f - J_f(u)}{2\rho - \int_\Omega |\nabla u(x)|^2 dx},$$

where

$$B_\rho = \left\{ u \in W_0^{1,2}(\Omega) : \int_\Omega |\nabla u(x)|^2 dx < 2\rho \right\},$$

the problem

$$\begin{cases} -\Delta u = \dfrac{1}{\lambda} f(x,u) & \text{in } \Omega \\ u_{|\partial\Omega} = 0 \end{cases}$$

has at least two weak solutions one of which lies in B_ρ.

PROOF. The proof follows directly from part (b) of Theorem 2.4 applied taking $X = W_0^{1,2}(\Omega)$, $\Psi = \Psi_2$ and $\Phi = -J_f$, once observed that $\sup_{B_\rho} J_f = \sup_{\overline{(B_\rho)}_w} J_f$ and that, by classical arguments, the assumptions imply that, for each $\lambda > 0$ the functional $u \to \lambda\Psi_2(u) - J_f$ satisfies the Palais-Smale condition and is unbounded below as well. □

THEOREM 7.2. *Let $p > n$, let $f : \mathbf{R} \to \mathbf{R}$ be a continuous function and let $\alpha \in L^1(\Omega)$, with $\inf_\Omega \alpha \geq 0$. Moreover, assume that there are sequences $\{r_k\}$ in $\mathbf{R}^+$, with $\lim_{n\to\infty} r_k = +\infty$, and $\{\xi_k\}$ in $\mathbf{R}$ such that, for each $k \in \mathbf{N}$, one has*

$$\frac{\text{meas}(\Omega)}{p}|\xi_k|^p < r_k \tag{7.2}$$

and

$$\int_0^{\xi_k} f(t)dt = \sup_{|\xi| \leq c^*(pr_k)^{\frac{1}{p}}} \int_0^\xi f(t)dt, \tag{7.3}$$

where c^* is given in (7.1). Finally, assume that

$$\limsup_{|\xi|\to+\infty} \frac{\int_\Omega \alpha(x)dx \int_0^\xi f(t)dt}{|\xi|^p} > \frac{\text{meas}(\Omega)}{p}. \tag{7.4}$$

Then, the problem

$$\begin{cases} -\mathrm{div}(|\nabla u|^{p-2}\nabla u) + |u|^{p-2}u = \alpha(x)f(u) & \text{in } \Omega \\ \dfrac{\partial u}{\partial \nu} = 0 & \text{on } \partial\Omega \end{cases}$$

admits an unbounded sequence of weak solutions in $W^{1,p}(\Omega)$.

PROOF. We are going to apply Theorem 2.4 (part (c)) taking $X = W^{1,p}(\Omega)$, $\Psi = \Psi_1$ and $\Phi = -J_f$. Using the notation of Theorem 2.4, we now show that $\gamma = 0$. To this end, since $\lim_{k\to\infty} r_k = +\infty$, it is enough to show that $\varphi(r_k) = 0$ for all $k \in \mathbf{N}$. So, fix $k \in \mathbf{N}$. Now, denote by w the constant function on Ω taking the value ξ_k. By (7.2), we have $\Psi(w) < r_k$. Moreover, for each $u \in X$ such that $\|u\|_1^p \le pr_k$, one has

$$\sup_{x\in\Omega} |u(x)| \le c^*(pr_k)^{\frac{1}{p}}$$

and so, by (7.3), it follows that

$$-\int_0^{u(x)} f(t)dt \ge -\int_0^{\xi_k} f(t)dt$$

for all $x \in \Omega$, and hence, multiplying by α and integrating over Ω, one has

$$\Phi(u) \ge \Phi(w).$$

This shows that

$$\Phi(w) = \inf_{\|v\|_1^p \le pr_k} \Phi(v),$$

and so $\varphi(r_k) = 0$. Now, observe that, by (7.4), we can choose $\eta > \frac{\mathrm{meas}(\Omega)}{p}$ and a sequence $\{s_k\}$ in $\mathbf{R}$, with $\lim_{k\to\infty} |s_k| = +\infty$, in such a way that

$$\int_\Omega \alpha(x)dx \int_0^{s_k} f(t)dt \ge \eta |s_k|^p$$

for all $k \in \mathbf{N}$. Denote by v_k the constant function on Ω taking the value s_k. One has

$$\Phi(v_k) + \Psi(v_k) = \frac{\mathrm{meas}(\Omega)}{p}|s_k|^p - \int_\Omega \alpha(x)dx \int_0^{s_k} f(t)dt \le \left(\frac{\mathrm{meas}(\Omega)}{p} - \eta\right)|s_k|^p.$$

Consequently, the functional $\Phi + \Psi$ is unbounded below. At this point, Theorem 2.4 (part (c)) ensures that there exists a sequence $\{u_k\}$ of critical points of $\Phi + \Psi$ such that $\lim_{k\to\infty} \Psi(u_k) = +\infty$. But, of course, Ψ is bounded on each bounded subset of X, and so the sequence $\{u_k\}$ is unbounded in X. This concludes the proof. □

THEOREM 7.3. *Let* $p > n$, *let* $f : \mathbf{R} \to \mathbf{R}$ *be a continuous function and let* $\alpha \in L^1(\Omega)$, *with* $\inf_\Omega \alpha \ge 0$. *Moreover, assume that there are sequences* $\{r_k\}$ *in*

$\mathbf{R}^+$, *with* $\lim_{n\to\infty} r_k = 0$, *and* $\{\xi_k\}$ *in* $\mathbf{R}$ *such that, for each* $k \in \mathbf{N}$, *conditions* (7.2) *and* (7.3) *are satisfied. Finally, assume that*

$$\limsup_{\xi\to 0} \frac{\int_\Omega \alpha(x)dx \int_0^\xi f(t)dt}{|\xi|^p} > \frac{\text{meas}(\Omega)}{p}. \tag{7.5}$$

Then, the problem

$$\begin{cases} -\text{div}(|\nabla u|^{p-2}\nabla u) + |u|^{p-2}u = \alpha(x)f(u) & \text{in } \Omega \\ \dfrac{\partial u}{\partial \nu} = 0 & \text{on } \partial\Omega \end{cases}$$

admits a sequence of non zero weak solutions which strongly converges to 0 *in* $W^{1,p}(\Omega)$.

PROOF. We take X, Φ, Ψ as in the proof of Theorem 7.2. After observing that $\inf_X \Psi = 0$, from the proof of Theorem 7.2, we know that $\varphi(r_k) = 0$ for all $k \in \mathbf{N}$. Then, since $\lim_{k\to\infty} r_k = 0$, we have $\delta = 0$. By (7.5), there are $\mu > \frac{\text{meas}(\Omega)}{p}$ and a sequence $\{\rho_k\}$ in $\mathbf{R} \setminus \{0\}$, converging to 0, such that

$$\int_\Omega \alpha(x)dx \int_0^{\rho_k} f(t)dt \geq \mu|\rho_k|^p$$

for all $k \in \mathbf{N}$. If we denote by w_k the constant function on Ω taking the value ρ_k, of course the sequence $\{w_k\}$ strongly converges to 0 in X, and one has

$$\Phi(w_k) + \Psi(w_k) \leq \left(\frac{\text{meas}(\Omega)}{p} - \mu\right) |\rho_k|^p < 0$$

for all $k \in \mathbf{N}$. Since $\Phi(0) + \Psi(0) = 0$, this means that 0 is not a local minimum of $\Phi + \Psi$. Then, since 0 is the only global minimum of Ψ, Theorem 2.4 (part (*d*)) ensures that there exists a sequence $\{u_k\}$ of critical points of $\Phi + \Psi$ such that $\lim_{k\to\infty} \Psi(u_k) = 0$, and the proof is complete. □

Here is a sample of application of Theorem 5.2.

THEOREM 7.4. *Let* $p > n$ *and let* $f : \mathbf{R} \to \mathbf{R}$ *be a continuous function for which there are* $\eta > 0$ *and* $\xi_1 \in \mathbf{R}$, *with* $\text{meas}(\Omega)|\xi_1|^p > p\eta$, *such that*

$$\text{meas}(\Omega) \sup_{|\xi|\leq c^*(p\eta)^{\frac{1}{p}}} \int_0^\xi f(t)dt < \frac{p\eta}{|\xi_1|^p} \int_0^{\xi_1} f(t)dt, \tag{7.6}$$

where c^* *is given in* (7.1). *Assume also that*

$$\limsup_{|\xi|\to+\infty} \frac{\int_0^\xi f(t)dt}{|\xi|^p} < \frac{1}{ap} \tag{7.7}$$

for some a *satisfying*

$$a > \frac{\eta}{\frac{p\eta}{|\xi_1|^p} \int_0^{\xi_1} f(t)dt - \text{meas}(\Omega) \sup_{|\xi|\leq c^*(p\eta)^{\frac{1}{p}}} \int_0^\xi f(t)dt}.$$

Then, there exist $r > 0$ and a non-empty open set $B \subseteq]0, a[$ with the following property: for each $\lambda \in B$ and for each Carathéodory function $g : \Omega \times \mathbf{R} \to \mathbf{R}$, with $\sup_{|\xi| \leq s} |g(\cdot, \xi)| \in L^1(\Omega)$ for all $s > 0$, there exists $\delta > 0$ such that, for every $\mu \in [0, \delta]$, the problem

$$\begin{cases} -\mathrm{div}(|\nabla u|^{p-2}\nabla u) + |u|^{p-2}u = \lambda f(u) + \mu g(x, u) & \text{in } \Omega \\ \dfrac{\partial u}{\partial \nu} = 0 & \text{on } \partial\Omega, \end{cases}$$

has at least three weak solutions whose norms in $W^{1,p}(\Omega)$ are less than r.

PROOF. Note that if $\Psi_1(u) \leq \eta$, then $\sup_\Omega |u| \leq c^*(p\eta)^{\frac{1}{p}}$, and so, by (7.6), if u_1 denotes the constant function in Ω taking the value ξ_1, we have

$$\sup_{u \in \Psi_1^{-1}(]-\infty, \eta])} J_f(u) \leq \mathrm{meas}(\Omega) \sup_{|\xi| \leq c^*(p\eta)^{\frac{1}{p}}} \int_0^\xi f(t)dt < \frac{p\eta}{|\xi_1|^p} \int_0^{\xi_1} f(t)dt = \eta \frac{J_f(u_1)}{\Psi_1(u_1)}.$$

Hence, due to the choice of a, by Proposition 1.3 of [**35**], for a suitable constant σ, we have

$$\sup_{\lambda \in \mathbf{R}} \inf_{u \in W^{1,p}(\Omega)} (\Psi_1(u) + \lambda(\sigma - J_f(u))) < \inf_{u \in W^{1,p}(\Omega)} \sup_{\lambda \in [0,a]} (\Psi_1(u) + \lambda(\sigma - J_f(u))).$$

Further, thanks to (7.7), we can fix γ satisfying

$$\limsup_{|\xi| \to +\infty} \frac{\int_0^\xi f(t)dt}{|\xi|^p} < \gamma < \frac{1}{ap}.$$

Then, it is easy to see that, for a suitable $\beta > 0$ and for each $\lambda \in [0, a]$, we have

$$\Psi_1(u) - \lambda J_f(u) \geq \left(\frac{1}{p} - a\gamma\right) \|u\|_1^p - \beta.$$

So, if we take $X = W^{1,p}(\Omega)$, $\Psi = \Psi_1$, $J = J_f$ and $I = [0, a]$, all the assumptions of Theorem 5.2 are satisfied, and the conclusion follows, taking, of course, $\Gamma = J_g$. □

Here is a sample of application of Theorem 5.3.

THEOREM 7.5. *Let $f \in \mathscr{A}$ be such that*

$$\max\left\{\limsup_{\xi \to 0} \frac{\sup_{x \in \Omega} F(x, \xi)}{|\xi|^p}, \limsup_{|\xi| \to +\infty} \frac{\sup_{x \in \Omega} F(x, \xi)}{|\xi|^p}\right\} \leq 0 \tag{7.8}$$

and

$$\sup_{u \in W_0^{1,p}(\Omega)} \int_\Omega F(x, u(x))dx > 0$$

where

$$F(x, \xi) = \int_0^\xi f(x, s)ds$$

for all $(x,\xi) \in \Omega \times \mathbf{R}$.

Then, if we set

$$\gamma = \frac{1}{p} \inf \left\{ \frac{\int_\Omega |\nabla u(x)|^p dx}{\int_\Omega F(x,u(x))dx} : u \in W_0^{1,p}(\Omega), \int_\Omega F(x,u(x))dx > 0 \right\},$$

for each compact interval $[a,b] \subset]\gamma,+\infty[$ *there exists* $r > 0$ *with the following property: for every* $\lambda \in [a,b]$ *and every* $g \in \mathscr{A}$ *there exists* $\delta > 0$ *such that, for each* $\mu \in [0,\delta]$, *the problem*

$$\begin{cases} -\mathrm{div}(|\nabla u|^{p-2}\nabla u) = \lambda f(x,u) + \mu g(x,u) & \text{in } \Omega \\ u = 0 & \text{on } \partial\Omega \end{cases}$$

has at least three weak solutions whose norms in $W_0^{1,p}(\Omega)$ *are less than* r.

Proof. We deal with the case $p \le n$ only, the proof being similar in the other case. We are going to apply Theorem 5.3 taking $X = W_0^{1,p}(\Omega)$, $\Psi = \Psi_2$ and $J(u) = J_f$. Fix $\epsilon > 0$. In view of (7.8), there exist ρ_1, ρ_2, with $0 < \rho_1 < \rho_2$, such that

$$F(x,\xi) \le \epsilon|\xi|^p \tag{7.9}$$

for all $(x,\xi) \in \Omega \times (\mathbf{R} \setminus ([-\rho_2,-\rho_1] \cup [\rho_1,\rho_2]))$. Then, since F is bounded on $[-\rho_2,-\rho_1] \cup [\rho_1,\rho_2])$, we can choose $c > 0$ and $q > p$ (with $q \le \frac{pn-n+p}{n-p}$ if $p < n$) in such a way that

$$F(x,\xi) \le \epsilon|\xi|^p + c|\xi|^q$$

for all $(x,\xi) \in \Omega \times \mathbf{R}$. So, by continuous embedding, for some $c_1 > 0$ (independent of ϵ), one has

$$J(u) \le c_1(\epsilon\|u\|_2^p + \|u\|_2^q)$$

for all $u \in X$. Hence, we have

$$\limsup_{u\to 0} \frac{J(u)}{\Psi(u)} \le pc_1\epsilon. \tag{7.10}$$

Further, by (7.7) again, for each $u \in X \setminus \{0\}$, we have

$$\begin{aligned} \frac{J(u)}{\|u\|_2^p} &= \frac{\int_{\Omega(|u|\le\rho_2)} F(x,u(x))dx}{\|u\|_2^p} + \frac{\int_{\Omega(|u|>\rho_2)} F(x,u(x))dx}{\|u\|_2^p} \\ &\le \frac{\mathrm{meas}(\Omega)\sup_{\Omega\times[-\rho_2,\rho_2]} F}{\|u\|_2^p} + c_2\epsilon, \end{aligned}$$

for a constant c_2 independent of ϵ and u. So, we get

$$\limsup_{\|u\|\to+\infty} \frac{J(u)}{\Psi(u)} \le pc_2\epsilon. \tag{7.11}$$

Since ϵ is arbitrary, (7.10) and (7.11) give

$$\max\left\{\limsup_{\|u\|_2\to+\infty}\frac{J(u)}{\Psi(u)},\limsup_{u\to 0}\frac{J(u)}{\Psi(u)}\right\}\leq 0.$$

Then, with the notation of Theorem 5.3, we have $\alpha = 0$. By assumption, we also have $\beta > 0$. Thus, all the hypotheses of Theorem 5.3 are satisfied. Clearly, $\gamma = \frac{1}{\beta}$. The current conclusion then follows taking $\Gamma = J_g$. □

8. Final comments and notes

In this final section, we make some comments on the results of the previous sections and provide a guide to the relevant literature.

The results of Section 2, but one, are taken from [**161**]. More precisely, parts (a), (b) and (c) of Theorem 2.1 correspond, under more refined assumptions, to Theorems 2.1, 2.2 and 2.3 of [**161**], respectively, while part (b) of Theorem 2.4 agrees with Theorem 6 of [**164**]. A non-smooth version of Theorem 2.4 was established by S. A. Marano and D. Motreanu in [**149**].

The results of Section 3, but Theorem 3.14 (which was proved in [**179**]), come from [**170**]. A very interesting complement to the theory developed in Section 3 was provided by X. L. Fan and S. G. Deng in [**95**] (see also [**84**]).

Theorem 4.1 is formally new although it was implicitly used to obtain Theorem 1 of [**169**] (see also [**174**]). The other results of Section 4 were proved in [**166**] (see also [**158**]).

The main differences between Theorem 2.1 (part (a)) and Theorem 4.1 are quite clear. Namely, under essentially the same assumptions, while Theorem 2.1 ensures the existence of a $\hat{\tau}_\Psi$-local minimum of $\Psi + \lambda\Phi$ for all $\lambda > 0$ small enough, Theorem 4.1 is able to guarantee the existence of a τ_s-local minimum of $\Psi + \mu_n\Phi$ for a suitable sequence $\{\mu_n\}$ in $\mathbf{R}^+$ tending to zero. In practice, it can happen that the topology $\hat{\tau}_\Psi$ is finer than τ_s, and so being τ_s-local minimum is better that $\hat{\tau}_\Psi$-local minimum. In view of this discussion, it would be of interest to answer the following question:

PROBLEM 8.1. Under the hypotheses of Theorem 4.1, is there a $\delta^* > 0$ such that, for each $\lambda \in]0, \delta^*]$, the function $\Psi + \lambda\Phi$ has a τ_1-local minimum lying in $\Psi^{-1}(]-\infty, \rho])$?

Theorems 5.1, 5.2 and 5.3 agree with Theorem 3 of [**160**], Theorem 1 of [**177**] and Theorem 2 of [**179**], respectively. Also, Theorem 5.9 was essentially obtained in [**173**], but with a different proof. When no perturbation is considered (that is, when $\Gamma = 0$), Theorem 5.2 was obtained in [**160**] as a direct consequence of Theorem 5.1, without the boundedness assumption on Ψ. It would be interesting to know whether also Theorem 5.2 remains valid without such an assumption. However, in [**160**] the space X was also supposed to be separable. A non-smooth version of this result was obtained by S.A. Marano and D. Motreanu in [**148**]. Non-smooth versions of Theorem 5.2 have been obtained by A. Iannizzotto in [**117**] and by A. Kristály, W. Marzantowicz and Cs. Varga in [**127**]. The crucial minimax inequality appearing in Theorem 5.2 was studied by G. Bonanno in [**33**], [**35**] (see also [**189**]) and by G. Cordaro in [**69**], [**73**]. A simple sufficient condition for its validity (with

$I = [0, +\infty[$ was first provided by Proposition 3.1 of [**159**]. Such a condition, for instance, is satisfied when

$$\sup_{x \in X} \min\{\Phi(x), J(x)\} > 0,$$

$$\lim_{r \to 0^+} \operatorname{diam}(\Phi^{-1}(]-\infty, r])) = 0,$$

and there is a strict local minimum x_0 of Φ such that $\Phi(x_0) = J(x_0) = 0$ and

$$\limsup_{x \to x_0} \frac{J(x)}{\Phi(x)} \leq 0.$$

It is clear, however, that in such a situation, it is not convenient to apply Theorem 5.2, due to the lack of further information on the set A. In this case, one rather has to use Theorem 5.3 which is able to ensure the validity of the conclusion of Theorem 5.2 for each compact interval A contained in $]a, +\infty[$, where

$$a = \inf_{x \in \Phi^{-1}(]0,+\infty[) \cap J^{-1}(]0,+\infty[)} \frac{\Phi(x)}{J(x)}.$$

We also mention a method to draw three critical points theorems on the basis of part (a) of Theorem 2.4. For the first time, it was implicitly applied by F. Faraci in [**100**]. Then, it was formalized by D. Averna and G. Bonanno in [**25**] and refined furthermore by G. Bonanno and P. Candito in [**40**] (see also [**24**]).

Theorem 6.1 was established in [**175**] (see also [**181**]).

Our final comments concern the applications to differential equations of the general results obtained in Sections 2–6.

Theorem 7.1 (obtained in [**164**]) is strictly related to the famous Theorem 3.10 of [**6**]. A problem on this matter, proposed in [**167**], was completely solved by G. Anello in [**13**]. In particular, from [**13**], it follows that Theorem 7.1, when $r = 0$, is a proper improvement of Theorem 3.10 of [**6**], since the assumption $\lim_{\xi \to 0} \frac{\sup_{x \in \Omega} |f(x,\xi)|}{\xi} = 0$ can be dropped at all.

Theorems 7.2 and 7.3 are taken from [**163**]. A major problem on Theorem 7.2, still open, is as follows:

PROBLEM 8.2. Let $p > n$, let $f : \mathbf{R} \to \mathbf{R}$ be a continuous function and let $\alpha \in L^1(\Omega)$, with $\inf_\Omega \alpha \geq 0$. Assume that

$$\limsup_{|\xi| \to +\infty} \frac{\int_\Omega \alpha(x)dx \int_0^\xi f(t)dt}{|\xi|^p} > \frac{\operatorname{meas}(\Omega)}{p}.$$

Moreover, suppose that there is a divergent sequence $\{b_k\}$ in $\mathbf{R}^+$ such that, for each $k \in \mathbf{N}$, one has

$$\int_0^{\xi_k} f(t)dt = \sup_{|\xi| \leq b_k} \int_0^\xi f(t)dt$$

for some ξ_k with $|\xi_k| < b_k$.

Under such hypotheses, does the conclusion of Theorem 7.2 hold?

A similar open problem about Theorem 7.3 can be formulated assuming that the sequence $\{b_k\}$ converges to 0 and that

$$\limsup_{\xi \to 0} \frac{\int_\Omega \alpha(x)dx \int_0^\xi f(t)dt}{|\xi|^p} > \frac{\text{meas}(\Omega)}{p}.$$

Adopting a technique introduced by J. Saint Raymond in [**183**] (see also [**19**]), partial answers to these problems were provided in [**20**] and [**109**].

Theorem 7.4 and Theorem 7.5 agree with Theorem 2 of [**177**] and Theorem 3 of [**179**], respectively.

At this point, we want to stress what we already said in the Introduction: the literature contains a large number of papers where some of the general results of Sections 2–5 (or, at least, some of the strictly related results mentioned above) is applied to study some differential, integral or discrete problem. The list of references below is very strict, in the sense that we have inserted only those papers which contain such applications. We also stress the broadness of the range of problems which are the object of these applications. So, besides the papers already quoted above, we find:

- boundary value problems for elliptic equations and systems involving the p-Laplacian (often with $p = 2$) or even the $p(x)$-laplacian, in bounded or unbounded domains ([**1**]–[**3**], [**7**]–[**12**], [**14**]–[**18**], [**21**], [**22**], [**25**]–[**32**], [**34**], [**36**]–[**38**], [**43**], [**44**], [**47**], [**51**], [**55**]–[**64**], [**68**], [**71**], [**74**]–[**76**], [**82**], [**86**], [**87**], [**88**], [**89**], [**90**]–[**92**], [**96**], [**101**], [**103**], [**108**], [**111**], [**115**]–[**116**], [**121**], [**123**]–[**126**], [**128**], [**131**]–[**133**], [**136**], [**138**], [**142**]–[**143**], [**146**], [**150**]–[**155**], [**162**], [**165**], [**168**], [**172**], [**178**], [**180**], [**184**], [**185**], [**186**], [**191**], [**193**], [**195**], [**199**])
- non-smooth problems ([**39**], [**40**], [**42**], [**46**], [**52**]-[**54**], [**65**], [**66**], [**77**]–[**81**], [**83**], [**106**], [**107**], [**120**], [**122**], [**134**], [**197**])
- periodic solutions for Hamiltonian systems ([**48**], [**49**], [**72**], [**99**], [**104**], [**105**], [**110**], [**187**], [**188**], [**190**], [**192**], [**194**])
- higher order problems ([**4**], [**45**], [**85**], [**141**], [**145**])
- Hammerstein nonlinear integral equations ([**23**], [**98**], [**102**], [**112**], [**144**])
- discrete problems ([**5**], [**41**], [**114**], [**118**], [**198**])
- problems on compact Riemann manifolds ([**129**], [**130**])
- problems involving Dirichlet forms ([**139**]).

We conclude by recalling two further open problems, keeping the notation of Section 7.

PROBLEM 8.3. Let $n \geq 3$, $p = 2$ and let τ be the weak topology of $W_0^{1,2}(\Omega)$. Is there some $\varphi \in \mathscr{A}$, with $q < \frac{n+2}{n-2}$, such that, for each $r \in \mathbf{R}$ and each $\lambda > 0$ small enough, the functional $\Psi_2 - \lambda J_\varphi$ is unbounded below and the set $(\Psi_2 - \lambda J_\varphi)^{-1}(r)$ has no isolated points with respect to the topology τ_s?

In the light of Theorem 4.1, the relevance of Problem 8.3 is clear. Actually, if f was answering Problem 8.3 in the affirmative, then, by Theorem 4.1, for a suitable sequence μ_k in $\mathbf{R}^+$ tending to 0, the functional $\Psi_2 - \mu_k J_\varphi$ would have infinitely many local minima in the topology τ_s. Consequently, problem $(DP_{\mu_k \varphi})$ would have infinitely many weak solutions for each $k \in \mathbf{N}$.

Let $L > 0$ be fixed. Denote by $\mathscr{C}_L$ the class of all Lipschitzian functions $\varphi : \mathbf{R} \to \mathbf{R}$, with $\varphi(0) = 0$, having Lipschitz constant L, such that

$$\sup_{\xi \in \mathbf{R}} \int_0^{\xi} \varphi(t)dt = 0.$$

For $\varphi \in \mathscr{C}_L$, denote by Λ_φ the set of all $\lambda > 0$ such that problem $(DP_{\lambda\varphi})$ has at least one nonzero classical solution.

Denote also by λ_1 the first eigenvalue of the problem

$$\begin{cases} -\Delta u = \lambda u & \text{in}\Omega \\ u_{|\partial\Omega} = 0. \end{cases}$$

In [**175**], using Theorem 6.1, it was pointed out that

$$\inf_{\varphi \in \mathscr{C}_L} \inf \Lambda_\varphi \geq \frac{3\lambda_1}{L}.$$

In [**94**], X. L. Fan proved that

$$\inf_{\varphi \in \mathscr{C}_L} \inf \Lambda_\varphi > \frac{3\lambda_1}{L}.$$

In [**97**], the following problem has been proposed:

PROBLEM 8.4. Let $n \geq 2$ and $\Omega = \{x \in \mathbf{R}^n : a < |x| < b\}$ with $0 < a < b$. Is there $\varphi \in \mathscr{C}_1$ for which $\Lambda_\varphi \neq 0$?

Bibliography

[1] G. A. Afrouzi and S. Heidarkhani, *Three solutions for a Dirichlet boundary value problem involving the p-Laplacian*, Nonlinear Anal., **66** (2007), 2281–2288.

[2] G. A. Afrouzi and S. Heidarkhani, *Three solutions for a quasilinear boundary value problem*, Nonlinear Anal., **69** (2008), 3330–3336.

[3] G. A. Afrouzi and S. Heidarkhani, *Existence of three solutions for a class of Dirichlet quasilinear elliptic systems involving the $(p_1, \ldots, p_n)$-Laplacian*, Nonlinear Anal., **70** (2009), 135–143.

[4] G. A. Afrouzi, S. Heidarkhani and D. O'Regan, *Existence of three solutions for a doubly eigenvalue fourth-order boundary value problem*, Taiwanese J. Math., to appear.

[5] K. Ai, *Multiplicity results for discrete eigenvalue problems*, Far East J. Dyn. Syst., **9** (2007), 181–192.

[6] A. Ambrosetti and P. H. Rabinowitz, *Dual variational methods in critical point theory and applications*, J. Funct. Anal., **14** (1973), 349–381.

[7] G. Anello, *A multiplicity theorem for critical points of functionals on reflexive Banach spaces*, Arch. Math. (Basel), **82** (2004), 172–179.

[8] G. Anello, *Existence of infinitely many weak solutions for a Neumann problem*, Nonlinear Anal., **57** (2004), 199–209.

[9] G. Anello, *Existence of two local minima for functionals on reflexive Banach spaces*, Nonlinear Anal., **61** (2005), 1179–1187.

[10] G. Anello, *On the Dirichlet problem for the equation $\Delta u = g(x, u) + \lambda f(x, u)$ with no growth conditions on f*, Taiwanese J. Math., **10** (2006), 1515–1522.

[11] G. Anello, *Existence and multiplicity of solutions for nonlinear boundary value problems*, J. Nonlinear Convex Anal., **7** (2006), 335–341.

[12] G. Anello, *Existence and multiplicity of solutions to a perturbed Neumann problem*, Math. Nachr., **280** (2007), 1755–1764.

[13] G. Anello, *A note on a problem by Ricceri on the Ambrosetti-Rabinowitz condition*, Proc. Amer. Math. Soc., **135** (2007), 1875–1879.

[14] G. Anello, *Existence of solutions for a perturbed Dirichlet problem without growth conditions*, J. Math. Anal. Appl., **330** (2007), 1169–1178.

[15] G. Anello, *Remarks on a multiplicity theorem for a perturbed Neumann problem*, J. Math. Anal. Appl. **346** (2008), 274–279.

[16] G. Anello and G. Cordaro, *Existence of solutions of the Neumann problem involving the p-Laplacian via a variational principle of Ricceri*, Arch. Math. (Basel), **79** (2002), 274–287.

[17] G. Anello and G. Cordaro, *An existence theorem for the Neumann problem involving the p-Laplacian*, J. Convex Anal., **10** (2003), 185–198.

[18] G. Anello and G. Cordaro, *An existence and localization theorem for the solutions of a Dirichlet problem*, Ann. Polon. Math., **83** (2004), 107–112.

[19] G. Anello and G. Cordaro, *Positive infinitely many and arbitrarily small solutions for the Dirichlet problem involving the p-Laplacian*, Proc. Royal Soc. Edinburgh Sect. A, **132** (2002), 511–519.

[20] G. Anello and G. Cordaro, *Infinitely many positive solutions for the Neumann problem involving the p-Laplacian*, Colloq. Math., **97** (2003), 221–231.

[21] G. Anello and G. Cordaro, *Three solutions for a perturbed sublinear elliptic problem in* $\mathbf{R}^N$, Glasg. Math. J., **47** (2005), 205–212.

[22] G. Anello and G. Cordaro, *Perturbation from Dirichlet problem involving oscillating nonlinearities*, J. Differential Equations, **234** (2007), 80–90.

[23] G. Anello and G. Cordaro, *Nonlinear equations involving non positive definite linear operators via variational methods*, J. Integral Equations Appl. **19** (2007) 1–12.

[24] D. Arcoya and J. Carmona, *A nondifferentiable extension of a theorem of Pucci and Serrin and applications*, J. Differential Equations **235** (2007), 683–700.

[25] D. Averna and G. Bonanno, *A three critical points theorem and its applications to the ordinary Dirichlet problem*, Topol. Methods Nonlinear Anal., **22** (2003), 93–104.

[26] D. Averna and G. Bonanno, *Three solutions for a quasilinear two point boundary value problem involving the one-dimensional p-Laplacian*, Proc. Edinb. Math. Soc., **47** (2004), 257–270.

[27] D. Averna and G. Bonanno, *Three solutions for a Neumann boundary value problem involving the p-Laplacian*, Matematiche (Catania), **60** (2005), 81–91.

[28] D. Averna and R. Salvati, *Three solutions for a mixed boundary value problem involving the one-dimensional p-Laplacian*, J. Math. Anal. Appl., **298** (2004), 245–260.

[29] Z. B. Bai and W. G. Ge, *Existence of three positive solutions for some second-order boundary value problems*, Comput. Math. Appl., **48** (2004), 699–707.

[30] G. Barletta and R. Livrea, *Existence of three periodic solutions for a non autonomous system*, Matematiche (Catania), **57** (2002), 205–215.

[31] G. Bilotta, *Existence of infinitely many solutions for a quasilinear Neumann problem*, Panamer. Math. J., **13** (2003), 1–18.

[32] G. Bonanno, *Existence of three solutions for a two point boundary value problem*, Appl. Math. Lett., **13** (2000), 53–57.

[33] G. Bonanno, *A minimax inequality and its applications to ordinary differential equations*, J. Math. Anal. Appl., **270** (2002), 210–229.

[34] G. Bonanno, *Multiple solutions for a Neumann boundary value problem*, J. Nonlinear Convex Anal., **4** (2003), 287–290.

[35] G. Bonanno, *Some remarks on a three critical points theorem*, Nonlinear Anal., **54** (2003), 651–665.

[36] G. Bonanno, *A critical points theorem and nonlinear differential problems*, J. Global Optim., **28** (2004), 249–258.

[37] G. Bonanno, *Multiple critical points theorem without the Palais-Smale condition*, J. Math. Anal. Appl., **299** (2004), 600–614.

[38] G. Bonanno and P. Candito, *Three solutions to a Neumann problem for elliptic equations involving the p-Laplacian*, Arch. Math. (Basel), **80** (2003), 424–429.

[39] G. Bonanno and P. Candito, *On a class of nonlinear variational-hemivariational inequalities*, Appl. Anal., **83** (2004), 1229–1244.

[40] G. Bonanno and P. Candito, *Non-differentiable functionals and applications to elliptic problems with discontinuous nonlinearities*, J. Differential Equations, **244** (2008), 3031–3059.

[41] G. Bonanno and P. Candito, *Infinitely many solutions for a class of discrete non-linear boundary value problems*, Appl. Anal., **88** (2009), 605–616.
[42] G. Bonanno and A. Chinnì, *Discontinuous elliptic problems involving the $p(x)$-Laplacian*, Math. Nachr., to appear.
[43] G. Bonanno and G. D'Aguì, *A Neumann boundary value problem for the Sturm-Liouville equation*, Appl. Math. and Comput., **208** (2009), 318–327.
[44] G. Bonanno and G. D'Aguì, *On the Neumann problem for elliptic equations involving the p-Laplacian*, J. Math. Anal. Appl., **358** (2009), 223–228.
[45] G. Bonanno and B. Di Bella, *A boundary value problem for fourth-order elastic beam equations*, J. Math. Anal. Appl., **343** (2008), 1166–1176.
[46] G. Bonanno and N. Giovannelli, *An eigenvalue Dirichlet problem involving the p-Laplacian with discontinuous nonlinearities*, J. Math. Anal. Appl., **308** (2005), 596–604.
[47] G. Bonanno and R. Livrea, *Multiplicity theorems for the Dirichlet problem involving the p-Laplacian*, Nonlinear Anal., **54** (2003), 1–7.
[48] G. Bonanno and R. Livrea, *Periodic solutions for a class of second order Hamiltonian systems*, Electron. J. Differential Equations, Vol. 2005 (2005), No. **115**, pp. 1–13.
[49] G. Bonanno and R. Livrea, *Multiple periodic solutions for Hamiltonian systems with not coercive potential*, J. Math. Anal. Appl., **363** (2010), 627–638.
[50] G. Bonanno and G. Molica Bisci, *Infinitely many solutions for a boundary value problem with discontinuous nonlinearities*, Bound. Value Probl., 2009, Art. ID 670675, 20 pp.
[51] G. Bonanno and G. Molica Bisci, *Infinitely many solutions for a Dirichlet problem involving the p-Laplacian*, Proc. Royal Soc. Edinburgh Sect. A, **140** (2010), 737–752.
[52] B. E. Breckner, A. Horváth and Cs. VARGA, *A multiplicity result for a special class of gradient-type systems with non-differentiable term*, Nonlinear Anal., **70** (2009), 606–620.
[53] B. E. Breckner and Cs. Varga, *A multiplicity result for gradient-type systems with non-differentiable term*, Acta Math. Hungar., **118** (2008), 85–104.
[54] B. E. Breckner and Cs. Varga, *Infinitely many solutions for a class of systems of differential inclusions*, Proc. Edinb. Math. Soc., to appear.
[55] F. Cammaroto and A. Chinnì, *Infinitely many solution for a two point boundary value problem*, Far East J. Math. Sci., **11** (2003), 41–51.
[56] F. Cammaroto, A. Chinnì and B. Di Bella, *Infinitely many solutions for the Dirichlet problem involving the p-Laplacian*, Nonlinear Anal., **61** (2005), 41–49.
[57] F. Cammaroto, A. Chinnì and B. Di Bella, *Infinitely many solutions for the Dirichlet problem via a variational principle of Ricceri*, in "Variational Analysis and Applications", F. Giannessi and A. Maugeri eds., 215–229, Springer, 2005.
[58] F. Cammaroto, A. Chinnì and B. Di Bella, *Some multiplicity results for quasilinear Neumann problems*, Arch. Math. (Basel), **86** (2006), 154–162.
[59] F. Cammaroto, A. Chinnì and B. Di Bella, *Multiple solutions for a two point boundary value problem*, J. Math. Anal. Appl., **323** (2006), 530–534.
[60] F. Cammaroto, A. Chinnì and B. Di Bella, *Multiple solutions for a quasilinear elliptic variational system on strip-like domains*, Proc. Edinb. Math. Soc., **50** (2007), 597–603.
[61] F. Cammaroto, A. Chinnì and B. Di Bella, *Multiplicity results for a perturbed nonlinear Schrödinger equation*, Glasg. Math. J., **49** (2007), 423–429.
[62] F. Cammaroto, A. Chinnì and B. Di Bella, *Multiple solutions for a Dirichlet problem involving the p-Laplacian*, Dynam. Systems Appl., **16** (2007), 673–679.
[63] F. Cammaroto, A. Chinnì and B. Di Bella, *Multiple solutions for a Neumann problem involving the $p(x)$-Laplacian*, Nonlinear Anal., **71** (2009), 4486–4492.
[64] P. Candito, *Existence of three solutions for a non-autonomous two point boundary value problem*, J. Math. Anal. Appl., **252** (2000), 532–537.
[65] P. Candito, *Infinitely many solutions to the Neumann problem for elliptic equations involving the p-Laplacian and with discontinuous nonlinearities*, Proc. Edinb. Math. Soc., **45** (2002), 397–409.
[66] G. Chang and Z. Shen, *Three solutions for an obstacle problem for a class of variational-hemivariational inequalities*, Appl. Math. Comput., **215** (2009) 2063–2069.
[67] G. Choquet, *Lectures on Analysis*, Vol. I, W. A. Benjamin, Inc., 1969.
[68] N. T. Chung, *Multiple solutions for quasilinear elliptic problems with nonlinear boundary conditions*, Electron. J. Differential Equations **2008**, No 165, 6 pp. (electronic).
[69] G. Cordaro, *On a minimax problem of Ricceri*, J. Inequal. Appl., **6** (2001), 261–285.

[70] G. Cordaro, *Three periodic solutions to an eigenvalue problem for a class of second order Hamiltonian systems*, Abstr. Appl. Anal., **18** (2003), 1037–1045.

[71] G. Cordaro, *A multiplicity result for a perturbation of a symmetric Dirichlet problem*, Appl. Anal., **83** (2004), 799–806.

[72] G. Cordaro, *Existence and location of periodic solutions to convex and non coercive Hamiltonian systems*, Discrete Contin. Dyn. Syst, **12** (2005), 983–996.

[73] G. Cordaro, *Further results related to a minimax problem of Ricceri*, J. Inequal. Appl., (2005), 523–533.

[74] G. Cordaro, *Multiple solutions to a perturbed Neumann problem*, Studia Math., **178** (2007), 167–175.

[75] G. Cordaro and G. Rao, *Three solutions for a perturbed Dirichlet problem*, Nonlinear Anal., **68** (2008), 3879–3883.

[76] G. Cordaro and G. Rao, *Three periodic solutions for perturbed second order Hamiltonian systems* J. Math. Anal. Appl., **359** (2009), 780–785.

[77] G. Dai, *Infinitely many solutions for a Neumann-type differential inclusion problem involving the $p(x)$-Laplacian*, Nonlinear Anal., **70** (2009), 2297–2305.

[78] G. Dai, *Three solutions for a Neumann-type differential inclusion problem involving the $p(x)$-Laplacian*, Nonlinear Anal., **70** (2009), 3755–3760.

[79] G. Dai, *Infinitely many solutions for a hemivariational inequality involving the $p(x)$-Laplacian*, Nonlinear Anal., **71** (2009), 186–195.

[80] G. Dai, *Infinitely many solutions for a differential inclusion problem in image involving the $p(x)$-Laplacian*, Nonlinear Anal., **71** (2009), 1116–1123.

[81] G. Dai, *Three symmetric solutions for a differential inclusion system involving the $(p(x), q(x))$-Laplacian*, Nonlinear Anal. **71** (2009), 1763–1771.

[82] G. Dai, *Infinitely many non-negative solutions for a Dirichlet problem involving $p(x)$-Laplacian*, Nonlinear Anal., **71**, (2009) 5840–5849.

[83] G. Dai and W. Liu, *Three solutions for a differential inclusion problem involving the $p(x)$-Laplacian*, Nonlinear Anal., **71** (2009), 5318–5326.

[84] S.-G. Deng, *A local mountain pass theorem and applications to a double perturbed $p(x)$-Laplacian equations*, Appl. Math. Comput., **211** (2009), 234–241.

[85] Y. Deng and H. Pi, *Multiple solutions for p-harmonic type equations*, Nonlinear Anal. **71** (2009), 4952–4959.

[86] A. G. Di Falco, *Infinitely many solutions of the Neumann problem for quasilinear elliptic systems*, Matematiche (Catania), **58** (2003), 117–130.

[87] A. G. Di Falco, *Infinitely many solutions to the Dirichlet problem for quasilinear elliptic systems*, Matematiche (Catania), **60** (2005), 163–179.

[88] Z. Du and L. Kong, *Existence of three solutions for systems of multi-point boundary value problems*, Electron. J. Qual. Theory Differ. Equ., Spec. Ed I, **2009**, No 10, 17 pp. (electronic).

[89] Z. Du, X. Lin and C. C. Tisdell, *A multiplicity result for p-Lapacian boundary value problems via critical points theorem*, Applied Math. Comput., **205** (2008), 231–237.

[90] S. El Manouni, *Multiplicity results for some elliptic problems of n-Laplace type*, preprint.

[91] S. El Manouni, *The study of nonlinear problems for p-Laplacian in* $\mathbf{R}^n$ *via a Ricceri's principle*, preprint.

[92] S. El Manouni and M. Kbiri Alaoui, *A result on elliptic systems with Neumann conditions via Ricceri's three critical point theorem*, Nonlinear Anal., **71** (2009), 2343–2348.

[93] X. L. Fan, *A remark on Ricceri's conjecture for a class of nonlinear eigenvalue problems*, J. Math. Anal. Appl., **349** (2009), 436–442.

[94] X. L. Fan, *On Ricceri's conjecture for a class of nonlinear eigenvalue problems*, Appl. Math. Letters, **22** (2009), 1386–1389.

[95] X. L. Fan and S. G. Deng, *Remarks on Ricceri's variational principle and applications to the $p(x)$-Laplacian equations*, Nonlinear Anal., **67** (2007), 3064–3075.

[96] X. L. Fan and C. Ji, *Existence of infinitely many solutions for a Neumann problem involving the $p(x)$-Laplacian*, J. Math. Anal. Appl., **334** (2007), 248–260.

[97] X. L. Fan and B. Ricceri, *On the Dirichlet problem involving nonlinearities with non-positive primitive: a problem and a remark*, Appl. Anal., to appear.

[98] F. Faraci, *Bifurcation theorems for Hammerstein nonlinear integral equations*, Glasg. Math. J., **44** (2002), 471–481.

[99] F. Faraci, *Three periodic solutions for a second order nonautonomous system*, J. Nonlinear Convex Anal., **3** (2002), 393–399.
[100] F. Faraci, *Multiplicity results for a Neumann problem involving the p-Laplacian*, J. Math. Anal. Appl., **277** (2003), 180–189.
[101] F. Faraci, *A bifurcation theorem for noncoercive integral functionals*, Comment. Math. Univ. Carolin., **45** (2004), 443–456.
[102] F. Faraci, *Existence and multiplicity results for a nonlinear Hammerstein integral equation*, in "Variational Analysis and Applications," F. Giannessi and A. Maugeri eds., 359–371, Springer, 2005.
[103] F. Faraci, *Multiple solutions for two nonlinear problems involving the p-Laplacian*, Nonlinear Anal., **63** (2005), e1017–e1029.
[104] F. Faraci *Multiple periodic solutions for second order systems with changing sign potential*, J. Math. Anal. Appl., **319** (2006), 567–578.
[105] F. Faraci and A. Iannizzotto, *A multiplicity theorem for a perturbed second order nonautonomous system*, Proc. Edinb. Math. Soc., **49** (2006), 267–275.
[106] F. Faraci, A. Iannizzotto, P. Kupán and Cs. Varga, *Existence and multiplicity results for hemivariational inequalities with two parameters*, Nonlinear Anal., **67** (2007), 2654–2669.
[107] F. Faraci, A. Iannizzotto, H. Lisei and Cs. Varga, *A multiplicity result for hemivariational inequalities*, J. Math. Anal. Appl., **330** (2007), 683–698.
[108] F. Faraci and A. Kristály, *One-dimensional field equations involving an oscillatory nonlinear term*, Discrete Contin. Dyn. Syst, **18** (2007), 107–120.
[109] F. Faraci and A. Kristály, *On an open question of Ricceri concerning a Neumann problem*, Glasg. Math. J., **49** (2007), 189–195.
[110] F. Faraci and R. Livrea, *Infinitely many periodic solutions for a second order nonautonomous system*, Nonlinear Anal., **54** (2003), 417–429.
[111] F. Faraci and R. Livrea, *Bifurcation theorems for nonlinear problems with lack of compactness*, Ann. Polon. Math., **82** (2003), 77–85.
[112] F. Faraci and V. Moroz, *Solutions of Hammerstein integral equations via a variational principle*, J. Integral Equations Appl., **15** (2003), 385–402.
[113] N. Ghoussoub and D. Preiss, *A general mountain pass principle for locating and classifying critical points*, Ann. Inst. H. Poincaré Anal. Non Linéaire, **6** (1989), 321–330.
[114] T. He and Y. Su, *On discrete fourth-order boundary value problems with three parameters*, J. Comput. Appl. Math., to appear.
[115] S. Heidarkhani and D. Motreanu, *Multiplicity results for a two-point boundary value problem*, Panamer. Math. J., **19** (2009), 69–78.
[116] A. Iannizzotto, *A sharp existence and localization theorem for a Neumann problem*, Arch. Math. (Basel), **82** (2004), 352–360.
[117] A. Iannizzotto, *Three critical points for perturbed nonsmooth functionals and applications*, Nonlinear Anal., **72** (2010), 1319–1338.
[118] L. Jiang and Z. Zhou, *Three solutions to Dirichlet boundary value problems for p-Laplacian difference equations*, Adv. Difference Equ., **2008**, Art. ID 345916, 10 pp.
[119] E. Klein and A. C. Thompson, *Theory of Correspondences*, John Wiley & Sons, 1984.
[120] A. Kristály, *Multiplicity results for an eigenvalue problem for hemivariational inequalities in strip-like domains*, Set-Valued Anal., **13** (2005), 85–103.
[121] A. Kristály, *Existence of two nontrivial solutions for a class of quasilinear elliptic variational systems on strip-like domains*, Proc. Edinb. Math. Soc., **48** (2005), 465–477.
[122] A. Kristály, *Infinitely many solutions for a differential inclusion problem in* $\mathbf{R}^N$, J. Differential Equations, **220** (2006), 511–530.
[123] A. Kristály, *A double eigenvalue problem for Schrödinger equations involving sublinear nonlinearities at infinity*, Electron. J. Differential Equations **2007**, No 42, 11 pp. (electronic).
[124] A. Kristály, *Multiple solutions for a sublinear Schrödinger equation*, NoDEA Nonlinear Differential Equations Appl., **14** (2007), 291–301.
[125] A. Kristály, *Infinitely many small and large solutions for elliptic systems*, preprint.
[126] A. Kristály, H. Lisei and Cs. Varga, *Multiple solutions for p-Laplacian type equations*, Nonlinear Anal., **68** (2008), 1375–1381.
[127] A. Kristály, W. Marzantowicz and Cs. Varga, *A non-smooth there critical points theorem with applications in differential inclusions*, J. Global Optim., **46** (2010), 49–62.

[128] A. Kristály, M. MihăIlescu and V. RĂDULESCU, *Two nontrivial solutions for a non-homogeneous Neumann problem: an Orlicz-Sobolev setting*, Proc. Roy. Soc. Edinburgh Sect. A, **139** (2009), 367–379.

[129] A. Kristály, N. S. PAPAGEORGIOU and Cs. VARGA, *Multiple solutions for a class of Neumann elliptic problems on compact Riemann manifolds with boundary*, Canad. Math. Bull., to appear.

[130] A. Kristály and V. RăDulescu, *Sublinear eigenvalue problems on compact Riemannian manifolds with applications in Emden-Fowler equations*, Studia Math., **191** (2009), 237–246.

[131] A. Kristály and Cs. Varga, *On a class of nonlinear eigenvalue problems in* $\mathbf{R}^N$, Math. Nachr., **278** (2005), 1756–1765.

[132] A. Kristály and Cs. Varga, *Multiple solutions for elliptic problems with singular and sublinear potentials*, Proc. Amer. Math. Soc., **135** (2007), 2121–2126.

[133] A. Kristály and Cs. Varga, *Multiple solutions for a degenerate elliptic equation involving sublinear terms at infinity*, J. Math. Anal. Appl., **352** (2009), 139–148.

[134] A. Kristály, Cs. Varga and V. Varga, *A nonsmooth principle of symmetric criticality and variational-hemivariational inequalities*, J. Math. Anal. Appl., **325** (2007), 975–986.

[135] K. Kuratowski and C. Ryll-Nardzewski, *A general theorem on selectors*, Bull. Acad. Polon. Sci. Sér. Sci. Math. Astronom. Phys., **13** (1965), 397–403.

[136] C. Li and C.-L. Tang, *Three solutions for a class of quasilinear elliptic systems involving the* (p,q)*-Laplacian*, Nonlinear Anal., **69** (2009), 3322–3329.

[137] C. Li and C.-L. Tang, *Three solutions for a Navier boundary value problem involving the p-biharmonic*, Nonlinear Anal., **72** (2010), 1339–1347.

[138] H. Lisei, G. MoroşAnu and Cs. Varga, *Multiplicity results for double eigenvalue problems involving the p-Laplacian*, Taiwanese J. Math., **13** (2009), 1095–1110.

[139] H. Lisei and Cs. Varga, *Multiple solutions for nonlinear equations involving Dirichlet forms*, Topics in mathematics, computer science and philosophy, 135–145, Presa Univ. Clujeană, Cluj-Napoca, 2008.

[140] H. Lisei, Cs. Varga and A. Horváth, *Multiplicity results for a class of quasilinear problems on unbounded domains*, Arch. Math. (Basel), **90** (2008), 256–266.

[141] H. Liu and N. Su, *Existence of three solutions for a p-biharmonic problem*, Dyn. Contin. Discrete Impuls. Syst. Ser. A Math. Anal., **15** (2008), 445–452.

[142] J. Liu and X. Shi, *Existence of three solutions for a class of quasilinear elliptic systems involving the* $(p(x),q(x))$*-Laplacian* Nonlinear Anal., **71** (2009), 550–557.

[143] Q. Liu, *Existence of three solutions for* $p(x)$*-Laplacian equations*, Nonlinear Anal., **68** (2008), 2119–2127.

[144] X.-L. Liu, *On a nonlinear Hammerstein integral equation with a parameter*, Nonlinear Anal., **70** (2009), 3887–3893.

[145] X.-L. Liu and W.-T. Li, *Existence and multiplicity of solutions for fourth-order boundary value problems with parameters*, J. Math. Anal. Appl., **327** (2007), 362–375.

[146] R. Livrea, *Existence of three solutions for a quasilinear two point boundary value problem*, Arch. Math. (Basel), **79** (2002), 288–298.

[147] S. A. Marano and D. Motreanu, *Existence of two nontrivial solutions for a class of elliptic eigenvalue problems*, Arch. Math. (Basel), **75** (2000), 53–58.

[148] S. A. Marano and D. Motreanu, *On a three critical points theorem for non-differentiable functions and applications to nonlinear boundary value problems*, Nonlinear Anal., **48** (2002), 37–52.

[149] S. A. Marano and D. Motreanu, *Infinitely many critical points of non-differentiable functions and applications to a Neumann type problem involving the p-Laplacian*, J. Differential Equations, **182** (2002), 108–120.

[150] I. I. Mezei, *Multiple radially symmetric solutions for a quasilinear eigenvalue problem*, Acta Univ. Sapientiae Math., **1** (2009), 109–120.

[151] I. I. Mezei and Cs. Varga, *Multiplicity result for a double eigenvalue quasilinear problem on unbounded domain* Nonlinear Anal., **69** (2008), 4099–4105.

[152] M. MihăIlescu, *Existence and multiplicity of solutions for a Neumann problem involving the* $p(x)$*-Laplace operator*, Nonlinear Anal., **67** (2007), 1419–1425.

[153] D. S. Moschetto, *Infinitely many solutions to the Dirichlet problem for quasilinear elliptic systems involving the* $p(x)$ *and* $q(x)$*-Laplacian*, Matematiche (Catania), **63** (2008), 223–233.

[154] D. S. Moschetto, *Infinitely many solutions to the Neumann problem for quasilinear elliptic systems involving the $p(x)$ and $q(x)$-Laplacian*, Int. Math. Forum, **4** (2009), 1201–1211.
[155] D. S. Moschetto, *A quasilinear Neumann problem involving the $p(x)$-Laplacian*, Nonlinear Anal., **71** (2009), 2739–2743.
[156] P. Pucci and J. Serrin, *A mountain pass theorem*, J. Differential Equations, **60** (1985), 142–149.
[157] B. Ricceri, *Some topological mini-max theorems via an alternative principle for multifunctions*, Arch. Math. (Basel), **60** (1993), 367–377.
[158] B. Ricceri, *A new method for the study of nonlinear eigenvalue problems*, C. R. Acad. Sci. Paris, Série I, **328** (1999), 251–256.
[159] B. Ricceri, *Existence of three solutions for a class of elliptic eigenvalue problems*, Math. Comput. Modelling, **32** (2000), 1485–1494.
[160] B. Ricceri, *On a three critical points theorem*, Arch. Math. (Basel), **75** (2000), 220–226.
[161] B. Ricceri, *A general variational principle and some of its applications*, J. Comput. Appl. Math., **113** (2000), 401–410.
[162] B. Ricceri, *Existence and location of solutions to the Dirichlet problem for a class of nonlinear elliptic equations*, Appl. Math. Lett., **14** (2001), 143–148.
[163] B. Ricceri, *Infinitely many solutions of the Neumann problem for elliptic equations involving the p-Laplacian*, Bull. London Math. Soc., **33** (2001), 331–340.
[164] B. Ricceri, *On a classical existence theorem for nonlinear elliptic equations*, in "Experimental, constructive and nonlinear analysis", M. Théra ed., 275–278, CMS Conf. Proc. **27**, Canad. Math. Soc., 2000.
[165] B. Ricceri, *A bifurcation theory for some nonlinear elliptic equations*, Colloq. Math., **95** (2003), 139–151.
[166] B. Ricceri, *A further improvement of a minimax theorem of Borenshtein and Shul'man*, J. Nonlinear Convex Anal., **2** (2001), 279–283.
[167] B. Ricceri, *Some research perspectives in nonlinear functional analysis*, Seminar on fixed point theory, Cluj-Napoca, **3** (2002), 99–110.
[168] B. Ricceri, *Three solutions for a Neumann problem*, Topol. Methods Nonlinear Anal., **20** (2002), 275–281.
[169] B. Ricceri, *Three topological problems about integral functionals on Sobolev spaces*, J. Global Optim., **28** (2004), 401–404.
[170] B. Ricceri, *Sublevel sets and global minima of coercive functionals and local minima of their perturbations*, J. Nonlinear Convex Anal., **5** (2004), 157–168.
[171] B. Ricceri, *Integral functionals on Sobolev spaces having multiple local minima*, in "Variational Analysis and Applications", F. Giannessi and . Maugeri eds., 953–961, Springer, 2005.
[172] B. Ricceri, *A multiplicity theorem for the Neumann problem*, Proc. Amer. Math. Soc., **134** (2006), 1117–1124.
[173] B. Ricceri, *Minimax theorems for limits of parametrized functions having at most one local minimum lying in a certain set*, Topology Appl., **153** (2006), 3308–3312.
[174] B. Ricceri, *Topological problems in nonlinear and functional analysis*, in "Open problems in topology II", E. Pearl ed., 585-593, Elsevier, 2007.
[175] B. Ricceri, *A remark on a class of nonlinear eigenvalue problems*, Nonlinear Anal., **69** (2008), 2964–2967.
[176] B. Ricceri, *Minimax theorems for functions involving a real variable and applications*, Fixed Point Theory, **9** (2008), 275–291.
[177] B. Ricceri, *A three critical points theorem revisited*, Nonlinear Anal., **70** (2009), 3084–3089.
[178] B. Ricceri, *On an elliptic Kirchhoff-type problem depending on two parameters*, J. Gobal Optim., **46** (2010), 543–549.
[179] B. Ricceri, *A further three critical points theorem*, Nonlinear Anal., **71** (2009), 4151–4157.
[180] B. Ricceri, *A note on the Neumann problem*, Complex Var. Elliptic Equ., **55** (2010), 593–599.
[181] B. Ricceri and C. ZăLinescu, *A class of non-contractive operators with a unique fixed point*, Fixed Point Theory, **7** (2006), 333–339.
[182] J. Saint Raymond, *On a minimax theorem*, Arch. Math. (Basel), **74** (2000), 432–437.
[183] J. Saint Raymond, *On the multiplicity of the solutions of the equation $-\Delta u = \lambda f(u)$*, J. Differential Equations, **180** (2002), 65–88.

[184] R. Salvati, *Multiple solutions for a mixed boundary value problem*, Math. Sci. Res. J., **7** (2003), 275–283.
[185] X. Shang and J. Zhang, *Three solutions for a perturbed Dirichlet boundary value problem involving the p-Laplacian* Nonlinear Anal., **72** (2010), 1417–1422.
[186] X. Shi and X. Ding, *Existence and multiplicity of solutions for a general $p(x)$-Laplacian Neumann problem*, Nonlinear Anal., **70** (2009), 3715–3720.
[187] L. K. Shilgba, *An application of a three critical points theorem to a system of differential equations*, Yokohama Math. J., **52** (2003), 73–80.
[188] L. K. Shilgba, *Multiplicity of periodic solutions for a boundary eigenvalue problem*, Dyn. Syst., **20** (2005), 223–232.
[189] L. K. Shilgba, *Commentary on a three critical points theorem*, Far East J. Appl. Math., **36** (2009), 223–236.
[190] Y. Tian and W. Ge, *Multiple solutions for certain nonlinear second-order systems*, J. Appl. Math. Comput., **25** (2007), 353–361.
[191] L. L. Wang, Y. H. Fan and W. G. Ge, *Existence and multiplicity of solutions for a Neumann problem involving the $p(x)$-Laplace operator*, Nonlinear Anal., **71** (2009), 4259–4270.
[192] X. Wu, *Saddle-point characterization and multiplicity of periodic solutions of non-autonomous second-order systems*, Nonlinear Anal., **58** (2004), 899–907.
[193] X. Wu and J. Chen, *Existence and multiplcity of solutions for elliptic equations involving the p-Laplacian*, NoDEA Nonlinear Differential Equations Appl., **15** (2008), 745–755.
[194] X. Wu, S. Chen and K. Teng, *On variational methods for a class of damped vibration problems*, Nonlinear Anal., **68** (2008), 1432–1441.
[195] Z. Yang, D. Geng and H. Yan, *Three solutions for singular p-Laplacian type equations*, Electron. J. Differential Equations **2008**, No 61, 12 pp. (electronic).
[196] E. Zeidler, *Nonlinear functional analysis and its applications*, vol. III, Springer-Verlag, 1985.
[197] G. Zhang and S. Liu, *Three symmetric solutions for a class of elliptic equations involving the p-Laplacian with discontinuous nonlinearities in $\mathbf{R}^N$*, Nonlinear Anal., **67** (2007), 2232–2239.
[198] G. Zhang, W. Zhang and S. Liu, *Multiplicity result for a discrete eigenvalue problem with discontinuous nonlinearities*, J. Math. Anal. Appl., **328** (2007), 362–375.
[199] L. Zhang and W. Ge, *Solvability of a kind of Sturm-Liouville boundary value problems with impulses via variational methods*, Acta Appl. Math., **110** (2010), 1237–1248.

Biagio Ricceri, Department of Mathematics, University of Catania, Viale A. Doria 6, 95125 Catania
E-mail address: ricceri@dmi.unict.it

The method of Nehari manifold

Andrzej Szulkin and Tobias Weth

ABSTRACT. We present a unified approach to the method of Nehari manifold for functionals which have a local minimum at 0 and we give several examples where this method is applied to the problem of finding ground states and multiple solutions for nonlinear elliptic boundary value problems. We also consider a recent generalization of this method to problems where 0 is a saddle point of the functional.

CONTENTS

1. Introduction

In the two papers [**33**] and [**34**] Nehari has introduced a method which turned out to be very useful in critical point theory and eventually came to bear his name. He considered a boundary value problem for a certain nonlinear second order ordinary differential equation in an interval (a, b) and showed that it has a nontrivial solution which may be obtained by constrained minimization. In [**34**] he also considered existence of solutions with a prescribed number of nodes in (a, b).

To describe Nehari's method in an abstract setting, let E be a real Banach space and $\Phi \in C^1(E, \mathbb{R})$ a functional. The Fréchet derivative of Φ at u, $\Phi'(u)$, is an element of the dual space E^*, and we shall denote $\Phi'(u)$ evaluated at $v \in E$ by $\Phi'(u)v$. Suppose $u \neq 0$ is a critical point of Φ, i.e., $\Phi'(u) = 0$. Then necessarily u is contained in the set

$$(1) \qquad \mathcal{N} := \{u \in E \setminus \{0\} : \Phi'(u)u = 0\}.$$

Key words and phrases. Nehari manifold, generalized Nehari manifold, minimax, elliptic boundary value problem, ground state, multiple solutions.

So $\mathcal{N}$ is a natural constraint for the problem of finding nontrivial (i.e., $\neq 0$) critical points of Φ. $\mathcal{N}$ is called the *Nehari manifold* though in general it may not be a manifold. Set

$$c := \inf_{u \in \mathcal{N}} \Phi(u). \tag{2}$$

Under appropriate conditions on Φ one hopes that c is attained at some $u_0 \in \mathcal{N}$ and that u_0 is a critical point.

Assume without loss of generality that $\Phi(0) = 0$. Assume that for each $w \in S_1(0) := \{w \in E : \|w\| = 1\}$ the function $\alpha_w(s) := \Phi(sw)$ attains a unique maximum s_w in $(0, \infty)$ and that $\alpha'_w(s) > 0$ whenever $0 < s < s_w$, $\alpha'_w(s) < 0$ whenever $s > s_w$ and $s_w \geq \delta$ for some $\delta > 0$ independent of $w \in S_1(0)$. Then $\alpha'_w(s_w) = \Phi'(s_w w)w = 0$. Hence $s_w w$ is the unique point on the ray $s \mapsto sw$, $s > 0$, which intersects $\mathcal{N}$. Moreover, $\mathcal{N}$ is bounded away from 0. It is easy to see that $\mathcal{N}$ is closed in E and there exists a radial bijection between $\mathcal{N}$ and $S_1(0)$. In Section 3.1 we shall see that if s_w is bounded on compact subsets of $S_1(0)$, then this bijection is in fact a homeomorphism. Clearly, c in (2), if attained, is positive. We shall also show in Section 3.1 that $u_0 \in \mathcal{N}$ is a critical point whenever $\Phi(u_0) = c$. Note that since $s \mapsto \alpha_w(s)$ is increasing for all $w \in S_1(0)$ and $0 < s < \delta$, 0 is a local minimum and hence a critical point of Φ. Since u_0 is a solution to the equation $\Phi'(u) = 0$ which has minimal "energy" Φ in the set of all nontrivial solutions, we shall call it a *ground state.*

Suppose in addition to the assumptions already made that E is a Hilbert space and $\Phi \in C^2(E, \mathbb{R})$. Then

$$\alpha''_w(s_w) = \Phi''(s_w w)(w, w) = s_w^{-2}\Phi''(u)(u, u) \leq 0, \quad \text{where } u = s_w w \in \mathcal{N}.$$

If $\Phi''(u)(u, u) < 0$ for all $u \in \mathcal{N}$, then, setting $G(u) := \Phi'(u)u$, we see that

$$G'(u)u = \Phi''(u)(u, u) + \Phi'(u)u = \Phi''(u)(u, u) < 0, \quad u \in \mathcal{N}.$$

Since $\mathcal{N} = \{u \in E \setminus \{0\} : G(u) = 0\}$, it follows from the implicit function theorem that $\mathcal{N}$ is a C^1-manifold of codimension 1 and $E = T_u(\mathcal{N}) \oplus \mathbb{R}u$ for each $u \in \mathcal{N}$. Hence in this case it is easily seen that any $u \in \mathcal{N}$ with $\Phi(u) = c$ (i.e., any minimizer of $\Phi|_{\mathcal{N}}$) satisfies $\Phi'(u) = 0$. More generally, a point $u \in E$ is a nonzero critical point of Φ if and only if $u \in \mathcal{N}$ and u is critical for the restriction of Φ to $\mathcal{N}$. In view of this property, one may apply critical point theory on the manifold $\mathcal{N}$ in order to find critical points of Φ.

Our goal in this survey is to present a unified approach to the method of Nehari manifold and to illustrate it with a number of examples where it can be applied in order to show the existence of solutions to nonlinear boundary value problems. Our approach is a little different from the usual and is taken from [**44**]. In particular, we do not need to make customary assumptions which imply that $\Phi \in C^2(E, \mathbb{R})$ and $\Phi''(u)(u, u) < 0$ on $\mathcal{N}$. Therefore our results will be somewhat more general than those which may be found in the literature. We shall also treat so-called indefinite problems in which 0 is a saddle point rather than a local minimum of Φ. Then $\mathcal{N}$ need neither be closed (its closure may contain the origin) nor does it need to intersect all rays $s \mapsto sw$, $s > 0$, $w \in S_1(0)$. In the applications we shall consider it turns out that in this case $c = 0$ and is not attained. In order to circumvent this difficulty we shall replace $\mathcal{N}$ by the generalized Nehari manifold $\mathcal{M}$ which has been introduced by Pankov in [**35**]. Ground states will be obtained by minimizing Φ over $\mathcal{M}$. Also here we follow the approach in [**44**].

The literature on the method of Nehari manifold is rather extensive. It would be impossible to cover all different aspects of this method or to provide a reasonably complete bibliography. Therefore we focus on a few topics which we think are representative and are in line with our interests and research experience. A particular topic which we leave out in order to keep this survey reasonably short is the so-called fibering method which has been introduced by Pohozaev. See [**12, 17, 24**] and the references there.

Although we assume the reader is somewhat familiar with critical point theory and its applications to nonlinear boundary value problems, we summarize some pertinent results in Chapter 2. More material may be found e.g. in [**3, 23, 32, 37, 42, 45, 46**].

The central part of this survey is Chapter 3 where we consider elliptic boundary value problems in bounded domains and in $\mathbb{R}^N$. We make assumptions which imply that the functionals Φ corresponding to these problems have a local minimum at 0 and show that $\inf_{\mathcal{N}} \Phi$ is attained and hence there exists a ground state. We also discuss the existence of sign-changing solutions, and of infinitely many solutions for even Φ.

In Chapter 4 we consider indefinite elliptic problems and obtain solutions by minimizing Φ over $\mathcal{M}$. More precisely, we have an orthogonal decomposition $E = E^+ \oplus E^0 \oplus E^-$ of a Hilbert space E and the functionals which correspond to our problems are of the form $\Phi(u) = \frac{1}{2}\|u^+\|^2 - \frac{1}{2}\|u^-\|^2 - I(u)$, where $u^\pm \in E^\pm$. Then the generalized Nehari manifold is defined by

$$\mathcal{M} := \{u \in E \setminus (E^0 \oplus E^-) : \Phi'(u)u = 0 \text{ and } \Phi'(u)v = 0 \text{ for all } v \in E^0 \oplus E^-\}.$$

So all critical points $u \notin E^0 \oplus E^-$ must be in $\mathcal{M}$ and one wants to show that $c := \inf_{\mathcal{M}} \Phi$ is positive, is attained and if $u_0 \in \mathcal{M}$, $\Phi(u_0) = c$, then u_0 is a critical point of Φ. Under some additional assumptions there are no nontrivial critical points in $E^0 \oplus E^-$, hence the minimizers on $\mathcal{M}$ are ground states.

We would like to mention that we have included a few results which seem to be new: Theorem 19 where existence of solutions for an equation involving the p-Laplacian is shown and Theorems 41, 42 where existence of a ground state is shown for an indefinite system of elliptic equations. The novelty is that in both cases we admit a nonlinearity satisfying a weak superlinearity condition at infinity instead of more common conditions of Ambrosetti-Rabinowitz type.

Notation. $B_\rho(p)$ and $S_\rho(p)$ will respectively denote the open ball and the sphere centered at p and having radius ρ. We also write S as a shorthand for $S_1(0)$. The sublevel set of c will be denoted by Φ^c, i.e.,

$$\Phi^c := \{u \in E : \Phi(u) \leq c\}.$$

The symbol "$\rightharpoonup$" will stand for weak convergence in E.

Acknowledgements: The second author likes to thank P. Bates for helpful discussions and for pointing out the paper [**14**].

2. Preliminaries

Below we summarize some facts from critical point theory and briefly describe a variational framework for problems we shall consider.

2.1. Critical point theory. Let E be a Banach space and $\Phi \in C^1(E, \mathbb{R})$ a functional. A point $u \in E$ is called *critical* if $\Phi'(u) = 0$. The corresponding value $c = \Phi(u)$ is a *critical value* or a *critical level.* A sequence $(u_n) \subset E$ is called a *Palais-Smale sequence* if $(\Phi(u_n))$ is bounded and $\Phi'(u_n) \to 0$. If $\Phi(u_n) \to c \in \mathbb{R}$ and $\Phi'(u_n) \to 0$, then (u_n) is a $(PS)_c$*-sequence.* The functional Φ is said to satisfy the *Palais-Smale condition* (or $(PS)_c$*-condition*) if each Palais-Smale sequence (or $(PS)_c$-sequence) has a convergent subsequence. It is clear that if a (subsequence of) a Palais-Smale sequence converges to u, then u is a critical point. The above definitions can be easily carried over to C^1-submanifolds of E (or more generally, to Finsler manifolds). The only difference is that Φ' must be considered as an element of the cotangent bundle of the manifold M, i.e., $\Phi'(u) \in T_u(M)^*$, the dual of the tangent space to M at u.

In order to formulate our results we shall need the notion of genus. Let A be a closed subset of $E \setminus \{0\}$ such that $A = -A$. The *genus* of A, denoted $\gamma(A)$, is the smallest integer k such that there exists an odd mapping $h \in C(A, \mathbb{R}^k \setminus \{0\})$. We also set $\gamma(\emptyset) := 0$ and $\gamma(A) := \infty$ if no h exists for any finite k. It can be shown that if A is homeomorphic to the unit sphere in $\mathbb{R}^k$ by an odd homeomorphism, then $\gamma(A) = k$. This implies that the unit sphere S in an infinite-dimensional Banach space E contains compact sets of any genus $k \geq 1$ and since $\gamma(A) \leq \gamma(B)$ whenever $A \subset B$, it follows that $\gamma(S) = \infty$. Properties of genus may be found e.g. in [**3, 37, 42**].

Let E be a Banach space such that the unit sphere S in E is a submanifold of class (at least) C^1 and let $\Phi \in C^1(S, \mathbb{R})$.

THEOREM 1. *If Φ is bounded below and satisfies the Palais-Smale condition, then $c := \inf_S \Phi$ is attained and is a critical value of Φ.*

PROOF. Let $(u_n) \subset S$ be a minimizing sequence for Φ. By Ekeland's variational principle [**25, 32, 45**] we may assume $\Phi'(u_n) \to 0$. So (u_n) is a Palais-Smale sequence and therefore $u_n \to u$ after passing to a subsequence. Hence $\Phi(u) = c$ and $\Phi'(u) = 0$. □

Let

$$\Gamma_j := \{A \subset S : A = -A, \ A \text{ is compact and } \gamma(A) \geq j\}$$

and

$$c_j := \inf_{A \in \Gamma_j} \sup_{u \in A} \Phi(u), \quad j = 1, 2, \ldots .$$

If E is infinite-dimensional, then $\gamma(S) = \infty$ and $\Gamma_j \neq \emptyset$ for any j. If Φ is bounded below, then we have $c_1 \leq c_2 \leq \cdots$ and $c_j < \infty$ for all j because the sets $A \in \Gamma_j$ are compact. The following holds:

THEOREM 2. *If E is infinite-dimensional, $\Phi \in C^1(S, \mathbb{R})$ is bounded below and satisfies the Palais-Smale condition, then Φ has infinitely many pairs of critical points.*

This result may be found in [**37, 42**] for S respectively in a Hilbert space and in a Banach space such that S is of class $C^{1,1}$, and in [**43**] for S in a Banach space such that $S \in C^1$. (For $S \in C^1$ Theorem 2 can also be easily deduced from [**19**]. One can in fact remove the compactness requirement in the definition of Γ_j; this requirement was essential for the argument in [**43**].)

We do not include the proof here but only point out that one shows all c_j are critical levels and if $c_j = \cdots = c_{j+p}$ for some $p \geq 0$, then $\gamma(K_{c_j}) \geq p+1$, where

$$K_{c_j} := \{u \in S : \Phi(u) = c_j \text{ and } \Phi'(u) = 0\}.$$

Hence the number of critical points is infinite regardless of whether the number of distinct c_j's is finite or not.

2.2. Differential equations and boundary value problems. Let $\Omega \subset \mathbb{R}^N$ be a bounded domain and consider the boundary value problem

$$\begin{cases} -\Delta u - \lambda u = f(x,u), & x \in \Omega \\ u = 0, & x \in \partial\Omega. \end{cases} \tag{3}$$

Here we assume that λ is a real constant and $f \in C(\Omega \times \mathbb{R}, \mathbb{R})$ satisfies the growth restriction

$$|f(x,u)| \leq a(1 + |u|^{q-1}) \quad \text{for some } a > 0 \text{ and } 2 < q < 2^*, \tag{4}$$

where $2^* := 2N/(N-2)$ if $N \geq 3$ and $2^* := \infty$ if $N = 1$ or 2. Recall that if $N \geq 3$, then 2^* is the critical exponent with respect to the Sobolev embedding $H^1(\Omega) \hookrightarrow L^q(\Omega)$. Let $H_0^1(\Omega)$ be the closure of $C_0^\infty(\Omega)$ with respect to the norm $\|u\|$ given by

$$\|u\|^2 := \int_\Omega |\nabla u|^2\, dx.$$

By Poincaré's inequality, on $H_0^1(\Omega)$ $\|\,.\,\|$ is equivalent to the usual $H^1(\Omega)$-norm. Let

$$\Phi(u) := \frac{1}{2}\int_\Omega |\nabla u|^2\, dx - \frac{1}{2}\int_\Omega \lambda u^2\, dx - \int_\Omega F(x,u)\, dx \equiv \frac{1}{2}\|u\|^2 - I_1(u) - I(u),$$

where

$$F(x,u) := \int_0^u f(x,s)\, ds.$$

By a solution of (3) we shall always mean a weak solution $u \in H_0^1(\Omega)$.

THEOREM 3. *Suppose that f is continuous and satisfies (4). Then:*

(i) $\Phi \in C^1(H_0^1(\Omega), \mathbb{R})$ *and* $\Phi'(u) = 0$ *if and only if* $u \in H_0^1(\Omega)$ *is a solution of (3).*

(ii) *The functionals I and I_1 are weakly continuous, i.e., if $u_n \rightharpoonup u$, then $I(u_n) \to I(u)$ and $I_1(u_n) \to I_1(u)$.*

(iii) *The operators I' and I_1' are completely continuous (or weak-to-strong continuous), i.e., if $u_n \rightharpoonup u$, then $I'(u_n) \to I'(u)$ and $I_1'(u_n) \to I_1'(u)$.*

(iv) *If* $f(x,u) = o(u)$ *uniformly in* x *as* $u \to 0$, *then* $I'(u) = o(\|u\|)$ *and* $I(u) = o(\|u\|^2)$ *as* $u \to 0$ *in* $H_0^1(\Omega)$.

(v) *If* $N = 1$, *then* f *need not satisfy any growth restriction.*

The proofs of (different parts of) the above result may be found in [**3, 23, 32, 37, 42, 45, 46**]. Here we only point out that weak continuity of I, I_1 and complete continuity of I', I_1' follow from the compactness of the embedding $H_0^1(\Omega) \hookrightarrow L^r(\Omega)$ for $1 \le r < 2^*$ (the Rellich-Kondrachov theorem) and the fact that the operator $u \mapsto f(x,u)$ is continuous from $L^q(\Omega)$ to $L^{q/(q-1)}(\Omega)$ (Krasnoselskii's theorem). (iv) is a consequence of the fact that $f(x,u) = o(u)$ implies that for each $\varepsilon > 0$ there is C_ε such that

$$|f(x,u)| \le \varepsilon |u| + C_\varepsilon |u|^{q-1}$$

and (v) follows by the compactness of the embedding $H_0^1(\Omega) \hookrightarrow C(\overline{\Omega})$.

REMARK 4. Consider the Newtonian system of differential equations

$$-\ddot{q} + q = W_q(q,t), \quad q \in \mathbb{R}^N, \ t \in \mathbb{R},$$

where V given by $V(q,t) := W(q,t) - \frac{1}{2}|q|^2$ is the Newtonian potential. If W and W_q are continuous and 2π-periodic in t, then the conclusions of Theorem 3 remain valid in the space $H^1(S^1, \mathbb{R}^N)$ consisting of 2π-periodic functions with the norm given by

$$\|q\|^2 := \int_0^{2\pi} (|\dot{q}|^2 + |q|^2)\, dt,$$

see e.g. [**32**]. In particular, no growth restriction on W is necessary. The functional here is

$$\Phi(q) := \frac{1}{2}\|q\|^2 - I(q), \quad \text{where} \quad I(q) := \int_0^{2\pi} W(q,t)\, dt.$$

Let $\Delta_p u := \operatorname{div}(|\nabla u|^{p-2}\nabla u)$, where $p \in (1,\infty)$. Δ_p is called the p-Laplacian and the problem corresponding to (3) is

$$\begin{cases} -\Delta_p u - \lambda |u|^{p-2}u = f(x,u), & x \in \Omega \\ u = 0, & x \in \partial\Omega. \end{cases} \tag{5}$$

Now the function f needs to satisfy the growth restriction (4) with 2^* replaced by $p^* := Np/(N-p)$ if $N > p$ and $p^* := \infty$ otherwise. Let

$$\Phi(u) := \frac{1}{p}\int_\Omega |\nabla u|^p\, dx - \frac{1}{p}\int_\Omega \lambda |u|^p\, dx - \int_\Omega F(x,u)\, dx \equiv \frac{1}{p}\|u\|^p - I_1(u) - I(u),$$
$$u \in W_0^{1,p}(\Omega).$$

THEOREM 5. *Suppose that f is continuous and satisfies (4) with 2^* replaced by p^* and 2 by p. Then:*

(i) $\Phi \in C^1(W_0^{1,p}(\Omega), \mathbb{R})$ *and* $\Phi'(u) = 0$ *if and only if* $u \in W_0^{1,p}(\Omega)$ *is a solution of (5).*
(ii) *The functionals I and I_1 are weakly continuous.*
(iii) *The operators I' and I_1' are completely continuous.*
(iv) *If $f(x,u) = o(|u|^{p-1})$ uniformly in x as $u \to 0$, then $I'(u) = o(\|u\|^{p-1})$ and $I(u) = o(\|u\|^p)$ as $u \to 0$ in $W_0^{1,p}(\Omega)$.*
(v) *If $N < p$, then f need not satisfy any growth restriction.*

The proofs of the above statements are very similar to the corresponding ones in Theorem 3 and are outlined in [**23**]. A comprehensive treatment of problems involving the p-Laplacian in a bounded domain may be found in [**22**].

Finally we consider the problem

$$(6) \qquad \begin{cases} -\Delta u + V(x)u = f(x,u), & x \in \mathbb{R}^N \\ u(x) \to 0, & |x| \to \infty, \end{cases}$$

where $V \in C(\mathbb{R}^N, \mathbb{R})$ is bounded and $f \in C(\mathbb{R}^N \times \mathbb{R}, \mathbb{R})$ satisfies the growth restriction

$$(7) \qquad |f(x,u)| \leq a(|u| + |u|^{q-1}) \quad \text{for some} \quad a > 0 \text{ and } 2 < q < 2^*.$$

This time we work in $H^1(\mathbb{R}^N)$, with the usual norm given by

$$\|u\|^2 := \int_{\mathbb{R}^N} (|\nabla u|^2 + u^2)\, dx,$$

and the functional is

$$\Phi(u) := \frac{1}{2}\int_{\mathbb{R}^N} (|\nabla u|^2 + V(x)u^2)\, dx - \int_{\mathbb{R}^N} F(x,u)\, dx \equiv \langle Lu, u\rangle - I(u).$$

Here $L : H^1(\mathbb{R}^N) \to H^1(\mathbb{R}^N)$ is bounded linear and $\langle .,. \rangle$ is the inner product in $H^1(\mathbb{R}^N)$.

THEOREM 6. *Suppose that V, f are continuous, V is bounded and f satisfies (7). Then:*

(i) $\Phi \in C^1(H^1(\mathbb{R}^N), \mathbb{R})$ *and* $\Phi'(u) = 0$ *if and only if* $u \in H^1(\mathbb{R}^N)$ *is a solution of (6).*
(ii) *If $f(x,u) = o(u)$ uniformly in x as $u \to 0$, then $I'(u) = o(\|u\|)$ and $I(u) = o(\|u\|^2)$ as $u \to 0$ in $H^1(\mathbb{R}^N)$.*

A proof that Φ is of class C^1 and $\Phi'(u) = 0$ if and only if $u \in H^1(\mathbb{R}^N)$ is a solution of (6) may be found in [**21**] or [**45**]. The fact that such solution must necessarily tend to 0 as $|x| \to \infty$ is discussed more in detail in [**18**] and [**36**] (in particular, if 0 is not in the essential spectrum of $-\Delta + V$, then $u(x) \to 0$ exponentially, see [**36**]). We note that in unbounded domains it is necessary to replace (4) by the stronger condition (7); however, the assumption that $f(x,u) = o(u)$ uniformly in x as $u \to 0$ and (4) imply (7). The proof of (ii) is exactly the same as that of (iv) in Theorem 3. Note that now one cannot expect I to be weakly continuous or I' to be completely continuous because the embedding $H^1(\mathbb{R}^N) \hookrightarrow L^r(\mathbb{R}^N)$ is not compact for any r.

Finally, we recall a result concerning the spectrum of $-\Delta + V$ for periodic V.

PROPOSITION 7. *Suppose that $V \in C(\mathbb{R}^N, \mathbb{R})$ is 1-periodic with respect to $x_1, \ldots, x_N$. Then the spectrum $\sigma(-\Delta + V)$ in $L^2(\mathbb{R}^N)$ is absolutely continuous, bounded below but not above and consists of disjoint closed intervals.*

A proof of this well-known fact may be found e.g. in [**27, 40**]. Note that we do not exclude the possibility that the spectrum consists of only one interval, i.e., $\sigma(-\Delta + V) = [a, \infty)$ for some $a \in \mathbb{R}$. Intervals (α, β) such that $\alpha, \beta \in \sigma(-\Delta + V)$ and $(\alpha, \beta) \cap \sigma(-\Delta + V) = \emptyset$ are called *spectral gaps.*

3. Nehari manifold

3.1. Abstract setting. In what follows we always assume that E is a uniformly convex real Banach space, $\Phi \in C^1(E, \mathbb{R})$ and $\Phi(0) = 0$. Recall that $S := S_1(0)$. A function $\varphi \in C(\mathbb{R}_+, \mathbb{R}_+)$ is said to be a *normalization function* if $\varphi(0) = 0$, φ is strictly increasing and $\varphi(t) \to \infty$ as $t \to \infty$. We shall need the following further assumptions:

(A_1) There exists a normalization function φ such that

$$u \mapsto \psi(u) := \int_0^{\|u\|} \varphi(t)\, dt \in C^1(E \setminus \{0\}, \mathbb{R}),$$

$J := \psi'$ is bounded on bounded sets and $J(w)w = 1$ for all $w \in S$.

(A_2) For each $w \in E \setminus \{0\}$ there exists s_w such that if $\alpha_w(s) := \Phi(sw)$, then $\alpha'_w(s) > 0$ for $0 < s < s_w$ and $\alpha'_w(s) < 0$ for $s > s_w$.

(A_3) There exists $\delta > 0$ such that $s_w \geq \delta$ for all $w \in S$ and for each compact subset $\mathcal{W} \subset S$ there exists a constant $C_{\mathcal{W}}$ such that $s_w \leq C_{\mathcal{W}}$ for all $w \in \mathcal{W}$.

J in (A_1) is called the duality mapping corresponding to φ, see [**22**] for a detailed discussion of this notion. An obvious sufficient condition for (A_1) to hold is that $\|\cdot\| \in C^1(E \setminus \{0\}, \mathbb{R})$ (then one can take $\varphi(t) = t$). Here we will be mainly interested in two cases: E a Hilbert space and E the Sobolev space $W_0^{1,p}(\Omega)$ with $\Omega \subset \mathbb{R}^N$ bounded and $p > 1$. In the first case we take $\varphi(t) := t$ - then J is the usual duality mapping between E and E^*. In the second case we put $\varphi(t) := t^{p-1}$. The associated functional ψ is given by $\psi(u) = \frac{1}{p}\|u\|^p$ and the duality mapping

$$J = \psi' : E \to E^*, \quad J(w)v = \int_\Omega |\nabla w|^{p-2} \nabla w \cdot \nabla v \, dx \tag{8}$$

is continuous and bounded on bounded sets, see [**22**] or Section 7.5A in [**23**]. It follows from (A_1) that S is a C^1-submanifold of E and the tangent space of S at w is

$$T_w(S) = \{z \in E : J(w)z = 0\}.$$

Recall that

$$\mathcal{N} := \{u \in E \setminus \{0\} : \Phi'(u)u = 0\}.$$

As has been shown in the introduction, (A_2) implies that $sw \in \mathcal{N}$ if and only if $s = s_w$. Moreover, by the first part of (A_3), $\mathcal{N}$ is closed in E and bounded away from 0. Define the mappings $\widehat{m} : E \setminus \{0\} \to \mathcal{N}$ and $m : S \to \mathcal{N}$ by setting

$$\widehat{m}(w) := s_w w \quad \text{and} \quad m := \widehat{m}|_S. \tag{9}$$

PROPOSITION 8. *Suppose Φ satisfies (A_2) and (A_3). Then:*

(a) *The mapping $\widehat{m}$ is continuous.*
(b) *The mapping m is a homeomorphism between S and $\mathcal{N}$, and the inverse of m is given by $m^{-1}(u) = u/\|u\|$.*

PROOF. (a) Suppose $w_n \to w \neq 0$. Since $\widehat{m}(tw) = \widehat{m}(w)$ for each $t > 0$, we may assume $w_n \in S$ for all n and it suffices to show that $\widehat{m}(w_n) \to \widehat{m}(w)$ after passing to a subsequence. Write $\widehat{m}(w_n) = s_n w_n$. By (A_2) and (A_3), (s_n) is bounded and bounded away from 0, hence, taking a subsequence, $s_n \to \overline{s} > 0$. Since $\mathcal{N}$ is closed and $\widehat{m}(w_n) \to \overline{s}w$, $\overline{s}w \in \mathcal{N}$. Hence $\overline{s}w = s_w w = \widehat{m}(w)$.

(b) This is an immediate consequence of (a).

□

We shall consider the functionals $\widehat{\Psi} : E \setminus \{0\} \to \mathbb{R}$ and $\Psi : S \to \mathbb{R}$ defined by

$$\widehat{\Psi}(w) := \Phi(\widehat{m}(w)) \quad \text{and} \quad \Psi := \widehat{\Psi}|_S. \tag{10}$$

Although we do not claim that $\mathcal{N}$ is a C^1-manifold, we shall show that $\widehat{\Psi}$ is of class C^1 and there is a one-to-one correspondence between critical points of Ψ and nontrivial critical points of Φ. In the context of a standard saddle point reduction with respect to subspaces, a similar observation has been known for a long time, see e.g. [**2, 14**]. Somewhat surprisingly, the variant below has not been used – up to our knowledge – before [**44**]. However, see [**6, 7**] where the reduction to a C^1-functional $\Psi : S \to \mathbb{R}$ has been made under stronger smoothness assumptions.

PROPOSITION 9. *Suppose E is a Banach space satisfying (A_1). If Φ satisfies (A_2) and (A_3), then $\widehat{\Psi} \in C^1(E \setminus \{0\}, \mathbb{R})$ and*

$$\widehat{\Psi}'(w)z = \frac{\|\widehat{m}(w)\|}{\|w\|}\Phi'(\widehat{m}(w))z \quad \textit{for all } w,\ z \in E,\ w \neq 0.$$

PROOF. Let $w \in E \setminus \{0\}$ and $z \in E$. Using the maximality property of $s_w w$ and the mean value theorem, we obtain

$$\begin{aligned}\widehat{\Psi}(w + tz) - \widehat{\Psi}(w) &= \Phi(s_{w+tz}(w + tz)) - \Phi(s_w w)\\ &\leq \Phi(s_{w+tz}(w + tz)) - \Phi(s_{w+tz} w)\\ &= \Phi'(s_{w+tz}(w + \tau_t tz))s_{w+tz}tz,\end{aligned}$$

where $|t|$ is small enough and $\tau_t \in (0,1)$. Similarly,

$$\begin{aligned}\widehat{\Psi}(w+tz) - \widehat{\Psi}(w) &\geq \Phi(s_w(w+tz)) - \Phi(s_w w) \\ &= \Phi'(s_w(w+\eta_t tz))s_w tz,\end{aligned}$$

where $\eta_t \in (0,1)$. Since the mapping $w \mapsto s_w$ is continuous according to Proposition 8, we see combining these two inequalities that

$$\lim_{t\to 0} \frac{\widehat{\Psi}(w+tz) - \widehat{\Psi}(w)}{t} = s_w \Phi'(s_w w)z = \frac{\|\widehat{m}(w)\|}{\|w\|}\Phi'(\widehat{m}(w))z.$$

Hence the Gâteaux derivative of $\widehat{\Psi}$ is bounded linear in z and continuous in w. It follows that $\widehat{\Psi}$ is of class C^1, see e.g. [**23, 45**]. □

COROLLARY 10. *Suppose E is a Banach space satisfying (A_1). If Φ satisfies (A_2) and (A_3), then:*

(a) *$\Psi \in C^1(S,\mathbb{R})$ and*

$$\Psi'(w)z = \|m(w)\|\Phi'(m(w))z \quad \text{for all } z \in T_w(S).$$

(b) *If (w_n) is a Palais-Smale sequence for Ψ, then $(m(w_n))$ is a Palais-Smale sequence for Φ. If $(u_n) \subset \mathcal{N}$ is a bounded Palais-Smale sequence for Φ, then $(m^{-1}(u_n))$ is a Palais-Smale sequence for Ψ.*

(c) *w is a critical point of Ψ if and only if $m(w)$ is a nontrivial critical point of Φ. Moreover, the corresponding values of Ψ and Φ coincide and $\inf_S \Psi = \inf_{\mathcal{N}} \Phi$.*

(d) *If Φ is even, then so is Ψ.*

PROOF. (a) follows from Proposition 9. Note only that since $w \in S$, we have $m(w) = \widehat{m}(w)$.

(b) We first note that by (A1) we have $E = T_w(S) \oplus \mathbb{R}w$ for every $w \in S$, and the projection $E \to T_w(S)$, $z + tw \mapsto z$ has uniformly bounded norm with respect to $w \in S$. Indeed, J is bounded on bounded sets and $J(w)(z+tw) = t$, so if $\|z+tw\| = 1$, then $|t| \leq C$ for some constant $C > 0$ and therefore

$$\|z\| \leq |t| + \|z+tw\| \leq (C+1)\|z+tw\| \qquad \text{for all } w \in S,\ z \in T_w S \text{ and } t \in \mathbb{R}.$$

Moreover, by (a) we have

$$\|\Psi'(w)\| = \sup_{\substack{z\in T_w(S)\\ \|z\|=1}} \Psi'(w)z = \|u\| \sup_{\substack{z\in T_w(S)\\ \|z\|=1}} \Phi'(u)z \tag{11}$$

with $u = m(w)$, and since $\Phi'(u)w = \Phi'(u)\frac{u}{\|u\|} = 0$, we conclude using (a) again that

$$\begin{aligned}\|\Psi'(w)\| \leq \|u\|\|\Phi'(u)\| &= \|u\| \sup_{\substack{z\in T_w(S),\, t\in\mathbb{R}\\ z+tw\neq 0}} \frac{\Phi'(u)(z+tw)}{\|z+tw\|} \\ &\leq (C+1)\|u\| \sup_{z\in T_w(S)\setminus\{0\}} \frac{\Phi'(u)z}{\|z\|} = (C+1)\|\Psi'(w)\|.\end{aligned}$$

Since $u \in \mathcal{N}$ and $\mathcal{N}$ is bounded away from 0, this two-sided estimate, together with the fact that $\Phi(u) = \Psi(w)$, easily yields the assertion.

(c) By (11), $\Psi'(w) = 0$ if and only if $\Phi'(m(w)) = 0$. The other part is clear.

(d) If Φ is even, then $s_w = s_{-w}$. Hence $\widehat{m}(-w) = -\widehat{m}(w)$ and the conclusion follows from the definition of Ψ. □

REMARK 11. We note that the infimum of Φ over $\mathcal{N}$ has the following minimax characterization:

$$c = \inf_{u \in \mathcal{N}} \Phi(u) = \inf_{w \in E \setminus \{0\}} \max_{s>0} \Phi(sw) = \inf_{w \in S} \max_{s>0} \Phi(sw).$$

Our next results give sufficient conditions for the existence of a ground state and of infinitely many critical points of Φ.

THEOREM 12. *Let E be a Hilbert space and suppose that $\Phi(u) = \frac{1}{2}\|u\|^2 - I(u)$, where*

(i) *$I'(u) = o(\|u\|)$ as $u \to 0$,*
(ii) *$s \mapsto I'(su)u/s$ is strictly increasing for all $u \neq 0$ and $s > 0$,*
(iii) *$I(su)/s^2 \to \infty$ uniformly for u on weakly compact subsets of $E \setminus \{0\}$ as $s \to \infty$,*
(iv) *I' is completely continuous.*

Then equation $\Phi'(u) = 0$ has a ground state solution. Moreover, if I is even, then this equation has infinitely many pairs of solutions.

In the following we say that Φ *satisfies the Palais-Smale condition on* $\mathcal{N}$ if every Palais-Smale sequence (u_n) for Φ with $u_n \in \mathcal{N}$ for all n contains a convergent subsequence.

PROOF. First we verify that (A_2) and (A_3) are satisfied. By (i) and (iii), $\alpha_w(s) > 0$ for $s > 0$ small and $\alpha_w(s) < 0$ for s large. Since

$$\alpha_w'(s) = \frac{d}{ds}\Phi(sw) = s(\|w\|^2 - s^{-1}I'(sw)w),$$

it follows from (ii) that there exists a unique s_w with $\alpha_w'(s_w) = 0$. Clearly, $s_w \geq \delta$ for some $\delta > 0$ and all $w \in S$ according to (i), and $s_w \leq C_{\mathcal{W}}$ for all w in a compact subset $\mathcal{W} \subset S$ according to (iii).

We shall demonstrate in Proposition 14 below that Φ satisfies the Palais-Smale condition on $\mathcal{N}$. Assuming this, let (w_n) be a Palais-Smale sequence for Ψ and set $u_n := m(w_n) \in \mathcal{N}$. Then (u_n) is a Palais-Smale sequence for Φ according to Corollary 10, hence $u_n \to u$ after passing to a subsequence and $w_n \to m^{-1}(u)$. It follows that Ψ satisfies the Palais-Smale condition.

Let (w_n) be a minimizing sequence for Ψ. By Ekeland's variational principle [**25, 32, 45**] we may assume $\Psi'(w_n) \to 0$, and by the Palais-Smale condition, $w_n \to w$ after passing to a subsequence. Hence w is a minimizer for Ψ and u is a ground state solution for the equation $\Phi'(u) = 0$, cf. Theorem 1 and Corollary 10.

Assume now that Φ is even. Then so is Ψ. Since $\inf_S \Psi > 0$, Ψ is bounded from below and the second conclusion follows from Corollary 10 and Theorem 2. □

In one of the problems considered in the next section we shall need an extension of the above result to Banach spaces. Let $I_0 \in C^1(E, \mathbb{R})$ be even, positively homogeneous of degree $p > 1$ (i.e., $I_0(su) = s^p I_0(u)$, $s > 0$) and such that

$$c_0 \|u\|^p \leq I_0(u) \leq c_0^{-1} \|u\|^p \tag{12}$$

for some constant $c_0 > 0$.

THEOREM 13. *Let E be a uniformly convex Banach space satisfying (A_1) and suppose that $\Phi(u) = I_0(u) - I(u)$, where*

(i) $I'(u) = o(\|u\|^{p-1})$ *as* $u \to 0$,
(ii) $s \mapsto I'(su)u/s^{p-1}$ *is strictly increasing for all* $u \neq 0$ *and* $s > 0$,
(iii) $I(su)/s^p \to \infty$ *uniformly for* u *on weakly compact subsets of* $E \setminus \{0\}$ *as* $s \to \infty$,
(iv) I' *is completely continuous,*
(v) I_0 *is weakly lower semicontinuous, positively homogeneous of degree* p, *satisfies (12) and*

$$(I_0'(v) - I_0'(w))(v - w) \geq c_1(\|v\|^{p-1} - \|w\|^{p-1})(\|v\| - \|w\|)$$

for some $c_1 > 0$ *and all* $v, w \in E$.

Then equation $\Phi'(u) = 0$ *has a ground state solution. Moreover, if* I *is even, then this equation has infinitely many pairs of solutions.*

PROOF. Using (12) and the homogeneity of I_0 it is easy to see that the argument of Theorem 12 goes through with obvious changes. □

Note that while in Theorem 12 the sphere S is a C^∞-manifold, in Theorem 13 S may be of class C^1 only. Therefore it is important to have a version of Theorem 2 which holds for such S. In the next section we shall encounter a problem where $S \in C^1$ but $\notin C^{1,1}$.

To complete the proofs of Theorems 12 and 13 we still need to show that Φ satisfies the Palais-Smale condition on $\mathcal{N}$. Since (v) of Theorem 13 is automatically satisfied if E is a Hilbert space and $I_0(u) = \frac{1}{2}\|u\|^2$, it suffices to show this for Theorem 13. It was first noted in [**31**] that (in the context of semilinear boundary value problems) the Palais-Smale condition for Φ on $\mathcal{N}$ is satisfied under weaker growth assumptions than needed when no restriction to $\mathcal{N}$ is made.

PROPOSITION 14. *The following holds under the assumptions of Theorem 13:*

(a) *If* $(u_n) \subset \mathcal{N}$ *is a sequence such that* $\sup_{n\in\mathbb{N}} \Phi(u_n) < \infty$, *then – passing to a subsequence – we have* $u_n \rightharpoonup u \neq 0$ *as* $n \to \infty$, *and there is* $s_u > 0$ *such that* $s_u u \in \mathcal{N}$ *and* $\Phi(s_u u) \leq \liminf_{n\to\infty} \Phi(u_n)$.
(b) $\Phi|_{\mathcal{N}}$ *is coercive, i.e.,* $\Phi(u_n) \to \infty$ *as* $u_n \in \mathcal{N}$, $\|u_n\| \to \infty$.
(c) Φ *satisfies the Palais-Smale condition on* $\mathcal{N}$.

The proof below is inspired by [**31**].

PROOF. (a) Let $(u_n) \subset \mathcal{N}$ be a sequence such that $\Phi(u_n) \leq d$ for all n. We first claim that (u_n) is bounded. Otherwise $\|u_n\| \to \infty$ and $v_n := u_n/\|u_n\| \rightharpoonup v$ in

E after passing to a subsequence. Suppose $v = 0$. Since $u_n \in \mathcal{N}$ and $v_n \rightharpoonup 0$, for each $s > 0$ we have

$$d \geq \Phi(u_n) = \Phi(s_{v_n} v_n) \geq \Phi(s v_n) \geq c_0 s^p - I(s v_n) \to c_0 s^p \tag{13}$$

because assumption (iv) of Theorem 13 implies that I is weakly continuous. This yields a contradiction upon choosing $s > (d/c_0)^{1/p}$. So $v \neq 0$ and hence

$$0 \leq \frac{\Phi(u_n)}{\|u_n\|^p} \leq \frac{1}{c_0} - \frac{I(\|u_n\| v_n)}{\|u_n\|^p} \to -\infty \tag{14}$$

as $n \to \infty$ by (iii) of Theorem 13, a contradiction again. It follows that (u_n) is bounded, so $u_n \rightharpoonup u$ after passing to a subsequence. If $u = 0$, we see as in (13) that

$$d \geq \Phi(u_n) \geq \Phi(s u_n) \geq d_0 s^p - I(s u_n) \to d_0 s^p \tag{15}$$

for all $s > 0$, where $d_0 = c_0 \inf_{\mathcal{N}} \|u\| > 0$, a contradiction. Hence $u \neq 0$. Moreover, as $I(s_u u_n) \to I(s_u u)$, again by the weak continuity of I, we have

$$\Phi(s_u u) \leq \liminf_{n\to\infty} \Phi(s_u u_n) \leq \liminf_{n\to\infty} \Phi(u_n)$$

since $u_n \in \mathcal{N}$.

(b) This follows directly from (a).

(c) Let $(u_n) \subset \mathcal{N}$ be a Palais-Smale sequence. By (a), $u_n \rightharpoonup u$ after passing to a subsequence. Moreover,

$$\Phi'(u_n) = I_0'(u_n) - I'(u_n) \to 0 \quad \text{and} \quad I'(u_n) \to I'(u) \ \text{ as } n \to \infty,$$

hence by (v) of Theorem 13,

$$o(1) = [I_0'(u_n) - I_0'(u)](u_n - u) \geq c_1(\|u_n\|^{p-1} - \|u\|^{p-1})(\|u_n\| - \|u\|).$$

If follows that $\|u_n\| \to \|u\|$ and therefore $u_n \to u$ by the uniform convexity of E. So Φ satisfies the Palais-Smale condition on $\mathcal{N}$. □

REMARK 15. It is easily seen that the proof of Proposition 14 still goes through if the condition that I_0 satisfies (v) of Theorem 13 is replaced by the more general condition that I_0 can be written as a sum of a functional satisfying (v) and a functional with completely continuous derivative. It will be convenient to use this observation in the next section.

3.2. Elliptic equations in bounded domains. Let $\Omega \subset \mathbb{R}^N$ be a bounded domain and consider the boundary value problem

$$\begin{cases} -\Delta u - \lambda u = f(x,u), & x \in \Omega \\ u = 0, & x \in \partial\Omega. \end{cases} \tag{16}$$

Here $\lambda < \lambda_1$, where λ_1 denotes the first Dirichlet eigenvalue of $-\Delta$ in Ω and $f \in C(\Omega \times \mathbb{R}, \mathbb{R})$ satisfies the growth restriction

$$|f(x,u)| \leq a(1 + |u|^{q-1}) \quad \text{for some } a > 0 \text{ and } 2 < q < 2^*. \tag{17}$$

Recall from Section 2.2 that $2^* := 2N/(N-2)$ if $N \geq 3$ and $2^* := \infty$ otherwise. Since $\lambda < \lambda_1$, we may introduce an equivalent norm in $E \equiv H_0^1(\Omega)$ by setting

$$\|u\|^2 := \int_\Omega (|\nabla u|^2 - \lambda u^2)\,dx.$$

Let

$$F(x,u) := \int_0^u f(x,s)\,ds$$

and

$$\Phi(u) := \frac{1}{2}\int_\Omega (|\nabla u|^2 - \lambda u^2)\,dx - \int_\Omega F(x,u)\,dx \equiv \frac{1}{2}\|u\|^2 - I(u).$$

By Theorem 3, $\Phi \in C^1(E,\mathbb{R})$, I' is completely continuous and critical points of Φ are solutions of (16). The following result on the existence and multiplicity of solutions of (16) is essentially contained in [**31**]. We note however that in [**31**] stronger assumptions on f were imposed for the multiplicity result in order to apply critical point theory on $\mathcal{N}$.

THEOREM 16. *Suppose that $\lambda < \lambda_1$, f satisfies (17) and*

(i) $f(x,u) = o(u)$ *uniformly in x as $u \to 0$,*
(ii) $u \mapsto f(x,u)/|u|$ *is strictly increasing on* $(-\infty,0)$ *and* $(0,\infty)$,
(iii) $F(x,u)/u^2 \to \infty$ *uniformly in x as $|u| \to \infty$.*

Then equation (16) has a ground state solution. Moreover, if f is odd in u, then (16) has infinitely many pairs of solutions.

PROOF. We want to use Theorem 12. It follows from (i) and Theorem 3 that $I'(u) = o(\|u\|)$ as $u \to 0$, and it is easy to see from (ii) that $s \mapsto I'(su)u/s$ is strictly increasing if $u \neq 0$ and $s > 0$. Next we verify (iii) of Theorem 12. Let $\mathcal{W} \subset E \setminus \{0\}$ be weakly compact and let $(u_n) \subset \mathcal{W}$. It suffices to show that if $s_n \to \infty$ as $n \to \infty$, then so does a subsequence of $I(s_n u_n)/s_n^2$. Passing to a subsequence, $u_n \rightharpoonup u \in E \setminus \{0\}$ and $u_n(x) \to u(x)$ a.e. Since $|s_n u_n(x)| \to \infty$ if $u(x) \neq 0$, an application of Fatou's lemma yields

$$\frac{I(s_n u_n)}{s_n^2} = \int_\Omega \frac{F(x, s_n u_n)}{(s_n u_n)^2} u_n^2\,dx \to \infty \quad \text{as } n \to \infty$$

(that $F \geq 0$ is an easy consequence of (i) and (ii) above). Finally, (iv) of Theorem 12 is ensured by Theorem 3.

Now Theorem 12 yields that equation (16) has a ground state solution and if f is odd in u, then (16) has infinitely many pairs of solutions. □

REMARK 17. One can show that each ground state solution u_0 is positive or negative. Indeed, if u is a sign-changing solution, then we let $u^+ := \max\{u,0\}$ and $u^- := \min\{u,0\}$ denote the positive and the negative part of u, respectively. Multiplying (16) by $u^\pm$ and integrating, we see that $u^\pm \in \mathcal{N}$. Hence $\Phi(u) = \Phi(u^+) + \Phi(u^-) \geq 2c$. It follows that $u_0 \geq 0$ or $u_0 \leq 0$ and by Harnack's inequality [**26**], $u_0 > 0$ or $u_0 < 0$ in Ω.

Suppose f satisfies (17), (i) of Theorem 16 and the following Ambrosetti-Rabinowitz condition [**4**]: There exist $\mu > 2$ and $R > 0$ such that

$$0 < \mu F(x,u) \leq f(x,u)u \quad \text{for all } |u| \geq R. \tag{18}$$

It is well known and may be found e.g. in [**3, 4, 37, 42, 45**] that under these assumptions (16) has a positive solution, and if in addition f is odd in u, then there are infinitely many pairs of solutions. A typical example of f satisfying both (18) and the assumptions of Theorem 16 is $f(x,u) = |u|^{q-2}u$, $2 < q < 2^*$. It is easy to show by integration that (18) implies $|F(x,u)| \geq a_1|u|^\mu - a_2$ for some $a_1, a_2 > 0$. On the other hand, there are functions which increase slower than that and yet satisfy (i)–(iii) of Theorem 16, e.g. $f(x,u) = ug(|u|)$, where $g(0) = 0$, g is strictly increasing and $g(s) \to \infty$, $g(s)/s^a \to 0$ for any $a > 0$ as $s \to \infty$. Of course, there also exist functions satisfying (18) but not (ii).

Next we add a result on the existence of least energy sign-changing solutions. For this we put

$$\mathcal{N}_{sc} := \{u \in E : u^\pm \in \mathcal{N}\} \quad \text{and} \quad c_{sc} := \inf_{u \in \mathcal{N}_{sc}} \Phi(u),$$

where as in Remark 17, u^+ and u^- are the positive and negative part of u. Since $u^\pm \in \mathcal{N}$, we see that every sign-changing solution of (16) lies in $\mathcal{N}_{sc}$.

THEOREM 18. *Under the assumptions of Theorem 16,* (16) *admits a least energy sign-changing solution, i.e. a solution* $u \in \mathcal{N}_{sc}$ *with* $\Phi(u) = c_{sc} \geq 2c$.

This result is due to Liu and Wang [**31**]. The existence of least energy solution was first asserted in [**41**] under somewhat stronger assumptions. However, the proof in [**41**] is not fully correct since, as in many other papers, it has been overlooked that the functionals

$$E \to \mathbb{R}, \qquad u \mapsto \int_\Omega |\nabla u^\pm|^2 \, dx$$

are *not* of class C^1, see [**9**, Section 3] for a discussion of this regularity problem. The first correct proof – under stronger assumptions – on the existence of a least energy sign-changing solution of (16) was given by Castro-Cossio-Neuberger [**15**].

PROOF OF THEOREM 18. Let $(u_n) \subset \mathcal{N}_{sc}$ be a sequence of functions such that $\Phi(u_n) \to c_{sc}$ as $n \to \infty$. Since $\Phi(u_n) = \Phi(u_n^+) + \Phi(u_n^-)$ and $\Phi(u_n^\pm) \geq c > 0$ for all n, we see that $\Phi(u_n^\pm)$ is bounded. Hence, by Proposition 14(a), $u_n^+ \rightharpoonup u_1 \neq 0$ and $u_n^- \rightharpoonup u_2 \neq 0$ and therefore $u_n \rightharpoonup \widetilde{u} := u_1 + u_2$ as $n \to \infty$. Since we may assume a.e. pointwise convergence along a subsequence, we find that $u_1 \cdot u_2 \equiv 0$. Moreover, for $u := s_{u_1}u_1 + s_{u_2}u_2 = u^+ + u^-$ we have, by Proposition 14(a),

$$2c \leq \Phi(u) = \Phi(s_{u_1}u_1) + \Phi(s_{u_2}u_2) \leq \liminf_{n\to\infty}[\Phi(u_n^+) + \Phi(u_n^-)] = c_{sc}.$$

Since $u \in \mathcal{N}_{sc}$, we conclude that $\Phi(u) = c_{sc}$.

Using the quantitative deformation lemma (see [**45**, Lemma 2.3]), we shall now prove that $\Phi'(u) = 0$. It follows from assumption (ii) of Theorem 16 that, for $s, t > 0$ and at least one of $s, t \neq 1$,

$$\Phi(su^+ + tu^-) = \Phi(su^+) + \Phi(tu^-) < \Phi(u^+) + \Phi(u^-) = c_{sc}. \tag{19}$$

If $\Phi'(u) \neq 0$, then there exist $\delta > 0$ and $\mu > 0$ such that

$$\|v - u\| \leq 3\delta \quad \Rightarrow \quad \|\Phi'(v)\| \geq \mu.$$

Let $D = \left[\frac{1}{2}, \frac{3}{2}\right] \times \left[\frac{1}{2}, \frac{3}{2}\right]$ and $g(s,t) = su^+ + tu^-$. It follows from (19) that $\Phi(g(s,t)) = c_{sc}$ if and only if $s = t = 1$ and $\Phi(g(s,t)) < c_{sc}$ otherwise. Hence

$$\beta := \max_{\partial D} \Phi \circ g < c_{sc}. \tag{20}$$

For $\varepsilon := \min\left\{\frac{c_{sc}-\beta}{4}, \frac{\mu\delta}{8}\right\}$, [**45**, Lemma 2.3] yields a deformation η such that

a) $\eta(1, v) = v$ if $v \notin \Phi^{-1}([c_{sc} - 2\varepsilon, c_{sc} + 2\varepsilon])$,
b) $\Phi(\eta(1, v)) \leq c_{sc} - \varepsilon$ for every $v \in E$ with $\|v - u\| \leq \delta$ and $\Phi(v) \leq c_{sc} + \varepsilon$.
c) $\Phi(\eta(1, v)) \leq \Phi(v)$ for all $u \in E$.

It is then clear that

$$\max_{(s,t)\in D} \Phi(\eta(1, g(s,t))) < c_{sc}.$$

We shall prove that $\eta(1, g(D)) \cap \mathcal{N}_{sc} \neq \emptyset$, contradicting the definition of c_{sc}. Let us define $h(s,t) := \eta(1, g(s,t))$ and

$$\begin{aligned}
\Psi_0(s,t) &:= (\Phi'(su^+)u^+, \Phi'(tu^-)u^-),\\
\Psi_1(s,t) &:= \left(\frac{1}{s}\Phi'(h^+(s,t))h^+(s,t),\ \frac{1}{t}\Phi'(h^-(s,t))h^-(s,t)\right).
\end{aligned}$$

Since $\Phi'(su^\pm)u^\pm > 0$ if $0 < s < 1$ and < 0 if $s > 1$, the product formula for the degree now yields $\deg(\Psi_0, D, 0) = 1$. It follows from (20) and the property a) of η that $g = h$ on ∂D. Consequently, $\Psi_1 = \Psi_0$ on ∂D and $\deg(\Psi_1, D, 0) = \deg(\Psi_0, D, 0) = 1$. Therefore $\Psi_1(s,t) = 0$ for some $(s,t) \in D$, hence $\eta(1, g(s,t)) = h(s,t) \in \mathcal{N}_{sc}$. We conclude that u is a critical point of Φ. □

Next we consider a related boundary value problem for the p-Laplacian. Let $\Omega \subset \mathbb{R}^N$ be a bounded domain. It is well known (see e.g. [**22**] or Section 7.5A in [**23**]) that

$$\lambda_1 := \inf_{\substack{u \in W_0^{1,p}(\Omega)\\ u \neq 0}} \frac{\int_\Omega |\nabla u|^p\, dx}{\int_\Omega |u|^p\, dx}$$

is attained and is the first Dirichlet eigenvalue of the p-Laplacian. Set

$$I_0(u) := \frac{1}{p}\int_\Omega (|\nabla u|^p - \lambda |u|^p)\, dx \equiv \frac{1}{p}\|u\|^p - I_1(u).$$

If $\lambda < \lambda_1$, then it is clear that I_0 is positively p-homogeneous and satisfies (12). However, $(pI_0)^{1/p}$ is not a norm and if $1 < p < 2$, then the unit sphere is not a $C^{1,1}$-submanifold of $E = W_0^{1,p}(\Omega)$. Our boundary value problem is now

$$\begin{cases} -\Delta_p u - \lambda |u|^{p-2}u = f(x,u), & x \in \Omega\\ u = 0, & x \in \partial\Omega. \end{cases} \tag{21}$$

Recall from Section 2.2 that $p^* := Np/(N-p)$ if $N > p$ and $p^* := \infty$ otherwise. The following result seems to be new.

THEOREM 19. *Suppose that $\lambda < \lambda_1$ and $f \in C(\Omega \times \mathbb{R}, \mathbb{R})$ satisfies (17) with 2^* replaced by p^* and 2 by p. Suppose further*

(i) *$f(x,u) = o(|u|^{p-1})$ uniformly in x as $u \to 0$,*
(ii) *$u \mapsto f(x,u)/|u|^{p-1}$ is strictly increasing on $(-\infty, 0)$ and $(0,\infty)$,*
(iii) *$F(x,u)/|u|^p \to \infty$ uniformly in x as $|u| \to \infty$.*

Then equation (21) has a ground state solution. Moreover, if f is odd in u, then (21) has infinitely many pairs of solutions.

PROOF. We shall apply Theorem 13 to the functional

$$\Phi(u) := \frac{1}{p}\int_\Omega |\nabla u|^p\,dx - \frac{1}{p}\int_\Omega \lambda |u|^p\,dx - \int_\Omega F(x,u)\,dx \equiv I_0(u) - I(u),$$

where

$$I_0(u) = \frac{1}{p}\|u\|^p - \frac{1}{p}\int_\Omega \lambda|u|^p\,dx \equiv \psi(u) - I_1(u)$$

and $I(u) = \int_\Omega F(x,u)\,dx$. It is clear that (A_1) holds with $\psi(u) = \frac{1}{p}\|u\|^p$ and $J = \psi'$ given by (8). Very similarly as in the proof of Theorem 16 we find that assumptions (i)–(iii) of Theorem 13 are satisfied. Moreover, both I' and I_1' are completely continuous by Theorem 5, so (iv) holds as well. In view of Remark 15 it therefore suffices to show that J satisfies

$$(J(v) - J(w))(v-w) \geq (\|v\|^{p-1} - \|w\|^{p-1})(\|v\| - \|w\|) \text{ for } v, w \in E,$$

but this property is well known, see e.g. [**22**] or (d) on p. 501 in [**23**]. □

The result is known if (ii) and (iii) are replaced by a condition of Ambrosetti-Rabinowitz type ($\mu > 2$ should be replaced by $\mu > p$ in (18), see [**22**]).

Also here one can show there exists a least energy sign-changing solution with $c_{sc} \geq 2c$ and each ground state solution satisfies $u \geq 0$ or $u \leq 0$.

3.3. Elliptic problems in $\mathbb{R}^N$.

Here we consider the problem

$$\begin{cases} -\Delta u + V(x)u = f(x,u), & x \in \mathbb{R}^N \\ u(x) \to 0, & |x| \to \infty. \end{cases} \tag{22}$$

If V is bounded, f continuous and satisfying (17), but with 1 replaced by $|u|$, then

$$\Phi(u) := \frac{1}{2}\int_{\mathbb{R}^N} (|\nabla u|^2 + V(x)u^2)\,dx - \int_{\mathbb{R}^N} F(x,u)\,dx$$

is of class C^1 on $E \equiv H^1(\mathbb{R}^N)$ and critical points of Φ correspond to solutions of (22), see Theorem 6.

THEOREM 20. *Suppose $V \in C(\mathbb{R}^N, \mathbb{R})$, $f \in C(\mathbb{R}^N \times \mathbb{R}, \mathbb{R})$ satisfies (17) and*
(i) *V, f are 1-periodic in $x_1, \dots, x_N$ and $V(x) > 0$ for all x,*
(ii) *$f(x,u) = o(u)$ uniformly in x as $u \to 0$,*
(iii) *$u \mapsto f(x,u)/|u|$ is strictly increasing on $(-\infty, 0)$ and $(0,\infty)$,*
(iv) *$F(x,u)/u^2 \to \infty$ uniformly in x as $|u| \to \infty$.*
Then equation (22) has a ground state solution.

This extends Theorem 2.1 in [**28**] where it was additionally assumed that f is of class C^1 with f_u satisfying a growth restriction. By (i), V is bounded and bounded away from 0, hence

$$\|u\|^2 := \int_{\mathbb{R}^N} (|\nabla u|^2 + V(x)u^2)\, dx \tag{23}$$

defines an equivalent norm and we have

$$\Phi(u) = \frac{1}{2}\|u\|^2 - \int_{\mathbb{R}^N} F(x,u)\, dx \equiv \frac{1}{2}\|u\|^2 - I(u).$$

(23) remains valid, and so does Theorem 20, if $V > 0$ is replaced by the weaker condition that $\sigma(-\Delta + V) \subset (0,\infty)$, where σ is the spectrum in $L^2(\mathbb{R}^N)$. As has been noticed in Section 2.2, (17) and (ii) imply that f satisfies the proper growth restriction which is (7). Hence $\Phi \in C^1(E, \mathbb{R})$. By the periodicity of V and f, if u is a solution of (22), then so is $u(\cdot - y)$ for any $y \in \mathbb{Z}^N$. Two solutions which are not translates of each other by an element of $\mathbb{Z}^N$ will be called *geometrically distinct.* If f is odd in u, then (22) has in fact infinitely many geometrically distinct solutions under the assumptions of Theorem 20. We do not include the lengthy proof of this fact here and refer the reader to [**44**]. We remark that under stronger conditions there exist infinitely many geometrically distinct solutions also for non-odd f, see [**21**].

Let us note that if V is a positive constant and $f = f(u)$, then $\Phi(u) = \Phi(u(\cdot - y))$ for all $y \in \mathbb{R}^N$ and any translate $u(\cdot - y)$ of a solution $u \neq 0$ is again a solution. Hence the correct notion of geometrically distinct solutions here is the requirement that they are not translates of each other by any $y \in \mathbb{R}^N$. It follows that existence of a single nontrivial solution automatically leads to the existence of infinitely many geometrically distinct ones in the $\mathbb{Z}^N$-sense. However, as we shall see in Section 3.5, it is not necessarily true that the number of those which are $\mathbb{R}^N$-distinct is infinite.

We shall need the following result whose simple proof is left to the reader:

LEMMA 21. *If f satisfies (ii) and (iii) of Theorem 20, then $F(x,u) > 0$ and $\frac{1}{2}f(x,u)u > F(x,u)$ for all $u \neq 0$.*

PROOF OF THEOREM 20. We cannot apply Theorem 12 directly. It is easy to verify that (i)–(iii) of this theorem still hold, with the same proof as in Theorem 16 (note only that $u_n \rightharpoonup u$ in E implies $u_n \to u$ in $L^2_{loc}(\mathbb{R}^N)$ and hence $u_n(x) \to u(x)$ a.e. after taking a subsequence). Therefore the results of Corollary 10 remain valid. However, I' is not completely continuous and Φ cannot satisfy the Palais-Smale condition on $\mathcal{N}$. Indeed, if $\Phi'(u) = 0$ and $u_n(x) := u(x - y_n)$, where $y_n \in \mathbb{Z}^N$, then (u_n) is a Palais-Smale sequence which converges weakly but not strongly to 0 if $|y_n| \to \infty$.

Let $(w_n) \subset S$ be a minimizing sequence for Ψ. By Ekeland's variational principle we may assume $\Psi'(w_n) \to 0$, hence also $\Phi'(u_n) \to 0$, where $u_n := m(w_n)$. We shall show that (u_n) is bounded. Suppose $\|u_n\| \to \infty$, then $v_n := u_n/\|u_n\| \rightharpoonup v$ in E and $v_n \to v$ a.e. after passing to a subsequence. By (ii) and (17), for each $\varepsilon > 0$ there is C_ε such that

$$|f(x,u)| \le \varepsilon |u| + C_\varepsilon |u|^{q-1}. \tag{24}$$

If $v = 0$ and $v_n \to 0$ in $L^q(\mathbb{R}^N)$, then for each $s > 0$, $\int_{\mathbb{R}^N} F(x, sv_n)\, dx \to 0$ according to (24) and therefore

$$d \ge \Phi(u_n) \ge \Phi(sv_n) = \frac{s^2}{2} - \int_{\mathbb{R}^N} F(x, sv_n)\, dx \to \frac{s^2}{2}, \tag{25}$$

a contradiction for $s > \sqrt{2d}$. So $v_n \not\to 0$ in $L^q(\mathbb{R}^N)$ and it follows from P.L. Lions' lemma [**29**], [**45**, Lemma 1.21] that

$$\int_{B_1(y_n)} v_n^2\, dx \ge \delta \tag{26}$$

for some $\delta > 0$, $y_n \in \mathbb{R}^N$ and almost all n. Since Φ and $\mathcal{N}$ are invariant by translations of the form $v \mapsto v(\cdot - y)$, $y \in \mathbb{Z}^N$, we may assume translating v_n if necessary that the sequence (y_n) is bounded. Since $v_n \to v$ in $L^2_{loc}(\mathbb{R}^N)$, (26) implies $v \neq 0$ and we get a contradiction since Fatou's lemma yields

$$0 \le \frac{\Phi(u_n)}{\|u_n\|^2} = \frac{1}{2} - \int_{\mathbb{R}^N} \frac{F(x,u_n)}{u_n^2} v_n^2\, dx \to -\infty \qquad \text{as } n \to \infty. \tag{27}$$

So (u_n) is bounded and we may assume $u_n \rightharpoonup u$ in E and $u_n \to u$ a.e. Hence u is a solution of (22), possibly the trivial one ($u = 0$). If $u_n \to 0$ in $L^q(\mathbb{R}^N)$, then $\int_{\mathbb{R}^N} f(x,u_n)u_n\, dx = o(\|u_n\|)$ by (24) and the Hölder and Sobolev inequalities. So

$$o(\|u_n\|) = \Phi'(u_n)u_n = \|u_n\|^2 - \int_{\mathbb{R}^N} f(x,u_n)u_n\, dx = \|u_n\|^2 + o(\|u_n\|) \tag{28}$$

and therefore $u_n \to 0$. However, this is a contradiction because $u_n \in \mathcal{N}$ and $\mathcal{N}$ is bounded away from 0. So $u_n \not\to 0$ in $L^q(\mathbb{R}^N)$ and applying P.L. Lions' lemma once more, we see that (26) holds with v_n replaced by u_n and as before we may assume translating u_n if necessary that $u_n \rightharpoonup u \neq 0$. Hence u is a nontrivial solution of (22) and in particular, $u \in \mathcal{N}$.

We still need to show that $\Phi(u) = c := \inf_{\mathcal{N}} \Phi$. Since we may assume passing to a subsequence that $u_n \to u$ a.e., Lemma 21 and Fatou's lemma imply

$$\begin{aligned} c + o(1) = \Phi(u_n) - \frac{1}{2}\Phi'(u_n)u_n &= \int_{\mathbb{R}^N} \left(\frac{1}{2} f(x,u_n)u_n - F(x,u_n)\right) dx \\ &\ge \int_{\mathbb{R}^N} \left(\frac{1}{2} f(x,u)u - F(x,u)\right) dx + o(1) \\ &= \Phi(u) - \frac{1}{2}\Phi'(u)u + o(1) = \Phi(u) + o(1). \end{aligned}$$

Hence $\Phi(u) \le c$. The reverse inequality is obvious. □

REMARK 22. Again, one sees by the argument of Remark 17 that $u > 0$ or $u < 0$ in $\mathbb{R}^N$ if u is a ground state solution.

REMARK 23. Also in the framework of Theorem 20 it is possible to replace $-\Delta$ by $-\Delta_p$ (and modify the other assumptions accordingly). A somewhat stronger result on the existence of a ground state for an equation involving the p-Laplacian has been obtained in [**30**] by means of different methods.

We add another result for a nonperiodic V corresponding to a potential well.

THEOREM 24. *Suppose* $V \in C(\mathbb{R}^N, \mathbb{R})$ *satisfies*

(i) $0 < \inf_{\mathbb{R}^N} V \le \sup_{\mathbb{R}^N} V = V_\infty < \infty$ *with* $V_\infty := \lim_{|x|\to\infty} V(x)$.

Suppose furthermore that the nonlinearity $f = f(u)$ *in (22) does not depend on* x, *is continuous and satisfies*

(ii) $f(u) = o(u)$ *as* $u \to 0$,

(iii) $u \mapsto f(u)/|u|$ *is strictly increasing on* $(-\infty, 0)$ *and* $(0, \infty)$,

(iv) $F(u)/u^2 \to \infty$ *as* $|u| \to \infty$.

Then equation (22) has a ground state solution.

This extends Theorem 2.1 in [**28**] where it was additionally assumed that f is of class C^1 with f_u satisfying a growth restriction. The proof we will sketch here is also somewhat simpler than the one in [**28**] since we do not use a bounded domain approximation of problem (22).

PROOF OF THEOREM 24. Since parts of the argument are similar to the proof of Theorem 20, we only give a brief sketch. We consider the associated limit energy functional

$$\Phi_\infty : E \to \mathbb{R}, \qquad \Phi_\infty(u) := \frac{1}{2}\int_{\mathbb{R}^N} (|\nabla u|^2 + V_\infty u^2)\, dx - \int_{\mathbb{R}^N} F(u)\, dx$$

and the corresponding minimax value

$$c_\infty := \inf_{w \in E\setminus\{0\}} \max_{s>0} \Phi_\infty(sw).$$

By assumption (i), we infer that $c \le c_\infty$. Moreover, by Theorem 20 and Remark 22, c_∞ is attained at a strictly positive or a strictly negative solution $u \in E$ of the equation $-\Delta u + V_\infty u = f(u)$. So if V is strictly smaller than V_∞ on a set of positive measure, we have $\Phi(su) < \Phi_\infty(su) \le c_\infty$ for all $s > 0$, and from this and assumption (iii) it easily follows that $c < c_\infty$. So we may assume without loss of generality that $c < c_\infty$ since otherwise we must have $V \equiv V_\infty$ and the assertion follows from Theorem 20. Now let (w_n) be a minimizing sequence for Ψ. Again we may assume $\Psi'(w_n) \to 0$, so that also $\Phi'(u_n) \to 0$ with $u_n := m(w_n)$. Arguing by contradiction, we suppose $\|u_n\| \to \infty$; then $v_n := u_n/\|u_n\| \rightharpoonup v$ in E and $v_n \to v$ a.e. after passing to a subsequence. If $v = 0$ and $v_n \to 0$ in $L^q(\mathbb{R}^N)$, then for each $s > 0$, $\int_{\mathbb{R}^N} F(sv_n)\, dx \to 0$ according to (24), and we obtain a contradiction as in (25). So $v_n \not\to 0$ in $L^q(\mathbb{R}^N)$ and therefore (26) holds again for some $\delta > 0$, $y_n \in \mathbb{R}^N$ and almost all n. If the sequence (y_n) is bounded, we obtain a contradiction as in the proof of Theorem 20. If, on the other hand, $|y_n| \to \infty$ for a subsequence, then $\widetilde{v}_n \to v \neq 0$ in $L^2_{loc}(\mathbb{R}^N)$ for the functions $\widetilde{v}_n = v_n(\cdot - y_n)$, and as in (27) Fatou's lemma yields the contradiction

$$0 \leq \frac{\Phi(u_n)}{\|u_n\|^2} = \frac{1}{2} - \int_{\mathbb{R}^N} \frac{F(u_n)}{u_n^2} v_n^2 \, dx = \frac{1}{2} - \int_{\mathbb{R}^N} \frac{F(\widetilde{u}_n)}{\widetilde{u}_n^2} \widetilde{v}_n^2 \, dx \to -\infty$$

as $n \to \infty$ with $\widetilde{u}_n := u_n(\cdot - y_n)$. So (u_n) is bounded. As in the proof of Theorem 20 we then find a new sequence of points $y_n \in \mathbb{R}^N$, $n \in \mathbb{N}$ such that $\widetilde{u}_n \rightharpoonup u \neq 0$ for the translated functions $\widetilde{u}_n := u_n(\cdot - y_n)$. We claim that the sequence (y_n) is bounded. Indeed, if $|y_n| \to \infty$ for a subsequence, the assumption on the asymptotic shape of V implies that u is a critical point of Φ_∞ and therefore

$$\begin{aligned} c + o(1) &= \Phi(u_n) - \frac{1}{2}\Phi'(u_n)u_n = \int_{\mathbb{R}^N} \left(\frac{1}{2} f(u_n)u_n - F(u_n)\right) dx \\ &= \int_{\mathbb{R}^N} \left(\frac{1}{2} f(\widetilde{u}_n)\widetilde{u}_n - F(\widetilde{u}_n)\right) dx \geq \int_{\mathbb{R}^N} \left(\frac{1}{2} f(u)u - F(u)\right) dx + o(1) \\ &= \Phi_\infty(u) - \frac{1}{2}\Phi'_\infty(u)u + o(1) = \Phi_\infty(u) + o(1) \geq c_\infty + o(1) \end{aligned}$$

which contradicts the inequality $c < c_\infty$. Hence the sequence (y_n) is bounded, so we may without loss of generality assume that $y_n = 0$ and therefore $\widetilde{u}_n = u_n$ for all n. Then we can conclude as in the proof of Theorem 20. □

We remark that in a similar way one can treat elliptic systems like

$$\begin{cases} -\Delta_p u = F_u(x, u, v), & x \in \Omega \\ -\Delta_p v = F_v(x, u, v), & x \in \Omega \\ u = v = 0, & x \in \partial\Omega, \end{cases}$$

where $\Omega \subset \mathbb{R}^N$ is bounded, F_u, F_v satisfy a suitable growth restriction and F, F_u, F_v satisfy suitably modified conditions (W_2)–(W_4) in Section 3.5. It is also possible to consider a system in $\mathbb{R}^N$, with $-\Delta_p$ replaced by $-\Delta u + u$ (and similarly for v) provided F is periodic in $x_1, \ldots, x_N$. The details are left to the reader.

3.4. A multiplicity result for a singularly perturbed problem with domain topology. In this section we revisit the classical work of Benci and Cerami [**11**] on the multiplicity of positive solutions of the problem

$$(29) \qquad \begin{cases} -\varepsilon^2 \Delta u + \mu u = f(u), & x \in \Omega, \\ u = 0 & x \in \partial\Omega. \end{cases}$$

Here $\Omega \subset \mathbb{R}^N$ is a smooth bounded domain, $\mu > 0$ is a fixed constant and the nonlinearity $f \in C(\mathbb{R}, \mathbb{R})$ satisfies (17) and (ii)–(iv) of Theorem 24. The following multiplicity result in terms of the Lusternik-Schnirelman category has been obtained by Benci and Cerami under stronger assumptions. In particular, they require the nonlinearity to be of class $C^{1,1}$ and such that the Ambrosetti-Rabinowitz condition (18) holds. We recall that the category $\mathrm{cat}_X(A)$ of a subset A of a topological space X is defined as the minimal $k \in \mathbb{N}$ such that A is covered by k closed subsets of X which are contractible in X. As usual, if no such k exists, we put $\mathrm{cat}_X(A) = \infty$. We also write $\mathrm{cat}(X) := \mathrm{cat}_X(X)$.

THEOREM 25. *Suppose that f satisfies (17), (ii)–(iv) of Theorem 24 and that the underlying domain Ω is topologically nontrivial, i.e., it is not contractible. Then there exists $\varepsilon_0 > 0$ such that, for any $\varepsilon \in (0, \varepsilon_0)$, problem (29) admits at least* $\operatorname{cat}(\Omega) + 1$ *positive solutions.*

In the following we briefly sketch the proof of this result and refer to the papers indicated below for the details. Let $E := H_0^1(\Omega)$ and consider the functional

$$\Phi : E \to \mathbb{R}, \qquad \Phi(u) := \frac{1}{2}\|u\|^2 - I(u)$$

with

$$\|u\|^2 = \int_\Omega (\varepsilon^2 |\nabla u|^2 + \mu u^2)\, dx \quad \text{and} \quad I(u) = \int_\Omega F(u^+)\, dx.$$

Consider also the subset $S_+ := \{u \in E : \|u\| = 1, u^+ \neq 0\}$ of the unit sphere in E, where as before, u^+ denotes the positive part of u. For the sake of simplicity, we do not use ε-dependent notation for the functions and sets introduced in this section. It is easy to see that Φ is a C^1-functional, and from the maximum principle it follows that critical points correspond to positive solutions of (29). As a substitute for the conditions (A_2) and (A_3) introduced in Section 3.1, we note the following properties.

(i) For each $w \in E$ with $w^+ \neq 0$ there exists s_w such that if $\Phi'(sw)w > 0$ for $0 < s < s_w$ and $\Phi'(sw)w < 0$ for $s > s_w$.
(ii) There exists $\delta > 0$ such that $s_w \geq \delta$ for all $w \in S_+$ and for each compact subset $\mathcal{W} \subset S_+$ there exists a constant $C_{\mathcal{W}}$ such that $s_w \leq C_{\mathcal{W}}$ for all $w \in \mathcal{W}$.
(iii) The map $m : S_+ \to \mathcal{N}$, $m(u) = s_u u$ is a homeomorphism between S_+ and $\mathcal{N}$, and the inverse of m is given by $m^{-1}(u) = u/\|u\|$.

Here, as before, $\mathcal{N} := \{u \in E \setminus \{0\} : \Phi'(u)u = 0\}$ is the Nehari manifold. Similarly as in Section 3.1 we now consider the functional $\Psi : S_+ \to \mathbb{R}$ defined by $\Psi(w) := \Phi(\widehat{m}(w))$. By the same argument as in the proof of Proposition 9, we see that Ψ is a C^1-functional on the open subset S_+ of the smooth manifold S. Moreover, we have

$$\Psi'(w)z = \|m(w)\|\Phi'(m(w))z \quad \text{for all } w \in S_+ \text{ and } z \in T_w(S_+). \tag{30}$$

Hence nontrivial critical points of Ψ are in 1-1-correspondence with nontrivial critical points of Φ. In order to apply variational methods to the functional Ψ, we need the following crucial observation:

LEMMA 26. (i) *Let $(u_n) \subset S_+$ be a sequence such that* $\operatorname{dist}(u_n, \partial S_+) \to 0$ *as $n \to \infty$ (where the distance is taken with respect to the norm $\|\cdot\|$). Then $\Psi(u_n) \to \infty$.*
(ii) *Ψ satisfies the Palais-Smale condition in S_+, i.e. every sequence (u_n) in S_+ such that $\Psi(u_n)$ is bounded and $\Psi'(u_n) \to 0$ as $n \to \infty$ contains a subsequence which converges strongly to some $u \in S_+$.*

PROOF. (i) Let $s > 0$, and note that

$$\Phi(su_n) = \frac{s^2}{2} - I(su_n),$$

where as a consequence of (17) and similar arguments as in the proof of [**8**, Lemma 3.1] we have

$$\begin{aligned}|I(su_n)| &= \left|\int_\Omega F(su_n^+)\,dx\right| \le C\Big(s^2\int_\Omega |u_n^+|^2\,dx + s^q\int_\Omega |u_n^+|^q\,dx\Big)\\ &\le \widetilde{C}\Big(s^2\operatorname{dist}^2(u_n,\partial S_+) + s^q\operatorname{dist}^q(u_n,\partial S_+)\Big) \to 0 \quad \text{as } n\to\infty.\end{aligned}$$

If follows that

$$\liminf_{n\to\infty}\Psi(u_n) \ge \liminf_{n\to\infty}\Phi(su_n) = \frac{s^2}{2} \quad \text{for every } s>0$$

and therefore $\Psi(u_n)\to\infty$, as claimed.

(ii) By the same argument as in the proof of Theorem 14, we see that Φ satisfies the Palais-Smale condition on $\mathcal{N}$. Combining this information with (i) and (30), we conclude that (ii) holds. □

Similarly as before, we now consider the least energy value

$$c = \inf_{u\in\mathcal{N}}\Phi(u) = \inf_{w\in E\setminus\{0\}}\max_{s>0}\Phi(sw) = \inf_{w\in S_+}\max_{s>0}\Phi(sw) = \inf_{w\in S_+}\Psi(w).$$

As a consequence of standard deformation arguments with respect to the flow of a pseudogradient vector field of Ψ on S_+ (see e.g. [**42**, Chapter 5]), we now derive an abstract multiplicity result for critical points of Ψ in terms of the Lusternik-Schnirelman category with respect to sublevel sets.

THEOREM 27. *If there exists $d \ge c$ and a compact set $K\subset S_+^d$ such that $\operatorname{cat}_{S_+^d}(K)\ge k$ for some $k\in\mathbb{N}$, where $S_+^d := \{u\in S_+ : \Psi(u)\le d\}$, then S_+^d contains at least k critical points of Ψ. If furthermore $k\ge 2$ and there exists $e>d$ such that K is contractible in S_+^e, then there exists another critical point of Ψ in $S_+^e\setminus S_+^d$.*

The proof uses standard minimax arguments for the Lusternik-Schnirelman category like e.g. in [**3, 42, 45**]. We note here that by the definitions of Ψ and m and property (III) above, the following corollary is immediate.

COROLLARY 28. *If there exists $d\ge c$ and a compact set $K\subset S_+^d$ such that $\operatorname{cat}_{\mathcal{N}^d}(K)\ge k$ for some $k\in\mathbb{N}$, where $\mathcal{N}^d := \{u\in\mathcal{N} : \Phi(u)\le d\}$, then $\mathcal{N}^d$ contains at least k critical points of Φ. If furthermore $k\ge 2$ and there exists $e>d$ such that K is contractible in $\mathcal{N}^e$, then there exists another critical point of Φ in $\mathcal{N}^e\setminus\mathcal{N}^d$.*

In view of Corollary 28, the proof of Theorem 25 is completed once we find, for each small enough $\varepsilon>0$, values $d=d(\varepsilon)\ge c$, $e=e(\varepsilon)>d$ and a compact set $K\subset\mathcal{N}^d$ such that

$$\operatorname{cat}_{\mathcal{N}^d}(K)\ge\operatorname{cat}(\Omega), \tag{31}$$

$$K \text{ is contractible in } \mathcal{N}^e. \tag{32}$$

While it is easy to see that (32) holds for a given compact set $K \subset \mathcal{N}$ and large enough e (see [**11**, p. 43]), property (31) requires more work and the details are technical. The idea is to use the fact that for ε positive and close to zero and values d slightly larger than c (depending on ε) the functions in $\mathcal{N}^d$ are concentrated around points in Ω, and therefore the *barycenter map*

$$u \mapsto \frac{\int_\Omega x|\nabla u|^2\,dx}{\int_\Omega |\nabla u|^2\,dx}$$

can be used to exploit the effect of the domain topology. The construction of a set K satisfying (31) has been carried out in detail in [**11**], and up to one point the arguments only rely on properties of the nonlinearity f guaranteed by (17) and (ii)–(iv) of Theorem 24. The point where additional information is needed is the analysis of Φ-minimizing sequences on $\mathcal{N}$ in the case where $\varepsilon = 1$ and the underlying domain is the exterior of a ball in $\mathbb{R}^N$, see [**11**, p. 38]. Here it is crucial to note that these sequences are bounded, but this can be seen precisely as in the proof of Theorem 20. For the sake of brevity we do not give more details in this survey. As a final remark we only note that the arguments in [**11**] can be somewhat simplified by replacing the barycenter map by a generalized barycenter as defined in [**16**] or [**10**].

3.5. Newtonian systems. In this section we briefly discuss the Newtonian system of ordinary differential equations

$$-\ddot{q} + q = W_q(q,t), \quad q \in \mathbb{R}^N, \ t \in \mathbb{R} \tag{33}$$

under the following assumptions on W:

(W_1) $W \in C(\mathbb{R}^N \times \mathbb{R}, \mathbb{R})$, $W_q \in C(\mathbb{R}^N \times \mathbb{R}, \mathbb{R}^N)$ and W is 2π-periodic in t,
(W_2) $W_q(q,t) = o(|q|)$ uniformly in t as $q \to 0$,
(W_3) $s \mapsto s^{-1}W_q(sq,t) \cdot q$ is strictly increasing for all $q \neq 0$ and $s > 0$,
(W_4) $W(q,t)/|q|^2 \to \infty$ uniformly in t as $|q| \to \infty$.

Since $W(q,t)$ may be replaced by $W(q,t) - W(0,t)$, we may also assume without loss of generality that $W(0,t) = 0$ for all t.

THEOREM 29. *Suppose W satisfies (W_1)–(W_4). Then system (33) has a 2π-periodic ground state solution. If W is even in q, then (33) possesses infinitely many pairs of 2π-periodic solutions.*

Let $E := H^1(S^1, \mathbb{R}^N)$, where $H^1(S^1, \mathbb{R}^N)$ is the space of 2π-periodic $\mathbb{R}^N$-valued functions with the norm given by

$$\|q\|^2 = \int_0^{2\pi} (|\dot{q}|^2 + |q|^2)\,dt.$$

The functional corresponding to (33) is

$$\Phi(q) := \frac{1}{2}\|q\|^2 - \int_0^{2\pi} W(q,t)\,dt \equiv \frac{1}{2}\|q\|^2 - I(q).$$

Since there is an embedding $E \hookrightarrow C(S^1, \mathbb{R}^N)$, no growth restriction on W_q is necessary, and since this embedding is compact, I' is a completely continuous operator, see Remark 4. Now it is easy to see that the proof of Theorem 16 may be adapted to the present situation.

A (non-periodic) solution q of (33) will be called *homoclinic* (to 0) if $q \not\equiv 0$ and $q(t), \dot{q}(t) \to 0$ as $|t| \to \infty$.

THEOREM 30. *Suppose W satisfies (W_1)–(W_4). Then system (33) has a homoclinic solution which is a ground state.*

Here $E := H^1(\mathbb{R}, \mathbb{R}^N)$ with the usual norm given by

$$\|q\|^2 := \int_{\mathbb{R}} (|\dot{q}|^2 + |q|^2)\, dt$$

and

$$\Phi(q) := \|q\|^2 - \int_{\mathbb{R}} W(q,t)\, dt.$$

Then $\Phi \in C^1(E, \mathbb{R})$ and critical points $q \neq 0$ of Φ are homoclinic solutions of (33), see e.g. [**20**]. The proof of Theorem 30 parallels that of Theorem 20. We omit the details.

Also now it would be possible to apply the arguments of [**44**] in order to show the existence of infinitely many geometrically distinct pairs of homoclinics if W is even in q, and also here it is known that for not necessarily even W there exist infinitely many homoclinics under some stronger assumptions, see e.g. [**5, 20**].

Suppose the system (33) is autonomous, i.e., $W = W(q)$, and W satisfies (W_1)–(W_4). Then there exists a homoclinic q and also all its translates $q(\cdot - \theta)$, $\theta \in \mathbb{R}$, are homoclinics. Consider now the single equation

$$-\ddot{q} + q = q^3, \quad q, t \in \mathbb{R}.$$

The potential $V(q) = \frac{1}{4}q^4 - \frac{1}{2}q^2$ is a W-shaped curve and it is easy to see by energy considerations that if q_0 is a homoclinic, then $A = \{\pm q_0(\cdot - \theta) : \theta \in \mathbb{R}\}$ is the set of all homoclinics. So there exist only two homoclinics which are geometrically distinct in the $\mathbb{R}$-sense (i.e., distinct in the real and not only formal sense) but infinitely many $\mathbb{Z}$-distinct ones, cf. the considerations preceding the proof of Theorem 20.

4. Generalized Nehari manifold

4.1. Abstract setting. In this section we assume E is a Hilbert space and $\Phi \in C^1(E, \mathbb{R})$. Moreover, we are given an orthogonal decomposition

$$E = E^+ \oplus E^0 \oplus E^- \equiv E^+ \oplus F$$

such that $\dim E^0 < \infty$. We shall write

$$u = u^+ + u^0 + u^- = u^+ + v, \quad u^\pm \in E^\pm,\ u^0 \in E^0,\ v \in F.$$

So in contrast to our previous notation, from now on $u^{\pm}$ do *not* denote the positive and the negative part of a function u. Let

$$S^+ := S \cap E^+ = \{u \in E^+ : \|u\| = 1\},$$

$$E(u) := \mathbb{R}u \oplus F \equiv \mathbb{R}u^+ \oplus F \quad \text{and} \quad \widehat{E}(u) := \mathbb{R}^+u \oplus F \equiv \mathbb{R}^+u^+ \oplus F, \tag{34}$$

where as usual, $\mathbb{R}^+ = [0, \infty)$. We make the following assumptions on Φ:

(B_1) $\Phi(u) = \frac{1}{2}\|u^+\|^2 - \frac{1}{2}\|u^-\|^2 - I(u)$, where $I(0) = 0$, $\frac{1}{2}I'(u)u > I(u) > 0$ for all $u \neq 0$ and I is weakly lower semicontinuous.

(B_2) For each $w \in E \setminus F$ there exists a unique nontrivial (i.e., $\neq 0$) critical point $\widehat{m}(w)$ of $\Phi|_{\widehat{E}(w)}$. Moreover, $\widehat{m}(w)$ is the unique global maximum of $\Phi|_{\widehat{E}(w)}$.

(B_3) There exists $\delta > 0$ such that $\|\widehat{m}(w)^+\| \geq \delta$ for all $w \in E \setminus F$, and for each compact subset $\mathcal{W} \subset E \setminus F$ there exists a constant $C_{\mathcal{W}}$ such that $\|\widehat{m}(w)\| \leq C_{\mathcal{W}}$ for all $w \in \mathcal{W}$.

The inequality in (B_1) may seem artificial and restrictive but it will be satisfied in our applications, see Lemma 21. Let

$$\mathcal{M} := \{u \in E \setminus F : \Phi'(u)u = 0 \text{ and } \Phi'(u)v = 0 \text{ for all } v \in F\}.$$

If $u \neq 0$ and $\Phi'(u) = 0$, then $\Phi(u) = \Phi(u) - \frac{1}{2}\Phi'(u)u = \frac{1}{2}I'(u)u - I(u) > 0$ while $\Phi \leq 0$ on F. Hence $\mathcal{M}$ contains all nontrivial critical points of Φ and $\widehat{E}(w) \cap \mathcal{M} = \{\widehat{m}(w)\}$ whenever $w \in E \setminus F$. Since $\widehat{m}(w) = \widehat{m}(tw)$ for all $t > 0$, it is easy to see that if $F = \{0\}$, then (B_2), (B_3) are equivalent to (A_2), (A_3) and $\mathcal{M} = \mathcal{N}$. We shall call the set $\mathcal{M}$ the *generalized Nehari manifold*. It has been introduced by Pankov in [**35**]. We also note that if $\Phi \in C^2(E, \mathbb{R})$ and the restriction of $\Phi''(\widehat{m}(w))$ to $E(w)$ is negative definite for all $w \in E \setminus F$, then $\mathcal{M}$ is a manifold of class C^1 as follows from the implicit function theorem. According to our definitions, we have

$$\widehat{m} : E \setminus F \to \mathcal{M} \quad \text{and} \quad m := \widehat{m}|_{S^+} : S^+ \to \mathcal{M}. \tag{35}$$

It is easy to see that m is a bijection whose inverse m^{-1} is given by

$$m^{-1}(u) = \frac{u^+}{\|u^+\|}.$$

Since by (B_3), $\|m(w)^+\| \geq \delta$ for all $w \in E \setminus F$, $\mathcal{M}$ is bounded away from F and closed. Let

$$c := \inf_{u \in \mathcal{M}} \Phi(u).$$

It follows from (B_2) that $c > 0$ if it is attained. We shall show that if $u_0 \in \mathcal{M}$ and $\Phi(u_0) = c$, then u_0 is a critical point of Φ and hence it must be a ground state for the equation $\Phi'(u) = 0$.

Let us mention here that a reduction of an indefinite functional to a functional on E^+ is well known under stronger differentiability conditions, see [**14**] and e.g. [**1, 13, 38, 39**] (and [**2**] for a related finite-dimensional reduction). In [**38, 39**] a reduction in two steps has been performed: first to E^+ and then to a Nehari manifold on E^+.

PROPOSITION 31. *Suppose* Φ *satisfies* (B_1)–(B_3). *Then:*

(a) *The mapping* $\widehat{m}$ *is continuous.*
(b) *The mapping* m *is a homeomorphism between* S^+ *and* $\mathcal{M}$.

PROOF. (a) Suppose $(w_n) \subset E \setminus F$, $w_n \to w \notin F$. Since $\widehat{m}(w) = \widehat{m}(w^+/\|w^+\|)$, we may assume without loss of generality that $w_n \in S^+$. It suffices to show that $\widehat{m}(w_n) \to \widehat{m}(w)$ after passing to a subsequence. Write $\widehat{m}(w_n) = s_n w_n + v_n$ ($v_n = v_n^0 + v_n^- \in F$). By (B_3), $(\widehat{m}(w_n))$ is bounded. So taking a subsequence, $s_n \to s$ and $v_n \rightharpoonup v_* = v_*^0 + v_*^-$. Setting $\widehat{m}(w) = sw + v$, it follows from (B_2) that

$$\Phi(\widehat{m}(w_n)) \geq \Phi(s_n w_n + v) \to \Phi(sw + v) = \Phi(\widehat{m}(w))$$

and hence, using the weak lower semicontinuity of the norm and I,

$$\begin{aligned}\Phi(\widehat{m}(w)) &\leq \lim_{n\to\infty} \Phi(\widehat{m}(w_n)) = \lim_{n\to\infty}\left(\frac{1}{2}s_n^2 - \frac{1}{2}\|v_n^-\|^2 - I(\widehat{m}(w_n))\right) \\ &\leq \frac{1}{2}s^2 - \frac{1}{2}\|v_*^-\|^2 - I(sw + v_*) \leq \Phi(\widehat{m}(w)).\end{aligned}$$

Hence the inequalities above must be equalities. It follows that (v_n^-) is strongly convergent and so is (v_n^0) because $\dim E^0 < \infty$. Consequently, $v_n \to v_*$ and by the uniqueness property (B_2), $v_* = v$.

(b) This is immediate from (a) and the fact that m^{-1} is continuous. □

Let

$$\widehat{\Psi} : E^+ \setminus \{0\} \to \mathbb{R}, \ \widehat{\Psi}(w) := \Phi(\widehat{m}(w)) \quad \text{and} \quad \Psi := \widehat{\Psi}|_{S^+}.$$

PROPOSITION 32. *Suppose* Φ *satisfies* (B_1)–(B_3). *Then* $\widehat{\Psi} \in C^1(E^+ \setminus \{0\}, \mathbb{R})$ *and*

$$\widehat{\Psi}'(w)z = \frac{\|\widehat{m}(w)^+\|}{\|w\|}\,\Phi'(\widehat{m}(w))z \quad \text{for all } w, z \in E^+,\ w \neq 0.$$

PROOF. Let $w \in E^+ \setminus \{0\}$, $z \in E^+$ and put $\widehat{m}(w) = s_w w + v_w$, $v_w \in F$. As in the proof of Proposition 9, we have

$$\begin{aligned}\widehat{\Psi}(w + tz) - \widehat{\Psi}(w) &= \Phi(s_{w+tz}(w + tz) + v_{w+tz}) - \Phi(s_w w + v_w) \\ &\leq \Phi(s_{w+tz}(w + tz) + v_{w+tz}) - \Phi(s_{w+tz} w + v_{w+tz}) \\ &= \Phi'(s_{w+tz} w + v_{w+tz} + \tau_t s_{w+tz} tz) s_{w+tz} tz\end{aligned}$$

for all $|t|$ small enough and some $\tau_t \in (0, 1)$, and also a similar reverse inequality holds. So continuing as in the above-mentioned proof, we obtain

$$\widehat{\Psi}'(w)z = s_w \Phi'(s_w w + v_w)z = \frac{\|\widehat{m}(w)^+\|}{\|w\|}\Phi'(\widehat{m}(w))z.$$

□

COROLLARY 33. *Suppose Φ satisfies (B_1)–(B_3). Then:*

(a) $\Psi \in C^1(S^+, \mathbb{R})$ *and*

$$\Psi'(w)z = \|m(w)^+\|\Phi'(m(w))z \quad \text{for all } z \in T_w(S^+).$$

(b) *If (w_n) is a Palais-Smale sequence for Ψ, then $(m(w_n))$ is a Palais-Smale sequence for Φ. If $(u_n) \subset \mathcal{M}$ is a bounded Palais-Smale sequence for Φ, then $(m^{-1}(u_n))$ is a Palais-Smale sequence for Ψ.*

(c) *w is a critical point of Ψ if and only if $m(w)$ is a nontrivial critical point of Φ. Moreover, the corresponding values of Ψ and Φ coincide and $\inf_{S^+} \Psi = \inf_{\mathcal{M}} \Phi$.*

(d) *If Φ is even, then so is Ψ.*

The argument is the same as in Corollary 10 except that the proof of (b) needs to be slightly modified. Now we have $E = T_w(S^+) \oplus E(w)$ for all $w \in S^+$ and setting $u = m(w)$, (11) is replaced by

$$\|\Psi'(w)\| = \sup_{\substack{z \in T_w(S^+) \\ \|z\|=1}} \Psi'(w)z = \|u^+\| \sup_{\substack{z \in T_w(S^+) \\ \|z\|=1}} \Phi'(u)z = \|u^+\| \, \|\Phi'(u)\|,$$

where the last equality follows because $\Phi'(u)v = 0$ for all $v \in E(w)$ and $E(w)$ is orthogonal to $T_w(S^+)$. Since $\|u^+\| \geq \delta > 0$ for $u \in \mathcal{M}$, this gives the conclusion.

REMARK 34. Similarly as in Remark 11, we now have the following variational characterization of the infimum of Φ over $\mathcal{M}$:

$$c = \inf_{u \in \mathcal{M}} \Phi(u) = \inf_{w \in E \setminus F} \max_{u \in \widehat{E}(w)} \Phi(u) = \inf_{w \in S^+} \max_{u \in \widehat{E}(w)} \Phi(u).$$

The following is an analogue of Theorem 12:

THEOREM 35. *Suppose Φ satisfies (B_1), (B_2) and*

(i) $I'(u) = o(\|u\|)$ *as* $u \to 0$,

(ii) *$I(su)/s^2 \to \infty$ uniformly for u on weakly compact subsets of $E \setminus \{0\}$ as $s \to \infty$.*

(iii) *I' is completely continuous.*

Then equation $\Phi'(u) = 0$ has a ground state solution. Moreover, if I is even, then this equation has infinitely many pairs of solutions.

PROOF. We show that (B_3) holds. By (B_1) and (i), we can find $\rho, \eta > 0$ such that $\Phi(w) \geq \eta$ whenever $w \in S_\rho(0) \cap E^+$. So for any $w \in E \setminus F$, $\Phi(\widehat{m}(w)) \geq \eta$. Since $\frac{1}{2}\|\widehat{m}(w)^+\| \geq \Phi(\widehat{m}(w))$, $\|\widehat{m}(w)^+\| \geq \delta$ for some $\delta > 0$. That $\|\widehat{m}(w))\| \leq C_{\mathcal{W}}$ for all w in a compact set $\mathcal{W} \subset S^+$ is an immediate consequence of (B_1), (ii) and the fact that $\Phi(\widehat{m}(w)) > 0$. Since $\widehat{m}(w) = \widehat{m}(w^+/\|w^+\|)$ for all $w \in E \setminus F$, this is also true if $\mathcal{W}$ is a compact subset of $E \setminus F$. We also note that $c = \inf_{\mathcal{M}} \Phi \geq \eta > 0$.

The rest of the argument is the same as in the proof of Theorem 12 taking Corollary 33 into account and assuming Proposition 36 below. □

We remark that it does not suffice to assume $s \mapsto I'(su)u/s$ is increasing for all $s > 0$ in order to ensure that (B_2) holds.

PROPOSITION 36. *Under the assumptions of Theorem 35, Φ satisfies the Palais-Smale condition on $\mathcal{M}$.*

PROOF. We modify some earlier arguments. Let $(u_n) \subset \mathcal{M}$ be a sequence such that $\Phi(u_n) \leq d$ for some $d > 0$ and $\Phi'(u_n) \to 0$. If (u_n) is unbounded, we set $v_n := u_n/\|u_n\|$. Passing to a subsequence, we may assume $\|u_n\| \to \infty$ and $v_n \rightharpoonup v$. It follows from (ii) that if $v \neq 0$, then

$$0 \leq \frac{\Phi(u_n)}{\|u_n\|^2} = \frac{1}{2}\|v_n^+\|^2 - \frac{1}{2}\|v_n^-\|^2 - \frac{I(\|u_n\|v_n)}{\|u_n\|^2} \tag{36}$$

and the right-hand side above tends to $-\infty$ (cf. (14)). Hence $v = 0$. By (36) and since $I \geq 0$, $\|v_n^+\| \geq \|v_n^-\|$. If $v_n^+ \to 0$, then also $v_n^- \to 0$ and therefore

$$\|v_n^0\|^2 = 1 - \|v_n^+\|^2 - \|v_n^-\|^2 \to 1.$$

Hence $v_n^0 \to v^0$ because E^0 is finite-dimensional. So $v \neq 0$, a contradiction. Therefore $v_n^+ \not\to 0$ and thus $\|v_n^+\| \geq \alpha$ for all n and some $\alpha > 0$, possibly after passing to a subsequence. We complete the proof of boundedness of (u_n) by noting that

$$d \geq \Phi(u_n) \geq \Phi(sv_n^+) \geq \frac{1}{2}\alpha^2 s^2 - I(sv_n^+) \to \frac{1}{2}\alpha^2 s^2 \tag{37}$$

for all $s > 0$, a contradiction again (cf. (15)). So (u_n) is bounded and

$$\Phi'(u_n) = u_n^+ - u_n^- - I'(u_n) \to 0.$$

Since I' is completely continuous and $\dim E^0 < \infty$, (u_n) has a convergent subsequence. □

4.2. Elliptic equations–the indefinite case. Now we return to the problem (16) in a bounded domain, but this time for $\lambda \geq \lambda_1$. Let $E = H_0^1(\Omega)$ and let $E = E^+ \oplus E^0 \oplus E^-$ be the orthogonal decomposition corresponding to the spectrum of $-\Delta - \lambda$ in E. More precisely, denote the Dirichlet eigenvalues of $-\Delta$ by $\lambda_1, \lambda_2, \ldots$ and a corresponding orthogonal (in E) set of eigenfunctions by $e_1, e_2, \ldots$. Suppose $\lambda_k < \lambda = \lambda_{k+1} = \cdots = \lambda_m < \lambda_{m+1}$, where $1 \leq k < m$. Then

$$E^- = \operatorname{span}\{e_1, \ldots, e_k\} \quad \text{and} \quad E^0 = \operatorname{span}\{e_{k+1}, \ldots, e_m\}.$$

We also admit the cases $k = 0$ and $k = m \geq 1$ which respectively correspond to $E^- = \{0\}$ and $E^0 = \{0\}$. Let $u = u^+ + u^0 + u^- \in E^+ \oplus E^0 \oplus E^-$. In E we introduce an equivalent norm such that

$$\int_\Omega (|\nabla u|^2 - \lambda u^2)\, dx = \|u^+\|^2 - \|u^-\|^2.$$

Then the functional Φ corresponding to (16) is given by

$$\Phi(u) = \frac{1}{2}\|u^+\|^2 - \frac{1}{2}\|u^-\|^2 - I(u), \tag{38}$$

where as before,

$$I(u) = \int_\Omega F(x, u)\, dx. \tag{39}$$

The following result is taken from [**44**] and is an extension of Theorem 16 to the case $\lambda \geq \lambda_1$:

THEOREM 37. *Suppose that $\lambda \geq \lambda_1$, f satisfies (17) and*

(i) *$f(x,u) = o(u)$ uniformly in x as $u \to 0$,*
(ii) *$u \mapsto f(x,u)/|u|$ is strictly increasing on $(-\infty, 0)$ and $(0, \infty)$,*
(iii) *$F(x,u)/u^2 \to \infty$ uniformly in x as $|u| \to \infty$.*

Then equation (16) has a ground state solution. Moreover, if f is odd in u, then (16) has infinitely many pairs of solutions.

PROOF. Since I is weakly continuous by Theorem 3, it follows from (38) and Lemma 21 that (B_1) holds, and so do (i), (ii) of Theorem 35 as has been shown during the course of the proof of Theorem 16. Since also (iii) holds according to Theorem 3, the conclusion will follow from Theorem 35 if we can verify (B_2). This will be done in Proposition 39 below. □

In Lemma 38 and Proposition 39 below we assume that the hypotheses of Theorem 37 are satisfied.

LEMMA 38. *Let u, s, v be real numbers such that $s \geq -1$ and let $w := su+v \neq 0$. Then*

$$f(x,u)\left[s\left(\frac{s}{2}+1\right)u + (1+s)v\right] + F(x,u) - F(x,u+w) < 0$$

for all $x \in \Omega$.

PROOF. We fix $x \in \Omega$ and $u, v \in \mathbb{R}$. For $s \geq -1$ we put $z = z(s) := (1+s)u+v$ (so $z = u + w$) and

$$g(s) := f(x,u)\left[s\left(\frac{s}{2}+1\right)u + (1+s)v\right] + F(x,u) - F(x,z).$$

We must show $g(s) < 0$ whenever $u \neq z$. Suppose $u = 0$. Then $z \neq 0$ and hence $g(s) = -F(x,z) < 0$ by Lemma 21. In what follows we assume $u \neq 0$. If $uz \leq 0$, we have, since $v = z - (1+s)u$,

$$\begin{aligned} g(s) &= f(x,u)\left[\left(\frac{s^2}{2}+s\right)u + (s+1)(z-(s+1)u)\right] + F(x,u) - F(x,z) \\ &< f(x,u)\left[\left(\frac{s^2}{2}+s\right)u + (s+1)(z-(s+1)u)\right] + \frac{1}{2}f(x,u)u - F(x,z) \qquad (40) \\ &= -\frac{1}{2}(s+1)^2 f(x,u)u + (s+1)f(x,u)z - F(x,z) \leq 0, \end{aligned}$$

where we have used Lemma 21 and the fact that $f(x,u)z \leq 0$ whenever $uz \leq 0$. Hence $g(s) < 0$ in this case. Suppose $uz > 0$. We note that

$$g(-1) = -\frac{1}{2}f(x,u)u + F(x,u) - F(x,v) < -F(x,v) \leq 0 \quad \text{and} \quad \lim_{s\to\infty} g(s) = -\infty$$

by Lemma 21. Moreover,

$$g'(s) = uz\left(\frac{f(x,u)}{u} - \frac{f(x,z)}{z}\right). \tag{41}$$

Suppose that g attains its maximum on $[-1,\infty)$ at some point s with $g(s) \geq 0$. Then $g'(s) = 0$ and $u = z$ by (41) and (ii) of Theorem 37, so $g(s) = -\frac{1}{2}s^2 f(x,u)u \leq 0$. It follows that $g(s)$ may be 0 if $u = z$ (i.e., $w = 0$) but must be negative otherwise. □

Recall the definition (34) of $\widehat{E}(u)$.

PROPOSITION 39. *(i) $\widehat{E}(w) \cap \mathcal{M} \neq \emptyset$ for any $w \in E \setminus (E^0 \oplus E^-) \equiv E \setminus F$.*
(ii) If $u \in \mathcal{M}$, then

$$\Phi(u+w) < \Phi(u) \quad \textit{whenever } u+w \in \widehat{E}(u), \ w \neq 0.$$

Hence u is the unique global maximum of $\Phi|_{\widehat{E}(u)}$.

PROOF. (i) Let $w \in E \setminus F$. Since $\widehat{E}(w) = \widehat{E}(w^+/\|w^+\|)$, we may assume $w \in S^+$. We claim that $\Phi \leq 0$ on $\widehat{E}(w) \setminus B_R(0)$ provided R is large enough. Arguing by contradiction, we find a sequence $(u_n) \subset \widehat{E}(w)$ such that $\|u_n\| \to \infty$ and $\Phi(u_n) \geq 0$. Setting $v_n := u_n/\|u_n\|$, we can use (36) and the argument following it to conclude that $v_n \rightharpoonup 0$ and at the same time $\|v_n^+\| = \|s_n w\| = s_n$ is bounded and bounded away from 0. But then $v_n^+ \to sw$, $s > 0$, a contradiction.

By (i) of Theorem 37, $\Phi(sw) = \frac{1}{2}s^2 + o(s^2)$ as $s \to 0$. Hence $0 < \sup_{\widehat{E}(w)} \Phi < \infty$. Since Φ is weakly upper semicontinuous on $\widehat{E}(w)$ and $\Phi \leq 0$ on $\widehat{E}(w) \cap F$, the supremum is attained at some point u_0 such that $u_0^+ \neq 0$. So u_0 is a critical point of $\Phi|_{\widehat{E}(w)}$ and hence $u_0 \in \mathcal{M}$.

Since $\dim F < \infty$, the argument above could be simplified. We have chosen not to do this because later on we shall encounter a situation where F is infinite-dimensional.

(ii) Let

$$B(v_1, v_2) := \int_\Omega (\nabla v_1 \cdot \nabla v_2 - \lambda v_1 v_2)\, dx, \quad v_1, v_2 \in E.$$

For $u \in \mathcal{M}$, let $u + w \in \widehat{E}(u)$. Then $u + w = (1+s)u + v$, where $s \geq -1$ and $v = v^0 + v^- \in F$. A calculation gives

$$\begin{aligned}
\Phi(u+w) - \Phi(u) &= \frac{1}{2}[B(u+w, u+w) - B(u,u)] + \int_\Omega (F(x,u) - F(x,u+w))\, dx \\
&= \frac{1}{2}[B((1+s)u + v, (1+s)u + v) - B(u,u)] \\
&\quad + \int_\Omega (F(x,u) - F(x,u+w))\, dx \\
&= \frac{1}{2}\Big([(1+s)^2 - 1]B(u,u) + 2(1+s)B(u,v) + B(v,v)\Big) \\
&\quad + \int_\Omega (F(x,u) - F(x,u+w))\, dx
\end{aligned}$$

$$
\begin{aligned}
&= -\frac{\|v^-\|^2}{2} + B\left(u, s\left(\frac{s}{2}+1\right)u + (1+s)v\right) \\
&\quad + \int_\Omega (F(x,u) - F(x,u+w))\,dx \\
&= -\frac{\|v^-\|^2}{2} + \int_\Omega \Big(f(x,u)\left[s\left(\frac{s}{2}+1\right)u + (1+s)v\right] \\
&\quad + F(x,u) - F(x,u+w)\Big)\,dx.
\end{aligned}
$$

In the last step we have used the fact that $z := s\left(\frac{s}{2}+1\right)u + (1+s)v \in E(u)$ and therefore

$$0 = \Phi'(u)z = B(u,z) - \int_\Omega f(x,u)z\,dx.$$

Since w is nonzero on a set of positive measure, the last integral above is negative according to Lemma 38 and hence $\Phi(u+w) < \Phi(u)$. □

We have completed the proof of Theorem 37. Next we turn to a result in [**44**] for the nonlinear Schrödinger equation (22) with V such that 0 is in a gap of the spectrum of $-\Delta+V$. In [**44**] it has been shown that (22) has a ground state solution and if, in addition, f is odd in u, then there are infinitely many geometrically distinct solutions. Here we only prove the existence of a ground state.

THEOREM 40. *Suppose* $V \in C(\mathbb{R}^N,\mathbb{R})$, $f \in C(\mathbb{R}^N \times \mathbb{R}, \mathbb{R})$ *satisfies (17) and*

(i) V, f *are 1-periodic in* $x_1,\dots,x_N$, $0 \notin \sigma(-\Delta+V)$ *and* $\sigma(-\Delta+V) \cap (-\infty,0) \neq \emptyset$,
(ii) $f(x,u) = o(u)$ *uniformly in* x *as* $u \to 0$,
(iii) $u \mapsto f(x,u)/|u|$ *is strictly increasing on* $(-\infty,0)$ *and* $(0,\infty)$,
(iv) $F(x,u)/u^2 \to \infty$ *uniformly in* x *as* $|u| \to \infty$.

Then equation (22) has a ground state solution.

PROOF. Let $E = H^1(\mathbb{R}^N)$. Since $0 \notin \sigma(-\Delta+V)$ and $\sigma(-\Delta+V)\cap(-\infty,0) \neq \emptyset$, $E^0 = \{0\}$, $\dim E^\pm = \infty$ and there is an equivalent norm in E such that

$$\int_{\mathbb{R}^N} (|\nabla u|^2 + V(x)u^2)\,dx = \|u^+\|^2 - \|u^-\|^2.$$

Hence (38) holds except that in (39) Ω should be replaced by $\mathbb{R}^N$. Since $F \geq 0$, I is weakly lower semicontinuous as follows easily from Fatou's lemma and the fact that if $u_n \rightharpoonup u$ in E, then $u_n \to u$ in $L^2_{loc}(\mathbb{R}^N)$ and hence a.e. after passing to a subsequence. This and Lemma 21 imply that (B_1) holds, and so does (B_2) by Proposition 39. (The only difference here is that integration is performed over $\mathbb{R}^N$ instead of Ω and $-\lambda v_1 v_2$ is replaced by $V(x)v_1v_2$ in the definition of the bilinear form B. Note also that in the setting of Theorem 37 we had $\dim E^- < \infty$ but as we have already mentioned in the proof of Proposition 39, this fact was not used.) That (B_3) is satisfied and $c = \inf_{\mathcal{M}} \Phi > 0$ follows as in the proof of Theorem 35.

Now it remains to combine the arguments of Theorem 20 and Proposition 36 as follows. We take a minimizing Palais-Smale sequence (w_n) for Ψ. Then (u_n), where $u_n := m(w_n)$, is a Palais-Smale sequence for Φ according to Corollary 33. Assuming $\|u_n\| \to \infty$ and setting $v_n := u_n/\|u_n\|$, we see as in (36) and the argument following it that $v_n \rightharpoonup 0$ after passing to a subsequence and $\|v_n^+\| \geq 1/\sqrt{2}$ (because $\|v_n^+\| \geq \|v_n^-\|$ and $\|v_n^+\|^2 + \|v_n^-\|^2 = 1$). If $v_n^+ \to 0$ in $L^q(\mathbb{R}^N)$, then using (24) we obtain a contradiction as in (37). Hence (26) holds with v_n replaced by v_n^+, and again we may assume, possibly after replacing v_n by $v_n(\cdot - y_n)$ for some $y_n \in \mathbb{Z}^N$, that (y_n) is a bounded sequence. But then $v^+ \neq 0$ and hence $v \neq 0$, a contradiction. We have shown that (u_n) is bounded. The rest of the argument is exactly the same as in Theorem 20 except that if $u_n \to 0$ in $L^q(\mathbb{R}^N)$, then (28) should read

$$o(\|u_n^+\|) = \Phi'(u_n)u_n^+ = \|u_n^+\|^2 - \int_{\mathbb{R}^N} f(x,u_n)u_n^+\,dx = \|u_n^+\|^2 + o(\|u_n^+\|).$$

Hence $u_n^+ \to 0$ in E and

$$\liminf_{n\to\infty} \Phi(u_n) = \liminf_{n\to\infty} \left(\frac{1}{2}\|u_n^+\|^2 - \frac{1}{2}\|u_n^-\|^2 - I(u_n)\right) \leq \frac{1}{2}\lim_{n\to\infty} \|u_n^+\|^2 = 0,$$

contradicting the fact that $\inf_{\mathcal{M}} \Phi > 0$. □

Below we formulate two results for systems of equations. Let $\Omega \subset \mathbb{R}^N$ be a bounded domain and consider the problem

$$\begin{cases} -\Delta u_1 = h(x,u_2), & x \in \Omega \\ -\Delta u_2 = g(x,u_1), & x \in \Omega \\ u_1 = u_2 = 0, & x \in \partial\Omega. \end{cases} \tag{42}$$

These systems, both in bounded and unbounded domains, have been considered in several recent papers, see e.g. [**38, 39**] and the references there. Our results seem to be new under the weak superlinearity conditions which we impose here.

Denote

$$G(x,u_1) := \int_0^{u_1} g(x,s)\,ds \quad \text{and} \quad H(x,u_2) := \int_0^{u_2} h(x,s)\,ds.$$

THEOREM 41. *Suppose g, h satisfy (17) and (i)–(iii) of Theorem 37. Then system (42) has a ground state solution. Moreover, if g is odd in u_1 and h odd in u_2, then (42) has infinitely many pairs of solutions.*

PROOF. Let $E := H_0^1(\Omega) \times H_0^1(\Omega)$ and

$$\Phi(u) := \int_\Omega \nabla u_1 \cdot \nabla u_2\,dx - \int_\Omega (G(x,u_1) + H(x,u_2))\,dx \qquad \text{for } u = (u_1,u_2) \in E.$$

Then $\Phi \in C^1(E,\mathbb{R})$ and critical points of Φ are solutions of (42). The quadratic form

$$u \mapsto \int_\Omega \nabla u_1 \cdot \nabla u_2\,dx$$

is indefinite and $E = E^+ \oplus E^-$, where

$$E^\pm = \{u \in E : u_2 = \pm u_1\}. \tag{43}$$

So $\dim E^{\pm} = \infty$ and each $u \in E$ may be represented as

$$u = u^+ + u^- = \frac{1}{2}(u_1 + u_2,\, u_1 + u_2) + \frac{1}{2}(u_1 - u_2,\, u_2 - u_1), \quad \text{where } u^{\pm} \in E^{\pm}.$$

Hence we can write

$$\Phi(u) = \frac{1}{2}\|u^+\|^2 - \frac{1}{2}\|u^-\|^2 - I(u)$$

with the norm $\|\cdot\|$ on E defined by $\|u\|^2 = \int_\Omega (|\nabla u_1|^2 + |\nabla u_2|^2)$ for $u = (u_1, u_2) \in E$. Now it remains to repeat the argument of Theorem 37. The only point which requires explanation is that, since Lemma 38 holds for both g and h, we obtain as in Proposition 39(ii):

If $u \in \mathcal{M}$, then $\Phi(u + w) < \Phi(u)$ for any $w \neq 0$ such that $u + w \in \widehat{E}(u)$.

Indeed, we can write

$$u + w = (1 + s)u + v \quad \text{with } s \geq -1 \text{ and } v \in E^-.$$

Hence a computation similar to the one in the proof of Proposition 39(ii) gives

$$\begin{aligned}
\Phi(u+w) - \Phi(u) = &-\frac{\|v\|^2}{2} + \int_\Omega \Big(\nabla u_1 \cdot \nabla \Big[s\Big(\frac{s}{2}+1\Big)u_2 + (1+s)v_2\Big]\, dx \\
&+ \int_\Omega \Big(\nabla u_2 \cdot \nabla \Big[s\Big(\frac{s}{2}+1\Big)u_1 + (1+s)v_1\Big]\, dx \\
&+ \int_\Omega (G(x,u_1) - G(x, u_1 + w_1) \\
&+ H(x,u_2) - H(x, u_2 + w_2))\, dx \\
= &-\frac{\|v\|^2}{2} + \int_\Omega \Big(g(x,u_1)\Big[s\Big(\frac{s}{2}+1\Big)u_1 + (1+s)v_1\Big] \\
&+ G(x,u_1) - G(x,u_1+w_1)\Big)\, dx \\
&+ \int_\Omega \Big(h(x,u_2)\Big[s\Big(\frac{s}{2}+1\Big)u_2 + (1+s)v_2\Big] \\
&+ H(x,u_2) - H(x,u_2+w_2))\, dx.
\end{aligned}$$

Since $w = (w_1, w_2) \neq 0$, at least one of the integrals above is negative and therefore $\Phi(u + w) < \Phi(u)$ as required. □

Finally we consider the problem

$$\begin{cases} -\Delta u_1 + u_1 = h(x, u_2), & x \in \mathbb{R}^N \\ -\Delta u_2 + u_2 = g(x, u_1), & x \in \mathbb{R}^N \\ u_1(x), u_2(x) \to 0, & |x| \to \infty. \end{cases} \tag{44}$$

THEOREM 42. *Suppose g, h satisfy (17) and (i)–(iv) of Theorem 40. Then system (44) has a ground state solution.*

Here $E = H^1(\mathbb{R}^N) \times H^1(\mathbb{R}^N)$,

$$\Phi(u_1, u_2) = \int_{\mathbb{R}^N} (\nabla u_1 \cdot \nabla u_2 + u_1 u_2)\, dx - \int_{\mathbb{R}^N} (G(x, u_1) + H(x, u_2))\, dx$$

and $E = E^+ \oplus E^-$, with $E^\pm$ given by (43). The proof is by inspection of the arguments of Theorems 40 and 41. It can also be seen by inspection of the arguments in [**44**] that if g, h are odd in u_1 and u_2 respectively, then there exist infinitely many geometrically distinct solutions. However, if g and h are independent of x, see our earlier comments in Section 3.3 concerning $\mathbb{Z}^N$-distinct vs. $\mathbb{R}^N$-distinct.

REMARK 43. It is easy to see that one can admit more general linear parts in (42) and (44). We do not pursue the details.

Bibliography

[1] N. Ackermann, *A nonlinear superposition principle and multibump solutions of periodic Schrödinger equations*, J. Funct. Anal. 234 (2006), 277–320.

[2] H. Amann, *Saddle points and multiple solutions of differential equations*, Math. Z. 169 (1979), 127–166.

[3] A. Ambrosetti and A. Malchiodi, *Nonlinear Analysis and Semilinear Elliptic Problems*, Cambridge Studies in Advanced Mathematics 104, Cambridge University Press, Cambridge, 2007.

[4] A. Ambrosetti and P.H. Rabinowitz, *Dual variational methods in critical point theory and applications*, J. Funct. Anal. 14 (1973), 349–381.

[5] G. Arioli, A. Szulkin and W. Zou, *Multibump solutions and critical groups*, Trans. Amer. Math. Soc. 361 (2009), 3159–3187.

[6] A. Bahri, *Topological results on a certain class of functionals and application*, J. Funct. Anal. 41 (1981), 397–427.

[7] A. Bahri and H. Berestycki, *A perturbation method in critical point theory and applications*, Trans. Amer. Math. Soc. 267 (1981), 1–32.

[8] T. Bartsch, Z.L. Liu and T. Weth, *Sign changing solutions of superlinear Schrödinger equations,* Comm. PDE 29 (2004), 25–42.

[9] T. Bartsch and T. Weth, *A note on additional properties of sign changing solutions to superlinear elliptic equations*, Topol. Meth. Nonl. Anal. 22 (2003), 1–14.

[10] T. Bartsch and T. Weth, *Three nodal solutions of singularly perturbed elliptic equations on domains without topology*, Ann. IHP Anal. Non Linéaire 22 (2005), 259–281.

[11] V. Benci and G. Cerami, *Multiple positive solutions of some elliptic problems via the Morse theory and the domain topology,* Calc. Var. PDE 2 (1994), 29–48.

[12] K.J. Brown and T.F. Wu, *A semilinear elliptic system involving nonlinear boundary condition and sign-changing weight function*, J. Math. Anal. Appl. 337 (2008), 1326–1336.

[13] B. Buffoni, L. Jeanjean and C.A. Stuart, *Existence of a nontrivial solution to a strongly indefinite semilinear equation*, Proc. Amer. Math. Soc. 119 (1993), 179–186.

[14] A. Castro, *Hammerstein integral equations with indefinite kernel*, Internat. J. Math. & Math. Sci. 1 (1978), 187–201.

[15] A. Castro, J. Cossio and J. M. Neuberger, *A sign-changing solution for a superlinear Dirichlet problem*, Rocky Mountain J. Math. 27 (1997), 1041 1053.

[16] G. Cerami and D. Passaseo, *The effect of concentrating potentials in some singularly perturbed problems*, Calc. Var. PDE 17 (2003), no. 3, 257–281.

[17] J. Chabrowski and D.G. Costa, *On a class of Schrödinger-type equations with indenite weight functions*, Comm. PDE 33 (2008), 1368–1394.

[18] J. Chabrowski and A. Szulkin, *On the Schrödinger equation involving a critical Sobolev exponent and magnetic field*, Topol. Meth. Nonl. Anal. 25 (2005), 3–21.

[19] J.N. Corvellec, M. Degiovanni and M. Marzocchi, *Deformation properties for continuous functionals and critical point theory*, Topol. Meth. Nonl. Anal. 1 (1993), 151–171.

[20] V. Coti Zelati and P.H. Rabinowitz, *Homoclinic orbits for second order Hamiltonian systems possessing superquadratic potentials*, J. Amer. Math. Soc. 4 (1991), 693–727.

[21] V. Coti Zelati and P.H. Rabinowitz, *Homoclinic type solutions for a semilinear elliptic PDE on R^n*, Comm. Pure Appl. Math. 45 (1992), 1217–1269.
[22] G. Dinca, P. Jebelean and J. Mawhin, *Variational and topological methods for Dirichlet problems with p-Laplacian*, Port. Math. 58 (2001), 339–378.
[23] P. Drábek and J. Milota, *Methods of Nonlinear Analysis. Applications to Differential Equations*, Birkhäuser, Basel, 2007.
[24] P. Drábek and S.I. Pohozaev, *Positive solutions for the p-Laplacian: application of the fibering method*, Proc. Royal Soc. Edinb. A 127 (1997), 703-726.
[25] I. Ekeland, *On the variational principle*, J. Math. Anal. Appl. 47 (1974), 324–353.
[26] D. Gilbarg and N.S. Trudinger, *Elliptic Partial Differential Equations of Second Order*, Springer-Verlag, Berlin, 1983.
[27] P. Kuchment, *Floquet Theory for Partial Differential Equations*, Birkhäuser, Basel, 1993.
[28] Y.Q. Li, Z.Q. Wang and J. Zeng, *Ground states of nonlinear Schrödinger equations with potentials*, Ann. IHP Analyse Non Linéaire 23 (2006), 829–837.
[29] P.L. Lions, *The concentration-compactness principle in the calculus of variations. The locally compact case*, Ann. IHP Analyse Non Linéaire 1 (1984), 109–145 and 223–283.
[30] S. Liu, *On ground states of superlinear p-Laplacian equations in R^N*, J. Math. Anal. Appl. 361 (2010), 48–58.
[31] Z. Liu and Z.Q. Wang, *On the Ambrosetti-Rabinowitz superlinear condition*, Adv. Nonl. Stud. 4 (2004), 563–574.
[32] J. Mawhin and M. Willem, *Critical Point Theory and Hamiltonian Systems*, Springer-Verlag, New York, 1989.
[33] Z. Nehari, *On a class of nonlinear second-order differential equations*, Trans. Amer. Math. Soc. 95 (1960), 101–123.
[34] Z. Nehari, *Characteristic values associated with a class of non-linear second-order differential equations*, Acta Math. 105 (1961), 141–175.
[35] A. Pankov, *Periodic nonlinear Schrödinger equation with application to photonic crystals*, Milan J. Math. 73 (2005), 259–287.
[36] A. Pankov, *On decay of solutions to nonlinear Schrödinger equations*, Proc. Amer. Math. Soc. 136 (2008), 2565–2570.
[37] P.H. Rabinowitz, *Minimax Methods in Critical Point Theory with Applications to Differential Equations*, CBMS Reg. Conf. Series Math. 65, Amer. Math. Soc., Providence, R.I., 1986.
[38] M. Ramos and H. Tavares, *Solutions with multiple spike patterns for an elliptic system*, Calc. Var. PDE 31 (2008), 1–25.
[39] M. Ramos and J.F. Yang, *Spike-layered solutions for an elliptic system with Neumann boundary conditions*, Trans. Amer. Math. Soc. 357 (2005), 3265–3284.
[40] M. Reed and B. Simon, *Methods of Modern Mathematical Physics, Vol. IV*, Academic Press, New York, 1978.
[41] M. Struwe, *Three nontrivial solutions of anticoercive boundary value problems for the pseudo-Laplace operator*, J. Reine Angew. Math. 325 (1981), 68–74.
[42] M. Struwe, *Variational Methods*, Springer-Verlag, Berlin, 1990.
[43] A. Szulkin, *Ljusternik-Schnirelmann theory on C^1-manifolds*, Ann. IHP Analyse Non Linéaire 5 (1988), 119–139.
[44] A. Szulkin and T. Weth, *Ground state solutions for some indefinite variational problems*, J. Func. Anal. 257 (2009), 3802–3822.
[45] M. Willem, *Minimax Theorems*, Birkhäuser, Boston, 1996.
[46] W. Zou and M. Schechter, *Critical Point Theory and Its Applications*, Springer-Verlag, New York, 2006.

Andrzej Szulkin, Department of Mathematics, Stockholm University, 106 91 Stockholm, Sweden
E-mail address: andrzejs@math.su.se

Tobias Weth, Institut für Mathematik, Goethe-Universität Frankfurt, 60054 Frankfurt, Germany
E-mail address: weth@math.uni-frankfurt.de

Solutions for elliptic problems with precise sign information

Zhitao Zhang*

ABSTRACT. In this paper, we give a survey over some results on precise sign information of solutions to elliptic problems with Dirichlet boundary condition, especially for existence of sign-changing solutions, nodal numbers and Morse index. By different means of topological tools, variational methods, Morse theory and bifurcation method, we show some recent results. Partial results on sign-changing solutions related with work of the author are shown here.

CONTENTS

1. Introduction

Many boundary value problems which arise in applications involve elliptic equations of second order

$$\begin{cases} -\triangle u = f(x,u) \text{ in } \Omega, \\ u = 0 \text{ on } \partial\Omega, \end{cases} \tag{1.1}$$

where $-\triangle$ is the Laplacian, Ω is a bounded domain in $\mathbb{R}^N, N \geq 1$ with smooth boundary $\partial\Omega$, $f : \bar{\Omega} \times \mathbb{R} \to \mathbb{R}$ is a nonlinear continuous function. It arises in numerous physical problems involving the steady state temperature distribution,

AMS Subject Classification (2000): 35J20, 35J25, 35J60, 35J70

Key words. Critical points, multiple solutions, sign-changing solution, Fucik spectrum, $p-$Laplacian, Lotka-Volterra competing system, Dirichlet Problems, bifurcation.

*Supported by National Natural Science Foundation of China (10671195, 10831005, 10971046).

in problems of chemical reactions, in the theory of stellar structures, and in problems of Riemannian geometry etc(see [**3**]).

In recent 20 years, besides the positive and negative solutions, people are concerned with sign-changing solutions of various differential equations, which especially play an important role in the study of dynamics of the Lotka-Volterra competing species system.(Related references are [**7**], [**8**], [**9**],[**12**], [**23**], [**26**],[**28**], [**30**], [**29**],[**37**], [**44**],[**64**] etc).

The Lotka-Volterra model of competing species describes the competition of a number of species in a fixed domain. Its general form is as following:

$$\begin{cases} \dfrac{\partial u_i}{\partial t} - \triangle u_i = f_i(u_i) - u_i \sum\limits_{j \neq i} b_{ij} u_j, & in\ \Omega \times (0, +\infty), \\ u_i = \phi_i, & on\ \Omega \times \{0\}, \end{cases} \tag{1.2}$$

where $b_{ij} > 0$ are constants and $1 \leq i, j \leq M$, and M is the number of the species and Ω is a bounded domain in $\mathbb{R}^N$ $(N \geq 1)$ with smooth boundary. The intra-species constraint term $f_i(u_i)$ is usually the logistic model:

$$f_i(u_i) = a_i u_i - u_i^2.$$

Usually we consider homogeneous Dirichlet or Neumann boundary condition. We are only concerned about nonnegative solutions, that is, $u_i \geq 0$ for all i. The study of this reaction-diffusion systems has a long history and there exist a great amount of works. Most of these works are concerned with the case of two species, and dynamics of the Lotka-Volterra competing species system with strong competition have good relation with sign-changing solutions of the limit equation of the stationary systems(see [**28**],[**30**]). This is one of the motivations to study the existence of sign-changing solutions and their properties. Moreover, solutions of the limit singular system of multiple species are related with sign-changing solutions of free boundary problems (see [**31**],[**57**]), which has a considerable attention in [**18**][**19**].

If $N = 1$ several methods are available which allow one to control the zero set of the solutions, for instance the shooting method and Nehari's methods of piecing together positive and negative solutions. But these techniques cannot be generalized to deal with (1.1), except the domain is radially symmetric (see [**6**])

In the case $N > 1$ essentially two methods are available to get existence of solution of (1.1), the fixed point approach and the variational approach. In the fixed point approach, it is a standard method of reducing (1.1) to an equivalent fixed point equation in an appropriate function space(e.g., in $L^p(\Omega), 1 < p < \infty, C(\bar{\Omega})$, or $C^1(\bar{\Omega})$). Namely we use the Green function k for the differential operator $-\triangle$ subject to Dirichlet boundary conditions to transfer (1.1) into a nonlinear Hammerstein integral equation

$$u(x) = \int_{\Omega} k(x, y) f(y, u(y)) dy, \quad x \in \Omega. \tag{1.3}$$

Usually for the use of partial order method, instead of the assumption that $f(x, \cdot)$ be increasing, it is possible to use the condition that there exists a nonnegative constant γ such that

$$f(x, \zeta) - f(x, \eta) \geq -\gamma(\zeta - \eta) \tag{1.4}$$

for all $x \in \bar{\Omega}$ and $\zeta, \eta \in \mathbb{R}$ satisfying $\zeta \geq \eta$. The boundary value problem is equivalent to

$$\begin{cases} -\triangle u + \gamma u = f(x,u) + \gamma u & \text{in } \Omega, \\ u = 0 & \text{on } \partial\Omega, \end{cases} \tag{1.5}$$

and $f_\gamma(x,\zeta) := f(x,\zeta) + \gamma\zeta$ is increasing for ζ and $-\triangle + \gamma$ is nonnegative satisfies the maximum principle, i.e., $K(u) := (-\triangle + \gamma)^{-1}(f(x,u) + \gamma u)$ maps the cone of positive functions to itself and is order preserving: $u \leq v$ implies $K(u) \leq K(v)$. This is the basis of the sub- and super-solution method (see Amann [**3**]). It is possible to obtain the existence of sign-changing solutions with the method by computing fixed point indices in positive and negative cones, then comparing these with some globalfixed point index. This requires a priori bounds on the set of fixed points which don't exist for superlinear problems like $f(x,u) = |u|^{p-2}u, p > 2$ (see [**6**]).

The variational approach is based on the fact that weak solutions of (1.1) are critical points of the functional

$$I(u) = \frac{1}{2}\int_\Omega |\nabla u|^2 dx - \int_\Omega F(x,u)dx,$$

where $F(x,t) = \int_0^t f(x,s)ds$ is a primitive of f. By Sobolev embedding Theorem, the growth conditions on $f(x,t)$ for $|t| \to \infty$ are required.

By Sobolev's work, we know that a "good" space to work with for finding critical points of the nonlinear functional I is the Sobolev space $H := H_0^1(\Omega)$ with the inner product

$$(u,v) = \int_\Omega \nabla u \cdot \nabla v$$

and the norm

$$\|u\| = \left(\int_\Omega |\nabla u|^2\right)^{\frac{1}{2}},$$

i.e., the space of functions whose (weak) derivative belongs to the space $L^2(\Omega)$. The functional I is well defined and continuous on H under some continuous and growth condition on $f(x,s)$, namely

$$|f(x,s)| \leq c + c|s|^p, \quad s \in \mathbb{R}.$$

$p \leq N^*, N^* = \frac{N+2}{N-2}, N \geq 3; N^* = \infty$ for $N = 1,2$. Then

$$F(x,s)| \leq c + c|s|^{2^*}, \quad s \in \mathbb{R}, \quad 2^* = \frac{2N}{N-2}, \quad N \geq 3.$$

Indeed, by the famous embedding theorem $H \hookrightarrow L^{2^*}(\Omega)$, $I(u)$ is well defined on H and the functional is continuous and differentiable.

We can also use invariant set of descent flow, bifurcation methods and Morse theory to obtain existence of solution especially sign-changing solutions for elliptic equations (see [**29**],[**44**],[**52**],[**55**],[**58**]).

We give a survey over some results on precise sign information of solutions to elliptic problems with Dirichlet boundary condition, especially for existence of sign-changing solutions, nodal numbers and Morse index. By different means of topological tools, variational methods, Morse theory, and bifurcation method, we show some recent results. There are a lot of papers on sign-changing solutions, only partial results on sign-changing solutions related with work of the author are shown here, e.g., Castro-Cossio-Neuberger's, Bartsch-Weth's, Dancer-Du's etc, and the author's.

2. Sign-changing solutions for superlinear Dirichlet problems

2.1. Nehari maniford and sign-changing solutions. Assume that $\varphi \in C^1(H,\mathbb{R})$ is such that $\varphi'(0) = 0$. A necessary condition for $u \in H$ to be a critical point of φ is that $(\varphi'(u), u) = 0$. This condition defines the Nehari manifold

$$S := \{u \in H : (\varphi'(u), u) = 0, u \neq 0\}.$$

We first introduce one result of Castro et al. [**12**]. They studied

$$\begin{cases} -\triangle u = f(u) & \text{in } \Omega, \\ u = 0 & \text{on } \partial\Omega, \end{cases} \tag{2.1}$$

where Ω be a smooth bounded region in $\mathbb{R}^N, N \geq 2$.

Assume that

$A_1)$ $f \in C^1(\mathbb{R},\mathbb{R})$ such that $f(0) = 0$. There exist constants $A > 0$ and $p \in (1, \frac{N+2}{N-2})$ such that $|f'(u)| \leq A(|u|^{p-1} + 1)$ for all $u \in \mathbb{R}$, resp. $p \in (2,\infty)$ in case $N = 2$.

$A_2)$ There exists $m \in (0,1)$ such that for all $u \in \mathbb{R}$

$$f(u)u - 2F(u) \geq mf(u)u,$$

where $F(u) = \int_0^u f(s)ds$.

$A_3)$ $f'(u) > \frac{f(u)}{u}, \forall u \neq 0$ and $\lim_{|u|\to\infty} \frac{f(u)}{u} = \infty$.

We know $A_1)$ implies that f is subcritical, i.e., there exists $B > 0$ such that $|f(u)| \leq B(|u|^p + 1)$; A3) implies that f is superlinear.

Let $0 < \lambda_1 < \lambda_2 \leq \lambda_3 \leq \cdots$ be the eigenvalues of $-\triangle$ with zero Dirichlet boundary condition in Ω. i.e. they are eigenvalues of the following problems

$$-\triangle u = \lambda u, \quad u|_{\partial\Omega} = 0.$$

Define the functional $J : H \to \mathbb{R}$ by

$$J(u) = \frac{1}{2}\int_\Omega |\nabla u|^2 dx - \int_\Omega F(u)dx. \tag{2.2}$$

By regularity theory for elliptic boundary value problems the weak solutions of (2.2)(i.e., the critical points of J) are classical solutions under the condition $f \in C^1$ (see [**33**]). They get

THEOREM 2.1. *(Castro-Cossio-Neuberger, Theorem 1.1 of* [**12**]*) Under assumptions* $A_1) - A_3)$*, if* $f'(0) < \lambda_1$*, then* (2.1) *has at least three nontrivial solutions:* $\omega_1 > 0$ *in* Ω*,* $\omega_2 < 0$ *in* Ω*, and* ω_3*. The function* ω_3 *changes sign exactly once in* Ω*, i.e.,* $(\omega_3)^{-1}(\mathbb{R} - \{0\})$ *has exactly two connected components. If nondegenerate, the one-sign solutions are Morse index* 1 *critical points of* J*, and the sign-changing solution has Morse index* 2*. Furthermore,*

$$J(\omega_3) \geq J(\omega_1) + J(\omega_2).$$

REMARK 2.2. *One can relax* $A_2)$ *to hold only for* $|u|$ *sufficiently large. This is necessary in order to consider nonlinearities* f *such that* $f'(0) > 0$.

REMARK 2.3. *If* $f'(0) > \lambda_1$*, then by multiplying* (2.1) *by an eigenfunction corresponding to* λ_1 *and integrating by parts, it is easily seen that* (2.1) *does not have one-signed solutions.*

Our assumptions on f imply that $J \in C^2(H, \mathbb{R})$, and that

$$J'(u)(v) = (J'(u), v) = \int_\Omega \nabla u \cdot \nabla v dx - \int_\Omega f(u) dx, \quad \forall v \in H. \tag{2.3}$$

Define $\gamma : H \to \mathbb{R}$ by $\gamma(u) = (\nabla J(u), u) = \int_\Omega [|\nabla u|^2 - f(u)u] dx$ and compute

$$\gamma'(u)(v) = (\nabla\gamma(u), v) = 2\int_\Omega \nabla u \cdot \nabla v dx - \int_\Omega f(u) v dx - \int_\Omega f'(u) u v dx, \quad \forall v \in H. \tag{2.4}$$

DEFINITION 2.4. *For* $u \in L^1(\Omega)$*, we define* $u^+(x) = \max\{u(x), 0\} \in L^1(\Omega)$*, and* $u^-(x) = \min\{u(x), 0\} \in L^1(\Omega)$*. If* $u \in H$ *then* $u^+, u^- \in H$*. We say that* $u \in L^1(\Omega)$ *changes sign if* $u^+ \neq 0$ *and* $u^- \neq 0$*. For* $u \neq 0$ *we say that* u *is positive* $(u > 0)$ *if* $u^- = 0$*, and similarly,* u *is negative* $(u < 0)$ *if* $u^+ = 0$.

We use S denote the Nehari manifold $S \subset H$ here and various subsets of S:

$$S = \{u \in H \setminus \{0\} : \gamma(u) = 0\}, \quad \hat{S} = \{u \in S : u^+ \neq 0, u^- \neq 0\},$$
$$S_1 = \{u \in \hat{S} : \gamma(u^+) = 0\}, \quad G^+ = \{u \in S : u > 0\}, \quad \hat{S}^+ = \{u \in S : \gamma(u^+) < 0\},$$
$$W^+ = G^+ \cup \hat{S}^+, \quad G^- = \{u \in S : u < 0\},$$
$$\hat{S}^- = \{u \in S : \gamma(u^+) > 0\}, \quad W^- = G^- \cup \hat{S}^-.$$

We have the disjoint unions $S = G^+ \cup \hat{S} \cup G^-$ and $\hat{S} = \hat{S}^+ \cup S_1 \cup \hat{S}^-$. We note that nontrivial solutions to (2.1) are in S, one-sign solution are in $G^+ \cup G^-$, and sign-changing solutions are in S_1. We define $S^\infty = \{u \in H : \|u\| - 1\}$.

LEMMA 2.5 (Lemma 2.2 of [**12**]). *Under the above assumptions we have:*

(a) 0 *is a local minimum of* J*. If* $u \in H \setminus \{0\}$*, then there exists a unique* $\bar{\lambda} = \bar{\lambda}(u) \in (0, \infty)$ *such that* $\bar{\lambda} u \in S$*. Moreover,* $J(\bar{\lambda} u) = \max_{\lambda > 0} J(\lambda u) > 0$*. If* $\gamma(u) < 0$ *then* $\bar{\lambda} < 1$*; and if* $\gamma(u) > 0$ *then* $\bar{\lambda} > 1$ *and* $J(u) > 0$.

(b) *The function* $\bar{\lambda} \in C^1(S^\infty, (0, \infty))$*. The set* S *is closed, unbounded, and a connected* C^1*-submanifold of* H *diffeomorphic to* S^∞.

(c) $u \in S$ *is a critical point of* J *on* $H \iff u$ *is a critical point of* $J|_S$.

(d) $J|_S$ *is coercive, i.e.,* $J(u) \to \infty$ *as* $\|u\| \to \infty$ *in* S. *Also,* $0 \notin S$ *and* $\inf_S J > 0$.

LEMMA 2.6. *(Lemma 2.3 of* [**12**]*) The function* $h : H \to H$ *defined by* $h(u) = u^+$ *is continuous. Also,* h *defines a continuous function from* $L^{p+1}(\Omega)$ *into itself.*

LEMMA 2.7 (Lemma 2.4 of [**12**]). *Given* $w \in \hat{S}$, *there exists a path* $r_\omega \equiv r \in C^1([0,1], S)$ *such that*

(a) $r(0) = aw^+ \in G^+$ *for some* $a > 0, r(1) = bw^- \in G^-$ *for some* $b > 0, r(\frac{a}{a+b}) = w$.
(b) $a > 1 \Leftrightarrow b < 1 \Leftrightarrow w \in \hat{S}^+, \gamma(r(t)^+) < 0 \Leftrightarrow t \in (0, \frac{1}{2}) \Leftrightarrow r(t) \in \hat{S}^+$.
(c) $r(\frac{1}{2}) = aw^+ + bw^- \in S_1, r([0,1]) \cap S_1 = \{r(\frac{1}{2})\}$.
(d) $J(r(0)) < J(r(t)) < J(r(\frac{1}{2}))$, *for* $t \in (0, \frac{1}{2})$.

LEMMA 2.8 (Lemma 2.5 of [**12**]). *The sets* $G^+, \hat{S}^+, W^+, S_1, W^-, \hat{S}^-$ *and* G^- *satisfy:*

(a) G^+, S_1 *and* G^- *are closed, and* G^+, G^- *are connected.*
(b) $\hat{S}$ *is open, and the subsets* $\hat{S}^+$ *and* $\hat{S}^-$ *are open and separated by* S_1.
(c) W^+ *and* W^- *are the only two components of* $S \setminus S_1$. *In particular,* G^+ *and* G^- *are separated by* S_1.
(d) *If* $w \in G^+$ *and* $J(w) = \min_{G^+} J$, *then* $J(w) = \min_{W^+} J$ *and* w *is a critical point of* J.

LEMMA 2.9 (Lemma 2.6 of [**12**]). *If* $w \in S_1$ *and* $J(w) = \min_{S_1} J$, *then* w *is a critical point of* J.

Now the proof of the main Theorem 2.1 is:

PROOF. We first establish the existence of one-sign solution $w_1 > 0$ (one finds $w_2 < 0$ similarly). Furthermore, we show that $J|_S$ has local minima at w_1 and w_2, and hence these two critical points are of Morse index 1(if nondegenerate).

We define $c_1 = \inf_{G^+} J$ and take $\{u_n\} \subset G^+$ with $\lim_{n\to\infty} J(u_n) = c_1$. The coercivity of J and the subcritical condition on f allow us to apply the Sobolev Imbedding Theorem, whence we find $\bar{u} \in H \subset L^{p+1}(\Omega)$ such that (without loss of generality)

$$u_n \rightharpoonup \bar{u} \text{ in } H, \quad u_n \to \bar{u} \text{ in } L^{p+1}(\Omega),$$
$$\int_\Omega u_n f(u_n)dx \to \int_\Omega \bar{u} f(\bar{u})dx, \quad \int_\Omega F(u_n)dx \to \int_\Omega F(\bar{u})dx.$$

That $\bar{u} \neq 0$ is evident, as $\int_\Omega \bar{u} f(\bar{u})dx = \lim_{n\to\infty} \int_\Omega u_n f(u_n)dx = \lim_{n\to\infty} \|u_n\|^2 > 0$ (See Lemma 2.5 (d)). By the continuity of $h : L^{p+1}(\Omega) \to L^{p+1}(\Omega)$, we see that $\bar{u} > 0$. We wish to show that $u_n \to \bar{u}$ in H. If we assume to the contrary that $u_n \not\to \bar{u}$ in H, then without loss of generality we may assume that $\|\bar{u}\| < \liminf \|u_n\|^2$. It follows that $\gamma(\bar{u}) < \liminf \gamma(u_n) = 0$, so by Lemma 2.5 (a) there exists $0 < \alpha < 1$ such that $\alpha\bar{u} \in G^+$. Consequentially, we get the following contradiction,

$$J(\alpha\bar{u}) < \liminf_{n\to\infty} J(\alpha u_n) \le \liminf_{n\to\infty} J(u_n) = \inf_{G^+} J = c_1.$$

We conclude that $u_n \to \bar{u}$ in H, $\alpha = 1, \bar{u} \in G^+$, and $J(\bar{u}) = \min_{G^+} J = c_1$. By Lemma 2.8 (d) we see that $w_1 = \bar{u} \in G^+$ is a critical point of J, and hence a positive solution to (2.1).

We can obtain the negative solution $w_2 \in G^- \subset W^-$ in the same fashion, whence we can define

$$c_2 = \inf_{G^-} J = \inf_{W^-} J = J(w_2).$$

Now we show that there exists a solution w_3 to (2.1) which changes sign exactly once. If the solution is a nondegenerate critical point, then it has Morse index 2.

Let $c_3 = \inf_{S_1} J$ and $\{u_n\} \subset S_1$ be such that $J(u_n) \to c_3$. Using $\gamma((u_n)^+) = 0$, we see that $\{(u_n)^+\} \subset G^+$ and $\{(u_n)^-\} \subset G^-$. Since $J|_S$ is coercive, $\{u_n\}$ is bounded. By the Sobolev Imbedding Theorem, without loss of generality, we can assume that there exist $u, v, w \in H$ such that

$$\begin{gathered} u_n \rightharpoonup u, \quad (u_n)^+ \rightharpoonup v, \quad (u_n)^- \rightharpoonup w \text{ in } H, \\ u_n \to u, \quad (u_n)^+ \to v, \quad (u_n)^- \to w \text{ in } L^{p+1}(\Omega). \end{gathered}$$

By Lemma 2.3 we know that $h : L^{p+1}(\Omega) \to L^{p+1}(\Omega)(u \to u^+)$ is a continuous transformation, so we see that $u^+ = v \geq 0$ and $u^- = w \leq 0$. Let us see that $u \in S_1$. Since $(u_n)^+ \to u^+$ in $L^{p+1}(\Omega)$ and f is subcritical,

$$\int_\Omega F((u_n)^+)dx \to \int_\Omega F(u^+)dx \quad \text{and} \quad \int_\Omega f((u_n)^+)(u_n)^+dx \to \int_\Omega f(u^+)u^+dx$$

By $(u_n)^+ \in S$ and Lemma 2.5 (d), we see that

$$\int_\Omega u^+ f(u^+)dx = \lim_{n\to\infty} \int_\Omega f((u_n)^+)(u_n)^+dx = \lim_{n\to\infty} \|(u_n)^+\|^2 > 0,$$

consequentially $u^+, u^- \neq 0$ and $u = u^+ + u^-$ is sign changing.

Let us see that $(u_n)^+ \to u^+$ in H. If we soppose not, then without loss of generality we may assume that $\|u^+\|^2 < \liminf_{n\to\infty} \|(u_n)^+\|^2$, whence

$$\begin{aligned} \gamma(u^+) &= \|u^+\|^2 - \int_\Omega u^+ f(u^+)dx \\ &< \liminf_{n\to\infty} \|(u_n)^+\|^2 - \lim_{n\to\infty} \int_\Omega (u_n)^+ f((u_n)^+)dx = 0. \end{aligned}$$

From Lemma 2.5 (a) we see that there exists $0 < \alpha < 1$ such that $\alpha u^+ \in G^+$, and similarly that there exists $0 < \beta \leq 1$ such that $\beta u^- \in G^-$. We conclude that $\alpha u^+ + \beta u^- \in S_1$. This provides a contradiction, since

$$\begin{aligned} J(\alpha u^+ + \beta u^-) &< \liminf_{n\to\infty} J(\alpha(u_n)^+ + \beta(u_n)^-) \\ &= \liminf_{n\to\infty} \{J(\alpha(u_n)^+) + J(\beta(u_n)^-)\} \\ &\leq \liminf_{n\to\infty} J((u_n)^+) + J((u_n)^-)\} \\ &= \liminf_{n\to\infty} J(u_n) = \inf_{S_1} J = c_3. \end{aligned} \tag{2.5}$$

Hence $(u_n)^+ \to u^+$ in H and $\alpha = 1$. Similarly, we conclude that $(u_n)^- \to u^-$ in H and $\beta = 1$, which proves that $u \in S_1, u_n \to u$ in H, and $J(u) = c_3$. Letting $w_3 = u$, we see that $J|_{S_1}$ attains its minimum at w_3. We find that w_3 is a critical point of J by Lemma 2.9, hence a solution to (2.1).

Let us see that w_3 changes sign exactly once. Since w_3 is of class C^2, hence continuous, $E = \{x \in \Omega : u(x) \neq 0\}$ is open. Suppose E has more than two components. Since w_3 changes sign, without loss of generality we can assume that there exist connected components A, B and C of E such that $u > 0$ in A and $u < 0$ in B. Let u_A, u_B and u_C be the zero extensions of $w_3|_A$, $w_3|_B$, and $w_3|_C$ to all of Ω. Since $\triangle w_3 + f(w_3) = 0$ on Ω it follows that $\gamma(u_A) = \gamma(u_B) = \gamma(u_C) = 0$. Hence $J > 0$ on S implies $J(u_A + u_B) < J(u_A + u_B + u_C) \leq J(w_3) = c_3$, a contradiction since $u_A + u_B \in S_1$. We conclude that E has exactly two components.

If w_3 is nondegenerate critical point, we see that w_3 has Morse index 2 in H by observing that $J''(w_3)(v, v) < 0$ for $v = (w_3)^+$ and $v = (w_3)^-$, $((w_3)^+, (w_3)^-) = 0$, and $J|_{S_1}$ has a minimum at w_3.

It is clear that

$$c_3 = J(w_3) = J(w_3^+) + J(w_3^-) \geq J(w_1) + J(w_2) = c_1 + c_2.$$

This completes the proof. □

2.2. Additional properties of sign-changing solutions to superlinear elliptic equations. Now we give Bartsch-Weth's results on additional properties of sign-changing solutions to superlinear elliptic equations (See [**9**])

$$\begin{cases} -\triangle u = f(x, u) & \text{for } x \in \Omega, \\ u|_{\partial\Omega} = 0 \end{cases} \tag{2.6}$$

on a smooth,bounded domain $\Omega \subset \mathbb{R}^N, N \geq 2$.

We make the follow assumptions similar to the above:

- (f_1) $f \in C^1(\Omega \times \mathbb{R}, \mathbb{R})$, $f(x, 0) = 0$ for all $x \in \Omega$.
- (f_2) There exist constants $p \in (2, \frac{2N}{N-2})$, resp. $p \in (2, \infty)$ in case $N = 2$, such that $|f'(x,t)| \leq C(|u|^{p-2} + 1)$ for all $x \in \Omega, t \in \mathbb{R}$, where $f' := \partial f/\partial t$.
- (f_3) $f'(x,t) > \frac{f(x,t)}{t}, \forall x \in \Omega, t \neq 0$.
- (f_4) There exists $R > 0$ and $\theta > 0$ such that for all $x \in \Omega, |t| \geq R$

$$0 < \theta F(x,t) \leq t f(x,t),$$

where $F(x,t) = \int_0^t f(x,s)ds$.

Clearly these assumptions hold for

$$f(x,t) = \sum_{i=1}^{d} a_i(x)|t|^{p_i - 2}t, \tag{2.7}$$

where $2 \leq p_1 < \cdots < p_d < 2N/(N-2), a_1, \ldots, a_d$ are bounded nonnegative C^1-functions and a_d is bounded from below by a positive constant. In fact, $a_i \in L^\infty$ is sufficient, since the differentiability of f with respect to x is not necessary in (f_1). They gain further information on sign changing solutions, in particular on the nodal structure, extremality properties and the Morse index with respect to the energy functional

$$\Phi(u) = \frac{1}{2}\int_\Omega |\nabla u(x)|^2 dx - \int_\Omega F(x, u(x))dx. \tag{2.8}$$

It is well known consequence of $(f_1), (f_2)$ that $\Phi \in C^2(H)$ and that critical points of Φ are weak solutions of (2.6).

Consider the set

$$M := \{u \in H : u^+ \neq 0, u^- \neq 0, \quad \Phi'(u)u^+ = \Phi'(u)u^- = 0\}.$$

The set M is not a manifold and we don't expect it to be a complete metric space if $\mu_1 < 0$ (where $\mu_1 < \mu_2 \leq \cdots$ are the Dirichlet eigenvalues of the operator $-\triangle - f'(x,0)$). The condition $\mu_1 > 0$ required in [**12**] implies that M is a closed subset of H, hence a complete metric space. Obviously, all sign-changing solutions of (2.6) are contained in M. Now put

$$\beta := \inf_{u\in M} \Phi(u). \tag{2.9}$$

We will show that $\Phi(\bar{u}) = \beta$, i.e. $\bar{u}$ is a least energy sign-changing solution. Thus we also obtain the infimum of Φ on M is achieved by a critical point of Φ. We then conclude by the following result.

THEOREM 2.10 (Bartsch-Weth, Theorem 1.3 of [**8**]). *Suppose $(f_1) - (f_4)$ hold. Then every weak solution $u \in M$ with $\Phi(u) = \beta$ has Morse index 2 and has precisely two nodal domains.*

$(f_1) - (f_4)$ can be weaken by the following hypotheses:

(A_1) $f : \Omega \times \mathbb{R} \to \mathbb{R}$ is a caratheodory and $f(x, 0) = 0$ for all $x \in \Omega$.

(A_2) There exist $p \in (2, 2N/(N-2)]$, resp. $p \in (2, \infty)$ in case $N = 2$, and $C > 0$ such that $f(x,t) \leq C(|t| + |t|^{p-1})$ for all $t \in \mathbb{R}$ and a.e. $x \in \Omega$.

(A_3) The function $t \to f(x,t)/|t|$ is nondecreasing on $\mathbb{R} \setminus \{0\}$ for a.e. $x \in \Omega$.

Stronger variant of (A_3):

$(\tilde{A}_3)$ The function $t \to f(x,t)/|t|$ is strictly increasing on $\mathbb{R} \setminus \{0\}$ for a.e. $x \in \Omega$.

Note that for nonlinearities of the form (2.7) condition (A_3) is a consequence of the sign condition

$$a_i(x) \geq 0 \quad \text{for } x \in \Omega, \quad i = 1, 2, \ldots, d.$$

In view of (A_2) the nonlinearity f may have critical growth at infinity. Nevertheless (A_2) ensures that every weak solution u of (2.6) is at least continuous in Ω. Indeed, combining (A_2) with Sobolev embeddings, we deduce that

$$\frac{f(\cdot, u(\cdot))}{u} \in L^{N/2}(\Omega).$$

Hence the Brezis-Kato Theorem yields $u \in L^q_{loc}(\Omega)$ for every $2 \leq q < \infty$. In particular there is a number $s > N/2$ such that $f(\cdot, u(\cdot)) \in L^s_{loc}(\Omega)$, thus u is continuous by elliptic regularity.

LEMMA 2.11 (Lemma 2.1 of [**8**]). *Suppose* $(A_1)-(A_3)$ *hold, and consider* $u \in H \setminus \{0\}$ *with* $\Phi'(u)u = 0$. *Then*

$$0 \leq \Phi(u) = \sup_{t \geq 0} \Phi(tu) \tag{2.10}$$

If in addition $(\tilde{A}_3)$ *is satisfied, then* $\Phi(u) > 0$.

PROOF. For $t \geq 0$ we define $h(t) := \Phi(tu)$. Then $h(0) = 0$, and

$$h'(t) = \Phi'(tu)u = t\int_\Omega \left(|\nabla u|^2 - \frac{f(x,tu)}{tu}u^2\right)$$

for $t > 0$. Hence (A_3) implies that $t \to \frac{h'(t)}{t}$ is nonincreasing on $(0, \infty)$, and thus the set $S := \{t > 0 : h'(t) = 0\}$ is a subinterval of $(0, \infty)$ which contains $t = 1$ by assumption. Let $b \leq \infty$ be the right endpoint of S. Then h is strictly decreasing on (b, ∞), whereas

$$0 \leq \max_{t \in [0,b]} h(t) \leq \max_{t \in S} h(t) = h(1).$$

This yields (2.10). If in addition $(\tilde{A}_3)$ holds, then $t \to h'(t)/t$ is strictly decreasing on $(0, \infty)$. Hence $S = \{1\}$, and $h'(t) > 0$ for $0 < t < 1$. We conclude that $\Phi(u) = h(1) > h(0) = 0$, as claimed. □

THEOREM 2.12 (Bartsch-Weth, Theorem 2.3 of [**8**]). *Suppose* $(A_1), (A_2)$ *and* $(\tilde{A}_3)$ *hold. Then every weak solution* $u \in M$ *of* (2.6) *with* $0 < \Phi(u) \leq \beta$ *has precisely* 2 *nodal domains.*

PROOF. Suppose in contradiction that u has at least three nodal domains. We choose nodal domains Ω_1, Ω_2 such that $u(x) \geq 0, x \in \Omega_1; u(x) \leq 0, x \in \Omega_2$. The associated functions $v_1, v_2 \in H$ are defined as

$$v_i(x) := \begin{cases} u(x), & \forall x \in \Omega_i, \\ 0, & \forall x \in \Omega \setminus \Omega_i. \end{cases} \tag{2.11}$$

Clearly $v_1 + v_2 \in M$ and the function $v := u - v_1 - v_2$ satisfies $\Phi'(v)v = 0$. This implies $\Phi(v) > 0$ by Lemma 2.11, hence $\beta \leq \Phi(v_1 + v_2) < \Phi(u)$, which contradicts the assumption. □

For the Morse index of sign-changing solutions with least energy, now we use the assumptions $(f_1)-(f_4)$. Consider the Hilbert space $H_1 := H \cap H^2(\Omega)$, endowed with the scalar product from $H^2(\Omega)$. Moreover, denote by $\|\cdot\|_1$ the induced norm. We need the following technical lemma concerning the functionals

$$Q_\pm : H \to \mathbb{R}, \quad Q_\pm(u) = \int_\Omega |\nabla u^\pm| dx = \int_\Omega \nabla u \cdot \nabla u^\pm dx,$$

$$\Psi_\pm : H \to \mathbb{R}, \Psi_\pm(u) = \int_\Omega f(x,u)u^\pm dx.$$

LEMMA 2.13 (Lemma 3.1 of [**8**]).
(a) *$Q_\pm$ is differentiable at $u \in H_1$ with derivative $Q'_\pm(u) \in H^{-1}$ given by*

$$Q'_\pm(u)v = \int_{\pm u>0} ((-\triangle u)v + \nabla u \nabla v)dx.$$

(b) $Q'_\pm|_{H_1} \in C^1(H_1)$.
(c) *$\Psi_\pm \in C^1(H)$ with derivative given by*

$$\Psi'_\pm(u)v = \int_\Omega f'(x,u^\pm)u^\pm v dx + \int_\Omega f(x,u^\pm)v dx.$$

LEMMA 2.14 (Lemma 3.2 of [**8**]). *The set $M \cap H_1$ is a C^1-manifold of codimension two in H.*

PROOF. Define

$$g_\pm : H \to \mathbb{R}, \quad g_\pm(u) = \Phi'(u)u^\pm,$$

so that $M \cap H_1 = \{u \in H_1 : u^+ \neq 0, u^- \neq 0, g_+(u) = 0 = g_-(u)\}$.
Lemma 2.13 implies that $g_\pm|_{H_1} \in C^1(H_1)$. For $u \in M \cap H_1$ we obtain

$$\begin{aligned} g'_+(u)u^+ &= \int_\Omega (|\nabla u^+|^2 - f'(x,u)(u^+)^2), \ g'_+(u)u^- = 0, \\ g'_-(u)u^- &= \int_\Omega (|\nabla u^-|^2 - f'(x,u)(u^-)^2), \ g'_-(u)u^+ = 0. \end{aligned}$$

Hence (f_3) yields $g'_+(u)u^+ < 0$ and $g'_-(u)u^- < 0$ for $u \in M \cap H_1$. Approximating u^+ and u^- by functions in H_1, we conclude that $(g'_+(u), g'_-(u)) \in \mathbb{L}(H_1, \mathbb{R}^2)$ is onto for every $u \in M \cap H_1$. From this the assertion follows. □

In the following, if $u \in H$ is a critical point of Φ, we denote by $m(u)$ the Morse index of u.

PROPOSITION 2.15 (Proposition 3.3 of [**8**]). *Let $u \in M$ be a critical point of Φ with $\Phi(u) = \beta$. Then $m(u) = 2$.*

PROOF. By (f_3) there holds

$$\Phi''(u)(u^\pm, u^\pm) = \int_\Omega (\nabla u^\pm|^2 - f'(x,u)(u^\pm)^2) < 0,$$

hence $m(u) \geq 2$. To show $m(u) \leq 2$, note first that $u \in H_1$ by elliptic regularity. Denote by $T \subset H_1$ the tangent space of the manifold $M \cap H_1$ at u. We show that

$$\Phi''(u)(v,v) \geq 0 \quad \text{for all } v \in T. \tag{2.12}$$

Indeed, by Lemma 2.14 there exists for every $v \in T$ a C^1-curve $\gamma : [-1,1] \to M \cap H$ such that $\gamma(0) = u$ and $\dot\gamma(0) = v$. Since $\Phi'(u)v = 0$, we calculate that $\Phi \circ \gamma : [-1,1] \to \mathbb{R}$ is even twice differentiable at $t = 0$ with derivative

$$\frac{\partial^2}{\partial t^2}\Phi \circ \gamma|_{t=0} = \Phi''(u)(v,v).$$

Recalling that $\Phi(u) = \min_{v \in M \cap H_1} \Phi(v)$, we infer that $\partial^2(\Phi \circ \gamma)/\partial t^2|_{t=0} \geq 0$, and hence (2.12) follows. Since $T \subset H_1$ has codimension two and H_1 is dense in H, we conclude $m(u) \leq 2$, as required. □

The proof of Theorem 2.10 follows from Theorem 2.12 and Proposition 2.15.

REMARK 2.16. *In* [**7**], *T. Bartsch, K.C. Chang, and Z.Q. Wang also have some results on the Morse indices of sign-changing solutions of nonlinear elliptic problems.*

3. Sign-changing solutions for jumping nonlinear problems

3.1. On limit equation of two species Lotka-Volterra competing equations system. The two species Lotka-Volterra competing equations system is

$$\begin{cases} -\triangle u = au - u^2 - cuv & \text{in } \Omega \\ -\triangle v = dv - v^2 - euv & \text{in } \Omega \\ u|_{\partial\Omega} = v|_{\partial\Omega} = 0. \end{cases} \tag{3.1}$$

Here $a, d > \lambda_1, c, e > 0$

Let λ_j be the j-th eigenvalue of $-\triangle$ with zero Dirichlet boundary data, $0 < \lambda_1 < \lambda_2 < \cdots < \lambda_k < \cdots$, denote by n_k the dimension of $ker(-\triangle - \lambda_k I)$ and let $\{\varphi_{k,1}, \varphi_{k,2}, \ldots, \varphi_{k,n_k}\}$ be an orthogonal basis of $ker(-\triangle - \lambda_k I)$ with the property that $\int_\Omega |\nabla \varphi_{k,j}|^2 = 1$ for $j = 1, 2, \ldots, n_k$ and $k = 1, 2, \ldots$. It is well known that $\lambda_1 > 0, \lambda_1$ is simple, i.e., $n_{1,1} = 1$, and $\varphi_{1,1} > 0$. We denote $\varphi_{1,1}$ by φ_1. i.e.,

$$-\triangle \varphi_1 = \lambda_1 \varphi_1, \quad \varphi_1|_{\partial\Omega} = 0.$$

As the interaction parameters c, e go to infinity, we get the following limiting equation

$$-\triangle u = h(u), \quad u|_{\partial\Omega} = 0, \tag{3.2}$$

where $h : \mathbb{R} \to \mathbb{R}$ is defined by

$$h(u) = \begin{cases} au - \alpha u^2, & u \geq 0, \\ du + \beta u^2, & u \leq 0, \end{cases}$$

and $\alpha \geq 0, \beta > 0$. In [**22**], Dancer and Du have shown that if (3.2) has a sign-changing solution u_0 which is isolated and has nonzero fixed point index, then for c, e both large with c/e close to α/β, (3.1) has a positive solution (u, v) which is close to $(u_0^+/\alpha, -u_0^-/\beta)$.

They also proved if $a, d > \lambda_2$, then (3.2) has at least one sign-changing solution, moreover if its nontrivial solution is isolated then (3.2) has a solution which changes sign and has fixed point index -1. In [**23**] they find the exact set of points (a, d) in $\mathbb{R}^2$ where the statement holds. It is completely determined by the homogeneous jumping nonlinear problem

$$-\triangle u = au^+ + du^-, \quad u|_{\partial\Omega} = 0. \tag{3.3}$$

Nontrivial solutions of (3.3) correspond to nontrivial critical points of the following functional on H

$$J_{(a,d)}(u) = \frac{1}{2}\int_\Omega |\nabla u|^2 dx - \frac{a}{2}\int_\Omega (u^+)^2 dx - \frac{d}{2}\int_\Omega (u^-)^2 dx. \tag{3.4}$$

We define

$$\Sigma = \{(a,d) \in \mathbb{R}^2 : (3.3) \text{ has a nontrivial solution}\}$$

and let

$$S_k = \{(a,d) \in \mathbb{R}^2, \lambda_k \le a, d \le \lambda_{k+1}, (a,d) \ne (\lambda_k, \lambda_k), (a,d) \ne (\lambda_{k+1}, \lambda_{k+1})\}.$$

By [**48**] we know $S_k \not\subset \Sigma$.

Define

$$D_k = \{(a,d) \in \mathbb{R}^2 : \text{there exists a path } \gamma(t) = (\gamma_1(t), \gamma_2(t)), t \in [0,1], \text{such that } (\gamma_1(t), \gamma_2(t)) \notin \Sigma \text{ and } \gamma(0) = (a,d), \gamma(1) \in S_k\}.$$

From the results cited in [**23**], there exists a continuous function $\eta(t)$ defined on $(\lambda_1, \lambda_2]$ with the properties that

(a) η is strictly decreasing, $\eta(\lambda_2) = \lambda_2$, $\lim_{\lambda \to \lambda_1 + 0} \eta(\lambda) = +\infty$;
(b) equation (3.3) has a nontrivial solution for $(a,d) = (a, \eta(a)), a \in (\lambda_1, \lambda_2]$ and $(a,d) = (\eta(d), d), d \in (\lambda_1, \lambda_2]$;
(c) equation (3.3) has no nontrivial solution for $\lambda_1 < d < \eta(a), a \in (\lambda_1, \lambda_2]$ or $\lambda_1 < a < \eta(d), d \in (\lambda_1, \lambda_2]$.

We denote by Γ the curve $\{(a, \eta(a)) : \lambda_1 < a \le \lambda_2\} \cup \{(\eta(d), d) : \lambda_1 < d \le \lambda_2\}$,

For convenience, we denote by $\tilde{S}$ the set of points (a,d) which are above Γ in the a-d plane.

We define

$$\triangle(a,d) = \{u \in H : J_{(a,d)}(u) < 0\}.$$

Let

$$S := \{(a,d) : a, d > \lambda_1, \triangle(a,d) \text{ is path-connected}\}.$$

It is proved in [**23**] that

$$\tilde{S} = S$$

LEMMA 3.1 (Dancer-Du, Lemma 2.1 of [**23**]). *$\triangle(a,d)$ with $a, d > \lambda_1$ is path-connected if and only if there exists a path in $\triangle(a,d)$ connecting φ_1 and $-\varphi_1$.*

THEOREM 3.2 (Dancer-Du, Theorem 2.5 of [**23**]). *If $(a,d) \in S$, then (3.2) has at least one sign-changing solution. Moreover, if each nontrivial solution of (3.2) is isolated, then there exists a sign-changing solution with fixed point index -1. Here the fixed point refers to that of the mapping $u \to (-\triangle)^{-1} h(u)$ from $W_0^{1,2}(\Omega)$ to itself.*

PROOF. We consider the cases $\alpha > 0$ and $\alpha = 0$ separately. Suppose first that $\alpha > 0$. Since $h(u)$ is concave on $(0, \infty)$, by a well-known result, one knows that (3.2) has a unique positive solution u^+. Similarly, (3.2) has a unique positive solution u^-. An easy upper and lower solution argument shows that $0 \le u^+ \le \alpha^{-1}a, -\beta d \le u^- \le 0$, and any other solution of (3.2) satisfies $-\beta d \le \alpha^{-1}a$. Moreover, except for the trivial solution $u \equiv 0$ and u^+ and u^-, any other solution of (3.2) changes sign. Therefore, to prove the theorem, we need only find a nontrivial solution other than u^+ and u^-.

Now we modify $h(u)$ outside $[-\beta^{-1}, \alpha^{-1}a]$ in such a way that the modified problem

$$-\triangle u = \tilde{h}(u), u|_{\partial\Omega} = 0 \tag{3.5}$$

has the same solution set as (3.2), that $\tilde{h}$ is smooth except at $u = 0$ and that $\tilde{h}$ is bounded on $\mathbb{R}^1$. It is easily seen that this is possible.

Define

$$H(u) = \int_0^u \tilde{h}(s)ds.$$

Then for $u \in [-\beta^{-1}d, \alpha^{-1}a]$,

$$H(u) = \begin{cases} \frac{1}{2}au^2 - \frac{1}{3}\alpha u^3, & 0 \le u \le \alpha^{-1}a, \\ \frac{1}{3}du^2 + \frac{1}{3}\beta u^3, & 0 \ge u \ge -\beta^{-1}d. \end{cases}$$

Define on H a functional I by

$$I(u) = \frac{1}{2}\int |\nabla u|^2 - \int H(u).$$

Then it is well known that I is C^1 and satisfies the PS condition, and critical points of I are solutions of (3.2). By the proof of Theorem 4.2 of [**22**], u^+, u^- are two strict local minima of I. Hence, define

$$c = \inf_{L \in \triangle} \max_{u \in L} I(u),$$

where $\triangle$ denotes the set of all continuous paths in H joining u^+ and u^-, it follows from the Mountain Pass Theorem that c is a critical value of I, and $c > \max\{I(u^+), I(u^-)\}$. Therefore, if we can show that $c \ne 0$, then c corresponds to a nontrivial critical point of I as $I(0) = 0$. This must be a sign-changing solution of (3.2) as it differs from $0, u^+$ and u^-.

Now we construct a path in H joining u^+ and u^-, and on which $I(u) < 0$. If this is done, then by the definition, $c < 0$ and (3.2) will have a sign-changing solution.

A simple calculation shows that

$$I(tu^+) = \frac{t^2}{2}\alpha\left(\frac{2}{3}t - 1\right)\int (u^+)^3 < 0 \text{ for } t \in (0, 1] \tag{3.6}$$

and

$$I(tu^-) = \frac{t^2}{2}\beta\left(1 - \frac{2}{3}t\right)\int (u^-)^3 < 0 \text{ for } t \in (0, 1]. \tag{3.7}$$

In particular,

$$J_{(a,d)}(tu^+) < I(tu^+) < 0, \quad J_{(a,d)}(tu^-) < I(tu^-) < 0, \ \ \forall t \in (0, 1].$$

Hence, u^+, u^- belong to $\triangle(a, d)$, and by the path connectedness of $\triangle(a, d)$, we can find a path $L_0 \subset \triangle(a, d)$ connecting u^+ and u^-. Moreover, since $(a, d) \in S$, L_0 can be chosen L^∞ bounded. We show that for $\varepsilon > 0$ small, $I(u) < 0$ for $u \in \varepsilon L_0 \equiv \{\varepsilon u : u \in L_0\}$. In fact, since L_0 is L^∞ bounded, there exists $\varepsilon_0 > 0$ such that

$$-\beta^{-1}d \leq \varepsilon u \leq \alpha^{-1}a \quad \text{for } u \in L_0 \text{ and } 0 < \varepsilon \leq \varepsilon_0.$$

Since L_0 is compact in H, there exists $\delta > 0$ such that

$$J_{a,d}(u) \leq -\delta \quad \text{for } u \in L_0,$$

and hence

$$J_{a,d}(u)(\varepsilon u) = \varepsilon^2 J(u) \leq -\varepsilon^2\delta \quad \text{for } u \in L_0 \text{ and } \varepsilon > 0.$$

Therefore,

$$\begin{aligned} I(\varepsilon u) &= J_{a,d}(u)(\varepsilon u) + \frac{1}{3}\alpha\varepsilon^3 \int (u^+)^3 - \frac{1}{3}\beta\varepsilon^3 \int (u^-)^3 \\ &\leq -\varepsilon^2\delta + \frac{1}{3}(\alpha + \beta)\varepsilon^3 \int |u|^3 < 0, \end{aligned} \tag{3.8}$$

for $u \in L_0$ and

$$0 < \varepsilon < \min\left\{\frac{3}{2(\alpha+\beta)} \cdot \delta \cdot (\max_{u \in L_0} \int |u|^3)^{-1}, \varepsilon_0\right\}.$$

Thus, by (3.6)–(3.8), for $\varepsilon > 0$ small, the path

$$\{tu^+ : 1 \geq t \geq \varepsilon\} \cup \varepsilon L_0 \cup \{tu^- : \varepsilon \leq t \leq 1\}$$

meets our requirement. This proves the first part of the theorem for the case $\alpha > 0$.

The second part of the theorem for case $\alpha > 0$ follows from the proof Theorem 4.2 of [**22**], (Note that, since $c < 0$, any solution with $I(u) = c$ is nonzero.)

The proof for the case $\alpha = 0$ is only a modification of the proof above. Therefore we only point out the differences. In this case, (3.2) has no positive solution, it has the trivial solution $u \equiv 0$ and a unique negative solution u^-, all other solutions change sign and satisfy

$$u \geq -\beta^{-1}d.$$

We now modify h outside $(-\beta^{-1}d, \infty)$ such that $\tilde{h}$ is bounded on $(-\infty, 0)$, is smooth except at $u = 0$ and that the modified problem (3.5) has the same solution set as

(3.2). Then u^- is a strict local minimum of I. Since $I(t\varphi_1) \to -\infty$ as $t \to \infty$, we can find that $t_0 > 0$ such that $I(t_0\varphi_1) < I(u^-)$. Therefore, if I satisfies the PS condition, then we are again in the mountain pass setting and a sign-changing solution is found once we can construct a path in H connecting $t_0\varphi_1$ and u^-. Then the path

$$\{t\varphi_1 : t_0 \geq t \geq \varepsilon\} \cup \varepsilon L_1 \cup \{tu^- : \varepsilon \leq t \leq 1\}$$

with $\varepsilon > 0$ small meets our requirements.

It remains to show that I satisfies the PS condition. This is a consequence of the fact that $\tilde{h}$ has the form $\tilde{h}(u) = au^+ + h^-(u)$, where h^- is bounded on $(-\infty, +\infty)$. We omit the details. This completes the the proof of the first part of the theorem for $\alpha = 0$. The second part in this case is proved in the same way as for the case $\alpha > 0$. The proof of Theorem 3.2 is now complete. □

REMARK 3.3. *In* [**30**], *dynamics of two species Lotka-Volterra competing equations system with diffusion and large interaction is studied by Dancer and Zhang, the solutions approach the stationary state as t tends to infinity. The sign-changing solution of limit equation is used.*

3.2. On general jumping nonlinear problem. Dancer and Zhang consider general jumping nonlinear elliptic boundary value problem with resonance at infinity:

$$\begin{cases} -\Delta u = f(x,u) \text{ in } \Omega, \\ u|_{\partial\Omega} = 0, \end{cases} \tag{3.9}$$

We assume that $f(x,0) = 0$, $f(x,u)u \geq 0$, and $f \in C(\bar{\Omega}\times\mathbb{R}^1)$ is locally Lipschitz continuous in u uniformly in x here. Furthermore, we assume $f(x,u)$ has jumping nonlinearities at zero or infinity,

(H_1) $\lim_{u\to 0^+} \dfrac{f(x,u)}{u} = a_1, \lim_{u\to 0^-} \dfrac{f(x,u)}{u} = d_1$, uniformly for $x \in \Omega$,

(H_2) $\lim_{u\to +\infty} \dfrac{f(x,u)}{u} = a_2, \lim_{u\to -\infty} \dfrac{f(x,u)}{u} = d_2$, uniformly for $x \in \Omega$.

Let

$$p(x,t) := f(x,t) - a_2t^+ - d_2t^-, P(x,t) = \int_0^t p(x,s)ds.$$

We assume that either

$$H(x,t) := 2P(x,t) - tp(x,t) \leq W(x) \in L^1(\Omega), \tag{3.10}$$

and

$$H(x,t) \to -\infty \text{ a.e. as } |t| \to \infty; \tag{3.11}$$

or

$$H(x,t) := 2P(x,t) - tp(x,t) \geq W(x) \in L^1(\Omega), \tag{3.12}$$

and

$$H(x,t) \to +\infty \text{ a.e. as } |t| \to \infty. \tag{3.13}$$

We sometimes assume:

(H_3) Problem (3.9) has a negative strict subsolution $\varphi(x)$ and a positive strict super solution $\psi(x)$, that is

$$-\triangle\varphi < f(x,\varphi), \quad \varphi < 0 \text{ in } \Omega, \varphi|_{\partial\Omega} = 0.$$
$$-\triangle\psi > f(x,\psi), \quad \psi > 0 \text{ in } \Omega, \psi|_{\partial\Omega} = 0.$$

Nontrivial solutions of (3.9) correspond to nontrivial critical points of the following functional on H

$$J(u) = \frac{1}{2}\int_\Omega |\bigtriangledown u|^2 dx - \int_\Omega F(x,u)dx, \tag{3.14}$$

where $F(x,u) = \int_0^u f(x,s)ds$.

We will find that the existence of solutions of (3.9) depends on the homogeneous jumping nonlinear problem (3.3), its nontrivial solutions correspond to nontrivial critical points of the following functional $J_{(a,d)}(u)$ (see (3.4)) on H.

We use the following "compactness condition": Let $\psi(t)$ be a positive non-increasing function on $(0,\infty)$ satisfying

$$\int_1^\infty \psi(t)dt = \infty. \tag{3.15}$$

We say that J satisfies $(\psi)_c$ if any sequence $\{u_k\} \subset H$ satisfying

$$J(u_k) \to c, \quad \frac{J'(u_k)}{\psi(\|u_k\|)} \to 0, \tag{3.16}$$

has a convergent subsequence. If this holds for every $c \in R^1$, we say that J satisfies (ψ). This reduces to the usual Palais-Smale condition for $\psi(t) \equiv 1$ and to a condition introduced by Cerami [**17**] for $\psi(t) = 1/(1+t)$.

We use the following lemma

LEMMA 3.4 (Lemma 5.1 of [**42**]). *If* (3.10), (3.11) *hold, then for any* $c \in R^1$,

$$J(u_k) \to c, \quad (1 + \|u_k\|)J'(u_k) \to 0$$

implies that $\{u_k\}$ *has a convergent subsequence, i.e., J satisfies the compactness condition* (ψ) *with* $\psi(t) = 1/(1+t)$.

REMARK 3.5. *As shown in* [**43**], *if* (3.12), (3.13) *hold, then for any* $c \in R^1$,

$$J(u_k) \to c, \quad (1 + \|u_k\|)J'(u_k) \to 0$$

implies that $\{u_k\}$ *has a convergent subsequence,i.e., J satifies the compactness condition* (ψ) *with* $\psi(t) = 1/(1+t)$ *too. In fact we can prove it easily as in the proof of Lemma 5.1 of* [**42**]. *(See Remark 3.17 in the following).*

Now we give some definitions : Since $f(x,u)$ is locally Lipschitz continuous for u, so J is a C^{2-0} functional defined on $C_0^1(\bar\Omega)$. Let $M = \{u \in H | J'(u) = 0\}$ and $X = C_0^1(\bar\Omega)$ with the usual norm $\|u\|_X = \max_{0\le|\alpha|\le 1} \sup_{x\in\bar\Omega} |D^\alpha u(x)|$. It is well known that $X \subset H$ is densely embedded into H, the critical points set $M \subset X$. Let $u(t,u_0), 0 \le t < \eta(u_0)$ ($\eta(u_0)$ is the maximum existence interval) be the forward strong solution of the initial value problem in X:

$$u' = -J'(u), \quad u(0) = u_0 \in X. \tag{3.17}$$

DEFINITION 3.6. *We call $N \subset X$ is an invariant set of descent flow of J if $\{u(t,u_0) | 0 \le t < \eta(u_0), u_0 \in N\} \subset N$. (It is similar to a definition in [**52**])*

Let $K = (-\triangle)^{-1}$, $\bar{F}(u) = f(x,u(x))$, for $x \in \bar{\Omega}, u \in H$. Then $J'(u) = u - K\bar{F}(u)$. The forward strong solution of the initial value problem (3.17) satisfies that

$$u(t,u_0) = e^{-t}[u_0 + \int_0^t e^s K\bar{F}(u(s,u_0))ds], (\text{for } t \ge 0). \tag{3.18}$$

Since $f(x,t)$ is local Lipschitz continuous in $t \in \mathbb{R}^1$ uniformly for x, $K : L^\infty(\Omega) \to C_0^1(\bar{\Omega})$ is a bounded linear operator, and $\bar{F} : C_0^1(\bar{\Omega}) \hookrightarrow L^\infty(\Omega)$ is local Lipschitz continuous, for every $u_0 \in X$, there exists $B_\delta(u_0)$, $l_1 > 0$ such that

$$\|K\bar{F}(u) - K\bar{F}(u_0)\|_X \le l_1 \|u - u_0\|_X, \quad \forall u \in B_\delta(u_0), \tag{3.19}$$

where $B_\delta(u_0) = \{u \in X : \|u - u_0\|_X < \delta, \delta > 0\}$. Thus the forward strong solution $u(t,u_0)$ of (3.17) exists and is unique in X.

Let $P_H := \{u \in H : u \ge 0$ almost everywhere$\}$, $P := P_H \cap X = \{u \in C_0^1(\bar{\Omega}) | u \ge 0\}$, P has nonempty interior $\overset{o}{P}$. Since $K(P) \subset P, K(-P) \subset (-P)$ and $P, -P$ are convex closed sets in X, by the theory of ordinary differential equations in Banach Spaces, we know $P, -P$ are both invariant sets of descent flow of J under the condition $f(x,u)u \ge 0$. Noticing $K\bar{F} : P \to \overset{o}{P}$ and (3.18), we know that $u(t,u_0) \in \overset{o}{P}$, $\forall t > 0$, for $u_0 \in P$, and $\overset{o}{P}$, $-\overset{o}{P}$ are both invariant sets of descent flow of J.

DEFINITION 3.7. *Suppose that $\Phi \in C^1(X, \mathbb{R}^1), a \in \mathbb{R}^1$. We say that Φ has the retracting property for a on X, if $\forall b > a$, $\Phi^{-1}[a,b] \cap X \cap M = \emptyset$, then $\Phi^a \cap X$ is a retract of $\Phi^b \cap X$, i.e. there exists $\eta : \Phi^b \cap X \to \Phi^a \cap X$ continuous in the topology of X, such that $\eta(\Phi^b \cap X) \subset \Phi^a \cap X, \eta|_{\Phi^a \cap X} = id|_{\Phi^a \cap X}$.*

LEMMA 3.8 (Lemma 1 of [**29**]). *Suppose that* (3.10),(3.11) *or* (3.12),(3.13) *are satisfied, then for any $a \in \mathbb{R}^1$, Then J has the retracting property for a on X.*

LEMMA 3.9 (Lemma 2 of [**29**]). *Suppose U is a bounded connected open set of $\mathbb{R}^2$, and $(0,0) \in U$, then there exists a connected component Γ' of the boundary of U, and each one sided ray l through the origin satisfies $l \cap \Gamma' \ne \emptyset$.*

THEOREM 3.10 (Dancer-Zhang, Theorem 1 of [**29**]). *Suppose that $(H_1), (H_2)$ are satisfied,* (3.10),(3.11) *(or* (3.12), (3.13)*) are satisfied, and $a_1, d_1 < \lambda_1$, $(a_2, d_2) \in \tilde{S}$, $(a_2, d_2) \in \Sigma$. Then* (3.9) *has at least one sign-changing solution, one positive solution, one negative solution.*

PROOF. Under these hypotheses (3.10),(3.11) (or (3.12), (3.13)), by Lemma 3.8, we know J has the retracting property for any $a \in \mathbb{R}^1$ in X.

First we prove that θ is a strictly local minima of J. (This is known, we only give this for completeness.)

We only give the proof for $\Omega \subset \mathbb{R}^N, N \geq 3$, (The proof is similar for $N \leq 2$). By (H_1) and $f(x,0) = 0, a_1, d_1 < \lambda_1$, we know that $\exists\ 0 < \varepsilon_0 < \lambda_1$ and $\delta > 0$ such that $|f(x,t)| < (\lambda_1 - \varepsilon_0)|t|, \forall |t| < \delta$, thus

$$|F(x,u)| \leq \int_0^u |f(x,t)|dt < \frac{1}{2}(\lambda_1 - \varepsilon_0)u^2, \text{ as } 0 < |u| < \delta. \tag{3.20}$$

By $(H_1), (H_2)$, we have $\exists b_1 > 0$ such that

$$|F(x,u)| \leq \int_0^u |f(x,t)|dt < \frac{1}{2}(\lambda_1 - \varepsilon_0)u^2 + b_1|u|^{\frac{2n}{n-2}}, \text{ for } -\infty < u < \infty. \tag{3.21}$$

Therefore, by the Poincaré inequality, we have that

$$\begin{aligned} |\int_\Omega F(x,u(x))dx| &\leq \int_\Omega |F(x,u(x))|dx \\ &< \frac{1}{2}(\lambda_1 - \varepsilon_0)\int_\Omega u^2(x)dx + b_1 \int_\Omega |u(x)|^{\frac{2n}{n-2}}dx \\ &\leq \frac{1}{2}\|u\|^2 - \frac{1}{2}\varepsilon_0 C^2(\Omega)\|u\|^2 + b_1 \int_\Omega |u(x)|^{\frac{2n}{n-2}}dx, \end{aligned} \tag{3.22}$$

and by the definition of J, we have that

$$J(u) > \frac{1}{2}\varepsilon_0 C^2(\Omega)\|u\|^2 - b_1 \int_\Omega |u(x)|^{\frac{2n}{n-2}}dx, \quad \forall u \in H. \tag{3.23}$$

Since $H \hookrightarrow L_{\frac{2n}{n-2}}(\Omega)$, we have that $\exists b_2 > 0$ such that

$$b_1 \int_\Omega |u|^{\frac{2n}{n-2}}(x)dx \leq b_2\|u\|^{\frac{2n}{n-2}},$$

Thus, we know

$$J(u) > \frac{1}{2}\varepsilon_0 C^2(\Omega)\|u\|^2 - b_2\|u\|^{\frac{2n}{n-2}}.$$

Therefore, there exists $b_3 > 0$ and an open neighbourhood U_{δ_0} of θ such that

$$J(u) > b_3\|u\|^2, \quad \forall u \in U_{\delta_0}, \quad \text{and} \quad f(0) < f(u), \quad \forall u \in U_{\delta_0}, u \neq 0. \tag{3.24}$$

i.e., θ is a strictly local minima of J, where $U_{\delta_0} = \{u \in H, \|u\| < \delta_0\}$.

Now we prove there exists a neighborhood of 0 in X, which is an invariant set of J.

Let $U_{\frac{1}{n}} = \{u \in U_{\delta_0} \cap X\,|\,J(u) < \frac{1}{n}\}, n = 1, 2, 3, \ldots$, then $U_{\frac{1}{n}}$ is an open neighborhood of θ in X for each n. Therefore, for sufficiently large n_0 such that

$$\frac{1}{n_0 b_3} < \delta_0^2,$$

and for $u_0 \in U_{\frac{1}{n_0}}$, the solution $u(t, u_0)$ of (12) satisfies the following formula:

$$b_3\|u(t,u_0)\|^2 < J(u(t,u_0)) \leq J(u_0) < \frac{1}{n_0}.$$

Thus we have that $U_{\frac{1}{n_0}}$ is an open, invariant set of descent flow of J.

Let $U_0 = U_{\frac{1}{n_0}}$.

Set

$$U_1 = \{u_1 \in X | \exists t' > 0 \text{ such that } u(t', u_1) \in U_0\}.$$

It is easy to know that U_1 is an open invariant set of descent flow of J in X. And since $u(t, u_1)$ has continuous dependence on the initial value u_1, we may prove that ∂U_1 is also an invariant set of descent flow of J in X if $\partial U_1 \neq \emptyset$. It is clear that $J(u) > 0$ on ∂U_1, and hence $J(u(t, u_0)) > 0$ for $u_0 \in \partial U_1$ by invariance.

Since $(a_2, d_2) \in \tilde{S}$, there exists (a_0, d_0) such that $(a_0, d_0) \in \Gamma$ and $a_2 > a_0, d_2 > d_0$. By Lemma 3.1, we get that there exists a path L_0 in $\triangle(a_2, d_2)$ connecting $\varphi_1, -\varphi_1$. And by the proof of Lemma 2.3 of [**23**],

$$\begin{aligned} L_0 = \{t\varphi_0^+ + (1-t)\varphi_1 : 0 \le t \le 1\} &\cup \{t\varphi_0^+ + (1-t)\varphi_0^- : 0 \le t \le 1\} \\ \cup \{t\varphi_0^- + (1-t)(-\varphi_1) &: 0 \le t \le 1\}, \end{aligned} \tag{3.25}$$

where φ_0 is one nontrivial solution of (3.3) with $(a, d) = (a_0, d_0)$. Since L_0 is compact, there is a ε neighborhood L_ε in H, such that $L_\varepsilon \subset \triangle(a_2, d_2)$. And in this neighborhood, we can choose a path L in X, such that $L \subset \triangle(a_2, d_2)$. In fact, we can choose $\varphi^1 \in X, \varphi^2 \in X$ such that

$$\|\varphi^1 - \varphi_0^+\| < \frac{\varepsilon}{2}, \quad \|\varphi^2 - \varphi_0^-\| < \frac{\varepsilon}{2}.$$

Define

$$\begin{aligned} L = \{t\varphi^1 + (1-t)\varphi_1 : 0 \le t \le 1\} &\cup \{t\varphi^1 + (1-t)\varphi^2 : 0 \le t \le 1\} \\ \cup \{t\varphi^2 + (1-t)(-\varphi_1) &: 0 \le t \le 1\}. \end{aligned}$$

We improve L by replacing it by $\frac{u}{\|u\|_X}$ for $u \in L$, and finally we can shrink the curve to remove intersections and get L with no self intersections on ∂B_1, where B_1 is the unit ball in X. (L can be chosen arc connected by Theorem 4.1 on p27 of Whyburn [**35**]).

(H_2) implies that for $\delta > 0, \delta\lambda_1^{-1} \sup\{\|u\| : u \in L\} < \inf\{|J_{(a_2,d_2)}(u)| : u \in L\}$, there exists $t_0 > 0$ such that

$$\begin{aligned} &|f(x,t) - a_2 t| < \delta t, \ \forall t > t_0, \\ &|f(x,t) - d_2 t| < \delta |t|, \ \forall t < -t_0. \end{aligned}$$

Thus

$$\begin{aligned} &|F(x,t) - \frac{a_2}{2}t^2| \le \frac{\delta}{2}t^2 + Ct, \ \forall t \in R^+, \\ &|F(x,t) - \frac{d_2}{2}t^2| \le \frac{\delta}{2}t^2 + C|t|, \ \forall t \in R^-. \end{aligned}$$

Therefore, for $u \in H$,

$$\begin{aligned} &\int_\Omega |F(x,u^+) - \frac{a_2}{2}(u^+)^2| + \int_\Omega |F(x,u^-) - \frac{d_2}{2}(u^-)^2| \\ &\le \delta \int_\Omega u^2 + C \int_\Omega |u| \le \delta\lambda_1^{-1}\|u\|^2 + C\|u\| \end{aligned}$$

Since

$$J(u) = J_{(a_2,d_2)}(u) - \int_\Omega (F(x,u^+) - \frac{a_2}{2}(u^+)^2) - \int_\Omega (F(x,u^-) - \frac{d_2}{2}(u^-)^2)$$

for $u \in H$ and since for $u \in L$ and $t \geq 0$, $J_{(a_2,d_2)}(tu) = t^2 J_{(a_2,d_2)}(u)$, we have

$$\begin{aligned} J(tu) &\leq t^2 J_{(a_2,d_2)}(u) + \delta\lambda_1^{-1} t^2 \|u\| + tC\|u\|, \\ &\leq t^2 \sup\{J_{(a_2,d_2)}(u) : u \in L\} + \delta\lambda_1^{-1} t^2 \sup\{\|u\| : u \in L\} \\ &\quad + tC \sup\{\|u\| : u \in L\}. \end{aligned} \tag{3.26}$$

Thus

$$J(tu) \to -\infty, (t \to +\infty) \text{ uniformly for } u \in L, \quad (\|u\|_X = 1). \tag{3.27}$$

Define the surface $Z := \{tu : t \geq 0, u \in L\}$. It is easily seen to be homeomorphic to a closed half space $\mathbb{R}^{2^+} := \{(x,y) \in \mathbb{R}^2 | y \geq 0\}$ in $\mathbb{R}^2$ by our careful choice of L.

It is easy to know $U_1 \cap Z$ is a bounded and relatively open set in Z, $\partial U_1 \cap Z \neq \emptyset$. We may assume that $U_1 \cap Z$ is connected. (Otherwise, we consider the connected component $U_1' \subset Z$ of $U_1 \cap Z$, with $(0,0) \in U_1'$, instead of $U_1 \cap Z$). By Lemma 3.9 there exists at least one connected component $\Lambda \subset (\partial U_1 \cap Z)$ such that $\Lambda \cap \overset{o}{P} \neq \emptyset, \Lambda \cap (-\overset{o}{P}) \neq \emptyset$, $\Lambda \cap [X \backslash (-P \cup P)] \neq \emptyset$. (Note that to apply Lemma 3.9, we consider the homeomorph of U_1 in the half space $\mathbb{R}^{2^+}$, and then add its reflection in the other half space to obtain an open set in $\mathbb{R}^2$).

Thus without loss of generality, we assume that $\partial U_1 \cap Z$ is connected, and ∂U_1 is connected.

Let

$$V_1 = \{h \in \partial U_1 |\ u' = -J'(u), u(0) = h, \exists t_0 \geq 0 \text{ such that } u(t_0,h) \in \overset{o}{P}\}, \tag{3.28}$$

$$V_2 = \{h \in \partial U_1 |\ u' = -J'(u), u(0) = h, \exists t_0 \geq 0 \text{ such that } u(t_0,h) \in -\overset{o}{P}\}. \tag{3.29}$$

Noticing (3.18) and $f(x,u)u \geq 0$, by the strongly order preserving property of K we know that $P \cap \partial U_1, (-P) \cap \partial U_1$ are both invariant sets of descent flow of J in X. V_1, V_2 are disjoint relatively open (in ∂U_1) invariant sets of descent flow of J, and V_1, V_2 are unchanged if we replace $\overset{o}{P}$ by P in (3.28)(3.29). By connectedness of ∂U_1, $\partial U_1 \backslash (V_1 \cup V_2)$ is not empty. Thus by the deformation lemma (Lemma 3.8), we have that every solution of (3.17) goes to negative energy or approaches a critical point. Regularity implies that it converges in X. So

$$\begin{aligned} &\forall u_0 \in P \cap \partial U_1, \quad u(t,u_0) \to u_1 \in M \cap P in X, \\ &\forall u_0 \in -P \cap \partial U_1, \quad u(t,u_0) \to u_2 \in M \cap (-P) in X, \\ &\forall u_0 \in \partial U_1 \backslash (V_1 \cup V_2), \quad u(t,u_0) \to u_3 \in M in X. \end{aligned}$$

Thus (1) has at least three solutions $u_1 \in P \cap \partial U_1, u_2 \in (-P) \cap \partial U_1, u_3 \in \partial U_1 \backslash (V_1 \cup V_2)$. And by the maximum principle, we know that $u_1 \in \overset{o}{P}, u_2 \in -\overset{o}{P}$. It is clear that u_3 is sign-changing, and $\exists \varepsilon_0 > 0$ such that $J(u_i) \geq \varepsilon_0, i = 1,2,3$. (If $u(t,u_0)$ approached a point in $P \cup (-P)$, $u_0 \in V_1 \cup V_2$.) □

THEOREM 3.11 (Dancer-Zhang, Theorem 2 of [**29**]). *Suppose that* (H_1), (H_2), (H_3) *are satisfied,* (3.10),(3.11) *(or* (3.12), (3.13)*) are satisfied,* $(a_1, d_1) \in D_k, k \geq 2, (a_2, d_2) \in \tilde{S}$, *and* $(a_2, d_2) \in \Sigma$, *Then* (3.9) *has at least* 7 *solutions, and three are sign-changing, two are positive, two are negative.*

PROOF. By a small modification of the proofs of Theorem 1 and Corollary 1 in [**26**], we get that there exists 4 nontrivial solutions $u_i, i = 1, 2, 3, 4$ in $[\varphi, \psi]$, u_1 is positive, u_2 is negative, and u_3, u_4 are sign-changing. In the proof of Theorem 1, we replace U_0 by the interior of $[\varphi, \psi]$ (Note the interior of $[\varphi, \psi]$ is connected). This is also an open invariant set of descent flow of J. (Note (3.17)). Much as before, we get that there exist three critical points u_5, u_6, u_7 outside $[\varphi, \psi]$, and $u_5 \in \partial U_1 \backslash (V_1 \cup V_2)$, $u_6 \in \overset{o}{P}, u_7 \in - \overset{o}{P}$. □

When $a_2 = d_2$, we have

COROLLARY 3.12. *If* $f \in C(\bar{\Omega} \times R)$ *satisfies*

$$\overline{\lim}_{t \to 0} \frac{f(x,t)}{t} < \lambda_1, \quad \lim_{|t| \to \infty} \frac{f(x,t)}{t} = \lambda_l, \ \ l > 2, \tag{3.30}$$

and

$$\lim_{|t| \to \infty} (2F(x,t) - tf(x,t)) = -\infty, \text{ uniformly in } \Omega; \tag{3.31}$$

or

$$\lim_{|t| \to \infty} (2F(x,t) - tf(x,t)) = +\infty, \text{ uniformly in } \Omega. \tag{3.32}$$

Then (3.9) *has at least one sign-changing solution, one positive solution, one negative solution.*

REMARK 3.13. *As in* [**43**], *we do not assume* $p(x,t) := f(x,t) - a_2 t^+ - d_2 t^-$ *grows sublinearly. For example* (3.10),(3.11) *(or* (3.12), (3.13)*) are satisfied if* $p(x,t) = -t/\ln|t|$ *(resp.* $t/\ln|t|$*) for* $|t|$ *large, even though there is no* $\sigma \in (0,1)$ *such that* $|p(x,t)| \leq C(|t|^\sigma + 1)$ *in this case.*

REMARK 3.14. (H_3) *is satisfied when* f *satisfies one of the following three conditions:*

- (F_1) $\limsup_{|t| \to +\infty} \frac{f(x,t)}{t} < \lambda_1$;
- (F_2) *there exists* $t_1 < 0, t_2 > 0$ *such that* $f(x,t_1) = f(x,t_2) = 0$;
- (F_3) *there is a number* $k > 0$ *such that* $|f(x,t)| < k$ *for* $t \in [-c, c]$, *where* $c = \max_{\Omega} e(x)$ *and* $e(x)$ *satisfies* $-\triangle e = k$ *in* $\Omega, e = 0$ *on* $\partial\Omega$.

REMARK 3.15. *It is easy to see that* $f(x,u)u \geq 0$ *is not necessary. Because* $f(x,u)$ *is asymptotically linear uniformly for* $x \in \Omega$, *there exists a positive number* m *such that* $g(x,u) = f(x,u) + mu$ *and* $g(x,u)u \geq 0$, *we can consider the following equation* $-\triangle u + mu = g(x,u)$, $u|_{\partial\Omega} = 0$ *instead of* (3.9).

REMARK 3.16. *If* $f \in C^1(\bar{\Omega} \times R^1)$, $f'_t(x,0) < \lambda_1$; $f'_t(x,t) \to \lambda_l, l > 2$ *as* $|t| \to +\infty$, *and* (3.31) *or* (3.32) *are satisfied, the conclusion of Corollary 3.12 is clearly valid.*

REMARK 3.17. *From the proof of Theorem 3.10, we know that* $J(u_i) \geq \varepsilon_0 > 0, i = 1, 2, 3$. *So in fact for the compact condition, we only use that* J *satisfies* $(\psi)_c$ *for any* $c > 0$, *and* J *has the retracting property for any* $c > 0$ *in* X. *From the proof of Lemma 5.1 of* [**43**], *we can use*

$$\begin{aligned} &H(x,t) \geq W_1(x) \in L^1(\Omega), \quad \forall t \in R^1, \\ &H(x,t) \geq W_2(x) \in L^1(\Omega) \text{ for large } |t|, \text{ and } \int_\Omega W_2(x) \geq 0 \end{aligned} \tag{3.33}$$

instead of (3.12),(3.13). *For the convenience of the reader, we give the proof here.*

PROOF. We prove for any $c > 0$,

$$J(u_k) \to c, \quad (1 + \|u_k\|)J'(u_k) \to 0 \tag{3.34}$$

implies that $\{u_k\}$ has a convergent subsequence, i.e., J satisfies $(\psi)_c$ for any $c > 0$.
By (3.34), we have

$$J_{(a_2,d_2)}(u_k) - \int_\Omega P(x, u_k) \to c, \tag{3.35}$$

$$(J'_{(a_2,d_2)}(u_k), v) - (p(x, u_k), v) \to 0, \quad \forall v \in H, \tag{3.36}$$

and

$$|(J'(u_k), u_k)| \leq \|u_k\| \cdot \|J'(u_k)\| \to 0.$$

Thus

$$(J'(u_k), u_k) = 2J_{(a_2,d_2)}(u_k) - \int_\Omega u_k p(x, u_k) \to 0. \tag{3.37}$$

Assume that $\rho_k = \|u_k\| \to +\infty$, and let $\tilde{u}_k = u_k/\rho_k$. Then $\|\tilde{u}_k\| = 1$. Thus there is a subsequence such that $\tilde{u}_k \to \tilde{u}$ weakly in H, strongly in $L^2(\Omega)$, and a.e. in Ω. By (3.37), we obtain

$$1 = \|\nabla \tilde{u}_k\|^2_{L^2(\Omega)} = a_2\|\tilde{u}_k^+\|^2_{L^2(\Omega)} + d_2\|\tilde{u}_k^-\|^2_{L^2(\Omega)} + \rho_k^{-2} \int_\Omega u_k p(x, u_k) + \varepsilon_k,$$

where $\varepsilon_k \to 0$ as $k \to +\infty$. Since $\tilde{u}_k \to \tilde{u}$ strongly in $L^2(\Omega)$ and $p(x,t)/t \to 0$, as $|t| \to +\infty$ uniformly in x, we can pass to the limit and find

$$1 = a_2\|\tilde{u}^+\|^2_{L^2(\Omega)} + d_2\|\tilde{u}^-\|^2_{L^2(\Omega)}. \tag{3.38}$$

This shows that $\tilde{u} \not\equiv 0$.
Moreover, (3.36) implies $J'_{(a_2,d_2)}(\tilde{u}) = 0$.
In fact, (3.36) implies

$$(J'_{(a_2,d_2)}(\tilde{u}_k), v) - (\frac{p(x, u_k)}{\rho_k}, v) \to 0, \quad \forall v \in H.$$

Taking the limit, we find $(J'_{(a_2,d_2)}(\tilde{u}), v) = 0$ for all $v \in H$, and hence $J'_{(a_2,d_2)}(\tilde{u}) = 0$.
Multiplying by $\tilde{u}$ gives

$$\|\tilde{u}\|^2 = a_2\|\tilde{u}^+\|^2_{L^2(\Omega)} + d_2\|\tilde{u}^-\|^2_{L^2(\Omega)}.$$

Hence by (3.38), $\|\tilde{u}\| = 1$. This show that $\tilde{u}_k \to \tilde{u}$ strongly in H. Combining (3.35) and (3.37), we obtain

$$\frac{1}{2}\int_\Omega H(x, u_k) \to -c < 0. \tag{3.39}$$

Let $\Omega_0 = \{x \in \Omega : \tilde{u}(x) = 0\}, \Omega_1 = \Omega \backslash \Omega_0$. Then Ω_0 has measure zero, $|u_k(x)| \to \infty$ for almost all $x \in \Omega_1$. By (3.33) and Fatou's Lemma, we have

$$\begin{aligned} \underline{\lim}_{k\to\infty} \int_\Omega H(x, u_k(x))dx &\geq \int_{\Omega_0} W_1(x)dx + \underline{\lim}_{k\to\infty} \int_{\Omega_1} H(x, u_k(x))dx \\ &\geq \int_{\Omega_1} \underline{\lim}_{k\to\infty} H(x, u_k(x))dx \\ &\geq \int_{\Omega_1} W_2(x) = \int_\Omega W_2(x) \\ &\geq 0, \end{aligned} \tag{3.40}$$

contradicting (3.39). Hence, the sequence $\{u_k\}$ is bounded in H. A standard argument now show that it has a convergent subsequence.

Under (3.10),(3.11), or (3.12),(3.13), we can obtain

$$\int_\Omega H(x, u_k) \leq \int_{\Omega_0} W(x) + \int_{\Omega_1} H(x, u_k) \to -\infty, \tag{3.41}$$

or

$$\int_\Omega H(x, u_k) \geq \int_{\Omega_0} W(x) + \int_{\Omega_1} H(x, u_k) \to +\infty, \tag{3.42}$$

Both (3.41) and (3.42) contradict that $\frac{1}{2}\int_\Omega H(x, u_k) \to -c$ for all $c \in \mathbb{R}^1$. □

3.3. Sign-changing solutions of p-Laplacian equations. We consider the Existence of multiple and sign-changing solutions of the problem

$$\begin{cases} -\Delta_p u = h(u) & \text{in } \Omega \\ u(x) = 0 & \text{on } \partial\Omega \end{cases} \tag{3.43}$$

where $-\Delta_p u = -div(|\nabla u|^{p-2}\nabla u)$ is the p-Laplacian, $1 < p < +\infty$, Ω is a smooth bounded domain in $\mathbb{R}^N (N \geq 1)$, $h(u)$ is local Lipschitz continuous, $h(0) = 0$, $h(u)u \geq 0$ and $h(u)$ has "jumping" nonlinearities at zero or infinity.

$$(H_1) \quad \lim_{u\to 0+} \frac{h(u)}{|u|^{p-2}u} = a_0, \qquad \lim_{u\to 0-} \frac{h(u)}{|u|^{p-2}u} = d_0;$$

$$(H_2) \quad \lim_{u\to +\infty} \frac{h(u)}{|u|^{p-2}u} = a_1, \qquad \lim_{u\to -\infty} \frac{h(u)}{|u|^{p-2}u} = d_1.$$

Zhang-Li [**61**] seems to be the first one to consider multiple and sign-changing solutions about p-Laplacian in the case of $N < p < \infty$.

In [**59**], we take the idea from [**61**] to construct carefully a pseudo-gradient vector field (in short, p.g.v.f.) in $C_0^1(\bar{\Omega})$, by which we obtain the positive and negative cones of $C_0^1(\bar{\Omega})$ are all the invariant sets of the descent flow of the corresponding functional, then we use differential equations theory in Banach space to obtain sign-changing and multiple solutions of (3.43).

Before giving our main result, let us recall a related homogeneous "jumping" nonlinear problem

$$(3.44) \qquad -\Delta_p u = a(u^+)^{p-1} - d(u^-)^{p-1} \quad \text{in} \quad \Omega, \qquad u|_{\partial\Omega} = 0,$$

where $u^+ = \max\{u, 0\}$, $u^- = \max\{-u, 0\}$. By [**20**], the Fučik spectrum of p-Laplacian on $W_0^{1,p}(\Omega)$ is defined as the set of Σ_p of those $(a, d) \in \mathbb{R}^2$ such that (3.44) has a nontrivial solution u. Also in [**20**], the authors construct the first nontrivial curve Γ in Σ_p and there exists a continuous function $\eta(t)$ defined on $(\lambda_1, \lambda_2]$ in ad-plane, such that

(a) (3.44) has a nontrivial solution for $(a, d) \in \Gamma = \{(a, \eta(a));\ \lambda_1 < a \leq \lambda_2\} \cup\{(\eta(d), d);\ \lambda_1 < d \leq \lambda_2\}$;
(b) (3.44) has no nontrivial solution for $\lambda_1 < d < \eta(a)$, $a \in (\lambda_1, \lambda_2]$ or $\lambda_1 < a < \eta(d)$, $d \in (\lambda_1, \lambda_2]$.

where λ_1 is the first eigenvalue of $-\Delta_p$ in $W_0^{1,p}(\Omega)$ with Dirichlet boundary condition, λ_1 is simple with eigenfunction $\phi_1(x) > 0$ in Ω and $\phi_1(x) \in C_0^1(\bar{\Omega})$ and λ_2 is defined as

$$\lambda_2 = \inf\{\lambda > \lambda_1;\ \lambda \text{ is an eigenvalue of } -\Delta_p \text{ in } W_0^{1,p}(\Omega)\}.$$

Now we denote by S the set of points (a, d) which are above Γ in the ad-plane, $\tilde{S} = S \cup \{+\infty\}$, and our main result read as follows.

THEOREM 3.18 (Zhang-Chen-Li, Theorem 1.1 of [**59**]). *Suppose that* (H_1), (H_2) *hold,* a_1, $d_1 < \lambda_1$, $(a_0, d_0) \in \tilde{S}$, *and*

(H_3) *$h(u)$ is nondecreasing in u and for any $u \in \mathbb{R}^1$ any $0 < t < 1$, there exist a continuous function $\eta(u, t) > 0$ such that $h(tu) \geq (1+\eta)t^{p-1}h(u), \forall u \geq 0$; $h(tu) \leq (1-\eta)t^{p-1}h(u), \forall u \leq 0$.*

Then (3.43) has at least three solutions, one unique positive, one unique negative, one is sign-changing solutions.

We can give an example here: $h(u) = a(u)|u|^{p-2}, a(u)u \geq 0$. $a(u)$ is local Lipschitz continuous.

$$\frac{a(u)}{u} \to a_0, \quad (u \to 0^+), \qquad \frac{a(u)}{u} \to d_0, \quad (u \to 0^-);$$
$$\frac{a(u)}{u} \to a_1, \quad (u \to +\infty), \qquad \frac{a(u)}{u} \to d_1, \quad (u \to -\infty);$$

and $\forall u \in \mathbb{R}^1, 0 < t < 1, \exists \eta(u, t) > 0$ such that

$$a(tu) \geq (1+\eta)t^{p-1}a(u), \quad \forall u \geq 0; \quad a(tu) \leq (1-\eta)t^{p-1}a(u), \quad \forall u \leq 0;$$

(i.e. $a(u)$ is sublinear.)

We don't give the proof here, if the readers are interested, please see [**59**].

4. Bifurcation and multiple sign-changing solutions

In [**58**], we consider the following semilinear elliptic equation:

$$\begin{cases} -\Delta u = \lambda f(u) \text{ in } \Omega, \\ u|_{\partial\Omega} = 0, \end{cases} \tag{4.1}$$

where $-\Delta$ is the Laplacian, Ω is a smooth bounded domain in $R^n, n \geq 1$, λ is a positive parameter. Let λ_k be the k-th eigenvalue of

$$\begin{cases} -\Delta \phi = \lambda \phi \text{ in } \Omega, \\ \phi|_{\partial\Omega} = 0, \end{cases} \tag{4.2}$$

Let ϕ_k be the corresponding eigenfunction with the property that $\int_\Omega \phi_k^2 = 1$ for $k = 1, 2, \ldots$. It is well known that $\lambda_1 > 0, \lambda_1$ is simple, $\phi_1 > 0$. Here we assume that all $\lambda_k s$ are simple eigenvalues, i.e. $0 < \lambda_1 < \lambda_2 < \cdots < \lambda_k < \cdots$. Moreover, we assume

(f_1) $f \in C^1(R, R), f(0) = 0, f'(0) > 0$;
(f_2) $\lim_{|u| \to +\infty} \frac{f(u)}{u} = f'(\infty) > 0$;
(f_3) $f(u)/u$ is increasing in $(0, \infty)$ and is decreasing in $(-\infty, 0)$;
(f_4) $f(u)/u$ is decreasing in $(0, \infty)$ and is increasing in $(-\infty, 0)$.

REMARK 4.1. *(f_3) and (f_4) are contradictory, they are used alternatively in the following.*

Nontrivial solutions of (4.1) correspond to nontrivial critical points of the following functional on H

$$J_\lambda(u) = \frac{1}{2} \int_\Omega |\nabla u|^2 dx - \lambda \int_\Omega F(u) dx, \tag{4.3}$$

where $F(u) = \int_0^u f(s) ds$.

Let u be an isolated critical point of the functional J, and let $c = J(u)$. We call

$$C_q(J, u) = H_q(J^c \cap V_u, (J^c \setminus \{u\} \cap V_u), G)$$

the qth critical group, with coefficient group G of J at u, $q = 0, 1, 2, \ldots$, where $J^c = \{z \in H : J(z) \leq c\}$, V_u is a neighborhood of u such that the critical set K satisfies $K \cap (J^c \cap V_u) = \{u\}$, and $H_*(X, Y; G)$ stands for the singular relative homology groups with the abelian cofficient group G.

Let $W(x) \in L^\infty(\Omega)$, consider an eigenvalue problem:

$$\begin{cases} -\Delta \phi = W(x)\phi + \mu_i(W)\phi \text{ in } \Omega, \\ \phi|_{\partial\Omega} = 0. \end{cases}$$

It is well known that, for $i = 1, 2, \ldots$,

$$\mu_i(W) = \min_i \max_i \frac{\int_\Omega (|\nabla z|^2 - W(x) z^2) dx}{\int_\Omega z^2 dx},$$

where $\max_i$ is over all $z(\neq 0) \in T_i$, and $\min_i$ is over all linear subspaces T_i of H of dimension i. For $W_1, W_2 \in L^\infty(\Omega)$ satisfying $W_2(x) \geq W_1(x)$ almost everywhere, $\mu_i(W_2) \leq \mu_i(W_1)$. If in addition $mes\,\{x \in \Omega : W_2(x) > W_1(x)\} > 0$, then $\mu_i(W_2) < \mu_i(W_1)$.

Let

$$(4.4)\quad \lambda_k^0 = \frac{\lambda_k}{f'(0)},\ \lambda_k^\infty = \frac{\lambda_k}{f'(\infty)},\ \lambda_k^m = \frac{\lambda_k}{\inf_{u\in R\backslash 0} f'(u)} \text{ and } \lambda_k^M = \frac{\lambda_k}{\sup_{u\in R\backslash 0} f'(u)}.$$

When (f_3) is satisfied, we have $f'(u) \geq f(u)/u$ for $u \in R\backslash 0$, and for $k \in N$, f satisfying $(f_1), (f_2), (f_3)$, $\lambda_k^M \leq \lambda_k^\infty < \lambda_k^0$. If in addition, $uf''(u) \geq 0$, then $\lambda_k^M = \lambda_k^\infty$. For $k \in N$, we define open intervals

$$I_k = (\lambda_k^\infty, \lambda_k^0),\ \tilde{I}_k = (\lambda_k^M, \lambda_k^0).$$

We define the Morse index $M(u)$ of a solution u to (4.1) to be the number of negative eigenvalues of the following problem:

$$(4.5)\qquad \begin{cases} -\Delta\phi = \lambda f'(u)\phi + \mu\phi \text{ in } \Omega, \\ u|_{\partial\Omega} = 0. \end{cases}$$

If u is a solution to (4.1), and 0 is not an eigenvalue of (4.5), then u is a non-degenerate solution, otherwise it is degenerate. In [**51**], the author gets the following main results (for simplicity, they assume $\lambda_k^M = \lambda_k^\infty$,i.e. $I_k = \tilde{I}_k$):

(a) If $\lambda \notin \bigcup_{j\in N} I_j$, then (4.1) has only the trivial solution $u = 0$;
(b) $\lambda \in I_k\backslash \bigcup_{j\neq k} I_j$, then (4.1) has exactly two nontrivial solutions which are non-degenerate and with Morse index $M(u) = k$;
(c) If $\lambda \in (I_k \bigcap I_{k+1})\backslash \bigcup_{j\neq k,k+1} I_j$, then there exists $\epsilon > 0$ such that (4.1) has exactly four nontrivial solutions for $\lambda \in (\lambda_{k+1}^\infty, \lambda_{k+1}^\infty + \epsilon) \cup (\lambda_k^0 - \epsilon, \lambda_k^0)$ (which is near the boundary of $I_k \cap I_{k+1}$), and all of them are non-degenerate with two of them have Morse index $M(u) = k$, the other two $M(u) = k + 1$.

4.1. Main results with parameter λ. When f satisfies $(f_1), (f_2), (f_3)$, for $u \in R\backslash 0$, we have

$$(4.6)\qquad f'(0) < \frac{f(u)}{u} < f'(u) \leq \sup_{u\in R} f'(u),\quad \frac{f(u)}{u} < f'(\infty).$$

If u is a nontrivial solution of (4.1), then

$$\begin{aligned} \mu_k(\lambda f'(0)) &> \mu_k(\lambda f(u)/u) > \mu_k(\lambda f'(u)) \\ &\geq \mu_k(\lambda \sup_{u\in R} f'(u)), \mu_k(\lambda f(u)/u) > \mu_k(\lambda f'(\infty)). \end{aligned}$$

Our main results are:

THEOREM 4.2 (Zhang, Theorem 2.1 of [**58**]). *Suppose that f satisfies (f_1), $(f_2), (f_3)$, $\lambda_k^0 \leq \lambda_{k+1}^M$, for $k = 1, 2, \ldots N$; Moreover, suppose that $uf''(u) \geq 0$ in a neighborhood of 0. Then for $k = 1, 2, \ldots, N$*

(a) *The branch of the solutions* ($\lambda \in (I_k \cup \{\lambda_k^0\}) \backslash \bigcup_{j\neq k} \tilde{I}_j$) *to (1) has the same critical groups* $C_q^{\pm}(J_\lambda, u^{\pm}(\lambda,\cdot)) = \delta_{qk}G$, *and the critical values are positive; let* $\Sigma_k^{\pm} = \{(\lambda, u_k^{\pm}(\lambda,\cdot)) : \lambda \in I_k\}$, *then* Σ_k^+ *and* Σ_k^- *join at* $(\lambda_k^0, 0)$, *and*

$$\lim_{\lambda\to(\lambda_k^\infty)^+} \|u_k^{\pm}(\lambda,\cdot)\| = \infty.$$

(b) *For* $\lambda \in I_k, k \geq 2$, *the nontrivial solutions are sign-changing; for* $\lambda \in I_1$, *the solutions are positive or negative.*

(c) *For* $s_1 < s_2, s_i \in I_1, i = 1, 2$, *the positive solutions* $u_{s_1} \not\leq u_{s_2}$, *and the negative solutions* $u_{s_1} \not\geq u_{s_2}$. ($u_1 \leq u_2 \Longleftrightarrow u_1(x) \leq u_2(x), \forall x \in \Omega$.)

PROOF. By Theorem 1.3 of [**50**], (1) has only trivial solution if $\lambda \notin \bigcup_{k\geq 1} I_k$, and has exactly two non-trivial solutions $u_k^{\pm}(\lambda,\cdot)$ if $\lambda \in I_k$. Moreover, $u_k^{\pm}(\lambda,\cdot)$ are non-degenerate and $M(u_k^{\pm}(\lambda,\cdot)) = k$; for $\lambda \in (\lambda_k^0, \lambda_k^\infty)$, all nontrivial solutions of (1) lie on two smooth curves $\sum_k^{\pm}$, Σ_k^+ and Σ_k^- join at $(\lambda_k^0, 0)$, and

$$\lim_{\lambda\to(\lambda_k^\infty)^+} \|u_k^{\pm}(\lambda,\cdot)\| = \infty.$$

(a) So on the branch $\Sigma_k^{\pm} \cup \{(\lambda_k^0, 0)\}$, for each non-trivial solution $u^{\pm}(\lambda,\cdot)$, we have the critical groups

$$C_q^{\pm}(J_\lambda, u^{\pm}(\lambda,\cdot)) = \delta_{qk}G.$$

Next for the point $(\lambda_k^0, 0)$ on the branch $\Sigma_k^{\pm}\cup\{(\lambda_k^0, 0)\}$, we use the Shifting-theorem (see Theorem 5.4 of [**16**],p50) to prove

$$C_q(J_{\lambda_k^0}, 0) = \delta_{qk}G. \tag{4.7}$$

First we know $(\lambda_k^0, 0)$ is degenerate and $M(u_k(\lambda_k^0, 0)) = k - 1$ since $-\triangle\phi_k = \lambda_k^0 f'(0)\phi_k$.

Let

$$Z = ker\{-\triangle - \lambda_k\} = \{t\phi_k : t \in R\}, \quad \tilde{J}_{\lambda_k^0}(u) := J_{\lambda_k^0}|_Z.$$

(since λ_k is simple, $dim Z = 1$). So

$$\tilde{J}_{\lambda_k^0}(t\phi_k) = \frac{1}{2}\int_\Omega |\bigtriangledown(t\phi_k)|^2 dx - \lambda_k^0 \int_\Omega F(t\phi_k)dx,$$

$$\frac{d\tilde{J}_{\lambda_k^0}(t\phi_k)}{dt} = t\lambda_k - \lambda_k^0 \int_\Omega f(t\phi_k)\phi_k dx,$$

$$\frac{d^2\tilde{J}_{\lambda_k^0}(t\phi_k)}{dt^2} = \lambda_k - \lambda_k^0 \int_\Omega f'(t\phi_k)\phi_k^2 dx.$$

Therefore,

$$\frac{d\tilde{J}_{\lambda_k^0}(t\phi_k)}{dt}|_{t=0} = t\lambda_k - \lambda_k^0 \int_\Omega f(t\phi_k)\phi_k dx|_{t=0} = 0,$$

$$\frac{d^2\tilde{J}_{\lambda_k^0}(t\phi_k)}{dt^2}|_{t=0} = \lambda_k - \lambda_k^0 \int_\Omega f'(t\phi_k)\phi_k^2 dx|_{t=0} = 0.$$

Noticing that $uf''(u) \geq 0$ in a neighborhood of 0, $\forall t \neq 0$ sufficiently small, we have

$$\frac{d^2 \tilde{J}_{\lambda_k^0}(t\phi_k)}{dt^2} = \lambda_k - \lambda_k^0 \int_\Omega f'(t\phi_k)\phi_k^2 dx < \lambda_k - \lambda_k^0 \int_\Omega f'(0)\phi_k^2 dx = 0. \tag{4.8}$$

Hence, we know that 0 is a local maximum of $\tilde{J}_{\lambda_k^0}$, and by the Shifting-theorem we get

$$C_q(J_{\lambda_k^0}, 0) = C_{q-(k-1)}(\tilde{J}_{\lambda_k^0}, 0) = \delta_{q,(k-1)+1}G = \delta_{qk}G.$$

Now we consider the critical values: Suppose u is a non-trivial solution of (4.1), $-\triangle u = \lambda f(u), u|_{\partial\Omega} = 0$, then

$$J_\lambda(u) = \frac{1}{2}\int_\Omega |\bigtriangledown u|^2 dx - \lambda \int_\Omega F(u)dx = \lambda \int_\Omega \Big(\frac{1}{2}f(u)u - F(u)\Big).$$

Since $f'(s) > \frac{f(s)}{s}$, for $s > 0$ we have $f'(s)s > f(s)$, and $\int_0^s f'(s)s\, ds > F(s)$, then $f(s)s > 2F(s)$; for $s < 0$ we have $f'(s)s < f(s)$ and $\int_s^0 f'(s)s\, ds < \int_s^0 f(s)\, ds$, then $f(s)s > 2F(s)$ too. Therefore, $J_\lambda(u) > 0$.

(b) Suppose u is a nontrivial solution of (4.1), then $\int_\Omega(-\triangle u)\phi_1 = \int_\Omega \lambda f(u)\phi_1$, and $\int_\Omega \lambda_1\phi_1 u = \int_\Omega \lambda f(u)\phi_1$. Thus by $mes\ \Omega_0 = 0$ (where $\Omega_0 = \{x \in \Omega : u(x) = 0\}$), we have that

$$\int_{\Omega\setminus\Omega_0} \Big(\lambda_1 - \lambda\frac{f(u)}{u}\Big) u\phi_1 = 0. \tag{4.9}$$

Noticing (4.6), we have $\lambda_1 - \lambda\frac{f(u)}{u} < \lambda_1 - \lambda f'(0) \leq 0$ when $\lambda \geq \frac{\lambda_1}{f'(0)}$. Therefore, by (4.9) we get u is sign-changing.

When $\lambda \in I_1$, it is obvious that there are only one positive and one negative solution for (4.1).

(c) Let $C_0(\Omega) = \{u \in C(\bar{\Omega})|u(x) = 0, \forall x \in \partial\Omega\}$, $P := \{u \in C_0(\Omega)|u(x) \geq 0\}$, and define operators $A_s(u) := (-\triangle)^{-1}sf(u), \forall u \in C_0(\Omega)$. Then P is a cone with nonempty interior points, A_s is compact and increasing for each $s > 0$. The solution u_s of

$$-\triangle u = sf(u), \quad u|_\Omega = 0, \ u > 0 \tag{4.10}$$

is equivalent to the fixed point of A_s in P.

Suppose $u_{s_1}, u_{s_2} (s_1 < s_2, s_i \in I_1)$ are the solutions of (4.10), then u_{s_2} is the super-solution of (4.10) with $s = s_1$, thus by lemma 4.1 of [**34**] (p294) the fixed point index

$$i(A_{s_1}, (0, u_{s_2}), P) = 1, \tag{4.11}$$

where $(0, u_{s_2}) := \{u \in C_0(\Omega)|0 < u(x) < u_{s_2}(x), \forall x \in \Omega\}$.

For $A_s, s \in I_1$, we prove its Fréchet differential operator at ∞ along P is $(-\triangle)^{-1}sf'(\infty)$.

$\forall \varepsilon > 0$, $\exists R > 0$ such that

$$|f(t)/t - f'(\infty)| < \varepsilon, \quad \forall t, \ |t| > R.$$

Thus $\forall u \in C_0(\Omega)$ with $\|u\|_{C_0(\Omega)} \geq R$, we denote $\Omega_0 = \{x \in \Omega : |u(x)| \leq R\}, \Omega_1 = \Omega \backslash \Omega_0$, then

$$\begin{aligned}
\|(-\triangle)^{-1}f(u) - sf'(\infty)(-\triangle)^{-1}u\|_{C_0(\Omega)} &\leq \|(-\triangle)^{-1}\| \cdot \|f(u) - sf'(\infty)u\|_{C_0(\Omega)} \\
&= \|(-\triangle)^{-1}\| \cdot \max_{x\in\bar{\Omega}} |f(u(x)) - sf'(\infty)u(x)| \\
&\leq \|(-\triangle)^{-1}\| \cdot (\max_{x\in\bar{\Omega}_0} |f(u(x)) - sf'(\infty)u(x)| \\
&\quad + \max_{x\in\bar{\Omega}_1} |f(u(x)) - sf'(\infty)u(x)|) \\
&\leq C + \varepsilon\|(-\triangle)^{-1}\| \cdot \|u\|_{C_0(\Omega)},
\end{aligned} \tag{4.12}$$

where $C > 0$ is a constant independent on u. Therefore, we have

$$\overline{\lim}_{u\in P, \|u\|_{C_0(\Omega)}\to\infty} \|(-\triangle)^{-1}f(u) - sf'(\infty)u\|_{C_0(\Omega)}/\|u\|_{C_0(\Omega)} \leq \varepsilon\|(-\triangle)^{-1}\|.$$

By the arbitrary of ε, we get

$$\lim_{\|u\|\to\infty} \|(-\triangle)^{-1}f(u) - sf'(\infty)u\|_{C_0(\Omega)}/\|u\|_{C_0(\Omega)} = 0.$$

Hence, A_s has its Fréchet differential operator at ∞ along P and

$$A_s'(\infty) = (-\triangle)^{-1}sf'(\infty).$$

Since $sf'(\infty) \in (\lambda_1, \lambda_2)$, we know that 1 is not an eigenvalue of $A_s'(\infty)$, and $A_s'(\infty)$ has $sf'(\infty)/\lambda_1$ as the first eigenvalue which is greater than 1 with positive eigenvalue ϕ_1. By lemma 4.10 of [**34**],p328, we know that $\exists\sigma > 0$ such that

$$i(A_s, P_{r_s}, P) = 0 \ (\forall\sigma \leq r_s < \infty). \tag{4.13}$$

By the additivity of the fixed point index, we know

$$i(A_{s_1}, P_{r_{s_1}} \backslash [0, u_{s_2}], P) = -1,$$

which implies that there is a fixed point $\bar{u}_{s_1}$ of A_{s_1} in $P_{r_{s_1}} \backslash [0, u_{s_2}]$, noticing that the uniqueness of the positive solution for each $s \in I_1$, we get $u_{s_1} = \bar{u}_{s_1}$.

So $u_{s_1} \not\leq u_{s_2}$.

Similarly we have the results for the negative solutions. □

REMARK 4.3. *$\lambda_k^0 \leq \lambda_{k+1}^M$ is equivalent to*

$$\lambda_{k+1}/\lambda_k \geq \sup_u f'(u)/f'(0).$$

Since $\lambda_{k+1}/\lambda_k \to 1$ as $k \to \infty$, then the λ-ranges of solution curves must have overlaps for large k.

THEOREM 4.4 (Zhang, Theorem 2.2 of [**58**]). *Suppose that f satisfies (f_1), $(f_2), (f_4)$, $\lambda_k^m \leq \lambda_{k+1}^0$, for $k = 1, 2, \ldots N$; Moreover, suppose that $uf''(u) \leq 0$ in a neighborhood of 0. Then for $k = 1, 2, \ldots, N$,*

(a) *The branch of the solutions* $(\lambda \in (I_k \cup \{\lambda_k^0\}) \setminus \bigcup_{j \neq k} \tilde{I}_j)$ *to* (4.1) *has the same critical groups* $C_q^{\pm}(J_\lambda, u(\lambda, \cdot)) = \delta_{q(k-1)}G$, *and the critical values are negative, where* $I_k = \left(\frac{\lambda_k}{f'(0)}, \frac{\lambda_k}{f'(\infty)}\right)$, $\Sigma_k^{\pm} = \{(\lambda, u_k^{\pm}(\lambda, \cdot)) : \lambda \in I_k\}$, Σ_k^+ *and* Σ_k^- *join at* $(\lambda_k^0, 0)$, *and*

$$\lim_{\lambda \to (\lambda_k^\infty)^+} \|u_k^{\pm}(\lambda, \cdot))\| = \infty.$$

(b) *For* $\lambda \in (\lambda_k/f'(0), \lambda_k/f'(\infty)), k \geq 2$, *the nontrivial solutions are sign-changing; for* $\lambda \in I_1$, *the solutions are positive or negative.*

(c) *For* $s_1 < s_2, s_i \in I_1, i = 1, 2$, *the positive solutions* $u_{s_1} \leq u_{s_2}$, *and the negative solutions* $u_{s_1} \geq u_{s_2}$.

PROOF. (a) When f satisfies $(f_1), (f_2), (f_4)$, we have for $u \neq 0$,

$$f'(0) > \frac{f(u)}{u} > f'(u) \geq \inf_{u \in R} f'(u), \quad \frac{f(u)}{u} > f'(\infty).$$

Similar to the proof of theorem 2.1, noticing that $uf''(u) \leq 0$ in a neighborhood of 0, now $\forall t \neq 0$ sufficiently small, (4.8) becomes

$$\text{(4.14)} \qquad \frac{d^2 \tilde{J}_{\lambda_k^0}(t\phi_k)}{dt^2} = \lambda_k - \lambda_k^0 \int_\Omega f'(t\phi_k)\phi_k^2 dx > \lambda_k - \lambda_k^0 \int_\Omega f'(0)\phi_k^2 dx = 0.$$

Hence, we know that 0 is a local minimum of $\tilde{J}_{\lambda_k^0}$, and by the Shifting-theorem we get

$$C_q(J_{\lambda_k^0}, 0) = C_{q-(k-1)}(\tilde{J}_{\lambda_k^0}, 0) = \delta_{q(k-1)}G.$$

And similarly we can prove that the critical values are negative.

(b) For (4.9), now we have $\lambda_1 - \lambda \frac{f(u)}{u} < \lambda_1 - \lambda f'(\infty) \leq 0$ when $\lambda \geq \frac{\lambda_1}{f'(\infty)}$. Therefore, by (4.9) we get u is sign-changing.

(c) Suppose $u_{s_1}, u_{s_2} (s_1 < s_2, s_i \in I_1)$ are the solutions of (4.10), then u_{s_2} is the super-solution of (11) with $s = s_1$. Noticing that $s_1 f'(0) > \lambda_1$, we have that $\varepsilon\phi_1$ ($\varepsilon > 0$ sufficiently small) is a sub-solution of (4.10) with $s = s_1$. So by the strongly increasing property of $(-\triangle)^{-1}$, we know the unique positive solution $u_{s_1} \in (\varepsilon\phi_1, u_{s_2})$, where $(\varepsilon\phi_1, u_{s_2}) := \{u \in C_0(\Omega) | \varepsilon\phi_1(x) < u(x) < u_{s_2}(x), \forall x \in \Omega\}$.

Now we consider super-linear problems. □

THEOREM 4.5 (Zhang, Theorem 2.3 of [**58**]). *Suppose that* f *satisfies* (f_1), $(f_2), (f_3)$, *and* $f'(\infty) = +\infty$. *Moreover, suppose that* $uf''(u) \geq 0$ *in a neighborhood of* 0; *suppose that* $|f'(t)| \leq C(1 + |t|^{\alpha - 1}), \alpha \in \left(1, \frac{n+2}{n-2}\right), n \geq 3$, *(if* $n < 2, \alpha$ *has no restriction) and* $\exists \theta > 2, M > 0$ *such that* $\theta F(t) \leq tf(t)$ *for* $|t| \geq M$. *Then*

(a) $\forall \lambda \in \left(0, \frac{\lambda_1}{f'(0)}\right)$, (4.1) *has at least one positive, one negative, one sign-changing solution.*

(b) (4.1) *has one signed nontrivial solution* $\Leftrightarrow \lambda \in \left(0, \frac{\lambda_1}{f'(0)}\right)$

(c) $\forall \lambda \in (0, \infty)$, (4.1) *has at least one sign-changing solution.*

(d) $\forall k \geq 2, \exists \varepsilon > 0$, *such that* $\forall \lambda \in \left(\frac{\lambda_k}{f'(0)} - \varepsilon, \frac{\lambda_k}{f'(0)}\right)$, (4.1) *has at least three sign-changing solutions.*

(e) *u is a nontrivial critical point of J_λ implies that $J_\lambda(u) > 0$.*

PROOF. (a) We can use similar (but simpler) proof of Theorem 3.10 to prove it (Here the condition is strong enough).

(b) By the proof of (b) of Theorem 4.2, and (a), we get it.

(c) We consider the critical points of the functionals $J_\lambda(u)$ of (4.3). Under the conditions of this theorem, we know $J_\lambda(u)$ satisfies the P.S. condition for each $\lambda \in (0, \infty)$.

First we compute the critical groups of 0.

As $\lambda \in \left(\frac{\lambda_k}{f'(0)}, \frac{\lambda_{k+1}}{f'(0)}\right), k = 1, 2, \ldots$, then 0 is a non-degenerate critical point of J_λ, the Morse index of 0 is k, so the critical group is $C_q(J_\lambda, 0) = \delta_{qk}G$.

As $\lambda \in \left(0, \frac{\lambda_1}{f'(0)}\right)$, the critical group $C_q(J_\lambda, 0) = \delta_{q0}G$.

As $\lambda = \frac{\lambda_k}{f'(0)}, k = 1, 2, \ldots$, by the proof (a) of Theorem 2.1, we get that the critical group is $C_q(J_\lambda, 0) = \delta_{qk}G$.

Following the proof of [**55**], since $f(t)t \geq 0$ and $F(t) > 0$ for any $|t| > M$, we have

$$\frac{f(t)}{F(t)} \geq \frac{\theta}{t}, \quad \forall t \geq M, \quad \frac{f(t)}{F(t)} \leq \frac{\theta}{t}, \quad \forall t \leq -M.$$

Hence

$$F(t) \geq C|t|^\theta, \quad \text{as} \quad |t| \geq M.$$

Thus $\forall u \in S^\infty$, the unit sphere in H,

$$J_\lambda(tu) \to -\infty \text{ as } t \to +\infty. \tag{4.15}$$

We now prove that $\forall a < 0$, $J_\lambda(tu) \leq a$ implies that $\frac{dJ_\lambda(tu)}{dt} < 0$.

If $J_\lambda(tu) = \frac{t^2}{2} - \lambda \int_\Omega F(tu(x))dx \leq a < 0$, then noticing that (f_3) implies that $sf(s) > 2F(s), \forall s \neq 0$, we have

$$\begin{aligned}\frac{dJ_\lambda(tu)}{dt} &= (dJ_\lambda(tu), u) = t - \lambda \int_\Omega f(tu(x))u(x)dx \\ &= \frac{2}{t}\left\{\frac{t^2}{2} - \frac{\lambda}{2}\int_\Omega tu(x)f(tu(x))\right\} \leq \frac{2}{t}\left\{\frac{t^2}{2} - \frac{2\lambda}{2}\int_\Omega F(tu(x))dx\right\} < 0.\end{aligned} \tag{4.16}$$

The implicit function theorem is employed to obtain a unique $T(u) \in C(S^\infty, R^1)$ such that

$$J_\lambda(T(u)u) = a, \quad \forall u \in S^\infty.$$

Since J_λ is continuous in 0, there is a open neighborhood V of 0, such that

$$J_\lambda(u) > \frac{a}{2}, \quad \forall u \in V.$$

Thus $\|T(u)\|$ possesses a positive lower bound $\delta > 0$.

Finally, let us define a deformation retract $\eta : [0,1] \times (H \backslash B_\varepsilon(0)) \to H \backslash B_\varepsilon(0)$, where $B_\varepsilon(0)$ is the $\varepsilon-$ball with center 0, by

$$\eta(s, u) = (1-s)u + sT(u)u, \quad \forall u \in H \backslash B_\varepsilon(0).$$

This proves that $H\backslash B_\varepsilon(0) \simeq J_\lambda^a$, i.e. $J_\lambda^a \simeq S^\infty$, where $J_\lambda^a = \{u \in H | J_\lambda(u) \leq a\}$. Thus $H_q(H, J_\lambda^a) \simeq H_q(H, S^\infty) \simeq 0$, the Betti numbers

$$\beta_q = 0, \quad \forall q = 0, 1, 2, \ldots. \tag{4.17}$$

Now $\forall \lambda \in \left[\frac{\lambda_k}{f'(0)}, \frac{\lambda_{k+1}}{f'(0)}\right), k = 1, 2, \ldots$, by the Morse inequality, we know that J_λ has at least one nontrivial critical point, i.e. (4.1) has at least one nontrivial solution, by (b) we know that it is sign-changing. Thus noticing (a) we get that $\forall \lambda \in (0, \infty)$, (4.1) has at least one sign-changing solution.

(d) By the bifurcation theory, we know that $\left(\frac{\lambda_k}{f'(0)}, 0\right), k = 1, 2, \ldots$ is a bifurcation point. Similarly to the proof of Lemma 2.6 of [**50**], we know $\lambda(s) < \frac{\lambda_k}{f'(0)}$ for $0 < |s| < \delta$, i.e. the curve turns to the left of $\frac{\lambda_k}{f'(0)}$.

So there is $\varepsilon > 0$ such that there is a curve $\sum_k^\pm$, and $\forall \lambda \in \left(\frac{\lambda_k}{f'(0)} - \varepsilon, \frac{\lambda_k}{f'(0)}\right)$, we know $u_1(\lambda, \cdot), u_2(\lambda, \cdot) \in \sum_k^\pm$.

Next we prove they are non-degenerate as $\varepsilon > 0$ sufficiently small.

Since $\|u_i(\lambda, \cdot)\|, i = 1, 2$ are sufficiently small as $\varepsilon > 0$ sufficiently small, by the regularity of the elliptic operator we know that $u_i(\lambda, \cdot) \in C^2(\Omega)$, and $\|u_i(\lambda, \cdot)\|_{C_0^1(\Omega)}$ are sufficiently small such that $\frac{\lambda_k f'(u_i(\lambda, \cdot))}{f'(0)} < \lambda_{k+1}$, then $\frac{\lambda f(u_i(\lambda, \cdot))}{u_i(\lambda, \cdot)} < \lambda f'(u_i(\lambda, \cdot)) < \lambda_{k+1}, \forall \lambda \in \left(\frac{\lambda_k}{f'(0)} - \varepsilon, \frac{\lambda_k}{f'(0)}\right)$. Now consider (4.5) with $u = u_i(\lambda, \cdot)$, we get

$$\begin{aligned} &\mu_{k-1}(\lambda f'(0)) < 0, \quad \mu_k(\lambda f'(0)) > 0; \\ &\mu_{k-1}(\lambda \frac{f(u_i)}{u_i}) < 0, \quad \mu_{k+1}(\lambda \frac{f(u_i)}{u_i}) > 0, \ \mu_k(\lambda \frac{f(u_i)}{u_i}) = 0. \end{aligned}$$

Thus $\mu_k(\lambda f'(u_i)) < 0$, $\mu_{k+1}(\lambda f'(u_i)) > 0$. So $u_i(\lambda, \cdot), \forall \lambda \in \left(\frac{\lambda_k}{f'(0)} - \varepsilon, \frac{\lambda_k}{f'(0)}\right)$ are non-degenerate, the critical groups are

$$C_q(J_\lambda, u_i(\lambda, \cdot)) = \delta_{qk} G.$$

By the proof of (c), we know $C_q(J_\lambda, 0) = \delta_{q(k-1)} G$. Suppose there were no more critical points of J_λ. The Morse type numbers over (H, J_λ^a) would be

$$M_{k-1} = 1, \quad M_k = 2, \quad M_q = 0, \quad q \neq k-1, \ k,$$

but the Betti numbers

$$\beta_q = 0, \quad \forall q = 0, 1, 2, \ldots.$$

By the Morse inequality, this is a contradiction. Thus there are at least 3 sign-changing solutions for $\lambda \in \left(\frac{\lambda_k}{f'(0)} - \varepsilon, \frac{\lambda_k}{f'(0)}\right)$.

(e) By the proof (a) of Theorem 4.2, we get it. □

REMARK 4.6. *If f satisfies the assumptions of Theorem 4.5, and f is odd, then* (4.1) *has infinitely many solutions; and when $\lambda \geq \frac{\lambda_1}{f'(0)}$,* (4.1) *has infinitely many sign-changing solutions.*

4.2. Equations without the parameter λ.

$$\begin{cases} -\triangle u = f(u) \text{ in } \Omega, \\ u|_{\partial\Omega} = 0. \end{cases} \tag{4.18}$$

THEOREM 4.7 (Zhang,Theorem 3.1 of [**58**]). *Suppose f satisfies $(f_1), (f_2), (f_3)$, and $\lambda_{k-1} < f'(0) \le \lambda_k \le f'(\infty) < \sup_{u\in R} f'(u) < \lambda_{k+1}$, as $k \ge 2$; $0 < f'(0) \le \lambda_1 \le f'(\infty) < \sup_{u\in R} f'(u) < \lambda_2$, as $k = 1$. (It is clear that the equalities don't hold simultaneously) Then*

(a) *If $f'(0) \ne \lambda_k, f'(\infty) \ne \lambda_k$, then (4.18) has exactly two nontrivial non-degenerate solutions with same critical groups $C_q(J, u_i) = \delta_{qk}G, i = 1, 2$, and the critical values are positive.*

If $f'(0) = \lambda_k, uf''(u) \ge 0$ in a neighborhood of 0, then (4.18) has only the trivial solution and $C_q(J, 0) = \delta_{qk}G$, where

$$J(u) = \frac{1}{2}\int_\Omega |\bigtriangledown u|^2 dx - \int_\Omega F(u)dx. \tag{4.19}$$

If $f'(\infty) = \lambda_k, \lambda_{k-1} < f'(0) < \lambda_k$, then (4.18) has only the trivial solution and $C_q(J, 0) = \delta_{q(k-1)}G$. (In fact, since $\lambda_{k-1} < f'(0) < \frac{f(u)}{u} < f'(\infty) = \lambda_k, \mu_{k-1}\left(\frac{f(u)}{u}\right) < 0, \mu_k\left(\frac{f(u)}{u}\right) > 0$, we get $-\triangle u = (f(u)/u)u$ in Ω, $u|_{\partial\Omega} = 0$ has no solution.)

(b) *For $k \ge 2$, the two nontrivial solutions are sign-changing; for $k = 1$, (4.18) has exactly one positive and one negative solution.*

THEOREM 4.8. (Zhang, Theorem 3.2 of [**58**]) *Suppose f satisfies $(f_1), (f_2), (f_4)$, $\lambda_{k-1} < f'(\infty) \le \lambda_k \le f'(0) < \lambda_{k+1}$, as $k \ge 2$, $0 < f'(\infty) \le \lambda_1 \le f'(0) < \sup_{u\in R} f'(u) < \lambda_2$, as $k = 1$.(It is clear that the equalities don't hold simultaneously) Then*

(a) *If $f'(\infty) \ne \lambda_k, f'(0) \ne \lambda_k$, then (4.18) has exactly two nontrivial non-degenerate solutions with the same critical groups $C_q(J, u_i) = \delta_{q(k-1)}G, i = 1, 2$, and the critical values are negative, (Now $C_q(J, 0) = \delta_{qk}G$).*

If $f'(0) = \lambda_k, f'(\infty) < \lambda_k$, $uf''(u) \le 0$ in a neighborhood of 0, then (4.18) has only the trivial solution and $C_q(J, 0) = \delta_{q(k-1)}G$.

If $f'(\infty) = \lambda_k, \lambda_k < f'(0) < \lambda_{k+1}$, then (4.18) has only the trivial solution and $C_q(J, 0) = \delta_{qk}G$.

(b) *For $k \ge 2$, the nontrivial solutions are sign-changing; for $k = 1$, the solutions are positive or negative.*

THEOREM 4.9 (Zhang, Theorem 3.3 of [**58**]). *Suppose that f satisfies (f_1), $(f_2), (f_3)$, $f'(\infty) = +\infty$. Moreover, suppose that $uf''(u) \ge 0$ in a neighborhood of 0 if $f'(0) = \lambda_k$; suppose that $\exists\theta > 2, M > 0$ such that $\theta F(t) \le tf(t)$ for $|t| \ge M$; $|f'(t)| \le C(1 + |t|^{\alpha-1}), \alpha \in \left(1, \frac{n+2}{n-2}\right), n \ge 3$, (if $n \le 2, \alpha$ has no restriction). Then*

(a) $f'(0) < \lambda_1$, *(4.18) has at least one positive, one negative, one sign-changing solution;*
(b) *(3) has no one-signed nontrivial solution if $f'(0) \ge \lambda_1$,*
(c) *If $f'(0) < +\infty$, (4.18) has at least one sign-changing solution.*

(d) $\forall k \geq 2, \exists \varepsilon > 0$ *such that* $f'(0) \in (\lambda_k - \varepsilon, \lambda_k)$ *implies that (3) has at least three sign-changing solutions.* $\left(-\Delta u = f'(0)\frac{f(u)}{f'(0)}\right)$

(e) u *is a nontrivial critical point of* J *implies that the critical value* $J(u) > 0$.

REMARK 4.10. *If* f *satisfies the assumptions of Theorem 4.9, and* f *is odd, then* (4.18) *has infinitely many solutions. If* $f'(0) \geq \lambda_1$, (4.18) *has infinitely many sign-changing solutions.(By the Morse inequality and the proof of Theorem 4.9*

5. Existence solutions of nonlocal elliptic problems

We study and obtain sign changing solutions of the problem

$$\begin{cases} -\left(a + b\int_\Omega |\nabla u|^2\right)\Delta u = f(x,u) & \text{in } \Omega \\ u = 0 & \text{on } \partial\Omega \end{cases} \tag{5.1}$$

where Ω is a smooth bounded domain in $\mathbb{R}^n$, $a, b > 0$, and $f(x,t)$ is locally Lipschitz continuous in $t \in \mathbb{R}$, uniformly in $x \in \overline{\Omega}$, and subcritical:

$$|f(x,t)| \leq C\left(|t|^{p-1} + 1\right) \quad \text{for some } 2 < p < 2^* = \begin{cases} \dfrac{2n}{n-2}, & n \geq 3 \\ \infty, & n = 1, 2 \end{cases} \tag{5.2}$$

where C denotes a generic positive constant.

This problem is related to the stationary analogue of the equation

$$u_{tt} - \left(a + b\int_\Omega |\nabla u|^2\right)\Delta u = g(x,t) \tag{5.3}$$

proposed by Kirchhoff [**36**] as an extension of the classical D'Alembert's wave equation for free vibrations of elastic strings. Kirchhoff's model takes into account the changes in length of the string produced by transverse vibrations. Some early classical studies of Kirchhoff equations were Bernstein [**10**] and Pohozaev [**46**]. However, equation (5.3) received much attention only after Lions [**39**] proposed an abstract framework to the problem. Some interesting results can be found, for example, in [**5, 13, 32**]. More recently Alves et al. [**2**] and Ma and Rivera [**40**] obtained positive solutions of such problems by variational methods. Similar nonlocal problems also model several physical and biological systems where u describes a process which depends on the average of itself, for example the population density, see [**1, 4, 14, 15, 54**].

5.1. Variational methods and invariant sets of descent flow.

We assume that

$$tf(x,t) \geq 0 \tag{5.4}$$

and consider three cases:

(i) 4-sublinear case: $p < 4$,

(ii) asymptotically 4-linear case:

$$\lim_{|t|\to\infty} \frac{f(x,t)}{b\,t^3} = \mu \quad \text{uniformly in } x, \tag{5.5}$$

(iii) 4-superlinear case:

$$\exists \nu > 4: \quad \nu\, F(x,t) \leq t f(x,t), \quad |t| \text{ large} \tag{5.6}$$

where $F(x,t) = \int_0^t f(x,s)ds$, which implies

$$F(x,t) \geq C\,(|t|^\nu - 1)\,. \tag{5.7}$$

By (5.2) and (5.5) (resp. (5.7)),

$$4 \leq p < 2^* \text{ (resp. } 4 < \nu \leq p < 2^*), \tag{5.8}$$

so $n = 1, 2$, or 3 in (ii) and (iii).

Case (ii) leads us to the nonlinear eigenvalue problem

$$\begin{cases} -\|u\|^2\,\Delta u = \mu\, u^3 & \text{in } \Omega \\ u = 0 & \text{on } \partial\Omega, \end{cases} \tag{5.9}$$

whose eigenvalues are the critical values of the functional

$$I(u) = \|u\|^4, \quad u \in S := \{u \in H = H_0^1(\Omega) : \int_\Omega u^4 = 1\}. \tag{5.10}$$

We will see in the next section that I satisfies the Palais-Smale condition (PS) and that the first eigenvalue $\mu_1 > 0$ obtained by minimizing I has an eigenfunction $\psi > 0$. We define a second eigenvalue $\geq \mu_1$ by

$$\mu_2 := \inf_{\gamma \in \Gamma} \max_{u \in \gamma([0,1])} I(u) \tag{5.11}$$

where Γ is the class of paths $\gamma \in C([0,1], S)$ joining $\pm\psi$ such that $\gamma \cup (-\gamma)$ is non-self-intersecting.

Using the Yang index we will construct an unbounded sequence of minimax eigenvalues $0 < \mu_1 \leq \mu_2 \leq \cdots$ of the nonlinear problem (5.9) for which the following theorem holds, where we set $\lambda_0 = \mu_0 = -\infty$ for convenience(See next subsection, (5.38)). Denote by $0 < \lambda_1 < \lambda_2 \leq \cdots$ the Dirichlet eigenvalues of $-\Delta$ on Ω.

$$\begin{cases} -\Delta u = \lambda\, u & \text{in } \Omega \\ u = 0 & \text{on } \partial\Omega \end{cases} \tag{5.12}$$

THEOREM 5.1 (Zhang-Perera, Theorem 1.1 of [**64**]). *Problem* (5.1) *has a positive solution, a negative solution, and a sign changing solution in the following cases:*

(i) $p < 4$ *and*

$$\exists \lambda > \lambda_2: \quad F(x,t) \geq \frac{a\lambda}{2}\,t^2, \quad |t| \text{ small}, \tag{5.13}$$

(ii) (5.5) *and* (5.13) *hold and* $\mu < \mu_1$,
(iii) (5.5) *holds,* $\mu > \mu_2$ *is not an eigenvalue of* (5.9), *and*

$$F(x,t) \le \frac{a\lambda_1}{2}\, t^2, \quad |t| \text{ small}, \tag{5.14}$$

(iv) (5.6) *and* (5.14) *hold.*

Recall that a function $u \in H$ is called a weak solution of (5.1) if

$$(a + b\|u\|^2) \int_\Omega \nabla u \cdot \nabla v = \int_\Omega f(x,u)\, v \quad \forall v \in H. \tag{5.15}$$

Weak solutions are the critical points of the C^{2-0} functional

$$\Phi(u) = \frac{a}{2}\, \|u\|^2 + \frac{b}{4}\, \|u\|^4 - \int_\Omega F(x,u), \quad u \in H. \tag{5.16}$$

They are also classical solutions if f is locally Lipschitz on $\overline{\Omega} \times \mathbb{R}$.
Let $X = C_0^1(\overline{\Omega})$, with the usual norm

$$\|u\|_X = \max_{0 \le |\alpha| \le 1} \sup_{x \in \overline{\Omega}} |D^\alpha u(x)|, \tag{5.17}$$

which is densely imbedded in H. By the elliptic regularity theory, the critical point set $K := \{u \in H : \Phi'(u) = 0\} \subset X$. Let $\widetilde{\Phi} = \Phi|_X$. A standard argument shows that $\widetilde{\Phi}$ has the retracting property (see, e.g.,[**29**], proof of Lemma 1).

Consider the initial value problem

$$\begin{cases} \dfrac{du}{dt} = -W(u), & t > 0 \\ u(0) = u_0 \in X \end{cases} \tag{5.18}$$

where

$$W(u) = \frac{\widetilde{\Phi}'(u)}{a + b\|u\|^2} = u - Au, \tag{5.19}$$

$A = KG$, $G : X \to L^\infty(\Omega)$, $u \mapsto \dfrac{f(x,u(x))}{a + b\|u\|^2}$, and $K = (-\Delta)^{-1} : L^\infty(\Omega) \to X$. Since $f(x,t)$ is locally Lipschitz in t, uniformly in $\overline{\Omega}$, G is locally Lipschitz, and K is a bounded linear operator, so W is locally Lipschitz on X. So a unique solution $u(t,u_0)$ of (5.18) satisfying

$$u(t,u_0) = e^{-t} \left(u_0 + \int_0^t e^s\, Au(s,u_0)\, ds \right) \tag{5.20}$$

exists in some maximal existence interval $[0, T(u_0))$, $T(u_0) \le \infty$. Since

$$\frac{d}{dt}\left(\widetilde{\Phi}(u(t,u_0))\right) = (\widetilde{\Phi}'(u), \frac{du}{dt}) = -\frac{\|\widetilde{\Phi}'(u)\|^2}{a + b\|u\|^2} \le 0, \tag{5.21}$$

$\widetilde{\Phi}$ is nonincreasing along the orbits.

Therefore, we can use the similar proof of Theorem 3.10 to prove Theorem 5.4.

Now, by the new information concerning functionals without Palais-Smale condition, we can study Kirchhoff type problems with 4-superlinear growth as $|u| \to +\infty$ and with weak conditions than (5.6). We know that

$$tf(x,t) \geq 0$$

which implies that

$$F(x,t) = \int_0^t f(x,s)ds \geq 0.$$

Let $\tilde{F}(x,u) := \frac{1}{4}f(x,u)u - F(x,u)$. We make the following assumptions:

$(S1)$ $\frac{F(x,u)}{u^4} \to +\infty$ as $|u| \to +\infty$ uniformly in $x \in \Omega$;
$(S2)$ $f(x,u) = o(|u|)$ as $|u| \to 0$ uniformly in $x \in \Omega$;
$(S3)$ $\tilde{F}(x,u) \to +\infty$ as $|u| \to +\infty$ uniformly in $x \in \Omega$, and there exists $\sigma > \max\{1, N/2\}$ such that $|f(x,u)|^\sigma \leq C\tilde{F}(x,u)|u|^\sigma$ for $|u|$ large.

We obtain:

THEOREM 5.2 (Mao-Zhang, Theorem 1.1 of [**41**]). *Assume* $(S1) - (S3)$ *hold. Then problem* (1.1) *has at least one nontrivial solution.*

Furthermore, we assume

$(S4)$ $f(x,t)$ is locally Lipschitz continuous in $t \in \mathbb{R}^1$, uniformly in $x \in \bar{\Omega}$.

THEOREM 5.3 (Mao-Zhang, Theorem 1.2 of [**41**]). *Assume* $(S1)$–$(S4)$ *hold. Then problem* (1.1) *has at least one positive solution, one NEGATIVE solution, and one sign-changing solution.*

The proofs of our results are based on combining the critical point theory, invariant set of descent flow and the minimax methods using the Cerami condition (See [**41**]).

Examples satisfying the 4-superlinear conditions $(S1)$–$(S3)$ are the following functions:

- *Example 1.* $f(x,u) = u^3 \ln(1+|u|)$;
- *Example 2.* $F(x,u) = |u|^\mu + (\mu-4)|u|^{\mu-\epsilon} \sin^2\left(\frac{|u|^\epsilon}{\epsilon}\right)$ where $N = 3$, $2^* > \mu > 4$ and $0 < \epsilon < \mu + N - \mu N/2$.

5.2. Yang index and critical groups to nonlocal problems. Perera and Zhang [**44**] obtain nontrivial solutions of a class of nonlocal quasilinear elliptic boundary value problems using the Yang index and critical groups.

THEOREM 5.4 (Perera-Zhang, Theorem 1.1 of [**44**]). *If* $\lambda \in (\lambda_l, \lambda_{l+1})$ *and* $\mu \in (\mu_m, \mu_{m+1})$ *is not an eigenvalue of* (5.9), *with* $l \neq m$, *then problem* (5.1) *has a nontrivial solution.*

Yang Index:

We briefly recall the definition and some basic properties of the Yang index. Yang [**56**] considered compact Hausdorff spaces with fixed-point-free continuous involutions and used the Čech homology theory, but for our purposes here it suffices to work with closed symmetric subsets of Banach spaces that do not contain the origin and singular homology groups.

Following [**56**], we first construct a special homology theory defined on the category of all pairs of closed symmetric subsets of Banach spaces that do not contain the origin and all continuous odd maps of such pairs. Let (X, A), $A \subset X$ be such a pair and $C(X, A)$ its singular chain complex with $\mathbb{Z}_2$ coefficients, and denote by $T_\#$ the chain map of $C(X, A)$ induced by the antipodal map $T(x) = -x$. We say that a q-chain c is symmetric if $T_\#(c) = c$, which holds if and only if $c = c' + T_\#(c')$ for some q-chain c'. The symmetric q-chains form a subgroup $C_q(X, A; T)$ of $C_q(X, A)$, and the boundary operator ∂_q maps $C_q(X, A; T)$ into $C_{q-1}(X, A; T)$, so these subgroups form a subcomplex $C(X, A; T)$. We denote by

$$Z_q(X, A; T) = \{c \in C_q(X, A; T) : \partial_q c = 0\}, \tag{5.22}$$

$$B_q(X, A; T) = \{\partial_{q+1} c : c \in C_{q+1}(X, A; T)\}, \tag{5.23}$$

$$H_q(X, A; T) = Z_q(X, A; T)/B_q(X, A; T) \tag{5.24}$$

the corresponding cycles, boundaries, and homology groups. A continuous odd map $f : (X, A) \to (Y, B)$ of pairs as above induces a chain map $f_\# : C(X, A; T) \to C(Y, B; T)$ and hence homomorphisms

$$f_* : H_q(X, A; T) \to H_q(Y, B; T). \tag{5.25}$$

Example: [1.8 of [**56**]] For the m-sphere,

$$H_q(S^m; T) = \begin{cases} \mathbb{Z}_2 & \text{for } 0 \le q \le m \\ 0 & \text{for } q > m. \end{cases} \tag{5.26}$$

Let X be as above, and define homomorphisms $\nu : Z_q(X; T) \to \mathbb{Z}_2$ inductively by

$$\nu(z) = \begin{cases} \operatorname{In}(c) & \text{for } q = 0 \\ \nu(\partial c) & \text{for } q > 0 \end{cases} \tag{5.27}$$

if $z = c + T_\#(c)$, where the index of a 0-chain $c = \sum_i n_i \sigma_i$ is defined by $\operatorname{In}(c) = \sum_i n_i$. As in [**56**], ν is well-defined and $\nu\, B_q(X; T) = 0$, so we can define the *index homomorphism* $\nu_* : H_q(X; T) \to \mathbb{Z}_2$ by $\nu_*([z]) = \nu(z)$.

Proposition 5.5 (2.8 of [**56**]). *If F is a closed subset of X such that $F \cup T(F) = X$ and $A = F \cap T(F)$, then there is a homomorphism $\Delta : H_q(X; T) \to H_{q-1}(A; T)$ such that $\nu_*(\Delta[z]) = \nu_*([z])$.*

Taking $F = X$ we see that if $\nu_*\, H_m(X; T) = \mathbb{Z}_2$, then $\nu_*\, H_q(X; T) = \mathbb{Z}_2$ for $0 \le q \le m$. We define the *Yang index* of X by

$$i(X) = \inf\{m \ge -1 : \nu_*\, H_{m+1}(X; T) = 0\}, \tag{5.28}$$

taking $\inf \emptyset = \infty$. Clearly, $\nu_* H_0(X;T) = \mathbb{Z}_2$ if $X \neq \emptyset$, so $i(X) = -1$ if and only if $X = \emptyset$.

Example: [3.4 of [**56**]] $i(S^m) = m$

PROPOSITION 5.6 (2.4 of [**56**]). *If $f : X \to Y$ is as above, then $\nu_*(f_*([z])) = \nu_*([z])$ for $[z] \in H_q(X;T)$, and hence $i(X) \leq i(Y)$. In particular, this inequality holds if $X \subset Y$.*

Recall that the *Krasnoselskii Genus* of X is defined by

$$\gamma(X) = \inf\{m \geq 0 : \exists \text{ a continuous odd map } f : X \to S^{m-1}\}. \tag{5.29}$$

By Example [3.4 of [**56**]] and Proposition 5.6,

PROPOSITION 5.7. $\gamma(X) \geq i(X) + 1$

PROPOSITION 5.8. *If $i(X) = m \geq 0$, then the reduced group $\widetilde{H}_m(X) \neq 0$.*

PROOF. By (5.28),

$$\nu_* H_q(X;T) = \begin{cases} \mathbb{Z}_2 & \text{for } 0 \leq q \leq m \\ 0 & \text{for } q > m. \end{cases} \tag{5.30}$$

We show that if $[z] \in H_m(X;T)$ is such that $\nu_*([z]) \neq 0$, then $[z] \neq 0$ in $\widetilde{H}_m(X)$. Arguing indirectly, assume that $z \in B_m(X)$, say, $z = \partial c$. Since $z \in B_m(X;T)$, $T_\#(z) = z$. Let $c' = c + T_\#(c)$. Then $c' \in Z_{m+1}(X;T)$ since $\partial c' = z + T_\#(z) = 2z = 0$ mod 2, and $\nu_*([c']) = \nu(c') = \nu(\partial c) = \nu(z) \neq 0$, contradicting $\nu_* H_{m+1}(X;T) = 0$. □

Variational eigenvalues of (5.9)**:**

Let $H = H_0^1(\Omega)$ be the usual Sobolev space, normed by

$$\|u\| = \left(\int_\Omega |\nabla u|^2\right)^{1/2}, \tag{5.31}$$

and let

$$I(u) = \|u\|^4, \quad u \in S := \{u \in H : \int_\Omega u^4 = 1\}. \tag{5.32}$$

By the Lagrange multiplier rule, $u \in S$ is a critical point of I if and only if

$$\|u\|^2 \int_\Omega \nabla u \cdot \nabla v = \mu \int_\Omega u^3 v \quad \forall v \in H \tag{5.33}$$

for some $\mu \in \mathbb{R}$, i.e., u is a weak solution of (5.9). Taking $v = u$ we see that the Lagrange multiplier μ equals the corresponding critical value $I(u)$. So the eigenvalues of (5.9) are precisely the critical values of the functional I. We use the customary notation

$$I^a = \{u \in S : I(u) \leq a\} \tag{5.34}$$

for the sublevel sets of I.

LEMMA 5.9. *I satisfies the Palais-Smale condition* (PS), *i.e., every sequence $\{u_j\}$ in H such that $I(u_j)$ is bounded and $I'(u_j) \to 0$, called a* (PS) *sequence, has a convergent subsequence.*

PROOF. Since $\|u_j\|$ is bounded, for a subsequence, u_j converges to some u weakly in H and strongly in $L^4(\Omega)$. Denoting by

$$P_j\, v = v - \left(\int_\Omega u_j^3\, v\right) u_j \tag{5.35}$$

the projection of $v \in H$ onto the tangent space to S at u_j, we have

(5.36)

$$\|u_j\|^2 \int_\Omega \nabla u_j \cdot \nabla(u_j - u) = I(u_j) \int_\Omega u_j^3\,(u_j - u) + \frac{1}{4}(I'(u_j), P_j\,(u_j - u)) \to 0,$$

so, passing to a subsequence, $u_j \to 0$ or u. □

Denote by $\mathcal{A}$ the class of closed symmetric subsets of S, let

$$\mathcal{F}_m = \{A \in \mathcal{A} : i(A) \geq m - 1\}, \tag{5.37}$$

and set

$$\mu_m := \inf_{A \in \mathcal{F}_m} \max_{u \in A} I(u). \tag{5.38}$$

PROPOSITION 5.10. *μ_m is an eigenvalue of* (5.9) *and $\mu_m \nearrow \infty$.*

PROOF. If μ_m is not a critical value of I, then there is an $\varepsilon > 0$ and an odd homeomorphism η of S such that $\eta(I^{\mu_m+\varepsilon}) \subset I^{\mu_m-\varepsilon}$ by the first deformation lemma. Taking $A \in \mathcal{F}_m$ with $\max I(A) \leq \mu_m + \varepsilon$, we have $A' = \eta(A) \in \mathcal{F}_m$ by Proposition 5.6, but $\max I(A') \leq \mu_m - \varepsilon$, contradicting (5.38).

Clearly, $\mu_{m+1} \geq \mu_m$. Since the sequence of Ljusternik-Schnirelmann eigenvalues $\widetilde{\mu}_m$ of (5.9) defined using the genus γ is unbounded (see, e.g., Struwe [**49**]) and $\mu_m \geq \widetilde{\mu}_m$ by Proposition 5.7, $\mu_m \to \infty$. □

When μ is not an eigenvalue of (5.9), 0 is the only critical point of the associated variational functional

$$I_\mu(u) = \|u\|^4 - \mu \int_\Omega u^4, \quad u \in H \tag{5.39}$$

and hence its critical groups at 0 are defined and given by

$$C_q(I_\mu, 0) = H_q(I_\mu^0, I_\mu^0 \setminus \{0\}) \tag{5.40}$$

(see, e.g., Chang [**16**]).

PROPOSITION 5.11. *If $\mu \in (\mu_m, \mu_{m+1})$ is not an eigenvalue of* (5.9), *then*

$$C_m(I_\mu, 0) \neq 0. \tag{5.41}$$

PROOF. Since I_μ is positive homogeneous, I_μ^0 is radially contractible to 0 and $I_\mu^0 \setminus \{0\}$ is homotopic to $I_\mu^0 \cap S$ via the radial projection onto S, so it follows from the long exact sequence of reduced homology groups of the pair $(I_\mu^0, I_\mu^0 \setminus \{0\})$ that

$$C_m(I_\mu, 0) = H_m(I_\mu^0, I_\mu^0 \setminus \{0\}) \cong \widetilde{H}_{m-1}(I_\mu^0 \cap S) = \widetilde{H}_{m-1}(I^\mu) \tag{5.42}$$

where the last equality follows from $I_\mu|_S = I - \mu$. Since $\mu > \mu_m$, there is an $A \in \mathcal{F}_m$ such that $A \subset I^\mu$, so $i(I^\mu) \geq i(A) \geq m - 1$ by Proposition 5.6. On the other hand, $I^\mu \notin \mathcal{F}_{m+1}$ since $\mu < \mu_{m+1}$, so $i(I^\mu) < m$. Hence $i(I^\mu) = m-1$, and $\widetilde{H}_{m-1}(I^\mu) \neq 0$ by Proposition 5.8. □

PROOF OF THEOREM 5.4. Recall that a function $u \in H$ is called a weak solution of (5.1) if

$$\left(a + b\|u\|^2\right) \int_\Omega \nabla u \cdot \nabla v = \int_\Omega f(x,u)\, v \quad \forall v \in H. \tag{5.43}$$

Weak solutions are the critical points of the C^1 functional

$$\Phi(u) = \frac{a}{2}\,\|u\|^2 + \frac{b}{4}\,\|u\|^4 - \int_\Omega F(x,u), \quad u \in H. \tag{5.44}$$

They are also classical solutions if f is locally Lipschitz on $\overline{\Omega} \times \mathbb{R}$. □

LEMMA 5.12. *If μ is not an eigenvalue of* (5.9), *Φ satisfies* (PS).

PROOF. As usual, it suffices to show that every (PS) sequence $\{u_j\}$ of Φ is bounded (see, e.g., Alves et al. [**2**], Lemma 1). Suppose that $\rho_j = \|u_j\| \to \infty$ for a subsequence. Setting $v_j = u_j/\rho_j$ and passing to a further subsequence, v_j converges to some v weakly in H, strongly in $L^4(\Omega)$, and a.e. in Ω. Passing to the limit in

$$\int_\Omega \nabla v_j \cdot \nabla w - \int_\Omega \frac{f(x,u_j)}{b\,u_j^3}\, \frac{v_j^3}{1 + a/b\rho_j^2}\, w = \frac{(\Phi'(u_j), w)}{\left(a + b\rho_j^2\right)\rho_j} \tag{5.45}$$

gives

$$\int_\Omega \nabla v \cdot \nabla w - \mu\, v^3\, w = 0 \tag{5.46}$$

for each $w \in H$, and passing to the limit with $w = v_j - v$ shows that $\|v\| = 1$, so μ is an eigenvalue of (5.9), contrary to assumption. □

Let

$$\Phi_0(u) = \frac{a}{2}\left(\|u\|^2 - \lambda \int_\Omega u^2\right), \qquad \Phi_\infty(u) = \frac{b}{4}\left(\|u\|^4 - \mu \int_\Omega u^4\right). \tag{5.47}$$

PROPOSITION 5.13. *If λ and μ are not eigenvalues of* (5.12) *and* (5.9), *respectively, then for all sufficiently small $\rho > 0$ and sufficiently large $R > 4\rho$ there is a* *functional* $\widetilde{\Phi} \in C^1(H, \mathbb{R})$ *such that*

$$\text{(i)}\ \widetilde{\Phi}(u) = \begin{cases} \Phi_0(u), & \|u\| \le \rho \\ \Phi(u), & 2\rho \le \|u\| \le R/2 \\ \Phi_\infty(u), & \|u\| \ge R, \end{cases}$$

in particular,

$$C_q(\widetilde{\Phi}, 0) = C_q(\Phi_0, 0), \qquad C_q(\widetilde{\Phi}, \infty) = C_q(\Phi_\infty, 0), \tag{5.48}$$

(ii) $u = 0$ *is the only critical point of* Φ *and* $\widetilde{\Phi}$ *with* $\|u\| \le 2\rho$ *or* $\|u\| \ge R/2$, *in particular, critical points of* $\widetilde{\Phi}$ *are the solutions of* (5.1),

(iii) $\widetilde{\Phi}$ *satisfies* (PS).

PROOF. Since λ and μ are not eigenvalues of (5.4) and (5.9), respectively, Φ_0 and Φ_∞ satisfy (PS) and have no critical points with $\|u\| = 1$, so

$$\delta_0 := \inf_{\|u\|=1} \|\Phi_0'(u)\| > 0, \qquad \delta_\infty := \inf_{\|u\|=1} \|\Phi_\infty'(u)\| > 0, \tag{5.49}$$

and

$$\inf_{\|u\|=\rho} \|\Phi_0'(u)\| = \rho\,\delta_0, \qquad \inf_{\|u} = R\|\|\Phi_\infty'(u) = R^3\,\delta_\infty \tag{5.50}$$

by homogeneity. Let

$$\Psi_0(u) = \frac{b}{4}\,\|u\|^4 + \int_\Omega \frac{a\,\lambda}{2}\,u^2 - F(x,u), \quad \Psi_\infty(u) = \frac{a}{2}\,\|u\|^2 + \int_\Omega \frac{b\,\mu}{4}\,u^4 - F(x,u). \tag{5.51}$$

By (5.2),

$$\sup_{\|u\|=\rho} |\Psi_0(u)| = o(\rho^2), \qquad \sup_{\|u\|=R} |\Psi_\infty(u)| = o(R^4) \tag{5.52}$$

and

$$\sup_{\|u\|=\rho} \|\Psi_0'(u)\| = o(\rho), \qquad \sup_{\|u\|=R} \|\Psi_\infty'(u)\| = o(R^3) \tag{5.53}$$

as $\rho \to 0$ and $R \to \infty$. Since $\Phi = \Phi_0 + \Psi_0 = \Phi_\infty + \Psi_\infty$, it follows from (5.50) and (5.53) that

$$\inf_{\|u\|=\rho} \|\Phi'(u)\| = \rho\,(\delta_0 + o(1))\,, \qquad \inf_{\|u\|=R} \|\Phi'(u)\| = R^3\,(\delta_\infty + o(1))\,. \tag{5.54}$$

Take smooth functions $\varphi_0, \varphi_\infty : [0,\infty) \to [0,1]$ such that

$$\varphi_0(t) = \begin{cases} 1, & t \le 1 \\ 0, & t \ge 2, \end{cases} \qquad \varphi_\infty(t) = \begin{cases} 0, & t \le 1/2 \\ 1, & t \ge 1 \end{cases} \tag{5.55}$$

and set

$$\widetilde{\Phi}(u) = \Phi(u) - \varphi_0(\|u\|/\rho)\,\Psi_0(u) - \varphi_\infty(\|u\|/R)\,\Psi_\infty(u). \tag{5.56}$$

Since

$$\|d(\varphi_0(\|u\|/\rho))\| = O(\rho^{-1}), \qquad \|d(\varphi_\infty(\|u\|/R))\| = O(R^{-1}), \tag{5.57}$$

(5.54) holds with Φ replaced by $\widetilde{\Phi}$ also, and (i) and (ii) follow.

As for (iii), $\|\widetilde{\Phi}'\|$ is bounded away from 0 for $\rho \leq \|u\| \leq 2\rho$ and for $\|u\| \geq R/2$ by construction, so every (PS) sequence for $\widetilde{\Phi}$ has a subsequence in $\|u\| < \rho$ or in $2\rho < \|u\| < R/2$, which is then a (PS) sequence for Φ_0 or for Φ, respectively. □

We are now ready to prove Theorem 5.4. Since $\lambda \in (\lambda_l, \lambda_{l+1})$ and $l \neq m$

$$C_m(\widetilde{\Phi}, 0) = C_m(\Phi_0, 0) = 0, \tag{5.58}$$

but

$$C_m(\widetilde{\Phi}, \infty) = C_m(\Phi_\infty, 0) = C_m(I_\mu, 0) \neq 0 \tag{5.59}$$

by Proposition 5.11, so $\widetilde{\Phi}$ must have a nontrivial critical point.

References

[1] C. O. Alves and F. J. S. A. Correa. On existence of solutions for a class of problem involving a nonlinear operator. *Comm. Appl. Nonlinear Anal.*, 8(2):43–56, 2001.

[2] C. O. Alves, F. J. S. A. Correa, and T. F. Ma. Positive solutions for a quasilinear ellitic equation of kirchhoff type. *Comput. Math. Appl.* 49 (2005), no. 1, 85–93.

[3] H. Amann, Fixed point equations and nonlinear eigenvalue problems in ordered Banach spaces, *SIAM Review* **18(4)**(1976),620-709.

[4] D. Andrade and T. F. Ma. An operator equation suggested by a class of nonlinear stationary problems. *Comm. Appl. Nonlinear Anal.*, 4(4):65–71, 1997.

[5] A. Arosio and S. Panizzi. On the well-posedness of the Kirchhoff string. *Trans. Amer. Math. Soc.*, 348(1):305–330, 1996.

[6] T. Bartsch, Critical point theory on partially ordered Hilbert spaces, *J. Functional Analysis* **186**(2001),117-152.

[7] T. Bartsch, K. C. Chang, and Z.Q. Wang, On the Morse indices of sign changing solutions of nonlinear elliptic problems. *Math. Z.* 233 (2000), no. 4, 655–677.

[8] T. Bartsch and Tobias Weth, A note on additional properties of sign changing solutions to superlinear elliptic equations, *Topol. Methods Nonlinear Anal.* 22 (2003), no. 1, 1–14.

[9] T. Bartsch and Zhi-Qiang Wang, On the existence of sign changing solutions for semilinear Dirichlet problems. *Topol. Methods Nonlinear Anal.* 7 (1996), no. 1, 115–131.

[10] S. Bernstein. Sur une classe d'équations fonctionnelles aux dérivées partielles. *Bull. Acad. Sci. URSS. Sér. Math. [Izvestia Akad. Nauk SSSR]*, 4:17–26, 1940.

[11] Siegfried Carl and Kanishka Perera, Sign-changing and multiple solutions for the p-Laplacian. *Abstr. Appl. Anal.* 7 (2002), no. 12, 613–625.

[12] A. Castro, J. Cossio, and J.M. Neuberger, A sign-changing solution for a superliner Dirichlet problem, *Rocky Mountain J. Math.* **27** (1997), no. 4, 1041-1053.

[13] M. M. Cavalcanti, V. N. Domingos Cavalcanti, and J. A. Soriano. Global existence and uniform decay rates for the Kirchhoff-Carrier equation with nonlinear dissipation. *Adv. Differential Equations*, 6(6):701–730, 2001.

[14] M. Chipot and B. Lovat. Some remarks on nonlocal elliptic and parabolic problems. *Nonlinear Anal.*, 30(7):4619–4627, 1997.

[15] M. Chipot and J.-F. Rodrigues. On a class of nonlocal nonlinear elliptic problems. *RAIRO Modél. Math. Anal. Numér.*, 26(3):447–467, 1992.

[16] Kung-ching Chang, Infinite dimentional Morse theory and multiple solutions problems, Birkhauser, 1993.

[17] G. Cerami, An existence criterion for the critical points on unbounded manifolds, *Istit. Lombardo Accad. Sci. Lett. Rend. A*, **112(2)**:332–336, 1978.

[18] M. Conti and V. Felli, Coexistence and segregation for strongly competing species in special domains. *Interfaces Free Bound.* 10 (2008), no. 2, 173–195.
[19] M. Conti, S. Terracini and G. Verzini, Neharis problem and competing species systems, *Ann. Inst. H. Poincare Anal. Non Lineaire* **19** (2002), no. 6, 871–888.
[20] M. Cuesta, D. de Figueiredo and J. P. Gosses, The begining of the Fučik spectrum for the p-Laplacian, *J. Diff. Equa.* **159**(1999), 212–238.
[21] E.N. Dancer, Competing species systems with diffusion and large interaction, *Rend. Sem. Mat. Fis. Milano 65* (1995), 23–33.
[22] E.N. Dancer and Y.H. Du, Competing species equations with diffusion, large interactions, and jumping nonlinearities, *J. Differential Equations 114* (1994), 434–475.
[23] E. N. Dancer and Yihong Du, Existence of changing sign solutions for some semilinear problems with jumping nonlinearities at zero, *Proceeding of the Royal Society of Edinburgh* 124 A, (1994), 1165–1176.
[24] E. N. Dancer and Yihong Du, Multiple solutions of some semilinear elliptic equations via generalized conley index, *Journal of Mathematical Analysis and Applications*, 189, 848–871, 1995.
[25] E. N. Dancer, Remarks on jumping nonlinearities, *Topics in Nonlinear Analysis*, edited by Escher and Simonett, Birkhauser, Basel, 101–116, 1999.
[26] E. N. Dancer and Yihong Du, On sign-changing solutions of certain semilinear elliptic problems, *Appl. Anal.*, 56, (1995), 193–206.
[27] E. N. Dancer and Yihong Du, Positive solutions for a three-species competition system with diffusionII: The case of equal birth rates, *Nonlinear Analysis,* 24(3), (1995), 359–373.
[28] E. N. Dancer and Z. M. Guo, Uniqueness and stability for solutions of competing species equations with large interactions, *Comm. Appl. Nonlinear Anal.* 1 (1994), 19–45.
[29] E. N. Dancer and Zhitao Zhang, Fucik spectrum, sign-changing, and multiple solutions for semilinear elliptic boundary value problems with resonance at infinity, *J. Math. Anal. Appl.* 250 (2000), no. 2, 449–464.
[30] E. N. Dancer and Zhitao Zhang, Dynamics of LotkaCVolterra Competition Systems with Large Interaction, *Journal of Differential Equations,* **182** (2002), no. 2, 470–489.
[31] E. N. Dancer and Kelei Wang, Zhitao Zhang, Dynamics of Strongly Competing Systems with Many Species. Submitted to Transaction of AMS.
[32] P. D'Ancona and S. Spagnolo. Global solvability for the degenerate Kirchhoff equation with real analytic data. *Invent. Math.*, 108(2):247–262, 1992.
[33] D. Gilbarg and N. Trudinger, Elliptic partial differential equations of second order, Berlin, New York: Springer-Verlag, 2001.
[34] Dajun Guo, Nonlinear Functional Analysis, Shandong Science and technology publishing house, Jinan, 1985. (In Chinese)
[35] Whyburn G.T., Topological Analysis. Princeton: Princeton University Press, 1958.
[36] G. Kirchhoff. *Mechanik.* Teubner, Leipzig, 1883.
[37] Shujie Li and Zhitao Zhang, Fucik Spectrum, sign-changing and multiple solutions for semilinear elliptic boundary value problems with jumping nonlinearities at zero and infinity, *Sci. in China*, 44(7), 856–866, 2001.
[38] Shujie Li and Zhitao Zhang, Sign-changing and multiple solutions theorems for semilinear elliptic boundary value problems with jumping nonlinearities, *Acta Math. Sin.* (Engl. Ser.) 16 (2000), no. 1, 113–122.
[39] J.-L. Lions. On some questions in boundary value problems of mathematical physics. In *Contemporary developments in continuum mechanics and partial differential equations (Proc. Internat. Sympos., Inst. Mat., Univ. Fed. Rio de Janeiro, Rio de Janeiro, 1977)*, volume 30 of *North-Holland Math. Stud.*, pages 284–346. North-Holland, Amsterdam, 1978.
[40] T. F. Ma and J. E. Muñoz Rivera. Positive solutions for a nonlinear nonlocal elliptic transmission problem. *Appl. Math. Lett.*, 16(2):243–248, 2003.
[41] Anmin Mao and Zhitao Zhang, Sign-changing and multiple solutions of Kirchhoff type problems without the P.S. condition, *Nonlinear Anal.* 70 (2009), no. 3, 1275–1287.
[42] K. Perera and Martin Schechter, Double resonance problems with respect to the Fucik spectrum, *Indiana Univ. Math. J.* 52 (2003), no. 1, 1–17.
[43] K. Perera and Martin Schechter, Computation of critical groups in Fucik resonance problems, *J. Math. Anal. Appl.* 279 (2003), no. 1, 317–325.

[44] Kanishka Perera and Zhitao Zhang, Nontrivial solutions of Kirchhoff-type problems via the Yang index, *J. Differential Equations* 221 (2006), no. 1, 246–255.
[45] Kanishka Perera and Zhitao Zhang, Multiple positive solutions of singular p-Laplacian problems by variational methods, *Bound. Value Probl.* 2005, no. 3, 377–382.
[46] S. I. Pohožaev. A certain class of quasilinear hyperbolic equations. *Mat. Sb. (N.S.)*, 96(138):152–166, 168, 1975.
[47] Bernhard Ruf, Superlinear elliptic equations and systems. Handbook of differential equations: stationary partial differential equations. Vol. V, 211–276, *Handb. Differ. Equ.*, Elsevier/North-Holland, Amsterdam, 2008.
[48] Martin Schecher. The Fucik spectrum. *Indiana University Mathematics Journal*, vol.43, No.4, 1139–1157, 1994.
[49] Michael Struwe, Variational Methods, Springer-Verlag, Berlin, 2000.
[50] Junping Shi and Junping Wang, Morse indices and exact multiplicity of solutions to semilinear elliptic problems, *Proc. Amer. Math. Soc.* 127(1999), No.12, 3685–3695.
[51] Junping Shi, Exact multiplicity of solutions to superlinear and sublinear problems, *Nonlinear Anal.* 50(2002), no. 5, Ser. A: Theory Methods, 665–687.
[52] J. Sun and Z. Liu. Calculus of variations and super- and sub-solution in reverse order, *Acta Mathematica Sinica*, vol.37, No.4, 512–514, 1994. (In Chinese)
[53] J. L. Vazquez, A strong maximum principle for some quasilinear elliptic equations, *Appl Math Optim* 12:191–202(1984).
[54] Carlos Frederico Vasconcellos. On a nonlinear stationary problem in unbounded domains. *Rev. Mat. Univ. Complut. Madrid*, 5(2–3):309–318, 1992.
[55] Z.Q. Wang, On a superlinear elliptic equation, *Ann Inst. H. Poincare Analyse Non Lineaire* 8(1991) 43–57.
[56] Chung-Tao Yang. On theorems of Borsuk-Ulam, Kakutani-Yamabe-Yujobô and Dyson. II. *Ann. of Math. (2)*, 62:271–283, 1955.
[57] Kelei Wang and Zhitao Zhang, Some New Results in Competing Systems with Many Species, *Ann. I. H. Poincare-AN* 27(2010), 739–761.
[58] Zhitao Zhang, On bifurcation, critical groups and exact multiplicity of solutions to semilinear elliptic boundary value problems, *Nonlinear Anal.* 58 (2004), no. 5–6, 535–546.
[59] Zhitao Zhang, Jianqing Chen, and Shujie Li, Construction of pseudo-gradient vector field and sign-changing multiple solutions involving p-Laplacian. *J. Differential Equations* 201 (2004), no. 2, 287–303.
[60] Zhitao Zhang, Marta Calanchi, and Bernhard Ruf, Elliptic equations in $\mathbb{R}^2$ with one-sided exponential growth. *Commun. Contemp. Math.* 6 (2004), no. 6, 947–971.
[61] Zhitao Zhang and Shujie Li, On sign-changing and multiple solutions of the p-Laplacian. *J. Funct. Anal.* 197 (2003), no. 2, 447–468.
[62] Zhitao Zhang, Shujie Li, Shibo Liu, and Weijie Feng, On an asymptotically linear elliptic Dirichlet problem. *Abstr. Appl. Anal.* 7 (2002), no. 10, 509–516.
[63] Zhitao Zhang and Xiaodong Li, Sign-changing solutions and multiple solutions theorems for semilinear elliptic boundary value problems with a reaction term nonzero at zero, *J. Differential Equations* 178 (2002), no. 2, 298–313.
[64] Zhitao Zhang and Kanishka Perera, Sign changing solutions of Kirchhoff type problems via invariant sets of descent flow, *J. Math. Anal. Appl.* 317 (2006), no. 2, 456–463.

Zhitao Zhang, Academy of Mathematics and Systems Science, Chinese Academy of Sciences, Beijing 100190, China

E-mail address: zzt@math.ac.cn

Index

Made in United States
Orlando, FL
04 November 2026

90200661R00378